# Novel Plant Bioresources

# Novel Plant Bioresources

## Applications in Food, Medicine and Cosmetics

Ameenah Gurib-Fakim
Center for Phytotherapy Research (CEPHYR), Mauritius

WILEY Blackwell

*Registered office:* John Wiley & Sons, Ltd, The Atrium, Southern Gate, Chichester, West Sussex, PO19 8SQ, UK

*Editorial offices:* 9600 Garsington Road, Oxford, OX4 2DQ, UK
The Atrium, Southern Gate, Chichester, West Sussex, PO19 8SQ, UK
111 River Street, Hoboken, NJ 07030-5774, USA

For details of our global editorial offices, for customer services and for information about how to apply for permission to reuse the copyright material in this book please see our website at www.wiley.com/wiley-blackwell.



***Library of Congress Cataloging-in-Publication Data***

Novel plant bioresources : applications in food, medicine and cosmetics / [edited by] Ameenah Gurib-Fakim.
pages cm
Includes bibliographical references and index.
ISBN 978-1-118-46061-0 (cloth)
1. Plant diversity. 2. Germplasm resources conservation–Economic aspects. 3. Germplasm resources, Plant–Economic aspects. 4. Plant biotechnology. I. Gurib-Fakim, Ameenah, editor of compilation.
QK46.5.D58N68 2014
333.95'3416–dc23

2013046825

A catalogue record for this book is available from the British Library.

Wiley also publishes its books in a variety of electronic formats. Some content that appears in print may not be available in electronic books.

Cover image: Close-up of neem leaves © istock/ Focal.Point
Cover design by Meaden Creative

Set in 9.25/11.5pt Minion by Laserwords Private Limited, Chennai, India
Printed and bound in Malaysia by Vivar Printing Sdn Bhd

1 2014

# Content

# List of contributors

**Abderrahman Aafi,** Forest Research Center, High Commission for Water, Forests and Combat Desertification, Agdal, Rabat, Morocco

**Azila Abdul-Aziz,** Institute of Bioproduct Development, Universiti Teknologi Malaysia Kuala Lumpur, Jalan Semarak, Kuala Lumpur, Malaysia

**Fatimah Corazon Abdullah,** Natural Products Discovery Laboratory, Institute of Bio-IT Selangor, Universiti Selangor, Malaysia

**Mohamed Aberchane,** Forest Research Center, High Commission for Water, Forests and Combat Desertification, Agdal, Rabat, Morocco

**M.O. Abukutsa-Onyango,** Jomo Kenyatta University of Agriculture and Technology, Kenya, Department of Horticulture, Nairobi, Kenya

**E.G. Achigan-Dako,** Laboratory of Plant Science, Department of Plant Production, Faculty of Agronomic Sciences (FSA), University of Abomey-Calavi, Cotonou, Republic of Benin

**Gauravi Agarkar,** Department of Biotechnology, Sant Gadge Baba Amravati University, Amravati (MS), India

**T. Balan,** Department of Biomedical Science, Faculty of Medicine and Health Sciences, Universiti Putra Malaysia, Serdang, Selangor, Malaysia

**Joanne Bero,** Pharmacognosy Research Group, Louvain Drug Research Institute, Université Catholique de Louvain, Bruxelles, Belgium

**John A. Beutler,** Molecular Targets Laboratory, Center for Cancer Research, NCI-Frederick, Frederick, MD, USA

**Anne Bialecki,** Laboratoire de Chimie des Substances Naturelles et des Sciences des Aliments, Faculté des Sciences et Technologies, Université de la Réunion, La Réunion, France

**Thomas Brendler,** Plantaphile Ltd, Collingswood (USA), Eastbourne (UK) and Berlin (Germany)

**Bruno Canard,** Laboratoire d'Architecture et de Fonction des Macromolécules Biologiques (AFMB-AMU-UMR7257), Marseille, France

**Kelly Chibale,** Department of Chemistry, University of Cape Town, Rondebosch, South Africa

**Nyaradzo T.L. Chigorimbo-Murefu,** Department of Chemistry, University of Cape Town, Rondebosch, South Africa

**Theresa Chimponda,** Biomolecular Interactions Analyses Group, Department of Biochemistry, University of Zimbabwe, Mt. Pleasant, Harare, Zimbabwe

**Elaine Chirisa,** Biomolecular Interactions Analyses Group, Department of Biochemistry, University of Zimbabwe, Mt. Pleasant, Harare, Zimbabwe

**Tariro Chitemerere,** Biomolecular Interactions Analyses Group, Department of Biochemistry, University of Zimbabwe, Mt. Pleasant, Harare, Zimbabwe

**Gordon M. Cragg,** Natural Products Branch, Developmental Therapeutics Program, NCI-Frederick, Frederick, MD, USA

**E. Dagne,** African Laboratory for Natural Products (ALNAP), Department of Chemistry, Addis Ababa University, Addis Ababa, Ethiopia

**Hugo de Boer,** Systematic Biology, Department of Organismal Biology, Evolutionary Biology Centre, Uppsala University, Uppsala, Sweden; Naturalis Biodiversity Center, Faculty of Science, Leiden, The Netherlands; Faculty of Science, Leiden University, Leiden, The Netherlands

**M.E. Dulloo,** Bioversity International, Rome, Italy

**J.N. Eloff,** Phytomedicine Programme, Department of Paraclinical Sciences, Faculty of Veterinary Science, University of Pretoria, Onderstepoort, South Africa

**C.A. Fatimah,** Institute of Bio-It Selangor, Universiti Selangor, Shah Alam, Selangor, Malaysia, Faculty of Sciences & Biotechnology, Universiti Selangor, Shah Alam, Selangor, Malaysia

**Ulrich Feiter,** Parceval Ltd, Cape Town, South Africa

**T.H. Fernandes,** Lúrio University, Nampula, Mozambique

**L.J. Ferrão,** Lúrio University, Nampula, Mozambique

**Mohamed Ghanmi,** Forest Research Center, High Commission for Water, Forests and Combat Desertification, Agdal, Rabat, Morocco

**Joyce Govinden-Soulange,** Faculty of Agriculture, University of Mauritius, Réduit, Mauritius

**Barbara Gravendeel,** Naturalis Biodiversity Center, National Herbarium Nederland – Leiden University, Leiden, The Netherlands

**Françoise Guéritte,** Centre de Recherche de Gif, Institut de Chimie des Substances Naturelles (ICSN), CNRS, LabEx LERMIT, Gif sur Yvette Cedex, France

**Lucia Guidi,** Department of Agriculture, Food and Environment, University of Pisa, Pisa, Italy

**Jean-Claude Guillemot,** Laboratoire d'Architecture et de Fonction des Macromolécules Biologiques (AFMB-AMU-UMR7257), Marseille, France

**Ameenah Gurib-Fakim,** Center for Phytotherapy Research (CEPHYR), Ebene, Mauritius

**Mariani Abdul Hamid,** Institute of Bioproduct Development, Universiti Teknologi Malaysia, Skudai, Johor, Malaysia

**Rosnani Hasham,** Institute of Bioproduct Development, Universiti Teknologi Malaysia, Skudai, Johor, Malaysia

**C.A. Houdegbe,** Laboratory of Plant Science, Department of Plant Production, Faculty of Agronomic Sciences (FSA), University of Abomey-Calavi, Cotonou, Republic of Benin

**D. Hunter,** Bioversity International, Rome, Italy, School of Agriculture and Wine Sciences (SAWS), Charles Sturt University, Orange, New South Wales, Australia

**Murray Hunter,** School of Business Innovation & Technoentrepreneurship, University Malaysia Perlis, Malaysia

**Abdul Latif Ibrahim,** Natural Products Discovery Laboratory, Institute of Bio-IT Selangor, Faculty of Sciences & Biotechnology, Universiti Selangor, Malaysia

**Mustapha Umar Imam,** Laboratory of Molecular Biomedicine, Institute of Bioscience, Universiti Putra Malaysia, Serdang, Selangor, Malaysia

**Maznah Ismail,** Laboratory of Molecular Biomedicine, Institute of Bioscience, Department of Nutrition and Dietetics, Faculty of Medicine and Health Sciences, Universiti Putra Malaysia, Serdang, Selangor, Malaysia

**Maurice Iwu,** Bioresource Development & Conservation Programme, Nigeria, Head Office, Abuja, FCT, Nigeria

**Yasmina Jaufeerally-Fakim,** Faculty of Agriculture, University of Mauritius, Reduit, Mauritius

**John F. Kamanula,** Chemistry Department, Mzuzu University, Luwinga, Mzuzu, Malawi

**F.H. Kamisan,** Department of Biomedical Science, Faculty of Medicine and Health Sciences, Universiti Putra Malaysia, Serdang, Selangor, Malaysia

**Ossy M.J. Kasilo,** World Health Organization, Regional Office for Africa, Brazzaville, Republic of the Congo

**David R. Katerere,** Department of Pharmaceutical Sciences, Faculty of Science, Tshwane University of Technology, Arcadia Campus, Pretoria, Republic of South Africa

**Abderrahim Khia,** Laboratory of Biotechnology, Environment and Quality, Faculty of Sciences, University Ibn Tofail, Kénitra, Morocco

**Salomé Kpoviessi,** Pharmacognosy Research Group, Louvain Drug Research Institute, Université Catholique de Louvain, Bruxelles, Belgium, Laboratory of Physic and Synthesis Organic Chemistry (LaCOPS), University of Abomey-Calavi (UAC), Faculty of Sciences and Technics (FAST), Cotonou, Benin, Laboratory of Pharmacognosy and Essential Oils (LAPHE), University of Abomey-Calavi (UAC), Faculty of Health Sciences (FSS), Faculty of Sciences and Technics (FAST), Cotonou, Benin

**Namrita Lall,** University of Pretoria, Department of Plant Science, Lynwood, Pretoria, South Africa

**Marco Landi,** Department of Agriculture, Food and Environment, University of Pisa, Pisa, Italy

**Mohammed Lamorde,** Infectious Diseases Institute, College of Health Sciences, Makerere University, Kampala, Uganda

**D. Leaman,** Canadian Museum of Nature, Ottawa, Canada

**Pieter Leyssen,** Rega Institute for Medical Research (KU Leuven), Leuven, Belgium

**Marc Litaudon,** Centre de Recherche de Gif, Institut de Chimie des Substances Naturelles (ICSN), CNRS, LabEx LERMIT, Gif sur Yvette Cedex, France

**M. Maaza,** UNESCO–UNISA Africa Chair in Nanosciences/Nanotechnology, College of Graduate Studies, University of South Africa (UNISA), Muckleneuk Ridge, Pretoria, South Africa, Nanosciences African Network (NANOAFNET), iThemba LABS–National Research Foundation, Somerset West, Western Cape Province, South Africa

**Mohamad Fawzi Mahomoodally,** Department of Health Sciences, Faculty of Science, University of Mauritius, Réduit, Mauritius

**Lucy Lynn Maliwichi,** University of Venda, Thohoyando, Republic of South Africa

**Cecilia Maliwichi-Nyirenda,** Indigenous Knowledge Centre, Blantyre, Malawi

**S.S. Mamat,** Department of Biomedical Science, Faculty of Medicine and Health Sciences, Universiti Putra Malaysia, Serdang, Selangor, Malaysia

**Rumbidzai Mangoyi,** Biomolecular Interactions Analyses Group, Department of Biochemistry, University of Zimbabwe, Mt. Pleasant, Harare, Zimbabwe

**L.J. McGaw,** Phytomedicine Programme, Department of Paraclinical Sciences, Faculty of Veterinary Science, University of Pretoria, Onderstepoort, South Africa

**Grace Mugumbate,** Department of Chemistry, University of Cape Town, Rondebosch, South Africa

**Stanley Mukanganyama,** Biomolecular Interactions Analyses Group, Department of Biochemistry, University of Zimbabwe, Mt. Pleasant, Harare, Zimbabwe

**S. N'Danikou,** Laboratory of Plant Science, Department of Plant Production, Faculty of Agronomic Sciences (FSA), University of Abomey-Calavi, Cotonou, Republic of Benin; Bioversity International, West and Central Africa Office, Cotonou, Benin

**David J. Newman,** Natural Products Branch, Developmental Therapeutics Program, NCI-Frederick, Frederick, MD, USA

**Jean-Baptiste Nikiema,** World Health Organization, Regional Office for Africa, Brazzaville, Republic of the Congo

**Norhayati Mohammad Noor,** Institute of Bioproduct Development, Universiti Teknologi Malaysia, Skudai, Johor, Malaysia

**Marco Nuno de Canha,** University of Pretoria, Department of Plant Science, Lynwood, Pretoria, South Africa

**Antonia Nyamukuru,** Sustainable Use of Plant Diversity (SUPD), Kampala, Uganda

**Christopher Okunji,** US Pharmacopeial Convention, Rockville, MD, USA

**Joseph Otieno,** Department of Medical Botany, Plant Breeding and Agronomy, Institute of Traditional Medicine, Muhimbili University of Health and Allied Sciences, Dar es Salaam, Tanzania

**Tony L. Palama,** Unité Mixte de Recherche – Peuplement Végétaux et Bioagresseurs en Milieu Tropical, Université de La Réunion, Saint-Denis, La Réunion, France

**L. Denzil Phillips,** Denzil Phillips International Ltd, Richmond (UK) and Sinsheim (Germany)

**Joëlle Quetin-Leclercq,** Pharmacognosy Research Group, Louvain Drug Research Institute, Université Catholique de Louvain, Bruxelles, Belgium

**Lida Rahimi,** Natural Products Discovery Laboratory, Institute of Bio-IT Selangor, Universiti Selangor, Malaysia

**Mahendra Rai,** Department of Biotechnology, Sant Gadge Baba Amravati University, Amravati (MS), India

**Philippe Rasoanaivo,** Institut Malgache de Recherches Appliquées (IMRA), Avarabohitra Itaosy, Antananarivo, Madagascar

**Dnyaneshwar Rathod,** Department of Biotechnology, Sant Gadge Baba Amravati University, Amravati (MS), India

**R. Rodzi,** Department of Biomedical Science, Faculty of Medicine and Health Sciences, Universiti Putra Malaysia, Serdang, Selangor, Malaysia

**Rokia Sanogo,** Faculté de Pharmacie, Université des Sciences, des Techniques et des Technologies de Bamako, Mali, Département de Médecine Traditionnelle, Institut National de Recherche en Santé Publique, Bamako, Mali

**Mohamad Roji Sarmidi,** Institute of Bioproduct Development, Universiti Teknologi Malaysia, Skudai, Johor, Malaysia

**Badr Satrani,** Forest Research Center, High Commission for Water, Forests and Combat Desertification, Agdal, Rabat, Morocco

**Smeetha Singh,** University of Pretoria, Department of Plant Science, Lynwood, Pretoria, South Africa

**Jacqueline Smadja,** Laboratoire de Chimie des Substances Naturelles et des Sciences des Aliments (LCSNSA), Faculté des Sciences et Technologies, Université de La Réunion, La Réunion, France

**C.N.A. Sossa-Vihotogbé,** Faculty of Agronomic Sciences, University of Abomey-Calavi (FSA/UAC, Benin), Cotonou, Benin

**John R.S. Tabuti,** Makerere University, College of Agricultural and Environmental Sciences (MUCAES), Kampala, Uganda

**D.A. Tchokponhoue,** Laboratory of Plant Science, Department of Plant Production, Faculty of Agronomic Sciences (FSA), University of Abomey-Calavi, Cotonou, Republic of Benin

**Tinde van Andel,** Naturalis Biodiversity Center, National Herbarium Nederland – Leiden University, Leiden, The Netherlands

**Sarina Veldman,** Systematic Biology, Department of Organismal Biology, Evolutionary Biology Centre, Uppsala University, Uppsala, Sweden

**R. Vihotogbé,** Faculty of Agronomic Sciences, University of Abomey-Calavi (FSA/UAC, Benin), Cotonou, Benin

**Susan A. Wren,** International Centre for Insect Physiology and Ecology (ICIPE), Nairobi, Kenya

**Harisun Yaakob,** Institute of Bioproduct Development, Universiti Teknologi Malaysia, Skudai, Johor, Malaysia

**F. Yahya,** Department of Biomedical Science, Faculty of Medicine and Health Sciences, Universiti Putra Malaysia, Serdang, Selangor, Malaysia

**Salah Eddine Bakkali Yakhlef,** Forest Research Center, High Commission for Water, Forests and Combat Desertification, Agdal, Rabat, Morocco

**Zainul Amiruddin Zakaria,** Department of Biomedical Science, Faculty of Medicine and Health Sciences, Universiti Putra Malaysia, Serdang, Selangor, Malaysia, Natural Products Discovery Laboratory, Institute of Bio-IT Selangor, Universiti Selangor, Malaysia

# Foreword

Biodiversity is the basis of life. At the dawn of a new millennium, one of the most pressing challenges of our time is the continuing, and at times irreversible, loss of biodiversity on our planet.

Global efforts to reduce biodiversity loss began in earnest with the establishment of the Convention on Biological Diversity (CBD) in 1993 at the Rio Earth Summit. Today, the CBD counts 193 parties (or governments) as members, representing a dramatic step forward in the conservation of biological diversity, the sustainable use of its components, and the fair and equitable sharing of benefits arising from the use of genetic resources. To maintain focus, the United Nations has decreed the period 2011–2020 as the Decade on Biodiversity.

Nowhere is the need for conservation and sustainable utilization of biodiversity greater than in sub-Saharan Africa, whose biodiversity wealth is uniquely important from a global conservation viewpoint. The African continent is home to between 40 000 and 60 000 plant species, of which at least 35 000 are found nowhere else. Africa's biodiversity wealth is not uniformly distributed. The Democratic Republic of Congo, Madagascar and South Africa have each been classified as 'megadiverse' countries, the world's 17 most biologically diverse countries that together account for nearly 70% of global species diversity. South Africa's Cape Floristic Region (CFR) represents less than 0.5% of the area of the African continent but is home to nearly 20% of its flora, containing nearly 3% of the world's plant species.

Despite its enormous natural wealth, sub-Saharan Africa faces daunting conservation challenges. Its flora and fauna are under unrelenting assault from a variety of threats: agricultural expansion, habitat loss and degradation, overexploitation, illegal hunting and trade, invasive alien species, rapid population growth, urbanization and climate change to name just a few. As argued in *Freedom to Innovate: Biotechnology in Africa's Development*, biodiversity loss in Africa has a significant impact on economic growth and social development. For Africa's rural citizens, it has the effect of removing key sources of food, fuel and medicines, as well as adversely affecting tourism and pharmaceuticals – from a reduction in the availability of medicinal plants. New knowledge, about conservation and whole plant utilization, is needed, not just to strengthen the conservation effort, but to harness Africa's unique patrimony of natural resources to foster economic development, reduce poverty and protect the environment.

Seen in this context, this volume, *Novel Plant Bioresources: Applications in Food, Medicine and Cosmetics* edited by Professor Ameenah Gurib-Fakim of Mauritius, is a major new contribution for promoting sustainable utilization and management of novel plant genetic resources in Africa and beyond. It advances our understanding of the increasingly crucial role that plants play in the economic, cultural, medical and social spheres of our lives. The volume's focus on underutilized plant species is welcome, and marks a first-of-its type effort to marshal, in one publication, novel uses of plants for food, cosmetics and medicines.

The volume includes contributions from a diverse range of scholars – a majority hailing from the African continent – who offer fresh, new insights on novel plants and a wide range of important, related topics, including conservation, discovery of new drugs, new molecular approaches, market standards, nutrition and commercial applications of medicinal plants, among others.

Sustainable utilization and management of plant genetic resources is a topic of contemporary significance. By marshalling the latest evidence and cutting-edge knowledge, this volume should find broad appeal among activists, business leaders, scientists, farmers, policymakers and all those who are committed to reducing biodiversity loss on our planet.

**Ismail Serageldin**
*Director*
*Bibliotheca Alexandrina, Egypt*

## Part One

# Novel Plant Bioresources: Applications in Medicine, Cosmetics, etc.

# 1 Plant Diversity in Addressing Food, Nutrition and Medicinal Needs

M.E. Dulloo[1], D. Hunter[1,2] and D. Leaman[3]

[1] Bioversity International, Rome, Italy
[2] School of Agriculture and Wine Sciences (SAWS), Charles Sturt University, Orange, New South Wales, Australia
[3] Canadian Museum of Nature, Ottawa, Canada

## 1.1 Introduction

The world presently still faces tremendous challenges in securing adequate food that is healthy, safe and of high nutritional quality for all, and doing so in an environmentally sustainable manner (Pinstrup-Andersen, 2009; Godfray *et al.*, 2010). With the growing demand of an expected 9 billion people by 2050, it remains unclear how our current global food system will cope (Foley *et al.*, 2011; Tilman *et al.*, 2011). Compounded with climate change, ecosystems and biodiversity under stress, ongoing loss of species and genetic diversity, increasing urbanization, social conflict and extreme poverty, there has never been a more urgent time for collective action to address food, nutrition security and health globally (Hunter and Fanzo, 2013). Currently, 868 million people suffer from hunger in spite of the target of Millennium Development Goal No. 1 to halve hunger by 2015, while micronutrient deficiencies, known as hidden hunger, undermine the growth and development, health and productivity of over 2 billion people (Micronutrient Initiative, 2009). At the same time, over 1 billion people, worldwide, are overweight (WHO, 2012), and as many as 80% of the world's people depend on traditional medicine (which involves the use of plants extracts or their active principles) for their primary health care needs (WHO *et al.*, 1993).

As we shall see in this chapter, plant diversity has a critical role to play in addressing the food and nutrition security and medicinal needs of people of this world. *The Plant List* (2010) reports that there are just over 1 million recorded scientific plant names at the species ranks, of which about 30% have accepted species names, 45% are synonyms and 25% are still unresolved. This reflects the estimations of the number of plant species that exist in the world as being between 250 000 and 400 000 (Govaerts, 2001; Bramwell, 2002). These numbers are most likely to change as new plants are being discovered and as taxonomists resolve the nomenclatures of recorded plant species. The plant diversity is not evenly distributed across the world and tends to be concentrated in specific diversity-rich areas. It is generally known that most diversity of species occurs within the warm regions of the tropics and less diversity exists in temperate and boreal regions of the world (Dulloo, 2013). Barthlott *et al.* (2005) has identified five centres that reach a species richness of more than 5000 plant species per 10 000 square kilometres (Costa Rica-Chocó, Atlantic Brazil, Tropical Eastern Andes, Northern Borneo, New Guinea). Most of the global centres are located in mountainous regions within the humid tropics, where suitable climatic conditions and high levels of geodiversity (i.e. the diversity of abiotic conditions) coincide (Barthlott *et al.*, 2005). Myers *et al.* (2000) noted that as many as 44% of all species of vascular plants and 35% of all species in four vertebrate groups are confined to 25 hotspots comprising only 1.4% of the land surface of the Earth, mostly located in tropical areas. Among crops plants, the Russian breeder Nicolai Vavilov identified eight centres of origin of cultivated plants including South Mexican and Central America, Southern America, Mediterranean centre, Middle East, Ethiopia, Central Asia, India and China (Vavilov, 1931).

As the primary producers of our planet, capturing sunlight energy that fuels life on Earth in the process of photosynthesis, plants are the most fundamental and essential resources for humankind. Besides this fundamental function, plant species provide us with sources of foods, medicines, clothes, ornamentals, building materials and other uses. Plants are also an intricate part of all our ecosystems and provide all the essential ecosystem services, including the provisioning, regulating, cultural and supporting services. Besides the obvious provisioning of food in ensuring that people are food and nutritionally secured, many plants contribute directly to our agriculture by providing valuable traits and genes that are used by modern-day breeders for crop improvement, in particular those plants which are closely related to crop plants, the so-called crop wild relatives (CWRs). In addition, all human societies use plants as medicines. Many plant species protect and enrich our soil: nitrogen-fixing bacteria in root nodules of leguminous plants fertilize the soil. They form an essential link in the biogeochemical cycles, including water, nitrogen and other nutrient cycles. Plants provide direct support for other life forms. For example, trees are habitats for many organisms, including providing nesting sites for birds and a harbour for many other animals. Mangroves protect our coasts and provide a breeding ground for many marine organisms. There is also an inextricable link between plants and culture (Posey, 1999). Many plant species play important cultural roles in the development of human cultures throughout the world. Indigenous, traditional and local communities have a deep knowledge about plants and their uses as medicines, in traditional customs and rituals and have sustainably used and conserved a vast diversity of plants. The use of biologically active materials from the natural environment as medicines to maintain

*Novel Plant Bioresources: Applications in Food, Medicine and Cosmetics*, First Edition. Edited by Ameenah Gurib-Fakim.

and restore health is an important human adaptation, as fundamental a feature of human culture as is use of fire, tools and speech (Alland, 1966; Johns, 1990). Having evolved over millennia, the knowledge, cultural traditions and medicinal resources of many human societies may be rapidly disappearing with the loss of cultural and biological diversity (Principe, 1991; Schultes, 1991).

In spite of this great diversity of plants on Earth and the fundamental role they play, the story of crops and humanity has shown an increasing reliance on a small proportion of plant species used by humans (Murphy, 2007). The beginnings of exploitation of plant diversity for food and nutrition are as old as humankind, and early hunter–gatherers in pre-agricultural times would have exploited their local environment for readily available fruits, berries, seeds, flowers, shoots, storage organs and fleshy roots to complement meat obtained from hunting.

Furthermore, the evolution of crop plants that began about 10 000 years ago resulted in an even greater reliance by humans on much-reduced plant diversity than was previously utilized for food supply. While the number of plant species used for food by pre-agricultural human societies is estimated at around 7000 (Wilson, 1992), another 70 000 are known to have edible parts (Kunkel, 1984). An estimated 50 000–70 000 plant species are used medicinally around the world (Schippmann *et al.*, 2002, 2006), of which relatively few are produced in cultivation (Mulliken and Inskipp, 2006). Prescott-Allen and Prescott-Allen (1990) calculated that the world's food comes just from 103 plant species based on calories, protein and fat supply; 30 crops provide 95% of the world food energy needs (FAO, 1998). However, only four crop species (maize, wheat, rice and sugar) supply almost 60% of the calories and proteins in the human diet (Palacios, 1998). Today, in population terms, 4 billion people rely on rice, maize or wheat as their staple food, while a further 1 billion people rely on roots and tubers (Millstone and Lang, 2008), and as these authors point out there are thousands of plant species with neglected potential utility for humans and which represent one of the most poorly underutilized and underappreciated food resources we have.

The great majority (70–90%) of the market demand for medicinal and aromatic plants is supplied through wild collection (Lange, 1998; Bhattacharya *et al.*, 2008), providing many rural communities with important sources of income. While some wild-sourced plants appear to be produced in a sustainable manner, others, particularly high-demand species in international trade, evidently are not sustainably sourced (Sheldon *et al.*, 1996; Oldfield and Jenkins, 2012). Moreover, medicinal plant species are likely to be threatened by loss of habitat, climate change and other factors contributing to the extinction of plants and other species worldwide (Vié *et al.*, 2009).

### 1.1.1 Threatened plants and crop varieties

There have been many studies and assessments undertaken at national, regional and global levels to show that plant diversity is globally threatened. Historically, our knowledge of the threatened plants stems from the pioneering work of Sir Peter Scott (then chairman of the International Union for Conservation of Nature (IUCN) Species Survival Commission) who initiated the compilation of a list of threatened plants which led to the publication of the IUCN Plant Red Data Book in 1978 (Lucas and Synge, 1978). This book provided the conservation status of 250 species (mainly European plants) of the 25 000 plant species estimated to be threatened at this time. This work encouraged other countries to develop their own lists of threatened plants and Plant Red Data Books (Gabrielyan, 1988; Strahm, 1989; Fu and Chin, 1992; Golding, 2002). The 1997 IUCN Red List of Threatened Plants was the first-ever published list of threatened vascular plants, including ferns and fern allies, gymnosperms and flowering plants, and listed 12.5% of the world's vascular flora (estimated at 270 000 at this time) as being threatened at the global scale, recognizing though that this assessment was based on incomplete data sets and the quality of data, which varied considerably depending upon regions and taxonomic groups (Walter and Gillett, 1998). The work did not take into account genetic erosion within the populations of species, which is important for plant genetic resources and wild relatives of cultivated plants.

IUCN (2001) has developed a uniform way of estimating the degree of threat to taxa. Taxa are listed in the IUCN Red List under categories that indicate the varying degrees of their probability of extinction. There are nine clearly defined IUCN categories under which every species (or lower taxonomic unit) in the world can be classified (Figure 1.1). Taxa are then classified to these categories, by assessment using five quantitative criteria and sub-criteria that take into account the population sizes, distribution range and degree of threats (IUCN, 2001). Such criteria, however, cannot be applied to cultivated plants, which require a different paradigm. Padulosi and Dulloo (2012) proposed a novel approach for monitoring cultivated plants that is based on assessing current trends and possible decline of its cultivation over time. The ultimate objective of monitoring cultivated species is to secure their effective use by people so as to meet sustainably their livelihood needs. This approach would allow us to 'raise the red flag' when such a decline goes below that level (compared with past use-trends) under which its benefits (nutritional, income generation, etc.) are no longer spread over the community. Thus, a four-cell framework has been proposed for assessing cultivated plants, mostly at varietal levels, based on the number of households and areas of cultivation (Padulosi and Dulloo, 2012) (Figure 1.2). Several countries have attempted to produce a red list of cultivated plants, including Romania (Antofie, 2011), Germany (Hammer and Khoshbakht, 2005; Meyer and Vögel, 2006) and Nepal (Joshi *et al.*, 2004).

This global concern about loss of plant diversity led botanists convened at the XVIth International Botanical Congress in St Louis Missouri, USA, in August 1999 to call for plant conservation as a global priority in biodiversity conservation. This in turn led to the development of The Global Strategy for Plant Conservation (GSPC), which was first adopted at the sixth meeting of the Conference of the Parties to the Convention on Biological Diversity (CBD) in April 2002 and was subsequently revised at the tenth Conference of Parties in 2010 in Nagoya, Japan. The GSPC established 16 outcome-oriented targets to halt the loss of plant diversity and provided a framework which facilitated harmony between existing initiatives aimed at plant conservation. It stimulated many countries to make progress to achieve the GSPC targets, including undertaking preliminary assessment of conservation status of their known species and lists of their threatened species. Recently, an ad hoc international expert group of ethnobotanists meeting at Missouri Botanic Garden (1–2 May 2013) called for a development global programme on the conservation of useful plants and associated knowledge for the successful implementation of the

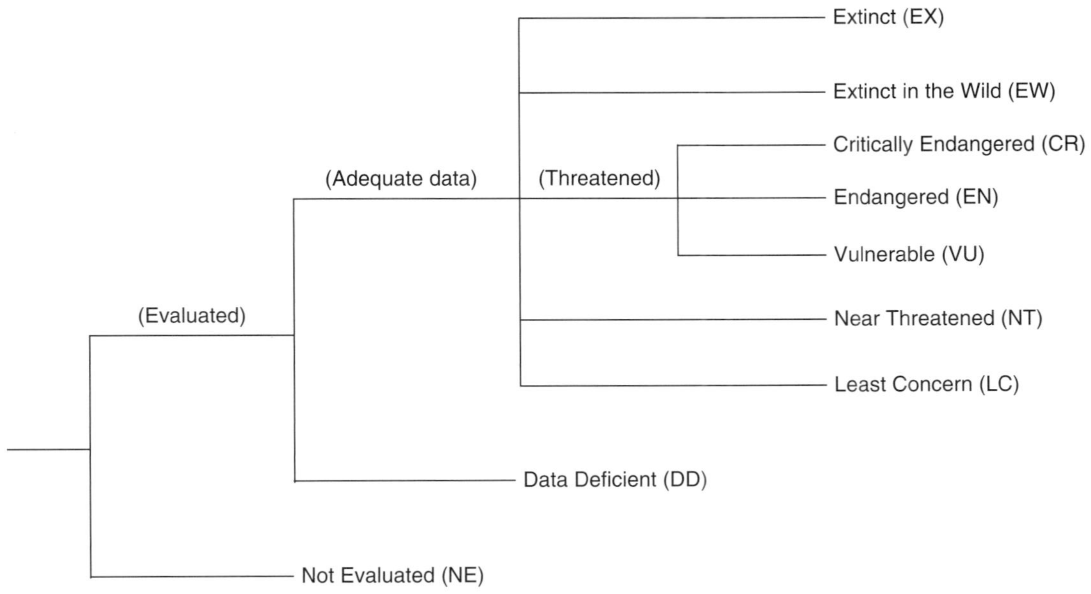


**Figure 1.1** The IUCN Red List categories

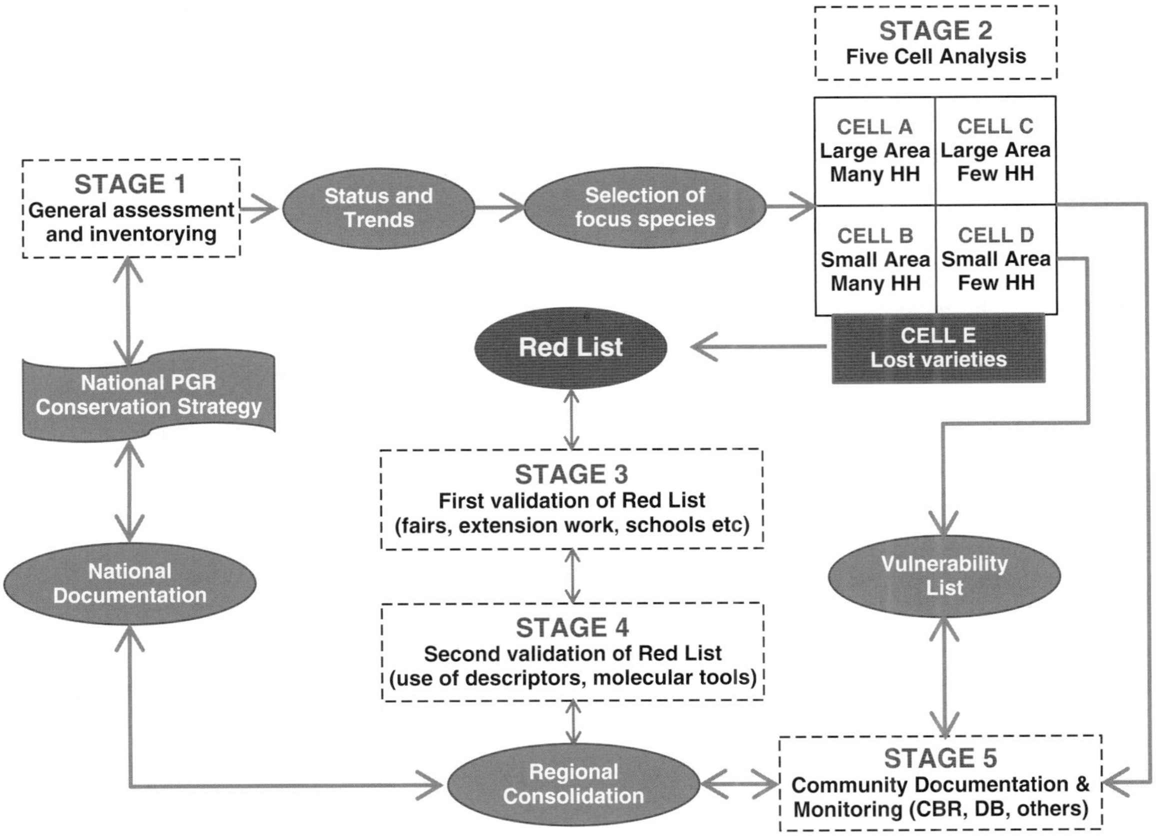


**Figure 1.2** Five-cell framework for assessing threatened cultivated plants. *Source*: Padulosi and Dulloo (2012)

GSPC objectives and targets by 2020 (Peter Wyse-Jackson, 2 May 2013, personal communication).

The Millennium Ecosystem Assessment (2005) identified five major direct drivers of biodiversity loss and ecosystem service changes. These are habitat change, climate change, invasive alien species, overexploitation and pollution. A study on the patterns of threats to the flora of an entire continent (South America) showed that population accessibility, expansion of agriculture and grazing pressure are also key drivers of immediate extinction risk of plant diversity (Ramirez-Villegas *et al.*, 2012).

With regard to plant biodiversity important for agriculture, the Food and Agriculture Organization of the United Nations (FAO) second State of the World Report on Plant Genetic Resources for Food and Agriculture (PGRFA; FAO, 2010) provided a review of the change in state of PGRFA since the first report was published in 1998 (FAO, 1998). By PGRFA we mean cultivated crops and their varieties as well as their wild relatives and wild food plants (i.e. wild harvested plants). PGRFA has an enormous contribution to make in ensuring food security, livelihood and resilience of the production system and in coping with climate change.

The FAO's recent publication on *Save and Grow* (FAO, 2011) informs us that 50% of food production growth actually comes from PGRFA, and consequently plays an important role in improving crop production. With the challenge of the need to increasing food production by 50–70% in order to meet the demand for food by 9.1 billion people by 2050 (Tomlinson, 2013), the FAO proposed a new paradigm of sustainable crop production intensification (SCPI) for producing more from the same land while conserving resources, reducing negative impacts on the environment and enhancing natural capital and flow of ecosystem services (FAO, 2011). To achieve this paradigm the FAO SCPI strategy quotes, 'Farmers will need a genetically diverse portfolio of improved crop varieties that are suited to a range of agro-ecosystem and farming practices and resilient to climate change' (FAO, 2011). In other words, the paradigm can only be realized if the production system is diversified. Jarvis *et al.* (2007) previously showed that the broader the diversity employed on farm, the more resilient will be the production system. In particular, local landraces, which are considered to be the reservoirs of adaptive variation in crops (Sthapit and Padulosi, 2012), will be key in sustaining on-farm production as well as providing raw materials for future plant breeding. Crop diversity also helps to reduce genetic vulnerability, whereby diversity within a field or within a production system helps to ensure stability in overall food production and thus reduces the risks to agricultural production. A more diverse cropping system helps to buffer against the spread of pests and diseases and the vagaries of weather, likely to occur in uniform monoculture cultivation. A bioversity project in Kitui, east of Kenya, showed that farmers who grew a wider range of crops on the farm coped better with drought conditions. While maize crops failed during April 2009, farmers who grew local drought-resistant crops such as *ngelenge* (a local type of lima bean, *Phaseolus lunatus* L.), cowpeas (*Vigna unguiculata* (L.) Walp.) and *mbumbu* (hyacinth bean, *Lablab purpureus* (L.) Sweet) and some forms of sorghum successfully gathered a good harvest (Bioversity International, 2009). In the global context, the phenomenon of genetic vulnerability represents a major risk with regard to the capacity of our agricultural systems to ensure sustainable food security, as well as the livelihoods of farmers.

With regard to medicinal plants, The World Bank has called on health officials, economists and other planner/decision-makers the world over to include the contribution of medicinal plants to national health and local economies in national resource accounting (Srivastava *et al.*, 1996; Lambert *et al.*, 1997). The contribution of medicinal plants to health and livelihoods is recognized directly and indirectly in international and regulatory policy frameworks focusing on the relationship between biodiversity conservation and human social, cultural, health, and economic security and development (see Box 1.1).

### Box 1.1. Relevant Targets on Agricultural Biodiversity

#### Millennium Development Goal

Target 7.B: Reduce biodiversity loss, achieving, by 2010, a significant reduction in the rate of loss.

#### Global Strategy on Plant Conservation (2011–2020)

Target 6: At least 75% of production lands in each sector managed sustainably, consistent with the conservation of plant diversity.
Target 7: At least 75% of known threatened plant species conserved *in situ*.
Target 8: At least 75% of threatened plant species in *ex situ* collections, preferably in the country of origin, and at least 20 % available for recovery and restoration programmes.
Target 9: 70% of the genetic diversity of crops including their wild relatives and other socio-economically valuable plant species conserved, while respecting, preserving and maintaining associated indigenous and local knowledge.
Target 11: No species of wild flora endangered by international trade.
Target 12: All wild-harvested plant-based products sourced sustainably.
Target 13: Indigenous and local knowledge, innovations and practices associated with plant resources, maintained or increased, as appropriate, to support customary use, sustainable livelihoods, local food security and health care.

*Source:* https://www.cbd.int/gspc

#### United Nations Strategic Plan for Biodiversity 2011–2020 (Aichi biodiversity targets)

Target 2: By 2020, at the latest, biodiversity values have been integrated into national and local development and poverty reduction strategies and planning processes and are being incorporated into national accounting, as appropriate, and reporting systems.
Target 4: By 2020, at the latest, governments, businesses and stakeholders at all levels have taken steps to achieve or have implemented plans for sustainable production and consumption and have kept the impacts of use of natural resources well within safe ecological limits.
Target 7: By 2020, areas under agriculture, aquaculture and forestry are managed sustainably, ensuring conservation of biodiversity
Target 12: By 2020, the extinction of known threatened species has been prevented and their conservation status, particularly of those most in decline, has been improved and sustained.
Target 13: By 2020, the genetic diversity of cultivated plants and farmed and domesticated animals and of wild relatives, including other socio-economically as well as culturally valuable species, is maintained and strategies have been developed and implemented for minimizing genetic erosion and safeguarding their genetic diversity.
Target 18: By 2020, the traditional knowledge, innovations and practices of indigenous and local communities relevant for the conservation and sustainable use of biodiversity, and their customary use of biological resources, are respected, subject to national legislation and relevant international obligations, and fully integrated and reflected in the implementation of the convention with the full and effective participation of indigenous and local communities, at all relevant levels.

*Source:* www.cbd.int/sp/targets

#### World Health Organization

The Alma-Ata Declaration (1978) urged countries and their governments to include traditional medicine in their primary health systems, and to recognize traditional medicine practitioners as health workers, particularly for primary health care at the community level.

International Consultation on Conservation of Medicinal Plants (Chiang Mai, Thailand), convened by WHO, IUCN and WWF in 1988, resulting in the 'Chiang Mai Declaration' calling for action to 'Save the Plants that Save Lives' (WHO *et al.*, 1993).

World Health Assembly resolution on medicinal plants (WHO, 1988), referring to the Chiang Mai Declaration, placed medicinal plants, their rational and sustainable use, and their conservation firmly in the arena of public health policy and concern.

*WHO traditional medicine strategy* (WHO, 2002a), included components to protect indigenous traditional medical knowledge aiming to promote their recording and documentation, and to protect medicinal plants aiming to promote their sustainable use and cultivation.

World Health Assembly resolution on traditional medicine (WHO, 2003a) requested the WHO to collaborate with other organizations of the UN system and nongovernmental organizations in various areas related to traditional medicine, including research, protection of traditional medical knowledge and conservation of medicinal plants resources.

*Guidelines on good agricultural and collection practices (GACP) for medicinal plants* (WHO, 2003b) provide general technical guidance on quality assurance and control of herbal medicines, including obtaining herbal materials of good quality for the sustainable production of herbal medicines.

## 1.2 Plant genetic resources for food and agriculture

In this section we provide a review of plant diversity within cultivated plants and their wild relatives, as well as wild harvested food plants. Plant and animal species for food have been collected, used, domesticated and improved through traditional systems of selection over many generations, resulting in even more intraspecific diversity developed by early farmers in terms of crop varieties and local landraces and breeds. Many of these varieties and their wild relatives are at risk from a wide range of drivers of biodiversity loss – changes in land use, replacement of traditional varieties by modern cultivars, agricultural intensification, increased population, poverty, land degradation and environmental change (including climate change) (van de Wouw *et al.*, 2009; FAO, 2010). The third report on the Global Biodiversity Outlook (GBO3) gives the example of the decline in the numbers of local rice varieties in China from 46 000 in the 1950s to slightly more than 1000 in 2006 (Secretariat of the Convention on Biological Diversity, 2010). The threats also threatened the attainment of global targets established by governments at the global level, such as those of the Millennium Development Goals, GSPC and CBD Aichi targets (see Box 1.1 for targets relevant to genetic diversity) and recently by the FAO Commission on Genetic Resources for Food and Agriculture (CGRFA) (CGRFA, 2013). With regard to genetic diversity, there is currently no monitoring mechanism in place to inform us of the status and trends of genetic diversity of PGRFA at the global level (Dulloo *et al.*, 2010; Pereira *et al.*, 2013), except for some 200–300 crop species for which it is thought that 70% of genetic diversity is conserved in genebanks (Secretariat of the Convention on Biological Diversity, 2010). For monitoring the status and trends of PGRFA and to take remedial actions to ensure both their conservation and use, biodiversity indicators have been developed by these global initiatives (Millennium Development Goals, GSPC, Strategic Plan for Biodiversity, Aichi Biodiversity targets and CGRFA) to monitor progress towards achievements of their respective targets.

### 1.2.1 Crop diversity

As mentioned earlier, of the 400 000 plant species, about 100 000 species are used by mankind, 30 000 species are edible, 7000 crop species are used as food at local levels, 120 crop species are important at the national scale, 30 crop species provide 90% of the world's calories and only 4 crops provide 60% of the calories and proteins globally (Prescott-Allen and Prescott-Allen, 1990; Wilson, 1992; FAO, 1998; Palacios, 1998). Food crops can be differentiated into major crops and minor crops, but the distinction can be very arbitrary depending on the criteria used to differentiate between them. Based on area harvested worldwide, FAOSTAT data for 2011 produces some 36 crops that are grown over more than 4 million hectares (http://faostat.fao.org/) (Table 1.1). The first State of the World Report on PGRFA (FAO, 1998) provides an overview of 30 crops that feed the world. Among the major crops, it lists wheat, rice, maize, millet, sorghum, potato, sugarcane, soybean, sweet potato, cassava, beans and banana/plantain as being crops that each supplies more than 5% of the plant-derived energy intake in one or more subregions (see Annex 2 in FAO (1998)).

### 1.2.2 Landrace diversity

Throughout history, farmers have subjected their domesticated plants to strong selection pressures, thereby developing a large diversity of morphologically recognizable traditional varieties or landraces as a result of selection, genetic drift or fragmentation of their populations (Harlan, 1992). Although these landraces are generally recognized, there is still a lack of a universally accepted definition of landrace. The on-farm Conservation and Management Taskforce of the European Cooperative Group on Genetic Resources defines a landrace as follows:

> A landrace of a seed-propagated crop is a variable population which is identifiable and usually have a name. It lacks formal crop improvement, is characterised by a specific adaptation to environment conditions of the area of cultivation, (tolerant to biotic and abiotic stresses of that area) and is closely associated with the use, knowledge habit, dialects and celebrations of the people who developed and continue to grow it.
>
> Vetelainen *et al.* (2009)

For each crop species there may exist thousands of landraces, but there is a lack of information on the number of extant landraces, although there have been attempts at making inventories of landraces in many European countries (Vetelainen *et al.*, 2009) and for some major crops. By definition, every crop that has been grown on a specific farm long enough for it to develop distinctive characteristics would make it a landrace in its own right. However, without a clear definition of a landrace and a nomenclature for landraces, it will be difficult to estimate the number of landraces globally. It should also be borne in mind that the evolutionary processes and selection forces that help shape diversity in agroecosystems is extremely dynamic and takes place at a much higher rate than in

**Table 1.1** World's major food crops (above 4 million ha, harvested)

| Crops | Area harvested (ha) | Production (million tonnes) |
|---|---|---|
| Wheat | 220 385 285 | 704 080 283 |
| Maize | 170 398 070 | 883 460 240 |
| Rice, paddy | 164 124 977 | 722 760 295 |
| Soybeans | 102 993 246 | 260 915 871 |
| Barley | 48 603 576 | 134 279 415 |
| Sorghum | 35 482 800 | 54 198 010 |
| Rapeseed | 33 645 342 | 62 454 482 |
| Millet | 31 929 408 | 27 705 271 |
| Beans, dry | 29 211 491 | 23 250 253 |
| Sunflower seed | 26 049 793 | 40 206 186 |
| Sugarcane | 25 436 924 | 1 794 359 190 |
| Groundnuts, with shell | 21 770 537 | 38 614 053 |
| Cassava | 19 644 071 | 252 203 769 |
| Potatoes | 19 248 586 | 374 382 274 |
| Oil palm fruit | 16 265 248 | 233 810 539 |
| Chick peas | 13 202 603 | 11 623 787 |
| Coconuts | 11 437 523 | 59 189 887 |
| Cow peas, dry | 10 426 698 | 4 928 280 |
| Cocoa beans | 10 003 270 | 4 395 657 |
| Oats | 9 679 190 | 22 504 708 |
| Olives | 9 634 576 | 19 845 300 |
| Sweet potatoes | 7 953 196 | 104 259 988 |
| Grapes | 7 086 022 | 69 654 926 |
| Sesame seed | 6 628 276 | 4 092 236 |
| Plantains | 5 496 411 | 38 901 406 |
| Bananas | 5 157 466 | 106 541 709 |
| Rye | 5 113 145 | 12 948 841 |
| Mangoes, mangosteens, guavas | 5 066 918 | 38 899 593 |
| Sugar beet | 5 061 732 | 271 644 917 |
| Yams | 4 882 306 | 56 613 722 |
| Apples | 4 766 775 | 75 635 283 |
| Tomatoes | 4 734 356 | 159 023 383 |
| Cashew nuts, with shell | 4 701 983 | 4 201 010 |
| Onions, dry | 4 290 645 | 85 375 125 |
| Lentils | 4 169 382 | 4 411 104 |
| Oranges | 4 039 313 | 69 605 815 |

*Source:* http://faostat.fao.org/, accessed 13 March 2013.

natural ecosystem, thus making an estimate of landrace diversity even more complicated.

However, some estimates of order of magnitude of landrace diversity and for specific crops are possible. For example, Delêtre *et al.* (2013) mentioned an estimated 200 000 or more landraces of rice (*Oryza sativa* L.) worldwide and about as many varieties of wheat (*Triticum aestivum* L. subsp. *aestivum*). There are about 47 000 varieties of sorghum, 30 000 varieties of common bean (*Phaseolus vulgaris* L.), chickpea (*Cicer arietinum* L.) and maize (*Zea mays* L.), approximately 20 000 varieties of pearl millet, 15 000 varieties of peanut (*Arachis hypogaea* L.), and between 7000 and 9000 varieties of manioc (*Manihot esculenta* Crantz) (FAO, 1998). In Nepal, over 132 local varieties of mango (Subedi *et al.*, 2008) and 2000 local varieties of rice (Gupta *et al.*, 1996) are known to exist, while in India there are about 1000 mango varieties and 5000 local rice varieties (B. Sthapit, 3 April 2013, personal communication). In China, there are just over 1000 varieties of rice, although it is believed that this number was more than 46 times higher 56 years ago (Secretariat of the Convention on Biological Diversity, 2010). In the Andean region of Peru, a main centre of domestication and diversification of crop plants of the world, Velasquez-Milla *et al.* (2011) found 1483 farmers' varieties of Andean tuber species maintained by households in the village Huánuco (Central Sierra region), while there were 507 in Cajamarca (Northern Sierra region). This difference was attributed to cultural identity, extent of cultivated land area, differences in the way of practicing traditional agricultural techniques and the levels of self-sufficiency of households. The Andean region of Peru is a highly specialized production system developed at high altitudes between 3300 and 4200 m and composed of many native tuber crops, including seven potato species (*Solanum ajanhuiri, S. chaucha, S. curtilobum, S. juzepczuckii, S. phureja, S. stenotomum* and *S. tuberosum*), with about 3000 varieties characterized by botanical descriptors: 'oca' (*Oxalis tuberosa*) with at least 50 technically described varieties, 'olluco' (*Ullucus tuberosus*) with 50–70 clones and 'mashua' (*Tropaeolum tuberosum*) with nearly 100 described varieties (Velasquez-Milla *et al.*, 2011).

### 1.2.3 Crop wild relatives

CWRs are another constituent of PGRFA that are important for food and agriculture. Generally, they are not directly used as food, although some are; for example, wild yams (*Dioscorea* spp.) (Hunter and Heywood, 2011) and many wild species of potatoes in Andean regions (Vellasquez-Milla *et al.*, 2011). Broadly speaking, CWRs are any species of the same genus as cultivated plants (i.e. crop) to which they are related (Maxted *et al.*, 2006). A more precise definition of a CWR refers to its degree of relationship with the crop and based on the Harlan and de Wet genepool concept or on a taxonomic relationship (Maxted *et al.*, 2006). The taxon group concept employs the taxonomic hierarchy as a proxy for taxon genetic relatedness, and thus crossability. Thus, a CWR is a wild plant taxon that has an indirect use derived from its relatively close genetic relationship to a crop; this relationship is defined in terms of the CWR belonging to genepools 1 or 2, or taxon groups 1 to 4 of the crop (Maxted *et al.*, 2006), where genepools 1 and 2 refer to primary and secondary genepools and taxon groups 1 to 4 are: taxon group 1a, crop; taxon group 1b, same species as crop; taxon group 2, same series or section as crop; taxon group 3, same subgenus as crop; taxon group 4, same genus; and taxon group 5, same tribe but different genus to crop.

The most comprehensive review of CWRs has been the work of Maxted and Kell (2009) that made an attempt at estimating the number of CWRs that may occur worldwide and described the centres of diversity for CWRs for 14 priority crops for food and agriculture. Based on the broad definition of a CWR, as any taxon belonging to the same genus as a crop, Maxted and Kell (2009) estimated that there are between 50 000 and 60 000 crop and CWR species. More recently, using genepool and/or taxon group concepts, Vincent *et al.* (2013) made the first global priority CWR inventory, along with analysis of its composition and geographic distribution, as a basis for future *in situ* and *ex situ* conservation and sustainable use. The inventory (www.cwrdiversity.org/checklist/) contains 1667 taxa from 37 families, 109 genera, 1392 species and

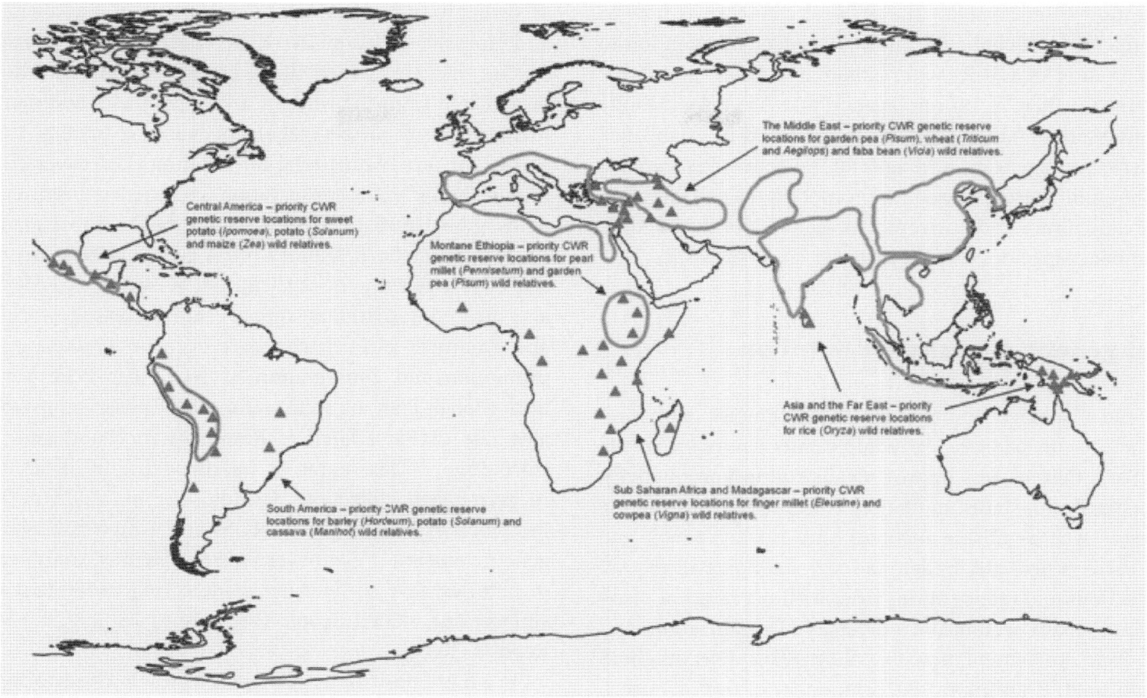


**Figure 1.3** Global priority genetic reserves locations for CWRs of 12 crops. *Source*: Maxted and Kell (2009). Reproduced with permission

299 sub-specific taxa and identified the regions with highest CWR presence in Western Asia (262), then China (222) and southeastern Europe (181). In addition, there are 203 countries that have at least one global priority native CWR (Vincent *et al.*, 2013). Maxted and Kell (2009) also identified priority locations to implement genetic reserves to conserve CWRs of 12 selected crops that may result in a global network of *in situ* conservation areas targeting CWRs and the development of guidelines for site selection and management of these resources (Maxted and Kell, 2009). For example, high-priority locations were identified in sub-Saharan Africa for the conservation of two high-priority wild relatives of finger millet (in Burundi, Democratic Republic of Congo, Ethiopia, Kenya, Rwanda, Uganda), one pearl millet wild relative (in Sudan), two garden pea wild relatives (in Ethiopia) and numerous cowpea wild relatives in several African countries (Maxted and Kell, 2009) (Figure 1.3).

CWRs represent an important reservoir of genetic resources for breeders (Maxted and Kell, 2009). Many useful traits from CWRs, such as pest and disease resistance, abiotic stress tolerance or quality improvements, have been introgressed in today's crops (Hajjar and Hodgkin, 2007). Like other wild plant species, CWRs are currently suffering genetic erosion – some entire CWR taxa are being lost, while for others their genetic diversity is being reduced or is shifting in response to environmental changes (Maxted *et al.*, 2012). There is an increasing awareness about the importance of CWRs and the role they play in ensuring food security. This has prompted IUCN to prioritize threat assessment of CWRs. The first extensive IUCN Red List assessment of CWR diversity has recently been published for European species (Kell *et al.*, 2012). In total, 571 native European CWRs of high-priority human and animal food-crop species were assessed: 313 (55%) were assessed as least concern, 166 (29%) as data deficient, 26 (5%) as near threatened, 22 (4%) as vulnerable, 25 (4%) as endangered and 19 (3%) as critically endangered. Many countries have also carried out inventories of CWRs (Idohou *et al.*, 2012). An analysis of the floras from the Arabian Peninsula shows that there are over 400 wild relatives of some 70 food and forage crops (Rao, 2013). The attention to the conservation of CWRs is also emphasized within the FAO second Global Plan of Action for the conservation and sustainable utilization of plant genetic resources for food and agriculture (FAO, 2012), which includes conservation of CWRs as a priority area, and Article 5 of the International Treaty on Plant Genetic Resources for Food and Agriculture (ITPGRFA) (FAO, 2002). The tenth Conference of the Parties to the Convention on Biological Diversity (CBD) also underlined the importance of CWRs in their Strategic Plan 2011–2020 agreed in Nagoya: 'Target 13: By 2020, the status of crop and livestock genetic diversity in agricultural ecosystems and of wild relatives has been improved.' Indicators to monitor the *in situ* conservation of CWRs and wild food plants have been developed (CGRFA, 2013).

Conservation interventions for CWRs have also gained importance. In Ethiopia, wild populations of *Coffea arabica* L. are conserved in the montane rainforest, and a study of their genetic diversity and economic value is being carried out in order to develop models for conserving *C. arabica* genetic resources both within and outside protected areas (Gole *et al.*, 2002). Bioversity International's UNEP/GEF project '*In situ* conservation of crop wild relatives through enhanced information management and field application' has made significant advances in promoting the *in situ* conservation of over 39 CWRs in protected areas in Armenia, Bolivia, Madagascar, Sri Lanka and Uzbekistan (Hunter and Heywood, 2011). Within *ex situ* collection, they represent only between 2 and 18% of accessions, depending on estimates (Khoury *et al.*, 2010). The Global Crop Diversity Trust in partnership with Royal Botanic Gardens Kew has embarked on an ambitious project funded by the Norway Government to prioritize CWR population for collecting and conservation in genebanks. The greatest challenge is how to ensure that relevant national authorities give adequate attention to *in situ* conservation of CWRs within their territories, given that in many countries there is no one agency that has responsibility for their conservation – they are outside the remit of established nature conservation agencies, and agricultural ministries have no conservation remit. There is a need for promoting the collaboration between the agriculture and environment sectors and for building local and national capacities for *in situ* conservation of CWRs.

## 1.3 Plant genetic diversity for nutrition

Addressing the challenge of malnutrition is complex and multifaceted. There is no panacea. It requires interventions that are interdisciplinary, that address improving agricultural productivity, dietary diversity, health status, sanitation, education, infrastructure and markets as well as raising incomes and moving people out of poverty. There is no silver bullet, and any solution will require appropriate actions and interventions from within a range of disciplines and across many sectors and a diversity of organizations and institutions, from implementation to policy. Plant diversity, both wild and cultivated, is an important component of our arsenal in efforts to improve nutrition and health. Staple crops such as wheat, maize and rice have been crucial in supporting global food security to the present day and will be equally important in the future. However, there is a wider plant diversity (neglected and underutilized species (NUSs)) that is largely untapped and which has high nutritional significance and is rich in proteins, fats, vitamins, minerals, antioxidants, nutraceuticals and other beneficial phytochemicals which, if made available and utilized effectively, could contribute significantly to enhancing dietary diversity, improving nutrition and health, as well as the livelihoods and well-being of millions of individuals in communities all over the world, both developed and developing. Unfortunately, there are numerous barriers that have hindered the effective and sustainable utilization of this wider plant diversity which has seen much of it relegated to a relatively minor role in agriculture, though important in local and regional food systems.

### 1.3.1 Neglected and underutilized plant diversity in addressing nutrition

Besides the major crops, there are also over 7000 crops species which are important at local and national levels. Many of those have historically been marginalized by mainstream agricultural research (Padulosi *et al.*, 2012). These so-called NUSs are known to play an important role in food security, nutrition, health, income generation and cultural practices (Jaenike and Hoeschle-Zeledon, 2006). Padulosi and Hoeschle-Zeledon (2004) described the main features that NUSs have in common. NUSs are an integral part of local culture, are present in traditional food preparations and are the focus of current trends to revive culinary traditions. They are highly adapted to agro-ecology niches and marginal areas, and are represented by ecotypes and landraces. Underutilized plant species are also cultivated and used by drawing on indigenous local knowledge. They are not well represented in *ex situ* genebanks and are characterized by a poor or non-existent local seed supply system and which renders them inaccessible (Padulosi and Hoeschle-Zeledon, 2004). NUSs encompass a variety of plant species that are farmed (local crops), reared (semi-domesticates) or gathered from the wild for a variety of uses and may contribute to nutrition (food, beverage), medicine, cosmetics, fodder, fibres or fuel, or provide material for building (Delêtre *et al.* 2013). The importance of NUS conservation and use has been recognized by many global initiatives, such as the FAO Global Plan of Action on Plant Genetic Resources (GPA) (FAO, 1996), Agenda 21 and the Global Forum for Agricultural Research (GFAR). There has also been international support for increased work on NUSs, and a number of initiatives have been undertaken. Although the promotion and conservation of NUSs has been part of FAO's first Global Plan of Action for the Conservation and Sustainable Use of Plant Genetic Resources for Food and Agriculture since 1996, NUSs represent less that 20% of all accessions held in germplasm collections (Padulosi *et al.*, 2002a). Inadequately described or characterized, NUSs are at high risk of cultural and genetic erosion (Vietmeyer, 1986). Further, only very limited inventories have so far been made owing to the lack of financial support and skilled people (Padulosi *et al.*, 2002a; FAO Country Reports, 2009).

On a more positive note, there is a growing body of evidence that many of the neglected and underutilized species and cultivars are nutritionally superior compared with their mainstream agriculture counterparts. Nutrient composition can differ dramatically between species and among cultivars of the same species. For example, sweet potato cultivars can vary in their carotenoid content by a factor of 200 or more (Box 1.2); protein content of rice cultivars can range from 5% to 13 % by weight; provitamin-A carotenoid content of bananas can be less than 1 μg/100 g for some cultivars and as high as 8500 μg/100 g for others (Box 1.3), meaning the consumption of one cultivar as opposed to another could be the difference between micronutrient deficiency and micronutrient adequacy (Burlingame *et al.*, 2009).

**Box 1.2. Orange-Fleshed Sweet Potatoes and Vitamin A Deficiency**

Intake of adequate vitamin A is critical for good health. Deficiency can limit growth, undermine immunity and lead to blindness, and increased mortality. Deficiency of vitamin A is particularly widespread among young children in sub-Saharan Africa. Food-based efforts to combat vitamin A deficiency aim to improve access to and intake of vitamin-A-rich foods. Among plant sources, orange-fleshed sweet potatoes (OFSP) have been demonstrated to have good to excellent amounts of beta-carotene, which is highly bioavailable. As little as 100 g of boiled or steamed OFSP can meet the daily-recommended intake levels of vitamin A of a child under 5 years old. Furthermore, unlike a number of vegetables, sweet potato also contains significant amounts of energy as well as vitamin A. In sub-Saharan Africa, the majority of cultivated landraces are white-fleshed, lacking in beta-carotene. One approach to reducing vitamin A deficiency in the region is the promotion of OFSP, and this intervention is based around working with households and communities to make changes in their sweet potato production and consumption practices to enhance the eating of orange-coloured cultivars instead of, or in conjunction with, white cultivars.

*Source:* Low *et al.* (2013)

**Box 1.3. Going Local: Promoting Nutritious Plant Genetic Resources in the Federated States of Micronesia**

The Federated States of Micronesia (FSM) has recently witnessed significant dietary shifts and increasing dependence on imported, often unhealthy, foods. Traditionally, the islands of FSM have achieved food and nutrition security through sustainable agriculture practices reliant on nutritious staples such as roots and tubers, breadfruit and banana. More recently, less nutritious, cheaper imported foods have contributed to meeting demand for more diversity of products; however, this is contributing to the poor health of the population. Nutritional and health indicators for FSM, though limited, are alarming. Vitamin A deficiency has been recorded among the highest in the world. Furthermore, over 70% of Pohnpei State adults are overweight (with around 43% obese) and about 32% of Pohnpei adults have diabetes. This despite the remarkable diversity of nutritious plant foods available in the country. Pohnpei alone has 133 cultivars of breadfruit, 55 bananas, 171 yams, 24 giant swamp taros, 9 tapiocas and many pandanus cultivars documented. To address the health and food system problems that have arisen, efforts were initiated in 1998 to identify local plant foods that could be promoted to alleviate the vitamin A deficiency problem. Local biodiversity experts highlighted the rare karat and other yellow-fleshed banana cultivars. Subsequent, compositional analyses demonstrated that karat, a cultivar traditionally given to infants, was rich in beta-carotene, the most important of the provitamin A carotenoids, with amounts much higher than in common white-fleshed bananas (a similar finding to that in yellow- and white-fleshed sweet potato, Box 1.1). The karat cultivar soon received international acclaim for its rich nutrient content, creating additional interest within FSM. Further studies have shown that there are many additional yellow-fleshed cultivars of banana, giant swamp taro, breadfruit and pandanus that are rich in beta-carotene and other carotenoids, nutrients and fiber and which could be an important part of food-based approaches to addressing nutritional problems in FSM, as well as other parts of the Pacific. The Island Food Community of Pohnpei, a national nongovernmental organization, has been working to promote the production, consumption and local marketing of local nutritious plant diversity through it's 'Let's Go Local' national campaign.

*Source:* Englberger and Johnson (2013)

The importance and plight of NUSs can be exemplified by the case study on the African leafy vegetable (ALV). More than 200 ALV species are used across the subSaharan Africa region (Chweya and Eyzaguirre, 1999). The Plant Resources of Tropical Africa (PROTA) reported an estimated 6376 useful indigenous African plants, of which 397 are vegetables. Guarino (1997) reported that there are more than 20 leafy vegetable species specific to Africa that are used in daily diets and are of nutritional importance. Being accessible to the low-income communities, ALVs play a crucial role in food security and in improving the nutritional status of poor families in many parts of Africa. Significant levels of provitamin A and vitamin C, proteins, several mineral micronutrients, other micronutrients and nutraceuticals are found in many ALVs (Box 1.4). Compared with more expensive and single nutrient-focused biofortification and supplementation interventions, these indigenous vegetables offer a more sustainable approach to supplying a range of nutrients, particularly for poor households and vulnerable groups such as pregnant and nursing mothers (Ojiewo *et al.*, 2013).

**Box 1.4. African Indigenous Vegetables to Improve Nutrition**

Indigenous vegetables like spiderplant, roselle and hair lettuce are known to be important sources of iron, while others like moringa (*Moringa oleifera*), African nightshade (*Solanum scabrum*) *and* jute mallow (*Corchorus olitorius*) are excellent sources of provitamin A. Studies have estimated that around half of all vitamin A requirements and about a third of iron requirements may be provided by the consumption of indigenous vegetables within poor households in Africa. Many of these indigenous vegetables also contain a rich diversity of beneficial nutraceuticals, such as allylic sulphides, flavonoids, isothiocyanates, lycopene and phenolic acids, some of which are antioxidants and play an important role in enhancing immunity and disease prevention. Furthermore, beneficial antiviral, antibacterial, anti-inflammatory and anti-mutagenic activities have been reported for some indigenous vegetables.

*Source:* Ojiewo *et al.* (2013), compiled from other sources

Edible aroids are grown globally throughout the humid tropics and include *Colocasia esculenta* (L.) Schott (taro), *Xanthosoma sagittifolium* (L.) Schott (cocoyam), *Cyrtosperma merkusii* (Hassk.) Schott (giant swamp taro) and *Alocasia macrorrhiza* (L.) G.Don (giant taro). They are very ancient crops and about half a billion people include them in their diets, as staples, subsistence foods or vegetables (Rao *et al.*, 2010). They represent an important source of energy, minerals and vitamins. Their fresh leaves are also consumed and are very rich in proteins. For example, in the case of taro, its corms are baked, roasted or boiled and the leaves are frequently eaten as a vegetable and represent an important source of vitamins, especially folic acid. The blades and petioles of taro leaves can be preserved or dried and are an important food in times of scarcity. Petioles and stolons are also eaten fried or pickled. The inflorescence is a delicacy in some food cultures of Asia and the Pacific. The corms and leaves are also used for medicinal purposes. Recent initiatives have targeted and identified nutrient-rich giant swamp taros which could be promoted to address serious nutrition deficiency problems in parts of Oceania. $\beta$-Carotene concentrations of 34 *Cyrtosperma merkusii* cultivars recently analysed varied from 50 to 4486 μg/100 g, with those yellow-fleshed cultivars largely containing higher carotenoid concentrations (Englberger *et al.*, 2003). *Colocasia* taro cultivars have also been analysed for carotenoid content and display high values (Champagne *et al.*, 2010).

Despite these good attributes, ALVs have been generally neglected by research and extension services, and this is one of the reasons why their diversity has become threatened. In addition, ALVs are associated with poor rural lifestyle and are thus regarded as a low-status food. Cultural changes and urbanization have further led to the neglect of these plants in many parts of Africa. ALVs are renowned as a crop being grown, processed and sold by women. In Cameroon, for instance, leafy vegetables constitute important sources of income for rural women who can be either farmers themselves or middle men (buyam–sellam) (Pouboum, 1999).

In India, minor millets comprise a group of small-seeded cereals represented by seven species; namely, finger millet (*Eleusine coracana* (L.) Gaertner), kodo millet (*Paspalum scrobiculatum* L.), foxtail millet (*Setaria italica* (L) Pal.), little millet (*Panicum sumatrense* Roth *ex* Roemer & Schultes), proso millet (*Panicum miliaceum* L.) and barnyard millet (represented by two species: *Echinochloa crus-galli* (L.) Beauv. and *E. colona* (L.) Link). Nutritionally, minor millets are comparable or even superior to other staple cereals, such as wheat and rice (Box 1.5) (Bergamini *et al.* 2013).

**Box 1.5. Minor Millets and Improved Nutrition in India**

It has been estimated that 100 g of cooked grain of foxtail millet contains nearly twice the protein of the equivalent amount of rice, finger millet over 38 times the amount of calcium, and little millet more than nine times the amount of iron. Minor millets are also rich in vitamins, sulphur-containing amino acids and certain phytochemicals and are sometimes referred to as *nutricereals*. They also lack gluten and are rich in non-starchy polysaccharides and dietary fibre, and their slow release of sugar on consumption means they are considered suitable for diabetics and for celiac-affected people. Processing, such as malting of finger millet, can improve the energy value of the grain and enhance the bioavailability of its protein, calcium and iron, as well as the content of vitamins such as niacin and folic acid. Preliminary studies into the nutritional impact of minor millets in school feeding programmes in Karnataka State in India show promise. After a 3-month project intervention, findings highlighted a significant improvement among school children with respect to weight and haemoglobin content in children fed on millets compared with a control group fed on rice.

*Source:* Bergamini *et al.* (2013), compiled from other sources

Forest-based plant diversity is equally important in supplying crucial nutrients and other components that are usually not supplied by staple foods alone, and many of the micronutrients provided by forest foods are nutrients commonly lacking from diets and important for health and development functions (Box 1.5). Furthermore, forest ecosystems seem to provide more diversified wild foods, compared with other land use types within the same region, and a recent study of village-based households in the East Usambara Mountains (Tanzania) has demonstrated that individuals using foods from forests had significantly more diverse and nutrient-dense diets than individuals who had not consumed forest foods (Powell *et al.*, 2011). The contribution of forest-based plant diversity to nutrition and sustainable diets has recently been reviewed by Vinceti *et al.* (2013), though Colfer *et al.* (2006) have highlighted that data on the nutrient content of many indigenous fruits are still either unavailable or unreliable. A review of the current knowledge of nutrient composition of selected indigenous fruits from sub-Saharan Africa has been prepared by Stadlmayr *et al.* (2013). Consumption of fruit in sub-Saharan Africa is estimated to be significantly below the recommended daily amount. Furthermore, about 60% of the population in sub-Saharan Africa, mostly women and children, suffer some form of undernutrition, with iron and vitamin A deficiency the most common manifestation. Indigenous fruit trees (IFTs) with substantial content of vitamins and minerals have considerable potential in addressing these problems by contributing to the supply of micronutrients to local households (Table 1.2). For example, consumption of just 40–100 g of the berries from the *Grewia tenax* (Forrsk.) Fiori tree could provide close to the daily iron requirement of an under-8-year-old child.

Despite their importance and contribution to nutrition security, wild foods have had limited acknowledgement in official statistics and food and nutrition policy (Bharucha and Pretty, 2010). A recent survey by these authors, summarizing information from 36 studies in 22 countries, highlights that wild biodiversity still plays an important role in local contexts with around 90–100 wild species known per location. Several of the studies reviewed by Bharucha and Pretty (2010) have demonstrated that wild plant foods are good sources of important micronutrients. For example, several edible desert plants are important providers of essential fatty acids, zinc, iron and calcium in the Sahel region. While in the arid Ferlo region of Senegal, about half of all edible plants have components that are considered edible and which are important sources of vitamins A, B2 and C. The plants commonly used by the Fulani in Nigeria during the dry season have been shown to be nutritionally superior in their energy and micronutrient content compared with species utilized during the wet season and can therefore play a critical role in year-round nutrition security. The use of wild plant diversity

**Table 1.2** Nutrient contents of some African indigenous and exotic fruits per 100 g edible portion (high values are highlighted in bold). RE is retinol equivalents

| Species | Energy (kcal) | Protein (g) (g) | Vit C (mg) | Vit A (RE) (μg–) | Iron (mg) | Calcium (mg) |
|---|---|---|---|---|---|---|
| **Indigenous fruits** | | | | | | |
| *Adansonia digitata L.* | 327 | 2.5 | 126–**509** | 0.03–0.06 | 6.2 | 275 |
| *Grewia tenax* (Forrsk.) Fiori | N.A. | 3.6 | N.A. | N.A. | **7.4–20.8** | **610** |
| *Sclerocarya birrea* Hochst. | 225 | 0.7 | 85–**319** | 0.035 | 3.4 | 35 |
| *Tamarindus indica* L. | 275 | 3.6 | 11–20 | 0.01–0.06 | 3.1 | 192 |
| *Ziziphus mauritiana* Lam. | 184 | 0.4 | 3–14 | 0.07 | 0.8 | 23 |
| **Exotic fruits** | | | | | | |
| Guava (*Psidium guajava* L.) | 68 | 2.6 | **228.3** | 0.031 | 0.3 | 18 |
| Mango (*Mangifera indica* L.) | 65 | 0.5 | 27.7 | 0.038 | 0.1 | 10 |
| Orange (*Citrus sinensis* L. Osbeck) | 47 | 0.9 | 53.0 | 0.008 | 0.1 | 40 |
| Pawpaw (*Carica papaya* L.) | 39 | 0.6 | 62.0 | **0.135** | 0.1 | 24 |

*Sources*: Kehlenbeck *et al.* (2013), Stadlmayr *et al.* (2013), Vinceti *et al.* (2013). Reproduced with kind permission from Earthscan, from Routledge and Barbara Vincenti.

in many societies forms an important component of indigenous knowledge systems (Slikkerveer, 1994) and contributes significantly to indigenous peoples' food systems (Box 1.6) (Kuhnlein *et al.*, 2009). A study recently conducted by Bioversity International has demonstrated the value and importance of the use of wild and neglected and underutilized foods in the diet of local populations of agro-pastoralists relying on supplementing their diet with food collected from neighbouring forests and fields in the eastern region of Baringo District in Kenya (Termote *et al.*, 2013). Five wild fruit and vegtable species (*Solanum nigrum* L., *Balanites aegyptiacus* (L.) Delile, *Ximenia americana* L., *Berchemia discolor* (Klotzsch) Hemsl. and *Ziziphus mauritiana* Lam.) based on the nutrient content were included in Save the Children's Cost of Diet tool, which estimated the lowest cost diet that meets the energy requirements and recommended nutrient intakes for mothers and children aged 6–24 months (LACON diet tool). The results of the study showed that adding the five selected wild foods to the LACON diet resulted in a reduction of the cost of the diet of up to 64% for children aged 12–24 months in the dry season, as well as meeting the recommended intakes for iron for women and children between 12 and 24 months in both seasons.

**Box 1.6. Indigenous Peoples' Food and Nutrition Systems**

A procedure for documenting indigenous people's food systems, both wild and domesticated, was developed by scientists working at the Centre for Indigenous Peoples' Nutrition and Environment (CINE) at McGill University, Canada, and the FAO. The approach was applied in case studies covering 12 indigenous communities in different global regions, including Ainu (Japan), Awajun (Peru), Baffin Inuit (Canada), Igbo (Nigeria), Ingano (Colombia), Karen (Thailand), Maasai (Kenya), Nuxalk (Canada) and Pohnpei (Federated States of Micronesia). Study highlights included documentation for the first time ever of the complexity of the Ingano diet, demonstrating the use of over 160 types of food ranging from roots to insects to palm tree products. The plants, *milpesos* palm, *yoco* liana, bitter cane and *cayamba* mushroom were found to be a particular priority for maintaining local health. In Pohnei, the study revealed a major diversity of local foods (381 items) and uncovered orange-fleshed local banana and pandanus cultivars very rich in carotenoids. The nutritional contribution of the Dalit food system was found to be reliant on many wild plant foods. Wild fruit species in particular were found to create a respect for the surrounding environment and its management. In all, a total of 329 plant species or cultivars providing food were found in the survey.

*Source:* Kuhnlein *et al.* (2009)

A comprehensive review of the contribution of agricultural biodiversity, including plant diversity, to nutrition and health, has recently been compiled by Heywood (2013), as has the role of underused plant food sources in supplying key nutrients by Kuyper *et al.* (2013).

### 1.3.2 Barriers to promoting plant diversity for improved nutrition

As highlighted earlier in this chapter, the use of wider plant diversity as part of food-based approaches to improving nutrition is certainly no panacea. There are many valid reasons why much of the plant diversity with nutritional potential is in fact underutilized. These barriers limiting the better promotion and use of plant diversity to address nutritional issues have recently been reviewed by many authors, including Hunter and Fanzo (2013), Heywood (2013), Kuyper *et al.* (2013) and Blasbalg *et al.* (2011). The more notable of these barriers are: (a) lack of innovative cross-sectoral and interdisciplinary approaches between agriculture and health and other sectors to better promote and mainstream plant diversity for improved nutrition (McEwan *et al.*, 2013); (b) continued neglect by international and national research and extension systems of the vast majority of nutritious plant diversity; (c) negative cultural perceptions and attitudes to local, traditional nutritionally rich plant diversity; (d) lack of reliable methods and information for linking plant diversity to dietary diversity and improved nutrition outcomes (Berti and Jones, 2013; Remans and Smukler, 2013); (e) poorly developed infrastructure and markets for the majority of plant diversity for food and nutrition; (f) inadequate agricultural and food security policies and strategies which promote major cereal staples at the expense of the dietary role of more nutritious plant diversity, such as minor millets, indigenous fruits and vegetables and root and tubers; (g) non-tariff trade barriers and strict food safety assessment regulations, such as the European Union's Novel Foods Regulation (NFR), which places a considerable burden of proof on those attempting to bring traditional plant diversity and foods, and their products to markets (Hermann, 2013); (h) lack of evidence demonstrating or comparing the most (cost-) effective methods and approaches for mobilizing plant diversity for delivering dietary and nutrition outcomes; and finally (i) poor information management of plant diversity for nutrition, where relevant information is often highly fragmented, scattered in various publications and reports or not easily accessible databases.

Just taking the last two barriers – approaches to mobilizing plant diversity for improved nutrition and poor information management – there is work currently underway which demonstrates that these barriers are by no means insurmountable and that progress can be made towards influencing policy and programmes about the benefits of food-based approaches using plant diversity to improve nutrition. Fanzo *et al.* (2013) recently carried out a global survey of current approaches and interventions to mobilize agricultural biodiversity, including plant diversity, for improving diets and nutrition that highlights many lessons learned, good practices and suggestions for scaling up activities and interventions. Innovative approaches promoting and mobilizing plant diversity for improving nutrition are also underway in relation to the development of sustainable diets (Fanzo *et al.*, 2012) and the use of innovative research tools such as 'Cost of Diet' to better understand the role of wild, neglected and underutilized plant foods in reducing the cost of a nutritionally adequate diet hold out much promise (Termote *et al.*, 2013).

Regarding the second highlighted barrier, though emerging scientific data on the nutritional and non-nutrient bioactive attributes of indigenous and traditional plant diversity indicate their immense nutritional value there still remain significant challenges related to data accessibility, accuracy and availability. Major gaps in data still also exist. Existing *food composition databases and tables*, which are usually country specific, provide information about nutritional characteristics of the main conventional foods. They often do not include wild, local and indigenous plant species and cultivars, and thus do not represent the diversity of local food systems and often omit significant variations in nutrient composition among different cultivars of plant species used for food. Filling these data gaps requires a systematic investigation of traditional, local and wild foods at the species and cultivar levels. In addition to the lack of nutritional composition data, there are also challenges related to the accessibility and accuracy of the existing composition data of much plant diversity (Nesbitt *et al.*, 2010).

Despite these limitations, significant efforts are underway to improve data management and availability of the nutritional composition of plant diversity. INFOODS is a global network of national food composition data centres, coordinated by the FAO, with the objective to improve the quality and availability of food analysis data. It provides an international framework aiming to generate and improve the usability of the data by addressing various issues related to the standardization of different aspects of data collection, analysis, compilation and publishing. INFOODS guides global activities related to data publishing, tool development, capacity building and data management at the global level. The national centres are grouped around 11 regional data centres covering all regions of the world. In 2012, the FAO in collaboration with the West African Health Organization (WAHO) of the Economic Community of West African States (ECOWAS) and Bioversity International, published the *West African Food Composition Table*, which contains details of the composition of 472 foods from the Western Africa region. The publication also includes nutrient data on the various ways foods are prepared (i.e. boiled, grilled and stewed) and genetic diversity information on local food cultivars (Stadlmayr *et al.*, 2012).

Furthermore, the FAO have also developed nutrition indicators for biodiversity on food composition and food consumption in an effort to encourage countries and scientists to generate, compile and disseminate cultivar-specific nutrient composition data and dietary consumption data. The indicators are a measure of biodiversity in food and are intended to become a tool to promote awareness of the importance of food biodiversity (wild, indigenous and traditional foods) to nutrition security. The nutrition indicator for biodiversity on food composition is a count of the number of foods with a sufficiently detailed description to identify genus, species, subspecies and cultivar/breed, and with at least one value for a nutrient or other bioactive component as found in the published and unpublished literature. The nutrition indicator for biodiversity on food consumption is a count of the number of foods being consumed and taxonomically described at genus, species, subspecies and cultivar/breed levels.

## 1.4 Plant diversity for medicines

Thousands of plants are or have been used as medicines in home remedies, ritual cures, poisons and tonics. In this section we provide a review of what is known of the diversity of plants used as medicines, and a summary of challenges to the ongoing use of medicinal plant biodiversity. A comprehensive list of all plant species used as medicines does not exist, notwithstanding many efforts over time to list and describe the uses of these species. Descriptions of plants and their medicinal uses, commonly called 'herbals' (Singer, 1927), are among the earliest books produced in ancient times (Stuart, 1979). In modern times, these references have been replaced by pharmacopoeias that provide specifications for identification and preparation of drugs derived from plant and non-plant sources, published by national, regional and international medical expert authorities (WHO, 2011). Among numerous efforts to identify and describe globally important medicinal plants (e.g. WHO, 1999, 2002b, 2007, 2009; van Wyk and Wink, 2004), the NAPRALERT® database, launched in 1975, provides a compendium of published information about natural products covering more than 250 plant genera, including more than 15 000 species with known medicinal uses and potential applications (University of Illinois Board of Trustees, 2012). Penso (1980) produced an inventory of 21 000 species for the World Health Organization (WHO); a revised list compiled by Penso and Proserpio (1997) included 12 305 taxa.

The number of medicinal plants in the world is a moving target, constantly adjusted by ongoing documentation of ethnomedicines, identification of new chemical constituents and applications, and discovery of known compounds in new taxa. Defining this number is further complicated by the lack of clear boundaries between medicines, foods and spices, and cosmetics as plant uses that contribute to health and well-being. The WHO defines a medicinal plant as '[a] plant (wild or cultivated) used for medicinal purposes' (WHO, 2003b) and defines herbal medicines broadly to include herbs, herbal materials, herbal preparations and finished herbal products (WHO, 2002a). Plants with aromatic properties overlap so extensively with those having medicinal properties that much relevant work on their diversity and use treats medicinal and aromatic plants as a single group (Leaman *et al.*, 1999; Padulosi *et al.*, 2002b).

Based on the NAPRALERT database and WHO inventories, Farnsworth and Soejarto (1991) estimated that 14–28% of the world's plant species is currently or has in the past been used in folk medicine. Extrapolating from the proportion of national floras included in 15 national pharmacopoeias, Schippmann *et al.* (2006) estimated that 50 000–70 000 species of higher plants are used as medicine worldwide, approximately 20% of the estimated global flora. A global checklist of medicinal plants is now being compiled from a wide range of standard references, pharmacopoeias and databases to support the work of the Medicinal Plant Specialist Group of the Species Survival Commission, IUCN (Biodiversity Indicators Partnership, 2010). The global checklist currently includes more than 27 000 accepted species names and more than 3500 synonyms linked to a medicinal use referenced in at least one publication. These species are distributed among five divisions, eight classes and more than 400 families in the plant kingdom. Addition of new sources is an ongoing process that is expected to substantially increase the number of taxa included.

This great diversity of taxa is indicative of the phytochemical diversity found in medicinal plants (Lewis and Elvin-Lewis, 1977; Evans, 1989). Selection of plants for use as medicine throughout human history is likely to reflect the distribution and diversity of disease, diverse cultural attitudes about treatment and response, as well as the diversity and availability of local flora (Etkin, 1988). The active principles of plant-based medicines are secondary metabolites, compounds that are biosynthetically derived from primary metabolites but have no apparent function in primary metabolic processes such as photosynthesis, mineral nutrition and carbohydrate metabolism. Biologically active secondary metabolites are believed to have evolved in plants as chemical defences against attack by herbivores and pathogens, and such compounds have been shown to have a broad range of inhibitory, pharmacological and toxic effects (Rosenthal and Janzen, 1979; Luckner, 1984; Harborne,

1988). Eldridge *et al.* (2002) speculated that each plant species may contain 500–800 different secondary metabolites.

The modern development of many important drugs currently in use is based on traditional use of active principles extracted from plants (Lewis and Elvin-Lewis, 1977; Farnsworth *et al.*, 1985). Farnsworth (1990) and Newman *et al.* (2003) have identified 135 drugs in current use that are derived from plants, the majority associated with long-established traditional uses. Research protocols for testing and improving the efficacy, quality and safety of traditional ; remedies (Bodeker, 1996; Dhawan, 1997; Walker, 1997) are supporting increased use of traditional remedies in formal health systems (Balasubramaniam, 1997; Eisenberg *et al.*, 1993; Fleurentin, 1998; Steinhoff, 1998).

Compounds from many other plant species used as traditional medicines have contributed indirectly to modern pharmacology as phytochemical leads for development of more complex drugs, as models for synthetic drugs and as taxonomic pointers to other plant sources of useful drugs (Oldfield, 1981; Plotkin, 1988; Principe, 1991). Over the last 50 years, technological advances in screening and analysis of biological activity from natural sources has enabled larger scale chemical exploration of the world's flora (Miller, 2011). Challenges contributing to a lower than anticipated success of these discovery efforts in identifying new candidates for drug development range from limitations in the preparation and handling of plant samples to the need for research institutions to develop and apply new ethical frameworks for access to plant genetic resources and sharing the benefits from their use as required by the United Nations Convention on Biological Diversity (Miller, 2011; Williams, 2011). However, considering that the vast majority of plant species have not yet been evaluated for their potential use in medicine, they remain a source of genetic diversity with great potential to meet current and future needs for medicines.

### 1.4.1 Conservation of medicinal plant diversity

Global conservation status assessment for plants lags behind assessment for animals. Of the estimated 320 000 existing species of vascular plants, 15 674 plant species (less than 5% of the estimated global flora) had been assessed using the current Red List categories and criteria (version 3.1) and included in the 2012 IUCN Red List, as compared with nearly 50 000 species of animals (IUCN, 2012). The IUCN Red List is contributing to a suite of indicators needed to measure progress towards the Millennium Development Goal to significantly reduce the rate of biodiversity loss (Butchart *et al.*, 2007; Biodiversity Indicators Partnership, 2010). A Red List Index on biodiversity used for food and medicine is included in the framework of global indicators of progress towards the 2010 biodiversity target agreed in CBD decision VIII/15 (CBD, 2006). Work on developing this indicator revealed that fewer than 3% of species included on the Global Checklist of Medicinal Plants had yet been assessed for conservation status (Biodiversity Indicators Partnership, 2010; Overgaard Larsen, 2011). Information about the distribution, abundance, ecology and genetic diversity of the great majority of medicinal plants is scarce and fragmented (Oldfield and Jenkins, 2012).

Greater attention to assessment of the conservation status of the world's plant species, including further development of a *Sampled Red List Index for Plants* (Royal Botanic Gardens, Kew, 2010) and a larger sample of assessments of medicinal plants, will make it possible to compare medicinal use with the impact of other factors that contribute to decline in conservation status of the world's plants. In South Africa, for example, 2062 indigenous plant species (10% of the total national flora) have been recorded as being used for traditional medicine (Williams *et al.*, 2013). A national assessment of South African plants completed in 2009 indicated that 4% of the medicinal flora are threatened with extinction at the national level. Calculation of a Red List Index based on the national assessment results suggests that traded species are more threatened than non-traded species, and that persistent commercial exploitation could increase the level of threat to several species currently at risk.

### 1.4.2 Ex situ conservation and cultivation of medicinal plants

Genebanks, botanic gardens and other *ex situ* conservation facilities can play a central role in conserving medicinal plant diversity by supporting inventory and conservation status assessment of medicinal plant species, preserving germplasm of rare, threatened, and other priority species, and by supplying plant material for propagation, species reintroduction, ecosystem restoration, agronomic improvement, research and education (Heywood, 1991; Leaman *et al.*, 1999). A small survey of major *ex situ* genebank collections of medicinal and aromatic plants undertaken to review and evaluate the status of *ex situ* conservation efforts for these species (Leaman *et al.*, 1999; Padulosi *et al.*, 2002b) suggested that, although most existing collections lack diversity in species and accessions, the primary focus of these collections is on species considered to be rare or at risk of extinction and species with high economic value. Botanic gardens also hold extensive *ex situ* plant collections, including seed, pollen and DNA banks, *in vitro* storage and live collections, in addition to herbarium collections. The PlantSearch database maintained by Botanic Garden Conservation International provides information and a means to access *ex situ* collection information for more than 140 000 plant taxa, of which at least 3000 are medicinal plant species (Hawkins, 2008).

An estimate that 3000 species of medicinal and aromatic plants are in international trade is acknowledged as a 'qualified guess' given the difficulty of deciphering the species-relevant information in available trade data (Schippmann *et al.*, 2006). Fewer than 1000 medicinal plant species in commercial trade appear to be cultivated (Mulliken and Inskipp, 2006). Cultivation of medicinal plants may reduce the extent and impacts of wild harvest, but may also contribute to conversion of important natural habitat for medicinal plants and other species, contribute to loss of genetic diversity of species in cultivation and reduce incentives to conserve and manage wild populations. A detailed summary of costs and benefits is provided by Schippmann *et al.* (2006).

### 1.4.3 In situ conservation and wild collection of medicinal plants

The majority of medicinal plants in international trade are wild collected, and many of those produced in cultivation are also wild collected (Mulliken and Inskipp, 2006). Medicinal plants are harvested from forests, alpine pastures and other multiple-use habitats. The general lack of information about the distribution and abundance of medicinal plant species makes it difficult to evaluate their presence in protected areas (Leaman *et al.*, 1999). Although there has not been a systematic effort to document the inclusion of medicinal plant species in protected areas, they are often mentioned in site reviews and inventories (WWF and IUCN, 1994, 1995, 1997). The high local value often placed on wild populations of medicinal plants is evident in the establishment of sacred groves and other types of community-designated protected areas (CEESP, 2008). These local values and their conservation benefits are being supported through initiatives such as the identification and conservation of important plant areas for medicinal plants (Hamilton and Radford, 2007).

A survey of the use of wild flora for improved rural livelihoods in a few high-biodiversity countries suggested that many wild plants important as livelihood resources appear to be used sustainably, but that information about the impacts of wild plant utilization is limited (Oldfield and Jenkins, 2012). Sustainable wild harvest of any resource requires a management system that is appropriate to the risk of overharvest (Peters, 1994; Cunningham, 2001). An analysis of this risk and the relevant resource assessment and monitoring requirements are being incorporated into new tools for evaluating the sustainability of wild collection of medicinal and other wild-harvested plants, such as the FairWild Standard (Kathe *et al.*, 2010; Kathe, 2011) and guidance on evaluating non-detriment of international trade to the survival of species listed on Appendix II of the Convention on International Trade in Endangered Species of Fauna and Flora (Leaman, 2008; CITES, 2010).

## Acknowledgements

DH would like to acknowledge the support of the UNEP/FAO/GEF Biodiversity for Food and Nutrition (BFN) Project and the country partners from Brazil, Kenya, Sri Lanka and Turkey. He would also like to thank Dunya Mijatovic for background research undertaken during the development of the BFN project and Barbara Vinceti for recent collaboration and discussions on the role of forests and trees in nutrition. Finally, he would like to express his sincere gratitude to all the authors who so kindly contributed to the recent Earthscan/Routledge publication *Diversifying Food and Diets*, which forms the basis of some of this chapter.

## References

Alland, A. (1966) Medical anthropology and the study of biological and cultural adaptation. *American Anthropologist*, **68**: 40–51.

Antofie, M.M. (2011) *The Red List of Crop Varieties for Romania*, Publishing House, Lucian Blaga University, Sibiu, Romania [in English and Romanian].

Balasubramaniam, K. (1997) Herbal medicine: consumer protection concerns, in *International Symposium on Herbal Medicine: A Holistic Approach. Documents, Proceedings, and Recommendations. 1-4 June 1997, Honolulu, Hawaii. San Diego, USA* (eds D.A. Wozniak, S. Yuen, M. Garrett and T.M. Schuman), International Institute for Human Resources Development, College of Health and Human Services, San Diego State University, San Diego, CA, pp. 169–223.

Barthlott, W., Mutke, J., Rafiqpoor, M.D. *et al.* (2005) Global centres of vascular plant diversity. *Nova Acta Leopoldina*, **92**: 61–83.

Bergamini, N., Padulosi, S., Bala Ravi, S. and Yenagi, N. (2013) Minor millets in India: a neglected crop goes mainstream, in *Diversifying Food and Diets: Using Agricultural Biodiversity to Improve Nutrition and Health* (eds J. Fanzo, D. Hunter, T. Borelli and F. Mattei), Earthscan, London.

Berti, P.R. and Jones, A.D. (2013) Biodiversity's contribution to dietary diversity: Magnitude, meaning and measurement, in *Diversifying Food and Diets: Using Agricultural Biodiversity to Improve Nutrition and Health* (eds J. Fanzo, D. Hunter, T. Borelli and F. Mattei), Earthscan, London.

Bharucha, Z. and Pretty, J. (2010) The roles and values of wild foods in agricultural systems. *Philosophical Transactions of the Royal Society B: Biological Sciences*, **365**, 2913–2926.

Bhattacharya, P., Bhattacharyya, R., Asokan, A. and Prasad, R. (2008) Towards certification of wild medicinal and aromatic plants in four Indian states. *Unasylva* **230** (59), 35–44.

Biodiversity Indicators Partnership (2010) *Biodiversity Indicators and the 2010 Target: Experiences and Lessons Learnt from the 2010 Biodiversity Indicators Partnership*, Secretariat of the Convention on Biological Diversity, Montréal, Canada, Technical Series No. 53.

Bioversity International (2009) Diverse diets fight malnutrition and more. Bioversity International Annual Report 2009, Rome Italy, pp 2–5, www.bioversityinternational.org/publications/search/results.html (accessed 9 May 2013).

Blasbalg, T.L., Wispelwey, B. and Deckelbaum, R.J. (2011) Econutrition and utilization of food-based approaches for nutritional health. *Food and Nutrition Bulletin* **32** (1, supplement), S4–S13.

Bodeker, G.C. (ed.) (1996) Traditional health systems and policy: the Gifts of Health reports. *The Journal of Alternative and Complementary Medicine*, **2** (3, special issue), 319–448.

Bramwell, D. (2002) How many plant species are there? *Plant Talk*, **28**, 32–34.

Burlingame, B., Charrondiere, R. and Mouille, B. (2009) Food composition is fundamental to the cross-cutting initiative on biodiversity for food and nutrition. *Journal of Food Composition and Analysis*, **22**, 361–365.

Butchart, S.H.M., Akçakaya, H.R., Chanson, J. *et al.* (2007) Improvements to the Red List Index. *PLoS ONE*, **2** (1), e140, doi:10.1371/journal.pone.0000140.

CBD (2006) Decisions adopted by the Conference of the Parties to the Convention on Biological Diversity at its Eighth Meeting, Curitiba, 20–31 March 2006. Secretariat of the Convention on Biological Diversity, Montreal, Canada.

CEESP (2008) Recognising and supporting indigenous and community conservation: ideas and experiences from the grassroots. CEESP Briefing Note 9, http://cmsdata.iucn.org/downloads/ceesp_briefing_note_9_iccas.pdf (accessed 29 April 2013).

CGRFA (2013) Targets and indicators for plant genetic resources for food and agriculture, Fourteenth Regular Session of the Commission on Genetic Resources for food and agriculture, Working document CGRFA/14/13/4.1.Rev.1, Rome, Italy, www.fao.org/nr/cgrfa/cgrfa-meetings/cgrfa-comm/fourteenth-reg/en/#c160710 (accessed 9 May 2013).

Champagne, A., Bernillon, S., Moing, A. *et al.* (2010) Carotenoid profiling of tropical root crop chemotypes from Vanuatu, South Pacific. *Journal of Food Composition and Analysis*, **23**, 763–771.

Chweya, J.A. and Eyzaguirre, P.B. (eds) (1999) *The Biodiversity of Traditional Leafy Vegetables*, International Plant Genetic Resources Institute, Rome.

CITES (2010) Non-detriment findings for timber, medicinal plants and agarwood, CoP15 Doc. 16.3, www.cites.org/eng/cop/15/doc/E15-16-03.pdf (accessed 29 April 2013).

Colfer, C.J.P., Sheil, D. and Kishi, M. (2006) Forests and human health: assessing the evidence, Occasional Paper No. 45, CIFOR.

Cunningham, A.B. (2001) *Applied Ethnobotany: People, Wild Plant Use and Conservation*, Earthscan, London.

Delêtre, M., Gaisberger, H. and Arnaud, E. (2013) Agrobiodiversity in perspective – a review of questions, tools, concepts and methodologies. Report prepared for Bioversity International and IRD-Sud Expert Plantes.

Dhawan, B.N. (1997) R & D issues and medical concerns in the development of herbal drugs, in *International Symposium on Herbal Medicine: a Holistic Approach. Documents, Proceedings, and Recommendations. 1–4 June 1997, Honolulu, Hawaii, USA* (eds D.A. Wozniak, S. Yuen, M. Garrett and T.M. Schuman), International Institute for Human Resources Development, College of Health and Human Services, San Diego State University, San Diego, CA, pp. 152–166.

Dulloo, M.E. (2013) Global challenges for agricultural biodiversity and international collaboration, in *Conservation of Tropical Plant Species* (eds M.N. Normah, H.F. Chin and B.M. Reed), Springer, New York, NY, pp 491–510.

Dulloo, M.E., Hunter, D. and Borelli, T. (2010) *Ex situ* and *in situ* conservation of agricultural biodiversity: major advances and research needs. *Notulae Botanicae Horti Agrobotanici Cluj-Napoca*, **38** (2), 123–153.

Eisenberg, D.M., Kessler, R.C., Foster, C. *et al.* (1993) Unconventional medicine in the United States: prevalence, costs, and patterns of use. *New England Journal of Medicine*, **328** (4), 246–252.

Eldridge, F.R., Vervoort, H.C., Lee, C.M. *et al.* (2002) High-throughput method for the production and analysis of large natural product libraries for drug discovery. *Analytical Chemistry*, **74**, 3963–3971.

Englberger, L. and Johnson, E. (2013) Traditional foods of the Pacific: go local, a case study in Pohnpei, Federated States of Micronesia, in *Diversifying Food and Diets: Using Agricultural Biodiversity to Improve Nutrition and Health* (eds J. Fanzo, D. Hunter, T. Borelli and F. Mattei), Earthscan, London.

Englberger, L., Aalbersberg, W., Ravi, P. *et al.* (2003) Further analysis on Micronesia banana, taro, breadfruit and other foods for provitamin A carotenoids and minerals. *Journal of Food Composition and Analysis*, **16**, 216–236.

Etkin, N.L. (1988) Ethnopharmacology: biobehavioral approaches in the anthropological study of indigenous medicines. *Annual Review of Anthropology*, **17**, 23–42.

Evans, C.E. (1989) *Trease and Evans' Pharmacognosy*, 13th edn, Ballière Tindall, London.

Fanzo, J., Cogill, B. and Mattei, F. (2012) *Metrics of Sustainable Diets and Food Systems*, Bioversity International, Rome.

Fanzo, J., Hunter, D., Borelli, T. and Mattei, F. (eds) (2013) *Diversifying Food and Diets: Using Agricultural Biodiversity to Improve Nutrition and Health*, Earthscan, London.

FAO (1996) *The Global Plan of Action for the Conservation and Sustainable Utilization of Plant Genetic Resources*, Food and Agriculture Organization of the United Nations, Rome.

FAO (1998) *The State of the World's Plant Genetic Resources for Food and Agriculture*, Food and Agriculture Organization of the United Nations, Rome.

FAO (2002) *The International Treaty on Plant Genetic Resources for Food and Agriculture*, Food and Agriculture Organization of the United Nations, Rome.

FAO (2010) *The Second Report on the State of the World's Plant Genetic Resources for Food and Agriculture*, Food and Agriculture Organization of the United Nations, Rome.

FAO (2011) *Save and Grow. A Policymaker's Guide to Sustainable Intensification of Smallholder Crop Production*, Food and Agriculture Organization of the United Nations, Rome.

FAO (2012) *The Second Global Plan of Action for the Conservation and Sustainable Utilization of Plant Genetic Resources for Food and Agriculture*, Food and Agriculture Organization of the United Nations, Rome.

FAO Country Reports (2009) Benin, Burkina Faso, Cameroon, Côte d'Ivoire, Djibouti, Guinea, Kenya, Madagascar, Mali, Namibia, Niger, Tanzania, Togo, Uganda, Zambia, Nigeria, Zimbabwe, www.fao.org/agriculture/crops/core-themes/theme/seeds-pgr/sow/en/ (accessed 31 March 2013).

Farnsworth, N.R. (1990) The role of ethnopharmacology in drug development, in *Bioactive Compounds from Plants* (eds D.J. Chadwick and J. Marsh), Ciba Foundation Symposium 154, John Wiley & Sons, Ltd, Chichester, pp. 2–11.

Farnsworth, N.R., and Soejarto, D.D. (1991) Global importance of medicinal plants, in *The Conservation of Medicinal Plants* (eds O. Akerele, V. Heywood and H. Synge), Cambridge University Press, Cambridge, pp. 25–27.

Farnsworth, N.R., Akerele, O., Bingel, A.S. *et al.* (1985) Medicinal plants in therapy. *Bulletin of the World Health Organization*, **63** (6), 965–981.

Fleurentin, J. (1998) French regulations concerning the use and commercialisation of medicinal plants, in *Medicinal Plant Trade in Europe: Conservation and Supply, Proceedings, First International Symposium on the Conservation of Medicinal Plants in Trade in EUROPE, 22–23 June 1998, Royal Botanic Gardens, Kew, UK*, TRAFFIC Europe, Brussels.

Foley, J.A., Ramankutty, N., Brauman, K.A. *et al.* (2011) Solutions for a cultivated planet. *Nature*, **478**, 337–342.

Fu, L.K. and Chin, C.-M. (1992) *China Plant Red Data Book: Rare and Endangered Plants*, vol. **1**, Science Press, Beijing.

Gabrielyan, E.Ts. (ed.) (1988) *Red Data Book of Armenia: Plants*, Yerevan (in Russian).

Godfray, H.C.J., Crute, I.R., Haddad, L. *et al.* (2010) The future of the global food system. *Philosophical Transactions of the Royal Society B: Biological Sciences*, **365**, 2769–2777.

Golding, J.S. (ed.) 2002. Southern African Plant Red Data Lists, Southern African Botanical Diversity Network Report No. 14, SABONET, Pretoria.

Gole, T.W., Denich, M., Teketay, D. and Vlek, P.L.G. (2002) Human impacts on the *Coffea arabica* genepool in Ethiopia and the need for its *in situ* conservation, in *Managing Plant Genetic Diversity* (eds J. Engels, V. Ramanatha Rao, A. Brown and M. Jackson), CAB International/International Plant Genetic Resources Institute (IPGRI), Wallingford/Rome, pp. 237–247.

Govaerts, R. (2001) How many species of plants are there? *Taxon*, **50**, 1085–1090.

Guarino, L. (ed.) (1997) *Traditional African Vegetables. Promoting the Conservation and Use of Underutilized and Neglected Crops 16. Proceedings of the IPGRI International Workshop on Genetic Resources of Traditional Vegetables in Africa. Conservation and Use*, ICRAF-HQ Nairobi.

Gupta, S.R., Upadhyay, M.P. and Katsumoto, T. (1996) Status of plant genetic resources in Nepal, in *19th Summer Crops Workshop, Regional Agricultural Research Station (RARS)*, 27–29 February, Parwanipur, Nepal.

Hajjar, R. and Hodgkin, T. (2007) The use of wild relatives in crop improvement: a survey of developments over the last 20 years. *Euphytica*, **156**, 1–13.

Hamilton, A.C. and Radford, E.A. (2007) *Identification and Conservation of Important Plant Areas for Medicinal Plants in the Himalaya*, Plantlife International/Ethnobotanical Society of Nepal, Salisbury/Kathmandu.

Hammer, K. and Khoshbakht, K. (2005) Towards a red list for crop plant species. *Genetic Resources and Crop Evolution*, **52** (3), 249–265.

Harborne, J.B. (1988) *Introduction to Ecological Biochemistry*, 3rd edn, Academic Press, Harcourt Brace Jovanovich, London.

Harlan, J. (1992) *Crops and Man*, 2nd edn, American Society of Agronomy, Madison, WI.

Hawkins, B. (2008) *Plants for Life: Medicinal Plant Conservation and Botanic Gardens*, Botanic Gardens Conservation International, Richmond, UK.

Hermann, M. (2013) Successes and pitfalls of linking nutritionally promising Andean crops to markets, in *Diversifying Food and Diets: Using Agricultural Biodiversity to Improve Nutrition and Health* (eds J. Fanzo, D. Hunter, T. Borelli and F. Mattei), Earthscan, London.

Heywood, V. (1991) Botanic gardens and the conservation of medicinal plants, in *The Conservation of Medicinal Plants: Proceedings of an International Consultation 21–27 March 1988 held at Chiang Mai, Thailand* (eds O. Akerele, V. Heywood and H. Synge), Cambridge University Press, Cambridge, pp. 213–228.

Heywood, V.H. (2013) Overview of agricultural biodiversity and its contribution to nutrition and health, in *Diversifying Food and Diets: Using Agricultural Biodiversity to Improve Nutrition and Health* (eds J. Fanzo, D. Hunter, T. Borelli and F. Mattei), Earthscan, London.

Hunter, D. and Fanzo, J. (2013) Agricultural biodiversity, diverse diets and improving nutrition, in *Diversifying Food and Diets: Using Agricultural Biodiversity to Improve Nutrition and Health* (eds J. Fanzo, D. Hunter, T. Borelli and F. Mattei), Earthscan, London.

Hunter, D. and Heywood, V.H. (eds) (2011) *Crop Wild Relatives. A Manual of In Situ Conservation*, Earthscan, London.

Idohou, R., Assogbadjo, A.E., Fandohan, B. *et al.* (2012) National inventory and prioritization of crop wild relatives: case study for Benin. *Genetic Resources and Crop Evolution*, **60** (4), 1337–1352.

IUCN (2001) IUCN Red List Categories and Criteria: Version 3.1, IUCN Species Survival Commission, IUCN, Gland/Cambridge.

IUCN (2012) The IUCN Red List of Threatened Species, 2012.2, www.iucnredlist.org (accessed 29 April 2013).

Jaenicke, H. and Hoschle-Zeledon, I. (eds) (2006) *Strategic Framework for underutilized Plant Species Research and Development with Special Reference to Asia and the Pacific and to Sub-Saharan Africa*, International Centre for Underutilized Crops/Global Facilitation for Underutilized Species, Colombo, Sri Lanka/Rome, Italy.

Jarvis, D.I., Padoch, C. and Cooper, D. (eds) (2007) *Managing Biodiversity in Agricultural Systems*, Columbia University Press/Bioversity International, New York, NY/Rome.

Johns, T. (1990) *With Bitter Herbs They Shall Eat It: Chemical Ecology and the Origins of Human Diet and Medicine*, University of Arizona Press Tucson, AZ.

Joshi, K.B., Upadhyay, M., Gauchan, D. *et al.* (2004) Red listing of agricultural crop species, varieties and landraces. *Nepal Agricultural Research*, **5**, 73–80.

Kathe, W. (2011) The new FairWild standard – a tool to ensure sustainable wild-collection of plants. *Medicinal Plant Conservation*, **14**, 14–17.

Kathe, W., Pätzold, B., Leaman, D. *et al.* (2010) *Wild for a Cure: Ground-Truthing a Standard for Sustainable Management of Wild Plants in the Field*, TRAFFIC International, Cambridge.

Kehlenbeck, K., Asaah, E. and Jamnadass, R. (2013) Diversity of indigenous fruit trees and their contribution to nutrition and livelihoods in sub-Saharan Africa: examples from Kenya and Cameroon, in *Diversifying Food and Diets: Using Agricultural Biodiversity to Improve Nutrition and Health* (eds J. Fanzo, D. Hunter, T. Borelli and F. Mattei), Earthscan, London, pp. 257–269.

Kell, S., Maxted, N., and Bilz, M. (2012) European crop wild relatives threat assessment: knowledge gained and lessons learnt, in *Agrobiodiversity Conservation: Securing the Diversity of Crop Wild Relatives and Landraces* (eds N. Maxted, M.E. Dulloo, B.V. Ford-Lloyd *et al.*), CABI Publishing, Wallingford.

Khoury, C., Laliberté, B. and Guarino, L. (2010) Trends in *ex situ* conservation of plant genetic resources: a review of global crop and regional conservation strategies. *Genetic Resources and Crop Evolution*, **57** (4), 625–639.

Kuhnlein, H.V., Erasmus, B. and Spigelski, D. (2009) *Indigenous Peoples' Food Systems: The Many Dimensions of Culture, Diversity and Environment for Nutrition and Health*, United Nations Food and Agriculture Organization, Rome, www.fao.org/docrep/012/i0370e/i0370e00.htm.

Kunkel, G. (1984) *Plants for Human Consumption*, Koeltz Scientific Books, Koenigstein, Germany.

Kuyper, E., Vitta, B. and Dewey, K. (2013) Novel and underused food sources of key nutrients for complementary feeding. *A&T Technical Brief*, Issue 6.

Lambert, J., Srivastava, J., and Vietmeyer, N. (1997) Medicinal plants: rescuing a global heritage, World Bank Technical Paper number 355, World Bank, Washington, DC.

Lange, D. (1998) *Europe's Medicinal and Aromatic Plants: Their Use, Trade and Conservation*, TRAFFIC International, Cambridge.

Leaman, D.J. (2008) The International Standard for the Sustainable Wild Collection of Medicinal and Aromatic Plants (ISSC-MAP): elements of the standard relevant to CITES NDF, in *International*

*Expert Workshop on CITES Non-Detriment Findings Perennial Plant Working Group (Ornamentals, Medicinal and Aromatic Plants)*, Cancun, Mexico.

Leaman, D.J., Fassil, H. and Thormann, I. (1999) Conserving medicinal and aromatic plant species: identifying the contributions of the International Plant Genetic Resources Institute (IPGRI), Unpublished report, IPGRI, Rome.

Lewis, W.H. and Elvin-Lewis, M.P.F. (1977) *Medical Botany: Plants Affecting Man's Health*, John Wiley & Sons, Inc., New York, NY.

Low, J., Arimond, M., Labarta, R. *et al.* (2013) The introduction of orange fleshed sweet potato in Mozambican diets: a marginal change to make a major difference, in *Diversifying Food and Diets: Using Agricultural Biodiversity to Improve Nutrition and Health* (eds J. Fanzo, D. Hunter, T. Borelli and F. Mattei), Earthscan, London.

Lucas, G. and Synge, H. (1978) *The IUCN Plant Red Data Book*, IUCN, Gland.

Luckner, M. (1984) *Secondary Metabolism in Microorganisms, Plants, and Animals*, 2nd edn, Springer-Verlag.

Maxted, N. and Kell, S.P. (2009) *Establishment of a Global Network for the In Situ Conservation of Crop Wild Relatives: Status and Needs*, FAO Commission on Genetic Resources for Food and Agriculture, Rome.

Maxted, N., Ford-Lloyd, B.V., Jury, S.L. *et al.* (2006) Towards a definition of a crop wild relative. *Biodiversity and Conservation*, **15** (8), 2673–2685.

Maxted, N., Kell, S., Ford-Lloyd, B. *et al.* (2012) Towards systematic conservation of global crop wild relative diversity. *Crop Science*, **52**, 1–12.

McEwan, M., Prain, G. and Hunter, D. (2013) Opening a can of *Mopane* worms; can cross sectoral partnerships leverage agricultural biodiversity for better quality diets?, in *Diversifying Food and Diets: Using Agricultural Biodiversity to Improve Nutrition and Health* (eds J. Fanzo, D. Hunter, T. Borelli and F. Mattei), Earthscan, London.

Meyer, A. and Vögel, R. 2006. Rote Liste für gefährdete Kulturpflanzen in Deutschland, Handlungsmöglichkeiten und ausgewählte Fallbeispiele aus der Region Brandenburg. A study by BMVEL, Bonn. Published as LUGV Fachbeiträge des Landesumweltamtes, No. 100, www.mugv.brandenburg.de/cms/media.php/lbm1.a.2320.de/fb_n100.pdf

Micronutrient Initiative (2009) *Investing in the Future: A United Call to Action on Vitamin and Mineral Deficiencies*, www.unitedcalltoaction.org/documents/Investing_in_the_future.pdf (accessed 5 November 2013).

Millennium Ecosystem Assessment (2005) *Ecosystems and Human Well-being: Biodiversity Synthesis*, World Resources Institute, Washington, DC.

Miller, J.S. (2011) The discovery of medicines from plants: a current biological perspective. *Economic Botany*, **65** (4), 396–407.

Millstone, E. and Lang, T. (2008) *The Atlas of Food: Who Eats What, Where and Why*, Earthscan, London.

Mulliken, T. and Inskipp, C. (2006) Medicinal plant cultivation – scope, scale and diversity: results from an initial analysis, in *Proceedings of the 1st IFOAM International Conference on Organic Wild Production, Teslic, Bosnia and Herzegovina, May 2006*, International Federation of Organic Agricultural Movements (IFOAM), Bonn, Germany.

Murphy, D.J. (2007) *People, Plants and Genes: The Story of Crops and Humanity*, Oxford University Press.

Myers, N., Mittermeier, R.A., Mittermeier, C.G. *et al.* (2000) Biodiversity hotspots for conservation priorities. *Nature*, **403**, 853–858.

Nesbitt, M., McBurney, R.P.H., Broin, M. and Beentje, H.J. (2010) Linking biodiversity, food and nutrition: the importance of plant identification and nomenclature. *Journal of Food Composition and Analysis*, **23**, 486–498.

Newman, D.J., Cragg, G.M. and Snader, K.M. (2003) Natural products as sources of new drugs over the period 1981–2002. *Journal of Natural Products*, **66**, 1022–1037.

Ojiewo, C., Tenkouano, C., Hughes, J. and Keatinge, J.D.H. (2013) Diversifying diets: using indigenous vegetables to improve profitability, nutrition and health in Africa, in *Diversifying Food and Diets: Using Agricultural Biodiversity to Improve Nutrition and Health* (eds J. Fanzo, D. Hunter, T. Borelli and F. Mattei), Earthscan, London.

Oldfield, M.L. (1981) Tropical deforestation and genetic resource conservation, in *Blowing in the Wind: Deforestation and Long-range Implications* (eds V.H. Sutlive, N. Altshuler and M.D. Zamora), Department of Anthropology, College of William and Mary, Williamsburg, VA.

Oldfield, S. and Jenkins, M. (2012) *Wild Flora for Improved Rural Livelihoods: Case Studies from Brazil, China, India and Mexico*, Botanic Gardens Conservation International, Richmond, UK.

Overgaard Larsen, H. (2011) Accessibility of wild products: biodiversity for food and medicine indicators partnership. *Medicinal Plant Conservation*, **14**, 24–29.

Padulosi, S. and Dulloo, M.E. (2012) Towards a viable system for monitoring agrobiodiversity on farm: a proposed new approach for red listing of cultivated plant species, in *On Farm Conservation of Neglected and Underutilized Species: Status, Trends and Novel Approaches to Cope with Climate Change. Proceedings of an International Conference, Frankfurt, 14–16 June 2011* (eds S. Padulosi, N. Bergamini and T. Lawrence), Bioversity International, Rome, pp. 171–197.

Padulosi, S. and Hoeschle-Zeledon, I. (2004) Underutilized plant species what are they? *Leisa Magazine*, (March), 5–6.

Padulosi S., Hodgkin T., Williams J.T. and Haq, N. (2002a) Underutilized crops: trends, challenges and opportunities in the 21st century, in *Managing Plant Genetic Diversity* (eds J.M.M. Engels, V. Ramanatha Rao, A.H.D. Brown and M.T. Jackson), CABI–IPGRI, pp. 323–338.

Padulosi, S., Leaman, D. and Quek, P. (2002b) Challenges and opportunities in enhancing the conservation and use of medicinal and aromatic plants. *Journal of Herbs, Spices, and Medicinal Plants*, **9** (4), 243–267.

Padulosi, S., Bergamini, N. and Lawrence, T. (eds) (2012) *On Farm Conservation of Neglected and Underutilized Species: Status, Trends and Novel Approaches to Cope with Climate Change. Proceedings of an International Conference, Frankfurt, 14–16 June 2011*, Bioversity International, Rome.

Palacios, X.F. (1998) Contribution to the estimation of countries' interdependence in the area of plant genetic resources. Background study paper no 7 rev 1, FAO Commission on Genetic Resources for Food and Agriculture, Rome, www.fao.org/nr/cgrfa/cgrfa-back/en/?no_cache=1 (accessed 1 April 2013).

Penso, G. (1980) *WHO Inventory of Medicinal Plants Used in Different Countries*, WHO, Geneva.
Penso, G. and Proserpio, G. (1997) *Index Plantarum Medicinalium Totius Mundi Eorumque Synonymorum*, 2nd edn, OEMF, Milan.
Pereira, H.M., Ferrier, S., Walters, M. *et al.* (2013) Essential biodiversity variables. *Science*, **339**, 277–278.
Peters, C. (1994) Sustainable harvest of non-timber forest plant resources in tropical moist forest. An ecological primer, WWF Biodiversity Support Program, Washington, DC.
Pinstrup-Andersen, P. (2009) Food security: definition and measurement. *Food Security*, **1**, 5–7.
Plotkin, M.J. (1988) Conservation, ethnobotany, and the search for new jungle medicines: pharmacognosy comes of age ... again. *Pharmacotherapy*, **8** (5), 257–262.
Posey, D.A. (1999) Cultural and spiritual values of biodiversity. A complementary contribution to the global biodiversity assessment, in *Cultural and Spiritual Values of Biodiversity* (ed. D.A. Posey), UNEP and Intermediate Technology Publications. London, pp. 1–19.
Pouboum, C.F.N. (1999) Cameroon, in *The Biodiversity of Traditional Leafy Vegetables* (eds J.A. Chweya and P.B, Eyzaguirre), International Plant Genetic Resources Institute, Rome.
Powell, B., Hall, J. and Johns, T. (2011) Forest cover, use and dietary intake in the East Usambara Mountains, Tanzania. *International Forestry Review*, **13** (3), 305–317.
Prescott-Allen, R. and Prescott Allen, C. (1990) How many plants feed the world? *Conservation Biology*, **4**, 365–374.
Principe, P.P. (1991) Valuing the biodiversity of medicinal plants, in *The Conservation of Medicinal Plants* (eds O. Akerele, V. Heywood and H. Synge), Cambridge University Press, Cambridge, pp. 79–124.
Ramirez-Villegas, J., Jarvis, A. and Touval, J. (2012) Analysis of threats to South American flora and its implications for conservation. *Journal for Nature Conservation*, **20** (6), 337–348.
Rao, N.K. (2013) Crop wild relatives from the Arabian Peninsula. *Genetic Resources and Crop Evolution*, **60** (5), 1709–1725, doi: 10.1007/s10722-013-9972-5.
Rao, R.V., Matthews, P.J., Eyzaguirre, P.B. and Hunter, D. (eds) (2010) *The Global Diversity of Taro: Ethnobotany and Conservation*, Bioversity International, Rome.
Remans, R. and Smukler, S. (2013) Linking biodiversity and research: research methodologies, in *Diversifying Food and Diets: Using Agricultural Biodiversity to Improve Nutrition and Health* (eds J. Fanzo, D. Hunter, T. Borelli and F. Mattei), Earthscan, London.
Rosenthal, G.A. and Janzen, D.H. (eds) (1979) *Herbivores: Their Interaction with Secondary Plant Metabolites*, Academic Press, New York, NY.
Royal Botanic Gardens, Kew (2010) Plants under pressure a global assessment. The first report of the IUCN Sampled Red List Index for Plants, Royal Botanic Gardens, Kew, UK, www.kew.org/science-research-data/directory/projects/SampledRedListIndexP.htm (accessed 29 April 2013).
Schippmann, U., Leaman, D. and Cunningham, A. (2002) *Impact of Cultivation and Gathering of Medicinal Plants on Biodiversity: Global Trends and Issues*, FAO, Rome.
Schippmann, U., Leaman, D. and Cunningham, A.B. (2006) Cultivation and wild collection of medicinal and aromatic plants under sustainability aspects, in *Medicinal and Aromatic Plants: Agricultural, Commercial, Ecological, Legal, Pharmacological and Social Aspects* (eds R.J. Bogers, L.E. Craker and D. Lange), Springer, Dordrecht.
Schultes, R.E. (1991) The reason for ethnobotanical conservation, in *The Conservation of Medicinal Plants* (eds O. Akerele, V. Heywood and H. Synge), Cambridge University Press, Cambridge, pp. 65–75.
Secretariat of the Convention on Biological Diversity (2010) Global Biodiversity Outlook 3, Montreal.
Sheldon, J.W., Balick, M.J. and Laird, S. (1996) *Medicinal Plants: Can Utilization and Conservation Coexist?*, Advances in Economic Botany, vol. 12, New York Botanical Garden, New York, NY.
Singer, C. (1927) The herbal in antiquity and its transmission to later ages. *The Journal of Hellenic Studies*, **47** (1), 1–52.
Slikkerveer, L. (1994) *Indigenous Agricultural Knowledge Systems in Developing Countries: A Bibliography*, Leiden Ethnosystems and Development Programme (LEAD), Institute of Cultural and Social Studies, Leiden.
Srivastava, J., Lambert, J. and Vietmeyer, N. (1996) Medicinal plants: an expanding role in development, World Bank Technical Paper 320, The World Bank, Washington, DC.
Stadlmayr, B., Charrondiere, U.R., Enujiugha, V.N. *et al.* (2012) *West African Food Composition Table/Table de Composition des Aliments d'Afrique de l'Ouest*, FAO, Rome.
Stadlmayr, B., Charrondiere, R., Eisenwagen, S. *et al.* (2013) Nutrient composition of selected indigenous fruits from sub-Saharan Africa. *Journal of the Science of Food and Agriculture*, **93**, 2627–2636.
Steinhoff, B. (1998) ESCOP and WHO monographs – a potential scientific basis for a rational assessment of herbal medicinal products in Europe under specific aspects of the regulatory situation, in *Medicinal Plant Trade in Europe: Conservation and Supply. Proceedings, First International Symposium on the Conservation of Medicinal Plants in Trade in Europe, 22–23 June 1998, Royal Botanic Gardens, Kew, UK*, TRAFFIC Europe, Brussels, pp. 67–73.
Sthapit, B. and Padulosi, S. (2012) On farm conservation of neglected and underutilized crops in the face of climate change, in *On Farm Conservation of Neglected and Underutilized Species, Status, Trends and Novel Approaches to Cope with Climate Change. Proceedings of an International Conference, Frankfurt, 14–16 June 2011* (eds S. Padulosi, N. Bergami and T. Lawrence), Bioversity International, Rome.
Strahm, W. (1989) *Plant Red Data Book of Rodrigues*, Koeltz Scientific Books for IUCN World Conservation Union, Koenigstein.
Stuart, M. (ed.) (1979) *Herbs and Herbalism*, Orbis, London.
Subedi, A., Bajracharya, J., Joshi, B.K. *et al.* (2008) Locating and managing the mango (*Mangifera indica* L.) genetic resources in Nepal. *Plant Genetic Resources Newsletter*, **155**, 52–61.
Termote, C., Cogill, B., Deptford, A. *et al.* (2013) Role of wild, neglected and underutilized foods in reducing the cost of a nutritionally adequate diet in the eastern region of Baringo District, Kenya, poster presented at Grand Challenges Exploration, Agriculture and Nutrition Meeting, 13–15 March 2013, Seattle, WA, Bioversity International, Rome.
*The Plant List* (2010) Version 1, www.theplantlist.org/ (accessed 1 April 2013).

Tilman, D., Balzer, C., Hill, J. and Befort, B.L. (2011) Global food demand and the sustainable intensification of agriculture. *Proceedings of the National Academy of Sciences of the United States of America*, **108** (50), 20260–20264.

Tomlinson, I. (2013) Doubling food production to feed the 9 billion: a critical perspective on a key discourse of food security in the UK. *Journal of Rural Studies*, **29**, 81–90.

University of Illinois Board of Trustees (2012) NAPRALERT. Program for collaborative research in the pharmaceutical sciences, College of Pharmacy, University of Illinois at Chicago, www.napralert.org (accessed 29 April 2013).

Van de Wouw, M., Kik, C., van Hintum, T. *et al.* (2009) Genetic erosion in crops: concept, research results and challenges. *Plant Genetic Resources Characterization and Utilization*, **8** (1), 1–15.

Van Wyk, B.-E. and Wink, M. (2004) *Medicinal Plants of the World*, Timber Press, Portland, OR.

Vavilov, N.I. (1931) The problem of the origin of world agriculture in the light of the latest investigations, in *Science at the Crossroads: Papers Presented to the International Congress of the History of Science and Technology Held in London from June 29th to July 3rd, 1931 by the Delegates of the U.S.S.R.* (ed. N.I. Bukharin), Kniga (England) Ltd, London, pp. 1–20 (reprint, 1971, Frank Cass, New York).

Velasquez-Milla, D., Casas, A., Torres-Guevara, J. and Cruz-Soriano, A. (2011) Ecological and socio-cultural factors influencing *in situ* conservation of crop diversity by traditional Andean households in Peru. *Journal of Ethnobiology and Ethnomedicine*, **7**, 40, doi:10.1186/1746-4269-7-40.

Vetelainen, M., Negri, V. and Maxted, N. (2009) European landraces: on-farm conservation, management and use, Bioversity Technical Bulletin No. 15, Bioversity International, Rome.

Vié, J.-C., Hilton-Taylor, C. and Stuart, S.N. (eds) (2009) *Wildlife in a Changing World – An Analysis of the 2008 IUCN Red List of Threatened Species*, IUCN, Gland.

Vietmeyer, N.D. (1986) Lesser-known plants of potential use in agriculture and forestry. *Science*, **232**, 1379–1384.

Vincent, H., Wiersema, J., Kell, S.P. *et al.* (2013) A prioritized crop wild relative inventory to help underpin global food security. *Biological Conservation*, **167**, 265–275.

Vinceti, B., Ickowitz, A., Powell, B. *et al.* (2013) The contribution of forests to sustainable diets, Background paper for the International Conference on Forests for Food Security and Nutrition, FAO, Rome, 13–15 May, 2013.

Walker, L.A. (1997) New therapies from natural sources: drug discovery at the National Center for the Development of Natural Products, in *International Symposium on Herbal Medicine: A Holistic Approach. Documents, Proceedings, and Recommendations. 1–4 June 1997, Honolulu, Hawaii* (eds D.A. Wozniak, S. Yuen, M. Garrett, and T.M. Schuman), International Institute for Human Resources Development, College of Health and Human Services, San Diego State University, San Diego, CA, pp. 144–151.

Walter K.S. and Gillett H.J. (eds) (1998) 1997 IUCN red list of threatened plants, Compiled by World Conservation Monitoring Centre/IUCN, The World Conservation Union, Gland/Cambridge.

WHO (1988) WHA41.19 Traditional medicine and medicinal plants, in *Forty-First World Health Assembly*, Geneva, 2–13 May.

WHO (1999) *WHO Monographs on Selected Medicinal Plants 1*, World Health Organization, Geneva.

WHO (2002a) *WHO Traditional Medicine Strategy: 2002–2005*. World Health Organization, Geneva (WHO/EDM/TRM/2002.1), www.who.int/medicines/publications/traditionalpolicy/en/ (last accessed 29 April 2013).

WHO (2002b) *WHO Monographs on Selected Medicinal Plants 2*, World Health Organization, Geneva.

WHO (2003a) WHA56.31, in *Fifty-Sixth World Health Assembly*, Geneva, 19–28 May.

WHO (2003b) *WHO Guidelines on Good Agricultural and Collection Practices (GACP) for Medicinal Plants*, World Health Organization, Geneva, www.who.int/medicines/publications/traditional/gacp2004/en/ (accessed 29 April 2013).

WHO (2007) *WHO Monographs on Selected Medicinal Plants 3*, World Health Organization, Geneva.

WHO (2009) *WHO Monographs on Selected Medicinal Plants 4*, World Health Organization, Geneva.

WHO (2011) Index of pharmacopoeias, Working document QAS/11.453, World Health Organization, Geneva.

WHO (2012) Obesity and overweight, WHO Fact Sheet No. 311, World Health Organization, Geneva.

WHO, IUCN and WWF (1993) *Guidelines on the Conservation of Medicinal Plants*, IUCN/WHO/WWF, Gland/Geneva/Gland.

Williams, C. (2011) Access and benefit sharing under the Convention on Biological Diversity – update from CoP 10 in Nagoya. *Medicinal Plant Conservation*, **14**, 17–21.

Williams, V.L, Victor, J.E. and Crouch, N.R. (2013) Red listed medicinal plants of South Africa: status, trends, and assessment challenges. *South African Journal of Botany*, **86**, 23–35.

Wilson, E.O. (1992) *The Diversity of Life*, Penguin, London.

WWF and IUCN (1994) *Centres of Plant Diversity: A Guide and Strategy for their Conservation. Vol. 1: Europe, Africa, South West Asia and the Middle East*, IUCN Publications Unit, Cambridge.

WWF and IUCN (1995) *Centres of Plant Diversity: A Guide and Strategy for their Conservation. Vol. 2: Asia, Astralasia and the Pacific*, IUCN Publications Unit, Cambridge.

WWF and IUCN (1997) *Centres of Plant Diversity: A Guide and Strategy for their Conservation. Vol. 3: The Americas*, IUCN Publications Unit, Cambridge.

# 2 World Health Organization Perspective for Traditional Medicine

Ossy M.J. Kasilo and Jean-Baptiste Nikiema

World Health Organization, Regional Office for Africa, Brazzaville, Republic of the Congo

## 2.1 Introduction

The World Health Organization (WHO) defines traditional medicine (TM) as including diverse health practices, approaches, knowledge and beliefs incorporating plant, animal and/or mineral-based medicines, spiritual therapies, manual techniques and exercises applied singularly or in combination to maintain well-being, as well as to treat, diagnose or prevent illness (WHO, 2002). The terms 'complementary' and 'alternative' medicine (CAM) are used to refer to a broad set of health care practices that are not part of a country's own tradition, or not integrated into its dominant health care system. Accordingly, in this document, 'TM' is used when referring to Africa, India and China, whereas 'CAM' is used when referring to Europe and/or North America and Australia.

The important role played by TM and its practitioners in health care delivery was recognized by the Alma-Ata declaration in its Conference on Primary Health Care (WHO, 1978). The relevant recommendations of WHO governing bodies also address the strategic options that are expected to help achieve health for all. Regional partners and agencies of the United Nations – for example, the African Union (AU) and its regional economic communities, African Development Bank, Regional Intellectual Property Organization (Decade of African TM 2011), Food and Agricultural Organization (De Silva, 1997), International Union of Conservation of Nature, United Nations Conference on Trade and Development (UNCTAD, 2000), United Nations Industrial Development Organization (Wijesekera, 1991), World Bank, World Intellectual Property Organization (http://www.wipo.int/ip-development/en/agenda/flexibilities/resources/tk_gr_tce_f.html) – have also been stressing the importance of TM. In particular, the Economic Community of West African States (ECOWAS)/West African Health Organization (WAHO) established a TM programme in 2007 and the Southern African Development Community (SADC) developed a TM strategy 2010–2014.

TM has been used by the majority of the population in the WHO African Region before the advent of modern medicine for the treatment of diseases. It is embedded in the culture of the African population. In some communities TM is the only source of health care that is available, affordable and accessible (Kasilo *et al.*, 2005). Indeed, the WHO currently estimates that as much as 80% of the people living in the African Region use TM for their health care needs. Some of the modern medicines, such as atropine, artemisinine, physiostigmine and vincristine and vinblastine, were isolated from medicinal plants.

There is growing recognition of the important role that TM plays in overall health care delivery in both developing and industrialized countries. In fact, trends in the use of TM and CAM are on the increase. For example, TM is used by 60% of the population in South Africa, 70% of the population in Benin, Côte d'Ivoire, Ghana, Mali and Rwanda, which is comparable to India; 80% in Burkina Faso, Democratic Republic of Congo and South Africa; and over 90% in Burundi and Ethiopia, comparable to China. Similarly, the use of CAM in industrialized countries is 40% in Belgium, 42% in the USA, 68.9% in Australia, 70% in Canada, 75% in France, 80% in Germany and 90% in the UK (WHO Regional Office for the Western Pacific, 2012; WHO Regional Office for Africa, 2013).

The economic potential of herbal medicines has increased over the past decade. The herbal medicine market has expanded tremendously and the total annual sale of herbal medicines is still growing. In 2005, the total annual sale of herbal medicines products in China alone totalled US$ 14 billion, and by 2011 the total global herbal market size was US$ 62.9 billion, with India's contribution of US$ 1 billion (Bodeker *et al.*, 2005). The magnitude of the African herbal medicine products market is not known. However, recorded sales in Burkina Faso increased from US$ 2.68 million in 1998 to US$ 5.37 million in 2000. In Madagascar, sales increased from US$ 2 million in 1999 to US$ 3.5 million in 2000. In 2008, it was estimated that trade in traditional medicinal plants and products in South Africa was worth ZAR 2.9 billion per year (WHO Regional Office for Africa, 2013). Algeria recorded 6% value growth in 2011 of herbal/traditional products (http://www.euromonitor.com/herbal-traditional-products-in-algeria/report) and Cameroon witnessed good growth in these products due to greater demand from consumers (http://www.euromonitor.com/herbal-traditional-products-in-cameroon/report).

Despite the use of TMs for many centuries, challenges regarding the safety, efficacy and quality of TM products exist, and the development of TM varies widely in different countries of the region. This chapter will provide information on some of the WHO policy orientation on TM, and will discuss how some of the challenges have been addressed through the implementation of the priority interventions of the regional strategy on promoting the role of TM in health systems. Going forward, the chapter will address the Second Decade of African TM (2011–2020), the updating of the regional TM strategy and the development of herbal pharmacopoeias.

*Novel Plant Bioresources: Applications in Food, Medicine and Cosmetics*, First Edition. Edited by Ameenah Gurib-Fakim.

## 2.2 Policies on traditional medicine

It is nearly 40 years since the WHO was mandated to deal with the promotion and development of TM with the introduction of the first paper on primary health care (PHC) to the Executive Board in January 1975. The paper called for harmonization of services organized by governments with those traditionally available to the people, such as traditional birth attendants and herbalists. A small consultative meeting was convened in January 1976, which proposed to the board to collect, analyse and study available data on traditional healers and indigenous systems of medicine, and suggests further action to be taken by member states and the WHO secretariat. Subsequently, the WHO programme on TM was then established at the WHO headquarters in 1976, and in the same year, the World Health Assembly (WHA), through 'Resolution WHA29.72 on Health manpower development', acknowledged the value of TM in PHC.

In May 2008, the sixty-first WHA adopted a global strategy and plan of action by its resolution WHA61.21 on public health innovation, and intellectual property, which charts the way forward for research and development of TM and protection of intellectual property. The WHO Congress on TM convened by the WHO in November 2008 adopted the Beijing Declaration, calling for further development based on research and innovation in line with the above-mentioned resolution. The declaration also underscored the need for governments to establish systems for the qualification, accreditation or licensing of TM practitioners. These later should upgrade their knowledge and skills based on national requirements. The declaration underscored the strengthening of communication between conventional and TM providers and establishment of appropriate training programmes for health professionals, medical students and relevant researchers, among others.

In 2009, the WHA resolution WHA62.13 on Traditional Medicine recalls the Declaration of Alma-Ata, which states, inter alia, that 'The people have the right and duty to participate individually and collectively in the planning and implementation of their health care' and 'Primary health care relies, at local and referral levels, on health workers, including physicians, nurses, midwives, auxiliaries and community workers as applicable, as well as traditional practitioners as needed, suitably trained socially and technically to work as a health team and to respond to the expressed health needs of the community'. The resolution also recognizes TM as one of the resources of PHC services that could contribute to improved health outcomes, including those in the Millennium Development Goals. This resolution urges member states to promote appropriate, safe and effective use of TM through formulation of national policies, regulations and standards, as part of comprehensive national health systems, develop TM based on research and innovation, assist traditional health practitioners (THPs) to upgrade their knowledge and skill in collaboration with relevant health providers and cooperate with each other in sharing knowledge and practices.

At regional level, the WHO Regional Committee for Africa adopted several resolutions on TM. For example, Resolution AFR/RC24/R14 of 1974 on the decision to include 'TM and its role in the development of health services in Africa' in the technical discussions at its twenty-sixth session in 1976; Resolution AFR/RC28/R3 of 1978 on the Use of essential medicines and the African pharmacopoeia; Resolution AFR/RC34/R8 of 1984 on Launching or development of a program of TM; and Resolution AFR/RC40/R8 of 1990 on Promotion of TM, development of TM systems and its role in the development of health services in Africa.

The ninety-ninth session of the WHO Regional Committee adopted Resolution AFR/RC49/R5 of 1999 on 'Essential drugs in the WHO African Region: situation and trend analysis', which requested the WHO Regional Office for Africa to support member states to carry out research on medicinal plants and promote their use in health systems. The committee in its deliberations requested the WHO Regional Office for Africa to develop a comprehensive strategy on African TM. The fifty-first session of the WHO Regional Committee for Africa, in 2000, approved the regional strategy (AFR/RC50/9), by its Resolution AFR/RC50/R3 on 'Promoting the role of TM in health systems: a strategy for the African Region' (WHO Regional Office for Africa, 2001) (hereafter referred to as Resolution AFRR/RC50/R3). The priority interventions of the regional strategy are policy formulation, research promotion, development of local production of TM products, including cultivation and conservation of medicinal plants, capacity building and protection of intellectual property rights as priority interventions.

In 2001, the AU Summit of Heads of State and Government declared the period 2001–2010 as the *Decade of African TM* and in 2003 adopted a plan of action for its implementation. The summit also in 2001 and the fifty-seventh Regional Committee for Africa in 2007 declared TM research as a priority. In 2008 the 'Algiers Declaration on Research for Health' recognized the need to promote research in TM and strengthen health systems, taking into account the sociocultural and environmental situation of the people. The year 2010 marked a decade since the adoption of the above-mentioned regional strategy on TM and the declaration on the Decade of African TM mentioned above.

In 2010, the regional office dedicated a special issue of the *African Health Monitor* to TM (WHO Regional Office for Africa, 2010). The issue covered the situation in the African Region and ECOWAS member states; regulation, clinical practices, collaboration between practitioners of TM and conventional medicine, research on medicinal plants used for hypertension, opportunistic infections, sickle-cell anaemia, herbal pharmacopeia, research networks, local production and protection of traditional knowledge. The AU Conference of African Ministers of Health held in Windhoek, from 17 to 21 April 2011, discussed the End-of-Decade Review report on African TM and renewed the Decade from 2011 to 2020. In July 2011, the African Heads of State and Government deliberated on, and endorsed, political declarations on the renewed Decade of African TM mentioned above and its implementation plan. In 2011, the sixty-first session of the WHO Regional Committee discussed the progress report (WHO Regional Office for Africa, 2011) on the implementation of the regional strategy and plan of action on the Decade of African Traditional TM (2001–2010), and recommended an update of the regional strategy.

## 2.3 Tools and guidelines

In 2001 the WHO established a regional expert committee on TM to support member states to effectively implement Resolution AFRR/RC50/R3 on Promoting the Role of Traditional Medicine in Health Systems: A Strategy for the African Region (WHO Regional Office for Africa, 2001) and monitor and evaluate the progress made in TM. The committee supported the regional office for the development of tools and guidelines on priority interventions of the regional strategy, some of which are reported below. Countries adapted the tools and guidelines to their situations.

### 2.3.1 Tools for institutionalization of traditional medicine in health systems

Institutionalization of TM in health systems is one of the guiding principles of the regional strategy document entitled 'Promoting the role of TM in health systems: a strategy for the African Region' (AFR/RC50/9). The term institutionalization is used here to express the formalization and official incorporation of TM into national health systems and services. Resolution AFR/RC50/R3 urges member states to actively promote, in collaboration with all other partners, the protection of intellectual property rights (IPRs) and indigenous knowledge in the field of TM.

The TM resolution AFR/RC50/R3 requested the WHO Regional Director to support member states to: develop guidelines for the formulation and evaluation of national policies on TM; advise countries regarding the relevant legislation for the practice of TM; the documentation of practices and medicines of proven safety, efficacy and quality; strengthen WHO collaborating centres and other research institutions to conduct research; and establish mechanisms for the protection of cultural and IPRs.

These orientations can be achieved through institutionalization of TM, which calls for a number of actions to be undertaken by member states and include:

1. The setting up or strengthening of structures for TM as essential for optimizing the use of TM and should be based on a thorough analysis of the prevailing systems with the involvement of THPs themselves and the communities.
2. The establishment of professional TM bodies.
3. The formulation of a code of ethics.
4. The development of norms and standards.
5. The establishment of mechanisms for the official recognition and support of TM, including the identification, registration and accreditation of qualified practitioners.
6. The development of mechanisms of collaboration between conventional health practitioners (CHPs) and THPs.

In order to support member states to institutionalize TM in health systems, the WHO Regional Office for Africa has developed documents to be used as tools for institutionalizing TM in health systems. These documents were for development, implementation, monitoring and evaluation of national policies on TM: a model legal framework for the practice of TM; a traditional health practitioner's bill; model codes of ethics and practice for THPs; and guidelines for the formulation of a national master plan for the development of TM. The development of all the documents involved member states and relevant stakeholders after which they were field tested and the country experiences shared through regional fora convened by the WHO on integration (WHO, 2003a) and institutionalization of TM (WHO, 2004a). Subsequently, the tools and guidelines were updated taking into account current trends and country experience. The four documents mentioned above were combined into one document for easy reference by countries as tools for institutionalizing TM in their health systems (WHO, 2004b), and these were adapted by countries to their specific situations.

The purpose of the tools for institutionalizing TM in health systems is to support member states wishing to institutionalize TM in their national health systems. The document is intended for use by competent national authorities responsible for the development and regulation of the practice of TM, experts in TM, professional health associations including THPs and traditional birth attendants, as well as other organizations involved in TM. Salient points related to the guidelines for national policies on TM and model codes of ethics and practice for THPs are indicated below as examples.

#### 2.3.1.1 Guidelines for national policies on traditional medicine

The content of a national policy on TM will vary from country to country owing to differences in the level of development of TM, the availability of resources and political commitment. However, the policy should address problems identified during the situation analysis. The role of the government includes identification of the need for a national policy, development of a policy statement, development of a strategic plan, implementation of the national policy, and evaluation and monitoring. In general, a national policy on TM should include a statement of the government's role in the development of TM in the health care delivery system. The policy should include the vision and mission as well as goals and objectives for TM policy. A national policy on TM may contain the government's mission statement and may include the following points:

1. regulation of the practice of THPs, provision of infrastructure, development of a code of ethics for THPs;
2. education and training of THPs and CHPs;
3. regulation of raw materials processed and finished products;
4. acceptable standards of safety and quality of TM practices and medicines;
5. promotion of the proper use of TM, particularly in PHC delivery systems;
6. need for new legislation or expansion of existing laws in the country;
7. capacity building to strengthen government expertise in TM;
8. conservation of medicinal plants and animals, as well as application of national or international conventions on biodiversity and endangered species;
9. promotion and advocacy of the role of TM in health care systems (including dissemination of information inside and outside governments);
10. research and development;
11. international technical cooperation and exchange; and
12. monitoring and evaluation of the implementation of the national policy on traditional medicine.

#### 2.3.1.2 Model codes of ethics and practice for traditional health practitioners

This chapter discusses the objectives of a code of ethics, which are to:

1. ensure a high standard of conduct and practice among THPs;
2. foster good relationships among THPs, patients and other practitioners; and

3. increase THPs' awareness of the existing rules and regulations governing the practice of TM.

The chapter explains relevant terminologies for an understanding of the basic foundation of ethical behaviour governing the conduct of THPs, based on moral values and professional principles in relation to their job, the patient, the public and other practitioners. It provides sample contents of a code of ethics. This is followed by a code of practice that outlines the rules of conduct governing the relationships between THPs and their clients and colleagues in performing duties within their area of competence. The next section addresses disciplinary procedures, and this is followed by the minimum standards for the practice of THPs.

Commitment to professional ethical standards starts with the swearing of a professional oath and acceptance of the professional code of ethics. The oath is usually brief, general and intended to oblige and inspire a member to abide by applicable laws, codified ethics and the dictates of conscience and religious principles. The details on the document can be found on the WHO website: http://intranet.who.int/afro/.

### 2.3.2 Training modules on traditional medicine of health science students

The purpose of these training modules is to facilitate the introduction of the teaching of TM in the curriculum of health science students, increase their awareness of the role played by THPs in the promotion and provision of PHC, encourage CHPs to be involved in TM research activities and foster collaboration between practitioners of both traditional and conventional medicine. The modules include 'From African TM to Western medicine', 'Basic concepts in African TM', 'Policy and regulatory framework for the practice of African TM', 'The use of medicinal plants in African TM', 'Establishment of ethnomedical information on African TM', 'Quality assurance of African TM products', 'The role of THPs in primary health care', 'TM in reproductive health care services', 'Apprenticeship in African TM practice', 'Introduction to the protection of IPRs' and 'TM knowledge (TMK)'.

For instance, the module related to basic concepts highlights the following concepts:

1. Definitions and terminologies used in TM.
2. African TM as distinct from CAM.
3. Categories of THPs, their training and apprenticeship. There are different categories of THPs, which include traditional herbal medicine practitioners or herbalists, bone setters, traditional midwives, ritualists and spiritualists and others. Some of these THPs have received a formal education and have also had a lot of experience. Others practice TM after several years of apprenticeship with established THPs.
4. The structure of African society, norms and taboos.
5. Cultural principles of African TM.
6. Medical anthropology in relation to African history.
7. The concept of health and wellness. Health in African TM is regarded as a balance between body, mental and spiritual well-being and includes harmony between humans and the social and metaphysical forces of the environment.
8. The concept of diseases and causation. Any disturbance in the function of any of the above factors is regarded as ill health.
9. The concept of diagnosis and treatment.
10. Cultural and traditional methods of providing health care services.
11. To understand the importance of rituals in an African cultural context.
12. The concept of prevention and the processes involved in preventive services. including mother and child (MCH) services, curative services, apprenticeship, counselling and ritual services.

The outline chapters of the module 'From African TM to Western medicine' include examples of modern medicines that have been derived from medicinal plants:

1. The isolation of quinine antimalarial from the *Cinchona* plant species and the synthesis of antimalarial molecules based on the quinine structure: chloroquine, amodiaquine, mefloquine, halofntrine.
2. The isolation of reserpine from *Rauwolfia vomitoria* and the use of this alkaloid that led to understanding of modern treatment of psychotic conditions.
3. The isolation of furocoumarins from the plant khellin (*Amni visnaga*) and its pharmacological studies that led to better understanding of bronchial asthma and the ultimate modelling and synthesis of chromoglycate (Intal).
4. The isolation and use of morphine and its 20 congener molecules from the tincture of morphine from *Papaver somnifera*, leading to the synthesis of pethidine and morphinans which revolutionized the treatment of pain, and the modernization of surgery and control of surgical pain in Western medicine.
5. The knowledge of the use of the bark extracts of *Prunus africana* for the management of prostate hyperplasia in men, which led to the understanding of the development of 5-alpha-reductase inhibitors for the management of this condition.
6. The knowledge of the African traditional use of the Madagascar periwinkle, *Catharanthus roseus* or *Vinca rosea*, for cancer therapy and the development of a battery of chemical compounds for cancer, starting from the alkaloid vincristine, which was first isolated from *Vinca rosea*.
7. The use of *Toddalia asciatica* and *Fagara zanthoxyllum* has led to the management of sickle cell anaemia.
8. There are many diseases whose treatment depends on TM; for example, cancer, spasm, pain, malaria.

The outline chapters of the module on the role of THPs, hereafter used interchangeably with the term 'traditional medicine practitioners' in PHC, include the Alma-Ata declaration on PHC, the role of THPs in PHC, the main PHC principles, which are appropriate technology, community involvement, intersectoral collaboration and equity, core services characteristics of traditional health practices, preventive and curative services, MCH services, counselling services.

### 2.3.3 Tools for continuing education of traditional health practitioners in primary health care

The general objectives are to enable the THPs to acquire awareness and appropriate skills, prevent preventable diseases and reduce morbidity and mortality due to communicable diseases and unhealthy behaviours, improve health conditions in the communities. Specifically, these tools aim to assist THPs to be able to understand the human body and how it works, understand the concept of good health, understand the role of the THP in PHC, promote health by giving appropriate advice concerning nutrition, safe food, safe water supply, personal hygiene and environmental health, immunization, family planning, organize campaigns for diseases prevention, provide first aid when needed, promote the preparation of safe medicinal plant products, promote rational use of TM products, conserve and cultivate medicinal plants, practice TM within the code of ethics and understand IPRs and indigenous medical knowledge.

The tool contains 28 chapters and has been divided into six sections. Each chapter has overall objectives, rationale, specific learning objectives, expected results, the substance of the chapter, a summary, practical exercises and follow-up and evaluation. The sections are as follows.

*Section A: Basic Principles.* This section contains a chapter on concepts used and their definitions, the overall objectives and specific learning objectives of the document. This chapter is followed by two chapters that describe anatomy, physiology and diagnostic methods that will serve as a bridge between the THP and the CHP to facilitate communication between these two stakeholders.

*Section B: Public Health.* This section has four chapters, which deal with hygiene and environmental health, food safety, safe water supply and nutrition. These are fundamental public health issues.

*Section C: Preventive Medicine.* This section contains six chapters, which provide orientation on preventive aspects of provision of health care.

*Section D: Communicable Diseases.* This section contains four chapters, which provide orientation on some of the most important communicable diseases that affect the people in the region. Two of them, HIV/AIDS and malaria, are among the five priority diseases of the TM programme.

*Section E: Non-Communicable Diseases.* This section contains five chapters, which give orientation on diseases that include three of the five priority diseases which the Health Systems and Services Development (DSD) Division (currently called Health Systems and Services Cluster) has selected for whose treatment research and development and local production must be accelerated. The three diseases are hypertension, diabetes and sickle cell disease.

*Section F: TM.* This section contains six chapters, which deal with all the various aspects of the practice of African TM. These chapters are written in such a way that the integration of the previous chapters can be facilitated.

It is expected that the use of the 28 chapters in this document will facilitate a gradual transition whereby the African THP and the mainstream or CHP will work together, hand in hand, to provide health care in the WHO African Region. For instance, the substance of the chapters related to anatomy and family planning are summarized in the following.

#### 2.3.3.1 Basic anatomy and physiology terminology with regard to modern medicine

It seems very inappropriate to train THPs in anatomy and physiology in the same way as medical students in modern medical science. THPs have their own way of teaching the anatomy and physiology courses to their trainees. As it is in modern medical sciences training, medical students in anatomy and physiology as part of the medical course take years to master the subjects. Training in traditional health practice is done using their own languages (ethnic group languages), of which the modern health practitioner, although they might belong to the very ethnic group, cannot understand. Therefore, in order to facilitate easy communication between the traditional and modern health practitioners it is wise for THPs to be oriented on some basic anatomy and physiology terminologies. This informative orientation to THPs on modern medical anatomy and physiology terminologies will assist referral of patients to modern health facilities.

Anatomy and physiology are two branches of sciences that help to understand the human body and how it functions. Anatomy is the study of body structures and the relationships that exist between them, whereas physiology is the study of how the different body structures are functioning. It is acknowledged that several THPs are known in their communities as being able to treat various diseases, including fractures, head injuries, broken legs, arms and backs, infertility, hypertension, diabetes and sickle cell anaemia. However, they are capable of knowing all of these in their own ethnic languages, which does not cconform to the modern scientific language, and hence a need to harmonize the two and have a common understanding.

A short, understandable explication of basic terminologies on selected topics, such as bone, thorax, skull, upper and lower extremities, hand, foot, joints, articulations, muscle, blood, fertilization of ovum, heart, respiratory apparatus, digestive apparatus, uro-genital apparatus and pancreas, might give some useful insight into communication by various THPs who are usually involved in bone setting and treating infertility, diabetes, hypertension and anaemia. This informal/non-academic training of THPs in basic terminologies of anatomy and physiology is not intended to give them an overloaded medical education to convert them into CHPs, but only to give them communication skills on human body organs in relation to disease and referral. To ease communication between the two, modern health practitioners would also undergo a self-initiated orientation of the basic anatomy and physiology terminologies in their own ethnic languages. This will enhance communication between them and facilitate documentation of TM practices.

#### 2.3.3.2 Family planning

Marriage and raising a family has been an important method of maintaining generations of human beings. The importance of a human being traditionally and culturally rests in the ability to reproduce, raise a family and maintain good health of both the mother and children produced. Family planning improves women's and children's health, and contributes to the reduction of maternal and child mortality. Family planning means adequate spacing of pregnancies and prevention of conception to protect and promote women's health. It is easier to undertake child spacing to women who start giving birth while they are 18–24 years than those of 30–35 years. It is known that too many close pregnancies and pregnancies to aged women weaken the mother's health, and her ability to take care of the family. In order to reduce maternal morbidity and mortality, and improve women's health, the promotion of family planning and early marriage (18–24 years) is important and encouraged. Large families may have problems in feeding, clothing, educating and general care of all their children. Doctors, midwives or health workers can advise on the most suitable family planning methods. No single method is perfect, since they all have advantages, disadvantages and contraindications.

There are different modern family planning methods: injectable contraceptive, sub-dermal implant (Norplant), intrauterine device, condom, diaphragm (cervical cap), spermicide, surgical contraception and natural method.

*Injectable contraceptive products.* These contain progestin, administered at intervals of 1, 2 or 3 months, depending on the product used. Their advantages are: they are very highly effective, do not reduce breast milk production and are appropriate for breast-feeding women; they protect against endometrial cancer and pelvic inflammatory disease; they prevent anaemia. They have disadvantages: they are not reversible until the end of the effective period; return to fertility may be delayed for 6–10 months; injections -require trained personnel.

*Subdermal implant (Norplant).* Six capsules containing progestin are inserted under the woman's skin (upper or lower arm). This stops ovulation by thickening the cervical mucus.

Advantages are that it is very effective, does not reduce breast milk production, it prevents anaemia, it is an alternative for women who cannot tolerate the side-effects of combined oral contraceptives, it lasts for 5 years and is not user dependent.

Disadvantages are that it-requires a small incision and a local anaesthetic for insertion and removal, the effectiveness drops after 2–3 years in women who weigh over 70 kg, return to fertility may be delayed for 6–10 months and it is not advisable for short contraception time (1 or 2 years).

*Intrauterine device (IUD).* This is a small plastic device (in different sizes and shapes) containing copper or progesterone. It is inserted into the uterus through the cervix and inactivates sperm, preventing its migration. It also causes changes in the uterus and fallopian tubes that prevent fertilization.

Advantages are that it is very highly effective, it is easily reversible, it does not reduce breast milk production, some types may last for 10 years, it is not user dependent and return to fertility is immediate upon removal.

Disadvantages are that it is not advisable for women who have not yet had a child, who are anaemic or who suffer from severe menstrual pains; insertion and removal require a trained person.

*Condom.* A latex or polyurethane sheath, worn over the penis (male condom) or inserted into the vagina (female condom) before sexual intercourse.

Advantages are that there are no systemic side effects, no medical contraindications, there is some degree of protection against sexually transmitted diseases, including AIDS, and it can be used as a back-up method. Disadvantages are a high failure rate, it must be discarded after use and there may be an allergic reaction to latex.

*Diaphragm and cervical cap.* This is a circular, dome-shaped rubber disc (in different sizes and styles) that is inserted into the vagina to cover the cervix; it prevents sperm from entering the uterus.

Advantages are that it is recommended for women who have sexual intercourse infrequently, it can be used as a back-up method and it can be washed, cleaned and reused.

Disadvantages are a high failure rate and it is not recommended for women who are prone to urinary tract infections.

*Spermicide.* Foam, suppository, tablet, jelly or cream that is placed inside the vagina and is often used with a diaphragm, cervical cap or a condom.

Advantages are that it can be used as a back-up method and it may protect against sexually transmitted infections.

Disadvantages are a high failure rate and it may cause irritation and/or allergic reactions.

*Natural method.* This is the prevention of fertilization by abstaining from sexual intercourse during a woman's fertile periods.

Advantages are that there are no systemic or long-term effects and it fosters communication between the couple.

Disadvantages are a high failure rate, it requires several days of abstinence and the reliable determination of woman's fertile and infertile periods is not easy.

*Surgical contraception.* This prevents fertilization by permanently blocking either the vas deferens in the male or the fallopian tubes in the female. It requires a small incision and anaesthesia.

Advantages are that it is very highly effective and it is suitable for individuals who already have the desired number of children or women who have medical contraindications to pregnancy.

Disadvantages are that it is considered permanent, as the reversal operation is usually not successful.

*Oral contraceptives – birth-control pills.* Pills containing hormones that occur naturally in the body. Advantages are that, when taken correctly, the pill is one of the most effective birth-control methods available. Disadvantages are the contraindication to some drugs (antibiotics, antiepileptics) and some health problems, such as hepatitis, cirrhosis and cancer of the womb or the breast.

Some of the traditional practices and products can be successfully used in the management of the family planning methods. Among traditional birth spacing methods are prolonged, regular breast-feeding or withdrawal, couple separation for at least 2 years, abstinence. Breast-feeding helps to space pregnancies since it delays the return of menstruation and so helps to postpone the next intended pregnancy. However, through collaboration between THPs and CHPs, patients can be referred to the care-giver who is best qualified to provide a specific health service. A good referral system can promote good collaboration and benefit the THP, CHP and the community. Such a system encourages the THP to make referrals to a health centre for problems they cannot manage, while the CHP refers patients back to the THP for follow up, after giving necessary advice and treatment.

### 2.3.4 Framework for regulation of traditional medicine practitioners, practices and products

This framework contains two parts:

- Part One – the guidelines for regulating traditional medical practices and TM products. It contains three chapters relating respectively to the licensing of practitioners, ethics code and licence to open and operate local clinics.
- Part Two – guidelines for regulating TM products. This part is subdivided into five chapters related to good manufacturing practices (GMPs), official certification of medicines, clinical trials, pharmacovigilance and advertisement control.

#### 2.3.4.1 Applicant procedure for licensing of traditional health practitioners

Any TM practitioner wishing to obtain a practitioner's licence must prepare an application. The procedure for such an application is summarized in the algorithm of practitioner licensing decision in Figure 2.1. The latter should consist of the following:

- a standard application form addressed to the competent authority and comprising the decisions of the administrative and health authorities of the locality in which the practitioner resides,
- a copy of birth certificate;
- a national identification certificate;
- a police criminal record of less than 3 months (variable validity period depending on the countries);
- an act of notoriety (issued by the administrative authority of the place of residence of the TM practitioner);
- a residence permit (issued by the administrative authority of the place of residence of the TM practitioner);
- an undertaking to abide by the medical code of ethics signed by the TM practitioner (available at the local first-level health authority);
- passport photographs;
- applicant information sheet (available at the local first-level health authority).

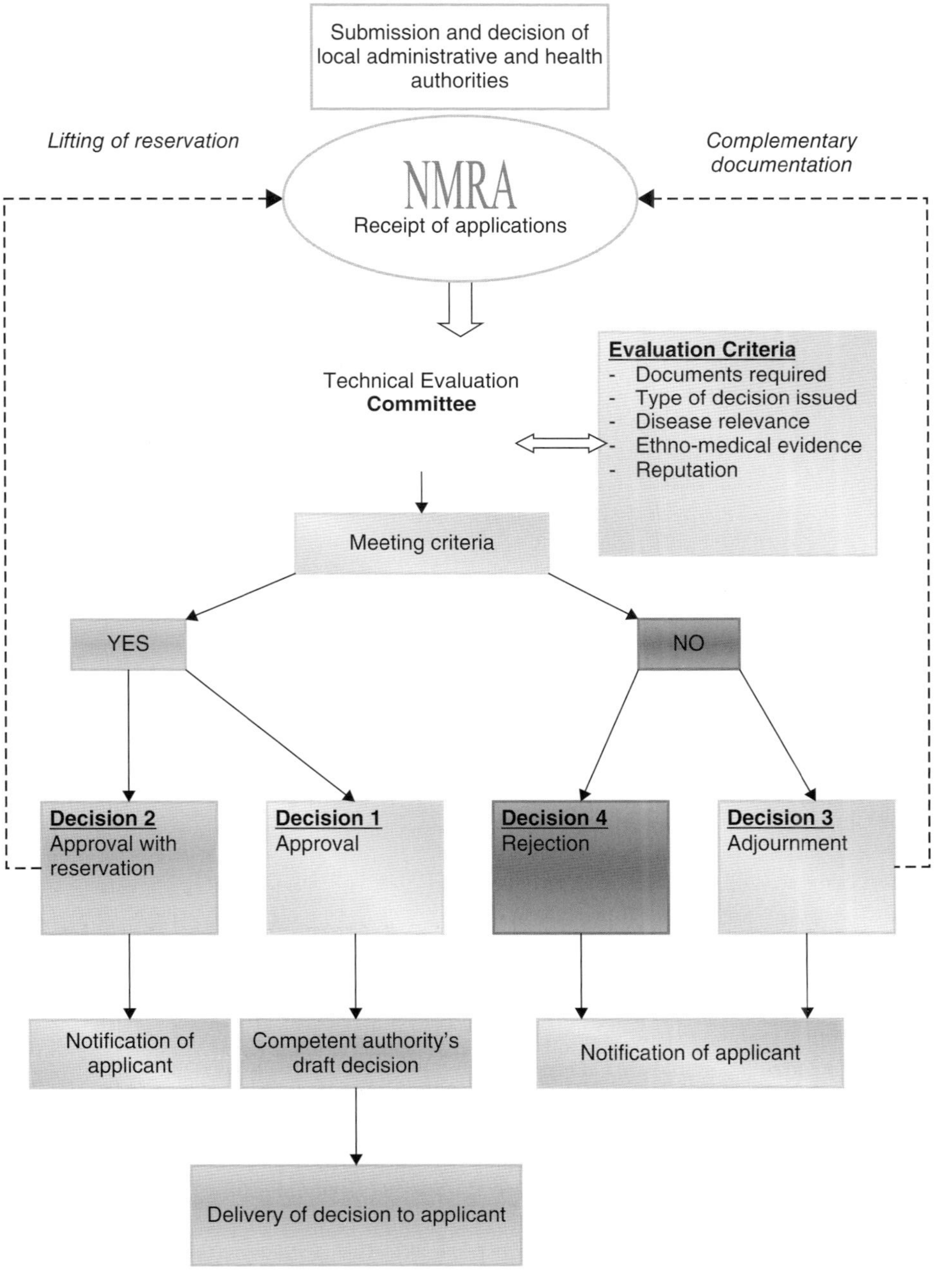


**Figure 2.1** Algorithm of practitioner licensing decision

Once the application is completed, it shall be submitted to the local first-level health authority to be forwarded to the competent national authority through the administrative chain of command.

### 2.3.4.2 Procedure for applying for registration of traditional medicine products

The granting of a marketing authorization (MA) by the competent authority must be supported by all the guarantees related to the pharmaceutical quality and safety as well as therapeutic efficacy. Consequently, the consideration of the documentation by a committee of experts and the decision of a national certification commission are therefore indispensable. The procedure for assessing an MA application is summarized in Figure 2.2 and must entail the following stages.

*Reception and administrative processing of application.* The application is received by the national pharmaceutical regulatory authority. A submission attestation is issued to the applicant. Upon receipt, an analysis and administrative assessment is conducted to determine the conformity of the application. The complete application is forwarded to the committee of experts.

*Committee of experts' assessment.* This seeks to ascertain the quality of the product submitted for certification. The technical expert, based on the assessment guide, undertakes a summary of the documentation received by highlighting the pharmaceutical, pharmaco-toxicological and clinical aspects. They compare the product with similar medicines that have already been certified. The committee members must be appointed by the competent authority.

*Decision of the commission.* The commission is responsible for validating the application. The decision of the commission relates to the following characteristics:

- therapeutic significance and efficacy;
- safety of the therapeutic doses;

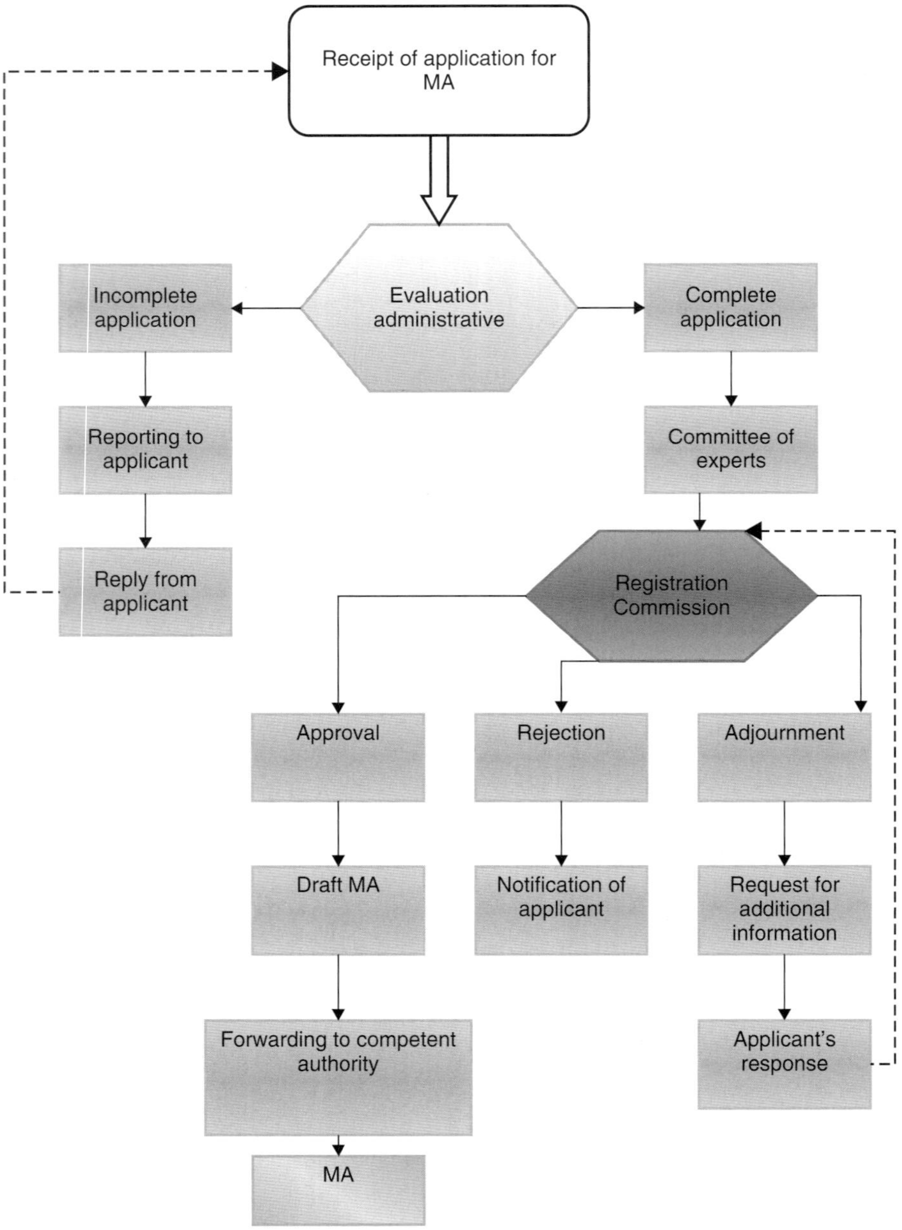


**Figure 2.2** Algorithm of decision for registration

- pharmaceutical quality (packaging and labelling);
- the wholesale price excluding taxes and costs of daily treatment and full treatment.

The characteristics as presented are compared with those of already certified similar medicines. Complementary analyses and controls may be requested by the certification commission.

*Authority's decision.* The marketing licence is granted or refused by the competent authority in accordance with the decision of the technical certification commission. A rejection must be justified and the applicant notified. The latter has the possibility of appealing and providing responses to the technical commission's objections.

### 2.3.4.3 Duration of validity and renewal of marketing authorizations

The marketing licence is granted by the competent authority for a 5-year duration. The renewal is subject to the submission of an application comprising:

- an application addressed to the competent authority;
- a certification indicating that no modification occurred in the elements produced in support of the certification application;a copy of the decision of the certification of the country of origin;
- proof of the payment of renewal fees.

*Decision of the commission.* The commission is responsible for validating the application. The decision of the commission relates to the following characteristics:

- therapeutic significance and efficacy;
- safety of the therapeutic doses;
- pharmaceutical quality (packaging and labelling);
- the wholesale price excluding taxes and costs of daily treatment and full treatment.

The characteristics as presented are compared with those of already certified similar medicines. Complementary analyses and controls may be requested by the certification commission.

*Authority's decision.* The marketing licence is granted or refused by the competent authority in accordance with the decision of the technical certification commission. A rejection must be justified and the applicant notified. The latter has the possibility of appealing and providing responses to the technical commission's objections.

*Duration of validity and renewal of MAs.* The marketing licence is granted by the competent authority for a 5-year duration. The renewal is subject to the submission of an application comprising:

- an application addressed to the competent authority;
- a certification indicating that no modification occurred in the elements produced in support of the certification application;
- a copy of the decision of the certification of the country of origin;
- proof of the payment of renewal fees.

### 2.3.4.4 Good manufacturing practices for traditional medicine products principles

According to the GMPs, the product must not only meet final specifications, but also be manufactured under the same conditions and procedures every time. The establishments, their systems, equipment and processes must be properly controlled to ensure the systematic manufacture of quality products. Thus, before starting work, it is always necessary to have the right instructions and to:

- follow the instructions to the letter and, in case of doubt, seek the right information;
- use the correct components;
- use the right equipment and materials;
- avoid contamination and adulteration;
- always pursue accuracy and precision;
- maintain cleanliness and order;
- watch out for risks of error and malpractice – any deviation should be reported immediately;
- prepare clear and accurate reports on the work done and the controls made.

#### *Rules of hygiene and cleanliness*

The basic rules of hygiene must be observed at all times. Precise cleaning and hygiene instructions must be applied to eliminate potential sources of contamination. These rules of hygiene and cleanliness must apply throughout the manufacturing process to the staff, premises, equipment and instruments, raw materials and anything that could be a source of contamination of the product.

#### *Staff hygiene and clothing*

Any person handling raw materials, semi-finished products and finished products must wash their hands and forearms with soapy water, wear gloves, a mask, a cap or a scarf, overall or suitable clothing for production. Change clothing on arrival and leaving the production venue. The clothing, gloves and mask must be cleaned after each use with soapy water every day, at the close of work. It is advisable to use one-off gloves or, where normal gloves are used, disinfect them in water and bleach mixture. It is also essential to avoid engaging in excessive talking during production. It is mandatory to conduct a medical examination prior to recruitment and, in the event of infectious diseases, the production activities must cease.

#### *Equipment sanitation*

Any equipment to be used for raw materials and finished products must be clean. The equipment must be cleaned after each use and immediately after handling. The clean and dry equipment must be kept at a clean and dust-free place. Each piece of equipment must have a specific and exclusive use.

### 2.3.4.5 Equipment cleaning procedure

#### *Premises sanitation*

The entire premises should be cleaned (floors and work surface tops) daily. The floors should be cleaned at least three times a week with bleach (diluted according to the directions on the packaging: floor and surface cleaning). Just before handling, clean all surfaces on which products are handled with water and bleach mixture. The same mop should not be used to wash floors and work surfaces. To facilitate the cleaning and avoid cross-contamination, specific steps, such as air extraction and use of specific premises, should be taken

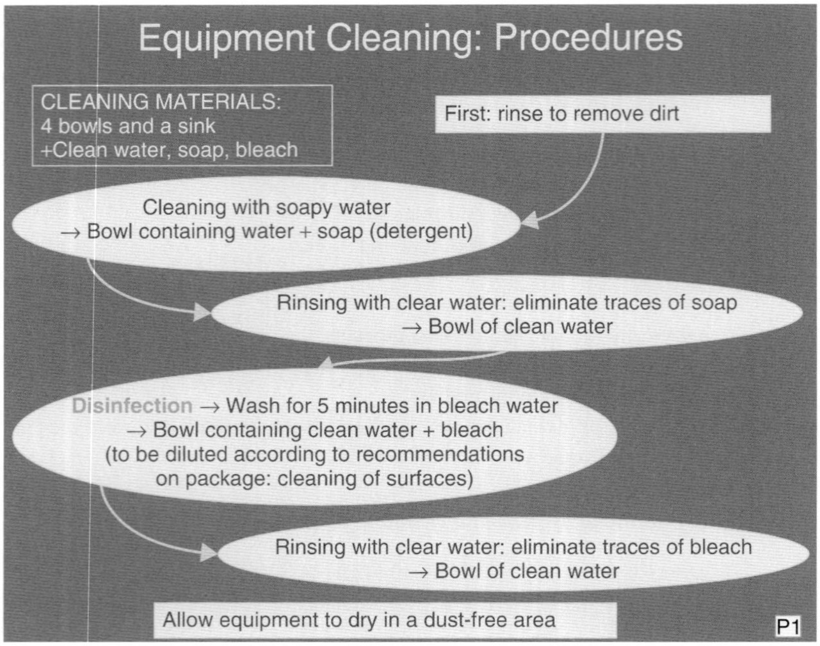


**Figure 2.3** Equipment cleaning procedure

during the removal of dust due to sampling, weighing, blending and processing of plants in the raw state. Figure 2.3 describes the equipment cleaning procedure.

***Site sanitation***

Waste bins must be regularly emptied and placed away from places of handling. It is not advisable to keep livestock on the same field or use pesticides or fertilizers on the site.

### 2.3.4.6 Organization

***Premises***

The premises must be suitable and sufficiently spacious. Depending on the activities, they may be divided into several compartments. It is essential to have shelves for keeping and storing various inputs and equipment. Inputs and equipment must be stored separately. This should be the same for storage and production areas. The premises and equipment must be located, designed, constructed, adapted and maintained in a manner that would suit the activities to be carried out. They should be maintained to prevent the proliferation of insects, microorganisms and animals (rodents). Their plan, arrangement, design and use must aim at minimizing risks of errors and facilitate effective cleaning and maintenance in order to avoid contamination, including cross-contaminations, dusts or dirt and, generally, any threat to the quality of the products. Lighting, temperature, moisture and ventilation must be appropriate in order not to directly or indirectly affect the medicines during their manufacture and storage or the smooth operation of the equipment.

***Staff***

Qualified and adequately trained staff is indispensable for the implementation of the GMP standards. A nominative organization chart of the establishment defining in a precise manner the tasks and responsibilities of each employee must be drawn up. Staff members who occupy positions of responsibility must have their specific tasks detailed in written job descriptions; they must be vested with the necessary authority to carry out their responsibilities. Their functions may be delegated to appointed alternates with suitable qualifications. There should not be any unexplained gap or overlap in the responsibilities of the staff concerned by the application of GMPs.

***Manufacturing process***

Any manufacturing process shall be clearly defined and systematically reviewed in light of experience; it must be demonstrated that the process is capable of reproducing the medicines to the required specification. A process entails several stages. All the production stages must be governed by procedures. The procedures are precise instructions for production and control. These enable:

- elimination of risks related to orally conveyed risks (erroneous or incomplete);
- traceability of the activities;
- limitation of contestations.

The procedures should be prepared in an appropriate style and clear and unambiguous vocabulary especially adapted to the means provided used. They must be used as reference by persons responsible for production and also used to train new staff members in the application of the method. The procedures will be updated in accordance with a review and revision time-table or when changes are made with regard to the method and equipment.

All the operators must receive training in order to implement the procedures correctly.

*Qualification and validation*

In accordance with GMP, each production structure must define the qualification and validation activities needed to prove that the critical aspects of the implementation of processes that are peculiar to it are well understood. Validation studies are a key element of GMPs and must be conducted in line with predefined and approved protocols. Manufacturing processes and procedures must be established on the basis of the results of the validations made. Responsibility for managing the validations must be clearly assigned. Qualification and validation must establish proof that: the premises, equipment and processes were designed in accordance with the requirements for applying GMPs. All aspects of the operation, including significant modifications made to the premises, installations, equipment or processes likely to directly or indirectly influence the quality of the product, must be qualified and validated.

*Documentation and archiving*

Plants used as raw materials must be documented and the necessary information for the purpose should comprise:

- botanical name;
- data on the origin of the plant, harvest period, harvesting method, possible use of pesticides;
- usable part;
- a description of the drying system;
- a description of the plant and its macroscopic and microscopic characteristics;
- identification trials;
- dosage method;
- appropriate methods for determining possible contamination by pesticides;
- trials to determine fungal and microbial contamination;
- trials to identify toxic metals, contaminants and possible fake products;
- trials to identify foreign substances.

Any treatment designed to reduce fungal or microbial contamination must be documented. The specifications comprising data on the treatment, trials and limitations of the residues must be available. Instructions related to the treatment must describe the various operations on the plant in the raw state, such as drying, crushing and sieving, as well as methods used to check the size of fragments or particles. They must also contain a description of the safety strainer or other methods for eliminating foreign substances. To ensure the traceability of the production, batch documents, monitoring of the settings during production, measurements and observations noted must be kept. It is equally advisable to have a physical trace, in the form of retained sample storage.

*Storage and transportation*

Plants in the raw state (unprocessed) must be stored in separate areas. The storage area must be well ventilated and be protected from insects and other animals, such as rodents. Effective steps must be taken to limit the proliferation of animal species and microorganisms introduced along with the raw plants in order to avoid cross-contamination. The storage of plants, extracts, stains and other products may require special conditions related to humidity, temperature and protection; these conditions must be met and verified.

*Quality control*

Quality control forms part of GMPs; it concerns sampling, actual control, as well as organizational, documentation and release procedures that ensure that the necessary and appropriate analyses have actually been conducted and that the raw materials, packaging items and finished products are not released for use, sale or supply without ensuring that they have first passed the quality test. In view of the often complex and variable nature of TMs, the control of the quality of raw materials and finished product is of capital importance.

Persons responsible for quality control must be specialized in the area of TMs in order to be able to undertake identification testing, recognize cases of spoilage, note the presence and development of microorganisms, detect infestations and lack of uniformity in the delivery of raw plant materials.

All control activities must follow written procedures. Sampling must be carefully carried out by staff with the requisite knowledge in order to eliminate errors related to the heterogeneity of TMs. Controls must follow written pharmacopoeia methods recognized by each country. The findings must be recorded and verified to ensure consistency. Documents related to sampling, control and recording, analytical reports and certificates, and outcomes of validation of control methods for each batch must be kept for 1 year after the expiry date and at least 5 years after the release of the batch.

### 2.3.4.7 Quality assessment

*Pharmaceutical assessment*

This assessment should cover all the important aspects of the quality assessment of the TMs. Where a pharmacopoeia has a monograph it should be consulted. Where it does not, one should be established using the same criteria as those for the monographs of the official pharmacopoeia. All the processes used should follow GMPs.

*Raw plant materials*

The botanical definition, namely the genus name, species and name of the author who described it first, should be provided for a correct identification of the plant. A definition and description of the portion of the plant used in manufacturing the medicine (e.g. leaf, flower, root) should be provided by indicating if the plant is used in the fresh state or dry state; or, after traditional processing, the active and characteristic ingredients must be specified. Foreign element and impurity content as well as microbial load should be defined or limited. Standard samples for each preparation batch should be authenticated by a qualified botanist and kept for a minimum of 10 years. A batch number should be assigned and mentioned on the label of the product.

*Plant preparations*

Plant preparations comprise plant materials in fragments or powder form, extracts, tinctures, fatty or essential oils, saps and preparations whose production involve fractioning, purification or concentration. The manufacturing method must be described in detail. Where a substance is added to the preparation during the manufacturing process to obtain a specific active ingredient content or characteristics, or for any purpose, that substance should be specified in the manufacturing method. An identification method, where possible for the titration of the preparation, should be indicated. Where it is not possible to identify the active ingredient, it should be sufficient to identify a substance or a mixture of characteristic substances (e.g. chromatographic profile) to guarantee the constant quality of the preparation.

*Finished product*

The manufacturing process and formula, including the quantity of excipients, should be described in detail. A finished product specification should be defined. A method of identification, and if possible of quantification of the plant preparation in the finished product, should be defined.

*Stability*

The physical and chemical stability of the product in the container intended for marketing should be assessed in specific storage conditions and the conservation period established.

## 2.3.5 Collaboration between traditional and conventional health practitioners

The policy orientation recommended in the regional strategy as an organizational arrangement for *institutionalization of TM* is development of mechanisms of collaboration between conventional and TM practitioners in areas such as referral of patients and information exchange at the local level. The strategy recommends that the ministry of health should also facilitate effective collaboration between THPs and CHPs. However, collaboration between THPs and CHPs is currently limited to prevention and care of people living with HIV/AIDS in sub-Saharan Africa (King, 2000) as demonstrated by studies in Cameroon, Central African Republic, South Africa, Malawi and Zambia. However, there are best practices of fruitful research-based collaborations and management of patients have been recorded in Mali, Senegal, South Africa and Uganda, and mechanisms for strengthening of collaboration between traditional health and conventional medicine practitioners have recently been proposed (Busia and Kasilo, 2010).

The recent economic crisis and globalization have resulted in malpractice in some countries, with charlatans and quacks practicing TM and mistrust between practitioners of the two systems of medicine, and this has hampered the efforts of some ministries of health to develop policy frameworks fostering collaboration between the two sectors. There is, therefore, a need for countries to accelerate the process of identification, licensing and accreditation of qualified practitioners, to develop medium- and long-term programmes for THPs' education, and well-defined frameworks for enhancing collaboration between practitioners of the two systems of medicine.

For effective collaboration, some key elements should be considered, and these include:

1. Development and implementation of national policies, legal and regulatory frameworks – member states should develop and implement national policies on TM as part of their comprehensive national health policies and strategic plans.
2. Governments should create enabling environments for the two systems to regularly engage in constructive dialogue to facilitate the development of well-defined national frameworks for effective collaboration.
3. Signing memorandum of understanding.
4. Stressing complementarities of both systems by referral from one health system to another.
5. Showing humility and respect.
6. Building mutual respect and trust.
7. Cultivating transparency through dialogue and negotiation.
8. Willingness to learn from one another.
9. Consistent exchange of information, materials and technology; for example, Involvement/participation of THPs and CHPs in joint research and development projects.
10. Existence of recognized, functional TM practitioners' associations for effective policy implementation and resource mobilization.
11. Establishment of collaborative health facilities.
12. Protection of indigenous medical knowledge of THPs.
13. Incentives to THPs.
14. Resource mobilization – TM is the least funded and developed health system in sub-Saharan Africa. In order to develop TM, the ministries of health should advocate and encourage partners and the private sector to invest more resources in TM.
15. Health insurance systems – the member states should establish a uniform charging and patient billing system and national health insurance systems for reimbursing TMs which have MAs and are included in the national essential medicines lists (NEMLs).
16. Acknowledgment of THPs in scientific publications. THPs should be cited in publications resulting from relevant information they provide to CHPs in research and academic institutions. Alternatively, they should at least be appropriately acknowledged. This will motivate them to share their knowledge more readily.

In spite of the huge benefits such collaborations and partnerships offer, considerable challenges still remain. For example, there is often a lack of transparency in the process of actualizing collaboration, resulting in a situation where the so-called 'collaboration process' is dominated by one group (often CHPs). The importance of collaboration between practitioners of the two systems of medicine cannot be overemphasized and is a necessary prerequisite for the ultimate institutionalization of TM in national health systems. The WHO Director General, Dr Margaret Chan, summarized the benefits of this collaboration very well in a speech delivered during the WHO Congress on TM held in Beijing, in November 2008:

> The two systems of traditional and Western medicine need not clash. Within the context of PHC, they can blend together in a beneficial harmony, using the best features of each system, and compensating for certain weaknesses in each. This is not something that will happen all by itself. Deliberate policy decisions have to be made. But it can be done successfully. Many countries have brought the two systems together

in highly effective ways. In several countries where health systems are organized around primary health care, TM is well integrated and provides a backbone of much preventive care and treatment of common ailments …

www.who.int/dg/speeches/2008/20081107/en/

## 2.4 Implementation of the regional strategy on traditional medicine

### 2.4.1 Policies, strategies and plans

Resolution AFR/RC5/R3 urges member states to formulate national policies on TM as part of their overall national health policies and to review existing legislation to conform to national policies. It also urges member states to establish a multidisciplinary and multisectoral mechanism to support the development and implementation of policies, strategies and plans. The resolution also requests the WHO to develop guidelines for the formulation and evaluation of national policies on TM, and advise countries regarding the relevant legislation for the practice of TM.

In implementing the above-mentioned resolution, the WHO developed tools and guidelines and organized regional and intercountry workshops to stimulate the development of national policies on TM mentioned above. The WHO also advises the countries on relevant legislation for the practice of TM. Countries adapted to their specific situations these tools for institutionalizing TM in health systems mentioned above for the formulation, implementation and monitoring of key components of national TM policies, strategies and plans, within the context of their national health policies.

The WHO 1999–2000 survey revealed that there were eight member states with national policies on TM, one with a THPs bill as a legal framework for the practice of TM, eight countries with laws and regulations and none with national codes of ethics for either THPs or national strategic plans. By 2012, 22 countries (Benin, Burkina Faso, Cameroon, Central African Republic, Chad, Congo, Democratic Republic of Congo, Equatorial Guinea, Gabon, Gambia, Ghana, Guinea, Guinea Bissau, Mali, Mozambique, Niger, Nigeria, Rwanda, South Africa, Tanzania, Uganda, Zimbabwe) mobilized additional financial resources as compared with 12 (Burkina Faso, Central Africa, Equatorial Guinea, Gabon, Ghana, Mali, Niger, Nigeria, Rwanda, Tanzania, Uganda, Zimbabwe) in 2000. Thirty-nine countries established national offices for TM and 24 established national programmes. In Ghana, herbal medicine clinics were established in 19 regional hospitals.

Forty countries (Angola, Benin, Botswana, Burkina Faso, Burundi, Cameroon, Central African Republic, Chad, Comoros, Congo, Côte d'Ivoire, Democratic Republic of Congo, Eritrea, Ethiopia, Equatorial Guinea, Gabon, Gambia, Ghana, Guinea Bissau, Guinea, Liberia, Madagascar, Malawi, Mali, Mauritania, Mauritius, Mozambique, Namibia, Niger, Nigeria, Rwanda, São Tomé and Príncipe, Senegal, Sierra Leone, South Africa, Tanzania, Togo, Uganda, Zambia and Zimbabwe) had established national TM policies, an increase of 87% from the baseline survey and 19 countries (Benin, Burkina Faso, Cameroon, Central African Republic, Congo, Côte d'Ivoire, Democratic Republic of Congo, Equatorial Guinea, Ghana, Guinea Bissau, Guinea, Malawi, Mali, Mozambique, Niger, Nigeria, Rwanda, Tanzania and Zimbabwe) had national TM strategic plans. In an effort to regulate, promote, develop and standardize the practice of African TM, country legal frameworks for TM practice, such as the Traditional Health Practioners Act, 2004 of South Africa (www.doh.gov.za/docs/legislation/acts/2004/act35.pdf), increased in number from one in 2000 to 29 in 2012. Similarly, the number of countries with codes of ethics and practice such as the Ghana Code of Ethics (GhanaWeb, 2004), rose from zero in 2000 to 19 in 2012. Six countries (Ethiopia, Ghana, Mali, South Africa, Tanzania and Zimbabwe) established national THPs' councils. Twenty-five countries established national expert committees for the development of TM, compared with 10 in 2000.

There is a need for enhancing efforts for implementation of the national policies through strategic plans within the context of national health policies, strategies and plans.

### 2.4.2 Cultivation, preservation and conservation of medicinal plants

Cultivation, preservation and conservation of medicinal plants is very important for ensuring sustainable raw materials for extemporaneous preparation of TMs, research and development and local production and domestic commercialization of medicines. The policy framework on the importance of medicinal plants is provided for by the Convention of Biological Diversity, the International Conservation of Nature, the Food Agricultural Organization and the WHO. Resolution AFR/RC5/R3 urges member states to actively promote, in collaboration with all other partners, the conservation of medicinal plants.

The WHO baseline survey of 2000 revealed that Botswana, Burkina Faso, Ghana and Lesotho had established policies related to conservation of medicinal plants, whereas Mali had a code on forests and legislation for the collection and conservation of medicinal plants. Seventeen countries (Burkina Faso, Cameroon, Ghana, Kenya, Madagascar, Malawi, Mali, Mauritania, Mozambique, Niger, Nigeria, Rwanda, São Tomé and Príncipe, Senegal, Seychelles, Zambia and Zimbabwe) (59%) indicated that they had a botanical garden or arboretum for *in situ* conservation of medicinal plants. By 2010, 17 countries (Benin, Botswana, Burkina Faso, Burundi, Cameroon, Congo, Democratic Republic of Congo, Ethiopia, Equatorial Guinea, Ghana, Guinea, Madagascar, Namibia, Mali, Mauritania, Swaziland and Zimbabwe) reported having established either commercial or community or individual cultivation of medicinal (e.g. *Artemisia annua*, chamomile, and *Moringa*) and aromatic plants (e.g. peppermint, shea and *Voacanga*). Thirteen countries (Benin, Burkina Faso, Cameroon, Congo, Chad, Democratic Republic of Congo, Ethiopia, Ghana, Nigeria, Rwanda, South Africa, Sierra Leone and Tanzania) reported having adopted the WHO guidelines on good agricultural and collection practices (GACP) and another 13 (Benin, Burkina Faso, Congo, Ethiopia, Ghana, Guinea, Madagascar, Mali, Mauritania, Rwanda, Senegal, South Africa, Swaziland) reported having adopted national policies on conservation of medicinal plants, while six (11%) had developed new medicinal plant varieties. In 2007, Cameroon developed guidelines related to the collection and conservation of medicinal plants.

By 2012, Benin reported to have established 14 medicinal plant gardens located in different ecological zones. The gardens were established by government in collaboration with the THPs' Association. The gardens are managed by THPs under supervision of the Ministry of Public Health. Other countries, such as Cameroon, Ghana, Mauritius, Senegal, South Africa and Zimbabwe, reported that they had developed national policies on conservation of medicinal plants. Twelve countries have reported ongoing activities in the promotion, conservation and cultivation of medicinal plants (Benin, Cameroon, Chad, Congo, Côte d'Ivoire, Democratic Republic of Congo, Gambia, Ghana, Madagascar, Nigeria, South Africa and Tanzania), while Kenya, Tanzania and Uganda reported large-scale cultivation of *Artemisia annua*.

Although countries have made limited progress in this area, there is still a need for additional effort in promoting home and botanical gardens, as well as selective use of modern biotechnology to develop improved medicinal plant varieties with higher yields of the bioactive constituents, integrated into the National Biotechnology Policy. There is also a need to promote practices which will guarantee consistent quality of the plant raw materials, training stakeholders, including THPs, and translating WHO guidelines on GACP into local languages. There is also a need for intersectoral collaboration and adequate information on well-researched medicinal plants required for commercial cultivation, among others.

### 2.4.3 Research and development of traditional medicine

Resolution AFR/RC5/R3 urges countries to produce inventories of effective practices as well as evidence on safety, efficacy and quality of TMs and undertake relevant research. The resolution requests the WHO to develop guidelines for the documentation of practices and medicines of proven safety, efficacy and quality and facilitate the exchange and utilization of this information by the countries. It also requests the WHO to strengthen WHO collaborating centres and other research institutions to carry out research, development of pharmacopoeia and monographs of medicinal plants and disseminate results on safety and efficacy of TMs.

The 1999–2000 WHO baseline survey revealed that 11 countries were conducting research, including ethno-medical surveys on TMs used for the treatment of malaria, opportunistic infections related to HIV/AIDS and sickle-cell disease. Mali had issued MAs for seven African TMs and included them in its NEML, including 'malarial'. In addition, Ghana, Guinea, Mozambique and Uganda had developed monographs on medicinal and aromatic plants in order to ensure safety, efficacy and quality of TMs. Equatorial Guinea and Guinea had national herbal pharmacopoeias, and the Organization of African Unity had published the African pharmacopoeia in 1985.

By 2012, there were 22 countries (Benin, Burundi, Burkina Faso, Cameroon, Côte d'Ivoire, Democratic Republic of Congo, Ethiopia, Ghana, Guinea, Kenya, Madagascar, Mali, Mozambique, Nigeria, Rwanda, Senegal, South Africa, Tanzania, Togo, Uganda, Zambia and Zimbabwe) which informed the WHO to conduct research on TMs used for the treatment of malaria, HIV/AIDS, sickle-cell anaemia, diabetes and hypertension using WHO guidelines. Seventeen countries (Benin, Cameroon, Chad, Equatorial Guinea, Ethiopia, Gambia, Guinea, Madagascar, Mali, Mauritius, Mozambique, Niger, Rwanda, Senegal, Seychelles, South Africa and Uganda) developed monographs on medicinal plants, whereas the Democratic Republic of Congo, Ghana and Nigeria and the Association for African Medicinal Plants Standards (AAMPS) developed national herbal pharmacopoeias and the African Herbal Pharmacopoeia respectively in order to set standards of safety, efficacy and quality. Cameroon, Chad, Côte d'Ivoire and Seychelles carried out national inventories of medicinal plants.

There is a lack of sound evidence on the safety, efficacy and quality of TM and weak dissemination and exchange of information on research results where this is available. There are also weak linkages between the TM research community, health services and policy makers hindering utilization of research results in policy making. There is inadequate equipment and infrastructure. There is limited collaboration between research institutions and the public; and very little public-private-partnerships (PPPs) between industries and universities. However, there is more collaboration with institutions and researchers outside the continent. There is also uncoordinated research into African TM, limited equipment and its maintenance, limited or absence of reference centers/laboratories for technical support to R&D; and inadequate research on TM practice. There are scarce documents on national standards for ensuring safety, efficacy and quality, such as monographs of medicinal and aromatic plants and herbal Pharmacopoeias to inform production of standardized TMs, large-scale cultivation of medicinal plants to ensure uninterrupted production of TMs.

Examples of common medicinal plants used for research and treatment of various diseases include *Calotropis procera* (Ait.) R. Br. (Asclepiadaceae) (Figure 2.4). The roots, latex and fruits of *Calotropis procera* (Ait.) R. Br. (Asclepiadaceae) are used for wound healing, antifungal (ringworm), dermatitis and boils. *Cassia occidentalis* L. (Caesalpiniaceae) (Figure 2.5). The leaves

**Figure 2.4** Fruits of *Calotropis procera* (Ait.) R. Br (Asclepiadaceae)

**Figure 2.5** The leaves of *Cassia* occidentalis L. (Caesalpiniaceae)

**Figure 2.6** *Adansonia digitata* L. (Bombacaceae) (Baobab)

**Figure 2.7** Fruits of *Carica papaya* (Caricaceae)

of *Cassia occidentalis* L. (Caesalpiniaceae) contain anthracene, flavonoids (antibacterial, antifungal and antiparasitic) and are used for vomiting, fever, malaria and joint pain. *Adansonia digitata L. (Bombacaceae)* (Figure 2.6), *Carica papaya* (Caricaceae) (Figure 2.7). The fruits of *Carica papaya* (Caricaceae) are used for skin ulcers, burns and jaundice. These plants and the summary on their use are shown below.

### 2.4.4 Local production and commercialization of traditional medicines

The WHO survey of 1999–2000 revealed that 16 member states locally produced small-scale TMs. There were eight member states with registration systems for African TMs in 1999–2000; however, less than half of the member states established herbal-medicine-related pharmacovigilance systems during the decade. By 2012, a total of 17 member states reported to have at least one to several manufacturing facilities dedicated to the production of standardized TMs. Other efforts include Ethiopia and Uganda, who adapted WHO guidelines to develop national registration documents and SADC for member states of its sub-region.

The member states' efforts in research and local production have not only resulted in contributing to improving access to some medicines used for communicable and non-communicable diseases, but also influenced policy change. However, the medicines

**Figure 2.8** Different forms of raw materials used for local production

included in national essential medicines lists are still not widely prescribed in health facilities even within the countries where they are commercialized. There is, therefore, the need to create enabling environments for the establishment and scaling up of post-market TM pharmacovigilance studies, production of regular supplies of plant raw materials (Figure 2.8), and local production, use and marketing of African TMs within sub-regions.

This would contribute to utilization of uninterrupted production capacity and registration of additional medicines from medicinal plants. There is persistent use of artisan techniques to manufacture TMs, weak production capacity, lacks of adequate structures, equipment, financial resources and investment by the pharmaceutical industries in local production of TMs, and limited trained personnel on local production of standardized African TMs; national registration procedures of TMs have not yet been put in place in some member states, and where they exist the criteria for the registration of TMs remain rigid, often in clear breach of WHO guidelines for registration of TMs.

### 2.4.5 Registration of traditional medicine products and inclusion in national essential medicines lists

TMs that meet the criteria of safety, efficacy and quality should be made available to the majority of the population through their registration by national medicines regulatory authorities. To facilitate this process the WHO convened two regional training workshops in South Africa in 2002 and in Spain in conjunction with the International Conference of Drug Regulatory Authorities in 2004 to review and contribute to improving the WHO draft guidelines on registration of TM products. In order to accelerate the process of registration, the WHO has proposed to member states the guidelines as a framework for facilitating the registration of TM products (WHO, 2004c). The proposed framework, which has a regional perspective, is intended to accelerate the registration and circulation of standardized African TM products within the WHO African Region, on the basis of criteria of pharmaceutical quality, safety of use and therapeutic efficacy.

Many countries in the WHO African Region are encountering problems in providing their people with equitable access to health care; at present, only about one half of the population in the region has access to conventional health care. Such considerations aside, TM products continue to be popular for both historical and cultural reasons. Countries have adopted the WHO guidelines on registration of TMs in the African Region, and by 2012 there were 13 countries (Burkina Faso, Cameroon, Congo, Côte d'Ivoire, Democratic Republic of Congo, Ghana, Madagascar, Mozambique, Niger, Nigeria, Tanzania and Zambia) that had used research results to issue MAs for TM products, ranging from three in Cameroon and Congo to over 1000 in Ghana and Nigeria. Eight countries (Burkina Faso, Cameroon, Democratic Republic of Congo, Ghana, Madagascar, Mali, Mozambique and Niger) included TM products into their NEMLs. Examples of TM products in the NEMLs include Saye (Traoré *et al.*, 2008) and N'dribala (Benoit-Vical *et al.*, 2003), used for malaria treatment in Burkina Faso. Similarly, Malarial (Diallo *et al.*, 2004) is used for malaria treatment in Mali and Faca (Nikiema *et al.*, 2010) is used for sickle-cell disease in Burkina Faso. Madeglucyl (Ratsimamanga, n.d.) for treatment of uncomplicated diabetes in Madagascar and Hepatisor used for hepatitis in Cameroon.

### 2.4.6 Protection of intellectual property rights and traditional medical knowledge

With regard to IPRs, Resolution AFR/RC5/R3 urges member states to actively promote, in collaboration with all other partners, the protection of IPRs and indigenous knowledge in the field of TM. The resolution requests the WHO to provide guidance to countries for the development of measures for the protection and preservation of traditional medical knowledge (TMK). In 1999–2000 no AU member states had developed a national policy or framework on IPRs for the protection of TMK. In 2000 the OAU Model Legislation for the Protection of Local Communities, Farmers, Breeders and Access to Biological Resources was adopted by the Council of Ministers.

IPRs are a relatively new area for which capacities need to be enhanced at all levels. As such, countries have not made much progress in this area. However, Kenya, Mozambique, South Africa and Zimbabwe have amended their legislation on industrial property to conform to the Trade-Related Aspects of Intellectual Property Rights (TRIPS) Agreement. Some countries have developed national policies and legislation, and documented African TM in the form of inventories, monographs, pharmacopoeias and recorded in the form of databases, among others. To provide guidance to countries developing measures for the protection and preservation of TMK, the WHO Regional Office for Africa has developed a WHO Regional Sui Generis Legislative 'Framework for the Protection of Indigenous Knowledge in African Traditional Medicine' and WHO 'Policy Guidance for the Protection of Indigenous Knowledge in African Traditional Medicine' for adaptation to the specific situations of member states. On its part, the African Regional Intellectual Property Organization (ARIPO) developed a draft legal instrument on the Protection of Traditional Knowledge and Expressions of Folklore in 2006.

By 2012, Botswana, Cameroon, Chad, Ghana, Nigeria and South Africa had adopted their national tools for protection of IPRs and TMK as compared to zero in 1999–2000. South Africa developed the indigenous knowledge systems (IKS) policy and a Biodiversity Act in 2004 and established a national office of IKS in 2006. Nigeria developed national legislation in 2006 and a Bill on IPRs in 2007. Nine countries (Burkina Faso, Congo, Democratic Republic of Congo, Ghana, Guinea, Madagascar, Mali, Rwanda and South Africa) had adopted the OAU Model Legislation, nine countries (Botswana, Chad, Eritrea, Ethiopia, Ghana, Nigeria, South Africa, Uganda and Zimbabwe) had adopted their national frameworks for the protection of IPRs and TMK, and eight countries (Benin, Cameroon, Democratic Republic of Congo, Ghana, Guinea, Mali, Senegal and South Africa) reported to have established databases on TM practitioners, TMK and access to biological resources; 17 countries (Burkina Faso, Cameroon, Central African Republic, Chad, Côte d'Ivoire, Equatorial Guinea, Gabon, Ghana, Guinea, Mali, Mozambique, Niger, Nigeria, Rwanda, Tanzania, Uganda and Zimbabwe) carried out inventories of medicinal plants.Mali and Tanzania organized a series of sub-regional and national sensitization workshops for the protection of TMK respectively. As a result, Tanzania is currently developing its national framework for the protection of indigenous knowledge. Furthermore, Cameroon, Chad, Cote d'Ivoire and Seychelles carried out national inventories of medicinal plants. With WHO support, 21 countries (Angola, Benin, Burkina Faso, Cameroon, Congo, Democratic Republic of Congo, Equatorial Guinea, Ethiopia, Gabon, Ghana, Kenya, Madagascar, Mali, Mozambique, Nigeria, Niger, Senegal, Seychelles, South Africa, Tanzania and Uganda) documented TM experiences and part of that information was used to develop audiovisual programmes, which were disseminated to all countries for use on the occasion of the celebration of the African TM Day.

There is a need for development of national TM policies by the ministries of health that would influence the development of IPR policies by relevant ministries. There is a need for one lead ministry to deal with issues of traditional knowledge pertaining to TMK and to coordinate the efforts of other ministries in this regard. The WHO will continue to work in close collaboration with organizations dealing with IPRs issues, particularly the AU, the ARIPO and the African Organization for Intellectual Property (OAPI) to disseminate information and to support countries to understand the implication of the TRIPS agreement as it relates to TM.

### 2.4.7 Capacity building

With regard to training, in 1999–2000, 17 countries had initiated training programmes to improve THPs' skills and PHC knowledge. Some of these countries provided training in TM for pharmacists as TM modules or through botany, phytotherapy, pharmacognosy and public health and for doctors (Ghana and Zimbabwe) and nurses (Ethiopia, Ghana, Mozambique and Zimbabwe). For example, Figure 2.9a and b shows health science students being trained in TM research and women THPs' exposure to the role of TM in health systems in South Africa. In order to facilitate development of curricula and training programmes in its member states, the WHO Regional Office for Africa developed training tools for health sciences in TM and in PHC, which were field-tested in 10 member states.

By 2012, 13 member states responded to have developed a total of 26 training programmes. The Kwame Nkrumah University of Science and Technology in Ghana established a Bachelor's Degree in Herbal Medicine in 2001, and as of 2012 had trained 80 graduate

**Figure 2.9** (a) Health science students and (b) women THPs' exposure to the role of TM in health systems

medical herbalists. A similar programme was established in Burkina Faso in 2011. A master's degree programme for health science students was established in Tanzania with enrolment of seven students, while Guinea and Sierra Leone are reported to be progressing this. Despite this progress, less than a quarter of the member states have developed curricula for health science students and training programmes for THPs.

There is a need for scaling up the efforts to promote and support provision of training to health sciences students, training and retraining of THPs, and the establishment of systems for the qualification, accreditation or licensing of THPs. The WAHO tools in TM adopted by ECOWAS deans of pharmacy and medical schools in October 2010, and the WHO training tools in TM for health sciences students and THPs in PHC which were adopted by experts, representatives from training institutions and associations of THPs, will go a long way to facilitate this process.

### 2.4.8 African Traditional Medicine Day in countries of the region for advocacy

Resolution AFR/RC50/R3 requests the WHO to institute in member states the African TM Day for advocacy. The WHO made this proposal in 2002, which was endorsed by heads of state and government in Maputo in 2003 for countries to commemorate the African TM Day on 31 August of every year. Only Ghana had instituted a TM Week in 1999. Since August 2003, all countries in the region have been commemorating the African TM Day on 31 August each year on specific themes. During the commemoration, THPs, CHPs, scientists, nongovernmental organizations (NGOs) and other stakeholders carried out joint activities such as exhibitions, debates, symposia, seminars, panel discussions and cultural shows.

These events have popularized TM by raising its awareness and the profile of TM, as well as promoted its institutionalization into national health systems, and this has resulted in the institution of a National TM Week in seven additional countries, namely Benin, Burkina Faso, Burundi, Congo, Mali, Senegal and Uganda, by 2012. In addition to celebrating the African TM Day, Burkina Faso and Mali celebrate an International Exhibition of African TM and International Week of TM respectively. THPs in these two and neighbouring countries participate in activities organized during these events, thereby fostering collaboration, networking and information exchange among associations of THPs. Box 2.1 reports an example of activities carried out during the first African TM Day that was commemorated in South Africa in conjunction with the fifty-third Regional Committee for Africa with the theme *African TM: Our Culture, Our Future* (WHO, 2003b).

**Box 2.1. Celebrating the first African TM Day, 31 August 2003, Pretoria, South Africa**

The first African TM Day was launched in Pretoria, South Africa, on 31 August 2003 with the theme 'TM: Our Culture, Our Future'. Henceforth, 31 August of each year has been designated as a day for advocacy on TM by the WHO Director.

The African TM Day was celebrated on the eve of the 53rd African Regional Committee for Africa (RC53), in conjunction with the launching of the South African National Reference Center for African TMs. Therefore, it was able to bring together ministers of health and other national health officials from the WHO African Region, the WHO staff who participated in RC53, the national authorities of South Africa, THPs and CHPs from South Africa and neighbouring countries, consumers and the press.

It was a very lively celebration, with speeches and presentations from representatives of associations of THPs and CHPs, national health authorities, the WHO Director General, the late Dr Lee Jong-wook, and the WHO Regional Director for Africa, Dr Ebrahim Malick Samba. The Director General called on African governments to accord formal recognition to TM, create an enabling environment for its practice and integrating the time-honoured system of medicine into their national health systems, as is the case in the People's Republic of China, the Republic of Korea and Vietnam. The WHO estimates that large and growing numbers of people in developed countries are turning to TM (sometimes called alternative or complementary medicine) for preventive or palliative care. The global market for traditional therapies stood at more than US$ 60 billion in 2000, and is steadily growing.

Dr Ebrahim Malick Samba stated that the theme 'TM: Our Culture, Our Future' was selected to highlight the recognition that TM forms the basis of the various cultural backgrounds of the peoples of Africa. With the collaboration of all stakeholders, Africans could benefit tremendously from the health and socio-economic advantages that TM offers for the modern world. He underscored the fact that African TM Day was a historic event, which must be accorded due recognition and support, and be commemorated by all Africans and stakeholders involved in the development of African TM.

The WHO leaders outlined other measures that should be taken by countries to promote the use of TM. These include investment in research to validate claims on the quality, safety and efficacy of TM products used to treat priority diseases, with a view to increasing access to treatment, large-scale cultivation and conservation of medicinal plants, and the protection of relevant IPRs.Box 2.2

Traditional medicines exhibited during the 10th African TM Day by one of the THPs in Ouagadougou, Burkina Faso, in 2012. The 10th African TM day was celebrated under the theme 'Decade of Institution of the African TM Day: What Impact?'

Box 2.2 is on TMs exhibited during the 10th African TM day by one of the THPs in Ouagadougou, Burkina Faso, in 2012. The celebration of the African TM Day continues to be annually with specific themes. Consequently, the 10th and 11th African TM Day were celebrated under the themes 'Decade of Institution of the African TM Day: What Impact?' and 'Traditional Medicine Research and Development', in 2012 and 2013 respectively, under the leadership of the current WHO Regional Director, Dr Luis Gomes Sambo.

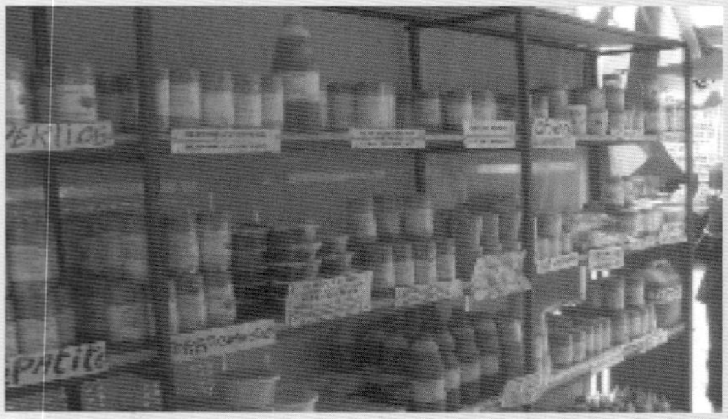

Traditional medicines exhibited during the 10th African Traditional Medicine day by one of the THPs in Ouagadougou, Burkina Faso, in 2012. The 10th African TM day was celebrated under the theme 'Decade of Institution of the African TM Day: What Impact?'

## 2.5 The way forward

### 2.5.1 Implemetation of the Second Decade of African Traditional Medicine plan of action (2011–2020)

The year 2010 marked a decade since the adoption of the above-mentioned regional strategy and the declaration of the first African TM Decade (African Union, 2005). The fourth Ordinary Session of the African Union Ministers held in Addis Ababa, Ethiopia, in May 2009 recommended that a final assessment of the End of Decade be undertaken by the African Union Commission and the WHO for its consideration of the report at its Fifth Session in 2011. The AU Conference of African Ministers of Health held in Windhoek from 17 to 21 April 2011 discussed the End-of-Decade Review report on African TM and renewed the decade from 2011 to 2020. There are four new elements of focus in the priority areas of intervention of the Second Decade of African TM plan of action: integration of TM into the public health system; community mobilization; research and development; and capacity strengthening. Effective implementation of the Second Decade of African TM plan of action (2011–2020) will go a long way to elevating TM in its rightful place.

### 2.5.2 Updating the regional strategy on promoting the role of traditional medicine in health systems

Since member states are at different stages of implementing the first regional strategy on *Promoting the role of TM in health systems: a strategy for the African region*, they need another decade to enhance implementation of the strategy to reach some of the goals set and address the challenges mentioned above. Therefore, the regional strategy adopted in 2000 is being updated for adoption by the WHO Regional Committee for Africa in 2013 to address the challenges that the member states have encountered. The promotion of positive aspects of practices and the use of TM products with proven efficacy and safety will supplement other efforts to contribute to the achievement of universal health coverage.

Promotion of the role of TM is critical to assisting countries to integrate it into their national health systems, as well as ownership of the strategy, development of country-specific strategies, utilization of research results for decision making, effective partnerships, and availability of medicines and sharing of information. The implementation of the updated strategy will require concerted collaboration among all the partners and stakeholders, and effective and rational mobilization of all resources available at the country and regional levels.

### 2.5.3 Development of African herbal pharmacopoeias for quality control of medicines

The WHO encourages member states to provide safe and effective traditional remedies and practices in public and private health services. Resolution AFR/RC50/R3, *Promoting the role of TM in health systems: a strategy for the African region*, requests the WHO to strengthen collaborating centres and other research institutions to develop pharmacopoeia and monographs of medicinal plants and disseminate results on safety and efficacy of TM products. The compendia most commonly used as standardized procedures for evaluating the quality control of medicinal products for human use are pharmacopoeia, pharmacopoeial formulary and dispensatory. A pharmacopoeia is used to set standards for the raw materials used in the preparation of the medicines, and a pharmacopoeial formulary gives the specifications of all raw materials and the quantities that are to be used to manufacture a specific medication. A dispensatory, on the other hand, combines some aspects of both the pharmacopoeia and the formulary.

The relevance of these compendia (pharmacopoeia, formulary and dispensatory) for the development of TM in countries of the African region cannot be overemphasized, as the only written documentation on African TM available now is all books published by individuals on plant medicines, which are useful publications. However, there is still a need for official compendia, which on approval by a member state will become a legal reference document for establishing the identity and quality of materials. In order to support member states in this regard, the WHO has published four volumes of *WHO monographs on selected medicinal plants* published in 1999 (WHO, 1999), 2003, 2007 and 2009 (WHO, 2009), and these available on the WHO website. The monographs aim to provide scientific information on the safety, efficacy and quality control of widely used medicinal plants; to provide models to support member states in developing their own monographs or formularies and to facilitate information exchange among member states.

WHO monographs are not intended to replace official compendia such as pharmacopoeias, formularies or legislative documents, but rather to provide comprehensive scientific references for national medicines regulatory authorities, pharmacists, physicians, research scientists, THPs, manufacturers and the general public. Monographs also promote harmonization in the use of herbal medicines with respect to levels of safety, efficacy and quality control. These aspects of herbal medicines depend largely on how the individual dosage form is prepared. For this reason, national regulatory authorities, experts, and health workers, as well as the scientific literature, should be consulted to determine whether a specific herbal preparation is appropriate for use in PHC.

The WHO will support member states wishing to develop such official compendia for the regulation of TM in the region, as has been done by the Democratic Republic of Congo (République Démocratique du Congo, 2009), Equatorial Guinea, Ghana (STEPRI, CSIR, 2007) and Nigeria (Federal Ministry of Health, 2008). In view of the geographical prevalence of medicinal plants, it may be necessary to develop such compendia on a regional basis, although some countries may wish to develop their own. However, it is important to state that it is the availability of such national or regional pharmacopoeial compendia that will facilitate the regulation of African TM. It is worth noting that the Organization of African Unity (OAU, now the AU) initiated the production of the African Pharmacopoeia in 1985 for this very reason. However, this document has limitations, mainly due to the attempt to cover medicinal plants common to all the countries of the continent.

Besides the African Pharmacopoeia, the WAHO in collaboration with the WHO Regional Office for Africa has, since 2008, been engaged in the development of monographs for the West African Herbal Pharmacopoeia, which will be an important regulatory tool towards improving the quality, safety and efficacy of medicinal plants in the ECOWAS member states. Also notable is the African Herbal Pharmacopoeia developed by the Association for African Medicinal Plants Standards (AAMPS) (Brendler *et al.*, 2010). Other efforts have been recorded in 19 countries (Benin, Burkina Faso, Cameroon, Chad, Equatorial Guinea, Ethiopia, Gambia, Ghana, Guinea, Madagascar, Mali, Mauritius, Mozambique, Niger, Rwanda, Senegal, Seychelles, South Africa and Uganda) which have developed monographs on medicinal plants in the African region.

Apart from the above-mentioned countries and AAMPS, AU and WAHO, which have developed monographs and published herbal pharmacopoeias, most countries and professional associations and networks have not published their work.

## 2.6 Conclusion

As part of the African culture, TM will continue to play an important role in health systems in countries of the region. For this to be realized, countries need to mobilize and allocate additional resources for research and development and capacity building. We believe that strong public–private and community partnerships and partners such as ministries of health and other relevant public and private institutions, NGOs, the AU, African Development Bank, United Nations International Development Organization, United Nations Conference on Trade and Development and the World Bank will be paramount for large-scale production and commercialization of TM products. This is the main goal of the updated regional TM strategy.

## References


African Union (2005) Plan of action on the implementation of the Decade of African TM (2001–2010), in *2nd Ordinary Session of the Conference of African Ministers of Health (CAMH2)*, Gaborone, Botswana, 10–14 October. Implementation of the Decision of the Lusaka Summit of Heads of State and Government, AHG/DEC.164 (XXXVII).

Benoit-Vical, F., Valentin, A., Da, B. *et al.* (2003). N'Dribala (*Cochlospermum planchonii*) versus chloroquine for treatment of uncomplicated *Plasmodium falciparum* malaria. *Journal of Ethnopharmacology*, **89** (1): 111–114.

Bodeker, G., Ong, C.K., Grundy, C. *et al.* (2005) *WHO Global Atlas of Traditional, Complementary and Alternative Medicine*, World Health Organization, Centre for Health Development, Kobe, Japan, pp. 3–12.

Brendler, T., Eloff, J.N., Gurib-Fakim, A. and Phillips, L.D. (eds) (2010) *African Herbal Pharmacopoeia*, Association for African Medicinal Plants Standards (AAMPS), Pretoria.

Busia, K. and Kasilo, O.M.J. (2010) Collaboration between traditional health practitioners and conventional health practitioners: some country experiences. *African Health Monitor*, (Special Issue 14), 40–46.

De Silva, T. (1997) Industrial utilization of medicinal plants in developing countries, in *Medicinal Plants for Forest Conservation and Health Care* (eds G.C. Bodeker, K.K.S. Bhat, J. Burley and P. Vantomme), FAO, Rome, pp. 34–44, http://www.fao.org/docrep/w7261e/W7261e07.htm (accessed 4 November 2013).

Diallo, D., Maiga, A., Diakité, C. and Wilcox, M. (2004) Malarial-5: development of an antimalarial phytomedicine in Mali, in *Traditional Herbal Medicines for Modern Times: Traditional Medicinal Plants and Malaria* (eds M. Wilcox, G. Bodeker and P.e Rasoanaivo), CRC Press, London, pp 117–130.

Federal Ministry of Health (2008) *Nigerian Herbal Pharmacopoeia*, 1st edn, Federal Ministry of Health/WHO, Abuja (ISBN: 978-978-088-020-0).

GhanaWeb (2004) Code of ethics for traditional medicine practitioners in Ghana, www.ghanaweb.com/GhanaHomePage/NewsArchive/artikel.php?ID=65168 (accessed 10 March 2013).

Kasilo, O.M.J., Soumbey-Alley, E., Wambebe, C. and Chatora, R. (2005) Regional overview: African region, in *WHO Global Atlas of Traditional, Complementary and Alternative Medicine* (eds G. Bodeker, C.-K. Ong, G. Burford et al.), World Health Organization, Geneva, pp. 3–12.

King, R. (2000) Collaboration with Traditional Healers in HIV/AIDS Prevention and Care in Sub-Saharan Africa. A Literature Review, Best practice collection, UNAIDS/00.29E, UNAIDS, Geneva.

Nikiema, J.B., Ouattara, B., Sembde, R. *et al.* (2010) Promotion de la médecine traditionnelle du Burkina Faso: essai de développement d'un médicament antidrépanocytaire. *African Health Monitor*, (Special Issue 14), 52–57.

Ratsimamanga, S.U. (n.d.) *Eugenia jambolana*: Madagascar, Malagasy Institute of Applied Research, Antananarivo, http://tcdc2.undp.org/GSSDAcademy/SIE/Docs/Vol7/Eugenia_Jambolana_Madagascar.pdf (accessed 4 November 2013).

République Démocratique du Congo (2009). *Pharmacopée Traditionnelle de la République Démocratique du Congo: Science et Tradition*, 1st edn, Ministère de la sante Publique, Kinshasa.

STEPRI, CSIR (2007) *Ghana Herbal Pharmacopoeia*, 2nd edn, Science and Technology Policy Research Institute, Council for Scientific and Industrial Research, Accra.

Traoré, M., Diallo, A., Nikièma, J.B. *et al.* (2008). *In vitro* and *in vivo* antiplasmodial activity of 'saye', an herbal remedy used in Burkina Faso TM. *Phytotherapy Research*, **22**, 550–555.

UNCTAD (2000) Systems and National Experiences for Protecting Traditional Knowledge, Innovations and Practices, United Nations Conference on Trade and Development, TD/B/COM.1/EM.13/2, http://www.unctad.org/en/Docs/c1em13d2.en.pdf (accessed 20 December 2012).

WHO (1978) Primary Health Care: Report of the International Conference on Primary Health Care, Alma-Ata, USSR, 6–12 September, 1978, Health for All Series No. 1, World Health Organization, Geneva.

WHO (1999) *WHO Monographs on Selected Medicinal Plants – Volume 1*, World Health Organization, Geneva.

WHO (2002) *WHO Traditional Medicine Strategy (2002–2005)*, WHO/EDM/TRM 2002.1, World Health Organization, Geneva.

WHO (2003a) The Regional Meeting on Integration of Traditional Medicine in Health Systems: Strengthening Collaboration between Traditional and Conventional Health Practitioners, Harare, Zimbabwe, 26–29 November 2001, Final Report, WHO Regional Office for Africa, Brazzaville, Republic of Congo, Document reference AFR/TRM/03.2.

WHO (2003b) African TM: Our Culture, Our Future. *African Health Monitor*, **4**.

WHO (2004a) Regional Meeting on Institutionalization of Traditional Medicine in Health Systems Held in Ouidah, Benin, 13–15 September 2004, WHO Regional Office for Africa, Brazzaville, Republic of Congo, Document reference AFR/TRM/04.6.

WHO (2004b) Tools for Institutionalizing Traditional Medicine in Health Systems of Countries in the WHO African Region, WHO Regional Office for Africa, Brazzaville, Republic of Congo, Document reference AFR/TRM/04.3.

WHO (2004c) Guidelines for Registration of TMs in the WHO African Region, WHO Regional Office for Africa, Geneva, Document reference AFR/TRM/04.1.

WHO (2009) *WHO Monographs on Selected Medicinal Plants – Volume 4*. World Health Organization, Geneva.

WHO Regional Office for Africa (2001) Document AFR/RC50/9 and Resolution AFR/RC50/R3, Promoting the Role of TM in Health Systems: A Strategy for the African Region.

WHO Regional Office for Africa (2010) Decade of African TM (2001–2010). *African Health Monitor*, (Special issue 14), 4–103.

WHO Regional Office for Africa (2011) Progress report on decade of traditional medicine in the African Region, Sixty-first session of WHO Regional Committee for Africa, Yamoussoukro, Cote d'Ivoire, 29 August–2 September, AFR/RC61/PR/2.

WHO Regional Office for Africa (2013) Enhancing the role of traditional medicine in health systems: a strategy for the African Region, Sixty-Third Session of WHO Regional Committee for Africa, Brazzaville, Republic of Congo, 2–6 September, AFR/RC63/6.

WHO Regional Office for the Western Pacific (2012) *The Regional Strategy for Traditional Medicine in the Western Pacific Region (2011–2020)*. WHO Regional Office for West Pacific Region, Manila (ISBN 9789290615590).

Wijesekera, R.O. (1991) Is there an industrial future for phytopharmaceutical drugs? An outline of UNIDO programmes in the sector. *Journal of Ethnopharmacology*, **32**, 217–224.

# 3 Cultivation of Novel Medicinal Plant Products and Associated Challenges

Ulrich Feiter
Parceval Ltd, Cape Town, South Africa

> Little did I know when, in 1995, I had, through some stroke of luck, secured an appointment with the managing director of DHU, the largest homoeopathic company in Germany. I was just 3 years into running my own very tiny company, growing a few herbs and manufacturing herbal tinctures in South Africa. Here I was, relatively inexperienced but eager to succeed. I placed some vials of sample extracts on the table. Dr Wartini looked at the samples and assured me that his colleagues would test them. 'But', he continued, 'you are coming from South Africa. We are using a plant called *Pelargonium sidoides* from your part of the world. Would you like to get involved in sourcing and cultivating this plant for us?' Excited as I was, I immediately agreed. We developed a basic framework for the first year of activities and off I went back to South Africa, full of enthusiasm and impatience. And that is where my personal journey into novel crops started. Little did I know how much sweat and toil I would encounter – but also the sweet taste of success! Here are some of the experiences I have gathered along the way to share with the valued reader.

## 3.1 Introduction

What is a novel plant? It is a plant that so far has not yet been commercially used, researched, cultivated, processed, and so on. And then, well, they are novel, which means that there is:

- *Nobody* you can ask for advice! Chances are that this is the first time ever that this plant has been cultivated. Nobody has probably tried or even has thought of cultivating – because, why? The plant in question may be little known, grows in remote places, has no ornamental value, is a slow grower or endemic to a specific climatic condition, comes from a country where production of food crops is more important than novel plants.
- *No* literature to look up anything specific about the plant. Often, botanical descriptions are all that can be found about a novel plant. And these may be difficult to find and then offer very limited information.
- *No* agricultural extension officer to help with seed sources, soil types, fertilizer requirements, irrigation needs, diseases, optimal growing conditions, and so on.
- *No* research to fall back on. No publications, no research projects, no previous experience anywhere available.

Therefore, *you* have to do it all yourself, which means:

- *Do* all the research, the search for background information, however small it may be. In the beginning, any snippet of information you can lay your hands on may contain important hints on how to approach the topic.
- *Go* to libraries, search the web, speak to colleagues, consult experts in cultivation.
- *Collect* plant material from the wild – in the form of seed, cuttings, whole plants. Nowadays, it is important to obtain all the relevant permits that may be applicable in that particular country.
- *Go* to where the plant in question grows and make your own observations about habitat, aspect, soil, altitude, climate, geology, plant communities, and so on.
- *Learn* to propagate and cultivate. Do all the trials of what time is the best for sowing and how much water, shade, fertilizer, and so on are needed.
- *Make* all the mistakes. In the early years of cultivation of a novel crop there is hardly a year where you will not adjust your cultivation techniques drastically by making and understanding your mistakes.
- *Carry* all the costs for the initial plant material, nursery space, trials, planting, weeding, and so on.
- *Be* very patient. The process of taking a novel crop into successful cultivation can take anything from 3 to 15 years. Whereby, if it only takes 3 years you are dealing either with an exceptionally easy plant or you are a genius already!

## 3.2 Basic principles of novel crop cultivation

Before looking at some case studies, it is important to ponder about some very basic principles behind novel crop cultivation.

### 3.2.1 Does your novel plant have real commercial potential?

With all due respect to everybody involved in the novel plant industry, at the beginning of a new project we gaze upon our discovery like young parents on their first-born child: it is the most beautiful

*Novel Plant Bioresources: Applications in Food, Medicine and Cosmetics*, First Edition. Edited by Ameenah Gurib-Fakim.

baby the world has ever set eyes upon! None like it before or after this one! All that potential!

It is useful to go through a few critical questions in the beginning to gain perspective on your novel plant first. After all, unless it is a real revolutionary breakthrough, it will have to compete against or, better, replace an existing product in the market place. This means that I have to spend time and a lot of money to communicate to people about the benefits of my product over what they are using for the same or similar purpose so far. Will this be successful? Therefore, explore these questions here. The product derived from my novel plant:

- is faster acting;
- is better tasting;
- is cheaper to produce;
- is longer lasting;
- is easier and more convenient to use;
- has less side effects;is more sustainable/green/environmentally friendly than products it is competing with;
- has some other compelling characteristic that will be highly desirable for a lot of people.

### 3.2.2 Botanical background

It is imperative to familiarize yourself with the novel plant in question. This is best done by collating and collecting all available source materials that may contain information about the plant. Initially it is wise to cast your net very wide – sometimes relevant information may be found in unlikely places. Typical sources to consult are:

- Libraries at botanical gardens, universities, herbaria – botanical books, old and new, may contain interesting bits about the plant! Beware, though, of that famous botanical pastime of taxonomists: changing botanical names! This can make the search for information very arduous, as a plant may have a string of synonyms or older names which all need to be investigated for relevant literature.
- Internet – there are many useful websites, especially if you are affiliated or have access to university networks. A word of caution here: although the Internet is a fantastic source of information, not all of it is relevant or well researched! Therefore, it is vital to check and cross-check again the data presented. This is particularly the case with regard to pictures of plants – they may be very pretty but may not necessarily depict the correct species.
- For many plants/plant groups/plant families there are specialist societies around; for example, The Succulent Society of South Africa, The International Geranium Society. These societies often have publications and offer exchange of plants, seed and advice between members. Your novel crop may have been cultivated by some enthusiasts as an ornamental already!
- Botanical societies in the country where the plant occurs.
- Go to where the plant occurs – local nurseries may by chance have a selection of indigenous plants – and your novel crop may be there. Or the proprietor may know where to turn to for more information.
- It cannot be overstressed that the plant itself is a most important source of information! Go to as many locations in its natural distribution range as you can muster. This is particularly instructive when the plant occurs over a wide range of habitats, altitudes or climatic zones. *Pelargonium sidoides*, for example, has a distribution from the coastal and subtropical lowlands of the Eastern Cape of South Africa to the alpine conditions of the Lesotho Highlands!
- Also, look 'left and right' from your novel plant! Are there other members of the genus or family that are already in cultivation – food, medicine or ornamental – somewhere? This may provide some pointers to start off with and save you time.
- Ask anyone you know who works with plants, whether academically or practically – the odds may have it that somebody there knows something or somebody who may know more.

### 3.2.3 Legal background

In today's context of the Convention for Biodiversity and the Nagoya Protocol it is essential to explore the legal framework of the country where the plant occurs. Especially within Africa, we find greatly varying levels of legislation implementing the spirit and letter of the convention. South Africa, for example, has a very advanced law, the National Environmental Management: Biodiversity Act (NEMBA, 2004) regulating the use of indigenous bioresources, whilst other countries have only very limited laws or none at all. Owing to the fact that environmental issues are often low on the agenda of a country, and especially so with developing countries, the use of indigenous plants may fall under more than one piece of legislation, which, with some luck, may even contradict each other. Also, the level of implementation of environmental legislation may be at an early stage with little or poor formal structures to put the act into practice. This may pose some obstacles, especially early on in the project. But lack of legislation or implementation in a country is not an excuse for ignoring this aspect – the international community is firmly moving in this direction and it is better to comply voluntarily, even under difficult circumstances, rather than being forced into compliance once one has been found to contravene local legislation of a provider country. It can be expected that within the next 3–5 years all signatory countries to the Nagoya Protocol will have some form of legislation in place. This will cover all aspects – from providing to trading with and receiving of bioresources.

Answer these questions for your novel plant project:

- Is my novel plant CITES (Convention for International Trade in Endangered Species – http://www.cites.org/) listed?
- Is there a local status under IUCN (International Union for Conservation of Nature http://www.iucn.org/) conservation criteria – protected/endangered/ … – in terms of a Red Data List if there is one available in that country.
- Do I know which local legislation covers biodiversity issues – prior informed consent, access and benefit sharing aspects, and so on? Which sort of aspects will I have to comply with?
- What is the current status of legislation and compliance with the Nagoya Protocol in my own country? What sort of hurdles must I expect to comply with when importing the material?
- Am I making use of traditional knowledge? If yes:
  - Can I clearly identify the holder of this knowledge?
  - Is the knowledge holder an individual, a group of people, a community, a university, … ?
  - Do I have a clear plan with regard to relevant agreements with individuals, groups or the country's government?
- Do I keep local people as well as the relevant government department in the loop? Do I maybe have to apply for a research permit

(as is the case under NEMBA in South Africa) before I conduct my initial research? Although it is said that 'asking for forgiveness is easier than asking for a permit', it is better practice to cooperate with the relevant parties as early as possible. This avoids unintended backlashes and delays in a project later on.

- Apart from local or international legislation, it is often also a general emotional atmosphere that will determine whether and what kind of opposition to the use of the resource is coming forth. The case of *Hoodia gordonii* – a plant used by the San people of southern Africa for quenching hunger and thirst during hunting trips – is a point in case. No nationally or internationally binding legislation existed at the time of development of the first commercial product and it was therefore impossible to comply with anything proper. However, the patenting and licensing out of the patent has been widely labelled as bio-piracy and used to demonstrate bio-piracy issues. The public international debate around *Hoodia* has been highly emotional most of the time – painting a picture of poor knowledge holders in the South being exploited by multinational giants in the North.

### 3.2.4 Knowledge management

A novel crop plant is only novel as long as there is only you or a few other people working with the plant. In the process of developing your crop you will spend a good amount of time, energy, money and sweat capital on learning how to cultivate the plant. Although sharing of knowledge for the wider benefit of mankind is important, it is equally important to protect the intricate bits of knowledge you gather over time. Understand which knowledge is key to your crop – all plants need water for instance, but only you know how often you need to water your plant, and this can make or break the success of your crop. Most of your cultivation knowledge is not so unique that it could be patented; therefore, the other equally powerful mechanism to protect knowledge – trade secrets – needs to be developed and preserved. Explore early in the process whether your novel crop has the potential for a commercial success. If yes, it is worth quietly beavering away at perfecting everything before you take a selection of information public.

### 3.2.5 Purpose of the novel crop

Depending on the final purpose and customer group, different demands will have to be covered by the plant material. As a basic rule, the higher you reach with the final purpose (e.g. 'my novel plant is the raw material for a prescription drug curing diabetes with the target market being the EU and the USA'), the more you will have to understand all possible dynamics around it. The lower you aim (e.g. 'my novel plant yields a compound that has a moisturizing effect on skin in a topical application to be used in a small and limited local product'), the less data will be needed. It is therefore useful to assess the future potential of the plant and prepare a project plan that reflects this (see Table 3.1).

### 3.2.6 Development cycles

What type of crop am I dealing with? Some plants or parts of a plant have very different time frames for their development period. Sometimes these factors compound and make the project very long term or may be so complex that the project as a whole may not be feasible. Table 3.2 shows various short-, medium- and long-term factors.

**Table 3.1** Low or high profile plant use determines the amount of input needed

| | High-profile use (e.g. prescription drug in country with high regulatory burden) | Low-profile use (e.g. topical use in cosmetics in country with low regulatory burden) |
|---|---|---|
| Base data about crop | Very important – from correct identity to suitability in cultivation | Very important |
| Cost effectiveness | Always important, but here even more so | Always important |
| Genetic variability | Must be understood, and any possible interference with quality to be ruled out | Good to know your plant, but genetic variability will have a much lesser impact on success |
| Quality parameters | Highly defined, well-understood, refined and validated test methods | Some general parameters to be known and assessed |
| Safety and efficacy data | Highly important – these are the cornerstones of a successful high-profile product. Expensive clinical, toxicological, safety, pharmacological trials have to be conducted | Important but not as critical – at this level, mostly no expensive trials are conducted but emphasis is on relying on 'traditionally safe' use |

### 3.2.7 Genetic variability

Botanically speaking, your crop plant is only one species. Within the natural populations, however, there may be variations in the genetic makeup of individual plants or groups of plants. This variation may be the stronger the larger the area of natural occurrence is – especially if a plant occupies many specific niches over different climatic zones. As mentioned above, *P. sidoides*, for instance, occurs over a spread of climatic zones from subtropical coastal to alpine montane habitats in its natural distribution range in South Africa and Lesotho. Also, the second case study plant, *Sutherlandia*, has a very wide geographical distribution right across South Africa – from the west coast (winter rainfall and Mediterranean climate) across the Karroo region (semi-arid to arid conditions) to central parts of South Africa (summer rainfall and temperate climate). This variability may express itself phenomenologically in different growth patterns, flowering times, leaf sizes, size of plant, and so on and may be reflected also in different ingredient profiles or higher/lower levels of ingredient. The latter one is, for obvious reasons, especially very attractive. Low levels of undesirable substances or high levels of active ingredients, for example, can provide for an exciting angle

**Table 3.2** Individual crop dynamics and final purpose are intimately linked

| Type of plant | Annuals | Perennials, shrubs | Trees |
|---|---|---|---|
| Plant parts | Leaves, flowers, flowering herb | Seed | Roots, barks, seed |
| Final purpose | Cosmetic, topical application, food with well-established use | Food, medicine, ingested materials, local traditional use | High activity, medicine, prescription drug |
| Toxicity | No toxicity – completely safe | Low toxicity | Profound toxicity |
| Activity | Low activity and limited use | Medium activity and wide spectrum of use | Highly active and very specific |
| Market | Limited and little regulated market, e.g. in a developing economy | Wider and somewhat regulated market | Large and highly regulated markets (EU, USA, Japan, etc.) with high requirements around product safety |
| Novelty of product | Well established and well-documented use, good data available | Some previous use restricted to a certain area, limited records and use data | Entirely new entity, no or little previous use data |

to your plant. The ornamental plant industry makes extensive use of this phenomenon and offers, for example, interesting flower shapes or colours as novel varieties.

Your novel crop may benefit by using some 'extreme' natural selections. It may be valuable to investigate populations which over a long time have been able to develop on their own, separated from the main body of the genetic pool of the species and possibly even under more extreme conditions. For this end, it is certainly useful to establish gene pool selections reflecting the natural distribution range of the plant in question. It needs to be noted here that the genetic makeup of a plant usually provides the plant with strategies to survive and thrive under specific natural and sometimes extreme conditions – droughts, floods, poor or rich soils, high or low natural radiation, extensive heat or extreme frost conditions, intense competition within plant communities, and so on. When moving plants from their natural (extreme) habitat into paradisiacal conditions of your nursery and greenhouse or what you consider as suitable growing conditions in other parts of the country of origin or in the world, some of these inert characteristics may disappear or exaggerate. Striking examples are the geraniums on the balconies of every Swiss chalet – they essentially are two species of *Pelargonium* from South Africa (*Pelargonium zonale* and *Pelargonium peltatum*) that have been developed from very humble origins and are by no means as spectacular and prolific in their growing and flowering patterns as they were in their old home country!

### 3.2.8 Record keeping

The modern world thrives on data and information. This is no different for your novel crop. Record keeping should cover:

- origin of your plant material, including all genetic variants (including GPS data, dates, herbarium specimen, photographs, maps, etc.);
- treatment of propagation material (seed/seedlings/cuttings/etc. with, for example, giberellic acid, fungicide);
- soil mixes and the origin of the mix components;
- nursery, greenhouse, and so on conditions;
- shade/no shade;time of year, weather conditions;
- propagation method used;quantities received, sown/planted, germinated, transplanted;
- soil type, soil analysis, treatments, application of fertilizer/compost/manure;
- plant distance, density per square metre or hectare;
- irrigation type, quantity, frequency, water quality, certificate of analysis;
- general observations, growth rate, diseases, pests, take relevant photographs frequently and file;
- harvest, harvesting method, quantity per plant, per square metre or per hectare
- manpower (man-hours or man-days) used for all activities;
- time from planting to harvesting;
- quality of harvested material, weather conditions;
- processing, description of, for example, cleaning, washing, cutting, drying, storing, extracting.

It is best to create standardized documents covering all activities. The data gathered will come in very handy at a later stage when the memory of what you actually have done with or to your plant has blurred or faded. It may provide crucial clues on which factors have produced certain outcomes and which are therefore vital to create consistent good-quality material.

Good agricultural and collection practice (GACP) guidelines give a good overview of the types of systems and documents to be created. Various GACP guidelines are available here:

- WHO version of 2003, http://whqlibdoc.who.int/publications/2003/9241546271.pdf;
- European Medicines Agency (EMEA) version of 2006, www.ema.europa.eu/docs/en_GB/document_library/Scientific_guideline/2009/09/WC500003362.pdf;
- European Herb Growers Association (EUROPAM) version 2010, www.europam.net/documents/gacp/EUROPAM_GACP_MAP_8.0.pdf.

In essence, they are very similar and cover the same topics.

Other standards of interest are:

- Sustainable Agriculture Network (SAN) together with the Rainforest Alliance, http://sanstandards.org/sitio/subsections/display/9;
- IFOAM (International Federation of Organic Agriculture Movements), www.ifoam.org/sites/default/files/page/files/ifoam_norms_version_august_2012_with_cover.pdf.

### 3.2.9 Project phases

Each novel plant project will have phases, with each phase having its own dynamic, priorities and outcomes. Here is an overview of the three main phases. The given time frames are largely dependent on the type of crop as well as the associated complexity – see Table 3.2 on time frames.

- Initiation phase:
  - first trials evaluating many parameters that may have an impact on the quality or quantity of the crop;
  - understanding basic techniques to be used and combined;
  - first cost and feasibility calculations;
  - time frame, 0–3 years.
- Experimental phase:
  - basic techniques have been agreed upon;
  - fine tuning and optimizing of all processes is now taking place;
  - scale up from initiation-phase size trials to field/plantation level;
  - solve scale-up-related challenges;
  - time frame, 2–5 years.
- Consolidation phase:
  - large-scale cultivation;
  - routine sets in;
  - few changes to the cultivation technique are required;
  - time frame, 2–5 years.

This overview can assist you in formulating your project proposal.

### 3.2.10 Propagation techniques

Early on in the game, the best propagation and cultivation technique will have to be worked upon. Expect frequent paradigm shifts in the beginning, when much will happen on a trial-and-error mechanism. As the knowledge base develops, deep changes will happen less frequently.

#### 3.2.10.1 Generative propagation

Genetically speaking, seed is always a mix of the parental genetic material. In most crops, this mix does not lead to major deviations of characteristic traits of the offspring plants. However, once clones have been chosen, generative propagation will not be possible anymore as this will lead to a split up into an infinite genetic variety. An exception to this is when clones have not been created by selection but through a long breeding programme.

- Seed:
  - a quick, cheap and mostly simple propagation;
  - method of choice when seed is easily obtainable;
  - seed is suitable when clone selections are not available or not necessary.
- Seed needs to be stored under controlled conditions (low temperature, humidity controlled and away from light).
- Viability of seed needs to be tested and a seed bank be established to ensure continuous availability of viable seed.
- Germination rate of seed can be improved by manifold treatments. Jelitto, a German seed company, has compiled a great variety of sowing instructions for the seed they offer in their catalogue. These instructions are very precise and constructive (https://www.jelitto.com/haupt_de.html). Typical methods are:
  - stratification (layering in sand and applying specific temperatures for an extended period of time);
  - cold/freezing/warmth treatment;
  - soaking in water/sulphuric acid/giberellic acid (at appropriate solution strength), and so on;
  - smoke treatment (e.g. www.sanbi.org/sites/default/files/documents/documents/order-form-smoke-plus-2010.pdf);
  - specific sowing medium mix;
  - heat treatment with boiling water;
  - controlled damaging or removal of the seed husk with sandpaper, knife, and so on;
  - some seeds take a long time (1–2 years!) to germinate – do not throw away seedling trays too early!
  - correct season – not all seed will germinate all year round even though the physical parameters may be optimal.

#### 3.2.10.2 Vegetative propagation

This is the preferred method when a specific genetic makeup of a plant needs to be preserved and multiplied, as this method results in identical plant material from the mother plants.

- Cuttings – a fast, reliable and cost-effective method of propagation. Useful especially when a clone selection process has taken place and elite plant material needs to be propagated. Not suitable for annual or biannual plants, most of which cannot be propagated by cuttings.
  - Top cuttings – top green to woody parts of branches or twigs.
  - Root cuttings – roots, rhizomes.
  - Leaf cuttings – whole leaves or parts thereof.
  - Correct season needs to be selected. Mostly spring or autumn provide good times for rooting of cuttings.
- Plant splitting – a very suitable method with perennial herbaceous plants to initially bulk up numbers. May also become the method of choice for plants with rhizomes (e.g. ginger and turmeric are propagated like this commercially).
- *In vitro* propagation – produces identical and very homogeneous plant material.
  - Initiation periods can be long, as a recipe for this particular plant will have to be developed (1–2 years).
  - Very costly method for low volumes of plants; more cost efficient when large volumes are required or when it is a long-term crop that can be harvested repeatedly.
  - For this method, a high level of understanding of your crop is necessary. A selection and breeding process for the best suitable clone(s) will have to precede *in vitro* propagation.
  - Not all plants respond successfully to *in vitro* propagation. Devil's claw (*Harpagophytum procumbens*), for example, does not develop a central tap root but only secondary roots. Although this is the part harvested, the lack of a tap root results in the complete destruction of the plant during the harvest.

### 3.2.11 Organic versus conventional cultivation

Organic, green, sustainable, carbon footprint, and so on – these are buzz words in today's modern world. One can have one's own preferences or doubt the organic approach. But – apart from all the other advantages of organic cultivation – one of them stands out: organic cultivation uses no harmful pesticides and fungicides. This also means that:

- there are no residues of pesticides and fungicides that could interfere, for instance, with analysis results;
- no extra testing for pesticide residues is necessary – avoiding huge cost implications, as these tests tend to be extremely costly.

It can therefore make economic sense to switch to or use organic methods right from the start. Organic certification agencies are active in most countries nowadays – either local agencies or international organizations. Worldwide, there are five significant accreditation schemes: IFOAM, JAS, ISO 65, the EU Regulation and US NOP. Within these schemes, individual certifying companies provide certification services. Globally operating companies such as Ecocert (France), Demeter (Germany), BCS (Germany), SGS (Switzerland), Soil Association (UK), IMO (Switzerland), to name a few, conduct inspections and issue organic certificates on compliance. A list of organic certifying agencies worldwide (status 2013) is available at http://www.grolink.se/docs/organic-certification-directory-2013.pdf.

Other beneficial side effects of organic farming have been observed:

- There can be a better ratio of ingredients to the produced biomass. This can be of specific interest if certain levels of ingredients have to be reached and may be influenced with cultivation techniques.
- Organic farming methods tend to be more observant of dynamics in nature; therefore, a closer match between a cultivated and a previously wild harvested plant quality can be achieved.

### 3.2.12 Choice of location

To some extent, each plant is a product of its environment. Plants adapt to a great variety of conditions of soil, climate, light availability, temperature ranges, drainage, and so on. In any horticultural or agricultural setup, specific conditions exist based on the location and its overall climate (temperate, subtropical, tropical, semi-arid, arid, alpine, coastal, etc.). These conditions can only be influenced within reason. It is usually easier to supplement a lack of a factor than reduce its impact. In a semi-arid area for instance, access to irrigation water can open up possibilities for crops. A striking example is the thriving grape industry in the Northern Cape of South Africa: in an area with normal rainfall of between 60 and 100 mm per annum, the Orange River provides the water for intense cultivation in a desert climate. This creates advantages such as low disease pressure (grapes are very susceptible to mildew when moderate temperatures go together with high humidity – the irrigation water rapidly evaporates in the dry desert air, providing the vineyards with sufficient moisture for growth but the humidity level in the leaf canopy stays too low for mildew to spread) or early access to markets because crops generally can be grown earlier due to the relative warm conditions after winter.

Other factors to be taken into account when choosing a location for your scale-up plantation are:

- General infrastructure – availability of good access roads, proximity to towns for all manner of supplies, sufficient electricity (e.g. for first-degree processing such as drying), availability of energy.
- Labour – some crops may be rather labour intensive. Access to labour (unskilled and sometimes, more importantly, skilled staff) at reasonable cost is very important.
- Favourable soil, agricultural and climatic conditions – the main 'comfort factors' should be within the determined range for your crop. Or, as the above example for grapes shows, mitigating measures are available.
- Sufficient size of free land available – often the best agricultural lands in favourable climates are already commercially used. Competition with existing high-input, and therefore high-value, commodities (e.g. fruit, wine, sugar) will drive the capital cost of land up as well as all general costs from labour to services. Make some wild projections of sales of your product and link them to yields per hectare and availability of land – you do not want to run out of land too quickly. Therefore, smallholdings generally are only useful in the initiation phase.
- Ownership of land – contract farming versus owning the agricultural entity.
  - Own agricultural entity: long-term security of land tenure; intellectual property around novel crop is easier to build up and protect; full control over all cost – overheads of land rest entirely on novel crop; suitable for medium- (3 years and longer) to long-term cycle crops (permanent crops); preferred for large-volume/long-term crops; ownership of land limits to given environmental conditions (altitude, climate, soil, water, etc.).
  - Contract farming: dependency on who owns the land – change of landownership results in loss of experience; intellectual property needs to be contractually protected; suitable for short- (annual or biannual plants) to medium-term cycle crops (3–4 years); preferred for smaller volume/short-term crops, as they can be moved to new locations; offers the possibility to choose best area without carrying the capital cost of setting up a horticultural or farming enterprise.

The ideal location for your novel crop is hard to find and may not exist at all. It is always necessary to weigh up all main factors of what your plant may need and balance them against the available options. Striking some form of compromise is inevitable!

### 3.2.13 Cost

One major factor concerning whether your novel crop will make it to a larger market is its cost effectiveness. Early evaluation and costing scenarios should be developed and continuously updated.

Major cost driving factors are:

- *Land – a capital cost.* Good agricultural land is usually planted with profitable crops already. The cost of land goes up the more fertile the land and the more profitable the crops. It is also hard to find vacant high-potential land.
- *Labour.* Nowadays, more and more countries have or are introducing strict labour laws, including minimum wages. In addition, unions and bargaining councils gain more and more weight even in remote and rural locations. Unskilled labour often is easy to employ – however, with novel crops, even more so than

with commodity crops, it is vital to have a pool of higher skilled personnel available. Novel crops have specific cultivation requirements which may well fall outside of any employee's experience spectrum.

- *Access to services.* This includes proximity to a variety of services usually located in larger centres or cities. This may vary from agricultural extension officers to specialized equipment dealers or repair workshops. It makes a huge difference, whether I need to spend 1 h to go and get a spare part for my equipment or a whole day. One is more likely to quickly drive 5 km to the nearest agricultural co-op than 50 km or more.
- *Infrastructure.* Here, we are talking about available infrastructure as well as on-farm or in the area. It is tempting to start on a vacant piece of land, but the cost of creating the necessary infrastructure, such as water supply (e.g. borehole, dam, wells), buildings, vehicles, implements, staff and housing to name a few, may be prohibitive and impact negatively on the production cost of the novel crop. The state of external infrastructure, such as access roads, is not to be discounted either – regular trips over a 50 km stretch on a rough road in rural areas will add dramatically to your overhead costs over time!
- *Favourable climates.* Favourable climatic conditions will have a negative impact on the cost of land, as competition with high-value agricultural commodities will be high.
- *Propagation cost.* It is vital to find ways of mass propagation that are cheap. It is therefore important to look at the dynamic of all steps along the value chain. Is my seed difficult to obtain (e.g. *P. sidoides* seed cannot be mechanically harvested – seed is collected over long periods by hand – this results in a seed harvesting cost of up to €2000/kg!)? I will not be able to afford to live with a low or erratic germination rate in, for example, direct field sowing. It will then be better to raise seedlings under controlled greenhouse conditions. *In vitro* propagation may be of interest, but essentially is a very expensive propogation method unless the cost can be counterbalanced by a high final crop value.
- *Cultivation cost.* It is important to mechanize as much as is possible during cultivation. This includes all activities from land preparation to final harvest and primary processing (e.g. drying) and must cover:
  - Soil and bed preparation – figure out the most effective way of planting – distances between plants within and between rows, numbers per square metre or hectare and prepare land accordingly.
  - Planting – can mechanical sowing machines or vegetable planters be used or adapted?
  - Weed control – choose plant distances and bed sizes to allow for a maximum of mechanical weeding. Consider techniques such as: mulching (plastic or organic matter) to suppress weed growth – especially with biannual or perennial type plants; specialized weeding machines for various crops can be obtained (e.g. www.kress-landtechnik.de/wEnglisch/index.shtml and www.hatzenbichler.com/); flame weeding – a technique used that is especially useful for the control of weeds during their early development – more information can be obtained at https://attra.ncat.org/attra-pub/summaries/summary.php?pub=110.
  - If hand weeding is unavoidable, try to use nifty tools that facilitate fast weeding without tiring the work force too much – for example, oscillating hand hoe (www.manufactum.de/schweizer-pendelhacke-p1409985/?a=56891&campaign=froogle&ia-pkpmtrack=31-83538353132363231313O3-105-103-191) or push hoe (www.valleyoaktool.com/wheel-hoe/).
  - Harvesting – the method of choice depends highly on the plant and the part that needs harvesting. Any mechanization must ensure that the integrity of the harvested goods is not jeopardized. Cutting, bruising, scratching, and so on is to be avoided – any damage to the plant tissue may lead to immediate reactions on the cellular level that may have an impact on the quality and ingredient composition of your crop. Think of the oxidation taking place within minutes when you cut an apple in half! Often, no machine is available for your specific crop, but machines from other crops may often be adjusted. Gingko leaves are harvested with adjusted cotton-picking equipment, root crops may be able to be lifted with potato harvesters, and so on.
- *Economy of scale.* Using smallholder farmers versus large-scale cultivation. Both systems have their merits and challenges; see Table 3.3.

### 3.2.14 Breeding programmes

Most novel crops have never been cultivated before and it may be necessary to embark on a breeding programme to improve certain characteristics of the plant in question. Traits that warrant improvement may be:

- agronomic yield – for example, kilograms/planted area;
- ingredient profiles may vary in different individual plants – select/breed for high-yielding strains;
- resistance to disease or insect attacks, general health characteristics;
- resistance to stress factors such as drought, flooding, and so on;
- ease of harvest of, for example, root crops – select for fewer larger roots (equals easy to harvest and fewer losses during harvesting) over many smaller roots (equals difficult to harvest with more potential losses);
- even size for ease of handling;
- purity of species in case where bastardization through, for example, cross-pollination with closely related species may occur;
- any breeding programme is time consuming and expensive, so a careful cost-to-benefit analysis needs to be undertaken to judge how much more potential can be unlocked from the plant.

The two main ways of crop improvement are breeding and selection.

- Breeding:
  - Involves usually quite a long period of selection and then cross-pollinating the selections until a stable seed that consistently carries the traits of the parents has been established.
  - Requires a time frame of 10 years and more, depending on whether this is an annual, biannual or perennial crop. Also, time to maturity and longevity (e.g. in fruit trees) may take a very long time to evaluate.

**Table 3.3** Pros and cons of large- and small-scale farming of novel crops

| | Large-scale cultivation | Multiple smallholder farmers |
|---|---|---|
| Economy of scale | Can be achieved with high level of mechanization and streamlining | Impossible to achieve – mostly manual labour made use of |
| Logistics | Easily organized – short distance between production area and processing | May be daunting due to long distance to processing facility, poor roads, quality losses due to the above |
| Capital cost of land | High, e.g. due to finance cost of land purchase | No direct cost of capital – smallholder farmers usually fully own their land without bond repayments |
| Infrastructure | Can be good and purpose made for the specific crop ensuring high quality | Usually no infrastructure for intermediate storage, etc. available |
| Socio-economic | Limited number of staff required and therefore limited socio-economic impact | Generates some income for a large number of individuals |
| Risk management | One site – in case of crop failure, everything is lost | Risk is mitigated by spread over many properties over a wider geographic area |
| Training and management | Is more controlled. Good training and management can be better achieved | Difficult to train many operators |
| Coordination | All activities from planting to harvest easily coordinated | Difficult to coordinate distribution of plants and planting as well as harvesting and transport to processing site |
| Crop quality | Much more easily controllable and achieved. Record keeping easier | Much poorer control on quality issues. Record keeping and batch-to-batch conformity poses big challenges |

- Results in a variety that can be protected under international breeding laws.
- Owing to the time involved, is very costly and rarely done by private companies unless massive trading volumes are achieved, as is the case with, for example, grains, vegetables and ornamental flowers.

- Selection:
  - Involves selecting individual plants with promising characteristics.
  - Is useful where plants can be propagated vegetatively or through *in vitro* methods to ensure the selection stays intact.
  - Is not suitable for annual or biannual plants.
  - Leads to success within less than 10 years.
  - Mostly cannot be protected by plant breeders rights because no new genetic variety has been created.

Both breeding and selection programmes are complex affairs: it is best to consult with relevant experts to assist at least in the design of the protocol. Failure to do so may lead to a waste of valuable time and money. Small aspects, forgotten to implement, may have a vast impact and may render many hours or even years useless. Examples of this are:

- no randomization of trials;
- forgetting about edge plants;
- incorrect statistical evaluations;
- non-standardized experimental conditions;
- flaws in the trial design;insufficient numbers of test plants;
- back-up of the tested genetic material in case the crop is lost;
- insufficient record keeping;
- frequent change of evaluating personnel.

### 3.2.15 Trials

It will be inevitable to set up and run a series of trials with your new crop. The extent of these trials depends much on what your final purpose of the crop will be (see Section 2.5). Initially, these trials will be small – a few hundred square metres may be involving a very limited number of plants. The purpose of an initial trial is to get a first handle on what challenges lie ahead when embarking on a larger scale project. Where do I obtain initial plant material? Is it botanically correct and verified? What soil is suitable? How much or little water? Fertilizer? And so on.

As the project progresses, more and more questions will arise that will need some form of answer. However, beware of setting up a trial the moment a question comes to mind! Evaluate with an exact and clear mind which questions really do matter. Too many trials will diffuse your energy, and what initially looks easy to control may soon turn into an avalanche of data collection and need for evaluation. This is time consuming and may not be productive.

Therefore:

- Identify and prioritize key questions.
- Keep trial numbers to a minimum.
- Eliminate, as much as possible, factors that will blur the final answer. If this is initially not possible, redesign the trial or break it up into smaller, and thus more manageable, trials.
- Keep trial design as simple as possible.
- Check trial design with experts in that field.
- Keep meticulous records. Three months or let alone 3 years later you will not remember a single detail unless it was clearly written down and documented.
- Have intermediate results and conclusions checked by experts not involved in the trials to avoid 'trial blindness'.

### 3.2.16 Enlist the help of experts

In the course of the development of a novel crop many different challenges have to be faced and solutions found. Even though, initially, there is limited knowledge in your specific crop around, it is advisable to engage with experts in fields that may border on or be applicable to your endeavour. You cannot know and do everything yourself! You are most likely the entrepreneur with the idea in your head – do not clutter it all the time with details! Keep an eye on the woods as well as on the trees!

## 3.3 Case study 1: *Pelargonium sidoides*

*P. sidoides* DC, Geraniaceae is endemic to South Africa and Lesotho. It spreads from the coastal and subtropical lowlands of the Southern and Eastern Cape north into the alpine zones of the Mountain Kingdom of Lesotho and further into the summer rainfall grassland habitats of the Transvaal and Limpopo regions of South Africa.

*P. sidoides* has a rosette-like cluster of greyish-green to green, heart- to kidney-shaped leaves. The small flowers are dark red to almost black – something found rarely in the plant kingdom (Figure 3.1). It has a strong tuberous storage root which enables it to survive adverse conditions such as drought and grass fires which regularly occur over much of its range. Whilst it is a very hardy plant, it will not survive cultivation and repeated ploughing. Its original range and occurrence probably was much larger than where it is found today.

Of medicinal interest is the tuberous root, which with increasing age displays a deep red colour. The bioactive ingredients in *P. sidoides* are the tri- and tetra-oxygenated coumarins, gallic acid and gallic acid methyl ester (polyphenols), various flavenoids, as well as significant levels of calcium and silica. Gallic acid and its methyl ester are present in large amounts and are identified as the immunomodulatory principle for this herbal medicine.

Macrophage activation was confirmed by an *in vitro* study based on *Leishmania* parasites (Kayser *et al.*, 2001). They also studied the antibacterial effect of the various coumarins and gallic acid compounds in *P. sidoides* and found that the potentially active compounds exhibited antibacterial activities with minimum inhibitory concentrations. This provides a scientific basis for the use and success of Umckaloabo® – a special extract from *Pelargonium*

**Figure 3.1** Close up of *Pelargonium sidoides* flowers. Note the dark red flowers. *Photo*: Ulrich Feiter, Parceval Pty, South Africa

marketed internationally under various trade names by Schwabe, Germany.

*Pelargonium* extract seems to have a four-way effect:

- antibacterial – it prevents bacteria attaching to the mucous membrane cells;
- antiviral – it prevents viruses attaching to mucous membrane cells;
- stimulates the immune system and prevents bacteria and viruses from multiplying;
- expectorant – allows the body to expel contaminated mucous, making conditions less suitable for the bacteria and viruses to multiply.

Schwabe, in Germany, a leading phytopharmaceutical group of companies with a global footprint, pioneered the modern use of *P. sidoides* by researching all aspects of toxicology, pharmacokinetics, and so on up to a series of clinical trials proving the efficacy of this humble plant from South Africa. Umckaloabo® is now a fully registered herbal medicine in Germany and is sold in more than 40 countries around the world.

Umckaloabo has been on the market quite a few decades now, but by the early 1990s it became clear that the use of wild harvested roots was, in the long run, not the best way forward, and thus Schwabe decided to embark on a long-term programme to propagate and cultivate this useful species.

Some members of the Geraniaceae family are widely used as ornamentals (*P. peltatum*, *P. zonale* and their hybrids) or for the distillation of rose geranium oil (e.g. *Pelargonium graveolens*). Therefore, *Pelargonium*, as such, is a well understood species from a horticultural point of view – literally millions and millions of them are propagated every year to grace the home gardens in central Europe and most notably the balconies of Swiss chalets. So why is *P. sidoides* included here as a novel plant? Precisely because, with all the knowledge around, taking *P. sidoides* into cultivation was expected to be an easy job but turned out to be a major challenge!

The initial task was to source plant material and propagate this to plant a first 1 ha with it. Seed was not obtainable from commercial sources and the quantities available from botanic gardens were negligible. *Pelargonium* usually produce three seeds per flower, which have long spiral and hairy tails that enable them to be carried by the wind for some distance. The seed pods open, once mature, usually in the course of the morning when the sun dries the pod and it cracks open, releasing the fluffy seeds which are carried away with the slightest whiff of wind. To find seed in the wild from plants scattered over large areas is a tedious exercise demanding much patience!

It was then decided to use cuttings as the initial method for propagation. The help of a specialist nursery in Port Elisabeth was enlisted who could locate plants in an area which was about to be developed into a road. A number of plants were removed to a nursery and a mother block for propagation purposes was established there. As the plants were growing, cuttings were taken for propagation. In parallel, a propagation set-up was established at the planned cultivation site near Cape Town in the Western Cape of South Africa.

This proved to be working alright but was no quick way to reach the targeted numbers within the targeted 12 month period. However, with a delay of some 6 months, the first 1 ha was planted. Intensive propagation trials continued so as to establish the best method for mass propagation should the cultivation become a success. Trials were conducted to establish best soil mixtures (sand, compost, vermiculite, bark compost, perlite, coco fibre, etc. were trialled in many combinations). Eventually, a mixture of sand, compost, bark compost and vermiculite provided the best solution.

Different type of cuttings were explored – taken from young plants or older specimens, different thicknesses of stems or number of leaves. Also, root cuttings were investigated. Again, did age play a role? Did the position of the root cutting (top, middle, bottom section of the root) have an influence? Leaf cuttings were considered, too. For this purpose, leaves were cut with sterilized knives, placed into an insulated box and whisked off to a specialist nursery where the cuttings were kept in pure perlite under high humidity. The leaves that survived grew into beautiful plants, However, the survival rate of leaves was less than 2% – hardly a successful and commercially viable method!

With 1 ha of plants established and growing, the possibility of collecting seed, however, presented itself. Teams of people walked the lines every morning and collected what they could find before the wind blew them into the horizon. Despite having a large number of plants, the yield per person per morning was measured in 2–3 g only. The cost of collecting a kilogram of seed came to around €2000! At a germination rate of around 35%, this was a costly but possible way forward.

Propagation containers were another issue looked at. The seeds germinated slowly, developed equally slow and needed an average 3 months from germination to be ready for planting out in the field. This is quite a long time, and the nutrition and space requirements for the seedlings had to be met. The locally available speedling trays proved to be unsuitable as they were too small or did not last very long. A special speedling tray was imported from Germany for the purpose that proved to be perfect for *Pelargonium*.

The duration of the cultivation was another interesting question. How long does a cultivated plant take to produce the same ingredient profile as that of a wild harvested plant? After all, it is vital that the cultivated plant matches to its wild cousin so as to produce the same clinical results. Another important aspect here is the commercial viability: the longer the plants grow, the more root mass they can produce. But would the annual addition of root mass stand in proportion to the cultivation cost? Eventually, after evaluating data from trial harvests over a period of 6 years it was found that a 3–4-year cultivation cycle gave the best ratio in terms of root mass to cultivation cost.

Plant spacing in field planting was looked at and it was found that different cultivation sites and specific soil and climatic conditions had a great influence on which spacing was used. In some instances, the approach of cultivating in raised beds proved of advantage, whilst in other locations straightforward row planting with distances to allow easy access with tractors and implements became the method of choice.

All in all it took some 10 years of earnest trials and errors that led to a cultivation monograph capturing all salient points and making cultivation under very different conditions and locations possible.

One of the most challenging observations was the fact that *Pelargonium* does not seem to follow a specific growth pattern. Plants could be found at any given time in a cultivation cycle to display all stages of emerging from a dormant period, being in full leaf, flowering, seeding and dying off to go into dormancy. This was puzzling at the best of times! Eventually, looking at the original habitats of the Eastern Cape of South Africa, it was concluded that the weather pattern there is so extreme that no rain may occur for long periods and rainfalls are not seasonally bound. *Pelargonium* seems to be able to adapt and react to this sort of pattern, growing and flourishing at any time of the year to ensure maximum survival of the species.

From the beginning, it was noted that yields per plants could vary dramatically – from producing a few meagre roots in some to bunches of healthy and strong roots in others. After researching factors such as nutrition and irrigation, the focus shifted to include genetic variability. At each harvest period, especially perfect specimens were selected and vegetatively propagated. In parallel, plant specimens were collected from different locations over the distribution area. In painstaking trials, these plants were evaluated for a host of factors ranging from number of roots and depth of rooting to development and yield per square metre. These trials had to run for at least 3 years at a time to reflect the cultivation cycle – so progress was extremely slow. One 3-year cycle was usually repeated because freak successes and failures due to unusual climatic conditions had to be ruled out. In parallel and most importantly, all agronomical trials had to be accompanied by full chemical analysis of each sample to ensure that the resulting plant material for extraction complied with the official pharmacopoeia as well as in-house specification. Eventually, the best-suited selections were propagated to take part in scaled-up field cultivation trials.

By now, *P. sidoides* is being successfully cultivated in South Africa, providing first-class and defined plant material for extraction, complying to the standards laid down for a successful plant-based medicinal product. Large-scale cultivation is the key to take the pressure off the natural resource by avoiding wild harvesting and thus assisting in the survival of *Pelargonium* in its natural habitat.

## 3.4 Case study 2: *Sutherlandia frutescens*

*Sutherlandia frutescens* (L.) R.Br. is a member of the Fabaceae and is endemic to southern Africa. Common names, such as balloon pea, refer to the puffed up seed pods of *Sutherlandia*. Other names refer very directly to its medicinal properties, such as Kankerbos (Afrikaans), Cancerbush (English), Umnwele – Spear of Light (Zulu and Xhosa),

*S. frutescens* (Figure 3.2) has been in use for hundreds of years as a medicinal plant in traditional healing in South Africa. The genus is named after the Scottish botanist James Sutherland (1639–1719) (Smith, 1966). According to some writers, *S. frutescens* is the most commonly used healing plant in South Africa. The range of traditional applications is broad and ranges from diabetes (*Sutherlandia* can help to regulate blood sugar levels), cancer, depression (the Zulu apparently used this plant to assist the grieving widows when their men did not return from war or hunting) or providing extra energy. During the great flu epidemic in 1918 many people successfully used *Sutherlandia* to fight the dreaded disease.

**Figure 3.2** *Sutherlandia frutescens. Photo*: Ulrich Feiter, Parceval Pty, South Africa

Nowadays, some astonishing case reports from clinics caring for large numbers of AIDS patients have unanimously testified to the considerable improvement in the overall health of patients living with HIV-Aids after taking *Sutherlandia* tablets. Although the original illness remained, there was an increase in body weight, an improved sense of wellbeing and consequently of quality of life.

A first Phase I clinical trial has been conducted showing the compatibility of *Sutherlandia* when taken by healthy adults (Johnson et al., 2007). The result of a follow-up clinical trial has as yet to be published.

The active ingredients found so far in *Sutherlandia* are l-canavanine, l-arginine, pinitol, $\gamma$-aminobutyric acid (GABA), asparagine and secondary plant metabolites such as saponins (triterpene glycosides) (Figure 3.3). The precise mechanism of action remains the subject of speculation. A study by Smith and Myburgh (2004) suggests that *S. frutescens* strengthens the immune system and activates physiological resources. To this extent, *Sutherlandia* can be regarded as an adaptogenic tonic. Trials have also been carried out *in vitro*, which reported anti-oxidative, anti-inflammatory and antibacterial effects. As a topical application it successfully treats minor wounds, eczema, rashes, insect bites, and so on.

There are various non-standardized *Sutherlandia* preparations on the South African market to date, and also elsewhere in the world. A fair amount of *Sutherlandia* products are traded over the Internet as well. Dosage forms are as powder and tablets for nutritional supplementation, for topical application in the form of gels or creams and as an immune system stimulating tea.

*S. frutescens* is endemic to southern Africa, where it occurs over a wide geographical area, occupying various habitats from coastal plains to succulent semi-arid zones. The plant is found from the Western Cape north into Namibia and Botswana and to the east into Lesotho and the Eastern Cape. It is absent in the tropical climates of the region. *Sutherlandia* is adapted to both summer and winter rainfall areas. In the semi-arid to arid plains of the Karroo it thrives along rivers and roads and is often browsed by sheep.

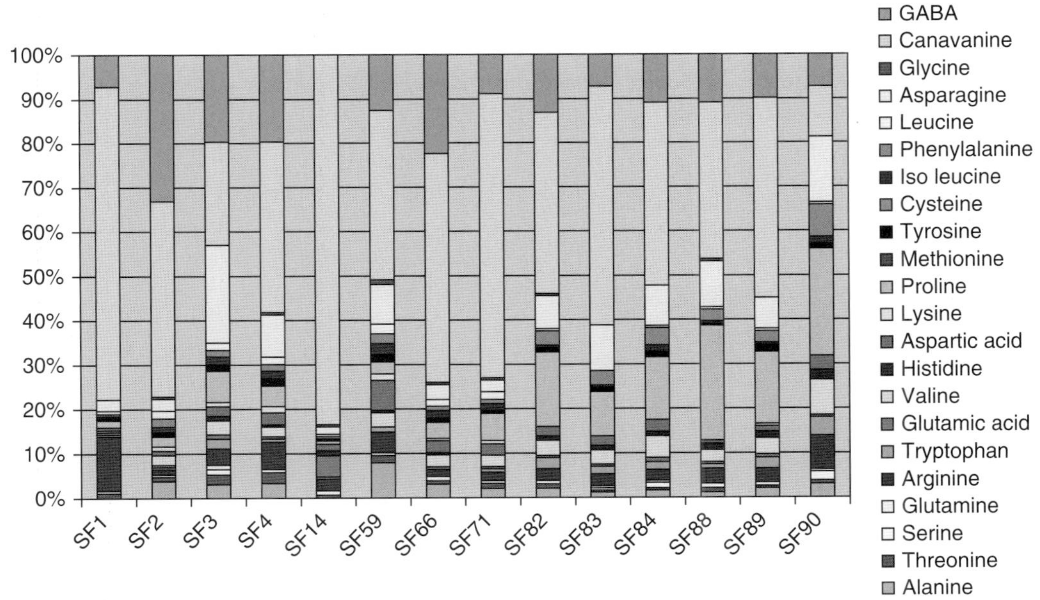


**Figure 3.3** Example of amino acid analysis of *Sutherlandia* – different part of plants and geographical distribution. *See plate section for colour version*

The distinct chemical variation displayed by geographical varieties from different locations spread over its range currently is the subject of a study conducted jointly by the University of Stellenbosch and Parceval Pty in South Africa.

The first *Sutherlandia* cultivation started sometime in the mid 1990s on a very small scale, initially just more for scientific and not for commercial purposes.

Being an annual to biannual to perennial plant depending on habitat and prevailing weather conditions, collecting of seed did not pose too great a challenge – *Sutherlandia* produces large numbers of seed pods on a regular basis. However, like many wild plants, how to get the seed to germinate needed to be understood. Various methods were tried out and two methods are now most commonly used:

- Boil water, let it cool down for a few seconds, pour it over the seed and let it stand for 24 h. Good seed will sink to the ground. After re-drying it can be sown. Floating seed is of inferior quality and should be discarded.
- Scrape seed between medium sandpaper to scratch the hard seed coat, soak in cold water for 24 h. After re-drying it can be sown.

Since the plant is occurring both in summer and winter rainfall areas, different optimal planting times resulted from this. Essentially, the best time is dependent on ambient climatic patterns: to have plants ready for planting with the onset of the relevant rainy season. If sufficient seed and irrigation are available, it is also possible to sow directly.

*Sutherlandia* usually grows naturally on well-drained soils but enjoys the additional water that is provided by run-off from roads or in areas where moisture is retained longer in the soil along rivers or low-lying areas. Where the planned cultivation site is not as well drained it will be best to provide additional drainage by planting on ridges and be careful with providing water so as to avoid 'wet feet'.

Harvesting (Figure 3.4) is usually performed by hand for a high-quality cut – approximately the top 20 cm of the plant are harvested. This gives the best ratio of material with active ingredients and does not contain too much woody components which otherwise dilutes the medicinal actions of the plant.

Overall, it can be said that owing to its nature as a relatively fast-growing shrub and being a pioneer plant, it was much easier to take into cultivation than the other case studies presented here.

**Figure 3.4** Harvesting *Sutherlandia frutescsens* is done by hand to achieve a high-quality raw material. *Photo*: Ulrich Feiter, Parceval Pty, South Africa

## 3.5 Case study 3: *Euphorbia resinifera*

*Euphorbia resinifera* O. Berg, Officinal Spurge or Resin Spurge (Engl.), is a plant occurring in North Africa, and is endemic to the Atlas Mountains of Morocco. It is a member of the Euphorbiaceae, which, with some 2100 species, is one of the largest plant families found on Earth. It is a succulent shrub growing some 60 cm tall. Its dark green erect stems branch prolifically into side stems, and thus form clumps of 1–2 m wide (Figure 3.5). Often mistaken as a cactus, it has short and sharp spines on the ridges of the four-sided stem.

It is one of the oldest medicinal plants recorded. Pliny the Elder reports that *Euphorbia* was named by King Juba II of Mauritania after Euphorbus, his Greek physician (Smith, 1849; Hodgkiss, 2007). King Juba II was a noted scientist and naturalist of his time and used this name to describe a plant he found on the slopes of Mount Atlas in today's Morocco. Its use in old times was as a purgative and emetic agent with very rapid and drastic efficacy.

The plant carries, like all Euphorbiaceae, a white milky latex, which is exuded when the plant is hurt. It runs down the stem and dries in the sun. Traditional harvesting methods include incising the stems and returning later to collect the dried bits and pieces of latex. Since this is a very invasive way of harvesting, many plant populations have been so much disturbed and decimated that *E. resinifera* was placed on the CITES list of protected plants.

The modern-day use of *E. resinifera* is usually restricted to the incorporation into homoeopathic medicines. This is probably due to its severity of action. Homoeopathic preparations such as Euphorbium comp.® (Biologische Heilmittel Heel) are indicated

**Figure 3.5** *Euphorbia resinifera* in a South African nursery – more than 30 years old! *Photo*: Ulrich Feiter, Parceval Pty, South Africa

and give fast relief in cases of rhinitis of various origins (viral, bacterial, allergic), rhinitis sicca, rhinitis hyperplastica and rhinitis atrophicans, hay fever, as well as acute and chronic sinusitis.

Since the plant is on the CITES list, all trade in the raw material is highly restricted. In order to secure a sustainable and legal supply of the dried resin, the German company Biologische Heilmittel Heel embarked on a project to produce sufficient raw material from cultivation. Heel is a pharmaceutical company that develops, manufactures and distributes medicines based on natural substances. Being the global leader in homeopathic combination preparations, the company is also a pioneer in the field of scientific research in natural health care. In cooperation with academic institutions, Heel actively fosters the concept of integrative medicine and is building the bridge between homeopathy and conventional medicine to improve patient care and health.

Initially, plants were obtained by a research-based nursery in Germany. To find plants as such was not too difficult as there are sufficient specialist nurseries or plant collections holding specimens of this succulent in many places. For a number of years, the emphasis was on propagation and increasing the numbers. Climatic conditions for the propagation and cultivation were though not too favourable in Germany, and so in 2010 it was decided to relocate the project to the company Parceval in South Africa. In order to do so, all necessary CITES requirements were observed and the cultivation site was registered with the local CITES authorities so as to be able to legally trade and export with the dried latex later. Additional specimens could be located in a few succulent nurseries in South Africa. Meticulous records were kept on origin of all plants, as well as what has since been propagated from the various origins. Plants are kept apart according to origin and age groups.

Initially, it was vital to obtain as much information as possible about the propagation of the plants as the emphasis in the first phase of the project was to increase the number of plants as fast as possible. Typically, no comprehensive guidelines were available, but in searching the Internet, contacting succulent growers, looking and following up any lead, a picture of how to propagate this species slowly emerged. Many trials followed under different conditions and at different times of the year until a successful protocol could be established.

Once the propagation protocol was established it was easy to start the roll out. Two years after the project started, the required amount of plants was propagated and happily growing. The number of plants needed was calculated by estimating a yield per plant per season and correlating this to the annual need of the client Biologische Heilmittel Heel in Germany. In order to not fall short of latex in the future, the basic number calculated/estimated was generously added to so as to allow for slow growth, disease (which so far has not been a problem) and other calamities.

*E. resinifera* is a semi-desert succulent growing under the harsh sun in Morocco. To our surprise we achieved interesting results by applying various degrees of shade during propagation as well as cultivation – it seems that *Euphorbia* enjoys life more with some shade!

Once the first plants were old enough to be harvested, an appropriate harvesting method had to be developed. The traditional harvesting method tends to be very invasive and often leads to the plants dying (hence it ending up on the CITES list!). Also, a lot of potential product is lost due to the juice running down the stems and disappearing to where it cannot be collected at a later stage. A more gentle as well as efficient approach was needed, so it was eventually decided to cut the stems in places and then capture the juice immediately with a specially designed suction gadget. The latex juice is collected in a small container and can easily be transported to the laboratory for controlled drying and further processing.

One important aspect of *Euphorbia* cultivation is the highly irritant nature of the latex. Great care is taken that no person handling the plant or harvesting the latex comes to harm. One of the key elements is to always work with the plant in large open or well-ventilated spaces – this will reduce the risk of a build up of potentially irritant fumes. Also, the usual protective clothing items (such as mouth mask, hair cover, long-sleeved gloves) are employed to protect the operators during handling and harvesting.

In parallel with the cultivation, trial harvests were conducted from time to time. The samples produced were sent to Biologische Heilmittel Heel for intense analysis and quality control. This was necessary to ensure that the material from wild harvest and cultivation was congruent in all critical parameters.

Over time, it was discovered that time of year, time of day as well as ambient temperature have a marked effect on the yield per plant. Today, the propagation, cultivation and production of commercial quantities of *Euphorbia* dried latex is well understood and the required volumes for the Heel Euphorbium comp.® range of homoeopathic products is covered from cultivation instead of wild harvest.

## 3.6 Conclusion

Taking a new crop into cultivation is and will always remain a daunting task. However, it can be done! Ensure the following:

- That sufficient expertise is available – green fingers work wonders!
- Allow for a minimum of 3–5 years experimenting before success.
- Allocate more cash than to any other cultivation.
- Expect curveballs frequently!
- Ask, ask, ask anyone who may have something to contribute – solutions often come from being humble!
- Spend lots of time with 'your plant' – in nature, on the field, in the greenhouse – its secrets are revealed at odd moments.
- Share your knowledge and others will share their's!
- Limit your trials to what is absolutely necessary – you will get lost in dozens of trials that may not really add to your knowledge.
- Do not give up! Be patient! A solution will come!
- One day, in years to come, you may be awarded the Black Belt of Novel Crop Growing!
- And then write the book!

## References

Hodgkiss, R.J. (2007) *Euphorbus*, http://www.succulent-plant.com/ephemera/euphorbus.html.

Johnson, Q., Syce, J., Nell, H. *et al.* (2007) A randomized, double-blind, placebo-controlled trial of *Lessertia frutescens* in healthy adults. *PLOS Clinical Trials*, **2** (4): e16. doi:10.1371/journal.pctr.0020016.

Kayser, O., Kolodziej, H. and Kiderlen, A.F. (2001) Immunomodulatory principles of *Pelargonium sidoides*. *Phytotherapy Research*, **15** (2), 122–126.

NEMBA (2004) National Environmental Management Biodiversity Act, Government Gazette, Republic of South Africa, vol. 467, No. 26436, http://innovation.mrc.ac.za/biodiversity.pdf.

Smith, C. and Myburgh, K.H. (2004) Treatment with *Sutherlandia frutescens* ssp. *microphylla* alters the corticosterone response to chronic intermittent immobilization stress in rats. *South African Journal of Science*, **100** (3–4), 229–232.

Smith, C.A. (1966) *Common Names of South African Plants*, Memoirs of the Botanical Survey of South Africa No. 35, The Government Printer, Pretoria.

Smith, W.G. (1849) *Dictionary of Greek and Roman Mythology*, Taylor and Watson, London.

## Further reading

Botanic Garden Conservation International, www.bgci.org/africa/abgn/.

The Succulent Society of South Africa, https://www.succulentsociety.co.za/.

National Environmental Management: Biodiversity Act, www.info.gov.za/gazette/acts/2004/a10-04.pdf.

List of organic certification authorities around the world, www.grolink.se/docs/organic-certification-directory-2013.pdf.

Convention for International Trade of Endangered Species (CITES), http://www.cites.org/.

Hager (2004) *Hagers Handbuch der Drogen und Arzneistoffe* [CD-ROM], Springer, Berlin.

International Union for Conservation of Nature (IUCN), http://www.iucn.org/.

PlantZAfrica Home page: www.plantzafrica.com/.

Raimondo, D., Von Staden, L., Foden, W. *et al.* (eds) (2009) *Red List of South African Plants. Strelitzia* 25. South African National Biodiversity Institute, Pretoria.

Van Der Walt J.J.A. and Vorster P.J. (1977) *Pelargoniums of Southern Africa*, vols 1 and 3, Juta, Cape Town.

Van Wyk, B.-E., Van Oudtshoorn, B. and Gericke, N. (1997) *Medicinal Plants of South Africa*, Briza Publications, Pretoria.

Van Wyk, B.-E. and Gericke, N. (2000) *People's Plants. A Guide to Useful Plants of Southern Africa*, Briza Publications, Pretoria.

# 4 Enabling Technologies to Facilitate Natural Product-Based Drug Discovery from African Biodiversity

Nyaradzo T. L. Chigorimbo-Murefu, Grace Mugumbate and Kelly Chibale

Department of Chemistry, University of Cape Town, Rondebosch, South Africa

## 4.1 Introduction

Biological diversity or biodiversity is defined as the 'variety of life' (Gaston and Spicer, 2004). It describes the breadth and depth in the variation of different species in a given space. Humans rely heavily on nature not only as a source of nutrition, but also as a means of transport, clothing and for health. As a result, the biodiversity of groups provides resources for different groups of people as well as opportunities for development. Africa's biodiversity has been hailed as one of the richest on the planet. African people have used natural products derived from that biodiversity for many millennia. The agricultural, pharmaceutical and resultant economic value of Africa's biodiversity is yet to be fully recognized and exploited. This chapter looks at how the African repository of natural products afforded by its biodiversity can be harnessed as a means to involve Africa in the global pharmaceutical industry. We focus specifically on the technologies that would enable the use of African natural products in modern drug discovery paradigms.

### 4.1.1 Drug discovery

Traditional medicines have been part of human civilizations well before methods of documenting them came into place and are still relied upon by over 80% of the current world population in this century according to the World Health Organization (WHO, 2005). African traditional medicine is quite extensive, with, for example, just the Zulus of South Africa making use of over 700 medicinal plants (Van Sertima, 1999). In order to clearly appreciate the contribution that African natural products can make to drug discovery we will look at a brief history of drug discovery, focusing on historical African players to identify strengths and weaknesses to build upon and mitigate respectively. In following with an Afrocentric approach to drug discovery we will briefly identify the gaps and suggest the technologies to enable afro-based drug discovery.

Humans have medicated themselves on plant-derived material since long before the concept of 'drugs' as we know them today (Newman *et al.*, 2000). The Ebers and Edwin Smith Papyri from Africa from about 1500 BC are the oldest ancient pharmaceutical records in which hundreds of medicines are described (Newsome, 1983). On the Asian continent, the Chinese first records date back to around 1100 BC and in the ancient Western world the Greek natural scientist Theophrastus around 300 BC made the first recorded Western contribution to medicine in his work, titled *History of Plants* (Newman *et al.*, 2000). This work was augmented and expanded by others and used in Roman medical schools, where pharmacy and medicine were taught. Knowledge from this period was preserved in European monasteries between 500 and 1200. It was the Arabs who opened the first pharmacies and further codified and incorporated Greco-Roman medicine with that of the other continents from knowledge gained from their travels (Newman *et al.*, 2000).

Before 1800, most drugs then (such as opium, hemlock and foxglove to name a few) did not have known chemical identities. They were water or water-based plant extracts with unknown active ingredients, very much like most of modern traditional medicines today (Raviña Rubira, 2011). The first isolation, extraction and purification of active herbal components and their use as pure drugs was of morphine by Merck in 1826 and started modern natural product-based drug discovery. The development of organic chemistry in the 19th century and subsequent advances in analytical chemistry were pivotal to the boom of the pharmaceutical industry. In 1806, Serturner, a pharmacology student, isolated morphine from opium using chemical methods and performed pharmacological studies in animals. He was the first investigator to take this approach, and it was used by many others to isolate pharmacologically active natural products (Raviña Rubira, 2011). Advances in the biological sciences have also been pivotal to the development of modern drug discovery and contribute significantly to the design of new drugs and understanding of their mechanisms and modes of action.

### 4.1.2 Modern drug discovery and the state of drug discovery in Africa

As described previously, the pharmaceutical industry and drug discovery were birthed from the advances made in organic and analytical chemistry. Together with the Industrial Revolution which provided the chemical feedstocks for this, the mass production of drugs was possible with the inception of Merck and Bayer in the 1800s. Developments in the disciplines of pharmacology and bacteriology ushered in the first period of drug discovery from 1820 to 1935, whose main scientific method was isolation and extraction of natural products, some chemical synthesis and assessing their pharmacology (Raviña Rubira, 2011). Modern drug discovery

*Novel Plant Bioresources: Applications in Food, Medicine and Cosmetics*, First Edition. Edited by Ameenah Gurib-Fakim.

paradigms rely heavily on methods developed during this period, but have largely departed from natural products following a bias towards synthetic drugs.

Many African traditional medicines have contributed to modern marketed herbal medicines. These include the anti-diarrhoeal kaolin used by the people of Mali, which is the active ingredient in Kaopectate used today, and reserpine the antipsychotic drug whose active ingredients are found in African traditional herbs used for those indications (Van Sertima, 1999). However, in spite of all the biodiversity in Africa and its application as traditional medicines, there is no newly marketed single chemical entity modern pharmaceutical product originating from this source. This is in spite of traditional medicines being used on the continent for thousands of years to treat and/or manage various ailments.

As shown, the seeds of drug discovery were sown by innovation in organic chemistry and analytical chemistry – particularly in natural product chemistry, which has a strong presence on the continent. In fact, a number of renowned natural product chemists have been trained from African institutions. The glaring disparity between the efforts above and the number of African-grown drug candidates, marketed drugs or pharmaceutical companies is caused by many reasons, ranging from a lack of effective coordination of continental efforts along the drug discovery and development value chain, to a lack of capacity and competency in several key areas, which are critical to the discovery of new medicines.

The historical strength of African natural product scientists in drug discovery has been in isolation, identification and characterization/structure elucidation of the bioactive components of traditional medicines. The lack of the attendant enabling technologies, including computational tools, medicinal chemistry and ADMET (absorption, distribution, metabolism, excretion, toxicology) integrated with pre-clinical pharmacology, has minimized the contribution of these efforts. Such an integrated approach is likely to result in a positive paradigm shift in the field of drug discovery in Africa and align it with trends in modern drug discovery that have yielded much success elsewhere outside the African continent. In this chapter, a short description of these enabling technologies is discussed including their importance and current status in Africa.

### 4.1.3 The modern drug discovery process

The process that a drug undergoes before it becomes a registered approved drug is a long capital-intensive process typically taking 10–15 years. A general summary of the process is provided in Figure 4.1. The process typically begins with the identification of a specific disease or condition that can be attenuated by small molecules. The cause of the disease is identified, whether it is an external pathogen or if it is caused by the dysfunction of cells or organs.

In some instances the molecular basis may be identified at protein level, and knowledge of the biology of the disease then enables the design of therapeutics against the causative agent. This is the target identification and validation step in the process (Figure 4.1).

The next step in the process involves the design, validation and implementation of disease assays that can be used to test for the effectiveness of molecules. Assays may target whole pathogens or critical molecular components, such as proteins. The screening process then involves the testing of molecules to ameliorate the disease phenotype and is often performed in high-throughput format *in vitro*. Thousands of molecules are typically screened in order to find hits that can effect a change on the target used in the assay. This marks the start of the drug discovery phase in which molecules are tested for activity to find hits. Hits generated, once validated, then undergo hit to lead and lead optimization, in which candidate molecules undergo adsorption, distribution, metabolism and excretion (ADME) pharmacokinetics (PK)-informed medicinal chemistry integrated with pre-clinical pharmacology. From this process, privileged drug candidates are identified, subjected to clinical assessment and production studies, and then marketed for use by patients. Many comprehensive reviews and books on this process have been written which the reader is referred to for a more in-depth representation of the process (Kennedy, 1997; Kubinyi, 2003; Raviña Rubira, 2011).

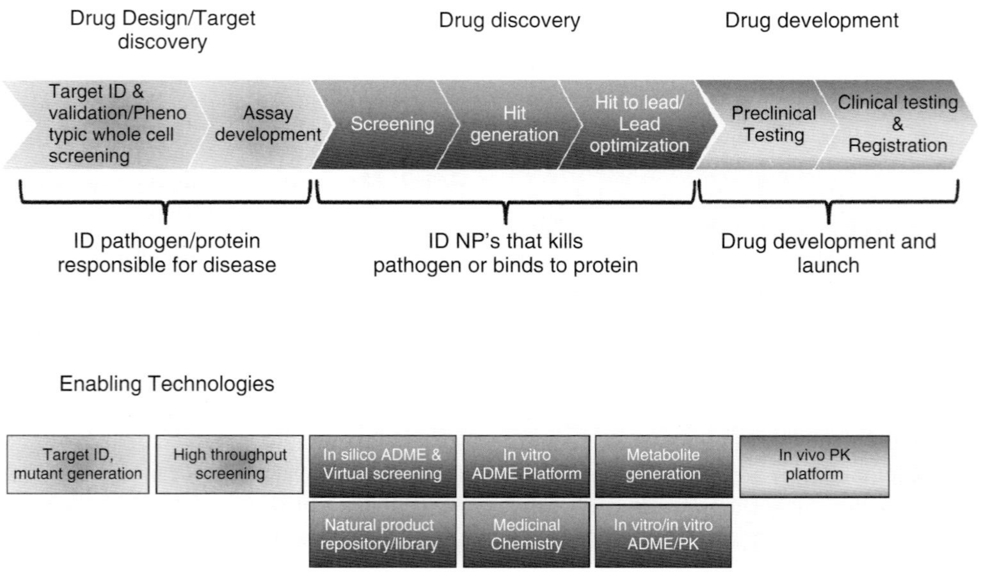


**Figure 4.1** Modern drug design, discovery and development pipeline with attendant enabling technologies. Adapted from Segall (2008)

## 4.2 Enabling-technology platforms

### 4.2.1 Medicinal chemistry

Medicinal chemistry is the study of the relationship between chemical structure and biological or pharmacological activity (Burger and Wolff, 1995). A definition given by Erhardt (2006) states that, 'medicinal chemistry uses physical organic principles to understand the interaction of small molecular displays with the biological realm'. Burger's medicinal chemistry and drug discovery book provides an excellent overview of the now distinct scientific discipline and describes its multidisciplinary nature (Burger and Wolff, 1995). Advances in biomolecular sciences enabled understanding of the correlation observed between biological activity and pharmacological mechanisms. Drug discovery has been dominated by rational drug design and medicinal chemistry integrated with pre-clinical pharmacology and ADME/PK and has departed significantly from the original natural product chemistry approaches to synthetic medicinal chemistry (Erhardt, 2006). Thus, synthesis of chemical libraries has been a hallmark of current drug discovery, and millions of molecules have been synthesized by medicinal chemists as drug discovery starting points. The process of screening chemical libraries to obtain bioactive molecules or 'hits' is the drug discovery starting point from which medicinal chemists typically get involved in drug discovery programmes (Figure 4.1). Obtaining drug leads from the hits described is the evolutionary process of lead development or optimization, in which chemical hits are selected and optimized for drug-like properties. In the process, improvements in specific properties, such as efficacy or stability of leads, by the iterative synthesis of analogues until a final drug candidate with the desired properties is obtained (Kubinyi, 2003).

#### 4.2.1.1 Medicinal chemistry of natural products

Farnsworth captured the allure of natural products to medicinal chemists in his statement, 'each plant [or any cellular organism for that matter] is a unique chemical factory capable of synthesizing an unlimited number of highly complex and unusual chemical substances whose structures could escape the imaginations of synthetic chemists forever' (Farnsworth, 1984). Natural products when used as chemical starting points for medicinal chemistry drug discovery could potentially shorten the drug discovery process. However, the difficulty of working with natural products is that they often do not lend themselves to chemical synthesis or are not available in sufficient quantities to facilitate the semi-synthesis of derivatives, even though various strategies, including biotransformations, can be employed to ameliorate these problems (Chigorimbo-Murefu *et al.*, 2012). As a result of these and other mitigating factors, as well as the unsurpassed chemical diversity that natural products offer, together with the failure of combinatorial chemistry to meet the anticipated higher numbers of drugs discovered, natural products are receiving renewed interest in drug discovery (Abel *et al.*, 2002). This development could strategically benefit an African natural product library proposed in the natural product repository section of this chapter.

### 4.2.2 Drug metabolism and pharmacokinetic–ADME

Modern drug discovery paradigms (post 1991) incorporate drug metabolism and PK (DMPK) assessment of drug candidates in the discovery stage of drug discovery and development. This is a result of the high drug candidate attrition rate that was prevalent when they were considered later in the pipeline (Kennedy, 1997; Eddershaw *et al.*, 2000; Kerns and Di, 2008). Hence, ADME studies are critical components of modern drug discovery projects (Kerns and Di, 2008; Nash *et al.*, 2010). *In vitro* ADME screens/assays are used to characterize a compound's 'drug-like' properties, including physico-chemical properties and metabolic stability, together with pharmacology assays (e.g. efficacy, bioavailability) to increase the progression of drug discovery compounds through the drug discovery and development value chain. Drug-like compounds are defined as 'compounds that have sufficiently acceptable ADME properties and acceptable toxicity properties to survive through the completion of Phase I clinical trial' (Lipinski, 2001). The strategy of earlier phase ADME assays has reduced the failure rate of drug candidates to only 10% after 2001 from the 40% failure rate before then (Figure 4.2).

Before 1991 ADME assays and screens were assessed only in the drug development stage and not the drug discovery stage. The

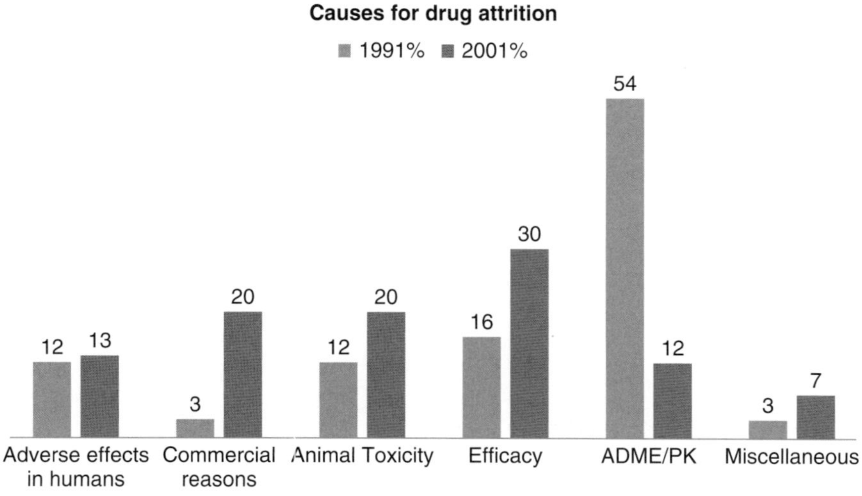


**Figure 4.2** Causes of drug attrition in 1991 and 2001 (adapted from PMA/FDA survey, 1991 and Pharmaceutical R&D Benchmarking Forum, General Metrics 2001)

inclusion of ADME assays in the drug discovery stage allowed compounds to 'fail fast and fail cheap'.

It then follows that not all biologically active natural products possess intrinsic drug-like properties and may require their modification or extraction of molecular information from them using medicinal chemistry tools. In most cases the drug-like properties of isolated active compounds from natural product extracts are not assessed. African drug discovery needs to move beyond the traditional culture of bioassay-guided fractionation of medicinally active plants, as was the case in early drug discovery, to modern integrated drug discovery paradigms incorporating pre-clinical pharmacology, ADME screens and lead development using medicinal chemistry. The relationship between medicinal chemistry, DMPK ADME and pharmacology in relation to drug discovery from African-sourced natural products is shown in Figure 4.3.

The workflow in the figure progresses drug leads sourced from African natural products identified using ADME screens which are then subjected to medicinal chemistry in which analogues are synthesized in order to obtain the structure–activity relationship and retested for pharmacological efficacy and ADME properties. This process is performed iteratively in order to obtain optimal drug candidates with acceptable drug-like properties and efficacy.

ADME screens are performed on compounds in both the discovery and development stages of the drug discovery process. Owing to the amounts of compounds being assessed, in quantity and availability, the assays are miniaturized and in high-throughput 96-well formats. As a result of this, the methods of detection for ADME assays usually are liquid chromatography–mass spectroscopy $(LC–MS)^n$ based, benefiting from advances in analytical chemistry and utilizing bioanalytical equipment such as ultra-performance LC–MS machines, for example. Table 4.1 gives a brief overview of the assays/screens routinely performed in current drug discovery programmes in the different stages. It is important to note that the ADME suite of assays is often tailored for the specific project, taking project specifications into consideration. For a more detailed and

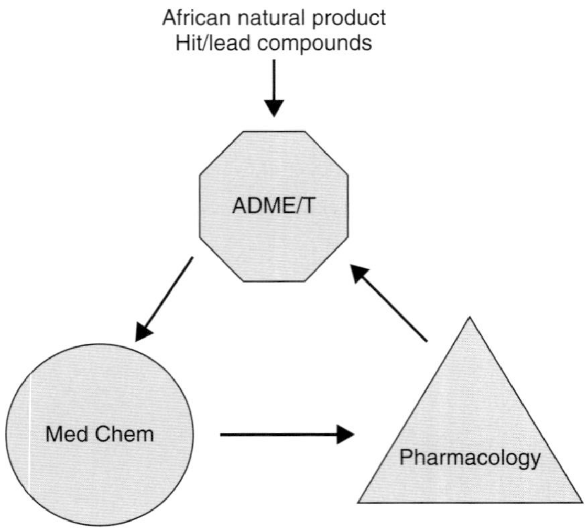


**Figure 4.3** The ADME-guided medicinal chemistry drug discovery workflow from African natural product hits or leads. Adapted from Masimirembwa *et al.* (2001)

**Table 4.1** Essential ADME profiling assays performed during the drug discovery and development stages

| Drug discovery stage screens | Drug development stage assays |
|---|---|
| *HTS lipophilicity assays (log D)* | |
| Important consideration that affects compound distribution, permeability and absorption. This may be determined computationaly or *in vitro* | Comprehensive assessment to accurately determine lead compound lipophilicity |
| *Solubility determination* | |
| Important marker for oral drug absorption and understanding solubility of the drug in other drug assays. Usually kinetic solubility assays performed at this stage | Thermodynamic/equilibrium solubility determination |
| *Membrane permeability* | |
| Important consideration for *in vivo* compound distribution, very important for drugs affecting the central nervous system. Non-cell-based assay, e.g. PAMPA at this stage | Permeability parameters determined in cell-culture-based assays at this stage usually, such as the Caco-2 cell line permeability |
| *Active transport* | |
| Important consideration for oral absorption and drug–drug interations | Permeability parameters determined in active transport expressing cell cultures MDCK cell lines |
| *Metabolic stability* | |
| Useful assays to determine bioavailability and drug clearance parameters using liver microsomes and/or hepatocytes | Drug metabolism studies performed in depth at this stage, where contribution of specific drug-metabolizing enzymes may be assessed, including drug metabolite identification |
| *CYP450 inhibition and induction* | |
| Important assays to determine possibility of drug–drug interactions | CYP450 inhibition and induction assays are usually confirmed at this stage |
| *Protein binding* | |
| Protein binding affects drug clearance, bioavailability and distribution | Lower throughput and more detailed protein binding assays may be commissioned at this stage |
| *Toxicology assays* | |
| For example, hERG and cytotoxicity | Detailed and more comprehensive project-specific toxicology assays are performed, e.g. phototoxicity or human kinase activity |

comprehensive exposition of ADME assays, Kerns and Di (2008) and Nash *et al.* (2010) are recommended to the reader.

#### 4.2.2.1 Recommendations

The inclusion of ADME/ADMET screens in African discovery programmes will enable the transition of African natural product drug discovery into modern drug discovery paradigms. The augmentation of current African drug discovery projects utilizing natural products with ADMET is likely to yield more African-produced drugs for African disease problems from African biodiversity. Figure 4.3 shows the movement of compounds through ADME screens and pharmacological testing, which can be done iteratively to obtain a suitable drug candidate derived from African natural product drug leads. The crucial role and interdependence of pharmacology, medicinal chemistry and ADME in drug discovery programmes as illustrated in Figure 4.3 reinforces the interdisciplinary nature of successful drug discovery. Current bottlenecks in Africa are in the ADME and medicinal chemistry areas, as some institutions have good synthetic organic chemistry capability, capacity, expertise and track record but have no access to ADME platforms and/or medicinal chemistry expertise. Currently, there are two ADME/PK platforms in Africa: one at the African Institute of Biomedical Sciences and Technology, AiBST, in Harare, Zimbabwe, and the other at the University of Cape Town's Drug Discovery and Development Centre (H3-D). The H3-D ADME platform has been established with funding from the South African Government's Department of Science and Technology (DST) and Technology Innovation Agency (TIA) to support integrated medicinal chemistry projects within H3-D and elsewhere in South Africa.

### 4.2.3 Repository of natural products

#### 4.2.3.1 Natural products in drug discovery

As shown in the previous sections, natural products have been relied upon as sources of bioactive molecules in the drug discovery process. The majority of natural products are derived from terrestrial plants, and up to a quarter of these possess bioactive properties such as antibacterial, antinematode, antiprotozoal, anticancer and anti-inflammatory activities (Bérdy, 2005; Penesyan *et al.*, 2010). The use of microbial natural products as therapeutic agents only began in the 20th century. However, 10% of all currently known bioactive natural products are of microbial origin and are the foundation of modern pharmaceuticals (Penesyan *et al.*, 2010). Nonetheless, in spite of the many successes of natural products in drug discovery, most pharmaceuticals departed from natural-product-based drug discovery. This was predominantly due to the advent of combinatorial chemistry, which promised an unlimited supply of synthetic molecules. Sadly, experience has now shown that combinatorial chemistry produced chemical libraries that are inferior to natural products for various reasons discussed in detail in various review articles. These reasons are varied and include the privileged nature of natural products in their ability to interact with protein targets, the diversity of natural product scaffolds (Abel *et al.*, 2002; Ortholand and Ganesan, 2004) and the potential to generate analogues through semi-synthesis and/or biotransformation (Chigorimbo-Murefu *et al.*, 2012).

Thus, a repository of African natural products as a chemical resource for drug screening would be beneficial to the drug discovery endeavours on the African continent. Such natural product libraries exist in academic institutions, and one such example is at the Queensland Compound Library (http://www.griffith.edu.au/science-aviation/queensland-compound-library) in Australia (Eskitis Institute, Griffiths University, Brisbane). Public–private partnerships with the same goal also exist, such as one at the Fundación MEDINA in Spain (www.medinaandalucia.es/en/). Such a model may be adapted in Africa. Currently, there is no African chemical library, and a natural product library would be attractive to many global players in the pharmaceutical industry and drug discovery centres at academic institutions.

This prospect has been realized (Guantai and Chibale, 2012) and is the impetus for the pan-African Natural Products Library pANPL initiative (Geary and Ubalijoro, 2012). This initiative proposes the creation of an African technology platform that will allow the exploration of the potential of African natural products to feed into modern paradigms of drug discovery in the form of an actual collection of African natural products. The platform also is proposed to include a virtual database that can be accessed for chemoinformatics. The benefits of the resource are as follows:

- it will provide a chemical resource for various screening platforms for African natural products that is inexpensive to operate and easy to transfer between research centres;
- it will facilitate the use of the emerging wealth of 'druggable' molecular targets of infectious diseases;
- it will facilitate the use of computational modelling to extract biologically active chemical properties of natural products;
- it will accelerate the development of sustainable infrastructure for natural-product-based drug discovery in Africa;
- it will have a positive impact on the preservation of endangered species and poverty alleviation in relevant communities once the value of the natural product is established;
- it will provide a means in which Africa can make a unique and innovative competitive contribution to drug discovery.

### 4.2.4 Computational (*in silico*) methods

Developments in medicinal chemistry techniques to date have seen large numbers of synthesized compounds generated that cannot be readily investigated experimentally, either *in vitro* or *in vivo*. Some of these molecules are stored in small-molecule databases (e.g. ZINC database; Irwin and Shoichet, 2005) that contains more than 20 million compounds. In addition, advances in genomic technologies, and hence molecular and structural biology, have enabled genome-wide DNA sequencing, gene expression analysis (Lenz *et al.*, 2000) and elucidation of protein structure, thereby providing large numbers of potential drug targets. These developments provide an opportunity to identify the best drug discovery targets that will lead to the best small-molecule drugs for many major diseases that have a high burden in Africa, such as malaria, tuberculosis and HIV/AIDS. On the other hand, drug discovery and development continues to be a lengthy and costly interdisciplinary endeavour. The development of fast computers brought some relief,

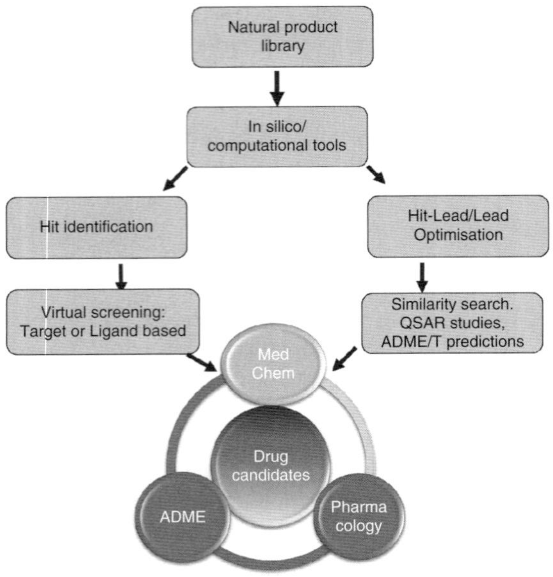


Figure 4.4 Computational approaches applied to natural product drug discovery

since computational, or *in silico*, methods have become important tools in drug discovery for a quick, low-cost delivery of potential novel drug candidates. This is achieved through the creation of computational models that help to predict aspects such as cell signalling and signal–response behaviour, clinical outcomes and ultimately identifying drug targets, as well as toxicity to different organs (Njuguna *et al.*, 2012). In drug design, computational tools may be used to facilitate the process at any of the following stages of drug discovery: hit identification, hit-to-lead optimization and lead optimization. Figure 4.4 summarizes the computational methods that can be used to identify novel drug entities from natural products in drug discovery.

### 4.2.4.1 Application of virtual screening to hit identification

The most important step in the discovery of drugs is a rapid identification of chemical entities that display activity against disease pathogens and are also synthetically tractable. Virtual screening, as the name implies, is an *in silico* technique that complements the collection of existing high-throughput screening in the search for novel chemical entities with the desired biological activity (Bajorath, 2002; McInnes 2007; Scior *et al.*, 2012). The approach is used to search large compound databases for a smaller number of compounds for biological evaluation, decide which compounds to synthesize and which ones to purchase from external suppliers (Leach and Gillet, 2007). Virtual screening can be divided into two major categories:

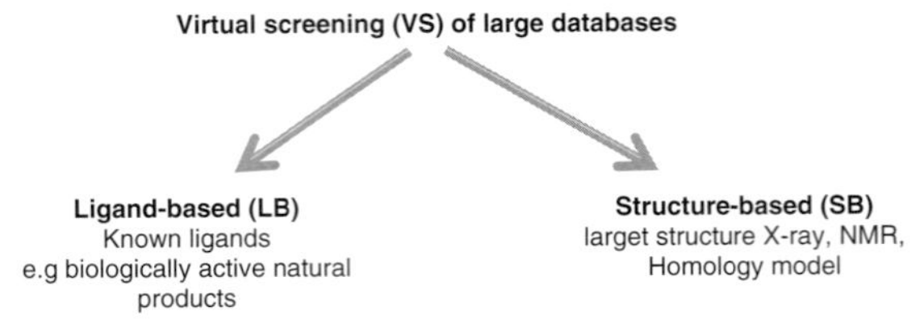


Figure 4.5 Virtual screening methods

structure-based and ligand-based virtual screening, as illustrated in Figure 4.5.

### 4.2.4.2 Structure-based virtual screening

Structure-based (or target-based) virtual screening utilizes available 3D structures of known drug targets generated experimentally from X-ray crystallography or nuclear magnetic resonance, or computationally via homology modelling. The method involves docking compounds from large small-molecule databases or in-house datasets into a protein target with the generation of different conformations (searching algorithms) of receptor–ligand complexes. This is followed by the application of a scoring function to estimate the affinity between the ligand and the receptor. A simplified protocol for structure-based virtual screening is given in Figure 4.6.

The process involves identification of a small-molecule database, preferably one with commercially available compounds, like the ZINC database (Irwin and Shoichet, 2005) or one that contains molecules with reported biological activity (e.g. ChEMBL database; Gaulton *et al.*, 2012). The database is filtered using suitable filtering criteria that could be based on drug-like properties. This can be achieved by utilising web-based filters or the OpenEye FILTER program (FILTER 2.0.2 OpenEye Scientific Software, Santa Fe), or through pipelining tools such as Pipeline Pilot from Accelrys (Hassan *et al.*, 2006). The filtered dataset is docked to a protein target of interest by making use of a suitable docking program. Among the most commonly used programs for high-throughput docking calculations and scoring are Glide with the GlideScore function from Schrodinger (Halgren *et al.*, 2004), Gold with the Goldscore function (Jones *et al.*, 1997), Autodock (Morris and Huey, 2009) and FlexX (Rarey and Kramer, 1996). High-ranked compounds (i.e. those that show high binding affinity to the targets and favourable binding modes) undergo biological assays to determine their biological activities. Within the context of natural products, Zhong *et al.* (2012) used structure-based virtual screening of the ZINC database of natural products containing 90 000 compounds and were able to identify the dipeptide-conjugate deoxyvasicinone derivative compound **1** (Figure 4.7) as an inhibitor of NEDD8-activating enzyme (NAE) in the micromolar range. NAE controls the specific degradation of proteins regulated by cullin-RING ubiquitin E3 ligase and is an

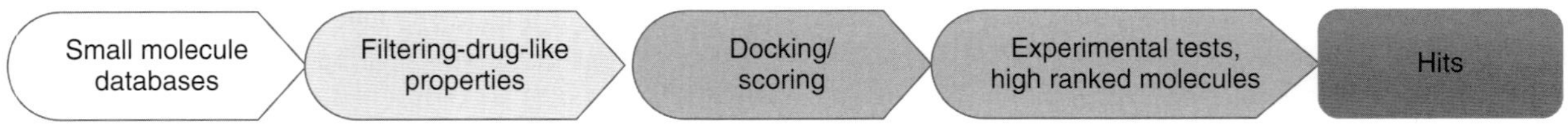


Figure 4.6 Simple protocol for structure-based virtual screening of large small-molecule databases

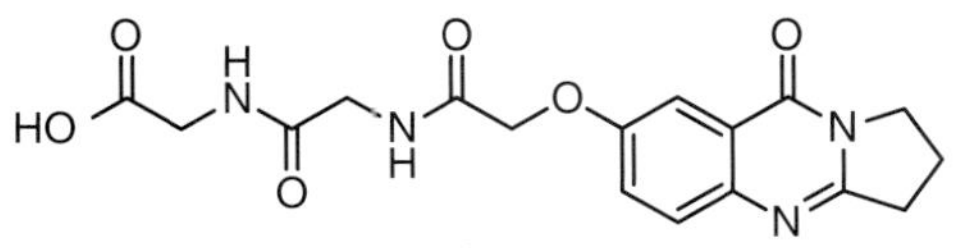

1

**Figure 4.7** Chemical structure of NEDD8-activating enzyme (NAE) inhibitor

attractive target of anti-cancer drugs. In addition to target validation, target–compound complexes with high binding affinity may also be used in structure-based modification and optimization of the natural product molecule.

#### 4.2.4.3 Ligand-based virtual screening

Based on the idea that increase in structural similarity increases biological similarity, ligand-based virtual screening ranks compounds based on their similarities to known active ligands. The top-ranked compounds are then selected for biological testing. In the case of several known biological active compounds, a 3D structural query or pharmacophore model is built. The query or model gives a summary of the important properties or molecular descriptors of the ligands and is used to identify molecules (e.g. from a selected database of natural products) that share some similarities or properties with the known compounds. When a number of actives and inactives are known, they are used to train machine-learning procedures, described by Leach and Gillet (2007), which can then be used in virtual screening (Leach and Gillet, 2007). The development of a variety of computational tools that can be used for molecular similarity analysis makes it possible to perform both 2D and 3D similarity searches.

Two-dimensional molecular similarity searches are normally based on molecular property descriptors, which are numerical values assigned to structures. These include physicochemical properties (e.g. molecular weight, log $P$ (where $P$ is the partition coefficient of a compound in octane and water), polar surface area, hydrogen-bond donors and acceptors, volume distribution), 2D properties (fingerprints, topological indices and maximum common substructure) and 3D properties that include molecular fields (Todeschini and Consonni, 2009). Similarity coefficients, like the commonly used Tanimoto coefficient, give a quantitative measure between two sets of molecular descriptors. Along with the most commonly used 2D similarity tools is the program Pipeline Pilot from Scitegic (Tresadern *et al.*, 2009). The program utilizes 2D fingerprints that describe the connectivity in molecules for example the extended connectivity fingerprint and the functional class-based extended connectivity fingerprint (López-Ramos and Perruccio, 2010).

Three-dimensional similarity search includes a 3D presentation of molecules and comparison of their 3D properties by overlaying them. One of the most commonly used methods is the rapid overlay of chemical structures (ROCS) method from OpenEye (ROCS, V 2.3. 1, OpenEye Scientific Software, Santa Fe), which performs shape-based similarity by aligning the whole molecules and the Tanimoto difference between their 3D steric fields is calculated. Currently, this approach is routinely applied within H3-D at the University of Cape Town, South Africa, to screen more than 13 million molecules from the drug-like subset of the ZINC database for compounds that show similarity to naturally occurring and biologically active compounds, such as gedunin. The molecular structure of gedunin is displayed in Figure 4.8a and its 3D representation, molecular descriptors and molecular volume/shape are presented in Figure 4.8b. For centuries, this natural product from the tree *Azadirachta indica* has been used in India and other parts of Asia to treat a variety of diseases because of its vast medicinal properties (Biswas and Chattopadhyay, 2002).

An overlay of Figure 4.8b with database compounds led to identification of compounds that were similar to gedunin as indicated by the shape Tanimoto scores greater than 0.7. Sticks in Figure 4.8c and d represent the 3D structures of two such compounds. As in any other screening, biological assays are carried out to evaluate activities of high-ranked compounds. Costa *et al.* (2008) successfully applied ligand-based virtual screening involving natural products to facilitate the discovery of novel antileishmanial agents: using two naturally occurring molecules as templates to search for novel antileishmanial entities from about 800 0000 compounds from the ZINC database, 14 were experimentally confirmed to possess significant *in vitro* antileishmanial activity.

### 4.2.5 Applications of other chemoinformatics aspects to drug design

Virtual screening is one aspect of chemoinformatics that has made a marked difference in identification of novel biologically active compounds in drug discovery. To tap into the immense quantities of knowledge that are generated by new drug discovery methods, other aspects of chemoinformatics, such as structure-based optimization, quantitative structure–activity relationship (QSAR) studies and *in silico* ADME/ADMET predictions play an important role in the hit-to-lead and lead optimization stages of drug discovery through to pre-clinical and clinical development.

#### 4.2.5.1 Structure-based optimization

For decades, the number of 3D structures of protein co-crystallized with substrates and/or inhibitors from X-ray crystallography, deposited in the Protein Databank, has increased drastically. This has paved ways for generation of structure-based libraries of modified and optimized biologically active compounds. LaMarche *et al.* (2012) reported a successful application of structure-guided design, synthesis, structure–activity relationships and biological evaluation of di- and tri-carboxylic acid analogues of the antibiotic natural product GE2270, shown in Figure 4.9. Optimization of the hit involved improving aqueous solubility and antibacterial activity of the natural product. Their work led to the commencement of clinical examination of the safety and efficacy of one analogue, LFF571 (Figure 4.9), for the treatment of *Clostridium difficile* infections in humans. In similar work, McCarren *et al.* (2012) applied molecular modelling, structure-based design, virtual screening and mechanism-based drug design to modify and optimize zinc binding in histone deacetylase. This work led to the identification of alternative zinc binding motifs for the metalloenzyme inhibitors.

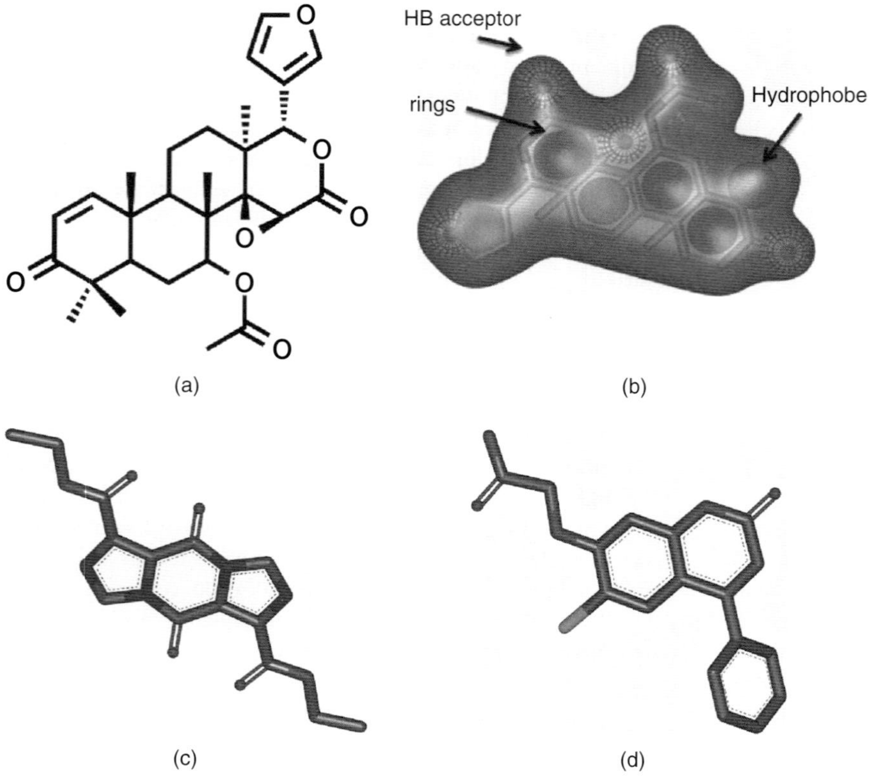


**Figure 4.8** (a) Structure of a natural product, gedunin; (b) 3D structure of gedunin showing volume space and molecular descriptors. Examples of high-ranked compounds from 3D shape analysis of the ZINC database using gedunin: (c) shape Tanimoto score of 0.72; (d) shape Tanimoto score of 0.70. *See plate section for colour version*

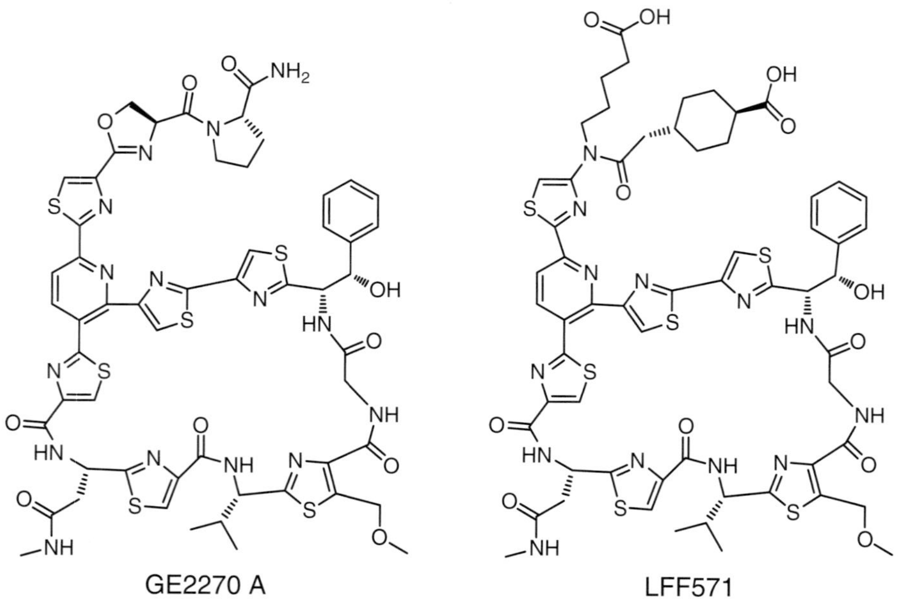


**Figure 4.9** Chemical structures of the antibiotic natural product GE2270A and its analogue LFF571

### 4.2.5.2 Quantitative structure–activity relationship studies

At each stage of drug design there is generation of huge volumes of data which are analysed to provide information and knowledge; for example, about the molecular structure, properties and activity of compounds. Optimization of hit compounds is dependent on the relationship between structural features and physicochemical properties, as well as between molecular structure and biological activity of compounds. To assist with this analysis, computational models, which enable the observed properties or activity to be related to the molecular structure, have been built (Leach and Gillet, 2007). QSAR models have been developed from, for example, metabolic stability, enzyme and metabolite identification, and cytochrome P450 inhibition data using recombinant enzymes (Masimirembwa and Thelingwani, 2012). A detailed account of more QSAR models is given by Leach and Gillet (2007). Co-crystallization of proteins and their substrates/inhibitors and the advancements in computational molecular modelling and docking calculations have collectively provided valuable information on the binding modes and the protein–ligand interactions that have led to generation of pharmacophore models used in structure-based drug optimization.

#### 4.2.5.3 *In silico* ADME/ADMET predictions

Prediction of physicochemical properties of a drug in the early stages is cost effective and important in reducing the number of drugs that fail at later stages of development.

Properties such as solubility, metabolic stability, permeability and toxicity are crucial for favourable (ADME/ADMET) properties of a drug (Kerns and Di, 2008). Mathematical and statistical models have been developed to predict, amongst other properties, solubility, Caco-2 permeability and volume of distribution. Prediction tools like Volsurf+ (Cruciani *et al.*, 2000) and Pentacle (Ermondi *et al.*, 2011) perform GRID-based calculations of a wide range of physicochemical properties that include lipophilicity (log *P* and log *D*), p$K_a$, molecular weight, volume distribution, solubility, Caco-2 permeability, and so on. Pipelining tools like Pipeline Pilot 8.5 perform atom-based calculations of these properties. Another approach to *in silico* prediction is the use of knowledge-based database systems that are capable of predicting toxicity of novel compounds based on their similarity to pre-existing substrates for which toxicity profiles have been well characterized (Njuguna *et al.*, 2012).

In summary, computational approaches have many advantages, which have resulted in the widespread adoption of these techniques to drug discovery design and development. For example, their application in Africa has been illustrated by the recent work reported by H3-D (Espinoza-Moraga *et al.*, 2013). In this reported work, a comparison was made between the chemical space of natural products active against different *Mycobacterium tuberculosis* strains and that of anti-tuberculosis drugs. An African computational platform employing these techniques in integrated drug discovery programmes would improve drug discovery initiatives on the continent. Such platforms exist at the H3-D centre (University of Cape Town, South Africa) and at AiBST in Harare, Zimbabwe, where they are integrated in drug discovery projects.

## 4.3 Natural product diversification and drug metabolite generation platform

One of the biggest detractions for the use of natural products in drug discovery is their often difficult or impossible chemical synthesis. This impediment is particularly pertinent in drug discovery programmes in which significant quantities of compounds are required for biological and other screens as well as analogues required for the elucidation of structure–activity relationships of compounds. This problem can be alleviated by the application of recent technological developments in the field of the biosynthesis and combinatorial biocatalysis.

Living cells normally produce various chemicals for their own requirements and can be engineered using genetic techniques to produce chosen products. Hence, biosynthesis enables the production of desired chemicals from living organisms such as microorganisms or plant cells. In its application, combinatorial biosynthesis is the genetic engineering of natural product biosynthetic pathways for desired products. It has been successfully used in the production of natural product drugs such as erythromycin, and also daptomycin, a natural product antibacterial (Staunton, 1998; Kirkpatrick *et al.*, 2003).

Biocatalysis involves the use of enzymes in the production of chemicals and is used in the synthesis of many pharmaceutical agents, such as the semi-synthesis of the cholesterol-lowering drug simvastin (Xie and Tang, 2007; Chigorimbo-Murefu *et al.*, 2012). Combinatorial biocatalysis involves the iterative use of enzymes for chemical synthesis and is usually used in the generation of natural product libraries of natural product analogues. This has been successfully used for the production of the bergenin natural product library (Krstenansky and Khmelnitsky, 1999). These technologies are not currently represented on the African continent, and a technology platform utilizing them would be important for the diversification of natural-productbased drug discovery.

Drug metabolite identification is an important consideration during lead optimization. Drug metabolites have been shown to also possess pharmacological activity and thus can contribute to the pharmacokinetic properties and toxicity of drugs. As a result, the US Food and Drug Administration regulations stipulate the profiling of drug metabolites. However, chemical synthesis of drug metabolites is usually difficult and tedious, resulting in the development and use of microbial models of drug metabolite generation (Azerad, 1999). Bioprocess engineering and bioreactor capabilities exist on the continent, for example, at the Center for Bioprocess Engineering and Research CeBER (www.ceber.uct.ac.za) and the H3-D centre drug metabolite platform in Cape Town, South Africa (www.H3D.co.za).

There are other platforms that are associated with drug discovery and development which have not been addressed in this chapter.

## 4.4 Conclusion

Nature has provided a rich repository for man to use for his developmental needs. Africa's biodiversity presents Africa and the world at large with a rich scientific research and development source that can be translated into life saving medicines. We have focused in this chapter on some of the technology platforms for the use of natural products in drug discovery and development. Although there are pockets of excellence in Africa in certain areas, the challenge of how to translate this into new potential medicines still remains. Africa's activities within the value chain of drug discovery, drug development and clinical testing of medicines are fragmented, under-funded with no absorptive capacity to attract and retain key skills. There is also a lack of continuity with respect to long term funding and sustained pipeline of projects. There is an urgent need to integrate the value chain and establish a new pharmaceutical industry in Africa that will grow the economy and create jobs while at the same time addressing the key health needs of the continent. To ensure sustainability, there is a need to build a pipeline of products and strategic alliances with international partners across the value chain. This is all while training a new generation of African scientists with the key skills needed for drug discovery – integrating medicinal chemistry with pre-clinical pharmacology as well as ADME/PK along with development of strong project management skills working to internationally recognized standards. Together this will bring a critical mass of new scientists to potentially develop drugs in various diseases, but with a unique focus on the issues facing sub-Saharan Africa.

## References


Abel, U., Koch, C., Speitling, M. and Hansske, F.G. (2002) Modern methods to produce natural-product libraries. *Current Opinion in Chemical Biology*, **6**, 453–458.

Azerad, R. (1999) Microbial models for drug metabolism. *Advances in Biochemical Engineering/Biotechnology*, **63**, 169–218.

Bajorath, J. (2002) Integration of virtual and high-throughput screening. *Nature Reviews. Drug Discovery*, **1**, 882–894.

Bérdy, J. (2005) Bioactive microbial metabolites. *The Journal of Antibiotics*, **58**, 1–26.

Biswas, K. and Chattopadhyay, I. (2002) Biological activities and medicinal properties of neem (*Azadirachta indica*). *Current Science*, **82**, 1336–1345.

Burger, A. and Wolff, M.E. (1995) *Burger's Medicinal Chemistry and Drug Discovery*, 5th edn, John Wiley & Sons, Inc., New York, NY.

Chigorimbo-Murefu, N.T., Njoroge, M., Nzila, A. *et al.* (2012) Biotransformation and biocatalysis: roles and applications in the discovery of antimalarials. *Future Medicinal Chemistry*, **4**, 2325–2336.

Costa, F., Nicoluci, R., Silva, M. *et al.* (2008) Natural products biological screening and ligand-based virtual screening for the discovery of new antileishmanial agents. *Letters in Drug Design & Discovery*, **5**, 158–161.

Cruciani, G., Pastor, M. and Guba, W. (2000) VolSurf: a new tool for the pharmacokinetic optimization of lead compounds. *European Journal of Pharmaceutical Sciences*, **11** (Suppl 2), S29–S39.

Eddershaw, P., Beresford, A. and Bayliss, M. (2000) ADME/PK as part of a rational approach to drug discovery. *Drug Discovery Today*, **5**, 409–414.

Erhardt, P.W. (2006) Using drug metabolism databases during drug design and development, in *Drug Discovery and Development. Volume 1: Drug Discovery* (ed. M.S. Chorghade), John Wiley & Sons, pp. 273–293.

Ermondi, G., Caron, G., Garcia Pintos, I. *et al.* (2011). An application of two MIFs-based tools (Volsurf+ and Pentacle) to binary QSAR: the case of a palinurin-related data set of non-ATP competitive glycogen synthase. *European Journal of Medicinal Chemistry*, **46**, 860–869.

Espinoza-Moraga, M., Njuguna, N.M., Mugumbate, G. *et al.* (2013). *In silico* comparison of antimycobacterial natural products with known antituberculosis drugs. *Journal of Chemical Information and Modeling*, **53**, 649–660 .

Farnsworth, N.R. (1984) The role of medicinal plants in drug development, in *Natural Products and Drug Development* (eds P. Krogsgaard-Larsen, S.B. Christensen and H. Kofod), Tindall and Cox, London, pp. 8–98.

Gaston, K.J. and Spicer, J.I. (2004) *Biodiversity: An Introduction*, 2nd edn, Wiley–Blackwell.

Gaulton, A., Bellis, L.J., Bento, A.P. *et al.* (2012) ChEMBL: a large-scale bioactivity database for drug discovery. *Nucleic Acids Research*, **40**, D1100–D1107.

Geary, G. and Ubalijoro, E. (2012) Searching for drugs that target multiple receptors for anthelmintics from African natural products, in *Drug Discovery in Africa – Impacts of Genomics, Natural Products, Traditional Medicines, Insights into Medicinal Chemistry and Technology Platforms in Pursuit of New Drugs* (eds K. Chibale, M. Davies-Coleman and C. Masimirembwa), Springer, Berlin, pp. 127–150.

Guantai, E. and Chibale, K. (2012) Natural product-based drug discovery in Africa: the need for integration into modern drug discovery paradigms, in *Drug Discovery in Africa – Impacts of Genomics, Natural Products, Traditional Medicines, Insights into Medicinal Chemistry and Technology Platforms in Pursuit of New Drugs* (eds K. Chibale, M. Davies-Coleman and C. Masimirembwa), Springer, Berlin, pp. 101–126.

Halgren, T. and Murphy, R. (2004) Glide: a new approach for rapid, accurate docking and scoring. 2. Enrichment factors in database screening. *Journal of Medicinal Chemistry*, **47**, 1750–1759.

Hassan, M., Brown, R.D., Varma-O'Brien, S. and Rogers, D. (2006) Cheminformatics analysis and learning in a data pipelining environment. *Molecular Diversity*, **10**, 283–299.

Irwin, J.J. and Shoichet, B.K. (2005) ZINC – a free database of commercially available compounds for virtual screening. *Journal of Chemical Information and Modeling*, **45**, 177–182.

Jones, G., Willett, P. and Glen, R. (1997) Development and validation of a genetic algorithm for flexible docking. *Journal of Molecular Biology*, **267**, 727–748.

Kennedy, T. (1997) Managing the drug discovery/development interface. *Drug Discovery Today*, **2**, 436–444.

Kerns, E. and Di, L. (2008) *Drug-like Properties: Concepts, Structure Design and Methods: from ADME to Toxicity Optimization*, Academic Press.

Kirkpatrick, P., Raja, A., LaBonte, J. and Lebbos, J. (2003) Daptomycin. *Nature Reviews. Drug Discovery*, **2**, 943–944.

Krstenansky, J.L. and Khmelnitsky, Y. (1999) Biocatalytic combinatorial synthesis. *Bioorganic & Medicinal Chemistry*, **7**, 2157–2162.

Kubinyi, H. (2003) *In search for new leads, in European Federation of Medicinal Chemistry Yearbook 2003*, LD Organisation sprl, Louvain-de-Neuve, pp. 14–28.

LaMarche, M.J., Leeds, J.A., Amaral, A. *et al.* (2012) Discovery of LFF571: an investigational agent for *Clostridium difficile* infection. *Journal of Medicinal Chemistry*, **55**, 2376–2387.

Leach, A. and Gillet, V. (2007) *An Introduction to Chemoinformatics*, Springer, Dordrecht.

Lenz, G., Nash, H. and Jindal, S. (2000) Chemical ligands, genomics and drug discovery. *Drug Discovery Today*, **5**, 145–156.

Lipinski, C.A. (2001) Drug-like properties and the causes of poor solubility and poor permeability. *Journal of Pharmacological and Toxicological Methods*, **44**, 235–249.

López-Ramos, M. and Perruccio, F. (2010) HPPD: ligand- and target-based virtual screening on a herbicide target. *Journal of Chemical Information and Modeling*, **50**, 801–814.

Masimirembwa, C. and Thelingwani, R. (2012) Application of *in silico*, *in vitro* and *in vivo* ADMET/PK platforms in drug discovery, in *Drug Discovery in Africa – Impacts of Genomics, Natural Products, Traditional Medicines, Insights into Medicinal Chemistry and Technology Platforms in Pursuit of New Drugs* (eds K. Chibale, M. Davies-Coleman and C. Masimirembwa), Springer, Berlin, pp. 151–192.

Masimirembwa, C.M., Thompson, R. and Andersson, T.B. (2001) *In vitro* high throughput screening of compounds for favorable metabolic properties in drug discovery. *Combinatorial Chemistry & High Throughput Screening*, **4**, 245–263.

McCarren, P., Hall, M.L. and Whitehead, L. (2012) The chemical tuning of a weak zinc binding motif for histone deacetylase using electronic effects. *Chemical Biology & Drug Design*, **80**, 203–214.

McInnes, C. (2007) Virtual screening strategies in drug discovery. *Current Opinion in Chemical Biology*, **11**, 494–502.

Morris, G. and Huey, R. (2009) AutoDock4 and AutoDockTools4: automated docking with selective receptor flexibility. *Journal of Computational Chemistry*, **30**, 2785–2791.

Nash, J., Beresford, A. and Yates, D. (2010) Intelligent ADME outsourcing. *European Biopharmaceutical Review* (Winter), 20–24.

Newman, D.J., Cragg, G.M. and Snader, K.M. (2000) The influence of natural products upon drug discovery (antiquity to late 1999). *Natural Product Reports*, **17**, 215–234.

Newsome, F. (1983) Black contributions to the early history of western medicine, in *Blacks in Science: Ancient and Modern*, (ed. I. Van Sertima), Transaction Books, New Brunswick.

Njuguna, N.M., Masimirembwa, C. and Chibale, K. (2012) Identification and characterization of reactive metabolites in natural products-driven drug discovery. *Journal of Natural Products*, **75**, 507–513.

Ortholand, J.-Y. and Ganesan, A. (2004) Natural products and combinatorial chemistry: back to the future. *Current Opinion in Chemical Biology*, **8**, 271–280.

Penesyan, A., Kjelleberg, S. and Egan, S. (2010) Development of novel drugs from marine surface associated microorganisms. *Marine Drugs*, **8**, 438–459.

Rarey, M. and Kramer, B. (1996) A fast flexible docking method using an incremental construction algorithm. *Journal of Molecular Biology*, **261**, 470–489.

Raviña Rubira, E. (2011) *The Evolution of Drug Discovery: From Traditional Medicines to Modern Drugs*, Wiley–VCH Verlag GmbH, Weinheim.

Scior, T., Bender, A., Tresadern, G. *et al.* (2012) Recognizing pitfalls in virtual screening: a critical review. *Journal of Chemical Information and Modeling*, **52**, 867–881.

Segall, M. (2008) Why is it still drug discovery? *European Biopharmaceutical Review* (Spring), 72–75.

Staunton, J. (1998) Combinatorial biosynthesis of erythromycin and complex polyketides. *Current Opinion in Chemical Biology*, **2**, 339–345.

Todeschini, R. and Consonni, V. (2009) *Molecular Descriptors for Chemoinformatics*, Wiley–VCH Verlag GmbH.

Tresadern, G., Bemporad, D. and Howe, T. (2009) A comparison of ligand based virtual screening methods and application to corticotropin releasing factor 1 receptor. *Journal of Molecular Graphics & Modelling*, **27**, 860–870.

Van Sertima, I. (1999) The lost sciences of Africa: an overview, in *African Renaissance: The New Struggle* (ed. M.W. Makgoba), Mafube/Tafelberg, Cape Town, pp. 305–330.

WHO (2005) National policy on traditional medicine and regulation of herbal medicines – Report of a WHO global survey. http://apps.who.int/medicinedocs/en/d/Js7916e/ (accessed 17 December 2012).

Xie, X. and Tang, Y. (2007) Efficient synthesis of simvastatin by use of whole-cell biocatalysis. *Applied and Environmental Microbiology*, **73**, 2054–2060.

Zhong, H., Ma, V.P.-Y. and Cheng, Z. (2012) Discovery of a natural product inhibitor targeting protein neddylation by structure-based virtual screening. *Biochimie*, **94**, 2457–2460.

# 5 Assessing Biodiversity: A Molecular Approach Using DNA Sequencing

Yasmina Jaufeerally-Fakim

Faculty of Agriculture, University of Mauritius, Reduit, Mauritius

## 5.1 Introduction

The world's biodiversity comprises all living entities from unicellular bacteria to the most complex of higher organisms. This richness forms the foundation of our natural environment upon which the survival of most species rely.

The plant kingdom is known to include more than 250 000 species classified according to established criteria for taxonomy (Spooner *et al.*, 2003). Their classification into families, genus and species is based on shared physical characteristics that allow individual groups to be distinguished. Such methodologies of species identification are critical for many applications in agriculture, ecology, germplasm assessment or conservation strategies. Plant bioresources represent an infinite source of biological material for humankind, not only for food but also for medicinal, fibre and bio-diesel use. It is imperative, therefore, to fully understand the diversity of living organisms and have means for precise identification at the species level. In addition, variations within populations provide an insight into the evolutionary history of their genetic changes. A wide range of molecular techniques developed over the past two decades have enabled a precise evaluation of such changes and have brought novel approaches for classification and measurement of relatedness among individuals. Most of the methods rely on the detection of sequence polymorphism at target sites within the genome. They differ in the complexity of the steps involved and in their resolution. The recent and most reliable way for biodiversity assessment is through DNA sequencing of either whole genomes or targeted loci in the genome. With the advent of next-generation sequencing methods, this new approach is rapidly overtaking others, mainly due to their high throughput and relative low costs. It is envisaged that species identification will soon be routinely done through bench-top sequencing machines linked to appropriate taxonomic classification software.

These new methodologies are discussed here to understand their applications for agriculture, biodiversity and crop wild relatives conservation.

## 5.2 Taxonomy and evolution

Plants form a kingdom of their own with 12 phyla within which classification is primarily based on seed, tissue structure and on reproductive characteristics. Within the phyla are the orders, classes and families. There are roughly 150 families, each of which group together several genera, with one genus comprising a collection of species. At every level of this classification, the grouping is built around shared characteristics, thus representing common evolutionary history. The class Monocotyledoneae has all the major crops, which form the main carbohydrate source for the human diet, with wheat, rice and corn. On the other hand, the Dicotyledoneae class includes all the major flowering plants.

Humankind is dependent on the world biodiversity for its survival. Its assessment will help to ensure a reliable food source for the future as well as contribute to finding new pharmacological compounds.

Understanding 'What is a species?' is very much centred around the notion of common characteristics. Even though this concept can be clearly defined for many organisms, it remains debatable for others. It is generally accepted that a species is a group of organisms sharing a number of morphological characteristics and that can breed among themselves. It implies that they also have the same genetic elements that will allow homologous chromosomes to pair with each other. This is a fundamental aspect of species because there cannot be the formation of a viable embryo unless homologues find each other at meiosis. The Darwinian concept of a species is one that describes it as a population or group of populations, distinct from other groups and that can evolve separately (Mayr and Bock, 2002). The post-Darwinian notion relies on evidence from molecular studies to point to a more integrated system of evolution where members of different species can interchange genetic material. The species remains a useful classification unit for understanding biodiversity.

Present-day biodiversity is the result of millions of years of evolution, which has shaped the natural world. The forces behind the changes come from the organisms' living environment and climatic conditions. Species, therefore, are not closed systems; rather, they are structures that allow outside forces to direct modifications inside. How extensive these modifications are determines how the organization of the populations is reshaped.

The aim of evolutionary biology studies is to reconcile data on genetic variation with the mechanisms of change that have accumulated over the years. Such data can be highly complex and require appropriate algorithms together with powerful computation for their analysis. The availability of DNA sequences at very low cost has transformed the way variation is measured within and among species. DNA sequences represent the highest resolution approach

*Novel Plant Bioresources: Applications in Food, Medicine and Cosmetics*, First Edition. Edited by Ameenah Gurib-Fakim.

for diversity measurement because each pair of nucleotides at a given locus represents one character. There are millions of such characters available in gene banks. This is unprecedented, given that scoring of morphological characteristics (which can be difficult to assess depending on the subjective analysis of the investigator) is limited to a few characters only.

Genomes are quite dynamic in the sense that alterations accumulate in DNA sequences at different rates for different loci. Such plasticity is an essential element of diversity and therefore represents a mechanism for species to evolve (Weber *et al.*, 2012). New technological systems available today allow for accurate comparisons of homologous sequences from many individuals through alignment. Several algorithms have been developed to address the need for precise alignment and phylogenetic inference (Zhu *et al.*, 1998).

## 5.3 Assessing diversity

Prior to the advent of molecular techniques, diversity among living organisms was only assessed through careful measurement of specific morphological characteristics. These characteristics follow Mendelian inheritance. Although very useful, such measurements can be influenced by the environment and other habitat-related factors. Above all, they are under either dominant or recessive expression, which require controlled crosses for identifying the heterozygotes. The phenotype of an individual results from the selective expression of genes through temporal and spatial control. Some genes (housekeeping) are constitutively expressed because they code for proteins that are always required in the cell, while many genes are under extremely tight regulation. The outcome of the mechanisms of control determines the phenotype of an organism at a given time or stage of development. This control is susceptible to outside influence, such as the presence of certain chemicals, which may alter gene expression.

### 5.3.1 Biochemical methods

Development in the methods for identification of alternative forms of the same enzyme (allozymes) emphasized the need to look closely into different forms of the same gene. Such biochemical markers have long contributed to measuring diversity (Zeidler, 2000). The presence of more than one form of the same enzyme indicates that there are alternative forms of the same gene. These, however, are limited as there are only few enzymes for which the assays can be revealing. The method also called for tedious handling of starch gel electrophoresis, which limits the number of samples that can be assayed in one experiment. This approach is only rarely used today.

The greatest impact on diversity studies has been the introduction of molecular techniques for measuring genetic variation. Since the early 1990s a large number of new techniques have been devised to target loci in the genomes of plants that are likely to reveal variation. The term 'molecular marker' was coined to indicate a DNA locus that shows polymorphism. Most of the methods are gel based with electrophoresis and/or utilizing fluorescent labels for detection purposes (Semagn *et al.*, 2006). For a marker to be used for diversity, it should be polymorphic within the populations under study and co-dominant.

The first few techniques were based on DNA–DNA hybridization, whereby homologous fragments would anneal to each other. This does not require that the homology be total. Fragments differing by 30–40% would still anneal under some conditions. With hybridization techniques, it is possible to assess for the presence of a homologous fragment of DNA in different genomes. It does not allow comparison of the exact sequences of the DNA fragment (Brown, 2001). Conventional hybridization methods are lengthy and produce limited information per assay.

The majority of molecular markers developed so far are based on the polymerase chain reaction (PCR), which amplifies a DNA segment through the use of appropriate primers. PCR is a powerful technique for *in vitro* DNA amplification and is simple to use. With this approach it is possible to obtain a large amount of a specific segment of DNA from a given genome. The many types of PCR-based markers differ in the way the amplification is effected and the loci that are targeted by the primers (De Giovanni *et al.*, 2004). There are, therefore, an infinite number of sites within a genome that can potentially reveal polymorphism among individuals of species or populations. A summary of some of the most widely used methods is provided here.

### 5.3.2 Amplification-based markers

PCR amplification requires the use of two primers, facing each other, annealing on the two different strands and which will act as anchors for the DNA polymerase to start synthesizing a new strand. It is an *in vitro* reaction that mimics the way DNA replication occurs in a cell. Denaturation of template DNA at high temperatures (94 °C) is required after each cycle of polymerization (Figure 5.1); in order for this to occur in the same tube, heat-stable polymerases such as Taq are used.

At the end of the reaction, the amplicon is present in large amounts, and this facilitates further manipulation.

The two primers have to anneal specifically, within a certain distance, at the sites of amplification, and hence they have to be properly designed so as to eliminate or minimize nonspecific amplification. Several computer applications are available from open source for the design of primers using a known DNA sequence.

#### 5.3.2.1 Random amplification of polymorphic DNA

This is a simple technique that uses a single primer for random amplification of genomic DNA. It uses random primers and only a single primer is used in the reaction. No sequence information is required prior to its application. Random amplification of polymorphic DNA (RAPD) primers became available from Operon Technologies Inc., Alameda, California, in the early days that they were developed (Williams *et al.*, 1990; Caetano-Anollés *et al.*,1992). This technique has been widely used for genetic variation assessment as well as for mapping of important traits for crop plants. Sequencing of specific RAPD fragments can also be used to obtain primers for more precise differentiation.

#### 5.3.2.2 Simple sequence repeats or microsatellite markers

This type of marker is based on the amplification of repeat sequences, usually four to six nucleotides long (Figure 5.2), that are commonly

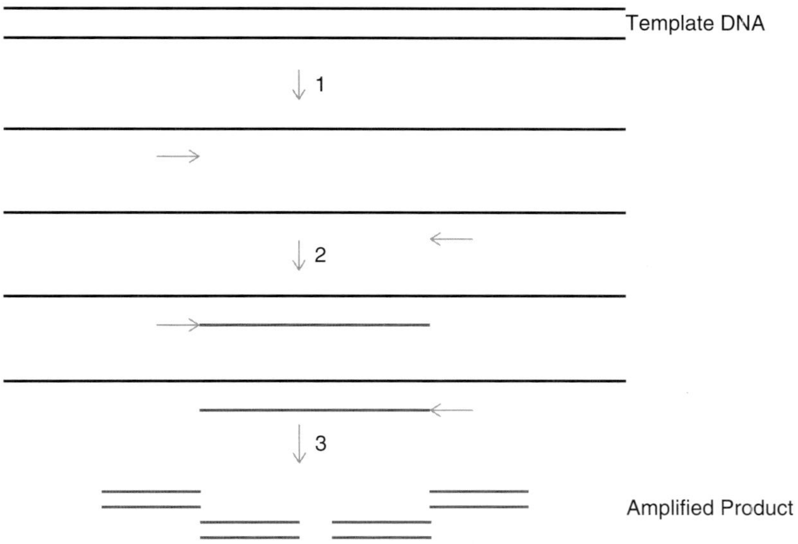


**Figure 5.1** PCR amplification of DNA: (1) DNA denaturation; (2) primer annealing; (3) amplification. Arrows indicate the primers. *See plate section for colour version*

found in plant genomes. The primers target the flanking regions of the repeat and the polymorphism is seen as the sizes of bands, which reflect the differences in number and types of repeats. This variation is the result of strand slippage during DNA replication, thus allowing extra nucleotides to be added. In order to use this method, the primer sequences have to be determined for a given species. The first step is to design the primers (Zane *et al.*, 2002) for a given species as the types of repeats differ among species. This can be done as follows:

1. First the genomic DNA is fragmented to a range of sizes and each fragment cloned into a vector. Once a whole library is obtained, each clone is then screened with a probe containing different types of repeats. The hybridization profile will reveal the types of repeats. Based on this, primers are then designed to amplify the region of repeats. Another approach for finding microsatellite sequences is through a PCR method using random primers. The rationale is that RAPD products often contain repeat regions. In both methods the clones are sequenced to obtain the exact DNA sequences.

2. Two alternative ways for isolating microsatellite sequences rely on the use of phage libraries of genomic DNA producing single-stranded DNA, which is then amplified with repeat-specific primers. The resulting clones containing repeats are then sequenced to design specific primers (Ostrander *et al.*, 1992; Paetkau, 1999).

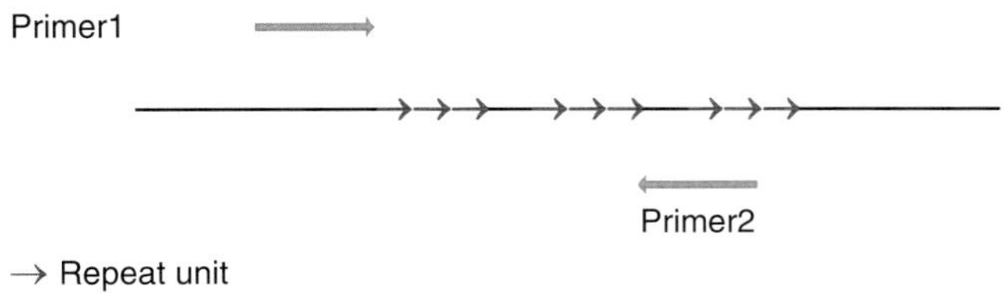


**Figure 5.2** Representation of a simple sequence repeat with three copies of a basic repeat unit of →→→. The primers on either side will amplify the region of repeats. Polymorphism is detected when there are different numbers or types of → repeats

Different types of oligonucleotide repeats are found in different organisms. Frequency and types of plant microsatellites sequences have been estimated using whole genome sequences of three monocots and three dicots (Sonah *et al.*, 2011). Comparison of the genomes of rice, sorghum, *Brachypodium* as monocots and *Arabidopsis*, *Medicago*, *Populus* as dicots, allowed an accurate study of the distribution, types and frequencies of the microsatellite sequences. Dicots seem to have more simple sequence repeat (SSRs) than monocots, while the distribution is not uniform. For both groups, the patterns are comparable: mononuceotide repeats are more dominant, while the frequency of SSRs decreases with increasing lengths. An A/T repeat is more common in plants and G/C is least commonly found. The most prevalent dinucleotide motif differs for each of the six species studied. Trinucleotide AGC/CGT, AGG/CCT and CCG/CGG are common in monocots, while dicots have more of AAC/GTT, AAG/CTT and AAT/ATT. Tetranucleotides are rarely found in plant genomes. This is an example of how whole plant genomes can be used for finding new markers for assessing biodiversity.

#### 5.3.2.3 Amplified fragment length polymorphism

This method (Vos *et al.*, 1995) relies on the amplification of a selective set of genomic fragments obtained after treatment with rare-cutting restriction enzymes, such as *Mse*1 and *Eco*R1. Adaptors are added to the fragments before amplification is performed with primers that anneal to the adaptor sequence. A pre-amplification is done with a primer that has one extra nucleotide added to the adaptor sequence. This is followed by a selective amplification with two or three extra nucleotides to the adaptor sequence. The polymorphism detected, therefore, reflects the nucleotide differences among the samples at the sites of primer annealing, which are the ends of the restriction fragments. The products are separated on acrylamide gels for viewing, or are detected in a genetic analyser if fluorescent labels are used. The fragments obtained differ in size and are scored accordingly. Amplified fragment length polymorphisms (AFLPs) are technically more demanding than the other amplification-based methods, but reveal a much higher level of polymorphism resulting from the loss or gain of nucleotides at the recognition sites of the restriction enzymes in use. Many markers are revealed in a few assays. No prior sequence information is required to use this method.

#### 5.3.2.4 Transposon-derived markers

The concept of mobile genetic elements was first described in maize by Barbara McClintock, at a time when no molecular evidence was yet possible. The variegation in kernel colour led to the hypothesis of DNA segments being capable of moving into genes and therefore changing their pattern of expression. Such elements consist of genes that allow them to break and reinsert into a different part of the genome. Transposable elements (TEs) can be either autonomous (i.e. they can move on their own) or non-autonomous and rely on the autonomous ones for transposition. Because of their abundance and ability to move and duplicate, they constitute a crucial element in the expansion and evolution of plant genomes. Many types of retrotransposons have been identified, and some are known to alter pattern and level of gene expression. Plant genomes have several classes of transposons classified on the basis of their origin.

Class I TEs include the retrotransposons for which transposition occurs when a copy of the RNA transcript is reverse transcribed before being inserted into a new site in the genome, thereby increasing the number of copies of that retrotransposon. The long terminal repeat (LTR)-retrotransposons comprise the Ty1-copia-like and Ty3-gypsy-like classes, while the non-LTR retrotransposons include the autonomous long interspersed elements and the non-autonomous short interspersed elements. LTR-retrotransposons possess terminal repeats flanking open reading frames (ORFs) for *gag*, *pol* and *env* genes (Figure 5.3). The *pol* gene encodes the protease, integrase, reverse transcriptase and RNaseH domains, the order of which is different between Ty1-copia and Ty3-gypsy (Suoniemi *et al.*, 1998).

Class II TEs are the DNA transposons which do not go through an RNA intermediate, but rather cut the TE and take it somewhere else in the genome. The *Ac*/*Ds* elements in maize are Class II TEs. The movement of non-autonomous element *Ds* requires it to be in the vicinity of the activator element *Ac*, which provides a functional transposase. The insertion of *Ds* into genes for transcription factors thus disrupts the expression of the anthocyanin pathway. Maize has a high diversity of TEs, making up 70% of its genome (Baucom *et al.*, 2010). It is worth discussing the proportion and diversity of TEs found in plant genomes, as recent evidence points to the significant contribution of these elements to species divergence.

The recent sequencing of several plant genomes has revealed the extensive occurrence of TEs, occupying between 15 and 75% of genome sequences. They constitute most of the repetitive sequences, given that they can duplicate and move without much restraint.

**Figure 5.3** Structure of LTR retrotransposon. *pol* contains IN (integrase), RT (reverse transciptase) and RH (RNase H)

Details of the barley genome (International Barley Genome Sequencing Consortium, 2012) indicated that it has as much as 84% of mobile and repetitive elements. Most of this is made up of LTR retrotransposons with the Ty-Gypsy (Suoniemi *et al.*, 1998) being more prevalent than Ty-copia. According to Tenaillon *et al.* (2010), there is a correlation between the sizes of plant genomes and the proportion of TEs they contain. This observation could explain the increase in size through expansion due to duplication and insertion of those elements throughout the genome. Although TEs can be seen as having 'invaded' plant genomes, they might be contributing to their stability by not causing any detrimental effect. Several studies have revealed the contribution of TEs to genetic polymorphism amongst populations by their dynamic restructuring of the genome (Baucom *et al.*, 2009). In situations of environmental stress and reduction of population sizes, this diversity could play a role in enhancing adaptation and survival (Hosid *et al.*, 2012). An assessment of LTR retrotransposons diversity among populations of the wild relative of wheat, *Aegilops speltoides*, showed that the inter-retrotransposon amplified polymorphism (IRAP) method is sensitive enough to reveal intra-specific variations. This approach uses primers targeting retrotransposons and facing each other to amplify the inter-TE regions. It has become a widely used method for studying genetic variation in wide-ranging genera. Primers are designed on known sequences of the retrotransposons, which can be found in the many databases devoted to retrotransposons, such as the GyDB, the Gypsy database or the Database of Retrotransposon Insertion Polymorphisms (dbRIP; http://dbrip.brocku.ca/). Several populations of the wheat wild relatives originating from the Iranian, Arabian to the Mediterranean, Euro-Siberian geographical regions were analysed by IRAP for four types of retrotransposons of the Ty1-copia and Ty3-gypsy families (Hosid *et al.*, 2012). The number and variations were shown to be significantly reduced in the marginal, stressed, micropopulations compared with the normal ones.

The heterogeneity of TEs is a very informative set of data for assessing intra-species diversity as there are acquisitions and losses over time. The genomes of maize B73 and *Zea luxurians* (teosinte), both of which have the same number of 20 chromosomes, were compared to investigate the content and types of TEs. This revealed that *Z. luxurians* has 1.35 times more TEs than maize, thus accounting for the overall difference in genome size (Tenaillon *et al.*, 2011). The distribution of the 20 families of TEs, however, was comparable between the two species.

Many other techniques based on amplification or membrane-based hybridization have become available over the last decade or so. They all target sites where changes in nucleotides can be identified by the approach used.

However, most amplification approaches require that PCR products be separated on gel media, thus representing a major constraint to upscaling the throughput. For many purposes, particularly for plant breeding, it is desirable to have a high number of markers that can be obtained in a direct, simple method. Hence, alternative approaches, such as those based on microarray hybridization or DNA sequencing, are more appropriate as they produce far more high-resolution data in one assay. Diversity arrays technologies are a microarray-based hybridization that can screen many thousands of clones at a time (Jaccoud *et al.*, 2001). The technique is based on identifying genomic restriction fragments produced using selected enzymes. First, a representative set of genomic clones from a species is prepared and spotted as probes on a slide. Then equivalent clones from the test samples (individuals from the same or related species) are similarly prepared and used as targets in the hybridization. These are Cy3 and Cy5 labelled and the hybridization signals are image captured as in other microarray methods. The approach is to reduce the complexity of fragments produced by restriction enzyme digestion so as to identify polymorphism. This polymorphism is indicative of nucleotide substitution at the restriction enzyme sites. First, genomic DNA is cut with the rare-cutting enzyme *Pst*1 and another frequent cutter, for example *Taq*I, *Bst*NI, *Msp*I, *Hpa*II, *Ban*II, *Mse*I or *Alu*I, which can be selected for a given species. Following ligation of *Pst*1 adaptors, amplification with *Pst*1 primers then produces a set of clones that forms a library which should be representative of all possible fragments of the species under study. Individual test samples are similarly prepared and used in the hybridization.

The results are obtained as presence or absence of a hybridization signal, or more precisely the restriction enzyme site is intact or modified. This method is genome wide and provides a large number of robust markers that are inherited in a Mendelian way (Sansaloni *et al.*, 2010).

## 5.4 DNA sequencing and barcoding

There is now massive evidence for the contribution of DNA sequencing in molecular phylogeny. The availability of sequence data has accelerated efforts for more precise definition of taxa. For a long time now, plant taxonomy and systematics have relied on exact genes or intergenic sequences of defined loci in the genomes. Chloroplastic DNA was first used to obtain sequence information for phylogeny; however, this is prone to hybridization and introgression, events that can interfere with the interpretation of the characters. In the early 1990s, nuclear ribosomal sequences provided enough evidence that these could be ideal targets for sequencing. Ribosomal genes in eukaryotes exist in tandem structures of the 18S-5.8S-26S genes.

There are hundreds or thousands of such copies throughout a plant's genome. Each unit of 18S-5.8S-26S (Figure 5.4) is transcribed with an intergenic transcribed spacer (ITS) that is then removed by further processing. Ribosomal genes are consitutively expressed as there is a perpetual need for the ribosomal RNAs in protein synthesis. The presence of multiple copies ensures a high level of expression. The genes are well conserved across genera, while the ITS region is more variable. Since the initial publication of Baldwin (1992) describing the use of ITS sequences for phylogenetic analysis, there has been a phenomenal number of papers published on the use of ITS for the same purpose. This can be explained by the fact that few loci with adequate sequence variability were then known to provide as many informative characters. The widespread use of

**Figure 5.4** Eukaryotic ribosomal genes organization. ETS: external transcribed spacer; ITS: internal transcribed spacer. The transcription unit includes the 18S, ITS1, 5.8S, ITS2 and 28S. The ITS regions are then removed during processing

ITS sequencing was fuelled by several properties of the technique. First, it was easy to use because universal primers had become available for several genera. Both parental alleles can be scored; the rDNA units, being present in multiple copies, can be amplified by PCR through conventional techniques of DNA extraction and amplification. This is less problematic than with single or low copy number genes.

However, as rightly discussed by Alvarez and Wendel (2003), use of ITS sequences for phylogenetic reconstruction poses several problems that have not been thoroughly debated. These considerations have been substantiated by whole-plant genomic sequences. Sequencing of PCR products, as such, does not ensure that all genomic loci of a particular gene are being assessed. Many genes are present in several copies in the genome and they originate from two main evolutionary processes. They could have come from ancestral copies (orthologues), present before speciation, or they could have appeared after speciation through gene duplication (paralogues). It is not possible to differentiate the two after a PCR. Orthologues are what should be used for phylogeny, and it often happens that paralogue sequences are used instead. This would seriously confound the resulting tree and lead to wrong classification.

Repeat units within genomes tend to homogenize through concerted evolution, which reduces variation among those units. rDNAs follow the same pattern and most of them have comparable sequences. However, it cannot be assumed that all the units in a given genome are always identical. Recombination and other genomic events can lead to alternative forms of the ITS sequences. Similarly, copies of genes can degenerate into pseudogenes, which are no longer expressed. They can be identified easily for protein coding genes by the loss of some amino acid codons or presence of stop codons, while for ribosomal products the change would lead to a possible structural alteration. Given the large number of rDNA units, it is likely that some of them are not expressed because of nucleotide substitutions.

Although ITS sequences have been extensively investigated for plant phylogeny, their use should be complemented with analysis of other nuclear genes. By now, there are millions of plant genes sequences as well as many whole genome sequences that can be tapped for designing new methods for phylogenetic reconstructions.

### 5.4.1 Plant DNA barcoding

The first initiative for using DNA barcoding came from the University of Guelph, in Canada, in 2003. The idea was simple, in the sense that it proposed to use DNA sequences for identification of species, which is what to some extent many studies were already trying to do with various types of sequences. But species identification has to be precise and reliable and the DNA to be used for this purpose must be identified first.

DNA barcoding allows accurate identification of specimens of all sorts, even those in a degraded state. It can complement the task of a conventional taxonomist. The major challenge that was faced by the barcoding scientists was to agree on which gene(s) to use for which class of organisms. This was successful for animals when 648 bp of the 5′ end of the cytochrome c oxidase I (*COI*) gene was shown to discriminate species in the majority of cases. *COI* is a mitochondrial gene coding for one of the 14 subunits that make up the last electron acceptor of the respiratory chain protein. It is well conserved within species and diverse enough for their differentiation.

A standard sequence library for all organisms under study was first constructed to host a maximum number of sequences (more than 1 million) from diverse species. This DNA-barcode-specific library is then used to match a query sequence for identification at the species level.

The evolution of *COI* in plants is too slow for it to be used for barcoding. Hence, many other genes have been sequenced across taxa to find the most suited for this purpose. So far, most of the ones that have been used are from plastid DNA.

The group II (G2) introns have been thoroughly investigated for plant barcoding. These are found in plant, fungal plastid DNA and some prokaryotic genomes. The mitochondrial and chloroplast G2 introns are of different origins (Kelchner, 2002). They are implicated in the mechanism for their own cis-splicing from the primary transcript of the tRNA gene. This is brought about by the folding of the RNA structure in such a way as to allow the proximity of an adenine residue with the G1 nucleotide at the intron–exon boundary, which causes a cleavage in the RNA at the 5′ intron–exon boundary through a nucleophilic attack. A trans-esterification reaction produces an intermediate molecule known as a lariat structure, which then undergoes another trans-esterification reaction at the 3′ intron–exon boundary. This then releases the intron and the rejoined RNA. Self-splicing is similar to spliceosome-mediated splicing, where several protein factors and enzymes are involved. G2 introns are organized into six helical domains (I–VI) and have a characteristic RNA folding pattern with a conserved secondary structure. Domain IV has an internal ORF for the enzyme maturase, which is required for the ribonucleoprotein (RNP) complex and is similar to reverse transcriptases. Toor *et al.* (2001) have classified the ORFs of G2 introns into six categories and found that most chloroplastic and mitochondrial G2 introns have lost their ORFs. Only two remain in higher eukaryotic organelles: *matK* in the *trnK* intron of chloroplast and *matR* in mitochondria. It is possible that the *matK* maturase can form RNP complexes with the other chloroplast G2 introns, thus compensating for the loss of the other G2 maturases (Kelchner, 2002). Prokaryotic and yeast G2 introns have functional reverse transcriptase and a zinc-finger motif. They have been shown to be mobile and undergo transposition. This is not the case for chloroplastic G2 introns. Other proteins are required for binding to G2 introns for splicing. CSR1 (chloroplast RNA splicing 1) has been shown to be required for the splicing of *atpF* intron in maize (Ostersetzer *et al.*, 2005). Interestingly, it is nuclear encoded and related to two other factors with similar function: CRS2 associated factor 1 (CAF1) and CAF2. They share the CRM (chloroplast splicing and ribosome maturation) common domain of 10 kDa.

Other genes have been identified for plant barcoding. Rubisco (ribulose bis-phosphate carboxylase) is the most abundant protein on Earth and is the most crucial enzyme for carbon fixation in photosynthesis. It is made up of eight chloroplast-encoded large subunits and eight small, nuclear-encoded subunits. This enzyme is very inefficient because of its ability to bind not only carbon dioxide, but also oxygen, thus resulting in a wasteful process of photorespiration. Prokaryotic rubisco is made up of a large subunit dimer. The catalytic activity is in the large subunit. Comparison of rubisco sequences of different origins has shown that rubisco is under positive selection in land plants and not in algae or cyanobacteria (Kapralov and Filatov, 2007). This was calculated using $d_N/d_S$ ratio with the maximum likelihood ratio test. Rates of non-synonymous to synonymous substitutions variation was indicative of positive selection for the rubisco gene; that is, that there is a tendency for this gene to change so as to produce novel forms that might be better adapted for its function.

A combination of *matK–rbcL* (de Vere, 2012) has been recommended for plant DNA barcoding, but this has not been discriminatory in some plant groups. For identification of species, the targeted region should have universal sequences for primer design as well as adequate sequence variation for interspecific differentiation. The *trnH–psbA* has recently been proposed as a supplementary locus that can provide additional data for closely related species. However, for many plant groups, amplification with the universal primers produces different sizes of products or multiple products. It was subsequently revealed that some *trnH–psbA* sequences contain inversions, which can complicate the alignment and lead to erroneous conclusions for species identification. The inversions are generally flanked by inverted repeats, which can form a hairpin structure commonly found in RNAs. A study of the *trnH–psbA* sequences of species of Gentianaceae has assessed the impact of inversions on phylogenetic inferences for conspecific sequences and those of closely related species. Figure 5.5 shows that the inverted sequence of 25 bp is flanked by palindromic regions of 18 bp. The authors used two matrices for alignment and tree building: one with the raw data and a second one in which the inversion is replaced by the complementary sequence, which is equivalent to inverting the loop. This was done to test how the presence of inversion within species can affect the phylogenetic inference. It was concluded that intraspecific inversions would make members of the same species seem more divergent and distant from each other, while closely related species can be mistakenly grouped together. This can pose serious problems if used generally for barcoding purposes.

A comprehensive evaluation of *trnH–psbA* has been done using available sequences from Genbank (Pang *et al.*, 2012). The sequences were thoroughly processed to exclude those of low quality, while inversions were replaced with the reverse complement to produce a list of 17 983 *trnH–psbA* sequences from 3595 species (Figure 5.6). This allowed an estimation of how effective *trnH–psbA* is for the precise species identification. Matching data sets were also made for *matK*, *rbcL* and ITS2 sequences from the same samples. The sequences were queried using the BLAST and BLAST+P distance method. In the latter, 100 hits of the BLAST were used for sequence alignment and distance calculation. The identity was made on the basis of the shortest distance of the neighbours with the query. The results showed that the BLAST+P distance performed better in the identification. At family level, *trnH–psbA* had a success rate of 6.3% for Begoniaceae and 86.2% for Celestraceae (Figure 5.7). The performance with the genera was low: 18% for three genera, including *Begonia*, and quite high (>88%) for *Ficus*, *Pedicularis*, *Primula* and *Solanum* species. The performance of *trnH–psbA* was compared with that of the other three loci (*matK*, *rbcL* and ITS2) and several combinations of all four. *trnH–psbA* gave the best identification results among the four individual ones for 36 genera.

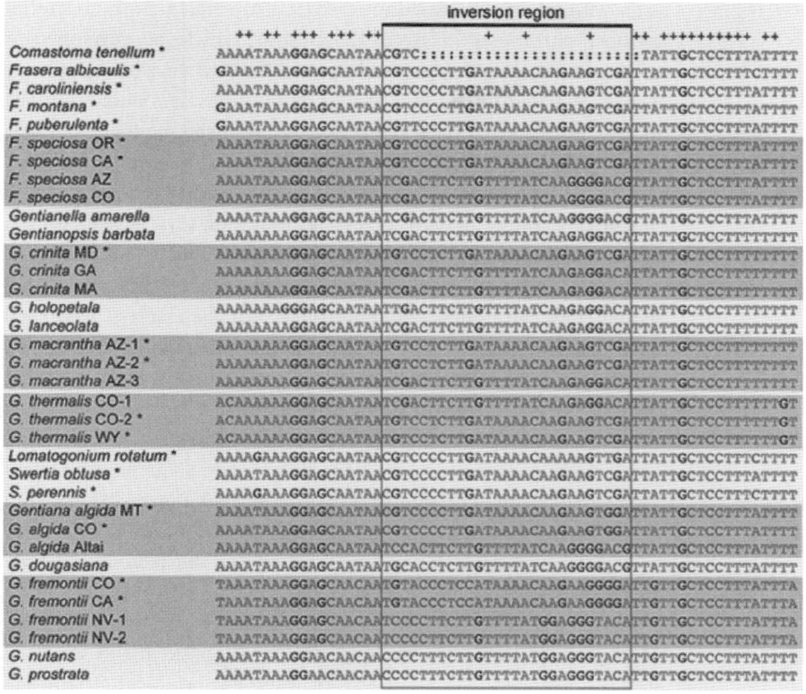


**Figure 5.5** Alignment of *trnH–psbA* sequences from Gentianaceae. Conspecific sequences are shaded and the box represents the 25 bp of the inversion region. *Source*: Whitlock et al. (2010). *See plate section for colour version*

Figure 5.7 illustrates the performance of the individual loci and their combinations. The *trnH–psbA*–ITS2 combination gave a higher discriminatory power than the *matK–rbcL* at both families and genera levels, including *Primula*, *Parnassia*, *Pedicularis* and *Rhododendron*.

Plant DNA barcoding has obvious applications in many areas, in particular for germplasm assessment and conservation strategies. In agriculture, development of new crop varieties depends on the introgression of traits of interest into commercial lines. These traits are often found in the crop wild relatives (CWRs) of the crop species. It is therefore of primary concern that such CWRs are identified and measures are taken for their conservation. Barcoding is an ideal method for this purpose. The rapidly increasing use of DNA sequences for identification purposes justifies the requirement for bioinformatics applications to analyse sequence data and enable reliable conclusions to be made. With the International Barcoding of Life project, the sequence libraries are continuously growing, with additional sequences being obtained from the many plant genera and families under study worldwide. It is therefore essential for users to have a platform for sequence analysis for proper identification and diversity assessment. In this context, the newly developed web portal for barcode analysis, iBarcode.org, provides the set-up for analysis and visualization of barcode data. It allows the user not only to identify the species, but also to compare the sequence with existing ones and highlight the differences in nucleotides. Such differences can identify new haplotypes. It calculates the $d_N/d_S$ ratios among all pairs of sequences within a data set, and this is represented in bar charts for within species or genera. Using a MEGA-formatted genetic distance matrix, a plot of maximum within-species divergence against minimum between-species divergence is obtained. This resource is a practical option for barcode analysis and visual displays.

## 5.5 Plant genomics

Plant diversity represents a reservoir of genes that can be used in for crop improvement and food security. Through precise species identification it is possible to evaluate biodiversity.

Genome sequence information is the ultimate high-resolution approach for diversity assessment. Plant genome sizes vary between 119 Mb for *Arabidopsis thaliana*, the plant model organism with five chromosomes, to 120 000 Mb in angiosperms. *Arabidopsis* was the first plant genome to be sequenced as it is among the smallest and a genetic map was available. Its gene content and metabolic pathways would reflect most existing plant genes (ArabidopsisGenomeInitiative, 2000). Rice (*Oryza*) was then chosen given its importance as a major crop species. It has $n = 12$ chromosomes and 400 Mb of DNA. It is a member of the grass family and comparison of its genome with other cereals has provided valuable information on the evolution of this group. Cereal genomics has shown that modern-day crops like rice, wheat, maize and sorghum have evolved from a common ancestor of five chromosomes ($n = 5$) which went through a first whole-genome duplication to produce $n = 10$ intermediate. Further chromosomal translocations and fusions produced two additional chromosomes as seen for rice. Reduction in the number of chromosomes and further whole-genome duplications then produced diploid maize ($n = 10$; 2500 Mb), diploid sorghum ($n = 10$; 700 Mb) and hexaploid wheat ($n = 7$; 17 000 Mb) (Wei *et al.*, 2007; Salse *et al.*, 2008). The common evolutionary processes undergone by cereal genomes are seen in the gene synteny and large common regions of DNA. The comparative genomics approach has confirmed the need for whole-genome sequences for precise evaluation of relatedness amongst different species. Although initially very costly, genomes sequencing is rapidly becoming affordable to many

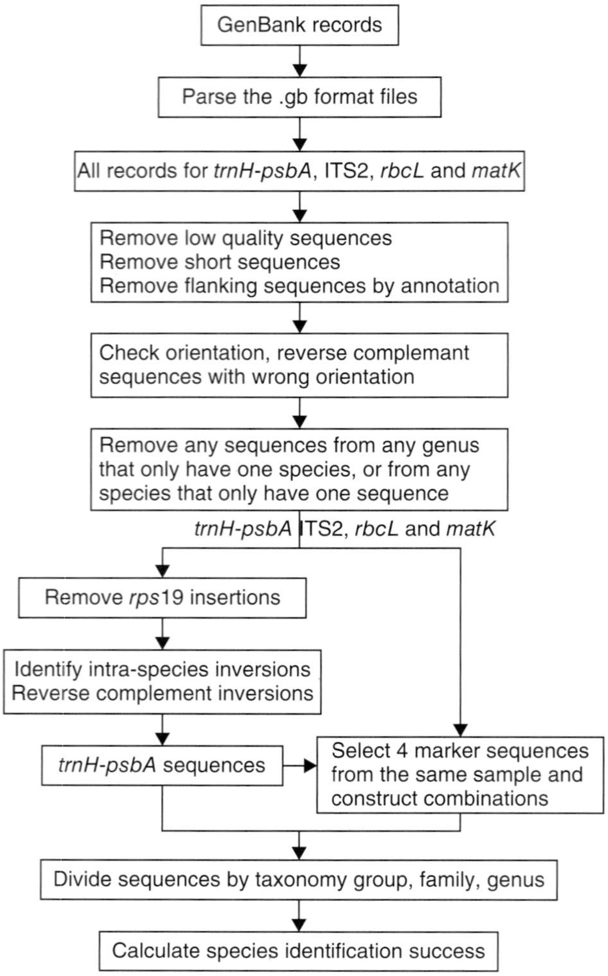


**Figure 5.6** Work flow used by Pang *et al.* (2012). *Source*: Pang *et al.* (2012)

institutions owing to new high-throughput technologies that have been developed.

### 5.5.1 DNA sequencing technologies: next-generation sequencing

Sanger dideoxynucleotide DNA sequencing was initially developed in the 1970s. The reaction is based on the polymerization of a new strand of DNA based on the template whose sequence is to be determined. The methodology worked well but was tedious and not very practical. Although it was used for many years as manual sequencing, it was later automated to allow reactions performed manually to be loaded into the sequencer for analysis. This was a major improvement as it bypassed the use of large polyacrylamide gels, which could take days to prepare and run. Samples used in the automated sequencer are labelled with fluorescent tags for detection. The use of four different labels means that all four reactions can be loaded together. The method has contributed considerably to DNA sequencing for decades; even the first human genome and the first plant genome were sequenced by this classical approach. Whole-genome sequencing is more complicated as the fragmented genomic DNA must first be cloned into vectors, which are transformed into bacterial cells. The latter are grown as individual clones, each with an insert. New innovative vectors, such as the bacterial artificial chromosome (BAC) or cosmids, had to be developed to accommodate longer inserts of up to 300 kb when whole genomes were sequenced. It still means that many thousands of clones have to be handled. Many bacterial genomes have been sequenced by this method. In one run, 700 bases of sequence is obtained from each well of a 96-well plate. Sequencing centres had several automated sequencers, and even with such facilities it took several years for the human genome to be completed.

Since 2003, there have been reports of new types of parallel sequencing techniques that could replace the Sanger method. They have the advantage of eliminating the step of cloning in vectors, which was a major advancement in the technology. They work on the basis of immobilizing single DNA strands onto a solid support, from where they are then amplified to produce a cluster of identical DNA strands. The latter are then subjected to sequencing. Detection depends on the chemistry of the sequencing reaction being used. Three important platforms, from Roche, Illumina and ABI, became the first commercial applications of the techniques. One approach was described using picolitre wells within a fibre-optic block (Margulies *et al.*, 2005): DNA is fragmented and ligated to adapters, which enables its capture on beads of 28 μm in size. The step is optimized to have one DNA fragment per bead. The beads are placed in an emulsion where the amplification takes place, thereby resulting in a cluster of identical DNA fragments. After the PCR, the emulsion is broken and each bead is placed in a picowell on a slide of 1.6 million wells, within which the sequencing reaction is performed. Additional smaller beads with the enzymes for reactions are added into the wells. The slides are in a flow chamber to which reagents are periodically added for the reaction to take place.

The sequencing reaction is based on pyrosequencing, where the signal for the incorporation of each nucleotide is recorded as a quantum of light proportional to the number of nucleotides added to the growing chain. Pyrosequencing is based on the activity of adenosine-triphosphate (ATP) sulfurylase to convert the pyrophosphate (PPi), generated with nucleotide addition during DNA synthesis, to ATP in the presence of adenosine 5′-phosphosulfate (APS):

$$\text{PPi} + \text{adenosine 5}'\text{-phosphosulfate} \xrightarrow{\text{ATP sulfurylase}} \text{ATP}$$

The ATP is used for the conversion of luciferin to oxyluciferin by the enzyme luciferase:

$$\text{luciferin} \xrightarrow{\text{ATP/luciferase}} \text{oxyluciferin}$$

The light thus produced is detected by a charge-coupled device (CCD) camera and recorded as peaks whose height is proportional to the number of nucleotides incorporated. A single nucleotide is added each time and has to be removed after the image capture. The enzyme apyrase is used at the end of each cycle to destroy the unused nucleotide and the ATP, so that the next nucleotide can be

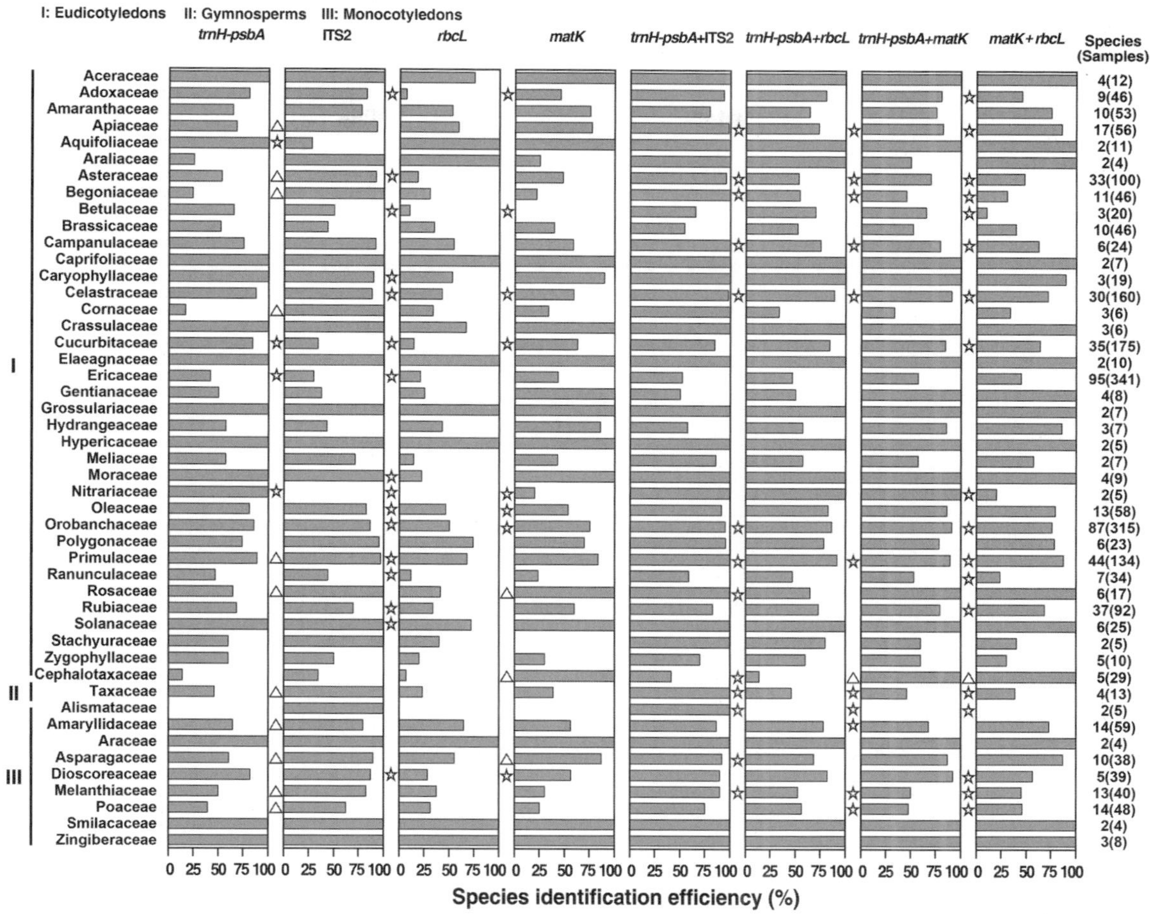


**Figure 5.7** Comparison of identification success rates with different markers for listed genera as indicated. *Source*: Pang *et al.* (2012)

added to the wells:

$$\begin{array}{c}\text{apyrase}\\ \downarrow\\ \text{dNTP} \rightarrow \text{dNDP} + \text{dNMP} + \text{phosphate}\end{array}$$

The above technology is used by the Roche 454 platform and has been modified since its first application.

The Illumina platform uses a different technique for amplification of DNA strands into clusters. The DNA is fragmented and repaired to form blunt ends and poly A nucleotides are added at the 3′ end. Adapters are ligated to both ends and the single-stranded fragments are immobilized onto a solid support to which the adapter strands have been fixed. The ends of the fragments, which are complementary to the adapter sequences on the support, will anneal to those and form a bridge type of structure. PCR reagents are added for the amplification with the fixed adapter sequence as primer for DNA polymerase. Several rounds of PCR will produce enough material for the sequencing reaction. Terminator, fluorescent nucleotides are used so that, after incorporation of each one, the image is captured by the CCD camera. The terminator group and the label are first removed before the next cycle of nucleotide addition. The resulting data are a series of images that reveal which nucleotide has been incorporated at each step and in each well. The nucleotide sequences are then assigned through an algorithm adapted to the reads. A quality check is also performed. Refer to www.illumina.com/documents/products/techspotlights/techspotlight_sequencing.pdf.

The third platform for high-throughput sequencing comes from Applied Biosystems and is the sequencing by oligonucleotide ligation and detection (SOLID). This approach is based on the sequential ligation of oligonucleotide to the template DNA and identifying which oligonucleotides allow the deduction of the actual template sequence.

The above technologies have considerably improved the output of research for measuring variation among organisms, in particular the use of DNA sequences as markers. Comparing DNA sequences for assessing variation can be cumbersome and expensive when a large number of samples are to be studied. Some methods for discriminating between DNA fragments use conformation polymorphism, which results from slight differences in nucleotides, on gel electrophoresis. Although quite reliable, this approach is limited owing to the small number samples that can be tested per assay. With the advent of next-generation sequencing systems, some modifications have been effected to the reactions so as to considerably increase the throughput for the identification of variants. This has been done by using nucletotide barcodes (Galan *et al.*, 2010), whereby the DNA strands are tagged. Short oligonucleotides are used to tag the ends of the DNA strands or added to the 5′ ends of the primers for sequencing. Thousands of samples can be tagged with various combinations of oligonucleotides. Figure 5.8 shows the

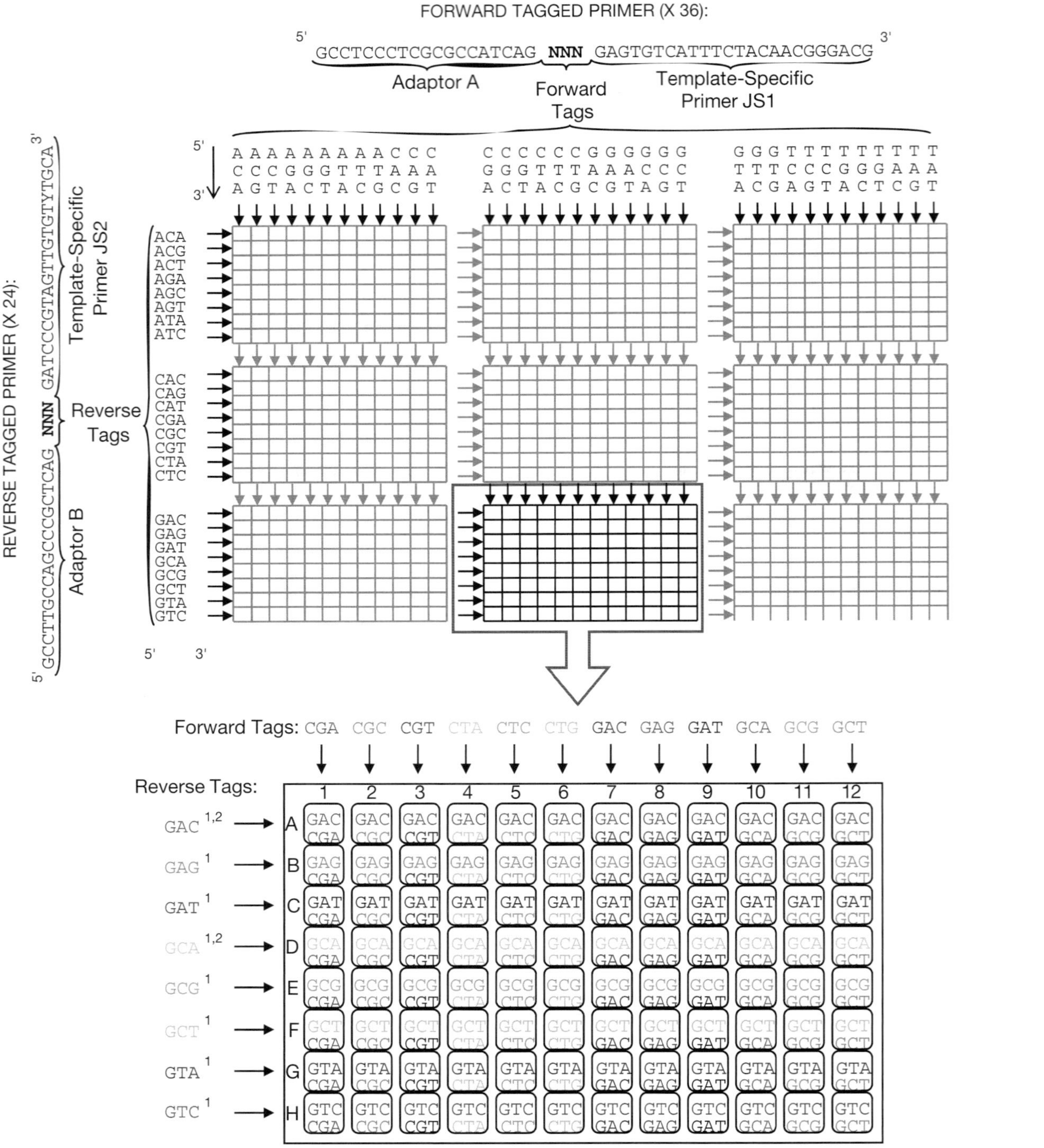


**Figure 5.8** Use of different six-nucleotide barcodes for sample identification in a 96-well plate with 36 forward and 24 reverse tags to have 864 combinations. *Source*: Galan *et al.* (2010). *See plate section for colour version*

distribution of different oliognucleotides across a plate horizontally, such that each row has a different one. The same is done vertically, resulting in a multiplexing effect. With 36 forward and 24 reverse tags in a 96-well plate, 864 combinations representing unique tags are possible. Multiplexing of tags within the same assay allows the genotyping of a very large number of individuals and is particularly effective when polymorphic genes are targeted.

It is now clear that the next-generation sequencing technologies are rapidly replacing capillary-based Sanger sequencing. This is very much due to the competitive costs and the high throughput of the methods. They have revolutionized the molecular detection of variants within and across species. Above all, it has contributed to an upsurge of genome sequencing projects, such that the 1000 human genomes project was set up to plough deep into the genotype variations. The human genome is diverse, consisting of thousands of combinations of rare and common variants: 38 million single-nucleotide polymorphisms, 1.4 million short insertions and deletions, and more than 14 000 larger deletions have been identified. Similarly, the 1000 plant genomes project initiated by the Beijing Genomics Institute was launched to sequence 100 plant transcriptomes in 2008.

## 5.6 Analysis of marker data

Molecular markers are of different types, depending on the technique being used, but are all based on variations in nucleotides at precise sites in genes, repeat regions or anywhere else in the genome. The aim of many studies using markers is for inferring relatedness amongst the individuals tested. As stated earlier, morphological traits are not only limited in the number of markers they can provide, but they are also susceptible to external factors which can influence their expression. The development of a large number of techniques for computing diversity among living organisms has resulted in more refined and precise definitions of phylogenetic relationships for many genera.

Gel-based markers, like RAPDs, AFLPs and many others, produce bands that are scored as present or absent for each primer. Each marker generates binary data, which is integrated into the calculation for phylogenetic inference. With direct sequence data, each site in the DNA molecule is a character that can have four states: A, T, G, C. The sequences are aligned using the many DNA alignment tools, and the comparison is done on the basis of similarity of nucleotides at each position. Nucleotides can be identical at a given position, or if they are not then this is the variation to be measured. The same type of analysis is done with protein sequences, except that there are 20 character states for all the amino acids, instead of the four nucleotides.

The variation represents nucleotide substitutions (if insertions and deletions are ignored) happening throughout the evolutionary history of the species. It is essential, therefore, to consider the types and possible models of substitutions. The four nucleotides of DNA are of two types: purines (C and T) with single six-membered rings and pyrimidines (A and G) with two fused rings. Each double strand of DNA consists of ladders of purine : pyrimidine pairs. The chemical structure of the nucleotides has an impact on the physical state of the molecule. A substitution in one strand is accompanied by the corresponding change in the other strand due to complementarity of the base pairing.

What are the forces that drive changes in nucleotides? Errors during DNA replication are normally repaired by specific enzymes, but if not they contribute significantly to the variation. Other external factors do also cause point mutations in DNA. Those changes accumulate at different rates for different groups of individuals. Similarly, rate of change varies with the region of the genome: the protein coding genes, non-coding, repeat elements and others.

Many statistical methods and algorithms have been developed to analyse data for evolution. These are based on different models of substitution. The Jukes–Cantor model assumes that all four nucleotides can replace each other with equal probability; that is, that an A–T change is as equally probable as an A–C or an A–G change is:

$$d = -\frac{3}{4} \ln \left[1 - \left(\frac{4}{3}\right) p\right]$$

where $d$ is the expected number of nucleotide substitutions per site and $p$ is the proportion of different nucleotides between two sequences. Although easy to implement, this assumption is not realistic. There is evidence for preferential substitutions.

The Kimura two-parameter model incorporates the higher rate of transitional (purine to purine or pyrimidine to pyrimidine) substitutions compared with transversional (purine to pyrimidine) substitution:

$$d = -\frac{1}{2} \ln(1 - 2P - Q) - \frac{1}{4} \ln(1-2Q)$$

where $P$ is the transition rate and $Q$ the transversion rate. Transversions are expected to occur four times more often than transitions.

The Feselstein model assumes the Kimura two-parameter model plus the unequal base frequencies.

The mathematical models of nucleotide substitutions calculate the evolutionary distances between sequences, and therefore between the taxa under study. Clustering methods are then applied for phylogeny inference so as to choose an appropriate tree. From a distance matrix, many tree topologies can be deduced; these are assessed so as to select one that represents at its best the original data set. The unweighted paired-group method with arithmetic mean algorithm clusters sequences through a series of steps which first chooses two operational taxonomic units (OTUs) with minimum distance. They are then considered as one OTU and the process is repeated. The distance between two clusters is the average distance between two sequences within one cluster.

The neighbour-joining approach finds the tree with the lowest total branch length. The maximum parsimony method identifies the most informative sites, which represents the minimum number of evolutionary changes, while maximum likelihood evaluates tree topology on the basis of which tree has data that match the highest likelihood for a given hypothesis for evolution.

## References


Alvarez, I. and Wendel, J.F. (2003) Ribosomal ITS sequences and plant phylogenetic inference. *Molecular Phylogenetics and Evolution*, **29**, 417–434.

ArabidopsisGenomeInitiative (2000) Analysis of the genome sequence of the flowering plant *Arabidopsis thaliana*. *Nature*, **408**, 796–815.

Baldwin, B.G. (1992) Phylogenetic utility of the internal transcribed spacers of nuclear ribosomal DNA in plants: an example from the Compositae. *Molecular Phylogenetics and Evolution*, **1**, 3–16.

Baucom, R.S., Estill, J.C., Chaparro, C. *et al.* (2009) Exceptional diversity, non-random distribution, and rapid evolution of retroelements in the B73 maize genome. *PLoS Genetics*, **5** (11), e1000732, doi:10.1371/journal.pgen.1000732.

Brown, T. (2001) Southern blotting. *Current Protocols in Immunology*, **6**, 10.6.1–10.6.13, doi: 10.1002/0471142735.im1006as06.

Caetano-Anollés, G., Bassam, B.J. and Gresshoff, P.M. (1992) DNA fingerprinting: MAAPing out a RAPD redefinition? *Bio/Technology*, **10** (9), 937.

De Giovanni, C., Dell'Orco, P., Bruno, A. *et al.* (2004) Identification of PCR-based markers (RAPD, AFLP) linked to a novel powdery mildew resistance gene (*ol-2*) in tomato. *Plant Science*, **166**, (1), 41–48.

De Vere, N., Rich, T.C.G., Ford, C.R. *et al.* (2012) DNA barcoding the native flowering plants and conifers of Wales. *PLoS ONE*, **7** (6), e37945, doi:10.1371/journal.pone.0037945.

Galan M., Guivier, E. Caraux, G. *et al.* (2010) A 454 multiplex sequencing method for rapid and reliable genotyping of highly polymorphic genes in large-scale studies. *BMC Genomics*, **11**, 296.

Hosid, E., Brodsky, L., Kalendar, R. *et al.* (2012) Diversity of long terminal repeat retrotransposon genome distribution in natural populations of the wild diploid wheat *Aegilops speltoides*. *Genetics*, **190**, 263–274.

International Barley Genome Sequencing Consortium (2012) A physical, genetic and functional sequence assembly of the barley genome. *Nature*, **491**, 711.

Jaccoud, D., Peng, K., Feinstein, D. and Kilian, A (2001) Diversity arrays: a solid state technology for sequence information independent genotyping. *Nucleic Acids Research*, **29** (4), E25.

Kapralov, M.V. and Filatov, D.A. (2007) Widespread positive selection in the photosynthetic rubisco enzyme. *BMC Evolutionary Biology*, **7**, 73, doi:10.1186/1471-2148-7-73.

Kelchner, S.A. (2002) Group II introns as phylogenetic tools: structure, function, and evolutionary constraints. *American Journal of Botany*, **89** (10), 1651–1669.

Margulies, M., Egholm, M., Altman, W.E. *et al.* (2005) Genome sequencing in microfabricated high-density picolitre reactors. *Nature*, **437**, 376–380.

Mayr, E. and Bock. W.J. (2002) Classifications and other ordering systems. *Journal of Zoology and Systematics Evolution Research*, **40**, 169–194.

Ostersetzer, O., Cooke, A.M., Watkins, K.P. and Barkan, A. (2005) CRS1, a chloroplast group II intron splicing factor, promotes intron folding through specific interactions with two intron domains. *Plant Cell*, **17**, 241–255.

Ostrander EA, Jong PM, Rine J, Duyk G (1992) Construction of small-insert genomic DNA libraries highly enriched for microsatellite repeat sequences. *Proceedings of the National Academy of Sciences of the USA*, **89**, 3419–3423.

Paetkau, D. (1999) Microsatellites obtained using strand extension: an enrichment protocol. *Biotechniques*, **26**, 690–697.

Pang, X., Liu, C., Shi, L. *et al.* (2012) Utility of the *trnH–psbA* intergenic spacer region and its combinations as plant DNA barcodes: a meta-analysis. *PLoS ONE*, **7** (11), e48833, doi:10.1371/journal.pone.0048833.

Salse, J., Bolot, S., Throude, M. *et al.* (2008) Identification and characterization of shared duplications between rice and wheat provide new insight into grass genome evolution. *Plant Cell*, **20**, 11–24.

Sansaloni, C.P., Petroli, C.D., Carling, J. *et al.* (2010) A high-density diversity arrays technology (DArT) microarray for genome-wide genotyping in *Eucalyptus*. *Plant Methods*, **6**, 16 (http://www.plantmethods.com/content/6/1/16).

Semagn, K, Bjørnstad, Å. and Ndjiondjop, M.N. (2006) An overview of molecular marker methods for plants. *African Journal of Biotechnology*, **5** (25) 2540–2568.

Sonah, H., Deshmukh, R.K., Sharma, A. *et al.* (2011) Genome-wide distribution and organization of microsatellites in plants: an insight into marker development in *Brachypodium*. *PLoS ONE*, **6** (6), e21298, doi:10.1371/journal.pone.0021298.

Spooner, D.M., Hetterscheid, W.L.A., van den Berg, R.G. and Brandenburg, W.A. (2003) Plant nomenclature and taxonomy: an horticultural and agronomic perspective, in *Horticultural Reviews*, vol. **28** (ed. J. Janick), John Wiley & Sons, Inc., New York, NY.

Suoniemi, A., Tanskanen, J., Schulman, A.H. (1998) *Gypsy*-like retrotransposons are widespread in the plant kingdom. *Plant Journal*, **13**, 699–705.

Tenaillon, M.I., Hollister, J.D. and Gaut, B.S. (2010) Plant transposable elements in three dimensions. *Trends in Plant Science*, **15**, 471–478.

Tenaillon, M.I., Hufford, M.B., Gaut B.S. and Ross-Ibarra J. (2011) Genome size and transposable element content as determined by high-throughput sequencing in maize and *Zea luxurians*. *Genome Biology and Evolution*, **3**, 219–229.

Toor, N., Hausner, G. and Zimmerly, S. (2001) Coevolution of group II intron RNA structures with their intron-encoded reverse transcriptases. *RNA*, **7**, 1142–1152.

Vos, P., Hogers, R., Bleeker, M. *et al.* (1995) AFLP: a new technique for DNA fingerprinting. *Nucleic Acids Research*, **23**, 4407–4414.

Weber, N., Halpin, C., Hannah, L.C. *et al.* (2012) Crop genome plasticity and its relevance to food and feed safety of genetically engineered breeding stacks. *Plant Physiology*, **160** (4), 1842–1853.

Wei, F., Coe, E., Nelson, W. *et al.* (2007) Physical and genetic structure of the maize genome reflects its complex evolutionary history. *PLoS Genetics*, **3**, e123.

Whitlock, B.A., Hale, A.M. and Groff, P.A. (2010) Intraspecific inversions pose a challenge for the *trnH–psbA* plant DNA barcode. *PLoS ONE*, **5** (7), e11533, doi:10.1371/journal.pone.0011533.

Williams, J.G.K., Kubelik, A.R., Livak, K.J. *et al.* (1990) DNA polymorphisms amplified by arbitrary primers are useful as genetic markers. *Nucleic Acids Research*, **18**, 6531–6535.

Zane, L, Bargelloni, L. and Patarnello, T. (2002) Strategies for microsatellite isolation: a review. *Molecular Ecology*, **11**, 1–16

Zeidler, M. (2000) Electrophoretic analysis of plant isozymes. *Acta Universitatis Palackianae Olomucensis Facultas Rerum Naturalium, Biologica*, **38**, 7–16.

Zhu, J., Liu, J.S. and Lawrence, C.E. (1998) Bayesian adaptive sequence alignment algorithms. *Bioinformatics*, **14** (1), 25–39.

# 6 Conservation of Endangered Wild Harvested Medicinal Plants: Use of DNA Barcoding

Sarina Veldman[1], Joseph Otieno[2], Barbara Gravendeel[3,4], Tinde van Andel[3,4] and Hugo de Boer[1,3,4]

[1]Systematic Biology, Department of Organismal Biology, Evolutionary Biology Centre, Uppsala University, Uppsala, Sweden
[2]Department of Medical Botany, Plant Breeding and Agronomy, Institute of Traditional Medicine, Muhimbili University of Health and Allied Sciences, Dar es Salaam, Tanzania
[3]Naturalis Biodiversity Center, Leiden, The Netherlands
[4]Faculty of Science, Leiden University, Leiden, The Netherlands

## 6.1 Wild harvested medicinal plants: background and challenges

Medicinal plants play an important role in society and have an impressive record of utilization. They have been used for thousands of years by many civilizations throughout the world (Halberstein, 2005). Even animals have been reported using herbs for self-medication (Huffman and Wrangham, 1994). Today, most people in developing countries continue using medicinal plants as a primary source for their health care. According to estimates of the World Health Organization (2003), this might even be up to 80% of the rapidly growing populations in some African and Asian countries. Reasons to keep relying on traditional medicine can be found in the fact that medicinal plant use has been part of culture and religion for many generations, but also in the high prices and availability of modern pharmaceuticals (Van Andel and Carvalheiro, 2013). An increased use of traditional medicine has been reported for developed countries as well, where people use them as alternatives to, or in addition to, modern medicine (Lambert *et al.*, 1997; Schippmann *et al.*, 2002). Although use of traditional medicine has become commercialized over the last decades, the majority of medicinal plants are still harvested from the wild. Several popular medicinal plant species are cultivated, but in most countries this is only a small percentage of the total number of species used in herbal remedies (Schippmann *et al.*, 2002; Hamilton, 2004). This increased commercialization puts a serious pressure on local plant populations and makes them susceptible to overharvesting, especially in combination with habitat destruction due to logging, agricultural clearance and other urban-associated development (Cunningham, 1993; Schippmann *et al.*, 2002). It is increasingly reported that species are becoming less abundant locally and more difficult to obtain in the wild (Bussmann *et al.*, 2007; Lee *et al.*, 2008; Williams *et al.*, 2013). Estimates of the number of medicinal plants that are threatened range from 4160 (Schippmann *et al.*, 2002) to 10 000 (Vorhies, 2000). For a number of species, conservation measures have been taken on different levels: through *in situ* and *ex situ* conservation, by means of research and education and also through national and international legislation. International conventions, such as the Convention on Biological Diversity (CBD) and the Convention on International Trade of Endangered Species (CITES), have been established to stimulate conservation by regulating trade in threatened species (Cunningham, 1993; Hamilton, 2004). However, implementation and monitoring of these international legislations can sometimes be rather difficult, as organisms are not always traded in morphological identifiable forms. In some cases it is possible to trace species based on their vernacular names, but one vernacular name may be used for several species or one species may have several vernacular names in the same language (Linares and Bye, 1987; Cunningham, 2001; Van't Klooster *et al.*, 2003; Kool *et al.*, 2012; Ouarghidi *et al.*, 2012). Identification is further complicated as medicinal plants can be traded in dried, shredded or powdered form or mixed with other species in pills, bitter tonics or other processed forms. Moreover, multiple parts of the plants can be marketed: the whole plant, leaves, seeds, fruits, twigs, bark, tubers, and so on, which can make it extremely difficult to determine which species is being traded (Van Andel *et al.*, 2008; Mati and de Boer 2011; Coghlan *et al.*, 2012). In this case it is necessary to rely on information of the informants only or, if present, on the information on the labels, as it is impossible to trace species on a morphological basis (Ouarghidi *et al.*, 2012). Species substitution and adulteration confound this process, and association of vernacular names may change over time due to disappearance of species through overharvesting and subsequent substitution (Ouarghidi *et al.*, 2012). Other causes of substitution can be unintentional due to insufficient knowledge, intentionally to avoid legislation or because of increasing wholesale prices due to declining abundance and availability of specific plants (Coghlan *et al.*, 2012; Ouarghidi *et al.*, 2012). A breakthrough to these identification challenges has been found in the emergent fields of DNA barcoding and molecular identification. Researchers have made major milestones in recent years: Wasser *et al.* (2007), using DNA

*Novel Plant Bioresources: Applications in Food, Medicine and Cosmetics*, First Edition. Edited by Ameenah Gurib-Fakim.

microsatellite markers, were able to trace the origin of the latest seizure of ivory to systematically unreported poaching in Zambia; Baker *et al.* (2010) have shown that illegally caught whale meat from Japan ended up in sushi restaurants in the USA; Eurlings *et al.* (2010) used DNA barcoding to identify threatened agarwood species from impounded samples; Kool *et al.* (2012) used the same technique to identify roots traded in markets in Morocco and uncovered that 25% of species were previously unknown; and Coghlan *et al.* (2012) used high-throughput next-generation sequencing to analyse the ingredients of traditional Chinese medicine (TCM) and discovered that the products contained more than 68 plant families and 47 animal species.

## 6.2 DNA barcoding general

The diversity of life underpins all biological studies, but its discrimination is no easy task. Few taxonomists can critically identify more than 0.01% of the estimated 10–15 million species (Hammond, 1992; Hawksworth and Kalin-Arroyo, 1995). The global scientific community lacks the expertise and continuity to identify all species diversity; but apart from this, morphological species identification has four significant limitations (Hebert *et al.*, 2003): (1) phenotypic plasticity and genetic variability in the characters employed for species recognition can lead to incorrect identifications; (2) morphologically cryptic taxa, which are common in many groups, can be overlooked (Knowlton, 1993; Jarman and Elliott, 2000; Burns *et al.*, 2008; Ragupathy *et al.*, 2009); (3) morphological keys are often effective only for a particular life stage or gender, and many individuals cannot be identified; (4) modern interactive keys represent a major advance, but the use of keys often demands such a high level of expertise that misdiagnoses are common. The limitations of morphology-based identification and the greying of the taxonomy community require a new approach to species detection and identification.

Advances in molecular biology have made it feasible and cost effective to use DNA sequences for species identification. The method of identifying living organisms to species level using DNA sequences has been coined DNA barcoding. It makes use of short (<1000 bp), agreed-upon regions of the genome (a 'barcode') that evolve fast enough to differ between closely related species (Kress *et al.*, 2005). When a barcode sequence has been retrieved from an unknown sample, an algorithm is used to compare it with a reference database containing barcodes from identified reference specimens, such as herbarium collections, thus enabling it to be identified. In other words, DNA barcodes function as molecular identifiers for individual species, in the same way as the machine-readable black-and-white barcodes are used in the retail industry to identify products.

The use of DNA sequences for species identification focused first on viruses, bacteria and protists. However, barcoding in its modern form was popularized by Hebert *et al.* (2003), who proposed to use the mitochondrial gene *CO1* as the standard barcode for all animals. This was readily adopted by the scientific community, and assessments have since shown that *CO1* can be used to distinguish over 90% of species in many animal groups; for example, spiders (Barrett and Hebert, 2005), birds (Kerr *et al.*, 2007), amphibians (Smith *et al.*, 2008), and butterflies (Burns *et al.*, 2008). DNA barcoding yields more and more data, especially with the development of high-throughput next-generation sequencing. At the same time, barcoding cannot replace morphology for identification and classification, and the surge of barcoding data increases the demand for taxonomic specialists, and the need for professional training of new expertise (Moritz and Cicero 2004; Will and Rubinoff, 2004; Rubinoff *et al.*, 2006). In recent years, the barcoding movement has grown substantially, and worldwide efforts coordinated by the Consortium for the Barcode of Life (CBOL) are now being put into retrieving barcode sequences from all organisms (Hollingsworth *et al.*, 2009; Hollingsworth, 2011).

## 6.3 DNA barcoding and species delimitation

Species delimitation has proven to be a challenge in the past and continues to be so today. Traditionally, species were identified, described and classified based mainly on their morphological characters, which proved successful in general, but more difficult when it came to cryptic, hybridizing or highly converged species. Nowadays, other characters, such as molecular data and behavior, are also used to distinguish species. The main reason that species delimitation is difficult lies in the fact that it remains unclear how a species should be defined (De Queiroz, 2007). A 2005 special issue of the journal *Science* contained a summary of the major unresolved questions in science, and 'What is a species?' was one of them. Not knowing what species really are is hard to believe when one considers the importance of the species concept in biology (Kennedy and Norman, 2005). Species concepts differ, and a single definition that fits all organisms has not been found. Different fields within biology look at species concepts from different angles, each putting an emphasis on what is considered most important to its related organisms. Ecologists are interested in occupation of distinct niches or adaptive zones, while phylogeneticists are interested in genetic differences, and others might look at unique evolutionary roles, phenotypic cohesion or intrinsic reproductive isolation (De Queiroz, 2007). However, even from a single point of view, delimiting species remains rather challenging.

Most species concepts agree on species being evolving metapopulation lineages, but delimiting species is often more problematic (De Queiroz, 2007). Species delimitation based on molecular data in the light of coalescent theory is being developed but requires many accessions as well as many loci (Knowles and Carstens, 2007). Initial studies on DNA barcoding suggested a significant barcoding 'gap' between intra- and interspecific variation (Hebert *et al.*, 2003; Barrett and Hebert, 2005; Hebert *et al.*, 2004b), but these studies have been criticized for undersampling both intraspecific and interspecific divergence (Meyer and Paulay, 2005). In an ideal situation, studies at population genetic level would have to be done for all species in a DNA barcoding database; but this is far from being achieved at present, and instead a more or less arbitrary cut-off value for sequence divergence is often used (Blaxter *et al.*, 2005; Lefébure *et al.*, 2006; Nielsen and Matz, 2006). The main methodological problem with DNA barcoding remains that it is often impossible to tell the difference between interspecific sequence variation and intraspecific sequence variation (Lahaye *et al.*, 2008; Chase and Fay 2009; Chen *et al.*, 2010). But notably, difficulties in distinguishing between intra- and interspecific variation are a widespread problem in morphological species delimitation as well.

Once a dataset of DNA sequences is generated, one can use several approaches to look at the interrelatedness of the different samples. The most common approaches nowadays are the use of similarity scores using BLAST (Altschul *et al.*, 1990), and phylogenetic approaches using maximum likelihood (Felsenstein, 1973), parsimony (Nixon, 1999) or Bayesian statistics (Huelsenbeck and Ronquist, 2001). These phylogenetic methods are implemented in respectively PAUP* and RAxML (Swofford, 2002; Stamatakis, 2006), PAUP* (Swofford, 2002) and MrBayes (Huelsenbeck and Ronquist, 2001). Some argue, however, that the use of tree-based methods for species delimitation has several drawbacks: (1) it is not possible to distinguish between recently diverged lineages, (2) incorrect results might occur when it comes to determining the closest relatives, (3) the classical approach and the barcoding approach will be difficult to unify, when the first utilizes character-based measures and the latter distance-based measures, and (4) the criteria used to delimit species are arbitrary, not objective (DeSalle *et al.*, 2005; Knowles and Carstens, 2007). Several suggestions have been made to minimize the effect of these drawbacks, such as a diagnostic system including other lines of evidence (DeSalle *et al.*, 2005), a probabilistic modelling approach (Knowles and Carstens, 2007), the use of dominant and codominant multilocus markers (Hausdorf and Hennig, 2010) and new heuristic methods without fixed species assignments (O'Meara, 2010), but still no consensus has been reached on this topic.

## 6.4 DNA barcodes for plants

Barcoding of plants has taken longer to develop, and it has been used less in applied research so far (Eurlings *et al.*, 2010; Chen *et al.*, 2011; Li *et al.*, 2011; Kool *et al.*, 2012). Early on, it became clear that the mitochondrial genome in plants evolves far too slowly to allow it to distinguish between species (Cho *et al.*, 2004). Various genes and non-coding regions in the plastid genome have been put forward as alternatives (Kress *et al.*, 2005; Kress and Erickson, 2007; Fazekas *et al.*, 2008, 2009; Ford *et al.*, 2009; Hollingsworth *et al.*, 2009; Burgess *et al.*, 2011). In addition to being sufficiently fast-evolving, a molecular barcode must also be flanked by conserved regions of the genome that can function as universal primer binding sites for polymerase chain reactions (PCRs). The ideal plant barcode needs to be amplifiable with only a single set of primers, so that it can be efficiently retrieved from any of the over 200 000 species of plants (Kress *et al.*, 2005). A single barcoding locus combining these two traits has not been found, and a combination of two or more, probably plastid, loci will almost certainly be required to approach the level of species discrimination and universality that *CO1* provides for animals (Kress and Erickson, 2007). Species identification of plants using DNA barcoding also faces many challenges due to frequent hybridization and the relatively young age of many lineages (Fazekas *et al.*, 2009). The high requirements of universality combined with the high variability among plants has led some to dub plant DNA barcodes a search for the Holy Grail (Rubinoff *et al.*, 2006).

Significant progress on plant DNA barcoding has been made. Kress *et al.* (2005) proposed to use the nuclear ribosomal internal transcriber spacer (nrITS) in combination with the plastid *trnH–psbA* spacer. They advocated the use of nrITS because it is the most frequently sequenced plant DNA marker, mainly owing to its use in phylogenetic evolutionary reconstructions. The *trnH–psbA* spacer was suggested because an analysis over plant DNA data showed it to be the most variable plastid marker. In addition, it is short, roughly 450 bp, and easy to amplify due to highly conserved flanking regions. In a later paper by Kress and Erickson (2007), the use of nrITS is dropped in favour of the plastid marker *rbcL*. nrITS has as disadvantage that it is present in multiple copies, that these copies can belong to different parent lineages in hybrids and that the PCR amplification success of these copies is unrelated to whether these copies are functionally transcribed or not, which in turn has an influence on the mutation rate of these sequences. The plastid marker *rbcL*, on the other hand, is widely available in NCBI GenBank, and it was one of the first widely applied plant DNA markers and crucial to pioneering research on the plant tree of life, the angiosperm phylogeny (cite APG I, II, III). As a marker, *rbcL* is easy to amplify but not very variable, which makes it ideal as an anchoring marker, allowing quick matching of an unknown plant sequence to a family or genus. Kress and Erickson (2007) had a two-step identification in mind with *rbcL* as the family anchor, and the *trnH–psbA* as the species identifier. In 2009, the CBOL Plant Working Group (Hollingsworth *et al.*, 2009) evaluated seven leading candidates with the aim of identifying a two-locus barcode for all plants. The result of their evaluation showed that three loci were the best, *matK*, *rbcL* and *trnH–psbA*, and they proposed the use of *matK+rbcL*. The marker *trnH–psbA* was rejected on the basis of problems obtaining high-quality bidirectional reads, despite its high amplification rate and 93% species identification rate among angiosperms. *MatK* was proposed as it is a rapidly evolving plastid locus making it suitable for species identification. Unfortunately, no one seems to be able to amplify it successfully without considerable effort (Kress and Erickson, 2007; Fazekas *et al.*, 2008; Kool *et al.*, 2012), rendering it a seemingly unsuitable universal barcode. Other markers that have been proposed include: *atpF–atpH* spacer (Hollingsworth *et al.*, 2009), *rpoB* gene (Hollingsworth *et al.*, 2009), *rpoC1* gene (Hollingsworth *et al.*, 2009; Kool *et al.*, 2012), *psbK–psbI* spacer (Hollingsworth *et al.*, 2009) and *trnL* (Taberlet *et al.*, 2007).

Recent research by Li *et al.* (2011) on the species discrimination success of nrITS, as well as the poor amplification success found for *matK* (Kress and Erickson, 2007; Fazekas *et al.*, 2008; Kool *et al.*, 2012), has led the CBOL Plant Working Group to adopt a more open strategy to plant barcoding and endorse the use of *trnH–psbA* and nrITS in addition to *rbcL* and *matK* (Hollingsworth, 2011). At the same time, the development of next-generation sequencing methods have reduced the need to have a limited set of markers, as online nucleotide repositories are flooded with useful data.

## 6.5 Examples of DNA barcoding of cryptic and prepared plant material

DNA barcoding is not only significant in the discovery of unknown species and for use in resolving taxonomic problems, but can also be of great value in the identification of cryptic and forensic plant material (see Box 6.1). Cryptic species can be difficult or impossible to identify using morphology alone, and can belong to species complexes consisting of multiple species sharing highly similar morphology in one or more life stages. Moreover, their morphological species circumscriptions are often disputed and regularly revised.

DNA barcoding can be used to shed more light on their evolutionary relations and support or invalidate taxonomic classifications made based on their morphology.

### Box 6.1. Identification of Traded Roots Using DNA Barcoding

Medicinal plant trade is important for local livelihoods, but destructive harvesting practices threaten the existence of some plant species. Harvesters of medicinal roots identify collected species according to their own ethnoclassifications, but once the dried or powdered roots enter the chain of commercialization, accurate identification becomes more challenging. de Boer and co-authors conducted a survey of morphological diversity among four root products traded in the medina of Marrakech and selected 51 root samples for molecular identification using three barcode markers, *trnH–psbA*, *rpoC1*, and nrITS. The sequences obtained were searched against a reference database of Moroccan medicinal plants and their closest relatives.

Using this method, the authors could identify the majority of the samples, but they also found that few of the barcoding identifications matched the scientific names that were given in the literature, including the most authoritative and widely cited pharmacopeia. Their finding has major implications for the monitoring of trade in endangered plant species, because it shows that morphology-based species identifications alone may not be accurate. As a result, trade in certain species may be overestimated, whereas the commercialization of other species may not be recorded at all.

*Source: Source*: de Boer *et al.* (2014)

Stech *et al.* (2013), for example, used two plastid markers (*rps4–trnT*, *trnT–trnL*) and two nuclear markers (nrITS1, nrITS2) to delimit (sub)species in the bryophyte species complex *Racomitrium canescens* (Grimmiaceae). Based on the combined sequence data of the plastid and nuclear markers, all eight species within the species complex appeared to be easily distinguishable. Their molecular findings supported the recent revision in morphological species circumscription of the *R. canescens* complex. Other examples of the use of DNA barcoding to elucidate cryptic species can be found for different groups of organisms – for example, skipper butterflies (Burns *et al.*, 2008; Hebert *et al.*, 2004a), birds (Hebert *et al.*, 2004b), spiders (Barrett and Hebert, 2005), fresh water fish (Lara *et al.*, 2010; April *et al.*, 2011), amphibians (Crawford *et al.*, 2012) and amoeba (Kosakyan *et al.*, 2012) – but also for cryptic medicinal species such as *Sphagnum cribrosum* (Johnson *et al.*, 2012) and grasses (Ragupathy *et al.*, 2009).

DNA barcoding is now also becoming an identification tool for traded and processed plant material. There is an increased demand for traceability systems in the trade of natural plant products, such as food, timber and herbal medicines, to ensure legality and safety of commercialized products in both national and international trade. Existing techniques, such as chromatography, microscopy, and immunological and electrophoresis assays, are frequently used to validate authenticity, but sequence-based methods such as DNA barcoding are becoming increasingly popular. One of the great advantages of DNA barcoding over other approaches is that DNA barcoding generates results for only small amounts of material, in addition to providing a reliably accurate identification. A recent study by Coghlan *et al.* (2012) showed the power of DNA barcoding in detecting species from highly processed material. They used high-throughput next-generation sequencing to screen 15 complex TCM samples, and generated over 49 000 sequence reads. These reads corresponded to 68 plant families, including two genera with potentially toxic species. Their screening revealed that some of the TCM samples contained traces of CITES-listed animal and plant genera, such as the Asiatic black bear (*Ursus thibetanus*), the Saiga antelope (*Saiga tatarica*) and a wild ginseng species (*Panax ginseng*). It should be noted, however, that reliable identification up to species level is still rather difficult and that non-organic toxic contaminants, such as heavy metals, are not detected with DNA barcoding.

Another example of the use of sequence-based methods in the validation of traded material is the research by Eurlings *et al.* (2010) on the identification of agarwood through polymorphic microsatellites. Asian agarwood (*Aquilaria* spp. and *Gyrinops* spp. – Thymelaeaceae) is used for traditional medicine and for the production of perfumes and incense, but destructive harvesting and overexploitation threaten the survival of both genera. Since the 13th CITES CoP meeting, all species in both genera are CITES appendix II listed. Eurlings *et al.* (2010) used microsatellite sequences obtained from leaf extracts of 27 *Aquilaria* trees harvested in the wild or produced in plantations in Vietnam, Thailand and China to create a molecular reference database for species identification. The system still needs to be expanded and tested on a larger sample of material confiscated from trade, but the results were promising as it was feasible to discriminate the samples on their geographic and wild versus cultivated origins.

Other research has shown that DNA barcoding can also be effective in tracing meat and seafood, either in fresh or processed form, in analysing the contents of dairy products and in identifying crop cultivars and other edible plants (Galimberti *et al.*, 2013); see Box 6.2. Possibilities for the use DNA barcoding in the identification of cryptic, forensic and processed plant material are myriad, and are likely to be of increasing importance as DNA sequencing becomes more affordable, reference databases increase in size and accuracy of identification, and DNA barcoding technology becomes more portable.

### Box 6.2. DNA Barcoding to Identify Illegal Orchid Trade

Chikanda is a traditional Zambian snack consisting mainly of processed orchid tubers (genera: *Disa*, *Satyrium*, *Habenaria*), and is commonly eaten in Zambia and southern Tanzania. All orchid species are CITES listed, and trade in orchids is regulated. Orchid tubers for chikanda are wild-harvested and traded through local markets. Many of the tubers originate from Tanzania, and substantial amounts of those are illegally collected in protected areas. Monitoring this trade and mapping species diversity is a major challenge, as orchid tubers lack distinctive characteristics for accurate species identification. The emergence of DNA barcoding provides a tool to tackle this challenge. Based on preliminary results and on results from other research, it is expected that DNA barcoding will enable successful identification of orchid tubers sold on the local markets and identification of species present in prepared chikanda. Moreover, DNA barcoding of orchids from protected areas will provide us with reference sequences. An ultimate goal will be to use DNA barcoding to trace the traded species back to local markets and collection sites.

Depending on the outcome, DNA barcoding might prove to be a powerful detection method for monitoring illegal orchid trade, enabling the research community to address conservation issues of importance to both orchid protection and sustainable local livelihoods.

*Source: Source*: Sarina Veldman, Uppsala University, research in progress.

## 6.6 Plant DNA authentication, verification and certification

DNA barcoding is perfectly suitable for purposes of verification of material in horticulture and the pharmaceutical industry. Molecular identification can be used to verify, authenticate or certify anything from potted plants to sterile tubers, and crude material to processed and packaged herbal medicines. In the pharmaceutical industry considerable effort is put into quality control in the production chain, and companies ensure that producers follow the United Nations Food and Agricultural Organization good agricultural practices (GAP). Even more important for the pharmaceutical industry are the World Health Organization good manufacturing practice (GMP) guidelines. Both GAP and GMP require accurate species knowledge. Molecular barcoding can reliably be used for identification of seed stocks, harvested, dried and processed plant material, and processed herbal products. Producers can use DNA certification to show the purity and accuracy of the species content of the produce, and importers, processors and manufacturers can use DNA authentication and verification to test the material delivered to them. Some pharmaceutical companies have developed their own barcoding laboratories, while others rely on commercial ventures such as AuthenTechnologies in San Francisco, USA (www.authentechnologies.com), Researgene in Uppsala, Sweden (www.researgene.com) or RIKILT in Wageningen, the Netherlands (http://wageningenur.nl).

Successful molecular identification depends on the accuracy, exhaustiveness and redundancy of the reference sequences with which the query sequence is compared. Development of a barcoding system for single species within a genus requires multiple accessions of the target species and accessions of all related species that might reasonably be substituted for the target species. The criterion of reasonable substitution can be based on species geographical distributions, related species in cultivation or reported adulterants or substitutes. An example of species geographical distributions is the reference library that Kool *et al.* (2012) used to identify traded roots in Morocco. Their library included all species found in Morocco within the same genus as the target species, but excluded those found elsewhere on the argument that such species would be unlikely to turn up in markets in southern Morocco. Another example that highlights some of the hurdles of species sampling is the research by Takundwa *et al.* (2012) that aimed to find a barcoding marker for the marama bean, *Tylosema esculentum* (Burchell) Schreiber, a species with nutritive seeds and tubers found in southern Africa. Their library included multiple accessions of *T. esculentum*, but only a single accession of another species within the same genus, *T. fassoglensis* (Schweinf.) Torre & Hillc. Their phylogenetic analysis shows *T. fassoglensis* to be nested within *T. esculentum*, but they are unable to conclude that the species should be synonymized, and thus also unable to identify unknown samples using their marker with certainty. In addition, the related *T. angolense* P.Silveira & S.Castro, which is found in the same geographical area, was not included in their study, which further undermines the resolution of their marker. On the other hand, the study by Feng *et al.* (2010) shows that careful species sampling can yield valuable insights into species adulteration using DNA authentication. Feng *et al.* used the nrITS marker to create a barcode system to identify *Angelica sinensis* (Oliv.) Diels, and based their reference library on previously reported adulterants within the same genus. Their selection included the target species, *A. sinensis*, and seven known adulterants. Testing their system using eight commercially available crude samples, they were able to show with reliable confidence that four samples consisted of two adulterants instead of the target species. In recent years, molecular identification sequences have been published for a large number of different species. A useful overview by Sucher and Carles (2008) shows the explosive progress in this field.

## 6.7 Future opportunities and challenges

In the decade or so that DNA barcoding for species identification has been advocated and used (Hebert *et al.*, 2003), great advances have been made in methodology, cost and reference collections. The acceptance of *CO1* as a universal barcode for animals (Hebert *et al.*, 2003), nrITS for fungi (Schoch *et al.*, 2012) and the use of several (nrITS, *matK*, *rbcL*, *trnH–psbA*, *rpoC1*) but limited markers for plants (Hollingsworth, 2011) has yielded an immense volume of reference sequences for all sorts of organisms.

Methodological advances in DNA extraction, PCR and sequencing have made it possible to obtain and sequence DNA from substrates such as highly processed samples (Galimberti *et al.*, 2013), processed and mixed samples (Coghlan *et al.*, 2012) and ancient material (van Geel *et al.*, 2008, 2011). These advances have been the result of a joint effort by the scientific community.

Despite this progress challenges do remain, but development and innovation in molecular biology keep taking leaps forward as new methods and insights emerge. The challenges of today could be the routines of tomorrow. First, and perhaps most importantly, DNA barcode reference databases will continue to expand; more species will be sequenced and samples of the same species originating from different geographical localities will be added, capturing both intraspecific and interspecific genetic variation (Fazekas *et al.*, 2009). This might enable future identification of unknown species throughout their distributional range (Kool *et al.*, 2012), and it will contribute to the public availability of more and more sequences through online databases such as NCBI GenBank, enabling scientists to easily access and use large amounts of sequence data.

The search for the most effective barcode for plants will continue: it should be easily obtained with a single primer pair, preferably suitable for bidirectional sequencing, it should need little to no manual sequence editing and, last but not least, it should be able to discriminate highly between species (Hollingsworth *et al.*, 2009). High-throughput sequencing techniques will keep improving, and enable more rapid generation of large amounts of sequence data at ever lowering costs. Ideally, advances in these techniques will lead to more accurate identifications up to species level from all sorts of material (Hollingsworth *et al.*, 2009). Advances in bioinformatics will make it easier to assign sequences to species with high accuracy, and automatically linking these identifications to secondary information such as distribution, medicinal properties, toxicity, IUCN red-list status, and CITES appendix status. A recently developed bioinformatics pipeline identifies sequences associated with CITES-listed plant species from next-generation sequence data (Lammers *et al.*, in preparation).

Technological advances are likely to integrate DNA sequencing and mobile phone technology into hand-held sequencing and molecular identification devices. At the forefront of development is the MinION USB DNA sequencer, a proof of concept device by UK firm Oxford Nanopore. A device like that could easily be connected to a smartphone by online reference sequence matching. Using a DNA barcode scanner to aid in the conservation of endangered wild harvested medicinal plants seems only a few steps away.

## Acknowledgements

We wish to thank the NWO-SIDA-COSTECH TASENE programme for the funding that enables this ongoing research project.

## References

Altschul, S.F., Gish, W., Miller, W. *et al.* (1990) Basic local alignment search tool. *Journal of Molecular Biology*, **215** (3), 403–410.

April, J., Mayden, R.L., Hanner, R.H. and Bernatchez, L. (2011) Genetic calibration of species diversity among North America's freshwater fishes. *Proceedings of the National Academy of Sciences of the United States of America*, **108** (26), 10602–10607.

Baker, C.S., Steel, D., Choi, Y. *et al.* (2010) Genetic evidence of illegal trade in protected whales links Japan with the US and South Korea. *Biology Letters*, **6** (5), 647–650.

Barrett, R.D.H. and Hebert, P.D.N. (2005) Identifying spiders through DNA barcodes. *Canadian Journal of Zoology*, **491**, 481–491.

Blaxter, M., Mann, J., Chapman T. *et al.* (2005) Defining operational taxonomic units using DNA barcode data. *Philosophical transactions of the Royal Society of London. Series B, Biological Sciences*, **360** (1462), 1935–1943.

Burgess, K.S., Fazekas, A.J., Kesanakurti, P.R. *et al.* (2011) Discriminating plant species in a local temperate flora using the *rbcL+matK* DNA barcode. *Methods in Ecology and Evolution*, **2** (4), 333–340.

Burns, J.M., Janzen, D.H., Hajibabaei, M., *et al.* (2008) DNA barcodes and cryptic species of skipper butterflies in the genus *Perichares* in Area de Conservacion Guanacaste, Costa Rica. *Proceedings of the National Academy of Sciences of the United States of America*, **105** (17), 6350–6355.

Bussmann, R.W., Sharon, D., Vandebroek, I. and Revene, Z. (2007) Health for sale: the medicinal plant markets in Trujillo and Chiclayo, northern Peru. *Journal of Ethnobiology and Ethnomedicine*, **3** (1), 37.

Chase, M.W. and Fay, M.F. (2009) Barcoding of plants and fungi. *Science*, **325** (5941), 682.

Chen, J., Li, Q., Kong, L. and Yu, H. (2011) How DNA barcodes complement taxonomy and explore species diversity: the case study of a poorly understood marine fauna. *PLoS ONE*, **6** (6), e21326.

Chen, S., Yao, H., Han, J. *et al.* (2010) Validation of the ITS2 region as a novel DNA barcode for identifying medicinal plant species. *PLoS ONE*, **5** (1), e8613.

Cho, Y., Mower, J.P., Qiu, Y.L. and Palmer, J.D. (2004) Mitochondrial substitution rates are extraordinarily elevated and variable in a genus of flowering plants. *Proceedings of the National Academy of Sciences of the United States of America*, **101** (51), 17741–17746.

Coghlan, M., Haile, J., Houston, J. *et al.* (2012) Deep sequencing of plant and animal DNA contained within traditional chinese medicines reveals legality issues and health safety concerns. *PLoS Genetics*, **8** (4), e1002657.

Crawford, A.J., Cruz, C., Griffith, E. *et al.* (2012) DNA barcoding applied to *ex situ* tropical amphibian conservation programme reveals cryptic diversity in captive populations. *Molecular Ecology Resources*, **13** (6), 1005–1018.

Cunningham, A.B. (1993) African medicinal plants: setting priorities at the interface between conservation and primary health care, *People and Plants Working Paper 1*, UNESCO, Paris.

Cunningham, A.B. (2001) *Applied Ethnobotany: People, Wild Plant Use and Conservation*, Earthscan/James & James.

De Boer, H.J., Ouarghidi, A., Abbad, A. *et al.* (2014) DNA barcoding reveals limited accuracy of identifications based on folk taxonomy. *PLOS ONE*. DOI: 10.1371/journal.pone.0084291.

De Queiroz, K. (2007) Species concepts and species delimitation. *Systematic Biology*, **56** (6), 879–886.

DeSalle, R., Egan, M.G. and Siddall, M. (2005) The unholy trinity: taxonomy, species delimitation and DNA barcoding. *Philosophical Transactions of the Royal Society B: Biological Sciences*, **360** (1462), 1905–1916.

Eurlings, M.C.M., Heuveling van Beek, H. and Gravendeel, B. (2010) Polymorphic microsatellites for forensic identification of agarwood (*Aquilaria crassna*). *Forensic Science International*, **197** (1–3), 30–34.

Fazekas, A.J., Burgess, K.S., Kesanakurti, P.R. *et al.* (2008) Multiple multilocus DNA barcodes from the plastid genome discriminate plant species equally well. *PLoS ONE*, **3** (7), e2802.

Fazekas, A.J., Kesanakurti, P.R., Burgess, K.S. *et al.* (2009) Are plant species inherently harder to discriminate than animal species using DNA barcoding markers? *Molecular Ecology Resources*, **9**, 130–139.

Felsenstein, J. (1973) Maximum likelihood and minimum-steps methods for estimating evolutionary trees from data on discrete characters. *Systematic Zoology*, **22** (3), 240–249.

Feng, T., Liu, S. and He, X. (2010) Molecular authentication of the traditional Chinese medicinal plant *Angelica sinensis* based on internal transcribed spacer of nrDNA. *Electronic Journal of Biotechnology*, **13** (1), 9–10.

Ford, C.S., Ayres, K.L., Toomey, N. *et al.* (2009) Selection of candidate coding DNA barcoding regions for use on land plants. *Botanical Journal of the Linnean Society*, **159** (1), 1–11.

Galimberti, A., De Mattia, F., Losa, A. *et al.* (2013) DNA barcoding as a new tool for food traceability. *Food Research International*, **50**, 55–63.

Halberstein, R.A. (2005) Medicinal plants: historical and cross-cultural usage patterns. *Annals of Epidemiology*, **15** (9), 686–699.

Hamilton, A.C. (2004) Medicinal plants, conservation and livelihoods. *Biodiversity & Conservation*, **13** (8), 1477–1517.

Hammond, P. (1992) Species inventory, in *Global Biodiversity: Status of the Earth's Living Resources* (ed. B. Groombridge), Springer, Houten, pp. 17–39.

Hausdorf, B. and Hennig, C. (2010) Species delimitation using dominant and codominant multilocus markers. *Systematic Biology*, **59** (5), 491–503.

Hawksworth, D.L. and Kalin-Arroyo, M.T. (1995) Magnitude and distribution of biodiversity. *Global Biodiversity Assessment* (ed. V.H. Heywood), Cambridge University Press, Cambridge, pp.107–191.

Hebert, P.D.N., Cywinska, A., Ball, S.L., deWaard, J.R., (2003) Biological identifications through DNA barcodes. *Proceedings of the Royal Society B*, **270**, 313–322.

Hebert, P.D.N., Penton, E.H., Burns, J.M. *et al.* (2004a). Ten species in one: DNA barcoding reveals cryptic species in the neotropical skipper butterfly *Astraptes fulgerator*. *Proceedings of the National Academy of Sciences of the United States of America*, **101** (41), 14812–14817.

Hebert, P.D.N., Stoeckle, M.Y., Zemlak, T.S. and Francis, C.M. (2004b) Identification of birds through DNA barcodes. *PLoS Biology*, **2** (10), e312.

Hollingsworth, P.M. (2011) Refining the DNA barcode for land plants. *Proceedings of the National Academy of Sciences of the United States of America*, **108** (49), 19451–19452.

Hollingsworth, P.M., Forrest, L.L., Spouge, J.L. *et al.* (2009) A DNA barcode for land plants. *Proceedings of the National Academy of Sciences of the United States of America*, **106** (31), 12794–12797.

Huelsenbeck, J.P. and Ronquist, F. (2001) MRBAYES: Bayesian inference of phylogenetic trees. *Bioinformatics*, **17** (8), 754–755.

Huffman, M.A. and Wrangham, R.W. (1994) Diversity of medicinal plant use by chimpanzees in the wild, in *Chimpanzee Cultures* (eds R.W. Wrangham, W.C. McGrew, F.B.M. de Waal and P. Heltne), Harvard University Press, Cambridge, MA, pp.129–148.

Jarman, S.N. and Elliott, N.G. (2000) DNA evidence for morphological and cryptic Cenozoic speciations in the Anaspididae, 'living fossils' from the Triassic. *Journal of Evolutionary Biology*, **13** (4), 624–633.

Johnson, M.G., Shaw, B., Zhou, P. and Shaw, A.J. (2012) Genetic analysis of the peatmoss *Sphagnum cribrosum* (Sphagnaceae) indicates independent origins of an extreme infra-specific morphology shift. *Biological Journal of the Linnean Society*, **106** (1), 137–153.

Kennedy, D. and Norman, C. (2005) What don't we know? *Science*, **309** (5731), 75.

Kerr, K.C.R., Stoeckle, M.Y., Dove, C.J. *et al.* (2007) Comprehensive DNA barcode coverage of North American birds. *Molecular Ecology Notes*, **7**, 535–543.

Knowles, L.L. and Carstens, B.C. (2007) Delimiting species without monophyletic gene trees. *Systematic Biology*, **56** (6), 887–895.

Knowlton, N. (1993) Sibling species in the sea. *Annual Review of Ecology and Systematics*, **24**, 189–216.

Kool, A., de Boer, H.J., Krüger, Å. *et al.* (2012) Molecular identification of commercialized medicinal plants in southern Morocco. *PLoS ONE*, **7** (6), e39459.

Kosakyan, A., Heger, T.J., Leander, B.S. *et al.* (2012) COI barcoding of nebelid testate amoebae (Amoebozoa: Arcellinida): extensive cryptic diversity and redefinition of the Hyalospheniidae Schultze. *Protist*, **163** (3), 415–434.

Kress, W.J. and Erickson, D.L. (2007) A two-locus global DNA barcode for land plants: the coding *rbcL* gene complements the non-coding *trnH-psbA* spacer region. *PloS ONE*, **2** (6), e508.

Kress, W.J., Wurdack, K.J., Zimmer, E.A. *et al.* (2005) Use of DNA barcodes to identify flowering plants. *Proceedings of the National Academy of Sciences of the United States of America*, **102** (23), 8369–8374.

Lahaye, R., van der Bank, M., Bogarin, D. *et al.* (2008) DNA barcoding the floras of biodiversity hotspots. *Proceedings of the National Academy of Sciences of the United States of America*, **105** (8), 2923–2928.

Lambert, J., Srivastava, J. and Vietmeyer, N. (1997) *Medicinal Plants: Rescuing a Global Heritage*, World Bank Publications.

Lara, A., Ponce de León, J.L., Rodríguez, R. *et al.* (2010) DNA barcoding of Cuban freshwater fishes: evidence for cryptic species and taxonomic conflicts. *Molecular Ecology Resources*, **10** (3), 421–430.

Lee, S., Xiao, C. and Pei, S. (2008) Ethnobotanical survey of medicinal plants at periodic markets of Honghe Prefecture in Yunnan Province, SW China. *Journal of Ethnopharmacology*, **117** (2), 362–377.

Lefébure, T., Douady, C.J., Gouy, M. and Gibert, J. (2006) Relationship between morphological taxonomy and molecular divergence within Crustacea: proposal of a molecular threshold to help species delimitation. *Molecular Phylogenetics and Evolution*, **40** (2), 435–447.

Li, D.-Z., Gao, L.-M., Li, H.-T. *et al.* (2011) Comparative analysis of a large dataset indicates that internal transcribed spacer (ITS) should be incorporated into the core barcode for seed plants. *Proceedings of the National Academy of Sciences of the United States of America*, **108** (49), 19641–19646.

Linares, E. and Bye, R.A. (1987) A study of four medicinal plant complexes of Mexico and adjacent United States. *Journal of Ethnopharmacology*, **19** (2), 153–183.

Mati, E. and De Boer, H.J. (2011) Ethnobotany and trade of medicinal plants in the Qaysari Market, Kurdish Autonomous Region, Iraq. *Journal of Ethnopharmacology*, **133**, 490–510.

Meyer, C.P. and Paulay, G., 2005. DNA barcoding: error rates based on comprehensive sampling. *PLoS Biology*, **3**(12), e422.

Moritz, C. and Cicero, C. (2004) DNA barcoding: promise and pitfalls. *PLoS Biology*, **2** (10), e354.

Nielsen, R. and Matz, M. (2006) Statistical approaches for DNA barcoding. *Systematic Biology*, **55** (1), 162–169.

Nixon, K.C. (1999) The parsimony ratchet, a new method for rapid parsimony analysis. *Cladistics*, **15** (4), 407–414.

O'Meara, B.C. (2010) New heuristic methods for joint species delimitation and species tree inference. *Systematic Biology*, **59** (1), 59–73.

Ouarghidi, A., Powell, B., Martin, G.J. *et al.* (2012) Species substitution in medicinal roots and possible implications for toxicity in Morocco. *Economic Botany*, **66**, 370–382.

Ragupathy, S., Newmaster, S.G., Murugesan, M. and Balasubramaniam, V. (2009) DNA barcoding discriminates a new cryptic grass species revealed in an ethnobotany study by the hill tribes of the Western Ghats in southern India. *Molecular Ecology Resources*, **9**, 164–171.

Rubinoff, D., Cameron, S. and Will, K. (2006) Are plant DNA barcodes a search for the Holy Grail? *Trends in Ecology & Evolution*, **21** (1), 1–2.

Schippmann, U., Leaman, D.J. and Cunningham, A.B. (2002) Impact of cultivation and gathering of medicinal plants on biodiversity: global trends and issues, in *Biodiversity and the Ecosystem Approach in Agriculture, Forestry and Fisheries*, FAO, Rome.

Smith, M.A., Poyarkov, N.A. and Hebert, P.D.N. (2008) *CO1* DNA barcoding amphibians: take the chance, meet the challenge. *Molecular Ecology Resources*, **8** (2), 235–246.

Stamatakis, A. (2006) RAxML-VI-HPC: maximum likelihood-based phylogenetic analyses with thousands of taxa and mixed models. *Bioinformatics (Oxford, England)*, **22** (21), 2688–2690.

Stech, M., Veldman, S., Larraín, J. *et al.* (2013) Molecular species delimitation in the *Racomitrium canescens* complex (Grimmiaceae) and implications for DNA barcoding of species complexes in mosses. *PLoS ONE*, **8** (1), e53134.

Sucher, N.J. and Carles, M.C. (2008) Genome-based approaches to the authentication of medicinal plants. *Planta Medica*, **74** (6), 603–623.

Swofford, D.L. (2002) PAUP*: phylogenetic analysis using parsimony (* and other methods). Version. 4, Sinauer Associates, Sunderland, MA.

Taberlet, P., Coissac, E., Pompanon, F. *et al.* (2007) Power and limitations of the chloroplast *trnL* (UAA) intron for plant DNA barcoding. *Nucleic Acids Research*, **35** (3), e14.

Takundwa, M., Chimwamurombe, P.M., Kunert, K. and Cullis, C.A. (2012) Developing DNA barcoding (*matK*) primers for marama bean [*Tylosema esculentum* (Burchell) Schreiber]. *African Journal of Biotechnology*, **11** (97), 16305-16312.

Van Andel, T.R. and Carvalheiro, L.G. (2013) Why urban citizens in developing countries use traditional medicines: the case of Suriname. *Evidence-Based Complementary and Alternative Medicine*, **2013**, article ID 687197.

Van Andel, T.R., Behari-Ramdas, J. Havinga, R. and Groenendijk, S. (2008) The medicinal plant trade in Suriname. *Ethnobotany Research & Applications*, **5**, 351–372.

Van Geel, B., Aptroot, A., Baittinger, C. *et al.* (2008) The ecological implications of a Yakutian mammoth's last meal. *Quaternary Research*, **69** (3), 361–376.

Van Geel, B., Guthrie, R.D., Altmann, J. *et al.* (2011) Mycological evidence of coprophagy from the feces of an Alaskan Late Glacial mammoth. *Quaternary Science Reviews*, **30** (17), 2289–2303.

Van't Klooster, C., Lindeman, J.C. and Jansen-Jacobs, M.J. (2003) *Index of Vernacular Plant Names of Suriname*, Blumea Supplements 15, Nationaal Herbarium Nederland, Universiteit Leiden branch, Leiden.

Vorhies, F. (2000) The global dimension of threatened medicinal plants from a conservation point of view, in *Medicinal Utilisation of Wild Species: Challenge for Man and Nature in the New Millennium* (eds S. Honnef and R. Melisch), WWF Germany/TRAFFIC Europe-Germany, Frankfurt am Main, pp. 26–29.

Wasser, S.K., Mailand, C., Booth, R. *et al.* (2007) Using DNA to track the origin of the largest ivory seizure since the 1989 trade ban. *Proceedings of the National Academy of Sciences of the United States of America*, **104** (10), 4228–4233.

Will, K.W. and Rubinoff, D. (2004) Myth of the molecule: DNA barcodes for species cannot replace morphology for identification and classification. *Cladistics*, **20**, 47–55.

Williams, V.L., Victor, J.E. and Crouch, N.R. (2013) Red listed medicinal plants of South Africa: status, trends, and assessment challenges. *South African Journal of Botany*, **86**, 23–35.

World Health Organization (2003) Traditional medicine, Factsheet No. 134, http://www.who.int/mediacentre/factsheets/2003/fs134/en/.

# 7 Market Entry, Standards and Certification

Susan A. Wren
International Centre for Insect Physiology and Ecology (ICIPE), Nairobi, Kenya

## 7.1 Sustainable utilization of indigenous plant products

### 7.1.1 International trade in indigenous medicinal and aromatic plants collected in the wild

There is a big transfer of natural ingredients from the developing world to the Western world. The global demand has increased significantly over the last 10 years. The markets for natural products are wide ranging and include culinary and herb teas, food flavourings, toiletries, cosmetics, baby products, fragrances, health care products, herbal supplements/nutraceuticals and other accessories. High quality, natural, design conscious, ethically/fair traded and environmentally friendly are the strongly apparent trends (Wren, 2007). An estimated 50 000–70 000 plant species are used in traditional and modern medicine throughout the world, as a wide variety of products. The great majority of these species are obtained through collection from the wild habitat. Plants harvested from natural populations continue to be the most important source of medicine to cover the primary health care needs of more than two-thirds of the world's population (Tolman *et al.*, 2012). The global industry is still dependent on wild collected plants as raw materials for food, medicine and cosmetics – over 90% of commercially used plant species are sourced from wild collection; this is over 70% of material trade. Only 500 plant species are commercially cultivated today (Kathe *et al.*, 2010).

At the international level, the natural products market has become both sophisticated and diverse, with a major trend towards natural ingredients and product certification (organic, ethical, etc). Consumers are looking for quality, authenticity and a feel-good factor. These market characteristics provide opportunities for developing world populations to benefit from enterprises based on indigenous plant products and to harness particular competitive advantages, such as pure and unpolluted source of materials, the lower cost of producing and, in some cases, processing the raw materials into finished products and the emotive storyline (community empowerment – incomes for women and families and the least advantaged, conservation through trade, etc). Recent trends show an increase in the use of herbal extracts in food formulations in the way of functional foods and also in cosmetics; these areas show a stronger growth compared with supplements. There is a general need for 'new' natural and botanical products (Wren, 2007).

### 7.1.2 Threats and challenges to the future of indigenous commercialized resources

About 15 000 wild collected plant species are threatened in at least part of their natural habitats (Schippmann *et al.* 2006). Not only overharvesting causes this threat, but also land conversion and other forms of habitat destruction or modification. In certain areas, wild collection secures the only income for rural households, though the collecting communities generally belong to the most impoverished and deprived sector of the society (Wren, 2012). Despite the fundamental importance of indigenous medicinal and aromatic plants to health and livelihoods, relatively little investment has been made in assessing the conservation status of most wild collected species or in developing more sustainable harvest and trade practices. With increasing demand for wild collected plant species, overharvesting and other threats to wild populations, the development of effective tools ensuring the sustainable use of the resources has increasing importance. Approaches to stimulate sustainable wild collection that engage governments as well as local, regional and international enterprises are urgently needed to help industry, resource managers, collectors and other stakeholders towards more sustainable and equitable sourcing practices (Twarog, 2006).

#### 7.1.2.1 Needs to reduce depletion of natural resources and mitigate impacts of climate change

Increasing pressure on resources and land degradation is common in the developing world and is exacerbated by the negative impacts of climate change (UNCED, 1992).

The social and economic impacts are severe. The domestication and sustainable wild harvesting of indigenous plant materials can provide a tangible incentive for communities to positively manage the natural resources into the future (Duerbeck, 2008). At the same time, agro-forestry and high-value cropping in the lands adjacent to watercourses can provide vital buffers to mitigate drought impacts. In human–wildlife conflict areas the cultivation of food crops is mostly unsuccessful. Sustainable wild harvesting and domestication of indigenous plant products may provide the only viable environmentally sustainable livelihood option. Such initiatives can also address important millennium goals (UN, 2000); however, so far, very few have been successfully developed.

*Novel Plant Bioresources: Applications in Food, Medicine and Cosmetics*, First Edition. Edited by Ameenah Gurib-Fakim.

#### 7.1.2.2 Sustainable wild harvesting

The development of sustainable wild harvest standards is imperative where products are sourced from indigenous resources. Sustainable wild harvest standards are devised to ensure that wild production comes from a clearly defined area and uses methods that meet international certification standards. Sustainable wild harvesting methods and certification systems are explained in Section 7.3.2.

#### 7.1.2.3 Domestication

For scaling up dryland plant product enterprises based on wild harvested plant products, the domestication of indigenous species should be encouraged where possible (Wren and Powys, 2008). This is particularly important for species that cannot be readily wild harvested at sustainable levels and for medicinal extracts that are derived from bark, wood and roots (Schippmann *et al.*, 2006).

## 7.2 Market entry

### 7.2.1 Achieving marketable quality: the example of essential oils

#### 7.2.1.1 Quality control at harvest

The plant material may vary considerably in quality depending on site, soil, weather and climatic conditions, time of harvest, genetic variations, plants health and quality.

- *Correct identification of indigenous species.* To prevent the harvesting of the wrong species it is important to develop simple pictorial guide sheets (laminated) and provide them to the harvesters and the grading teams. Training in correct species identification should be carried out at the beginning and re-enforced through ongoing training in sustainable wild harvesting techniques, handling, and so on.
- *Contamination with other species.* Contamination with related or non-related species is a common problem where the required species are growing in mixed stands. The collectors should have basic training in plant identification.
- *Ensuring that the right part of the plant is harvested at the correct season.* There is wide variation in essential oil content of the required product according to plant part and season and weather conditions. If flowers are harvested, the product may be in its optimum state during the late dry season when there has been no rain. Time of day can also be a factor: late afternoon during the dry season will reduce the water content of the flower heads (Tisserand and Balacs, 1995). Harvesting information should be assessed using distillation results and comparative gas chromatograph (GC) data.

#### 7.2.1.2 Potential sources of contamination

- *Plants.* Use of pesticides and other chemicals, altitude, soil conditions and rainfall, plant species and varieties.
- *Processing.* This may include: poor methods of equipment cleaning and standards of hygiene leading to contamination from degraded oils, detergents, and so on; poor processing techniques, such as allowing entry of foreign matter or any plant material other than that from the intended essential oil species; inclusion of parts of the essential oil plant that are not accepted under the industry standards/norms for that specific essential oil species (Tisserand and Balacs, 1995).
- *Packaging and handling.* Chemical degradation can occur with exposure to heat, light or oxygen.
- *Storage.* Contamination from poor or inadequate storage is a common fault. Essential oils should be stored in tightly closed, darkened glass containers in a cool place to ensure lasting quality. Oxidation rates vary, but if stored in dark, cool conditions most essential oils do not show signs of degrading within 1–2 years or more.

#### 7.2.1.3 Methods of evaluation

There are many aspects that affect the quality of essential oils from the growing conditions of the plants to their harvesting, distillation, manufacture, distribution and storage. Qualitative testing determines what individual constituents are present in an essential oil, whilst quantitative testing provides information on how much of each component is present.

- *Sensory evaluation.* The first steps of essential oil testing begin with sensory evaluation. The sensory capacity of a trained person can identify inferior oils quickly, based on the viscosity, colour, clarity and odour of an essential oil. An odour test can help to detect certain adulterants and can be used to detect whether the sample has been sourced from a more or less desirable place of origin.
- *Physical parameters.* Once an essential oil sample passes the sensory tests the next stage is to test the physical parameters. This is achieved by measuring the specific gravity, optical rotation and refractive index. This will confirm or reject the authenticity of an essential oil's botanical species and country of origin, whilst possibly revealing or confirming the presence of any adulteration by foreign substances. If the first two stages are successfully passed the oil is then tested using GC–mass spectrometry (Tisserand and Balacs, 1995).

### 7.2.2 Statutory regulations and standards

Essential oils are sourced from many regions of the world, and suppliers or companies usually obtain oils from farmers or wholesalers whose practices and integrity they have come to trust over time. However, over time a number of statutory bodies and certifying organizations have developed to regulate quality and safeguard the overall integrity of the industry and the welfare of consumers at large (Buckle, 2003).

Essential oil suppliers are required by the industry to provide the following information:

- Latin name and country of origin (where the essential oil was produced),
- statement about purity,
- physical tests,
- certification status.

#### 7.2.2.1 Essential oils regulations around the world

Various agencies abroad have been proactive in setting standards for essential oil quality. Examples of these agencies include the following.

- *AFNOR (Association Française de Normalisation).* AFNOR's 'New Approach' provides directives and standards for members of the European Union states to facilitate intra-community trade. All companies wishing to exchange goods within Europe are therefore obliged to comply with their directives. For essential oils, they provide guidelines and information on various topics, including determination of water content, chromatographic profiles, determination of acid value, phenols content, and so on. For more information, see www.afnor.org/en.
- *ISO (International Organization for Standardization).* The ISO mission is to promote the development of standardization in the areas of intellectual, scientific, technological and economic activity. For essential oils, they provide guidelines for packaging, conditioning, storage, labelling, sampling, testing, and so on. ISO also provides, for a fee, quality standards for individual essential oils. ISO International Standards ensure that products and services are safe, reliable and of good quality. For business, they are strategic tools that reduce costs by minimizing waste and errors and increasing productivity. ISO does not provide certification or conformity assessment.

Relevant ISO standards for essential oil producers and processors are:

- ISO 9000 Quality management
- ISO 14000 Environmental management
- ISO 3166 Country codes
- ISO 26000 Social responsibility
- ISO 50001 Energy management
- ISO 31000 Risk management
- ISO 22000 Food safety management
- ISO 4217 Currency codes
- ISO 639 Language codes

For more information, see www.iso.org.

#### 7.2.2.2 United States 'Federal Food, Drug, and Cosmetic Act' and 'Dietary Supplement Health and Education Act'

The Food and Drug Administration (FDA) has responsibility for regulating foods, food additives, drugs, cosmetics and dietary supplements. The FDA considers essential oils as either cosmetics or drugs, depending on their intended use, and makes decisions concerning the regulation of essential oils on a case-by-case basis. The vast majority of essential oils are handled as cosmetics, foods and flavouring agents. Claims made in advertising, but not on product labelling, are regulated by the Federal Trade Commission. Room fragrance systems (deodorizers, odour control) are the responsibility of the Consumer Product Safety Commission. For more information, see the FDA website: www.fda.gov/cosmetics/productandingredientsafety/productinformation/ucm127054.htm.

For the United States Federal Food, Drug, and Cosmetic Act and the Dietary Supplement Health and Education Act, see www.health.gov/dietsupp/ch1.htm and www.fda.gov/RegulatoryInformation/Legislation/FederalFoodDrugandCosmeticActFDCAct. Source of information: Office of Cosmetics and Colors Fact Sheet, March 13, 2000: Aromatherapy.

#### 7.2.2.3 Customs, duties and taxes

Customs policies can vary from country to country. The importer, as well as the exporter, is responsible for making sure that the product/quantities are permitted within the importing country. International orders may be subject to customs duties, import taxes and fees specified by each importing country.

#### 7.2.2.4 Good agricultural and collection practices for medicinal plants

Good agricultural and collection practices (GACP) for medicinal plants are the first step in quality assurance, on which the safety and efficacy of herbal medicinal products directly depend. These practices also play an important role in protecting natural resources of medicinal plants for sustainable use. As part of its Essential Drugs and Medicine Policy website, the World Health Organization (WHO) has created a series of important guidelines related to GACP for medicinal plants. This is also related to a broader policy agenda within the WHO that is committed to protecting such plants, along with promoting their sustainable use and cultivation. Under the overall context of quality assurance and control of herbal medicines, the WHO developed the guidelines to provide general technical guidance on obtaining medicinal plant materials of good quality for the sustainable production of herbal products classified as medicines. The guidelines concern the cultivation and collection of medicinal plants and include certain post-harvest operations, and are also related to WHO's work on the protection of medicinal plants, aiming promotion of sustainable use and cultivation of medicinal plants.

The main objectives of these guidelines are to:

- contribute to the quality assurance of medicinal plant materials used as the source for herbal medicines to improve the quality, safety and efficacy of finished herbal products;
- guide the formulation of national and/or regional GACP guidelines and GACP monographs for medicinal plants and related standard operating procedures;
- encourage and support the sustainable cultivation and collection of medicinal plants of good quality in ways that respect and support the conservation of medicinal plants and the environment in general.

Further information can be obtained at www.who.int/medicines/library/trm/medicinalplants/agricultural.shtml.

### 7.2.3 Other relevant guidelines

TRAFFIC, the International Union for Conservation of Nature (IUCN) and the World Wide Fund for Nature (WWF) have engaged a wide variety of stakeholders in the development of tools such as the Convention on Biological Diversity (CBD) Global Strategy for Plant Conservation, WHO/IUCN/WWF/TRAFFIC Guidelines on the Conservation of Medicinal Plants and the FairWild Standard to reduce overexploitation.

The International Standard for Sustainable Wild Collection of Medicinal and Aromatic Plants (ISSC-MAP) has been developed by the Medicinal Plant Specialist Group (MPSG) of the Species Survival Commission (SSC), IUCN, on behalf of a steering group consisting of the MPSG, Bundesamt für Naturschutz (German Federal Agency for Nature Conservation; BfN), WWF Germany and TRAFFIC. An international advisory group of more than 150 experts provided guidance in drafting the ISSC-MAP. The ISSC-MAP is designed to help those involved in the harvest, management, trade, manufacture and sale of wild-collected medicinal and aromatic plant (MAP) resources to understand and comply with the conditions under which the sustainable collection of these resources can take place.[1]

The ISSC-MAP formed the recommendations of the 1993 WHO/IUCN/WWF Guidelines on the Conservation of Medicinal Plants (WHO *et al.*, 1993, 2013) and the WHO Guidelines on Good Agricultural and Collection Practices (GACP) for Medicinal Plants (WHO, 2003). These guidelines provide general recommendations for the development of a global framework of practice standards for MAP. The group, WHO, IUCN, WWF and TRAFFIC, through an international consultation process incorporated broader guidance and principles related to sustainable use of biological diversity, access and benefit sharing, and fair business practices. ISSC-MAP information can be found at http://www.floraweb.de/map-pro.

#### 7.2.3.1 The European Plant Conservation Strategy

The targets of the European Plant Conservation Strategy (EPCS) translate the internationally accepted Global Strategy for Plant Conservation (GSPC) into a European context. The EPCS provides individuals, institutions, organizations and governments across Europe with a unique opportunity to take action to secure the future of wild plants and fungi and the ecosystem services and livelihoods that depend upon them.

ISSC-MAP is named in the EPCS document as a valuable instrument contributing to achieve Objective 3: Using plant diversity sustainably.

#### 7.2.3.2 Access and benefit-sharing requirements

The fair and equitable sharing of the benefits arising out of the utilization of genetic resources is one of the CBD's three objectives. Parties to the CBD recognize that access to these resources must be subject to prior informed consent (PIC) and has to be based on mutually agreed terms (MAT), including fair and equitable sharing of any resulting benefits. The CBD also requires that access to related traditional knowledge (TK) occurs with the approval of the holders of that knowledge, who must also participate equitably in the resulting benefits. However, awareness of access and benefit sharing (ABS) remains limited at the community, government, civil society and private sector levels. Furthermore, policies that help in developing and implementing ABS remain a significant challenge. It is useful to note that the FairWild Standard includes ABS elements as important constituents of sustainable use practices (see Section 7.3.3.1).

### 7.2.4 Export compliance

Export procedure describes the documents required for exporting to most countries/economic trading blocks. However, additional or alternative documents may be required for certain types of products or destinations, such as a quality control inspection certificate from the Export Inspection Agency. Some food, aromatic and pharmaceutical products may require a health or sanitary certificate for export.

#### 7.2.4.1 Standard shipping documents

Most exporters use the services of a shipping agent to arrange the shipping documents and ensure that export compliances have been met. It is necessary for all exporters, however, to be aware of the shipping required for each commercial contract:

The *shipping bill/bill of export* is the main document required by the customs authority. The export documents required for the processing of the shipping bill include: packing list stating the contents; quantity; gross and net weights; invoices. which contain all relevant particulars, like quantity, unit rate, total value, correct and full description of goods and so on; purchase order of the overseas buyer; and the inspection certificate.

Documents required for most export consignments include the following:

- Customs declaration form and dispatch note.
- Commercial invoice. Issued by the seller for the consignment, as per trade term.
- Consular invoice. Needed for the countries like Kenya, Uganda, Tanzania, Mauritius, New Zealand, Burma, Iraq, Australia, Fiji, Cyprus, Nigeria, Ghana, Zanzibar and so on. It must be prepared in the prescribed format and be signed/certified by the counsel of the importing country located in the country of export.
- Customs invoice. Needed for the countries like USA, Canada, and so on. It is prepared on a special form being presented by the customs authorities of the importing country. Facilitates entry of goods in the importing country at preferential tariff rate.
- Certified invoice. Required when the exporter needs to certify on the invoice that the goods are of a particular origin or manufactured/packed at a particular place and in accordance with a specific contract.
- Packing list. Shows the details of goods contained in each parcel/ shipment.
- Certificate of inspection. Shows that goods have been inspected before shipment.
- Weight note. Required to confirm the packets or bales or other form are of a stipulated weight.
- Manufacturer's/supplier's quality/inspection certificate. Required in addition to the certificate of origin for few countries to show

[1] The ISSC-MAP defines a framework for the sustainable use of biological diversity. The UN CBD provides both global and national contexts for these efforts. Under the CBD, specific guidance for ecological, socio-economic and equity basis for conservation and sustainable use of biodiversity is articulated in the *Ecosystem Approach* (CBD, 2000), the *Global Strategy for Plant Conservation* (CBD, 2002a), the *Bonn Guidelines on Access to Genetic Resources and Fair and Equitable Sharing of Benefits Arising out of their Utilization* (CBD, 2002b) and *Principles and Guidelines for the Sustainable Use of Biodiversity* (CBD, 2004)

that the goods shipped have actually been manufactured and are available.

- Certificate of chemical analysis. Required to ensure the quality and grade of certain items, such as metallic ores, pigments, and so on.
- Certificate of shipment. Signifies that a certain lot of goods has been shipped.
- Health/sanitary certification – required for export of foodstuffs.
- Certificate of conditioning. Issued by the competent office to certify compliance of humidity factor, dry weight, and so on.
- Shipping order. Issued by the shipping (conference) line, which intimates the exporter about the reservation of space of shipment of cargo through the specific vessel from a specified port and on a specified date.
- Shipping advice. Prepared in an aligned document to be used to inform the overseas customer about the shipment of goods.
- Phytosanitary certificate. Most countries permit imports of plants and plant products only on the basis of a phytosanitary certificate. This must be issued for each consignment by the national authority of the exporting country.

#### 7.2.4.2 Material safety data sheet

The material safety data sheet (MSDS) is required under the Occupational Safety and Health Administration (OSHA) Hazard Communication Standard and describes the physical and chemical properties of the product. The MSDS is a detailed informational document prepared by the manufacturer or importer of a hazardous chemical. Essential oils are defined as a hazardous chemical. The MSDSs contain useful information such as flash point, toxicity, procedures for spills and leaks, and storage guidelines. The MSDS must be compiled to comprise the information listed in Table 7.1.

The US version of the MSDS is used to comply with OSHA's Hazard Communication Standard, 29 CFR 1910.1200. The standard must be consulted for specific requirements. Standard forms are provided by the US Department of Labor Occupational Safety and Health Administration.

#### 7.2.4.3 Organic labelling

On 20 June, the European Commission began the implementation of Regulation No. 508/2012, setting out detailed rules for dealing with recognition of equivalence in certification of organic products. The certification bodies recognized for providing the necessary certification are listed in the regulation, including locally accredited bodies and EU-based certification agencies. Also, on 1 July 2012, the regulation related to the new organic logo entered into force. The aim of the new logo, which is to be used throughout the EU, is to create more transparency on packaged and certified organic products. The new labelling requirements do not apply to bulk fresh fruit and vegetables. The correct labelling for organic products and probable restrictions at import will be enforced and checked at customs. The newly given EU code number (143) must be used to avoid difficulties during customs clearing. It will be necessary to check the current status of labelling requirements with the EU-based importer(s) prior to shipping (EC, 2012).

**Table 7.1** Stipulated information for the MSDS form

| |
|---|
| Section 1. Chemical product and company identification. Links the MSDS to the material. Identifies the supplier of the MSDS. Identifies a source for more information. Include manufacturer's name. |
| Section 2. Composition/information on ingredients. Lists the OSHA hazardous components. May also list significant nonhazardous components. |
| Section 3. Hazards identification, including emergency overview. Provides information on the potential adverse human health effects and symptoms that might result from reasonably foreseeable use and misuse of the material. |
| Section 4. First aid measures. Provides instructions to be taken if accidental exposure requires immediate treatment. |
| Section 5. Fire fighting measures. Provides basic fire fighting guidance, including appropriate extinguishing media. Describes other fire and explosive properties useful for avoiding and fighting fires involving the material, such as flash point or explosive limits. |
| Section 6. Accidental release measures. Describes actions to be taken to minimize the adverse effects of an accidental spell, leak or release of the material. |
| Section 7. Information on appropriate practices for safe handling and storage. |
| Section 8. Exposure controls/personal protection. Information on practices, or equipment, or both, that are useful in minimizing worker exposure. |
| Section 9. Physical and chemical properties. Provides additional data that can be used to help characterize the material and design safe work practices. |
| Section 10. Stability and reactivity. Describes the conditions to be avoided or other materials that may cause a reaction that would change the intrinsic stability of the material. |
| Section 11. Toxicological information. May be used to provide background toxicological information on the material, its compounds, or both. |
| Section 12. Ecological information. May be used to provide information on the effects the material may have on plants or animals and on the material's environment. |
| Section 13. Disposal considerations. |
| Section 14. Transport information. May provide basic shipping classification information. |
| Section 15. Regulatory information. May be used to provide any additional information on regulations affecting the material. |
| Section 16. Other relevant information. |

## 7.3 Certification

### 7.3.1 Organic certification

Organic certification addresses a growing worldwide demand for organic food. It is intended to assure quality and prevent fraud, and to promote commerce through the process of third-party regulatory certification. For organic producers, certification identifies suppliers of products approved for use in certified operations. For consumers, 'certified organic' serves as a product assurance. Certification is essentially aimed at regulating and facilitating the sale of organic products to consumers.

Organic certification, as well as fair-trade certification, has the potential to directly and indirectly contribute to the achievement of the Millennium Development Goal (MDG) targets. A study by Setboonsarng (2008) reveals that organic certification substantially contributes to MDG1 (poverty and hunger) and MDG7 (environmental sustainability) by way of premium prices and better market access, among others. This study concludes that for this market-based development scheme to broaden its poverty impacts, public sector support in harmonizing standards, building up the capacity of certifiers, developing infrastructure development and innovating alternative certification systems will be required (www.adbi.org/discussion-paper/2008/08/25/2675.organic.fairtrade.certification/organic.agriculture.and.certification.an.overview/).

In some countries, organic standards are formulated and overseen by the government. The USA, the EU, Canada and Japan have comprehensive organic legislation, and the term 'organic' may be used only by certified producers. Internationally, equivalency negotiations are underway, and some agreements are already in place, to harmonize certification between countries, facilitating international trade. There are also international certification bodies, including members of the International Federation of Organic Agriculture Movements (IFOAM), working on harmonization efforts. Where formal agreements do not exist between countries, organic product for export is often certified by agencies from the importing countries, which may establish permanent foreign offices for this purpose. In 2011, IFOAM introduced the 'IFOAM Family of Standards', which attempts to simplify harmonization.

EU countries acquired comprehensive organic legislation with the implementation of the EU-Eco-regulation 1992. Supervision of certification bodies is handled on the national level. The European Commission promotes a Europe-wide label for organic food, which culminated in 2009 in a European organic logo. However, most countries are still using their own nationally promoted labelling schemes and logos.

In the USA, the Secretary of Agriculture has established the National Organic Program (NOP). The final rule was published in the Federal Register in 2000. It restricts the use of the term 'organic' to certified organic producers (excepting growers selling under $5000 a year, who must still comply and submit to a records audit if requested, but do not have to formally apply). Certification is handled by state, non-profit and private agencies that have been approved by the US Department of Agriculture (USDA). In Canada, certification was implemented at the federal level on 30 June 2009. Mandatory certification is required for agricultural products represented as organic in import, export and inter-provincial trade.

Organic compliance has to be verified and certified by an accredited certification body. It is best to seek advice from the buyer/importer as to the choice of certification body/agency, according to the product type, production system, locations and types of target market, and so on.

See www.ifoam.com for listing of IFOAM-accredited certification bodies and agencies.

### 7.3.2 Sustainable wild harvest and certification

As expounded in many conservation publications, it is well established that there is a growing need to safeguard the future sustainable use of biodiversity. Certification of sustainable collection methods can provide incentives for conservation of important habitats and strengthens local economies. In most countries there are currently no regulatory mechanisms for the wild harvesting of indigenous plant products, although some national governments have developed regulations and licensing for the collection of specific Convention on International Trade in Endangered Species (CITES)-listed species. If such mechanisms align with an internationally endorsed system that is market recognized, then taking the legitimate route can enable commercial actors to reach premium international markets. However, if it acts as a trade barrier, then the illegal market will be further stimulated.

A growing number of companies are realizing that investing in sustainable management of wild harvests and associated supply chains makes good business sense. As well as avoiding the need to discontinue or reformulate products, investment in sustainable sourcing, particularly when coupled with 'fair trade', can enhance the competitive advantage of a product. This is reinforced by growing consumer awareness of and demand for 'ethical' products. Governments are also increasingly recognizing the economic importance of wild-product trade, and so are keen to address sustainability issues and ensure product quality through more transparent supply chains. Groups promoting 'alternative livelihoods' based on commercialization of non-timber forest products have also come to realize that success relies on stable supplies, as well as stable or growing markets (Kathe *et al.*, 2010).

The development of sustainable wild harvest standards is imperative where products are sourced from indigenous resources. Sustainable wild harvest standards are devised to ensure that wild production comes from a clearly defined area and using methods that meet international certification standards (Duerbeck, 2008).

### 7.3.3 Certification options for indigenous plant products

#### 7.3.3.1 EU ISO65 sustainable wild harvest protocols

Most organic products on the market originate from cultivation, which has to comply with the respective organic production requirements. But a considerable number of organic products are not cultivated, but collected where they grow naturally. For such wild collection products the principles of the organic regulations apply as much as for 'cultivated' crops. However, the different situation concerning the risk of contamination, ownership of the land, vast size of the collecting areas and sustainability implies that there are some differences regarding requirements for organic certification of wild collected plants. Therefore, most organic standards contain a small chapter of organic wild collection.

The minimum requirements for organic wild collection according to the organic regulations of the EU (Regulations (EEC) 834/07 and 889/08) and the US standard (USDA NOP) are explained in a document compiled by the Institute for Marketecology (IMO) and the Swiss Import Promotion Programme (SIPPO), which can be found at www.imo.ch/portal/pics/IMO_CH/documents/organic_wild_collection_guidance_manual_version2.pdf.

The EU definition for 'wild' collection is:

1. The collected plants grow naturally in an area which has not been treated with prohibited inputs (according to the respective organic regulations) for at least 3 years.
2. The collection areas are not owned by the company itself (public land) and/or are of vast size.

3. The collected plants must grow and regenerate naturally without any agricultural measures.
4. Certified are *plants* grown in an approved (by an accredited certifier) area. The area (*land*) itself is not certified.

Generally, neglected old plantations of perennial plants such as orchards, and so on, cannot be considered as 'wild' as usually the ownership is clearly defined and the plants did not grow spontaneously. The owner has a clear interest to increase the production of such a plantation; consequently, the risk that prohibited inputs are used is higher than with land that belongs to someone else. Additionally, for an area that is vast and badly accessible, it is far less likely that someone actually uses prohibited inputs, such as fertilizers or pesticides, to increase the yield of the wild growing plants.

In that case the 'normal' (cultivation) farm inspection system applies, which takes the low-key production system into account. In cases where the situation is not clear, it is up to the certification body to decide if the project can be considered as 'wild collection'.

Plants that are not indigenous but are collected in the wild (naturally regenerated) are considered as 'wild'. In organic agriculture, fields (*land*) and *crops* are certified 'organic'. In wild collection projects, only the *plants* are certified organic. The land needs to be *approved*, but is not *certified*. This is an important difference to organic agriculture, as land used as a collection area cannot automatically be used for organic agriculture without a conversion period.

### 7.3.3.2 FairWild standard

The FairWild standard is an international established label that aims at ecological and economic sustainability by offering a comprehensive standard including management systems, resource assessments, access rights, and so on, as well as providing economic support by fair-trade systems, labelling and marketing promotion. Integrating with an international certification standard such as this will enable governments in the Arid and Semi-Arid Lands (ASAL) region to develop a regulatory framework in a way that ensures the commercial operators that benefits can be achieved and secured through compliance.

FairWild standard development was led by TRAFFIC, WWF, IUCN, BfN and others, and involved extensive consultation with representatives from government, the private sector, academia, non-governmental organizations (NGOs) and certification agencies. The FairWild standard now provides a reference and best-practice framework for sustainable wild collection and trade. The ISSC-MAP standard was tested for applicability to community resource management in pilot projects in six countries: Bosnia and Herzegovina, Brazil, Cambodia, India, Nepal and Lesotho. Lessons learned were published in a report, *Wild for a Cure*, launched in September 2010. Under the auspices of the FairWild Foundation (www.fairwild.org), the ISSC-MAP and the FairWild standards are implemented to assure buyers that products are produced in a socially and ecologically sound manner. The standard for wild collection is based on existing knowledge and appropriate resource management, focusing on:

- promoting sustainable resource management and conservation of wild plants;
- establishing a system for certification of sustainable wild collection and fair trade practices;
- providing advice on the application of standards for sustainable and fair trade in conservation and trade policy and regulations.

The standard forms the basis of a third-party audited certification system, complementing organic and fair trade certification processes, which typically lack mechanisms for certifying that wild harvest levels are sustainable. The FairWild standard's principles and approach are being used by a growing number of herbal product, food and other companies dependent on wild-sourced botanicals, and by governments keen to bring or maintain wild plant harvests and trade within sustainable levels. The FairWild standard is maintained by the FairWild Foundation, a Swiss-based charity. Specifically designed to be relevant to the private sector, civil society organizations and governments alike, the standard and associated guidance tools (see FairWild Standard in Practice at www.fairwild.org).

The initial and successive annual inspections are conducted by the accredited inspection body, IMO, a Swiss certification company accredited to provide inspection to the EU ISO65 organic standards and the FairWild standard. (For further information see www.imo.ch.)

#### *Fair sourcing and trade practices*

The standard can be implemented by wild-collection operations and government agencies, such as forest authorities, in a variety of ways:

- species-specific certification;
- as a guide for community-based plant collection and resource management;
- as a guide for resource management through government authorities

For more information on fair sourcing and trade practices, please visit the Fairwild Foundation website at www.fairwild.org/international-legal-agreements.

#### *National legislation and policies*

Governments can use the FairWild standard as a model for the development of national laws and regulations governing fair trade practices, the conservation of biodiversity or the management of plant collection from the wild. FairWild has helped shape biodiversity and resource management legislation in countries such as Bosnia-Herzegovina. [2] The National Medicinal Plants Board of India has included parts of FairWild's ecological module (the former ISSC-MAP) in its *Guidelines for Good Field Collection Practices of Medicinal Plants* (in press). If you are interested in following this development, or related matters, please visit the FairWild website at http://www.fairwild.org/trade-practices.

[2] Lessons learnt from a FairWild implementation project in the Brazilian Amazon were presented at a CBD meeting on finalizing the text of an international regime on ABS in Colombia in April 2010. The project promotes a better comprehension of and guidance on implications of the use of biological resources and associated TK, aiming to establish agreements on conservation and benefit sharing with different partners, including the private sector.

While the entire standard needs to be considered and implemented successfully in order to achieve FairWild certification, it is possible to focus only on specific parts of the standard to help guide national or regional legislative and policy processes.

#### *FairWild and CITES*

The FairWild standard is also proving instrumental for the implementation of existing regulatory frameworks provided by national resource management systems as well as by international conventions such as CITES.

Countries exporting plant and animal species listed in Appendix II of CITES are required to demonstrate the levels of export are not detrimental to the survival of the species concerned. This is achieved through the compilation and issuing of a so-called non-detriment finding (NDF) by the CITES scientific authority of the country concerned.

Currently, there is little detailed technical guidance on how to compile NDFs or how to make them comparable between different countries. CITES is currently preparing such guidance through its Animals and Plants Committees.

The FairWild standard has been useful in this process. Members and partners of the FairWild Foundation have ensured that the relevant ecological criteria (formerly the ISSC-MAP standard) of the FairWild standard were considered at international CITES expert meetings in Mexico and elsewhere and a background paper on supporting the implementation of ISSC-MAP in CITES through the NDF process prepared. All elements relevant to CITES in ISSC-MAP were used in the resultant Conference of Parties (CoP)15 document.

The ISSC-MAP standard is also referenced in another CoP15 document, which states: 'the aim of ISSC-MAP is to bridge the gap between existing broad conservation guidelines and management plans developed for specific local conditions'.

The implementation of an adaptive management scheme based on regular monitoring is an important consideration in the ongoing NDF evaluation process discussion. FairWild management planning guidance can also provide guidance to this.

#### *FairWild and the Convention on Biological Diversity*

FairWild has clear links to the CBD's core aim of conservation of biological resources, including their sustainable use and fair sharing of benefits resulting from such use. The FairWild Standard provides both public and private sectors with a means to achieve these CBD objectives within operations focusing on wild plant collection.

#### *Global Strategy for Plant Conservation*

The strategy aims to contribute to poverty alleviation and sustainable development through inclusion of its targets into government policy, as national biodiversity strategies or targeted national plant conservation strategies and into private-sector guidelines. The comprehensiveness of the FairWild standard and the flexibility in its implementation make the FairWild Foundation an ideal partner for governments, companies, NGOs and community-based plant sourcing operations for working towards fulfillment of the GSPC targets. ISSC-MAP, whose criteria form the ecological part of the FairWild standard, is cited as a tool for implementing specific GSPC targets in the Plant Conservation Report that was presented at CBD CoP9. This report was developed by the CBD Secretariat in collaboration with the Global Partnership for Plant Conservation and other relevant organizations and stakeholders.

## 7.3.4 Defining the structure of the commercial operation/project

All sites where wild plants are collected, stored or processed need to comply with the organic regulation and must be inspected by the certification body. All parties involved define their cooperation by contracts.

All collecting areas must be known to the certification body. Purchase locations and main processing and trading locations are inspected by the certification body every year. Collectors' homes are spot-checked. Certification takes place all along the complete supply chain. Each step from the collection to the shelf needs to be known to the certifier.

To achieve the necessary transparency, all wild collection activities need to be organized in a well-structured 'project'. Such projects usually consist of three major parties: the collectors, the purchase centres and the main processor/trader, which is the actual wild project operator. The main processing/trading unit is usually responsible for the organization of the project. The purchase centres mostly belong to the main company, but can also be independent dealers with contracts both with the main company and with the organic certifier. The collectors are contracted by the main company (or the purchaser). Simple processing like drying or cleaning is often done at the collectors' homes or the purchasing centres. Further processing, packing and trading is done at the facilities of the project operator, and such central processing units are regarded by the certifier as dedicated handling units. Sometimes there are no separate purchase centres but all goods are directly transported from the collectors to the project operators' facilities.

## 7.3.5 Protocol for sustainable wild harvested of selected plants

### 7.3.5.1 Mapping and zoning

The map is the first tool for developing sustainable wild harvest systems. The map should identify the main ecosystems and predominant botanical species. The species that is to be the subject of wild harvesting is then marked clearly in the regions where it is most abundant. The areas where wild harvesting would damage the health of the species /habitat/natural diversity should be marked in red as non-intervention zones (blocked from collection).

This initial mapping stage has been assessed only in general terms during this assignment, and a detailed mapping will need to be carried out by local botanists familiar with the area, preferably using global positioning system plotting methodology. The plotted areas are then incorporated into operation maps, clearly marked as harvesting zones for each specific species. From this point, a detailed botanical survey of the selected plant species growing in these zones should take place using transects, as described in this document, together with assessment of the overall plant stand. This information gives rise to the harvesting capacity and more exact harvesting thresholds of the selected species within the identified zones. From this position, the rotation of harvesting across and within the zones over the entire mapped areas can be developed. This is the foundation for the sustainable wild harvesting protocols.

### 7.3.5.2 Land use history

Current land use must be recorded as part of the mapping process; land history is also of importance when the area is to be certified under the international statutory standards requirements (i.e. 5-year history). This should include any current or past fertilizer use in the areas where the species are to be harvested, and run-off pollution from towns over the harvesting area and dump sites within the harvesting area. Harvesting should be carried out at least 30 m from agricultural fields, roadsides and dumps, and at least 500 m from urban settlements.

### 7.3.5.3 Harvesting methodology

Once the mapping component is in place, the harvesting methodology has to be developed. This is again a process of monitoring and evaluation. The economic aspects need to be included in the harvesting methodologies, as well as ensuring the continued health of the plant. As an example, for the sustainable collection of indigenous tree seed, it is recommended that the seed is harvested in a plastic hessian sack from the area beneath the tree, removing up to a maximum of 50% of the fallen seed within the season.

### 7.3.5.4 Harvesting levels (quantity)

The botanical assessment must be conducted by a specialist ethnobotanist and recommendations for best sustainable harvesting practice provided. The results of a botanical baseline survey of the wild plant materials to be harvested are mapped and the status of each plant species is defined. Before the first harvest, a calculation must be carefully made to set the threshold of the amount of the plant material harvested from the zoned area and the whole collecting area via a certain number of selected control expanses. The result is mapped.

To calculate the amount to be harvested, the types of harvested parts of plants and their time of reconstitution have to be taken into account. Different formulas can be used for the calculation. The second harvest may be conducted at the same place only after the calculated time of reconstitution. The possible amount to be harvested in a sustainable way is again calculated and compared with the first one. If needed, corrections will be proposed. Successive harvests should be carried out according to former propositions. After a certain number of harvesting periods, recommended by trained botanists employed to monitor the health of the species and habitat, this has to be carefully calculated again and compared with the first calculation. This may lead again to corrections of the harvest. The specific approach for the calculation of the amount to be harvested is decided by the commissioned botanists.

### 7.3.5.5 Noncompliance action

Noncompliances must be adhered to. For example, for the sustainable collection of indigenous tree seed. The removal of seeds still attached to the tree is not permitted; supply containing any seed that is not sufficiently ripe to have fallen naturally from the tree will be rejected and no payment or reimbursement of costs will be given to the supplier. Any damage caused to the trees from harvesting the seed will result in immediate removal of the entire group from the contract agreement with the project.

### 7.3.5.6 Establishing the risk analysis and critical control points

The risk analysis and critical control points are designed to prevent overharvesting or poor harvesting practices that threaten survival of the plant. This aspect is a central component of the sustainability protocols and the monitoring methodology, and is set up as a verification and validation mechanism of these sustainability protocols.

### 7.3.5.7 Verification system

The risk-based management system is a practical tool that attempts to address the conservation concerns and the requirements of the internal control system (ICS) implementation and those of external certification systems. Within this system the protocols must be constructed in a way that can be clearly interpreted by the communities involved. Using critical points along the production chain, the quality of the final product can be improved if extension and management focus is directed at key critical points along the production chain, in 'critical control production points'.

At that point along the chain, if corrective measures are not planned for in advance by the farmer or farmer group, then deterioration of quality can be observed, which directly affects the overall returns to the farmer. When harvesters understand the complete chain they have the ability to plan in advance for possible eventualities or risks and establish certain standards that should be followed, and thus have a better control of their production process and reward.

### 7.3.5.8 Developing the mapping and zoning of the harvesting areas

Further standards and criteria are to be developed by individual wild harvester operations to exclude or minimize possible non-natural influences to the plants to be harvested. Special attention is to be given to distances to roads, industrial areas, floodplains, and so on. The further activities, such as processing, transportation, packing, and so on, have separate standards requirements equal to existing regulations of organic agriculture.

The botanic database and maps, the guidelines for collection, and the knowledge and experiences of the collector groups are a good basis for the further development of a sustainable use of wild plant resources in the region concerned.

### 7.3.5.9 Definition and description of collection areas

The definitions and description of the selected harvesting areas must accord with those officially recognized and established, and include the local names and references known by the resident population/ harvesters.

### 7.3.5.10 Designated harvesting areas and harvesters

There must be a strong relation between the area size and the number of collectors to assure that sustainable harvesting levels can be established without negatively impacting the overall plant population in a given area. A continuity of supply must be assured while avoiding overharvesting.

After the approved map is put on the wall at the registered collection centres, the collectors sign that each delivery is harvested within the approved areas. In order not to disturb the overall ecology of the harvesting area, the harvesters should use certain identified routes to reach the harvesting areas, and risk management protocols should be established to govern movement within these areas.

#### 7.3.5.11 Recording systems

The process starts with harvesting and ends with the market. Working from the market perspective, quality attributes (in all dimensions) that consumers are interested in are identified before production starts. The recording sheets must be maintained by each harvester. The project extension staff must ensure that the harvesters are able and willing to use the sheet. Follow-up extension activities will be necessary to identify and address deficiency (illiteracy or lack of willingness/diligence, non-compliance of the harvesters). The activity calendar is a useful management tool, but also a requirement for certification and must be maintained and updated by the project manager.

Other documentation includes:

- a detailed list of harvester groups;
- critical control points and sanctions;
- internal inspectors' audits;
- harvest area and species lists;
- wild harvest record sheets;
- annual harvesting and activity calendar.

### 7.3.6 Developing sustainable wild harvest system with collector groups using the internal control system

#### 7.3.6.1 Basic requirements for group certification

Organic certification for the wild harvesting groups implies that the:

- Cooperative (central body) is responsible for ensuring the group's compliance with organic standards.
- Organic certification is registered for the entire group and the products identified and managed by the ICS.
- Group members are operating under contractual or binding membership requirements, which specify the commitment to comply with organic standards.
- ICS is managed by the cooperative. Files are maintained, identifying all members of the group.
- Each operation is inspected at least once a year.
- List of all individual member producers is available.
- Non-compliances are dealt with according to set procedures and sanctions.

#### 7.3.6.2 Conversion period

The conversion period for sustainably wild harvested plant materials can be less than 1 year. In the case of non-agricultural enterprises, conversion time is based more on the assessment of the smooth and competent operation of the ICS. The standard 2-year conversion period for organic certification of agricultural systems is thus lowered depending on the adaptability of the producers/harvesters and the ICS staff to manage system adequately.

#### 7.3.6.3 Internal control system development

***Relevance of the internal control system to the management of sustainable wild harvesting protocols***

Under the organic certification mechanism, the harvesters will be responsible for maintaining the ICS, which is a statutory requirement of the organic certification standards form producer groups. The ICS is based on risk assessment, risk management and internal policing of the agreed risk management criteria. There are specific sustainable wild harvest standards under organic certification. In general terms it follows the points made above. They differ slightly between the organic certification bodies, so it will be necessary to check the exact requirements of the standards once the certification body has been selected.

The ICS is a system that guards the integrity of the organic quality of the products generated by the registered producers/harvesters. It is a system in which all persons dealing with the product (harvesters/producers, agents, store keepers and processors, etc.) are identified, registered, instructed on the requirements for organic certification and contracted to ensure compliance. The activities of these persons are then monitored in a system of regular visits and documented control. Besides this, the persons involved are made aware of their common responsibility for the product, which implies a certain social control. In case of deviation from the standards, the ICS has defined rules on action to be taken. The responsibility for the implementation of the ICS lies with association/cooperative. The ICS defines specific the responsibilities of each level in the project. It should provide for transparency, which is clearly documented so that the external inspection agency can easily understand and evaluate the functioning of the system. The personnel involved must be training to perform their duties accurately. The ICS can utilize the traditional community leadership and existing cooperative system. Community headmen and cooperative staff can supervise the ICS in the key roles, explained below. The documentation required for the ICS is the basis for the traceability system for organic certification and for other purposes, as mentioned above. This type of recording system is also a fundamental requirement for managing any commercial supply chain.

#### 7.3.6.4 Documentation

The basic documents, registration form, contract and annual inspection form, are filled in by hand and kept in the hand-written form. The harvester lists are stored on computer and the computer generates forms for the harvester lists and buying records. These are updated prior to every harvesting season. The field officers and the buying officers receive a copy of these lists.

***Yield estimates***

Each season, an average yield estimate is made for each harvesting group depending on the preceding season yields, weather patterns and flowering for the season in question.

*Code numbering*

Each harvester has a unique code number that consists of three sets of letters indicating the division, the group and a three-digit number allocated to him or her within the group.

*Example documentation*

- Internal regulation, which includes:
  - organic production standards equivalent to those in country of product destination;
  - rules of participation for new members, including procedure for conversion to organic;
  - procedure for excluding individual operators in case of violation of standards;
  - all producers must be informed of the internal regulations.
- Structure of producer organization:
  - organization chart.
  - constitution.
  - other activities of the organization.

*Producer lists*

Includes producer's code, full name of harvester, location, entrance date, total surface area of organic harvesting, amount delivered in past year, harvest estimation for next year, name of internal inspector, date of internal inspection, result of internal inspection:

- names, qualifications and responsibilities of ICS staff;
- list of members being sanctioned;
- completely filled in internal inspection report forms;
- updated maps;
- product flow – traceability from land to final product.

The following documents must be available for each member:

- formal commitment of harvester members to fulfil internal standard (written contracts);
- member entrance form – harvesting data sheet;
- harvesting records – harvested quantities, post harvest procedures;
- maps;
- annual inspection checklist.

*Internal inspection*

Each contracted harvester is visited once a year by the field officer during the harvesting season and the visit is recorded in the internal farm inspection forms. The farmers are inspected for compliance to organic standards. In case of any deviations, these farmers are reported to the sanction committee for sanctions.

*Project staff*

- Project manager
- Project supervisor
- Sanction committee
- Executive committee
- Processing manager
- Sales manager
- Field officer
- Secretary
- Treasurer
- Groups representatives.

*Flow chart showing the internal control system structure*

This is required as a visual reference. It identifies the main actors and flow of responsibility for the management of documentation and risk assessment within the ICS structure.

*Traceability at the operations office*

The following files must be kept which relate to the ICS:

- the map showing all harvesting areas;
- copies of entrance form;
- harvester's list;
- copy of harvester's contracts;
- internal inspection reports;
- store registration forms;
- store declaration forms;
- buying records;
- sales records;
- staff contracts;
- organization's operational manual.

Apart from these, the bookkeeping of the project operating costs must also be kept at the project office, as well as a file with the minutes of management committee meetings and a file of correspondence with the certification body.

*Traceability responsibilities of the field officer*

The field officers are responsible for maintaining the following:

- copies of the maps of their villages;
- copies of the harvesters list.;
- buying records (during buying season).

#### 7.3.6.5 Risk management

The chain of custody and associated possible risks are listed in Table 7.2, and Table 7.3 lists the major potential risks.

*Sanctions for deviation*

A harvester who is found to have violated the protocols due to intent is expelled from the project and their contract revoked. A harvester found guilty of violating the organic standards due to intent will predispose all of their group to the risk of exclusion from the project. Harvesters are well informed of the potential risks leading to their expulsion from the project and are encouraged to report any anomaly that is taking place or will take place within their group members. Mild sanctions, such as extending the conversion period, are made to harvesters who could have violated the organic standards by ignorance.

Table 7.2 Chain of custody and possible risk

| Harvesting | Farm storage → | Transport → | Buying centre → |
|---|---|---|---|
| Contamination<br>Pests and predators<br>Equipments/tools<br>Less skills<br>Overharvesting | Contamination<br>Poor or lack of labelling<br>Pests/predators | Commingling<br>Spillage<br>Contamination<br>Adulteration | Poor records<br>Low-skilled buyers<br>Contamination<br>Labelling and coding<br>Lots separation<br>Inadequate ICS staff |
| **Market** | **Warehouse ←** | **Processor ←** | **Transport ← ←** |
| Poor prices<br>Non-compliance with industrial standards (GC)<br>Loss of organic certification compliance<br>Loss of buyer<br>Thefts<br>Adulteration or reduction in quality, loss of certification due to poor transportation environment/methods | Mixing<br>Odour and physical contamination<br>Lack of labelling<br>Lack of epoxy-resin-lined containers<br>Poor storage conditions<br>Batch lots mixing<br>Improper records<br>Spillage<br>Oxidization | Contamination<br>Poor records<br>Inadequate ICS staff<br>Inadequate equipment/tools<br>Poor distillation techniques<br>Poor cleaning methods<br>Grading<br>Contamination | Mixing<br>Spillage<br>Lack of labelling<br>Contamination<br>Delays in delivery |

Table 7.3 Major potential risks

| Possible risks | Solutions |
|---|---|
| Harvesting in the parks and restricted areas | Restricting the membership of the groups in only the designated areas<br>Product rejection at the buying centres if harvest from outside the designated areas |
| Contaminated harvesting tools | Use clean harvesting equipment; field officer to do spot checks during harvesting season; sanctions |
| Use of inappropriate bags | Use of only recommended bags allowed under the certification system<br>Product rejection at the buying centres if presented in inappropriate bags |
| Unsuitable, unsustainable harvesting methods | Retrain harvesters in the sustainability protocols<br>Only stipulated procedures will be certified<br>Field officers to do spot checks during harvesting. |
| Contamination during storage | Clean storage area; field officer to do spot inspection<br>Training on proper honey handling |
| Adulteration | Intensify inspection and penalties<br>Product rejection at the buying centres if adulterated<br>Sensitization on importance of harvesting, storage and handling protocols |

#### 7.3.6.6 Due diligence procedures

The orientation and training needs of the communities/harvesting groups involved must be first assessed before the sustainable harvesting programme is started. Management and organizational support areas will include:

- organizing the community into self-managed and well-organized producer groups;
- developing the technical capacity of communities for sustainable harvesting according to the protocols developed;
- developing ethical community benefit sharing methodology;
- developing the recording systems indicated below;
- arranging the transparent financial handling mechanism to accord with cooperative principles and ethical trade requirements.

#### 7.3.6.7 Utilizing the internal control system traceability systems

Due diligence can be greatly enhanced through developing the ICS management capacity. The elaborate but simple systems provide excellent transparency and traceability for organic and fair-trade certification, for monitoring and evaluation of the conservation impact of the sustainable wild harvest systems, as well as meeting the traceability requirements of international buyers.

*Harvesting procedure*

The harvesting procedures to be followed are:

- Maximum volume of material to be harvested is estimated for all the harvesters prior to the start of every harvesting period and overseen by the field officers.
- All harvesting groups have received the necessary level of training.
- All harvesting groups have received information guides (laminated pictorial sheets).
- Before harvesting is done, harvesting materials, including bags and harvesting kits, are cleaned thoroughly. This is confirmed by field officers after doing spot checks.

*Buying procedures*

Buying days are announced by the field officers. On the buying day, the following must be at the buying centre:

- resident field officer;
- executive committee member.

The membership card shows the harvesters name, group name and code number.

The purchase officer and the committee member(s) confirm that the delivered amount of the plant material is reasonable by comparing the actual delivery with the estimate (noted in the buying record by the internal inspector at latest 1 week before delivery). If there is any doubt about the quantity, the plant material is kept apart until the ICS manager has checked with the respective harvester and has indicated whether or not to allow the produce into the organic supply chain.

The quantity of plant material delivered is recorded in the buying record with a statement.

A cash voucher (receipt) is completed. The vouchers will be stamped 'organic' or 'in-conversion'. The quantity delivered, the number of the receipt and the date of delivery are noted down on the buying record and later entered in a computerized produce control sheet. All plant material delivered by the harvester to the collection centre must be confirmed by themselves by signing against their submission and receipts on that date. Monitoring deliveries will be based on yield estimate and confirmed by the executive committee members and resident field officer. Extra submission up to 10% of estimated yield will be tolerated; beyond that, they must receive confirmation from the field officer and excess deliveries forms filled in. The form will then be submitted to the project manager.

*Storage and handling procedures*

Organic plant material will be stored separately at all times from non-certified plant materials and any other other products, to ensure that there is no possibility to mix conventional and organic material at any one time.

The field officer checks all storage areas and ensures that there is no possibility of contamination. When the fresh plant material is harvested it is taken to the designated depot centre within the same day. This depot is registered and laid down in the depot store declaration form as complying with the handling procedures. Containers/bags used for transportation of the plant material must not be used for any other purpose, to prevent contamination. Wild thyme and helichrysum can be packaged in bags designed to reduce condensation inside the bag and the consequent risk of fungal or bacterial growth (i.e. plastic polypropylene/hessian bags are most suitable). The best method of post-harvest handling of wild thyme will be as for cultivated thyme.

*Processing procedures*

All processing is done at the designated depot/central processing centre. Harvesters do not at any one time do any form of processing outside of these controlled environments.

*Financial handling*

This has to be conducted transparently through the use of recording systems. These have been well designed by fair trade organizations and companies such as Traidcraft, Share Interest and FLO and may provide a useful tool even if fair trade certification is not sought. Government auditing assistance is very helpful for third-party verification, and is usually conducted annually.

### 7.3.6.8 Training and extension

Achieving sustainable harvesting of indigenous/wild plant species can only be achieved if the harvesters understand the need for the systems, gain benefit from them and receive sufficient training and extension assistance in keeping them up. The endorsement of these systems by third-party verification, such as organic certification, enabling the producers to access more rewarding markets, is also a fundamental part of ensuring that the sustainable wild harvesting protocols are well managed and maintained into the future. A full training programme will need to be developed which will be enacted as a specific training session over the year and as an ongoing facility for the wild harvesting groups. The training will include all relevant staff and extension and training officers.

1. Ensure that the harvesters are fully aware of the species that is required:
   i. correct identification of plant;
   ii. correct part of the plant to harvest;
   iii. correct time of year to harvest;
   iv. mapping, zones and managing the zonal area;
   v. defining boundaries, non-intervention and buffer areas;
   vi. practical sustainable techniques for harvesting;
   vii. practical negative impact of damages to ecosystem on environment, ecology and livelihoods.
2. Organic certification:
   i. purpose of organic certification and market relationship;
   ii. certification requirements for sustainable wild harvesting under the EU/US-NOP regulations;
   iii. sustainable harvesting protocols and certification;
   iv. developing the ICS;
   v. managing the ICS;
   vi. preparation for certification.

3. Product quality and market entry:
   i. quality management system (harvesting, handling, storage) and associated protocols;
   ii. managing the recording systems to meet buyer traceability requirements;
   iii. correct processing techniques.

These training areas will need to be developed progressively as time-framed activities, combining practical demonstration, lectures and group work, demonstration and learning visits to other similar commercial sustainable wild harvesting programmes.

The training will need to be supported by sensitization/educational /training materials, including simple laminated pictorial leaflets explaining the sustainable wild harvesting methods for each of the selected species, posters and educational videos.

## 7.4 Developing indigenous plant-based enterprises as viable businesses with developing country communities

### 7.4.1 Success through partnership

To realistically address the longer term livelihoods development, developing plant-based enterprises into viable businesses that are both environmentally and socially supportive cannot be a single-handed operation; rather, it requires a tightly bound partnership approach, endorsed through contractual agreements with its research, development and commercial partners that add value and bring in different skills, experiences and competences. Development of trained extension services in plant-based products and value-addition technologies is in its infancy in the ASAL. Government and NGO extension staff in the rural areas are potentially powerful agents for supporting viable agri-business development. Not only in creating enabling legal and policy environment for smallholder market development, they can also play supportive roles in agricultural extension, business development services, market information and development, technical and management training, facilitating formation and legal recognition of marketing entities, access to credit and other public services (Wren, 2012).

Private-sector actors play a key role in supply chain and market development. Their presence increases the confidence of the producer; they can provide precise market information that can be translated in technical and management training and guidance; some provide market guarantees and access to trade finance.

### 7.4.2 Building lasting capacity

The first important investment is to raise awareness of all relevant actors and stakeholders of the benefits of conservation/bio-enterprises from a livelihood and environmental point of view, what is involved in building the capacity of the producer groups in relation to production, quality control, business and management. If communities and producers are assisted to go through the process of understanding and making business decisions, they will drive forward the enterprise as their own, and therefore its future is more secure.

Developing a pilot phase of indigenous plant product enterprise support will enable the correct and appropriate facilities and equipment to be installed/constructed, planting materials of the correct genotype to be sourced and the producer groups, extension staff and operators to be sufficiently trained before the enterprises embark on the full business development. Commercial partners should be engaged under transparent and equitable agreements and operational partnerships secured with relevant NGOs and government departments.

Through feasibility studies and business planning processes, the plant products identified as potentially viable should be selected on the basis of their commercial value, their availability, suitability for sustainable wild harvesting, potential for domestication, conservation value to the region, and so on. Commercial trials must be undertaken for the domesticated species before full commercial production is considered.

As all the indigenous plant products business activities have to be environmentally and socially supportive, the structures and systems not only need to ensure efficient and effective running of the operations, but also compliance with the transparency and traceability requirements of international certification standards and of the buyer/market. Producer groups will need to establish ICSs that ensure transparency and traceability of the products. These aspects are fundamental if products are to reach premium markets based on organic and ethical trade certifications. The assessment of the sustainability threshold for wild harvested plant products, according to certification protocols, the organization and structuring of the producer groups and the education of the wild harvesters as to the criteria of a sustainable commercial operation are also focal areas of indigenous-plant-based enterprises (Wren, 2012).

Depot centres will be required in strategic positions for the collection, payment and semi-processing of the raw materials. A central processing facility is necessary for achieving uniform products that meet the market criteria and can maintain consistent market supply; and also for bulking and final processing, packing, branding, marketing and exporting the finished products. This facility will provide final quality control, enabling the product to reach international premium markets.

Monitoring tools that have been established by a coalition of international development organizations, such as the Cristal, Community-based Risk Screening Tool – Adaptation & Livelihoods, provides a useful stating position. Baseline data should be established so that monitoring and qualitative and quantitative evaluation will be conducted on an annual basis. Monitoring and evaluation methods that incorporate communities involvement is shown to have positive results, leading to greater empowerment and responsibility at community level. Take-up of directly related activities that address concerns arising from data evidence is stronger.

### 7.4.3 Key factors in enabling appropriate selection of indigenous plant product enterprises for producer groups

The enterprise types recommended for future development will need to have potential to provide low risk and medium to high return for the participating communities and attractive returns within a relatively short term from inception. Viable opportunities for developing sustainable wild harvest operations for rural communities are also focal areas for investment in providing tangible reasons for these communities to maintain the future welfare of these natural resources. The majority of indigenous-plant-based product enterprises do not require high capital cost to equip and operate, and involve value-addition activities that enable equitable income-generating activities for both genders at the rural community level.

Initiatives to assist the development of indigenous plant product enterprises must be based on business principles, substantial enough to stimulate sufficient economies and scale for viable businesses to operate and achieve sustainable livelihoods impact. The main relevant business principles are identified as follows.

#### *Competitive advantage*

The main factors of competing in the market are the price and the quality. Well-presented and processed/formulated products, the price setting, capitalizing on the intrinsic condition, branding and marketing the storyline also provide competitive advantage.

#### *Economies of scale*

In order to enter the marketplace, producers need to first identify the initial minimum supply levels. It is then necessary to achieve the economies of scale that maintain the market position/open up greater markets, to ensure equipment is efficiently utilized and the overhead, loan servicing and investment requirements of operating a developing business are covered by adequate turnover.

#### *Ability to meet market forces*

To ensure the viability of the enterprise (i.e. annual profit made over the variable costs incurred in the season) the fostering of long-term trading relationships is a central factor, from the large-scale to the small-scale operator. To succeed, producer groups must be in a position to make at least a basic business plan. For the producer to reach the position of export and then to gain a foothold in the international market, the two central requirements are consistent quality and quantity. Business support to producer groups will be necessary; that is, business assessment, feasibility studies, cost price calculations, development of recording and accounting systems, including ICSs. This will enable producer group managers to make informed decisions and run more efficient businesses. By providing specific and tailored technical training, advice and information, the production and primary processing expertise of producer and harvester groups will improve, as well as their income-generating potential. The key success of these enterprises is that they hold the interest and drive of participation communities, to be low risk and achieve sufficiently attractive returns to producers within a short time from inception.

#### *Producer/operator capacity*

A careful assessment must be made of the organizational and management ability of the producer group/association/cooperative and commercial operator. The majority of the project effort will need to be placed in developing well-organized, structured and technically competent groups/cooperatives, and then in the relevant technical skills development. This input is not technically demanding, but requires significant manpower and time. A programme of training is necessary to enable the producers to operate efficiently, sustainably and viably to achieve market entry and sustain the business, and for replicable models to emerge (including securing affordable loans for the expansion of the bio-enterprises).

#### *Management challenges of organizing producer groups and community-owned micro-enterprises*

These are clearly demonstrated by the studies conducted by the programme partners and have highlighted the importance of active and well-focused extension facilities. The associated cost can be significant; and these risks need to be reduced for the private sector to be attracted to invest in bio-enterprises in such remote regions and develop viable and sustainable business operations. The programme will therefore provide key services (training, extension, certification, etc.) and some of the infrastructural costs necessary to kick-start the bio-enterprises and to reach a viable point where commercial partners can viably engage.

#### *Investment capital*

Developing new enterprises, particularly community based, means that producers need to secure necessary investment capital to fulfil the infrastructure, training and management needs of the start-up phase, which is critical to developing a successful and replicable bio-enterprise model.

### 7.4.4 Summary of key activities required to develop indigenous plant product enterprises with producers groups

- *Ensuring understanding of local context and information needed before decisions can be made*:
  - exchange visits;
  - field days.
- *Central methodologies*:
  - Pilot period.
  - Demonstration and trials; capacity-building the skills and awareness of extension network and producer group leaders.
  - Developing the network of physical infrastructure; adequately equipped.
  - Utilizing the ICS as a method of creating a vertically and horizontally strong management and communication network. A highly adaptive value-chain mechanism, as well as an international requirement for producer group organic and ethical trade certification.
  - Integral financial handling systems, trade finance, rural micro-savings and credit facilities.
- *How to start the process of identifying indigenous plant products for enterprises*. Feasibility studies and business plans to assess potential plant-based enterprises on realistic commercial grounds and against conservation and social livelihood development goals. This should include the main 'opportunity cost' areas:
  - extension, demonstration and training;
  - infrastructure and physical value addition;
  - non-physical value addition – certification;

- strengthen supply-chain efficiency – including encouraging high community participation and promoting gender mainstreaming.
- *How to assist communities to decide which enterprises and why?*
  - cost–benefit and business planning exercises;
  - production trials.
- *How to assist communities identify ways to overcome hurdles?*
  - ICS risk assessment approach.
- *How to assist communities to mobilize themselves.* Training and extension support to capacity-building producer groups in:
  - organization;
  - technical skills;
  - business and financial management;
  - development of area management units.
- *Quality control and monitoring*:
  - ICS, as explained above;
  - sustainable wild harvest protocols;
  - trainer extension teams, lead farmers;
  - technical information materials in indigenous languages, training courses, demonstration, field days.
- *Actions to engage partners*:
  - Regional seminar–workshop to discuss challenges, opportunities, realistic requirements for supporting and engaging in alternative livelihood approaches.
  - Introduction and sensitization meetings held with all potential stakeholders and actors in the target areas.
  - Key partners involved in discussion making processes through the area management units, in training sessions. Information materials distributed.
  - Contractual agreements/memoranda of understanding to form strategic alliances with research, development, commercial partners.
- *Monitoring and evaluation parameters to consider*:
  - level and speed of uptake of the business skills and technologies;
  - impact on the household food security and livelihoods of the activities;
  - increase in environmental resilience to climate change;
  - increase in social and livelihood resilience to climate change;
  - increase women's and other groups' empowerment;
  - adapting activities to issues that arise.

### 7.4.5 Cross-cutting factors in the developing indigenous plant product enterprises

Examples of cross-cutting factors in the developing indigenous plant product enterprises are listed in Table 7.4.

**Table 7.4** Examples of cross-cutting factors in the developing indigenous plant product enterprises

| Business criteria | Main factors | Actions |
|---|---|---|
| Economically sound | Critical partnerships<br>Competitive advantage<br>Economies of scale<br>Assessing export viability<br>Sustaining the market position<br>Producer/operator capacity | Baseline data collection<br>Professional input on opportunity assessment (product development, branding, marketing)<br>Feasibility study and basic business plan<br>Assessment of the potential for certification<br>Micro-credit for individual investment in scaling up<br>Grant and loan support to the producer group to develop purchasing and processing capacity, and trade finance<br>Memoranda of understanding between the producer groups/association/cooperatives and the development, research, government and commercial partners |
| Economic advancement of women | Micro-credit and savings facility<br>Ongoing training and extension services<br>Organization and business training<br>Sensitization of communities to the role of women in business<br>Sustainable wild harvesting business opportunities<br>Value-addition involving processing of raw materials<br>Organized trading associations | Should be proactively promoted and encouraged at the community level as well as being endorsed as part of the commercial business approach<br>Developing the constitution and bylaws of the producer groups working with the enterprises<br>Training in improved techniques that are appropriate for women's participation and adequate to achieve the quality expected in target markets<br>Women retain their incomes for their efforts through small accounts where sale can be directly placed |
| Pro-poor | Micro-credit and savings facility<br>Ongoing training and extension services<br>Organization and business training<br>Organized trading associations inclusive of all sections of society<br>High-value cropping that can be grown on small land sizes<br>Sustainable wild harvesting business opportunities<br>Value-addition involving processing raw material | Micro-credit and savings facility<br>Training of producer groups in technical, management and financial aspects<br>Management and organizational capacity building<br>Sustainable wild harvest protocols and regulation through agreement based on compliance with these protocols<br>Construction of the depot and processing centres<br>Developing demonstration sites<br>Trade finance and capital to develop the business |

## Acknowledgements

This chapter has been compiled as a result of invaluable interaction with AMMPS, FairWild and many indigenous plant products initiatives within developing countries and within international markets. I am extremely grateful for the opportunities and the insights that I have gained, and to the development agencies and NGOs that have invited me to assist indigenous plant products collectors/producers, processors, traders and exporters over the past 20 years. I give my sincere thanks to Dr Ameenah Gurib-Fakim of the Centre for Phytotherapy Research, Mauritius, for her invitation to contribute to this publication. I greatly admire her undiminished enthusiasm and energy in enhancing the world knowledge on medicinal plants utilization and conservation. Much of the material for this chapter has been drawn from published data and, as such, I would specifically like to acknowledge the formative and highly influential work of the IMO, the FairWild Foundation and TRAFFIC in the development of guidelines, standards and certification systems for the sustainable utilization of indigenous plants across the world.

## References

Buckle, J. (2003) *Clinical Aromatherapy: Essential Oils in Practice*, 2nd edn, Churchill Livingstone, Edinburgh.

CBD (2000) *Ecosystem approach*. Secretariat of the Convention on Biological Diversity, Montreal, Canada, COP5 Decision V/6, www.biodiv.org/decisions/default.aspx.

CBD (2002a) *Global strategy for plant conservation*. Secretariat of the Convention on Biological Diversity, Montreal, Canada, COP6 Decision VI/9, www.biodiv.org/decisions/default.aspx

CBD (2002b) *Access and benefit sharing as related to genetic resources*. Secretariat of the Convention on Biological Diversity, Montreal, Canada, COP6 Decision VI/24, www.biodiv.org/decisions/default.aspx.

CBD (2004) *Sustainable use (Article 10)*. Secretariat of the Convention on Biological Diversity, Montreal, Canada, COP7 Decision VII/12, www.biodiv.org/decisions/default.aspx.

Duerbeck, K. (2008) Sustainable wild collection: requirements for resource management and resource use for natural ingredients. Abstracts of The World Congress on Medicinal and Aromatic Plants, Cape Town, November. *African Journal of Traditional, Complementary and Alternative Medicines (AJTCAM)*, http://journals.sfu.ca/africanem/index.php/ajtcam/article/view/670.

EC (2012) Commission implementing regulation EU 508/2012, amending Regulation (EC) No. 1235/2008 laying down detailed rules for implementation of Council Regulation (EC) No. 834/2007 as regards the arrangements for imports of organic products from third countries, *Official Journal of the European Union*, 21 June 2012, http://eur-lex.europa.eu/LexUriServ/LexUriServ.do?uri=OJ:L:2012:162:0001:0045:EN:PDF.

Kathe, W., Pätzold, B., Leaman, D. *et al.* (2010) *Wild For A Cure: Ground-Truthing, a Standard for Sustainable Management of Wild Plants in the Field*, Traffic International, Cambridge, UK.

Schippmann, U., Leaman, D. and Cunningham, A.B. (2006) A comparison of cultivation and wild collection of medicinal and aromatic plants under sustainability aspects, in *Medicinal and Aromatic Plants: Agricultural, Commercial, Ecological, Legal, Pharmacological and Social Aspects* (eds R.J. Bogers, L.E. Craker and D. Lange), Wageningen UR Frontis Series, Vol. 17, Springer, http://library.wur.nl/frontis/medicinal_aromatic_plants/06_schippmann.pdf.

Setboonsarng, S. (2008) Can ethical trade certification contribute to the attainment of the millennium development goals? A review of organic and fair-trade certification, Discussion Paper 115, ADBI, Tokyo.

Tisserand, R. and Balacs, T. (1995) *Essential Oil Safety: A Guide for Health Professionals*,. Churchill Livingstone, Edinburgh.

Tolman, B., Nguyen, M. and Timoshyna, A. (2012) TRAFFIC pilots sustainable medicinal and aromatic plant harvesting project in Viet Nam, www.iucn.org/about/union/commissions/sustainable_use_and_livelihoods_specialist_group/sulinews/issue_2/sn2_vietnam (accessed 5 November 2013)

Twarog, S. (2006) Organic agriculture: a trade and sustainable development opportunity for developing countries, in *Trade and the Environment Review, 2006*, UNCTAD/DITC/TED/2005/12, United Nations, New York, pp. 141–180.

UN (2000) United Nations Global Millennium Development Goals 2000, www.unaids.org/en/aboutunaids/unitednationsdeclarationsandgoals/2000millenniumdevelopmentgoals/

UNCED (1992) *Agenda 21*, United Nations action plan on sustainable development. Outcome of the United Nations Conference on Environment and Development (UNCED) held in Rio de Janeiro, Brazil, in 1992, http://sustainabledevelopment.un.org/content/documents/Agenda21.pdf.

WHO (2003) *WHO Guidelines on Good Agricultural and Collection Practices (GACP) for Medicinal Plants*, World Health Organization, Geneva.

WHO, IUCN and WWF (1993) *Guidelines on the Conservation of Medicinal Plants*, IUCN, Gland.

WHO, IUCN, WWF and TRAFFIC (2013) *Revised Guidelines on the Conservation of Medicinal Plants*, in press.

Wren, S.A. (2007) Conservation non timber forest products opportunity assessment of the Samburu Heartland, AWF report.

Wren, S.A. and Powys, A. (2008) A feasibility study to establish the potential for bio-enterprise development in Laikipia. LWF report.

## Further reading

EPA definition of certified organic: www.epa.gov/agriculture/torg.html, www.epa.gov/nrmrl/std/lca/resources.html, www.ifoam.org/about_ifoam/pdfs/PGS_PDFs/Studies_Book_Web_20091030ILB.pdf.

IFOAM family of standards: www.ifoam.org/about_ifoam/standards/family_of_standards/familiy_of_standards.html.

European Commission. Organic farming: questions and answers: www.ifad.org/evaluation/public_html/eksyst/doc/agreement/pl/organic.htm.

TRAFFIC, www.traffic.org/home/2011/10/19/conservation-of-medicinal-plants-top-of-who-agenda.html.

Wren, S.A and Speranza, C.I. (2010) The struggle to diversify rural livelihoods: bio-enterprise development initiatives and their impacts on agro-/pastoral communities in the drylands of Kenya. *European Journal of Development Research*, Special edition, 'New avenues for pastoral development in sub-Saharan Africa', **22**, 751–769.

Wren, S.A. (2012) Socio-economic and livelihood impacts of environmentally supportive bio-enterprise development for the agro-/pastoral communities in Samburu Heartland, Kenya, University of Plymouth, UK, http://hdl.handle.net/10026.1/1093.

# 8 European Union Market Access Categories and Regulatory Requirements for Novel Natural Products

Thomas Brendler[1] and L. Denzil Phillips[2]

[1]Plantaphile Ltd, Collingswood (USA), Eastbourne (UK), Berlin (Germany)
[2]Denzil Phillips International Ltd, Richmond (UK), Sinsheim (Germany)

## 8.1 Introduction

While Asia may have the longest tradition in herbal medicine, Europe is still the largest consumer of commercially manufactured herbal medicinal products (HMPs) in the world. More than 300 different plant species are regularly used in European herbal products today. Europe has been formulating and manufacturing HMPs since the Middle Ages. At this time monasteries were not only centres of knowledge about herbal medicine but also places were treatments were provided and herbal remedies produced. Modern companies like Germany's Kneipp Werke and Klosterfrau were built around such monastic traditions.

Europe is also a global leader in the formulation, extraction and manufacture of phytomedicines and plant-based fragrances and flavours and cosmetics. Many of the world's leading brands and manufacturers of herbal medicines, such as Schwabe, Bioforce, and Arkopharma, are European, and the same holds for natural cosmetics, where, for instance, Yves Rocher, Body Shop and Dr Hauschka are global players. Furthermore, some of the most important 'alternative' systems of medicine, such as homeopathy, naturopathy and anthroposophical therapies, were invented by Europeans.

This long and extensive tradition of the manufacture and use of herbal medicines and natural products has also given rise to a long and intensive interest in the safety and efficacy of such products. As a result, a well-established and complex set of rules and regulations have existed in many European countries in connection with the manufacture, distribution and sale of medicines, cosmetics and food products.

The formation and expansion of the European Union in the late 20th century into what is now the world's largest trade block posed enormous challenges to regulators of food, medicines and beauty care products in Europe. Most sovereign nations in Europe still believe that the health and safety of its citizens, like military defence, is a national matter and not one that can be handed over in its entirety to the European Union (EU) in Brussels. Most of the recent legislations that are outlined in this chapter have been the result of sometimes acrimonious struggles between those who wish to see uniformity within the EU and those who want to ensure that national rights concerning health and safety are maintained. Moreover, the speed by which each EU nation has adopted such legislation differs widely, with the result that regulatory approval is often much faster and easier in some member states than in others.

The complex set of regulations that now apply throughout Europe to regulate the manufacture and sale of herbal products are often difficult for outsiders to understand and conform to. These regulations are often considered as a barrier to trade with non-EU member countries. Once all the presently planned regulations for drugs, foods and beauty care products are fully operational throughout the EU, they, however, provide a logical and comprehensive set of rules to protect the health and safety of EU citizens in the 21st century.

## 8.2 Raw materials

Less than 10% of the 750 000 plant species recorded are used for health or beauty care purposes. While the World Health Organization (WHO) estimates that 20 000 medicinal plants are used in herbal medicine, only about 700 of these are used commercially; 75% of these are found in Europe or Asia (see Figure 8.1).

Some of the 70 most important plant materials traded globally for medicinal and beauty care purposes are listed in Table 8.1.

### 8.2.1 Plant material and extracts, essential oils, gums and resins

#### 8.2.1.1 Introduction

Botanical raw materials can be traded/exported/imported without specific requirements to safety and quality. It is only when botanical ingredients are processed into finished products (see below) that stricter requirements apply. These are, however, specific to the category of finished product and are laid out further below.

Intermediate products of herbal origin intended for further processing by an authorized manufacturer or for the preparation of a

*Novel Plant Bioresources: Applications in Food, Medicine and Cosmetics*, First Edition. Edited by Ameenah Gurib-Fakim.

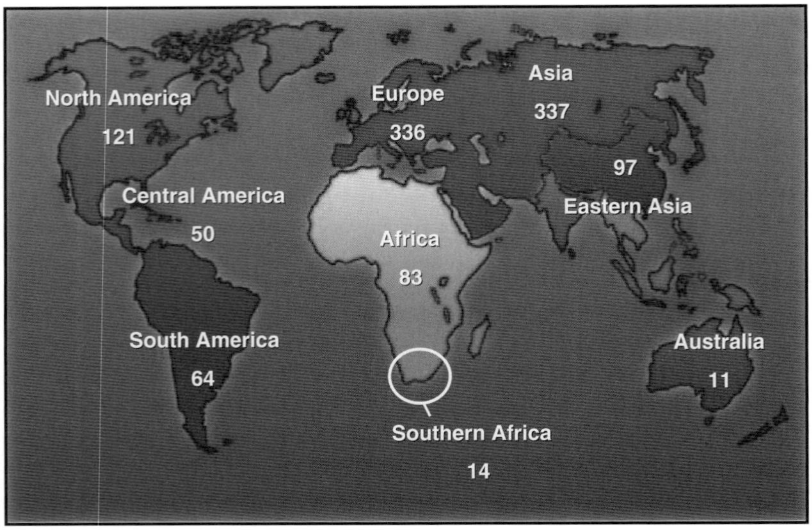


Figure 8.1 Continental diversity of commercialized medicinal plants. Courtesy Ben Erik Van Wyk, WOCMAP 2009

Table 8.1 Selection of the 70 most important medicinal plants traded globally

| | | | |
|---|---|---|---|
| Acacia | Emblica | Licorice | Schisandra |
| Pelargonium | Ephedra | Milk thistle | Senna |
| Aloe | Garlic | Olive | St John's wort |
| American ginseng | Gentian | Peony | Tamarind |
| Angelica | Ginger | Psyllium | Taxus |
| Artemisia | Ginkgo | Rauvolfia | Valerian |
| Astralagus | Grape (seed) | Rhubarb | Warburgia |
| Calendula | Hawthorn | Rosemary | Withania |
| Cinnamon | Horse chestnut | Salvia | Willow |
| Echinacea | Korean ginseng | Saw palmetto | |

formula do not fall under the definition of a finished product and are, therefore, not subject to marketing authorization as such. Nevertheless, in order to guarantee their constant quality, these intermediate products have to be controlled. The regulations in member states are described as follows.

Intermediate products are handled like all other raw materials intended for medicinal, food or cosmetics products. Good manufacturing practices (GMPs) rules must be followed. Specific storage conditions must be maintained according to the nature and origin of the material (e.g. proper temperature, well-ventilated area, dry, regular control for animals and insects).

The assessment of products based on plants non-native to Europe must comply with the European Community legislation. If preparations of plants non-native to Europe are marketed as finished products, they are regarded as 'new entities' and must fulfil the complete requirements concerning quality, safety and efficacy, particularly in terms of toxicological data and clinical documentation.

### 8.2.1.2 Category definitions

*Herbal substances:* all mainly whole, fragmented or cut plants, plant parts, algae, fungi, lichen in an unprocessed, usually dried, form, but sometimes fresh. Certain exudates that have not been subjected to a specific treatment are also considered to be herbal substances. Herbal substances are precisely defined by the plant part used and the botanical name according to the binomial system (genus, species, variety and author).

*Herbal preparations:* these are obtained by subjecting herbal substances to treatments such as extraction, distillation, expression, fractionation, purification, concentration or fermentation. These include comminuted or powdered herbal substances, tinctures, extracts, essential oils, expressed juices and processed exudates.

### 8.2.1.3 Overlapping documentation requirements

The overlapping documentation requirements are shown in Figure 8.2.

#### *Phytosanitary certificates*

Phytosanitary certificates are issued to indicate that consignments of plants, plant products or other regulated articles meet specified

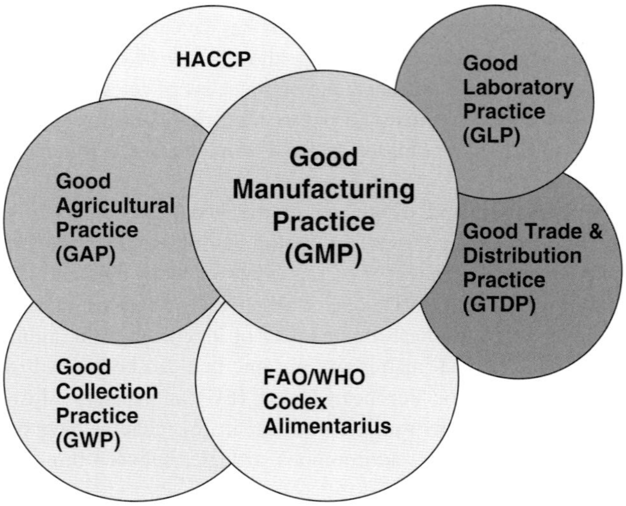


Figure 8.2 Overlapping document requirements

phytosanitary import requirements and are in conformity with the certifying statement of the appropriate model certificate.

Importing countries should not require phytosanitary certificates for plant products that have been processed in such a way that they have no potential for introducing regulated pests, or for other articles that do not require phytosanitary measures.

#### *Convention on International Trade in Endangered Species of Wild Fauna and Flora*

The Convention on International Trade in Endangered Species of Wild Fauna and Flora (CITES) is an international agreement between governments. Its aim is to ensure that international trade in specimens of wild animals and plants does not threaten their survival. CITES works by subjecting international trade in specimens of selected species to certain controls. All import, export, re-export and introduction from the sea of species covered by the convention has to be authorized through a licensing system. Each party to the convention must designate one or more management authorities in charge of administering that licensing system and one or more scientific authorities to advise them on the effects of trade on the status of the species.

A specimen of a CITES-listed species may be imported into or exported (or re-exported) from a state party to the convention only if the appropriate document has been obtained and presented for clearance at the port of entry or exit. There is some variation of the requirements from one country to another and it is always necessary to check on the national laws that may be stricter, but the basic conditions that apply are described below.

#### *Appendix I plants*

An import permit issued by the management authority of the state of import is required. This may be issued only if the specimen is not to be used for primarily commercial purposes and if the import will be for purposes that are not detrimental to the survival of the species. In the case of a live animal or plant, the scientific authority must be satisfied that the proposed recipient is suitably equipped to house and care for it. An export permit or re-export certificate issued by the management authority of the state of export or re-export is also required. This is issued only if the specimen was legally obtained, the trade will not be detrimental to the survival of the species and an import permit has already been issued.

A re-export certificate may be issued only if the specimen was imported in accordance with the provisions of the convention and, in the case of a live animal or plant, if an import permit has been issued.

#### *Appendix II plants*

An export permit or re-export certificate issued by the management authority of the state of export or re-export is required. An export permit may be issued only if the specimen was legally obtained and if the export will not be detrimental to the survival of the species. A re-export certificate may be issued only if the specimen was imported in accordance with the convention. In the case of a live animal or plant, it must be prepared and shipped to minimize any risk of injury, damage to health or cruel treatment. No import permit is needed unless required by national law.

The convention allows or requires parties to make certain exceptions to the general principles described above, notably in the following cases:

- for specimens in transit or being transhipped;
- for specimens that were acquired before CITES provisions applied to them;
- for specimens that are personal or household effects;
- for plants that were 'artificially propagated';
- for specimens that are destined for scientific research.

There are special rules in these cases and a permit or certificate will generally still be required.

#### *Major medicinal plants listed under CITES Appendix II*

| | |
|---|---|
| • *Adonis vernalis* | Adonis |
| • *Aloe ferox* | Cape Aloes |
| • *Aquilera malaccensis* | Agarwood |
| • *Bletilla striata* | Bletilla hyacinth |
| • *Cibotium barometz* | Vegetable lamb |
| • *Dendrobium nobile* | Dendrobium |
| • *Discorea deltoidea* | Wild yam |
| • *Gastrodia elata* | Gastrodia |
| • *Guaiacum officinale* | Pockwood tree |
| • *Hydrastis canadensis* | Goldenseal |
| • *Nardostachys grandifolia* | Spikenard |
| • *Picrorhiza kurroa* | Kutki |
| • *Panax quinquefolius* | Ginseng |
| • *Podophyllum hexandrum* | Himalayan may apple |
| • *Prunus africana* | Pygeum |
| • *Pterocarpus santalinus* | Sandalwood |
| • *Rauvolfia serpentina* | Rauwolfia |
| • *Taxus wallichiana* | Yew. |

#### *Hazard Analysis Critical Control Points*

The hazard analysis critical control point (HACCP) is a systematic preventive approach to food safety and pharmaceutical safety that addresses physical, chemical and biological hazards as a means of prevention rather than finished product inspection. HACCP is used in the food industry to identify potential food safety hazards, so that key actions, known as critical control points (CCPs), can be taken to reduce or eliminate the risk of the hazards being realized. The system is used at all stages of food production and preparation processes, including packaging, distribution, and so on. HACCP has been increasingly applied to industries other than food, such as cosmetics and pharmaceuticals. This method, which in effect seeks to plan out unsafe practices, differs from traditional 'produce and test' quality control methods which are less successful and inappropriate for highly perishable foods. Good hygiene practice (GHP) is established in the implementation of HACCP (see above). Specifically, potential sources of contamination from the environment need to be considered. Thus, processing should not be carried on in areas where the presence of potentially harmful substances would lead to an unacceptable level of such substances in the finished product.

***Good agricultural practices/good agricultural and collection practices/good agriculture and sourcing practice***

Good agricultural practices (GAPs) are a collection of principles applied to on-farm production and postproduction processes, resulting in safe and healthy agricultural products, while taking into account economical, social and environmental sustainability. Good agricultural and collection practices (GACPs) additionally include the collection and harvesting of wild plants, fungi and animals. All EU member states have a formal obligation to develop GAP/GACP guidelines as part of their national rural development plans. In this context, GAPs/GACPs generally focus on social and environmental issues. For companies supplying traditional herbal products or phytomedicines, the definition and purpose of GACPs are more narrowly focused. There are a number of different abbreviations used in connection with good agricultural and sourcing protocols. These include:

- good agricultural practice GAP
- good farming practice GFP
- good agriculture and hygiene practice GAHP
- good agricultural and collection practice GACP
- good agriculture and sourcing practice GASP
- good wild crafting practice GWP
- good sourcing practice GSP
- good field collection practice GFCP

The European Medicines Agency (EMEA), in its *Guidelines on Good Agricultural and Collection Practice for Starting Materials of Herbal Origin* (EMEA, 2006), states:

> to ensure appropriate and consistent quality of medicinal plant/herbal substances it is necessary to establish good agricultural and collection practices for herbal starting materials. The concept of good manufacturing practice for the manufacture, processing, packing and storage of active pharmaceutical ingredients also applies to medicinal plants/herbal substances.

***Good manufacturing practices***

GMPs are part of a quality system covering the manufacture and testing of active pharmaceutical ingredients, diagnostics, foods, pharmaceutical products and medical devices. GMPs are guidelines that outline the aspects of production and testing that can impact the quality of a product. The EU has legislated that manufacturers of medicinal, food and cosmetic products must follow GMP procedures, and have created their own GMP guidelines that correspond with their legislation. For example, Volume 4 of 'The rules governing medicinal products in the European Union' contains guidance for the interpretation of the principles and guidelines of GMPs for medicinal products for human and veterinary use laid down in Commission Directives 91/356/EEC, as amended by Directive 2003/94/EC, and 91/412/EEC respectively.

#### 8.2.1.4 Certifications

There are many different types of certification with many different objectives:

- ecologically responsible forest management standards (e.g. FSC) that assess water and soil conservation, preservation of wildlife and habitat, and maintenance of forest structure, function and processes;
- fair trade certification programs that assure equitable sharing of profits with producers, worker's rights and decent working conditions;
- organic certification standards that assure pesticide-free agricultural production (and are occasionally applied to agro-forestry and forestry production systems).

***Organic certification***

Organic certification is a certification process for producers of organic food and other organic agricultural products. In general, any business directly involved in food production can be certified, including seed suppliers, farmers, food processors, retailers and restaurants. Requirements vary from country to country, and generally involve a set of production standards for growing, storage, processing, packaging and shipping that include:

- avoidance of most synthetic chemical inputs (e.g. fertilizer, pesticides, antibiotics, food additives), genetically modified organisms, irradiation and the use of sewage sludge;
- use of farmland that has been free from synthetic chemicals for a number of years;
- keeping detailed written production and sales records (audit trail);
- maintaining strict physical separation of organic products from non-certified products;
- undergoing periodic on-site inspections.

The European organic food label has been mandatory throughout the EU since July 2010.

Countries wishing to export organic products to North America, Japan or the EU must obtain the certificate of the different respective trading block as they is no global recognition process.

The International Federation of Organic Agriculture Movements is the worldwide umbrella organization for the organic movement, uniting more than 750 member organizations in 116 countries (www.ifoam.org).

***Kosher***

Procedural details and requirements for kosher certification are available from the respective authorities. The 'kosherness' of a hechshered product is not absolute; its formulation and manufacture will conform broadly to the appropriate Jewish laws which govern kosher foods, but some kosher certifiers are more stringent, while others tend towards leniency in the interpretation of those laws. In applying for kosher certification of a product the stringency/leniency of a certifying agency may be important in the market where the product will be offered for sale. Agencies can be found at www.kashrut.com/agencies/.

### *Halal*

Halal is an Arabic word meaning lawful or permitted. The opposite of halal is haram, which means unlawful or prohibited. Halal and haram are universal terms that apply to all facets of life. However, we will use these terms only in relation to food products, meat products, cosmetics, personal care products, pharmaceuticals, food ingredients and food contact materials. While many things are clearly halal or haram, there are some things which are not clear:

- swine/pork and its by-products;
- animals improperly slaughtered or dead before slaughtering;
- alcoholic drinks and intoxicants;
- carnivorous animals, birds of prey and certain other animals;
- foods contaminated with any of the above products.

Foods containing ingredients such as gelatine, enzymes, emulsifiers and flavours are questionable (mashbooh), because the origin of these ingredients is not known.

The Islamic Food Council of Europe is one of the certification agencies in Europe, and the Islamic Food and Nutrition Council of America is one of the leading halal-food certification organizations in North America and can be found at http://www.ifanca.org /index.php.

### *Fairtrade*

Fair trade is an organized social movement and market-based approach that aims to help producers in developing countries obtain better trading conditions and promote sustainability. The movement advocates the payment of a higher price to producers as well as social and environmental standards. Although no universally accepted definition of fair trade exists, fair trade labelling organizations most commonly refer to a definition developed by FINE, an informal association of four international fair trade networks: fair trade is a trading partnership, based on dialogue, transparency and respect, that seeks greater equity in international trade.

*Fairtrade International* is a non-profit, multi-stakeholder association involving 25 member and associate member organizations (labelling initiatives and producer networks), traders and external experts. In 2004 it split into two organizations: *Fairtrade Labelling Organizations International (FLO)*, which develops and reviews Fairtrade standards and assists producers in gaining and maintaining Fairtrade certification and capitalizing on market opportunities, and *FLO-CERT*, which inspects and certifies producer organizations and audits traders.

The *European Fair Trade Association*, created in 1990, is a network of European alternative trading organizations that import products from some 400 economically disadvantaged producer groups in Africa, Asia and Latin America (www.european-fair-trade-association.org/).

## 8.3 Finished products

### 8.3.1 Herbal medicines

#### 8.3.1.1 Introduction

Within all EU states the basis of all legislation regarding any pharmaceutical product is European directives. Each directive contains subsections called articles. Some articles have been amended by later directive(s). Therefore, there is more than one place to look when considering all the current legislation.

A medicinal product is defined as:

- any substance or combination of substances presented as having properties for treating or preventing disease in human beings;
- any substance or combination of substances which may be used in or administered to human beings either with a view to restoring, correcting or modifying physiological functions by exerting a pharmacological, immunological or metabolic action, or to make a medical diagnosis.

An HMP may consist of one or more herbal ingredients of different levels of modification (processing). The conditioning, handling or processing of the herb is, per definition, responsible for whether a product is a herbal substance or a herbal preparation.

A plant part used without a specific processing is defined as a herbal substance. This could be the whole, fragmented or cut plant or parts of the plants as well as certain exudates. The herbal substance is characterized by the plant part used and the botanical name according to the binomial system and may be specified, for example, whether it is fresh or dried (e.g. dried, whole or fragmented bark).

It is possible to obtain a full marketing authorization for herbal medicines where the required levels of efficacy as well as safety and quality can be demonstrated. Following European Commission (EC) guidelines, a marketing authorization for HMPs can be obtained in different ways depending on the substance, indication and intended claim. Thus, the possibilities to obtain marketing authorization for HMPs are:

- full marketing authorization – procedure with full documentation with new tests and trials;
- marketing authorization of well-established-use HMPs – full bibliographic documentation;
- mixed application – including bibliographic evidence, supplemented by own research;
- registration as a traditional HMP (THMP).
- registration as a homoeopathic or anthroposophic medicine;
- registration as a medical device.

#### 8.3.1.2 Pharmacopoeias

The WHO undertook a survey in 2005 of 138 member states to find out whether a national pharmacopoeia that includes herbal medicines was in existence. If member states indicated that a national pharmacopoeia existed, the survey asked for bibliographical information about it and asked about its legal status. If member states indicated that they lacked a national pharmacopoeia, they were asked whether one was being developed and, further, whether a pharmacopoeia from another country or region was used instead.

Only 24% (34 countries) of the responding countries indicated that a national pharmacopoeia existed and was in use. Of the 104 countries lacking such a national pharmacopoeia, 25% (26 countries) indicated that such a document was in preparation.

Of those 104 member states lacking a national pharmacopoeia, 56% (58 countries) indicated that another pharmacopoeia was in use. Many countries reported the use of several different pharmacopoeias. Finally, 31 countries (30%) reported not using any pharmacopoeia.

The *European pharmacopoeia* is used most frequently where no national pharmacopoeia is available, followed by the *British pharmacopoeia* and the *United States pharmacopoeia*.

In 85% of the 34 countries with a national pharmacopoeia, the information it contains is legally binding. And in 59% of the 58 countries using another country or region's pharmacopoeia, the information contained in the other pharmacopoeia is also legally binding.

### 8.3.1.3 Monographs

***Use of monographs globally***

The WHO traditional medicine survey of 2005 included a section on the use of monographs in member states. Of the responding member states, 46 (33%) reported that they had national monographs on herbal medicines. Of the 84 countries that reported not having national monographs, 25 (28%) indicated that national monographs were in development.

Eighty-four countries reported not having national monographs. Instead, many countries reported the use of multiple monographs. The *WHO monographs* series was reported as being used by the largest number of countries, followed by the *European pharmacopoeia* and the European Scientific Cooperative on Phytotherapy monographs. Of the 46 countries with their own national monographs, 52% (24 countries) reported that their monographs were legally binding. Of the 34 countries that reported using other monographs, 44% (15 countries) reported that such texts are legally binding.

***European Union Community Monographs/Community List***

An EU community herbal monograph comprises the scientific opinion of the Committee on Herbal Medicinal Products (HMPC) on safety and efficacy data concerning a herbal substance and its preparations intended for medicinal use. The HMPC evaluates scientifically all available information, including non-clinical and clinical data, but has also documented long-standing use and experience in the community. Community Monographs are divided into two columns: well-established use (marketing authorization) and traditional use (simplified registration):

- well-established use – demonstrated with sufficient safety and efficacy data,
- traditional use – accepted on the basis of sufficient safety data and plausible efficacy.

A final Community Monograph can be used in application reference material by a marketing authorization applicant (well-established use part) and by a traditional use registration applicant (traditional use part).

In contrast to the community herbal monographs, the 'Community List of herbal substances, preparations and combinations thereof for use in traditional herbal medicinal products' is legally binding to applicants and competent authorities in the member states in so far as:

- an applicant will not be required to provide evidence of the safe and traditional use of a medicinal product for which he seeks a traditional use registration if he demonstrates that the proposed product and related claims in the application comply with the information contained in the Community List;
- competent authorities will not have the opportunity to require additional data to assess the safety and the traditional use of the product.

The Community List is developed gradually through 'list entries'. When a draft list entry for a given herbal substance or preparation has been produced by the Committee, it is released for public consultation on this website for a period of three months. The list entry is then finalized by the HMPC from a scientific point of view, before being submitted for approval by the EC. Once approved by the EC, the Community List is published and becomes a legal binding document throughout the EU.

### 8.3.1.4 Common technical document

The common technical document (CTD) is an internationally agreed format for the preparation of applications to be submitted to regulatory authorities in the three International Conference on Harmonisation (ICH) regions of Europe, the USA and Japan. It is intended to save time and resources and to facilitate regulatory review and communication. The CTD gives no information about the content of a dossier and does not indicate which studies and data are required for a successful approval. Regional requirements may affect the content of the dossier; therefore, the dossier will not necessarily be identical for all regions. The CTD indicates an appropriate format for the data that are required in an application. Applicants should not modify the overall organization of the CTD. However, in the non-clinical and clinical summaries, applicants may modify individual formats if needed in order to provide the best possible presentation of the technical information.

The CTD is organized into five modules. Module 1 is region specific. Modules 2–5 are intended to be common for all regions:

- Module 1 – administrative information and prescribing information;
- Module 2 – CTD summaries;
- Module 3 – quality;
- Module 4 – non-clinical study reports;
- Module 5 – clinical study reports.

### 8.3.1.5 Summary of product characteristics

The summary of the product characteristics shall contain, in strict order, key product information including composition, pharmaceuticals form, therapeutic indications, side effects, pre-clinical safety data, shelf life and storage requirements, and so on.

#### 8.3.1.6 Patient information leaflet

The inclusion in the packaging of all medicinal products of a package leaflet shall be obligatory unless all the information required is directly conveyed on the outer packaging or on the immediate packaging.

The package leaflet shall be drawn up in accordance with the summary of the product characteristics; it shall include, in the following order:

- the name of the medicinal product, followed by its strength and pharmaceutical form;
- the pharmaco-therapeutic group or type of activity;
- the therapeutic indications;
- information that is necessary before the medicinal product is taken (warnings, etc.);
- instructions for proper use, in particular the dosage and posology;
- a description of the adverse reactions that may occur under normal use;
- reference to the expiry date indicated on the label, and special storage instructions;
- the full qualitative (in active substances and excipients) and quantitative composition in active substances, using common and scientific names;
- list of the names authorized in each member state;
- date on which the package leaflet was last revised.

The package leaflet shall reflect the results of consultations with target patient groups to ensure that it is legible, clear and easy to use.

#### 8.3.1.7 Packaging

The following particulars shall appear on the outer packaging of medicinal products or, where there is no outer packaging, on the immediate packaging:

- The name of the medicinal product, followed by its strength and pharmaceutical form, and, if appropriate, whether it is intended for babies, children or adults. Where the product contains up to three active substances, the international non-proprietary name (INN) shall be included, or, if one does not exist, the common name.
- A statement of the active substances expressed qualitatively and quantitatively per dosage unit or according to the form of administration for a given volume or weight, using their common names.
- The pharmaceutical form and the contents by weight, by volume or by number of doses of the product.
- A list of those excipients known to have a recognized action or effect. However, if the product is injectable, or a topical or eye preparation, all excipients must be stated.
- The method of administration and, if necessary, the route of administration. Space shall be provided for the prescribed dose to be indicated.
- A special warning that the medicinal product must be stored out of the reach and sight of children.
- A special warning, if this is necessary for the medicinal product.
- The expiry date in clear terms (month/year).
- Special storage precautions, if any.
- Specific precautions relating to the disposal of unused medicinal products or waste derived from medicinal products, where appropriate, as well as reference to any appropriate collection system in place.
- The name and address of the marketing authorization holder and, where applicable, the name of the representative appointed by the holder to represent them.
- The number of the authorization for placing the medicinal product on the market.
- The manufacturer's batch number.
- In the case of non-prescription medicinal products, instructions for use.

#### 8.3.1.8 Herbal medicine: regulatory categories

A 2005 WHO survey of the use of traditional and herbal medicine in more than 150 member states included questions concerning the different regulatory categories under which herbal medicines were sold in the respective member states. Detailed descriptions of seven possible regulatory categories for herbal medicines were given on the survey form. These definitions are presented below.

1. *Prescription medicines:* medicines/drugs that can only be purchased with a prescription (i.e. a physician's order). In some countries, the legal framework allows traditional practitioners to prescribe medicines.
2. *Over-the-counter medicines:* medicines/drugs that can be purchased without a prescription from a physician.
3. *Self-medication only:* medicines/drugs permitted for self-medication purposes only.
4. *Dietary supplements:* a dietary supplement is a substance that contains, for instance, a vitamin, a mineral, a herb or other botanical or an amino acid. A dietary supplement may be intended to increase the total daily intake of a concentrate, metabolite, constituent, extract or combination of these ingredients.
5. *Health food:* health foods could be products that are presented with specific health claims and therefore regulated differently from other foods.
6. *Functional foods:* like health foods, functional foods may be products which are offered with specific health claims, and therefore regulated differently from other foods.
7. *Other:* products classified differently from the above-mentioned categories.

Table 8.2 lists the categories of the 2005 WHO survey and the number of respondents of the 131 member countries surveyed.

Thus, herbal medicines and herbal substances can be categorized as food supplements in one country and hence subject to food safety legislation while in a neighbouring country the same product or products can be considered over-the-counter medicines and are subject to drug safety laws.

#### 8.3.1.9 European Union drug legislation

The body of EU legislation in the pharmaceutical sector is compiled in Volume 1 and Volume 5 of the publication 'The rules governing medicinal products in the European Union'.

**Table 8.2** Results of the 2005 WHO survey

| Category | Respondent countries |
|---|---|
| Over the counter medicine | 97 |
| Prescription medicine | 50 |
| Dietary supplements | 47 |
| Self-medication only | 40 |
| Special herbal medicine category | 25 |
| No status | 23 |
| Health food | 15 |
| Other | 12 |
| Functional foods | 9 |
| No answer | 4 |

The basic legislation is supported by a series of guidelines that are also published in the other volumes of 'The rules governing medicinal products in the European Union'.

Medicinal products classed as for paediatric use, as orphans, as HMPs and as advanced therapies are governed by specific rules. For more information refer to 'EU Legislation – Eudralex'. These are the most important regulations, the details of which can be found in Eudralex:

### *Category definitions*

The principle of the EU drug regulatory approval mechanism can be seen as a pyramid (Figure 8.3). At the peak of the pyramid are drugs with full market authorization (complete application) making strong proven claims. Below this are drugs regulated under their well-established use or under the Traditional Herbal Medicinal Products Directive. Below this are products that either make no claim at all or those that attempt fraudulently to make claims outside of this regulatory framework.

### *Full market authorization*

A product can be authorized after evaluation of a marketing authorization application consisting of only product-specific safety and efficacy data ('full dossier'). As a result, the product is granted a marketing authorization by a member state or by the EMEA via the centralized procedure if all requirements are met.

### *Full dossier documentation requirements*

The application shall be accompanied by the following particulars and documents, submitted in accordance with the requirements for CTD: quantitative and qualitative particulars of constituents, therapeutic indications, posology, results of clinical trials and description of pharmacovigilance needs. The documents and information concerning the results of the pharmaceutical and pre-clinical tests and the clinical trials (where applicable) shall be accompanied by detailed summaries.

### *Market authorization based on well-established use*

A product can be classified under well-established medicinal use provisions ('well-established use'). This is demonstrated with sufficient safety and efficacy data. As a result, the product is granted a marketing authorization usually by a member state or by the EMEA under certain conditions.

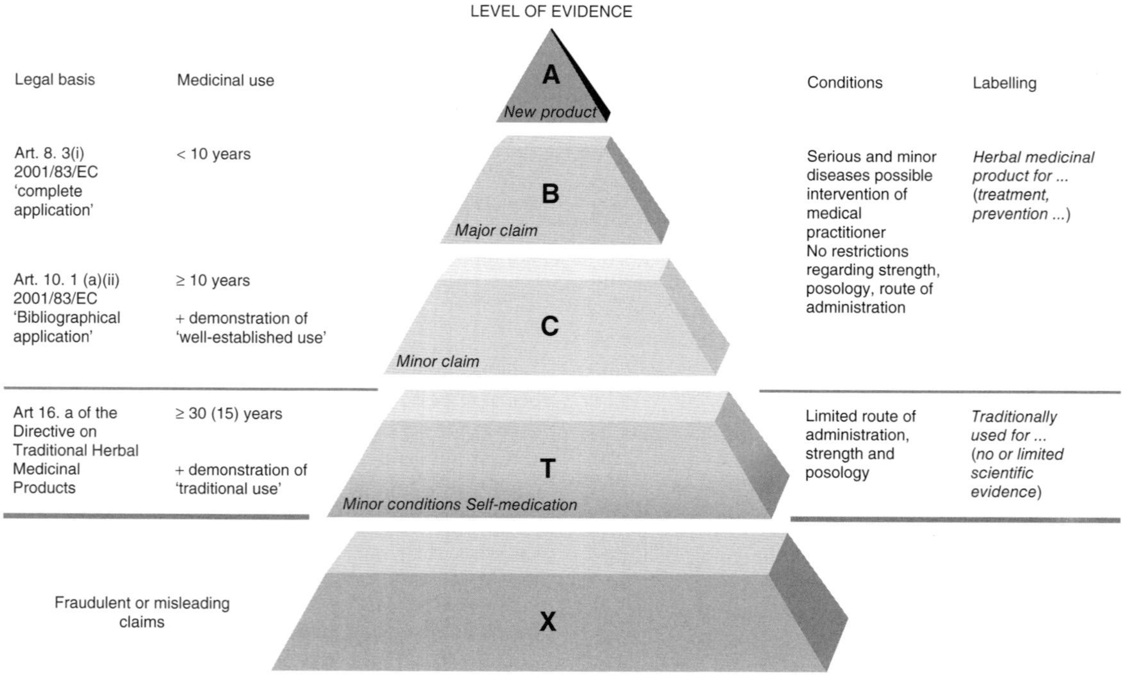


**Figure 8.3** Principle of the EU drug regulatory approval mechanism.
*Source:* A. J. Vlietinck (2009). Reproduced with permission

Definition of and guidance on well-established medicinal use are given in Annex 1 Part II of Council Directive 2003/63/EC (25.06.2003) as follows.

(a) Factors important to establish whether a product should be classified under the well-established medicinal use category are:
  - the time over which a substance has been used, not less than 10 years from first and systematic use as a medicinal product;
  - quantitative aspects of the use of the substance;
  - the degree of scientific interest in the use of the substance;
  - the coherence of scientific assessments.

  Therefore, different periods of time may be necessary for establishing well-established use of different substances. In any case, however, the period of time required for establishing a well-established medicinal use of a constituent of a medicinal product must not be less than one decade from the first systematic and documented use of that substance as a medicinal product in the European Community.
(b) The bibliography should cover all aspects of the safety and/or efficacy assessment and must include or refer to a review of the relevant literature, taking into account pre- and post-marketing studies and published scientific literature concerning experience in the form of epidemiological studies and in particular of comparative epidemiological studies. All documentation, both favourable and unfavourable, must be communicated.
(c) Concerning missing information, justification must be given why demonstration of an acceptable level of safety and/or efficacy can be supported although some studies are lacking.
(d) Non-clinical and/or clinical overviews must explain relevance of any data submitted which concern a product different from the product intended from marketing.
(e) Post-marketing experience with other products containing the same constituents is of particular importance and applicants should put a special emphasis on this issue.

### *Documentation requirements*

The application shall be accompanied by the following particulars and documents, submitted in accordance with the requirements for CTDs: quantitative and qualitative particulars of the constituents, therapeutic indications, posology, results of clinical trials and description of pharmacovigilance needs. The documents and information concerning the results of the pharmaceutical and pre-clinical tests and the clinical trials (where applicable) shall be accompanied by detailed summaries.

### *Traditional herbal medicinal products, traditional Chinese medicine, kampo, ayurveda, unani, homoeopathy, anthroposophy and others*

*Homeopathic medicinal product.* Any medicinal product prepared from substances called homeopathic stocks in accordance with a homeopathic manufacturing procedure described by the European pharmacopoeia or, in the absence thereof, by the pharmacopoeias currently used officially in the member states. A homeopathic medicinal product may contain a number of principles.

*Anthroposophic medicinal product.* The anthroposophic medicinal products described in an official pharmacopoeia and prepared by a homeopathic method are to be treated, as regards registration and marketing authorization, in the same way as homeopathic medicinal products.

*HMP.* Any medicinal product, exclusively containing as active ingredients one or more herbal substances or one or more herbal preparations, or one or more such herbal substances in combination with one or more such herbal preparations.

*THMP.* A product can be classified under traditional medicinal use provisions ('traditional use') accepted on the basis of sufficient safety data and plausible efficacy: the product is granted a traditional use registration (simplified registration procedure) by a member state. Under Directive 2004/24/EC on THMPs, applicants are required to produce bibliographic or expert evidence of traditional use. Essentially, the medicinal product (or corresponding products(s) characterized by having the same active ingredients, the same or similar intended purpose, equivalent strength and posology and the same or similar route of administration) need to have been in medicinal use throughout a period of at least 30 years preceding the date of application. At least 15 of the 30 years of use must relate to the EU. The requirement to prove medicinal use for a period of 30 years can also be satisfied if number or quantity of ingredients has been reduced over that period of time. Whenever the product complies with a Community Monograph as established by the HMPC (see below), the requirement to demonstrate traditional use is lifted.

### *Application*

*Homoeopathic medicinal products.* Only homeopathic medicinal products which satisfy all of the following conditions may be subject to a special, simplified registration procedure:

- they are administered orally or externally;
- no specific therapeutic indication appears on the labelling of the medicinal product or in any information relating thereto;
- there is a sufficient degree of dilution to guarantee the safety of the medicinal product; in particular, the medicinal product may not contain either more than one part per 10 000 of the mother tincture or more than 1/100th of the smallest dose used in allopathy with regard to active substances whose presence in an allopathic medicinal product results in the obligation to submit a doctor's prescription.

An application for special, simplified registration may cover a series of medicinal products derived from the same homeopathic stock or stocks.

*THMPs.* A simplified registration procedure is established for HMPs that fulfil all of the following criteria:

- they have indications exclusively appropriate to THMPs which, by virtue of their composition and purpose, are intended and designed for use without the supervision of a medical practitioner for diagnostic purposes or for prescription or monitoring of treatment;
- they are exclusively for administration in accordance with a specified strength and posology;
- they are an oral, external and/or inhalation preparation;
- the period of traditional use as laid down in Directive 2004/24/EC has elapsed;

- the data on the traditional use of the medicinal product are sufficient; in particular, the product proves not to be harmful in the specified conditions of use and the pharmacological effects or efficacy of the medicinal product are plausible on the basis of long-standing use and experience.

The presence in the HMP of vitamins or minerals for the safety of which there is well-documented evidence does not prevent the product from being eligible for registration, provided that the action of the vitamins or minerals is ancillary to that of the herbal active ingredients regarding the specified claimed indication(s).

HMPs from other traditional medicinal paradigms, such as traditional Chinese medicine, kampo, ayurveda, unani, and so on may qualify for registration as a THMP if they fulfil the aforementioned requirements; otherwise, licensing in another medicinal products or finished products category would be required.

For the purpose of assisting with the registration of THMPs, regulators are providing Community Monographs and a Community List of herbal medicines (see above).

### *Traditional herbal medicinal product and homeopathic documentation requirements*

A product dossier needs to be submitted in CTD format to the relevant authorities, comprising of the following information.

*Homoeopathic medicinal product.* The following documents shall be included with the application in order to demonstrate, in particular, the pharmaceutical quality and the batch-to-batch homogeneity of the products concerned:

- scientific name or other name given in a pharmacopoeia of the homeopathic stock or stocks, together with a statement of the various routes of administration, pharmaceutical forms and degree of dilution to be registered;
- dossier describing how the homeopathic stock or stocks is/are obtained and controlled, and justifying its/their homeopathic use, on the basis of an adequate bibliography;
- manufacturing and control file for each pharmaceutical form and a description of the method of dilution and potentization;
- manufacturing authorization for the medicinal product concerned;
- copies of any registrations or authorizations obtained for the same medicinal product in other member states;
- one or more mock-ups of the outer packaging and the immediate packaging of the medicinal products to be registered;
- data concerning the stability of the medicinal product.

*THMPs.* The application shall be accompanied by the following particulars and documents:

- Name or corporate name and permanent address of the applicant and, where applicable, of the manufacturer.
- Name of the medicinal product.
- Qualitative and quantitative particulars of all the constituents of the medicinal product, including the reference to its INN recommended by the WHO, where an INN for the medicinal product exists, or a reference to the relevant chemical name.
- Evaluation of the potential environmental risks posed by the medicinal product. This impact shall be assessed and, on a case-by-case basis, specific arrangements to limit it shall be envisaged.
- Description of the manufacturing method.
- Therapeutic indications, contraindications and adverse reactions.
- Posology, pharmaceutical form, method and route of administration and expected shelf life.
- Reasons for any precautionary and safety measures to be taken for the storage of the medicinal product, its administration to patients and for the disposal of waste products, together with an indication of potential risks presented by the medicinal product for the environment.
- Description of the control methods employed by the manufacturer.
- A summary of the product characteristics, a mock-up of the outer packaging, and of the immediate packaging of the medicinal product, together with a package leaflet.
- A document showing that the manufacturer is authorized in their own country to produce medicinal products.
- The results of pre-clinical (toxicological and pharmaceutical) tests.
- In case of combinations, the information relating to the combination as such; if the individual active ingredients are not sufficiently known, the data shall also relate to the individual active ingredients.
- Any authorization or registration obtained by the applicant in another member state, or in a third country, to place the medicinal product on the market, and details of any decision to refuse to grant an authorization or registration, whether in the European Community or a third country, and the reasons for any such decision.
- Bibliographical or expert evidence to the effect that the medicinal product in question, or a corresponding product, has been in medicinal use throughout a period of at least 30 years preceding the date of the application, including at least 15 years within the European Community. At the request of the member state where the application for traditional-use registration has been submitted, the HMPC shall draw up an opinion on the adequacy of the evidence of the longstanding use of the product, or of the corresponding product. The member state shall submit relevant documentation supporting the referral.
- A bibliographic review of safety data together with an expert report, and where required by the competent authority, upon additional request, data necessary for assessing the safety of the medicinal product.

A corresponding product is characterized by having the same active ingredients, irrespective of the excipients used, the same or similar intended purpose, equivalent strength and posology and the same or similar route of administration as the medicinal product applied for.

### *Efficacy/proof of tradition*

For THMPs the requirements for showing efficacy have been replaced by showing proof for traditional use. The requirement to show medicinal use throughout the period of 30 years is satisfied even where the marketing of the product has not been based on a specific authorization. It is likewise satisfied if the number or quantity of ingredients of the medicinal product has been reduced during that period. Where the product has been used in the European Community for less than 15 years, but is otherwise eligible for simplified registration, the member state where the application for traditional-use registration has been submitted shall refer the product to the HMPC. The member state shall submit relevant documentation supporting the referral.

## 8.3.2 Food products

### 8.3.2.1 Introduction

The harmonization of EU food law has been in progress for 40 years. From the early 1960s to 1985 most of the focus was on vertical legislation with 'recipe' or compositional criteria established for product categories. Of the original programme of over 30 'recipe' directives, only nine were agreed to and adopted during the first 20 years of the European Community. In 1985 a white paper from the EC to the Council of the EU recommended a complete overhaul with respect to the approach to food legislation and that the original concept of restrictive vertical legislation be replaced by a system of horizontal legislation.

There are complex laws on flavourings that control the source materials. In addition, flavourings produced from natural botanical sources must comply with requirements that impose low limits on biologically active principles (BAPs), which occur naturally in plants. BAPs subject to strict control include coumarin, safrole and berberine. Companies hoping to export products to Europe are advised to ensure that their suppliers certify that the composition of the flavours is in conformance with European law.

### 8.3.2.2 Category definitions

***Food***

'Food' (or 'foodstuff') means any substance or product, whether processed, partially processed or unprocessed, intended to be, or reasonably expected to be, ingested by humans. 'Food' includes drink, chewing gum and any substance, including water, intentionally incorporated into the food during its manufacture, preparation or treatment. 'Food' does not include:

- feed;
- live animals, unless they are prepared for placing on the market for human consumption;
- plants prior to harvesting;
- medicinal products;
- cosmetics;
- tobacco and tobacco products;
- narcotic or psychotropic substances;
- residues and contaminants.

***Food additives***

Food additives are substances added intentionally to foodstuffs to perform certain technological functions; for example, to colour, to sweeten or to preserve.

Food additives are defined in European Community legislation as

> any substance not normally consumed as a food in itself and not normally used as a characteristic ingredient of food whether or not it has nutritive value, the intentional addition of which to food for a technological purpose ... results ... in it or its by-products becoming directly or indirectly a component of such foods.

For the full definition see: Article 1(2) of Directive 89/107/EEC.

The European Community legislation on food additives is based on the principle that only those additives that are explicitly authorized may be used. Most food additives may only be used in limited quantities in certain foodstuffs. If no quantitative limits are foreseen for the use of a food additive, it must be used according to GMP; that is, only as much as necessary to achieve the desired technological effect.

The use of food additives must always be labelled on the packaging of food products by their category (antioxidant, preservative, colour, etc.) with either their name or E-number. Detailed rules on labelling of additives in foodstuffs, and on additives sold as such to food producers and consumers, are laid down in European Community legislation (Directive 2000/13/EC, Regulation 50/2000/EC and Directive 89/107/EEC).

***Food flavourings***

Flavourings are substances used to give taste and/or smell to food. Community legislation defines different types of flavourings, such as natural, natural-identical or artificial flavouring substances, flavouring preparations of plant or animal origin, process flavourings which evolve flavour after heating and smoke flavourings (for the full definition, see Article 1(2) of Directive 88/388/EEC).

### 8.3.2.3 Novel foods and genetically modified foods

If a herbal product is to be placed on the market as food or food ingredients it must be ensured that it is not a novel or genetically modified (GM) food. Otherwise it has to undergo novel food approval. Ingredients that do not have a history of significant consumption within the EU prior to 15 May 1997 may be subject to the controls of Novel Food Regulation (EC) 258/97. All novel foods are subject to a premarket safety assessment under Novel Food Regulation (EC) 258/97. The Novel Food Regulation is presently undergoing substantial revision. The main objective is to separate foods that have been traditionally used, but not within the EU, from foods that are entirely new in origin, such as GM products. The regulation requires that a full dossier on the substance or process be reviewed by a competent authority in the EU country of intended first sale.

### 8.3.2.4 Good practices for food production and manufacturing

Food manufacturing, processing and marketing are governed by the following good practices:

- *Good hygienic practices* – all practices regarding the conditions and measures necessary to ensure the safety and suitability of food at all stages of the food chain.

- *Codex HACCP system* – a system that identifies, evaluates and controls hazards which are significant for food safety, described in the Annex to the Codex General Principles of Food Hygiene.
- *HACCP-based system* – a system that is consistent with the seven principles of HACCP but does not conform to the layout or steps of the Guidelines for the Application of the Codex HACCP system.
- *Food safety management system* – a holistic system of controls that manage food safety in a food business. This includes good hygienic practices, the HACCP system, management policies and traceability/recall systems.
- *GMPs* – these are part of a quality system covering the manufacture and testing of foods. GMPs are guidelines that outline the aspects of production and testing that can impact the quality of a product.

### 8.3.2.5 Food product laws and regulations

The food and consumer goods legislation regulates, amongst other things, the manufacture and handling of food and consumer goods.

The Codex Alimentarius Commission (CAC) develops food standards that serve as a reference for international food trade. More than 180 countries are members of the CAC and more than 200 international non-governmental and intergovernmental organizations have the observer status. The World Trade Organization (WTO) Agreement on the Application of Sanitary and Phytosanitary Measures (SPS Agreement) considers that WTO members applying the Codex Alimentarius Standards meet their obligations under this agreement.

The 27 member states of the EU are all members of the CAC. In 2003, the European Community (now the EU) also became a full member of the CAC and shares the competence with its member states on the basis of the level of harmonization of the relevant legislation. Since the entry into force of the Treaty of Lisbon on 1 December 2009 the EU replaced the European Community. The EU and its member states elaborate EU position papers on issues discussed in the CAC, the various codex committees and task forces.

On 28 January 2002 the European Parliament and the Council of the EU adopted Regulation (EC) 178/2002 laying down the general principles and requirements of food law.

The regulation contains general provisions for traceability (applicable from 1 January 2005) that cover all food and feed, all food and feed business operators, without prejudice to existing legislation on specific sectors such as beef, fish, genetically modified organisms (GMOs), and so on. Importers are similarly affected as they will be required to identify from whom the product was exported in the country of origin. Unless specific provisions for further traceability exist, the requirement for traceability is limited to ensuring that businesses are at least able to identify the immediate supplier of the product in question and the immediate subsequent recipient, with the exemption of retailers to final consumers (one step back–one step forward).

Directive 2002/46/EC of the European Parliament and Council of the EU of 10 June 2002 on the approximation of the laws of member states relating to food supplements establishes harmonized rules for the labelling of food supplements and introduces specific rules on vitamins and minerals in food supplements. The aim is to harmonize the legislation and to ensure that these products are safe and appropriately labelled so that consumers can make informed choices.

Annex II of Directive 2002/46/EC is a list of permitted vitamin or mineral preparations that may be added for specific nutritional purposes in food supplements. It has been amended by Commission Directive 2006/37/EC and Commission Regulation (EC) 1170/2009 to include additional substances. The trade of products containing vitamins and minerals not listed in Annex II has been prohibited since 1 August 2005.

With respect to flavouring substances, European Parliament and Council of the EU Regulation 2232/96/EC sets out the basic rules for the use of these substances in or on foodstuffs in the EU. Under this regulation, the member states have informed the EC which flavouring substances are currently authorized for use in foodstuffs at national level. This information has been compiled by the EC in a register of about 2800 substances, adopted as Commission Decision 1999/217/EC, and amended by Commission Decision 2000/489/EC. The register forms the basis for a 5-year evaluation programme of these flavouring substances as adopted by Commission Regulation 1565/2000/EC. After completion of the evaluation programme, a European Community-wide positive list of flavouring substances for use in foodstuffs shall be established.

### 8.3.2.6 Documentation requirements

Food and feed imported into the European Community for placing on the market within the Community shall comply with the relevant requirements of food law or conditions recognized by the Community to be at least equivalent thereto or, where a specific agreement exists between the Community and the exporting country, with requirements contained therein. Unless food is for a specific (dietetic or medicinal) purpose no specific documentation will need to be provided

#### *Food safety*

Food that complies with specific European Community provisions governing food safety shall be deemed to be safe insofar as the aspects covered by the specific European Community provisions are concerned.

A two-level tiered conceptual framework for safety assessment consisting of a safety assessment based on available knowledge and a subsequent level in which further testing and/or data are required is proposed by the scientific committee. Botanicals or botanical preparations for which an adequate body of knowledge exists can benefit from a "presumption of safety" without any need for further testing. Based on reasonable evidence, they can then be assumed to be safe, sometimes under certain restrictions.

Botanicals and botanical preparations for which a presumption of safety is not possible should be subject to a more extensive safety assessment, based on additional data to be provided according to the methodology described by European Food Safety Authority (EFSA).

For a safety assessment, EFSA expects the following data to be compiled:

- Identity and nature of the source material (scientific name, synonym(s), common names, geographical origin, growth and harvesting conditions).

- Manufacturing process (method(s) of manufacturing, substances entering the manufacturing process, standardization criteria).
- Chemical composition (compounds classified according to their chemical structure, constituents which characterize quality and/or biological activity, compounds of potential chemical, physiological or toxicological concern).
- Specifications. (These may be based on nutritional or biologically active components or, when these are not known, on selected chemical markers. Limits for or absence of specific undesirable/toxic substances should be specified. The specifications should include concentrations of major groups of constituents present in the botanical preparation including, for example, amino acids, lipids, polysaccharides, volatile oil, inorganic ions, polyphenols, alkaloids, terpenes, alkenylbenzenes, lignin, saponins, as well as the major constituents within these classes. In addition, information on maximum levels for possible contaminants including, for example, heavy metals, mycotoxins, pesticide residues, and polycyclic aromatic hydrocarbon residues should be provided.)
- The stability of the botanical ingredient should be demonstrated over the shelf-life time. Any information concerning possible degradation should also be provided.
- Proposed uses and use levels (intended uses and recommended intakes with respect to food product categories and especially with regard to population groups with specific uses (e.g. children).
- Information on existing assessments.
- Information on the extent and duration of exposure.
- Toxicological data (studies should be carried out according to the principles of good laboratory practice (GLP) described in Council Directive 87/18/EEC and accompanied by a statement of GLP compliance).

## 8.3.3 Food supplements, foods for particular nutritional uses, fortified foods

### 8.3.3.1 Category definitions

*Food supplements*

Food supplements are concentrated sources of nutrients or other substances with a nutritional or physiological effect whose purpose is to supplement the normal diet. They are marketed 'in dose' form; that is, as pills, tablets, capsules, liquids in measured doses, and so on.

Supplements fall into four categories:

- Supplements containing only vitamins and minerals.
- Supplements containing vitamins, minerals and other non-botanical actives. In this category, non-botanical actives include ingredients such as glucosamine and CoQ10. The vitamin and mineral levels in such combinations still have to comply with the same rules as category 1.
- Supplements containing herbs and non-herbal active ingredients, including vitamins and minerals. The legislation on herbs sold under food law is very strict across Europe. In EU countries, most products in this category cannot be sold as dietary supplements under food law and can only be sold as registered medicines. The requirements for each country are so complex that every product is subjected to expert regulatory review.
- Supplements containing only herbs or botanicals.

*Foods for particular nutritional uses*

Foods for particular nutritional uses, or dietetic foods, are specially manufactured products intended to satisfy the particular nutritional requirements of specific groups of people. These could be baby foods, foods for people suffering from gluten or lactose intolerance, foods for special medical purposes, and so on. Dietetic food must fulfil strict criteria to meet the required health-related standards of specific groups of persons with special nutritional needs like babies or diabetics. For this reason, the competent authorities have to be aware of the products circulating on the market. Specific dietetic food (e.g. for special medical purposes or for pregnant and breastfeeding women) must therefore be reported.

*Fortified foods*

Fortified foods are foods or food products to which extra nutrients have been added. Vitamins and minerals are usually added to foods for three main reasons:

- They may replace some of the nutrients lost during manufacture and storage. For example, iron and several types of vitamin B are added to wheat flour after processing. This is known as restoration.
- They can be added to a 'substitute' food in order to compensate for a missing nutrient present in the 'traditional' alternative. The best known example of this is margarine, where vitamins A and D are added during production so it will have a similar vitamin level to butter.
- Nutrients may also be added to foods to enrich them further, even if the vitamins and minerals added are not normally present in that food. Calcium is often added to fruit juices, and this can provide an important source for people who do not eat dairy products.

### 8.3.3.2 Food supplement: documentation requirements

While national laws still apply for a number of matters concerning the sale of food supplements, especially (a) the issue of upper and lower safe limits for vitamins and minerals (where EU laws have not yet been finally agreed) and (b) the decision whether a product is considered a food supplement or medicinal product (where national authorities may have different views), there are an increasing number of EU-wide rules that apply to the distribution and sales of food supplements. There are four main EU directives pertaining to food supplements:

1. EC 202/46/EC of 10 June 2002
2. EC 2006/37/EC of 30 March 2006
3. EC 2008/100/EC of 28 October 2008
4. EC 1170/2009/EC of 30 November 2009.

Directive 202/46 requires supplement producers to indicate on the box and label:

- the names of the categories of the nutrients or substances that characterize the product or an indication of the nature of those nutrients or substances;

- the portion of the product recommended for daily consumption and a warning not to exceed the stated recommended daily dose;
- a declaration to the effect that the supplement is not a substitute for a varied diet;
- that the name, label and information should not suggest (labelling, presentation and advertising) and must not imply this can prevent, treat or cure a human disease;
- that information on vitamins and minerals shall also be expressed as a percentage of the reference values mentioned, as per Annex 1 of EC Directive 2008/100/EC.

The information must be presented together in one place in tabular form, with the numbers aligned if space permits. Where space does not permit, the information shall be presented in linear form. It shall be printed in legible and indelible characters in a conspicuous place.

#### 8.3.3.3 Health claims

The following documentation requirements apply particularly to foods for particular nutritional uses and novel foods. They apply to health claims related to the consumption of a food category, a food or its constituents (including a nutrient or other substance, or a combination of nutrients/other substances):

- Information on the characteristics of the food/constituent for which a health claim is made. Where applicable, this information should contain aspects considered pertinent to the claim, such as the composition, physical and chemical characteristics, manufacturing process, stability, and bioavailability.
- A proposal for the wording of the health claim, including, as appropriate, the specific conditions of use. The following should be specified, with a rationale: the target population for the intended health claim; where appropriate, a statement addressed to persons who should avoid using the food/constituent for which the health claim is made; the quantity of the food/constituent and pattern of consumption required to obtain the claimed effect, and whether this quantity could reasonably be consumed as part of a balanced diet; a warning for a food/constituent that is likely to present a health risk if consumed to excess; any other restrictions of use; directions for preparation and/or use.

The application must also contain all pertinent scientific data (published and unpublished, data in favour and not in favour) identified that form the basis for substantiation of the health claim. Data from studies in humans addressing the relationship between the consumption of the food/constituent and the claimed effect will be required for substantiation of a health claim; because of the scientific uncertainties in extrapolating non-human data to humans, data from studies in animals or model systems may be included only as supporting evidence (e.g. to explain the mechanism underlying the claimed effect of the food/constituent).

A comprehensive review of the data from human studies addressing the specific relationship between the food/constituent and the claimed effect is required. This review, and the identification of data considered pertinent to the claim, should be performed in a systematic and transparent manner in order to demonstrate that the application reflects adequately the balance of all the evidence available.

In cases where any of the required data does not apply for a particular application, reasons/justification must be given for the absence of such data in the application.

Health claims should be substantiated by taking into account the totality of the available scientific data and by weighing the evidence, subject to the specific conditions of use. In particular, the evidence should demonstrate the extent to which:

- the claimed effect of the food/constituent is relevant for human health;
- a cause and effect relationship is established between the consumption of the food/constituent and the claimed effect in humans (such as the strength, consistency, specificity, dose–respons, and biological plausibility of the relationship);
- the quantity of the food/constituent and pattern of consumption required to obtain the claimed effect could reasonably be achieved as part of a balanced diet;
- the specific study group(s) in which the evidence was obtained is representative of the target population for which the claim is intended.

It should be noted that, to date, the vast majority of health-claim dossiers submitted to EFSA under the Food Supplements Directive have been rejected on the ground of insufficient evidence; 1700 claims have been reviewed 170 have been approved. The latest batch – the third and penultimate – consisted of 808 claims, of which 733 (91%) were rejected. In the main, the positive opinions were for essential vitamins and minerals, with just a handful relating to other substances (botanicals).

#### 8.3.3.4 Packaging, labelling, declaration

There are very detailed pan-European labelling requirements for products, which include the declaration of all ingredients in descending order by weight, open expiry date coding and storage conditions, the name of the manufacturer, packer or importer established within the EU, a generic description of the product and usage instructions. Additionally, there are rules for the labelling of ingredients and additives derived from GMOs.

The EC is also separately taking forward the amendment of Council Directive 90/496/EEC on the nutrition labelling of foodstuffs. The EC plans to revise the list of vitamins and minerals and their recommended daily allowances, update energy conversion factors and include a definition for fibre.

The EC issued a proposal for a new Food Information Regulation on 4 February 2008. This proposal follows an EU-wide review of both general food and nutrition labelling legislation, which began in 2004.

#### 8.3.3.5 Food supplement claims

With the advent of the health claims regulation, the European regulatory landscape has become unified throughout the various member states. Whereas before each country could more or less decide independently on the stringency, or lack thereof, of its national claim regulations, today there is a firm European-wide framework in place. Most important is the definition of two types of claims: 'nutrition claims' and 'functional claims.' The latter are subdivided into 'well-accepted claims,' 'innovative claims' and 'risk-reduction claims.' There are lists of accepted claims for both nutrition and well-accepted claims.

## 8.3.4 Cosmetics

### 8.3.4.1 Category definitions

'Cosmetic product' means any substance or mixture intended to be placed in contact with the external parts of the human body (epidermis, hair system, nails, lips and external genital organs) or with the teeth and the mucous membranes of the oral cavity with a view exclusively or mainly to cleaning them, perfuming them, changing their appearance, protecting them, keeping them in good condition or correcting body odours;

### 8.3.4.2 Cosmetic governing laws and regulations

There are the following directives and regulations:

- Cosmetics Directive 76/768/EC as amended.
- Regulation (EC) No 1223/2009 of the European Parliament and of The Council of 30 November 2009 on cosmetic products.
- Regulation (EC) No 1272/2008 of the European Parliament and of The Council of 16 December 2008 on classification, labelling and packaging of substances and mixtures, amending and repealing Directives 67/548/EEC and 1999/45/EC, and amending Regulation (EC) No 1907/2006.
- Regulation (EC) No 1907/2006 of the European Parliament and of the Council of 18 December 2006 concerning the Registration, Evaluation, Authorisation and Restriction of Chemicals (REACH) and establishing a European Chemicals Agency.

A comprehensive list of substances that are not permitted for use in cosmetics or for which restrictions apply can be found in the appendices to Regulation EC 1223/2009. There are more than 35 plants that are prohibited for use in cosmetics, such as belladona, *Datura* and *Aristolochia*, all of which contain highly toxic compounds.

### 8.3.4.3 Cosmetic documentation requirements

When a cosmetic product is placed on the market, the responsible person shall keep a product information file for it. The product information file shall be kept for a period of 10 years following the date on which the last batch of the cosmetic product was placed on the market. The product information file shall contain the following information and data, which shall be updated as necessary:

- The qualitative and quantitative composition of the product; in the case of perfume compositions and perfumes, the name and code number of the composition and the identity of the supplier.
- The physical and chemical and microbiological specifications of the raw materials and the finished product and the purity and microbiological control criteria of the cosmetic product.
- The method of manufacture complying with the GMP laid down by European Community law or, failing that, laid down by the law of the member state concerned; the person responsible for manufacture or first importation into the European Community must possess an appropriate level of professional qualification or experience in accordance with the legislation and practice of the member state which is the place of manufacture or first importation;
- Assessment of the safety for human health of the finished product. To that end, the manufacturer shall take into consideration the general toxicological profile of the ingredients, their chemical structure and their level of exposure. It shall take particular account of the specific exposure characteristics of the areas on which the product will be applied or of the population for which it is intended. There shall be *inter alia* a specific assessment for cosmetic products intended for use on children under the age of 3 years and for cosmetic products intended exclusively for use in external intimate hygiene. Should the same product be manufactured at several places within European Community territory, the manufacturer may choose a single place of manufacture where that information will be available. In this connection, and when so requested for monitoring purposes, it shall be obliged to indicate the place so chosen to the monitoring authority or authorities concerned. In this case this information shall be easily accessible.
- The name and address of the qualified person or persons responsible for the assessment referred to in the previous bullet point. That person must hold a diploma as defined in Article 1 of Council Directive 89/48/EEC in the field of pharmacy, toxicology, dermatology, medicine or a similar discipline.
- Existing data on undesirable effects on human health resulting from use of the cosmetic product.
- Proof of the effect claimed for the cosmetic product, where justified by the nature of the effect or product.
- Data on any animal testing performed by the manufacturer, their agent or suppliers, relating to the development or safety evaluation of the product or its ingredients, including any animal testing performed to meet the legislative or regulatory requirements of non-member countries.

### 8.3.4.4 Cosmetic safety

A cosmetic product made available on the market shall be safe for human health when used under normal or reasonably foreseeable conditions of use, taking account, in particular, of the following:

- presentation, including conformity with Directive87/357/EEC;
- labelling;
- instructions for use and disposal;
- any other indication or information provided by the responsible person.

The safety of a cosmetic product is assessed by the Scientific Committee on Consumer Products (SCCP). In general, the safety evaluation of cosmetic ingredients by the SCCP is based upon the principles and practice of the risk assessment process usually applied for ingredients in medicinal products, pesticides, food additives, and so on. This risk assessment procedure is subdivided in four parts:

1. hazard identification;
2. dose–response assessment;
3. exposure assessment;
4. risk characterization.

Cosing is the European Commission database with information on cosmetic ingredients contained in the:

- 'Cosmetics Directive' 76/768/EEC (Cosmetics Directive), as amended;
- Inventory of Cosmetic Ingredients, as amended; and
- opinions on cosmetic ingredients of the Scientific Committee for Consumer Safety.

#### 8.3.4.5 Cosmetic efficacy

Cosmetics Europe provides guidelines for the evaluation of the efficacy of cosmetic products. These guidelines aim to assist the cosmetics industry to comply with the applicable European regulations for the efficacy evaluation of cosmetic products. Because methodologically sound research is essential for the efficacy evaluation, these guidelines provide an overview of established testing methodologies.

#### 8.3.4.6 Cosmetic packaging, labelling, declaration

The container and packaging of cosmetic products should bear the following information:

- the name or registered name and address of responsible person;
- country of origin;
- the nominal content at the time of packaging, given by weight or by volume;
- the date until which the cosmetic product, stored under appropriate conditions,
- particular precautions to be observed in use;
- the batch number of manufacture
- the function of the cosmetic product, unless it is clear from its presentation;
- a list of ingredients.

#### 8.3.4.7 Cosmetic claims

In the labelling, making available on the market and advertising of cosmetic products, text, names, trademarks, pictures and figurative or other signs shall not be used to imply that these products have characteristics or functions that they do not have. The label may state that no animal tests on the finished cosmetic product have been carried out, or its prototype, or any of the ingredients contained in it, or used any ingredients that have been tested on animals by others for the purpose of developing new cosmetic products.

## 8.4 Summary

Access to the European market for producers of herbals requires a clear understanding of a complex set of overlapping regulations that cover the manufacture and sale of herbal remedies, foodstuffs, food supplements and beauty care products. The potential exporter must first find out which regulations apply to their product and which do not. In some cases it is easy to decide whether a product is a herbal medicine, a foodstuff, a supplement or a beauty care product. In other cases it is difficult to determine and will require detailed discussion and debate with regulatory staff specializing in what are known as 'borderline products'. Furthermore, not all EU countries will take an identical view on the matter!

Once a clear decision has been reached as to what category best fits the product, the task of preparing (and submitting) a dossier to the appropriate regulatory agency in the prescribed format begins. This is almost always a labour-intensive and time-consuming activity requiring very close attention to detail and rigorous scientific analysis. Materials will need to be translated into the language of the country in which the product is to be marketed. In some cases (as with most food supplements), the regulations will lay down precisely the type and quantity of the ingredients that can be used and the exact wording of claims that can be made on the packaging and promotional material. In other situations – as with most THMPs – it is possible to negotiate with the regulatory authorities on everything from the colour of the label to the composition and quantity of ingredients used.

As a general rule, regulatory controls are more stringent on medicines than they are on foods and cosmetics. This is clearly apparent to any cosmetic manufacturer who moves away from producing simple decorative topical products to trying to launch a therapeutic skin care range requiring a full set of clinical studies and a dossier no different in depth and content to an over-the-counter medicine.

For those companies wishing to market products under the Traditional Herbal Medicinal Products Directive the regulatory situation is by now fairly well known and researched. The Traditional Herbal Medicinal Products Directive has been in force for some years in most EU countries, and many hundreds of dossiers have been submitted and reviewed by regulatory authorities. In the case of the EU Food Supplements Directive, the regulations are still far from being fully applied Europe-wide. EFSA has rejected more than 80% of the proposed health claims that companies have submitted for approval, and while recommendations for safe levels of vitamin and mineral supplements have been agreed upon, dialogue with industry as to recommended levels of botanical ingredients in supplements has not even started.

Certification is another extremely important topic for those involved in marketing herbals in Europe. While organic or fair trade certification is not compulsory, consumers increasingly want to learn more about exactly what goes into the products they buy and where and under what conditions the products are grown and processed. Some of these issues are already covered under EU legislation – THMPs, for instance, will only get regulatory approval if the producer can show that the raw materials are grown according to EU-approved GACPs. There is, however, no independent inspection process to ensure that these rules are being complied with, and that is where organic and fair trade certifications come into play. These certificates are issued by professional certifying bodies whose job it is to monitor operating procedures and production methods and undertake annual inspections. Unfortunately, some of the certification processes suffer from problems similar to those of EU regulations. They are not always of uniform format and quality; for example, EU organic certificates are not accepted in Japan.

Entering the EU herbal market is not always straightforward as there are many regulatory issues that need to be considered. The market is, however, large and sophisticated and, barring the recent European recession, has been growing at a very fast pace throughout the 21st century.

## Reference

EMEA (2006) *Guidelines on Good Agricultural and Collection Practice for Starting Materials of Herbal Origin*. European Medicines Agency, London.

## Further reading

### Useful literature on raw materials

EC (2005) Guidelines on HACCP, GMP and GHP for ASEAN Food SMEs, http://ec.europa.eu/food/training/haccp_en.pdf.

EHIA (n.d.) Guidelines for good agricultural and hygiene practices for raw materials used for herbal infusions (GAHP) approved on 03/06/2005, www.ehia-online.org.

Europam (2003) Guidelines for good wild crafting practice (GWP) of medicinal and aromatic plants, www.europam.net/documents/gacp/EUROPAM_GACP_MAP_8.0.pdf.

Europam (2006) Guidelines for good agricultural practice (GAP) of medicinal and aromatic plants, http://www.europam.net/documents/gacp/EUROPAM_GACP_MAP_8.0.pdf.

FAO Committee on Agriculture (2003) Development of a framework for good agricultural practices. ftp://ftp.fao.org/docrep/fao/010/a1193e/a1193e00.pdf.

FAO/WHO (2006) FAO/WHO guidance to governments on the application of HACCP in small and/or less-developed food businesses, FAO Food and Nutrition Paper 86, ftp://ftp.fao.org/docrep/fao/009/a0799e/a0799e00.pdf.

Kabelitz, L. and Sievers, H. (2004) Contaminants of Medicinal and Food Herbs with a View to EU Regulations. *Innovations in Food Technology*, **11**, 25.

Lange, D. (1998) Europe's Medicinal and Aromatic Plants: Their Use, Trade and Conservation. TRAFFIC, Cambridge, UK.

Leaman, D.J. and Salvador, S. (2005) International Standard for Sustainable Wild Collection of Medicinal and Aromatic Plants ISSC-MAP. Available online at www.floraweb.de/map-pro/Standard_Version1_0.pdf.

Scholten, W.K. (2003) The new European Union good agricultural and collection practice rules. Are good practices good enough if they do not result in batch-to-batch consistency? *Therapeutic Innovation & Regulatory Science*, **37**, 321–327.

United Nations Food and Agriculture Organization. Good Agricultural Practice Framework. Available online at http://www.fao.org/DOCREP/MEETING/006/Y8704e.htm.

WHO (2003) WHO Guidelines on Good Agricultural and Collection Practices (GACP) for Medicinal Plants. World Health Organization, Geneva.

WHO Guidelines on Good Agricultural and Collection Practices (GACP). Available online at whqlibdoc.who.int/publications/2003/9241546271.pdf.

### Useful literature on herbal medicines

Brendler, T., Phillips, L.D. and Spiess, S. (2009) *A Practical Guide to Licensing Herbal Medicinal Products*, Pharmaceutical Press, London.

EMEA (2004) EMEA/CPMP/HMPWP/1156/03. Final concept paper on the implementation of different levels of scientific evidence in core data for herbal medicinal drugs. European Medicines Agency, London.

Medicine and Healthcare Products Regulatory Agency (MHRA) (2001) Guidance on requesting a review of a provisional determination issued by the borderline section of the medicines control agency.

Medicine and Healthcare Products Regulatory Agency (MHRA) (2005) List of herbal ingredients which are prohibited or restricted in medicines.

Medicine and Healthcare Products Regulatory Agency (MHRA) (2007a) Guidance Note 8: A guide to what is a medicinal product.

Medicine and Healthcare Products Regulatory Agency (MHRA) (2007b) Traditional herbal medicines registration scheme – guidance for retailers, wholesalers, importers and manufacturers on the requirements of the THMRS.

### Useful literature on food supplements, novel foods

AFSSA (2003) Framework for the evaluation of the safety, the effect and the claims of foodstuff made from plants for the human diet, http://www.anses.fr/sites/default/files/documents/NUT-Ra-PlantesEN.pdf.

Codex Alimentarius (1997)

Hazard analysis and critical control point (HACCP) system and guidelines for its application. Annex to CAC/RCP 1–1969, Rev-3 Council of Europe, 2005.

Council of Europe (2005) Guidelines on the safety of plant-based food supplements and consumer interest in those supplements. Public Health Committee, Committee of Experts on Nutrition, Food Safety and Consumer Health, Ad-hoc Group on Food Supplements, 24 May.

EFSA (2005) Opinion of the scientific committee on a request from EFSA related to a harmonised approach for risk assessment of substances which are both genotoxic and carcinogenic. *EFSA Journal*, **282**, 1–31.

EFSA (2006) Guidance document of the Scientific Panel on Genetically Modified Organisms for the risk assessment of genetically modified microorganisms and their derived products intended for food and feed use. *EFSA Journal*, **374**, 1–115.

EFSA Scientific Committee (2009) Guidance on safety assessment of botanicals and botanical preparations intended for use as ingredients in food supplements. *EFSA Journal*, **7** (9), 1249.

### Useful literature on cosmetics

Colipa (2001) *Guidelines for the Evaluation of the Efficacy of Cosmetic Products*, second edition.

Cosmetic Product Information Requirements in the European Union – Amended Guidelines for the Cosmetic Industry Based on Article 7a of the 6th Amendment to the EU Cosmetics Directive, http://ec.europa.eu/enterprise/cosmetics/html/cosm_borderline_docs.htm.

Declaration of Helsinki, adopted by the 18th World Medical Assembly, Helsinki, Finland, June 1964; amended by the 29th World Medical Assembly, Tokyo, Japan, October 1975, the 35th World Medical Assembly, Venice, Italy, October 1983 and by the 41st World Medical Assembly, Hong Kong, September 1989.

EC (2006) The SCCP's Notes of Guidance for the Testing of Cosmetic Ingredients and their Safety Evaluation, 6th Revision.

IFSCC (1987) *Principles of Product Evaluation: Objective Sensory Methods*, Monograph No. 1, Micelle Press.

Rogiers, V. (2001) EEMCO guidance for the assessment of transepidermal water loss in cosmetic sciences. *Skin Pharmacology and Applied Skin Physiology*, **14**, 117–128.

# 9 Nutrition, Health and Food Security: Evidence and Priority Actions

L.J. Ferrão and T.H. Fernandes
Lúrio University, Nampula, Mozambique

## 9.1 Introduction

Among human rights, proper nutrition and health are considered fundamental. During the last two decades, all world governments, gathered in several World Food Summits, pledged to eradicate one of the worst scourges weighing on society's collective conscience: hunger.

One of the Millennium Development Goals (MDGs) was to halve the number of people undernourished in the world by the year 2015. However, presently in 2013, it is realized that the objective will not be attained. Indeed, the world nations are confronted with the sad reality that virtually no progress has been made towards reducing hunger and improving well-being, while there has been only a meager reduction of some 3 million undernourished people in few Asian countries, but not in Africa (UNSCN, 2010). But sadly and wrongly, malnutrition is the only focus, and only malnutrition is considered in the MDGs. Nutrition-related chronic diseases are not explicitly addressed in the MDGGs, and this omission will no doubt be corrected when they are revised.

While medicine is quite an old discipline, nutrition is quite a recent one, and most developing countries do not have qualified nutritionists in sufficient numbers. This lack of specialized human resources strongly impairs the fight against hunger and nutritional diseases.

The world population is around 7 billion people. If one virtually allocated people shoulder to shoulder, the space needed for the whole world population would be some 1295 km$^2$ (i.e. the area of Los Angeles) or three world populations in the Emirate of Dubai. This reflects clearly that the main problem is not demographic explosion but namely distribution of available food.

A smaller percentage of the populations of developing countries are undernourished today in relation to a decade ago; however, chronic malnutrition figures for children continue to be very high, such as the 50% recorded in Mozambique presently.

Human health is considerably affected by availability of foods, with nutritional quality maintaining and strengthening physiological and metabolic needs improving the immune system. A strong component is otherwise related to environmental health, which may in turn affect crop yield, production and nutritive quality of many agricultural products.

Nutrition is a cornerstone that affects and defines the health of all people, rich and poor. It paves the way for us to grow, develop, work, play and resist infection and aspire to realization of our fullest potential as individuals and societies. Conversely, malnutrition makes us all more vulnerable to disease and premature death.

Poverty is a major cause and consequence of ill-health worldwide. Poverty, hunger and malnutrition stalk one another in a vicious circle, compromising health and wreaking havoc on the socioeconomic development of whole countries, entire continents. Nearly 30% of humanity, especially those in developing countries – infants, children, adolescents, adults and older persons – bear this triple burden (Cafiero, 2012). Most developing countries have enjoyed remarkable rates of growth during recent decades. High growth rates of GDP per capita are a key factor in reducing food insecurity and malnutrition.

However, economic growth per se does not guarantee success. Furthermore, since the development of the wheel and fire, humankind has attempted to reduce the effort involved with activities both at home and away from the home. The striving to increase the tastefulness of the diet has been equally important. Fat and sugar are two of the more pleasurable elements of the diet in terms of taste preferences. Even in poor countries there is a negative shift in dietary patterns involving reductions in fibre and whole-grain intakes, increases in refined carbohydrate intakes (particularly in sweeteners and sweet beverages), and increases in intakes of animal protein and partially hydrogenated fats (FAO, WFP and IFAD, 2012).

Health and sustainable human development are equity issues. In our globalized 21st century, equity must begin at the bottom, hand in hand with healthy nutrition. It must also be realized that resources allocated to preventing and eliminating disease will be effective only if the underlying causes of malnutrition – and their consequences – are successfully addressed. One should even define strategies that will prevent disease and malnutrition from occurring in the first place, but this is not so easy in developing countries.

## 9.2 Well-being and nutrition

Malnutrition, the single most important risk factor for disease, means 'badly nourished', but it is more than a measure of what one eats, or fails to eat. Clinically, malnutrition is characterized by inadequate intake of protein, energy and micronutrients and by frequent infections or disease. Nutritional status is the result of the complex interaction between the food eaten, the overall state of health and the environment in which one lives – in short, food, health and caring, the three 'pillars of well-being'. Malnutrition

*Novel Plant Bioresources: Applications in Food, Medicine and Cosmetics*, First Edition. Edited by Ameenah Gurib-Fakim.

casts long shadows, affecting close to 800 million people – 20% of all people in the developing world.

Malnutrition kills, maims, cripples and blinds on a massive scale worldwide, affecting one in every three people worldwide, afflicting all age groups and populations, especially the poor and vulnerable. It plays a major role in half of the 10.4 million annual child deaths in the developing world; it continues to be a cause and consequence of disease and disability in the children who survive. It should be noted that malnutrition is not only medical; it is also a social disorder rooted in poverty and discrimination having economic ripple effects that can jeopardize development.

African governments have finally realized this link, and special investment has been located in establishing new nutrition and health degrees that may tackle the problem by graduating nationals that will spread the impact of balanced nutrition on preventing illness and guide the necessary changes, namely in rural areas. However, malnutrition is as much a technical problem as political, and only with an interdisciplinary approach can its effects in human health be minimized.

The impact of nutrition on general health is mainly reflected on the malnutrition-related diseases, which include iron deficiency/anaemia, iodine deficiency disorders, vitamin A deficiency, intrauterine growth retardation, obesity and protein-energy malnutrition (Popkin, 2006). Being a chronic development, malnutrition has its roots and present dimension largely linked to past underdevelopment and discrimination. Exacerbated by poverty, malnutrition combines with disease, both chronic and infectious, to form a deadly duo that together can deal a lethal blow to development. The consequences include death, disability, stunted mental and physical growth and, as a result, delayed national development. All approaches should be human rights based, target the poor, promote gender equality, enhance long-term resilience and allow sustainable graduation out of poverty.

Healthy eating and regular physical activity play a substantial role in preventing chronic diseases, including heart disease, cancer and stroke, the three leading causes of death among adults aged over 18 years in developed countries. In Africa, eating habits avoid civilization diseases, while people are used to walking for long distances in a rural environment.

Micronutrient deficiencies ('hidden hunger') still affect over 30% of the world's population, causing increased morbidity and mortality, impaired cognitive development and reduced learning ability and productivity, reduced work capacity in populations due to high rates of illness and disability, and tragic loss of human potential. Overcoming micronutrient malnutrition is a precondition for ensuring development. The world is increasingly faced with a double burden of malnutrition, whereby undernutrition, especially among children, coexists with overweight and diet-related chronic diseases and micronutrient malnutrition.

Poor diet and physical inactivity among younger persons can lead to an increased risk for certain chronic health conditions, including high blood pressure, type 2 diabetes and obesity. The world is shifting rapidly toward a diet linked with non-communicable diseases. In Africa, these civilization malfunctions have been slowly introduced with fast food and drink in major cities. South African women in urban areas are considered the most obese in the world (Mendez *et al.*, 2005). Neighbouring countries follow a similar trend of 'globesity' – a swelling global tidal wave of obesity. This means a huge difference of nutritional status among urban and rural populations and, therefore, different approaches for the authorities.

A major problem in rural areas concerns the lack of variation of main foods and the predominance of fewer staple diets. Maize and cassava are good sources of energy, and supplemented with a protein source such as beans, eggs, meat and fish may constitute a balanced diet. However, the latter are not so immediately available in all provinces of many countries. A further problem, indirectly related, is the availability of quality water free of pathogens. Less than 1% of the world's water is liquid fresh water, and having clean water is not enough if people do not have access to it.

Nutrient-deficient diets provoke health problems; malnutrition increases susceptibility to disease. The spiral of poverty and illness can end in death. Strategic objectives should be pursued within an integrated agriculture–nutrition–health framework. Improving food security and nutrition is about more than just increasing the quantity of energy intake; it is also about improving the quality of the food in terms of dietary diversity, variety, nutrient content and safety. Measures to achieve greater dietary diversity and adequate intake of micronutrients may include the judicious use of targeted supplementation for the poorest until the cost of a diversified diet becomes affordable. Overnutrition should be addressed through changes in lifestyle and healthier diets.

## 9.3 Traditional food cultures

There is a compelling need for prevention of chronic diseases, as well as of nutritional deficiencies. Adequate and healthful food supplies, which often have much in common with long-established food cultures, need to be protected, although their dietary diversity may need to be improved.

In order to halt the progression of obesity, diabetes and other chronic diseases in least-resourced and impoverished countries, the causes of these diseases that can be changed need to be confronted (Smith *et al.*, 2006) . These include patterns of diet and physical activity.

As mentioned, long-established traditional food cultures, often higher in vegetables, fruits and dietary fibre than industrialized food supplies, should be protected and revived, and regular physical activity should again become the social norm.

Food systems and ways of life typical of industrialization in its modern and current forms are by and large determined by forces of economic globalization, privatization of public services and deregulation, especially as this has led to the vast expansion of transnational corporations.

Food-based dietary guidelines that take account of local and regional food cultures should be compiled, published and made widely available. For those already diagnosed with diabetes, health professionals should be available to advise how to eat healthily and to make appropriate changes in diet. All this means that many more adequately trained nutrition professionals are needed in sub-Saharan Africa.

## 9.4 Nutrition in pregnancy and infancy

In Africa most of all, improved nutrition of women and infants is crucial. This improves reproductive health and child survival in the short term. It also protects against obesity and chronic diseases in the longer term and for following generations. The time of pregnancy and the first 2 years of life are now well known as the 'window of opportunity' to reduce risk of disease in infants and young children.

Maternal malnutrition, or obesity at the other end of the spectrum, gestational diabetes or pregnancy-induced hypertension all put infants at risk. Nutrition is often a neglected component of maternal and child health programmes. Most likely this is because those who should provide prenatal care are not aware of the importance of nutrition, or do not have the time for it, or do not know about nutrition. Similarly, the first 2 years of life are critical not only for optimal growth and development, but also to reduce the long-term risk associated with low as well as high birth weight.

In the World Health Organization's (WHO's) review of key interventions for maternal and child health (PMNCH, 2011), nutrition interventions identified as essential at community and primary levels are primarily micronutrient supplementation (iron, folate, calcium) in pre-pregnancy and pregnancy. It is disappointing to find no mention of nutrition monitoring and counselling. Nutrition is not positioned as part of the antenatal care package, except for anaemia. The WHO guide lists 'advice and counsel on nutrition and self-care', but without making links, for example, to haemoglobin level, weight or weight gain, current dietary practices or household access to food. There is no mention either of nutritional management of gestational diabetes, which is fairly common now in low-income countries. Nutrition counselling only appears for mothers in the post-natal period, but here again no link is made with poor nutrition and future health risks for mother and child.

## 9.5 Health and nutrition education is central for development

Investing in nutrition education is crucial for health, for food sovereignty, for human rights, for economic development, for social protection and thus for development of a country. The World Bank positions nutrition as central to development. But while progress has been noted in most regions, sub-Saharan Africa has stagnated. In Africa in 2010, well over one-third (38%) of under-5 children were stunted. According to available data, the prevalence of anaemia is highest in Africa, with nearly a half (47.5 %) of under-5 children and two-thirds (67.6%) of women affected. An estimated 2% of under-5 African children and 10% of African pregnant women experience night blindness, dangerous especially in rural areas, and the first clinical sign of vitamin A deficiency (World Bank, 2008).

Chronic (non-communicable) diseases are currently responsible for nearly two-thirds of world deaths, and four-fifths of these occur in low- and middle-income countries. For diabetes alone, the prevalence in sub-Saharan Africa ranges between 3 and 7.7% in cities and from 1 to 3.5% in rural areas, and is rising rapidly, from a total of an estimated 12 million in 2010 to a projected 24 million in 2030 (WHO, 2012).

Based on the recommended needs of 100–500 BSc- (Licentiates), 10–50 MSc- and 5–25 PhD-level nutritionists per 5 million people, the minimum number of nutrition graduates that are needed every year can be estimated. Competent nutritionists require several years of theoretical and practical training (Hughes *et al.*, 2012). Furthermore, the very field of nutrition should be better defined; currently, its specificity and boundaries are hazy.

Nutrition encompasses the various dimensions of the relationship of human beings with their food: biological, of course, but also psychological, social, political, economic and environmental. When these dimensions are ignored the discipline is no longer nutrition; it is more likely really to be, say, food science, biochemistry or agriculture. So why in Africa especially are there so few nutritional professionals and specialists working in the health sector? And why is this fact not recognized as an acute problem? One reason may be that the scale and width of nutrition as a discipline and in practice, engaged with social, economic and environmental as well as biological and behavioural aspects of life, is not yet well understood.

In Africa, the relevance of current nutrition training programmes has to be questioned. Competency-based training is a need and a must. But changes from objectives to actual real practised competency will only take place progressively in training programmes. Objectives in terms of knowledge, attitudes and skills will no doubt continue to be pursued, but it is primarily in assessment methods that objectives and competencies differ. Improved childhood nutrition and access to education can improve cognitive development and thereby raise levels of income when those children become adults – with benefits at the individual level as well as for society as a whole.

In several low-income countries of Africa, nutritionists are trained to address malnutrition, but not chronic diseases. Nutrition training programmes tend to focus on food production, composition and technology. This expertise is vital in the context of food insecurity and child malnutrition, but is not sufficient. Nutrition is interdisciplinary. It is at the crossroads of agriculture and health.

High-income countries can afford specialists in public health nutrition. Low-income countries cannot. What Africa needs now is a strong cadre of professional nutritionists trained in general nutrition practice at the undergraduate level, representing the base of the higher education pyramid. For this general practice, they require a thorough knowledge and understanding of food composition, food systems, food behaviour, nutrition requirements and recommended intakes, in addition to public and clinical nutrition (Mazzocchi *et al.*, 2012).

In low-income countries there should be no marked separation between public health and clinical nutrition. Treatment of severe malnutrition cannot be rationally dissociated from management of moderate malnutrition and its prevention. There is wide consensus on prevention of chronic diseases, but nutritional management of those already affected, say, with diabetes, cardiovascular disease or hypertension, and of the sick and hospitalized, at high nutritional risk, is also needed. This is the rationale for training professional nutritionists for general practice, before specializing some of them in public health nutrition at MSc level.

The promotion of sound eating practices and the nutritional management of diseases are among the tasks that professional nutritionists should have. Competent specialists are also needed, to train other professionals and field workers, for advocacy with governments, to design and evaluate nutrition programmes and to interact with the food and pharmaceutical industry.

## 9.6 Research and development

The current research and development (R&D) system is driven by market forces, not health needs, and relies overwhelmingly on the patent system to recoup R&D costs by charging high prices for the medical tools that reach the market. This creates two key problems (UN, 2012).

First, the needs of people in wealthy countries trump the needs of people in poor countries. When people affected by a given disease are too few or too poor to compete with markets in wealthy countries, medical challenges go unaddressed. Since wealthy country needs drive innovation, tools primarily designed for high-infrastructure, resource-rich environments are only subsequently rolled out for use in resource-limited settings – even if they may not be practical to use or are not designed for the disease burden, or indeed the specific disease strains in developing countries.

Second, when medical tools do exist – because the diseases affect rich and poor countries alike and the population of patients in wealthy countries is enough of a market pull – they are often priced out of reach. This holds true for new HIV/AIDS and cancer (e.g. human papillomavirus) drugs and vaccines, for example. This has led to bitter disputes as those seeking to enforce patent-protected monopolies collide with those seeking to secure the widest and most affordable access to new medical and food additive tools.

Most African countries do not conduct enough R&D activities in the nutritional domain, and only in the last decade has better attention been given. Research is scarce, and when conducted it is not published in peer review journals; therefore, information gets lost. Although new and regular funding has been available in many African countries from philanthropic institutions, NGOs and governments, innovation on nutrition and health is meagre. There is also a vast gap on innovation in Africa while new funding from philanthropic foundations and governments has been forthcoming. Efforts to stimulate R&D have resulted in some urgently needed improvements. However, on the whole, these efforts have been ad hoc, inadequate and are not part of a sustainable system designed to meet today's developing country nutrition and health challenges (Delisle, 2012).

It has long been also argued that professionals should be trained in their own countries, that training overseas may not be appropriate for the competencies needed and that there is a brain-drain favouring wealthy countries. But for Africans there are more opportunities to train abroad, and very little support for university studies in Africa.

The private sector is far more effective at pursuing and applying innovation than government could ever be. However, there are key areas in which government policy must strengthen the ability of the private sector to innovate effectively. Although, generally, there is no robust government funding for research in Africa, the focus should be for government resources towards research programmes that advance the development of knowledge, and on technologies with widespread application and potential to serve as the foundation for private sector innovation and commercialization.

There is an urgent need for a complementary system to drive and fund innovation for people in developing countries. Such a new system needs to speed up the development of traditional drugs, vaccines, diagnostics, nutraceuticals and pharmanutrients designed for the nutritional and medical needs of people in developing countries. But who can finance innovation initiatives in Africa? Many good ideas may occur, but these are later developed in wealthy countries and the product being sold to the original source. The science should be an input to the public policy decision.

As previously mentioned, we live today in a world of food abundance but with unbalanced distribution, transport and conservation, through lack of communication roads, cold enterprises, storage places and also due to economic interests and the greediness of big international groups. Short-term profits for a few are long-term losses for many. Improved health and increasing agricultural productivity in the past have catalysed unprecedented social and economic transformations. Today there is more than enough food for all, but it is concentrated in rich and productive countries. Despite abundance, hunger hovers; despite progress, poverty persists.

It is necessary to have active public policies to ensure that the fruits of economic growth are widely shared, and also requires making good use of the public revenue generated by fast economic growth for social services, especially for public healthcare and public education. As in other nutrition interventions, evidence of impact of nutrition monitoring and counselling in pregnancy may be lacking, due to lack of human resources and funds for this type of research. Here, there is a vicious circle: no research, no scientific evidence, no evidence-based nutrition practice; no research, no evidence, …

## 9.7 Role of agricultural growth on reducing poverty, hunger and malnutrition

Agriculture is still the dominant source of employment in the economies of many low-income countries, and the urban poor spend most of their income on food (FAO, 2009). Agricultural growth is particularly effective in reducing hunger and malnutrition. Most of the extreme poor depend on agriculture and related activities for a significant part of their livelihoods.

Agricultural growth involving smallholders, especially women, will be most effective in reducing extreme poverty and hunger when it increases returns to labour and generates employment for the poor. Moreover, for the foreseeable future, the majority of the poor and hungry will continue to live in rural areas and depend directly or indirectly on investments in rural infrastructure and smallholder-based agriculture to improve their livelihoods. Therefore, more investment is required in agriculture, food chains, markets and commerce of products from farmers.

All stakeholders must promote practical solutions that aim to promote sustainable intensification of food production systems, ensure a strong involvement of smallholder farmers and other rural poor, and preserve natural resources – including by minimizing post-harvest losses and waste throughout the food chain. Higher prices of agricultural commodities provide positive incentives for increased investment in agriculture. However, the largest commodity in the world, the US dollar, has a strong impact in the value of everything, rendering all countries dependent on the capacity to obtain that currency in order to improve their balance of payments.

Production growth in agriculture during the past 20 years was accounted for primarily by increased area harvested, with relatively little deriving from higher yields. A growth pattern based on land expansion raises questions of sustainability (FAO, 2006). Better policy responses and improved governance are needed to ensure sustainability and to address the effects of increased price volatility and of higher costs of the food basket for the poor, most of who are net food buyers or live at subsistence levels.

Economic and agricultural growth should be 'nutrition sensitive'. Growth needs to result in better nutritional outcomes through enhanced opportunities for the poor to diversify their diets; improved access to safe drinking water and sanitation; improved access to health services; better consumer awareness regarding adequate nutrition and childcare practices; and targeted distribution of supplements in situations of acute micronutrient deficiencies. Good nutrition, in turn, is key to sustainable economic growth.

## 9.8 Concluding remarks

Healthy people feel stronger, can work better and may have more earning opportunities to gradually lift them out of both poverty and malnutrition. Healthier, more productive societies are a potential outcome. The number of nutrition graduates must be increased in parallel with other health professionals and the approach must be multidisciplinary in the search of solutions for well-known problems.

Good African governance must link with communities, farmers, agricultural cooperatives and associations, industries, universities and state agencies to develop specific guidance for the commodities most often associated with food-borne illness outbreaks. With advanced research and continued scientific breakthroughs, state-of-the-art monitoring and a collaborative instead of combative relationship between regulators, producers and businesses, major developments were achieved in developed countries but with weak immediate impact in developing countries. The latter may reveal that strong economic growth is necessary but not sufficient to accelerate reduction of hunger and malnutrition.

Each country must safeguard its food supply, prevent water and food-borne illnesses and improve consumers' knowledge about the food they eat. Administration demands numerous steps for approval of new drugs and systems that are stifling medical and nutritional innovation. Nutrient-deficient diets provoke health problems; therefore, attention to local foods and medicinal plants, and possibly the creation of a specific food pyramid with local indigenous foods, could be a way to improve nutritional status.

In nutrition as well as in other health disciplines, human resources are needed, for intervention, policy and research. The importance of generating new knowledge locally through research and to apply research results in policy and programming, even in low-income countries, is now widely recognized.

Developments are only possible by the combined effects of increased attention to world hunger, overall economic and agricultural growth, and targeted policy interventions. It is imperative to reinstate, protect, preserve and, as needed, adjust long-established traditional food systems. New approaches are required to equip African societies for the health challenges of the 21st century.

## References

Cafiero, C. (2012) Advances in hunger measurement: traditional FAO methods and recent innovations, in *The 1st International Conference on Global Food Security*, Noordwijkerhout, The Netherlands, www.fao.org/fileadmin/templates/ess/ess_test_folder/Food_security/Cafiero_Global_Food_Security.pdf.

Delisle, H. (2012) Empowering our profession in Africa. *World Nutrition*, **3** (6), 269–284.

FAO (2006) *The State of Food Insecurity in the World 2006: Eradicating World Hunger – Taking Stock Ten Years after the World Food Summit*, Food and Agriculture Organization of the United Nations, Rome.

FAO (2009) *The State of Food Insecurity in the World: Economic Crisis – Impacts and Lessons Learned*, Food and Agriculture Organization of the United Nations, Rome.

FAO, WFP and IFAD (2012) *The State of Food Insecurity in the World 2012. Economic Growth is Necessary but not Sufficient to Accelerate Reduction of Hunger and Malnutrition*, Food and Agriculture Organization of the United Nations, Rome.

Hughes, R., Shrimpton, R., Recine, E. and Margetts, B. (2012) Empowering our profession. *World Nutrition*, **3** (2), 33–56.

Mazzocchi, M., Shankar, B. and Traill, B. (2012) *The Development of Global Diets since ICN 1992: Influences of Agri-food Sector Trends and Policies*, Food and Agriculture Organization of the United Nations, Rome.

Mendez, M.A., Monteiro, C.A. and Popkin, B.M. (2005) Overweight exceeds underweight among women in most developing countries. *American Journal of Clinical Nutrition*, **81**, 714–721.

PMNCH (2011) *Essential Interventions, Commodities and Guidelines for Reproductive, Maternal, Newborn and Child Health*, The Partnership for Maternal, Newborn and Child Health, World Health Organization, Geneva.

Popkin, B.M. (2006) Global nutrition dynamics: the world is shifting rapidly toward a diet linked with non-communicable diseases. *American Journal of Clinical Nutrition*, **84**, 289–298.

Smith, L.C., Alderman, H. and Aduayom, D. (2006) Food insecurity in sub-Saharan Africa. New estimates from household expenditure surveys, Research Report 146, International Food Policy Research Institute, Washington, DC.

UN (2012) *The Future We Want*, Outcome document adopted at Rio+20, United Nations Conference on Sustainable Development, www.uncsd2012.org/content/documents/727The%20Future%20We%20Want%2019%20June%201230pm.pdf [accessed 12 November 2013].

UNSCN (2010) *Nutrition and the MDGs: Accelerating Progress Towards 2015*, United Nations Standing Committee on Nutrition, Geneva.

WHO (2012) Global Health Observatory (GHO) data repository, www.who.int/gho/en/ [accessed 12 November 2013].

World Bank (2008) *World Development Report 2008: Agriculture for Development*, World Bank, Washington, DC.

## Part Two

# Medicine (Plants as Medicine: Humans and Animal Health)

# 10 Anticancer Potential of African Plants: The Experience of the United States National Cancer Institute and National Institutes of Health[1]

John A. Beutler[1], Gordon M. Cragg[2], Maurice Iwu[3], David J. Newman[2] and Christopher Okunji[4]

[1]Molecular Targets Laboratory, Center for Cancer Research, NCI-Frederick, Frederick, MD, USA
[2]Natural Products Branch, Developmental Therapeutics Program, NCI-Frederick, Frederick, MD, USA
[3]Bioresource Development & Conservation Programme, Nigeria, Head Office, Abuja, FCT, Nigeria
[4]US Pharmacopeial Convention, Rockville, MD, USA

## 10.1 Introduction

Nature has been, and continues to be, an indispensable source of novel lead molecules for the development of effective drugs for the treatment of many of the diseases afflicting humankind. A recent analysis indicated that 64% of the 1073 small-molecule drugs approved for the treatment of all diseases worldwide during the period 1981 to 2010 could be considered as naturally derived or inspired, leaving 36% as truly synthetic (i.e. totally devoid of natural inspiration). In the cancer treatment area, close to 80% can be regarded as naturally derived or inspired, while approximately 68% of anti-infectives (anti-bacterial, -fungal, -parasitic and -viral) fall into this category (Newman and Cragg, 2012).

The effective discovery and development of novel drugs always requires multidisciplinary collaboration. In the area of natural products, the bioactive compound isolated from the natural source often is not suitable as such for development into a drug, but may be regarded as a lead molecule for structural optimization, a process that requires significant input from biochemists and medicinal and synthetic chemists. In addition, preclinical development always requires close collaboration with pharmacologists and toxicologists in order to adjust pharmacodynamic and toxicological properties to advance the agent into clinical trials with human patients.

An additional, important requirement in natural product bioactive lead discovery and development, however, is close international collaboration. In the following discussions, features of the policies and collaborative agreements established by the United States National Cancer Institute (NCI) and Fogarty International Center at the National Institutes of Health (FIC/NIH) ensuring 'fair and equitable' collaboration with selected, biodiversity-rich African countries through government agencies and/or appropriate organizations are presented. These policies adhere to the objectives established by the Convention on Biological Diversity (CBD) for 'access and benefit-sharing' associated with the use of genetic resources, or the use of traditional knowledge pertaining to the resources of biodiversity-rich countries, at the United Nations Conference on Environment and Development held in Rio de Janeiro (the Rio 'Earth Summit') in June, 1992, as well as guidelines and protocols developed at subsequent meetings of the Conference of the Parties (www.biodiv.org/convention/articles.asp) in Bonn in 2002 (www.cbd.int/doc/publications/cbd-bonn-gdls-en.pdf) and in Nagoya in 2010 (www.cbd.int/abs/text/). The requirements and implications of these guidelines and protocols are discussed in detail in a recent review by Cragg *et al.* (2012).

## 10.2 The United States National Cancer Institute programme

### 10.2.1 Background

The NCI (www.nci.nih.gov) was established in 1937, 'to provide for, foster and aid in coordinating research related to cancer.' Since 1955, NCI has provided resources for preclinical screening and clinical development of compounds and materials submitted by public and private scientists and institutions worldwide, and has played a major role in the discovery and development of many commercial and investigational anticancer agents. During this period, more than 500 000 chemicals, both synthetic and pure natural products, have been screened for antitumour activity using a variety of *in vitro* and *in vivo* screening methods. The success of this effort has depended on close collaboration with organizations worldwide, and interna-

[1]The opinions expressed in this chapter are those of the authors, not necessarily those of the US Government.

*Novel Plant Bioresources: Applications in Food, Medicine and Cosmetics*, First Edition. Edited by Ameenah Gurib-Fakim.

tional collaboration continues to be an important feature of the NCI programmes (Cragg and Newman, 2005).

In the early days of the programme from 1955 to 1982, most of the materials screened were pure synthetic compounds, but it was soon appreciated that natural products were an excellent source of complex chemical structures with a wide variety of biological activities. From 1960 to 1982, random collections of over 35 000 plant samples, mainly from temperate regions, were performed by the US Department of Agriculture (USDA) through an interagency agreement with NCI, while marine invertebrates were generally collected by academic investigators, mainly funded through NCI grants. Microbial samples were obtained from pharmaceutical companies and research institutes, partially funded by contracts with the NCI. Thus, from 1960 to 1982 (i.e. more than 30 years before the CBD to 10 years before), over 180 000 microbial-derived, some 16 000 marine-organism-derived and over 114 000 plant-derived extracts were screened for antitumour activity, and, as mentioned above, a number of clinically effective chemotherapeutic agents were developed (Cragg and Newman, 2009).

After a lapse in funding from 1982 to 1986, the systematic collection of marine invertebrates and terrestrial plants resumed, coordinated by the NCI Natural Products Branch (NPB; http://dtp.nci.nih.gov/branches/npb/index.html). The collection of marine organisms began with the Caribbean and Australasia, but has now expanded worldwide through a contract with the Coral Reef Research Foundation, which is based in Palau in Micronesia. Terrestrial plant collections have been carried out in over 25 countries in tropical and subtropical regions worldwide, with collections in sub-Saharan Africa and Madagascar performed through a contract with the Missouri Botanical Garden (MBG). Over 60 000 plant samples have been collected worldwide, with 17 420 being from sub-Saharan countries, and the permanent repository of over 120 000 extracts is intended to be a resource of potential agents for the treatment of all human diseases. All collections are performed subject to the terms of the NCI Letter of Collection (LOC; http://dtp.nci.nih.gov/branches/npb/agreements.html), which states NCI's willingness to collaborate with local scientists and/or authorities in the discovery and development of novel drugs from organisms (plants, marine invertebrates, microbes) collected in their country and/or territorial waters. Agreements based on the LOC have been signed with 14 countries, including Gabon, Ghana, Madagascar and Tanzania, and it is notable that the first agreement with any country worldwide was signed with Centre National D'Applications des Recherches Pharmaceutiques, Antananarivo, in 1990, 2 years prior to the signing of the CBD. Collections by MBG were performed in Cameroon and the Central African Republic without the finalization of a formal LOC agreement, but the authorities in these two countries were fully aware of the terms of the LOC and granted the necessary permits without requiring a formal agreement. In this respect, the NCI is totally committed to the terms of the LOC irrespective of whether or not a formal agreement has been signed (Kaufman, 1993). From 1989 to 2001, 18 officials and scientists from African countries visited NCI one or more times for periods of 1–2 weeks, either to discuss participation in NCI contract collections or for direct collaboration in the drug discovery process. In carrying out these collections, NCI contractors work closely with qualified source country (SC) organizations (SCOs), and SCO scientists collaborate in field collection activities and taxonomic identification. Voucher specimens of each species collected are deposited in the national herbarium or repository, and NCI contractors provide training opportunities for local personnel through conducting workshops and presentation of lectures, both in-country and at the contractor's US facilities. In addition, NCI has invited 10 African scientists nominated by SCOs to visit its facilities, or equivalent facilities in other approved US organizations, for 1–12 months to participate in collaborative natural products research involving the screening and bioassay-directed fractionation of extracts. The LOC also dictates benefit sharing and use of source-country resources in the event of the licensing and development of a promising drug candidate. Successful licensees are required to negotiate agreements with source country government agencies or SCOs dictating terms of collaboration and compensation.

Particularly in the area of plant-related studies, SC scientists and governments are becoming increasingly committed to performing more of the drug discovery operations in-country, as opposed to simply exporting raw materials. Thus, NCI has established collaborative agreements – memoranda of understanding (MOUs), http://dtp.nci.nih.gov/branches/npb/agreements.html – with over 20 SCOs worldwide qualified to perform in-country processing. These include Rhodes University, South Africa (1998), the Council for Scientific and Industrial Research (CSIR), Division of Food, Biological and Chemical Technologies (1996) and the Zimbabwe National Traditional Healers Association (1994). NCI undertakes to abide by the same policies of collaboration and compensation as specified in the LOC, and depending on available resources, NCI also assists the SCOs in establishing their own drug-discovery programmes through training in the processes of antitumour screening and natural product isolation. Since 1988, NCI has sponsored long-term visiting scientists from 18 countries worldwide, including three from southern Africa. Under the terms of the MOU, the discovery of novel anticancer drugs is undertaken by the SCO at its own expense, and NCI assists with the provision of free secondary *in vitro* and *in vivo* testing. All results from such secondary testing are considered the sole intellectual property of the SCO, and can be used by the SCO to apply for patents covering promising inventions. NCI will also commit its resources to collaborating with the SCO in the preclinical and clinical development of any SCO-discovered drug that meets the NCI selection criteria, and agrees to transfer any knowledge, expertise and technology developed during such collaboration to the SCO, subject to the provision of mutually acceptable guarantees for the protection of intellectual property associated with any patented technology.

As noted above, NCI has played a major role in the discovery and development of many of the available commercial and investigational anticancer agents. Organizations or individuals wishing to have pure compounds tested free of charge in the NCI screening programme may submit their compounds through an online submission process at http://dtp.nci.nih.gov/docs/misc/common_files/submit_compounds.html. A screening agreement (http://dtp.nci.nih.gov/docs/misc/common_files/canagr.html) may be signed with the NCI which includes terms stipulating confidentiality and levels of collaboration in the drug development process.

Further details of the NCI collaboration policies may be found in Cragg *et al.* (2012) and Beutler *et al.* (2012).

### 10.2.2 African plant-derived anticancer agents: recent developments

While over 3000 plant species have reportedly been used in the treatment of cancer, in many instances the 'cancer' is undefined, or reference is made to conditions such as 'hard swellings', abscesses, calluses, corns, warts, polyps or tumours (Hartwell, 1982). Thus, many of these claims should be viewed with scepticism because cancer, as a specific disease entity, is likely to be poorly defined in folkloric and traditional medicine (Cragg *et al.*, 1994). This is in contrast to other plant-based therapies used in traditional medicine for the treatment of afflictions such as malaria and cutaneous fungal infection, which are more easily defined, and which are prevalent in the regions where traditional medicine systems are used extensively.

Some of the most effective plant-derived anticancer agents in clinical use are the so-called vinca alkaloids, vinblastine (VLB) and vincristine (VCR), isolated from the Madagascar periwinkle, *Catharanthus roseus*, used by various cultures for the treatment of diabetes. They were first discovered during an investigation of the plant as a source of potential oral hypoglycaemic agents, so their discovery may be indirectly attributed to the observation of an unrelated medicinal use of the source plant. Though the plant was originally endemic to Madagascar, the samples used in the discovery of VLB and VCR were collected in Jamaica and the Philippines. More recent semi-synthetic analogues of these agents are vinorelbine and vindesine. Other important plant-derived, clinically used agents are the semi-synthetic derivatives of epipodophyllotoxin, etoposide and teniposide; the taxanes, paclitaxel (Taxol®) and docetaxel (Taxotere); topotecan (hycamptamine) and irinotecan (CPT-11), semi-synthetically derived from camptothecin; and homoharringtonine. These and other important anticancer agents in clinical and preclinical development have been comprehensively reviewed, and interested readers should consult a review for details (Cragg and Newman, 2009).

#### 10.2.2.1 The combretastatins: models for combinatorial chemistry

The combretastatins were isolated from the South African 'bush willow', *Combretum caffrum* (Eckl. & Zeyh.) Kuntze, collected as part of a random collection programme in southern Africa performed in the 1970s for NCI by the USDA, working in collaboration with the Botanical Research Institute of South Africa. The combretastatins are stilbene derivatives that act as anti-angiogenic agents, causing vascular shutdown in tumours, leading to tumour necrosis (Pinney *et al.*, 2012). Two water-soluble analogues, combretastatin A-4 phosphate (CA4P; fosbretabulin disodium; Zybrestat) and combretastatin A-1 phosphate (CA1P; OXi4503) (Figure 10.1) have advanced into clinical trials under the sponsorship of Oxigene, Inc.

To date, CA4P has been in nine clinical trials against a variety of cancers, mostly used in combination with other agents such as

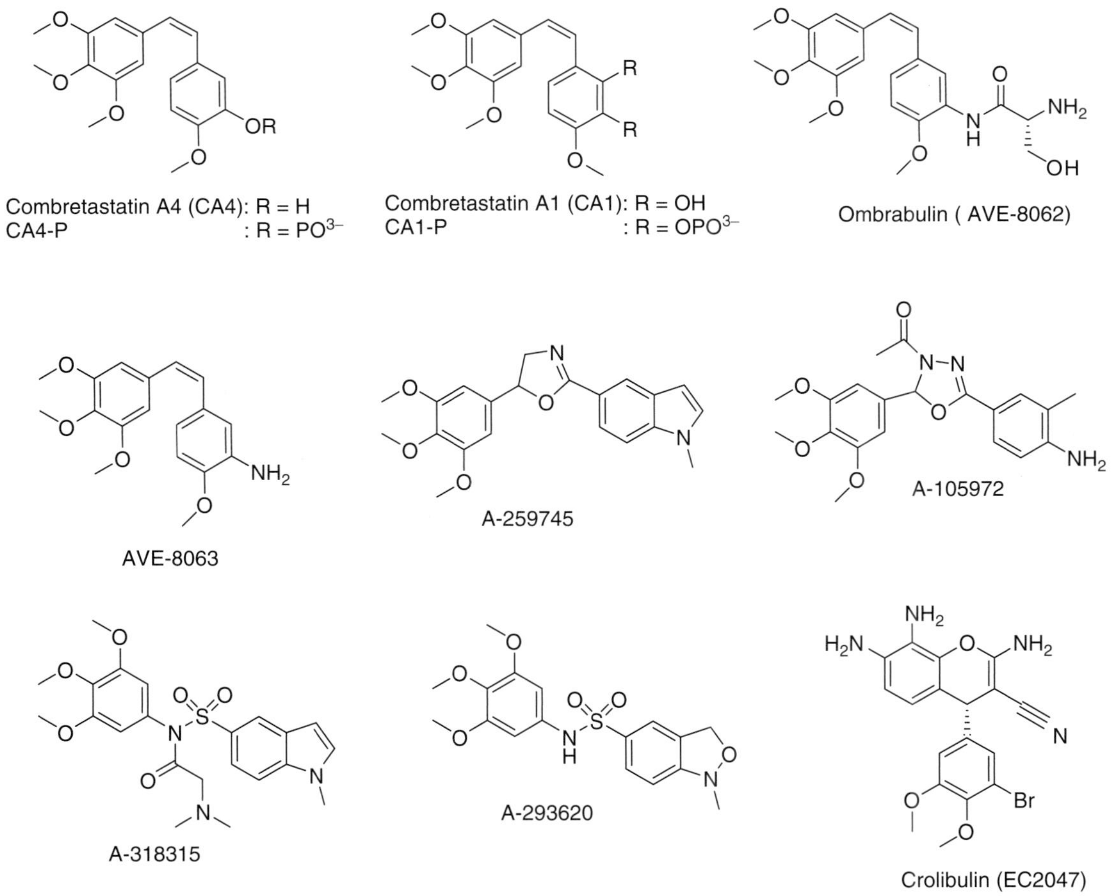


Figure 10.1 Some combretastatin analogues and mimics

carboplatin, paclitaxel or bevacizumab, as well as radiation treatment, and recruitment for two other trials is ongoing (November 2012). In March 2012, Zybrestat was granted orphan drug status by the US Food and Drug Administration (FDA) and the European Medicines Agency for the treatment of anaplastic thyroid cancer. The clinical experience with CA4P has been reviewed (Siemann *et al.*, 2009), and a significant partial response rate (29%) was observed in patients with platinum-resistant ovarian cancer when CA4P was used in combination with carboplatin and paclitaxel. CA4P has also shown promising activity against ophthalmological diseases such as myopic macular degeneration, and clinical trials using CA4P for the treatment of exudative age-related macular degeneration are also in progress (Eichler *et al.*, 2006; Pinney *et al.*, 2012).

Combretastatin A1P, like CA4P, shows excellent activity in preclinical studies, and a Phase I safety trial has been completed with patients having hepatic carcinomas; in addition, a Phase I trial against acute myelogenous leukemia (AML) or myelodysplastic syndrome is currently recruiting patients, and in late 2012 it was announced that it had received orphan drug status from the FDA for the treatment of AML.

The combretastatin chemical class has served as a model for the synthesis of a host of analogues containing the essential trimethoxy aryl moiety linked to substituted aromatic moieties through a variety of two- or three-atom bridges, including heterocyclic rings and sulfonamides, thereby providing an impressive display of the power of a relatively simple natural product structure to spawn a prolific output of medicinal and combinatorial chemistry. Some are of these are shown in Figure 10.1. One of these analogues, ombrabulin (AVE8062), is currently in a Phase II clinical trial against advanced soft-tissue sarcoma. A structural variant, crolibulin (EC2407), in combination with cisplatin, is advancing into Phase I/II clinical trials against anaplastic thyroid cancer and other solid tumours. Further information on the status of these clinical trials may be found at www.clinicaltrials.gov/ct2/results?term=cancer+combretastatins.

#### 10.2.2.2 Maytansine: targeting toxic natural products

Frequent liabilities of natural products, at least in the area of cancer chemotherapy, are that, while many are potent cytotoxins, they have limited solubility in aqueous solvents and exhibit considerable toxicity, often resulting in a narrow therapeutic index. As a result, a number of pure natural products have failed as promising leads, including the plant-derived agents bruceantin and maytansine. An alternative approach to using such agents is the investigation of their potential as 'warheads' attached to monoclonal antibodies specifically targeting epitopes on the tumour of interest. Such agents are called antibody drug conjugates (ADCs). The promise of this approach to cancer therapy has been the subject of several reviews, and readers are referred to them for further details of developments in this area (Senter, 2009; Teicher, 2009; Alley *et al.*, 2010; Caravella and Lugovskoy, 2010; Teicher and Chari, 2011).

A promising candidate for such an approach is maytansine (Figure 10.2), which was isolated in extremely low yield ($2 \times 10^{-5}$% based on plant dry weight) in the early 1960s from the Ethiopian plant *Maytenus serrata* (Hochst. Ex A. Rich.) Wilczek (Celastraceae family) collected for the NCI as part of a random collection programme performed in the early 1960s through a collaboration with the USDA (Lambert, 2010; Yu *et al.*, 2012). The novel structure and very potent *in vitro* activity of maytansine prompted great interest in pursuing further preclinical, and possible clinical, investigation. Following an extensive survey of other species of *Maytenus* and related members of the Celastraceae family, *Maytenus buchanii*

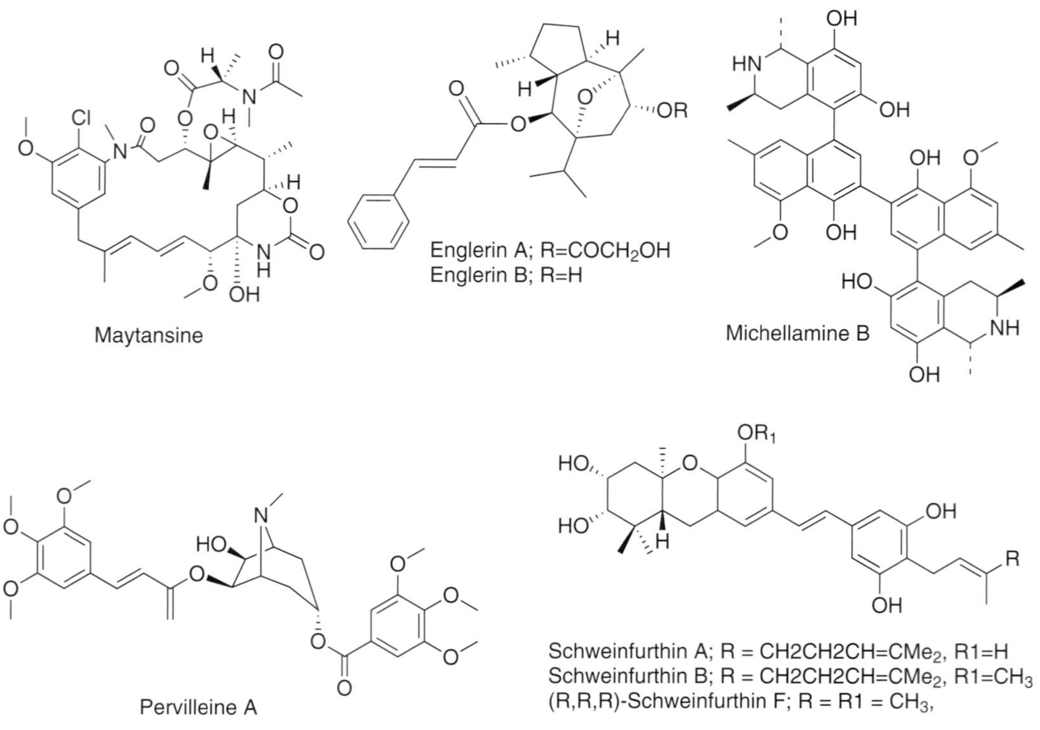


**Figure 10.2** Maytansine, englerins, schweinfurthins, pervilleine A, michellamine B

(Loes) R. Wilczek collected in Kenya was identified as a more productive source, and a large collection (about 9000 kg) in 1976 gave sufficient quantities for pursuit of preclinical and clinical development. NCI expressed its appreciation of the collaboration shown by the Kenyan authorities in the collection process by donating supplies of the antileukaemic drugs VLB and VCR, and preliminary studies were initiated aimed at the cultivation of *M. buchanii* (personal communication, Dr Robert Perdue (deceased, July, 2011)). Despite the promising activity shown in preclinical animal testing of maytansine, insignificant efficacy was observed in human clinical trials, and studies were terminated in the early 1980s.

The subsequent isolation of closely related compounds, the ansamitocins, from a microbial source, the actinomycete *Actinosynnema pretiosum*, posed the question as to whether the maytansines are truly plant products, or are produced through an association between a microbial symbiont and the plant. While this remains a topic of continued interest, the microbial production of the ansamitocins has allowed for production of larger quantities of this class of compounds, and this factor, together with their extreme potency, stimulated heightened interest in pursuing their development.

The maytansanoid derivatives DM1 and DM4 can be conjugated through either thioether or disulfide linkages with various monoclonal antibodies (mAbs) targeting a variety of cancers (Lambert, 2010; Yu *et al.*, 2012). Conjugates prepared with the mAb huC242 target cancers expressing *muc*1, such as pancreatic, biliary, colorectal and gastric cancers, and huC242-DM4 has undergone a Phase I clinical trial for the treatment of CanAg-expressing tumours, such as carcinomas of the colon and pancreas (Mita *et al.*, 2007). DM4-conjugated anti-Cripto monoclonal antibody BIIB015 was in a Phase I study in patients with relapsed/refractory solid tumours; Cripto, a member of the EGF-CFC family of growth factor-like molecules, is overexpressed in carcinomas, such as those of the breast, ovary, stomach, lung and pancreas, while not expressed in normal tissues (www.cancer.gov/drugdictionary/?CdrID=596550). As of November 2012, the most complete listing of the preclinical and clinical status of these agents through the end of 2009 is reported in the review by Lambert (2010). In summary, there have been 14 DM1- or DM4-linked agents that have been reported as either preclinical or in clinical trials. A recent review by Teicher and Chari (2011) gives the clinical status of these and several other ADCs.

The most advanced of the 14 DM-linked conjugates, known as T-DM1 or trastuzumab emtansine, is currently in 10 clinical trials (November 2012), either as a single agent or in combination with other agents, such as docetaxel or paclitaxel (www.cancer.gov/clinicaltrials/search/results?protocolsearchid=8186237). It has shown significant efficacy in the treatment of patients with advanced or metastatic HER2-positive breast cancer who have failed at least two treatments with the currently approved drugs, trastuzumab and the tyrosine kinase inhibitor lapatinib. On February 22, 2013, the FDA approved Kadcyla (ado-trastuzumab emtansine) as a new therapy for patients with HER2-positive, late-stage (metastatic) breast cancer (www.fda.gov/NewsEvents/Newsroom/PressAnnouncements/ucm340704.htm). Details of the agent and results to the end of 2009 were reported by Lambert (2010) and in more detail by Niculescu-Duvaz (2010). Further detailed discussions are given by Yu *et al.* (2012) and LoRusso *et al.* (2011), and a less technical commentary is given in the article by Hughes (2010).

#### 10.2.2.3 Englerins

The englerins (Figure 10.2) are novel guaiane sesquiterpenes isolated from the Tanzanian plant *Phyllanthus engleri* Pax collected by Z. Mwambo and MBG staff in 1989. The extract was prepared at NCI in 1992, and testing of the extract in the NCI 60-cancer cell line screen in 1997 showed selective inhibition of growth of the renal cancer cell lines; however, the extract was not pursued at that point. Bioassay-guided fractionation in 2007 yielded englerins A and B, and englerin A showed 1000-fold selectivity against six of eight renal cancer cell lines with GI(50) values ranging from 1 to 87 nM (Ratnayake *et al.*, 2009). These compounds are being developed in collaboration with scientists from the Institute of Traditional Medicine of Muhimbili University of Health and Allied Services in Dar Es Salaam, Tanzania. While the original LOC with the Institute of Traditional Medicine had expired at the time of the discovery, an updated LOC was negotiated in 2009 which permitted a multi-kilogram recollection of plant material to be made. This single collection yielded gram quantities of englerin A for preclinical development at NCI, which is ongoing. The outstanding activity of englerin A in the NCI 60 cell line screen has led many synthetic chemistry groups to attempt the total synthesis of englerin A. The first total synthesis was reported by the Christmann group from Dortmund, Germany (Willot *et al.*, 2009). This work also established the absolute configuration of the natural product. Other groups have since reported different synthetic strategies, including Nicolaou (Nicolaou *et al.*, 2010), Echavarren (Molawai *et al.*, 2010), Ma (Zhou *et al.*, 2010), Lin (Sun *et al.*, 2011; Wang et al., 2013) and Parker (Lee and Parker, 2012). Recently, a more efficient process has been developed by Chain, in which englerin A was prepared in eight steps and 20% overall yield from readily available starting materials (Li *et al.*, 2011). A review by Pouwer *et al.* (2012) has encapsulated the varied synthetic approaches to date.

Structure–activity relationships have been explored for the englerins by both the Chen (Chan and Chen, 2011) and Christmann groups (Radtke *et al.*, 2011). Notably, replacement of the cinnamate ester with a naphthoate moiety led to improved selectivity (Radtke *et al.*, 2011), while a reverse glycol ester and a lactate ester were worthy of note (Chan and Chen, 2011). 9-Deoxy-englerin A showed significantly reduced activity (Ushakov *et al.*, 2011). A series of chlorinated analogues with potency and selectivity have also been reported (Akee *et al.*, 2012).

The mechanism of englerin A's effect on renal cancer has been proposed to involve protein kinase C agonism, with inhibitory effects on the insulin pathway (Sourbier *et al.*, 2013). Given the dependence of renal cancers on glycolysis, this effect of englerin A may lead to its observed experimental therapeutic effects (Sourbier *et al.*, 2013). A recent report established that englerin A does not act via apoptotic mechanisms; rather, it promotes cellular necrosis (Sulzmaier *et al.*, 2012). The availability of labelled compounds (e.g. fluorescent or biotinylated derivatives) derived from synthetic approaches to englerins will likely shed more light on the mechanistic questions.

#### 10.2.2.4 Schweinfurthins

The schweinfurthins (e.g. schweinfurthins A, B and F; Figure 10.2) were isolated in 1996 from the African plant *Macaranga schweinfurthii* Pax, collected in Cameroon in 1987 by MBG. They displayed significant selective activity against central nervous system, renal and breast cancer cell lines in the NCI 60 cell line anticancer assay (Beutler *et al.*, 2006). The spectrum of anticancer activity does not match that of any currently used agent, indicating that these compounds may be acting at a previously unrecognized target or through a novel mechanism. Thus far, 10 schweinfurthins (A to J) have been isolated from nature by the NCI and Kingston/Virginia Tech groups (Thoison *et al.*, 1992; Beutler *et al.*, 1998; Yoder *et al.*, 2006). While successful scale-up recollections and isolation were undertaken in collaboration with colleagues in Cameroon in 1998–1999 and thereafter (Klausmeyer *et al.*, 2010), the isolation of sufficient quantities of the schweinfurthins from the natural source remains challenging, and synthetic strategies have been developed to provide a reliable source of natural schweinfurthins and synthetic analogues for further biological testing (Neighbors *et al.*, 2005, 2006). In the case of schweinfurthin F, total synthesis of the (*R,R,R*)- and (*S,S,S*)-enantiomers and comparisons of spectral data, optical rotations and bioassay data with those reported for the natural product have resulted in assignment of the natural material as the (*R,R,R*)-isomer (Mente *et al.*, 2007). These synthetic efforts are continuing, and most of the naturally occurring schweinfurthins have now been obtained by total synthesis (Mente *et al.*, 2008; Kuder *et al.*, 2009; Topczewski *et al.*, 2009, 2011; Ulrich *et al.*, 2010; Topczewski and Wiemer, 2011).

Investigations into the mechanism of action of schweinfurthins have yet to identify a proximate molecular target; however, in glioblastoma cell lines, it appears that a defective neurofibromatosis type 1 (NF1) pathway confers sensitivity (Turbyville *et al.*, 2010). Other groups have identified oxysterol-binding proteins and sterol metabolism as possible targets of schweinfurthins (Burgett *et al.*, 2011; Holstein *et al.*, 2011).

#### 10.2.2.5 The pervilleines: potential multidrug resistance inhibitors

A serious problem encountered in cancer chemotherapy is the resistance developed by many cancer patients to treatment with standard anticancer agents (Fojo and Bates, 2003). Repeated exposure of a cell population to treatment with a drug may induce resistance to that drug, and the cell population may subsequently show broad cross-resistance to other anticancer agents, even though it has never been exposed to those agents. This phenomenon known as multidrug resistance (MDR) has been linked to the presence of the *MDR1* gene which encodes P-glycoprotein (Pgp) which effectively pumps the drugs out of the cell, thereby precluding their antitumour actions. Despite the large number of compounds discovered which reverse this effect *in vitro* in cell line studies (so-called MDR inhibitors), their effectiveness in the clinic has been mostly disappointing. Thus, there is a continuing search for more effective MDR inhibitors.

The pervilleines, isolated from *Erythroxylum pervillei* Baillon and collected in Madagascar in 2003, have shown promising MDR activity both *in vitro* and *in vivo* (Silva *et al.*, 2001; Mi *et al.*, 2003; Chin *et al.*, 2006a). Preclinical development of pervilleine A (Figure 10.2) proceeded through a collaboration between an NCI-supported National Cooperative Drug Discovery Group and the Institut Malgache de Recherches Appliquées, and the weak cholinergic and adrenergic receptor-mediated activities exhibited by racemic pervilleine A hydrochloride were considered advantageous for its further development as a new adjuvant in cancer chemotherapy (Chin *et al.*, 2007).

#### 10.2.2.6 Other novel agents discovered through direct National Cancer Institute support

Several other novel, antitumour agents discovered through projects directly supported by NCI, either through the early programme (1955–1982) or the more recent NCI LOC programme (Section 10.2.1) have been mentioned in a recent review by Magadula and Erasto (2009). Sansevistatins 1 (Figure 10.3) and 2, new cytotoxic spirostanol saponins, were isolated from *Sansevieria ehrenbergii*, initially collected for the NCI in Kenya in 1966 by the USDA (Section 10.2.1) (Pettit *et al.*, 2005). Several novel cytotoxic agents have been isolated from Tanzanian plants collected by the NCI contractor, MBG (Section 10.2.1). In addition to the englerins (Figure 10.2) discussed in Section 10.2.2.3, these include new agarofuran sesquiterpenes, reissantins A–E (E; Figure 10.3), isolated from *Reissantia buchananii* collected in 1989 (Chang *et al.*, 2003) and rautandiols A (Figure 10.3) and B, pterocarpans isolated from *Neorautanenia mitis* collected in 1993 (Sakurai *et al.*, 2006). While the rautandiols did not exhibit significant cytotoxicity, several known rotenone derivatives were isolated as the cytotoxic constituents. Visiting scientists to the University of Illinois at Chicago from Tanzania, sponsored by the NCI, isolated two new compounds, (3α,12α)-dihydroxy-ent-8(14),15-isopimaradien-18-al (Figure 10.3) and (−)-*trans*-9-acetyl-4,90-di-*O*-methyl-30-de-*O*-methyldehydrodiconiferyl alcohol, from *Euphorbia quinquecostata* collected by Z. Mbwambo in 1999 (Su *et al.*, 2002), while *Berchemia discolor*, collected in 1999, yielded new prenylated flavonoids, including (3*S*)-discoloranone (Figure 10.3) (Chin *et al.*, 2006b).

### 10.2.3 African marine-derived anticancer agents

Marine invertebrates are proving to be an increasingly valuable source of potential anticancer agents, and interested readers are referred to several reviews for details (Newman and Cragg, 2004a,b). Regarding drug discovery from African marine resources, it is interesting to note that the first collections of *Dollabella auricularia*, source of the dolastatins (Flahive and Srirangam, 2012), were made by Professor G. R. Pettit of Arizona State University off the coast of Mauritius in the early 1960s. Likewise, the first member of the hemiasterlin family of compounds was isolated by Professor Yoel Kashman of Tel Aviv University from the sponge *Hemiasterella minor* (Kirkpatrick), collected in Sodwana Bay, South Africa (Andersen and Roberge, 2012). Both the dolastatins and the hemiasterlins have given rise to anticancer agents that have progressed to clinical trials (Andersen and Roberge, 2012; Flahive and Srirangam, 2012). Notably, brentuximab vedotin (Adcetris), prepared by Seattle Genetics, Inc. by conjugating the humanized anti-CD30 monoclonal antibody SGN-30 to the cytotoxic agent monomethyl auristatin E (Katz *et al.*, 2012), a synthetic analogue of dolastatin 10, received accelerated approval from the FDA in August 2011 for the treatment of Hodgkin lymphoma and a rare systemic anaplastic large-cell lymphoma. Collections of the marine worm *Cephalodiscus gilchristii* by the Pettit group off the southeastern shores of South Africa in 1988 yielded cephalostatin 1 (Pettit *et al.*,

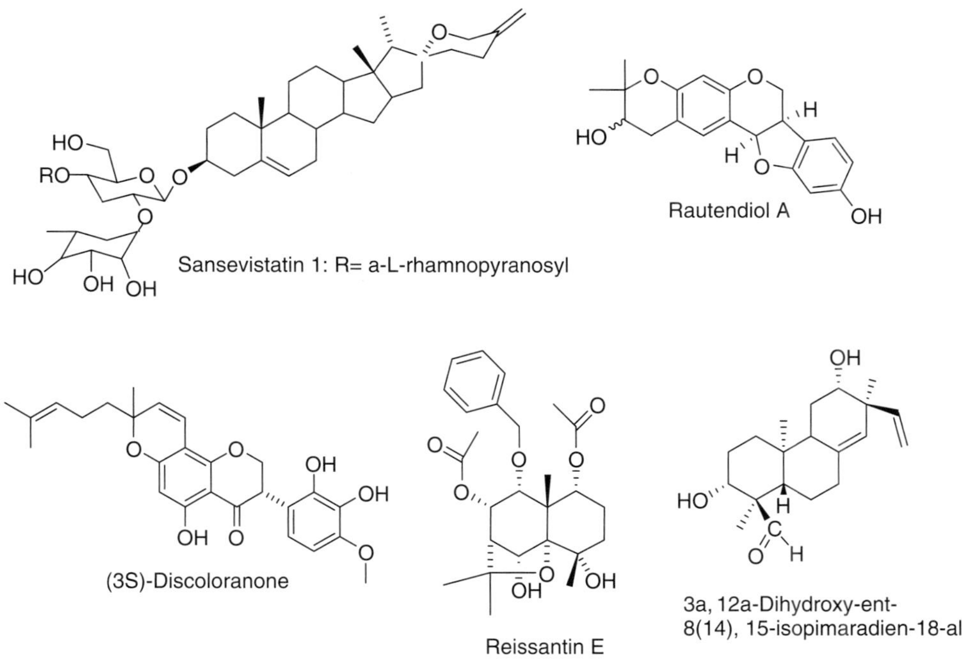


**Figure 10.3** Other novel antitumour agents

1988), and a further 18 members of this family were isolated from a large-scale recollection. These and the closely related ritterazines have been reviewed by Moser (2008). Exploring the potential of African marine invertebrates as a source of anticancer agents has continued through collections off the coasts of eastern and southern Africa through the NCI contract with Coral Reef Research Foundation (see Section 10.2.1). Collections off the southern African coast have been performed in close collaboration with Professor Michael Davies-Coleman through an MOU with Rhodes University.

### 10.2.4 Other discoveries: michellamine B, a potential anti-HIV agent

From 1987 to 1996, the NCI tested over 60 000 extracts of natural origin in an *in vitro* cell-based anti-HIV screen which determined the degree of HIV-1 replication in treated infected lymphoblastic cells versus that in untreated infected control cells. Several plant-derived natural products have shown *in vitro* activity, and of these, michellamine B (Figure 10.2), isolated from *Ancistrocladus korupensis* D. Thomas and Gereau, collected in the Korup region of southwestern Cameroon, is the only anti-HIV agent isolated from an African source to advance into preclinical development at the NCI. Despite *in vitro* activity against an impressive range of HIV-1 and HIV-2 strains, the difference between the toxic dose level and the anticipated level required for effective antiviral activity (the therapeutic index) was too small, and NCI discontinued further studies aimed at clinical development. However, the discovery of novel antimalarial agents, the korupensamines, from the same species added further promise for this species (Hallock *et al.*, 1994, 1997). For more details on the development of michellamine B, interested readers are referred to Beutler *et al.* (2012).

## 10.3 The International Cooperative Biodiversity Groups programme

### 10.3.1 Background

The International Cooperative Biodiversity Groups (ICBG) programme originally adopted several key elements of the NCI approach described above, but the programme is broader in concept, having multiple goals (www.icbg.org/). The programme was initiated in 1993 as a US Government (USG) interagency initiative led by the FIC/NIH, having as its aims the discovery of new bioactive molecules from nature while providing incentives for sustainable biodiversity conservation. Currently, the programme is jointly sponsored by five USG agencies, NIH (including NCI and several other institutes), National Science Foundation (NSF) Directorates of Biological Sciences and Geosciences, USDA National Institute of Food and Agriculture, Department of Energy (DOE) Office of Biological and Environmental Research, and the National Oceanographic and Atmospheric Administration Oceans and Human Health Initiative. The ICBG programme, like NCI programmes, is one of international collaboration that pays particular attention to access and benefit-sharing concerns. However, the ICBGs differ in structure and scope, in that they are a competitive grants programme (http://grants.nih.gov/grants/guide/rfa-files/rfa-tw-08-010.html). International research consortia are assembled in response to a solicitation to build partnerships comprising at least one US academic research institution and at least one non-US host organization in biodiversity-rich, low- and middle-income countries. Consortia typically include between three and ten organizations with different missions and expertise, all collaborating toward joint, interdependent goals.

Since 1993 there have been seven planning grants and 14 full ICBGs, several of which have been competitively renewed for multiple 5-year cycles. Multi-year collaborative efforts have operated in 20 countries, including Cameroon, Madagascar and Nigeria. While the earlier ICBGs focused largely on tropical plants, the projects have expanded to involve marine sources and microorganisms. Viable agreements for collections, research and the sharing of benefits with organizations in the 20 collaborating countries have been negotiated, and each agreement is vetted and approved by the appropriate national and local authorities. The projects are also governed by legal agreements negotiated individually by the multiple participants in each consortium, sometimes including participants from three or more countries.

Despite the challenges of negotiating and maintaining complex research partnerships, the ICBGs have been very successful in generating exciting new science and providing benefits both locally and nationally to host countries. From 1993 to 2009, over 83 000 extracts from approximately 10 630 uniquely identified species (of which about 10% were of marine origin) and 50 000 isolated microbial strains have been tested in over 78 therapeutic and agrochemical assay areas. Over 1400 unique bioactive molecules have been isolated, many of which are new to science, and some 30 molecules are currently considered active leads for potential new therapeutics or agrochemicals. These achievements have been described in over 600 publications. Significant contributions have been made to building research capacity in host countries, with over 4000 individuals trained in biodiversity and bio-discovery sciences. These include 350 long-term (1 year or longer) training opportunities, 70% of which involved scientists from host countries, and resulted in the award of 135 advanced degrees. Collaborating laboratories in host countries have been upgraded through provision of chemistry- and bioassay-related equipment, nuclear magnetic resonance technology, spectrometers, microscopes, polymerase chain reaction machines, computers and others. In some instances, the ICBG projects have contributed to the creation of national laboratories for chemistry and biological screening.

Contributions to biodiversity conservation have been substantial, involving the deposition of voucher specimens in national and international repositories, publication in print or on the web of national and regional taxonomy and identification guides, as well as descriptions of medicinal plants in the public domain. ICBG efforts have led to the creation of new protected areas in Panama, Laos, Suriname and Madagascar and have strengthened existing reserves in many other countries, leading to the formation of local and regional collaborations and new economic activities in most of these countries.

Readers interested in learning further details of the ICBG process and achievements are referred to the article by Cragg *et al.* (2012).

### 10.3.2 Biodiversity conservation and drug discovery in Madagascar

The ICBG programme in Madagascar, led by Dr David Kingston of Virginia Tech, is in its fourth 5-year cycle. The ICBG Consortium includes Conservation International (CI), Madagascar's Centre National d'Application et des Recherches Pharmaceutiques (CNARP), Centre National de Recherches sur l'Environnement (CNRE) and Centre National de Recherches Oceanographiques (CNRO), MBG, the University of Maryland Center for Environmental Studies (UMCES) and industrial partners Eisai Inc. (Eisai) and Dow Agrosciences (DAS).

Madagascar is the world's fourth largest island and is noted for its unique biodiversity. The interest in exploring the potential of these unique biological resources as sources of novel bioactive compounds stemmed from the isolation of the plant-derived anticancer agents VLB and VCR (Section 10.2.2). The ICBG efforts in this regard are focused on the discovery of novel drugs and agrochemicals. Plants indigenous to Madagascar are collected by botanists from MBG, CI and CNARP, marine organisms by teams from CNRO and microbial strains are isolated from soil samples by scientists from CNRE with training from DAS and from UMCES. Extracts of these biological materials are then tested in assays related to cancer, immunology, malaria, neurological disease, tuberculosis and agrochemistry. The isolation of 270 new natural products isolated from Madagascar higher plants, liverworts and marine organisms during the period 1991–2009 has been reviewed by Hou and Harinantenaina (2010).

The ICBG project also undertakes economic development and biodiversity conservation initiatives. Among the areas set aside as protected areas as a result of assessments carried out by the ICBG partners, CI and MBG, and preparation of dossiers for the Malagasy Government, are the Montagne des Français and the Ambodivahibe Bay Marine Reserve, which form parts of the larger Ramena protected area that also includes the Oronjia Reserve. The Montagne des Français is a limestone massif on the northern tip of Madagascar which is home to a large number of plant, mammalian, amphibian and reptilian species, some of which are critically endangered. The Ambodivahibe Bay Marine Reserve was selected after a series of consultations with the local residents of Ambodivahibe, and is expected to become a tourist destination. Economic development has included training for local scientists in botany, bioassays and microbiology, and for local residents around the collection sites, training in animal husbandry, the improvement in the efficiency of vegetable farming, and for women, in Western cooking standards to enable them to act as hosts to ecotourists. The highly detrimental environmental effects of charcoal production from valuable natural resources have been offset by the planting of fast-growing eucalyptus trees to provide a renewable source of charcoal. In all these conservation and economic development efforts, the ICBG was generously provided with substantial upfront funds by the industrial partners. Further details of these aspects are found in the review by Kingston (2011).

#### 10.3.2.1 Plant-derived antitumour agents

The unique Madagascar plant biodiversity has been a prolific source of *in vitro* active antitumour compounds, many of which possess novel structures. Generally, antitumour activity was determined against the A2780 human ovarian cancer cell line. The classes of antitumour compounds and their plant sources are listed in Table 10.1.

**Table 10.1** Classes of antitumour compounds derived from Madagascar plants

| Chemical class | Plant source | Reference |
|---|---|---|
| Acetogenins | *Uvaria* sp. | Dai *et al.* (2012) |
| Alkaloids | *Tabernaemontana calcarea* | Chaturvedula *et al.* (2003a) |
| | *Vepris punctate* | Chaturvedula *et al.* (2003b) |
| | *Ambavia gerrardii* | Pan *et al.* (2011b) |
| Alkylresorcinols | *Oncostemon bojerianum* | Chaturvedula *et al.* (2002a) |
| Anthraquinones | *Scutia myrtina* | Hou *et al.* (2009a) |
| Benzophenones | *Ochrocarpos punctatus* | Chaturvedula *et al.* (2002b) |
| Butanolides | *Tambourissa* sp. | Yoder *et al.* (2007a) |
| Butenolides | *Artabotrys madagascariensis* | Murphy *et al.* (2008b) |
| Cardenolides | *Pentopetia androsaemifolia* Decne | Adou *et al.* (2010) |
| | *Elaeodendron* sp. | Cao *et al.* (2007a) |
| | *Elaeodendron alluaudianum* | Hou *et al.* (2009b) |
| | *Roupellina (Strophanthus) boivinii* | Karkare *et al.* (2007) |
| | *Leptadenia madagascariensis* | Pan *et al.* (2011a) |
| Coumarins | *Ochrocarpos punctatus* | Chaturvedula *et al.* (2002b) |
| | *Astrotrichilia* sp. | Harinantanaina *et al.* (2011) |
| Diphenylpropanes | *Bussea sakalava* | Pan *et al.* (2010b) |
| Diterpenes | *Cassipourea madagascariensis* | Chaturvedula *et al.* (2006) |
| | *Cordyla madagascariensis* ssp. *madagascariensis* | Hou *et al.* (2008) |
| | *Casearia lucida* | Prakash *et al.* (2002) |
| | *Podocarpus madagascariensis* | Reynolds *et al.* (2006) |
| | *Casearia nigrescens* | Williams *et al.* (2007) |
| | *Cyphostemma greveana* | Cao *et al.* (2011) |
| Flavanoids | *Mundulea chapelieri* | Cao *et al.* (2004a) |
| | *Physena madagascariensis* | Cao *et al.* (2006a) |
| | *Schizolaena hystrix* | Murphy *et al.* (2005) |
| | *Pongamiopsis amygdaina* | Harinantenaina *et al.* (2010) |
| Guttiferones | *Rheedia calcicola* | Cao *et al.* (2007b) |
| Liminoids | *Malleastrum* sp. | Murphy *et al.* (2008a) |
| Long-chain fatty acids | *Schizolaena hystrix* | Murphy *et al.* (2006) |
| Naphthaquinones | *Mendoncia cowanii* | Williams *et al.* (2006) |
| Phlorglucinols | *Protorhus thouvenotii* | Cao *et al.* (2004b) |
| Pterocarpans | *Pongamiopsis amygdaina* | Harinantenaina *et al.* (2010) |
| Sesquiterpenes | *Vernonia pachyclada* | Williams *et al.* (2005) |
| | *Cyphostemma greveana* | Cao *et al.* (2011) |
| | *Athroisma proteiforme* | Pan *et al.* (2011c) |
| Steroids | *Helmiopsis sphaerocarpa* | Cao *et al.* (2009) |
| Stilbenes | *Macaranga alnifolia* | Yoder *et al.* (2007b) |
| Triterpenes | *Acridocarpus vivy* | Cao *et al.* (2004c) |
| | *Albizia gummifera* | Cao *et al.* (2007c) |
| | *Helmiopsis sphaerocarpa* | Cao *et al.* (2009) |
| | *Brachylaena ramiflora* | Chaturvedula *et al.* (2002c) |
| | *Vepris punctate* | Chaturvedula *et al.* (2004) |
| | *Cassipourea lanceolata* | Hou *et al.* (2010) |
| Triterpenoid saponins | *Albizia gummifera* | Cao *et al.* (2007c) |
| | *Terminalia tropophylla* | Cao *et al.* (2010) |
| | *Polyscias amplifolia* | Chaturvedula *et al.* (2003c) |
| | *Pittosporum viridiflorum* | Seo *et al.* (2002) |
| Xanthones | *Terminalia calcicola* | Cao *et al.* (2007d) |
| | *Psorospermum molluscum* | Leet *et al.* (2008) |
| | *Symphonia pauciflora* | Pan *et al.* (2010a) |

The structures of a selection of novel *in vitro* active antitumour compounds belonging to the chemical classes and their source plants listed in Table 10.1 are shown in Figure 10.4.

#### 10.3.2.2 Plant-derived anti-plasmodial compounds

The following compound classes exhibiting activity against the drug-sensitive HB3 strain of *Plasmodium falciparum* were isolated: phenols from *Sloanea rhodantha* (Baker) Capuron var. *rhodantha* (Cao *et al.*, 2006b); anthraquinones from *Scutia myrtina* (Hou *et al.*, 2009a); and pseudoguaianolides of *Athroisma proteiforme* (Pan *et al.*, 2011c). Recently, the exciting finding of potent activity in

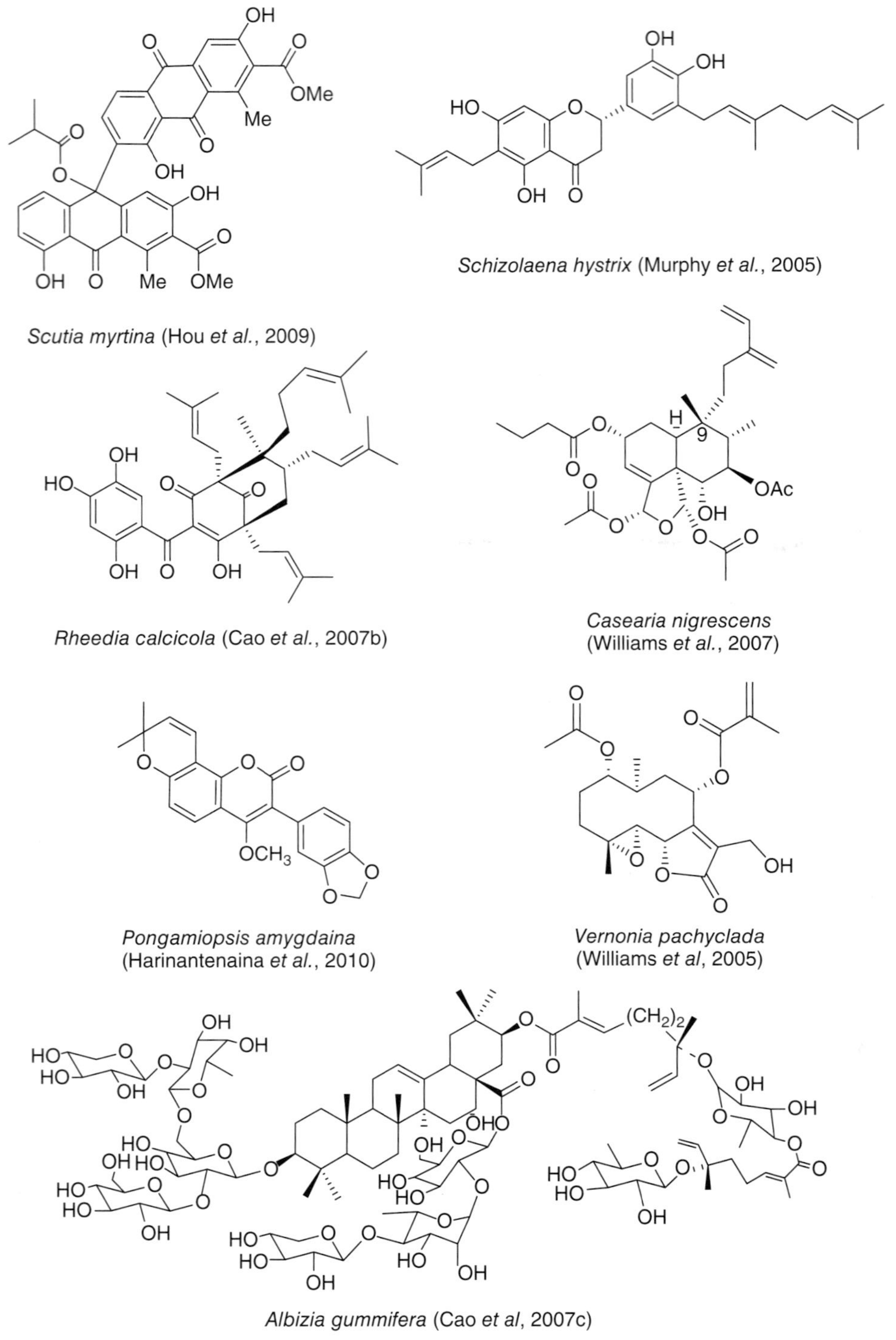


**Figure 10.4** Some novel *in vitro* active antitumour compounds from Madagascar plants

some simple phloroglucinols from *Mallotus oppositifolius* has been reported (Harinantenaina *et al.*, 2013).

#### 10.3.2.3 Marine-derived antitumour compounds

Antitumour bistramides (lipopeptides) were isolated from *Trididemnum cyclops* (Murphy *et al.*, 2009).

### 10.3.3 Drug development and conservation of biodiversity in West and Central Africa

The main goal of the ICBG on drug development and conservation of biological diversity in West and Central Africa was the development of an effective platform for the sustainable management of natural resources to promote biodiversity conservation and drug discovery that addressed unmet human health needs and economic development of rural communities in the host countries. The central approach was through building partnerships between various local stakeholders and members of the project, as well as strengthening existing collaborative programmes and institutions. The overriding goal was to increase the net worth of West–Central African forests as living resources, and to demonstrate the feasibility of using drug development as an incentive for the conservation of biological diversity. The ICBG was also envisioned as an integrated conservation and development project, which retained the three core elements of the programme (biodiversity conservation, drug discovery and economic development) in both the design and in the monitoring and assessment of the effectiveness of the project.

#### 10.3.3.1 Organizational structure of the West and Central Africa International Cooperative Biodiversity Groups

The ICBG was a collaboration of the Walter Reed Army Institute of Research, Washington, DC (WRAIR), Bioresources Development & Conservation Programme (BDCP), the Smithsonian Tropical Research Institute, and nine other institutions in Cameroon, Nigeria and the USA. The organizational structure provided for a consortium relationship between the three main partners, WRAIR, BDCP and the Smithsonian Institution, through a Cooperative Research and Development Agreement (CRADA). Other partner institutions of the ICBG were linked to the programme through a research agreement with one of the three principal CRADA partners. Contract screening agreements were also executed between the ICBG and other institutions as the need arose. By this arrangement, the principal partner institutions, WRAIR, BDCP and the Smithsonian, were involved at all phases of the ICBG programme. BDCP was an integral institution in the coordination of this ICBG. BDCP is an African-based not-for-profit, non-governmental organization with an International Liaison Office in Washington, DC, and branches in Nigeria and Cameroon.

The ICBG was an integrated biodiscovery and sustainable development programme that was formulated into five associate programmes as follows:

- *AP-1. Biodiversity conservation (biodiversity conservation programmes and training):* Smithsonian Tropical Research Institute; Smithsonian Conservation Training Council; Smithsonian Man and Biosphere Program; BDCP – Nigeria and Cameroon, Korup National Park, Cameroon, Cross-River National Park, Okwangwo, Nigeria.
- *AP-2. Drug development A. Non-parasitic diseases (antiviral/anticancer/cytotoxicity/central nervous system (CNS) activity, etc.):* University of Dchang, Cameroon; Center for Drug Discovery, University of Utah, Salt Lake City; International Centre for Ethnomedicine and Drug Development (InterCEDD), Nsukka, Nigeria; University of Miami.
- *AP-3. Phytomedicine development, phytochemistry and preliminary bioassays:* InterCEDD, Nsukka; University of Dschang; Howard University, Washington, DC; University of Mississippi.
- *AP-4. Drug development B. Malaria and other parasitic diseases:* Division of Experimental Therapeutics, WRAIR; InterCEDD, Nsukka; University of Dschang.
- *AP-5. Ethnobiology and economic botany:* BDCP Nigeria and Cameroon;, InterCEDD, Nsukka.

#### 10.3.3.2 Plant selection strategy

Ethnomedical plant selection and screening methods were utilized to increase the chance to identify leads with therapeutic activity. The ICBG developed a customized approach that involved a carefully designed ethnomedical survey, followed by information analysis and, finally, integration of the results to produce a highly selective priority list of plant samples which constituted a Database of African Medicinal Plants (AfricMed). This proprietary database contains inventories of plants used in traditional medicine in West and Central Africa for the treatment of a wide range of diseases, ranging from malaria and infectious diseases to skin problems and diseases of the respiratory system. The collected plant materials were catalogued and voucher specimens deposited at the source country herbaria. The database was created in 1992 following extensive ethnobotanical surveys of various localities in Nigeria and Cameroon. The searchable database entries continue to grow through the collaborative efforts of traditional herbalists in the InterCEDD network. An extract library was developed to supply extracts to different ICBG screening programmes. Plant samples selected for the anticancer drug discovery project were from the AfricMed database.

These samples were processed according to the protocol described earlier by Schuster *et al.* (1999). Briefly, plant specimens consisting of different plant parts were air dried in the field and finely ground into a powder. The powder (200–500 g) was extracted with organic solvents, as shown in Table 10.2. Finally, the resultant marc was extracted with water and then lyophilized to yield the aqueous extract. The plant extracts were provided by InterCEDD, Nsukka, and the University of Dschang for anticancer testing. These consisted of extracts of all parts of the plant (bark, stems, wood, pods, leaves, fruits, fruit-pulp, roots, rhizomes, flowers and seeds). The samples were logged in and the data were stored in a computerized Chemical Information System (CIS) database at WRAIR. A unique identification number referred to as the SU-number was assigned to each plant sample. Samples were submitted to collaborating ICBG institutions for evaluation for potential antiparasitic, antifungal, antiplasmodial, antiviral, cytotoxic and CNS activity. A total of 160 samples were sent to the University of Utah for anticancer and cytotoxicity tests, and active extracts were subjected to bioassay-guided fractionation.

#### 10.3.3.3 Antitumour plants from West and Central Africa

West and Central African forests have been shown by various studies to be a rich source of potential anticancer agents (Iwu, 1993; Kuete *et al.*, 2011; Sawadogo *et al.*, 2012). Plant selection based on prior ethnomedical use has been fairly predictive of the possibility of identifying active leads (Kuete *et al.*, 2011; Sawadogo *et al.*, 2012). It has been suggested that ethnopharmacological usages such as immune and skin disorders, and inflammatory ailments, should be considered when selecting plants to treat cancer, since these reflect disease states bearing relevance to cancer or cancer symptoms (Kuete and Efferth, 2011). For the ICBG, a total of 160 plant samples used in

**Table 10.2** Cytotoxicity of Central and West African plants. Activity: $IC_{50} = 10\,\mu g/mL$

| Family | Genus/species | Plant part[a] | Extract solvent[b] |
|---|---|---|---|
| Ancistrocladaceae | *Ancistrocladus barteri* Scott-Elliot | StBk | MeOH |
| Annonaceae | *Cleistopholis staudtii* Engl. & Diels | StBk | EtOH |
| | *Enantia chlorantha* Oliv | Bk | MeOH |
| | *Uvaria chamae P. Beauv* | Rt | Chf; PE |
| Apocynaceae | *Picralima nitida* (Stapf) T. Durand & H. Durand | Frp | Chf |
| Asparagaceae | *Dracaena mannii* Baker | Frp | Chf |
| Bignoniaceae | *Kigelia africana* (Lam.) Benth. | Fr | MeOH |
| Capparaceae | *Crateva adansonii* DC | Lf | MeOH |
| Euphorbiaceae | *Alchornea cordifolia* (Schumach. & Thonn.) Müll. Arg. | Lf | Hexane |
| Fabaceae | *Albizia ferruginea* (Guill. & Perr.) Benth. | Pl | Hexane |
| Lamiaceae | *Hoslundia opposita* Vahl | Px | DCM |
| Meliaceae | *Khaya grandifoliola* A. Juss. | StBk | Acetone |
| Monimiaceae | *Xymalos monospora* (Harv.) Baill. ex Warb | Bk | DCM |
| Moraceae | *Dorstenia barteri* Bureau | Wd | DCM |
| Plantaginaceae | *Scoparia dulcis* L | Px | DCM |
| Rubiaceae | *Araliopsis tabouensis* Aubrév. & Pellegr. | StBk | Chf/MeOH (1/1) |
| | *Diodia scandens* Hutch. & Dalziel | Hb | MeOH |
| | *Morinda lucida* Benth. | Lf | DCM; PE |
| Rutaceae | *Fagara lemairei* De Wild. | RtBk | MeOH |
| Siparunaceae | *Glossocalyx brevipes* Benth. | StBk | PE |
| Zingiberaceae | *Aframomum melegueta* (Roscoe) K. Schum. | Sd | Hexane |
| | *Kaempferia aethiopica* Benth. | Rt | DCM |

[a]Bk, bark; Fr, fruit; Frp, fruit pulp; Hb, herb; Lf, leaves; Pl, whole plant; Px, aerial parts; Rt, roots; RtBk, root bark; Sd, seeds; StBk, stem bark; Wd, wood.
[b]Chf, chloroform; DCM, dichloromethane; PE, petroleum ether.

traditional medicine in Nigeria and Cameroon were screened for cytotoxicity in the HCT116, KB and CCRF-CEM human cancer cell lines. These medicinal plant species were distributed across 43 families and 93 species; of these, 75 (46.9%) of the extracts were cytotoxic at 100 μg/mL or less, 18 (11.2%) demonstrated weak activity and 67 (41.8%) showed no activity. At a concentration of 10 μg/mL, the 22 samples (14%) listed in Table 10.2 were found to be significantly cytotoxic and were selected for further studies.

Among the plants showing marginal cytotoxicity ($IC_{50} = 100\,\mu g/mL$) were *Lannea acida* A. Rich. and *Spondias mombin* L. (Anacardiaceae); *Cleistopholis patens* (Benth.) Engl. & Diels (Annonaceae); *Gongronema latifolium* Benth., *Pleiocarpa pycnantha* (K. Schum.) Stapf, and *Rauvolfia vomitoria* Afzel (Apocynaceae); *Dracaena arborea* (Willd.) Link (Asparagaceae); *Vernonia amygdalina* Delile (Asteraceae); *Trema guineense* (Schum. & Thonn.) Ficalho (Cannabaceae); *Hippocratea welwitschii* Oliv (Celastraceae); *Combretum glutinosum* Perr. (Combretaceae); *Discoglypremna caloneura* (Pax) Prain and *Neoboutonia velutina* Prain (Euphorbiaceae); *Cassia siamea* Lam., *Cylicodiscus gabunensis* Harms, *Erythrina senegalensis* DC, and *Mezoneurum benthamianum* Baill. (Fabaceae); *Hyptis suaveolens* (L.) Poit. (Lamiaceae); *Napoleonaea imperialis* P. Beauv (Lecythidaceae); *Triumfetta tomentosa* Bojer and *Waltheria indica* L. (Malvaceae); *Guarea thompsonii* Sprague & Hutch and *Khaya senegalensis* (Desr.) A. Juss. (Meliaceae); *Olax viridis* Oliv. (Olacaceae); *Peperomia* sp. (Piperaceae); *Mitracarpus scaber* Zucc. (Rubiaceae); *Chrysophyllum albidum* G. Don (Sapotaceae); *Aframomum daniellii* (Hook. f.) K. Schum. and *Renealmia polypus* Gagnep. (Zingiberaceae).

Several extracts were identified with specified effects on the cell cycle. The most striking effect was that of hydrophobic extracts of *Uvaria chamae* roots (SU-799), which causes aneuploidy in CHO cells.

#### 10.3.3.4 Plant-derived antiplasmodial compounds

A major component of this ICBG was the emphasis on the evaluation of plants from Nigeria and Cameroon for the treatment of tropical diseases and new leads in the search for better drugs for parasitic infections, such as malaria, leishmaniasis and trypanosomiasis. Plants collected as described above which were used in traditional medicine for the treatment of protozoan, fungal and viral infections were screened for antimalarial activity and cytotoxicity. Empirical evidence suggested that a possible correlation exists between the observed antimalarial properties and antitumour activities of medicinal plants. Duffy *et al.* (2012) have reported that 14 out of 15 nature-derived antimalarials (93%) referenced by the World Health Organization and 146 of 235 antimalarial natural species (62%) from a defined MEDLINE search have anticancer activity. Natural products with antimalarial activity, therefore, provide a feasible platform for much needed leads in anticancer drug discovery.

Five hundred extracts from plants used in traditional medicine for the treatment of malaria were screened *in vitro* against *P. falciparum*; 70% showed activity with $LC_{50}$ values of 100 μg/mL or less, and 18 plants were selected for bioassay-guided isolation of the active compounds. This led to the identification and selection of some chemotypes such as *Picralima* alkaloids and related betacarboline indoles, cryptolepine analogues, enantin and related isoquinolines, labdane diterpenes, and morindone and other anthraquinones as templates for the semi-synthesis of new antimalarials (Schuster, 2002; Okunji *et al.*, 2005).

It is noteworthy that one of the plants investigated, *Reneilmia cincinnata* (Family, Zingiberaceae), used only in steam-baths for the traditional treatment of fevers and malaria, exhibited strong ($IC_{50}$ = 4–25 μg/mL) antiplasmodial activity (Tchuendem *et al.*, 1999). Bioassay-guided fractionation and isolation of the active compounds yielded the sesquiterpenoids 6,7,10-trihydroxyisodaucane, oplodiol, opiopanone, 5*E*,10(14)-germacradien-1β,4β-diol, 1(10)*E*,5*E*-germacradien-4α-ol and eudesman-1,4,7-triol as the active constituents responsible for the antimalarial property of the Reneilmia steam-bath (Tchuendem *et al*, 1999).

## 10.4 Conclusions

African biodiversity has proved to be the source of several promising anticancer drugs. Three plant-derived agents – the combretastatin derivatives CA1P and CA4P and the antibody-conjugated maytansanoid derivative trastuzumab emtansine – as well as the marine-derived antibody-conjugated monomethyl auristatin E derivative brentuximab vedotin have been approved for clinical use in the treatment of a variety of cancers. Several other products are in clinical trials or various stages of preclinical development, while a substantial number of naturally derived compounds showing significant antitumour activity have been identified. While natural products are a proven source of novel, bioactive molecules, the actual compound isolated from the natural source often may not be suitable for development into an effective drug, but it may serve as a lead molecule that can provide the basis for further chemical or biochemical modification. The discovery and development of promising bioactive molecules always involve close multidisciplinary collaboration. In the case of natural products, however, such collaboration is most often international in nature, and through the signing of collaborative agreements (LOCs or MOUs) with qualified research groups and communities worldwide, the NCI and NIH ICBG drug discovery programmes are committed to policies of 'fair and equitable' collaboration.

## Acknowledgements

We gratefully acknowledge the collaboration of the permitting authorities in source countries where collections for the NCI and NIH ICBGs were performed. These include Cameroon, Central African Republic, Ethiopia, Gabon, Ghana, Kenya, Madagascar, Nigeria, South Africa, Tanzania and Zimbabwe. We thank Dr David Kingston for helpful comments and suggestions. Support for the ICBG programmes through awards from USG agencies led by the FIC/NIH is gratefully acknowledged (U01 TW00313/Madagascar; 5U01TW000327-09 and 1U01TW001023-01/West-Central Africa).

## References

Adou, E., Miller, J.S., Ratovoson, F. *et al.* (2010) Antiproliferative cardenolides from *Pentopetia androsaemifolia* Decne. from the Madagascar rain forest. *Indian Journal of Experimental Biology*, **48**, 248–257.

Akee, R.K., Ransom, T.T., Ratnayake, R. *et al.* (2012) Chlorinated englerins with selective inhibition of renal cancer cell growth. *Journal of Natural Products*, **75**, 459–463.

Alley, S.C., Okeley, N.M. and Senter, P.D. (2010) Antibody–drug conjugates: targeted drug delivery for cancer. *Current Opinion in Chemical Biology*, **14**, 529–537.

Andersen, R.J. and Roberge, M. (2012) A synthetic analog of the antimitotic natural product hemiasterlin, in *Anticancer Agents from Natural Products*, 2nd edn (eds G.M. Cragg, D.G.I. Kingston and D.J. Newman), CRC Taylor & Francis, Boca Raton, FL, pp. 347–362.

Beutler, J.A., Shoemaker, R.H., Johnson, T. and Boyd, M.R. (1998) Cytotoxic geranyl stilbenes from *Macaranga schweinfurthii*. *Journal of Natural Products*, **61**, 1509–1512.

Beutler, J.A., Jato, J.G., Cragg, G.M. *et al.* (2006) The schweinfurthins. Issues in development of a plant-derived anticancer lead, in *Medicinal and Aromatic Plants* (eds J. Bogers, L.E. Craker and D. Langa), Springer, Amsterdam, pp. 301–309.

Beutler, J.A., Cragg, G.M. and Newman, D.J. (2012) The National Cancer Institute and natural product-based drug discovery in Africa, in *Drug Discovery in Africa* (eds K. Chibale, M. Davies-Coleman and C. Masimirembwa), Springer-Verlag, Berlin, pp. 29–52.

Burgett, A.W., Poulsen, T.B., Wangkanont, K. *et al.* (2011) Natural products reveal cancer cell dependence on oxysterol-binding proteins. *Nature Chemical Biology*, **7**, 639–647.

Cao, S.G., Schilling, J.K., Miller, J.S. *et al.* (2004a) Cytotoxic compounds from *Mundulea chapelieri* from the Madagascar rainforest. *Journal of Natural Products*, **67**, 454–456.

Cao, S.G., Schilling, J.K., Randrianasolo, A. *et al.* (2004b) New cytotoxic alkyl phloroglucinols from *Protorhus thouvenotii*. *Planta Medica*, **70**, 683–685.

Cao, S.G., Guza, R.C., Miller, J.S. *et al.* (2004c) Cytotoxic triterpenoids from *Acridocarpus vivy* from the Madagascar rain forest. *Journal of Natural Products*, **67**, 986–989.

Cao, S.G., Norris, A., Miller, J.S. *et al.* (2006a) Cytotoxic compounds of *Physena madagascariensis* from the Madagascar rain forests. *Natural Product Research*, **20**, 1157–1163.

Cao, S.G., Ranarivelo, L., Ratsimbason, M. *et al.* (2006b) Antiplasmodial activity of compounds from *Sloanea rhodantha* (Baker) Capuron var. *rhodantha* from the Madagascar rain forest. *Planta Medica*, **72**, 1438–1440.

Cao, S.G., Brodie, P.J., Miller, J.S. *et al.* (2007a) Antiproliferative cardenolides of an *Elaeodendron* sp from the Madagascar rain forest. *Journal of Natural Products*, **70**, 1064–1067.

Cao, S., Brodie, P.J., Miller, J.S. *et al.* (2007b) Guttiferones K and L, antiproliferative compounds of *Rheedia calcicola* from the Madagascar rain forest. *Journal of Natural Products*, **70**, 686–688.

Cao, S.G., Norris, A., Miller, J.S. *et al.* (2007c) Cytotoxic triterpenoid saponins of *Albizia gummifera* from the Madagascar rain forest. *Journal of Natural Products*, **70**, 361–366.

Cao, S., Brodie, P.J., Miller, J.S. *et al.* (2007d) Antiproliferative xanthones of *Terminalia calcicola* from the Madagascar rain forest. *Journal of Natural Products*, **70**, 679–681.

Cao, S., Brodie, P., Miller, J.S. *et al.* (2009) Antiproliferative compounds of *Helmiopsis sphaerocarpa* from the Madagascar rain forest. *Natural Product Research*, **23**, 638–643.

Cao, S., Brodie, P.J., Callmander, M. *et al.* (2010) Saponins and a lignan derivative of *Terminalia tropophylla* from the Madagascar dry forest. *Phytochemistry*, **71**, 95–99.

Cao, S., Hou, Y., Brodie, P. *et al.* (2011) Antiproliferative compounds of *Cyphostemma greveana* from a Madagascar dry forest. *Chemistry & Biodiversity*, **8**, 643–650.

Caravella, J. and Lugovskoy, A. (2010) Design of next-generation protein therapeutics. *Current Opinion in Chemical Biology*, **14**, 520–528.

Chan, K.P. and Chen, D.Y. (2011) Chemical synthesis and biological evaluation of the englerin analogues. *ChemMedChem*, **6**, 420–423.

Chang, F.R., Hayashi, K., Chen, I.H. *et al.* (2003) Antitumor agents. 228. Five New agarofurans, reissantins A–E, and cytotoxic principles from *Reissantia buchananii*. *Journal of Natural Products*, **66**, 1416–1420.

Chaturvedula, V.S.P., Schilling, J.K., Miller, J.S. *et al.* (2002a) New cytotoxic bis 5-alkylresorcinol derivatives from the leaves of *Oncostemon bojerianum* from the Madagascar rain forest. *Journal of Natural Products*, **65**, 1627–1632.

Chaturvedula, V.S.P., Schilling, J.K. and Kingston, D.G.I. (2002b) New cytotoxic coumarins and prenylated benzophenone derivatives from the bark of *Ochrocarpos punctatus* from the Madagascar rain forest. *Journal of Natural Products*, **65**, 965–972.

Chaturvedula, V.S.P., Schillin, J.K., Miller, J.S. *et al.* (2002c) Two new triterpene esters from the twigs of *Brachylaena ramiflora* from the Madagascar rain forest. *Journal of Natural Products*, **65**, 1222–1224.

Chaturvedula, V.S.P., Sprague, S., Schilling, J.K. and Kingston, D.G.I. (2003a) New cytotoxic indole alkaloids from *Tabernaemontana calcarea* from the Madagascar rain forest. *Journal of Natural Products*, **66**, 528–531.

Chaturvedula, V.S.P., Schilling, J.K., Miller, J.S. *et al.* (2003b) New cytotoxic alkaloids from the wood of *Vepris punctata* from the Madagascar rain forest. *Journal of Natural Products*, **66**, 532–534.

Chaturvedula, V.S.P., Schilling, J.K., Miller, J.S. *et al.* (2003c) New cytotoxic oleanane saponins from the infructescences of *Polyscias amplifolia* from the Madagascar rain forest. *Planta Medica*, **69**, 440–444.

Chaturvedula, V.S.P., Schilling, J.K., Miller, J.S. *et al.* (2004) New cytotoxic terpenoids from the wood of *Vepris punctata* from the Madagascar rain forest. *Journal of Natural Products*, **67**, 895–898.

Chaturvedula, V.S.P., Norris, A., Miller, J.S. *et al.* (2006) Cytotoxic diterpenes from *Cassipourea madagascariensis* from the Madagascar rain forest. *Journal of Natural Products*, **69**, 287–289.

Chin, Y.W., Jones, W.P., Waybright, T.J. *et al.* (2006a) Tropane aromatic ester alkaloids from a large-scale recollection of *Erythroxylum pervillei* stem bark obtained in Madagascar. *Journal of Natural Products*, **69**, 414–417.

Chin, Y.W., Mdee, L.K., Mbwambo, Z.H. *et al.* (2006b) Prenylated flavonoids from the root bark of *Berchemia discolor*, a Tanzanian medicinal plant. *Journal of Natural Products*, **69**, 1649–1652.

Chin, Y.W., Kinghorn, A.D. and Patil, P.N. (2007) Evaluation of the cholinergic and adrenergic effects of two tropane alkaloids from *Erythroxylum pervillei*. *Phytotherapy Research*, **21**, 1002–1005.

Cragg, G.M. and Newman, D.J. (2005) International collaboration in drug discovery and development from natural sources. *Pure and Applied Chemistry*, **77**, 1923–1942.

Cragg, G.M. and Newman, D.J. (2009) Nature: a vital source of leads for anticancer drug development. *Phytochemistry Reviews*, **8**, 313–331.

Cragg, G.M., Boyd, M.R., Cardellina II, J.H. *et al.* (1994) Ethnobotany and drug discovery: the experience of the US National Cancer Institute, in *Ethnobotany and the Search for New Drugs* (eds J. Chadwick and J. Marsh), Ciba Foundation Symposium Vol. 185, John Wiley & Sons, Ltd, Chichester, pp. 178–196.

Cragg, G.M., Katz, F., Newman, D.J. and Rosenthal, J. (2012) The impact of the United Nations Convention on Biological Diversity on natural products research. *Natural Product Reports*, **29**, 1407–1423.

Dai, Y., Harinantenaina, L., Brodie, P.J. *et al.* (2012) Two antiproliferative acetogenins from a *Uvaria* sp. from the Madagascar dry forest. *Journal of Natural Products*, **75**, 479–483.

Duffy, R., Wade, C. and Chang, R. (2012) Discovery of anticancer drugs from antimalarial natural products: a MEDLINE literature review. *Drug Discovery Today*, **17** (17–18), 942–953.

Eichler, W., Yafai, Y., Wiedemann, P. and Fengler, D. (2006) Antineovascular agents in the treatment of eye diseases. *Current Pharmaceutical Design*, **12**, 2645–2660.

Flahive, E. and Srirangam, J. (2012) The dolastatins: novel antitumor agents from *Dolabella auricularia*, in *Anticancer Agents from Natural Products*, 2nd edn (G.M. Cragg, D.G.I. Kingston and D.J. Newman), CRC Taylor & Francis, Boca Raton, FL, pp. 263–289.

Fojo, T. and Bates, S. (2003) Strategies for reversing drug resistance. *Oncogene*, **22**, 7512–7523.

Hallock, Y.F., Manfredi, K.P., Blunt, J.W. *et al.* (1994) Korupensamines A–D, novel antimalarial alkaloids from *Ancistrocladus korupensis*. *Journal of Organic Chemistry*, **59**, 6349–6355.

Hallock, Y.F., Manfredi, K.P., Dai, J.R. *et al.* (1997) Michellamines D–F, new HIV-inhibitory dimeric naphthylisoquinoline alkaloids, and korupensamine E, a new antimalarial monomer, from *Ancistrocladus korupensis*. *Journal of Natural Products*, **60**, 677–683.

Harinantenaina, L., Brodie, P., Slebodnick, C. *et al.* (2010) Antiproliferative compounds from *Pongamiopsis amygdaina* from the Madagascar dry forest. *Journal of Natural Products*, **73**, 1559–1562.

Harinantenaina, L., Brodie, P.J., Callmander, M. *et al.* (2011) Astrotricoumarin, an antiproliferative 4′-hydroxy-2′,3′-dihydroprenylated methylcoumarin from an *Astrotrichilia* sp. from the Madagascar dry forest. *Natural Product Communications*, **6**, 1259–1262.

Harinantenaina, L., Bowman, J.D., Brodie, P.J. *et al.* (2013) Antiproliferative and antiplasmodial dimeric phloroglucinols from *Mallotus oppositifolius* from the Madagascar dry forest. *Journal of Natural Products*, **76**, 388–393.

Hartwell, J.L. (1982) *Plants Used Against Cancer*, Quarterman, Lawrence, MA.

Holstein, S.A., Kuder, C.H., Tong, H. and Hohl, R.J. (2011) Pleiotropic effects of a schweinfurthin on isoprenoid homeostasis. *Lipids*, **46**, 907–921.

Hou Y. and Harinantenaina, L. (2010) New and bioactive natural products isolated from Madagascar plants and marine organisms. *Current Medicinal Chemistry*, **17**, 1191–1219.

Hou, Y.P., Cao, S.G., Brodie, P. *et al.* (2008) Biodiversity conservation and drug discovery in Madagascar, part 32. Antiproliferative cassane diterpenoids of *Cordyla madagascariensis* ssp madagascariensis from the Madagascar rain forest. *Journal of Natural Products*, **71**, 150–152.

Hou, Y., Cao, S., Brodie, P.J. *et al.* (2009a) Antiproliferative and antimalarial anthraquinones of *Scutia myrtina* from the Madagascar forest. *Bioorganic & Medicinal Chemistry*, **17**, 2871–2876.

Hou, Y.P., Cao, S.G., Brodie, P. *et al.* (2009b) Antiproliferative cardenolide glycosides of *Elaeodendron alluaudianum* from the Madagascar rain forest. *Bioorganic & Medicinal Chemistry*, **17**, 2215–2218.

Hou, Y.P., Cao, S.G., Brodie, P.J. *et al.* (2010) Euphane triterpenoids of *Cassipourea lanceolata* from the Madagascar rain forest. *Phytochemistry*, **71**, 669–674.

Hughes, B. (2010) Antibody–drug conjugates for cancer: poised to deliver? *Nature Reviews Drug Discovery*, **9**, 665–667.

Iwu, M.M. (1993) *Handbook of African Medicinal Plants*, CRC Press, London.

Karkare, S., Adou, E., Cao, S. *et al.* (2007) Cytotoxic cardenolide glycosides of *Roupellina (Strophanthus) boivinii* from the Madagascar rain forest. *Journal of Natural Products*, **70**, 1766–1770.

Katz, J., Janik, J.E. and Younes, A. (2012) Brentuximab vedotin (SGN-35). *Clinical Cancer Research*, **17**, 6428–6436.

Kaufman, D. (1993) *Botany 2000-Asia Newsletter*, **2**, 6.

Kingston, D.G.I. (2011) Modern natural products drug discovery and its relevance to biodiversity conservation. *Journal of Natural Products*, **74**, 496–511.

Klausmeyer, P., Van, Q.N., Jato, J.G. *et al.* (2010) Schweinfurthins I and J from *Macaranga schweinfurthii*. *Journal of Natural Products*, **73**, 479–481.

Kuder, C.H., Neighbors, J.D., Hohl, R.J. and Wiemer, D.F. (2009) Synthesis and biological activity of a fluorescent schweinfurthin analogue. *Bioorganic & Medicinal Chemistry*, **17**, 4718–4723.

Kuete, V. and Efferth, T. (2011) Pharmacogenomics of Cameroonian traditional herbal medicine for cancer therapy. *Journal of Ethnopharmacology*, **137**, 752–766.

Kuete, V., Krusche, B., Youns, M. *et al.* (2011) Cytotoxicity of some Cameroonian spices and selected medicinal plant extracts. *Journal of Ethnopharmacology*, **134**, 803–812.

Lambert, J.M. (2010) Antibody–maytansinoid conjugates: a new strategy for the treatment of cancer *Drugs of the Future*, **35**, 471–480.

Lee, J. and Parker, K.A. (2012) A formal synthesis of (−)-englerin A by RRCM and transannnular etherification. *Organic Letters*, **14**, 2682–2685.

Leet, J.E., Liu, X., Drexler, D.M. *et al.* (2008) Cytotoxic xanthones from *Psorospermum molluscum* from the Madagascar rain forest. *Journal of Natural Products*, **71**, 460–463.

Li, Z., Nakashige, M. and Chain, W.J. (2011) A brief synthesis of (−)-englerin A. *Journal of the American Chemical Society*, **133**, 6553–6556.

LoRusso, P.M., Weiss, D., Guardino, E. *et al.* (2011) Trastuzumab emtansine: a unique antibody–drug conjugate in development for human epidermal growth factor receptor 2-positive cancer. *Clinical Cancer Research*, **17**, 6437–6447.

Magadula, J.J. and Erasto, P. (2009) Bioactive natural products derived from the East African flora. *Natural Product Reports*, **26**, 1535–1554.

Mente, N.R., Wiemer, A.J., Neighbors, J.D. *et al.* (2007) Total synthesis of (*R,R,R*)- and (*S,S,S*)-schweinfurthin F: differences of bioactivity in the enantiomeric series. *Bioorganic & Medicinal Chemistry Letters*, **17**, 911–915.

Mente, N.R., Neighbors, J.D. and Wiemer, D.F. (2008) $BF_3{\cdot}Et_2O$-mediated cascade cyclizations: synthesis of schweinfurthins F and G. *Journal of Organic Chemistry*, **73**, 7963–7970.

Mi, Q., Cui, B., Silva, G.L. *et al.* (2003) Characterization of tropane alkaloid aromatic esters that reverse the multidrug-resistance phenotype. *Anticancer Research*, **23**, 3607–3616.

Mita, M.M., Ricart, A.D., Mita, A.C. *et al.* (2007) A phase I study of a CanAg-targeted immunoconjugate, huC242-DM4, in patients with Can Ag-expressing solid tumors. *Journal of Clinical Oncology*, **25** (18S), 3062.

Molawai, K., Delpont, N. and Echavarren, A.M. (2010) Enantioselective synthesis of (−)-englerins A and B. *Angewandte Chemie International Edition*, **49**, 3517–3519.

Moser, B.R. (2008) Review of cytotoxic cephalostatins and ritterazines: isolation and synthesis. *Journal of Natural Products*, **71**, 487–491.

Murphy, B.T., Cao, S.G., Norris, A. *et al.* (2005) Cytotoxic flavanones of *Schizolaena hystrix* from the Madagascar rain forest. *Journal of Natural Products*, **68**, 417–419.

Murphy, B.T., Cao, S., Norris, A. *et al.* (2006) Cytotoxic compounds of *Schizolaena hystrix* from the Madagascar rain forest. *Planta Medica*, **72**, 1235–1238.

Murphy, B.T., Brodie, P., Slebodnick, C. *et al.* (2008a) Antiproliferative limonoids of a *Malleastrum* sp from the Madagascar rain forest. *Journal of Natural Products*, **71**, 325–329.

Murphy, B.T., Cao, S., Brodie, P.J. *et al.* (2008b) Antiproliferative compounds of *Artabotrys madagascariensis* from the Madagascar rain forest. *Natural Product Research*, **22**, 1169–1175.

Murphy, B.T., Cao, S.G., Brodie. P. *et al.* (2009) Antiproliferative bistramides from *Trididemnum cyclops* from Madagascar. *Journal of Natural Products*, **72**, 1338–1340.

Neighbors, J.D., Beutler, J.A. and Wiemer, D.F. (2005) Synthesis of nonracemic 3-deoxyschweinfurthin. *Journal of Organic Chemistry*, **70**, 925–931.

Neighbors, J.D., Salnikova, M.S., Beutler, J.A. and Wiemer, D.F. (2006) Synthesis and structure–activity studies of schweinfurthin B analogs: evidence for the importance of a D-ring hydrogen bond donor in expression of differential cytotoxicity. *Bioorganic & Medicinal Chemistry*, **14**, 1771–1784.

Newman, D.J. and Cragg, G.M. (2004a) Advanced preclinical and clinical trials of natural products and related compounds from marine sources. *Current Medicinal Chemistry*, **11**, 1693–1714.

Newman, D.J. and, Cragg, G.M. (2004b) Marine natural products and related compounds in clinical and advanced preclinical trials. *Journal of Natural Products*, **67**, 1216–1238.

Newman, D.J. and Cragg, G.M. (2012) Natural products as sources of new drugs over the 30 years from 1981 to 2010. *Journal of Natural Products*, **75**, 211–235.

Nicolaou, K.C., Kang, Q., Ng, S.Y. and Chen, D.Y. (2010) Total synthesis of englerin A. *Journal of the American Chemical Society*, **132**, 8219–8222.

Niculescu-Duvaz, I. (2010) Trazuzumab emastine, an antibody–drug conjugate for the treatment of HER2+ metastatic breast cancer. *Current Opinion in Molecular Therapeutics*, **12**, 350–360.

Okunji, C.O., Iwu, M.M., Ito, Y. and Smith, P.L. (2005) Preparative separation of indole alkaloids from the rind of *Picralima nitida* (Stapf) T. Durand & H. Durand by pH-zone-refining countercurrent chromatography. *Journal of Liquid Chromatography and Related Technologies*, **28**, 775–783.

Pan, E.D., Cao, S.G., Brodie, P.J. *et al.* (2010a) An antiproliferative xanthone of *Symphonia pauciflora* from the Madagascar rain forest. *Natural Product Communications*, **5**, 751–754.

Pan, E.D., Harinantenaina, L., Brodie, P.J. *et al.* (2010b) Four diphenylpropanes and a cycloheptadibenzofuran from *Bussea sakalava* from the Madagascar dry forest. *Journal of Natural Products*, **73**, 1792–1795.

Pan, E.D., Harinantenaina, L., Brodie, P.J. *et al.* (2011a) Antiproliferative cardenolides of *Leptadenia madagascariensis* from the Madagascar dry forest. *Bioorganic & Medicinal Chemistry*, **19**, 422–428.

Pan, E.D., Cao, S., Brodie, P.J. *et al.* (2011b) Alkaloids of *Ambavia gerrardii* from the Madagascar dry forest and the synthesis of 9-hydroxyeupolauridine. *Journal of Natural Products*, **74**, 1169–1174.

Pan, E.D., Gorka, A., Alumasa, J. *et al.* (2011c) Antiplasmodial and antiproliferative pseudoguaianolides of *Athroisma proteiforme* from the Madagascar dry forest. *Journal of Natural Products*, **74**, 2174–2180.

Pettit, G.R., Inoue, M., Kamano, Y. *et al.* (1988) Antineoplastic agents. 147. Isolation and structure of the powerful cell growth inhibitor cephalostatin 1. *Journal of the American Chemical Society*, **110**, 2006–2007.

Pettit, G.R., Zhang, Q., Pinilla, V. *et al.* (2005) Antineoplastic agents. 534. Isolation and structure of sansevistatins 1 and 2 from the African *Sansevieria ehrenbergii*. *Journal of Natural Products*, **68**, 729–733.

Pinney, K.G., Pettit, G.R., Trawick, M.L. *et al.* (2012) The discovery and development of the combretastatins, in *Anticancer Agents from Natural Products*, 2nd edn (eds G.M. Cragg, D.G.I. Kingston and D.J. Newman), CRC Taylor & Francis, Boca Raton, FL, pp. 27–63.

Pouwer, R.H., Richard, J.A., Tseng, C.C. and Chen, D.Y. (2012) Chemical synthesis of the englerins. *Chemistry – An Asian Journal*, **7**, 22–35.

Prakash, C.V.S., Hoch, J.M. and Kingston, D.G.I. (2002) Biodiversity conservation and drug discovery in Madagascar, part 2. Structure and stereochemistry of new cytotoxic clerodane diterpenoids from the bark of *Casearia lucida* from the Madagascar rain forest. *Journal of Natural Products*, **65**, 100–107.

Radtke, L., Willot, M., Sun, H. *et al.* (2011) Total synthesis and biological evaluation of (−)-englerin A and B: synthesis of analogues with improved activity profile. *Angewandte Chemie International Edition*, **50**, 1–6.

Ratnayake, R., Covell, D., Ransom, T.T. *et al.* (2009) Englerin A, a selective inhibitor of renal cancer cell growth, from *Phyllanthus engleri*. *Organic Letters*, **11**, 57–60.

Reynolds, M., Chaturvedula, V.S.P., Ratovoson, F. *et al.* (2006) Cytotoxic diterpenoids from *Podocarpus madagascariensis* from the Madagascar rain forest. *Natural Product Research*, **20**, 606–610.

Sakurai, Y., Sakurai, N., Taniguchi, M. *et al.* (2006) Rautandiols A and B, pterocarpans and cytotoxic constituents from *Neorautanenia mitis*. *Journal of Natural Products*, **69**, 397–399.

Sawadogo, W.R., Schumacher, M., Teiten, M-H. *et al.* (2012) Traditional West African pharmacopeia, plants and derived compounds for cancer therapy. *Biochemical Pharmacology*, **84**, 1225–1240.

Schuster, B.G. (2002) Development of antimalarial agents and drugs for parasitic infections based on leads from traditional medicine: the Walter Reed experience, in *Ethnomedicine and Drug Discovery* (eds M.M. Iwu and J. Wootton), Elsevier, Amsterdam, pp. 163–171.

Schuster B.G., Jackson J.E., Obijiofor C.N. *et al.* (1999) Drug development and conservation of biodiversity in West and Central Africa: a model for collaboration with indigenous people. *Pharmaceutical Biology* **37** (Suppl), 1–16.

Senter, P.D. (2009) Potent antibody drug conjugates for cancer therapy. *Current Opinion in Chemical Biology*, **13**, 235–244.

Seo, Y., Berger, J.M., Hoch, J. *et al.* (2002) Biodiversity conservation and drug discovery in Madagascar. Part 1. A new triterpene saponin from *Pittosporum viridiflorum* from the Madagascar rain forest. *Journal of Natural Products*, **65**, 65–68.

Siemann D.W., Chaplin D.J. and Walicke P.A. (2009) A review and update of the current status of the vasculature-disabling agent combretastatin-A4 phosphate (CA4P). *Expert Opinion on Investigational Drugs*, **18**, 189–197.

Silva, G.L., Cui, B., Chávez, D. *et al.* (2001) Modulation of the multidrug-resistance phenotype by new tropane alkaloid aromatic esters from *Erythroxylum pervillei*. *Journal of Natural Products*, **64**, 1514–1520.

Sourbier, C., Scroggins, B.T., Ratnayake, B. *et al.* (2013) Englerin A stimulates PKC theta to inhibit insulin signaling while simultaneously activating HSF1: a case of pharmacologically induced synthetic lethality. *Cancer Cell*, **23**, 228–237.

Su, B.N., Park, E.J., Mbwambo, Z.H. *et al.* (2002) New chemical constituents of *Euphorbia quinquecostata* and absolute configuration assignment by a convenient Mosher ester procedure carried out in NMR tubes. *Journal of Natural Products*, **65**, 1278–1282.

Sulzmaier, F.J., Li, Z., Nakashige, M.L. *et al.* (2012) Englerin A selectively induces necrosis in human renal cancer cells. *PLoS ONE*, **7**, e48032.

Sun, B.F., Wang, C.L., Ding, R. *et al.* (2011) Concise approach to the core of englerin A via an organocatalytic [4+3] cycloaddition reaction. *Tetrahedron Letters*, **52**, 2155–2158.

Tchuendem, M.H.K., Mbah, J.A., Tsopmo, A. *et al.* (1999) Anti-plasmodial sequiterpenoids from the African *Reneilmia cincinnata*. *Phytochemistry*, **52**, 1095–1099.

Teicher, B.A. (2009) Antibody–Drug conjugate targets. *Current Cancer Drug Targets*, **9**, 982–1004.

Teicher, B.A. and Chari, R.V. (2011) Antibody conjugate therapeutics: challenges and potential. *Clinical Cancer Research*, **17**, 6389–6397.

Thoison, O., Hnawia E., Guéritte-Voegelein, F. and Sévenet, T. (1992). Vedelianin, a hexahydroxanthene derivative isolated from *Macaranga vedeliana*. *Phytochemistry*, **31**, 1439–1442.

Topczewski, J.J. and Wiemer, D.F. (2011) First total synthesis of (+)-vedelianin, a potent antiproliferative agent. *Tetrahedron Letters*, **52**, 1628–1630.

Topczewski, J.J., Neighbors, J.D. and Wiemer, D.F. (2009) Total synthesis of (+)-schweinfurthins B and E. *Journal of Organic Chemistry*, **74**, 6965–6972.

Topczewski, J.J., Kodet, J.G. and Wiemer, D.F. (2011) Exploration of cascade cyclizations terminated by tandem aromatic substitution: total synthesis of (+)-schweinfurthin A. *Journal of Organic Chemistry*, **76**, 909–919.

Turbyville, T.J., Gürsel, D.B., Tuskan, R.G. *et al.* (2010) Schweinfurthin A selectively inhibits proliferation and Rho signaling in glioma and neurofibromatosis type 1 tumor cells in an NF1-GRD dependent manner. *Molecular Cancer Therapeutics*, **9**,1234–1243.

Ulrich, N.C., Kodet, J.G., Mente, N.R. *et al.* (2010) Structural analogues of schweinfurthin F: probing the steric, electronic and hydrophobic properties of the D-ring substructure. *Bioorganic & Medicinal Chemistry*, **18**, 1676–1683.

Ushakov, D.B., Navickas, V., Ströbele, M. *et al.* (2011) Total sythesis and biological evaluation of (−)-9-deoxy-englerin A. *Organic Letters*, **13**, 2090–2093.

Wang, J., Chen, S.G., Sun, B.F. *et al.* (2013) Collective total synthesis of englerin A and B, orientalol E and F, and oxyphyllol: application of the organocatalytic [4+3] cycloaddition reaction. *Chemistry - A European Journal*, **19**, 2539–2547.

Williams, R.B., Norris, A., Slebodnick, C. *et al.* (2005) Cytotoxic sesquiterpene lactones from *Vernonia pachyclada* from the Madagascar rain forest. *Journal of Natural Products*, **68**, 1371–1374.

Williams, R.B., Norris, A., Miller, J.S. *et al.* (2006) Two new cytotoxic naphthoquinones from *Mendoncia cowanii* from the rainforest of Madagascar. *Planta Medica*, **72**, 564–566.

Williams, R.B., Norris, A., Miller, J.S. *et al.* (2007) Cytotoxic clerodane diterpenoids and their hydrolysis products from *Casearia nigrescens* from the rain forest of Madagascar. *Journal of Natural Products*, **70**, 206–209.

Willot, M., Radtke, L., Könning, D. *et al.* (2009) Total synthesis and absolute configuration of the guaiane sesquiterpene englerin A. *Angewandte Chemie International Edition*, **48**, 9105–9108.

Yoder, B., Norris, A., Miller, J.S. *et al.* (2006) Cytotoxic prenylated stilbenes and flavonoids from *Macaranga alnifolia* from the Madagascar rain forest. *Journal of Natural Products*, **70**, 342–346.

Yoder, B.J., Cao, S.G., Norris, A. *et al.* (2007a) Tambouranolide, a new cytotoxic hydroxybutanolide from a *Tambourissa* sp (Monimiaceae). *Natural Product Research*, **21**, 37–41.

Yoder, B.J., Cao, S.G., Norris, A. *et al.* (2007b) Antiproliferative prenylated stilbenes and flavonoids from *Macaranga alnifolia* from the Madagascar rainforest. *Journal of Natural Products*, **70**, 342–346.

Yu, T.-W., Floss, H.G., Cragg, G.M. and Newman, D.J. (2012) Ansamitocins (maytansinoids), in *Anticancer Agents from Natural Products*, 2nd edn (eds G.M. Cragg, D.G.I. Kingston and D.J. Newman), CRC Taylor & Francis, Boca Raton, FL, pp. 407–427.

Zhou, Q., Chen, X., and Ma, D. (2010) Asymmetric, protecting-group-free total synthesis of (−)-englerin A. *Angewandte Chemie International Edition*, **49**, 3513–3515.

# 11 Biodiversity as a Source of Potent and Selective Inhibitors of Chikungunya Virus Replication

Pieter Leyssen[1], Jacqueline Smadja[2], Philippe Rasoanaivo[3], Ameenah Gurib-Fakim[4], Mohamad Fawzi Mahomoodally[5], Bruno Canard[6], Jean-Claude Guillemot[6], Marc Litaudon[7] and Françoise Guéritte[7]

[1]Rega Institute for Medical Research (KU Leuven), Leuven, Belgium
[2]Laboratoire de Chimie des Substances Naturelles et des Sciences des Aliments (LCSNSA), Faculté des Sciences et Technologies, Université de La Réunion, La Réunion, France
[3]Institut Malgache de Recherches Appliquées (IMRA), Avarabohitra Itaosy, Antananarivo, Madagascar
[4]Center for Phytotherapy Research (CEPHYR), Ebene, Mauritius
[5]Department of Health Sciences, Faculty of Science, University of Mauritius, Réduit, Mauritius
[6]Laboratoire d'Architecture et de Fonction des Macromolécules Biologiques (AFMB-AMU-UMR7257), Marseille, France
[7]Centre de Recherche de Gif, Institut de Chimie des Substances Naturelles (ICSN), CNRS, LabEx LERMIT, Gif sur Yvette Cedex, France

## 11.1 The epidemiology of chikungunya virus

Chikungunya virus (CHIKV) was isolated and identified in the mid 1950s from the serum of a febrile patient during a dengue-like epidemic in Tanzania, East Africa. The virus and associated disease were named by the clinical symptoms it caused: 'chikungunya' means 'stooped walk' in Makonde language, and refers to a long-lasting painful arthralgia. Other clinical symptoms are similar to those of dengue fever, including sudden and high fever, nausea, myalgia and rash. For this reason, chikungunya has long been confused with dengue, especially during the acute phase, although its long-lasting symptoms are more characteristic, especially the joint pain (Pialoux *et al.*, 2007).

CHIKV is an emerging mosquito-borne alphavirus that belongs to the family Togaviridae. The main vectors of chikungunya are *Aedes* mosquitoes from the Culicideae family. Until recently, CHIKV in Africa was spread by different species of *Aedes* such as *A. furcifer*, *A. luteocephalus*, *A. taylori* and *A. africanus*. Introduction of the virus into Asia happened alongside with its adaptation to transmission by *A. aegypti*. During the epidemics in the Indian Ocean, it was established that it further adapted to transmission by a new vector, *A. albopictus* (Pialoux *et al.*, 2007). Because of the presence of these latter mosquito species in most tropical and subtropical regions worldwide, CHIKV has become a pathogen with pandemic potential.

However, unlike dengue viruses, and for yet unknown reasons, CHIKV still displays a re-emergence pattern that is linked with long-distance migration. Ever since its discovery, CHIKV has been somewhat ignored by scientists except when it has caused large human epidemics. In the 1960s and 1970s, few major chikungunya disease outbreaks were reported in India and Southeast Asia, but in 2004 a large outbreak began on the coast of Kenya. The epidemic then reached many islands in the Indian Ocean, India and various parts of Southeast Asia, and in many other countries across Asia, Europe and North America, cases were reported, mainly travellers that returned from areas with ongoing CHIKV activity (Enserink, 2007). At the end of 2013, an outbreak of the virus was detected in the Carribean islands, which may again herald a massive outbreak, this time in Central and/or South America.[1]

### 11.1.1 The chikungunya virus genome

The CHIKV genome (Kahn *et al.*, 2002; Li *et al.*, 2012) is a single-stranded, positive-sense, capped and polyadenylated RNA molecule, which upon infection is translated in the cytoplasm of the host cell into two large polyproteins. The largest one, which is encoded by almost two-thirds of the viral genome, is processed co-translationally into the four polyvalent non-structural proteins that are essential for virus replication (nsp1, nsp2, nsp3, nsp4 and combinations of nsp1–4 cleavage products), whereas the smallest one, which is located in the downstream one-third of the viral genome, gives rise to the structural proteins, and which will make up the virus particle (Figure 11.1; Li *et al.*, 2012).

The nsP1 protein carries an alphavirus unique methyltransferase able to methylate a guanosine 5′-triphosphate (GTP) molecule at its N7 position. Nsp1 might be the still elusive alphavirus guanylyltransferase involved in RNA capping of viral mRNAs. It carries, however, a N7-guanine methyltransferase acting first on GTP and mechanistically coupled to covalent attachment of N7-GMP at conserved residue His38 (Semliki Forest numbering). Nsp1 is also a target of mycophenolic acid and ribavirin treatment, as resistance mutations can be mapped onto nsp1 regarding these antiviral

[1]http://www.who.int/csr/don/2013_12_10a/en/index.html. (Accessed 21 January 2014)

*Novel Plant Bioresources: Applications in Food, Medicine and Cosmetics*, First Edition. Edited by Ameenah Gurib-Fakim.

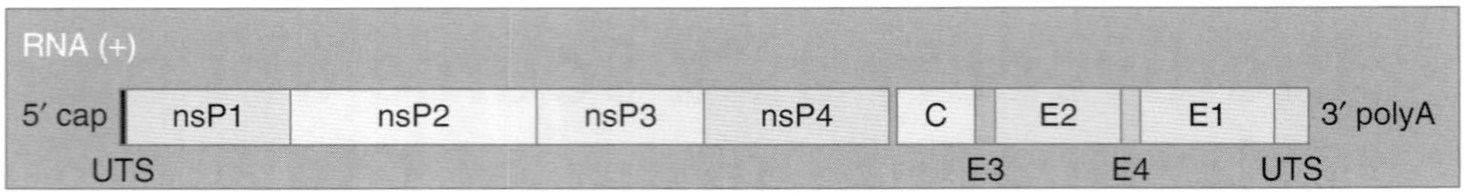


**Figure 11.1** Chikungunya virus genome. *Source*: Deng *et al.* (2012). Reproduced with permission of Elsevier

agents affecting the GTP metabolism. The nsp1 protein carries an amphipathic helix, which is required as a membrane anchor, and nsp1 MTase activity is only observed in the presence of membranes, implying that membrane binding triggers activation through a conformational change. The nsP2 protein is a bifunctional protein that appears to play a central role in RNA synthesis. The 798aa CHIKV nsP2 is a superfamily I RNA helicase. It contains nucleoside and RNA triphosphatase, as well as RNA helicase activities in its 470aa N-terminal part, while the C-terminal part carries a papain-like protease, responsible for processing the viral polyprotein into the four non-structural proteins 1 to 4. The nsP3 protein is composed of three distinct domains. The N-terminal part is a *macro* domain, also called domain X, the second is conserved and unique to alphaviruses, and the C-terminal third domain is highly phosphorylated and hypervariable. *Macro* proteins domains are found in many proteins involved in poly-ADP-ribose metabolism (e.g. poly(ADP-ribose) polymerases). The nsP4 protein is the RNA-dependent RNA polymerase (RdRp). It is the most conserved alphavirus protein. In association with other nsps, it promotes genomic and subgenomic RNA synthesis.

### 11.1.2 Inhibitors of the chikungunya virus replication

Currently, there is still no vaccine or any antiviral drug on the market for the prophylaxis or treatment of CHIKV infections. Patients are given extra fluids and are recommended or even forced by the disease to rest in bed because of severe arthralgia. Drugs, like ibuprofen, naproxen, acetaminophen, or paracetamol, are administered to alleviate the symptoms of fever and aching. As in the case of dengue fever virus infection, aspirin should be avoided (www.cdc.gov/chikungunya/symptoms/).

The years 2005–2006 witnessed a massive outbreak in the Indian Ocean region. Chikungunya was before considered to be an obscure disease because of its low prevalence and limited geographical distribution. It therefore did not attract a lot of attention. At that period of time, virus-cell-based assays were available for many viruses and were first used for the exploration of antiviral discovery. A logical approach was to evaluate compounds with reported antiviral activity for a potential conjunct effect on CHIKV replication. As a consequence, in 2004, one study reported on the selective *in vitro* anti-CHIKV activity of ribavirin (**1**) and 6-azauridine (**2**) (Figure 11.2), as well as glycyrrhizin and INFα2b (Briolant *et al.*, 2004). These results were modestly surprising as these compounds have a mechanism of action that targets a metabolic or signal transduction cascade in the host cell on which the virus depends for its replication. For ribavirin in particular, the pleiotropic and still-debated mechanism of action is mainly the result of inhibition of the cellular enzyme inosine monophosphate dehydrogenase (IMP-DH). IMP-DH inhibition results in depletion of intracellular GTP pools, which has an indirect antiviral effect. Indeed, GTP shortage decreases viral RNA genome synthesis. Mycophenolic acid (MPA, **3**; Figure 11.2), a ribavirin unrelated drug documented to be a potent IMP-DH inhibitor, also selectively inhibits CHIKV replication (Khan *et al.*, 2011). Sindbis virus, which belongs to the same virus genus as CHIKV, acquires resistance to treatment with either ribavirin or MPA, giving insight into actual antiviral mechanism: resistance mutations map in the guanylyltransferase nsp1 confering an increased affinity for GTP (Scheidel and Stollar, 1991).

Ribavirin was approved long ago for clinical use by the US Food and Drug Administration for specific uses other than for chikungunya. Results obtained with this compound *in vitro* against chikungunya may have spurred the undertaking of a small trial with a limited number of patients to investigate whether or not ribavirin

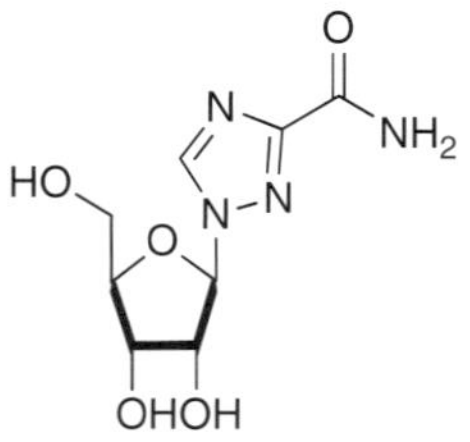

**1**

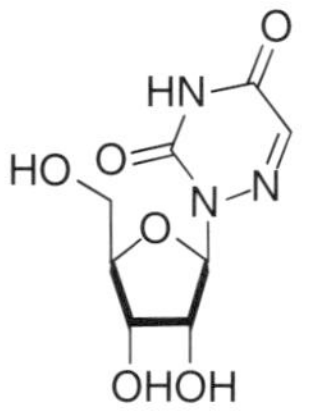

**2**

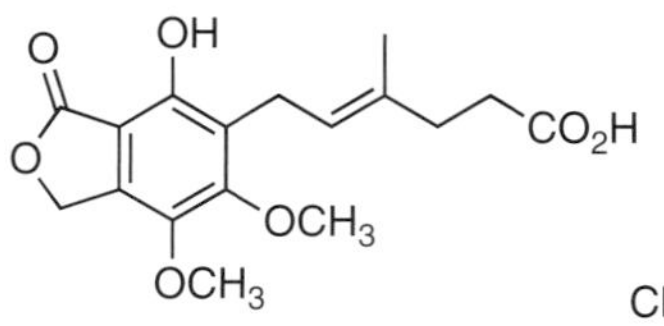

**3**

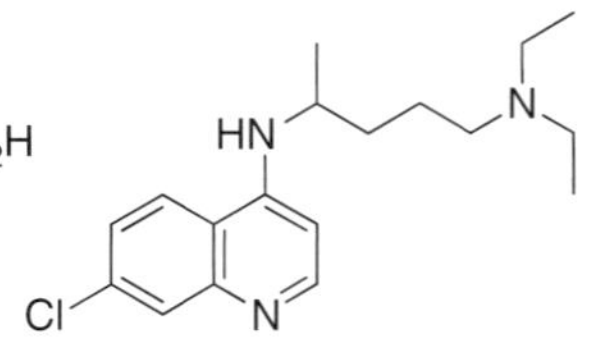

**4**

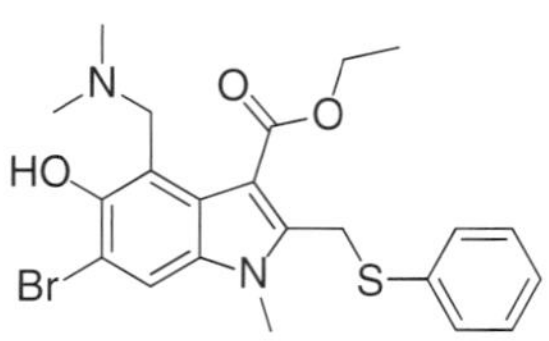

**5**

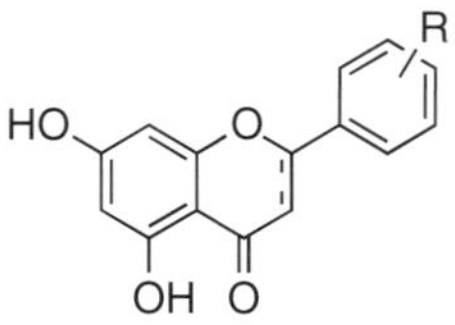

**6**

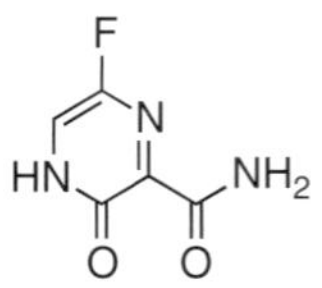

**7**

**Figure 11.2** Compounds inhibiting the CHIKV replication

has a beneficial effect on chikungunya-induced arthritis. The conclusion of this study was, 'Ribavirin may have a direct antiviral property against chikungunya leading to faster resolution of joint and soft tissue manifestations' (Ravichandran and Manian, 2008). This, however, may not have been the right conclusion to draw from this study as the patients only received treatment at least 2 weeks after the febrile episode. In humans, no active CHIKV replication has been reported beyond the acute stage of the infection, and the persistent arthralgia is to be attributed to a derailed immune response rather than to ongoing virus replication. The amelioration of symptoms that is observed in this study, therefore, may be the result of the immunomodulatory properties of ribavirin rather than to a direct antiviral effect (Ogbomo *et al.*, 2010). To our knowledge, no large-scale clinical study has yet been undertaken with ribavirin in the context of a CHIKV infection.

Following the rapid emergence of CHIKV worldwide and, as a consequence, the pronounced change in interest, the quest for novel and selective inhibitors of this viral pathogen was launched on a larger scale. Because of the urgency in this outbreak situation to have a drug available for the treatment or prevention of the disease, a library of compounds that were already available for clinical use was first screened for antiviral activity in a CHIKV-cell-based assay.

Chloroquine (**4**; Figure 11.2), a widely used drug in the treatment of malaria, significantly decreased *in vitro* CHIKV replication in a dose-dependent manner and with a considerable window of antiviral selectivity (i.e. the difference in concentration of compound that causes the antiviral effect and the concentration which has an adverse effect on the host cell). Addition of chloroquine before or at the same time of virus infection drastically reduced CHIKV replication. However, the drug was ineffective once the virus successfully passed the initial stages of infection (Khan *et al.*, 2010). Chloroquine is a weak base and specifically inhibits the acidification of intracellular endosomes, a mechanism of action that is already well known for this compound (Savarino *et al.*, 2003). As CHIKV, like many other viruses, depends on a decrease of the pH inside the endocytotic vesicles for successful infection of the host cell, it is therefore not unexpected that this compound also has an antiviral effect against this virus (Gay *et al.*, 2012). This compound, like ribavirin, does not specifically inhibit CHIKV replication. It affects virus replication by acting on the host cell enzymes and pathways. Because of its previously approved clinical use, its efficacy was assessed in a small double-blind trial involving 54 chikungunya patients that were randomly distributed between a placebo and short-course drug treatment group. Even though the drug was administered in the acute phase at relatively high doses, no beneficial effect was observed either on febrile arthralgia or virus titers. At day 200, long-lasting arthralgia was even more prevalent in the group that received treatment, indicating there is no justification for use of chloroquine in the context of acute CHIKV infection (De Lamballerie *et al.*, 2008; Delogu and de Lamballerie, 2011).

Another compound that acts in the early stages of virus infection and that has broad-spectrum antiviral activity *in vitro* as well as *in vivo* is arbidol (**5**; Figure 11.2) (Shi *et al.*, 2007). More particularly, arbidol interferes at the level of fusion of the virus particle with host cell membranes. In the case of influenza virus, stabilization of the virion may be the precise mechanism of action, while interaction with the E protein and/or modulation of the endocytotic membrane by intercalation is put forward as possible mechanism in the case of hepatitis C virus (Boriskin *et al.*, 2008). Recently, arbidol was also reported to selectively inhibit CHIKV infection. The identification of a G407R mutation in the envelope protein of CHIKV that acquired resistance to the antiviral effect of the compound indicates a similar mechanism of action (Delogu *et al.*, 2011). No clinical studies with this compound against CHIKV have been performed so far.

In recent years, more and more publications are becoming available in which the antiviral activity of compounds on the replication of CHIKV is reported. Harringtonine, a cephalotaxine alkaloid that inhibits an early stage of (alpha)virus replication, was identified in an immunofluorescence-based screening campaign of a natural product compound library (Kaur *et al.*, 2013). Similarly, a screening campaign of other libraries in a CHIKV replicon system as well as a Sindbis virus-cell-based assay (a model system quite frequently used instead of the more pathogenic CHIKV) yielded hits such as chlorpromazine and 5,7-dihydroxyflavones (**6**; Figure 11.2) (Pohjala *et al.*, 2011).

All together, it is striking that most of the compounds that have been reported so far do not have a high specificity for CHIKV replication as such but rather have a quite broad spectrum of antiviral activity. This indicates there are still many targets in the replication cycle of this virus to be explored and exploited. However, considering the high costs of drug development and the small market potential in the case of CHIKV infections, it is unlikely that major strides forward will be made in the near future. A breakthrough is only to be expected when new drugs, that are approved for the treatment of other viral diseases, also prove to inhibit CHIKV replication. Nucleoside analogues in development for the treatment of hepatitis C or dengue fever virus infections may be used against CHIKV as well and possibly could be used off-label for the treatment of this disease as soon as they reach the market. Also, T-705, favipiravir (**7**; Figure 11.2), which currently is in clinical development for the treatment of influenza virus infections, has shown potential as an anti-alphavirus agent (Julander *et al.*, 2009).

### 11.1.3 Chikungunya virus enzymes as targets for antiviral drugs

Among the non-structural proteins, the CHIKV nsP2 protein is the only one so far which has been studied *in vitro* and *in silico* as a potential target for CHIKV inhibitors.

A virtual screening campaign was performed using a database of 350 000 compounds. These simulations have allowed the finding of a potent inhibitor of this enzyme, but no compound structure was reported (Bassetto *et al.*, 2011). Also, another study investigated the *in silico* interaction of small molecules with the nsP2 protease protein. One compound *N*-butyl-9-(3,4-dipropoxy-5-

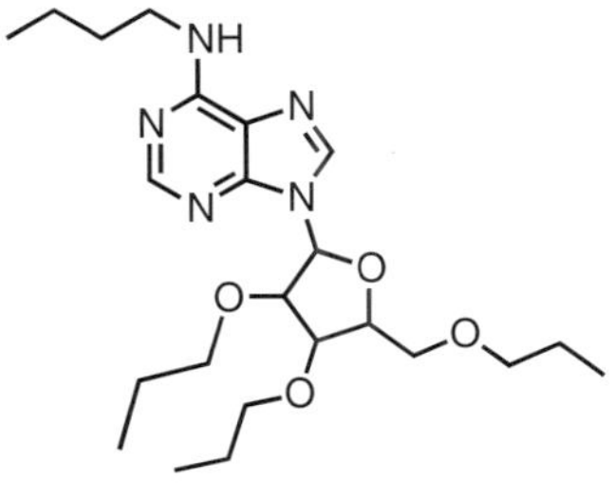


**8**

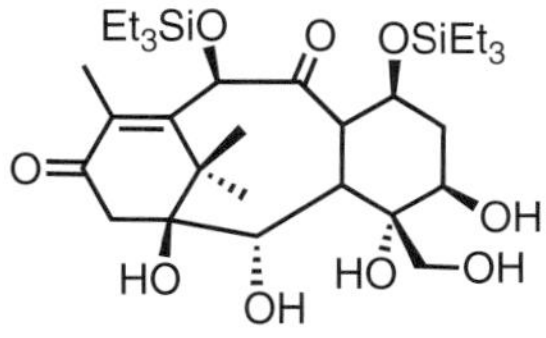


**9**

**Figure 11.3** Compounds inhibiting the CHIKV replication *in silico* and *in vitro* (propoxymethyl)oxolan-2-yl)purin-6-amine (**8**; Figure 11.3) with absorption, distribution, metabolism, excretion and toxicity properties was found to bind efficiently to the protein (Bora *et al.*, 2012).

Recently, the screening of a chemical library using a phenotypic assay targeting the non-structural protein nsP2 led to the identification of one compound (Lucas-Hourani *et al.*, 2013). This compound (**9**; Figure 11.3) belongs to the taxoid family, and inhibits also CHIKV replication *in vitro*.

As new purified enzymes and proteins become available, drug design programmes should emerge provided that associated phenotypes are easily monitored.

## 11.2 The PHYTOCHIK programme for the discovery of natural compounds active against chikungunya virus

The project PHYTOCHIK, selected by the Centre de Recherche et de Veille des maladies émergentes dans l'Océan Indien (CRVOI) for financial support, began in March 2009 and ended in December 2011.

This project, entitled 'Biodiversity and emerging viruses in the Indian Ocean: selection of drug candidates targeting the Chikungunya virus', had as its overall objective the discovery and characterization of selective antiviral compounds derived from the vast Indian Ocean plant biodiversity and capable of fighting emerging viral diseases such as Chikungunya. To this end, the project was conducted through north–south collaborative network of natural product chemists and biologists, including (i) Indian Ocean Islands institutes/laboratories (namely, Institut Malgache de Recherches Appliquées (Madagascar), Laboratoire de Chimie des Substances Naturelles et des Sciences des Aliments (La Réunion) and the Department of Health Sciences, Faculty of Science (Mauritius)), (ii) two French laboratories (Institut de Chimie des Substances Naturelles, CNRS Gif-sur-Yvette and Laboratoire d'Architecture et Fonction des Macromolécules Biologiques, Marseille) and (iii) one Belgian laboratory (Laboratory of Virology and Chemotherapy, Rega Institute for Medical Research, KU Leuven, Leuven).

### 11.2.1 Plants of the Indian Ocean as a source of bioactive compounds

Throughout history, plants have been the first and traditional source of active substances for most medical treatments. At the present time, about 50% of the drugs approved for clinical use are natural products or inspired by natural products (Newmann *et al.*, 2012). This is not surprising, because natural compounds offer structural diversity combined with surprising/unexpected biological properties, which cannot be rivalled by the creativity and synthetic ingenuity of medicinal chemists. In addition, they provide scaffold molecules for a much greater percentage of semi-synthetically and synthetically produced molecules. The advent of today's modern techniques in drug discovery, such as combinatorial chemistry, combined with medium- to high-throughput screening, bioinformatics, 'omics' methods, and so on, has led to the neglect of plants as a source of bioactive compounds (Koehn and Carter, 2005). However, technological progress for the chemical and biological screening of natural extracts, together with improvements in natural product isolation, has led to a renewed interest for the discovery of anti-infective natural products (Clardy and Walsh, 2004; Cos *et al.*, 2004).

Small Indian Ocean islands (Comoros, Mauritius, Reunion, Rodrigues, Seychelles) and Madagascar (Figure 11.4) pack a great

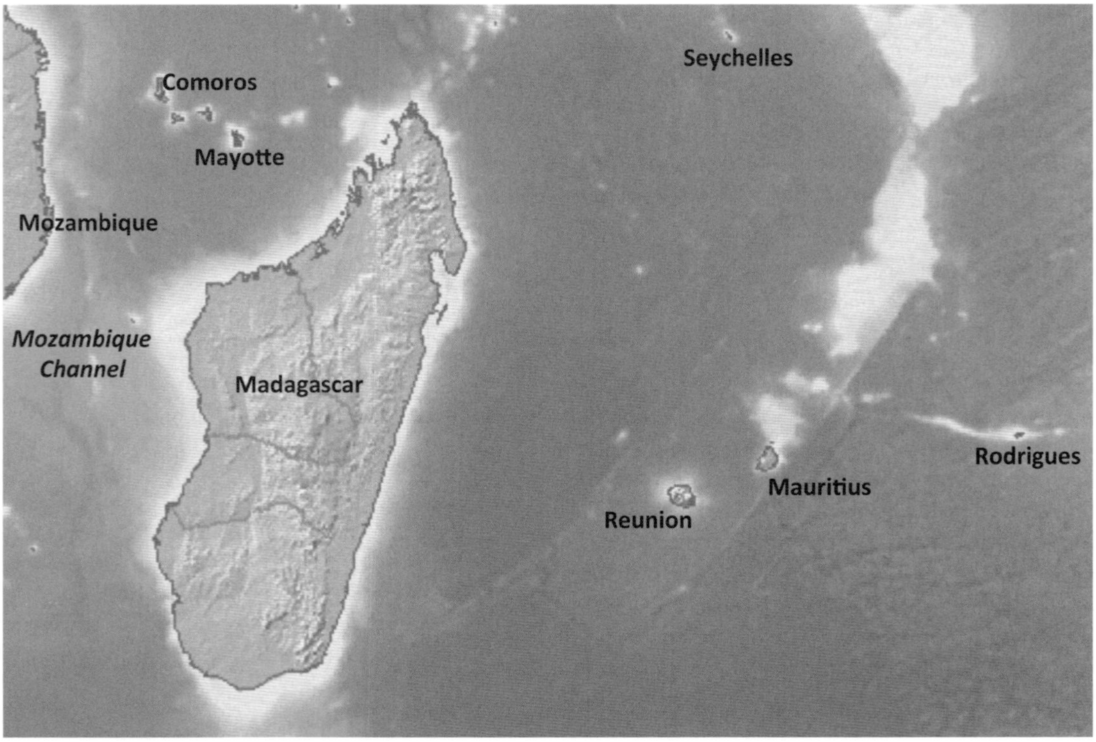


**Figure 11.4** Madagascar and small Indian Ocean islands (Comoros, Mauritius, Reunion, Rodrigues and Seychelles)

deal of biodiversity into a relatively small area of 594 221 $km^2$. In terms of the original extent of their native habitats, Madagascar and the small Indian Ocean islands represents the 10th largest of the 25 biodiversity hotspots that have been identified by Conservation International. They rank eighth amongst hotspots retaining an intact habitat, according to the most recent estimates of tropical forest cover. They harbour 11 endemic plant families, at least 310 endemic genera and 8900–10 500 endemic species.

### 11.2.2 The assays: inhibition of chikungunya virus replication and activity on chikungunya virus enzymes

#### 11.2.2.1 Selection of extracts active on chikungunya virus replication

When starting from plant extracts, which are composed of numerous different molecules, a reliable and rigorous screening strategy is required. CHIKV causes a pronounced cytopathic effect in, for example, Vero cells, which served as a starting point for the optimization of a virus-cell-based screening assay employing the MTS (3-(4,5-dimethylthiazol-2-yl)-5-(3-carboxymethoxyphenyl)-2-(4-sulfophenyl)-*2H*-tetrazolium, inner salt) method. Following microtitre plate readout, assay conditions can be quickly and automatically sorted as inactive/toxic (no increase in optical density (OD) at 493 nm) or with potential antiviral activity (a greater than 50% increase in OD compared with the control conditions). However, and this was expected based on prior experience with the screening of complex samples, a high initial positives rate was obtained (up to 10%). To discriminate true hits from false positives, a microscopic quality-control step was introduced: only samples that completely inhibited virus-induced cell death without causing an adverse effect on the host cell and monolayer morphology were retained for further study. The same assay was used to guide the fractionation process, in the end yielding selective inhibitors of CHIKV replication.

#### 11.2.2.2 Chikungunya enzymes

Very few proteins and enzymes of alphaviruses in general have been made available as targets for purified compounds. In fact, alphaviruses perhaps represent the last most important challenge in terms of biochemistry and structural biology of (+) RNA virus replication enzymes. This is, in part, due to technical difficulties associated with the purification of membrane-associated macromolecular complexes. The nsp1–4 proteins are known to require specific proteolytic cleavages in order to associate to each other into functionally defined complexes. As an example nsp4, the replicative RNA-dependent polymerase, has different template specificities when it is associated with uncleaved nsp123 than when associated with nsp1, nsp2 and nsp3, indicating that protein–protein contacts are important in order to obtain fully functional enzymes.

The only purified and crystallized enzyme domain of chikungunya enzymes is an nsp3 domain, the macro domain (Malet *et al.*, 2009). It is an ADP-ribose and RNA binding module of unknown function. In coronaviruses, it is dispensable for replication but plays a role as a scaffold protein; that is, it can be mutated but not deleted. In order to use it as a drug design target, a phenotype should first be determined, not to say a possible function. Based on the comparison with other coronaviruses (Eriksson *et al.*, 2008), it might play a role in counteracting the innate immune response of the infected cell rather than playing a direct role in the replication of viral RNA.

Nsps from other alphaviruses (Sindbis, VEEV) have been studied, and should provide interesting guidelines for chikungunya enzymes. For example, the nsp2 protease from VEEV has been crystallized (Russo *et al.*, 2006) and should provide an interesting target that could potentially speed up design of inhibitors against chikungunya nsp2. Likewise, the nsp4 RNA-dependent RNA polymerase from Sindbis virus has been purified and characterized as a truncated enzyme exhibiting RNA terminal transferase activity (Rubach *et al.*, 2009). This result indicates that a putative processivity factor must exist, and its discovery should allow robust RdRp activity assays useful for drug screening or characterization.

### 11.2.3 Collect, extraction and screening

Overall, 972 plants were collected in Madagascar, Reunion and Mauritius for the PHYTOCHIK project. For each plant, an ethyl acetate extract was performed on small amounts of plant material. In the case of species selected on ethnopharmacological data, an aqueous extract was prepared. The ethyl acetate extracts were then submitted to a small-cartridge polyamide filtration to remove tannins and then weighed and stored in appropriate vials. These extracts, once filtered through polyamide cartridges, were evaluated for their antiviral activity in the CHIKV-cell-based assay as well as for their cytotoxicity. After this screening, extracts were selected on the basis of their antiviral effect, taking into account their cytotoxicity and a microscopic examination, as mentioned earlier.

#### 11.2.3.1 Plant extracts from Madagascar

In Madagascar, 682 plant species were collected in different ecological regions, ranging from the xerophytic forests of the south to the evergreen rainforests of the east. They were distributed in 117 families and 414 genera. Thirty-three extracts (4.83%) were shown to be active in the primary screening, of which five extracts, from *Echitella lusiantiflora* (Apocynaceae), *Delonix boiviniana* (Fabaceae), *Grevea madagascariensis* (Montiniaceae), *Olea lancea* (Oleaceae) and *Gaertneria obovata* (Rubiaceae), were found to be particularly promising. The 28 most active extracts were then submitted to a second round of testing for further reconfirmation. Twelve extracts only, from *Flacourtia ramontchi* (Salicaceae), *Croton chapelieri*, *Argomuellera perrieri* (Euphorbiaceae), *Pandanus* cf. *aridus* (Pandanaceae), *Anacolosa pervilleana* (Olacaceae), *Gonioma malagasy*, *Alafia thouarsii*, *Roupellina boivinii* (Apocynaceae), *Doratoxylon chouxii* (Sapindaceae), *Menabea venenata* (Asclepiadaceae), *Canarium madagascariensis* (Burseraceae) and *Chassalia bojeri* (Rubiaceae), were found to be active. Ten of them were collected in the transition forest of the Sakaraha region. This unique transition forest between the xerophytic plants (arid region of the south) and the deciduous plants (seasonally dry of the west) holds among the highest endemic species in Madagascar.

#### 11.2.3.2 Plant extracts from Reunion Island

In April 2009, Reunion lab work was begun first by the collection of plants through a cooperative network of botanist collaborators.

The harvests were made with best regard to the biological patrimony presented by the exceptional biodiversity of the island (Pitons, Cirques and Remparts of Reunion Island are listed in UNESCO World Heritage). The plant species were selected partly for their known ethnopharmacological properties (against Chikungunya and other antiviral diseases). One hundred and seventy-one species, taxonomically authenticated and classified in a photographic library, were collected in 1 year as well on highlands and at the seaside of Reunion island at different sites, such as Colorado, Plaine des Palmistes, Forêt de Bébour, Forêt de Bélouve, les Avirons, Forêt de Mare Longue, Langevin, Saint-Philippe and Grand-Etang. The selected plant parts (leaves and trunk bark) were oven dried and ground following a standard process. The 171 species belonging to 57 families representing 117 genera and corresponding to four batches were stored in powder flasks in air-conditioned rooms. Among the families, some of them included several species such as the Rubiaceae family with 16 species, the Euphorbiaceae family with 15 species and the Myrtaceae with 13 species evaluated. The primary screening on antiviral activity against CHIKV replication in cell culture showed that eight families gave the best response, including 13 species (eight leaves extracts and five bark extracts). The 13 active extracts were submitted to further testing for confirmation. Finally, eight plant extracts (four leaf and four bark extracts) were then found to be active. The selected plants are *Aphloia theiformis*, indigenous (Flacourtiaceae); *Croton mauritianus*, endemic to Reunion (Euphorbiaceae); *Indigofera ammoxylum*, endemic to Reunion (Fabaceae); *Phyllanthus phyllireifolius*, endemic to Reunion (Phyllanthaceae); *Securinega durissima*, indigenous to Reunion, Mauritius and Madagascar (Phyllanthaceae); *Stillingia lineata*, endemic to Reunion and Mauritius (Euphorbiaceae); *Agarista buxifolia*, endemic to Madagascar and Mascarenes; *Doratoxylon apetalum*, indigenous (Sapindaceae) (Girard-Valenciennes *et al.*, 2010; Leyssen *et al.*, 2011). Two examples of active species from Reunion belonging to the Euphorbiaceae family are presented in Figure 11.5.

#### 11.2.3.3 Plant extracts from Mauritius

In Mauritius, 119 plants were collected and around 160 crude filtered extracts were screened for possible cytotoxicity and antiviral evaluation. The samples were collected following field trips from conservation management areas in Mauritius and selection of plant samples was based mainly on (1) ethnopharmacology/traditional use and (2) on chemotaxonomic markers. Interestingly, 12 extracts proved to be potent against CHIKV, namely, *Elaeodendron orientale* (Celastraceae), twigs and leaves; *Cassine orientalis* (Celastraceae), leaves; *Dodonaea viscosa* (Sapindaceae), twigs; *Pleurostylia leucocarpa* (Celastraceae), leaves; *Allophyllus cobbe* (Sapindaceae), leaves; *Diospyros tessellaria* (Ebenaceae), leaves; *Azadirachta indica* (Meliaceae), leaves; *Dombeya acutangula* (Sterculiaceae), leaves; *Tarenna borbonica* (Rubiaceae), leaves; *Mimusops elengi* (Sapotaceae), leaves.

*For all these results*, it was nevertheless frustrating to note in most cases the loss of the antiviral activities after test reconfirmation of the active extracts, or during bioassay-directed fractionation of the promising extracts.

(a) (b)

**Figure 11.5** Two active species from Reunion belonging to the Euphorbiaceae family. (a) *Croton mauritianus*. *Photo*: Emmanuelle Girard-Valenciennes 2009 © LCSNSA. Reproduced with permission. (b) *Stillingia lineata*. *Photo*: Sophie Techer 2009 © LCSNSA. Reproduced with permission.

### 11.2.4 Bioguided fractionation and isolation of pure compounds

Several extracts prepared from plants from Madagascar, Reunion and Mauritius have shown inhibitory activity in the CHIKV-cell-based assay, but only few of them led to active compounds.

The stem bark of *Flacourtia ramontchi*, collected in Madagascar, led to several new phenolic glucosides named flacourtosides, but none of them showed activity against CHIKV (Bourjot *et al.*, 2012b). However, modest activity of flacourtosides was observed in a dengue virus polymerase assay, and betulinic acid 3-β-caffeate, isolated alongside the phenolic glucosides, showed significant inhibition in this assay (Bourjot *et al.*, 2012b).

Similarly, bioguided study of *Anacolosa pervilleanna* allowed the isolation of a series of polyacetylene compounds, which again were not found active on CHIKV. However, the triterpenes lupenone (**10**) and β-amyrone (**11**) isolated from this plant (Figure 11.6) showed a moderate antiviral effect on CHIKV, with $EC_{50}$ of 77 μM and 86 μM respectively (Bourjot *et al.*, 2012c).

The anti-CHIKV activity of lupenone (**10**) and β-amyrone (**11**) is consistent with previously described involvment of triterpenes with lupane or oleanane skeletons, against alphavirus replication (Pohjala *et al.*, 2009; Khan *et al.*, 2011).

The eight plant extracts from Reunion possessing selective antiviral activity were submitted to bioassay-guided fractionation in order to identify some active molecules.

From *Aphloia theiformis* and *Phyllanthus phyllireifolius*, at this time no new chemical compounds have been detected. Only triterpenes such as 28-*O*-β-D-glucopyranosyl-6-D-glucopyranosyl-6-β-hydroxy tormentic acid (moderate activity) from *Aphloia theiformis* and sitosterol (**12**; Figure 11.6) (moderate activity) from *Phyllanthus phyllireifolius* have been isolated. Once again, we note the antiviral activity, however small, of compounds having a pentacyclic triterpene skeleton.

From *Stillingia lineata*, four compounds were isolated with two known molecules (tonantzitlolone and 4′-acetoxytonantzitlolone (**13**; Figure 11.6), respectively non-active and active) and two new molecules that are under investigation.

The bioassay-guided fractionation of the five other hits (*Agarista buxifolia*, *Indigofera ammoxylum*, *Croton mauritianus*, *Securinega durissima* and *Doratoxylon apetalum*) is still in progress.

Only five plants from Mauritius (*Cinnamomum camphora*, leaves; *Cassine orientalis*, leaves; *Dodonaea viscosa*, stem and leaves; *Pleurostylia leucocarpa*, leaves; *Allophyllus cobbe*, leaves) were initially selected for further fractionation, phytochemical analysis and

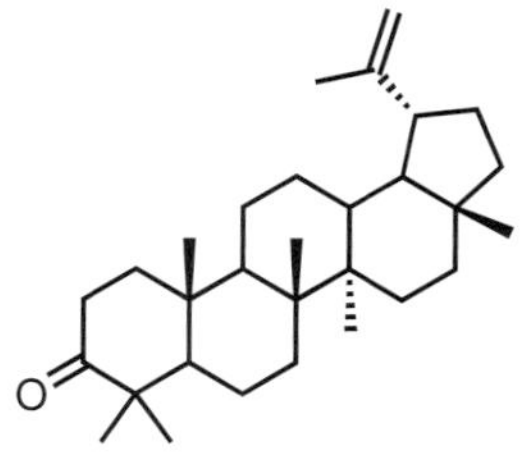

10

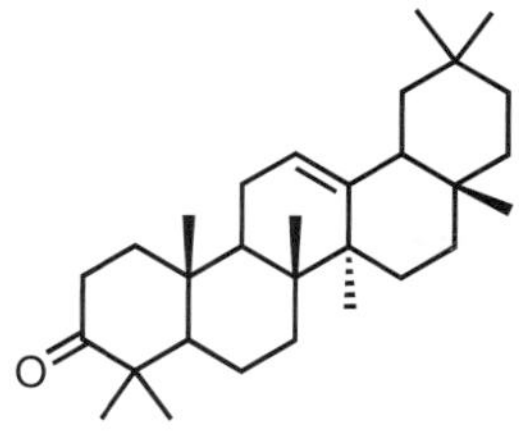

11

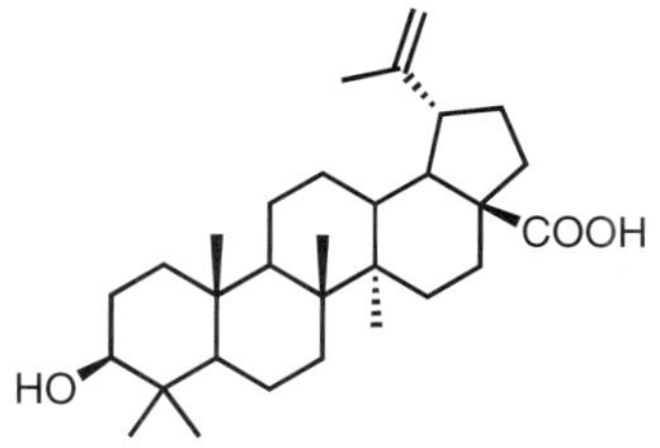

12

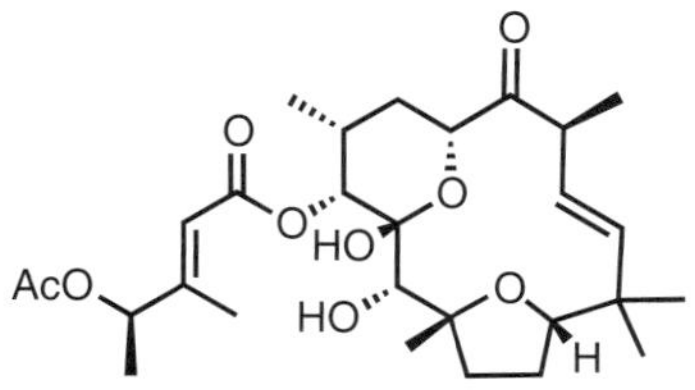

13

**Figure 11.6** Four pure compounds (**10**, **11**, **12**, **13**) with inhibitory activity in the CHIK virus-cell-based assay

anti-CHIKV evaluation. It was found that fractions of the leaves and stem of *Cassine orientalis* were active as well as fractions of *Dodonaea viscosa* (stem) and that of *Pleurostylia leucocarpa* (leaves). The bioguided phytochemistry of these plants is in progress.

*As for the extracts*, it has often been disappointing to lose antiviral activity during fractionation and purification. Indeed, most of the extracts studied that possess a potent anti-CHIKV activity did not lead to active products, and only a few of them gave known products with a moderate antiviral activity. Today, it is difficult to explain this loss of activity during the confirmations of activity or purifications. However, other plants from Madagascar, Mauritius and Reunion are being studied and will perhaps lead to potent inhibitors of CHIKV replication, as has been the case in a specific study of the antiviral activity of Euphorbiaceae.

**Figure 11.7** *Trigonostemon cherrieri*. (a) Tree. *Photo*: Poullain 2009. Reproduced with permission. (b) Flower. *Photo*: Dumontet 2009. Reproduced with permission

## 11.3 Euphorbiaceae, abundant source of anti-chikungunya virus compounds

The PHYTOCHIK programme allowed us to build a tight network of virologists and chemists of natural products. We therefore decided to continue our work on the search for anti-CHIKV compounds from plants collected in other countries than in the Indian Ocean.

Taking advantage of the results from a screen on the dengue polymerase NS5 protein assay, we isolated a series of highly oxygenated daphnane diterpenoid orthoesters (DDOs) bearing an uncommon chlorinated moiety. These compounds, named trigocherrins and trigocherriolides, were purified from *Trigonostemon cherrieri* ethyl acetate bark and wood extracts. This species is an endemic Euphorbiaceae of New Caledonia (Figure 11.7), and is to date the only representative of the *Trigonostemon* genus present in this overseas country. The antiviral potential of trigocherrins and trigocherriolides was evaluated in a virus-cell-based assay against three members of the genus alphavirus: CHIKV, Sindbis virus (SINV) and Semliki forest virus (SFV). Some of these compounds were shown to possess significant inhibitory activity on the replication of CHIKV, SINV and SFV, although the activity was slightly lower for the last two viruses, as shown in Table 11.1 (Allard *et al.*, 2012a,2012b).

From these two series of compounds, trigocherrins (**14a–c**) were shown to be the most potent anti-CHIKV molecules, since their selectivity index ($CC_{50}$ Vero/$EC_{50}$ CHIKV) was, on average, 10 times higher than that of the trigocherriolides (**15a–c**) (Figure 11.8). Also, trigocherrins (**14a–c**) were found to be more potent than chloroquine, which was included as reference compound used in this assay.

As mentioned before, trigocherrins (**14a–c**) and trigocherriolides (**15a–c**) belong to a large and structurally diverse group of DDOs. This chemical family was studied recently in particular for their anti-HIV activity (Huang, L. *et al.*, 2011; Huang, S.Z. *et al.*, 2012; Vidal *et al.*, 2012), and one highly oxygenated daphnane diterpenoid isolated from *Trigonostemon thyrsoideum* was shown to prevent the cytopathic effects induced by HIV-1 (Zhang *et al.*, 2010).

Seeking new daphnanes with potential antiviral properties, we undertook to study another species of the same genus, *Trigonostemon howii* (Figure 11.9), whose ethyl acetate extract of the bark showed a significant anti-CHIKV activity. This time, no active DDO was isolated from this plant collected in central Vietnam; only one compound belonging to the tigliane series, trigowiin A (**16**), presented moderate antiviral activity. Given the structural similarity between this compound and the well-known tiglianes prostratin

**Table 11.1** Anti-metabolic and antiviral activities of trigocherrins **14a–14c** and trigocherriolides **15a–15c** (data in micrograms)

| Compound | $CC_{50}$ Vero cells | $EC_{50}$ (CHIKV) | SI (CHIKV) | $EC_{50}$ (SINV) | $EC_{50}$ (SFV) |
|---|---|---|---|---|---|
| **14a** | 35 ± 8 | 1.5 ± 0.6 | 23 | 7.7 ± 1.8 | 16 ± 3 |
| **14b** | 93± 3 | 2.6 ± 0.7 | 35 | 21 ± 3 | 45 ± 0.6 |
| **14c** | 23.1 ± 0.6 | 3.0 ± 1.2 | 7.7 | 5.6 ± 0.6 | 18 |
| **15a** | 5.3 ± 0.2 | 2.5 ± 0.3 | 2.1 | 3.3 ± 0.2 | 4.2 |
| **15b** | 4.6 ± 0.8 | 1.9 ± 0.6 | 2.4 | 3.2 ± 0.1 | 4.0 |
| **15c** | 10.5 ±0.1 | 3.9 ± 1.0 | 2.7 | 5.3 ± 1.2 | 5.1 ± 0.4 |
| Chloroquine (**4**) | 100 ± 25 | 10 ± 5 | 10 | 11.0 ± 2.1 | 13.8 ± 2.3 |

SI: selectivity index.

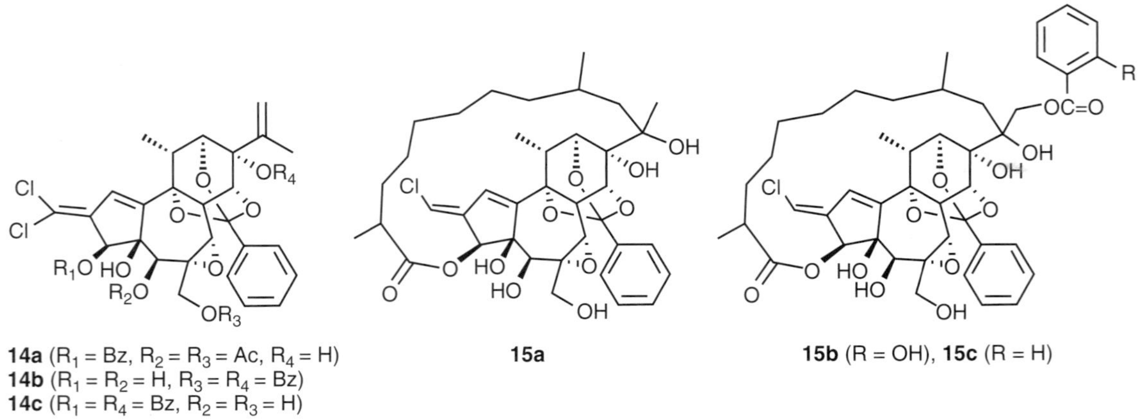


**Figure 11.8** Potent anti-CHIKV compounds belonging to the daphnane series

**Figure 11.9** *Trigonostemon howii. Photos*: Dao Dinh Cuong 2005. Reproduced with permission

(**17**) and phorbol (**18**), these compounds were studied together with 12-*O*-tetradecanoylphorbol-13-acetate (**19**, TPA) and its 4α-epimer (**20**, 4α-TPA) for their potential anti-CHIKV activity (Figure 11.10, Table 11.2) (Bourjot *et al.*, 2012a).

As shown in Table 11.2, prostratin (**17**) and TPA (**19**) proved to be the most potent inhibitors of CHIKV replication, as apparent from their lower $EC_{50}$ values and higher selectively indices (SI calculated as $CC_{50}$ Vero/$EC_{50}$ CHIKV, a value that is representative for the window of antiviral selectivity). Furthemore, TPA (**19**) proved to be 65 times more potent than prostratin (**17**) and 1000 times more potent than its 4α-epimer (**20**) (SI of 1965, 30.3 and 1.9 respectively), while phorbol (**18**) was found to be completely inactive. None of the compounds were active against SFV, while only TPA (**19**) exhibited antiviral activity against SINV, although it was far less potent than the effect observed on the replication of CHIKV. TPA (**19**) as well as prostratin (**17**) could be the starting point for further chemical optimization and more profound biological studies.

Besides, the anti-HIV activity of daphnane and tigliane diterpenoids is often linked to their action on protein kinase C. Whether or not this is so, the mechanism of action of these compounds in the context of CHIKV has yet to be investigated.

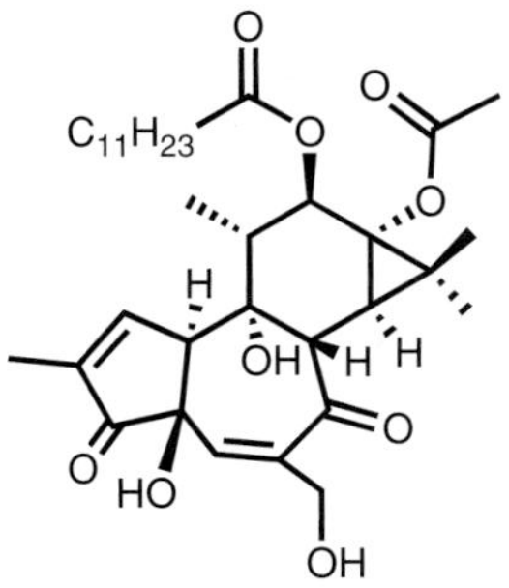

**16**

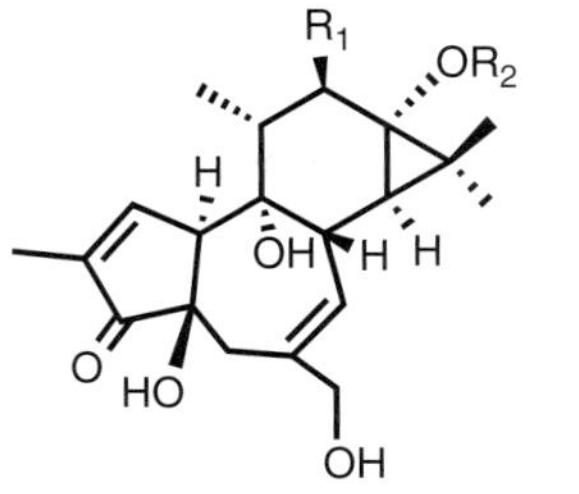

**17** ($R_1$ = H, $R_2$ = Ac)
**18** ($R_1$ = OH, $R_2$ = OH)
**19** ($R_1$ = $OCOC_{13}H_{27}$, $R_2$ = Ac)
**20** ($R_1$ = $OCOC_{13}H_{27}$, $R_2$ = Ac, 4$\alpha$-OH)

**Figure 11.10** Potent anti-CHIKV compounds belonging to the tigliane series

**Table 11.2** Antimetabolic and antiviral activities of trigowiin (**16**), prostratin (**17**), phorbol (**18**), TPA (**19**) and 4α-TPA (**20**) (data in micrograms)

| Compound | $CC_{50}$ Vero cells | $EC_{50}$ (CHIKV) | SI (CHIKV) | $EC_{50}$ (SINV) | $EC_{50}$ (SFV) |
|---|---|---|---|---|---|
| **16** | >100 | 43.5 ± 12.8 | >2.3 | – | – |
| **17** | 79.0 ± 17.4 | 2.6 ± 1.5 | 30.3 | inactive | inactive |
| **18** | inactive | inactive | | inactive | inactive |
| **19** | 5.7 ± 1.7 | 0.0029 ± 0.0003 | 1965 | 2.2 ± 0.2 | inactive |
| **20** | 5.3 ± 0.6 | 2.8 ± 0.5 | 1.9 | n.d. | n.d. |
| Chloroquine (**4**) | 89 ± 28 | 11 ± 5 | 8.1 | 11.0 ± 2.1 | 14 ± 2 |

SI: selectivity index; n.d.: not determined.

## 11.4 Conclusion

As part of a project dedicated to the discovery of natural inhibitors of CHIKV, our efforts were initially focused on the antiviral screening of plants from the islands of Madagascar, Reunion and Mauritius. The bioguided phytochemical studies of the active extracts were often disappointing after the loss of the antiviral activity during fractionation and did not result in the isolation of very powerful antiviral compounds. However, we have shown the inhibitory activity of pentacyclic triterpenes on the replication of chikungunya. It is interesting to note that some of these triterpenes are known to possess antiviral properties on other alphavirus such as Sindbis and Semliki Forest viruses.

Looking for more powerful inhibitors of chikungunya replication, we investigated species belonging to the Euphorbiaceae family, collected in Vietnam and New Caledonia. New unique products belonging to the daphnane and tigliane chemical series were isolated and were shown to possess significant activity against chikungunya. Finally, screening diterpenoids structurally related to tiglianes has led to the identification of the potent activity of prostratin and 12-O-tetradecanoyl-phorbol-13-acetate on the CHIKV *in cellulo*.

Much remains to be done to determine and understand the exact mechanism of action of these active compounds, to design and synthesize structural analogues and select the best candidates for *in vivo* studies.

## Acknowledgements

Part of this research was supported by CRVOI. We gratefully acknowledged Professor Koussay Dellagi, Director of CRVOI (www.crvoi.org).

The bioguided phytochemical studies of *Trigonostemon* species are part of the PhD thesis of Pierre-Marie Allard (2011, http://www.theses.fr/14826140X) and Mélanie Bourjot (2012, http://www.theses.fr/s82153). All the results concerning the bioactive plants from Reunion Island are part of the PhD thesis of Sophie Techer (April 2013).

M. Litaudon and F. Guéritte thank D.D. Cuong and Q.B. Nguyen (VAST, Hanoi, Vietnam) for the collect of *Trigonostemon howii*, C. Poullain (Laboratoire des Plantes Médicinales, ICSN-CNRS, Nouméa, Nouvelle-Calédonie) and V. Dumontet (ICSN, CNRS, Gif-sur-Yvette) for that of *Trigonostemon cherrieri*.

The LCSNSA thanks the Regional Council of Île de La Réunion and the Environment-Biodiversity-Health Federation of Reunion University (FEBS-UR) for financial support. The researchers of the LCSNSA are grateful to the 'Amis des plantes et de la Nature (APN)', the 'Office National des forêts (ONF)', the 'Parc National de La Réunion (PNR)' and the 'UMR C_53 PVBMT' for collection and identification of the plants. They wish to express in particular their gratefulness to Dominique Strasberg, Hermann Thomas, Max Felicite, Jean-Maurice Tamon and Eloi Boyer for their valuable help in plant material collection.

## References


Allard, P.M., Leyssen, P., Martin, M.T. *et al.* (2012a) Antiviral chlorinated daphnane diterpenoid orthoesters from the bark and wood of *Trigonostemon cherrieri*. *Phytochemistry*, **84**, 160–168.

Allard, P.M., Martin, M.T., Tran Huudau, M.E. *et al.* (2012b) Trigocherrin A, the first natural chlorinated daphnane diterpene orthoester from *Trigonostemon cherrieri*. *Organic Letters*, **14** (1), 342–345.

Bassetto, M., de Burghgraeve, T., Leyssen, P. *et al.* (2011) Computer-aided design and evaluation of novel anti-CHIKV compounds. *Antiviral Research*, **90** (2), A67.

Bora, L. (2012) Homology modeling and docking to potential novel inhibitor for chikungunya (37997) protein nsP2 protease. *Journal of Proteomics & Bioinformatics*, **5** (2), 054–059.

Boriskin, Y.S., Leneva, I.A., Pécheur, E.I. and Polyak, S.J. (2008) Arbidol: a broad-spectrum antiviral compound that blocks viral fusion. *Current Medicinal Chemistry*, **15**, 997–1005.

Bourjot, M., Delang, L., Nguyen, V.H. *et al.* (2012a) Prostratin and 12-*O*-tetradecanoylphorbol 13-acetate are potent and selective inhibitors of chikungunya virus replication. *Journal of Natural Products*, **75**, 2183–2187.

Bourjot, M., Delmotte, S., Leyssen, P. *et al.* (2012b) Flacourtosides A–F, phenolic glycosides isolated from *Flacourtia ramontchi*, potential inhibitors of emerging viruses. *Journal of Natural Products*, **75**, 752–758.

Bourjot, M., Leyssen, P., Eydoux, C. *et al.* (2012c) Chemical constituents of *Anacolosa pervilleana* and their antiviral activities. *Fitoterapia*, **83** (6), 1076–1080

Briolant, S., Garin, D., Scaramozzino, N. *et al.* (2004) *In vitro* inhibition of chikungunya and Semliki Forest viruses replication by antiviral compounds: synergistic effect of interferon-α and ribavirin combination. *Antiviral Research*, **61** (2), 111–117.

Clardy, J. and Walsh, C. (2004) Lessons from natural molecules. *Nature*, **432**, 829–837.

Cos, P., Vlietinck, A.J., Vanden Berghe, D. and Maes, L. (2006) Anti-infective potential of natural products: how to develop a stronger in vitro “proof-of-concept”. *Journal of Ethnopharmacology*, **106**, 290–302.

De Lamballerie, X., Boisson, V., Reynier, J.C. *et al.* (2008) On chikungunya acute infection and chloroquine treatment. *Vector-borne and Zoonotic Diseases*, **8**, 837–839.

Delogu, I. and de Lamballerie, X. (2011) Chikungunya disease and chloroquine treatment. *Journal of Medical Virology*, **83**, 1058–1059.

Delogu, I., Pastorino, B., Baronti, C. *et al.* (2011) *In vitro* antiviral activity of arbidol against chikungunya virus and characteristics of a selected resistant mutant. *Antiviral Research*, **90** (3), 99–107.

Enserink, M. (2007) Chikungunya: no longer a Third World disease. *Science*, **318**, 1860–1861.

Eriksson, K.K., Cervantes-Barragán, L., Ludewig, B. and Thiel, V. (2008) Mouse hepatitis virus liver pathology is dependent on ADP-ribose-1″-phosphatase, a viral function conserved in the alpha-like supergroup. *Journal of Virology*, **82** (24), 12825–12834.

Gay, B., Bernard, E., Solignat, M. *et al.* (2012) pH-dependent entry of chikungunya virus into *Aedes albopictus* cells. *Infection, Genetics and Evolution*, **12** (6), 1275–1281.

Girard-Valenciennes, E., Clerc, P., Grondin, I. *et al.* (2010) Medicinal plant potentialities from Reunion Island, in *VIII International Symposium on Ethnobotany Disciplines*, 3–8 October, Lisbon, Portugal.

Huang, L., Ho, P., Yu, J. *et al.* (2011) Picomolar dichotomous activity of gnidimacrin against HIV-1. *PLOS ONE*, **6**, e26677.

Huang, S.Z., Zhang, X.J., Li, X.Y. *et al.* (2012) Daphnane-type diterpene esters with cytotoxic and anti-HIV activities from *Daphne acutiloba* Rehd. *Phytochemistry*, **75**, 99–107.

Julander, J.G., Smee, D.F., Morrey, J.D. and Furuta, Y. (2009) Effect of T-705 treatment on western equine encephalitis in a mouse model. *Antiviral Research*, **82**, 169–171.

Kaur, F., Thiruchelvan, M., Lee, R.C. *et al.* (2013) Inhibition of chikungunya virus replication by harringtonine, a novel antiviral that suppresses viral protein expression. *Antimicrobial Agents and Chemotherapy*, **57**, 155–167.

Khan, A.H., Morita, K., Parquet, M. *et al.* (2002) A Complete nucleotide sequence of Chikungunya Virus and evidence for an internal polyadenylation site. *Journal of General Virology*, 3075–3084.

Khan, M., Santhosh, S. R., Tiwari, M. *et al.* (2010) Assessment of *in vitro* prophylactic and therapeutic efficacy of chloroquine against chikungunya virus in Vero cells. *Journal of Medical Virology*, **82** (5), 817–824.

Khan, M., Dhanwani, R., Patro, I.K. *et al.* (2011) Cellular IMPDH enzyme activity is a potential target for the inhibition of chikungunya virus replication and virus induced apoptosis in cultured mammalian cells. *Antiviral Research*, **89**, 1–8.

Koehn, F.E. and Carter, G.T. (2005) The evolving role of natural product in drug discovery. *Nature Reviews Drug Discovery*, **4**, 206–220.

Leyssen, P., Litaudon, M., Guillemot, J.C. *et al.* (2011) PHYTOCHIK: biodiversity as a source of selective inhibitors of CHIKV replication. *Antiviral Research*, **90** (2), A36–A37.

Li, X.F., Jiang, T., Deng, Y. *et al.* (2012) Complete genome sequence of a chikungunya virus isolated in Guangdong, China. *Antiviral Research*, **86** (16), 8904–8905.

Lucas-Hourani, M., Lupan, A., Desprès, P. *et al.* (2013) A phenotopic assay to identify chikungunya virus inhibitors targeting the non structural protein nsP2. *Journal of Biomolecular Screening*, **18**, 172–179.

Malet, H., Coutard, B., Jamal, S. *et al.* (2009) The crystal structures of chikungunya and Venezuelan equine encephalitis virus nsP3 macro domains define a conserved adenosine binding pocket. *Journal of Virology*, **83** (13), 6534–6545.

Newmann, D.J. and Cragg, G.M. (2012) Natural products as source of new drugs over the 30 years from 1981 to 2010. *Journal of Natural Products*, **75**, 311–335.

Ogbomo, H., Michaelis, M., Altenbrandt, B. *et al.* (2010) A novel immunomodulatory mechanism of ribavirin in suppressing natural killer cell function. *Biochemical Pharmacology*, **79**, 188–197.

Pialoux, G., Gauzere, B.A., Jaureguiberry, S. and Strobe, M. (2007) Chikungunya, an epidemic arbovirosis. *Lancet Infectious Diseases*, **7**, 319–327.

Pohjala, L., Alakurtti, S., Ahola, T. *et al.* (2009) Betulin-derived compounds as inhibitors of alphavirus replication. *Journal of Natural Products*, **72**, 1917–1926.

Pohjala, L., Utt, A., Varjak, M. *et al.* (2011) Inhibitors of alphavirus entry and replication identified with a stable chikungunya replicon cell line and virus-based assays. *PLOS ONE*, **6** (12), e28923.

Ravichandran, R. and Manian, M., 2008. Ribavirin therapy for chikungunya arthritis. *Journal of Infection in Developing Countries*, **2**, 140–142.

Rubach, J.K., Wasik, B.R., Rupp, J.C. *et al.* (2009) Characterization of purified Sindbis virus nsP4 RNA-dependent RNA polymerase activity *in vitro*. *Virology*, **384** (1), 201–208.

Russo, A.T., White, M.A. and Watowich, S.J. (2006) The crystal structure of the Venezuelan equine encephalitis alphavirus nsP2 protease. *Structure*, **14** (9), 1449–1458.

Savarino, A., Boelaert, J.R., Cassone, A. *et al.* (2003) Effects of chloroquine on viral infections: an old drug against today's diseases. *Lancet Infectious Diseases*, **3** (11), 722–727.

Scheidel, LM. and Stollar, V. (1991) Mutations that confer resistance to mycophenolic acid and ribavirin on Sindbis virus map to the nonstructural protein nsP1. *Virology*, **181**, 490–499.

Shi, L., Xiong, H., He, J., *et al.* (2007) Antiviral activity of arbidol against influenza A virus, respiratory syncytial virus, rhinovirus, coxsackie virus and adenovirus *in vitro* and *in vivo*. *Archives of Virology*, **152**, 1447–1455.

Vidal, V., Potterat, O., Louvel, S. *et al.* (2012) Library-based discovery and characterization of daphnane diterpenes as potent and selective HIV inhibitors in *Daphne gnidium*. *Journal of Natural Products*, **75**, 414–419.

Zhang, L., Luo, R.H., Wang, F. *et al.* (2010) Daphnane diterpenoids isolated from *Trigonostemon thyrsoideum* as HIV-1 antivirals. *Phytochemistry*, **71**, 1979–1883.

# 12 Using African Plant Biodiversity to Combat Microbial Infections

J.N. Eloff and L.J. McGaw

Phytomedicine Programme, Faculty of Veterinary Science, University of Pretoria, Onderstepoort, South Africa

## 12.1 Introduction and problem statement

Microbial infections have a major effect on human health, especially in rural parts of Africa where antibiotics are not freely available or too expensive. They also have a very important effect on the quality of life though microbial infections of animals and plants produced for food.

People have been using plants to treat ailments for many years. In a 60 000-years-old Neanderthal Shanidar cave burial site in present-day Iraq, pollen was found of eight plant species; seven of these species, including an *Ephedra* species, are still being used as medicinal plants around the world (Solecki, 2007). In South Africa, Khoi San rock paintings include eight clearly defined medicinal plant species, including *Aloe* and *Harpagaphyton* spp. For thousands of years, plants were the only resources that people had to combat diseases. In the training of doctors, knowledge of plants was so important that Galen once stated: 'The doctor who does not know his plants should quit the profession'.

In the 16th century, Paracelsus started using chemicals, and for a long time two systems were in use. In days when people still accepted the spontaneous development of life, it was very difficult to understand certain diseases until microorganisms were discovered and Louis Pasteur developed the germ theory of infectious diseases. A giant step in the battle between plant medicine and chemical medicine was taken when Paul Ehrlich discovered Salvarsan as a 'magic bullet' against organisms causing syphilis. This was an enormous improvement compared with the mercury salt treatment of syphilis that led to horrible side effects, such as losing hair and teeth. The most remarkable change away from using herbal medicines, however, came after the discovery and commercialization of penicillin.

It is not widely known that many ancient cultures, in Greece and India, have already used moulds to treat infections. There have been other people who have discovered antibiotic activity of fungi on bacteria before the well-known serendipitous discovery of Alexander Fleming in 1928 that a *Penicillium notatum* infection on one of his *Staphylococcus* plates inhibited the growth of the bacteria. The penicillin in his culture was unstable and the yield was very low (1 ppm).

Ten years later, the biochemists Florey, Chang and Heatly in England were able to produce a stable penicillin. During the Second World War, scientists in the USA succeeded in getting higher yielding *Penicillium* strains and developing fermentation technology to produce large quantities of penicillin in time to save thousands of lives of soldiers with infected wounds. The discovery of antibiotics had a major influence on the development of the pharmaceutical industry and the downgrading of the herbal medicine industry.

Very soon after the discovery of penicillin, the development of resistant bacteria was discovered in London (Levy and Marshall, 2004). Owing to misuse of antibiotics, there is such a development of resistance that some authors have warned that we are entering the post-antibiotic era. Recently, Britain's chief medical officer, Sally Davies, stated that the issue should be added to the list of national emergencies (Kåhrström, 2013). The World Health Organization has identified several levels of drug-resistant *Mycobacterium tuberculosis* strains; that is, multiple drug-resistant, extensively drug-resistant and even total drug-resistant (TDR) strains discovered in India, Italy and Iran (Daley and Carminero, 2013). Dr Shelly Batra, President of a New-Delhi based non-governmental organization that fights tuberculosis (TB) stated that "We are on the brink of another epidemic, and it has no treatment. If TDR spreads, we will go back to the Dark Ages".

Despite substantial research efforts by the pharmaceutical industry, no new class of broad-spectrum antibiotics were developed after the fluoroquinolones nearly 50 years ago (Lewis and Ausubel, 2006). There are many problems in finding new antibiotics active against Gram-negative bacteria. Abad *et al.* (2007) made the following statement:

> "In the past few decades, a worldwide increase in the incidence of fungal infections has been observed as well as a rise in the resistance of some species of fungus to different fungicidals used in medicinal practice. Fungi are one of the most neglected pathogens, as demonstrated by the fact that the amphotericin B, a polyene antibiotic discovered as long ago as 1956, is still used as a 'gold standard' for antifungal therapy. The last two decades have witnessed a dramatic rise in the incidence of life threatening systemic fungal infections".

Many pharmaceutical products are based on plant products. In a survey in the 1970s, it was found that 25% of prescription medicines in the USA were based on compounds originally discovered from plants (Farnsworth and Morris, 1976). Some of the most important anti-malarial pharmaceuticals are plant based and several anticancer pharmaceuticals are based on compounds isolated from plants.

There have been many review papers written on this topic. Some of the papers we consulted to prepare this chapter are shown in Table 12.1.

*Novel Plant Bioresources: Applications in Food, Medicine and Cosmetics*, First Edition. Edited by Ameenah Gurib-Fakim.

**Table 12.1** Some of the papers consulted in preparing this chapter

| Reference | Title |
|---|---|
| Cowan (1999) | Plant products as antimicrobial agents |
| Gibbons (2004) | Anti-staphylococcal plant natural products |
| Levy and Marshall (2004) | Antibacterial resistance worldwide:causes, challenges and responses |
| Rios and Recio (2005) | Medicinal plants and antimicrobial activity |
| Cos *et al.* (2006) | Anti-infective potential of natural products: How to develop a stronger in vitro 'proof of concept' |
| Lewis and Ausubel (2006) | Prospects for plant-derived antibacterials |
| Eloff and McGaw (2006) | Plant extracts used to manage bacterial, fungal and parasitic infections in southern Africa |
| Abad *et al.* (2007) | Active antifungal substances from natural sources |
| Van Vuuren (2008) | Antimicrobial activity of South African medicinal plants |
| Svetaz *et al.* (2010) | Value of ethnomedicinal information for the discovery of plants with antifungal properties. A survey among seven Latin American countries |
| Kuete *et al.* (2011) | Antibacterial activity of some natural products against bacteria expressing a multi-resistant phenotype |
| Queiros *et al.* (2012) | Modern approaches in the search for new active compounds from crude extracts of natural sources |
| Sortino *et al.* (2012) | The role of natural products in the discovery of new anti-infective agents with emphasis on antifungal compounds |
| Cragg *et al.* (2012) | Natural products in drug discovery: recent advances |

## 12.2 Commercial use of African medicinal plants in the herbal medicine industry

Single chemical entities as pharmaceuticals have taken the place of herbal medicines for the treatment of many serious diseases since the widespread use of antibiotics in the 1950s. There has, however, been a resurgence in the use of formal herbal medicines to treat less-serious illnesses and for maintaining health. Africa has missed out in the growth in this industry because the trade in African medicinal plants to the developed world is very low. Therefore, this limits job and wealth creation by growing, beneficiating and exporting of herbal medicines.

One of the major constraints for trade in African medicinal plants was identified at the Medicinal Plants Forum for Commonwealth Africa held in Cape Town in 2000 as the lack of suitable technical specifications and quality control standards (Brendler *et al.*, 2010). This conclusion was justified when van Wyk and Wink (2004) analysed the origin of commercialized herbal medicines. The percentage of commercialized medicinal plants was much lower from southern Hemisphere countries, where the traditional knowledge was transferred orally, than from Europe, India and Asia, where traditional knowledge has been written down for many centuries. Africa, including the Indian Ocean islands, contains about 60 000 plant species (Klopper *et al.*, 2007), and only 83 species are used as herbal medicines in the developed world compared with the 434 herbal medicines commercialized from the 13 600 species occurring naturally in Europe. This clearly shows that there is a tremendous opportunity to develop many more herbal medicines from Africa.

## 12.3 Why is there such a difference in product development for antimicrobials versus other medicinal applications?

A search on Google Scholar using the terms 'Africa antimicrobial plant' without citations or patents since 2000 yields about 16 600 hits. Even using such a rough measure it is clear that many papers have been written investigating plants for antimicrobial activities. In a very large proportion of the publications the aim of the authors was probably to eventually identify compounds that could be used as new antibiotics. Many papers written unfortunately had limited, if any, value because the methods used were not acceptable and do not yield comparable results.

A common error bedevilling many publications in this field is that statements are made that a certain plant extract or compound has antibacterial or antifungal activity. If there is no quantitative measure attached, such a statement is meaningless. If the concentration is high enough, probably all compounds or extracts would be toxic to microbes. Nobody would think of sucrose as an antibiotic, but a high enough concentration will certainly kill microorganisms.

In many of these 16 000 papers there are serious problems with the methods used. For example if 'agar' is included in the search terms, Google Scholar yields 14 800 hits. This may mean that 89% of papers delivered since 2000 probably used agar diffusion techniques. As discussed in Section 12.4.3, this makes comparisons difficult, if not impossible.

## 12.4 Methods used in developing useful products

### 12.4.1 Extraction of plant material

Plants contain at least 100 000 small compounds (Lewis and Ausubel, 2006) and many of these compounds would not readily be soluble in different extractants. One way of classifying these compounds is by the polarity. Non-polar or lipophilic compounds are soluble in non-polar solvents such as hexane or dichloromethane. The more polar or hydrophilic compounds are mainly soluble in polar solvents such as water or methanol. By using the wrong solvent, one may therefore miss active compounds. Some authors use mixtures of solvents to extract as many different compounds as possible in searching for plants with promising activities. Another approach is to use an intermediate polarity extractant that would extract compounds over a wide range of polarities. Among the commonly available solvents, acetone would fulfil this role. The ability of acetone to mix with polar and non-polar solvents is important. Several extractants were investigated on a 5-point scale and given different weights (in brackets) for the quantity extracted (3) from *Combretum erythrophyllum* and *Anthocleista grandiflora*, the rate of extraction (3), number of compounds extracted (5), number of antibacterial compounds extracted (5), the toxicity to pathogens in subsequent bioassays (4) and ease of removal (5) and hazardous to use (2). The average values out of a potential 130 were acetone 102, methanol:chloroform:water (12:5:3) 81, methylene dichloride 79, methanol 71, ethanol 68 and water 47 (Eloff, 1988a). If only one solvent is used, acetone is by far the best extractant based on these results.

Acetone also had the lowest toxicity for four fungi with the following minimum inhibitory concentration (MIC): acetone (51%) dimethylsulfoxide (45%), methanol (16–44%) and ethanol (30%) (Eloff *et al.*, 2007). Acetone is also very useful when determining MICs of volatile oils or non-polar fractions during bioassay-guided fractionation because it dissolves non-polar compounds and mixes with aqueous growth mediums.

When a series of extractants were used to extract dried leaves of *Combretum microphyllum* (Kotze and Eloff, 2002) or *Combretum woodii* (Eloff *et al.*, 2005) it was clear that the most polar (water) and most non-polar solvents (hexane) did not extract antimicrobial compounds (Figure 12.1).

Hardly any other plant water extracts in our laboratory had any antimicrobial activity. In a review of the literature for 1989–1999 only 14/157 water extracts were listed (Srivastava *et al.*, 2000). In another survey on the antibacterial activity of South African plants used for medicinal purposes only 4/27 water plant extracts were active (Rabe and van Staden, 1997). In our experience, water extracts

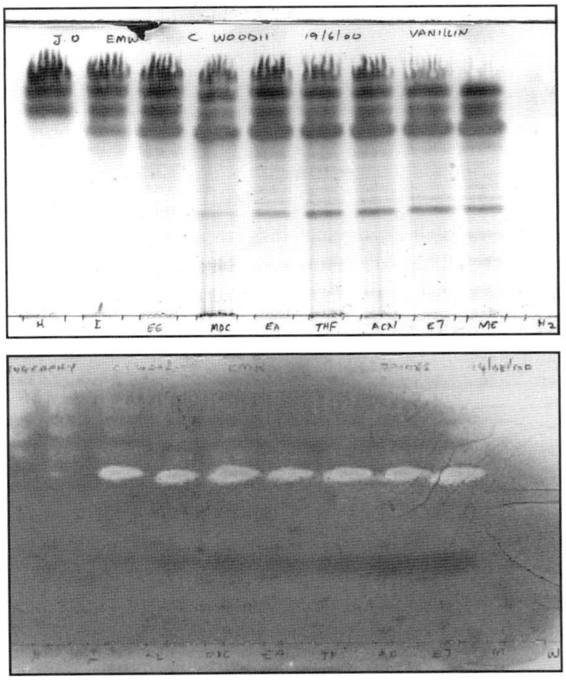


**Figure 12.1** Separation of 100 µg of compounds extracted from dried leaves of *Combretum woodii* by hexane, di-isopropylether, diethyl ether, methylene dichloride, ethyl acetate, tetrahydrofuran, acetone–ethanol, methanol and water separated with ethyl acetate:methanol:water and sprayed with vanillin sulfuric acid (top) and with a culture of *Staphylococcus aureus*, incubated overnight and then sprayed with tetrazolium violet. Clear areas indicate where bacterial growth is inhibited. *Source*: Eloff *et al.* (2005). Reproduced with permission of Academic Journals

with antimicrobial activity may contain tannins that would inhibit microbial growth but would not be useful potential antibiotics or prophylactic growth enhancers.

### 12.4.2 Selection of plant material to investigate

There are different ways to select the plants that are to be investigated. Cos *et al.* (2006) identified the following: random selection of species and then chemical screening, random selection of species for bioassays, following up on publications on biological activities and using ethnomedicinal leads. Other approaches could be to search on a taxonomic basis investigating families or genera with high activity (Eloff, 1998c).

Because traditional healers usually only have water available as extractant and the antimicrobial activities of aqueous extracts are very low, this means that traditional leads may not be useful to find compounds that can become antibiotics.

The Phytomedicine Programme has determined the MIC of acetone leaf extracts of trees that have been used traditionally based on information on herbarium sheets (Arnold *et al.*, 2002). There were no differences in antibacterial activity between trees used traditionally and randomly selected. There were no differences in the average for all the pathogens with MICs of 0.94 and 0.95 mg/mL. For the trees not used medicinally, 67 out of 93 extracts (i.e. 9.0%) had average MICs <0.16 mg/mL against six bacteria and two fungi. For the trees used medicinally the MIC for 30 out of 53 (i.e. 6.3%) extracts had an average MIC of <0.16 mg/mL against the same bacteria and fungi. The distribution in activity against four important nosocomial pathogens indicated that nearly all extracts had an MIC of 2.5 mg/mL or lower (Figure 12.2).

It must be stressed that there were no indications for what trees were used. If we investigated only trees with an established traditional use of the leaves against microbial infections, the results may have differed. Svetaz *et al.* (2010) investigated the antifungal activity of 327 plant species from seven Latin American countries against 11 human opportunistic and pathogenic fungi. They found a much higher hit rate among extracts of plants used to treat fungal infections than of plants not used traditionally. The number of hits was also higher against dermatophyte infections that could easily be diagnosed by traditional healers than against yeast or *Aspergillus* infections, where diagnosis would not be as easy.

### 12.4.3 Determining antimicrobial activity

Many scientists have used agar diffusion techniques to determine antibacterial activity. This technique is relatively simple and is useful if the activity of a known compound, such as an antibiotic, is to be determined. Briefly, microbes are freshly inoculated into the growth medium in a Petri dish and then either a hole is made into the agar and filled with the test solution or a disk containing the test solution is placed on the agar. The active compound(s) diffusing from the area applied leads to growth inhibition along a concentration gradient. There is a linear relationship between the logarithm of the concentration of the antimicrobial compound and the zone of microbial inhibition (Hewitt and Vincent, 1998). The following factors influence the inhibition zone size: concentration of antibiotic, volume of test solution, density of the inoculum, duration

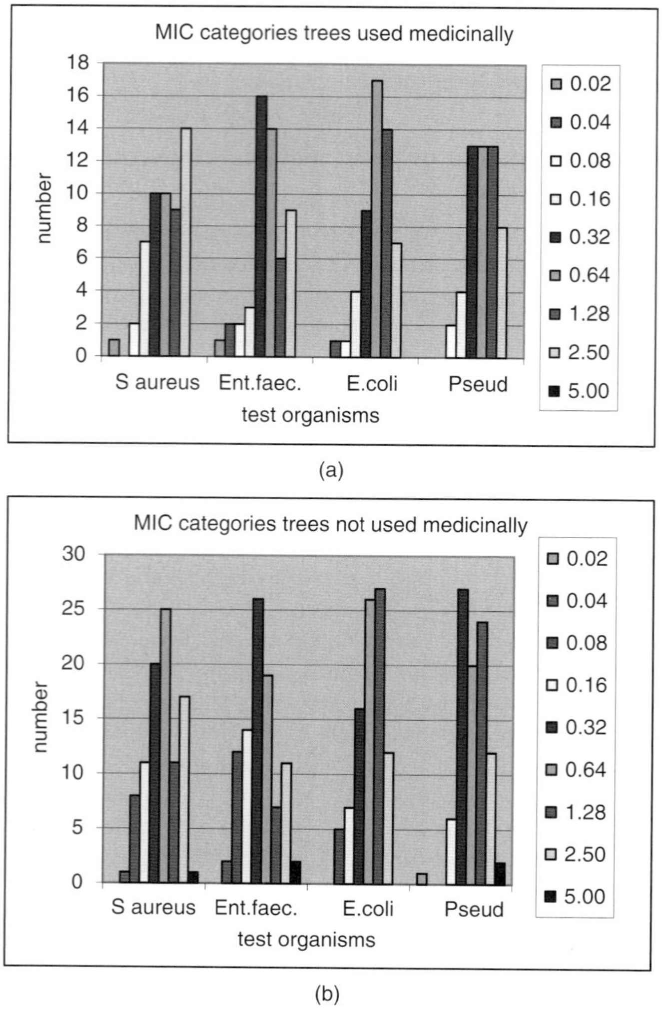


**Figure 12.2** The number of extracts that had MICs of 0.02–5 mg/mL against *Staphylococcus aureus*, *Enterococcus faecalis*, *Escherichia coli* and *Pseudomonas aeruginosa* for trees that were used medicinally (a) and trees not used medicinally (b)

and temperature of diffusion phase before incubation, thickness of the agar medium, composition of the medium and incubation temperature (Hewitt and Vincent, 1998). Most authors apparently never addressed these factors.

What is even more important is that if there are compounds with different polarities in the extract, the non-polar compounds will diffuse very slowly or even precipitate in the aqueous agar medium. It therefore becomes impossible to compare activities between different plant extracts accurately. As shown in Section 12.4.2, 89% of papers identified by Google Scholar have the word 'agar' somewhere in the text. When the same terms (Africa antimicrobial plant) are searched in all databases of Web of Science only 26% (54/206) of listed manuscripts contained the word 'agar'. Manuscripts containing the word 'agar' do not necessarily mean that agar diffusion techniques were used, but it probably shows that journals accredited by Thomson Reuters do not accept papers using agar diffusion techniques in plant antimicrobial studies. Van Vuuren (2008) evaluated

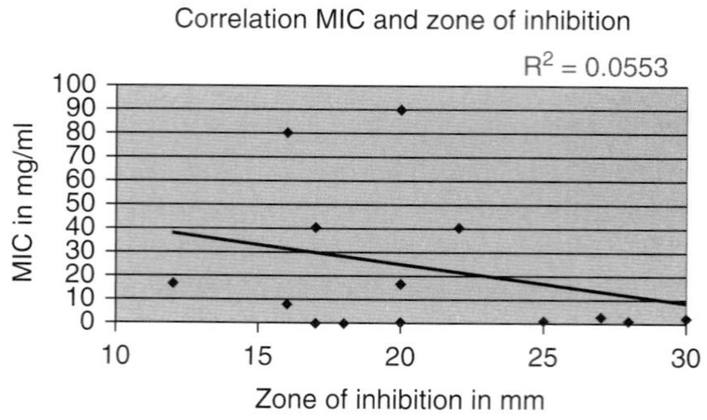


**Figure 12.3** The correlation between MIC and zone of inhibition of methanol and water extracts of *Dichrostachys cineraria* on different bacteria. Data from of Eisa *et al.* (2000)

the incidence of using different methods of determining antimicrobial activity.

Eisa *et al.* (2000) investigated the antibacterial activity of *Dichrostachys cineraria* of methanol and water extracts on eight different bacteria using agar diffusion (results in millimetres diameter) and serial dilution (results in minimum inhibitory activity in milligrams per millilitre) assays. We analysed their data and found that there is no predictability between the two sets of results with a correlation coefficient of 0.0553 (Figure 12.3). This may explain why high-quality journals do not accept agar diffusion assays for plant extracts. Agar dilution assays where different concentrations of the extract, fraction or compound are incorporated in the agar growth medium overcome the diffusion problems. These assays are, however, much less sensitive than serial dilution assays (Eloff, 1998b).

A microplate dilution assay was developed that gave robust and reproducible results without the need for sophisticated apparatus, required a small quantity of extract, and has been used worldwide with 809 citations (January 2014) by Google Scholar (Eloff, 1998b). A slight modification was required to apply this method to fungi (Masoko *et al.*, 2007)

In essence, extracts are made up to a concentration of 10 mg/mL in acetone and twofold serial dilutions are made in 100 μL of water. To this 100 μL, an actively growing bacterial culture is added. Because this is such a high inoculum there is no lag phase in the growth phase and the MIC results are insensitive to the cell concentration of the bacteria. There was no difference when a 1% inoculum of *Staphylococcus aureus* cultures in Müller–Hinton broth was incubated at 37 °C for 1, 3, 6 and 24 h and a 50% inoculum used in the assay. The microplates were sealed and grown overnight under 100% humidity. It was not easy to determine the turbidity of the cells with a microplate reader because cells clumped at the bottom of the well with some test organisms. Precipitation of compounds present in extracts of certain plants and the green colour of the extracts at high concentrations also made it difficult to see unambiguously where growth occurred with a microplate reader. To detect growth, addition of 40 μL of 0.2 mg/mL tetrazolium violet (INT) led to the best results because bacterial changed the colourless INT to a red formazan (Eloff, 1998b). By registering the lowest concentration where growth was inhibited (i.e. where there was a decrease in the colour) gave the MIC (Figure 12.4).

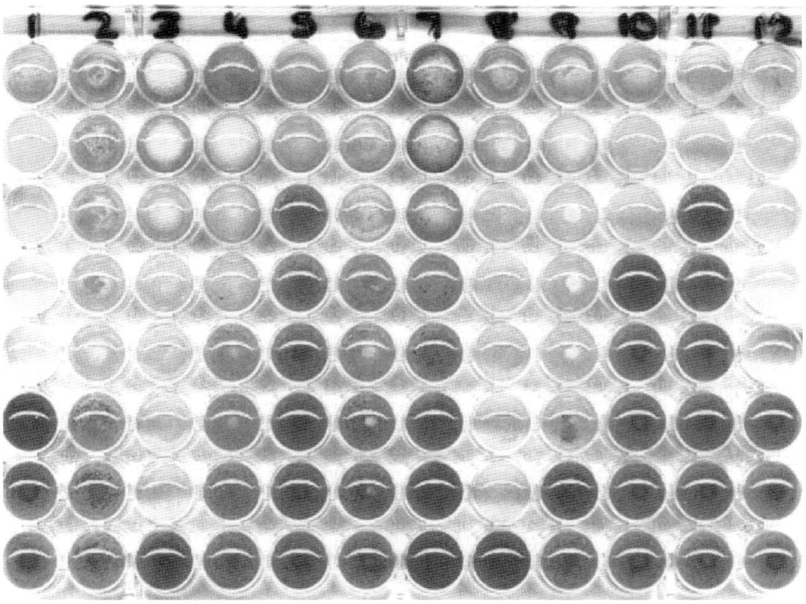


Figure 12.4 Illustration of using tetrazolium violet to determine MIC of different plant extracts in a serial dilution assay, with highest concentration at the top. Extracts in lanes 3 and 8 were the most active and extracts in lanes 5 and 11 the least active. The extracts where microbial growth is inhibited led to a reduction or absence in colour. *See plate section for colour version*

One should take into account that adding the 100 μL of the bacterial culture to the 100 μL in the series of diluted wells also decreases the concentration the cells were subjected to. If one started with a 10 mg/mL extract, the MIC in lanes 5 and 11 would be 0.125 mg/mL.

### 12.4.4 Determining number of antibacterial compounds

In examining plant extracts for the discovery of compounds with interesting activity it is also very useful to know how many different compounds are present in different plant extracts or fractions. Bioautography detects biological activity of separated compounds on a chromatogram. There are three ways in which this can be done. The first is by the agar overlay technique, where an inoculated agar solution is poured over a chromatogram and then incubated. In theory, one should be able to see where microbial growth is inhibited. In our hands, this is a very messy procedure. In particular, the separated polar compounds tended to dissolve in the agar growth medium and move it from its original position. Another technique is to blot the chromatogram with sterile filter paper and then place this on an agar culture of the microorganism. This also did not work well in our experience.

Our approach was based on that of Begue and Klein (1972). We remove the solvents from chromatograms and then spray the chromatogram with a 10 times concentrated culture of the organism in growth medium directly on the thin-layer chromatogram. After incubating overnight under 100% relative humidity, the chromatogram is sprayed with 2 mg/mL INT in methanol. This technique worked well with many microbes on condition that the eluents used for the thin-layer chromatography were volatile enough to be removed by a cold stream of air within a short period. We have developed eluents (Kotze and Eloff, 2002) that worked well. With a slight variation the method also worked very well for fungal pathogens (Masoko and Eloff, 2005). An example of a bioautogram is presented in Figure 12.1.

## 12.5 Results of random screening of large number of species

### 12.5.1 Problems with results in the literature

Many scientists and funding agencies have become disillusioned with screening plants for antimicrobial activities. If one compares the number of publications on antibacterial activities of plants with the number of successful products that have resulted, it seems screening for antimicrobial compounds in plants is an exercise in futility. Much has been promised and little delivered.

Many wide screening processes in the past have failed miserably to deliver worthwhile new pharmaceutical or herbal products, especially in the field of antimicrobial activity. The failure may be due to one or more of the following reasons:

1. The traditional agar diffusion bioassay method to determine antimicrobial activity does not work well with plant extracts; consequently, comparable quantitative results are seldom produced.
2. Scientists have been focusing on the wrong plant species by investigating plants used traditionally to treat infections. Because all aqueous plant extracts that we have tested to date have very little activity in our assays, we currently believe that these plants may have been effective through other mechanisms and not by inhibiting microbial growths directly.
3. Because relatively simple methods were involved, many inexperienced scientists jumped on the bandwagon.
4. Inefficient extractants may have been used.
5. There has been very limited follow up on results of screening publications because it requires substantial expertise of a different kind.
6. Journals have not been critical enough in evaluating manuscripts before publication, even stating that an extract with an MIC of 7 mg/mL was active (Fabry *et al.*, 1998). There appears to be reasonable consensus that only MICs of 100 μg/mL or lower

should be considered active for extracts (Eloff, 2004) and MICs of 10 μg/mL should be considered active for pure compounds (Rios and Recio, 2005).

## 12.6 Our approach to random screening

### 12.6.1 Selection of plants to work on

In our experience, plants used traditionally to treat infections are not more active than randomly selected plant species. With the new Benefit Sharing and Access Control legislation in South Africa to prevent biopiracy, there are administrative complications in investigating plants used traditionally. We therefore selected random screening of plants as our approach.

It is practically impossible to screen all the approximately 24 000 southern African angiosperm species. Even collecting and screening one representative of each of the approximately 2300 genera (Arnold and de Wet, 1993) would be a mammoth task, but it may be feasible to screen good representative members of the 224 families. One spin off of such an approach would be that it would be possible to determine which plant families or genera had the highest activity. High activity in a family should then be followed up by investigating representatives of genera and eventually species.

The most demanding part of this project would be to collect the plant material. If roots, bark, leaves and flowers are to be sampled, this would vastly complicate the project. It would quadruple the number of analyses; it would require getting permission from nature conservation and landowners to collect plant material; it would require collecting during the flowering season; it would rule out the possibility of collecting plants in a botanical garden.

To limit this project to feasible dimensions, we decided to focus on tree species and to examine only leaves and twigs, because this is a renewable resource. All of our plant materials were accessed from botanical gardens because it is easy to collect leaves from the same tree at a later stage to confirm results or expand work. This may also reduce the onerous task of collecting voucher specimens if the botanical garden's herbarium contains voucher specimens.

According to Coates Palgrave (2005), there are 112 families containing 236 genera and 2839 species of trees in southern Africa. This list includes trees occurring south of the Zambezi and Cunene rivers as well as alien species that have become naturalized in southern Africa. We planned to screen at least one species of each genus, subgenus or section of the trees available in botanical gardens and to analyse all the data.

For practical reasons we decided to use only properly dried material. In some cases at least, dried material retains its biological activity for many decades (Eloff, 1999). Furthermore, we ensure that material is properly collected to limit the effect that microbial infection of leaves could have. Leaves are only picked when they are dry from dew, collected in open mesh orange bags hung in the shade to limit photo-oxidative damage. Leaves are ground to a fine powder and stored in closed glass containers.

### 12.6.2 Novelty of our approach

Many scientists have focused on looking for compounds that can be used as single-substance pharmaceuticals. We have found substantial evidence for synergistic antimicrobial effects in plant extracts. We believe that there is a reasonable chance to develop anti-infective extracts rather than isolate single compounds that can be patented. The novelty of our approach is that:

1. We have a good basis for collecting plants in association with the National Botanical Gardens of the South African National Biodiversity Institute.
2. We use validated techniques for a variety of bioassays with a wide range of expertise within our group.
3. We focus on extracts with a higher percentage of success. This does not rule out the less likely possibility that novel pharmaceuticals may be discovered, but that is not our main aim.
4. We try to develop low-level technology that can be applied where it is needed most.

### 12.6.3 Methods used

The procedures used are explained in more detail elsewhere (Eloff and McGaw, 2006). Leaf powder was extracted with acetone, made up to 10 mg/mL and then the antibacterial activity is determined (see Section 12.4). The test organisms used were the four most important nosocomial pathogens, *S. aureus* (ATCC 29213), *Escherichia coli* (ATCC 27853), *Enterococcus faecalis* (ATCC 29212) and *Pseudomonas aeruginosa* (ATCC 25922), as well as methicillin-resistant *S. aureus* (MRSA) and *Mycobacterium smegmatis* (a non-pathogenic species of the TB-causing genus). Two fungi, *Candida albicans* and *Cryptococcus neoformans*, were also used.

### 12.6.4 Some preliminary results

Many plant extracts with exciting activities were discovered. Many extracts had exciting activities against *M. smegmatis* (Table 12.2) and *C. neoformans* (Table 12.3). In both cases, several extracts had MICs lower than 0.08 mg/mL and no extract had an activity higher than 2.5 mg/mL.

Before the current serial dilution microplate assays were developed, Vlietinck *et al.* (1995) investigated 100 Rwandese medicinal plants used by traditional healers to treat infections using agar diffusion and agar dilution assays. They found that 45% of the extracts were active against *S. aureus*, 2% against *E. coli*, 16% against *P. aeruginosa*, 7% against *C. albicans*, 80% against *Microsporum canis* and 60%

**Table 12.2** Minimal inhibitory activities of acetone leaf extracts of 324 tree species against *Mycobacterium smegmatis*

| MIC | No. | Cumulative | %Total (%) |
|---|---|---|---|
| 0.02 | 15 | 15 | 5 |
| 0.03 | 7 | 22 | 7 |
| 0.04 | 16 | 38 | 12 |
| 0.08 | 27 | 65 | 20 |
| 0.12 | 13 | 78 | 24 |
| 0.16 | 46 | 124 | 38 |
| 0.32 | 54 | 178 | 55 |
| 0.64 | 60 | 238 | 73 |
| 1.28 | 36 | 274 | 85 |
| 2.5 | 50 | 324 | 100 |

**Table 12.3** Minimal inhibitory activities of acetone leaf extracts of 707 tree species against *Cryptococcus neoformans*

| MIC | No. | Cumulative | Total (%) |
|---|---|---|---|
| 0.02 | 4 | 4 | 1 |
| 0.04 | 22 | 26 | 4 |
| 0.08 | 77 | 103 | 15 |
| 0.16 | 131 | 234 | 33 |
| 0.31 | 147 | 381 | 54 |
| 0.63 | 139 | 520 | 74 |
| 1.25 | 78 | 598 | 85 |
| 2.5 | 109 | 707 | 100 |

against *Trichophyton mentagrophytes*. It is a pity that this important work published in an excellent journal cannot really be compared with MIC data. It does, however, show that the 80% ethanol-in-water extracts had lower activity against Gram-negative organisms.

It is interesting that in the excellent paper of Svetaz *et al.* (2010) investigating the MICs of 80% ethanol extracts of 327 plant species against 11 fungal pathogens only 10 out of the 1067 MICs reported in tables had MICs lower than 62.5 µg/mL. Only about 1% of the assays led to MICs lower than 62.5 µg/mL compared with 54% of our extracts against *C. neoformans*. This may be due to a high sensitivity of the *C. neoformans* strain we used, to the effect of different extractants or to other differences in the methods used.

## 12.7 Activity of compounds isolated against *Staphylococcus aureus*

Gibbons (2004) wrote an excellent review on the use of compounds isolated from plants to combat *S. aureus* infections. There is no single chemical entity derived from a plant that is used clinically as an antibacterial. Lewis and Ausubel (2006) stated that many, if not most, of the 100 000 small-molecule compounds in plants have antimicrobial activity. Many of these compounds have very weak activity, frequently orders of magnitude lower than antibiotics obtained from fungi or bacteria.

Gibbons' review of more than 150 papers published between 1995 and 2003 shows that 116 single chemical entities have been isolated with MICs lower than 64 µg/mL against different staphylococci. He provided the structures of 116 different compounds that have been isolated during this period and discussed it under headings of monoterpenes, sesquiterpenes, diterpenes, triterpenes, phenylpropoanoids and stilbenoids, simple phenols and tropolones, flavonoids, alkaloids, polyketides and polyynes, sulfur-containing products and acylphloroglucinols. Many of these compounds had activities lower than 1 µg/mL. Some were more active against MRSA than against *S. aureus*.

The question arises as to why have many of these compounds not been developed to become new antimicrobials, especially since the first vancomycin-resistant *S. aureus* was discovered in 2002. One reason may be that the scientists working in this field may not determined the cytotoxicity of the isolated compounds. If a compound is active against many diverse organisms, it may contain a general metabolic toxin that would be toxic to human or animal cells as well. A second stumbling block is the absorption, distribution, metabolism and excretion pharmacokinetic parameters that affect the clinical usefulness require a totally different expertise. Many compounds may be too difficult to synthesize *de novo*, and partial synthesis from a plant-based precursor may be too expensive if it depends on isolation from plant material that is not widely available. The most important reason is probably that pharmaceutical companies have developed and invested heavily in fermentation technology and developing products from microbial sources.

## 12.8 Discovering antifungal compounds from natural products

Plants were the most important group investigated for antifungal activity in *Journal of Natural Products* during the last decade with 41% of publications, followed by fungi (34%), marine organisms (5%) and bacteria (5%) (Sortino *et al.*, 2012). There has been an increase in the number of papers published on antifungal compounds in two of the leading journals in this field, the *Journal of Ethnopharmacology* and *Planta Medica*. Between 1981 and 1990 there were 40 papers, followed by 139 papers in the next decade and 223 papers published between 2001 and 2010 (Sortino *et al.*, 2012).

Sortino *et al.* (2012) also analysed the antifungal studies registered in the NAPRALERT database between 1975 and 2005 to determine the families to which the approximately 2000 species investigated belong. The most-studied families were Asteraceae, Fabaceae, Lamiaceae, Euphorbiaceae, Ranunculaceae and Myrtaceae. They concluded that the selection of species was correlated with the abundance of species in the families and the accessability. What was more interesting is that the families with the largest percentage of active species from species investigated were Solanaceae (57%), Poaceae (56%), Rutaceae (38%), Brassicaceae (37.5%) and Lamiaceae (35%). Sortino *et al.* (2012) also provided the structures of 89 antifungal compounds isolated.

## 12.9 Review papers focusing on antimicrobial activity of plants from Africa

Vlietinck *et al.* (1995) investigated the antimicrobial activity of 100 plant species from Rwanda. The methods used at that stage were still agar diffusion and agar dilution assays. In 2008, an issue of the *Journal of Ethnopharmacology* was dedicated to review papers research done in South Africa. A paper by van Vuuren (2008) listed the contributions of many authors in this area and also discussed methods and evidence for investigating synergism. Eloff *et al.* (2008) provided information on the compounds isolated from members of the Combretaceae in different countries in Africa and the biological activities. Eloff and McGaw (2006, 2008) provided some results obtained in the Phytomedicine Programme and focused on methods used, including methods to be applied in determining cellular toxicity and *in vivo* animal experiments. In Cameroon, an excellent paper evaluated the activity of some natural products against bacteria expressing a multidrug-resistant phenotype (Kuete *et al.*, 2011). They found that some natural products are substrates of efflux pumps acting in resistant strains of four bacteria, but that terpenoids in general were not substrates.

## 12.10 Promising new approaches

There has been a strong development of new methods that can aid in the identification of active compounds. Queiros *et al.* (2012) provide an excellent summary of recent developments, especially in hyphenated techniques and metabolomics. In an older paper, Lewis and Ausubel (2006) discussed some interesting possibilities; for example, using extracts from plants that were stimulated to synthesize antibacterials by pathogen attack or treatment with immune elicitors such as salicylic acid. They also discussed the hypothesis that plants use a different strategy for controlling infections by acting in combination. This echoes our conclusion after isolating many active compounds from active fractions only to discover that the activity is much lower than what we expected.

One way in which such a synergistic combination can work is by resistance-modifying agents (RMAs). Examples of these RMAs are methicillin-resistance reversing agents and modulators of multidrug resistance (MDR) where efflux transporters are involved. It appears that most MDR inhibitors are large molecules with a high degree of lipophilicity (Gibbons, 2004). We had thought that the efficacy of aqueous extracts in healing patients of traditional healers may be related to other compounds enhancing the activity of antibacterial compounds with a low activity or that some compounds may affect the efflux pumps in bacteria. The high lipophilicity of these compounds makes this idea less attractive.

Another idea that extracts may target bacterial virulence rather than bacterial growth (Lewis and Ausubel, 2006) could be an attractive explanation for the apparent efficacy of aqueous extracts. These authors identified three potential aspects to be considered for future research.

### 12.10.1 Find a viable new lead

It is important that a viable lead must have specificity to be selectively toxic against the pathogen. The drug industry has developed excellent new antibiotics against Gram-positive species; for example, Zyvox, Synercid (two protein synthesis inhibitors) and Cubicin (a membrane-acting inhibitor) (Lewis and Ausubel, 2006).

Gram-negative bacteria have developed a sophisticated permeability barrier with a hydrophilic lipopolysaccharide outer membrane. This limits the entry of most drugs that are hydrophobic or amphipathic compounds. The drug industry has failed to find new broad-spectrum compounds against Gram-negative species. The fluoroquinolines, the last class of broad spectrum antibiotics, were discovered 40 years ago.

It therefore seems a viable option to search for natural products with activity against Gram-negative bacteria. As stated above, there is also an urgent need to find new antifungal drugs.

In both cases it appears to be important to determine the mechanism of activity. If a novel mechanism of activity is found, then the compound may have much promise for use in combination with existing antibiotics.

The second approach would be to search for MDR inhibitors that could be used to inhibit the Gram-negative resistance – nodulation-cell division efflux pump. This excretes amphipathic compounds from Gram-negative bacteria. If such compounds that are non-toxic are found, it would be a great advantage.

The third possibility is to find compounds that would block pathogen virulence. This would be a challenging exercise. An interesting model has been developed to infect the free-living nematode *Caenorhabditis elegans* with bacteria and then see if a compound or extract can cure the nematode from the bacteria by decreasing the virulence.

There is another possibility; that is, by investigating the effect of plant extracts on the adherence of bacteria (Katsikogianni and Missirlis, 2004). If compounds in a plant extract inhibit the adherence of microorganisms, this may explain why aqueous plant extracts that have hardly any direct antibacterial activity may be effective when used by traditional healers to treat patients.

## 12.11 The potential of using African medicinal plants as extracts

It appears that much of the work done to date has stopped before determining the safety of extracts or compounds or the efficacy of the product in animal or field experiments. Furthermore, in the human herbal product market it appears that marketing is much more important than activity of the product. This has led to frustration, where the human herbal medicine market was not interested in developing products with higher activity.

Consequently, we decided to focus on developing products that could be used in the animal and plant production industry. We have found that there is slower development of resistance against complex plant extracts than against single-compound antibiotics. Additionally, there is a growing market for organically produced foods. It therefore seems as if there is a good motivation for investigating the use of plant extracts in animal and plant production. The increase in activity by potentizing extracts and an application in plant production and another in animal production will be discussed.

There are challenges in applying plant extracts as therapeutic products. The use of the plant material should be on a sustainable basis. That is why we focus only on using tree leaves. There may be variation in activity based on the season, genetic factors and environmental factors. Quality control could be difficult. The long term use would probably involve selection of chemotypes with a higher activity and then production of the trees as a source of the leaves to ensure consistent quality. The use of leaf extracts may be cheaper than using chemical control and it would probably be safer to the environment and may deliver "organically grown" products.

### 12.11.1 Increasing activity of extracts by low cost potentizing of extracts

By investigating the chromatogram and bioautogram in Figure 12.1, it is clear that the hexane and di-isopropylether and water extracts did not extract the active compounds from leaves of *Combretum woodii*. By extracting leaves, with hexane or water, prior to extracting it with acetone, it is possible to remove a number of inactive compounds, leading to a higher activity per mass unit in the extracts.

The *C. woodii* extract also had excellent antioxidant activity, and the activity against coccidiosis was evaluated in a poultry experiment (Naidoo *et al.*, 2008). Unfortunately, the extract was toxic to the poultry.

By using such an approach to remove inactive components, we have been able to develop a grape seed extract with nearly double the antioxidant activity of the best product on the market, leading to a South African patent and a commercial product, Biooxidin.

### 12.11.2 Example of a plant extract with higher activity than commercial fungicides against plant fungal pathogens

Eloff *et al* (2006) found that an extract of *Melianthus comosus* Vahl growing widely in southern Africa had excellent activity against animal fungal pathogens, but the toxicity to animals would have complicated the development of a product.

The extracts also had an excellent activity against 10 plant fungal pathogens investigated (*Rhizoctonia solani, Fusarium oxysporum, Penicillium janthinelum, Penicillium expansum, Colletotrichum glocosponicales, Trichoderma harzianum, Pythium ultimum, Phytophthora nicotiana, Aspergillus niger* and *Aspergillus parasiticus*). The extract contained one major antifungal compound, and this compound was isolated and characterized as 3-hydroxy-12-oleanen-30-oic acid.

By selective extraction and solvent fractionation, the activity of the extract could be increased, leading to an extract with an average MIC of 0.066 mg/mL against all 10 fungal pathogens. If the MIC values of 0.16 mg/mL against *P. expansum* and *A. niger* are ignored, the average MIC for the other fungi is 0.04 mg/mL. The acetone extract did not lose activity after storage at room temperature for a month. The dried extract was slightly soluble in water and ethanol, reasonably soluble in ethyl acetate and highly soluble in acetone. The potentized extract had a higher antifungal activity than six commercially used fungicides against some important plant fungal pathogens. In a limited field trial it gave a much better result than a commercial fungicide even though it was used at a quarter of the dose of the commercial fungicide.

This shows that plant extracts have the potential to deliver products that can control fungal infections.

### 12.11.3 Example of a plant extract with similar activity to a commercial fungicide against the animal fungal pathogen Aspergillus fumigatus

After determining the MIC of many tree leaf extracts against *C. neoformans*, we investigated the activity of extracts with good activity against *Aspergillus fumigatus*, which is a fungal pathogen causing serious problems in the poultry industry and also in immune-compromised humans. The acetone extract of one of the species *Loxostylus alata* had good *in vitro* activity against *A. fumigatus* (Suleiman *et al.*, 2010). The toxicity of this extracts to poultry was determined. At a dose of 300 mg/kg there were signs of toxicity, but not at concentrations up to 200 mg/kg. Young birds were infected with *A. fumigatus* and treated with 50, 100 and 200 mg/kg as well as with 60 mg/kg ketoconazole, the best therapeutic drug. By measuring the pathology and presence of *A. fumigatus*, a dose-related protection was obtained with the plant extract. The highest concentration tested gave the same level of protection as the positive control, ketoconazole (Suleiman *et al.*, 2012).

These results indicate that there is a good possibility of developing useful products based on a plant extract and not a single chemical entity. Registering products may still lead to complications if the use has environmental toxicity implications.

## 12.12 Conclusions

There have been a large number of publications investigating the antimicrobial activity of plant extracts. Up to about one or two decades ago the methods used did not make it possible to compare the activity of different plants.

### 12.12.1 Work on Gram-positive bacteria

Since newer methods have been developed there has been much work done, especially on Gram-positive bacteria, and it is surprising that, with so many highly active compounds discovered, so little progress has been made in the development of these single chemical entities to new drugs. One limitation in existing work on Gram-positive organisms is that, apparently, not much work has been done on determining the safety of these compounds and in doing *in vivo* efficacy studies. It may also be very useful to evaluate the mechanism of activity of these promising compounds. This work would require a multidisciplinary approach.

Development of resistance in Gram-positive bacteria is frequently associated with efflux pumps that remove antibiotics from the microbe cells. There have been encouraging results in identifying compounds that can inhibit the efflux pump to such a level that the microbe can be controlled. It is encouraging that there is much evidence of MDR compounds than can reverse the development of resistance in Gram-positive bacteria. Combination of older drugs with plant extracts may reverse the resistance of many Gram-positive bacteria. There have been some new drugs with very good activity developed recently.

The situation is not that positive with the *Mycobacterium* species causing TB in many organisms. It is not clear if they are closer to the Gram-positive or Gram-negative organisms. Some have classified them as acid-fast Gram-positive bacteria owing to their lack of an outer cell membrane, while other authors consider them closer to Gram-negative bacteria. The latter have an unusual, waxy coating on their cell surface (primarily mycolic acid), which makes the cells impervious to Gram staining.

### 12.12.2 Gram-negative bacteria

It appears that there are many more difficulties in developing new drugs from plants against Gram-negative bacteria. This has to do with the hydrophilic outer membrane of Gram-negative bacteria. Because most antibacterial compounds are non-polar or amphipathic, transport through the hydrophilic polysaccharide is difficult. A search for compounds that would enhance the uptake of antibacterial compounds by Gram-negative bacteria appears to be a useful endeavour.

One place to start would be to investigate plant extracts that have a good activity against Gram-negative bacteria, especially if the isolated active compounds have a low activity.

Investigating factors that would decrease virulence of these bacteria may be a difficult, but promising option (Sortino *et al.*, 2012).

### 12.12.3 Antifungal compounds

From this review it is clear that there is a great need to develop antifungal compounds, especially with the increase of fungal infections of immunocompromised patients.

The application of plant antifungal extracts to control fungal infections affecting plant or animal productivity appears to be a promising area to investigate. Here, evaluating the safety of products and the mechanism of activity should be priority areas requiring a multidisciplinary and collaborative approach.

## Acknowledgements

This research was funded by the National Research Foundation and the Medical Research Council of South Africa. The curators of several national botanical gardens of the South African National Biodiversity Institute allowed us to collect plant material in their botanical gardens.

## References

Abad, M.J., Ansuategui, M. and Bermejo, P. (2007) Active antifungal substances from natural sources. *Archive of Organic Chemistry*, **2007** (7), 116–145.

Arnold, T. and de Wet, C. (1993) *Plants of Southern Africa: Names and Distribution*, Memoirs of the Botanical Survey of South Africa 62, National Botanical Institute, Pretoria.

Arnold, T.H., Prentice, C.A., Hawker, L.C. *et al.* (2002) *Medicinal and Magical Plants of Southern Africa: An Annotated Checklist*, Strelitzia 13, National Botanical Institute, Pretoria.

Begue, W.J. and Klein, R.M. (1972). The use of tetrazolium salts in bioautographic procedure. *Journal of Chromatography*, **88**, 182–184.

Brendler, T., Eloff, J.N., Gurib-Fakim, A. and Phillips, D. (eds) (2010) *African Herbal Pharmacopoeia*, AAMPS Publishing, Mauritius (ISBN 9789990389098).

Coates Palgrave, M. (2005) *Keith Coates Palgrave Trees of Southern Africa*, 3rd edn, Struik Publishers, Cape Town.

Cos, P., Vlietinck, A.J., Vanden Berghe, D. and Maes, L. (2006) Anti-infective potential of natural products: how to develop a stronger *in vitro* 'proof of concept'. *Journal of Ethnopharmacology*, **106**, 290–302.

Cowan, M.M. (1999) Plant products as antimicrobial agents. *Clinical Microbiology Reviews*, **12**, 564–582.

Cragg, M.G., Grothaus, P.G. and Newman, D.J. (2012) Natural products in drug discovery: recent advances, in *Plant Bioactives and Drug Discovery: Principles, Practice, and Perspectives* (ed. V. Cachinel-Filho), John Wiley & Sons, Inc., Hoboken, NJ, pp. 1–42.

Daley, C.L. and Caminero, J.A. (2013) Management of multidrug resistant tuberculosis. *Seminars in Respiratory and Critical Care Medicine*, **34** (01), 044–059.

Eisa, M.M., Almagboul, A.Z., Omer, M.E.A. and Elegami A.A. (2000) Antibacterial activity of Dichrostachyscinerea. *Fitoterapia*, **71**, 324–327.

Eloff, J.N. (1998a) Which extractant should be used for the screening and isolation of antimicrobial components from plants? *Journal of Ethnopharmacology.* **60**, 1–8.

Eloff, J.N. (1998b) A sensitive and quick microplate method to determine the minimal inhibitory concentration of plant extracts for bacteria. *Planta Medica*, **64**, 711–714.

Eloff, J.N. (1998c) Conservation of medicinal plants: selecting medicinal plants for research and gene banking, in *Conservation of Plant Genes III: Conservation and Utilization of African Plants* (eds R.P. Adams and J.E. Adams), Monographs in Systematic Botany from the Missouri Botanical Garden 71, Missouri Botanical Garden Press, St. Louis, MO, pp. 209–222.

Eloff, J.N. (1999) It is possible to use herbarium specimens to screen for antibacterial components in some plants. *Journal of Ethnopharmacology*, **67**, 355–360.

Eloff, J.N. (2004) Quantifying the bioactivity of plant extracts during screening and bioassay-guided fractionation. *Phytomedicine*, **11**, 370–371.

Eloff, J.N. and McGaw, L.J. (2006) Plant extracts used to manage bacterial, fungal and parasitic infections in southern Africa, in *Modern Phytomedicine: Turning Medicinal Plants into Drugs* (eds I. Ahmad, F. Aqil and M. Owais), Wiley–VCH Verlag GmbH, Weinheim, pp. 97–121.

Eloff, J.N. and McGaw, L.J. (2008) Application of plant extracts and products in veterinary medicine, in *New Strategies Combating Bacterial Infection* (eds I. Ahmad and F. Aqil), Wiley–VCH Verlag GmbH, Weinheim, pp. 205–228.

Eloff, J.N., Famakin, J.O. and Katerere, D.R.P. (2005) *Combretum woodii* (Combretaceae) leaf extracts have high activity against Gram-negative and Gram-positive bacteria. *African Journal of Biotechnology*, **4**, 1161–1166.

Eloff, J.N., Angeh, I. and McGaw, L.J. (2006) A plant antifungal product from *Melianthus comosus* (Melianthaceae) leaf extracts. *Planta Medica*, **72**, 982.

Eloff, J.N., Picard, J. and Masoko, P. (2007) Resistance of animal fungal pathogens to solvents used in bioassays. *South African Journal of Botany*, **73**, 667–669.

Eloff, J.N., Katerere, D.R. and McGaw L.J. (2008) The biological activity and chemistry of the southern African Combretaceae. *Journal of Ethnopharmacology*, **119**, 686–699.

Fabry, W., Okemo, P.O. and Ansorg, R. (1998) Antibacterial activity of East African medicinal plants. *Fitoterapia*, **60**, 79–84.

Farnsworth, N.R. and Morris, R.W. (1976) Higher plants – the sleeping giant of drug development. *American Journal of Pharmaceutical Education*, **148**, 46–52.

Gibbons, S. (2004) Anti-staphylococcal plant natural products. *Natural Products Reports*, **21**, 263–276.

Hewitt, W. and Vincent, S. (1989) *Theory and Application of Microbiological Assay*, Academic Press, San Diego, CA.

Kåhrström, C.T. (2013) Entering a post-antibiotic era? *Nature Reviews Microbiology*, **11**, 146.

Klopper, R.R., Gautier, L., Chatelain, C. *et al.* (2007) Floristics of the angiosperm flora of sub-Saharan Africa: an analysis of the African Plant Checklist and Database. *Taxon*, **56**, 201–208.

Kotze, M. and Eloff, J.N. (2002) Extraction of antibacterial compounds from *Combretum microphyllum* (Combretaceae). *South African Journal of Botany*, **68**, 62–67.

Kuete, V., Alibert-Franco, S., Eyong, K.O. *et al.* (2011) Antibacterial activity of some natural products against bacteria expressing a multidrug-resistant phenotype. *International Journal of Antimicrobial Agents*, **37**, 156–161.

Levy, S.B. and Marshall, B. (2004) Antibacterial resistance worldwide: causes, challenges and responses. *Nature Medicine*, **10**, 122–129.

Lewis, K. and Ausubel, F.M. (2006) Prospects for plant-derived antibacterials. *Nature Biotechnology*, **24**, 1504–1507.

Katsikogianni, M. and Missirlis, Y.F. (2004) Concise review of mechanisms of bacterial adhesion to biomaterials and of techniques used in estimating bacteria–material interactions. *European Cells and Materials*, **8**, 37–57.

Masoko, P. and Eloff, J.N. (2005) The diversity of antifungal compounds of six South African *Terminalia* species (Combretaceae) determined by bioautography. *African Journal of Biotechnology*, **4**, 1425–1431.

Masoko, P., Picard, J. and Eloff, J.N. (2007) The antifungal activity of twenty-four South African *Combretum* species (Combretaceae). *South African Journal of Botany*, **73**, 173–183.

Naidoo, V., McGaw, L.J., Bisschop, S.P.R. *et al.* (2008) The value of plant extracts with antioxidant activity in attenuating coccidiosis in broiler chickens. *Veterinary Parasitology*, **153**, 214–219.

Queiroz, E.F., Hostettmann, K. and Wolfender, J.-L. (2012) Modern approaches in the search for new active compounds from crude extract of natural products, in *Plant Bioactives and Drug Discovery: Principles, Practice, and Perspectives* (ed. V. Cachinel-Filho), John Wiley & Sons, Inc., Hoboken, NJ, pp. 43–80.

Rabe, T. and van Staden, J. (1997) Antibacterial activity of South African plants used for medicinal purposes. *Journal of Ethnopharmacology*, **56**, 81–87.

Rios, J.L. and Recio, M.C. (2005) Medicinal plants and antimicrobial activity. *Journal of Ethnopharmacology*, **100**, 80–84.

Solecki, R.S. (1977) The implications of the Shanidarcave Neanderthal flower burial. *Annals of the New York Academy of Sciences*, **293**, 114–124.

Sortino, M., Derita, M., Svetaz, L. *et al.* (2012) The role of natural products in discovery of new anti-infective agents with emphasis on antifungal compounds, in *Plant Bioactives and Drug Discovery: Principles, Practice, and Perspectives* (ed. V. Cachinel-Filho), John Wiley & Sons, Inc., Hoboken, NJ, pp. 205–240.

Srivastava, A., Shukla, Y.N. and Kumar, S. (2000) Recent development in plant derived antimicrobial constituents: a review. *Journal of Medicinal and Aromatic Plant Sciences*, **22**, 349–405.

Suleiman, M.M., McGaw, L.J., Naidoo, V. and Eloff, J.N. (2010) Evaluation of several tree species for activity against the animal fungal pathogen *Aspergillus fumigatus*. *South African Journal of Botany*, **76**, 64–71.

Suleiman, M.M., Duncan, N., Eloff, J.N. and Naidoo, V. (2012) A controlled study to determine the efficacy of *Loxostylis alata* (Anacardiaceae) in the treatment of aspergillosus in a chicken (*Gallus domesticus*) model in comparison to ketoconazole. *BMC Veterinary Research*, **8**, 210.

Svetaz, L., Zuljan, F., Derita, M. *et al.* (2010) Value of the ethnomedical information for the discovery of plants with antifungal properties. A survey among seven Latin American countries. *Journal of Ethnopharmacology*, **127**, 137–158.

Van Vuuren, S.F. (2008) Antimicrobial activity of South African medicinal plants. *Journal of Ethnopharmacology*, **119**, 462–472.

Van Wyk B.-E. and Wink, M. (2004) *Medicinal Plants of the World: An Illustrated Scientific Guide to Important Medicinal Plants and their Uses*, Timber Press, Portland, OR.

Vlietinck, A.J., Van Hoof, L., Tott, J. *et al.* (1995) Screening of hundred Rwandese medicinal plants for antimicrobial and antiviral properties. *Journal of Ethnopharmacology*, **46**, 31–47.

# 13 Plant Medicines Used in the Treatment of Malaria

John R.S. Tabuti[1], Antonia Nyamukuru[2] and Mohammed Lamorde[3]

[1]Makerere University, College of Agricultural and Environmental Sciences (MUCAES), Kampala, Uganda
[2]Sustainable Use of Plant Diversity (SUPD), Kampala, Uganda
[3]Infectious Diseases Institute, College of Health Sciences, Makerere University, Kampala, Uganda

## 13.1 Introduction

Globally, malaria ranks among the top three communicable killer diseases. Each year, malaria causes 781 000 deaths and illnesses in an estimated 225 million people worldwide (WHO, 2010). Most deaths occur in sub-Saharan Africa (WHO, 2008), with young children and pregnant women at greatest risk for severe forms of malaria and death (WHO, 2008). Uganda has one of the highest burdens of malaria globally, with an estimated 70 000–110 000 children dying of malaria each year (Lynch *et al.*, 2005).

At the household and at the national level, malaria exerts significant social and economic costs (Sachs and Malaney, 2002). These costs include direct medical costs to treat the disease and also lost income resulting from inability to work or time taken off work to care for relatives with malaria. Malaria in children impacts negatively on school attendance and decreases savings at the household level. At the national level, governments incur the costs of prevention and treatment of malaria. Furthermore, the economic impact of malaria endemicity includes lower trade, tourism and foreign direct investment (Sachs and Malaney, 2002).

Allopathic medicines are widely used to treat malaria, but traditional medicines are also used (Tabuti, 2008). Historically, allopathic medicines lose effectiveness periodically due to emergence of drug-resistant strains of *Plasmodium* spp., the causative pathogen of malaria. Consequently, widely accessible and affordable medicines, including chloroquine and sulfadoxine–pyrimethamine, are no longer used in the Uganda malaria treatment programme (MOH, 2005). These agents have been replaced by highly efficacious combination therapies containing artemisinin derivatives. For uncomplicated malaria, the World Health Organization recommends artemisinin combination therapy (ACT) containing a rapidly acting and highly potent artemisinin derivative combined with a longer acting less potent partner drug that prevents malaria relapse and may contribute to preventing the emergence of resistance to the artemisinins. Despite this strategy, resistance to artemisinin has begun to appear (Phyo *et al.*, 2012).

Consequently, there is an urgent need to identify and develop new therapies that will control malaria. Research for new drugs for the treatment of malaria is highly justified, especially in Africa, because access to efficacious allopathic medicines like ACTs is limited, and were it not because of their being highly subsidized by the donor community or the government, in this case the Ugandan Government, many Ugandans would not have any access at all to ACTs because they are costly.

One of the most effective ways of identifying efficacious and safe medicines is to document indigenous knowledge (IK) associated with traditional medicine in ethnobotanical studies (Balick, 1990; Prance, 1991). Indeed, some of the most effective antimalarials have been developed from traditional medicines, including quinine from the bark of *Cinchona* spp. and artemisinin from *Artemisia annua* L. (annual wormwood). In this chapter we review the plant materials that have been documented so far for the treatment of malaria in Uganda and what evidence exists with regard to the efficacy and safety of these medicinal plants.

## 13.2 Approach used in the review

We conducted a literature search for published articles on research conducted in Uganda. We included articles in which materials for treatment of malaria were documented. We included articles that were specifically focused on malaria, as well as general ethnobotanical surveys in which malaria treatments were documented alongside other material for treating other diseases. We prioritized the reported medicines using two criteria. First, species mentioned from three and/or more regions in Uganda; second, species which were highly ranked by the author(s). The information on rankings was not uniform, and some authors did not rank the species. We found three reports that reported directly on malaria and seven general ethnobotanical papers that were not explicitly focused on malaria but reported it among other traditional medicines. Most studies aimed at drug discovery for malaria focused on documenting known materials, and only a handful of pharmacological studies have been conducted to evaluate the efficacy and safety of plant species used to treat malaria. Only one validation study was identified, and this focused on the efficacy of *Cardiospermum halicacabum* and *Momordica foetida* (Waako *et al.*, 2005). After developing a select species list, we expanded our review to include international articles that reported on the safety and efficacy of plants that had been documented in the Ugandan studies.

*Novel Plant Bioresources: Applications in Food, Medicine and Cosmetics*, First Edition. Edited by Ameenah Gurib-Fakim.

## 13.3 Plant species commonly used to treat malaria in Uganda

In our review, we identified 129 species used for malaria treatment. Some plant species are used for malaria treatment in more than one region in Uganda (Table 13.1). The finding that these species are known in different ethnic and geographical groups may indicate that these species are efficacious. Using this criterion, we identified 12 species from the 129 species as priority plants for further study (Table 13.1). The remedies are prepared by boiling hard plant parts, like the bark or the root, to prepare decoctions or by crushing leaves and mixing in water (infusions). We found most therapies are prepared from single species; however, some concoctions are prepared from mixtures of species; for example, *Vernonia amygdalina* plus *Momordica foetida*; *Cannabis* sp. plus *Chenopodium opulifolium*; or *Mangifera indica* plus *Tamarindus indica* (Tabuti, 2008).

Studies reported that therapies are administered orally in variable doses; for example, 100–500 mL for adults and older children, or 1–3 tablespoons for children younger than 5 years one to three times a day either for short duration (1–3 days) or until the patient's condition has improved (Tabuti, 2008; Stangeland *et al.*, 2011).

For five of the species in Table 13.1 we found articles describing efficacy and safety. This information is summarized below and suggests that some of these species may be active against malaria.

*V. amygdalina* is used to treat fevers and malaria in several African countries, including Cameroon (Betti, 2004), Nigeria (Tor-anyiin *et al.*, 2003), Rwanda (Hutchings *et al.*, 1996) and Ethiopia (Karunamoorthi and Tsehaye, 2012). Widespread use of this plant for malaria treatment is further evidence of efficacy and/or safety. In studies reporting the activity of leaf extracts of this plant against *Plasmodium falciparum*, the half maximal inhibitory concentration ($IC_{50}$), which represents the concentration of the extract at which 50% of parasite is inhibited, ranged from 2.5 to 9.7 μg/mL, depending on solvent used in the extraction (Tona *et al.*, 2004; Chenniappan and Kadarkarai, 2010). A review by Magadula and Erasto (2009) indicates that *V. amygdalina* has active sesquiterpenoids, and that the sesquiterpene lactones (STLs) vernodalin, vernodalol, vernolide and hydroxyvernolide have moderate antiplasmodial activity against the multidrug-resistant K-1 strain of *P. falciparum*, with vernodalin being the most active compound with an $IC_{50}$ of 4 μg/mL. Young leaves of *V. amygdalina* contain higher concentrations of vernodalin than the other STLs. This suggests that the observed antimalarial efficacy found especially in leaf extracts of the species may be partly due to vernodalin. In addition, leaves of *V. amygdalina* contain steroidal saponins of the class vernonioside A1, A2, A3, A4 and B1. These saponins were seen to have weak antiplasmodial activity against the multidrug-resistant K-1 strain (Ohigashi *et al.*, 1994; Magadula and Erasto, 2009).

*Azadirachta indica* is used in Tanzania (Gessler *et al.*, 1995), Ghana (Abbiw, 1996; Asase *et al.*, 2005) Kenya (Njoroge and Bussmann, 2006; Nguta *et al.*, 2010), and Nigeria (Sofowora, 1993; Tor-anyiin *et al.*, 2003). Antiplasmodial activity of this plant has been demonstrated clinically and experimentally (Sofowora, 1993).

**Table 13.1** Plant species most frequently mentioned in literature from Ugandan studies for the treatment of malaria. The table also shows the number of reports encountered in the study and the geographical regions from which the reviewed studies were conducted

| Species | No. of reports | Uganda geographical region[a] | Source |
|---|---|---|---|
| *Vernonia amygdalina* Delile | 8 | a, b, c, d | Tabuti *et al.* (2003), Ssegawa and Kasenene (2007), Galabuzi (2008), Tabuti (2008), Oryema *et al.* (2010), Stangeland *et al.* (2011) |
| *Azadirachta indica* A. Juss. | 4 | b, c, d | Tabuti *et al.* (2003), Galabuzi (2008), Tabuti (2008), Stangeland *et al.* (2011) |
| *Pseudarthria hookeri* Wight &Arn | 5 | a, c, d | Ssegawa and Kasenene (2007), Galabuzi (2008), Oryema *et al.* (2010), Stangeland *et al.* (2011) |
| *Mangifera indica* L. | 3 | a, b | Tabuti *et al.* (2003), Tabuti (2008) |
| *Momordica foetida* Schumch. Et Thonn | 4 | b, c | Tabuti *et al.* (2003), Tabuti (2008), Oryema *et al.* (2010), Stangeland *et al.* (2011) |
| *Cajanus cajan* (L.) Millsp. | 3 | a, b, c | Tabuti (2008), Stangeland *et al.* (2011) |
| *Zanthoxylum chalybeum* Engl. | 3 | a, b, d | Ssegawa and Kasenene (2007), Galabuzi (2008), Tabuti (2008), Oryema *et al.* (2010) |
| *Markhamia lutea* (Benth.) K. Schum. | 3 | c, d | Galabuzi (2008), Stangeland *et al.* (2011) |
| *Maesa lanceolata* Forssk | 3 | c, d | Galabuzi (2008), Stangeland *et al.* (2011) |
| *Microglossa pyrifolia* Kuntze (TS) | 3 | c, d | Galabuzi (2008), Stangeland *et al.* (2011) |
| *Senna didymobotrya* (Fresen.) H.S. Irwin & Barneby | 3 | c, d | Ssegawa and Kasenene (2007), Stangeland *et al.* (2011) |
| *Vernonia lasiopus* O. Hoffm. | 3 | c, d | Ssegawa and Kasenene (2007), Galabuzi (2008), Stangeland *et al.* (2011) |

[a](a) Northern Uganda, includes the Oyam and Erute counties; (b) eastern Uganda, includes studies from Kaliro and Budiope districts close to Lake Kioga; (c) western Uganda, includes Kiohima, Mbarara and Nyakayojo; (d) south western Uganda has Sango Bay region. In the eastern region the study was conducted among people of the Basoga ethnic group, in south western region among the Banyankole and in the northern region among the Acholi.

The active principles are reported to be nimbolide and gedunin (Sofowora, 1993).

*Mangifera indica* is widely used for the treatment of malaria (Willcox and Bodeker, 2004). In Africa it is used in Ghana (Asase *et al.*, 2005), Nigeria (Tor-anyiin *et al.*, 2003), Tanzania (Gessler *et al.*, 1995) and Rwanda (Hutchings *et al.*, 1996). The bark shows moderate to low antiplasmodial activity ($IC_{50}$ 6.76–11.50 μg/mL) depending on solvent used in the extraction (Chenniappan and Kadarkarai, 2010).

*Momordica foetida* water and ethyl acetate extracts were found to have weak *in vitro* antiplasmodial activity with $IC_{50}$ values greater than 28.00 μg/mL. *In vivo* studies of water extracts of *M. foetida* given orally in the dose range 200 and 1000 mg/kg twice daily prolonged survival of mice infected with *Plasmodium berghei* from 7.0 ± 1.8 to 17.9 ± 1.8 days (Waako *et al.*, 2005).

*Zanthoxylum chalybeum* is used in different countries, including Tanzania (Gessler *et al.*, 1995) and Kenya (Njoroge and Bussmann, 2006; Nguta *et al.*, 2010), for the treatment of malaria. According to Gansané *et al.* (201), alkaloids from the stem bark are active against *P. falciparum* W2 strain ($IC_{50}$ of 1.2 μg/mL). However, the alkaloids were toxic with a selectivity index (SI) >20. Gessler *et al* (1995) reported high *in vitro* antiplasmodial activity for *Z. chalybeum* root bark obtained from Tanzania (Gessler *et al.*, 1994) with $IC_{50}$ of 0.7 μg/mL and 0.43 μg/mL in the ethyl acetate fraction and water fraction respectively. However, the level of activity depended on the plant part and region from which the material was collected. Root bark extracts were more effective than stem bark, while *Z. chalybeum* root bark obtained from the Kagera region was three times more active than root bark samples obtained from Dar es Salaam.

*Cajanus cajan* (L.) Millsp. is a promising species for malaria treatment. It has been reported for use as an antimalarial in several countries, including Kenya (Njoroge and Bussmann, 2006), Nigeria (Ajaiyeoba *et al.*, 2006) and in South America (see references in Duker-Eshun *et al.* (2004)). Crude extracts of this plant contain compounds that have been shown to have antimalarial activity. Duker-Eshun *et al.* (2004) reported $IC_{50}$ values ranging from 19 to 34 μM for stilbenes, longistylin A and C, and betulinic acid, while cajachalcone, 2′,6′-dihydroxy-4-methoxy chalcone is reported to have an $IC_{50}$ value of 2.0 μg/mL (7.4 μM) (Ajaiyeoba *et al.*, 2013). The species appears to be non-toxic (Ajaiyeoba *et al.*, 2006).

Use of the same species for the same use is one demonstration of the importance of a species, and in this particular case indirect evidence that the species may be effective against malaria. It is also evident that little research has been done to validate plant materials used to treat malaria. There is a need, therefore, to undertake scientific studies to build up scientific evidence for these species with positive implications for drug development.

## 13.4 Conclusions and recommendations

In this review we found information on many plants known to treat malaria in Uganda. Some species are used in several parts of Uganda, as well as other parts of the world, and this provides initial evidence that these plants may be efficacious for the treatment of malaria. However, for the foreseeable future, allopathic medicines are likely to remain the mainstay of malaria treatment in Uganda and other developing countries. Artemisinins have $IC_{50}$ values in the nanomolar range, representing much greater potency than what is reported for the Ugandan medicinal plants (Basco and Le Bras, 1993). In a Ugandan clinical trial, very high levels of effectiveness were demonstrated (96% cure rate; Achan *et al.*, 2009). With the incorporation of ACTs into national treatment programmes and their free distribution to patients diagnosed with malaria, counter-intuitively a significant barrier may exist, to research into medicinal plants for malaria. Although, the limited validation data available from these plants suggest that they are unlikely to be as potent as artemisinins, the combination approach with ACTs suggests that less potent compounds can play an important role in malaria treatment by preventing the emergence of resistant parasites or by synergistic action with existing drugs.

The history of malaria treatment demonstrates that research on medicinal plants can result in the development of highly efficacious, life-saving medicines. It is therefore surprising that so limited published data on medicinal plants are available for malaria treatment in Uganda, where malaria is endemic. Indeed, where these plants are documented, limited information exists with regard to the efficacy or safety of the extracts and more studies are required to confirm their viability as antimalarial agents. We recommend further study of the 12 species that were prioritized in this review. National databases containing information of identity, safety and efficacy of medicinal plants for malaria will facilitate future research in this field.

## References

Abbiw, D.K. (1990) *Useful Plants of Ghana: West African Uses of Wild and Cultivated Plants*, Intermediate Technology Publications and The Royal Botanic Gardens, Kew, London.

Achan, J., Tibenderana, J.K., Kyabayinze, D. *et al.* (2009) Effectiveness of quinine versus artemether–lumefantrine for treating uncomplicated falciparum malaria in Ugandan children: randomised trial. *BMJ*, **339**, b2763.

Ajaiyeoba, E.O., Abiodun, O.O., Falade, M.O. *et al.* (2006) *In vitro* cytotoxicity studies of 20 plants used in Nigerian antimalarial ethnomedicine. *Phytomedicine*, **13**, 295–298.

Ajaiyeoba, E.O., Ogbole, O.O., Abiodun, O.O. *et al.* (2013) Cajachalcone: an antimalarial compound from *Cajanus cajan* leaf extract. *Journal of Parasitology Research*, **2013**, 703781.

Asase, A., Oteng-Yeboah, A.A., Odamtten, G.T. and Simmonds, M.S. (2005) Ethnobotanical study of some Ghanaian anti-malarial plants. *Journal of Ethnopharmacology*, **99** (2), 273–279.

Balick, M.J. (1990) Ethnobotany and the identification of therapeutic agents from the rainforest, in *Bioactive Compounds from Plants* (eds D.J. Chadwick and J. Marsh), John Wiley & Sons, Ltd, Chichester, pp. 22–39.

Basco, L. and Le Bras, J. (1993) *In vitro* activity of artemisinin derivatives against African isolates and clones of *Plasmodium falciparum*. *American Journal of Tropical Medicine and Hygiene*, **49** (3), 301–307.

Betti, J.L. (2004) An ethnobotanical study of medicinal plants among the Baka Pygmies in the Dja Biosphere Reserve, Cameroon. *African Study Monographs*, **25** (1), 1–27.

Chenniappan, K. and Kadarkarai, M. (2010) *In vitro* antimalarial activity of traditionally used Western Ghats plants from India and their interactions with chloroquine against chloroquine-resistant *Plasmodium falciparum*. *Parasitology Research*, **107** (6), 1351–1364.

Duker-Eshun, G., Jaroszewski, J.W., Asomaning, W.A. í (2004) Antiplasmodial constituents of *Cajanus cajan*. *Phytotherapy Research*, **18**, 128–130.

Galabuzi, C. (2008) Conservation and local utilization of key anti-malarial medicinal plants in the Sango Bay area, southern Uganda. MSc Dissertation, Makerere University.

Gansané, A., Sanon, S., Ouattara, L.P. *et al.* (2010) Antiplasmodial activity and toxicity of crude extracts from alternatives parts of plants widely used for the treatment of malaria in Burkina Faso: contribution for their preservation. *Parasitology Research*, **106** (2), 335–340.

Gessler, M.C., Nkunyab, M.H.H., Mwasumbic, L.B. *et al.* (1994) Screening Tanzanian medicinal plants for antimalarial activity. *Acta Tropica*, **56**, 65–77.

Gessler, M.C., Msuya, D.E., Nkunya, M.H. *et al.* (1995) Traditional healers in Tanzania: the treatment of malaria with plant remedies. *Journal of Ethnopharmacology*, **48** (3), 131–144.

Hutchings, A., Scott, A.H., Lewis, G. and Cunningham, A. (eds) (1996) *Zulu Medicinal Plants: An Inventory*, University of Natal Press, Scottsville.

Karunamoorthi, K. and Tsehaye, E. (2012) Ethnomedicinal knowledge, belief and self-reported practice of local inhabitants on traditional antimalarial plants and phytotherapy. *Journal of Ethnopharmacology*, **141**, 143–150.

Lynch, M., Koek, I., Beach, R. *et al.* (2005) *Rapid Assessment Report – Uganda*. President's Malaria Initiative, www.fighting-malaria.gov/countries/mops/assessments/uganda_assessment.pdf (accessed 12 March 2013).

Magadula, J.J. and Erasto, P. (2009) Bioactive natural products derived from the East African flora. *Natural Product Reports*, **26**, 1535–1554.

MOH (2005) *National Policy on Malaria Treatment*, Malaria Control Programme, Ministry of Health Kampala, Uganda.

Nguta, J.M., Mbaria, J.M., Gakuya, D.W. *et al.* (2010) Antimalarial herbal remedies of Msambweni, Kenya. *Journal of Ethnopharmacology*, **128**, 424–432.

Njoroge, G.N. and Bussmann, R.W. (2006) Diversity and utilization of antimalarial ethnophytotherapeutic remedies among the Kikuyus (central Kenya). *Journal of Ethnobiology and Ethnomedicine*, **2**, 8.

Ohigashi, H., Huffman, M.A., Izutsu, D. *et al.* (1994) Toward the chemical ecology of medicinal plant use in chimpanzees: the case of *Vernonia amygdalina*, a plant used by wild chimpanzees possibly for parasite-related diseases. *Journal of Chemical Ecology*, **20**, 541–553.

Oryema, C., Ziraba, R.B., Omagor, N. and Opio, A. (2010) Medicinal plants of Erute county, Lira district, Uganda with particular reference to their conservation. *African Journal of Ecology*, **48**, 285–298.

Phyo, A.P., Nkhoma, S., Stepniewska, K. *et al.* (2012) Emergence of artemisinin-resistant malaria on the western border of Thailand: a longitudinal study. *Lancet*, **379** (9830), 1960–1966.

Prance, G.T. (1991) What is ethnobotany today? *Journal of Ethnopharmacology*, **32**, 209–216.

Sachs, J. and Malaney, P. (2002) The economic and social burden of malaria. *Nature*, **415**, 680–685.

Sofowora, A. (1993) *Medicinal Plants and Traditional Medicine in Africa*, Spectrum Books Ltd, Ibadan, Nigeria.

Ssegawa, P. and Kasenene, J.M. (2007) Plants for malaria treatment in southern Uganda: traditional use, preference and ecological viability. *Journal of Ethnobiology*, **27** (1), 110–131.

Stangeland, T., Alele, P.E., Katuura, E. and Lye, K.A. (2011) Plants used to treat malaria in Nyakayojo sub-county, western Uganda. *Journal of Ethnopharmacology*, **137**, 154–166.

Tabuti, J.R.S. (2008) Herbal medicines used in the treatment of malaria in Budiope county, Uganda. *Journal of Ethnopharmacology*, **116** (1), 33–42.

Tabuti, J.R.S., Dhillion, S.S. and Lye, K.A. (2003) Traditional herbal drugs of Bulamogi, Uganda: plants, use and administration. *Journal of Ethnopharmacology*, **88**, 19–44.

Tona, L., Cimanga, R.K., Mesia, K. *et al.* (2004) *In vitro* antiplasmodial activity of extracts and fractions from seven medicinal plants used in the Democratic Republic of Congo. *Journal of Ethnopharmacology*, **93**, 27–32.

Tor-anyiin, T.A., Sha'ato, R. and Oluma, H.O.A. (2003) Ethnobotanical survey of anti-malarial medicinal plants among the Tiv people of Nigeria. *Journal of Herbs, Spices & Medicinal Plants*, **10** (3), 61–74.

Waako, P.J., Gumede, B., Smith, P. and Folb, P.I. (2005) The *in vitro* and *in vivo* antimalarial activity of *Cardiospermum halicacabum* L. and *Momordica foetida* Schumch. Et Thonn. *Journal of Ethnopharmacology*, **99**, 137–143.

WHO (2008) *World Malaria Report 2008*, World Health Organization, Geneva.

WHO (2010) *World Malaria Report 2010*, World Health Organization, Global Malaria Programme, Geneva.

Willcox, M.L. and Bodeker, G. (2004) Traditional herbal medicines for malaria. *BMJ*, **329**, 1156–1159.

# 14 Multiple Anti-Infective Properties of Selected Plant Species from Zimbabwe

Rumbidzai Mangoyi, Tariro Chitemerere, Theresa Chimponda, Elaine Chirisa and Stanley Mukanganyama

Biomolecular Interactions Analyses Group, Department of Biochemistry, University of Zimbabwe, Mt. Pleasant, Harare, Zimbabwe

## 14.1 Introduction

Natural products have been described as the birth place of antibiotics, and with the emergence of drug resistance in pathogens there is the need for new classes of these drugs (Upadhyay *et al.*, 2010). There is also a need for identifying new resistance factors as microbes evolve new resistance mechanisms. Infectious diseases account for 50% of all deaths in tropical countries. There have been epidemics due to drug-resistant microorganisms; therefore, these microbes pose enormous public health concerns (Gupta *et al.*, 2009). Plants are good sources of phytomedicines as they offer great promise in the treatment of intractable infectious diseases, including opportunistic AIDS anti-infective agents (Buwa and Afolayan, 2009). Since time immemorial, mankind has always sought to extract subsistence and medicines from their environment. It is the accumulation of this experience that constitutes the basis of therapeutics today. Plants have served as a source of new pharmaceutical products and inexpensive starting materials for the synthesis of many known drugs. Natural products and their derivatives represent more than 50% of drugs in clinical use in the world (Cowan, 1999).

Traditional medicine refers to the knowledge, skills and practices based on the theories, beliefs and experiences indigenous to different cultures, used in the maintenance of health and in the prevention, diagnosis, improvement or treatment of physical and mental illness (WHO, 2002). According to WHO (2002), traditional medicine continues to provide health coverage for over 80% of the world's population, especially in the developing world. This can be attributed mainly due to its affordability, availability and accessibility. Considering the great number of chemicals that have been derived from plants as medicine, scientific evaluation of plants used traditionally for the treatment of microbial infection is a logical step of exploiting the medicinal compounds that may be present in plants. Plant-based antimicrobials represent a vast untapped source of medicines with enormous therapeutic potential (Cowan, 1999).

It has been shown that there is now an increase in antibiotic resistance in nosocomial and community-acquired infections (Ozhak-Baysan *et al.*, 2012). Over the years, prolonged selective pressure by different drugs has resulted in organisms bearing additional kinds of resistance mechanisms that led to multidrug resistance. These mechanisms include novel penicillin-binding proteins, enzymatic mechanisms of drug inactivation and modification, mutated drug targets, enhanced efflux pump expression, and altered membrane permeability (Alekshun and Levy, 2007) towards hydrophilic compounds. Enhanced drug efflux has now been shown to occur in bacterial, mycobacterial, fungal and even human cancer cells. Once identified, one of the strategies is to counter efflux-mediated resistance (Poole and Lomovskaya, 2006). Therefore, renewed strategies on treatment and prevention of drug resistance in infectious organisms are now required. A multi-pronged approach is required that should include prevention (such as vaccination), improved monitoring and the development of new treatments. The development of new treatments encompasses the use of plants as sources of herbal medicines. The use of plants for drug discovery provides for new drug leads, many of which, if approved, may undergo clinical trials against cancer, bacterial infections, fungal infections and for inflammatory conditions (Conforti *et al.*, 2008; Prasad *et al.*, 2011).

Inflammatory diseases are a major cause of morbidity in the world (Shah *et al.*, 2011). Reactive oxygen species play a crucial role in the inflammatory response and include active species such as nitric oxide (NO), superoxide anion, hydrogen peroxide ($H_2O_2$) and hydroxyl radical, as well as enzymes such as inducible nitric oxide synthase (iNOS), which is important for the production of the free radicals (Jeong *et al.*, 2011). iNOS is highly expressed in macrophages; its activation leads to organ destruction in some inflammatory diseases when the enzyme is improperly regulated (Yoon *et al.*, 2009; Zhao *et al.*, 2009). Oxidative stress, chronic inflammation and infections stimulate iNOS, which consequently enhances the generation of NO (El-Desouky and Gamal-Eldeen, 2009). Pro-inflammatory genes are activated via signal transduction pathways that lead to the production of pro-inflammatory parameters such as NO (Wu, 2012). The activities of pro-inflammatory mediators and cytokines that include inflammatory iNOS and cyclooxygenase-2 are regulated by NF-κβ (Khan *et al.*, 2011). The potential of a plant to act as an antioxidant can be demonstrated by the reduction of the production of NO in a cell. RAW macrophages are an *in vitro* cell line model system that can be used to study the production of reactive oxygen species and moderation of the process by exogenous compounds.

A number of Zimbabwean plants are used in traditional medicines (Table 14.1) and it is useful to validate these as regards a scientific basis for their use. In this study, we report our findings on various

*Novel Plant Bioresources: Applications in Food, Medicine and Cosmetics*, First Edition. Edited by Ameenah Gurib-Fakim.

**Table 14.1** Zimbabwean medicinal plants evaluated for infective properties

| Plant material (crude extract) | | Part used | Activity/ethnomedicinal information |
|---|---|---|---|
| Scientific | Local | | |
| *Xeroderris stuhlmannii* | Murumanyama | Leaves | Mastitis and backache (Ruffo, 1991) |
| *Parinari curatellifolia* | Muhacha | Roots | Skin rashes, tuberculosis, chronic diarrhoea, herpes zoster, herpes simplex |
| *Parinari curatellifolia* | Muhacha | Leaves | Skin rashes, tuberculosis, chronic diarrhoea, herpes zoster, herpes simplex |
| *Vernonia adoensis* | Musikavakadzi | Leaves | Tuberculosis (Kisangau *et al.*, 2007) |
| *Rhus lancea* | Muchokochiana | Flowers | |
| *Rhus lancea* | Muchokochiana | Leaves | – |
| *Syzigium cordatum* | Mukute | Leaves | Herpes zoster, herpes simplex, skin rashes |
| *Faurea* spp. | Mutsatsati | Leaves | – |
| *Combretum zeyheri* | Muruka, mupembere-kono, muchenja | Leaves | Coughs, diarrhoea, rectal prolapse, snake bites and stomach ache (Ruffo, 1991) |
| *Oxalis lutifolia* | Gungwe, kahungwarara | Leaves | – |
| *Oxalis lutifolia* | Gungwe, kahungwarara | Roots | – |
| *Cossonia natalensis* | Mutobvi, mufenje, mushondya | Leaves | – |
| *Croton* sp. | Gunukira, Mufandemengemugugu, mubvukuta | Leaves | – |
| *Syzigium cordatum* | Mukute | Bark | Herpes zoster, herpes simplex, skin rashes (Chigora *et al.*, 2007) |
| *Aloe ferox* | Gavakava | Leaves | – |
| | Mutsvairachuru | Leaves | – |
| *Catunaregum spinosa* | Murovaduri | Leaves | Intestinal worms, gonorrhoea and syphilis (Ruffo, 1991) |
| *Brachystegia boehmii* | Mupfuti | Leaves | – |
| *Uapaca kirkiana* | Muzhanje | Fruit | – |
| *Syzigium cumini* | Mukoyo | Leaves | – |
| *Bidens pilosa* | Tsine | Leaves | Anti-inflamatory, anti-rheumatic (Wang *et al.*, 2003); wounds and relapsing fevers in children, oral candidiasis (Ruffo, 1991) |
| *Lippia javanica* | Zimbani | Leaves | – |
| *Salons delagoense* | Nhundurwa | Fruit | Scabies (Chigora *et al.*, 2007) |
| *Abelmoschus esculentus* | Derere | Fruit | – |
| *Mangifera indica* | Mumango | Leaves | Astringent, gonorrhoea, asthma, prolongs ejaculation, anthelmintic (Kadavul and Dixit, 2009) |
| *Prunus cerasoides* | – | Leaves | – |
| *Asclepias fruticosa* | – | Leaves | – |
| *Lantana camara* | Mbarambati | Leaves | Ring worm infections |
| *Zyziphus mucronata* | Muchecheni | Leaves | Snake bites and stomach ache (Ruffo, 1991) |
| *Garcinia huillensis* | Mutunduru | Leaves | Treatment of cryptococcal meningitis |
| *Flacourtia indica* | Munhunguru | Leaves | Coughs, snake bites, infertility and stomach ache (Ruffo, 1991) |
| *Gymnosporia senegalensis* | Chizhuzhu, musosawafa | Leaves | – |
| *Conyza sumatrensis* | – | leaves | Puff adder (*Bitis arietans*) bites and stomach ache |
| *Solanum mauritianum* | – | Leaves | – |
| *Cajanus cajan* (Druce) | | N8E7 | Stomach ailments (Iwawela *et al.*, 2007) |
| *Callistemon citrinus* (Curtis Skeels) | | UZ2E7 | Haemorrhoid treatment (Oyedeji *et al.*, 2009) |
| *Terminalia pruniodes* (Lawson) | Mudziyashe | N6E7 | Diarrhoea (Ruffo *et al.*, 1991) |
| *Aloe barbadensis* (Mill.) | Gavakava | N11E7 | Sap is used to treat skin rashes and the leaves are prepared and used to treat tuberculosis (Gelfand *et al.*, 1985) |
| *Combretum apiculatum* (L) | Muruka | C9E7 | Coughs, diarrhoea, snake bites stomach ache (Ruffo, 1991) |
| *Cussonia natalensis* (Sond) | Mutobvi | UZ9E7 | Diarrhoea (Ruffo, 1991) |
| *Croton gratissimus* (Burch) | Gunukira | UZ13E7 | Malaria, rabies, gonorrhoea, wounds, ascariasis, internal worms (Iwawela *et al.*, 2007) |
| *Euphorbia tiraculli* (L) | | N10E7 | Removal of benign moles using the latex (Iwawela *et al.*, 2007), rabies treatment |
| *Parinari curatellifolia* (Planch ex Benth) | Muhacha | C6E7 | Facilitates conception in women (Chigora *et al.*, 2007) |
| *Syzigium guineense* (Will D.C) | Mukute | C12E7 | Tuberculosis, fevers (Chigora *et al.*, 2007) |
| *Rhus lancea* (Barkely) | Muchokochiana | C11E7 | Stomach ailments, fevers (Gelfand *et al.*, 1985) |
| *Xeroderris stuhlmanni* (Mend) | Murumanyama | C4E7 | Stomach ailments (Iwawela *et al.*, 2007) |

*Source*: Chitemerere and Mukanganyama (2011). Reproduced with permission of Global Science Books.
Key: E7 – leaves; N – Norton samples; C – centenary samples.

plants from Zimbabwe that we have found to have significant anti-infective medicinal properties, which include antiproliferative activity (*Combretum platypetalum*, *Parinari curatellifollia*), antimycobacterial activity (*Parinari curatellifolia*, *Combretum imberbe*, *C. platypetalum*), antifungal activity (*Combretum zeyheri*, *C. imberbe*, *Combretum apiculatum*), antibacterial (*Callistemon citrinus*, *Vernonia adoensis*, *C. zeyheri*, *Mangifera indica*) and anti-inflammatory activity (*C. platypetalum*, *P. curatellifolia*).

## 14.2 Preparation of plant extracts

Plant samples were collected from three provinces of Zimbabwe: Mashonaland West Province (Norton), Mashonaland Central (Centenary) and Harare (University of Zimbabwe). The plant samples were authenticated by Mr Christopher Chapano, a taxonomist at the National Herbarium and Botanic Gardens (Harare, Zimbabwe). Roots, leaves, flowering parts and stems of plants were separated from the plant, dried at 50 °C in an oven (Memmert, Schwabach, SRG Germany) and then ground to powder using a two-speed blender (Cole Parmer Instrument Company, Connecticut, USA). Powdered samples were then extracted using different solvents: distilled water, methanol, ethanol and acetone.

**Table 14.2** Minimum inhibitory concentrations and minimum fungicidal concentrations of extracts towards the *Candida* species

| Plant species (crude extract) | | *Candida albicans* | *Candida krusei* |
|---|---|---|---|
| *Combretum zeyheri* (leaves) | Zone of inhibition[a] | 18.5 ± 0.7 mm | 18 ± 0.2 mm |
| | MIC | 0.08 mg/mL | 0.16 mg/mL |
| | MFC | 0.31 mg/mL | 0.31 mg/mL |
| *Combretum molle* (leaves) | Zone of inhibition | 17 ± 0.1 mm | 15 ± 0.3 mm |
| | MIC | 0.31 mg/mL | 0.31 mg/mL |
| | MFC | 0.63 mg/mL | 1.25 mg/mL |
| *Cussonia natalensis* (leaves) | Zone of inhibition | 16 ± 1.4 mm | – |
| | MIC | 0.31 mg/mL | |
| | MFC | 1.25 mg/mL | |
| *Syzigium cordatum* (leaves) | Zone of inhibition | 15 ± 0.1 mm | 12 ± 0.1 mm |
| | MIC | 0.63 mg/mL | 0.63 mg/mL |
| | MFC | 1.25 mg/mL | 2.5 mg/mL |
| *Syzigium cordatum* (bark) | Zone of inhibition | 15 ± 0.1 mm | 12 ± 0.6 mm |
| | MIC | 0.63 mg/mL | 0.63 mg/mL |
| | MFC | 1.25 mg/mL | 2.5 mg/mL |
| Positive control (miconazole) | Zone of inhibition | 20 ± 0.8 mm | 22.5 ± 0.7 mm |
| | MIC | 0.31 mg/mL | 0.63 mg/mL |
| | MFC | 0.31 mg/mL | 0.63 mg/mL |
| Negative control (DMSO) | Zone of inhibition | 6 mm | 6 mm |

*Source*: Mangoyi and Mukanganyama (2011). Reproduced with permission of Global Science Books.
MIC: minimum inhibition concentration; MFC: minimum fungicidal concentration; DMSO: dimethylsulfoxide.
[a]Results are the average (±SD) of two separate antifungal susceptibility tests (each antifungal susceptibility test was followed by a disk diffusion assay done in quadruplicate).

### 14.2.1 Antifungal effects

The antifungal activity of ethanol extracts of 38 selected plants were investigated against a clinical strain of *Candida krusei* and *Candida albicans* strain ATCC 10231 using the agar disc diffusion method. Briefly, the test material equivalent to 500 µg, dissolved in appropriate solvent, was applied on sterile paper discs (6 mm diameter, cartridge susceptibility discs, Mast Diagnostics, Mast Group Ltd, Merseyside, UK). The solvent was allowed to evaporate from filters deposited on 96-well plates at room temperature. The discs were then deposited on the surface of the inoculated agar plates. Plates were incubated at 4 °C for 2 h and then at 37 °C in a Labcon incubator (England) overnight. The inhibition zone around each of the discs was measured and recorded. Inhibition zone was measured as the diameter of inhibition of growth around each of the disc plus the diameter of the disc. Of these 38 tested samples, 19 plant extracts showed profound antifungal

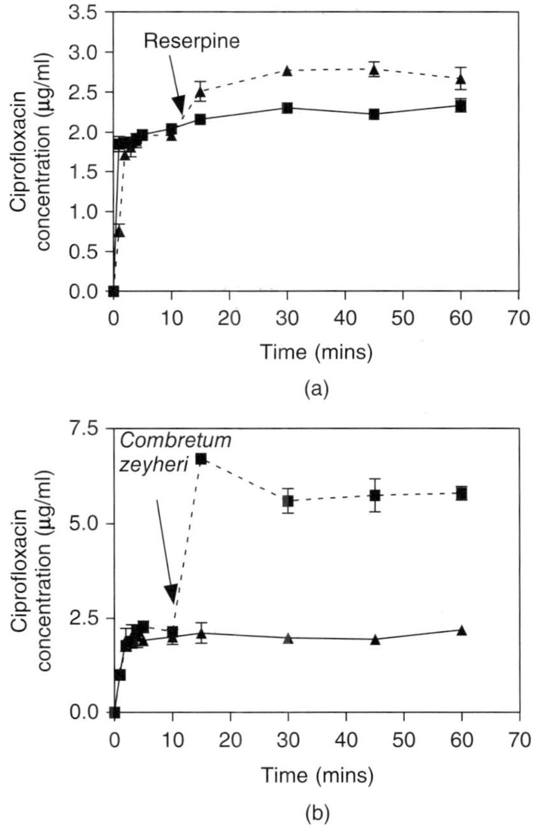


**Figure 14.1** Inhibition of drug efflux pumps in *Candida albicans* by (a) a standard influx pump inhibitor, reserpine, and (b) crude ethanolic extracts of *Combretum zeyheri*. The tubes containing either *C. albicans* cells and ciprofloxacin (20 mg/L) (control) or *C. albicans* cells, ciprofloxacin and the plant extract (100 µmol/L) (test sample) were incubated at 37 °C. The active efflux of ciprofloxacin was then examined at time intervals in all tubes. The arrow points at the time when the plant extract was added to the cells incubated in the presence of ciprofloxacin. *Source*: Mangoyi and Mukanganyama (2011). Reproduced with permission of Global Science Books

effects when compared with the positive control, miconazole. The minimum inhibition concentration (MIC) was determined for plant extract that showed high efficacy against the tested microorganisms using 3-(4,5-dimethylthiazol-2-yl)-2,5-diphenyltetrazolium bromide assay. The minimum fungicidal concentration (MFC) of the extract was determined on respective solid media. The MFC assay showed that 0.31 mg/mL of *C. zeyheri* was able to completely kill both *C. albicans* and *C. krusei* (Table 14.2).

The ciprofloxacin accumulation assay was carried out to investigate if these plant extracts could inhibit the activity of active drug efflux pumps in both *Candida* species. To determine the effect of plant extracts on fungal efflux pumps, accumulation of ciprofloxacin was measured by the method of Peiyuan *et al.* (2006) with modifications. The active efflux of ciprofloxacin was determined at the excitation and emission wavelengths of 270 nm and 452 nm respectively using an RF-1501 Shimadzu spectrofluorimeter (Shimadzu Corporation, Japan). The active efflux of ciprofloxacin was examined as well in the presence of reserpine, a standard drug efflux pump inhibitor. *C. zeyheri*, which had the most potent antifungal effects, also inhibited the *Candida* efflux pumps (Figure 14.1).

It is well known that fungal infections, especially systemic ones, are difficult to diagnose, and even after the diagnosis there are fewer classes of antifungal drugs to use compared with antibacterial drugs (Cannon *et al.*, 2009). The results of this study provide an ethnopharmacological basis for the use of plant extracts for infectious diseases, especially against fungal infections. Not only are these plant extracts able to inhibit fungal growth, but they are also able to enhance the accumulation of drugs inside *Candida*. This activity may contribute to the suppression of drug resistance to existing antifungals. Therefore, the antifungal activity of these plants may be due to direct effects on growth as well as indirect effects to do with inhibition of the efflux of compounds from these cells. This study has also identified *C. zeyheri* plant as a plant with potential antifungal activity and partial isolation of active agents has shown that it can be used as a source for the isolation of active compounds that may serve as lead compounds in antifungal drug development (Mangoyi *et al.*, 2012). These results also confirm the potential of plants used by traditional healers in Zimbabwe as a source of bioactive compounds.

### 14.2.2 Antibacterial effects

A total of 19 plants of different families were selected to assess their antibacterial potential against five bacterial species: *Staphylococcus aureus* (ATCC 9144), *Pseudomonas aeruginosa* (ATCC 27853), *Escherichia coli* (ATCC 11229), *Bacillus cereus* (ATCC 11778) and *Bacillus subtilis* (ATCC 6633) using the disc diffusion method. Thirteen extracts exhibited antibacterial activity against the test microorganisms. Extracts were not strain specific and showed antibacterial activity for all five bacterial species with the exceptions from members of the Asteraceae family that showed no activity against all test bacteria except *B. subtilis*. The five most potent extracts showing considerably good antibacterial activity for each test organism were

**Table 14.3** A summary table for the minimum inhibitory concentration (MIC) and minimum bactericidal concentration (MBC) assays

| Organism | G+/G− | Plant species | MIC (μg/mL) | MBC (μg/mL) | MIC amp | MBC amp |
|---|---|---|---|---|---|---|
| *Escherichia coli* | G− | *Callistemon citrinus* | 125 | 250 | 8 | 16 |
| | | *Vernonia adoensis* | 188 | 500 | | |
| | | *Parinari curatellifolia* (stems) | 188 | 500 | | |
| | | *Parinari curatellifolia* | 300 | 500 | | |
| | | *Mangifera indica* | 47 | 250 | | |
| *Pseudomonas aeruginosa* | G− | *Vernonia adoensis* | 188 | 1000 | 16 | 125 |
| | | *Callistemon citrinus* | 63 | 250 | | |
| | | *Parinari curatellifolia* (stems) | ND | >1000 | | |
| | | *Parinari curatellifolia* | 188 | 500 | | |
| | | *Mangifera indica* | 46 | 250 | | |
| *Staphylococcus aureus* | G+ | *Vernonia adoensis* | 94 | 1000 | <2 | 8 |
| | | *Callistemon citrinus* | 47 | 250 | | |
| | | *Parinari curatellifolia* | 375 | 500 | | |
| | | *Parinari curatellifolia* (stems) | 250 | 500 | | |
| | | *Faurea* sp | ND | >1000 | | |
| *Bacillus subtilis* | G+ | *Callistemon citrinus* | 24 | 63 | 4 | 8 |
| | | Asteraceae family (flowers) | 292 | 1000 | | |
| | | *Vernonia adoensis* | 188 | 1000 | | |
| | | *Faurea* sp | 500 | >1000 | | |
| | | *Mangifera indica* | 40 | 125 | | |
| *Bacillus cereus* | G+ | *Vernonia adoensis* | 188 | 1000 | 4 | 8 |
| | | *Callistemon citrinus* | 63 | 125 | | |
| | | *Parinari curatellifolia* (stems) | 179 | 500 | | |
| | | *Lantana camara* | ND | >1000 | | |
| | | *Parinari curatellifolia* | 375 | >1000 | | |

*Source*: Chitemerere and Mukanganyama (2011). Reproduced with permission of Global Science Books.
ND: no activity detected.

selected to determine MIC values. From the MIC values obtained (0.06–0.5 mg/mL), extracts were shown to be potent bacterial growth inhibitors (Table 14.3). However, ampicillin proved to be a more potent antibacterial agent than all the extracts, with MIC values ranging from 0.002 to 0.008 mg/mL. The effect of plant extracts on ABC drug efflux pumps in bacteria was investigated by determining the concentration of rhodamine (R6G) that accumulated in the bacteria in the presence of plant extracts. Absorbance of R6G was measured at 527 nm using a Shimadzu 1601 UV spectrophotometer (Shimadzu, Tokyo, Japan).

Figure 14.2 shows the effects of plant extracts on the efflux of R6G from bacteria from *S. aureus* and *P. aeruginosas*. Reserpine is an antihypertensive plant alkaloid that was first isolated from the roots of *Rauwolfia vomitoria* and is used as a standard inhibitor of efflux pumps. All the test bacteria in this study responded to the effects of reserpine by increasing R6G influx by 70–90%. In most cases the plant extracts used in this assay were more effective in blocking drug efflux pumps than reserpine was (Figure 14.2).

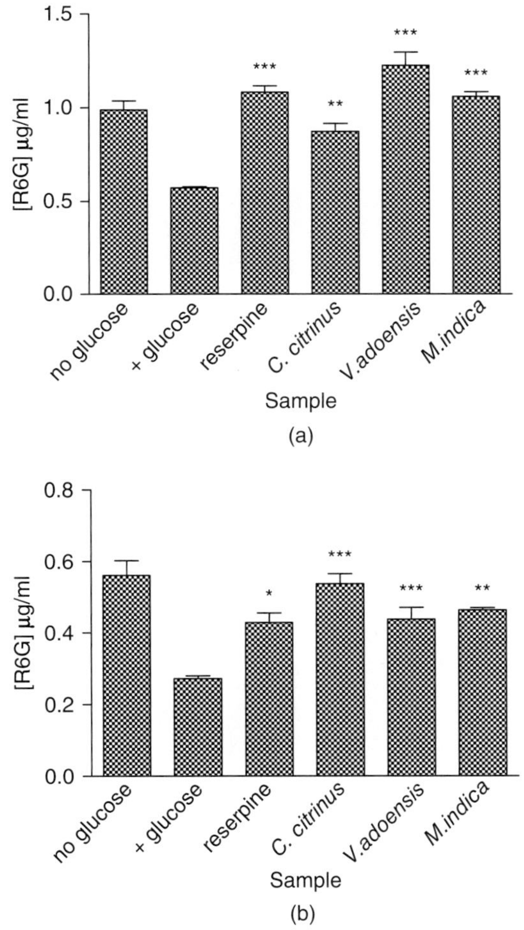


**Figure 14.2** The accumulation of R6G in bacterial cells over time. The graphs are a plot of R6G concentration inside the cell after 30 min of incubation against the sample used (plant extract, ampicillin or reserpine). Glucose was used to provide energy for the efflux pumps in the form of ATP. (a) *Staphylococcus aureus*; (b) *Pseudomonas aeruginosa*. Error bar denotes the standard deviation from mean ($n = 4$). The asterisks indicate statistical significant differences with the control (+glucose) ($p < 0.05$). *Source*: Chitemerere and Mukanganyama (2011). Reproduced with permission of Global Science Books

From this study, *M. indica*, *C. citrinus* and *V. adoensis* extracts were a potential source of compounds that had efflux pump inhibitory activity. This was shown by an increase in the accumulation of R6G within the cell in the presence of these extracts. These plant extracts were more effective in blocking drug efflux pumps than reserpine was. In another study, *M. indica* was found to contain alkaloids and glycosides, giving it great pharmacological importance (Nwinuka *et al.*, 2008). Although the antibacterial activities of *V. adoensis*, *M. indica* and *C. citrinus* observed in this study confirm the findings of other previous similar studies, it is worth noting that these extracts also showed activity as efflux pump inhibitors, an interesting finding that has not been reported in previous studies investigating the antibacterial activity of these three extracts (Chitemerere and Mukanganyama, 2011). These extracts, hence, have a promising future for the development of effective efflux pump inhibitors, which

**Table 14.4** Minimum inhibitory concentrations and minimum bactericidal concentrations of extracts towards the mycobacterial species

| Plant extract | [a]Zone of inhibition (mm) at 500 µg/disc<br>[b]MIC (µg/disc)<br>[c]MBC (µg/disc)<br>*M. aurum* | Plant extract | Zone of inhibition (mm) at 500 µg/disc<br>MIC (µg/disc)<br>MBC (µg/disc)<br>*C. glutamicum* |
|---|---|---|---|
| *Vernonia adoensis* leaves | [a]28 ± 1<br>[b]31<br>[c]250 | *Mangifera indica* (leaves) | 17 ± 1<br>250<br>>500 |
| *Faurea* sp. leaves | 18 ± 1<br>250<br>>500 | *Vernonia adoensis* leaves<br>roots | 15 ± 1<br>–<br>250<br>>500 |
| *Syzigium cumini* leaves | 19 ± 2<br>–<br>– | *Parinari curatellifolia* stems | 13 ± 1<br>125<br>>500 |
| *Xerroderris stuhlmannii* leaves | 18 ± 1<br>31<br>125 | *Zyziphus mucronata* stems | 13 ± 1<br>125<br>500 |
| *Parinari curatellifolia* stems | 18 ± 1<br>8<br>63 | *Lippia javanica* roots | 12 ± 1<br>32<br>>500 |
| *Parinari curatellifolia* leaves | 18 ± 1<br>63<br>125 | | |
| Rifampicin | 29 ± 2<br>2<br>16 | Rifampicin | 40 ± 1<br><1<br>63 |

*Source*: Chimponda and Mukanganyama (2010). Reproduced with permission of Malaysian Society of Parasitology and Tropical Medicine. [a]Results are the average (±SD) of two separate antibacterial susceptibility tests (each antibacterial susceptibility test was followed by a disk diffusion assay done in quadruplicate). The zone of inhibition being determined at a concentration of 500 µg/disc. [b]MIC: minimum inhibition concentration; [c]MBC: minimum bactericidal concentration.

would augment the antibacterial activities of standard antibiotics (Van Bambeke *et al.*, 2010).

### 14.2.3 Antimycobacterial effects

The antimycobacterial activities of 19 Zimbabwean medicinal plant extracts against two model non-pathogenic actinomycetales species, *Corynebacterium glutamicum* (ATCC 13022) and *Mycobacterium aurum* ($A^+$), were investigated using the disc diffusion assay. *V. adoensis* had the highest zone of inhibition ($28 \pm 1$ mm), which was comparable to that of rifampicin (positive control) ($29 \pm 2$ mm) against *M. aurum*, while *M. indica* was the most potent against *C. glutamicum* (Table 14.4). The MIC and MBC were determined using the agar disc diffusion method. *P. curatellifolia* had the lowest MIC of 8 μg/disc, while rifampicin had an MIC of 2 μg/disc when using *M. aurum* as the test organism. *Lippia javanica* had the lowest MIC of 12 μg/disc when using *C. glutamicum*, while rifampicin had a value that was less than 1 μg/disc.

The plant extract and rifampicin were combined and the effects assessed on the growth of the bacteria. *P. curatellifolia*, *Xerroderris stuhlmannii* (Figure 14.3) and *Ziziphus mucronata* showed an antagonistic activity between the plant extract and rifampicin. *L. javanica* had no effect on the activity of rifampicin. Lastly, the ciprofloxacin accumulation assay was carried out. Overexpression of ATP binding cassette (ABC) transporters has been proposed as a major mechanism contributing to the innate drug resistance in *Mycobacterium tuberculosis* and related mycobacteria. The extract from *P. curatellifolia* had the highest drug efflux pump inhibitory effect against both *C. glutamicum* and *M. aurum*. *V. adoensis* and *Faurea saligna* leaf extracts also inhibited the action of the efflux pumps (Figure 14.4).

The intrinsic resistance *of M. tuberculosis* to drugs has been linked to reduced entry of the drugs into the cell (De Rossi *et al.*, 2006; Nguyen and Thompson, 2006). Since its discovery in 1882, *M. tuberculosis* infections have remained a major global problem (Mohamad *et al.*, 2011). The plant extracts in this study were not only able to inhibit mycobacterial growth of *C. glutamicum* and

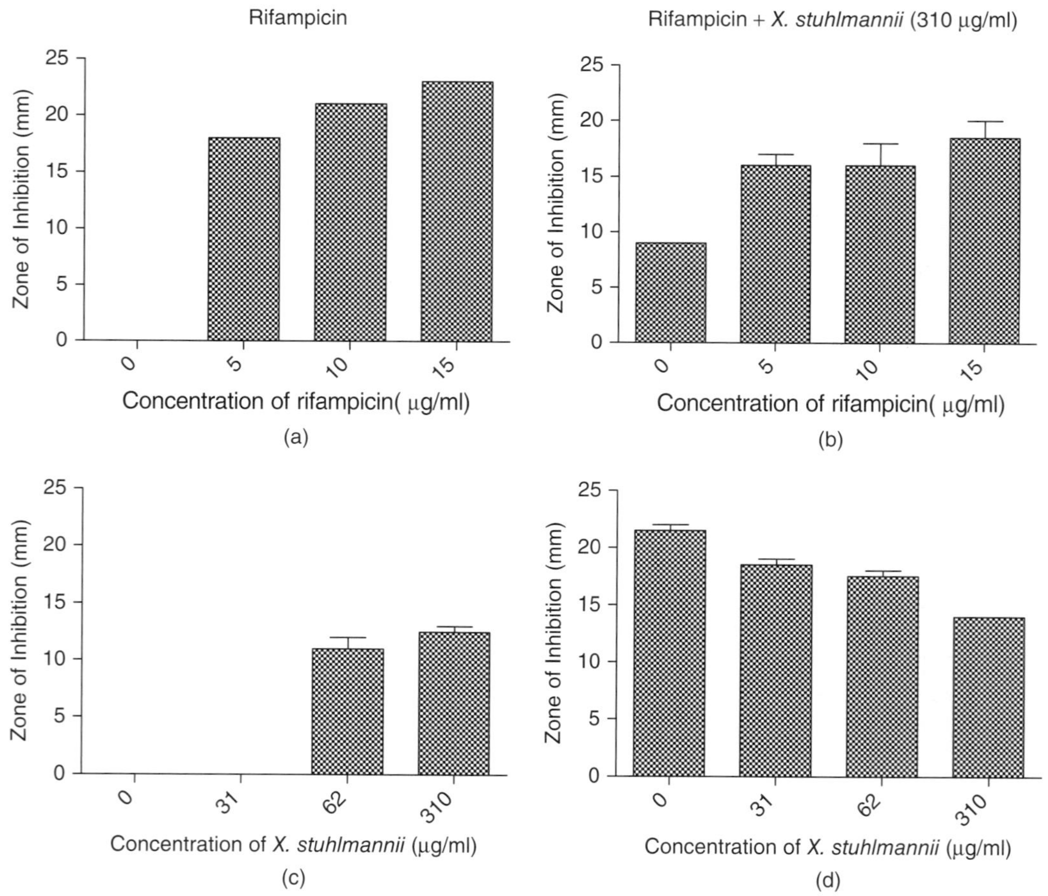


**Figure 14.3** The effects of combining rifampicin and the extract from *Xerroderris stuhlmannii* on the growth of *Mycobacterium aurum*. Rifampicin and the extract were kept constant (a, c) or varied (b, d). The combined effects of the drug and extract were assessed on the growth inhibition. Data are the mean +SD for $n = 4$. ** Values differ significantly ($P < 0.01$). *Source*: Chimponda and Mukanganyama (2010). Reproduced with permission of Malaysian Society of Parasitology and Tropical Medicine

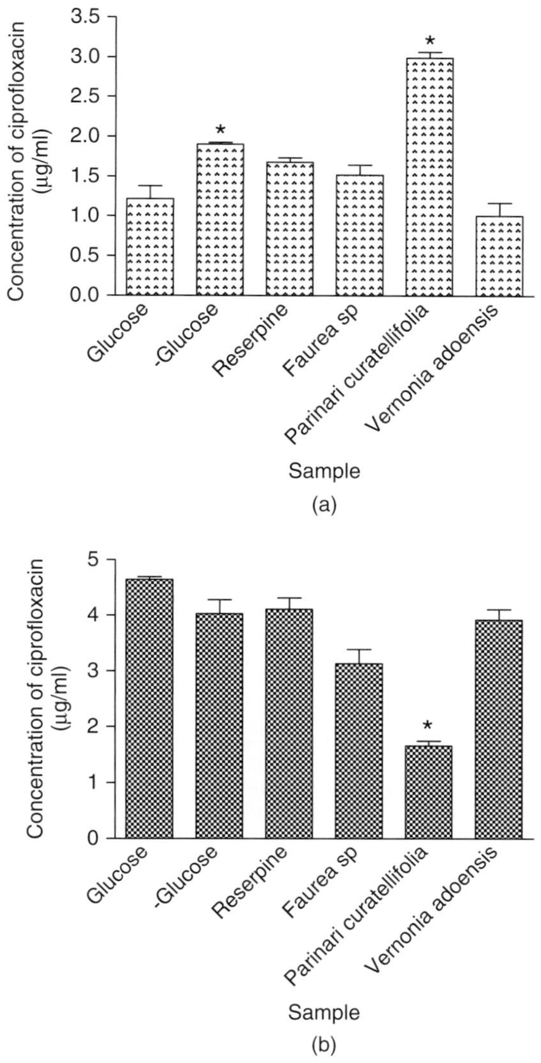


**Figure 14.4** The effects of the plant extracts on the accumulation and efflux of ciprofloxacin in *Mycobacterium aurum*. Cells were loaded with ciprofloxacin and then incubated with the plant extract. The accumulation (a) was measured by determining the fluorescence of ciprofloxacin in the lysed cells. The efflux (b) was determined from the amount of ciprofloxacin in the supernatant of cells before lysis. Experiments were performed in duplicate and each measurement was done twice. Values differ significantly (***P* < 0.01;**P* < 0.05). All values are compared with the value of the sample with glucose. *Source*: Chimponda and Mukanganyama (2010). Reproduced with permission of the Malaysian Society of Parasitology and Tropical Medicine

*M. aurum*, but were also able to enhance the accumulation of drugs inside the bacteria (Chimponda and Mukanganyama, 2010). This activity may contribute to the suppression of drug resistance to the existing antimycobacterial drugs when used in combination. These results also indicate the potential of plants used by traditional healers in Zimbabwe as a source of possible bioactive antimycobacterial compounds.

### 14.2.4 Antiproliferative effects

The antiproliferative effects of plant extracts were investigated on EUACC strain Jurkat E6 (T cell lymphocytic cell line) and an ATCC strain Wil2 (B cell lymphocytic cell line). The Jurkat cells that were used had a viability of greater than 90% using the trypan blue assay. The effects of extracts from *Combretum* species on growth of Jurkat cells was determined by incubating cells in the presence of extracts and then counting live cells using the haemocytometer. Doxorubcin and 1,3-bis(2-chloroethyl)-1-nitrosourea (BCNU), were used as positive controls in these assays. The results were determined after 24, 48 and 72 h of incubation. *Combretum* species were found to be active in the inhibition of growth of cancer cell lines, with *C. platypetalum* extract showing the most inhibition (Figure 14.5) with a $GI_{50}$ of 26 µg/mL (Table 14.5). Using Wil2 cells, *P. curatellifolia* extract and *Aloe barbadensis* were also shown to have antiproliferative effects (Figure 14.6a and b). *P. curatellifolia* leaf extracts at 50 µg/mL reduced cancer cell proliferation by more than 50% compared with the standard anticancer drug doxorubicin, which reduced proliferation by 25%. Leaf extracts from *Croton gratissimus* and *V. adoensis* similary reduced cell proliferation, but not to the same extent as *P. curatellifolia* and *A. barbadensis* (Figure 14.6c and d).

Therefore, widespread use of *Combretum* species in Africa for different ailments justifies their investigations into other possible biological activities of this group of plants (McGaw *et al.*, 2001). To discover lead compounds against cancer, there is a need to look at ethno-medicinal knowledge as regards toxicity. Using a human leukaemic cancer cell line, Jurkat T cells, we investigated the effects of other *Combretum* species from Zimbabwe and the results from this study (Table 14.5) showed that *C. platypetalum* was the most effective extract to inhibit the proliferation of the cells with a $GI_{50}$ close to that of BCNU, a standard anticancer drug. It is possible that phenanthrenes and stilbenes that have been found in other *Combretum* species are responsible for the antiproliferation activities observed with the species from Zimbabwe. The solvent of extraction was also important, as *Combretum molle* from Mashonaland central province of Zimbabwe extracted with ethanol or methanol produced different effects on Jurkat T cells. The extract from *P. curatellifolia* had a $GI_{50}$ of 93 µg/mL against Wil2 cells after 1 h of incubation (Mukanganyama *et al.*, 2011). Extracts from *P. curatellifolia* have been found to have moderate antiproliferative effects against breast cancer cell lines (Fouche *et al.*, 2009). This property could be attributed to the high levels of ent-kaurene terpenoids that have been isolated from *P. curatellifolia* leaves (Rundle *et al.* 2001). The root bark from *P. curatellifolia* collected from Zimbabwe was found to have cytotoxic ent-kaurene diterpenoids 15-oxozoapatlin, 13-methoxy-15-oxozoapatlin and 13 hydroxy-15-oxozoapatlin that were found to have anticancer effects (Lee *et al.* 1996). *A. barbadensis* was found to have a $GI_{50}$ of 115 µg/mL. *A. barbadensis* is traditionally used for easing labour and abortion as it has anti-inflammatory activity (Iwawela *et al.*, 2007). Aloe emodin is an anthraquinone that

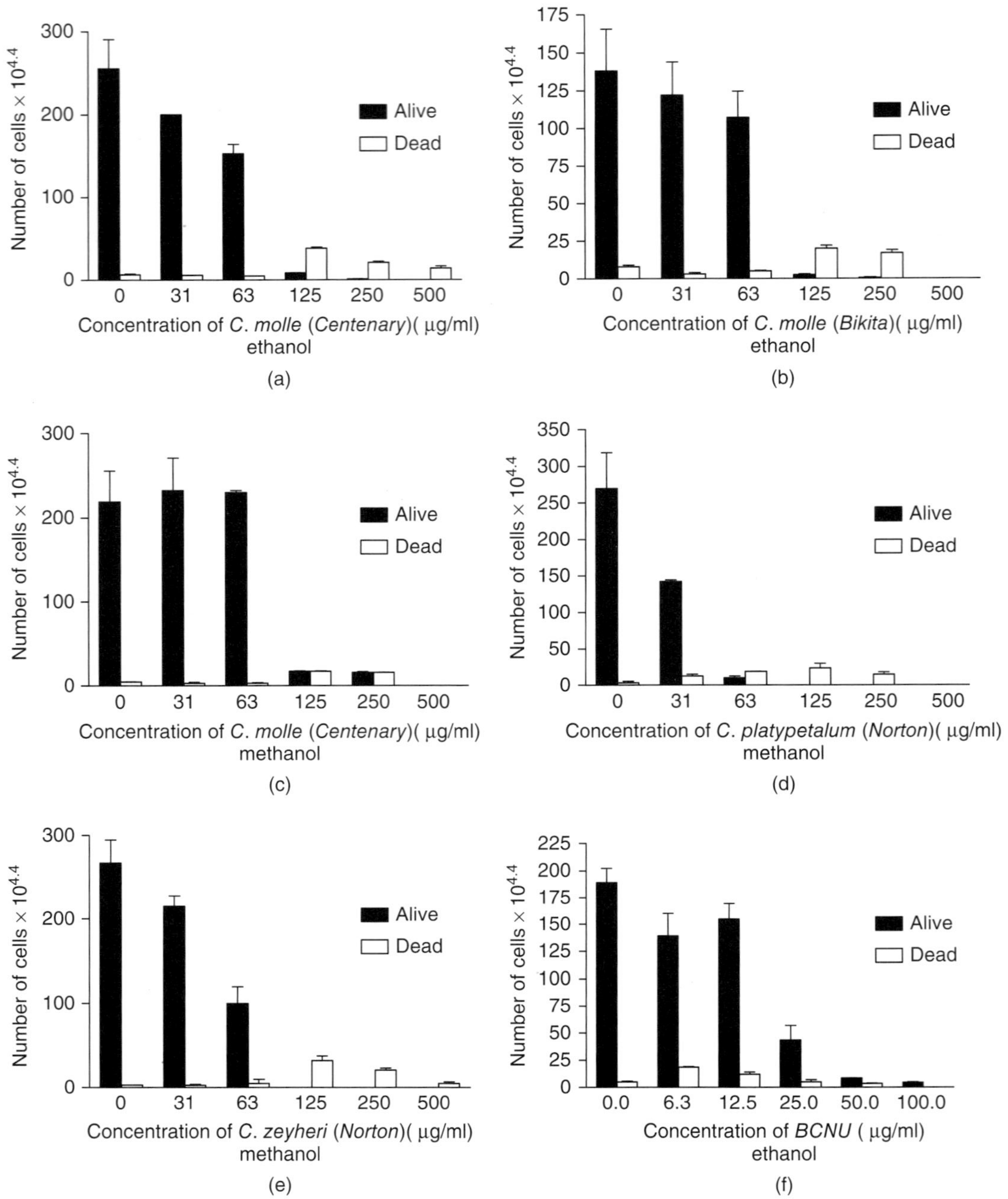


**Figure 14.5** The antiproliferative and cytotoxic effect of extracts from *Combretum* species on Jurkat T cells in comparison with anticancer drug BCNU (f). Number of readings ($n = 2$) and each point represents the mean plus/minus standard deviation

has been isolated from *A. barbadensis* (Ni *et al.*, 2004) and could be involved in the inhibition of proliferation, as these compounds have been reported to bring about cell cycle arrest in cervical cancer cells (Ni *et al.*, 2004).

### 14.2.5 Anti-inflammatory effects

The anti-inflammatory effects of plant extracts were investigated on RAW cells, and this was done by determining the effect of the plant extracts on the production of NO by cells in the presence of menadione or $H_2O_2$. The reaction and standard curve samples were read in a SpectraMax® microplate reader (Molecular Devices, Sunnyvale, USA) at 540 nm. Menadione was shown to increase the nitrite concentration when incubated with RAW 264.7 cells compared with other agents that can cause oxidative stress in the cell (Figure 14.7). So the RAW 264.7 cells when incubated would act as a model system for the mimicking of oxidative stress in the body. Combination of menadione and the *C. platypetalum* ethanol extract led to an increase in the nitrite concentration when compared with the baseline and with the concentration of menadione alone (Figure 14.8a). The nitrite concentration increased when *C. platypetalum* and *C. zeyheri* ethanol extracts were combined with

**Table 14.5** Differential inhibition of Jurkat T cell growth by plant leaf extracts from different geographical locations of Zimbabwe

| Plant species | Province of Zimbabwe | Solvent | $GI_{50}$ (μg/mL)[a] | $GI_{90}$ (μg/mL)[a] |
|---|---|---|---|---|
| *Combretum molle* | Mashonaland Central (Centenary) | Ethanol | 64 | 175 |
| *Combretum molle* | Masvingo (Bikita) | Ethanol | 81 | 205 |
| *Combretum molle* | Mashonaland Central (Centenary) | Methanol | 125 | 279 |
| *Combretum platypetalum* | Mashonaland West (Norton) | Methanol | 26 | 81 |
| *Combretum zeyheri* | Mashonaland West (Norton) | Methanol | 50 | 144 |
| BCNU | | Ethanol | 18 | 51 |

[a] $GI_{50}$ and $GI_{90}$ are concentrations of the extract or drug that causes 50% and 90% inhibition of growth of the cancer cells respectively.

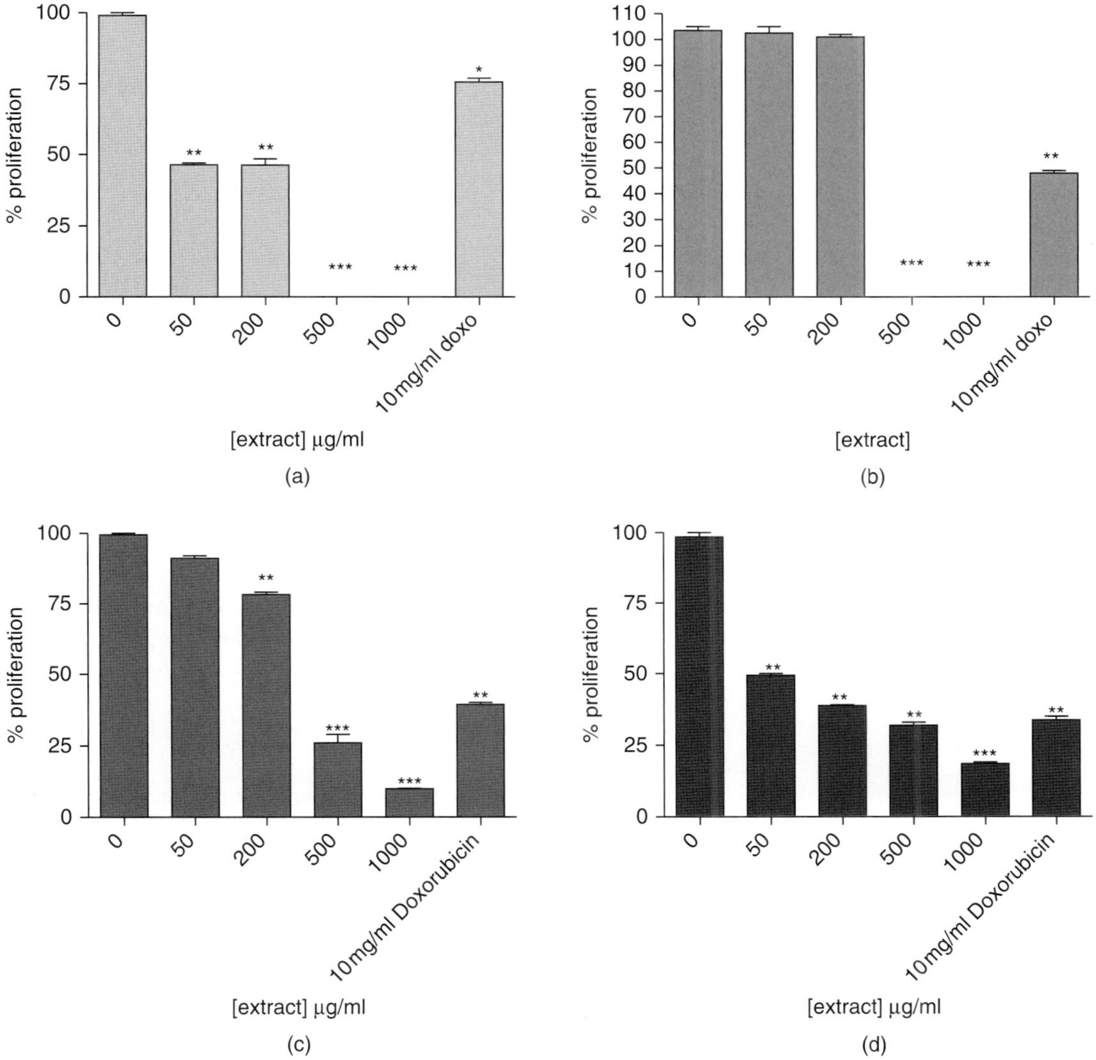


**Figure 14.6** The antiproliferative and cytotoxic effect of plant extracts: (a) *Parinari curatellifolia*, (b) *Aloe barbadensis*, (c) *Croton gratissimus* and (d) *Vernonia adoensis* on Wil2 cells in comparison with anticancer drug doxorubicin. $^{*}p < 0.05$, $^{**}p < 0.01$, $^{***}p < 0.001$. Number of readings ($n = 3$) and each point represents the mean plus/minus standard deviation. *Source*: Mukanganyama *et al.* (2012). Reproduced with permission of Global Science Books

$H_2O_2$. However, $H_2O_2$ when combined with *P. curatellifolia* and *C. molle* had similar effects on the nitrite concentration as it decreased in the presence of the extracts (Figure 14.8b).

The increase in NO production observed in our study when menadione was incubated with RAW cells could have been due to the ability of this compound to induce oxidative stress in cells. Menadione has been shown to be able to induce the generation of free radical species in the Hep G2 cell line and pancreatic acinar cells *in vitro* (Chen and Cederbaum, 1997; Criddle *et al.*, 2006). The combination of menadione and the *C. platypetalum* ethanol extract led to an increase in the nitrite concentration, and increase in the concentration could be attributed to some

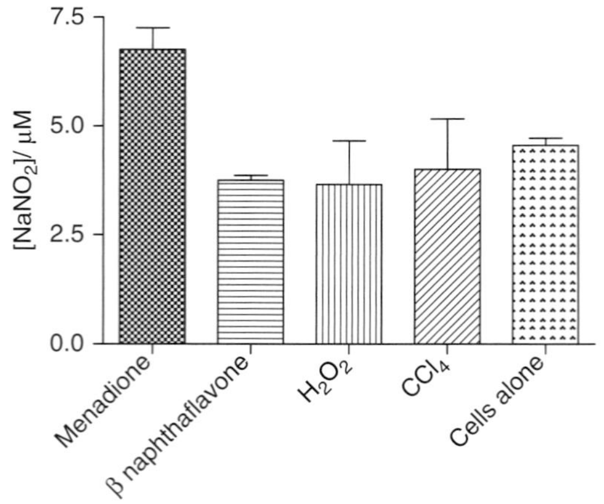


**Figure 14.7** The effects of selected compounds on production of nitrite ions in RAW 264.7 cells. The compounds were selected on their ability to induce oxidative stress on cells

of the phytoconstituents present in *C. platypetalum* acting as pro-oxidants. Flavonoids isolated from *C. platypetalum* increased the nitrite concentration in the cells. Phytoconstituents, flavonoids in particular, have been found to act as pro-oxidants when in high concentrations (Yordi *et al.*, 2011). Combining the *C. platypetalum*, *C. zeyheri*, *C. molle* and *P. curatellifolia* extracts with menadione showed that the extracts increased nitrite concentration compared with the RAW cells alone in the presence of menadione. Nitrite concentration increased when *C. platypetalum* and *C. zeyheri* ethanol extracts were combined with $H_2O_2$. However, $H_2O_2$ when combined with *P. curatellifolia* and *C. molle* had similar effects on the nitrite concentration as it decreased in the presence of the extracts. The effect of the extracts could have been due to the quenching of the peroxide molecule before oxidative stress could be induced. Pro-oxidant activity of plant extracts has been shown to be one of the mechanisms of their antiproliferative effects. However, in inflammatory conditions this would lead to an exacerbation of oxidative stress, which in turn can increase the extent of inflammation. However, the pro-oxidant activity of these extracts could be responsible for their antibacterial, antimycobacterial, antifungal and antiproliferative effects in the model systems used in this study.

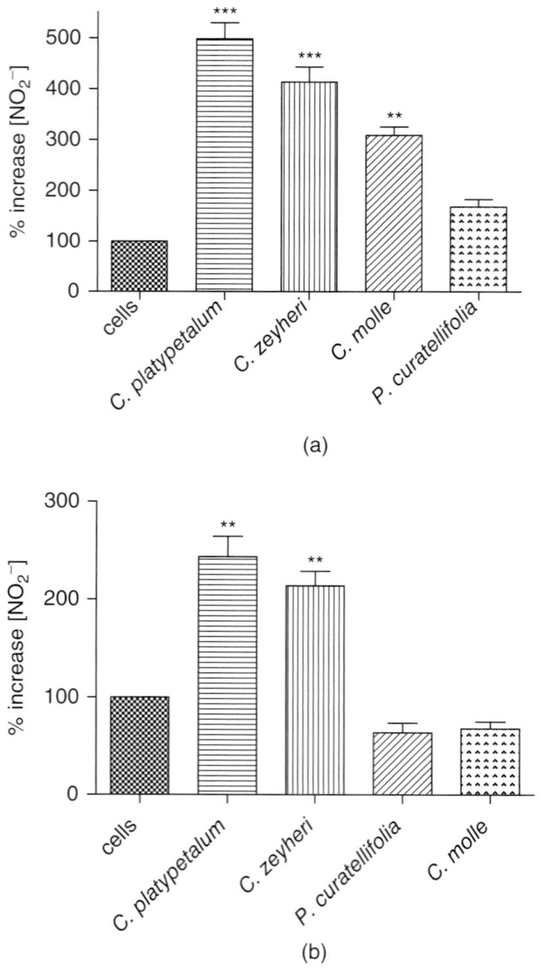


**Figure 14.8** The effects of combining menadione (a) or $H_2O_2$ (b) with plant leaf extracts on the production of nitrite ions in RAW 264.7 macrophage cells

## 14.3 Conclusions

The extracts of plants from Zimbabwe have been found to have significant anti-infective medicinal properties, which included antiproliferative activity (*C. platypetalum*, *P. curatellifollia*), antimycobacterial activity (*P. curatellifolia*, *C. imberbe*, *C. platypetalum*), antifungal activity (*C. zeyheri*, *C. imberbe*, *C. apiculatum*), antibacterial (*C. citrinus*, *V. adoensis*, *C. zeyheri*, *M. indica*) and anti-inflammatory activity (*C. platypetalum*, *P. curatellifolia*). This study has further confirmed the anecdotal evidence as provided by healers when they use these plants for treatment. Further work is required to isolate the lead compounds from the active plants so that they can be further developed as potent phytomedicines.

## Acknowledgements

This study was sponsored by the International Foundation in Sciences (IFS), Stockholm, Sweden, grant number F/3413-03F. Support from the International Science Programs (ISP) through the International Program in the Chemical Sciences (IPICS: ZIM01), Uppsala University, Uppsala, Sweden, and the support from the University of Zimbabwe Research Board RB72/11 (Harare, Zimbabwe) is also acknowledged.

## References

Alekshun, M.N. and Levy, S.B. (2007) Molecular mechanisms of antibacterial multidrug resistance. *Cell*, **128**, 1037–1050.

Buwa, L.V. and Afolayan, A.J. (2009) Antimicrobial activity of some medicinal plants used for the treatment of tuberculosis in the Eastern Cape Province, South Africa. *African Journal of Biotechnology*, **8**, 6683–6687.

Cannon, R.D., Lamping, E., Holmes, A.R. *et al.* (2009) Efflux-mediated antifungal drug resistance. *Clinical Microbiology Reviews*, **22**, 291–321.

Chen, Q. and Cederbaum, A.I. (1997) Menadione cytotoxicity to Hep G2 cells and protection by activation of nuclear factor-κB. *Molecular Pharmacology*, **52**, 648–657.

Chigora, P., Masocha, R. and Mutenheri, F. (2007) The role of indigenous medicinal knowledge (IMK) in the treatment of ailments in rural Zimbabwe: the case of Mutirikwi communal lands. *Journal of Sustainable Development in Africa*, **9**, 26–43.

Chimponda, T. and Mukanganyama, S. (2010) Antimycobacterial activities of selected medicinal plants from Zimbabwe against *Mycobacterium aurum* and *Corynebacterium glutamicum*. *Tropical Biomedicine*, **27**, 595–610.

Chitemerere, T.A. and Mukanganyama, S. (2011) *In vitro* activity of selected medicinal plants from Zimbabwe. *African Journal of Plant Science and Biotechnology*, **5** (1), 1–7.

Chokunonga, E., Borok, M.Z., Chirenje, Z.M. and Nyakabau, A.M. (2010) *Pattern of Cancer in Zimbabwe*, Zimbabwe National Cancer Registry, 2006 Annual Report.

Conforti, F., Ioele, G., Statti, G.A. *et al.* (2008) Antiproliferative activity against human tumor cell lines and toxicity test on Mediterranean dietary plants. *Food and Chemical Toxicology*, **46**, 3325–3332.

Cowan, M. (1999) Plant products as antimicrobial agents. *Clinical Microbiology*, **12**, 564–582.

Criddle, D.N., Gillies, S., Baumgartner-Wilson, H.K. *et al.* (2000) Antibacterial and antifungal drug discovery. *Nature Biotechnology*, **18**, 24–26.

Criddle, D.N., Gillies, S., Baumgartner-Wilson, H.K. *et al.* (2006) Menadione induced reactive oxygen species generation via redox cycling promotes apoptosis of murine pancreatic acinar cells. *Journal of Biological Chemistry*, **281**, 40485–40492.

De Rossi, E., Aínsa, J.A. and Riccardi, G. (2006) Role of mycobacterial efflux transporters in drug resistance: an unresolved question. *FEMS Microbiology Reviews*, **30**, 36–52.

El-Desouky, S.K. and Gamal-Eldeen, A.M. (2009) Cytotoxic and anti-inflammatory activities of some constituents from the floral buds of *Syringa patula*. *Pharmaceutical Biology*, **47**, 872–877.

Fouche, G., Senabe, J., Khorombi, E. *et al.* (2009) High throughput screening of South African plants for anticancer properties. *CISR Biosciences Journal*, **119**, 455–461.

Gelfand, M., Mavi, S., Drummond, R.B. and Ndemera, B. (1985) *The Traditional Medical Practitioner in Zimbabwe*, Mambo Press, Gweru, Zimbabwe, p. 296.

Gupta, A., Bhakta, S., Kundu, S. *et al.* (2009) Fast-growing, non-infectious and intracellularly surviving drug-resistant *Mycobacterium aurum*: a model for high-throughput antituberculosis drug screening. *Journal of Antimicrobial Chemotherapy*, **64**, 774–781.

Iwawela, O.E., McGaw, L.J., Naidoo, V. and Eloff, J.N. (2007) Inflammation: the foundation of disease and disorders: a review of phytomedicines of southern Africa origin used to treat pain and inflammatory conditions. *Journal of African Biotechnology*, **6**, 2868–2885.

Jeong, J., Ryu, Suk, D. and Lee, D. (2011) Anti-inflammatory effects of ethanol extract from *Orostachys japonicus* on modulation of signal pathways in LPS-stimulated RAW 264.7 cells. *BMB Reports*, **44** (6), 399–404.

Kadavul, K. and Dixit, A.K. (2009) Ethnomedicinal studies of the woody species of Kalrayan and Shervarayan Hills, Eastern Ghats, Tamil Nadu. *Indian Journal of Traditional Knowledge*, **4**, 592–597.

Khan, S., Shin, E.M., Choi, R.J. *et al.* (2011) Suppression of LPS-induced inflammatory and NF-κB responses by anomalin in RAW 264.7 macrophages. *Journal of Cellular Biochemistry*, **112**, 2179–2188.

Kisangau, D.P., Hosea, K.M., Joseph, C.C. and Lyaruu, H.V.M. (2007) *In vitro* antimicrobial assay of plants used in traditional medicine in Bukoba rural district, Tanzania. *African Journal of Traditional, Complimentary and Alternative Medicines*, **4**, 510–523.

Lee, I.S., Shamon, L.A., Chai, H.B. *et al.* (1996) Cell cycle specific cytotoxicity mediated by rearranged ent-kaurene diterpenoids isolated from *Parinari curatellifolia*. *Chemico-Biological Interactions*, **99**, 193–204.

Mangoyi, R., Mafukidze, W., Marobela, K. and Mukanganyama, S. (2012) Antifungal activities and preliminary phytochemical investigation of *Combretum* species from Zimbabwe. *Journal of Microbial and Biochemical Technology*, **4**, 037–044.

McGaw, L.J., Rabe, T., Sparg, S.G. *et al.* (2001) An investigation on the biological activity of *Combretum* species. *Journal of Ethnopharmacology*, **75**, 45–50.

Mohamad, S., Zin, N.M., Wahab, H.A. *et al.* (2011) Antituberculosis potential of some ethnobotanically selected Malaysian plants. *Journal of Ethnopharmacology*, **133**, 1021–1026.

Mukanganyama, S., Dumbura, S.C. and Mampuru, L. (2011) Anti-proliferative effects of plant extracts from Zimbabwean medicinal plants against human leukaemia cell lines. *African Journal of Plant Science and Biotechnology*, **6**, 14–20.

Ni, Y., Yates, K.M. and Tizard, I.R. (2004) Aloe polysaccharides, in *Aloes: The Genus Aloe* (ed. T. Reynolds), CRC Press, Boca Raton, FL, pp. 75–87.

Nguyen, L. and Thompson, C.J. (2006) Foundations of antibiotic resistance in bacterial physiology: the mycobacterial paradigm. *Trends in Microbiology*, **14**, 304–312.

Nwinuka, N.M., Monanu, M.O. and Nwiloh, B.I. (2008) Effects of aqueous extract of *Mangifera indica* L. (mango) stem bark on haematological parameters of normal albino rats. *Pakistan Journal of Nutrition*, **7**, 663–666.

Oyedeji, O., Lawal, O.A., Shode, F.O. and Adebola, O. (2009) Chemical composition and antibacterial activity of the essential oils of *Callistemon citrinus* and *Callistemon viminalis* from South Africa. *Molecular Journal*, **14**, 1990–1998.

Ozhak-Baysan, B., Ofunc, D., Colak, D. *et al.* (2012) Distribution and antifungal susceptibility of *Candida* species casuing nosocomial candiduria. *Medical Mycology*, **50**, 529–532.

Poole, K. and Lomovskaya, O. (2006) Can efflux inhibitors really counter resistance? *Drug Discovery Today: Therapeutic Strategies*, **3**, 145–152.

Prasad, R., Sharma, M. and Rawal, M.K. (2011) Functionally relevant residues of Cdr1p: a multidrug ABC transporter of human pathogenic *Candida albicans*. *Journal of Amino Acids*, **25**, 1360–1312.

Ruffo, C.K. (1991) A survey of medicinal plants in Tabora region, Tanzania, in *Proceedings of International Conference on Traditional Medicinal Plants*, Dar Es Salaam University Press, Dar Es Salaam.

Rundle, N.T., Xu, L., Anderson, R.J. and Roberge, M. (2001) $G_2$ DNA damage checkpoint inhibition and antimitotic activity of 13-hydroxy-15-oxozoapatlin. *Journal of Biological Chemistry*, **45**, 160–164.

Shah, B.N., Seth, A.K. and Maheshwari, K.M. (2011) A review on medicinal plants as a source of anti-inflammatory agents. *Research Journal of Medicinal Plant*, **5**, 101–115.

Upadhyay, R.K., Ahmad, S., Tripathi, R. *et al.* (2010) Screening of antimicrobial potential of extracts and pure compounds isolated from *Capparis deciduas*. *Journal of Medicinal Plants Research*, **4**, 439–445.

Van Bambeke, F., Pagès, J.M. and Lee, V.J. (2010) Inhibitors of bacterial efflux pumps as adjuvants in antibacterial therapy and diagnostic tools for detection of resistance by efflux. *Frontiers in Anti-Infective Drug Discovery*, **1**, 138–175.

Wang, N., Yao, X., Ishii, R. and Kitanaka, S. (2003) Bioactive sucrose esters from *Bidens parviflora*. *Phytochemistry*, **62**, 741–746.

WHO (2000) General guidelines for methodologies on research and evaluation of traditional medicine, WHO/EDM/TRM/2000.1, World Health Organization, Geneva.

Wu, W. (2012) Evaluation of anti-inflammatory effects of *Broussonetia papyrifera* stem bark. *Indian Journal of Pharmacology*, **44**, 26–40.

Xia, P., Feng, P., Zhong, L. *et al.* (2002) Accumulation of ciprofloxacin and lomefloxacinin fluoroquinolone-resistant strains of *Escherichia coli*. *Chinese Medical Journal*, **115**, 31–35.

Yoon, W., Ham, Y.M., Kim, S. *et al.* (2009) Suppression of pro-inflammatory cytokines, iNOS, and COX-2 expression by brown algae *Sargassum micracanthum* in RAW 264.7 macrophages. *EurAsian Journal of BioSciences*, **3**, 130–143.

Yordi, E., Pérez, E.M., Matos, M. and Villares, E.U. (2011) Structural alerts for predicting clastogenic activity of pro-oxidant flavonoid compounds: quantitative structure–activity relationship study. *Journal of Biomolecular Screening*, **17**, 216–224.

Zhao, F., Wang, L. and Liu, K. (2009) *In vitro* anti-inflammatory effects of arctigenin, a lignan from *Arctium lappa* L., through inhibition on iNOS pathway. *Journal of Ethnopharmacology*, **122**, 457–462.

# 15 Development of Phytodrugs from Indigenous Plants: The Mali Experience

Rokia Sanogo

Faculté de Pharmacie, Université des Sciences, des Techniques et des Technologies de Bamako, Mali
Département de Médecine Traditionnelle, Institut National de Recherche en Santé Publique, Bamako, Mali

## 15.1 Introduction

African traditional medicine (TM) is a cultural heritage and a reservoir of knowledge still largely untapped. It offers possibilities of effective, available, accessible, acceptable and affordable treatments for the common diseases in African rural communities. The World Health Organization (WHO) estimates that the vast majority of rural populations in developing countries use the resources of TM for their health needs (WHO, 2001). The use of plant-based systems continues to play an essential role in health care. It has been estimated that approximately 80% of the population in developing countries depend on TM for their primary health care (Hostettman and Marston, 2002). In Ghana, Mali, Nigeria and Zambia, the first line of treatment for 60% of the children with high fevers, resulting from malaria, is the use of herbal medicines at home (WHO, 2003). Since the declaration of Alma Ata in 1978, the WHO has recommended the inclusion of TM and the pharmacopoeia in primary health care (WHO, 1978).

In 2000, WHO Africa reaffirmed the usefulness and desirability of the inclusion in the armamentarium of traditional remedies, which gave evidence of safety, efficacy and quality (WHO, 2000). In 2001, the Abuja Declaration of Heads of State and Government of the African Union focused on the research about the resources of TM for supporting Malaria, HIV AIDS, tuberculosis and other priority diseases treatments. The same year, the WHO Africa launched the African Decade of Traditional Medicine (2001–2010). In 2008, 30 years after Alma Ata, the WHO reaffirmed the need of the revival of primary health care (WHO, 2008). The fifth session of the Conference of Ministers of Health in Windhoek, Namibia, in April 2011, approved the report of the end of decade review (African Union, 2010, 2011) and recommended the renewal of the decade and the development of the plan of action for implementation of the 2nd Decade of African Traditional Medicine (2011–2020). An interesting point of the plan of action of this new decade is to continue research and development of TM and medicinal plants to produce evidence on the safety, efficacy and quality of new phytodrugs.

In Mali, the sanitary situation is characterized by a predominance of many endemic and epidemic diseases, together with a lack of qualified health workers, medicines and equipment. To improve their state of health, people use both conventional and TM. TM, being a significant element in the cultural patrimony, still remains the main recourse for a large majority of people for treating health problems. A high level of Malian Government commitment supported the research and development on TM, with the creation of the first institute for the study on medicinal plants in 1968. Today, this institute is the Department of Traditional Medicine (DMT) within the National Institute for Research on Public Health (INRSP) of the Ministry of Health of Mali. This institute is a collaborating centre of the WHO in the research on TM. The main objectives of the DMT are to assure the collaboration between traditional and conventional medicines and to produce medicines from local resources, in particular medicinal plants (MS/INRSP/DMT, 2005).

The main activity of the DMT is the research and development of phytodrugs from local plants, named improved TMs (MTAs; *medicaments traditionnels ameliores*). The aim for promoting the local production of phytodrugs is to provide cost-effective medicines to populations, particularly those with limited access to quality health care.

The research methodology of phytodrugs from herbal remedies passes through a long process. The different steps are as follows:

- Ethnobotanical surveys in order to collect information about herbal remedies and plants used in the traditional treatment of the specific disease. The approach consists of selecting remedies and medicinal plants in accordance with the indication of traditional health practitioners in certain contexts of use. The indication is generally related to symptoms and signs of the diseases. Other data, like remedies preparation, allegations of use, including dosage, are also collected.
- Literature review of preclinical and clinical literature on plants mainly used in the management of the same disease.
- Selection of herbal remedies and medicinal plants for further studies.
- Preclinical study, including extraction, characterization of chemical constituents, toxicity determination, and evaluation of the effectiveness of plant extracts. The preclinical investigation

*Novel Plant Bioresources: Applications in Food, Medicine and Cosmetics*, First Edition. Edited by Ameenah Gurib-Fakim.

permits one to demonstrate the safety and efficacy of the extracts, and to identify the dosage and form suitable for the clinical studies.
- Clinical study, conducted according to a protocol approved by scientific and ethic committees of the INRSP of Ministry of Health.

The results obtained at the different steps are used to prepare a dossier with data on quality, safety and efficacy (Sanogo *et al.*, 2000). Since 1979, the DMT has developed many phytodrugs or MTAs. MTAs are indeed drugs from TM and the 'improvement' lies in the pharmacologic evidence of safety and efficacy, in the standardized dose and in the quality control procedures. For each MTA, this dossier is submitted to the scientific committee of INRSP and the national regulatory authorities to obtain the marketing authorization (autorisation de mise sur le marché; AMM) of the MTAs.

The Malian regulation authority classifies TMs in four categories:

1. TM prepared by a traditional health practitioners for an individual patient with fresh or dried raw materials, with a short life;
2. TM currently used in the community, prepared in advance, and composed of crude raw materials;
3. standardized extracts prepared in advance following scientific research;
4. molecules purified from TMs following scientific research.

After obtaining the AMM, the DMT undertakes the large-scale production and marketing of MTAs.

The requirements for the dossier vary according to the category (Willcox *et al.*, 2012). Most MTAs are in category 2, for which clinical trials are not an absolute requirement.

Since 1990, seven MTAs have been included in the Essential Drugs List of Mali, and in the Malian National Formulary (MSSPA/DNSP, 1998), alongside conventional drugs, and are distributed through pharmacies and community health centres. These MTAs are a safe, effective, locally produced and less expensive alternative to the imported medicines.

Monographs for all the plants used to prepare the seven improved TMs were available during the ICS MAPs programme from 2005 to 2008. This programme has now stopped; see http://folk.uio.no/ingvilau/ITM.html. The seven MTAs developed and produced by DMT are as follows.

### 15.1.1 Balembo®, antitussive syrup

This is prepared with powder of the fruit of *Crossopteryx febrifuga* Benth. (Rubiaceae). *Balembo* is the local name of *C. febrifuga*; it is a small tree 5–6 m high with small round fruit, which become black when ripe (Figure 15.1).

In Africa, in local medicine, the different parts of the plant are used in the treatment of fever. The decoction of the roots is used in the treatment of coughs and gastrointestinal complaints (Dalziel, 1937). In Mali, the powder of the fruits is used as a sedative in the treatment of cough (Adjanohoun *et al.*, 1981).

The fruits are rich in flavonoids and polysaccharides. The saponins (crossoptin A and crossoptin B) with anti-inflammatory, mucolytic and anti-oedemic activities were isolated from roots and bark of the plant (Foresta *et al.*, 1988). The seeds have anti-inflammatory properties (Maiga *et al.*, 2006).

**Figure 15.1** *Crossopteryx febrifuga* (Rubiaceae) black fruits fruits are used to make a cough syrup. *Photo*: Sergio Giani © Aidemet. Reproduced with permission

At the DMT, the powder of fruits is boiled in water to produce syrup (Kodio, 1986). There was very low toxicity at a dose of 30 g/kg administrated orally in mice (Dolo, 1991). *C. febrifuga* was the most effective of several antitussive remedies tested (Occhiuto *et al.*, 1999; Sutovska *et al.*, 2009). In the model of cough in guinea-pigs provoked by nebulized citric acid, it reduced the number of coughs by 63% at a dose of 250 mg/kg, and by 77% at a dose of 1 g/kg (compared with 76% by codeine at 10 mg/kg). The lower dose of 50 mg/kg was not effective (Occhiuto *et al.*, 1999). The remedy at the oral dose of 1 g/kg also reduced antigen-induced bronchoconstriction in guinea-pigs by 54%, compared with a reduction of 78% by disodium cromoglicate at a dose of 10 mg/kg (Occhiuto *et al.*, 1999). However, it had no effect on histamine-induced bronchoconstriction (Occhiuto *et al.*, 1999). Aqueous extracts were not active against bacteria commonly causing respiratory infections (Sanogo *et al.*, 1998a).

Actually, in Mali, extracts of fruits are used for the production of 'Improved Traditional Prescription' Syrup Balembo for adults and children (Figure 15.2).

In 1986, in a clinical trial, Balembo syrup was used by 32 patients suffering cough, excluding tuberculosis and infectious cough. The results demonstrated an effective activity against cough after

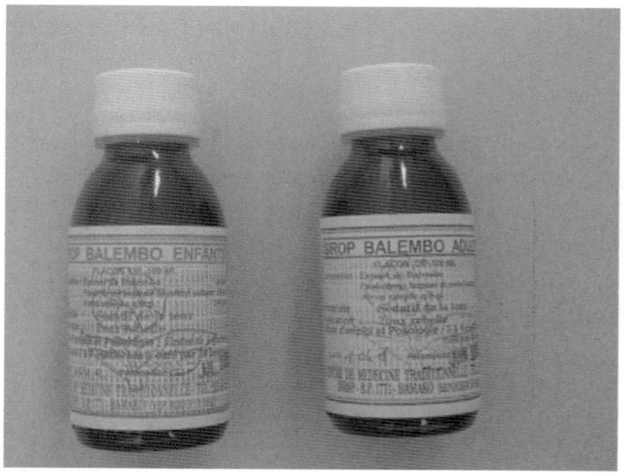


**Figure 15.2** Balembo syrup (child, adult). *Photo*: Sergio Giani © Aidemet. Reproduced with permission

7 days of treatment. The efficacy of the syrup was demonstrated by expectoration fluidization, the diminution of cough frequency and stopping emesis (Kodio, 1986).

The market studies, conducted in Bamako, demonstrated that Balembo syrup is the first MTA for pharmacy clients and patients (Konipo, 2001). Now, Balembo syrup is the most frequently prescribed MTA; it is prescribed by up to 76% of biomedical health workers, and is also the most widely known by patients (Diallo *et al.*, 2010).

### 15.1.2 Dysenteral®, antidysentérique

This is prepared with powder of the aerial parts of *Euphorbia hirta* L, Euphorbiaceae).

*Demba sinji* or *Bojara*, local names of *E. hirta*, is a common pantropical weed (Figure 15.3), widely used in many African countries for the treatment of dysentery (Adjanohoun *et al.*, 1981; Neuwinger, 1996; Hirt and M'Pia, 2008). The plant has a reputation as an analgesic in severe headache, rheumatism, pains in pregnancy, and so on. The plant is regarded as medication for gastrointestinal disorders, particularly intestinal parasitosis, amoebic dysentery, diarrhoea and ulcer. An aqueous decoction is used for the treatment of acute enteritis and dysentery. The plant is also used in bronchial and respiratory disorders, including asthma, bronchitis and hay fever.

*E. hirta* is active *in vitro* against several different microorganisms that cause diarrhoea or dysentery (Willcox *et al.*, 2012). All studies show a good activity against *Entamoeba histolytica*, and the whole plant is more active than the leaves alone (Tona *et al.*, 1998, 2000).

Furthermore, *E. hirta* contains tannins (such as ellagic acid and gallic acid) and flavonoids (such as quercitin), which have antidiarrhoeal and antiamoebic activity (Tona *et al.*, 1998; Galvez *et al.*, 1993).

The water extract has spasmolytic activity on guinea-pig ileum (Dhar *et al.*, 1968; Neuwinger, 1996) and reduces castor-oil-induced diarrhoea in mice (Hore *et al.*, 2006).

The pharmacology, chemistry and toxicology of the plant have been studied extensively (Ridet and Chartol, 1964; Neuwinger, 1996; Ogbulie *et al.*, 2007). The aerial parts contain shikimic acid and the roots and sap contain phorbol esters, which are potentially carcinogenic (Neuwinger, 1996; Diallo *et al.*, 2004), although in an experimental model *E. hirta* ether extracts had almost no tumour-promoting activity compared with other *Euphorbia* species (Ito *et al.*, 1984). Roots should not be used, and it is said that aerial parts should be dried, which removes the carcinogenic effect (Diallo *et al.*, 2005).

However, one study found that drying diminishes the *in vitro* antiamoebic activity (Duez *et al.*, 1991). Rats fed up to 5% powdered whole plant for up to 97 days in their diet showed no symptoms of poisoning, and no gross pathology on autopsy (Neuwinger, 1996).

Although, hypotensive effects were found in cats and dogs (Dhar *et al.*, 1968). A daily dose of alcohol extract (corresponding to 3 g/kg of dried plant) was tested in rats for up to 27 days with no observed toxic effects (Ridet and Chartol, 1964). Aqueous extracts produced no toxicity when given orally to rats at doses of up to 30 g/kg.

Uncontrolled clinical trials of several different extracts have been carried out. The lyophilized decoction (three doses of 10 g) was effective in treating a series of 10 patients with amoebic dysentery in Senegal (Neuwinger, 1996). A tincture of fresh aerial parts (1:2) was tested in a series of 40 cases of acute amoebic dysentery or diarrhoea at the Centre Muraz, Bobo-Dioulasso, Burkina Faso (Ridet and Chartol, 1964). Thirty-eight patients were treated with a dose of 10 ml four times daily for 9 days, with normalization of the stools within 2–6 days and no reported adverse effects (Ridet and Chartol, 1964). A second case series (using the same extract in tablet form, with a daily dose of 7.2 g of dried plant material, for 8 days) resulted in 125 cures out of 150 patients treated (Martin *et al.*, 1964). In this series, the adverse effects reported were 12 cases of hypotension, two cases of nausea/vomiting and one allergic reaction, none of which were severe. An unpublished clinical trial was conducted in Mali concerning Dysenteral treatment at a dose of 10 g boiled in 500 ml water by day for 3 days. Trophozoites of *E. histolytica* disappeared from the stool after 2 days, which was equivalent to treatment with metronidazole (Diallo *et al.*, 2005).

Actually, in Mali the powder of aerial parts is used for the production of the 'improved traditional prescription' Dysenteral, used in the treatment of dysenteric disease. It is presented as a box of nine bags of 10 g and given at a dose of one bag (10 g) three times per day for 3 days (Figure 15.4).

**Figure 15.3** *Euphorbia hirta* L. flowering plant and branch used to make Dysenteral. *Photo*: Sergio Giani © Aidemet. Reproduced with permission

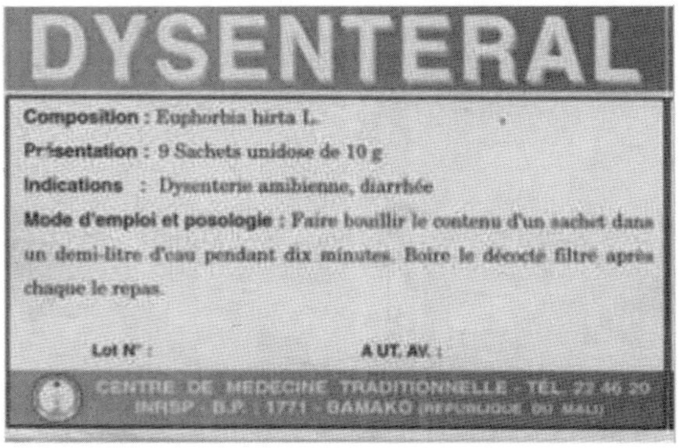


**Figure 15.4** Dysenteral notice. *Photo*: Sergio Giani © Aidemet. Reproduced with permission

### 15.1.3 Gastrosedal®, anti-gastric product

This is prepared with the powder of roots of *Vernonia kotschyana* Sch. Bip. Ex Walp Asteraceae. *Buaye*, the local name of *V. kotschyana*, is a herbaceous plant that grows to a height of about 1 m (Figure 15.5).

The roots are widely used in Mali for the treatment of gastrointestinal disorders and wound healing (Adjanohoun *et al.*, 1985). The healers recommend different preparations of the drug: powder, macerate and infusion. Its thick roots are powdered and used (often mixed with hot water) for indigestion and stomach pains in Mali and Nigeria (Nergard *et al.*, 2004; Pousset, 2004).

The roots contain steroid glycosides called vernoniosides D1–D3, F1 and F2 (Figure 15.6) (Sanogo *et al.*, 1998b, 2002), and different types of polysaccharides, such as inulin, pectins and arabinogalactans, some of which have anti-inflammatory and immunomodulatory properties (Nergard *et al.*, 2004, 2005; Austarheim *et al.*, 2012).

The gastro-protective properties of *V. kotschyana* root showed effect against chemical-agents-induced injury *in vivo*. The aqueous extract is active on experimental ulcers induced by very different mechanisms of action, like restraint and cold, ethanol and indomethacin. The aqueous extract does not act on histamine receptors or gastric secretions. The *n*-butanol fraction protects gastric mucosa also in particular stressing conditions, such as indomethacin administration followed by pyloric ligature while aqueous fraction is inactive (Germano *et al.*, 1996; Sanogo *et al.*, 1996). Aqueous extracts are effective in preventing ethanol and stress-induced gastric ulcers in rats, equivalent to 50 mg/kg of ranitidine (Sanogo *et al.*, 1996). Recently, inulin-rich fractions from *V. kotschyana* roots showed anti-ulcer activity in mice (Austarheim *et al.*, 2012). In that study, Vk50-I (98% inulin) and Vk100-I (83% inulin) from *V. kotschyana* significantly inhibited the formation of gastric lesions in mice at 100 mg/kg. No immunomodulating activities or toxicity were found. These data show that inulin is probably partly responsible for the anti-ulcer activity of Gastrosedal. It is possible that water-soluble polysaccharides (mainly inulin) have an indirect impact on the general health of the gastric injury (Austarheim *et al.*, 2012).

Two uncontrolled clinical trials have been carried out in patients with gastric ulcers. In the first, 80% (of 47 patients) reported symptomatic improvement (Diallo *et al.*, 1990). In the second, 16 patients were followed up after 30 days of taking *V. kotschyana* root tablets (6 g daily). Half of the patients had symptomatic improvement, and the ulcers had healed in six patients (Toure, 1989).

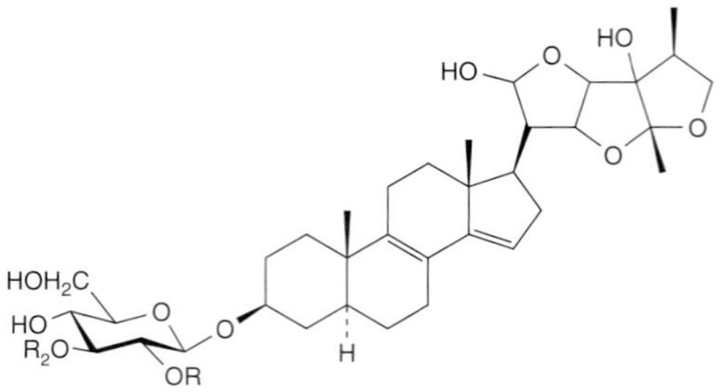


Vernonioside $D_1$ R=$R_1$=H

Vernonioside $D_2$ R=H $R_1$=β-D-xylopyranosyl

Vernonioside $D_3$ R=β-D-glucopyranosyl $R_1$=H

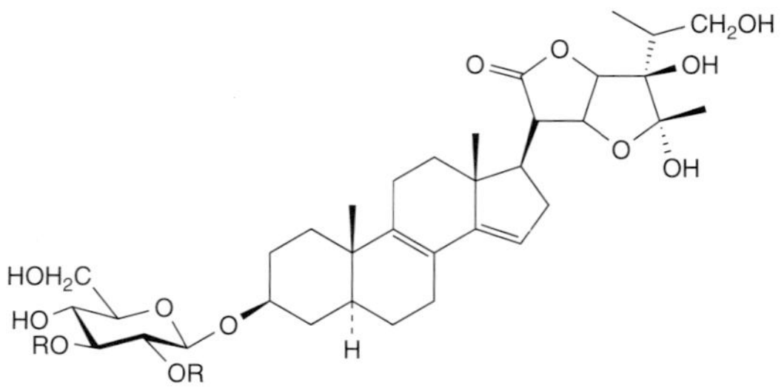


Vernonioside $F_1$ R=H

Vernonioside $F_2$ R=β-D-xylopyranosyl

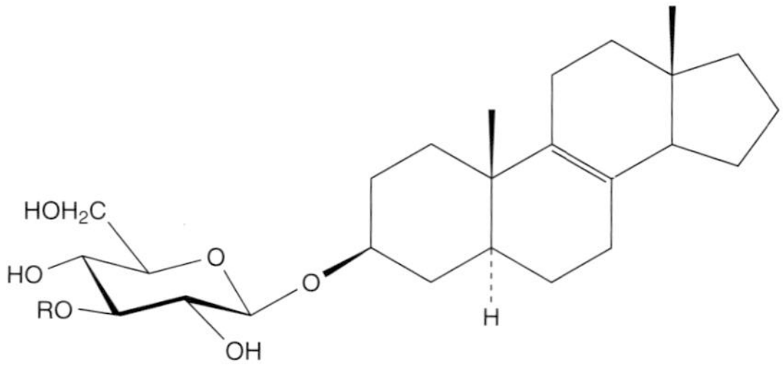


**5** R=β-D-glucopyranosyl

**Figure 15.6** Vernoniosides and androstane glycosides from the roots of *Vernonia kotschyana*. *Source*: Sanogo *et al.* (1998b). Reproduced with permission of Elsevier

**Figure 15.5** *Vernonia kotschyana* Sch. Bip. Ex Walp. (Asteraceae) rhizome is used to make a phytodrug effective against indigestion and gastric ulcers. *Photo*: Sergio Giani © Aidemet. Reproduced with permission. *See plate section for colour version*

Actually, in Mali, the powder of root is used as Gastrosedal, an MTA used in the treatment of gastritis and gastro duodenal ulcers (Figure 15.7).

It is used at a dose of one bag (5 g) three times per day for 45 days.

### 15.1.4 Hepatisane®, choleretic drug

This is prepared with the powder of leaves of *Combretum micranthum* G. Don, Combretaceae. The leaves of this plant yield a drug called 'Kinkeliba', which has been listed in the French, Spanish and British *Extra Pharmacopoeia* under supplementary drugs and indicated for black-water fever and bilious gastrointestinal conditions. It was added to the *French Pharmacopoeia* in 1937 and to the *African Pharmacopoeia* in 1985 (OAU, 1985; Pousset, 2004).

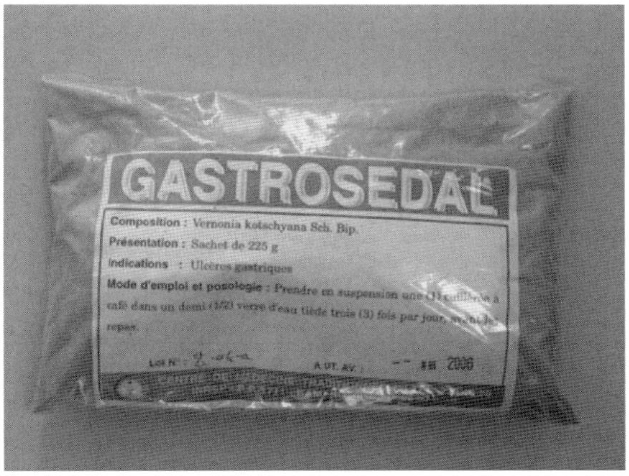


**Figure 15.7** Gastrosedal tea. *Photo*: Sergio Giani © Aidemet. Reproduced with permission

*C. micranthum* (Combretaceae), locally called *N'golobé*, is a small tree, common on the poor soil of the Sahel savannah. Twigs with leaves are commonly sold in markets by the name of Kinkeliba (Pousset, 2004). Like other *Combretum* species, it has characteristic four-winged fruit (Figure 15.8).

The plant is used in West Africa for the treatment of different diseases (Irvine, 1961; Kerharo and Adam, 1974; Adjanohoun, *et al.*, 1985; Ake Assi and Guinko, 1991). The leaves are diuretic, and leaf infusion is used in the treatment of bilious fevers with haematuria. The decoction is taken as a beverage against chronic constipation (Pousset, 2004). It has a wide variety of other medicinal uses in West Africa, particularly as a cholagogue.

Chemical constituents include potassium nitrate, flavonoids (such as vitexin), organic acids, tannins, coumarins, sterols, terpenoids, carbohydrates (inositol, mannitol, sorbitol) and alkaloids (choline, stachydrine) (Ogan, 1972; Bassene *et al.*, 1981, 1986; Pousset, 2004; Diallo *et al.*, 2005).

The leaves are often used to prepare a refreshing tea, but tea is also used for jaundice and hepatitis (Adjanohoun *et al.*, 1981; Arbonnier, 2002). *C. micranthum* extracts are not only interesting sources for antimicrobial activities, but also potential sources of phenolic antioxidants (Karou *et al.*, 2005).

Extracts obtained from *C. micranthum* and others plants commonly used in Cote d'Ivoire by native healers for the treatment of malaria were tested on two strains of *Plasmodium falciparum*: FcB1-Colombia (chloroquine-resistant) and F32-Tanzania (chloroquine-sensitive). The extract from *C. micranthum* was among those extracts that had an apparent inhibitory effect on *P. falciparum* growth *in vitro*, whereas others seemed to be less efficient (Benoit *et al.*,1996). Methanol extract of *C. micranthum* leaves was studied for anti-inflammatory activity in rats and mice using the carrageenan-induced rat paw oedema and the acetic-acid-induced vascular permeability in mice. The extract (50, 100 mg/kg) significantly ($P < 0.05$) inhibited oedema production induced by carrageenan in rats. Increased vascular permeability caused by acetic acid injection was also inhibited by the extract, within the same dose range. *C. micranthum* extract (100 mg/kg) inhibited granuloma formation in rats to a similar degree as indomethacin (5 mg/kg). These results provide evidence for the anti-inflammatory property of *C. micranthum* leaves (Olajide *et al.*, 2003).

In Mali, patients with non-obstructive jaundice have been treated with Hepatisane and their bilirubin and transaminases returned to normal within 2–3 weeks of starting the treatment (Diallo *et al.*, 2005). About 50 asymptomatic patients with chronic viral hepatitis B have been treated with Hepatisane in a clinical trial, but there was no clearance of the hepatitis B surface antigen (J. Falquet, personal communication). However, the herbal treatment was well tolerated, there were no adverse effects and compliance was good. Further clinical trials are needed to determine whether this treatment helps patients with symptomatic hepatitis, or whether it helps to prevent long-term consequences of chronic viral hepatitis.

Hepatisane is an 'improved traditional prescription' produced with of powder of leaves of *C. micrathum* leaves. It is presented in a packet of 14 bags of 10 g of powder of leaves of the plant. The content of one bag (10 g) is boiled in 0.5 l of water for 10 min. The decoction is taken orally; the operation is repeated twice daily for 7 days (Figure 15.9).

Hepatisane is used in symptomatic treatment of hepatic insufficiency, in colds, in hepatitis, dyspepsia and constipation. It is also diuretic. Hepatisane is not toxic and can be safely consumed when used appropriately.

**Figure 15.8** *Combretum micranthum* G. Don (Combretaceae) leaves are used to make Hepatisane tea. *Photo*: Sergio Giani © Aidemet. Reproduced with permission

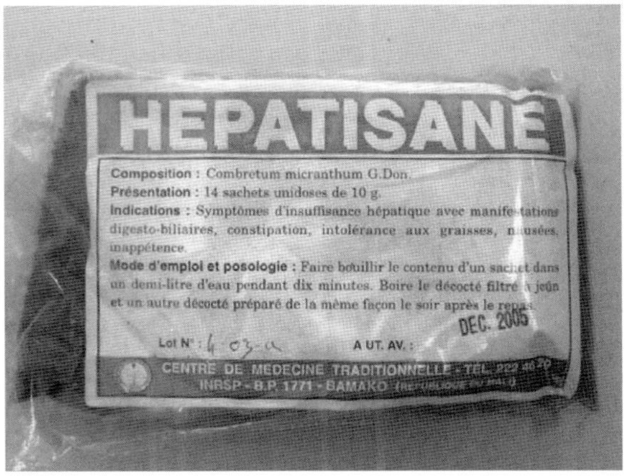


**Figure 15.9** Hepatisane tea. *Photo*: Sergio Giani © Aidemet. Reproduced with permission

### 15.1.5 Laxa-Cassia®, laxative drug

This is prepared with the powder of leaves of *Cassia italica* Mill, Cesalpiniaceae. *Senna italica* Mill (synonym *C. italica*), locally called *Balibali*, is a widespread shrub in the dry regions of Africa (Figure 15.10). In West Africa, the plant is used as laxative and in hepatobiliary affections (Kerharo and Adam, 1974). It is the local equivalent of senna (which in Europe is made from *Cassia senna* and *Cassia angustifolia*). Research has confirmed that its chemical composition is the same as that of *C. angustifolia* (Neuwinger, 1996).

The leaves and fruit are rich in sennosides (anthraquinone glycosides) (Kazmi *et al.*, 1994), which are broken down in the colon and caecum by the gut flora to the active rhein anthrone. This stimulates peristalsis (Heinrich *et al.*, 2004). A common side-effect of overstimulation is colic; this can be reduced by using dry rather than fresh plant material (Neuwinger, 1996). Recently, studies reported myocardial infarction (MI) has been reported in a 45-year-old female patient who sustained acute inferior MI 8 h following *C. italica* ingestion. Subsequently, the cardiac catheterization showed damage in the arteries of the patient. It is possible that, in this case, MI may be partly related to herbal ingestion (El-Menyar *et al.*, 2006). It is not to be used in pregnancy or by lactating mothers, nor in children less than 12 years old.

Constipation is surprisingly common in Malian villages, attributable to insufficient fibre in the diet (which often consists of little more than maize-meal gruel) and insufficient water intake. As everywhere, it is important to remember that dietary advice (increasing fibre intake and water intake) is the first-line treatment, and laxatives are a second-line treatment to be used only when absolutely necessary, because of the risk of side-effects. However, it is very sensible for African countries to manufacture their own senna from local plants, rather than to import European senna at a higher price.

Laxa-Cassia (Figure 15.11) is an 'improved traditional prescription' produced with *C. italica* leaves powder. Laxa-Cassia is used in the treatment of constipation. It is presented as a box of four bags of 5 g and given at the dose of one bag (5 g) once a night for 2–3 days.

Figure 15.10 *Cassia italica* (Miller) Lamarck leaves are used for the production of laxative. *Photo*: Sergio Giani © Aidemet. Reproduced with permission

Figure 15.11 Laxa-Cassia tea. *Photo*: Sergio Giani © Aidemet. Reproduced with permission

### 15.1.6 Malarial®, antimalarial drug

This is prepared with the mixture of powder of leaves of *Senna occidentalis* L. (Cesalpiniaceae), *Lippia chevalieri* Moldenke (Verbenaceae) and flowers of *Acmella oleracea* L. (Jansen) (Asteraceae). Malarial is the first MTA formulated by the DMT. It is based on a recipe prepared with three plants and produced as a standardized phytomedicine (Willcox *et al.*, 2012).

*S. occidentalis* (L.) Link. (synonymous with *Cassia occidentalis*, Caesalpinaceae) is a pantropical plant (Figure 15.12). In Mali, the leaves are considered as a diuretic, against oedema; a decoction is made for malaria, fevers in children and pregnant women, yellow fever and headache (Adjanohoun *et al.*, 1985). The plant is widely used for the treatment of malaria (Diallo *et al.*, 2004) and is active *in vitro* against malaria parasites (Iwalewa *et al.*, 1990). The aqueous-ethanolic extract of leaves of *C. occidentalis* produced significant hepatoprotection (Jafri *et al.*, 1999). The antimalarial

Figure 15.12 *Cassia occidentalis* (Caesalpinaceae) leaves are major constituent of Malarial. Photo: Sergio Giani © Aidemet. Reproduced with permission

activity of *C. occidentalis* crude extracts was confirmed. The plant showed more than 60% inhibition of the parasites growth *in vitro* at a test concentration of 6 μg/mL (Tona *et al.*, 1999, 2001, 2004). Studies indicated that *C. occidentalis* was not genotoxic and exerted no other toxic signs and symptoms in treated animals (Sharma *et al.*, 1999). Ingestion of large doses of senna laxatives may expose people to the risk of hepatotoxicity (Borrelli *et al.*, 2005; Vanderperren *et al.*, 2005). The monograph of the plant is *Ghana Herbal Pharmacopoeia* (PORSPI, 1992).

*L. chevalieri* Mold. (Verbenaceae) is an aromatic herb that is used in West Africa to flavour tea and treat fevers (Kerharo and Adam, 1974) (Figure 15.13).

The antimalarial activities of infusion and decoction of *Lippia multiflora* were demonstrated on two strains of *P. falciparum*: FcB1-Colombia (chloroquine-resistant) and F32-Tanzania (chloroquine-sensitive) growth *in vitro* (Benoit *et al.*, 1996). The essential oil extracted from dried leaves of *L. multiflora* exhibited larvicidal and viscidal activities against eggs of *Anopheles gambiae*. *L. multiflora* showed the highest activity against *A. gambiae* eggs and *Anophles aegypti* larvae (Bassole *et al.*, 2003, 2005).

*A. oleracea* (L.) R.K. Jansen (synonymous with *Spilanthes oleracea* L., Asteraceae) is a sprawling plant with yellow capitules (Figure 15.14), having a variety of uses in TM, including as a local anaesthetic for toothache and as an antipyretic. It contains spilanthol, which is effective against *P. falciparum* (Jellal *et al.*, 1998). Malarial was evaluated against malaria parasites *in vitro* and in mice. It was not very active *in vitro* ($IC_{50}$ = 470–600 μg/mL) but prolonged the survival of malaria-infected mice by 2–3 days compared with the untreated controls (Gasquet *et al.*, 1993). It was also non-toxic to mice.

Three clinical studies were carried out to evaluate the safety and efficacy of Malarial. The first took place in Baguineda in 1984–1985. The second was a randomized controlled trial comparing it with chloroquine (Kéita *et al.* 1990; Koita, 1990). There were 53 patients included, of whom 36 were randomized to Malarial and 17 to chloroquine. Follow-up to day 21 was completed by 75% of the Malarial group and 59% of the chloroquine group. Fever clearance was similar in both groups, but parasite clearance was better in the chloroquine group. Malarial was better tolerated than chloroquine (Willcox *et al.*, 2012). It was felt that the amount of *A. oleracea* (4%) present in this formulation of Malarial was insufficient for a truly effective schizonticidal activity.

**Figure 15.13** *Lippia chevalieri* Mold. (Verbenaceae) leaves are the second constituent of Malarial. *Photo*: Sergio Giani © Aidemet. Reproduced with permission

**Figure 15.14** *Spilanthes oleracea* (*Acmella oleracea*), Asteraceae. Flowerheads are an antimalarial component of Malarial. *Photo*: Sergio Giani © Aidemet. Reproduced with permission

According to this result, it was therefore decided to increase the amount of *A. oleracea* in Malarial to 6%, and this new formulation was tested in an observational cohort study on patients with uncomplicated malaria (Doumbia, 1997).

Thirty patients were included, aged 5 years or above, with a temperature of >37.5 °C and a parasitaemia of >3000/mcl *P. falciparum*. There was no control group. Parasitaemia declined and symptoms improved. Parasitaemia after at day 7 of treatment remained higher in patients aged 8–19 than in older patients. This suggests that patient immunity was playing a role in clearing the parasites.

In the phytomedicine, *C. occidentalis* and *L. chevalieri* extracts contribute to the antiplasmodial activity and the management of the symptoms of malaria such as fever, while *A. oleracea* is the component of Malarial with effective schizonticidal activity.

Malarial 'improved traditional prescription' is a combination of three plants, leaves of *C. occidentalis* (64%), leaves of *L. chevalieri* (32%) and inflorescence of *A. oleracea* (6%).

Malarial is presented as a confection of 11 bags of 10 g of the mixture. The content of one bag is boiled in 0.5 l of water for 10 min. The decoction is taken at a dose of two bags two times per day for 4 days and one bag a day for 3 days. For children, the dose is half the adult dose (Figure 15.15).

### 15.1.7 Psorospermine®, anti-eczema ointment

This is prepared with the roots of *Psorospermum guineense* Hochr. (Hypericaceae). *P. guineense* Hochr., called Karijakuma, is a West African shrub (Figure 15.16). A decoction of its bark, roots or

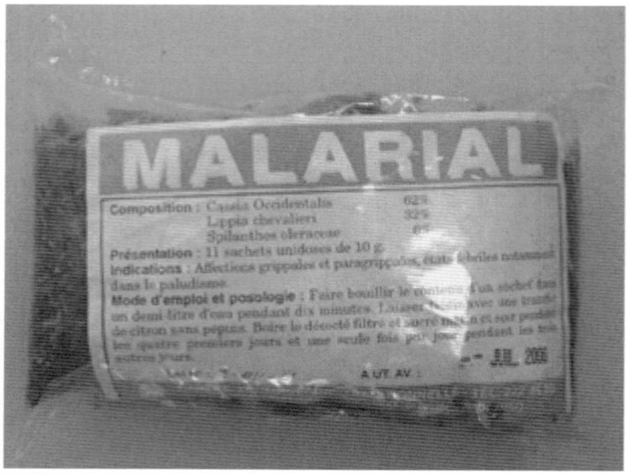


**Figure 15.15** Malarial tea made with the mixture of three plants. *Photo*: Sergio Giani © Aidemet. Reproduced with permission

**Figure 15.16** *Psorospermum guineense* root extract is used to make an ointment for eczema. *Photo*: Sergio Giani © Aidemet. Reproduced with permission

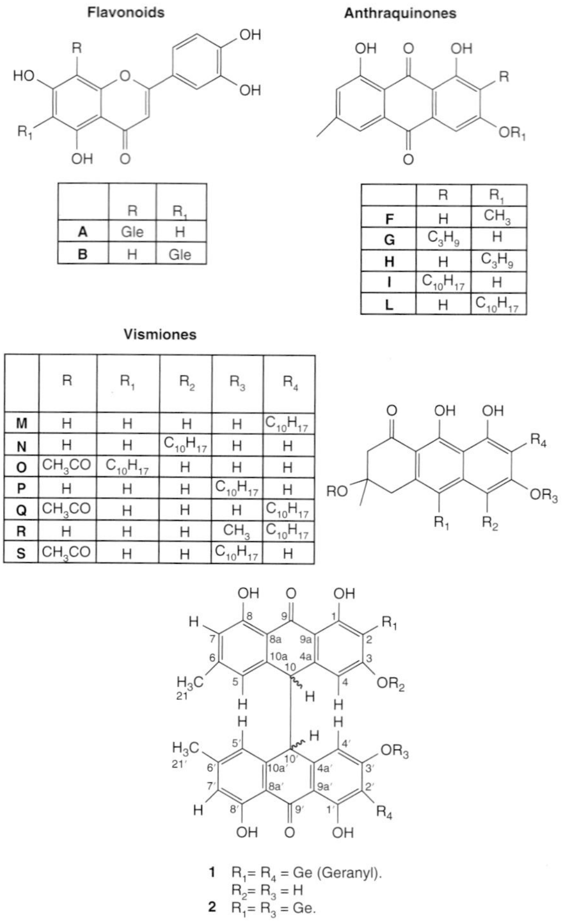


Flavonoids

| | R | $R_1$ |
|---|---|---|
| A | Gle | H |
| B | H | Gle |

Anthraquinones

| | R | $R_1$ |
|---|---|---|
| F | H | $CH_3$ |
| G | $C_3H_9$ | H |
| H | H | $C_3H_9$ |
| I | $C_{10}H_{17}$ | H |
| L | H | $C_{10}H_{17}$ |

Vismiones

| | R | $R_1$ | $R_2$ | $R_3$ | $R_4$ |
|---|---|---|---|---|---|
| M | H | H | H | H | $C_{10}H_{17}$ |
| N | H | H | $C_{10}H_{17}$ | H | H |
| O | $CH_3CO$ | $C_{10}H_{17}$ | H | H | H |
| P | H | H | H | $C_{10}H_{17}$ | H |
| Q | $CH_3CO$ | H | H | H | $C_{10}H_{17}$ |
| R | H | H | H | $CH_3$ | $C_{10}H_{17}$ |
| S | $CH_3CO$ | H | H | $C_{10}H_{17}$ | H |

**Figure 15.17** Chemical structures of major constituents of extract of the root bark from Malian samples *of Psorospermum guineense*. *Source*: Politi *et al.* (2004). Reproduced with permission of John Wiley & Sons

branches with leaves is widely used for skin conditions such as eczema, psoriasis (Diarra, 1990), scabies, cold sores and leprosy (Pousset, 2004). It contains tannins, anthraquinones and xanthones (Bilia *et al.*, 2000; Politi *et al.*, 2004) (Figure 15.17). A dichloromethane extract of root bark is active against the intracellular forms of *Leishmania major*. Extract from *P. guineense* was found to be significantly active against both extracellular and intracellular forms of *Leishmania major* (Ahua *et al.*, 2007).

A double-blind, randomized, controlled clinical trial was conducted in Mali comparing Psorospermine (60 patients) with Shea butter (50 patients) for the treatment of eczema in patients aged 3 years and above. It was applied twice daily after washing with soapy water. After 15 days, the treatment failure rate was 18% in the Psorospermine group, compared with 86% in the Shea butter group ($p < 0.0001$). It was more effective for acute than for chronic eczema (12% and 33% of treatment failures respectively). It was more effective than Shea butter for the different symptoms of eczema: pruritus, erythema, vesicles, exudation and lichenification. Of the patients treated with Psorospermine, 82% rated it as good or excellent, compared with 14% of patients treated with Shea butter (Traoré, 1995). Shea butter comes from the nut of the African Shea tree, *Vitellaria paradoxa* (Sapotaceae).

Psorospermine anti-eczema ointment is prepared with the roots of *P. guineense* (Figure 15.18).

## 15.2 Development of new phytodrugs

On the increasing trend in the use of herbal preparations as remedies for major diseases, there is also a growing concern about their efficacy, safety and control. For this, according to the WHO recommendations, we are using the WHO 'Research Guidelines for Evaluating the Safety and Efficacy of Herbal Medicines' for

**Figure 15.18** Psorospermine notice. *Photo*: Sergio Giani © Aidemet. Reproduced with permission

investigation of traditional remedies used in the management of main illnesses such as malaria, hypertension, diabetes, sickle cell anaemia and HIV/AIDS.

Actually, our institute is working to the development of new phytodrugs from other plants used by traditional health practitioners in Mali. Our activities are focused on the research on new, more effective MTAs against malaria, and against hypertension.

### 15.2.1 Improved traditional medicines against malaria

Until 2002, research in Mali had been ongoing to produce a new phytodrug for malaria that is more effective than Malarial, which is the least frequently used MTA.

After a long process, *Argemone mexicana* (Papaveraceae) decoction was selected (Adjobimey *et al.*, 2004; Diallo *et al.*, 2006) and has already undergone clinical trials (Willcox *et al.*, 2007, 2011a,b; Graz *et al.*, 2010).

It is anticipated that it will soon be approved as a recommended MTA for malaria. The name of this new phytomedicine will be Sumafoura Tiémoko Bengaly, in honour of the traditional healer who mainly used *A. mexicana* to treat malaria. Actually, other researches are ongoing to prepare syrup with decoction and hydro-ethanolic extract of *A. mexicana* (Sanogo *et al.*, 2008, 2012). This syrup can contribute to a better management of uncomplicated malaria in children. We are actually finalizing the investigations and completing the dossier of this new phytodrug.

The antimalarial drug in tea or syrup prepared with the areal parts without seed of *A. mexicana* L. (Papaveraceae) is a new candidate to be listed in the Malian Formulary Therapeutic Guidelines of the Minister of Health.

### 15.2.2 A new phytodrug against hypertension

Despite availability and usage of a wide array of antihypertensive drugs, blood pressure has remained poorly controlled in most health care settings, particularly in sub-Saharan Africa. The treatment of hypertension with conventional medicines is too expensive, leading many patients to use TMs. At the DMT in Bamako, different plants used in TM have actually been investigated for their diuretic activities. Diuretics cause increase in the rate of urine flow, and are thus employed in hypertension and other cardiovascular disorders.

Stem barks of *Sclerocarya birrea* and *Vitex doniana* are used to prepare the phytodrug called Nitrokundang® by the traditional health practitioner Mr Salif Traoré. The diuretic activity of the aqueous extract was demonstrated in mice. The infusion prepared according to the traditional health practitioner recommendations presented the best diuretic effect. The urinary output was 170% with the extemporary infusion; in the same conditions, furosemide at a dose of 20 mg/kg presented 184%. The infusion showed also an important output of sodium but not potassium (Sanogo *et al.*, 2009). In our previous study, the antioxidant activity of phenolic compounds from *S. birrea* was demonstrated as beneficial for the new phytodrug (Braca *et al.*, 2003).

## 15.3 Discussion

Medicinal plants have played a key role in world health care, with about 80% of Africans depending on phytodrugs, which have shown a wide range of uses in the treatment of diseases, especially priority diseases of Africa, such as HIV/AIDS, malaria, sickle-cell anaemia, diabetes and hypertension. These medicinal plants have more beneficial effects than their synthetic counterparts, and are safer, acceptable, affordable, culturally compatible and suitable for chronic treatments. Some African phytodrugs are well known in the international market, and so supply economic benefit for the producing countries. The use of phytodrugs should be integrated into the health agenda, since effective health care cannot be achieved in Africa by using orthodox medicine alone. This can be achieved by adopting the WHO memorandum and guidelines for the institutionalization of phytodrugs into the health sector (WHO, 2004). However, there are some problems limiting the development of phytodrugs, such as lack of standardization, efficacy and quality control of plants used, extinction of some plant species, lack of funds and others. If these problems can be fully addressed, this will help in the future harmonization and development of phytodrugs.

To ensure sustainability and wide availability of phytodrugs, a standardized mode of production, which meets modern pharmaceutical manufacturing standards, must be developed. It will then be possible to make these phytodrugs available through direct sale or by prescription, depending on the registration category assigned to the product by the National Medicines Regulatory Authority.

These 'improved phytodrugs' have many advantages. The first advantage is that they are culturally accepted because they are already used by TM systems.

The second advantage is that they can be developed much faster and more inexpensively than new modern drugs. This is because their use, preparation and safety are already understood in traditional knowledge systems, and so preclinical development can be greatly accelerated. WHO guidelines (WHO, 1996) state: 'If the product has been traditionally used without demonstrated harm, no specific restrictive regulatory action should be undertaken unless new evidence demands a revised risk–benefit assessment.' The WHO maintains the position that there is no requirement for preclinical toxicity testing, but rather that evidence of traditional use or recent clinical experience is sufficient (WHO, 1998). This is reflected in the Malian regulations, which only require toxicology and clinical trials for extracts, not for crude traditional preparations (Willcox *et al.*, 2012).

The third advantage is that the end product is more widely available and affordable to patients in resource-poor settings than many pharmaceuticals are. MTAs are now considered part of the essential drugs list in Mali (MSSPA/DNSP, 1998). In terms of sales, the most successful have been Balembo, Gastrosedal and Hepatisane, because there is no affordable conventional equivalent for the treatment of jaundice or hepatitis. Malarial was the least preferred because of its lower efficiency and price compared with equivalent conventional drugs used in the management of malaria. It is for this reason that the DMT has conducted research for the development of a new phytodrug that is more effective against malaria. A new phytodrug for malaria, Sumafoura Tiemoko Bengaly tea, recommended for the first-line treatment for malaria, is currently being developed (Willcox *et al.*, 2012). Further studies are underway to improve the use of the tea and to produce antimalarial syrup produced with extracts of *A. mexicana*.

The fourth advantage is that prescribers and patients perceive most MTAs to be effective. For this, many patients choose self-medication with MTAs. This was the main reason for choosing to prescribe them in 85% of health workers surveyed in Kadiolo district (Diallo *et al.*, 2010); interestingly, only 38% of prescribers and 29% of patients quoted the lower price as their primary reason for choosing to use MTAs. The most popular MTAs with prescribers were Gastrosedal and Laxa-Cassia, and the most popular with patients were Laxa-Cassia and Hepatisane. Another previous survey in Bamako classed in first place Balembo syrup (Konipo, 2001).

Medical students are taught about MTAs during their training, and the DMT has conducted training for doctors on the prescription of MTAs.

In general, it seems that practising doctors only prescribe MTAs if they have received an additional training session. As in other parts of the world, it seems that some patients and doctors integrate traditional and modern medicines according to their perceived effectiveness.

One disadvantage of the production of standardized phytodrugs (as opposed to teaching people to grow and produce their own herbal medicines) is that their distribution is hindered by infrastructure problems, which also affect the distribution of conventional medicines. In the survey in Kadiolo district, phytodrugs were out of stock on an average of 78 days per year (21% of the time) (Diallo *et al.*, 2010). Research on phytodrugs has been constrained by limited financial resources and the need to strengthen research capacity. In particular, many of the clinical trials have been small and uncontrolled, and remained unpublished.

For the development of TM and medicinal plants to produce phytodrugs, it is important to reinforce and to give more support to the DMT. For this, we need pragmatic issues, requiring several considerations, include funding, facilities and involvement of more properly trained research personnel and traditional medical practitioners. A lot of preclinical data exist on many plants that can be used to develop new phytodrugs for the management of, among others, chronic diseases such as diabetes, hypertension, sickle-cell anaemia, liver diseases, benign prostatic hypertrophy, AIDS and HIV, and so on. For the development of these new phytodrugs, we need to conduct clinical research that must be carried out under conditions of ensuring adequate safety for the subjects in collaboration with selected institutions with adequate facilities. DMT needs human resources trained on process technology and development of herbal remedies industries.

The sustainable production of phytodrugs in Mali also needs institutional and financial support to cultivate the plants used for the production of phytodrugs. Effectively, an important limiting factor in the local production of TM is the availability of adequate plant resources. Concerted efforts should therefore be made towards the conservation and cultivation of medicinal plants, while the harvesting and collection methodologies of the main plants should be carried out according to the good collection practices of the WHO. Special efforts should be made to engage the traditional health practitioners and farmers in the conservation, cultivation, harvesting and post-harvest processing of medicinal plants. It is anticipated that the economic viability of these plants will encourage investors to continue to cultivate them so that the activity becomes sustainable.

For the development of phytodrugs, special attention should be paid also to intellectual property rights and issues related to equitable benefit sharing emanating from the commercial exploitation of TMs.

## 15.4 Conclusion

African medicinal plants, in addition to their traditional and cultural values, hold great public and medical interest worldwide as sources of standardized phytodrugs or novel lead compounds for drug development.

Phytodrugs of DMT of Mali are important in the availability of essential medicines used by the majority of the population. The successes of these phytodrugs open new opportunities for patients and stimulate research on medicinal plants with direct relevance to human health care.

Actually, many interesting results are recorded with crude extracts showing the rationale behind plant usage in Mali, but it is not enough when we consider the large amount of plants that have not been screened for their biochemical composition or for their pharmacological properties. The research is underway to complete preclinical and clinical data for the development of new phytodrugs for human health care application. This integration of phytodrugs into the health system should be developed in such a way as to bring harmony between the traditional and modern systems of health care with minimum threat to each other.

Actually, in Mali, appropriate legislation and regulations and a functional research institute exist. Moreover, the sustainable production of phytodrugs in Mali requires a stronger political will, more trained human resources, more leadership and more effective partnerships between traditional health practitioners and researchers. It is also necessary to develop partnership arrangements between the traditional health practitioners and researchers, as well as with the public and private pharmaceutical industry for strengthening the local production of phytodrugs. For this, considerable efforts are needed in order to inspire the private and public sectors to take up the challenge of the production of phytodrugs in Mali, from the perspective of a regional collaboration. Such partnerships will be based on fair play, honesty, equality, mutual respect and equitable sharing of benefits.

## References

Adjanohoun, E.J., Ake Assi, L., Floret, J. *et al.* (1981) *Médecine Traditionnelle et Pharmacopée: Contribution aux Études Ethnobotaniques et Floristiques au Mali*, Agence de Coopération Culturelle et Technique, Paris.

Adjanohoun, E.J., Ake Assi, L., Floret, J.J. *et al.* (1985) *Médecine Traditionnelle et Pharmacopée: Contribution aux Études Ethnobotaniques et Floristiques au Mali*, Edition Agence de Coopération Culturelle et Technique (ACCT), Paris.

Adjobimey, T., Edayé, I., Lagnika, L. *et al.* (2004) Activités antiplasmodiales in vitro de quelques plantes antipaludiques de la Pharmacopeé béninoise. *Comptes Rendus Chimie*, **4** (7), 1023–1027.

African Union (2010) End of Decade Review Tool for Implementation of the Plan of Action of the 1st AU Decade of Traditional Medicine (2001–2010).

African Union (2011) Fifth Ordinary Session of the African Union Conference of Ministers of Health, Windhoek, Namibia, 17–21 April 2011 (CAMH/Min/Draft/Decl.(V)).

Ahua, K.M., Ioset, J.R., Ioset, K.N. *et al.* (2007) Antileishmanial activities associated with plants used in the Malian traditional medicine. *Journal of Ethnopharmacology*, **110** (1), 99–104.

Ake Assi, L. and Guinko, S. (1991) *Plants Used in Traditional Medicine in West Africa*, Editions Roche.

Arbonnier, M. (2002) *Arbres, Arbustes et Lianes des Zones Sèches d'Afrique de l'Ouest*, CIRAD, Museum National d'Histoire Naturelle, Paris.

Austarheim, I., Nergard, C.S., Sanogo, R. *et al.* (2012) Inulin-rich fractions from *Vernonia kotschyana* roots have anti-ulceractivity. *Journal of Ethnopharmacology*, **144** (1), 82–85.

Bassene, E., Olschwang, D. and Pousset, J.L. (1981) African medical plants. I: characterization of inositol and sorbitol, probable active principles of kinkeliba (*Combretum micranthum* G. Don). *Dakar Medical*, **26** (2), 219–225.

Bassene, E., Olschwang, D. and Pousset, J.L. (1986) African medicinal plants. Alkaloids of *Combretum micranthum* G. Don (Kinkeliba). *Annales Pharmaceutiques Françaises*, **44** (3), 191–196.

Bassole, I.H., Guelbeogo, W.M., Nebie, R. *et al.* (2003) Ovicidal and larvicidal activity against *Aedes aegypti* and *Anopheles gambiae* complex mosquitoes of essential oils extracted from three spontaneous plants of Burkina Faso. *Parassitologia*, **45** (1), 23–26.

Bassole, I.H.N., Nebie, R., Savadogo, A. *et al.* (2005) Composition and antimicrobial activities of the leaf and flower essential oils of *Lippia chevalieri* and *Ocimum canum* from Burkina Faso. *African Journal of Biotechnology*, **4** (10), 1156–1160.

Benoit, F., Valentin, A., Pelissier, Y. *et al.* (1996) *In vitro* antimalarial activity of vegetal extracts used in West African traditional medicine. *American Journal of Tropical Medicine and Hygiene*, **54** (1), 67–71.

Bilia, A.R., Yusuf, A.W., Braca, A. *et al.* (2000) New prenylated anthraquinones and xanthones from *Vismia guineensis*. *Journal of Natural Products*, **63** (1), 6–21.

Borrelli, F., Capasso, R., Aviello, G. *et al.* (2005) Senna and the formation of aberrant crypt foci and tumors in rats treated with azoxymethane. *Phytomedicine*, **2** (6–7), 501–505.

Braca, A., Politi, M., Sanogo, R. *et al.* (2003) The chemical composition and antioxidant activity of phenolic compounds from wild and cultivated *Sclerocarya birrea* (Anacardiaceae) leaves. *Journal of Agricultural and Food Chemistry*, **51**, 6689–6695.

Dalziel, J.M. (1937) *The Useful Plants of West Tropical Africa*, Crown Agents for Overseas Government Administration, London.

Dhar, M.L., Dhar, M.M., Dhawan, B.N. *et al.* (1968) Screening of Indian plants for biological activity: part I. *Indian Journal of Experimental Biology*, **6**, 232–247.

Diallo, D., Coumaré, A. and Koita, N. (1990) Etude préliminaire d'une plante médicinale du Mali: *Vernonia kotschyana* (Sch. Bip), in *Cahier Spécial de l'INRSP*, no. 1, INRSP, Bamako, Mali, pp. 52–56.

Diallo, D., Maiga, A., Diakite, C. and Willcox, M. (2004) Malarial-5: development of an antimalarial phytomedicine in Mali, in *Traditional Medicinal Plants and Malaria* (eds M. Willcox, G. Bodeker and P. Rasoanaivo), CRC Press, Boca Raton, FL.

Diallo, D., Paulsen, B.S. and Hveem, B. (2005) Les produits de la medecine traditionnelle, in *Les Plantes Sauvages du Sahel Malien* (eds G. Berge, D. Diallo and B. Hveem), Editions Karthala, Paris, pp. 135–141.

Diallo, D., Graz, B., Falquet, J. *et al.* (2006) Malaria treatment in remote areas of Mali: use of modern and traditional medicines, patient outcome. *Transactions of the Royal Society of Tropical Medicine and Hygiene*, **100** (6), 515–520.

Diallo D., Diakité, C., Diawara, A. *et al.* (2010) Study of the consumption of the improved traditional phytomedicines in the Health District of Kadiolo (region of Sikasso, Mali). *Mali Médical*, **25** (1), 5–13.

Diarra I.I. (1990) Contribution to the study of skin diseases and their traditional treatments in Mali. PharmD thesis, Ecole Nationale de Médecine et de Pharmacie au Mali, Université de Bamako, Bamako, Mali (in French).

Dolo A. Contribution to the study of the toxicology of Malian medicinal plants. PharmD thesis, Ecole Nationale de Médecine et de Pharmacie (ENMP), Bamako, Mali (in French).

Doumbia, S. (1997) Study of antimalarial plants in Mali. Pharm D thesis, Faculté de Médecine, Pharmacie et Odonto-Stomatologie, Université de Bamako, Bamako, Mali (in French).

Duez, P., Livaditis, A., Guissou, P.I. *et al.* (1991) Use of an *Amoeba proteus* model for *in vitro* cytotoxicity testing in phytochemical research. Application to *Euphorbia hirta* extracts. *Journal of Ethnopharmacology*, **34** (2–3), 235–246.

El-Menyar, A.A., Helmy, A.H., Mubarak, N.M. and Arafa, S.E. (2006) Acute myocardial infarction with patent epicardial coronary vessels following *Cassia italica* ingestion. *Cardiovascular Toxicology*, **6** (2), 81–84.

Foresta, P., Ghirardi, O., Gabetta, B. and Cristoni, A. (1988) Triterpene saponins having anti-inflammatory, mucolytic and antiedemic activities, process for the preparation thereof and pharmaceutical compositions containing them. European Patent Application EP 87–109117 19870625.

Galvez, J., Crespo, M.E., Jimenez, J. *et al.* (1993) Antidiarrhoeic activity of quercitrin in mice and rats. *Journal of Pharmacy and Pharmacology*, **45** (2), 157–159.

Gasquet, M., Delmas, F., Timon-David, P. *et al.* (1993) Evaluation *in vitro* and *in vivo* of traditional antimalarial, 'Malarial 5'. *Fitoterapia*, **64** (5), 423–426.

Germano, M.P., De Pasquale, R., Iauk, L. *et al.* (1996) Antiulcer activity of *Vernonia kotschyana* Sch. Bip. *Phytomedicine*, **2** (3), 229–233.

Graz, B., Willcox, M.L., Diakite, C. *et al.* (2010) *Argemone mexicana* decoction versus artesunate/amodiaquine for the home-based management of malaria in Mali. Policy and public health implications. *Transactions of the Royal Society of Tropical Medicine & Hygiene*, **104**, 33–41.

Heinrich, M., Barnes, J., Gibbons, S. and Williamson, E.M. (2004) *Fundamentals of Pharmacognosy and Phytotherapy*, Churchill Livingstone, Edinburgh.

Hirt, H., and M'Pia B. (2008) *Natural Medicine in the Tropics I: Foundation Text*, Anamed, Winnenden.

Hore, S.K., Ahuja, V., Mehta, G. *et al.* (2006) Effect of aqueous *Euphorbia hirta* leaf extract on gastrointestinal motility. *Fitoterapia*, **77** (1), 35–38.

Hostesttmann, K. and Marston, A. (2002) Twenty years of research into medicinal plants: results and perspectives. *Phytochemistry Reviews*, **1**, 275–285.

Irvine, F.R. (1961) *Woody Plants of Ghana with Special References to their Uses*, Oxford University Press.

Ito, Y., Tokuda, H., Ohigashi, H. and Koshimizu, K. (1984) Distribution and characterization of environmental promoter substances as assayed by synergistic Epstein-Barr virus-activating system, in *Cellular Interactions by Environmental Tumor Promoters* (ed. H. Fujiki), Japan Science Society Press, Tokyo.

Iwalewa, E.O., Lege-Oguntoye, L., Rai, P.P. *et al.* (1990) *In vitro* antimalarial activity of leaf extracts of *Cassia occidentalis* and *Guiera senegalensis* in *Plasmodium yoelii nigeriensis*. *West African Journal of Pharmacology and Drug Research*, **9** (Suppl), 19–21.

Jafri, M.A., Jalis Subhani, M., Javed, K. and Singh, S. (1999) Hepatoprotective activity of leaves of *Cassia occidentalis* against paracetamol and ethyl alcohol intoxication in rats. *Journal of Ethnopharmacology*, **66** (3), 355–361.

Jellal, A., Lemerre, S., Michot, P. *et al.* (1998) *Le Spilanthes*. Etablissement National d'Enseignement Supérieur Agronomique (ENESA), Dijon.

Karou, D., Dicko, M.H., Simpore, J. and Traore A.S. (2005) Antioxidant and antibacterial activities of polyphenols from ethnomedicinal plants of Burkina Faso. *African Journal of Biotechnology*, **4** (8), 823–828.

Kazmi, M.H., Malik, A., Hameed, S. *et al.* (1994) An anthraquinone derivative from *Cassia italica*. *Phytochemistry*, **36** (3), 761–763.

Kéita, A., Doumbo, O., Koïta, N. *et al.* (1990) Recherche expérimentale sur un anti-malarique traditionnel. 'Etude préliminaire sur la faisabilité d'un protocole d'essai clinique'. *Bulletin de Médecine Traditionnelle et Pharmacopée*, **4** (2), 139–146.

Kerharo, J. and Adam, J.G. (1974) *La Pharmacopée Sénégalaise Traditionnelle: Plantes Médicinales et Toxiques*, Vigot Frères, Paris.

Kodio, A. (1986) Contribution to the study of the consumption of cough syrups in the Republic of Mali: evaluation of the need and improvement of Balembo syrup. PharmD thesis, Ecole Nationale de Médecine et de Pharmacie (ENMP), Bamako, Mali (in French).

Koita, N. (1990) A comparative study of the traditional remedy 'Suma-Kala' and chloroquine as treatment for malaria in the rural areas, in *Proceedings of an International Conference of Experts from Developing Countries on Traditional Medicinal Plants* (eds K.E. Mshigeni, M.H.H. Nkunya, V. Fupi *et al.*), Dar Es Salaam University Press, Dar Es Salaam, pp. 68–82.

Konipo, A. (2001) Etude de marché des médicaments traditionnels améliorés et mise au point d'une pommade dermique à base de *Mitracarpus scaber* Zucc. (Rubiaceae). Thèse de Pharmacie, Faculté de Médecine de Pharmacie et d'Odontostomatologie, Université de Bamako.

Maiga, A., Malterud, K.E., Diallo, D. and Paulsen, B.S. (2006) Antioxidant and 15-lipoxygenase inhibitory activities of the Malian medicinal plants *Diospyros abyssinica* (Hiern) F. White (Ebenaceae), *Lannea velutina* A. Rich (Anacardiaceae) and *Crossopteryx febrifuga* (Afzel) Benth. (Rubiaceae). *Journal of Ethnopharmacology*, **104** (1–2), 132–137.

Martin, M., Ridet, J., Chartol, A. *et al.* (1964) Action therapeutique de l'extrait de *Euphorbia hirta* dans l'amibiase intestinale. *Medecine Tropicale*, **24** (3), 250–261.

MS/INRSP/DMT (2005) Politique Nationale de Médecine Traditionnelle du Mali.

MSSPA/DNSP (1998) *National Therapeutic Formulary*, Ministry of Health, Elderly People and Solidarity, Bamako (in French).

Nergard, C.S., Diallo, D., Michaelsen, T.E. *et al.* (2004) Isolation, partial characterisation and immunomodulating activities of polysaccharides from *Vernonia kotschyana* Sch. Bip. ex Walp. *Journal of Ethnopharmacology*, **91** (1), 141–152.

Nergard, C.S., Matsumoto, T., Inngjerdingen, M. *et al.* (2005) Structural and immunological studies of a pectin and a pectic arabinogalactan from *Vernonia kotschyana* Sch. Bip. ex Walp. (Asteraceae). *Carbohydrate Research*, **340** (1), 115–130.

Neuwinger, H.D. (1996) *African Ethnobotany: Poisons and Drugs*, Chapman & Hall, London.

OAU (1985) *African Pharmacopoeia*, Vol. 1, Organization of African Unity, Scientific Technical & Research Commission, Lagos.

Occhiuto, F., Sanogo, R., Germano, M.P. *et al.* (1999) Effects of some Malian medicinal plants on the respiratory tract of guinea-pigs. *Journal of Pharmacy and Pharmacology*, **51** (11), 1299–1303.

Ogan, A.U. (1972) The alkaloids in the leaves of *Combretum micranthum*. Studies on West African medicinal plants. *Planta Medica*, **21** (2), 210–217.

Ogbulie, J.N., Ogueke, C.C., Okoli, I.C. and Anyanwu, B.N. (2007) Antibacterial activities and toxicological potentials of crude ethanolic extracts of *Euphorbia hirta*. *African Journal of Biotechnology*, **6** (13), 1544–1548.

Olajide, O.A., Makinde, J.M. and Okpako, D.T. (2003) Evaluation of the anti-inflammatory property of the extract of *Combretum micranthum* G. Don (Combretaceae). *Inflammopharmacology*, **11** (3), 293–298.

Politi, M., Sanogo, R., Ndjoko, K. *et al.* (2004) HPLC-UV/PAD and HPLC-MS(*n*) analyses of leaf and root extracts of *Vismia guineensis* and isolation and identification of two new bianthrones. *Phytochemical Analysis*, **15** (6), 355–364.

PORSPI (1992) *Ghana Herbal Pharmacopoeia*, Policy Research and Strategic Planning Institute, Accra.

Pousset, J.-L. (2004) *Plantes Medicinales d'Afrique*, Edisud, Aix-en-Provence.

Ridet, J. and Chartol, A. (1964) Les proprietes antidysenteriques de *Euphorbia hirta*. *Medecine Tropicale*, **24** (2), 119–143.

Sanogo, R., De Pasquale, R., Germano, M.P. *et al.* (1996) *Vernonia kotschyana* Sch. Bip.: tolerability and gastroprotective activity. *Phytotherapy Research*, **10** (Suppl), S169–S171.

Sanogo, R., Crisafi, G., Germanò, M.P. *et al.* (1998a) Evaluation of Malian traditional medicines: screening for antimicrobial activity. *Phytotherapy Research*, **12**, S154–S156.

Sanogo, R., Germano, M.P., De Tommasi, N. *et al.* (1998b) Vernonioisides and an androstane glycoside from *Vernonia kotschyana*. *Phytochemistry*, **47** (1), 73–78.

Sanogo, R., Keita, A. and De Pasquale, R. (2000) Pharmacognosie et pharmacodynamie de plantes utilisées en médecine traditionnelle au Mali, in *Des Sources du Savoir aux Médicaments du Futur – From the Source of Knowledge to the Medicine of Future, Proceedings of the 4th European Congress on Ethnopharmocology* (eds J. Fleurentin, J.M. Pelt and G. Mazars), IRD Editions, Montpellier, pp. 383–387.

Sanogo, R., Giani, S., Braca, A. *et al.* (2002) Pharmacochemical investigation on some Malian medicinal plants. *Current Topics in Phytochemistry*, **5**, 225–234.

Sanogo, R., Djimdé, A., Guirou, C. *et al.* (2008) Etude de la toxicité sub-chronique du décocté de *Argemone mexicana* L. *Pharmacopée et Médecine Traditionnelles Africaines*, **15**, 26–31.

Sanogo, R., Karadji, A.H., Dembélé, O. and Diallo, D. (2009) Activité diurétique et salidiurétique d'une recette utilisée en médecine traditionnelle pour le traitement de l'hypertension artérielle. *Mali Médical*, **24** (4), 1–6.

Sanogo, R., Traoré, F., Djimdé, A. *et al.* (2012) Formulation de sirops antipaludiques à base d'extraits de *Argemone mexicana*, in *Pharmacopée et Médecine et Traditionnelles Africaines*, vol. 16, http://publication.lecames.org/index.php/pharm/article/view/22 (accessed 14 November 2013).

Sharma, N., Trikha, P., Athar, M. and Raisuddin, S. (1999) Protective effect of *Cassia occidentalis* extract on chemical-induced chromosomal aberrations in mice. *Drug and Chemical Toxicology*, **22** (4), 643–653.

Sutovska, M., Franova, S., Priseznakova, L. *et al.* (2009) Antitussive activity of polysaccharides isolated from the Malian medicinal plants. *International Journal of Biological Macromolecules*, **44** (3), 236–239.

Tona, L., Kambu, K., Ngimbi, N. *et al.* (1998) Antiamoebic and phytochemical screening of some Congolese medicinal plants. *Journal of Ethnopharmacology*, **61** (1), 57–65.

Tona, L., Ngimbi, N.P., Tsakala, M. *et al.* (1999) Antimalarial activity of 20 crude extracts from nine African medicinal plants used in Kinshasa, Congo. *Journal of Ethnopharmacology*, **68** (1–3), 193–203.

Tona, L., Kambu, K., Ngimbi, N., *et al.* (2000) Antiamoebic and spasmolytic activities of extracts from some antidiarrhoeal traditional preparations used in Kinshasa, Congo. *Phytomedicine*, **7** (1), 31–38.

Tona, L., Mesia, K., Ngimbi, N.P. *et al.* (2001) *In-vivo* antimalarial activity of *Cassia occidentalis, Morinda morindoides* and *Phyllanthus niruri. Annals of Tropical Medicine and Parasitology*, **95** (1), 47–57.

Tona, L., Cimanga, R.K., Mesia, K., *et al.* (2004) *In vitro* antiplasmodial activity of extracts and fractions from seven medicinal plants used in the Democratic Republic of Congo. *Journal of Ethnopharmacology*, **93** (1), 27–32.

Toure, I.A.K. (1989) Evaluation of the therapeutic efficacy of an improved traditional recipe (Gastrosédal) in the treatment of gastritis. MD thesis, Faculté de Médecine, Pharmacie et Odonto-Stomatologie, Université de Bamako, Bamako, Mali (in French).

Traoré, B. (1995) Contribution to research on a traditional treatment for eczema: preliminary study on the feasibility of a clinical protocol. MD thesis, Faculté de Médecine, Pharmacologie et Odonto-Stomatologie, Université de Bamako, Bamako, Mali (in French).

Vanderperren, B., Rizzo, M., Angenot, L. *et al.* (2005) Acute liver failure with renal impairment related to the abuse of senna anthraquinone glycosides. *Annals of Pharmacotherapy*, **39** (7–8), 1353–1357.

WHO (1978) *Primary Health Care: Report of the International Conference on Primary Health Care of Alma-Ata*, World Health Organization, Geneva.

WHO (1996) *Expert Committee on Specifications for Pharmaceutical Preparations: Thirty-fourth Report*, World Health Organization, Geneva.

WHO (1998) *Guidelines for the Appropriate Use of Herbal Medicines*, WHO Regional Office for the Western Pacific, Manila.

WHO (2000) *Regional Committee for Africa – Fiftieth session*, www.who.int/director-general/speeches/2000/english/20000828_afro.html (accessed 14 November 2013).

WHO (2001) *Promotion du Rôle de la Médecine Traditionnelle dans le Système de Santé: Stratégie de la Région Africaine*, AFR/RC50/9, Bureau Régional pour l'Afrique, Harare, Zimbabwe.

WHO (2003) *Traditional medicine*, Fact Sheet no. 134, www.who.int/mediacentre/factsheets/fs134/en/ (accessed 14 November 2013).

WHO (2004) *Tools for Institutionalizing Traditional Medicine in Health Systems of Countries in the WHO African Region*, AFR/TRM/04.3, WHO Regional Office for Africa, Brazzaville, Republic of Congo.

WHO (2008) *The Ouagadougou Declaration on Primary Health Care and Health Systems in Africa: Achieving Better Health for Africa in the New Millennium*, http://ahm.afro.who.int/issue12/pdf/AHM12Pages10to21.pdf (accessed 14 November 2013).

Willcox, M.L., Graz, B., Falquet, J. *et al.* (2007) *Argemone mexicana* decoction for the treatment of uncomplicated falciparum malaria. *Transactions of the Royal Society of Tropical Medicine & Hygiene*, **101** (12), 1190–1198.

Willcox, M.L., Graz, B., Diakite, C. *et al.* (2011a). Is parasite clearance clinically important after malaria treatment in a high transmission area? A 3-month follow-up of home-based management with herbal medicine or ACT. *Transactions of the Royal Society of Tropical Medicine and Hygiene*, **105**, 23–31.

Willcox, M.L., Graz, B., Falquet, J. *et al.* (2011b) A new methodology for developing antimalarial phytomedicines. *Malaria Journal*, **10** (Suppl 1), S8, www.malariajournal.com/content/10/S1/S8 (accessed 14 November 2013).

Willcox, M., Sanogo, R., Diakité, C. *et al.* (2012) Improved traditional medicines in Mali. *Journal of Alternative and Complementary Medicine*, **18**, 212–220.

# 16 Healing Aloes from the Mascarenes Islands

Joyce Govinden-Soulange

Faculty of Agriculture, University of Mauritius, Réduit, Mauritius

## 16.1 Introduction

Since time immemorial, mankind has been relying on plants for fulfilling vital needs, including food and primary health care. In spite of all the advancement made in synthetic organic chemistry and technology, plants still constitute a fundamental source of food, drugs, cosmetics and raw materials. Alkaloids, cardiac glycosides, essential oils and organic rubber are but a few commercial products isolated from plants. In fact, about 255 of the drugs prescribed worldwide come from plants and a considerable number of synthetic drugs are obtained from natural precursors (Rates, 2001). A few of the countless examples include digoxin from *Digitalis* spp., quinine and quinidine from *Cinchona* spp., vincristine and vinblastine from *Catharanthus roseus*, and morphine and codeine from *Papaver somniferum*. In recent years, there has been increasing interest in unconventional therapies and therapeutic use of natural products, especially those derived from plants. The inaccurate or abusive use of synthetic drugs, including antibiotics, often results in undesirable effects, and populations are increasingly inclined towards the use of natural products for health care and food.

## 16.2 The Asphodelaceae

One plant family that has sustained interest for the past decade is the aloe group belonging to the Asphodelaceae. The latter is a subfamily of the monocot family Xanthorrhoeaceae in the order Asparagales with at least 17 genera and 814 species distributed in Asia, Europe, Africa, Australia and the USA. The Asphodelaceae subfamily name is derived from the generic name of the type genus, Asphodelus. It includes the *Aloëceae* (*Aloeceae*, *Aloeaceae*). The *Aloe* genus has approximately 400 species, is by far the largest genus in Alooideae and also the most widespread (Reynolds, 1966; Viljoen, 1999; Glen and Hardy, 2000; Klopper and Smith, 2007). It is distinguished from other Alooideae genera by many features, including morphology, growth form and distribution.

The plants are mostly perennial or annual herbs, or shrubs, or 'arborescent' (with some woody forms with trunks up to several metres high). They are succulent or non-succulent perennial plants commonly with a basal aggregation of leaves, or with terminal aggregations of leaves, when woody. Their leaves are usually small to very large, alternate, 'herbaceous', or leathery, or fleshy, or leathery and fleshy (Watson and Dallwitz, 1992). The leaves are simple, and the leaves' sheaths have free margins, with laminar neither inverted nor twisted through 90°, linear, lanceolate or ovate. The mesophyll contains mucilage cells (with raphides) and calcium oxalate crystals. The stem can possess secondary thickening, from a single cambial ring, or can be absent or anomalous (e.g. *Aloë*) (Watson and Dallwitz, 1992). The inflorescence consists of a raceme or panicle, simple or branched with pedicellate flowers persistent at the base of pedicels, branches and sometimes along branches. The flowers are either actinomorphic or sometimes zygomorphic. The flowers occur as aggregates in the inflorescence, in racemes, or in spikes. The fruit is usually a dehiscent capsule, mostly non-fleshy; or fleshy (e.g. *Lomatophyllum*) with endospermic seeds. The plants of the Asphodelaceae family are normally hermaphrodites with floral nectarines and nectar secretion from the gynoecium (via septal nectarines).

Asphodelaceae may be distinguished from other lilioid monocot groups by a combination of characters: general presence of anthraquinones, lack of steroidal saponins, simultaneous microsporogenesis, atypical ovular morphology and presence of an aril. Each of these occurs in other asparagoid groups, so it may be only the combination that is distinctive. All Alooideae have caps of 'aloin' cells at their phloem poles, which are thought to be involved in secondary metabolite production and are absent in most, but not all, Asphodeloideae (Beaumont *et al*., 1985).

There is still an ongoing debate as to whether aloes and their close relatives should be placed in the Asphodelaceae, Liliaceae or Aloaceae family. Dahlgren *et al.* (1985) claimed that both the *Aloe* and *Lomatophyllum* genera are in the same newly identified family of Asphodelaceae and are classified in a common subfamily, Alooideae, hence agreeing with the APG I (Angiosperm Phylogeny Group, 1998) classification (Bremer *et al*., 2004). Moreover, the inclusion of *Lomatophyllum* within the *Aloe* genus is also supported by molecular data. Diversity assessed by molecular markers is independent of the influence of environment and estimated using DNA from any growth stage. Molecular systematics makes use of data on the molecular constitution of biological organisms, namely nucleic acids (DNA, RNA) and proteins, in order to resolve questions in systematics such as their correct classification or taxonomy. As gene sequencing becomes easier and cheaper, molecular systematics is being applied to more and more groups of living organisms. Chase *et al.* (2000) sequenced the plastid *rbcL* gene for 21 taxa to evaluate the placement of Asphodelaceae in the asparagoid lilies, particularly their separation from Anthericaceae. Molecular systematics

*Novel Plant Bioresources: Applications in Food, Medicine and Cosmetics*, First Edition. Edited by Ameenah Gurib-Fakim.

including gene sequencing has been used to enlighten the phylogenetic relationships within the Asphodelaceae and compare their patterns of molecular evolution. Using parsimony analysis, Chase *et al.* (2000) have grouped *Lomatophyllum* and *Aloe* in the same clade within the subfamily Alooideae.

Various methods can be used for the detection of polymorphisms in trees, which reveal significant DNA sequence polymorphisms and detect genetic variation among cultivars and within populations. These include numerical and structural chromosomal changes: restriction fragment length polymorphism (RFLPs) of nuclear and organellar genomes (Rani and Raina, 2000), random oligonucleotide fingerprinting patterns, inter simple sequence repeats (ISSRs; Martin *et al.*, 2004), random amplified polymorphic DNA (RAPD; Santos *et al.*, 2008), amplified fragment length polymorphism (ALFP; Pang *et al.*, 2006), single sequence repeats (SSRs; Ahmad *et al.*, 2004) and inter-retrotransposon amplified polymorphism (IRAP; Leigh et *al.*, 2003). DNA markers are superior to traditional phenotypic markers because they are less affected by age, physiological condition of samples and the environment; they are not tissue specific and can be detected at any developmental phase of the organism. Comparing DNA sequences will help to accurately resolve phylogenetic relationships between the two genera and eradicate the aforementioned confusion that surrounds them.

Polymerase chain reaction (PCR) has been used as a technique for the recognition of aloes and identification of their plant origin. Three DNA markers, specific for *Aloe* and reproducible at low concentrations were developed. Adams *et al.* (2000) investigated the physical organization of 18S-5.8S-26S and 5S ribosomal DNA (rDNA) using fluorescent *in situ* hybridization in 13 *Aloe* species. The organization was compared with a phylogenetic tree of 28 species constructed by sequence analysis of the internal transcribed spacer (ITS) of 18S-5.8S-26S rDNA. In a phylogenetic study carried out by Grace (2009), new sequences of the ribosomal ITS, chloroplast *trnL* intron, *trnL-f* spacer and the *matk* gene were acquired from 29 species of the *Aloe* section *Pictae*. Chase *et al.* (2000) acquired indels pointing to the closeness of *Aloe* and *Lomatophyllum*. Darokar *et al.* (2003) carried out a molecular assessment of germplasm diversity in *Aloe* species using RAPD and AFLP analysis, which revealed considerable inter- and intra-specific variation. Tripathi *et al.* (2011) have confirmed the high discriminative capacity of AFLPs to assess polymorphisms within closely related accessions of *Aloe vera*.

Within the subfamily Alooideae, *Lomatophyllum* comprises 212 species (Marais and Coode, 1978; Rauh, 1998; Pailler *et al.*, 2002), out of which 145 are from Madagascar, one from Pemba (Misali Island, Tanzania), one from Mayotte (Comoros archipelago), one from Aldabra (Seychelles archipelago), one from Réunion, two from Mauritius and one from Rodrigues (Mascarenes archipelago). '*Lomatophyllum*' has been reported to be a generic concept very slightly supported from the morphological, anatomical, cytological and palynologic points of view (Schill, 1973; Rauh, 1995, 1998; Rowley, 1996; Steyn and Smith, 2001). Van Wyk *et al.* (1995) have used chemotaxonomy to elucidate the dilemma though the characterization of root anthraquinones. Consequently, the common presence of anthraquinones chrysophanol (Figure 16.1) and asphodelin (Figure 16.2), known to be present in *Aloe* was also detected in all seven *Lomatophyllum* species studied. Van Wyk *et al.*'s (1995) postulates are also supported by the fact that anthraquinone precursors

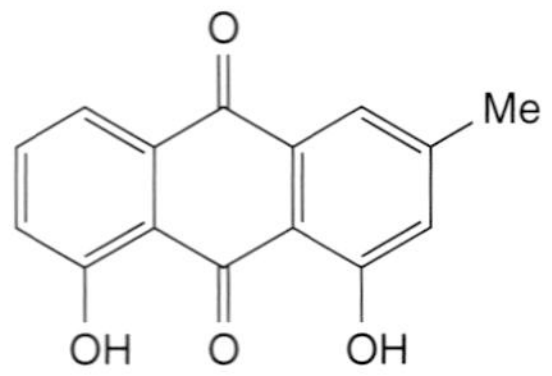


**Figure 16.1** Chrysophanol

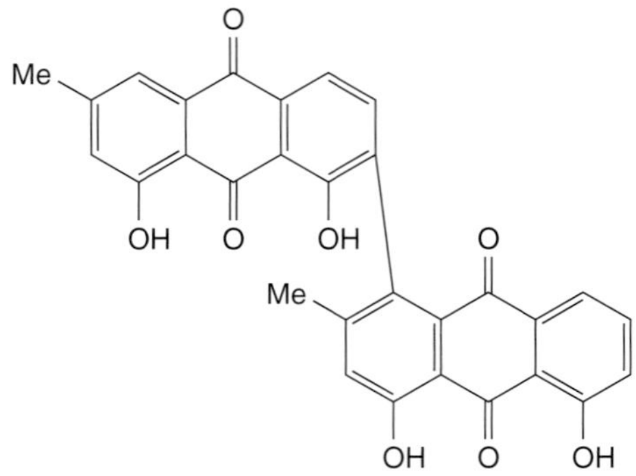


**Figure 16.2** Asphodelin

such as aloechrysone, identified in many *Aloes*, was also detected in six of the seven *Lomatophyllum* species investigated.

*Aloe tormentorii* (Marais) Newton & Rowley and *Aloe purpurea* Lam. are species endemic to Mauritius (Marais and Coode, 1978; Gurib-Fakim, 2003), while *Aloe macra* Haw. is endemic to Réunion Island (Bosser *et al.*, 1976; Marais and Coode, 1978; Pailler *et al.*, 2000). A taxonomic confusion between *A. purpurea* and *A. macra* still needs to be elucidated on the grounds of their morphological features. Marais (1974, 1975) has described some of the genus *Aloe* of the western Indian Ocean islands and Marais and Coode (1978) have tried to review these aloes. Thus, *A. purpurea* has a distinct reddish leaf-margin and larger inflorescences, whereas *A. macra* is slender and has smaller inflorescences. Conversely, these dissimilar morphological features are not sufficient to distinguish between *A. purpurea* and *A. macra* upon examination of a greater number of individuals of both species (Sanmukhiya *et al.*, 2010).

### 16.2.1 Healing aloes

Aloes are indigenous to Africa and the famous *A. vera* (*Aloe barbadensis* Mill.) (Figure 16.3), native to the Arabian Peninsula, is extensively cultivated, and there are over 250 species of aloe grown around the world (Grace *et al.*, 2008). *A. barbadensis* and *Aloe arborescens* are also cultivated in the USA in the Rio Grande valley of south Texas, Florida and southern California. Internationally, the plants can be found in Mexico, the Pacific rim countries, India, South America, the Caribbean, Australia and Africa. Aloes in South Africa (approximately 155) can be found in almost all vegetation types, from the arid to semi-arid regions, the winter rainfall fynbos region, succulent thickets, grassland and savanna, and even in comparatively high-rainfall forests (Gutterman and Chauser-Volfson, 2007). Most aloe species in southern Africa are used in ethnomedicine to treat microbial and parasitic infections of humans and animals,

Figure 16.3 *Aloe vera*

digestive disorders, pregnancy, labour and postnatal care, skin complaints, pain, respiratory system, genito-urinary system, poisoning and circulatory system, among others (Grace *et al.*, 2008). Several aloe species have also ben reported for their use in magical and ritual applications in South Africa.

Most aloes have small or fairly limited distribution ranges, and are often associated with only dry-land vegetation regions. Aloes are well adapted to the arid conditions found in their natural environment. Their succulent leaves enable them to survive prolonged periods of drought. Sharp thorns and spines, as well as usually bitter leaf sap, are good deterrents for many herbivores, although certain animals seem completely undaunted by these defence mechanisms of aloes, and the juicy leaves are often eaten. Aloes are generally drought resistant, and the mode of propagation is either by offsets or seeds. Aloes range in size from diminutive pot plants to large trees and massive clumps. The plants are characterized by rosettes of fleshy succulent leaves, which can grow close to the ground, or on branched and unbranched stems. There are variations: *Aloe plicatilis*, is distichous, while the unusual *Aloe polyphylla* from Lesotho grows in spirals. Some species, such as *A. vera*, have medicinal value and substantial cash value as crops. Aloes are also valued horticultural subjects, making stunning additions to gardens in drier climes. Aloe flowers vary tremendously in size, shape, colour and seasonal occurrence. The tubular flowers tend to be orange, red or yellow and appear on naked stalks from the centre of the leaf rosettes.

Mature aloes often produce offshoots, small rosettes that appear at the base of the plant. To produce a new plant, remove these offshoots gently and pot in crushed lava rock, sand, charcoal or pumice, mixed with a small amount of potting soil. Aloes are usually propagated from leaf cuttings from the mid-portion of the main rosette, where leaves are actively growing. To accelerate the rooting process, the end of the leaf can be treated with a rooting hormone and left to stand at room temperature for 2–3 days to allow the cut end to dry, forming a callus. The cutting can then be transferred into a planting mix, watered and kept in indirect light with daily misting and watering only lightly when the soil is dry. Rooting usually occurs in several weeks. However, *A. vera* plants are usually vegetatively propagated since their flowers are often male sterile (Natali *et al.*, 1990). Aloes flower in spring in warm tropical areas and cannot survive freezing temperatures. In young plants, the leaves appear at ground level, but the stem can grow up to 30 cm long in mature plants.

Aloes have been utilized to treat ailments in many countries of the world since, or probably long before, Roman times (Crosswhite and Crosswhite, 1984). These plants have a very long history of medicinal, cosmetic and cultural uses. Many early civilizations were familiar with the healing properties of *A. vera* in particular. This species likely originates from the Arabian Peninsula, but is today cultivated almost worldwide and has become naturalized in many areas. Over the past decades the use of *Aloe ferox* and *A. arborescens*, indigenous to South Africa, in the traditional pharmaceutical and cosmetic industry has increased significantly. Likewise, the leaves of *A. arborescens* can also be consumed as food (Gutterman and Chauser-Volfson, 2007). *Aloe excelsa* is one of the tree aloes found mostly in Zimbabwe, but species are also present on the southern African continent. Traditional healers from Zimbabwe have been using mainly the leaf gel exudates for treatment of various medicinal disorders (Coopoosamy and Magwa, 2007).

Aloe leaves contain a clear gel that is often used as a topical ointment. The green part of the leaf that surrounds the gel can be used to produce a juice or a dried substance (called latex) that is taken by mouth. The leaves of *Aloe* species are used both in traditional and modern medicine of many countries, including Africa, India, China, Mexico and in modern medicine of the USA and Europe, in the cosmetic industry and also as a food supplement; aloes are now also widely used in landscaping projects. Historically, aloe was used topically to heal wounds and for various skin conditions, and orally as a laxative. Today, in addition to these uses, aloe is used as a folk or traditional remedy for a variety of conditions, including diabetes, asthma, epilepsy and osteoarthritis. It is also used topically for osteoarthritis, burns, sunburns and psoriasis. *A. vera* gel can be found in hundreds of skin products, including lotions and sunblocks. The US Food and Drug Administration has approved *A. vera* as a natural food flavouring.

The first written account of the uses of *A. vera* is found in *De Materia Medica*, written in AD 41 by the Greek physician Dioscorides. From then till now, scientists have been researching to explore the uses of *A. vera*; it has a variety of uses, from beauty products, to medicinal uses, to commercial uses, to ornamental uses, to cooking. Aloe first gained popularity in the USA in the 1930s with reports of its success in treating X-ray burns (Moghaddasi and Verma, 2011). Aloes have been used as a multipurpose folk remedy, and are found in many pharmacopoeias. The bitter leaf exudate (drug aloes) is used worldwide as a laxative. The aloe gel, a thin clear jelly-like substance obtained from the parenchymal tissue, contains many bioactive ingredients, including amino acids, minerals, vitamins, enzymes and glycoproteins. This gel is known to have many therapeutic attributes. Many species of Asphodelaceae may be distinguished from members of the Liliaceae family by the presence of anthraquinones (Chase *et al.*, 2000). The phytochemistry and bioactivity of *Aloe* species have been widely reported since the late nineteenth century, when the trade in drug aloes prepared for the leaf exudate bloomed. Nowadays, the principal source of aloe-based products is the *A. vera* (L.) Burm. f., whose leaf parenchyma is used to produce the renowned 'aloe gel'. Diverse natural products have been isolated from the leaves of *A. vera*, including barbaloin (Manitto *et al.*, 1990; Bhavsar *et al.*, 1991), tetranitroaloeemodin (Djerassi, 1994), aloesin, isoaloesin (Yuan, 1993), *p*-sitosterol (Ando and Yamaguchi, 1990), gibberellins (Oden, 1991), polysaccharides, including carrisyn (McDaniel and McAnalley, 1987) and aloeferon (Madis, 1989), some alkanes, fatty acids and alkyl benzenes (Yamaguchi *et al.*, 1993). Anthraquinones are also well reputed for their biological attributes. Recent work has reported the presence of 12 of these phenolic compounds found exclusively in the plant sap, with the important ones being aloin and aloe emodin (Atherton, 1998; Sanmukhiya *et al.*, 2010). The presence of anthraquinones has also been reported in the roots of *Aloe saponaria* (Yagi *et al.*, 1974), *Aloe acutissima* (Iriki *et al.*, 1986), *Aloe berhana* (Dagne *et al.*, 1992) and *Aloe graminicolu* (Dagne *et al.*, 1994), *Lomatphyllun* (syn. *Aloe*): *Aloe aldabrense*, *Aloe antsingyense*, *Aloe lomatophylloides*, *Aloe macra*, *Aloe occidentale* and *Aloe purpureum* (Van Wyk *et al.*, 1995). Characteristic anthraquinone and preanthraquinone constituents from the roots of *Aloe* species have been described by Dagne *et al.* (1994) as being chrysophanol, asphodelin, aloechrysone, aloesaponarin, laccaic acid D-methyl ester, aloesaponol I and aloesaponol II.

Anthraquinones are known to act as laxatives and painkillers and to possess important antibacterial and antifungal effects. Chen et *al.* (2007) and Chiu et *al.* (2009) suggest that aloe-emodin represents a suitable and novel chemotherapeutic drug candidate for the treatment of human gastric carcinoma and human tongue squamous cancer. Moreover, emodin has been reported to have antiviral activities to some kinds of viruses, such as human cytomegalovirus, herpes simplex virus type 1 and poliovirus. The anthraquinone content of *A. vera* leaf extract measured as barbaloin has been reported to be of the order of 4.24% (Sanmukhiya *et al.*, 2010).

Aloe has also been used for the treatment of ailments associated with ageing, high blood pressure, diabetes, heart condition, liver, rheumatoid arthritis, kidney and digestive disorders. Yoo *et al.* (2008) have reported that ethanol extracts of *Aloe* may be useful as a drug against reactive-oxygen-species-mediated diseases.

Saponins are soapy substances forming about 3% of the *A. vera* gel and capable of cleansing and having antiseptic properties. The presence of salicylic acid has also been linked to the anti-inflammatory and antibacterial properties. These act powerfully as antimicrobials against bacteria, viruses, fungi and yeast (Atherton, 1998). The anti-arthritic activity of the anthraquinone present in the aloe gel and its anti-osteoarthrosis activity have also been reported (Davis *et al.*, 1986). In Mauritius, leaf extracts of aloes are applied on the breasts to help sever young infants from the mother (Gurib-Fakim, 2003).

More extensive work has been reported on the composition of the *A. vera* leaf gel whose main constituents are water (99.1%) and mesophyll cells (0.9% dry matter), which can be divided into three distinct fractions: cell wall, microparticles and liquid gel (Grundmann, 2012). The chief carbohydrate is mannose-6-phosphate, followed by other sugars in varying concentrations depending on the fraction. Overall, the five neutral sugars (i.e. arabinose, xylose, mannose, galactose and glucose) account for 69.2% of the total sugars in the gel. Mucopolysaccharides are mainly present as acemannan – a highly acetylated β-1-4-linked polysaccharide (>1 kDa) made mainly of mannose – with various side-chain glycosylation patterns. Acemannan (Figure 16.4) is the complex carbohydrate polymer material responsible for the immunostimulating activity ascribed to the *A. vera* leaf gel (Womble and Helderman, 1988; Tizard *et al.*, 1989).

The anthraquinone content should be below 50 ppm and is considered an impurity from the leaf extract of *A. vera* (Grundmann, 2012). Other ingredients include various amino acids, enzymes and vitamins, which have not been quantified, polysaccharides and glucomannan. Carboxypeptidase, magnesium, zinc, calcium, glucose, cholesterol, salicylic acid, prostaglandin precursors (gammalinolenic acid), vitamins A, C, E, lignins, saponins, plant sterols

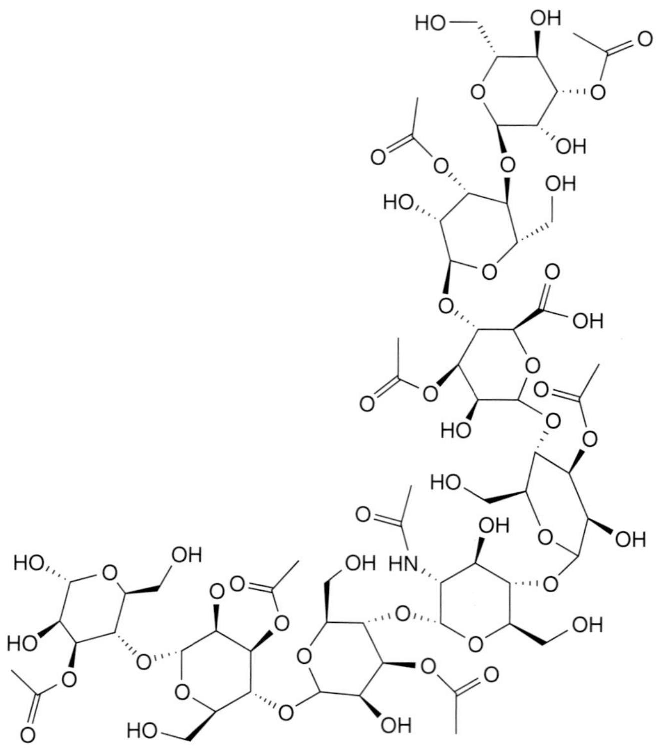

**Figure 16.4** Acemannan (Yates, 2012)

and amino acids have also been identified. From the latex leaf lining there are anthraquinone glycosides: aloin, aloe-emodin and barbaloin (Moghaddasi and Verma, 2011).

According to the Convention on International Trade of Endangered Species of Wild Fauna and Flora (CITES) checklist, there are over 450 species in the genus *Aloe*. Aloes are faced with many of the same threats that affect biodiversity worldwide: agricultural and livestock activities, mining and hydroelectric projects (and their associated road building), urban expansion, competition from invasive and exotic plants, and harvesting and collecting for medicinal and economic uses, including as ornamentals. Many African nations have plant conservation programmes and protection laws for aloes and other plants.

In 2012, the worldwide market for *A. vera* products was estimated to have reached $13 billion (IASC, 2012). The trade in *Aloe* is dominated by two markets: the cosmetic/medicinal industry and horticulture. Within the African continent, South Africa is the major exporter of *Aloe* plants both conventionally and micropropagated. South Africa exports important quantities of *A. ferox* for the cosmetic/medicinal industry, including an array of other wild *Aloe* taxa, in much smaller quantities, for the horticultural market. South Africa almost exclusively supplies the demand for wild *Aloe* leaf extracts (mainly *A. ferox*) for the markets in North America, Europe and Asia. Although not a major exporter of *Aloe*, Madagascar has also been a source of a range of wild *Aloe* plants for the ornamental, cosmetic and medicinal trades.

Canada, the Republic of Korea and Spain have exported the largest quantities of artificially propagated *A. vera* plants. The major markets for live plants are the USA, China and Switzerland.

Most *Aloe* species, endemic to the Mascarene Islands, are edging to the brink of extinction, being ruthlessly invaded in their habitat by exotic flora and owing to the innate dormancy or sterility of their seeds. In addition, harvesting of the natural products for research and industry has resulted in all species of *Aloe*, with the exception of *A.vera*, being protected by national as well as international conventions, namely CITES (Grace *et al.*, 2008).

### 16.2.2 Mascarene aloes

The Mascarene Islands (French: Îles des Mascareignes) collectively comprise the islands of Réunion, Mauritius and Rodrigues, which are situated in a line along a submarine ridge, the Seychelles–Mauritius Plateau, 640–800 km northeast from southern Madagascar in the western Indian Ocean. All these islands are volcanic in origin. The name Mascarene is taken from the 16th-century Portuguese explorer Pedro de Mascarenhas. All aloes growing in the Mascarene Indian Ocean belong to the section *Lomatophyllum*, 12 in all (Smith and Van Wyk, 1991) and most probably originate from Madagascar, having been spread by marine currents (Castillon and Castillon, 2010).

Consequently, the 12 Lomatophyllum aloes of the Mascarenes have been reported as *Aloe olygophylllum*, *A. antsingyense*, *A. lomatophylloides*, *A. macra*, *A. occidentale*, *Aloe orientale*, *A. purpurea*, *Aloe prostratum*, *A. tormentorii*, *Aloe roseum*, *Aloe sociale* and *Aloe viviparum*. Both *A. vera* and the *Aloe* section *Lomatophyllum* belong to the same subfamily Alooidae and family Asphodelaceae.

*A. tormentorii* (Marais) Newton & Rowley and *A. purpurea* Lam. are species endemic to Mauritius (Marais and Coode, 1978; Gurib-Fakim, 2003), while *A. macra* Haw. is endemic to Réunion Island and *A. lomatophylloides* is native to Rodrigues (Bosser *et al.*, 1976; Marais and Coode, 1978; Pailler *et al.*, 2000). However, a taxonomic confusion exists between *A. purpurea* and *A. macra* even if the several authors have described the two species as different (Sanmukhiya *et al.*, 2010). Marais (1974, 1975) has described some of the genus *Aloe* of the western Indian Ocean islands and Marais and Coode (1978) reported differences in the morphology of these aloes. Thus, *A. purpurea* has a distinct reddish leaf-margin and larger inflorescences, whereas *A. macra* is slender and has smaller inflorescences.

*A. macra*, *A. purpurea* and *A. tormentorii*, in spite of having some morphological differences, are often confused among themselves or with other members of the *Aloe* genus (Sanmukhiya *et al.*, 2010), particularly with the introduced species *A. vera*, locally named as 'mazambron'. Aloes indigenous to the Mascarenes are vernacularly known as 'mazambron marron', 'mazambron sauvage' or 'ananas marron'. *A. purpurea* has 1 m long dark green leaves with pink or red margins and orange–red to pinkish orange flowers. *A. tormentorii* has 60 cm × 15 cm rigid and straight, pale green or bluish leaves with a marginal pink border and is found on Round Island and Gunner's Coin Island (Mauritius). *A. lomatophylloides* (ananas marron) is endemic to Rodrigues; it has narrow and light green–yellow leaves up to 75 cm × 8 cm long (Bosch, 2008). *A. macra*, endemic to Réunion Island, looks like *A. purpurea* but it consists of a short stem and has smaller inflorescences (Sanmukhiya *et al.*, 2010).

### 16.2.3 *Aloe lomatophylloides* Balf.f

*A. lomatophylloides* (Figure 16.5) is a stemless succulent herb up to 40 cm tall with bright green leaves arranged in a rosette, erect to spreading without stipules or petiole. The leaf blade is lanceolate, up to 75 cm × 8 cm in size with short teeth 5–15 mm apart at the margin. The inflorescence has two to three branches and can reach up to 30 cm with small pinkish orange flowers. The fruits are multi-seeded berries. The plant usually thrives in dry habitats.

The plant is reputed for its use in the folk medicine of Rodrigues Island. The crushed leaves are applied as a poultice as a muscle pain reliever. A decoction of the leaves is also taken to increase

**Figure 16.5** *Aloe lomatophylloides*. *Photo*: Lobine 2013. Reproduced with permission

the menstrual flow (Gurib-Fakim, 2003). The plant is mainly cultivated as an ornamental. *A. lomatophylloides* is no longer found in the wild as its population has been decimated due to habitat destruction and overharvesting. Some 30 plants are conserved in the Grande Montagne Reserve and at Cascade Victoire in the north of Rodrigues Island (Gurib-Fakim *et al.*, 1994). *A. lomatophylloides* is now a famous ornamental in botanical gardens and is actively being propagated and, therefore, is not heavily threatened with extinction. The phytochemistry of *A. lomatophylloides* has not been fully described but is thought to be composed of flavones, isovitexin (apigenin-6-C-glucoside) occurring together with the anthronealoin (Schmelzer and Gurib-Fakim, 2008). Van Wyk *et al.* (1995) have reported the presence of eight anthraquinones common to *A. vera* in the roots of *A. lomatophylloides*; namely, chrysophanol, asphodelin, aloechrysone, aloesaponarin, laccaic acid D-methyl ester, aloesaponol I and aloesaponol II, thereby providing additional evidence for the inclusion of *Lomatophyllum* into the *Aloe* genus. The molecular, phytochemical and bioactive profiles of *A. lomatophylloides* have yet to be described.

### 16.2.4 *Aloe macra* Haw.

*A. macra* (Figure 16.6), traditionally known as mazambron marron or mazambron sauvage, is native to Réunion Island (France) and is found only in the dry to semi-dry ecosystems of the island, including the rocky cliffs and dry forests of Grand Bassin, Mafate, Grande Chaloupe (Castillon and Castillon, 2010). In the folk medicine of Réunion Island, the leaves of *A. macra* are used to treat minor infections, boils, constipation and otherwise as a healing substance for external use. These plants have also been used as a painkiller and as a laxative (Marais and Coode, 1978).

*A. macra* is quite variable in morphology and can have a short compact stem and small canaliculated leaves, sometimes up to 30 cm long or much longer. Species with long stems have also been reported on Réunion Island, hence the confusion with *A. purpurea* from Mauritius. *A. macra* yields small reddish orange flowers. The molecular, bioactivity and phytochemical profile of *A. macra* has been described by Sanmukhiya *et al.* (2010). These studies have revealed that, in addition to the common compounds reported in all aloes (namely, anthraquinones, carbohydrates, flavonoids, steroids, sugars and polysaccharides), other flavonoids, such as apigenin, isoorientin and isovitexin, have been identified in *A. macra* (Sanmukhiya *et al.*, 2010). However, *A. macra* has been found to have

**Figure 16.6** *Aloe macra* Haw. *Photo*: Lavergne 2013. Reproduced with permission

inferior barbaloin content compared with other Mascarene aloes (Sanmukhiya *et al.*, 2010). RAPD markers have disclosed more than 28 polymorphic bands between *A. vera* and the Mascarene *Aloe* species, showing the marked difference which exists between species of different origin. Consequently, authors have confirmed that *A. macra* and *A. purpurea* are genetically distinct species.

The population of *A. macra* has been declining due to the destruction of its habitat, by alien herbivores and predators (IUCN, 1997). Classified as endangered in Réunion Island, it is among the most damaged of all native vegetation communities in this country, having been decimated by introduced giant African land snails. The other main threat is predation by rats, which have also been introduced to Réunion Island (Cadet, 1984). Propagation efforts have been highly successful, but attempts to reinforce wild populations have often been met with difficulties owing to the impact of the alien herbivores and predators. The challenge is now to successfully re-establish viable populations in the original habitat, which will depend on the ability to curtail damage by both rats and snails.

*A. macra* has been reported to contain steroid and possess antiviral activity against herpes simplex type 1 virus (HSV-1; Fortin *et al.*, 2002). Sanmukhiya *et al.* (2010) have also reported the antibacterial activity of different fractionated extracts of *A. macra* against *Escherichia coli*, *Klebsiella pneumoniae* and *Staphylococcus aureus*, hence validating its use in the folk medicine of Réunion Island. Interestingly, the crude extracts of *A. macra* from Réunion Island demonstrated a more significant antimicrobial activity against the tested microorganisms than those from other Mascarene aloes and even *A. vera*. This potent antimicrobial activity has been attributed to the presence of isoorientin, a luteolin-6-C-glucoside, a flavone not detected in other Mascarene aloes or even the *A. vera*.

### 16.2.5 *Aloe purpurea* Lam.

Traditionally referred to as mazambron marron, mazambron sauvage and socotrine du pays, *A. purpurea* (Figure 16.7), is native to Mauritius but is also thought to exist in Réunion. *A. purpurea* is the only Mascarene aloe that forms a shrub, with a solitary trunk up to 3 m tall. The leaves are deep green or brownish green with pink or red margins with short spines, close to each other at the base and spaced at the tip. Flowers are borne in panicles and are reddish orange in colour. The fruit, a 2-cm-sized berry, is bigger than fruits of *A. macra*. *A. purpurea* is considered endangered and can be found growing in nature parks such as the Yemen forest and the Black River Gorges and in private gardens around the island.

The anthraquinone content of the roots of *A. purpurea* (then known as *Lomatophyllum purpurea*) has been portrayed by Van Wyk *et al.* (1995) and has been shown to contain six anthraquinones common to aloes; namely, chrysophanol, asphodelin, aloechrysone, laccaic acid D-methyl ester, aloesaponol I and aloesaponol II. Preliminary phytochemical screening of *A. purpurea* leaves from Mauritius has revealed the presence of alkaloids, anthraquinones, terpenes, phenols, flavonoids, coumarins, saponins and tannins (Sanmukhiya *et al.*, 2010). Nonetheless, *A purpurea* leaf extracts have been found to contain the highest anthraquinone (1.13% barbaloin) levels compared with the other Mascarene aloes *A. macra* and *A. tormentorii*. Except for the common presence of reddish orange flowers, morphologically, *A. macra* can be distinguished from *A. purpurea* by a spindlier character and the absence of red margin on the leaf borders

(Marais, 1975). Molecular profiling through the use of RAPD markers has confirmed the disparity between *A. macra* and *A. purpurea* at the genetic level (Sanmukhiya *et al.*, 2010).

Reports on the biological activities of *A. purpurea* are limited. Crude leaf methanolic extracts of *A. purpurea* from Mauritius have been shown to possess a higher antimicrobial activity against *E. coli*, *K. pneumoniae* and *S. aureus* than *A. vera*. This significant bioactivity of *A. purpurea* crude leaf extracts has been attributed to the difference in its flavonoid profile compared with *A. vera* (Govinden-Soulange *et al.*, 2009; Sanmukhiya *et al.*, 2010).

### 16.2.6 *Aloe tormentorii* (Marais) Newton & Rowley

This aloe (Figure 16.8) is commonly known as 'mazambron vert' and grows on two small rocky islands in the north of Mauritius: Round Island and Gunner's Coin Island. This plant is also used as ornamental in botanical gardens and resorts. It is a stemless or short-stemmed plant, not erect, but lying with a few suckers and thick, compact rosette-type leaves that thrives well in dry, sunny areas. Leaves are rigid and straight, triangular, pale green or greenish blue and brown when exposed to the sun. At maturity, the leaf borders turn pinkish and have a spiny border. It has branched inflorescences with flowers borne in compact panicles, which are salmon pink in colour and hermaphrodite. The fruit is a 2.5-mm-sized berry.

The phytochemical and bioactive profiles of *A. tormentorii* have been described. Consequently, alkaloids, anthraquinones, terpenes, phenols, flavonoids, coumarins, saponins and tannins have been detected in leaf extracts of *A. tormentorii*, and the flavonoids detected in this species have been described to belong to the apigenin

**Figure 16.7** *Aloe purpurea* Lam. *Photo*: Lobine 2013. Reproduced with permission

**Figure 16.8** *Aloe tormentorii* (Marais) Newton & Rowley. *Photo*: Pynee 2013. Reproduced with permission

family. *Aloe* species from Mauritius are richer in anthraquinone (0.95% expressed as barbaloin for *A. tormentorii* leaf extracts) than those from Réunion Island (Sanmukhiya *et al.*, 2010). DNA profiling of these Mascarene aloes has revealed that *A. purpurea* is genetically closer to *A tormentorii* as shown by PCR amplifications (Govinden-Soulange *et al.*, 2009; Sanmukhiya *et al.*, 2010). The antimicrobial activity of *A. tormentorii* against *E. coli*, *K. pneumoniae* and *S. aureus* has been portrayed as significantly superior than the antimicrobial activity of *A. vera* extracts against the same bacteria but weaker than that of *A. purpurea* (Govinden-Soulange *et al.*, 2009; Sanmukhiya *et al.*, 2010). The higher barbaloin concentration of *A. purpurea* (1.13% versus 0.95%) was attributed to this more significant antimicrobial effect.

## 16.3 Prospects and research avenues

The *A. vera* plant, its juice and derivatives are highly reputed in Mauritius and in the south west Indian Ocean Islands. Highly reputed as the most common medicinal plant in this decade, it is found in almost every garden of the Mascarene inhabitants. In addition, its products are marketed in the cosmetic, medicinal and the food industries. This review has highlighted the potential of Mascarene aloes, which can be exploited provided further research is carried out in order to validate their properties.

The principal inference from this review highlights the fact that the aloes of the Mascarenes have not been comprehensively studied. In view of the interesting preliminary screening reports with respect to phytochemistry and bioactivity of Mascarene aloes, there is much prospect for advanced research with respect to the identification of phytochemicals using sophisticated chromatographic techniques (gas chromatography, high-performance liquid chromatography) for the identification and characterisation of bioactive molecules. The pharmacological potential of Mascarene aloes still needs to be further examined especially with respect to the antioxidant, antiviral and anticancer attributes. Metabolomic profiling of the Mascarene aloes through the use of nuclear magnetic resonance spectroscopy is another thrust area for establishing the relationship between the bioactivity of extracts and phytochemical profile. This technology coupled with isolation and purification of bioactive molecules will certainly target and identify lead molecules with a view to pharmacological exploitation.

The application of molecular systematics by using data on the molecular structure of organisms, namely nucleic acids (DNA, RNA) and proteins, in order to resolve questions in systematics such as their correct classification or taxonomy is also of relevance in the context of Mascarene aloes. As gene sequencing becomes easier and cheaper, molecular systematics is being applied to more and more groups of living organisms. Germplasm diversity of Mascarene *Aloe* species in relation to other aloes still remains to be elucidated. Molecular markers (RFLP, AFLP, ISSR and IRAP) can be considered to characterize and to establish genetic relationships by revealing DNA sequence polymorphisms and detect genetic variation among Mascarene aloes.

*In vitro* propagation is one of the key modern technologies for plant conservation *ex situ*. It provides a means for generating collections of endangered species with enhanced multiplication rates, which are genetically identical to ensure homogeneity. Given that most aloes of the Mascarene are heavily endangered, it would be judicious to investigate the application of micropropagation technology through tissue culture to rescue and restore these endangered species.

## References


Adams, S.P., Leitch, I.J., Bennett, M.D. *et al.* (2000) Ribosomal DNA evolution and phylogeny in *Aloe* (Asphodelaceae). *American Journal of Botany*, **86**, 985–990.

Ahmad, R., Potter, D. and Southwick, S.M. (2004) Genotyping of peach and nectarine cultivars with SSR and SRAP molecular markers. *Journal of the American Society for Horticulture Science*, **129**, 204–210.

Ando, N. and Yamaguchi, I. (1990) β-Sitosterol from *Aloe vera* [*Aloe vera* (L.) Burm. f.] gel. *Kenkyu Kiyo-Tokyo Kasei Daigaku*, **30**, 15–20 [*Chemical Abstracts* (1990), **113**, 148952].

Angiosperm Phylogeny Group (1998) An ordinal classification for the families of flowering plants. *Annals of the Missouri Botanical Garden*, **85** (4), 531–553.

Atherton, P. (1998) *Aloe vera*: magic or medicine? *Nursing Standards*, **41**, 49–50.

Beaumont, J., Cutler, D.F., Reynolds, T. and Vaughan, J.G. (1985) The secretory tissue of aloes and their allies. *Israel Journal of Botany*, **34**, 265–282.

Bhavsar, G.C., Chauhan, M.G. and Upadhyay, U.M. (1991) Evaluation of Indian aloes. *Indian Journal of Natural Products*, **6**, 11–13 [*Chemical Abstracts*, **116**, 11091].

Bosch, C.H. (2008). *Aloe lomatophylloides* Balf. f., in *Prota 11(1): Medicinal Plants/Plantes Médicinales 1* [CD-ROM] (eds G.H. Schmelzer and A. Gurib-Fakim), PROTA, Wageningen.

Bosser, J., Cadet, Th., Gueho, J. *et al.* (eds) (1976) *Flore des Mascareignes: La Reunion, Maurice, Rodrigues*, Mauritius Sugar Industry Research Institute/L'Office de la Recherche Scientifique et Technique Outre-Mer/Royal Botanic Gardens, Kew, Port Louis/Paris/London.

Bremer, K., Mark, W.C. and Peter, F.S. (2004) An ordinal classification for the families of flowering plants. *Annals of the Missouri Botanical Gardens*, **85**, 531–553.

Cadet, T. (1984) *Plantes rares ou remarquables des Mascareignes*, Agence de Cooperation Vulturelleet Technique, Paris.

Castillon, J.B. and Castillon, J.P. (2010) *The Aloes of Madagascar*, Etang Sale/MSM Ltd, Réunion/Mauritius.

Chase, M.W., de Bruijn, A., Cox, A.V. *et al.* (2000) Phylogenetics of Asphodelaceae (Asparagales): an analysis of plastid *rbcL* and *trnL-F* DNA sequences. *Annals of Botany*, **86**, 935–951.

Chen, S.H., Lin, K.Y., Chang, C.C. *et al.* (2007). Aloe-emodin-induced apoptosis in human gastric carcinoma cells. *Food and Chemical Toxicology*, **45** (11), 2296–2303.

Chiu, T.H., Lai, W.W., Hsia, T.C. *et al.* (2009). Aloe-emodin induces cell death through S-phase arrest and caspase-dependent pathways in human tongue squamous cancer SCC-4 cells. *Anticancer Research*, **29**, 4503–4511.

Coopoosamy, R.M. and Magwa, M.L. (2007) Traditional use, antibacterial activity and antifungal activity of crude extract of *Aloe excelsa*. *African Journal of Biotechnology*, **6**, 2406–2410.

Crosswhite, F.S. and Crosswhite, C.D. (1984) *Aloe vera*, plant symbolism and the threshing floor. *Desert Plants*, **6**, 43–50.

Dagne, E., Casser, I. and Steglich, W. (1992) Aloechrysone, a dihydroanthracenone from *Aloe berhana*. *Phytochemistry*, **31**, 1791–1793.

Dagne, E., Yenesew, A., Asmellash, S. *et al.* (1994) Anthraquinones, pre-anthraquinones and isoeleutherol in the roots of *Aloe* species. *Phytochemistry*, **35**, 401–406.

Dahlgren, R.M.T., Clifford, H.T. and Yeo, P.F. (1985) *The Families of the Monocotyledons, Structure, Evolution and Taxonomy*, Springer-Verlag, Berlin.

Darokar, M.P., Rai, R., Gupta, A.K. *et al.* (2003) Molecular assessment of germplasm diversity in *Aloe* species using RAPD and AFLP analysis. *Journal of Medicinal and Aromatic Plant Sciences*, **25**, 354–361.

Davis, R.H., Agnew, P.S. and Shapiro, E. (1986) Antiarthritic activity of anthraquinones found in *Aloe* for podiatric medicine. *Journal of the American Podiatric Medical Association*, **76**, 61–66.

Djerassi, C. (1994) *Dictionary of Natural Products*, Chapman and Hall, London.

Fortin, H., Vigor, C., Lohezie-Le Devehat, F. *et al.* (2002). *In vitro* antiviral activity of thirty six plants from La Réunion Island. *Fitoterapia*, **73**, 346–350.

Glen, H.F. and Hardy, D.S. (2000) Aloaceae: *Aloe*, in *Flora of Southern Africa*, vol. 1, pt. 1, fasc. 1, *Aloaceae (First Part): Aloe* (ed. G. Germishuizen), National Botanical Institute, Pretoria, pp. 1–16.

Govinden-Soulange, J., Khoyratty, S., Ranghoo-Sanmukhiya, M. *et al.* (2008) DNA and antimicrobial fingerprinting of *Aloe* species from the Mascarene Islands, in *African Journal of Traditional, Complementary and Alternative Medicines (AJTCAM), Abstracts of the World Congress on Medicinal and Aromatic Plants, Cape Town, November 2008*, pp. 313–314.

Grace, O.M. (2009) Contributions to the systematics and biocultural value of *Aloe* L. (Asphodelaceae). Thesis, University of Pretoria.

Grace, O.M., Simmonds, M.S.J., Smith, G.F. and van Wyk, A.E. (2008) Therapeutic uses of *Aloe* L. (Asphodelaceae). *Journal of Ethnopharmacology*, **119**, 604–614.

Grundmann, O. (2012) *Aloe vera* gel research review. *Natural Medicine Journal*, **4** (9), www.naturalmedicinejournal.com/article_content.asp?edition=1&section=2&article=356.

Gurib-Fakim, A. (2003) *An Illustrated Guide to the Flora of Mauritius and the Indian Ocean Islands*, Caractère Ltée, Baie du Tombeau, Mauritius.

Gurib-Fakim, A., Guého, J., Sewra, M.D. and Dulloo, E. (1994) *Plantes Médicinales de l'Île Rodrigues*, Editions de l'Océan Indien, Rose-Hill, Mauritius.

Gutterman, Y. and Chauser-Volfson, E. (2007) Secondary phenol metabolites, distribution and content of some *Aloe* species, originated from the arid zones of South Africa: a review. *American Journal of Food Technology*, **2**, 555–569.

IASC (2012) International Aloe Science Council, Summit Proceedings, Las Vegas, NV, November.

Iriki, Y., Iizuka, K., Furuse, M. *et al.* (1986) Oligosaccharides of *Aloe acutissma* (Part I). *Nippon Nogei Kagaku Kaishi*, **60**, 111–113 [*Chemical Abstracts* (1986), **105**, 75866].

IUCN (1997) *Red List of Threatened Plants* (ed. S.K. Malher), World Conservation Monitoring Centre.

Klopper, R.R. and Smith, G.F. (2007) The genus *Aloe* L. (Apshodelaceae: Alooideae) in Namaqualand, South Africa. *Haseltonia*, **13**, 1–13.

Leigh, F., Kalendar, R., Lee, V. *et al.* (2003) Comparison of the utility of barley retro-transposons families for genetic analysis by molecular markers techniques. *Molecular Genetics and Genomics*, **269**, 464–474.

Madis, V.H., Omar, M.M. and Madis, V. (1989) Aloeferon isolation, manufacturing and its applications, US Patent 4,861,761 [*Chemical Abstracts*, (1990), **112**, 164948].

Manitto, P., Monti, D. and Speranza, G. (1990) Studies on aloe. Part 6. Conformation and absolute configuration of aloins A and B and related 10-C-glucosyl-9-anthrones. *Journal of the Chemical Society, Perkin Transactions* **1** (5), 1297–1300.

Marais, W. (1974) The extra-Madagascan species of *Lomatophyllum* (Liliaceae). *Kew Bulletin*, **29**, 721–723.

Marais, W. (1975) A further note on *Lomatophyllum* (Liliaceae). *Kew Bulletin*, **30**, 601–602.

Marais, W. and Coode, M.J.E. (1978) 183. Liliacées, in *Flore des Mascareignes: La Réunion, Maurice, Rodrigues. 177. Iridacées á 188. Joncacées* (eds J. Bosser, T. Cadet, J. Guého *et al.*), Mauritius Sugar Industry Research Institute/Office de la Recherche Scientifique et Technique Outre-Mer/Royal Botanic Gardens, Kew, Port Louis/Paris/London, pp. 1–41.

Martin, M., Sarmento, D. and Oliveira, M.M. (2004) Genetic stability of micropropagated almond plantlets, as assessed by RAPD and ISSR markers. *Plant Cell Reports*, **23**, 492–496.

McDaniel, H.R. and McAnalley, B.H. (1987) Evaluation of polymannoacetate (carrysin) in the treatment of AIDS. *Clinical Research*, **35**, 483A.

Moghaddasi, M. and Verma, S. (2011) *Aloe vera* their chemicals composition and applications: *International Journal of Biological and Medical Research*, **2**, 466–471.

Natali, L., Castorena Sanchez, I. and Cavallini, A. (1990) *In vitro* culture of *Aloe barbadensis* Mill. Micropropagation from vegetative meristems. *Plant Cell Tissue and Organ Culture*, **20**, 71–74.

Oden, P. (1991) The use of gibberellins for the treatment of prostatitis and psoriasis, PCT Int. Appl. WO 199108751 A1 [*Chemical Abstracts*, **115**, 270698].

Pailler, T., Picot, F., Lavergne, C. and Strasberg, D. (2000) Statut démographique et biologie de la reproduction d'une espèce endémique menacée de l'Île de La Réunion: *Lomatophyllum macrum* (Haw.) Salm-Dyck (Liliacées). *Revue d'Écologie (Terre et Vie)*, **55**, 201–214.

Pang, X.M., Wen, X.P., Hu, C.G. and Deng, X.X. (2006) Genetic diversity of *Poncirus* accessions as revealed by amplified fragment length polymorphism (AFLP). *Journal of Horticultural Science & Biotechnology*, **81** (2), 269–275.

Rani, V. and Raina, S.N. (2000) Genetic fidelity of organized meristem derived micropropagated plants: a critical reappraisal. *In Vitro Cellular & Developmental Biology*, **36**, 319–330.

Rates, S.M.K. (2001) Plants as sources of drugs. *Toxicon*, **39**, 603–613.

Rauh, W. (1995) *Succulent and Xerophytic Plants of Madagascar*, vol. **1**, Strawberry Press, Mill Valley, CA.

Rauh, W. (1998) Three new species of *Lomatophyllum* and one new *Aloe* from Madagascar. *Bradleya*, **16**, 92–100.

Reynolds, G.W. (1966) *The Aloes of Tropical Africa and Madagascar*, The Aloes Book Fund, Mbabane, Swaziland.

Rowley, G.D. (1996) The berried aloes: *Aloe* section *Lomatophyllum*. *Excelsa*, **17**, 59–62.

Sanmukhiya, M.R., Soulange, J.G., Lavergne, C. *et al.* (2010) Molecular biology, phytochemistry and bioactivity of three endemic *Aloe* species from Mauritius and Réunion islands. *Phytochemical Analysis*, **21**, 566–574.

Santos, C.A.F., Lima-Neto, F.P., Rodrigues, M.A. and Costa, J.G. (2008) Similaridade genética de acessos de mangueira de diferentes origens geográficas avaliadas por marcadores AFLP. *Revista Brasileira de Fruticultura*, **30**, 736–740.

Schill, R. (1973) Studien zur systematischen Stellung der Gattung *Lomatophyllum* Willd. *Beitrage zur Biologie der Pflanzen*, **49**, 273–289.

Schmelzer, G.H. and Gurib-Fakim, A. (2008) *Prota 11. Plant Resources of Tropical Africa: Medicinal Plants 1* (eds G.H. Schmelzer and A. Gurib-Fakim), PROTA, Wageningen.

Smith, G.F. and van Wyk, B.E. (1991) Geberic relationships in the Alooideae (Asphodelaceae). *Taxon*, **40**, 557–581.

Steyn, E.M.A. and Smith, G.F. (2001). Are ovules and seeds in *Lomatophyllum* Willd. (*Aloe* Sect. *Lomatophyllum* sensu Auct.) anatropous and exarillate? *Bothalia*, **31**, 237–240.

Tizard, I., Busbee, D., McAnalley, B. and Kemp, M. (1989) The biological activities of mannans and related complex carbohydrates. *Molecular Biotherapy*, **1**, 290–296.

Tripathi, N., Saini, N. and Tiwari, S. (2011) Assessment of genetic diversity among *Aloe vera* accessions using amplified fragment length polymorphism. *International Journal of Medicinal and Aromatic Plants*, **1**, 115–121.

Van Wyk, B.E., Yenesew A, and Dagne, E. (1995) The chemotaxonomic significance of root anthraquinones and pre-anthraquinones in the genus *Lomatophyllum* (Asphodelaceae). *Biochemical Systematics and Ecology*, **23**, 806–808.

Viljoen, A.M. (1999) A chemotaxonomic study of phenolic leaf compounds in the genus *Aloe*. PhD thesis, Rand Afrikaans University, Johannesburg.

Watson, L. and Dallwitz, M.J. (1992) *The families of flowering plants – Asphodelaceae Juss.*, www.biologie.uni-hamburg.de/b-online/delta/angio/www/asphodel.htm.

Womble, D.J. and Helderman, J. (1988) Enhancement of allo-responsiveness of human lymphocytes by acemannan (Carrisynk). *International Journal of Immunopharmacology*, **10**, 967–974.

Yagi, A., Makino, K. and Nishioka, I. (1974) Studies on the constituents of *Aloe sapnaria* Haw. I. The structures of tetrahydroanthracene derivatives and the related anthraquinones. *Chemical and Pharmaceutical Bulletin*, **22**, 1159–1166.

Yamaguchi, I., Mega, N. and Sanada, H. (1993) Components of the gel of *Aloe vera* (L.) Burm. f. *Bioscience, Biotechnology and Biochemistry*, **57**, 1350 [*Chemical Abstracts* (1993), **119**, 278474.].

Yates, K.M. (2012) *Acemannan Review*, CelaCare Technologies, Inc. report, www.celacare.com/Images/acemannanreviewwebversion 120803.pdf (accessed 14 November 2013).

Yoo, E.A., Kim, S.D., Lee, W.M. *et al.* (2008) Evaluation of antioxidant, antinociceptive, and anti-inflammatory activities of ethanol extracts from *Aloe saponaria* Haw. *Phytotherapy Research* **22**, 1389–1395.

Yuan, A.X. (1993) The molecular structure of iso-aloesin isolated from the leaves of *Aloe vera* L. var. *chinensis* (Haw.) Berge. *Zhongguo Zhong Yao Za Zhi*, **18**, 609–611 [*Chemical Abstracts*, (1994), **120**, 101981].

# 17 Pharmacological Activities of Some of the Neglected and Underutilized Tropical Plants in Malaysia

Z.A. Zakaria[1,2], F. Yahya[1], T. Balan[1], S.S. Mamat[1], R. Rodzi[1], F.H. Kamisan[1], C.A. Fatimah[2,3] and A.L. Ibrahim[2,3]

[1] Department of Biomedical Science, Faculty of Medicine and Health Sciences, Universiti Putra Malaysia, Serdang, Selangor, Malaysia
[2] Institute of Bio-It Selangor, Universiti Selangor, Shah Alam, Selangor, Malaysia
[3] Faculty of Sciences & Biotechnology, Universiti Selangor, Shah Alam, Selangor, Malaysia

## 17.1 Introduction

Plants have been an important source of medicine for many years. According to the World Health Organization, it was estimated that up to 80% of people from Asian and African countries rely mainly on traditional remedies for primary healthcare (WHO, 2003). Approximately, one-quarter of prescribed drugs today contain active ingredients obtained or modelled from plant substances. Industrially, many health care products have been developed from plants based on their traditional claims. Thus, the search for new pharmacologically active agents from natural sources has led to the discovery of many clinically useful drugs that are important in the treatment of human diseases. For example, aspirin, the most popular analgesic, was originally derived from species of *Salix* and *Spiraea*. Some of the most valuable anticancer agents, such as paclitaxel and vinblastine, are also derived solely from plant sources (Katzung, 1995; Pezzuto, 1996).

Malaysia abounds in botanical riches, and this wide diversity of plants provides us not only with food but also a wide range of medicinal values. Medicinal plants usually are interrelated to herbal plants, but there are also non-herbal plants that may have their very own medicinal values. These non-herbal plants are mostly underutilized and are paid less attention to owing to their lack of traditional use in treating diseases, or possibly also due to their habitat. A scientific approach may help to bring out and prove their valuable pharmacological properties, which may contribute to our modern medicine. Issues on the interest and conducting research and studies on neglected or underutilized plant species have been debated with some intensity in recent years (Withers, 2005). Numerous studies are being done by researchers nowadays in an attempt to discover new leads from these plant species. Therefore, this review is intended to summarize the studies that have been done by our team on some of the neglected/underutilized plants in Malaysia in order to prove their medicinal values via research and a scientific approach.

## 17.2 Muntingia calabura

*Muntingia calabura* L. (Elaeocarpaceae) is a flowering plant commonly known as Jamaica cherry and 'Kerukup Siam' among Malays (Chin, 1989). *M. calabura* is native to the American continent and is widely cultivated in warm areas of Asian region, including Malaysia. According to the traditional claims in Peru folklore, the leaves of *M. calabura* can be used to provide relief from swelling of the prostate gland and reducing gastric ulcers; meanwhile, the barks are commonly used for lower extremities washing to reduce swelling. The flowers are used to treat abdominal cramps and spasms and are also consumed to relief headache and cold (Morton, 1987).

### 17.2.1 Antinociceptive activity

Zakaria *et al.* (2006c, 2007b, 2008c) have earlier reported on the antinociceptive profiles of aqueous extract of *M. calabura* (AEMC) leaves collected from its natural habitat in Shah Alam, Selangor, Malaysia. AEMC (27, 135 and 270 mg/kg) was administered subcutaneously (s.c.) to rats assessed using the formalin-induced paw-licking test. In this study, all doses of AEMC were found to be significantly ($p < 0.05$) effective in reducing the number of paw lickings in a dose-dependent manner in the early phase, whereas in the late phase, except 135 mg/kg, significant antinociceptive action was demonstrated by the other two doses. Morphine (5 mg/kg), which was used as a standard reference drug, exhibited a significant result of antinociceptive activity in this test compared with all doses of AEMC.

In a recent study Zakaria *et al.* (2008c), AEMC that was given s.c. at the concentration of 5, 50 and 100% prior to the abdominal constriction test demonstrated antinociceptive activity in a dose-dependent manner. The 50% concentration of AEMC showed equi-effective antinociceptive action to that of morphine (0.8 mg/kg) or acetylsalicylic acid (ASA; 100 mg/kg), whereas a 100% concentration of AEMC demonstrated a highly significant ($p < 0.05$) reduction of abdominal constriction with analgesia of 68.81%.

*Novel Plant Bioresources: Applications in Food, Medicine and Cosmetics*, First Edition. Edited by Ameenah Gurib-Fakim.

AEMC at 50% concentration was further studied to determine the involvement of various receptor antagonists and effects of enzyme and pH in the antinociceptive activity. Pretreatment of AEMC against 10 mg/kg naloxonazine, 10 mg/kg pindolol and 5 mg/kg atropine showed significant ($p < 0.05$) reduction of activity, but not against 10 mg/kg funaltreaxamine, 10 mg/kg naltrindole, 10 mg/kg phenoxybenzamine, 10 mg/kg bicuculine or 5 mg/kg mecamylamine. AEMC demonstrated significant ($p < 0.05$) increase in antinociceptive activity after pretreatment at pH 9 and 11 (alkaline pH) and was able to maintain the activity under both the alkaline (pH 13) and acidic (pH 2) conditions.

Another study on treatment of AEMC was carried out in male Balb/c mice to examine the possible involvement of L-arginine/nitric oxide/cyclic guanosine monophosphate (L-arginine/NO/cGMP) pathway. The extract, which was administered s.c. in doses of 27, 135 and 270 mg/kg after 5 min pretreatment of distilled water (d$H_2O$) was found to exert potential in a dose-dependent manner. Pretreatment with L-arginine (20 mg/kg) blocked the antinociceptive activity of AEMC only at the dose of 270 mg/kg; meanwhile, pretreatment with $N^G$-nitro-L-arginine methyl ester (L-NAME; 20 mg/kg) caused significant ($p < 0.05$) enhancement of antinociceptive activity at the low concentration but inhibited the antinociception activity at the high concentration. On the other hand, methylene blue (MB; 20 mg/kg) was found to significantly ($p < 0.05$) enhance antinociception of AEMC at all concentrations. Co-treatment of L-NAME reversed the effect of L-arginine insignificantly when given alone and significantly when administered with low concentration of AEMC; meanwhile, co-treatment with MB was found to significantly ($p < 0.05$) reverse the effect of L-arginine when given alone or with 27 mg/kg of AEMC. These results demonstrated the involvement of the L-arginine/NO/cGMP pathway in antinociceptive activity conferred by AEMC (Zakaria *et al.*, 2006c).

Zakaria *et al.* (2007c) also reported on the effect of temperature and the involvement of opioid receptor in the antinociceptive activity of AEMC. Abdominal constriction and a hot plate test were used to determine the antinociception effect of AEMC at concentrations of 1, 5, 10, 50 and 100%. All the doses were found to show significant ($p < 0.05$) activity in both the tests. In the abdominal constriction test, AEMC caused reduction in the number of abdominal constriction in a concentration-dependent manner, but not in the hot plate test. A 50% concentration of AEMC in the abdominal constriction assay exerted a similar strength in antinociception when compared against 100 mg/kg of ASA with $IC_{50}$ approximately 72.53 mg/kg, and 48% analgesic effect; meanwhile, the 10% concentration of AEMC was equi-effective to 0.8 mg/kg morphine with $IC_{50}$ approximately 0.88 mg/kg, and 67% analgesic effect. Administration of AEMC at 50% concentration, pretreated with naloxone (10 mg/kg) caused significant ($p < 0.05$) inhibition of antinociceptive activity but not in lower dosage of naloxone (2 mg/kg). In the hot plate test, the 1% concentration of AEMC showed activity at 120 min, and the 10% concentration of AEMC has a high latency effect as early as 60 min after treatment. All concentrations of AEMC exhibited antinociceptive activity that lasted until the end of the experiment (180 min). In this test, 5 mg/kg morphine, which was used as a positive control, showed onset activity at 30 min but lost its activity 180 min later. The temperatures (40, 60 and 100 °C) enhanced the activity of AEMC in the hot plate test. Pretreatment of naloxone (2 and 10 mg/kg) blocked the extract's antinociceptive action, indicating the involvement of opioid receptor in the mechanism of action shown by AEMC.

A recent study by Mohd Sani *et al.* (2012) using the methanol extract of *M. calabura* (MEMC) leaves aimed to determine its antinociceptive activity using acetic-acid-induced abdominal constriction and the formalin-, capsaicin-, glutamate-, serotonin-induced paw licking test and hot plate test. MEMC (100, 250 and 500 mg/kg) that was given orally 60 min prior to the respective tests demonstrated significant ($P < 0.05$) antinociceptive action in all assays. In the acetic-acid-induced abdominal constriction test, MEMC showed dose-dependent activity, where 250 mg/kg MEMC was equi-effective to the standard control (100 mg/kg ASA). MEMC at the highest dose (500 mg/kg) showed an increase ($P < 0.001$) of latency against heat discomfort in the hot plate test. In the formalin-induced paw licking test, only 250 and 500 mg/kg of MEMC exhibited antinociceptive action in first phase, whereas all doses of MEMC caused a significant result in late phase in a dose-dependent manner. In comparison with MEMC, 100 mg/kg ASA and 5 mg/kg morphine produced antinociception in the first phase and second phase of the formalin test. All doses of MEMC showed dose-dependent inhibition in the capsaicin- and glutamate-induced paw licking test with analgesic effect ranging from 20 to 62% and 35 to 72%, respectively. Interestingly, the result was reversed after pre-challenge with the treatment of 5 mg/kg of naloxone, indicating that the antinociceptive activity observed was modulated partly via the opioid receptor system.

### 17.2.2 Anti-inflammatory properties

The anti-inflammatory profile of s.c. administered AEMC (27, 135 and 270 mg/kg) that was subjected to the carrageenan-induced paw oedema test has been reported by Zakaria *et al.* (2007b). The AEMC exhibited concentration-independent anti-inflammatory activity, where the highest dose, 270 mg/kg, demonstrated lower activity than the other doses. Interestingly, the activity exhibited by 27 mg/kg and 135 mg/kg of AEMC was significantly greater than 100 mg/kg of ASA between the time intervals of 180 and 240 min.

### 17.2.3 Antipyretic properties

Intraperitoneally (i.p) administered AEMC was reported to exert antipyretic activity in a dose-independent manner when assessed at 0, 30, 60, 120, 180, 240, 300, 360, 420 and 480 min using the Brewer's yeast (BY)-induced pyrexia test. At a dose of 270 mg/kg, no activity was observed, whereas 27 mg/kg and 135 mg/kg of AEMC were found to exert significant antipyretic action after 240 min administration of the respective treatment, which lasted until the end of the experiment. However, 100 mg/kg of ASA was reported to have more effective antipyretic activity than AEMC.

### 17.2.4 Antioxidant activity

Zakaria *et al.* (2011b) have investigated the *in vitro* antioxidant activity of various extracts of *M. calabura*, namely methanol, aqueous and chloroform extracts (MEMC, AQMC and CEMC). All extracts of *M. calabura* (20, 100 and 500 µg/mL) possessed antioxidant activity when tested using the di(phenyl)-(2,4,6-trinitrophenyl) iminoazanium (DPPH) and superoxide scavenging assays. MEMC demonstrated the highest DPPH scavenging activity, approximately between 92.1–99.9% and 85.7–89.0% of superoxide scavenging activity, followed by AEMC (30.7–94.9 and 79.8–77.9%) and CEMC (27.2–99.9 and 60.0–87.2%). Furthermore, the total phenolic contents of the AEMC, MEMC and CEMC were 2970.4 ± 6.6 mg, 1279.9 ± 6.1 mg and 2978.1 ± 4.3 mg, respectively, per 100 g gallic acid. From the phytochemical screening, AEMC was reported to contain saponins, flavonoids and condensed tannins, but not alkaloids, triterpenes and steroids. The CEMC contained flavonoids, condensed tannins, triterpenes and steroids, but not alkaloids and saponins, whereas saponins, flavonoids, condensed tannins and steroids, but not the alkaloids and triterpenes, were found in the MEMC.

A study was done by Zakaria (2007) to investigate the free-radical scavenging properties of some plants found in Malaysia, and one of them was *M. calabura*. The free-radical scavenging activity of AEMC was evaluated by DPPH and xanthine/xanthine oxidase superoxide (XOD) scavenging assay. As for the DPPH scavenging assay, the reaction mixtures contained test samples in methanolic solution and 200 μM DPPH dissolved in ethanolic solution were incubated at 37 °C for 30 min before the percentage of inhibition was calculated. The XOD scavenging assay was based on the removal rate of XOD-generated superoxide by measuring the reduction of nitro blue tetrazolium (NBT). AEMC, which was assayed against the DPPH and XOD scavenging assays, was found to exhibit remarkable radical scavenging activity, with percentage inhibitions of 94.80 ± 1.14% and 83.70 ± 2.05%, respectively. Interestingly, the phytochemical screening of AEMC demonstrated the presence of flavonoids, triterpenes, saponins, tannins and a high content of steroids but no traces of alkaloids. Besides that, the high-performance liquid chromatography (HPLC) profile of AEMC was compared against standard flavonoids, which were rutin (retention time $R_T = 14.402$ min) and catechin ($R_T = 21.786$ min); interestingly, AEMC showed an $R_T$ that was almost similar with the standard catechin, $R_T = 21.792$ min.

### 17.2.5 Antimicrobial activity

The possible antibacterial activity of AEMC, MEMC, and CEMC against several bacterial strains by using the *in vitro* disc diffusion method was investigated by Zakaria *et al.* (2006b). The extracts in a volume of 20 μL with different concentrations (10 000, 40 000, 70 000 and 100 000 ppm) were loaded into sterilized blank discs. Chloramphenicol (30 μg/μL), the commercial antibiotic discs, were used for comparison. AEMC, MEMC and CEMC were tested against *Corneybacterium diphtheria*, *Staphylococcus aureus*, *Bacillus cereus*, *Proteus vulgaris*, *Staphylococcus epidermis*, *Kosuria rhizophila*, *Shigella flexneri*, *Escherichia coli*, *Aeromonas hydrophila* and *Salmonella typhi*. Bacterial cultures (0.1 mL, $10^6$ bacteria/mL) were distributed homogeneously onto Mueller Hinton agar in Petri dishes and the discs, which were loaded with the extracts, were positioned on the agar. The result showed that CEMC was less effective than AEMC and MEMC, in terms of number of bacterial strains inhibited. AEMC was effective against *S. aureus* and *K. rhizophila* at all concentrations, while MEMC was effective against *S. flexneri*, *B. cereus*, *S. aureus*, *P. vulgaris*, *A. hydrophila* and *K. rhizophila*. At the highest concentration (100 000 ppm), AEMC produced maximum inhibitory effect (≥20.0 mm) against *C. diphteriae* and showed inhibition (≥13.0 mm) against *C. diphteriae*, *S. aureus* and *S. epidermidis*. The MEMC was found to show inhibition (≥13.0 mm) against *C. diphteriae*, *S. aureus*, *P. vulgaris*, *K. rhizophila* and *S. flexneri*. At 100 000 ppm concentration, MEMC exhibited inhibition (≥20.0 mm) against *S. aureus* and *S. flexneri*. Extract of *M. calabura* demonstrated a potential antibacterial property, which was comparable to the standard antibiotics used.

Zakaria *et al.* (2007a) also investigated the antistaphylococcal activity of several neglected plants in Malaysia, which included extract of *M. calabura* against various strains of *S. aureus*, namely *S. aureus* 29213α, *S. aureus* 33591, *S. aureus* 700699, vancomycin-intermediate *S. aureus* (VISA) and vancomycin-resistant *S. aureus* (VRSA). The single screening *in vitro* microtitre plate dilution method had been used to screen the extract for its antibacterial activity as well as to determine the minimal inhibitory concentration (MIC) and minimum bactericidal concentration (MBC). The results showed that the extracts produced antibacterial activity against *S. aureus* at a concentration of 5 μg/μL. Interestingly, all the *M. calabura* extracts were found to be only effective against the normal *S. aureus* strains which were *S. aureus* 29231α, *S. aureus* 33591 and *S. aureus* 700699. Furthermore, the range of MIC for AEMC, MEMC and CEMC were 5.000 μg/μL, 1.250 μg/μL and 1.250–2.500 μg/μL respectively, while the MBC values, determined only for MEMC and CEMC, were 2.500 μg/μL.

In another study, Zakaria *et al.* (2010b) had reported the *in vitro* antimicrobial activity of various extracts, partitions and fractions of *M. calabura* leaves against *S. aureus* ATCC 25923 (MSSA), *S. aureus* ATCC 33591 (a multidrugs-resistant *S. aureus* (MRSA) isolate), *E. coli* ATCC 10536, *Pseudomonas aeruginosa* ATCC 27853, *Candida albicans* ATCC 10231 and *Microsporum canis* ATCC 36299. Antimicrobial activity of AEMC, MEMC and CEMC were evaluated using the liquid micro-dilution method. The bacterial assays were employed by standard mycorrhiza helper bacteria. MEMC was reported to show moderate inhibitory effect against MSSA (MIC = 1250 μg/mL; MBC = 1250 μg/mL) and MRSA (MIC = 2500 μg/mL; MBC = 2500 μg/mL) and none of the crude extracts were effective against the fungi tested, *C. albicans* and *M. canis*. The MEMC was considered to be the most effective extract and was further partitioned sequentially using petroleum ether, ethyl acetate and water. Based on the result, the ethyl acetate partition (EAMC) exhibited the most effective antibacterial activities, with MIC and MBC values of 156 μg/mL and 313 μg/mL, respectively, against *S. aureus* 33591. The EAMC (10 g) was subjected to fractionation by vacuum liquid chromatography using silica gel 60 to give 35 fractions. These fractions were then pooled together according to their chromatogram similarity to yield 15 fractions, labelled as A1–A15 and subjected to antimicrobial testing against *S. aureus* 25923 and *S. aureus* 33591. From the findings, it was reported that only fractions A9 until A15 exhibited antimicrobial activity against the bacterial strain with the MIC/MBC values ranging from 78 to 2500 μg/mL. The most effective fraction was fraction A10, which showed the maximum inhibition, MIC and MBC at 78 μg/mL.

### 17.2.6 Antiproliferative activity

In addition, Zakaria *et al.* (2011b) have reported the *in vitro* antiproliferative activity of MEMC, AEMC and CEMC, which were assessed using 3-(4,5-dimethylthiazol-2-yl)-2,5-diphenyltetrazolium bromide (MTT) assay against 3T3 (normal mouse fibroblast), MCF-7 (oestrogen-dependent human breast adenocarcinoma), HeLa (human cervical adenocarcinoma), HT-29 (human colon cancer), HL-60 (acute promyelocytic leukaemia), K-562 (chronic myelogenous leukaemia) and MDA-MB-231 (human breast carcinoma). Dimethylsulfoxide (DMSO) 1% was used as the negative control, while Tamoxifen was used for comparison of $IC_{50}$ with the extracts. The viability of the cells was determined with tryphan blue reagent. DMSO alone was reported to be ineffective in exhibiting antiproliferative activity, as indicated by its failure to produce $IC_{50}$ against all the cell lines. The AEMC inhibited the proliferation of MCF-7, HeLa, HT-29 and K-562 cancer cells which produce $IC_{50}$ values of approximately 18 μg/mL, 52 μg/mL, 16 μg/mL, and 18 μg/mL, respectively. MEMC inhibited the proliferation of MCF-7, HeLa, HT-29, HL-60 and K-562 with $IC_{50}$ values of 22 μg/mL, 23 μg/mL, 46 μg/mL, 7 μg/mL and 39 μg/mL, respectively, while the chloroform extract only inhibited the proliferation of MCF-7, HeLa, HL-60 and K-562 cancer cells, which produce $IC_{50}$ values of 98 μg/mL, 22 μg/mL, 29 μg/mL and 42 μg/mL, respectively. All extracts of *M. calabura* leaves failed to inhibit the proliferative activity of the MDA-MB-231 cells.

## 17.3 Dicranopteris linearis

*Dicranopteris linearis*, locally known as 'Resam', which belongs to the family Gleicheniaceae, is one of the neglected plants found in Malaysia. This plant is commonly found in secondary forests and grows well on poor clay soils (Zakaria *et al.*, 2008b). *D. linearis* has plenty of traditional uses, such as reducing body temperature, controlling fever, treating asthma, women's sterility, external wounds, ulcers and boils, and also to get rid of intestinal worms (Chin, 1992; Vasudeva, 1999). *D. linearis* is also taken by the Malays as a cooling drink to reduce body temperature during fever (Chin, 1992). The leaves are squeezed in water and the water then drunk or the wet leaves applied on the body as a poultice (Zakaria *et al.*, 2008b). This plant has been proven scientifically to possess several medicinal values, depending on several different extraction types. Based on the previous studies, the plant has antinociceptive, anti-inflammatory, antipyretic (Zakaria *et al.*, 2006a, 2008b), *in vitro* anticancer and antioxidant effects (Zakaria *et al.*, 2011c) anti-staphylococcal (Zakaria *et al.*, 2007a) and antiproliferative activity (Mamat *et al.*, 2013). *D. linearis* has been reported to contain different flavonoids of the flavonol 3-*O*-glycosides type. Depending on the varieties of fern used, *D. linearis* contained afzelin, quercitrin, isoquercitrin, astragarin, rutin, kaempferol 3-*O*-(4-*O*-*p*-coumaroyl-3-*O*-α-L-rhamnopyranosyl)-α-L-rhamnopyranosyl-(1→6)-β-D-glucopyranoside, and (6*S*,13*S*)-6-[6-*O*-acetyl-β-D-glucopyra nosyl-(1-4)-α-L-rhamnopyranosyloxy]-13-[α-L-rhamnopyranosyl-(1→4)-β-D-fucopyranosyloxy]cleroda-3,14-diene (Zakaria *et al.*, 2008b).

### 17.3.1 Antinociceptive activity

Zakaria *et al.* (2012a) have carried out a study to determine the effect of various receptor antagonists on the antinociceptive activity of aqueous extracts of *D. linearis* (AEDL) leaves using the acetic-acid-induced writhing test. The powdered leaves of the plants were soaked separately in $dH_2O$ for 72 h at room temperature, filtered and then subjected to the freeze-drying process. The crude extracts (300–1000 mg/kg), prepared by dissolving in $dH_2O$, were administered (s.c.) into mice 30 min prior to subjection to the relevant assay. The most significant dose (500 mg/kg) was then chosen for the next study, where the extract was administered (s.c.) 10 min following the administration of the respective antagonist (3 and 10 mg/kg naloxone, 5 mg/kg atropine, 5 mg/kg methysergide, 1 mg/kg haloperidol, 5 mg/kg yohimbine and 2 mg/kg reserpine). The results obtained demonstrated that this plant possessed a dose-independent antinociceptive activity. Antagonists pretreatment showed that the activity of AEDL was blocked by naloxone, atropine and yohimbine. In conclusion, the antinociceptive activity of *D. linearis* involved modulation of the opioids, muscarinic and $\alpha_2$-adrenergic receptors.

Another study by Zakaria *et al.* (2012a) was aimed to determine the effect of various pain receptor antagonists and pain mediator blockers on the peripheral antinociceptive activity of AEDL using the abdominal constriction test as the nociceptive model. The crude dried aqueous extract (300, 500 and 1000 mg/kg), $dH_2O$ or 100 mg/kg ASA were administered (s.c.) into mice and 30 min later the animals were subjected to the abdominal constriction test in an attempt to establish their antinociceptive profiles. From the antinociceptive profile, a single dose (500 mg/kg) of the extract was chosen for further study on their mechanisms of peripheral antinociception. After 10 min pretreatment (s.c.) with various pain receptor antagonists or pain mediator blockers, the test solutions were administered (s.c.), followed 30 min later by the antinociceptive assay. In the mechanisms of the antinociception study, the antinociceptive activity of AEDL was attenuated by naloxone, atropine and yohimbine. It can be concluded that the antinociceptive activity of AEDL involves activation of the opioids, muscarinic, $\alpha_2$-adrenergic and adenosine receptors.

The abdominal constriction, hot plate and formalin tests were the three tests used to evaluate the antinociceptive effect of AEDL as reported by Zakaria *et al.* (2008b). The extract was prepared in doses of 13.2, 66.0, 132.0, and 660.0 mg/kg and administered (s.c.) 30 min before subjecting the animals to the assays mentioned above. Generally, from the study, the extract, at all doses used, was found to have significant ($p < 0.05$) concentration-independent antinociception.

Zakaria *et al.* (2006a) also reported the antinociceptive properties of chloroform leaves extract of *D. linearis* (CEDL). Abdominal constriction, formalin and hot plate tests were used to measure the antinociceptive activity. The extraction process was done by soaking the sample for 72 h and was subjected to evaporation under reduced pressure to dryness. DMSO 10% was used to prepare the extracts in doses of 20, 100 and 200 mg/kg. These extracts were then administered (s.c) 30 min prior to the assays mentioned. All the doses used exhibited significant ($p < 0.05$) antinociceptive activity in a dose-dependent manner. Thus, it was concluded that CEDL possesses antinociceptive activity.

### 17.3.2 Anti-inflammatory and antipyretic activity

Zakaria *et al.* (2008b) investigated the anti-inflammation and antipyretic activity using carrageenan-induced paw oedema and the BY-induced pyrexia test. AEDL was prepared in doses of 13.2, 66.0, 132.0, and 660.0 mg/kg and administered (s.c.) 30 min before subjecting the animals to the assays mentioned above. From the studies, the extract was found to have significant ($p \leq 0.05$) concentration-independent anti-inflammatory and antipyretic activity at all the doses tested.

Th carrageenan-induced paw oedema test was done to measure the anti-inflammatory activity of CEDL (Zakaria *et al.*, 2006a). The extraction process was done by soaking the sample for 72 h and was subjected to evaporation under reduced pressure to dryness. DMSO 10% was used to dissolve the extracts to the doses of 20, 100 and 200 mg/kg. These extracts were then administered (s.c.) 30 min prior to the assays mentioned. The extract exhibited significant ($P < 0.05$) antinociceptive activity in a dose-dependent manner, whereas for the anti-inflammatory assay the CEDL demonstrated significant ($P < 0.05$) activity in a dose-independent manner.

### 17.3.3 Antioxidant activity

Antioxidant activity of AEDL, CEDL and MEDL were investigated using DPPH and superoxide scavenging assays. All the extracts in different doses (20, 100 and 500 μg/mL) were found to exert antioxidant activity; with MEDL reported for its highest activity, followed by AEDL and CEDL in both the assays. As for the total phenolic content, MEDL, AEDL and CEDL were recorded at 3417.3 ± 4.7 mg, 3112.1 ± 6.7 mg and 1012.7 ± 5.3 mg, respectively, per 100 g gallic acid (Zakaria *et al.*, 2011c).

### 17.3.4 Antimicrobial activity

In order to determine the antimicrobial activity of AEDL, CEDL and MEDL, the micro-broth dilution method was used. *S. aureus* 25923, *S. aureus* 33591 (a methicillin-resistant *S. aureus* isolate), *S. aureus* 700699 (a VISA isolate), *S. aureus* 156 (a VRSA isolate), *E. coli* 35218 and *C. albicans* 10231 were the target strains used. The results obtained showed that the methanol extract showed the most effective antimicrobial activity, with MIC/MBC values of 625 μg/mL. Thus, MEDL was further fractionated and a total of 11 fractions were obtained (B1–B11). From these fractions, only B5, B6 and B11 fractions were found to be effective against *S. aureus* 33591 and *S. aureus* 25927, with MIC/MBC values ranging between 1250 and 2500 μg/mL (Zakaria *et al.*, 2010a). In conclusion, *D. linearis* possesses mild antibacterial activity against the selected panel of microbes. This explains the lack of claims made on the antimicrobial activity of the plant.

### 17.3.5 Anticancer and cytotoxic activity

Attempts to determine the anticancer activity of AEDL, CEDL and MEDL were also made by Zakaria *et al.* (2008a), *in vitro* against MCF-7 and HT-29 cancer cell lines using the MTT assay. Either $dH_2O$ or DMSO was used to dissolve the crude extracts. The starting concentration was 100 μg/mL and the extract was further serially diluted in a twofold manner. Only CEDL was effective against the MCF-7 cell line. Potential cytotoxic activity against the MCF-7 cell line was demonstrated by CEDL and MEDL, with $IC_{50}$ values of 95.5 μg/mL and 85 μg/mL, respectively. The $IC_{50}$ values of this extract were not reached even though the AEDL and CEDL showed a remarkable decrease in the percentage of HT-29 cells viability, approximately 65% and 78%, respectively. From this test, it can be deduced that the extract alone did not show any cytotoxic effect against the normal cell line (3T3).

Zakaria *et al.* (2011c) have also studied the cytotoxicity of AEDL, CEDL and MEDL leaves. By using the MTT assay, the cytotoxic effect was determined against 3T3 and cancer cell lines (MCF-7, HeLa, HT-29, HL-60, K-562 and MDA-MB-231). AEDL was not effective against any of the cancer cells studied; meanwhile, CEDL was effective only against MCF-7 and HeLa. MEDL was effective against all the cancer cells used. Interestingly, in terms of safety, all extracts failed to produce a cytotoxic effect against the normal cell line.

### 17.3.6 Hepatoprotective activity

Ismail *et al.* (personal communication) have studied the hepatoprotective activity of *D. linearis* leaf aqueous extract (DLAE) using two models of liver injury in rats. Rats were divided into 10 groups. The rats received $dH_2O$ (negative control), 200 mg/kg silymarin (positive control) or DLAE (50, 250 and 500 mg/kg) orally once daily for seven consecutive days. On the eighth day, the rats were subjected to hepatotoxic induction either using carbon tetrachloride ($CCl_4$) or paracetamol (PCM). The blood and liver were collected and subjected to biochemical and microscopic analysis. From the data obtained, only the highest dose of DLAE significantly ($p < 0.05$) reduced the alkaline phosphatase (ALP), alanine aminotransferase (ALT) and aspartate aminotransferase (AST) levels in $CCl_4$- and PCM-induced hepatotoxic rats, while the other doses caused significant ($p < 0.05$) reduction only in the levels of ALT and AST. The histological results showed the hepatoprotection occurred in a dose-dependent manner, which was in line with the biochemical analysis that demonstrated a reduction in formation of necrosis, steatosis and inflammation in the $CCl_4$- and PCM-induced tissue.

Besides that, hepatoprotective activity of methanol extract of *D. linearis* (MEDL) was investigated by Mamat *et al.* (2013) against two different inducers: PCM and $CCl_4$. All doses of MEDL showed significant ($p < 0.05$) reduction of ALT and AST levels in $CCl_4$-induced hepatotoxic rats, while only the 500 mg/kg MEDL caused significant ($p < 0.05$) reduction in the level of both the enzymes in the PCM-induced liver toxicity model. On the other hand, the histological results obtained were in line with the biochemical analysis.

## 17.4 Bauhinia purpurea

This is another neglected plant that has been investigated for its pharmacological properties. *Bauhinia purpurea*, known as 'Pokok Tapak Kerbau' in Malaysia, belongs to family Fabaceae (Zakaria *et al.*, 2012b). This plant has been traditionally used in India, Sri Lanka and Pakistan, but there is no documentation showing the use of *B. purpurea* as a traditional medicine in Malaysia (Zakaria *et al.*, 2009). *B. purpurea* has been used to treat wounds, ulcers, stomach tumours, glandular swellings, diarrhoea and fever (Zakaria *et al.*, 2007d); it also has been claimed to possess astringent effects and used as a laxative, while the flowers of *B. purpurea* have been used as a purgative in Pakistan (Zakaria *et al.*, 2009).

### 17.4.1 Antinociceptive activities

Zakaria *et al.* (2009) investigated the antinociceptive activity of chloroform extracts of *B. purpurea* (CEBP) leaves that were collected from their natural habitat in Shah Alam, Selangor, Malaysia. The study demonstrated that, in the abdominal writhing test, CEBP was found to produce significant ($p < 0.05$) reduction in the number of abdominal constrictions at all doses tested. In terms of the percentage of antinociception, all doses of CEBP demonstrated above 90% antinociceptive activity compared with the 100 mg/kg ASA that showed approximately 50% activity. In the hot plate test, the temperature was maintained at 50 ± 2 °C. Any responses shown by the animal placed in the Perspex cylinder with the heated surface were recorded. The animal may experience discomfort, indicated by either flipping or licking the paws, or jumping off the surface. All the concentrations of CEBP (20, 100 and 200 mg/kg) demonstrated antinociceptive activity in a concentration-independent manner. The 20 and 50 mg/kg CEBP exhibited their significant ($p < 0.05$) activity at intervals of 4–5 and 3–5 h after the administration of the test solution. However, the 100 mg/kg CEBP exhibited a significant ($p < 0.05$) activity after 2 h of its administration to mice and this activity lasted for another 3 h (interval of 5 h) before being completely diminished at the end of the experiment. The formalin test of CEBP was found to exhibit antinociceptive activity in a concentration-independent manner. The extract, at all doses tested, exhibited significant ($p < 0.05$) antinociceptive activity in the early and late phases of the test. Interestingly, all doses of CEBP, which produced an insignificant activity when compared against each other, exhibited an activity when compared with 5 mg/kg morphine in the early phase.

In another Zakaria and co-workers' study (Abdul Hisam *et al.*, 2012) the aqueous extract of *B. purpurea* (AEBP) (6.0, 30.0 and 60.0 mg/kg) was found to cause insignificant change in the number of abdominal constrictions at all doses, indicating its failure to inhibit the acetic-acid-induced abdominal constriction test. In the hot plate test, AEBP exhibited significant ($p < 0.05$) antinociceptive activity at 2 h after subjection of the mice to the test, at all of the doses used. The activity reached its maximum effect, indicated by prolonged latency of discomfort, at the interval of 3 h, with only the activity of 60.0 mg/kg remaining to be seen until the end of the experiment. However, the activity of AEBP was less in magnitude when compared with the 5 mg/kg morphine at all time. As for the formalin test, all of the doses used exhibited significant ($p < 0.05$) activity in the early and late phases of the test. In both the phases, the activity occurred in a dose-independent manner, with all doses of the extract exhibiting insignificant activity when compared together. The 6.0 mg/kg AEBP exhibited the highest activity, which was equi-effective when compared with 5 mg/kg morphine in the early phase but with 100 mg/kg ASA in the late phase.

### 17.4.2 Anti-inflammatory activities

Zakaria *et al.* (2009) reported that CEBP (20, 100 and 200 mg/kg) possessed anti-inflammatory activity in a non-concentration-dependent manner in the carrageenan-induced paw oedema test in rats. The result showed CEBP reached its peak after 3 h of administration and gradually decreased until the end of the experiment. Unexpectedly, the 100 mg/kg CEBP produced a less remarkable activity with significant ($p < 0.05$) activity observed only at the interval of 1 h. Interestingly, the 20 and 200 mg/kg CEBPs were found to exhibit significant ($p < 0.05$) anti-inflammatory activity in a similar degree, indicated by their insignificant effect when compared against each other. This activity was observed at the interval of 1 h and lasted for another 5 h before being completely diminished. On the other hand, 100 mg/kg ASA produced significant ($p < 0.05$) activity after 3 h of its administration and this activity lasted until the end of the experiment.

However, in another study done by Zakaria *et al.* (2007d), AEBP (6.0, 30.0 and 60.0 mg/kg) exhibited significant ($p < 0.05$) anti-inflammatory activity only at a dose of 30.0 mg/kg. Interestingly, the 30.0 mg/kg AEBP showed an equi-effective activity when compared with the 100 mg/kg ASA only between the interval times of 1 and 4 h, after which the ASA activity was found to be greater than the extract until the end of the experiment.

### 17.4.3 Antioxidant activities

Phytochemical screening of CEBP contained flavonoids, saponins, triterpenes and steroids, but not alkaloids and tannins. In addition, the HPLC profile of CEBP was compared against the standard compounds catechin (200 nm, $R_T$ = 6.386 min; or 280 nm, $R_T$ = 8.164 min), rutin (280 nm, $R_T$ = 17.515) and epigallocatechin gallate (200 nm, $R_T$ = 9.180 min; or 280 nm, $R_T$ = 12.081 min) and did not match any of the peaks obtained at 200 nm or 280 nm, indicating the absence of these compounds from CEBP.

On the other hand, phytochemical screening of AEBP demonstrated the presence of flavonoids, saponins, triterpenes and steroids, but not alkaloids and tannins. In addition, the HPLC profile of AEBP was compared against standard flavonoids at 200 nm, which were catechin ($R_T$ = 6.742 min), rutin ($R_T$ = 14.402 min) and fisetin ($R_T$ = 21.786 min) and interestingly AEBP possessed RT that was almost similar to the standard, fisetin ($R_T$ = 21.786 min).

### 17.4.4 Antipyretic activities

The antipyretic profile of AEBP was assessed using BY-induced pyrexia in rats. Pyrexia was induced by i.p. injection of 10% (w/v) BY suspension (10 mL/kg) 30 min after the administration of the test agents. The AEBP exhibited significant ($P < 0.05$) antipyretic activity. A dose-independent antipyretic activity was observed only at the concentrations of 6.0 and 30.0 mg/kg, with the former showing remarkable activity even when compared with 100 mg/kg ASA. AEBP was effective only at the interval of 1 h after the BY administration, before it lost its activity until the end of the experiment. There was a remarkable decrease in the rat's rectal temperature in the first 2 h after the BY administration, before the temperature was seen to increase gradually until it reached the normal rectal temperature at the end of the experiment (Zakaria *et al.*, 2007d).

### 17.4.5 Antiproliferative activities

According to Zakaria *et al.* (2011e), AEBP, MEBP and CEBP had been reported for their antiproliferative potential as indicated by their ability to inhibit in cell lines of the American Type Culture Collection, namely 3T3, MCF-7, HeLa, Caov-3, HL-60, CEMss and MDA-MB-231. Comparison of the antiproliferative activity of those

**Table 17.1** Comparison of the $IC_{50}$ values between tamoxifen and the *B. purpurea* extracts

| Compound/extract | $IC_{50}$ (μg/mL) | | | | | | |
|---|---|---|---|---|---|---|---|
| | 3T3 | MCF-7 | MDA-MB-231 | Caov-3 | HL-60 | CEMss | HeLa |
| Tamoxifen (mM/mL) | ND | 13 | 8 | 9 | ND | 6 | ND |
| *B. purpurea* extracts (μg/mL) | | | | | | | |
| aqueous | ND | 9 | 17 | 16 | 12 | 63 | 24 |
| chloroform | ND | 23 | 19 | 25 | 13 | 18 | 21 |
| methanol | ND | 83 | 75 | 41 | 14 | ND | 84 |

ND: not detected because the concentration of extracts/drugs required to exhibit $IC_{50}$ value were above 100 μg/mL (the highest concentration used).

extracts of *B. purpurea* leaves was made against tamoxifen using their $IC_{50}$ values (Table 17.1). Generally, tamoxifen exhibited potential antiproliferative activity when compared against the respective extract, as indicated by its lower $IC_{50}$ values. The study found that the AEBP and CEBP successfully inhibited the proliferation of all cancer cells, while the MEBP inhibited the proliferation of all cells except the CEMss cells when assessed using the MTT assay. The AEBP was effective against MCF-7, MDA-MB 231 and Caov-3, while the CEBP was highly effective against the CEMss and HeLa. However, the MEBP was highly effective only against the HL-60 cell lines. Interestingly, all extracts did not inhibit the proliferation of 3T3 cells, which did not affect the percentage of cells' viability when tested against the normal cell line (3T3), suggesting their non-cytotoxic properties.

### 17.4.6 Antiulcer activity

Studies had been done by Zakaria *et al.* (2011a) and Abdul Hisam *et al.* (2012) to scientifically prove the gastroprotection activity conferred by *B. purpurea*. Chloroform, methanol and aqueous extract of *B. purpurea* leaves (BPAE, MEBP and CEBP) have been tested for their antiulcer activity in rats.

BPAE, MEBP and CEBP were administered in doses of 100, 500 and 1000 mg/kg orally to 48-h-fasted rats ($n = 6$). The extracts were subjected to the ethanol- and indomethacin-induced gastric ulcer model and also pyloric ligation assay. Omeprazole (30 mg/kg) was used as the standard control and 8% Tween 80 (10 mL/kg) was used as the vehicle for CEBP pretreated rats, whereas saline (10 mL/kg) was the negative control for MEBP pretreated rats. In the ethanol-induced ulcer model, ulceration was induced in the rats by administrating absolute ethanol (1 mL/200 g) 30 min after the treatment and they were sacrificed after 15 min. As for the indomethacin-induced ulcer model, the rats received 100 mg/kg indomethacin 30 min after treatment in order to induce the ulceration in the stomach. Four hours later, they were euthanized by exposure to diethyl ether. Stomachs were removed and opened along the greater curvature to determine the gastric lesion. The ulcer area ($mm^2$) was determined and the protection percentage was calculated accordingly. Stomachs were fixed in 10% formalin and further subjected to histopathological study.

On the other hand, pylorus ligation was performed on anaesthetized rats where a midline incision was made in order to ligate the pylorus. Stomach was placed back carefully, abdomen was sutured and the rats were allowed to recover. After 4 h, the animals were sacrificed, stomachs were removed and the content was subjected to centrifugation in order to obtain the gastric secretion. The pH and total acidity of the gastric secretion were determined. The stomachs were used to determine gastric wall mucus content.

The study done by Zakaria *et al.* (2011a) had clearly revealed the significant activity of BPAE in preventing stomach ulceration. Pretreatment of BPAE at doses of 500 and 1000 mg/kg in the ethanol-induced ulcer model demonstrated significant ($p < 0.05$) reduction of gastric lesion, contributing to 74.2% and 75.3% protection respectively. As for the indomethacin-induced ulcer groups, administration of BPAE exhibited significant ($p < 0.05$) decrease in total ulcer area, with the percentage protection being 63.8%, 55.1% and 39.1% for the 100 mg/kg, 500 mg/kg and 1000 mg/kg respectively. Histopathological analysis supported these findings, where mild lesions to almost normal architecture of the mucosa with moderate to mild haemorrhage and oedema had been observed in stomach tissues of rats pretreated with BPAE in both the models. Gastroprotection in a dose-dependent manner had been observed in the stomach tissues of rats induced using ethanol. In the pylorus ligation assay, only 500 mg/kg BPAE demonstrated a significant ($p < 0.05$) reduction in the total ulcer area when compared with the control group with the percentage protection of 93.3%. BPAE was found to exert protection in a dose-independent manner in this group, as the highest dose (1000 mg/kg) of BPAE failed to reduce the ulcer area significantly ($p < 0.05$). However, it did not elevate the pH nor decrease the total acidity, as BPAE significantly ($p < 0.05$) reduced the pH and increased the total acidity without altering the gastric volume when compared with the control group. Furthermore, pretreatment with 500 mg/kg of BPAE increased gastric wall mucus content significantly ($p < 0.05$).

Abdul Hisam *et al.* (2012) have also demonstrated the antiulcer effect possessed by *B. purperea*, where oral administration of MEBP exerted significant ($p < 0.05$) antiulcer activity in all the models tested. Macroscopic analysis showed that only the 500 and 1000 mg/kg doses of MEBP reduced gastric lesions significantly ($p < 0.05$), with the protection reported to be 70% and 80% respectively in the ethanol-induced ulcer model. As for the indomethacin-induced ulceration, MEBP demonstrated significant ($p < 0.05$) reduction in total ulcer area, where the percentage of protection recorded was 31.6%, 50.8% and 47.4% for the three different doses. Microscopic analysis supported these findings by showing a reduction in haemorrhage and oedema formation, the preservation of normal mucosal architecture and the absence of leukocyte infiltration. In the pylorus ligation assay, MEBP was able to reduce the total ulcer area significantly ($p < 0.05$) in the range of approximately 40–83%. Significant increase ($p < 0.05$) in the volume of gastric juice released was recorded in the MEBP-treated groups, together with a slight increment in its total acidity, which contributes to a reduction in pH. Furthermore, MEBP increased the gastric wall mucus content significantly ($p < 0.05$), which was by twofold compared with the control group.

On the other hand, Abdul Hisam *et al.* (2012) proved the gastroprotection activity of the plant in chloroform extract. CEBP showed significant ($p < 0.05$) dose-dependent antiulcer activity in the ethanol-induced model. The highest dose exerted 100% protection, whereas 500 mg/kg and 100 mg/kg doses showed 86% and 50% inhibition, respectively. Furthermore, CEBP also demonstrated significant ($p < 0.05$) gastroprotection in the indomethacin-induced ulcer model in a dose-independent manner, where the protection was 31%, 24% and 58% from the lowest to the highest doses. Both these findings were supported by microscopic analysis, which showed a reduction of haemorrhage, oedema, inflammatory exudates formation and leukocytes infiltration in the treated groups. As for the pylorus ligation assay, the extract significantly ($p < 0.05$) increased the gastric wall mucus production and pH of gastric content, while significantly ($p < 0.05$) reducing the total volume and total acidity of the gastric content.

## 17.5 Melastoma malabathricum

*Melastoma malabathricum* L. (Melastomataceae) known to the Malays as 'senduduk' is one of the 22 species found in the Southeast Asian region, including Malaysia. Considered as native to tropical and temperate Asia and the Pacific Islands, this commonly found small shrub has gained herbal status in the Malay folklore belief as well as the Indian, Chinese and Indonesian folk medicines (Mohd Joffry *et al.*, 2012). Traditionally, the plant has been used by practitioners in Malayasia to treat various types of ailments, such as inflamed wounds, bleeding (Zakaria *et al.*, 2006d), pox scars, diarrhoea (Institute of Medical Research, 2002), and haemorrhoids (Jaganath and Ng, 2000). Scientific findings also revealed the wide pharmacological actions of various parts of *M. malabthricum*, such as antinociceptive, anti-inflammatory, wound healing, antidiarrhoeal, cytotoxic and antioxidant activities (Mohd Joffry *et al.*, 2012).

### 17.5.1 Antinociceptive activities

The antinociceptive activity of aqueous extract of *M. malabathricum* (AEMM) (Zakaria *et al.*, 2006d) and chloroform extract *M. malabathricum* (CEMM) (Zakaria *et al.*, 2006e), administered s.c. in different concentrations (10%, 50%, and 100%) were determined using the acetic-acid-induced abdominal writhing, hot plate and formalin tests.

In the abdominal writhing test, Zakaria *et al.* (2006d) reported that the AEMM, at all concentrations used exhibited significant ($p < 0.05$) antinociceptive activity in a concentration-independent manner. Interestingly, the equi-effective activity was observed in all concentrations of AEMM when compared with the 100 mg/kg ASA. In the hot plate test, the temperature was maintained at $50 \pm 2\,°C$. Any response by the animal placed in the Perspex cylinder with heated surface were recorded. The animal may experience discomfort, indicated by either flipping or licking of the paws, or jumping off the surface. In this study, all the concentrations of AEMM exhibited a significant ($p < 0.05$) antinociceptive activity in a concentration-dependent manner, at least for the first two concentrations (10% and 50% concentrations) in the time interval of 1–3 h. The 50% and 100% concentrations of AEMM, but not the 10% concentration, exhibited the antinociceptive activity. Although there was a significant ($p < 0.05$) antinociceptive activity in the 100% concentration of AEMM, this activity was lower than that of the 50% concentration extract. In the formalin test, pain was induced by injecting 50 μL of 5% formalin in the subplantar region of the left hind paw. The results were recorded in two phases: the early phase was measured at 0–5 min, while the late phase was measured at 15–30 min after formalin injection. In the early phase, concentration-independent activity was observed, while concentration-dependent activity was observed in the late phase. Interestingly, the AEMM showed an antinociceptive activity that was similar in manner to morphine, in which the latter was found to give a latency of discomfort at $16.20 \pm 0.98$ s and $4.29 \pm 0.97$ s for the early and late phases, respectively.

Zakaria *et al.* (2006e) also reported that the CEMM exhibited significant ($p < 0.05$) antinociceptive activity in a concentration-independent manner for all concentrations used in acetic-acid-induced abdominal writhing test. For CEMM, the 50% concentration produced a more effective activity than that of the 100% concentration CEMM when compared to the control group, with the former causing an approximately sixfold decrease in the number of abdominal constrictions, while the latter caused only a twofold decrease. The 10% and 100% concentrations of CEMM produced an equi-effective activity when compared with the 100 mg/kg ASA. The hot plate test tested using CEMM showed that all the concentrations used exhibited a significant ($p < 0.05$) antinociceptive activity in a concentration-independent manner. The 10% concentration CEMM was found to produce constant antinociceptive activity throughout the experimental time, while the other two concentrations showed somewhat inconsistent activity between the interval of 1 and 2 h. The 50% concentration CEMM lost its activity at the interval time of 1 h, while the 100% concentration CEMM lost its activity at the interval time of 2 h after administration. Interestingly, the activity of both concentrations of the extract were found to increase significantly ($p < 0.05$) after that and maintained this until the end of the experiment. Throughout the study, the 5 mg/kg morphine's antinociceptive activity was found to be greater than the extract except at the last interval time (5 h), where its activity was significantly ($p < 0.05$) lower than the 50% concentration CEMM. The antinociceptive profile of CEMM assessed using the formalin test demonstrated that all the concentrations used exhibited significant ($p < 0.05$) antinociception in both phases of nociception, with a more effective activity seen in the late phase. The extract was found to block the early phase nociception only at the 100% concentration used, whereas the inhibition of the late phase was observed in all of the concentrations used. Interestingly, the 100% concentration CEMM caused almost complete inhibition of the nociception in the early and late phases, respectively.

### 17.5.2 Anti-inflammatory activities

Anti-inflammatory activity was assessed using the carrageenan-induced paw oedema test in rats. Based on studies conducted by Zakaria *et al.* (2006d), acute inflammation was produced by subplantar injection of 0.1 mL of 1% carrageenan suspension in $dH_2O$ in the right hind paw of the rats, 30 min after the administration (s.c.) of test agents. The study showed that AEMM exhibited significant ($p < 0.05$) anti-inflammatory activity only at the 50% and 100% concentrations activity 1 h after administration, which lasted until the end of the experiment (Zakaria *et al.*, 2006d). The CEMM exhibited significant ($p < 0.05$) activity at all of the concentrations used in a concentration-independent manner. The activity can be seen as early as 1 h after the extract administration and lasted for the first 6 h. Except for the 50% concentration CEMM, which still maintained its significant ($p < 0.05$) activity, the other concentrations of CEMM were found to lose their anti-inflammatory activity. Except at the interval time of 4 h, where it lost the activity, the 100 mg/kg ASA was found to produce significant ($p < 0.05$) anti-inflammatory activity for the first 6 h before the activity diminished for the next 2 h. Throughout the experimental time, the 100 mg/kg ASA's anti-inflammatory activity was found to be lower than the extract.

### 17.5.3 Antipyretic activities

Zakaria *et al.* (2006d) investigated the antipyretic profile of AEMM as assessed using the BY-induced pyrexia test in rats. Pyrexia was induced by intraperitoneal injection of 10% (w/v) BY suspension (10 mL/kg) 30 min after the administration of the test agents. The AEMM exhibited significant ($p < 0.05$) antipyretic activity at all concentrations used, which lasted for the first 6 h after BY administration. Interestingly, there was a rapid decrease in the antipyretic activity of the 100% concentration of AEMM 3 h after BY administration when compared with the other concentrations of AEMM, with completely diminished activity observed for the last 2 h of the experiment.

### 17.5.4 Antioxidant assays

The antioxidant activities of the AEMM, CEMM and MEMM were assessed using the DPPH radical scavenging and superoxide scavenging assays. The 20, 100 and 500 mg/mL MEMM produced approximately 13–98% of DPPH radical scavenging activities, while the AEMM followed by the CEMM produced 7–70% and 0–18% radical scavenging activities, respectively. For the superoxide scavenging activity, the 20, 100 and 500 mg/mL MEMM produced approximately 91–99% effect, while the AEMM and CEMM caused approximately 80–96% and 28–53% scavenging activities, respectively. In addition, AEMM and MEMM showed the highest phenolic contents, which were approximately 3344 mg and 3055 mg respectively per 100 g gallic acid. However, total phenolic content in CEMM was approximately 92 mg per 100 g of gallic acid. From the phytochemical screening, AEMM was found to contain only flavonoids, CEMM contained flavonoids and steroids, while MEMM was found to contain saponins, flavonoids, condensed tannin and steroids (Zakaria *et al.*, 2011d). However, another study done by Zakaria *et al.* (2006d) showed flavonoids, triterpenes, tannins, saponins and steroids, but no alkaloids were detected in AEMM. The HPLC profile of AEMM was compared with the standard flavonoids, namely catechin, rutin and fisetin. Interestingly, the AEMM bioactive compounds were best separated at 330 nm, as indicated by the increase in the number of peaks detected.

### 17.5.5 Antiproliferative activities

Zakaria *et al.* (2011d) carried out a study to investigate the possible antiproliferative potential of AEMM, MEMM and CEMM as indicated by their ability to inhibit several cell lines of the American Type Culture Collection, namely 3T3, MCF-7, HeLa, Caov-3, HL-60, CEM-SS and MDA-MB-231. Interestingly, all extracts did not inhibit the proliferation of 3T3 cells, thus indicating their non-cytotoxic properties. The AEMM exhibited antiproliferative activity against Caov-3 that produced an $IC_{50}$ value of 34 mg/mL and HL-60 cell lines, while the CEMM inhibited the proliferation of Caov-3 by producing an $IC_{50}$ value of 41 mg/mL, HL-60, and CEM-SS cell lines with an approximate $IC_{50}$ value of 22 mg/mL. The MEMM demonstrated antiproliferative activity against more cell lines, including the MCF-7 with an estimated $IC_{50}$ value of 87 mg/mL, HeLa that produced an $IC_{50}$ value of approximately 88 mg/mL, Caov-3 with an $IC_{50}$ value of 58 mg/mL, HL-60, CEM-SS with an approximate $IC_{50}$ value of 30 mg/mL, and MDA-MB-231 cancer cell lines that produced an approximate $IC_{50}$ value of 59 mg/mL.

### 17.5.6 Hepatoprotective activities

Based on the study to determine the potential of hepatoprotective activity, rats were divided into 10 groups ($n = 6$) and received $dH_2O$ or 10% DMSO (negative control), 200 mg/kg silymarin (positive control) or extract in different dosages (50, 250 and 500 mg/kg) orally once daily for 7 days and on the eighth day subjected to hepatotoxic induction by using either $CCl_4$ or PCM. The blood and liver samples were collected and subjected to biochemical and microscopic analysis.

In AEMM, as reported by Yahya *et al.* (2012), the extract did not cause significant changes in normal behaviour pattern, and there were no signs and symptoms of mortality and toxicity observed. Only the 500 mg/kg of AEMM exhibited significant ($p < 0.05$) hepatoprotective activity against the $CCl_4$- and PCM-induced liver toxicity models by reducing the plasma levels of AST and ALT when compared with the control group. The blood parameter findings supported the histological observations.

On the other hand, Kamisan *et al.* (2013) reported that MEMM also did not cause significant changes in normal behaviour pattern, and no signs and symptoms of mortality and toxicity were observed. The extract exhibited a significant ($p < 0.05$) hepatoprotective activity against the $CCl_4$- and PCM-induced liver models of toxicity by improving liver function, as indicated by the reduction in levels of the liver enzymes AST and ALT compared with the control group. Histopathologic studies of the livers removed from the PCM- and $CCl_4$-induced rats pretreated with 10% DMSO revealed significant ($p < 0.05$) damage to the architecture of the liver, with severe necrosis of hepatocytes in the parenchymal region and haemorrhage. These observations were attenuated in the groups pretreated with MEMM or silymarin. The group induced with vehicle (10% glycerol) did not show any changes in the level of liver enzymes and, histologically, did not exert any structural changes.

### 17.5.7 Antiulcer activities

In an antiulcer study carried out by Wan Zainulddin *et al.* (personal communication) rats were divided into 10 groups and received 10% DMSO (negative control), 100 mg/kg ranitidine (positive control) or AEMM (50, 250 and 500 mg/kg) orally for 7 days and, on the eighth day subjected to gastric ulcer induction by using either ethanol or indomethacin. The stomachs were collected and subjected to macroscopic and microscopic analysis.

This study demonstrated the gastroprotective effect of AEMM, where the extract, at all dosages used, exhibited a significant ($p < 0.05$) antiulcer activity in a dose-dependent manner against the ethanol, but not in the indomethacin-induced gastric ulcer model. The percentage of antiulcer activity for the 50–500 mg/kg AEMM ranged between 51 and 82% compared to the negative control group (Wan Zainulddin *et al.*, personal communication). On the other hand, Zabidi *et al.* (2012) reported that MEMM exhibited a significant ($p < 0.05$) antiulcer activity in a dose-dependent manner at all dosages used against the ethanol-induced gastric ulcer model, but not for the indomethacin-induced gastric ulcer model. The percentage of 50–500 mg/kg MEMM ranged between 3 and 75 (Zabidi *et al.*, 2012). Gross observations were supported by histological findings for the both extracts.

## 17.6 Conclusion

From the detailed review, it can be concluded that various extracts of *M. calabura*, *D. linearis*, *B. purpurea* and *M. malabathricum* were found to show potential pharmacological activities in several aspects, even though these plants are categorized under the neglected/underutilized tropical plants. Therefore, these findings would serve as a primary investigation, which may contribute to the discovery of new bioactive compounds that may be a breakthrough in the field of drug discovery.

## References


Abdul Hisam, E.E., Zakaria, Z.A., Mohtaruddin, N. *et al.* (2012). Antiulcer activity of the chloroform extract of *Bauhinia purpurea* leaf. *Pharmaceutical Biology*, **50** (12), 1498–1507.

Chin, W.Y. (1989) *A Guide to the Wayside Trees of Singapore*, BP Singapore Science Centre, Singapore, p. 145.

Chin, W.Y. (1992) *A Guide to Medicinal Plants*, Singapore Science Centre, p. 24.

Institute of Medical Research (2002) *Compendium of Medicinal Plants used in Malaysia*, 2nd edn, Herbal Medicine Research Centre, Institute for Medical Research, Kuala Lumpur, pp. 131–132.

Jaganath, I.B. and Ng, L.T. (2000) *Herbs: the Green Pharmacy of Malaysia*, Vinpress, Kuala Lumpur, pp. 58–59.

Kamisan, F.H., Yahya, F., Ismail, N.A. *et al.* (2013) Hepatoprotective activity of methanol extract of *Melastoma malabathricum* leaf in rats. *Journal of Acupuncture and Meridian Studies*, **6** (1), 52–55.

Katzung, B.G. (1995) *Basic and Clinical Pharmacology*, 6thedn, Prentice Hall International, London.

Mamat, S.S., Kamisan, F.H., Wan Zainulddin, W.N. *et al.* (2013) Effect of methanol extract of *Dicranopteris linearis* leaves against paracetamol- and carbon tetrachloride ($CCl_4$)-induced liver toxicity in rats. *Journal of Medicinal Plants Research*, **7** (19), 1305–1309.

Mohd Joffry, S., Yob, N. J., Rofiee, M. S. *et al.* (2012) *Melastoma malabathricum* (L.) Smith ethnomedicinal uses, chemical constituents, and pharmacological properties: a review. *Evidence-Based Complementary and Alternative Medicine*, **2012**, article ID 258434.

Mohd Sani, M.H., Zakaria, Z.A., Balan, T. *et al.* (2012) Antinociceptive activity of methanol extract of *Muntingia calabura* leaves and the mechanisms of action involved. *Evidence-Based Complementary and Alternative Medicine*, **2012**, article ID 890361.

Morton, J.F. (1987) Jamaica cherry, in *Fruits of Warm Climates* (ed. J.F. Morton), Julia F. Morton, Miami, FL, pp. 65–69.

Pezzuto, J. (1996) Taxol production in plant cell culture comes of age. *Nature Biotechnology*, **14**, 1083.

Vasudeva, S.M. (1999) Economic importance of pteridophytes. *Indian Fern Journal*, **16** (1–2), 130–152.

Withers, L. (2005) *Strategic approaches for funding work on underutilized species*, The Global Facilitation Unit for Underutilized Species, Italy, www.underutilized-species.org/documents/Publications/underutilized_species_funding%20strategy.pdf (accessed 4 February 2013).

WHO (2003) *Traditional medicine*, WHO Fact Sheet No. 134, revised 2003, World Health Organization, Geneva, www.who.int/mediacentre/factsheets/fs134/en/ (accessed 2 January 2013).

Yahya, F., Ismail, N.A., Kamisan, F.H. *et al.* (2012) Effect of aqueous extract of *Melastoma malabathricum* leaves against paracetamol- and $CCl_4$-induced liver toxicity in rats. *African Journal of Pharmacy and Pharmacology*, **6** (37), 2682–2685

Zabidi, Z., Wan Zainulddin, W.N., Mamat, S.S. *et al.* (2012) Antiulcer activity of methanol extract of *Melastoma malabathricum* leaves in rats. *Medical Principle and Practice*, **21** (5), 501–503.

Zakaria, Z.A. (2007) Free radical scavenging activity of some plants available in Malaysia. *Iranian Journal of Pharmacology and Therapeutics*, **6** (1), 87–91.

Zakaria, Z.A., Abdul Ghani, Z.D.F., Raden Mohd Nor, R.N. *et al.* (2006a) Antinociceptive and anti-inflammatory activities of *Dicranopteris linearis* leaves chloroform extract in experimental animals. *Yakugaku Zasshi*, **126** (11), 1197–1203.

Zakaria, Z.A., Fatimah, C.A., Mat Jais, A.M., *et al.* (2006b) The *in vitro* antibacterial activity of *Muntingia calabura* extracts. *International Journal of Pharmacology*, **2** (3), 290–293.

Zakaria, Z.A., Sulaiman, M.R., Mat Jais, A.M. *et al.* (2006c) The antinociceptive activity of *Muntingia calabura* aqueous extract and the involvement of L-arginine/nitric oxide/cyclic guanosine monophosphate pathway in its observed activity in mice. *Fundamental & Clinical Pharmacology*, **20** (4), 365–372.

Zakaria, Z.A., Raden Mohd Nor, R.N., Hanan Kumar, G. *et al.* (2006d) Antinociceptive, anti-inflammatory and antipyretic properties of *Melastoma malabathricum* leaves aqueous extract in experimental animals. *Canadian Journal of Physiology and Pharmacology*, **84** (12), 1291–1299.

Zakaria, Z.A., Raden Mohd Nor, R.N., Sulaiman, M.R. *et al.* (2006e) Antinociceptive and anti-inflammatory properties of *Melastoma malabathricum* leaves chloroform extract in experimental animals. *Journal of Pharmacology and Toxicology*, **1** (4), 337–345.

Zakaria, Z.A., Mat Jais, A.M., Mastura, M. *et al.* (2007a) *In vitro* antistaphylococcal activity of the extracts of several neglected plants in Malaysia. *International Journal of Pharmacology*, **3** (5), 428–431.

Zakaria, Z.A., Mohd Nor Hazalin, N.A., Mohd Zaid, S.N.H. *et al.* (2007b) Antinociceptive, anti-inflammatory and antipyretic effects of *Muntingia calabura* aqueous extract in animal models. *Journal of Natural Medicine*, **61** (4), 443–448.

Zakaria, Z.A., Mustapha, S., Sulaiman, M.R. *et al.* (2007c) The antinociceptive action of aqueous extract from *Muntingia calabura* leaves: the role of opioid receptors. *Medical Principles and Practice*, **16** (2), 130–136.

Zakaria, Z.A., Wen, L.Y., Abdul Rahman, N.I. *et al.* (2007d) Antinociceptive, anti-inflammatory and antipyretic properties of the aqueous extract of *Bauhinia purpurea* leaves in experimental animals. *Medical Principle and Practice*, **16** (6), 443–449.

Zakaria Z.A., Mohamed A.M., Mohd Jamil N.S. *et al.* (2008a) *In vitro* anticancer activity of various extracts of neglected Malaysian plants (*Muntingia calabura* and *Dicranopteris linearis*) against MCF-7 and HT-29 cancer cell lines. *e-PharmaNexus*, **1** (1), 10–17.

Zakaria, Z.A., Abdul Ghani, Z.D.F., Raden Mohd Nor, R.N.S. et al. (2008b) Antinociceptive, anti-inflammatory, and antipyretic properties of an aqueous extract of *Dicranopteris linearis* leaves in experimental animal models. *Journal of Natural Medicines*, **62** (2), 179–187.

Zakaria, Z.A., Somchit, M.N., Sulaiman, M.R. et al. (2008c) Effects of various receptor antagonists, pH and enzymes on *Muntingia calabura* antinociception in mice. *Research Journal of Pharmacology*, **2** (3), 31–37.

Zakaria, Z.A., Abdul Rahman, N.I., Wen, L.Y. *et al.* (2009) Antinociceptive and anti-inflammatory activities of the chloroform extract of *Bauhinia purpurea* L. (Leguminosae) leaves in animal models. *International Journal of Tropical Medicine*, **4** (4), 140–145.

Zakaria, Z.A., Mat Desa, A., Ramasamy, K. *et al.* (2010a) Lack of antimicrobial activities of *Dicranopteris linearis* extracts and fractions. *African Journal of Microbiology Research*, **4** (1), 071–075.

Zakaria, Z.A., Sufian, A.S., Ramasamy, K. *et al.* (2010b) *In vitro* antimicrobial activity of *Muntingia calabura* extracts and fractions. *African Journal of Microbiology Research*, **4** (4), 304–308.

Zakaria, Z.A., Abdul Hisam, E.E., Rofiee, M.S. *et al.* (2011a) *In vivo* antiulcer activity of the aqueous extract of *Bauhinia purpurea* leaf. *Journal of Ethnopharmacology*, **137** (2), 1047–1054.

Zakaria, Z.A., Mohamed, A.M., Mohd Jamil, N.S. *et al.* (2011b) *In vitro* antiproliferative and antioxidant activities of the extracts of *Muntingia calabura* leaves. *American Journal of Chinese Medicine*, **39** (1), 183–200.

Zakaria, Z.A., Mohamed, A.M., Mohd Jamil, N.S. *et al.* (2011c) *In vitro* cytotoxic and antioxidant properties of the aqueous, chloroform and methanol extracts of *Dicranopteris linearis* leaves. *African Journal of Biotechnology*, **10** (2), 273–282.

Zakaria, Z.A., Rofiee, M.S., Mohamed, A.M. *et al.* (2011d) *In vitro* antiproliferative and antioxidant activities and total phenolic contents of the extracts of *Melastoma malabathricum* leaves. *Journal of Acupuncture and Meridian Studies*, **4** (4), 248–256.

Zakaria, Z.A., Rofiee, M.S., Teh, L.K. *et al.* (2011e) *Bauhinia purpurea* leaves' extracts exhibited *in vitro* antiproliferative and antioxidant activities. *African Journal of Biotechnology*, **10** (1), 65–74.

Zakaria, Z.A., Sodri, N.H., Hassan, H. and Anuar, K. (2012a) Effects of various receptor antagonists on the peripheral antinociceptive activity of aqueous extracts of *Dicranopteris linearis*, *Melastoma malabathricum* and *Bauhinia purpurea* leaves in mice. *International Journal of Genuine Traditional Medicine*, **2** (4), 38.1–38.6 (http://dx.doi.org/10.5667/tang.2012.0017).

Zakaria, Z.A., Abdul Hisam, E.E., Norhafizah, M. *et al.* (2012b) Methanol extract of *Bauhinia purpurea* leaf possesses anti-ulcer activity. *Medical Principle and Practice*, **21** (5), 476–482.

# 18 Multiple Applications of Endophytic *Colletotrichum* Species Occurring in Medicinal Plants

Mahendra Rai, Gauravi Agarkar and Dnyaneshwar Rathod

Department of Biotechnology, Sant Gadge Baba Amravati University, Amravati (MS), India

## 18.1 Introduction

Endophytes are microorganisms that survive within the plant tissues without causing any observable symptoms of disease. They offer remarkable potential for the exploitation of novel and eco-friendly secondary metabolites used in medicine, pharmaceutical industry and agriculture. Endophytes attack the tissues of living plants and cause unapparent and asymptomatic infections entirely within plant tissues without showing any symptoms of disease. Petrini *et al.* (1992) reported that there are chances of more than one type of endophytic fungus to be found within one plant. Herre *et al.* (2007) demonstrated that endophytic fungi play a very important mutualistic role by augmenting host defence responses against pathogens. Endophytes could be providing protection to the host and increasing the expression of intrinsic host defence mechanisms and providing additional sources of defence, extrinsic to those of the host. The production of antibiotics by *Pseudomonas*, such as 2-4-diacethylphloroglucinol, phenazines, pyrrolnitrin, pyoluteorin and hydrogen cyanine antibiotics show antifungal, antibacterial, antihelminthic and phytotoxic activity.

There has been an immense interest in the prospecting of these microbial endophytes as a source of novel bioactive natural products. Endophytes thus have the ability to show much chemical diversity, including alkaloids, peptides, steroids, terpenoids, isocoumarins, quinones, phenylpropanoids and lignans, phenols, phenolic acids, aliphatic compounds, lactones and others. Among these compounds, many have interesting biological activity. Schwarz *et al.* (2004) optimized the culture conditions of *Phoma* species and reported highest nematicidal activity in yeast malt glucose medium.

It is well known that after the discovery of taxol producer *Taxomyces andreanae*, which was isolated from *Taxus brevifolia*, the interest in endophyte research has increased. Endophytic fungi are an exciting group of microbes connected with the healthy tissues of plants. Endophytic fungal diversity is higher in tropical and subtropical plants than in other climatic zones (Banerjee, 2011). Karsten *et al.* (2007) reported herbicidal and algaecidal activity in ethyl acetate extract of an endophytic *Phoma* sp. isolated from *Fagonia cretica*. Randa *et al.* (2010) isolated a mycoendophyte (*Botryosphaeria rhodina*) from the stem of the medicinal plant *Bidens pilosa* (Asteraceae) that showed anti-inflammatory, antiseptic and antifungal effects. Santiago *et al.* (2012) reported that an endophytic fungus isolated from the plant *Cinnamomum mollissimum* showed antifungal activity against *Aspergillus niger* and also anticancerous activity. Budhiraja *et al.* (2012) isolated *Aspergillus* and *Penicillium* spp. from *Gloriosa superba*. They studied the antimicrobial activity of these endophytes against seven standard pathogenic strains, like *Staphylococcus aureus*, *Bacillus subtilis*, *Escherichia coli*, *Pseudomonas aeruginosa*, *Salmonella typhimurium*, *Saccharomyces cerevisiae* and *Candida albicans*. Vaz *et al.* (2012) isolated an endophyte *Colletotrichum* from leaves of *Myrciaria floribunda* and *Alchornea castaneifolia*, and an endophyte *Mycosphaerella* from *Eugenia* aff. *bimarginata*. Endophytic fungi are being more and more accepted as an ecological group of microorganisms that may offer sources for new secondary metabolites with useful biological activities. A group of active principles has been isolated and characterized from endophytes, and many of these have diverse bioactivities (anticancerous, antioxidants, antifungal, antibacterial, antiviral, anti-insecticidal and immune suppressants). Isolation of endophytic fungi from coffee plants (*Coffea arabica* and *Coffea robusta*) demonstrated antimicrobial activity against various human pathogenic bacteria (Sette *et al.*, 2006).

The genus *Colletotrichum* (Glomerellaceae, Sordariomycetidae, Sordariomycetes, Ascomycota) was described in 1831 by Corda. *Colletotrichum* species with dark setae and curved conidia are known as anthracnose pathogens of a number of economically important hosts and are often identified as *Colletotrichum dematium*. *C. dematium* has been reported to be synonymous with many species, including the type of the genus, *Colletotrichum lineola* (Damm *et al.*, 2009). *Colletotrichum acutatum* causes economically significant losses of temperate, subtropical and tropical crops. *Colletotrichum* is one of the most common and important genera of plant-pathogenic fungi. Virtually every crop grown throughout the world is susceptible to one or more species of *Colletotrichum*. *Colletotrichum* causes post-harvest rots, anthracnose spots and blights of aerial plant parts. Members of this genus cause major economic losses, especially of fruits, vegetables, and ornamentals. *Colletotrichum* is very damaging to important food crops, including bananas, cassava, sorghum and pulses, grown by subsistence farmers in developing countries throughout the tropics and subtropics. In addition to their considerable economic importance, the *Colletotrichum* is highly significant as experimental models in

*Novel Plant Bioresources: Applications in Food, Medicine and Cosmetics*, First Edition. Edited by Ameenah Gurib-Fakim.

studies of many aspects of fungal development, infection processes, host resistance, signal transduction and the molecular biology of plant–pathogen interactions. Many laboratories around the world are studying the biology and pathology of various species of *Colletotrichum*.

The main aim of this chapter is to discuss various applications of endophytic *Colletotrichum* species as a source of natural products, occurring in medicinal plants, with special reference to their importance in medicine and pharmaceutical industries.

## 18.2 Diversity of endophytic *Colletotrichum* sp. in medicinal plants

Medicinal plants are reported to contain a diverse flora of endophytes (Strobel, 2002) that provide protection to their host from pathogens and environmental stresses. Many of these endophytes have the potential of producing metabolites useful in pharmaceutical and agricultural industries (Petrini *et al.*, 1992). *Colletotrichum* is a ubiquitous endophyte and have been reported from several plant hosts (Brown *et al.*, 1998; Suryanarayanan *et al.*, 2002). The species of genus *Colletotrichum* are reported frequently as endophytes from various host plant tissues (Larran *et al.*, 2001, 2002; Hata *et al.*, 2002; Santamaría and Bayman, 2005; Chareprasert *et al.*, 2005; Lin *et al.*, 2007; Osono, 2007; Verma *et al.*, 2007; Joshee *et al.*, 2009; Vega *et al.*, 2010; Higgins *et al.*, 2011) in general and medicinal plants in particular.

*Colletotrichum gloeosporioides* had the maximum colonization frequency among the endophytes isolated from *Lippia sidoides*, a medicinal plant used as an antiseptic in the northeast of Brazil (de Siqueira *et al.*, 2011). Huang *et al.* (2008) isolated fungal endophytes from 29 traditional Chinese medicinal plants, out of which 23 host plants were colonized by *Colletotrichum* sp. and it was a relatively high occurrence in 13 of the host plants, especially in *Artemisia lactiflora* (Table 18.1). Thalavaipandian *et al.* (2011) reported that *Colletotrichum* sp. isolated from medicinal plants showed the highest relative frequency among all other endophytes. *Colletotrichum* sp. colonized more in leaves than other parts of the plant. Wang *et al.* (2008) reported *C. gloeosporioides* as an endophyte of *Taxus mairei* which showed high colonization in leaf. *Glomerella cingulata* (anamorph of *C. gloeosporioides*) was one of the dominant endophytes of *Tripterygium wilfordii* (Kumar and Hyde, 2004). Suwannarach *et al.* (2012) studied the diversity of endophytic fungi associated with wild cinnamon tree, *Cinnamomum bejolghota* and found that *C. gloeosporioides* was the most common fungal endophyte.

*Colletotrichum* sp. isolated from *Orthosiphon spiralis* (Shobana, 2011), *Piper hispidum* (Orlandelli *et al.*, 2012), and *Centella asiatica* (Rakotoniriana *et al.*, 2008) were the second most dominant genera possessing high colonization frequencies in their respective host medicinal plants. Seasonal variations also affect the occurrence of foliar endophytes as the *Colletotrichum* sp. was the dominant endophyte during the wet period (Suryanarayanan and Thennarasan, 2004). *C. gloeosporioides* was dominant in northeast monsoon and winter seasons and was most frequently isolated from leaves (Thalavaipandian *et al.*, 2011). *Colletotrichum dematium* and *Colletotrichum linicola* were isolated from the stem, leaves, petiole and roots of *Tinospora cordifolia*, whereas *C. linicola* occurred completely in winter season (Mishra *et al.*, 2012).

Some of the chemical compounds isolated from endophytic *Colletotrichum* sp. and their chemical structures are presented in Figure 18.1 and Figure 18.2.

## 18.3 Biomedical applications

Different endophytic *Colletotrichum* species isolated from medicinal plants have been studied for their applications in the biomedical field. Although several drugs are available to combat often-deadly bacterial infections, many of these pathogens have acquired multidrug resistance (Arivudainambi *et al.*, 2011). So it is inevitable to search for new compounds to overcome the problem of drug resistance. There has been increasing interest in screening of endophytes for valuable bioactive compounds. Arivudainambi *et al.* (2011) reported that the endophytic fungus *C. gloeosporioides* isolated from the medicinal plant *Vitex negundo* L. is a potential resource for the production of metabolites against multidrug-resistant *S. aureus*. Further, they studied the antimicrobial activity of extract *C. gloeosporioides* against multidrug-resistant *S. aureus*, *B. subtilis*, *E. coli*, *P. aeruginosa* and *C. albicans*.

Many researchers have reported that endophytic *Colletotrichum* sp. shows antibacterial and antifungal activity against human pathogens (Gond *et al.*, 2011). *Nyctanthes arbor-tristis* is a well-known medicinal plant of India used for its antibacterial, antifungal, antileishmanial and cytotoxic activity. The endophytic *C. dematium* isolated from *N. arbor-tristis* exhibited a broad range of antibacterial activity against *Shigella flexnii*, *S. boydii*, *Salmonella enteritidis*, *S. paratyphi*, and *P. aeruginosa* (Gond *et al.*, 2011). Zou *et al.* (2000) extracted a new antimicrobial metabolite colletotric acid, from a liquid culture of endophytic *C. gloeosporioides* isolated from *Artemisia mongolica* (Table 18.2). Colletotric acid inhibited the growth of *B. subtilis*, *S. aureus* and *Sarcina lutea*. In another case, endophytic *Colletotrichum* sp. isolated from *Tectona grandis* showed antibacterial activity against *S. aureus* and *B. subtilis* (Chareprasert *et al.*, 2005).

Vaz *et al.* (2012) reported an endophytic *C. gloeosporioides* isolated from the leaves of *Myrciaria floribunda* and *Alchornea castaneifolia* collected from Brazilian Cerrado ecosystems. Moreover, they also studied antifungal activity against *C. albicans*, *C. krusei*, *C. parapsilosis*, *C. glabrata*, *Cryptococcus neoformans* and antibacterial activity against *S. aureus*, *S. typhimurium* and *B. cereus*. Fungal endophytes *Colletotrichum siamense* and *Colletotrichum* sp. isolated from *Rafflesia cantleyi* showed the antifungal activity against *C. albicans* (Refaei *et al.*, 2011). Mishra *et al.* (2012) reported endophytic *C. dematium* and *C. linicola* isolated from *Tinospora cordifolia*, which exhibited significant antibacterial activity against human pathogens like *S. flexnii*, *S. enteritidis* and *S paratyphi*. All these studies strongly support the view that the endophytic *Colletotrichum* sp. isolated from different host species inhabiting different ecosystems are promising sources of natural antimicrobial compounds.

A variety of metabolites were extracted from the culture of endophytic *Colletotrichum* sp. isolated from stem of *Artemisia annua*. After characterization they confirmed the compounds like 6-isoprenylindole-3-carboxylic acid, 3β,5α-dihydroxy-6β-acetoxy-ergosta-7,22-diene and 3β,5α-dihydroxy-6β-phenylacetyloxy-ergosta-7,22-diene (Figure 18.1). Further, they also studied the antibacterial and fungal activity against *B. subtilis*, *S. aureus*, *S. lutea* and *Pseudomonas* species and against *C. albicans* and *A. niger* respectively (Lu *et al.*, 2000).

In addition, the species of genus *Colletotrichum* have been demonstrated to produce some economically important bioactive metabolites, such as taxol. Taxol is a potent anticancer drug isolated for the first time from the plant *T. brevifolia*; it has powerful antimitotic activity and is widely used in the treatment of a variety

**Table 18.1** List of endophytic *Colletotrichum* species isolated from different medicinal plants

| Sr. no. | Host medicinal plant | Endophytic *Colletotrichum* species | Explant | Reference |
|---|---|---|---|---|
| 1 | *Artemisia annua* | *Colletotrichum* sp. | Stem | Lu *et al.* (2000) |
| 2 | *Artemisia mongolica* | *C. gloeosporioides* | Stem | Zou *et al.* (2000) |
| 3 | *Tripterygium wilfordii* | *Colletotrichum* sp. | Twig bark | Kumar and Hyde (2004) |
| 4 | *T. wilfordii* | *C. musae* | Leaves | Kumar and Hyde (2004) |
| 5 | *T. wilfordii* | *Colletotrichum* sp. | Twig xylem | Kumar and Hyde (2004) |
| 6 | *Plumeria rubra* | *Colletotrichum* sp. | Leaves | Suryanarayanan and Thennarasan (2004) |
| 7 | *Brucea javanica* | *Colletotrichum* sp. | Stems and leaves | Choi *et al.* (2005) |
| 8 | *Brucea javanica* | *Colletotrichum* sp. | Branches or leaves | Choi *et al.* (2005) |
| 9 | *Tectona grandis* | *Colletotrichum* sp. | Leaves | Chareprasert *et al.* (2005) |
| 10 | *Samanea saman* | *Colletotrichum* sp. | Leaves | Chareprasert *et al.* (2005) |
| 11 | *Calotropis gigantea* | *C. gloeosporioides* | Leaves | Devarajan and Suryanarayanan (2006) |
| 12 | *Cryptocarya mandioccana* | *C. gloeosporioides* | Leaves | Inácio *et al.* (2006) |
| 13 | *Ocimum basilicum* | *C. falcatum* | Leaves | Gangadevi and Muthumary (2007) |
| 14 | *O. basilicum* | *C. gloeosporioides* | Leaves | Gangadevi and Muthumary (2007) |
| 15 | *Centella asiatica* | *C. higginsianum* | Leaves | Rakotoniriana *et al.* (2008) |
| 16 | *Taxus mairei* | *C. gloeosporioides* | Leaves | Wang *et al.* (2008) |
| 17 | *Pteromischum* sp. | *C. dematium* | Stem | Ren *et al.* (2008) |
| 18 | *Allamanda cathartica* | *Colletotrichum* sp. | Stem | Huang *et al.* (2008) |
| 19 | *Alstonia scholaris* | *Colletotrichum* sp. | Stem | Huang *et al.* (2008) |
| 20 | *Alyxia sinensis* | *Colletotrichum* sp. | Stem | Huang *et al.* (2008) |
| 21 | *Catharanthus roseus* | *Colletotrichum* sp. | Stem | Huang *et al.* (2008) |
| 22 | *Cerbera manghas* | *Colletotrichum* sp. | Stem | Huang *et al.* (2008) |
| 23 | *Melodinus suaveolens* | *Colletotrichum* sp. | Stem | Huang *et al.* (2008) |
| 24 | *Nerium oleander* | *Colletotrichum* sp. | Stem | Huang *et al.* (2008) |
| 25 | *Plumeria rubra* | *Colletotrichum* sp. | Stem | Huang *et al.* (2008) |
| 26 | *Strophanthus divaricatus* | *Colletotrichum* sp. | Stem | Huang *et al.* (2008) |
| 27 | *Tabernaemontana divaricata* | *Colletotrichum* sp. | Stem | Huang *et al.* (2008) |
| 28 | *Thevetia peruviana* | *Colletotrichum* sp. | Stem | Huang *et al.* (2008) |
| 29 | *Trachelospermum jasminoides* | *Colletotrichum* sp. | Stem | Huang *et al.* (2008) |
| 30 | *Asclepias curassavica* | *Colletotrichum* sp. | Stem | Huang *et al.* (2008) |
| 31 | *Graphistemma pictum* | *Colletotrichum* sp. | Stem | Huang *et al.* (2008) |
| 32 | *Hoya carnosa* | *Colletotrichum* sp. | Stem | Huang *et al.* (2008) |
| 33 | *Toxocarpus wightianus* | *Colletotrichum* sp. | Stem | Huang *et al.* (2008) |
| 34 | *Tylophora ovate* | *Colletotrichum* sp. | Stem | Huang *et al.* (2008) |
| 35 | *Artemisia indica* | *Colletotrichum* sp. | Stem | Huang *et al.* (2008) |
| 36 | *A. lactiflora* | *Colletotrichum* sp. | Stem | Huang *et al.* (2008) |
| 37 | *Scutellaria indica* | *Colletotrichum* sp. | Stem | Huang *et al.* (2008) |
| 38 | *Polygonum capitatum* | *Colletotrichum* sp. | Stem | Huang *et al.* (2008) |
| 39 | *Pavetta hongkongensis* | *Colletotrichum* sp. | Stem | Huang *et al.* (2008) |
| 40 | *Cestrum nocturnum* | *Colletotrichum* sp. | Stem | Huang *et al.* (2008) |
| 41 | *Justicia gendarussa* | *C. gloeosporioides* | Leaves | Gangadevi and Muthumary (2008) |
| 42 | *Dendrobium nobile* | *Colletotrichum* sp. | Leaves | Yuan *et al.* (2009) |
| 43 | *Hiptage benghalensis* | *Colletotrichum* sp. | Leaves, bark | Theantana *et al.* (2009) |
| 44 | *Calamus kerrianus* | *C. gloeosporioides* | Petiole, vein, intervein | Lumyong *et al.* (2009) |
| 45 | *Wallichia caryotoides* | *C. gloeosporioides* | Petiole, vein, intervein | Lumyong *et al.* (2009) |
| 46 | *Plumeria acutifolia* | *C. gloeosporioides* | Leaves | Nithya and Muthumary (2009) |
| 47 | *Elaeagnus pungens* | *C. gloeosporioides*<br>*C. montemartinii* | Leaves, branches | Liu *et al.* (2010) |
| 48 | *Mahonia bealei* | *C. gloeosporioides* | Leaves, branches | Liu *et al.* (2010) |
| 49 | *Rohdea japonica* | *C. gloeosporioides*<br>*C. montemartinii* | Branches | Liu *et al.* (2010) |
| 50 | *Sabina procumbens* | *C. gloeosporioides* | Leaves, branches | Liu *et al.* (2010) |
| 51 | *Eriobotrya japonica* | *C. montemartinii* | Leaves, branches | Liu *et al.* (2010) |
| 52 | *Camptotheca acuminata* | *Colletotrichum* sp. | Leaves | Ding *et al.* (2010) |
| 53 | *Couroupita guianensis* | *C. gloeosporioides* | Roots | Pria Dharsini *et al.* (2010) |
| 54 | *Vateria indica* | *Colletotrichum* sp. | Bark | Ruma *et al.* (2011) |
| 55 | *Artocarpus hirsutus* | *Colletotrichum* sp. | Bark, twig | Ruma *et al.* (2011) |
| 56 | *Salacia chinensis* | *C. gloeosporioides* | Stem, leaves | Bhagya *et al.* (2011) |
| 57 | *Nyctanthes arbor-tristis* | *C. dematium* | Leaves | Gond *et al.* (2011) |
| 58 | *Orthosiphon spiralis* | *Colletotrichum* sp. | Leaves | Shobana *et al.* (2011) |

Table 18.1 (*Continued*)

| Sr. no. | Host medicinal plant | Endophytic *Colletotrichum* species | Explant | Reference |
|---|---|---|---|---|
| 59 | *Rafflesia cantleyi* | *C. siamense, Colletotrichum* sp. | Flower | Refaei *et al.* (2011) |
| 60 | *Cephalotaxus hainanensis* | *Colletotrichum* sp. | Phloem | Chen *et al.* (2011) |
| 61 | *Piper ornatum* | *Colletotrichum* sp. | Phloem | Tianpanich *et al.* (2011) |
| 62 | *Huperzia serrata* | *Colletotrichum* sp. | Stems, leaves and roots | Chen *et al.* (2011) |
| 63 | *Ficus benghalensis* | *Colletotrichum* sp.<br>*C. gloeosporioides* | Leaf, bark | Maheswari and Rajagopal (2011) |
| 64 | *F. religiosa* | *Colletotrichum* sp.<br>*C. gloeosporioides* | Leaf, bark | Maheswari and Rajagopal (2011) |
| 65 | *Lippia sidoides* | *C. gloeosporioides* | Leaves, stem | de Siqueira *et al.* (2011) |
| 66 | *Cassia auriculata* | *Colletotrichum* sp. | Leaf, stem, bark | Thalavaipandian *et al.* (2011) |
| 67 | *Morinda tinctoria* | *Colletotrichum* sp. | Leaf, stem, bark | Thalavaipandian *et al.* (2011) |
| 68 | *Phyllanthus emblica* | *Colletotrichum* sp. | Leaf, stem, bark | Thalavaipandian *et al.* (2011) |
| 69 | *Pongamia glabra* | *Colletotrichum* sp. | Leaf, stem, bark | Thalavaipandian *et al.* (2011) |
| 70 | *Thespesia populnea* | *Colletotrichum* sp. | Leaf, stem, bark | Thalavaipandian *et al.* (2011) |
| 71 | *Tinospora cordifolia* | *C. linicola*<br>*C. crassipes* | Bark | Mishra *et al.* (2012) |
| 72 | *Panax ginseng* | *C. ignotum* | Petiole | Park *et al.* (2012) |
| 73 | *P.ginseng* | *C. pisi* | Root, stem, petiole | Park *et al.* (2012) |
| 74 | *Myrciaria floribunda* | *C. gloeosporioides* | Leaves | Vaz *et al.* (2012) |
| 75 | *Alchornea castaneifolia* | *C. gloeosporioides* | Leaves | Vaz *et al.* (2012) |
| 76 | *Lawsonia inermis* | *C. carssipes* | Stem, leaves | Amirita *et al.* (2012) |
| 77 | *L. inermis* | *C. falcatum* | Stem, leaves | Amirita *et al.* (2012) |
| 78 | *L. inermis* | *C. gleosporioides* | Stem, leaves | Amirita *et al.* (2012) |
| 79 | *Costus igneus* | *C. gleosporioides* | Stem, leaves | Amirita *et al.* (2012) |
| 80 | *Piper hispidum* | *Colletotrichum* sp. | Leaves | Orlandelli *et al.* (2012) |
| 81 | *P. hispidum* | *C. gloeosporioides* | Leaves | Orlandelli *et al.* (2012) |
| 82 | *P. hispidum* | *C. theobromicola* | Leaves | Orlandelli *et al.* (2012) |
| 83 | *P. hispidum* | *C. boninense* | Leaves | Orlandelli *et al.* (2012) |
| 84 | *P. hispidum* | *Glomerella cingulata* | Leaves | Orlandelli *et al.* (2012) |
| 85 | *Cinnamomum bejolghota* | *C. acutatum* | Vein, stem | Suwannarach *et al.* (2012) |
| 86 | *C. bejolghota* | *C. coccodes* | Vein | Suwannarach *et al.* (2012) |
| 87 | *C. bejolghota* | *C. gloeosporioides* | Vein, intervein, stem | Suwannarach *et al.* (2012) |

of cancers (Wani *et al.*, 1971). Taxol is produced by all plants of the yew species, but in extremely low amounts (Vidensek *et al.*, 1990; Georg *et al.*, 1994). Nowadays, all taxol in the world market is originated from *Taxus* sp. because the chemical synthesis is too expensive for commercialization. However, Stierle *et al.* (1993, 1994) reported that endophytic fungus *Taxomyces andreanae* produces taxol (Figure 18.1, **D**). Later on, a number of fungal endophytes were demonstrated to produce taxol (Li *et al.*, 1996; Strobel *et al.*, 1996; Wang *et al.*, 2000). Gangadevi and Muthumary (2008) reported *C. gloeosporioides* isolated from *Justicia gendarussa*, which produced taxol. Moreover, they also studied the cytotoxic activity against BT 220, H116, Int 407, HL 251 and HLK 210 human cancer cells *in vitro*. They reported that the endophyte can serve as a potential anticancerous agent to improve the production of taxol by fungal engineering (Gangadevi and Muthumary, 2008). Nithya and Muthumary (2009) studied the taxol-producing endophytic *C. gloeosporioides* isolated from the latex-bearing medicinal plant *Plumeria acutifolia*. The endophytic *Colletotrichum* species isolated from *Couroupita guianensis*, which exhibited antibiotic, antifungal, antiseptic and analgesic activity, also produces derivatives of taxol (Pria Dharsini *et al.*, 2010).

Some *Colletotrichum* species isolated from medicinal plants have been shown to produce novel bioactive metabolites. A novel peptide antimycotic, colutellin, was recovered from endophytic *C. dematium* isolated from a *Pteromischum* species growing in a tropical forest in Costa Rica (Ren *et al.*, 2008). Further, they studied the activity of colutellin which showed IL-2 production by activated CD4+ T-cells (Ren *et al.*, 2008). Tianpanich *et al.* (2011) reported five known isocoumarin derivatives – monocerin, its demethylated derivative, fusarentin 6,7-dimethyl ether, fusarentin 6-methyl ether and a fusarentin derivative – and a new phthalide named colletotrialide extracted from endophytic *Colletotrichum* species isolated from *Piper ornatum* (Table 18.2). Similarly, they analysed the cytotoxic activity of the monocerin derivative against the HepG2 cell line, which showed a similar activity to that of the standard anticancer drug etoposide (Figure 18.2).

The enzyme asparginase is used as a chemotherapeutic agent in the treatment of human cancers (Nakamura and Wilkinson, 1999; Long and Shu, 2002; Stams and den Boer, 2003; Lorenzi and Reinhold, 2006). Theantana *et al.* (2009) reported that endophytic *Colletotrichum* sp. E5T9 isolated from a Thai medicinal plant showed potential as a source of asparginase, which demonstrated 100% inhibition of the proliferation of CaCO-2 human Caucasian

**Table 18.2** Bioactivities of metabolites isolated from endophytic *Colletotrichum* species

| Host medicinal plant | Endophytic *Colletotrichum* sp. | Metabolites | Bioactivity of metabolite(s) | Reference |
|---|---|---|---|---|
| *Artemisia mongolica* | *Colletotrichum gloeosporioides* | Colletotric acid | Antimicrobial activity | Zou *et al.* (2000) |
| *A.annua* | *Colletotrichum* sp. | 6 Isoprenylindole-3-carboxylic acid | Antimicrobial activity | Lu *et al.* (2000) |
| | | 3β,5α-Dihydroxy-6β-acetoxy-ergosta-7,22-diene | Antimicrobial activity | |
| | | 3β,5α-Dihydroxy-6β-phenylacetyloxy-ergosta-7,22-diene | Antimicrobial activity | |
| | | 3β-Hydroxy-ergosta-5-ene | Fungistatic to plant pathogens | |
| | | 3-Oxo-ergosta- 4,6,8(14),22-tetraene | Fungistatic to plant pathogens | |
| | | Indole-3-acetic acid (IAA) | Plant growth hormone | |
| *Pteromischum* sp. | *C. dematium* | Colutellin A (peptide) | Immunosuppressive and Antimycotic activity | Ren *et al.* (2008) |
| *Justicia gendarussa* | *C. gloeosporioides* | Taxol | Anticancer | Gangadevi and Muthumary (2008) |
| *Plumeria acutifolia* | *C. gloeosporioides* | Taxol | Anticancer | Nithya and Muthumary (2009) |
| *Betula alnoides* | *Colletotrichum* sp. | Asparginase | Anticancer | Theantana *et al.* (2009) |
| *Couroupita guianensis* | *Colletotrichum sp.* | Taxol | Anticancer | Pria Dharsini *et al.* (2010) |
| *Piper ornatum* | *Colletotrichum* sp. | Monocerin, | Antioxidant activity | Tianpanich *et al.* (2011) |
| | | demethylated derivative of monocerin | Cytotoxic and Antioxidant activity | |
| | | Fusarentin 6,7-dimethyl ether, | Antioxidant activity | |
| | | Fusarentin 6-methyl ether, | Antioxidant activity | |
| | | Colletotrialide (phthalide) | | |

colon adenocarcinoma and inhibited proliferation of HepG2 human Caucasian hepatocyte carcinoma by 70%. The authors concluded that asparginase would be a potential source of natural drug for cancer antiproliferation.

## 18.4 Agriculture applications

The majority of all global agricultural products are lost each year due to various diseases and pests (Bajwa *et al.*, 2003), of which 20% of the losses are the result of pathogenic fungal diseases (Agrios, 2000). Although chemicals are effective in control of diseases, they are hazardous to the environment, and pathogens also have tendency to develop resistance against them. Thus, the natural and biological control of these disease-causing microorganisms has gained much attention in recent years, as a way of reducing the use of harmful chemical products in agriculture. Fungal endophytes have been reported to protect their plant host from pathogen attack, and in adverse environmental conditions by different strategies (Liu *et al.*, 2010).

Different *Colletotrichum* species isolated as endophytes from medicinal plants have been inhibitory to plant pathogens. Endophytic *Colletotrichum* sp. isolated from a Chinese medicinal plant exhibited antifungal activity against phytopathogenic fungi such as *C. gloeosporioides*, *Scopulariopsis* sp., *Trichoderma viride*, *Fusarium* sp., *Phytophthora nicotianae*, and *Verticillium* species (Li *et al.*, 2005). The endophytic fungus *Colletotrichum* sp. isolated from *Camptotheca acuminata* showed antifungal activity against pathogenic fungi like *Rhizoctonia solani*, *Gibberella fujikuroi*, *Gibberella zeae*, *Pyricularia grisea* and *Fusarium oxysporum* f. sp. *vasinfectum* and antibacterial activity against *Pseudomonas solanacearum* (Ding *et al.*, 2010). Endophytic *C. dematium* recovered from *N. arbor-tristis* showed antifungal activity against *Alternaria alternata*, *Cladosporium cladosporioides*, *Fusarium udum*, *Microsprum gypseum*, *Curvularia lunata* (Gond *et al.*, 2011). Further, they also concluded that antifungal activity of endophytic *C. dematium* signifies the potential of extracting antifungal compounds. The antifungal bioactive compound colletotric acid extracted from endophytic *Colletotrichum* sp. isolated from *Artemisia mongolica* showed antifungal activity against *Helminthosporium sativum* (Zou *et al.*, 2000) (Table 18.2).

*C. gloeosporioides* isolated from *Sabina procumbens* showed significant antifungal activity against the phytopathogens like *F. oxysporum* and *A. alternata* (Liu *et al.*, 2010). Moreover, they also studied the activity of *Colletotrichum montemartinii* against phytopathogens such as *R. solani*, *F. oxysporum*, *Fusarium graminearum*, *A. alternata*, *Phytophthora capsici*, *Glomeralla glycines*. Lastly, they also concluded that endophytic *Colletotrichum* sp. may play an important role in the protection of plants from diseases as well as being an excellent resource for natural antifungal compounds. The endophytic *Colletotrichum* sp. has the ability to synthesize antifungal metabolites that could be used as biocontrol agents in the control of plant diseases and might contribute for stress-resistance by reducing disease symptoms.

Lu *et al.* (2000) extracted different bioactive compounds, like 6-isoprenylindole-3-carboxylic acid, 3β,5α-dihydroxy-6β-acetoxy-

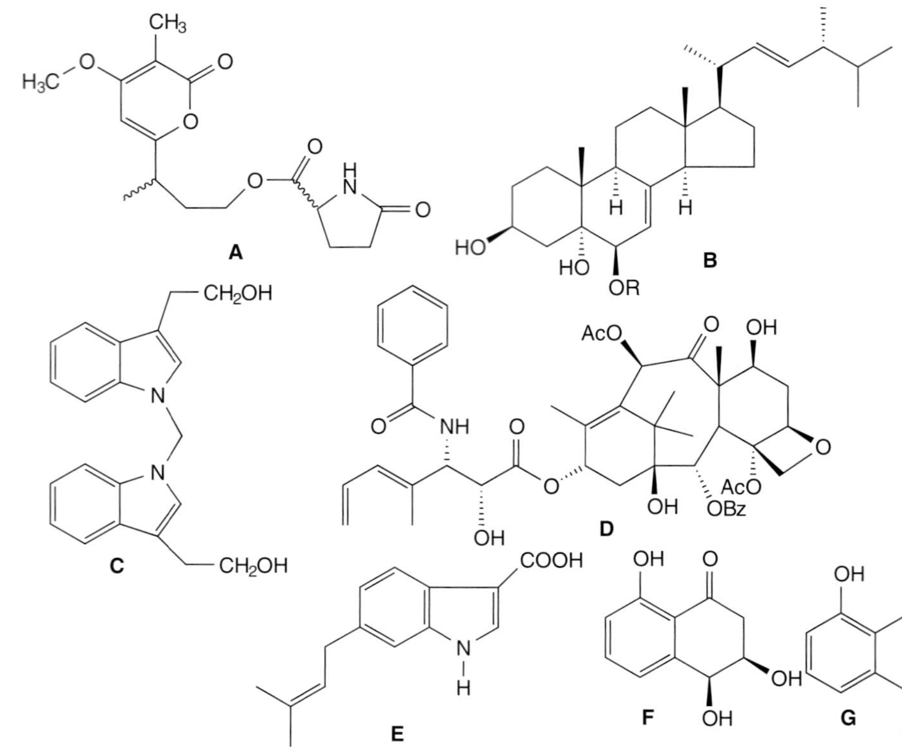


**Figure 18.1** 3-(4-Methoxy-3-methyl-2-oxo-2*H*-pyran-6-yl) butyl pyroglutamate (**A**); 3β,5α-dihydroxy-6β-acetoxy-ergosta-7,22-diene and 3β,5α-dihydroxy-6β-phenylacetyloxy-ergosta-7,22-diene (**B**); 2,20-(1,10-methylenebis(1*H*-indole-3,1-diyl))diethanol (**C**); taxol (**D**); 6-isoprenylindole-3-carboxylic acid (**E**); cis-4-hydroxy-6-deoxyscytalone (**F**); (4*R*)-4,8-dihydroxy-α-tetralone (**G**)

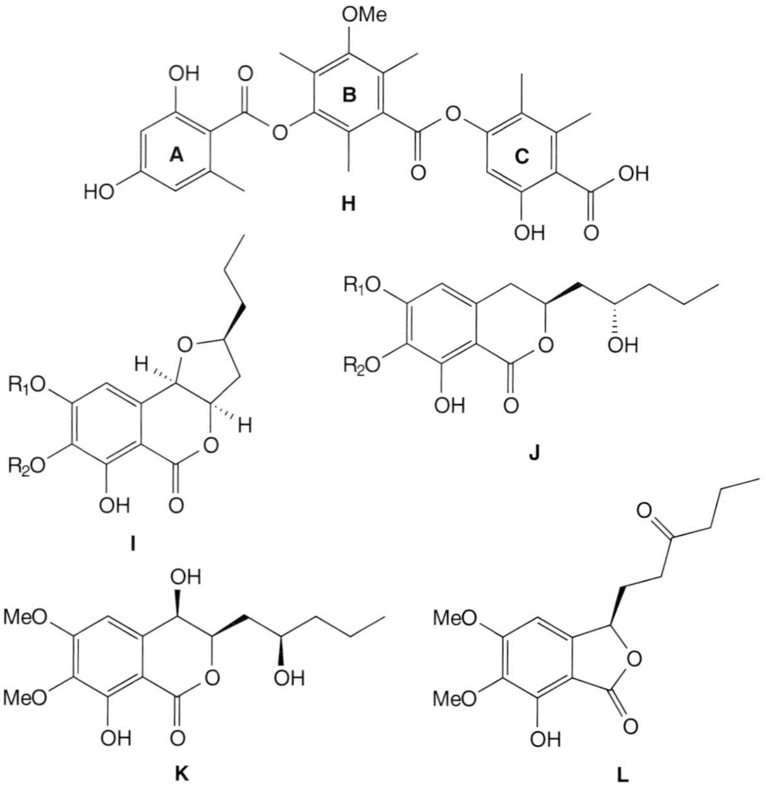


**Figure 18.2** Structure of colletotric acid (**H**); monocerin and its derivative (**I**); fusarentin derivatives (**J, K**); phthalide (**L**)

ergosta-7,22-diene and 3β,5α-dihydroxy-6β-phenylacetyloxy-ergosta-7,22-diene and 3β-hydroxy-ergosta-5-ene, from the endophytic *Colletotrichum* sp. recovered from stem of *A. annua*. They also studied the fungistatic activity against plant pathogenic fungi like *Gaeumannomyces graminis* var. *tritici*, *Rhizoctonia cerealis*, *H. sativum* and *P. capisici*. The endophytic *Colletotrichum* species isolated from *A. annua* produce plant growth regulators indole-3 acetic acid (IAA) can regulate the growth of the host and is presumably involved in the superior adaptability and competitiveness of *A. annua* in nature (Lu *et al.*, 2000). Mejía *et al.* (2008) suggested that the endophytic *C. gloeosporioides* can reduce the damage associated with pathogens in a variety of different ways in plants. Later on, they reported that endophytic *C. gloeosporioides* which was isolated from *Theobroma cacao* tissues was reinoculated on *T. cacao* leaves. It showed that treatment with *C. gloeosporioides* significantly decreased pod loss due to black pod rot disease. From the above results they concluded that *C. gloeosporioides* has restricted the proliferation of pathogens by competitive inhibition and can potentially be used as an effective biocontrol agent.

## 18.5 Industrial applications

Endophytes can also be exploited for the production industrially important compounds such as enzymes. Enzymes are biocatalysts, which have a variety of applications in pharmaceutical, food, textile and leather industries. For example, the enzyme amylases have a wide spectrum of applications, such as in the textile industry, paper and cellulose, leather, detergents, beer, liquor, bread, children's cereals, liquification and conversion of starch to sugar, animal chow, the fermentation industry (vitamins, amino acids, antibiotics), and the chemical and pharmaceutical industries (Pandey *et al.*, 1999; Butzen and Haefele, 2008) (Figure 18.3). Fungal enzymes are more stable than enzymes that are obtained from plants and animals (Maria *et al.*, 2005). The screening and identification of filamentous fungi capable of secreting extracellular enzymes with biotechnological potential has great importance (Onofre *et al.*, 2011). The selected species having the potential for enzyme secretion can be improved by optimizing environmental conditions favouring maximum production of the product for industrial use. The microbial capacity of hydrolysing different organic compounds, natural and synthetic, as well as inorganic compounds, has great economical importance (Onofre *et al.*, 2011). Therefore, the search for new chemical compounds with biological activities, particularly from endophytic fungi, is an area of interest of current research.

Assis *et al.* (2010) reported that *C. gloeosporioides* has a large potential for producing the α-amylase and glucoamylase enzymes. Choi *et al.* (2005) claimed that endophytic *Colletotrichum* sp. isolated from *Brucea javanica* produces extracellular cellulase, intracellular amylase, ligninase, pectinase and xylanase. In addition, they also studied endophytic *Colletotrichum* sp. that showed extracellular or intracellular production of amylase, cellulase and xylanase (Choi *et al.*, 2005). Onofre *et al.* (2011) evaluated the capacity of the endophytic *C. gloeosporioides* producing the amylolytic enzymes α-amylase and glucoamylase. The results provided evidence that the *C. gloeosporioides* showed the highest peaks of enzymatic production at 96 h, giving 563.32 U/g of α-amylase and 345.30 U/g of glucoamylase. From all observations they concluded that the fungus *C. gloeosporioides* has great capacity for producing amylolytic enzymes (α-amylase and glucoamylase) through fermentation in a rice-based solid state without any supplementation. Three endophytic *Colletotrichum* species isolated from the medicinal plants *Lawsonia inermis* and *Costus igneus* were tested for their ability to produce extracellular enzymes (i.e. amylase, cellulase, laccase, lipase and protease) by qualitative assays. Extracellular

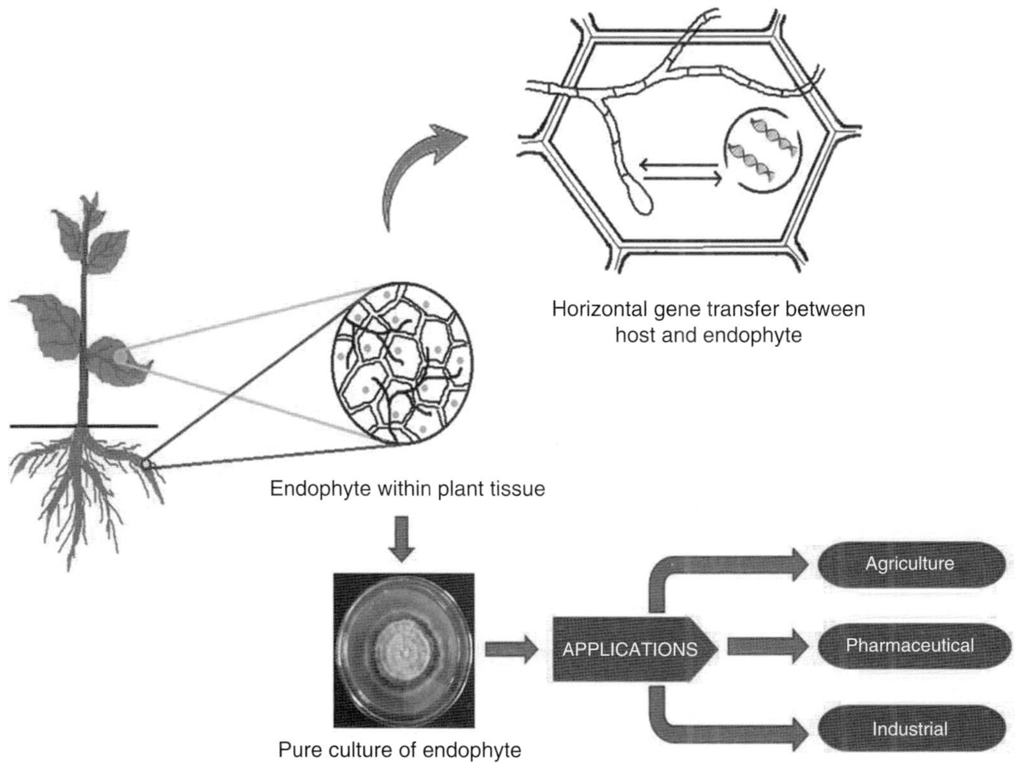


**Figure 18.3** Hypothetical mechanism of host–endophyte interactions and their broad applications

enzymes production was shown by *Colletotrichum carssipes* (amylase and protease), *Colletotrichum falcatum* (lipase and protease), *C. gleosporioides* (amylase and protease) (Amirita *et al.*, 2012).

## 18.6 Perspectives

Researchers from all over the world are engaged in the search for new antibiotics, chemotherapeutic agents and agrochemicals that are highly effective, possess low toxicity and have minor environmental impact. This search is driven by the development of resistant infectious microorganisms (e.g. species of *Mycobacterium*, *Streptococcus* and *Staphylococcus*). Furthermore, new diseases, like AIDS and respiratory syndrome, need the development of novel active drugs to fight them. Similarly, new drugs are needed for immunocompromised cancer patients and those receiving organ transplants, who are at high risk of opportunistic pathogens, such as *Aspergillus* species, *Cryptococcus* species and *Candida* species.

Apart from the above reasons, there is pressing need to search new and novel drug molecules to overcome many problems like scarcity of medicines, resistance in commercially available drugs, to minimize side effects of some drugs, to control plant pathogens, and so on. By keeping in mind the capabilities of fungi for the production of bioactive metabolites, researchers should be encouraged to isolate and screen endophytic fungi from diverse habitats and environments to search for novel bioactive metabolites. It has been reported by many researchers that fungal endophytes present in the host plants produce bioactive compounds similar to plants. Experts tried to find out the mechanism behind this and hypothesized that there are possibilities of transfer of genetic material or particular genes responsible for the production of bioactive compounds from plants to endophytes. It was assumed that horizontal gene transfer may play an active role in this mechanism.

## 18.7 Conclusion

The need for new bioactive compounds to overcome the growing problems of drug resistance in microorganisms and the appearance of new diseases is of increasing importance. The capability of fungi to produce bioactive metabolites has encouraged researchers to isolate and screen fungi from diverse habitats and environments to search for novel bioactive metabolites. Therefore, endophytes have proved an outstanding source of both novel and bioactive natural products, which have an enormous potential for the development of new drugs and agricultural products. Consequently, endophytes are known to be a rich and reliable source of biologically active compounds with potential benefits in medicinal, industrial and agricultural applications. In contrast, the development of techniques such as combinatorial chemistry and equipment such as peptide synthesizers gives rise to exciting opportunities and expectations for the synthesis of biological active compounds. Additionally, as most endophyte research has been conducted on cultivated species, the diversity of endophytes studied has been further restricted by limitations in the ability to cultivate endophytes in the laboratory.

In spite of the increased number of reports in the last decade, endophytes are still a relatively poorly investigated group of microorganisms. Therefore, the research focusing on endophytes is a promising field in the chemistry and biological properties of natural products. To overcome infectious diseases, there is need for a variety of novel antimicrobial compounds of biological origin. The endophytes hold enormous potential as sources of antimicrobials. These endophytes may open new vistas for the development of new drugs and agricultural products. The multidrug resistance problem in microbes underscores the need for further research on novel metabolites obtained from endophytes. Finally, it can be concluded that the *Colletotrichum* species have proved to be an outstanding novel bioactive compound producer of endophytes.

## References


Abdou, R., Scherlach, K., Dahse, H.-M. *et al.* (2010) Botryorhodines A–D, antifungal and cytotoxic depsidones from *Botryosphaeria rhodina*, an endophyte of the medicinal plant *Bidens pilosa*. *Photochemistry*, **71** (1), 110–116.

Agrios, G.N. (2000) Significance of plant diseases, in *Plant Pathology*, Academic Press, London, pp. 25–37

Amirita, A., Sindhu, P., Swetha, J. *et al.* (2012) Enumeration of endophytic fungi from medicinal plants and screeening of extracellular enzymes. *World Journal of Science and Technology*, **2** (2), 13–19.

Arivudainambi, U.S.E., Anand, T.D., Shanmugaiah, V. *et al.* (2011) Novel bioactive metabolites producing endophytic fungus *Colletotrichum gloeosporioides* against multidrug-resistant *Staphylococcus aureus*. *FEMS Immunology and Medical Microbiology*, **61**, 340–345.

Assis, T.C., Menezes, M., Andrade, D.E.G.T. *et al.* (2010) Estudo comparativo de isolados de *Colletotrichum gloeosporioides* quanto ao efeito da nutrição de carboidratos no crescimento, esporulação e patogenicidade em frutos de três variedades de mangueira. *Summa Phytopathologica*, **27**, 208–212.

Bajwa, R., Khalid, A. and Cheema, T.S. (2003) Antifungal activity of allelopathic plant extracts III: growth response of some pathogenic fungi to aqueous extract of *Parthenium hysterophorus*. *Journal of Plant Pathology*, **2**, 145–156.

Banerjee, D. (2011) Endophytic fungal diversity in tropical and subtropical plants. *Research Journal of Microbiology*, **6**, 54–62.

Bhagya, N., Sheik, S., Sharma, M.S. and Chandrashekar, K.R. (2011) Isolation of endophytic *Colletotrichum gloeosporioides* Penz. from *Salacia chinensis* and its antifungal sensitivity. *Journal of Phytology*, **3** (6), 20–22.

Brown, K.B., Hyde, K.D. and Guest, D.I. (1998) Preliminary studies on endophytic fungal communities of *Musa acuminata* species complex in Hong Kong and Australia. *Fungal Diversity*, **1**, 27–51.

Budhiraja, A., Nepali, K., Kaul, S. and Dhar, K.L. (2012) Antimicrobial and cytotoxic ativities of fungal isolates of medicinal plant *Gloriosa superba*. *International Journal of Recent Advances in Pharmaceutical Research*, **2** (1), 37–45.

Butzen, S. and Haefele, D. (2008) Dry-grind ethanol production from corn. *Crop Insights*, **18**, 1–5.

Chareprasert, S., Piapukiew, J., Thienhirun, S. *et al.* (2005) Endophytic fungi of teak leaves *Tectona grandis* L. and rain tree leaves *Samanea saman* Merr. *World Journal of Microbiology & Biotechnology*, **22**, 481–486.

Chen, G., Dai, H.F., Sha, Y. and Pei, Y.H. (2011) Two new compounds from the endophytic fungus *Colletotrichum* sp. L10 of *Cephalotaxus hainanensis*. *Journal of Asian Natural Products Research*. **13** (11), 1042–1046.

Choi, Y.W., Hodgkiss, I.J. and Hyde, K.D. (2005) Enzyme production by endophytes of *Brucea javanica*. *Journal of Agricultural Technology*, **1**, 55–66.

Damm, U., Woudenberg, J.H.C., Cannon, P.F. and Crous, P.W. (2009) *Colletotrichum* species with curved conidia from herbaceous hosts. *Fungal Diversity*, **39**, 45–87.

De Siqueira, V.M., Conti, R., Araújo, J.M., and Souza-Motta, C.M. (2011) Endophytic fungi from the medicinal plant *Lippia sidoides* Cham. and their antimicrobial activity. *Symbiosis*, **53**, 89–95.

Devarajan, P.T. and Suryanarayanan, T.S. (2006) Evidence for the role of phytophagous insects in dispersal of non-grass fungal endophytes. *Fungal Diversity*, **23**, 111–119.

Ding, T., Jiang, T., Zhou, J. *et al.* (2010) Evaluation of antimicrobial activity of endophytic fungi from *Camptotheca acuminata* (Nyssaceae). *Genetics and Molecular Research*, **9** (4), 2104–2112.

Gangadevi, V. and Muthumary, J. (2007) Endophytic fungal diversity from young, mature and senescent leaves of *Ocimum basilicum* L. with special reference to taxol production. *Indian Journal of Science and Technology*, **1** (1), 1–12.

Gangadevi, V. and Muthumary, J. (2008) Isolation of *Colletotrichum gloeosporioides*, a novel endophytic taxol-producing fungus from the leaves of a medicinal plant, *Justicia gendarussa*. *Mycologia Balcanica*, **5**, 1–4.

Georg, G.I., Chen, T.T., Ojima, I. and Vyas, D.M. (eds) (1994) *Taxane Anticancer Agents: Basic Science Current Status*, American Chemical Society, Symposium Series No. 583, American Chemical Society, Washington, DC.

Gond, S.K., Mishra, A., Sharma, V.K. *et al.* (2011) Diversity and antimicrobial activity of endophytic fungi isolated from *Nyctanthes arbor-tristis*, a well-known medicinal plant of India. *Mycoscience*, **53**, 113–121.

Hata, K., Atari, R. and Sone, K. (2002) Isolation of endophytic fungi from leaves of *Pasania edulis* and their within-leaf distributions. *Mycoscience*, **43**, 369–373.

Herre, E.A., Mejia, L.C., Kyllo, D.A. *et al.* (2007) Ecological implications of anti-pathogen effects of tropical fungal endophytes and mycorrhizae. *Ecology*, **88** (3), 550–558.

Higgins, H.L., Coley, P.D., Kursar, T.A. and Arnold, A.E. (2011) Culturing and direct PCR suggest prevalent host generalism among diverse fungal endophytes of tropical forest grasses. *Mycologia*, **103**, 247–260.

Huang, W.Y., Cai, Y.Z., Hyde, K.D. *et al.* (2008) Biodiversity of endophytic fungi associated with 29 traditional Chinese medicinal plants. *Fungal Diversity*, **33**, 61–75.

Inácio, M.L., Silva, G.H., Teles, H.L. *et al.* (2006) Antifungal metabolites from *Colletotrichum gloeosporioides*, an endophytic fungus in *Cryptocarya mandioccana* Nees (Lauraceae). *Biochemical Systematics and Ecology*, **34**, 822–824.

Joshee, S., Paulus, B.C., Park, D. and Johnston, P.R. (2009) Diversity and distribution of fungal foliar endophytes in New Zealand Podocarpaceae. *Mycological Research*, **113**, 1003–1015.

Karsten, K., Umar, F., Ulrich, F. *et al.* (2007) Secondary metabolites isolated from an endophytic *Phoma* sp. absolute configuration of tetrahydropyrenophorol using the solid-state TDDFT CD methodology. *European Journal of Organic Chemistry*, 3206–3211.

Kumar, D.S.S. and Hyde, K.D. (2004) Biodiversity and tissue-recurrence of endophytic fungi from *Tripterygium wilfordii*. *Fungal Diversity*, **17**, 69–90.

Larran, S., Mónaco, C. and Alippi, H.E. (2001) Endophytic fungi in leaves of *Lycopersicon esculentum* Mill. *World Journal of Microbiology & Biotechnology*, **17**, 181–184.

Larran, S., Perelló, A., Simón, M.R. and Moreno, V. (2002) Isolation and analysis of endophytic microorganisms in wheat (*Triticum aestivum* L.) leaves. *World Journal of Microbiology & Biotechnology*, **18**, 683–686.

Li, H., Qing, C., Zhang, Y. and Zhao, Z. (2005) Screening for endophytic fungi with antitumour and antifungal activities from Chinese medicinal plants. *World Journal of Microbiology & Biotechnology*, **21**, 1515–1519.

Li, J.Y., Strobel, G.A., Sidhu, R. *et al.* (1996) Endophytic taxol-producing fungi from bald cypress, *Taxodium distichum*. *Microbiology*, **142**, 2223–2226.

Lin, X., Lu, C., Huang, Y., *et al.* (2007) Endophytic fungi from a pharmaceutical plant, *Camptotheca acuminata*: isolation, identification and bioactivity. *World Journal of Microbiology & Biotechnology*, **23**, 1037–1040.

Liu, C., Liu, T., Yuan, F. and Gu, Y. (2010) Isolating endophytic fungi from evergreen plants and determining their antifungal activities. *African Journal of Microbiology Research*, **4**, 2243–2248.

Long, G.Q. and Shu, W.M. (2002) Comparison of antitumor effect of recombinant L-asparginase with wild type one *in vitro* and *in vivo*. *Acta Pharmacologica Sinica*, **23**, 946–951.

Lorenzi, P.L. and Reinhold, W.C. (2006) Asparginase synthetase as a causal, predictive biomarker for L-asparginase activity in ovarian cancer cells. *Molecular Cancer Therapeutics*, **5**, 2613–2623.

Lu, H., Zou, W.X., Meng, J.C. *et al.* (2000) New bioactive metabolites produced by *Colletotrichum* sp. an endophytic fungus in *Artemisia annua*. *Plant Science*, **151**, 76–73.

Lumyong, S., Techa, W., Lumyong, P. *et al.* (2009) Endophytic fungi from *Calamus kerrianus* and *Wallichia caryotoides* (Arecaceae) at Doi Suthep-Pui National Park, Thailand. *Chiang Mai Journal of Science*, **36** (2), 158–167.

Maheswari, S. and Rajagopal, K. (2011) Biodiversity of endophytic fungi associated with *Ficus religiosa* and *F. benghalensis*. *Mycologia Balcanica*, **8**, 169–172.

Maria, G.L., Sridhar, K.R. and Raviraja, N.S. (2005) Antimicrobial and enzyme activity of mangrove endophytic fungi of southwest coast of India. *Journal of Agricultural Technology*, **1**, 67–80.

Mejía, L.C., Rojas, E.I., Maynard, Z. *et al.* (2008) Endophytic fungi as biocontrol agents of *Theobroma cacao* pathogens. *Biological Control*, **46** (1), 4–14.

Mishra, A., Gond, S.K., Kumar, A. *et al.* (2012) Season and tissue type affect fungal endophyte communities of the Indian medicinal plant *Tinospora cordifolia* more strongly than geographic location. *Microbial Ecology* , **64**, 388–398.

Nakamura, C.T. and Wilkinson, R. (1999) Pancreatitis and perotitis following therapy with L-asparginase. *International Pediatrics*, **14**, 25–27.

Nithya, K. and Muthumary, J. (2009) Growth studies of *Colletotrichum gloeosporioides* (Penz.) Sacc. – a taxol producing endophytic fungus from *Plumeria acutifolia*. *Indian Journal of Science and Technology*, **2** (11), 14–19.

Onofre, S.B., Steilmnn, P., Bertolini, J. *et al.* (2011) Amylolytic enzymes produced by the fungus *Colletotrichum gloeosporioides* in rice semi-solid fermentation. *Journal of Yeast and Fungal Research*, **2** (3), 28–32.

Orlandelli, R.C., Alberto, R.N., Rubin Filho C.J. and Pamphile, J.A. (2012) Diversity of endophytic fungal community associated with *Piper hispidum* (Piperaceae) leaves. *Genetics and Molecular Research*, **11** (2), 1575–1585.

Osono, T. (2007) Endophytic and epiphytic phyllosphere fungi of red-osier dogwood (*Cornus stolonifera*) in British Columbia. *Mycoscience*, **48**, 47–52.

Pandey, A., Benjamin, S., Soccol, C.R. *et al.* (1999) The realm of microbial lipases in biotechnology. *Biotechnology and Applied Biochemistry*, **29**, 119–113.

Park, Y.H., Lee, S.G., Ahn, D.J. *et al.* (2012) Diversity of fungal endophytes in various tissues of *Panax ginseng* Meyer cultivated in Korea. *Journal of Ginseng Research*, **36** (2), 211–217.

Petrini, O., Sieber, T.N., Toti, L. and Viret, O. (1992) Ecology, metabolite production, and substrate utilization in endophytic fungi. *Natural Toxins*, **1**, 185–196.

Pria Dharsini, K., Jenefar, S., Malathi, S. *et al.* (2010) Molecular and morphological identification of *Colletotrichum gloeosporioides* Penz. on Cannon ball tree. *Journal of Biosciences Research*, **1** (2), 65–69.

Rakotoniriana, E.F., Munaut, F., Decock, C. *et al.* (2008) Endophytic fungi from leaves of *Centella asiatica*: occurrence and potential interactions within leaves. *Antonie van Leeuwenhoek*, **93**, 27–36.

Refaei, J., Jones, E.B.G., Sakayaroj, J. and Santhanam, J. (2011) Endophytic fungi from *Rafflesia cantleyi*: species diversity and antimicrobial activity. *Mycosphere*, **2** (4), 429–447.

Ren, Y., Strobel, G.A., Graff, J.C. *et al.* (2008) Colutellin A, an immunosuppressive peptide from *Colletotrichum dematium*. *Microbiology*, **154**, 1973–1979.

Ruma, K., Shailasree, S., Sampath, K. K. K., *et al.* (2011) Diversity of fungal endophytes from two endemic tree species *Artocarpus hirsutus* Lam. and *Vateria indica* Linn. of Western Ghats, India. *World Journal of Agricultural Sciences*, **7** (5), 577–582.

Santamaría, J. and Bayman, P. (2005) Fungal epiphytes and endophytes of coffee leaves (*Coffea arabica*). *Microbial Ecology*, **50**, 1–8.

Santiago, C., Fitchett, C., Munro, M.H.G. *et al.* (2012) Cytotoxic and antifungal activities of 5-hydroxyramulosin, a compound produced by an endophytic fungus isolated from *Cinnamomum mollisimum*. *Evidence-Based Complementary and Alternative Medicine*, **2012**, article ID 689310.

Schwarz, M., Köpcke, B., Weber, R.W.S. *et al.* (2004) 3-Hydroxypropionic acid as a nematicidal principle in endophytic fungi. *Phytochemistry*, **65**, 2239–2245.

Sette, L.D., Passarini, M.R.Z., Delarmelina, C. *et al.* (2006) Molecular characterization and antimicrobial activity of endophytic fungi from coffee plants. *Journal of Microbiology and Biotechnology*, **22**, 1185–1195.

Shobana, G., Rather, R.A. and Kathiravan, G. (2011) Seasonal recurrence of endophytic fungi of *Orthosiphon spiralis*. *International Journal of Science and Nature*, **2** (4), 723–726.

Stams, W.A.G. and den Boer, M.L. (2003) Sensitivity to L-asparginase is not associated with expression levels of asparagine synthetase in t(12;21)$^+$ pediatric ALL. *Blood*, **101** (7), 2743–2747.

Stierle, A., Stierle, D., Strobel, G. *et al.* (1994) Bioactive metabolites of the endophytic fungi of Pacific yew, *Taxus brevifolia*, in *Taxane Anticancer Agents: Basic Science Current Status* (eds G.I. Georg, T.T. Chen, I. Ojima and D.M. Vyas), American Chemical Society, Symposium Series No. 583, American Chemical Society, Washington, DC, pp. 81–97.

Stierle, A., Strobel, G.A. and Stierle, D. (1993) Taxol and taxane production by *Taxomyces andreanae*, an endophytic fungus of Pacific yew. *Science*, **260**, 214–216.

Strobel, G.A. (2002) Microbial gifts from rain forests. *Canadian Journal of Plant Pathology*, **24**, 14–20.

Strobel, G.A., Hess, W.M., Ford, E. *et al.* (1996) Taxol from fungal endophytes and the issue of biodiversity. *Journal of Industrial Microbiology*, **17**, 417–423.

Suryanarayanan, T.S. and Thennarasan, S. (2004) Temporal variation in endophyte assemblages of *Plumeria rubra* leaves. *Fungal Diversity*, **15**, 197–204.

Suryanarayanan, T.S., Murali, T.S. and Venkatesan, G. (2002) Occurrence and distribution of fungal endophytes in tropical forests across a rainfall gradient. *Canadian Journal of Botany*, **80**, 818–826.

Suwannarach, N., Bussaban, B., Nuangmek, W. *et al.* (2012) Diversity of endophytic fungi associated with *Cinnamomum bejolghota* (Lauraceae) in northern Thailand. *Chiang Mai Journal of Science*, **39** (3), 389–398.

Thalavaipandian, A., Ramesh, V., Bagyalakshmi *et al.* (2011) Diversity of fungal endophytes in medicinal plants of Courtallam hills, Western Ghats, India. *Mycosphere*, **2** (5), 575–582.

Theantana, T., Hyde, K.D. and Lumyong, S. (2009) Asparginase production by endophytic fungi from Thai medicinal plants: cytoxicity properties. *International Journal of Integrative Biology*, **7** (1), 1–8.

Tianpanich, K., Prachya, S., Wiyakrutta, S. *et al.* (2011) Radical scavenging and antioxidant activities of isocoumarins and a phthalide from the endophytic fungus *Colletotrichum* sp. *Journal of Natural Products*, **74**, 79–81.

Vaz, A.B.M., Brandão, L.R., Vieira, M.L.A. *et al.* (2012) Diversity and antimicrobial activity of fungal endophyte communities associated with plants of Brazilian savanna ecosystems. *African Journal of Microbiology Research*, **6** (13), 3173–3185.

Vega, F.E., Simpkins, A., Aime, M.C. *et al.* (2010) Fungal endophyte diversity in coffee plants from Colombia, Hawai'i, Mexico and Puerto Rico. *Fungal Ecology*, **3**, 122–138.

Verma, V.C., Gond, S.K., Kumar, A. *et al.* (2007) The endophytic mycoflora of bark, leaf, and stem tissues of *Azadirachta indica* A. Juss (Neem) from Varanasi (India). *Microbial Ecology*, **54**, 119–125.

Vidensek, N., Lim, P., Campbell, A. and Clarson, C. (1990) Taxol content in bark, wood, root, leaf, twig and seedlings from several *Taxus* species. *Journal of Natural Products*, **53**, 1609–1610.

Wang, J.F., Li, G.L., Lu, H.Y. *et al.* (2000) Taxol from *Tubercularia* sp. strain TF5, an endophytic fungus of *Taxus mairei*. *FEMS Microbiol. Letters*, **193**, 249–253.

Wang, Y.T., Lo, H.S. and Wang, P.H. (2008) Endophytic fungi from *Taxus mairei* in Taiwan: first report of *Colletotrichum gloeosporioides* as an endophyte of *Taxus mairei*. *Botanical Studies*, **49**, 39–43.

Wani, M.C., Taylor H.L., Wall, M.E. *et al.* (1971) Plant antitumor agents. VI. The isolation and structure of taxol, a novel antileukemic and antitumor agent from *Taxus brevifolia*. *Journal of the American Chemical Society*, **93** (9), 2325–2327.

Yuan, Z.L., Chen Y.C. and Yang, Y. (2009) Diverse non-mycorrhizal fungal endophytes inhabiting an epiphytic, medicinal orchid (*Dendrobium nobile*): estimation and characterization. *World Journal of Microbiology and Biotechnology*, **25** (2), 295–303.

Zou, W.X., Meng, J.C., Lu, H. *et al.* (2000) Metabolites of *Colletotrichum gloeosporioides*, an endophytic fungus in *Artemisia mongolica*. *Journal of Natural Products*, **63** (11), 1529–1530.

# 19 African Plants with Potential for Development into Ethnoveterinary Products

L.J. McGaw and J.N. Eloff

Phytomedicine Programme, Department of Paraclinical Sciences, Faculty of Veterinary Science, University of Pretoria, Onderstepoort, South Africa

## 19.1 Introduction

Africa is exceptionally rich in possessing a great abundance and variety of living species, constituting a unique biodiversity. With its assortment of ecosystems and habitats situated in a broad range of climates, the range of plant species occurring in this continent is astounding. Ecosystems generally occur in a latitudinal pattern (White, 1983), with more species found towards the equator (Mutke and Barthlott, 2005), although numbers of plant species are high in the Mediterranean, winter-rainfall climate regions of northern Africa and the southern Cape (Cowling *et al.*, 1996). Between these two regions are the subtropical deserts, which commonly possess lower diversity. Smaller areas of rich biodiversity occur in many regions, particularly in tropical montane areas (Rahbek, 1995). Climatic factors (such as water availability) and other factors (such as topography) play a role in the richness of species diversity. It has been reported that 32 424 angiosperm taxa occur in tropical Africa and 22 755 taxa are found in southern Africa (Klopper *et al.*, 2006). The South African Cape Floristic Region, a relatively small area of 90 000 $km^2$, has one of the richest floras in the world, being home to around 9000 indigenous plant species, of which almost 69% are endemic (Goldblatt and Manning, 2000). The sustainability of this precious resource is an important issue requiring efficient and realistic solutions. African herbal medicine relies more on wild harvested plants than any other continent (Brendler *et al.*, 2010). With the highest rate of deforestation in the world, many medicinal plants stand to become extinct before they are documented (Brendler *et al.*, 2010).

For centuries, plants have been used as sources of medicines for treating humans suffering from innumerable ailments. As well as use in human medicine, plants have been used to cure diseases occurring in livestock and companion animals. In almost every culture, there exists a treasure of information relating to traditional healing systems, and in some countries, such as China and India, this has developed into a sophisticated system of diagnosis, preparation of medicine and treatment (Houghton, 2010). Natural products constitute a significant resource of novel chemicals for drug development and discovery (Newman and Cragg, 2007). Natural substances constitute many active isolated chemicals that have been incorporated into familiar pharmaceutical dosage forms, such as tablets, capsules, injections, creams and ointments (Houghton, 2010). Pharmaceutical chemistry has enhanced the utility of natural compounds as drugs as they can be used as lead compounds for chemical manipulation of their structures to improve efficacy and reduce toxicity (Houghton, 2010). The use of traditional plant-based extracts dwindled over time in many countries, but a rising interest surrounding natural remedies has been experienced in recent decades. The growth of interest in the use of herbal medicines, both for humans and animals, has been fuelled in part by the pharmaceutical industry searching for new lead compounds for the treatment of various diseases, and also by an emerging interest and pride in traditions evidenced by many countries moving away from a colonial history (Houghton, 2010). This research substantiates traditional claims and may also, if efficacy and safety can be demonstrated, generate substitutes for expensive Western drugs (Houghton, 2010). A scientific study of the traditional remedies used may lead to improvements in the quality and effectiveness of the treatments.

## 19.2 What is ethnoveterinary medicine?

In many countries, animal health is a major constraint to livestock production, and veterinary services are often not readily available, being inaccessible or too expensive for marginal farmers and pastoralists (ITDG and IIRR, 1996). Traditional animal health care practices, known as 'ethnoveterinary medicine', offer low cost and easily available alternatives to orthodox medicine. In addition to the use of plant-based remedies, ethnoveterinary medicine covers traditional practices embedded in local tradition (Schillhorn van Veen, 1997). Such traditional practices may be useful in contributing to overall disease control, and hence integrated disease management, as they include methods relating to diagnosis, herd-, grazing- and pasture-management as well as treatment (Schillhorn van Veen, 1997). Traditional animal health care is generally unregulated and may be affected by quackery, and a large proportion of conventional practitioners are sceptical of its benefits; however, acknowledging the value of traditional knowledge, with some regulation for safety and to avoid abuse, empowers local livestock-keepers to attempt to solve animal disease problems cost effectively (Schillhorn van Veen, 1997). Regulation would also assist with the appropriate use of ethnoveterinary medicine in relation to companion animals.

*Novel Plant Bioresources: Applications in Food, Medicine and Cosmetics*, First Edition. Edited by Ameenah Gurib-Fakim.

In industrially developed countries, intensive farming is commonly practised. It was recommended as long ago as 1969 that antibiotics used to treat people and animals should not be used as growth promoters in animals (Swann, 1969). The banning of these antibiotic feed additives in many countries globally, particularly in Europe, as well as a growing interest of consumers in organic farming, has promoted the possible use of plant-based alternatives to animal treatments. The widespread use of antibiotics as feed additives to promote growth and as prophylactic treatments has led to concerns over the development of resistance in pathogens affecting humans and animals. Traditional medicines with different mechanisms of action may produce substitutes for these antibiotic growth promoters.

In most rural societies, animals are important as sources of wealth as well as food and transport. They are suppliers of dairy products, meat, wool and other products, which all stand to be lost when an animal becomes sick and dies (ITDG and IIRR, 1996; Houghton, 2010). In African countries particularly, livestock are often a major source of food and income for millions of people, and for many the animals may be the sole source of livelihood (ITDG and IIRR, 1996). In addition to the use of traditional remedies on livestock, a surge of interest in alternative treatments for companion animals has led to a growth in the herbal products industry led by demand from more affluent members of society willing to spend considerable sums of money on their pets (Houghton, 2010). Since around the late 1970s, holistic veterinarians have renewed interest in the study of medicinal plants and plant-derived drugs (Wynn and Fougère, 2007). Occurring at the same time as the growing importance of organic agriculture, opportunities for exploration of biologically active plant extracts have expanded. Research needs to continue into the efficacy, safety and appropriate applications, as well as the costs and benefits, of herbal medicines (Wynn and Fougère, 2007). The value of herbal medicine, and in particular its approach to chronic diseases, cannot be underestimated (Wynn and Fougère, 2007).

Ethnoveterinary medicine should not be seen as an alternative to orthodox veterinary medicine. Under field conditions, standard veterinary medicine and traditional practices should be complementary (ITDG and IIRR, 1996). Vaccinations provided by standard veterinary practices prevent infectious disease outbreaks, and certain diseases are very effectively treated using antibiotics. Conventional drugs are often more effective and convenient to use than the ethnoveterinary remedies (ITDG and IIRR, 1996). On the other hand, traditional ethnoveterinary treatments are locally available and often cheaper, and the owner of the animals can prepare and administer the remedy instead of relying on expensive professionals to provide this service (ITDG and IIRR, 1996). There are few veterinarians covering rural areas of many countries in Africa, so traditional remedies are often the only choice (ITDG and IIRR, 1996).

If the traditional use of plant extracts can be scientifically validated, it may lead to additional products that can be used in commercial farming or in the use of plant extracts in rural settings. There has been a strong growth in publications of scientists investigating the traditional use of animal remedies (McGaw and Eloff, 2008; Katerere and Luseba, 2010). In addition to preclinical experiments on animals for plant products intended for human use, there is also a growth in *in vivo* experiments using plant products for application in animal health.

## 19.3 Ethnoveterinary medicine in Africa

In many regions of Africa, the people traditionally rely on a wide range of indigenous practices to treat their livestock when sick, as well as to keep their animals healthy (ITDG and IIRR, 1996). This ethnoveterinary medicine, which has developed over centuries, includes the use of medicinal plants, surgery techniques and also management practices to treat or prevent various diseases (ITDG and IIRR, 1996). Some ethnoveterinary practices are common knowledge among livestock keepers (for example, those used to treat the more widespread ailments), while others are known only to a few healers (ITDG and IIRR, 1996). A manual describing a range of ethnoveterinary practices in Kenya and among the Somali community in neighbouring Ethiopia, intended as a practical guide for local people working with livestock and covering the most important livestock diseases and problems, has been produced (ITDG and IIRR, 1996). Such a collection of traditional indigenous practices is a valuable addition to the pharmacopoeia of a country. Not only is it a useful reference book where standard veterinary medicine is not available or affordable, it is intended that the information therein complements standard veterinary practice. Works such as these generate interest and stimulate research on validating traditional practice as well as documenting this valuable information.

McCorkle and Mathias-Mundy (1992) reviewed African ethnoveterinary medicine. More recent literature demonstrates a proliferation of papers reporting on surveys of ethnoveterinary usage of plants, mirroring the growth in information available on uses of plants to treat human ailments. There are still vast gaps in knowledge, however, and several authors have pointed out that much information remains to be recorded from current indigenous knowledge holders. A brief summary of some of the published ethnoveterinary uses of plants from regions in Africa will be presented in this section. Much of the information appears to have been gathered using structured questionnaires and general conversation with farmers and herbalists.

### 19.3.1 Southern Africa

The ethnoveterinary use of southern African plants and studies evaluating their medicinal properties were reviewed by McGaw and Eloff (2008). It was concluded that, although a large proportion of the population uses traditional remedies to treat common diseases of livestock, only a small percentage of these medicines have been evaluated for biological activity or toxicity, and that studies on plant species used in ethnoveterinary medicine could provide promising leads for development of standardized medications to be used on a commercial basis (McGaw and Eloff, 2008). Information on the bioactivities of southern African ethnoveterinary plants summarized by McGaw and Eloff (2008) is presented in Table 19.1.

Plants used in southern African ethnoveterinary medicine have also been reviewed by Maphosa *et al.* (2010a), with the families to which species belong being analysed in more depth. The region covered 11 countries, stretching from South Africa to Tanzania. In further ethnoveterinary research, Maphosa and Masika (2010) conducted a survey to document information on medicinal plants used by farmers to control internal parasites in goats in the Eastern Cape Province. This revealed that 28 plant species from 20 families are

**Table 19.1** Plants used in South Africa for ethnoveterinary purposes, and bioactivity of tested species

| Family | Species | Indication | Plant part used | Chemical constituents | Screened for activity |
|---|---|---|---|---|---|
| Acanthaceae | *Hypoestes aristata* (Vahl) Soland. Ex Roem. & Schult. | Used to drench calves suffering from a condition referred to as white scours | Whole plant infusions | | |
| Alliaceae | *Agapanthus praecox* Willd. | Diarrhoea in sheep and goats (Dold and Cocks, 2001) | Roots | | |
| Amaranthaceae | *Exomis microphylla* (Thunb.) Aellen | Endometritis and vaginitis (Dold and Cocks, 2001) | Leaf decoction | | |
| Amaryllidaceae | *Ammocharis coranica* (Ker-Gawl.) Herb. | Used medicinally for cattle (Gerstner, 1938) | Unspecified parts | Alkaloids, organic acid and haemolytic saponin (Watt and Breyer-Brandwijk, 1962) | |
| Amaryllidaceae | *Boophane disticha* (L.f.) Herb. | Redwater in cattle, constipation in cattle, used to facilitate healing of broken limbs (Dold and Cocks, 2001); abortion (van der Merwe *et al.*, 2001) | Bulb, root, bulb scales | | |
| Amaryllidaceae | *Crinum delagoense* Verdoorn | Used medicinally for cattle (Gerstner, 1939) | Unspecified parts | Unknown | |
| Amaryllidaceae | *Crinum moorei* Hook. F. | Used medicinally for cattle (Gerstner, 1939) | Unspecified parts | Lycorine, cheryrilline, crinamidine, crinidine, dihydrocrinidine and powelline and phenols (Watt and Breyer-Brandwijk, 1962) | |
| Amaryllidaceae | *Haemanthus albiflos* Jacq. | Healing of broken limbs (Dold and Cocks, 2001) | Bulb | | |
| Anacardiaceae | *Ozoroa paniculosa* (Sond.) R. & A. Fernandes | Abdominal problems in animals (Hutchings *et al.*, 1996); diarrhoea, redwater, sweating sickness (van der Merwe *et al.*, 2001) | Bark, root bark | Volatile oil (Watt and Breyer-Brandwijk, 1962) | |
| Anacardiaceae | *Protorhus longifolia* (Bernh.) Engl. | Heartwater and diarrhoea in cows (Dold and Cocks, 2001) | Bark | | |
| Anacardiaceae | *Rhus incisa* L. f. | Roots given to livestock as treatment for shock after an accident, bark given to cows for diarrhoea (Dold and Cocks, 2001) | Root and bark decoctions | | |
| Anacardiaceae | *Rhus lancea* L.f. | Diarrhoea, gallsickness (van der Merwe *et al.*, 2001) | Roots, bark | | Antibacterial, anthelmintic, brine shrimp toxicity (McGaw *et al.*, 2007) |
| Anacardiaceae | *Sclerocarya birrea* (A. Rich.) Hochst. | Diarrhoea, fractures (van der Merwe *et al.*, 2001) | Bark | | Antibacterial, anthelmintic, brine shrimp toxicity (McGaw *et al.*, 2007) |
| Apiaceae | *Heteromorpha trifoliata* (Wendl.) Eckl. & Zeyh. | Zulus use bark for colic, scrofula and vermifuge for horses (Gerstner, 1938); Xhosa use roots for threadwork in horses (Watt and Breyer-Brandwijk, 1962); redwater, gallsickness (Masika *et al.*, 2000) | Bark, roots | Falcarindiol and sarisan (antifungal) (Villegas *et al.*, 1988) | |

(*continued*)

**Table 19.1** (*Continued*)

| Family | Species | Indication | Plant part used | Chemical constituents | Screened for activity |
|---|---|---|---|---|---|
| Apocynaceae | *Acokanthera oppositifolia* (Lam.) Codd | Heartwater in goats and sheep, redwater in cattle, snakebite, anthrax, tapeworm, swollen limbs (Dold and Cocks, 2001) | Leaves, roots | | |
| Apocynaceae | *Secamone filiformis* (L.f.) J.H.Ross | Infectious diseases in cattle (van der Merwe, personal communication) | Aerial parts | | Antibacterial, anthelmintic, brine shrimp toxicity (McGaw *et al.*, 2007) |
| Apocynaceae | *Strophanthus speciosus* (Ward & Harv.) Reber | Given to cattle for snakebite (Hutchings *et al.*, 1996) | Unspecified | Cardiac glycosides stropeside and christyoside (Watt and Breyer-Brandwijk, 1962) | |
| Araceae | *Zantedeschia albomaculata* (Hook.) Baill. | Medicine for cattle (Jacot Guillarmod, 1971) | Unspecified parts | | |
| Araliaceae | *Cussonia spicata* Thunb. | Leaves applied in hot fomentations to goats paralyzed in their hind quarters (Palmer and Pitman, 1972); bark used for retained placenta in stock, leaves used to treat endometritis and/or vaginitis in cows, bark decoction for gallsickness in cattle (Dold and Cocks, 2001; Masika *et al.*, 2000) | Leaves, bark | | Antibacterial, anthelmintic, brine shrimp toxicity (McGaw *et al.*, 2007); antibacterial, anti-inflammatory, mutagenic (Luseba *et al.*, 2007) |
| Arecaceae | *Phoenix reclinata* Jacq. | Wash to treat footrot in sheep and goats (Dold and Cocks, 2001) | Roots | | |
| Asclepiadaceae | *Sarcostemma viminale* (L.) R. Br. | Stems used to encourage lactation in cows, galactagogue in cows (Dold and Cocks, 2001); wounds and maggots (Luseba and van der Merwe, 2006) | Stems, aerial parts | | Antibacterial, anti-inflammatory, mutagenic (Luseba *et al.*, 2007) |
| Asclepiadaceae | *Secamone filiformis* (L. f.) J.H. Ross | Diarrhoea in cattle (Dold and Cocks, 2001) | Stem | | |
| Asparagaceae | *Asparagus laricinus* Burch. | Sores, redwater, uterine infections (van der Merwe *et al.*, 2001) | Tubers | | |
| Asparagaceae | *Asparagus setaceus* (Kunth) Oberm. | Used to treat livestock for shock after an accident (Dold and Cocks, 2001) | Roots | | |
| Asparagaceae | *Asparagus suaveolens* (Burch.) Oberm. | Retained placenta in cows (Dold and Cocks, 2001); sores, redwater, uterine infections (van der Merwe *et al.*, 2001) | Roots, tubers | | |
| Asparagaceae | *Protasparagus virgatus* (Bak.) Oberm. | Anthelmintics for animals and humans (Watt and Breyer-Brandwijk, 1962) | Root infusions or decoctions | | |
| Asphodelaceae | *Aloe arborescens* Mill. | Used to drench sick calves (Hutchings *et al.*, 1996) | Leaf decoctions | Aloin, barbaloin, aloe emodin, aloenin, polysaccharides, lectins and other compounds (Hutchings *et al.*, 1996) | |

| | | | | | |
|---|---|---|---|---|---|
| Asphodelaceae | *Aloe cooperi* Bak. | Used to protect cattle from the ill effects of eating improper food (Watt and and Breyer-Brandwijk, 1962) | Smoke from burning leaves | | |
| Asphodelaceae | *Aloe ferox* Mill. | Typhoid, ticks and lice in poultry, redwater in cattle (Dold and Cocks, 2001); redwater, intestinal worms (Masika *et al.*, 2000) | Leaves, juice from leaves | | |
| Asphodelaceae | *Aloe greatheadii* var. *davyana* (Schönland) H.F.Glen & D.S.Hardy | Burns, general ailments, blood cleansing, internal parasites, eye infections (van der Merwe *et al.*, 2001) | Leaves, roots, whole plant | | |
| Asphodelaceae | *Aloe maculata* All. | Used for 'blood scours' in calves and enteritis and indigestion in poultry (Hutchings *et al.*, 1996) | Leaf infusions | | |
| Asphodelaceae | *Aloe marlothii* Berger | Newcastle disease in chickens (Luseba and van der Merwe, 2006); gallsickness, parasites, diarrhoea, constipation, retained placenta, dystocia, maggots (van der Merwe *et al.*, 2001) | Leaves | | Antibacterial, anti-inflammatory, mutagenic (Luseba *et al.*, 2007); anti-rickettsial (Naidoo *et al.*, 2006); anti-babesial (Naidoo *et al.*, 2005) |
| Asphodelaceae | *Aloe tenuior* Haw. | Retained placenta in cows, tapeworm, redwater, intestinal parasites (Dold and Cocks, 2001) | Leaves | | |
| Asphodelaceae | *Aloe zebrina* Baker | Wounds and maggots (Luseba and van der Merwe, 2006); burns, general ailments, blood cleansing, internal parasites, eye infections (van der Merwe *et al.*, 2001) | Fresh leaves, roots, whole plant | | |
| Asphodelaceae | *Bulbine alooides* (L.) Willd. | Redwater in cattle (Dold and Cocks, 2001) | Roots | | |
| Asphodelaceae | *Bulbine asphodeloides* (L.) Willd. | Used to treat sick cattle and goats (Hutchings *et al.*, 1996) | Unspecified parts | | |
| Aspidiaceae | *Dryopteris athamantica* (Kunze) Kuntze | Retained placenta in cows by the Sotho (Jacot Guillarmod, 1971) | Rhizome decoctions | | |
| Aspidiaceae | *Polystichum* sp. | Administered to horses with bots (Hutchings *et al.*, 1996) | Rhizome decoctions | | |
| Asteraceae | *Arctotus arctotoides* (L.f.) O.Hoffm. | Heartwater in goats (Dold and Cocks, 2001) | Whole plant | | |
| Asteraceae | *Bidens pilosa* L. | Equine anthelmintics (Hutchings *et al.*, 1996) | Unspecified | The polyacetylene phenylheptatriyene and chalcones (Graham *et al.*, 1980; Hoffman and Hoelzl, 1988) | |
| Asteraceae | *Brachylaena discolor* DC. | Anthelmintics for calves, sheep and goats (Hutchings *et al.*, 1996) | Dried leaf milk infusions | Onopordopicrin (Zdero and Bohlmann, 1987) | |

(*continued*)

**Table 19.1** (*Continued*)

| Family | Species | Indication | Plant part used | Chemical constituents | Screened for activity |
|---|---|---|---|---|---|
| Asteraceae | *Brachylaena elliptica* (Thunb.) DC. | Used for treating calves (Gerstner, 1939) | Roots | | |
| Asteraceae | *Brachylaena ilicifolia* (Lam.) Phill. & Schweick. | Diarrhoea in lambs (Dold and Cocks, 2001) | Leaves | | |
| Asteraceae | *Callilepis laureola* DC. | Used to kill maggots in cattle (Watt and Breyer-Brandwijk, 1962) | Root paste | Atractyloside and its aglycone, atractyligenin (Candy *et al.*, 1977) | |
| Asteraceae | *Dicoma anomala* Sond. | Gallsickness in stock animals; powdered plants used for sores and wounds on horses (Watt and Breyer-Brandwijk, 1962) | Root decoctions | Germacranolides (Hutchings *et al.*, 1996) | |
| Asteraceae | *Microglossa mespilifolia* (Less.) B.L. Robinson | Tonics for stock animals (Hutchings *et al.*, 1996) | Infusions from leaves and stems | Epi-friedelinol and $C_{17}$ acetylenic compounds (Bohlmann and Fritz, 1979) | |
| Asteraceae | *Mikania capensis* DC. | Plants used for horse sickness (Gerstner, 1939) | Unspecified | | |
| Asteraceae | *Printzia pyrifolia* Less. | Used for treating calves (Gerstner, 1939) | Roots | Matricaria ester and *p*-coumarate (Bohlmann and Zdero, 1978a) | |
| Asteraceae | *Schkuhria pinnata* (Lam.) Thell. | Eye infections, pneumonia, diarrhoea, heartwater (van der Merwe *et al.*, 2001) | Aerial parts | | Antibacterial, anthelmintic, brine shrimp toxicity (McGaw *et al.*, 2007); antibacterial, anti-inflammatory, mutagenic (Luseba *et al.*, 2007) |
| Asteraceae | *Senecio oxyriifolius* DC. | Swellings in animals (Hutchings *et al.*, 1996) | Leaves | A tricyclic sesquiterpenoid, angeloyl and bisabolenes (Bohlmann and Zdero, 1978b) | |
| Asteraceae | *Senecio tamoides* DC. | Anthrax and 'quarter evil' in cattle (Gerstner, 1939) | Unspecified parts | | |
| Asteraceae | *Vernonia mespilifolia* Less. | Heartwater in goats (Dold and Cocks, 2001) | Stems | | |
| Asteraceae | *Vernonia neocorymbosa* Hilliard | Used by Zulus to treat calves (Gerstner, 1939); root decoctions administered by Lobedu as anthelmintics to donkeys (Hutchings *et al.*, 1996); pounded leaf and root infusions administered by Vhavenda as anthelmintics to domestic animals (Mabogo, 1990) | Roots, leaves | Squalene, vernolide and vernodalin from aerial parts, and 13-hydroxybisabol-2,10-dien-1-one and small amounts of onopordopicrin in roots (Bohlmann *et al.*, 1983) | |
| Balanitaceae | *Balanites maughamii* | Diarrhoea in cattle (Luseba and van der Merwe, 2006) | Leaves | | |
| Boraginaceae | *Ehretia rigida* (Thunb.) Druce | Gallsickness in cattle (Hutchings *et al.*, 1996); fractures (van der Merwe *et al.*, 2001) | Roots | | |
| Capparaceae | *Capparis sepiaria* L. | Used by Xhosa for gallsickness in stock (Watt and Breyer-Brandwijk, 1962) | Root decoctions | | |

| | | | | | |
|---|---|---|---|---|---|
| Capparaceae | *Capparis tomentosa* | Root ash paste applied to sore teats in cows; root infusions used for stomach ailments in animals, particularly diarrhoea in cattle (Watt and Breyer-Brandwijk, 1962; Pujol, 1990), root decoction for gallsickness in stock (Dold and Cocks, 2001) | Paste made from root ashes, root infusions and decoctions | The alkaloids stachydrine (*Dictionary of Natural Products on CD-ROM*, 1996) and 3-hydroxy-4-methoxy-3-methyl-oxindole (Dekker *et al.*, 1987) | |
| Celastraceae | *Cassine aethiopica* Thunb. | Used by Zulus to drench worm-infested calves (Watt and Breyer-Brandwijk, 1962) | Milk or whey bark infusions | | |
| Celastraceae | *Cassine transvaalensis* (Burtt Davy) Codd | Diarhoea (van der Merwe *et al.*, 2001) | Bark | | |
| Celastraceae | *Maytenus heterophylla* (Eckl. & Zeyh.) N.K.B.Robson | Administered by Zulus to stock animals for diarrhoea (Watt and Breyer-Brandwijk, 1962) | Bark and leaf infusions | Dulcitol, a spermidine alkaloid, celacinnine, triterpenoids, maytansine (Hutchings *et al.*, 1996) | |
| Celastraceae | *Mystroxylon aethiopicum* (Thunb.) Loes. | Heartwater in cattle, worms in calves, intestinal parasites (Dold and Cocks, 2001) | Bark | | |
| Chenopodiaceae | *Chenopodium album* L. | Decoctions made from plants mixed with *C. ambrosioides* administered to goats and sheep for anaemia (Hutchings *et al.*, 1996) | Unspecified | Hydrocyanic acid, potassium oxalate, ascorbic acid, sitosterol, oleanic acid (Watt and Breyer-Brandwijk, 1962; Hutchings *et al.*, 1996) | |
| Chenopodiaceae | *Chenopodium ambrosioides* L. | Decoctions made from plants mixed with *C. album* administered to goats and sheep for anaemia (Hutchings *et al.*, 1996) | Unspecified | Saponins (Watt and Breyer-Brandwijk, 1962) flavonoids, quercetin, oxalic, malic and succinic acids, triterpenoid glycosides, chenopodoside A and B, amino acids, ascaridole (Hutchings *et al.*, 1996) | |
| Colchicaceae | *Gloriosa superba* L. | Used to kill lice, for skin eruptions, tick infections and screw-worm on cattle (Gerstner, 1939; Roberts, 1990) | Corms | Colchicine, chelidonic acid various alkaloids and other constituents (Hutchings *et al.*, 1996) | |
| Combretaceae | *Combretum caffrum* (Eckl. & Zeyh.) Kuntze | Conjunctivitis (Masika *et al.*, 2000) | Drops from squeezed leaves used | | Antibacterial, antifungal (Masika and Afolayan, 2002) |
| Combretaceae | *Combretum erythrophyllum* (Burch.) Sond. | Small doses administered as fattening tonics to dogs (Watt and Breyer-Brandwijk, 1962) | Roots | | |
| Combretaceae | *Combretum paniculatum* Vent. | Fertility problems (Luseba and van der Merwe, 2006) | Root bark | | |
| Combretaceae | *Terminalia sericea* Burch. Ex DC. | Wounds (Luseba and van der Merwe, 2006); diarrhoea (van der Merwe *et al.*, 2001) | Leaves, roots | | |
| Convolvulaceae | *Seddera sufruticosa* Hallier f. | Fractures (van der Merwe *et al.*, 2001) | Roots | | |
| Cornaceae | *Curtisia dentata* (Burm. F.) C.A.Sm. | Heartwater in cows (Dold and Cocks, 2001) | Bark | | |

*(continued)*

**Table 19.1** (*Continued*)

| Family | Species | Indication | Plant part used | Chemical constituents | Screened for activity |
|---|---|---|---|---|---|
| Cucurbitaceae | *Cucumis africanus* L. f. | Used as animal medicines by the Xhosa (Hutchings *et al.*, 1996) | Unspecified | Toxic cucurbitacins (Hutchings *et al.*, 1996) | |
| Dioscoreaceae | *Dioscorea dregeana* (Kunth) Dur. & Schinz | Sores and wounds in animals and humans by Xhosa (Watt and Breyer-Brandwijk, 1962) | Water heated in scooped out tuber | An alkaloid and organic acids (Watt and Breyer-Brandwijk, 1962) | |
| Dioscoreaceae | *Dioscorea sylvatica* (Kunth) Eckl. | Swollen udders and uterine problems in cows (Watt and Breyer-Brandwijk, 1962) | Lotions from boiled crushed inner parts of tubers | Diosgenin (Watt and Breyer-Brandwijk, 1962) | |
| Dracaenaceae | *Sansevieria hyacinthoides* (L.) Druce | Fresh leaf sap applied to eyes of sheep and goats for conjunctivitis (Dold and Cocks, 2001) | Leaf sap | | |
| Ebenaceae | *Diospyros mespiliformis* Hochst. Ex A. DC. | For milk production (Luseba and van der Merwe, 2006) | Bark | | |
| Euphorbiaceae | *Clutia pulchella* L. | Drenches for griping pains in calves (Hutchings *et al.*, 1996) | Milk infusions of leaves, stems and roots | | |
| Euphorbiaceae | *Croton gratissimus* Burch. var. *gratissimus* | Pneumonia, tonic, fertility enhancement (van der Merwe *et al.*, 2001) | Leaves, roots | | |
| Euphorbiaceae | *Euphorbia cooperi* N.E.Br. ex A.Berger | Blackquarter (Luseba and van der Merwe, 2006) | Aerial parts | | |
| Euphorbiaceae | *Jatropha curcas* L. | Drench for constipation in cattle and goats (Luseba and van der Merwe, 2006) | Seeds | | |
| Euphorbiaceae | *Jatropha zeyheri* Sond. | General ailments, diarrhoea (Luseba and van der Merwe, 2006) | Roots | | Antibacterial, anti-inflammatory, mutagenic (Luseba *et al.*, 2007) |
| Euphorbiaceae | *Phyllanthus burchellii* Müll.Arg. and *P. parvulus* Sond. | Eye infections (van der Merwe *et al.*, 2001) | Aerial parts | | |
| Euphorbiaceae | *Ricinus communis* L. | Administered as a purgative to calves refusing to suckle (Hutchings *et al.*, 1996); constipation, internal parasites (van der Merwe *et al.*, 2001) | Powdered seed | Seeds contain a fixed oil, ricin, lipases and ricinine (Trease and Evans, 1983) | Antibacterial, anthelmintic, brine shrimp toxicity (McGaw *et al.*, 2007); antibacterial, anti-inflammatory, mutagenic (Luseba *et al.*, 2007) |
| Euphorbiaceae | *Spirostachys abbrevia* Sond. | Sap applied to cattle sores to kill maggots (Hutchings *et al.*, 1996); sweating sickness (van der Merwe *et al.*, 2001) | Sap, wood | | |
| Euphorbiaceae | *Synadenium cupulare* (Boiss.) L.C. Wheeler | Eye infection, blackquarter (Luseba and van der Merwe, 2006) | Milky latex | | Antibacterial, anthelmintic, brine shrimp toxicity (McGaw *et al.*, 2007) |
| Fabaceae | *Acacia decurrens* Willd. | Hastens oestrus (Masika *et al.*, 2000) | Bark decoction | | |
| Fabaceae | *Acacia karroo* Hayne | Diarrhoea in goats, intestinal parasites in goats, sheep, poultry and pigs (Dold and Cocks, 2001); fractures and diarrhoea (van der Merwe *et al.*, 2001) | Bark, leaves | | |

| | | | | | |
|---|---|---|---|---|---|
| Fabaceae | *Acacia tortilis* (Forssk.) Hayne | Diarrhoea (van der Merwe *et al.*, 2001) | Branch tips | | |
| Fabaceae | *Adenopodia spicata* (E. Mey.) Presl | Powdered roots used by Zulus to fatten goats (Hutchings *et al.*, 1996); Mfengu use bark for colds in horses (Watt and Breyer-Brandwijk, 1962) | Roots, bark | Saponins (Watt and Breyer-Brandwijk, 1962) | |
| Fabaceae | *Calpurnia aurea* (Ait.) Benth. | Zulus use plant to destroy maggots in sores (Bryant, 1966) | Unspecified parts | Alkaloids (Hutchings *et al.*, 1996) | |
| Fabaceae | *Calpurnia villosa* Harv. | The Sotho use plant infusions topically on maggot-infested sores on cattle (Gerstner, 1939) | Unspecified parts | The alkaloid oroboidine (Hutchings *et al.*, 1996) | |
| Fabaceae | *Cassia abbreviata* Oliv. | Drench for worm infestations (Luseba and van der Merwe, 2006) | Bark | | |
| Fabaceae | *Elephantorrhiza elephantina* (Burch.) Skeels | The Xhosa use roots for diarrhoea and dysentery in cattle, horses and humans (Watt and Breyer-Brandwijk, 1962), root given to cows for mange (Dold and Cocks, 2001); heartwater, blackquarter, appetite stimulant or tonic (Luseba and van der Merwe, 2006); diarrhoea, heartwater, coughing, pneumonia (van der Merwe *et al.*, 2001) | Roots, aerial parts and bulb | Tannin (Watt and Breyer-Brandwijk, 1962) | Anti-rickettsial (Naidoo *et al.*, 2006); anti-babesial (Naidoo *et al.*, 2005) |
| Fabaceae | *Erythrophleum lasianthum* Corbishley | Lung sickness in cattle and abortions in dogs (Hutchings *et al.*, 1996) | Bark | Seeds and bark contain erythrophleine (Watt and Breyer-Brandwijk, 1962) | |
| Fabaceae | *Indigofera frutescens* L. f. | Anthelmintics in animals and humans, especially roundworm (Watt and Breyer-Brandwijk, 1962) | Root bark decoctions | | |
| Fabaceae | *Indigofera sessilifolia* DC. | Diarrhoea in calves (Dold and Cocks, 2001) | Roots | | |
| Fabaceae | *Macrotyloma axillare* (E. Mey.) Verdc. | Administered to cows with swollen udders after calving (Hulme, 1954) | Warm water leaf and stalk infusions | | |
| Fabaceae | *Peltophorum africanum* Sond. | Tonic, diarrhoea (van der Merwe *et al.*, 2001) | Bark, root bark | | Antibacterial, antioxidant, anthelmintic (Bizimenyera *et al.*, 2005, 2006a,b) |
| Fabaceae | *Pterocarpus angolensis* DC. | General illness, gallsickness, intestinal worms, blackquarter (Luseba and van der Merwe, 2006) | Bark | | Antibacterial, anthelmintic, brine shrimp toxicity (McGaw *et al.*, 2007); antibacterial, anti-inflammatory, mutagenic (Luseba *et al.*, 2007) |
| Fabaceae | *Schotia brachypetala* Sond. | Infectious diseases in cattle (van der Merwe, personal communication) | | | Antibacterial, anthelmintic, brine shrimp toxicity (McGaw *et al.*, 2007) |
| Fabaceae | *Schotia latifolia* Jacq. | Redwater in cattle (Dold and Cocks, 2001) | Bark decoction | | Antibacterial, antifungal (Masika and Afolayan, 2002) |

*(continued)*

**Table 19.1** (*Continued*)

| Family | Species | Indication | Plant part used | Chemical constituents | Screened for activity |
|---|---|---|---|---|---|
| Fabaceae | *Senna italica* Mill. | Diarrhoea and gallsickness (Luseba and van der Merwe, 2006); gallsickness, intestinal diseases, heartwater, anthrax, pneumonia (van der Merwe *et al.*, 2001) | Bark, roots | | |
| Fabaceae | *Tephrosia kraussiana* Meisn. | Plants used by Zulus for protecting cattle against quarter-evil and other diseases (Doke and Vilakazi, 1972) | Unspecified parts | Roots may contain a saponin (Watt and Breyer-Brandwijk, 1962) | |
| Fabaceae | *Tephrosia macropoda* (E. Mey.) Harv. | Roots and seeds used for killing vermin on animals and humans (Gerstner, 1941); leaf extracts used as anthelmintics for cattle (Bryant, 1966) | Roots, seeds, leaves | Roots contain toxicarol and deguelin (Watt and Breyer-Brandwijk, 1962) | |
| Geraniaceae | *Monsonia emarginata* (L. f.) L'Hérit | Stomach ailments in calves, lambs and humans (Watt and Breyer-Brandwijk, 1962) | Unspecified parts | Phloroglucin tannin (Watt and Breyer-Brandwijk, 1962) | |
| Geraniaceae | *Pelargonium luridum* (Andr.) Sweet | Administered to sick calves (Hutchings *et al.*, 1996) | Leaf infusions | Coumarins in roots (Wagner and Bladt, 1975) | |
| Geraniaceae | *Pelargonium reniforme* Curtis | Diarrhoea in goats and cows, heartwater in cattle, liver disorders in cattle and sheep (Dold and Cocks, 2001) | Root decoction | | |
| Geraniaceae | *Pelargonium sidoides* DC. | Used as anthelmintics for calves with *Ziziphus zeyheriana* Sond. (Watt and Breyer-Brandwijk, 1962) | Decoctions of unspecified parts | Coumarins in roots (Wagner and Bladt, 1975) | |
| Gesneriaceae | *Streptocarpus prolixus* C.B. Cl. | Administered by the Zulus as purgatives to cows (Hulme, 1954) | Cold water leaf infusions | | |
| Gunneraceae | *Gunnera perpensa* L. | Used to facilitate expulsion of afterbirth in animals and women (Gerstner, 1939) | Roots | Bitter principle, celastrin (Watt and Breyer-Brandwijk, 1962) | |
| Hyacinthaceae | *Albuca* sp. | Purgative and vermifuge for animals and humans (Gerstner, 1938) | Unspecified | | |
| Hyacinthaceae | *Ledebouria cooperi* (Hook. F.) Jessop | Administered to cows to ensure a succession of calves of the same gender (Watt and Breyer-Brandwijk, 1962) | Unspecified | | |
| Hyacinthaceae | *Ledebouria revoluta* (L. f.) Jessop | Gallsickness in animals by the Xhosa (Watt and Breyer-Brandwijk, 1962); bulb infusion for diarrhoea in goats, leaf decoction for gallsickness (Dold and Cocks, 2001) | Bulbs, leaves | | |
| Hyacinthaceae | *Scilla natalensis* Planch. | Administered to cattle with lung sickness (Hutchings *et al.*, 1996) | Unspecified | | |

| | | | | | |
|---|---|---|---|---|---|
| Hyacinthaceae | *Scilla nervosa* (Burch.) Jessop | Used as purges for calves (Gerstner, 1941) | Unspecified | | |
| Hyacinthaceae | *Urginea altissima* (L. f.) Baker | Intestinal parasites in cattle, retained afterbirth (Dold and Cocks, 2001) | Bulb decoction | | |
| Hyacinthaceae | *Urginea physodes* (Jacq.) Bak. | 'Itch' in goats (Gerstner, 1941) | Unspecified | | |
| Hyacinthaceae | *Urginea sanguinea* Schinz | General ailments, intestinal diseases, internal parasites, gallsickness, heartwater, redwater, sores, retained placenta (van der Merwe *et al.*, 2001) | Bulbs | | Anti-babesial (Naidoo *et al.*, 2005) |
| Hypoxidaceae | *Hypoxis hemerocallidea* Fisch. & C.A. Mey. [and *Hypoxis rigidula* Baker] | Fertility enhancement, general ailments, heartwater, abortion (van der Merwe *et al.*, 2001) | Corms | | |
| Icacinaceae | *Apodytes dimidiata* E. Mey. Ex Arn. | Purgatives for calves (Gerstner, 1938); worms in cattle (Hutchings *et al.*, 1996) | Bark, leaves | | |
| Iridaceae | *Crocosmia paniculata* (Klatt) Goldbl.. | Used for bovine diarrhoea by the Sotho (Watt and Breyer-Brandwijk, 1962) | Unspecified | | |
| Iridaceae | *Dietes iridioides* (L.) Sweet ex Klatt | Tonics for goats (Hulme, 1954); used by the Xhosa to prevent or treat stomach ailments in goats and sheep (Hutchings and Johnson, 1986) | Ground rhizomes | | |
| Iridaceae | *Watsonia densiflora* Bak. | Diarrhoea in calves by the Sotho (Watt and Breyer-Brandwijk, 1962) | | | |
| Lamiaceae | *Leonotis leonurus* (L.) R. Br. | Pounded roots and leaves are added to drinking water to prevent sickness in poultry and are used for gallsickness in cattle (Hulme, 1954); eye inflammation (Masika *et al.*, 2000) | Roots, leaves, drops used from squeezed leaf for eyes | Volatile oil and diterpenoids (labdane type lactones) for example marrubin (*Dictionary of Natural Products on CD-ROM*, 1996) | |
| Lamiaceae | *Leonotis ocymifolia* (Burm. F.) Iwarsson | Pounded roots and leaves are added to drinking water to prevent sickness in poultry and are used for gallsickness in cattle (Hulme, 1954) | Roots, leaves | | |
| Lamiaceae | *Leucas capensis* (Benth.) Engl. | Gallsickness in stock (Dold and Cocks, 2001) | Leaves | | |
| Lamiaceae | *Marrubium vulgare* L. | Gallsickness in stock (Dold and Cocks, 2001) | Leaves | | |
| Lamiaceae | *Plectranthus laxiflorus* Benth. | Drenches for animals (Watt and Breyer-Brandwijk, 1962) | Powdered aerial parts | | |

(*continued*)

**Table 19.1** (*Continued*)

| Family | Species | Indication | Plant part used | Chemical constituents | Screened for activity |
|---|---|---|---|---|---|
| Lamiaceae | *Tetradenia riparia* | Used for gallsickness and fevers in cattle (Hutchings *et al.*, 1996) | Leaves | A diterpene diol, ibozol, and related diterpenoids, large amounts of α-pyrones (*Dictionary of Natural Products on CD-ROM*, 1996) | |
| Lamiaceae | *Teucrium africanum* Thunb. | Gallsickness in cattle, heartwater in goats and sheep, bloat in goats, anthrax (Dold and Cocks, 2001) | Leaves | | |
| Loganiaceae | *Strychnos decussata* (Pappe) | Roundworm in cows (Dold and Cocks, 2001) | Bark infusion | | |
| Loganiaceae | *Strychnos henningsii* Gilg. | Heartwater and diarrhoea in cattle (Dold and Cocks, 2001) | Bark infusion | | |
| Malvaceae | *Hibiscus malacospermus* E.Mey. ex Harv. & Sond. | Retained placenta, intestinal worms (Masika *et al.*, 2000) | Root decoction | | |
| Melianthaceae | *Bersama tysoniana* Oliv. | Used by the Xhosa for gallsickness in cattle (Watt and Breyer-Brandwijk, 1962) | Bark decoctions | | |
| Moraceae | *Ficus ingens* (Miq.) Miq. | Administered to cows to increase milk production by Zulus (Watt and Breyer-Brandwijk, 1962) and Vhavenda (Mabogo, 1990) | Bark decoctions | Tannin (Watt and Breyer-Brandwijk, 1962) | |
| Moraceae | *Ficus sur* Forssk. | Zulus use leaf and bark infusions as bovine galactagogues (Hutchings *et al.*, 1996); Vhavenda use root decoctions for retained placenta in cows (Watt and Breyer-Brandwijk, 1962) | Leaves, bark, roots | Bark may contain tannin (Hutchings *et al.*, 1996) | |
| Myrsinaceae | *Rapanea melanophloeos* (L.) Mez | Heartwater in cows (Dold and Cocks, 2001) | Bark | | |
| Myrtaceae | *Heteropyxis natalensis* Harv. | Drench for stock animals (Watt and Breyer-Brandwijk, 1962) | Powdered leaves | Essential oils from ground dried leaves contain many constituents (Hutchings *et al.*, 1996) | |
| Olacaceae | *Ximenia americana* L. var. *microphylla* | Internal parasites (van der Merwe *et al.*, 2001) | Roots | | |
| Oleaceae | *Olea europaea* L. | Leaves used for endometritis and vaginitis in cows, bark infusion for diarrhoea in goats, gallsickness in cattle, eye lotion for animals and humans (Dold and Cocks, 2001) | Leaves, bark | | |
| Orchidaceae | *Eulophia speciosa* (R. Br. Ex Lindl.) H. Bol. | Emetics for animals and humans (Gerstner, 1941) | Root infusions | | |

| | | | | | |
|---|---|---|---|---|---|
| Pedaliaceae | *Dicerocaryum eriocarpum* (Dcne.) J.Abels [and *D. senecioides* (Kltzsch.) J.Abels] | Dystocia, drench for retained placenta (Luseba and van der Merwe, 2006; van der Merwe *et al.*, 2001) | Aerial parts, roots, whole plant | | Antibacterial, anthelmintic, brine shrimp toxicity (McGaw *et al.*, 2007); antibacterial, anti-inflammatory, mutagenic (Luseba *et al.*, 2007) |
| Pedaliaceae | *Harpagophytum procumbens* DC. | Retained placenta (van der Merwe *et al.*, 2001) | Fruit | | |
| Poaceae | *Cymbopogon marginatus* (Steud.) Stapf ex Burtt Davy | Gall sickness in animals (Jacot Guillarmod, 1971) | Unspecified | | |
| Phytolaccaceae | *Phytolacca heptandra* Retz. | The Xhosa use roots for lung sickness in cattle (Watt and Breyer-Brandwijk, 1962) | Roots | | |
| Phytolaccaceae | *Phytolacca octandra* L. | Lung sickness in cattle (Watt and Breyer-Brandwijk, 1962) | Root infusions | Triterpenoid saponins, known as yiamoloside B (Moreno and Rodriguez, 1981) | |
| Pittosporaceae | *Pittosporum viridiflorum* Sims | Gallsickness (Masika *et al.*, 2000) | Bark decoction | | |
| Plumbaginaceae | *Plumbago auriculata* Lam. | Diarrhoea in cows (Dold and Cocks, 2001) | Roots | | |
| Plumbaginaceae | *Plumbago zeylanica* L. | Pneumonia (van der Merwe *et al.*, 2001) | Roots | | |
| Podocarpaceae | *Podocarpus falcatus* (Thunb.) R. Br. Ex Mirb. | Distemper in dogs (Dold and Cocks, 2001) | Leaf decoction | | |
| Podocarpaceae | *Podocarpus latifolius* (Thunb.) R. Br. Ex Mirb. | Distemper in dogs, gallsickness in cattle (Dold and Cocks, 2001; Masika *et al.*, 2000) | Leaf, root or bark decoction | | |
| Polygalaceae | *Polygala hottentotta* Presl | Anthrax (Jacot Guillarmod, 1971) | Unspecified parts | | |
| Polygonaceae | *Emex australis* Steinh. | Threadworm in horses (Hutchings *et al.*, 1996) | Leaf decoctions | Anthraquinones (Watt and Breyer-Brandwijk, 1962) | |
| Polygonaceae | *Rumex lanceolatus* Thunb. | Used with *Euclea coriacea* A. DC. to treat gallsickness in stock animals (Jacot Guillarmod, 1971) | Unspecified parts | Chrysophanic acid, emodin and volatile oil (Watt and Breyer-Brandwijk, 1962) | |
| Proteaceaae | *Protea caffra* Meisn. | Enemas given to calves with bloody diarrhoea (Hutchings *et al.*, 1996) | Root bark decoctions | | |
| Proteaceae | *Protea welwitschii* Engl. | Dysentery and diarrhoea in calves and humans (Watt and Breyer-Brandwijk, 1962) | Decorticated root infusions | | |
| Ptaeroxylaceae | *Ptaeroxylon obliquum* (Thunb.) Radlk. | Anthrax remedy, for ticks in cattle (Hutchings *et al.*, 1996) | Wood | Powdered wood is irritating and induces sneezing (Hutchings *et al.*, 1996); timber has high oil and resin content (Watt and Breyer-Brandwijk, 1962) | |
| Ranuncluaceae | *Clematis brachiata* Thunb. | Vermifuge and for bots in horses (Hutchings *et al.*, 1996) | Infusions from shoots and leaves | Contains anemonol (Watt and Breyer-Brandwijk, 1962) | |

(*continued*)

**Table 19.1** (*Continued*)

| Family | Species | Indication | Plant part used | Chemical constituents | Screened for activity |
|---|---|---|---|---|---|
| Rhamnaceae | *Berchemia zeyheri* (Sond.) Grubov | Infectious diseases in cattle (van der Merwe, personal communication) | Bark | | Antibacterial, anthelmintic, brine shrimp toxicity (McGaw *et al.*, 2007) |
| Rhamnaceae | *Ziziphus mucronata* Willd. | Fertility enhancement, sores, burns (van der Merwe *et al.*, 2001) | Roots, leaves | | Antibacterial, anthelmintic, brine shrimp toxicity (McGaw *et al.*, 2007); antibacterial, anti-inflammatory, mutagenic (Luseba *et al.*, 2007) |
| Rhamnaceae | *Ziziphus zeyheriana* Sond. | Diarrhoea, internal parasites, general ailments (van der Merwe *et al.*, 2001) | Root-stock | | |
| Rosaceae | *Prunus persica* (L.) Batsch. | Leaf decoction for diarrhoea in lambs and kid goats, roots for broken bones (Dold and Cocks, 2001) | Leaf decoctions, roots | | |
| Rubiaceae | *Pentanisia prunelloides* (Klotzsch ex Eckl. & Zeyh.) Walp. subsp. *prunelloides* | Used by Xhosa to aid expulsion of retained animal or human placenta (Watt and Breyer-Brandwijk, 1962) | Root decoctions | | |
| Rutaceae | *Clausena anisata* (Willd.) Hook. F. Ex Benth. | Dysentery in cattle (Hutchings *et al.*, 1996) | Bark infusions | Terpenoid hydrocarcbons, alkaloids, coumarins and many other compounds (Hutchings *et al.*, 1996) | |
| Rutaceae | *Zanthoxylum capense* (Thunb.) Harv. | Gallsickness in stock (Dold and Cocks, 2001) | Leaves, root decoction | | |
| Rutaceae | *Zanthoxylum davyi* (Verdoorn) Waterm. | Administered as tonics to animals and humans (Hutchings *et al.*, 1996) | Root decoctions | | |
| Salicaceae | *Salix* L. spp. | Retained placenta (Masika *et al.*, 2000) | Decoction or infusion of unspecified parts | | Antibacterial, antifungal (Masika and Afolayan, 2002) |
| Salvadoraceae | *Azima tetracantha* Lam. | Dystocia in cows (Dold and Cocks, 2001) | Root | | |
| Sapindaceae | *Hippobromus pauciflorus* (L. f.) Radlk. | Leaf and root infusions used to clear mucus from noses of sheep and goats (Watt and Breyer-Brandwijk, 1962); root infusions given to stock animals with coughs (Hutchings *et al.*, 1996); leaf sap used for inflamed eyes in animals and humans (Watt and Breyer-Brandwijk, 1962; Masika *et al.*, 2000); bark used for heartwater and diarrhoea in cattle (Dold and Cocks, 2001) | Leaf and root infusions or decoctions, leaf sap, bark | Resinous and oily substances in wood and leaves (Watt and Breyer-Brandwijk, 1962) | Antibacterial, anthelmintic, brine shrimp toxicity (McGaw *et al.*, 2007) |
| Sapindaceae | *Pappea capensis* Eckl. & Zeyh. | Used medicinally by Zulus for calves (Gerstner, 1939) | Unspecified parts | | |
| Sapotaceae | *Englerophytum magalismontanum* Krause | Fertility enhancement (van der Merwe *et al.*, 2001) | Roots | | |

| | | | | | |
|---|---|---|---|---|---|
| Sapotaceae | *Sideroxylon inerme* L. | Preparations administered by Zulus as tonics to calves and goats (Hutchings *et al.*, 1996); Xhosa use bark for gallsickness in stock (Watt and Breyer-Brandwijk, 1962); bark decoction used for redwater in cattle (Dold and Cocks, 2001) | Unspecified parts; bark | Cinnamic acid, kaempferol and leucanthocyanidins from bark (Hutchings *et al.*, 1996) | |
| Solanaceae | *Datura stramonium* L. | Powdered leaves are applied by Zulus to animal and human bruises and wounds to draw out inflammation and pus (Watt and Breyer-Brandwijk, 1962) | Powdered leaves | Alkaloids including hyoscyamine and hyoscine (Oliver-Bever, 1986) | |
| Solanaceae | *Nicotiana tabacum* L. | Eye infections (van der Merwe *et al.*, 2001) | Leaves | | |
| Solanaceae | *Solanum aculeastrum* Dun. | Ringworm in cattle and horses and also for anthrax (Hutchings *et al.*, 1996) | Fruit | Solanine (Watt and Breyer-Brandwijk, 1962) | |
| Solanaceae | *Solanum capense* L. | Fruit pulp used by Xhosa for warts and ringworm in animals and humans, fruit sap for sores, distemper and sore eyes in dogs (Watt and Breyer-Brandwijk, 1962) | Fruit | | |
| Solanaceae | *Solanum hermannii* Dun. | Fruit sap and leaf paste used for sores on sheep and horses by Xhosa and Sotho (Watt and Breyer-Brandwijk, 1962) | Fruit, leaves | Solanin, solanidine, azosolanidin, quinhydrone solasonine, solasodine, solasodamine from fruit (Watt and Breyer-Brandwijk, 1962) | |
| Solanaceae | *Solanum incanum* L. | Sores (van der Merwe *et al.*, 2001) | Roots | | |
| Solanaceace | *Solanum panduriforme* E. Mey. | Diarrhoea (van der Merwe *et al.*, 2001) | Fruit sap | | |
| Solanaceae | *Solanum lichtensteinii* Willd. | Respiratory problems (Luseba and van der Merwe, 2006) | Aerial parts | | |
| Solanaceae | *Solanum mauritianum* Scop. | Dystocia in cows (Dold and Cocks, 2001) | Roots | | |
| Solanaceae | *Withania somnifera* (L.) Dun. | Used to stimulate milk production in cows (Gerstner, 1941); roots used for black gallsickness in cattle (Hutchings *et al.*, 1996); diarrhoea (van der Merwe *et al.*, 2001) | Unspecified parts, roots | Many compounds including choline, tropanaol, glycowithanolides, withanolides, withaferine and withasomnine (Hutchings *et al.*, 1996 and references cited therein) | |
| Sterculiaceae | *Dombeya rotundifolia* (Hochst.) Planch. | Newcastle disease in chickens (Luseba and van der Merwe, 2006); infectious diseases in cattle (van der Merwe, personal communication) | Leaves and flowers | | Antibacterial, anthelmintic, brine shrimp toxicity (McGaw *et al.*, 2007) |
| Thymelaeaceae | *Gnidia capitata* L.f. | Heartwater in cows, anthrax (Dold and Cocks, 2001) | Root decoction | | Antibacterial, anthelmintic, brine shrimp toxicity (McGaw *et al.*, 2007) |

(*continued*)

**Table 19.1** (*Continued*)

| Family | Species | Indication | Plant part used | Chemical constituents | Screened for activity |
|---|---|---|---|---|---|
| Thymelaeaceae | *Gnidia kraussiana* Meisn. | Preparations injected near the site of fractured limbs of stock animals by the Sotho (Jacot Guillarmod, 1971) | Unspecified parts | Flavone heteroside from roots, toxic diterpenoid fraction, polysacchardies daphnane orthoesters (Hutchings *et al.*, 1996) | |
| Tiliaceae | *Grewia flava* DC. | Fertility enhancement (van der Merwe *et al.*, 2001) | Roots | | |
| Tiliaceae | *Grewia occidentalis* L.f. | Gallsickness in stock (Dold and Cocks, 2001) | Leaves | | |
| Tiliaceae | *Triumfetta sonderi* L. | Retained placenta (van der Merwe *et al.*, 2001) | Root bark | | |
| Typhaceae | *Typha capensis* (Rohrb.) N.E. Br. | Decoctions taken or applied externally to aid expulsion of afterbirth in animals and humans (Roberts, 1990) | Unspecified parts | Quercetin 3′-dimethyl ether 4′-glucoside from leaf (Hutchings *et al.*, 1996) | |
| Urticaceae | *Pouzolzia mixta* Solms | Retained placenta, bloat, vaginal discharge (van der Merwe *et al.*, 2001) | Roots, leaves, stems | | Antibacterial, anthelmintic, brine shrimp toxicity (McGaw *et al.*, 2007) |
| Verbenaceae | *Clerodendrum glabrum* | Unspecified parts are used as purgatives for calves (Hutchings *et al.*, 1996); the Sotho and Swazi use topical leaf decoctions to prevent parasites developing in animal wounds (Watt and Breyer-Brandwijk, 1962); the Tswana use leaf infusions and bark scrapings as anthelmintics for dogs, calves and donkeys (Roberts, 1990) | Leaves, bark | | |
| Verbenaceae | *Lantana rugosa* Thunb. | Pastes or infusions used for animal and human eye complaints (Watt and Breyer-Brandwijk, 1962) | Leaves | Volatile oil and the alkaloid lantanin (Watt and Breyer-Brandwijk, 1962). | |
| Verbenaceae | *Vitex zeyheri* Sond. ex Schauer | Eye infections (van der Merwe *et al.*, 2001) | Leaves | | |
| Vitaceae | *Cissus quadrangularis* L. | Used by Zulus as a drench for sick horses (Watt and Breyer-Brandwijk, 1962), aerial parts used as poultice for wounds, lumpy skin disease and as tick repellent (Luseba and van der Merwe, 2006) | Aerial parts | A steroidal mixture and triterpenoids (Hutchings *et al.*, 1996) | Antibacterial, anthelmintic, brine shrimp toxicity (McGaw *et al.*, 2007); antibacterial, anti-inflammatory, mutagenic (Luseba *et al.*, 2007) |
| Vitaceae | *Cyphostemma natalitium* (Szyszyl.) J. V. D. Merwe | Used by Zulus for colic in cattle (Gerstner, 1939) | | | |

| | | | | | |
|---|---|---|---|---|---|
| Vitaceae | *Rhoicissus digitata* (L. f.) Gilg & Brandt | Cattle diseases (Hutchings *et al.*, 1996) | Tubers | | |
| Vitaceae | *Rhoicissus tomentosa* (Lam.) Wild & Drum. | Anthelmintics for calves (Watt and Breyer-Brandwijk, 1962) | Roots | | |
| Vitaceae | *Rhoicissus tridentata* (L. f.) Wild & Drum. | Cattle diseases (Pujol, 1990), diarrhoea in goats and sheep (Dold and Cocks, 2001); heartwater, redwater, internal parasites, general ailments, abortion (van der Merwe *et al.*, 2001) | Tubers | | Anti-babesial (Naidoo *et al.*, 2005) |
| Zamiaceae | *Stangeria eriopus* (Kunze) Baill. | Internal parasites in livestock (Dold and Cocks, 2001) | Rootstock | | |
| Zingiberaceae | *Siphonochilus aethiopicus* (Schweinf.) B.L. Burtt | Administered to horses as prophylactics against horse sickness (Watt and Breyer-Brandwijk, 1962) | Rhizome infusions | Volatile oil with a characteristic sesquiterpenoid, α-terpineol and other monoterpenoids (Van Wyk *et al.*, 1997) | |
| Zygophyllaceae | *Tribulus terrestris* L. | Retained placenta, bloat (van der Merwe *et al.*, 2001) | Whole plant, aerial parts | | |

commonly used in such treatments. Species belonging to the Asphodelaceae family were frequently used (21.4% of plants used), with *Aloe* the most utilized species (50%). Leaves were the most common plant parts used (45.9%), and decoctions were the most popular means of preparing remedies (70%). Combinations of medicinal plants were generally employed, but some plants were used on their own. The plant remedies against helminths were administered orally, mainly by using bottles. Some of the species described had anthelmintic properties according to literature reports, while others possessed other activities. It was concluded that if their safety and efficacy could be confirmed, these plant species could form an alternative cost-effective strategy to manage helminth infections (Maphosa and Masika, 2010).

Medicinal plants used to treat wounds and retained placenta in livestock in South Africa were investigated by Luseba *et al.* (2007). Dichloromethane extracts had antibacterial and anti-inflammatory activity, and *Cissus quadrangularis* stem and *Jatropha zeyheri* root extracts were selectively inhibitory against cyclooxygenase-2 in the anti-inflammatory assay. None of the extracts was mutagenic in the Ames test against *Salmonella typhimurium* strain TA98 without metabolic activation. Although water is traditionally the most commonly available solvent used to prepare medicinal extracts, the activity of organic extracts need not be disregarded. In livestock wound treatment, the whole-plant material is often processed and applied locally, and with ailments such as retained placenta where the treatment mixtures are given orally (and are probably unfiltered), active ingredients may be released (Luseba *et al.*, 2007).

### 19.3.2 East Africa

The ethnoveterinary use of plants used in the region of East Africa, covering Kenya, Uganda, Tanzania, Eritrea, Ethiopia, Somalia and Djibouti, was reviewed by Githiori and Gathumbi (2010). It was noted that, in this area, livestock farming is central to farming; for example, in Kenya about 69% of the land can only be used for livestock farming rather than for growing crops (Githiori and Gathumbi, 2010). Pastoralists dominate the farming system, which also includes smallholder farmers keeping a few animals. Most information on ethnoveterinary plants used to treat various conditions was available from Kenya and Uganda, with 161 plant species recorded out of a total of 193 species documented. A total of 64 plant families were used to prepare the different treatments. A large number of treatments were available, with the highest number for a single condition of 19 in Kenya for treatment of wounds. It was held that the ethnoveterinary practices documented were expected to be similar throughout the subcontinent because of shared customs and traditions (Githiori and Gathumbi, 2010).

An ethnobotanical study was conducted in Bale Mountains National Park, Ethiopia, on veterinary medicinal plants, leading to the observation that 74 plant species belonging to 64 genera and 37 families were used (Yineger *et al.*, 2007). Roots were the most frequently used plant parts, followed by leaves, and fresh material was preferred (Yineger *et al.*, 2007). The route of administration was most commonly oral, followed by dermal (Yineger *et al.*, 2007). Zorloni (2007) evaluated the activity of plants used to combat tick infestations of camels in Ethiopia.

Tabuti *et al.* (2003) reported that 38 plant species, distributed in 37 genera and 28 families, are used to treat common cattle diseases in Bulamogi county, Uganda, and roots are most frequently used, followed by leaves. The species used to treat cattle were also used to treat human ailments (Tabuti *et al.*, 2003). In the pastoral Karamoja region of northern Uganda, people rely mainly on ethnoveterinary knowledge to control common livestock health problems (Gradé *et al.*, 2009). In documenting the remedies used to treat livestock diseases, Gradé *et al.* (2009) presented information on 209 plant species belonging to 116 genera and 54 families. Bark and underground parts were used more frequently than other plant parts, and most remedies employed a single ingredient; route of administration was primarily oral, followed by topical application (Gradé *et al.*, 2009).

Among the Kikuyus of central Kenya, it was reported that 40 plant species in 26 families were used in the traditional management of various cattle diseases (Njoroge and Bussmann, 2006). Two plant families, namely the Asteraceae and Lamiaceae, were particularly frequently used, and the most commonly utilized plant species were *Synadenium compactum* (Euphorbiaceae), *Solanecio mannii* (Asteraceae) and *Senna didymobotrya* (Caesalpinaceae) (Njoroge and Bussmann, 2006).

### 19.3.3 West Africa

In West Africa, most livestock belongs in the care of traditional herdsmen, among whom ethnoveterinary health care remains a major component of disease management (Okoli *et al.*, 2010 and references cited therein). In many rural West African communities, traditional animal health care, or ethnoveterinary medicine, has remained popular because it is a readily available, low-cost alternative to modern veterinary services (Okoli *et al.*, 2010). The amount of published information available on diversity and efficacy of traditional animal health care is limited considering its popularity among stock raisers (Okoli *et al.*, 2010). In south-eastern Nigeria, indigenous farmers and healers used 24 plant species in the treatment of common livestock ailments such as diarrhoea, retained placenta and ecto- and endo-parasites (Okoli *et al.*, 2002). Methods of preparing and administering treatments included feeding of the plant parts, drenching with aqueous decoctions or external application of plant juice to affected parts (Okoli *et al.*, 2002).

A number of ethnoveterinary medicine practices, as well as herbal remedies, used by different West African communities have been reviewed by Okoli *et al.* (2010). It was noted that biological activity has in some cases been tested using *in vitro* studies but that little phytochemical work had been conducted on the species listed (Okoli *et al.*, 2010). It was also highlighted that a single remedy or plant extract may be used in treating more than one disease, but also that combinations of various plant extracts for broad-spectrum therapy were common (Okoli *et al.*, 2010). The Fulani herdsmen in north-eastern Nigeria rely on traditional knowledge to cure common herd diseases, and popular species used include *Adansonia digitata* (used for diarrhoea and skin disorders), *Zingiber officinale* (used as a laxative, appetizer and anti-bloat agent), *Parkia filicoides* (used against skin infections, wounds and worms) and *Nicotiana tabacum* (used against myiasis, hoof infections and ectoparasites) (Okoli *et al.*, 2010). It was concluded that a wealth of knowledge exists concerning traditional ethnoveterinary practices of West African people, and that there is a need to investigate and integrate traditional animal health care practices into primary animal health

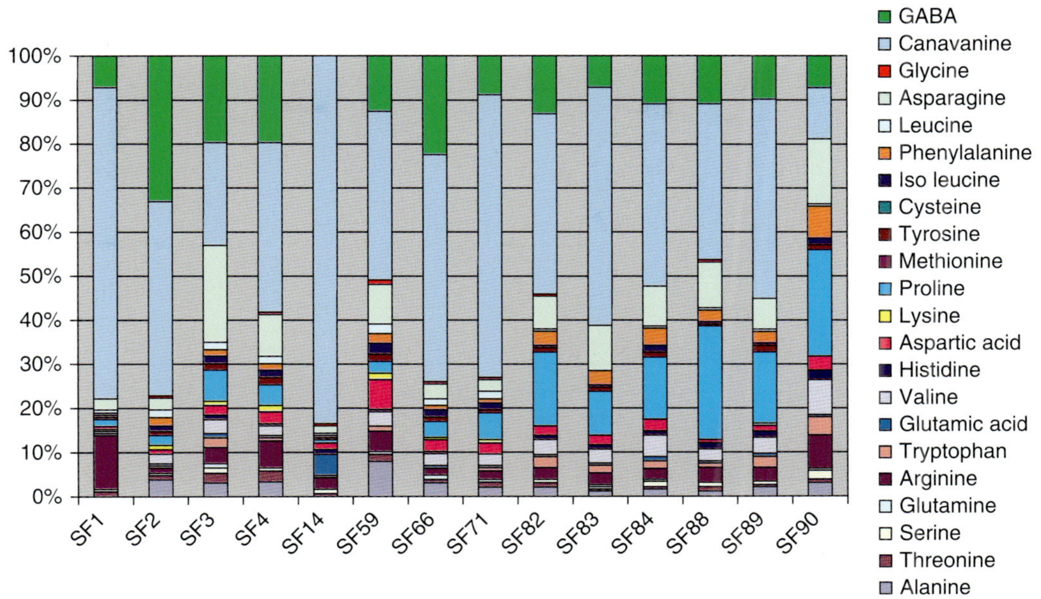


**Figure 3.3** Example of amino acid analysis of *Sutherlandia* – different part of plants and geographical distribution

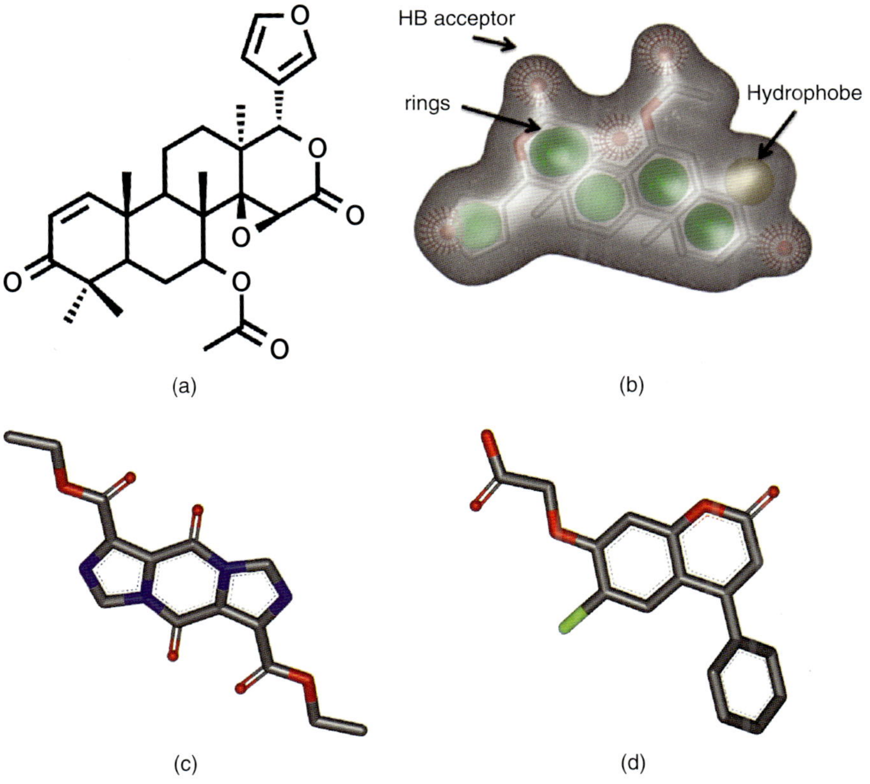


**Figure 4.8** (a) Structure of a natural product, gedunin; (b) 3D structure of gedunin showing volume space and molecular descriptors. Examples of high-ranked compounds from 3D shape analysis of the ZINC database using gedunin: (c) shape Tanimoto score of 0.72; (d) shape Tanimoto score of 0.70

*Novel Plant Bioresources: Applications in Food, Medicine and Cosmetics*, First Edition. Edited by Ameenah Gurib-Fakim

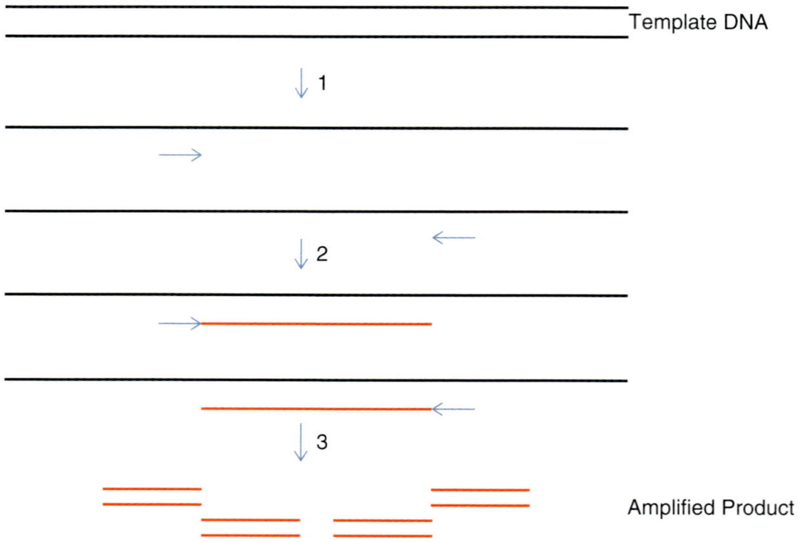


**Figure 5.1** PCR amplification of DNA: (1) DNA denaturation; (2) primer annealing; (3) amplification. Arrows indicate the primers

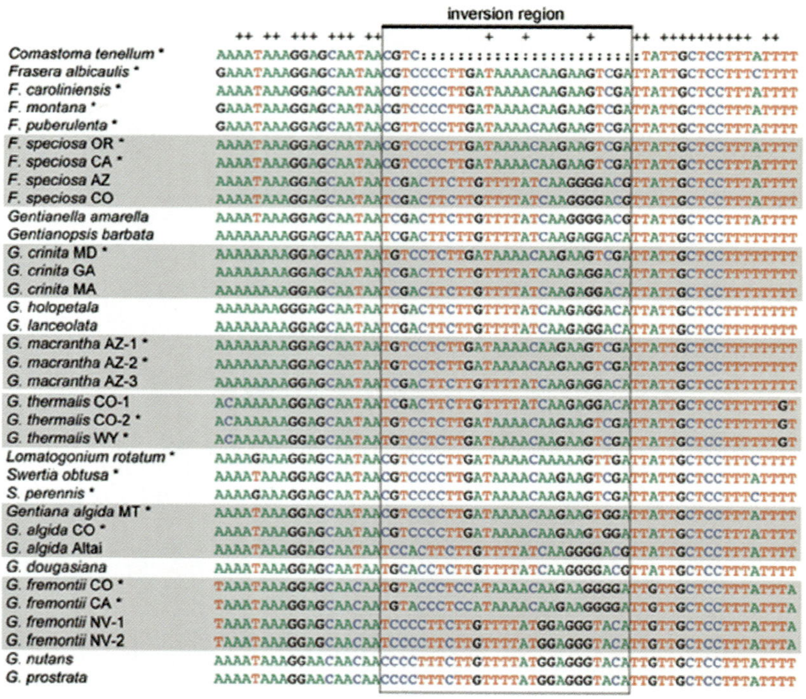


**Figure 5.5** Alignment of *trnH–psbA* sequences from Gentianaceae. Conspecific sequences are shaded and the box represents the 25 bp of the inversion region. *Source*: Whitlock *et al.* (2010)

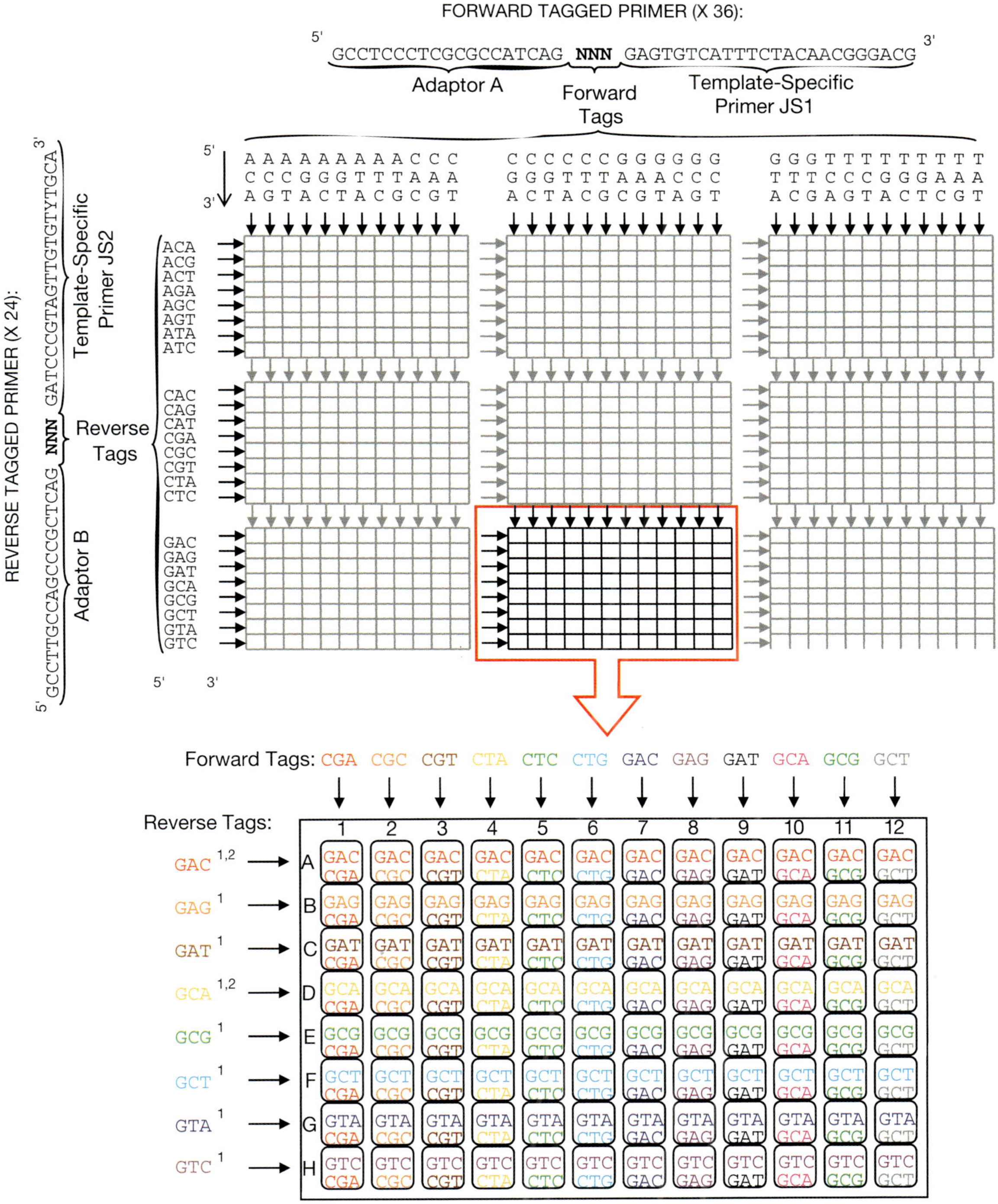


**Figure 5.8** Use of different six-nucleotide barcodes for sample identification in a 96-well plate with 36 forward and 24 reverse tags to have 864 combinations. *Source*: Galan *et al.* (2010)

**Figure 12.4** Illustration of using tetrazolium violet to determine MIC of different plant extracts in a serial dilution assay, with highest concentration at the top. Extracts in lanes 3 and 8 were the most active and extracts in lanes 5 and 11 the least active. The extracts where microbial growth is inhibited led to a reduction or absence in colour

**Figure 15.5** *Vernonia kotschyana* Sch. Bip. Ex Walp. (Asteraceae) rhizome is used to make a phytodrug effective against indigestion and gastric ulcers. *Photo*: Sergio Giani ©Aidemet. Reproduced with permission

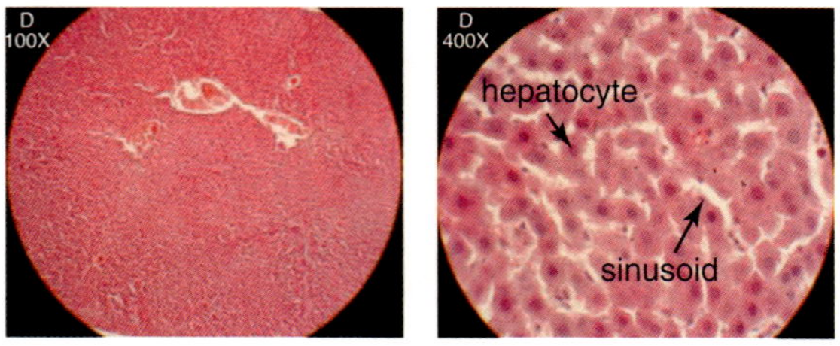


**Figure 26.2** Photomicrographs of haematoxylin-and-eosin (H&E)-stained liver sections of male rats treated with MMAPP. Liver sections from other rat groups showed a similar histoarchitecture

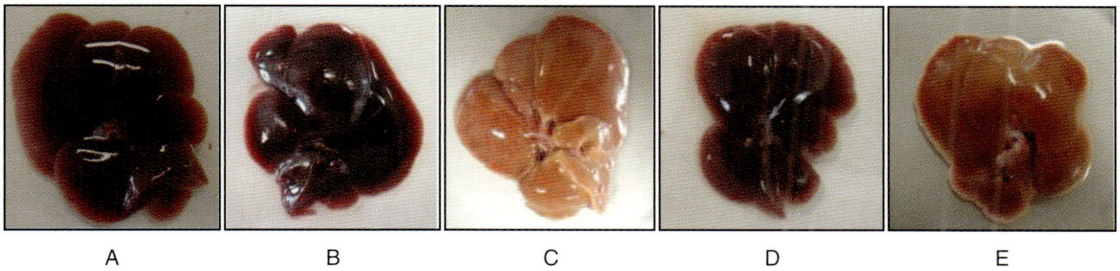


(A) Dark brick-red of normal liver (control); (B) Dark brick-red of SMN + PCM-induced liver(positive control); (C) Pale brownish-red of saline + PCM-induced liver (negative control); (D) Dark brick-red of 400 mg/kg MMAPL + PCM-induced liver; (E) Pale brownish-red of liver 400 mg/kg MMAPP + PCM-induced liver

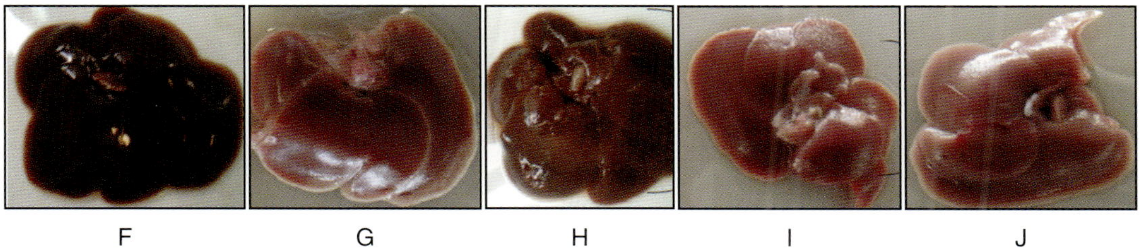


(F) Dark brick-red of normal liver (control); (G) Bright red of SMN+$CCl_4$-induced liver(positive control); (H) Brown of saline + $CCl_4$-induced liver (negative control); (I) Bright red of 400 mg/kg MMAPL + $CCl_4$-induced liver; (J) Bright red of 400 mg/kg MMAPP + $CCl_4$-induced liver

**Figure 26.3** Colour change observed in the livers isolated from the control and treated groups. Pale brownish-red colour reflects toxicity level, hepatic leakage and lipid accumulation. Red spots indicate hepatocellular necrosis and inflammation. (a) Dark brick-red cof normal liver (control); (b) dark brick-red of SMN + PCM-induced liver (positive control); (c) pale brownish-red of saline + PCM-induced liver (negative control); (d) dark brick-red of 400 mg/kg MMAPL + PCM-induced liver; (e) pale brownish-red of liver 400 mg/kg MMAPP + PCM-induced liver; (f) dark brick-red of normal liver (control); (G) bright red of SMN + $CCl_4$-induced liver(positive control); (h) brown of saline + $CCl_4$-induced liver (negative control); (i) bright red of 400 mg/kg MMAPL + $CCl_4$-induced liver; (j) bright red of 400 mg/kg MMAPP + $CCl_4$-induced liver

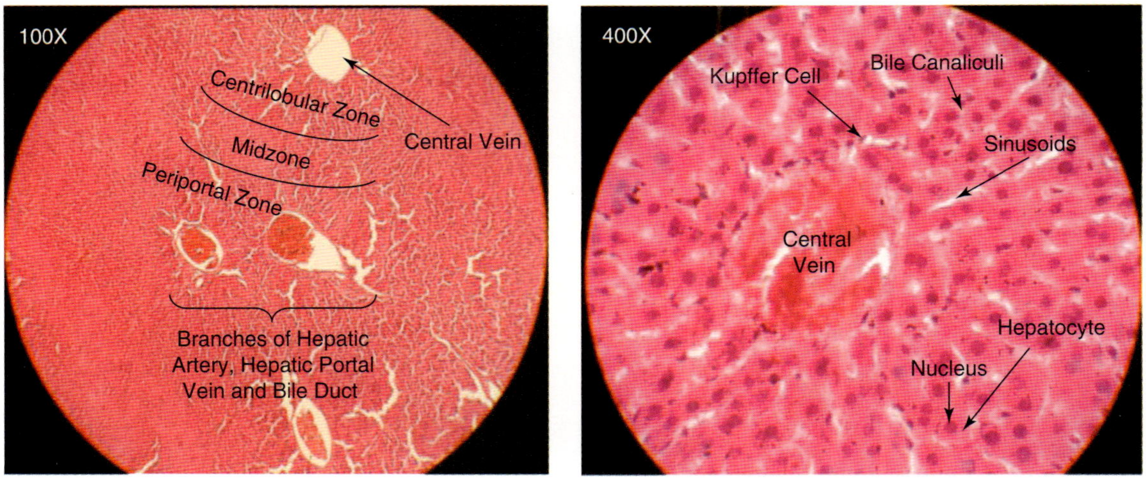


**Figure 26.4** Photomicrographs under light microscope of H&E-stained section from normal liver group

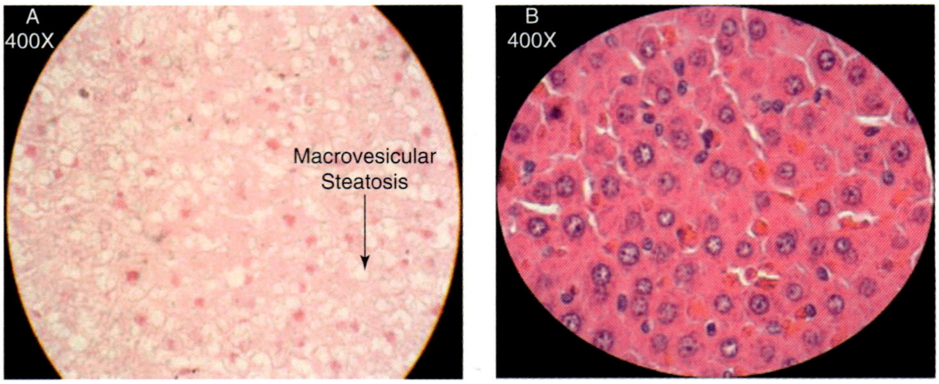


**Figure 26.5** (a) Negative control group; (b) positive control group

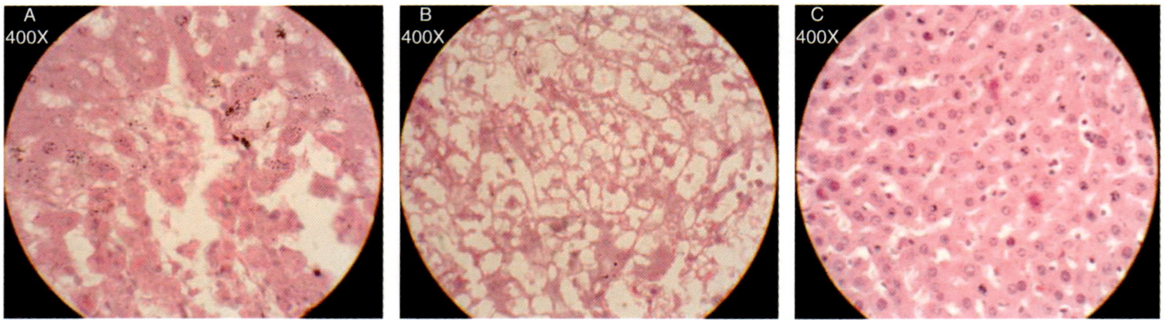


**Figure 26.6** Pretreated with (a) 100 mg/kg, (b) 200 mg/kg and (c) 400 mg/kg MMAPL

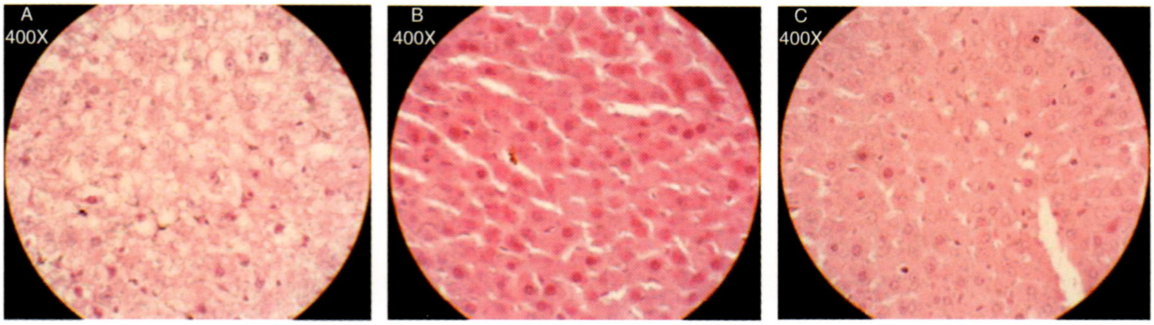


**Figure 26.7** Pretreated with (a) 100 mg/kg, (b) 200 mg/kg and (c) 400 mg/kg MMAPP

**Figure 34.2** Flower of *V. planifolia*. Photograph by T.L. Palama

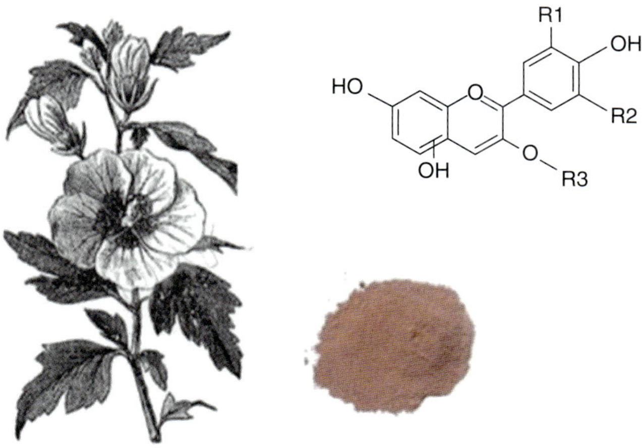


**Figure 35.2** Structure of *Hibiscus sabdariffa* anthocyanins: delphinidin 3-*O*-(2-*O*-D-xylopyranosyl)-D-glucopyranoside (R = OH, **1**), cyanidin 3-*O*-(2-*O*-D-xylopyranosyl)-D-glucopyranoside (R = H, **2**)

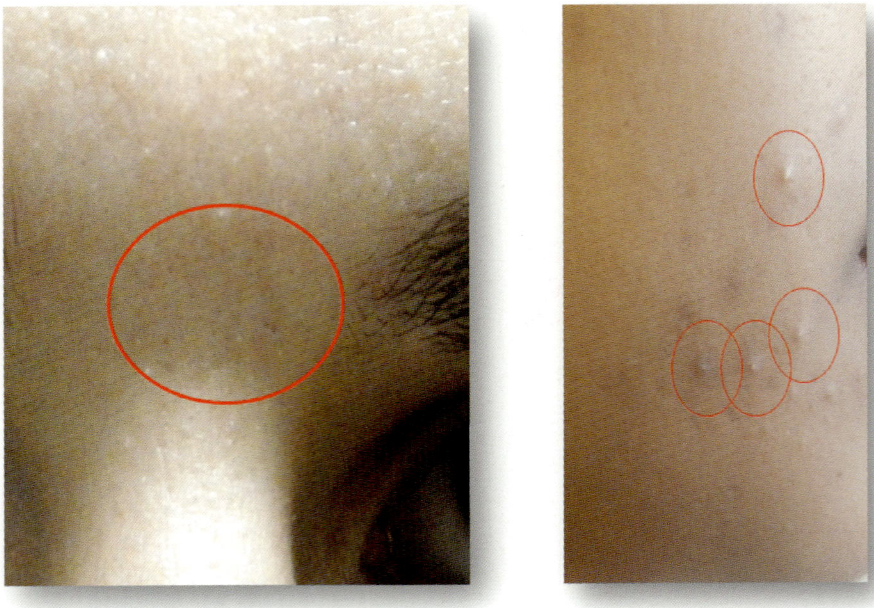

(a) (b)

**Figure 36.3** (a) Blackheads on the forehead and (b) inflammatory lesions on the face. *Source*: Photographs were taken by author M.N. de Canha.

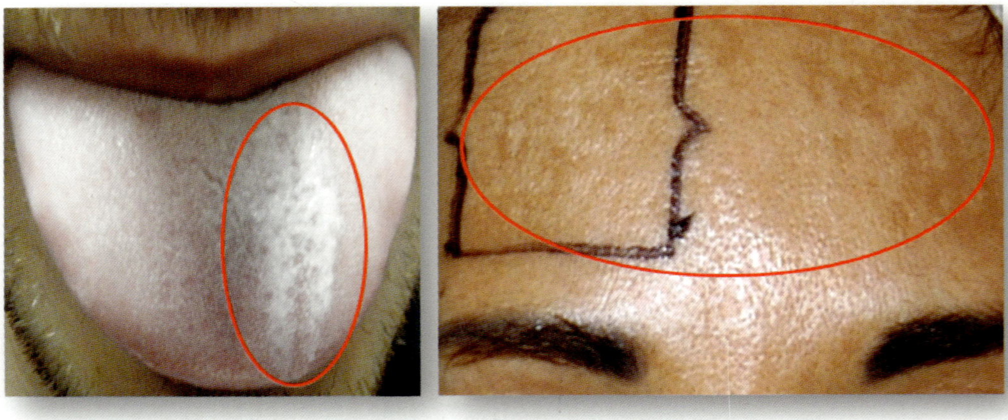

(a) (b)

**Figure 36.5** (a) Fissured tongue and (b) hypopigmentation of the forehead. *Source*: Lowenstein & Lowenstein (2011). Reproduced with permission of Elsevier.

care delivery systems in the region (Okoli *et al.*, 2010). The review highlighted the importance of traditional animal health caregivers in livestock health care systems in West Africa.

The potential role of medicinal plants in livestock production in Nigeria was reviewed by Adewumi (2004). He demonstrated that indigenous livestock keepers were competent in diagnosing various animal diseases, and have diverse methods of preparing medicinal plants to treat their animals. It was concluded that the toxicity of plants containing toxic components, such as pyrrolizidine alkaloids, should be made known to livestock keepers (Adewumi, 2004). Alawa *et al.* (2002) questioned 50 herdsmen and village livestock producers in northern Nigeria, where livestock in the country are concentrated, and found that many indigenous plant-based treatments are used and administered to animals for a variety of ailments. Taking into account the combination of ingredients used, it was held to be likely that additive, synergistic and nutritional effects could be involved in alleviating disease conditions in the animals (Alawa *et al.*, 2002). Koné and Kamanzi Atindehou (2008) undertook an ethnobotanical and ethnopharmacological study of medicinal plants used by rural communities of northern Côte d'Ivoire, and reported on the use of 55 medicinal recipes employing 44 plant species belonging to 40 genera and 30 families. Ethnoveterinary medicine in the northern provinces of Cameroon was reviewed and documented by Nfi *et al.* (2001). It was noted that the farmers interviewed had a rich history of successful agropastoralism with extensive knowledge of the prevention, control and treatment of cattle diseases using ethnoveterinary practices (Nfi *et al.*, 2001).

### 19.3.4 Northern Africa

Geographically, this region spans all the countries north of the Sahara, which is semi-arid to arid, but also includes areas of forest, mountains and fertile riverine valleys (Watson *et al.*, 1997). Watson *et al.* (1997) claim that this area, in spite of its aridity, contains 10% of the world's species of higher plants, of which half are endemic to the area. Boukraa *et al.* (2010) reviewed herbal medicines used in animal health in the Middle East and North Africa region and listed the most popular medicinal plants used. Some of the most common and important plant species used include *Urtica dioica* (which has anti-inflammatory, diuretic, antidiabetic, anticancer, antiviral, astringent and wound-healing effects), *Artemisia herba-alba* (with purgative, leishmanicidal, antifungal and antioxidant effects), *Thymus capitatus* (used to treat respiratory problems and wounds, with anti-inflammatory, antispasmodic and antifungal properties) and *Juniperus phoenicea* (used to treat diabetes, diarrhoea and as an antiseptic and antiparasitic).

## 19.4 African plants as sources of commercial remedies

African traditional medicine is the oldest and possibly the most diverse of medicinal systems globally, although the various cultural healing practices remain poorly recorded (Van Wyk and Wink, 2004). The importance of the commercial potential of Africa's diversity is highlighted by the fact that over 5400 African medicinal plant taxa with more than 16 300 medicinal uses have been recorded (Neuwinger, 2000). This implies that 10.8% of the African flora is used for traditional medicinal applications (Van Wyk, 2011). In southern Africa, it has been estimated that 13.8% (Van Wyk and Gericke, 2000) or 13.5% (Arnold *et al.*, 2002) of plant species are used in traditional medicine, percentages based on the figure of 21 817 angiosperm species occurring in southern Africa (Germishuizen and Meyer, 2003). It has been recorded that 83 African medicinal plants have been commercialized as branded products, with only about 14 southern African species viewed as being partially or fully commercialized compared with 336 species in Europe (Van Wyk and Wink, 2004). Brendler *et al.* (2010) noted that less than 8% of the 1100 medicinal plants commercialized internationally originate in Africa. Van Wyk (2011) published a list of promising indigenous South African species, indicating at least 90 medicinal plant species that could be considered for further development and commercialization. As an indication of the value of medicinal plants, Van Wyk (2011) described *Aloe vera* (of North African origin) as possibly being the most important medicinal plant, as fresh *Aloe vera* gel products had an annual retail value of approximately US$ 125 million in 2004 (Afolayan and Adebola, 2006). Van Wyk (2008) reported on the recent upsurge in research and development of new medicinal products and crops, evidenced by a rapid increase in the number of scientific publications and patents. From this, it is apparent that there are vast potential resources available from Africa for research and development of new products. However, the difficulties associated with bringing a product to the market should not be underestimated. Although there are potentially fewer constraints to developing an ethnoveterinary medicine than a botanical medication intended for human internal use, issues such as raw material supply, identification of active compounds, extract development, absorption studies, formulation development, clinical safety and efficacy studies, and intellectual property protection require attention (Gericke, 2011).

A major constraint to the growth of a modern industry based on plant medicines in Africa was identified by the Commonwealth Medicinal Plants Business Forum held in 2000 as the lack of suitable technical specification and quality control standards (Brendler *et al.*, 2010). It is known that some plants display marked regional variation in genetic, chemical and morphological characters that need to be thoroughly investigated (Van Wyk, 2008). The emerging commercialization process should be supported by the availability of basic biological information that may facilitate selection of superior clones, development of new cultivars and standardization of raw materials (Van Wyk, 2008). National and international buyers need to be in a position to evaluate the safety and efficacy of plants and extracts and should also be able to compare batches of a certain product originating from different places or different years (Brendler *et al.*, 2010). In Europe and Asia this is not the case, as traditional methods and formulations are recorded and evaluated, facilitating world trade in medicinal plants (Brendler *et al.*, 2010). In light of this, a project was established to identify the most important plant species in African medicine and to prepare a set of monographs or quality control standards for selected species. The major African species currently traded regionally or internationally, in addition to plants considered to have long-term future potential, were included. The African Association of Medicinal Plant Standards was established to prepare a comprehensive pharmacopoeia and a living database that would allow the continual upgrading and addition of relevant information. The first set of approximately 50 monographs was published in 2010 (Brendler *et al.*, 2010), and work is ongoing to upgrade the information available on African medicinal plants with commercial potential (www.aamps.org).

The commercialization of traditional ethnoveterinary remedies based on plant species is only one aspect to be considered, as ethnoveterinary medicine is a complex process. Promotion of all available resources that work, including people's animal healthcare practices which form part of ethnoveterinary medicine, is important to improve animal healthcare services in rural areas (Mathias, 2007). Various approaches have been proposed to validate remedies used in ethnoveterinary medicine such as literature studies, laboratory and clinical tests used in modern medicine, social science methods (e.g. requesting livestock keepers to rank treatments by how effective they are), and studying the influence of the remedy on animal production and economic parameters (Mathias, 2004). Laboratory methods can be used to identify candidates with promising biological activities and provide leads for isolation of active compounds (McGaw *et al.*, 2007), but further laboratory and clinical studies are needed to comply with legal requirements (Mathias, 2007). Alternative approaches include testing ethnoveterinary practices in the field, such as combining literature study, interviews and participatory methods with mortality and morbidity studies, veterinary diagnostics and keeping records of ethnoveterinary treatments (Ghotge *et al.*, 2002). It is important to note again that medicinal plant use is just one aspect of ethnoveterinary medicine as it is a complex system combining treatments with management practices that help prevent diseases (Mathias, 2007). The complexity of ethnoveterinary medicine with its emphasis on preventing disease complicates the study of the efficacy of individual aspects (Mathias, 2007). In exploring the potential of ethnoveterinary medicine, it should be kept in mind that intellectual property rights should be respected, and endangered plants need to be conserved (Mathias, 2007).

## 19.5 Examples of African medicinal plants used for ethnoveterinary purposes with scope for commercialization

In the Phytomedicine Programme at the Department of Paraclinical Sciences, University of Pretoria (www.up.ac.za/phyto), a major research focus targets the investigation of the pharmacological efficacy and potential toxicity of ethnoveterinary remedies. Some of these endeavours have culminated in the development of patentable products. This section of the chapter constitutes a brief review of some examples of the studies our research group has undertaken in recent years to provide an indication of the potential of ethnoveterinary plant remedies in producing commercial products.

Most of the studies involve *in vitro* investigations of antimicrobial and antiparasitic activity of extracts and purified plant-derived compounds as these are more affordable and easier to conduct. Once highly promising candidates are identified in laboratory studies, animal efficacy and safety trials are carried out. It should be kept in mind that lack of efficacy *in vitro* may be attributed to various factors, such as the complexity of ethnoveterinary practices where combinations of plant extracts may be used, and the remedies may function rather to enhance the immune system than to have a direct effect on the disease-causing organism (Eloff and McGaw, 2009).

### 19.5.1 Antimicrobial plants (antibacterial and antifungal)

Plants used to treat infectious diseases in cattle were screened for antibacterial activity against *Escherichia coli*, *Enterococcus faecalis*, *Pseudomonas aeruginosa* and *Staphylococcus aureus* using a broth microdilution assay (McGaw *et al.*, 2007). Hexane, methanol and water extracts were tested and found to be most active against the Gram-positive organisms, namely *E. faecalis* and *S. aureus*. In this study, 33% of plant extracts had minimum inhibitory concentration (MIC) values less than 1 mg/mL, and these levels of antibacterial activity may contribute towards rationalizing the use of the plants in ethnoveterinary medicine (McGaw *et al.*, 2007). *Ziziphus mucronata* (Rhamnaceae) had very good antibacterial activity, and 2,3-dihydroxyl-up-20-en-28-oic acid and zizyberanalic acid were isolated from the leaves of this species (Moloto, 2004). The first compound was highly active against *S. aureus*, providing rationale for the use of a *Z. mucronata* leaf paste in treating bacterial infections in animals and humans.

A similar technique was used to identify good antibacterial activity against *S. aureus* and *P. aeruginosa* in *Peltophorum africanum* (Fabaceae) extracts (Bizimenyera *et al.*, 2005). Pastoralists and farmers use the root and bark extracts to treat diarrhoea and dysentery, among other stomach ailments, in cattle, and the antibacterial activity of the species validates its traditional usage against bacterial infections.

*Gunnera perpensa* (Gunneraceae) is used to treat endometritis and retained placenta in cattle, and was investigated to discover whether antibacterial efficacy could be the reason for this reported use (McGaw *et al.*, 2005). Only weak antibacterial activity was shown by the *G. perpensa* extracts against the four bacterial species mentioned above, and so antibacterial efficacy was unlikely to be the reason for its widespread use.

Upon investigating antifungal activity of plant extracts against clinical isolates of the most common and important disease-causing fungi in animals, Masoko *et al.* (2005) reported significant antifungal activity in several *Terminalia* species (Combretaceae). The fungal pathogens represented various morphological forms of fungi, including yeasts (*Candida albicans* and *Cryptococcus neoformans*), moulds (*Aspergillus fumigatus*) and thermally dimorphic fungi (*Sporothrix schenckii*). The acetone extracts displayed the highest antifungal activity of the hexane, dichloromethane, acetone and methanol extracts of *Terminalia* leaves (Masoko *et al.*, 2005).

### 19.5.2 Antiparasitic activity

Antibabesial activity of plant extracts was evaluated in a cell culture-based assay (Naidoo *et al.*, 2005). The acetone extract of *Elephantorrhiza elephantina* rhizome was active against *Babesia caballi* cultures. *E. elephantina* is a popular species used in ethnoveterinary remedies. *Urginea sanguinea*, *Rhoicissus tridentata* and *Aloe marlothii* acetone extracts were not active (Naidoo *et al.*, 2005). Two of the same plant species were tested for antirickettsial activity against *Ehrlichia ruminantium*, namely *E. elephantina* and *A. marlothii* (Naidoo *et al.*, 2006). *E. ruminantium* cultures were incubated with leaf acetone extracts of the plants, which both had anti-ehrlichial activity with $EC_{50}$ values of 111.4 and 64.5 mg/mL and $EC_{90}$ values of 228.9 and 129.9 mg/mL respectively. The corresponding $EC_{50}$ and $EC_{90}$ for oxytetracycline, the positive control, were 0.29 mg/mL and

0.08 mg/mL. The plant extracts were postulated to produce their inhibitory activity by a similar mechanism, unrelated to that of the tetracyclines (Naidoo *et al.*, 2006).

Diseases in cattle and other livestock caused by helminth parasites are common, especially in rural areas, and anthelmintic remedies are important components of ethnoveterinary medicine. Most laboratory evaluations of anthelmintic efficacy of ethnoveterinary plant species are *in vitro* trials, as *in vivo* tests are expensive and time consuming. Parasitic nematodes are difficult to maintain in culture in laboratory conditions. However, two tests are commonly used: the egg hatch assay, where test samples incubated with fresh nematode eggs may inhibit hatching, and the larval development assay, where test substances prevent the development of the nematodes into infective larvae. The two assays used together can provide a reasonable indication of anthelmintic activity of plant extracts or natural products. The popular and widely used ethnoveterinary plant *Peltophorum africanum* is used to treat helminth diseases, and the acetone extracts of the leaves, bark and roots were tested for activity against *Haemonchus contortus* and *Trichostrongylus colubriformis* in the egg hatch and larval development assays (Bizimenyera *et al.*, 2006a, 2006b). The extracts were active in the assays at a concentration of 0.2 mg/mL, lending support to the use of this plant species in traditional medicine. However, *in vivo* evidence of effectiveness and lack of toxicity is necessary before further development of this active plant extract can take place. Several other studies revealing useful results were undertaken on anthelmintic efficacy of plant extracts, demonstrating the potential that plants have to combat helminth parasites of cattle as well as humans (Shai *et al.*, 2009; Maphosa *et al.*, 2010b; Ademola and Eloff, 2010, 2011a–d; Vatta *et al.*, 2011; Nchu *et al.*, 2012; Adamu *et al.*, 2013). It was shown by Adamu *et al.* (2012) that some plant extracts used for treatment of parasitic helminth infections also have good antifungal activity, which may facilitate bioassay-guided fractionation of active compounds, as antifungal bioassays are much easier than anthelmintic tests.

Tick-borne diseases cause major problems for livestock keepers. Promising results have been obtained from studies of the repellent and toxic effects of plant extracts against ticks. In one study, Nchu *et al.* (2005) investigated the repellent effects and direct toxicity of *Allium* species against adult *Hyalomma marginatum rufipes*. A high repellency index (65–79.48%) was shown by the *Allium porrum* acetone extract and the *Allium sativum* dichloromethane extract was toxic to 100% of ticks within 1 h of exposure. In further work, essential oils from aerial parts of *Lippia javanica* and *Tagetes minuta* showed concentration-dependent repellence on ticks (Nchu, 2004). *T. minuta* essential oil delayed molting of the *H. marginatum rufipes* ticks to the adult stage of 60% of engorged nymphs. A similar study reported that ethyl acetate extracts of *Senna italica* ssp. *arachoides* had an increasing acaricidal activity against *H. marginatum rufipes* with increasing concentration (Thembo, 2006). An *in vivo* study showed that the aqueous extracts of *S. italica* ssp. *arachoides* ingested by guinea-pigs and rabbits may interfere with the feeding performance of adult *H. marginatum rufipes* ticks (Thembo, 2006). Some plants used traditionally as arthropocides in South Africa, namely *Eucalyptus globoidea* and *Lavendula angustifolia*, are effective tick repellents (Mkolo, 2008). Zorloni (2007) recorded plant species used in Ethiopia for ethnoveterinary tick control, and analysed 28 of these plants for repellent and toxic effects against adult *Rhipicephalus pulchellus* ticks. Organic solvent extracts of many species showed promising repellent activities, and *Calpurnia aurea* had the best results in the tick toxicity assay (Zorloni, 2007; Zorloni *et al.*, 2010). Interestingly, *C. aurea* extracts, in contrast to many other plant species evaluated, had slight attractant rather than repellent properties and was toxic to the ticks at the concentrations tested (Zorloni *et al.*, 2010). It was held that development of this extract could lead to a product that may be used commercially to protect animals against tick infestations (Zorloni *et al.*, 2010).

Coccidiosis, caused by *Eimeria* species, is one of the most important diseases of poultry, and economically viable control of the coccidian parasite is desirable. Antioxidant compounds could be useful in controlling *Eimeria* infections because of the association of coccidial infection with lipid peroxidation of the intestinal mucosa. Four antioxidant plant extracts were screened for their anticoccidial activity in chickens with toltrazuril as the positive control (Naidoo *et al.*, 2008). *Combretum woodii* at a concentration of 160 mg/kg was extremely toxic to the birds, but treatment with *Tulbaghia violacea* (35 mg/kg), *Vitis vinifera* (75 mg/kg) and *Artemisia afra* (150 mg/kg) resulted in feed conversion ratios similar to toltrazuril, and higher than that of the untreated control. *T. violacea* also significantly lowered oocyst production in the chickens. It was concluded that antioxidant-rich plant extracts have potential benefits in treating coccidial infections (Naidoo *et al.*, 2008).

Myiasis is an infestation of the mammalian skin by various fly species belonging to the order Diptera. Mukandiwa *et al.* (2012) investigated the *in vitro* effect on blowfly larvae of four plant species (*Aloe zebrina*, *Clausena anisata*, *Erythrina lysistemon* and *Spirostachys africana*) which are used traditionally in southern Africa to treat myiasis. Larval behaviour, larval development and emergence of the adult flies were assessed after exposing the larvae to meat treated with acetone or aqueous extracts of the plant species. Increasing the concentration of the acetone plant extracts caused decreases in the ingestion of meat by larvae, pupal mass and adult emergence rates (Mukandiwa *et al.*, 2012). In the case of *C. anisata* and *S. africana* extracts, the increasing concentration was also associated with emerging adult flies being smaller. It was concluded that the plant species may contain compounds that interfere with larval feeding and the neuroendocrine control mechanisms in the blowfly. Mukandiwa *et al.* (2012) stated that some of the plant species tested have the potential to develop a product that can be used to control myiasis.

### 19.5.3 Antiviral plants

Plant-based antiviral agents appear to have much promise for providing therapeutic possibilities (Eloff and McGaw, 2009). Many different compounds isolated from plants have antiviral activity, including flavonoids, terpenoids, polyphenolics, saponins, proteins, alkaloids, polysaccharides and essential oils (Vlietinck and Vanden Berghe, 1991). Several of these phytochemicals have potentially complementary mechanisms of action, including antiviral effects by inhibiting the formation of viral DNA or RNA or by inhibiting the activity of viral reproduction (Jassim and Naji, 2003). Some plant compounds have unique antiviral mechanisms and are interesting candidates for further clinical research (Vlietinck and Vanden Berghe, 1991). It is interesting that, unlike the case with antibacterial or antifungal plant compounds, several antiviral plant chemicals have shown competitive *in vitro* and *in vivo* antiviral activities with those of synthetic antiviral drugs (Vlietinck and Vanden Berghe, 1991).

In a recent publication, Bagla *et al.* (2012) screened extracts of six plants traditionally used against infections for antiviral activity using virucidal and attachment assays. The test viruses included canine distemper virus (CDV), canine parainfluenza virus-2, feline herpesvirus-1 (FHV-1) and lumpy skin disease virus (LSDV). The best inhibition was observed with the acetone and methanol extracts of *Podocarpus henkelii* against CDV and LSDV, which inhibited replication of the viruses by more than 75% at 3 mg/mL with selectivity index (SI) values ranging between 12 and 45. Promising activity was found with the hexane extracts of *Plumbago zeylanica* and *Carissa edulis* against CDV, with the extracts lowering viral-induced cytopathogenic effect by 50% and 75% respectively. The hexane extract of *C. edulis* had some activity against FHV-1 with $EC_{50} < 70$ mg/mL and $SI < 2$ (Bagla *et al.*, 2012).

## 19.6 Toxicity

When used as directed and under the supervision of knowledgeable individuals, most herbal remedies are generally safe, but the potential for adverse effects or intoxications exists (Poppenga, 2007). In addition, because many herbal remedies contain multiple biologically active constituents, possible interactions with conventional drugs are a concern (Poppenga, 2007). Most traditional medicinal plants have never been subjected to exhaustive toxicological tests as is routine for modern pharmaceutical compounds, but based on their long-term traditional use are often assumed to be safe (Verschaeve and Van Staden, 2008; Street *et al.*, 2008). However, it has been shown that many plants which are used as food ingredients or in traditional medicine are mutagenic *in vitro*, or have toxic and carcinogenic properties (Verschaeve and Van Staden, 2008). The absence of regulation of the medicinal plant trade in terms of collection, processing and storage leads to concerns of the safety of the plants, and factors such as environmental pollution, misidentification and adulteration are further issues for consideration (Street *et al.*, 2008). Screening for toxicity as well as efficacy and stability of compounds or medicinal preparations derived from plants should be incorporated into comprehensive programmes to ensure a complete understanding of these plants (Fennell *et al.*, 2004).

Poisoning of an animal can occur in various ways, including using a remedy containing a known toxin, administration of a misidentified plant or contamination of commercially prepared herbal remedies with toxic plants (Poppenga, 2007). Additionally, many naturally occurring constituents of herbal remedies cause induction of one or more liver P-450 metabolizing enzymes which can cause altered metabolism of other chemicals and hence enhanced or diminished drug efficacy or toxicity (Poppenga, 2007). An issue of concern to veterinarians is the possibility of varying reactions encountered in different species; for example, differing susceptibilities to the toxic effects of herbal constituents (Poppenga, 2007).

It was noted in a recent review of southern African ethnoveterinary remedies that many of the plant species used are known to be toxic, belonging to families such as the Amaryllidaceae, Hyacinthaceae and Solanaceae (Maphosa *et al.*, 2010a). Although it was recognized that fatalities among animals owing to administration of these remedies would be difficult to pinpoint, it was proposed that the users are sufficiently knowledgeable about doses and methods of preparation to limit toxicity while optimizing therapeutic benefits (Maphosa *et al.*, 2010a).

## 19.7 Conclusions

Africa possesses an extraordinary diversity of plant species, many of which are used in preparing remedies for the treatment of a range of ailments in humans, as well as in livestock and companion animals. Natural products derived from plants are an important source of novel chemicals which may be useful in drug discovery and development. A growing interest in the use of plants for treating numerous ailments is evidenced by an increasing number of publications reviewing and documenting various species used in human and animal health care. Pharmacological and toxicological studies of potentially useful plants are being undertaken, although there is much room for research to be done in these fields. In many cases, the activity of ethnopharmacological or ethnoveterinary remedies has been confirmed, and in a few cases *in vivo* experiments have substantiated efficacy. Some products have been patented, although it does not appear as if many have been commercialized to date.

The study of ethnoveterinary plant treatments is a growing area, and there have been sessions on ethnoveterinary medicine at many recent international conferences relating to medicinal plants. The challenge is to spark the interest of more companies supplying animal medicines, or other agencies which can assist in the development of products useful in improving animal health and productivity. Such products may find markets in livestock animal production or in the health of companion animals. It may be easier to market products for the animal health sector where premiums may be paid for organically raised livestock, providing incentives for their use. Developing countries also stand to benefit from the production and distribution of affordable remedies based on plant extracts where efficacy and safety have been verified.

## References

Adamu, M., Naidoo, V. and Eloff, J.N. (2012) Some southern African plant species used to treat helminth infections in ethnoveterinary medicine have excellent antifungal activities. *BMC Complementary and Alternative Medicine*, **12**, 213.

Adamu, M., Eloff, J.N. and Naidoo, V. (2013) Efficacy and toxicity of thirteen plant leaf acetone extracts used in ethnoveterinary medicine in South Africa on egg hatching and larval development of *Haemonchus contortus*. *BMC Veterinary Research*, **9**, 38.

Ademola, I.O. and Eloff, J.N. (2010) *In vitro* anthelmintic activity of *Combretum molle* (R. Br. ex G. Don) (Combretaceae) against *Haemonchus contortus* ova and larvae. *Veterinary Parasitology*, **169**, 198–203.

Ademola, I.O. and Eloff, J.N. (2011a) Anthelminthic activity of acetone extract and fractions of *Vernonia amygdalina* against *Haemonchus contortus* eggs and larvae. *Tropical Animal Health and Production*, **43**, 521–527.

Ademola, I.O. and Eloff, J.N. (2011b) *In vitro* anthelmintic effect of *Anogeissus leiocarpus* (dc.) Guill. & Perr. leaf extracts and fractions on developmental stages of *Haemonchus contortus*. *African Journal of Traditional, Complementary and Alternative Medicines*, **8**, 134–139.

Ademola, I.O. and Eloff, J.N. (2011c) Ovicidal and larvicidal activity of *Cassia alata* leaf acetone extract and fractions on *Haemonchus contortus*: *In vitro* studies. *Pharmaceutical Biology*, **49**, 539–544.

Ademola, I.O. and Eloff, J.N. (2011d) Anthelmintic efficacy of cashew (*Anacardium occidentale* L.) on *in vitro* susceptibility of the ova and larvae of *Haemonchus contortus*. *African Journal of Biotechnology*, **10**, 9700–9705.

Adewumi, C.O. (2004) The potential role of ethnoveterinary pharmacy in livestock production in Nigeria. *Vom Journal of Veterinary Science*, **1**, 125–137.

Afolayan, A.J. and Adebola, P.O. (2006) *Aloe vera* (L.) Burm.f. [Internet] record from protabase, in *PROTA (Plant Resources of Tropical Africa / Ressources végétales de l'Afrique tropicale)* (eds G.H. Schmelzer, and A. Gurib-Fakim), PROTA, Wageningen, http://database.prota.org/search.htm (accessed 14 July 2011).

Alawa, J.P., Jokthan, G.E. and Akut, K. (2002) Ethnoveterinary medical practice for ruminants in the subhumid zone of northern Nigeria. *Preventive Veterinary Medicine*, **54**, 79–90.

Arnold, T.H., Prentice, C.A., Hawker, L.C. *et al.* (2002) *Medicinal and Magical Plants of Southern Africa: An Annotated Checklist*, Strelitzia 13, National Botanical Institute, Pretoria.

Bagla, V.P., McGaw, L.J. and Eloff, J.N. (2012) The antiviral activity of six South African plants traditionally used against infections in ethnoveterinary medicine. *Veterinary Microbiology*, **155**, 198–206.

Bizimenyera, E.S., Swan, G.E., Chikoto, H. and Eloff, J.N. (2005) Rationale for using *Peltophorum africanum* (Fabaceae) extracts in veterinary medicine. *Journal of the South African Veterinary Association*, **76**, 54–58.

Bizimenyera, E.S., Githiori, J.B., Swan, G.E. and Eloff, J.N. (2006a) *In vitro* ovicidal and larvicidal activity of the leaf, bark and root extracts of *Peltophorum africanum* Sond. (Fabaceae) extracts on *Haemonchus contortus*. *Journal of Animal and Veterinary Advances*, **5**, 608–614.

Bizimenyera, E.S., Githiori, J.B., Eloff, J.N. and Swan, G.E. (2006b) *In vitro* activity of *Peltophorum africanum* (Fabaceae) extracts on the egg hatching and larval development of the parasitic nematode *Trichostrongylus colubriformis*. *Veterinary Parasitology*, **142**, 336–343.

Bohlmann, F. and Fritz, U. (1979) New flavones from *Heteromorpha simplicifolium*. *Phytocemistry*, **18**, 1081.

Bohlmann, F. and Zdero, C. (1978a) Naturally occurring terpene derivatives. Part 132. Diterpenes with a new carbon skeleton from *Printzia laxa*. *Phytochemistry*, **17**, 487–489.

Bohlmann, F. and Zdero, C. (1978b) Naturally occurring terpene derivatives. Part 150. A new type of sesquiterpene from *Senecio oxyriifolius*. *Phytochemistry*, **17**, 1669–1671.

Bohlmann, F., Ates, N. and Jakupovic, J. (1983) Naturally occurring terpene derivatives. Part 464. Hirsutinolides from South African *Vernonia* species. *Phytochemistry*, **22**, 1159–1162.

Boukraa, L., Benbarek, H. and Benhanifia, M. (2010) Herbal medicines for animal health in the Middle East and North Africa (MENA) region, in *Ethnoveterinary Botanical Medicine: Herbal Medicines for Animal Health* (eds D.R. Katerere and D. Luseba), CRC Press, Boca Raton, FL, pp. 303–319.

Brendler, T., Eloff, J.N., Gurib-Fakim, A. and Phillips, L.D. (2010) Foreword, in *African Herbal Pharmacopoeia* (eds T. Brendler, J.N. Eloff, A. Gurib-Fakim and L.D. Phillips), Graphic Press Ltd, Mauritius.

Bryant, A.T. (1966) *Zulu Medicine and Medicine-Men*, C. Struik, Cape Town.

Candy, H.A., Pegel, K.H., Brookes, B. and Rodwell, M. (1977) The occurrence of atractyloside in *Callilepis laureola*. *Phytochemistry*, **16**, 1308–1309.

Cowling, R.M., Rundel, P.W., Lamont, B.B. *et al.* (1996) Plant diversity in Mediterranean-climate regions. *Trends in Ecology and Evolution*, **11** (9), 362–366.

Dekker, T.G., Fourie T.D., Matthee, E. and Snyckers, F.O. (1987) An oxindole from the roots of *Capparis tomentosa*. *Phytochemistry*, **26**, 1845–1846.

Dictionary of Natural Products on CD-ROM (1996) release 4.2, Chapman & Hall, London.

Doke, C.M. and Vilakazi, B.W. (1972) *Zulu–English Dictionary*, 2nd edn, Witwatersrand University Press, Johannesburg.

Dold, A.P. and Cocks, M.L. (2001) Traditional veterinary medicine in the Alice district of the Eastern Cape Province, South Africa. *South African Journal of Science*, **97**, 375–379.

Eloff, J.N. and McGaw, L.J. (2009) Application of plant extracts and products in veterinary infections. in *New Strategies Combating Bacterial Infections* (eds I. Ahmad and F. Aqil), Wiley–VCH Verlag GmbH, Weinheim.

Fennell, C.W., Lindsey, K.L., McGaw, L.J. *et al.* (2004) Assessing African medicinal plants for efficacy and safety: pharmacological screening and toxicology. *Journal of Ethnopharmacology*, **94**, 205–217.

Gericke, N. (2011) Muthi to medicine. *South African Journal of Botany*, **77**, 850–856.

Germishuizen, G. and Meyer, N.L. (2003) *Plants of Southern Africa: An Annotated Checklist*, Strelitzia 14, National Botanical Institute, Pretoria.

Gerstner, J. (1938) A preliminary checklist of Zulu names of plants with short notes. *Bantu Studies*, **12** (3), 215–236; 12 (4), 321–342.

Gerstner, J. (1939) A preliminary checklist of Zulu names of plants with short notes. *Bantu Studies*, **13** (1), 49–64; 13 (2), 131–149; 13 (4), 307–326.

Gerstner, J. (1941) A preliminary checklist of Zulu names of plants with short notes. *Bantu Studies*, **15** (3), 277–301; 15 (4), 369–383.

Ghotge, N.S., Ramdas, S.R., Ashalata, S. *et al.* (2002) Social approach to the validation of traditional veterinary remedies – the Anthra project. *Tropical Animal Health and Production*, **34**, 121–143.

Githiori, J.B. and Gathumbi, P.K. (2010) Ethnoveterinary plants used in East Africa, in *Ethnoveterinary Botanical Medicine: Herbal Medicines for Animal Health* (eds D.R. Katerere and D. Luseba), CRC Press, Boca Raton, FL, pp. 289–302.

Goldblatt, P. and Manning, J.C. (2000) *Cape Plants: A Conspectus of the Cape Flora of South Africa*, Strelitzia 9, National Botanical Institute, Pretoria.

Gradé, J.T., Tabuti, J.R.S. and Van Damme, P. (2009) Ethnoveterinary knowledge in pastoral Karamoja, Uganda. *Journal of Ethnopharmacology*, **122**, 273–293.

Graham, K., Graham, E.A. and Towers, G.H.N. (1980) Cercaricidal activity of phenyl heptatriyne and α-terthienyl naturally occurring compounds in species of Asteraceae Compositae. *Canadian Journal of Zoology*, **58**, 1955–1958.

Hoffman, B. and Hoelzl, J. (1988) New chalcones from *Bidens pilosa*. *Planta Medica*, **54**, 52–54.

Houghton, P. (2010) Foreword, in *Ethnoveterinary Botanical Medicine: Herbal Medicines for Animal Health* (eds D.R. Katerere and D. Luseba), CRC Press, Boca Raton, FL pp. ix–x.

Hulme, M.M. (1954) *Wild Flowers of Natal*, Shuter & Shooter, Pietermaritzburg.

Hutchings, A. and Johnson, C.T. (1986) Glimpses of a Xhosa herbal. *Veld & Flora*, **72**, 59–62.

Hutchings, A., Scott, A.H., Lewis, G. and Cunningham, A.B. (1996) *Zulu Medicinal Plants: An Inventory*, University of Natal Press, Pietermaritzburg.

ITDG and IIRR (1996) *Ethnoveterinary Medicine in Kenya: A Field Manual of Traditional Animal Health Care Practices*, Intermediate Technology Development Group and International Institute of Rural Reconstruction, Nairobi, Kenya.

Jacot Guillarmod, A. (1971) *Flora of Lesotho*, Cramer, Lehr.

Jassim, S.A.A. and Naji, M.A. (2003) Novel antiviral agents: a medicinal plant perspective. *Journal of Applied Microbiology*, **95**, 412–427.

Katerere, D.R. and Luseba, D. (eds) (2010) *Ethnoveterinary Botanical Medicine: Herbal Medicines for Animal Health*, CRC Press, Boca Raton, FL.

Klopper, R.R., Chatelain, C., Bänninger, V. *et al.* (2006) Checklist of the flowering plants of sub-Saharan Africa. An index of accepted names and synonyms, South African Botanical Diversity Network Report No. 42, SABONET, Pretoria.

Koné, W.M. and Kamanzi Atindehou, K. (2008) Ethnobotanical inventory of medicinal plants used in traditional veterinary medicine in northern Côte-d'Ivoire (West Africa). *South African Journal of Botany*, **74**, 76–84.

Luseba, D. and van der Merwe, D. (2006) Ethnoveterinary medicine practices among Tsonga speaking people of South Africa. *Onderstepoort Journal of Veterinary Research*, **73**, 115–122.

Luseba, D., Elgorashi, E.E., Ntloedibe, D.T. and Van Staden, J. (2007) Antibacterial anti-inflammatory and mutagenic effects of some medicinal plants used in South Africa for the treatment of wounds and retained placenta in livestock. *South African Journal of Botany*, **73**, 378–383.

Mabogo, D.E.N. (1990) The ethnobotany of the Vhavenda. MSc Thesis, University of Pretoria.

Maphosa, V. and Masika, P.J. (2010) Ethnoveterinary uses of medicinal plants: a survey of plants used in the ethnoveterinary control of gastro-intestinal parasites of goats in the Eastern Cape Province, South Africa. *Pharmaceutical Biology*, **48**, 697–702.

Maphosa, V., Tshisikhawe, P., Thembo, K. and Masika, P. (2010a) Ethnoveterinary medicine in southern Africa, in *Ethnoveterinary Botanical Medicine: Herbal Medicines for Animal Health*. (eds D.R. Katerere and D. Luseba), CRC Press, Boca Raton, FL, pp. 257–288.

Maphosa, V., Masika, P.J., Bizimenyera, E.S. and Eloff, J.N. (2010b) *In vitro* anthelminthic activity of crude aqueous extracts of *Aloe ferox*, *Leonotis leonurus* and *Elephantorrhiza elephantina* against *Haemonchus contortus*. *Tropical Animal Health and Production*, **42**, 301–307.

Masika, P.J. and Afolayan, A.J. (2002) Antimicrobial activity of some plants used for the treatment of livestock disease in the Eastern Cape, South Africa. *Journal of Ethnopharmacology*, **83**, 129–134.

Masika, P.J., van Averbeeke, W. and Sonandi, A. (2000) Use of herbal remedies by small-scale farmers to treat livestock diseases in central Eastern Cape Province, South Africa. *Journal of the South African Veterinary Association*, **71**, 87–91.

Masoko, P., Picard, J. and Eloff, J.N. (2005) Antifungal activities of six South African *Terminalia* species (Combretaceae). *Journal of Ethnopharmacology*, **99**, 301–308.

Mathias, E. (2004) Ethnoveterinary medicine: Harnessing its potential. *Veterinary Bulletin*, **74**, 27–37.

Mathias, E. (2007) Ethnoveterinary medicine in the era of evidence-based medicine: mumbo-jumbo. or a valuable resource? *Veterinary Journal*, **173**, 241–242.

McCorkle, M.C. and Mathias-Mundy, E. (1992) Ethnoveterinary medicine in Africa. *Journal of the International African Institute, London*, **62**, 59–93.

McGaw, L.J. and Eloff, J.N. (2008) Ethnoveterinary use of southern African plants and scientific evaluation of their medicinal properties. *Journal of Ethnopharmacology*, **119**, 559–574.

McGaw, L.J., Gehring, R., Katsoulis, L. and Eloff, J.N. (2005) Is the use of *Gunnera perpensa* extracts in endometriosis related to antibacterial activity? *Onderstepoort Journal of Veterinary Research*, **72**, 129–134.

McGaw, L.J., Van Der Merwe, D. and Eloff, J.N. (2007) *In vitro* anthelmintic, antibacterial and cytotoxic effects of extracts from plants used in South African ethnoveterinary medicine. *Veterinary Journal*, **173**, 366–372.

Mkolo, N.M. (2008) Anti-tick properties of some of the traditionally used plant-based products in South Africa. MSc thesis, University of Limpopo.

Moloto, M.P. (2004) Isolation and characterization of antibacterial, anthelmintic, antioxidant and cytotoxic compounds present in *Ziziphus mucronata*. MSc thesis, Medical University of South Africa.

Moreno, M. and Rodriguez, V.M. (1981) Medicinal pre-Hispanic plants. Part 9. Yiamoloside B, a fungistatic saponin of *Phytolacca octandra*. *Phytochemistry*, **20**, 1446–1447.

Mukandiwa, L., McGaw, L.J., Eloff, J.N. and Naidoo, V. (2012) Extracts of four plant species used traditionally to treat myiasis influence pupation rate, pupal mass and adult blowfly emergence of *Lucilia cuprina* and *Chrysomya marginalis* (Diptera: Calliphoridae). *Journal of Ethnopharmacology*, **143**, 812–818.

Mutke, J. and Barthlott, W. (2005) Patterns of vascular plant diversity at continental to global scales. *Biologiske Skrifter*, **55**, 521–537.

Naidoo, V., Zweygarth, E., Eloff, J.N. and Swan, G.E. (2005) Identification of anti-babesial activity for four ethnoveterinary plants *in vitro*. *Veterinary Parasitology*, **130**, 9–13.

Naidoo, V., Zweygarth, E. and Swan, G.E. (2006) Determination and quantification of the *in vitro* activity of *Aloe marlothii* (A. Berger) subsp. *marlothii* and *Elephantorrhiza elephantina* (Burch.) Skeels acetone extracts against *Ehrlichia ruminantium*. *Onderstepoort Journal of Veterinary Research*, **73**, 175–178.

Naidoo, V., McGaw, L.J., Bisschop, S.P.R. *et al.* (2008) The value of plant extracts with antioxidant activity in attenuating coccidiosis in broiler chickens. *Veterinary Parasitology*, **153**, 214–219.

Nchu, F. (2004) Developing methods for the screening of ethnoveterinary plants for tick control. MSc thesis, Medical University of South Africa.

Nchu, F., Magano, S.R. and Eloff, J.N. (2005) *In vitro* investigation of the toxic effects of the extracts of *Allium sativum* bulbs on adults of *Hyalomma marginatum rufipes* and *Rhipicephalus pulchellus*. *Journal of the South African Veterinary Association*, **76**, 99–103.

Nchu, F., Githiori, J.B., McGaw, L.J. and Eloff, J.N. (2012) Anthelmintic and cytotoxic activities of extracts of *Markhamia obtusifolia* Sprague (Bignoniaceae). *Veterinary Parasitology*, **183**, 184–188.

Neuwinger, H.D. (2000) *African Traditional Medicine. A Dictionary of Plant Use and Applications*, Medpharm Scientific Publishers, Stuttgart.

Newman, D.J. and Cragg, G.M. (2007) Natural products as sources of new drugs over the last 25 years. *Journal of Natural Products*, **70**, 461–477.

Nfi, A.N., Mbanya, J.N., Ndi, C. *et al.* (2001) Ethnoveterinary medicine in the northern provinces of Cameroon. *Veterinary Research Communications*, **25**, 71–76.

Njoroge, G.N. and Bussmann, R.W. (2006) Herbal usage and informant consensus in ethnoveterinary management of cattle diseases among the Kikuyus (Central Kenya). *Journal of Ethnopharmacology*, **108**, 332–339.

Okoli, I.C., Nwokeocha, J.R., Anyanwu, G.A. and Okeudo, N.J. (2002) Analysis of Imo State abattoir records (1995–1999) II: assessment of estimated volumes and values of slaughter animals. *International Journal of Environmental Health and Human Development*, **3**, 16–20.

Okoli, I.C., Tamboura, H.H. and Hounzangbe-Adote, M.S. (2010) Ethnoveterinary medicine and sustainable livestock management in West Africa, in *Ethnoveterinary Botanical Medicine: Herbal Medicines for Animal Health* (eds D.R. Katerere and D. Luseba), CRC Press, Boca Raton, FL, pp. 321–351.

Oliver-Bever, B. (1986) *Medicinal Plants in Tropical West Africa*, Cambridge University Press, Cambridge.

Palmer, E. and Pitman, N. (1972). *Trees of Southern Africa*, vol. 3, Balkema, Cape Town.

Poppenga, R.H. (2007) Herbal medicine: potential for intoxication and interactions with conventional drugs, in *Veterinary Herbal Medicine* (eds S.G. Wynn and B.J. Fougère), Mosby Elsevier, St. Louis, MO, pp. 183–207.

Pujol, J. (1990) *NaturAfrica – the Herbalist Handbook*, Jean Pujol Natural Healers' Foundation, Durban.

Rahbek, C. (1995) The elevational gradient of species richness: a uniform pattern? *Ecography*, **18**, 200–205.

Roberts, M. (1990) *Indigenous Healing Plants*, Southern Book Publishers, Halfway House.

Schillhorn van Veen, T.W. (1997) Sense or nonsense? Traditional methods of animal parasitic disease control. *Veterinary Parasitology*, **71**, 177–194.

Shai, L.J., Bizimenyera, E.S., Bagla, V. *et al.* (2009) *Curtisia dentata* (Cornaceae) leaf extracts and isolated compounds inhibit motility of parasitic and free living nematodes. *Onderstepoort Journal of Veterinary Research*, **76**, 249–256.

Street, R.A., Stirk, W.A. and Van Staden, J. (2008) South African traditional medicinal plant trade –challenges in regulating quality, safety and efficacy. *Journal of Ethnopharmacology*, **119**, 705–710.

Swann, M. (1969) *Report of the Joint Committee on the Use of Antibiotics in Animal Husbandry and Veterinary Medicine*, Cmnd 4190, HMSO, London.

Tabuti, J.R.S., Dhillion, S.S. and Lye, K.A. (2003) Ethnoveterinary medicines for cattle (*Bos indicus*) in Bulamogi county, Uganda: plant species and mode of use. *Journal of Ethnopharmacology*, **88**, 279–286.

Thembo, M.K. (2006) The anti-tick effects of *Senna italic* subsp. *arachoides* extracts on adults of *Hyalomma marginatum rufipes*. MSc thesis, University of Limpopo.

Trease, G.E. and Evans, W.C. (1983) *Pharmacognosy*, 12th edn, Bailliere Tindall, London.

Van der Merwe, D., Swan, G.E. and Botha, C.J. (2001) Use of ethnoveterinary medicinal plants in cattle by Setswana-speaking people in the Madikwe area of the North West Province of South Africa. *Journal of the South African Veterinary Association*, **72**, 189–196.

Van Wyk, B.-E. (2008) A broad review of commercially important southern African medicinal plants. *Journal of Ethnopharmacology*, **119**, 342–355.

Van Wyk, B.-E. (2011) The potential of South African plants in the development of new medicinal products. *South African Journal of Botany*, **77**, 812–829.

Van Wyk, B.-E. and Gericke, N. (2000) *People's Plants: A Guide to Useful Plants of Southern Africa*, Briza Publications, Pretoria.

Van Wyk, B.-E. and Wink, M. (2004) *Medicinal Plants of the World*, Briza Publications, Pretoria.

Van Wyk, B.-E., Van Oudtshoorn, B. and Gericke, N. (1997) *Medicinal Plants of South Africa*, Briza Publications, Pretoria.

Vatta, A.F., Kandu-Lelo, C., Ademola, I.O. and Eloff, J.N. (2011) Direct anthelmintic effects of *Cereus jamacaru* (Cactaceae) on trichostrongylid nematodes of sheep: *in vivo* studies. *Veterinary Parasitology*, **180**, 279–286.

Verschaeve, L. and Van Staden, J. (2008) Mutagenic and antimutagenic properties of extracts from South African traditional medicinal plants. *Journal of Ethnopharmacology*, **119**, 575–587.

Villegas, M., Vargas, D., Msonthi, J.D. *et al.* (1988) Isolation of the antifungal compounds falcarindiol and sarisan from *Heteromorpha trifoliata*. *Planta Medica*, **54**, 36–39.

Vlietinck, A.J. and Vanden Berghe, D.A. (1991) Can ethnopharmacology contribute to the development of antiviral drugs? *Journal of Ethnopharmacology*, **32**, 141–153.

Wagner, H. and Bladt, S. (1975) Coumarins from South African *Pelargonium* spp. *Phytochemistry*, **14**, 2061–2064.

Watson, R.T., Zinyowera, M.C. and Moss, R.H. (eds) (1997) Middle East and Arid Asia, in *IPCC Special Report on the Regional Impacts of Climate Change: An Assessment of Vulnerability*, Cambridge University Press, Cambridge, www.grida.no/climate/ipcc/regional/index.htm (accessed 15 November 2013).

Watt, J.M. and Breyer-Brandwijk, M.G. (1962) *The Medicinal and Poisonous Plants of Southern and Eastern Africa*, 2nd edn, Livingstone, London.

White, F. (1983) Vegetation of Africa – a descriptive memoir to accompany the Unesco/AETFAT/UNSO vegetation map of Africa, Natural Resources Research Report No. 20, United Nations Educational, Scientific and Cultural Organization, Paris.

Wynn, S.G. and Fougère, B.J. (2007) The roots of veterinary botanical medicine, in *Veterinary Herbal Medicine* (eds S.G. Wynn and B.J. Fougère), Mosby Elsevier, St. Louis, MO, pp. 33–49.

Yineger, H., Kelbessa, E., Bekele, T. and Lulekal, E. (2007) Ethnoveterinary medicinal plants at Bale Mountains National Park, Ethiopia. *Journal of Ethnopharmacology*, **112**, 55–70.

Zdero, C. and Bohlmann, F. (1987) Sesquiterpene lactones from the genus *Brachylaena*. *Phytochemistry*, **26**, 2597–2601.

Zorloni, A. (2007) Evaluation of plants used for the control of animal ectoparasitoses in southern Ethiopia (Oromiya and Somali regions). MSc thesis, Phytomedicine Programme, University of Pretoria.

Zorloni, A., Penzhorn, B.L. and Eloff, J.N. (2010) Extracts of *Calpurnia aurea* leaves from southern Ethiopia attract and immobilise ticks. *Veterinary Parasitology*, **168**, 160–164.

# 20 African Plant Biodiversity in Pest Management

S. N'Danikou[1,2], D.A. Tchokponhoue[1], C.A. Houdegbe[1] and E.G. Achigan-Dako[1]

[1]Laboratory of Plant Science, Department of Plant Production, Faculty of Agronomic Sciences (FSA), University of Abomey-Calavi, Cotonou, Republic of Benin
[2]Bioversity International, West and Central Africa Office, Cotonou, Benin

## 20.1 Introduction

Secoy and Smith (1983) defined pest as

> including household insects, external human parasites, external and irritating arthropods of domestic animals, wound flies and maggots, insects and vertebrates which attack field crops and orchards, and, destroyers of stored foods and other materials. The term also includes other forms, such as wolves, scorpions, snakes and crocodilians, that threaten shepherds and fishermen. However, they did not include scabies and mange, although caused by parasitic arthropods, since any recorded use of plants seems to have been to alleviate the symptoms rather than to kill or repel the mites.

In the context of this chapter, a crop pest is considered as an organism causing damage to crops. This includes all vertebrates and invertebrates, plants and microorganisms that are damaging to crops, either at the farm (e.g. insects, rabbits, weeds, fungi) or in storage (e.g. insects, rodents). Thus, the review is limited to crop pests and does not include livestock pests and human disease vectors.

Biodiversity is defined as the variety and variability of life at all levels of organization; that is, genetic, species and ecosystem levels. Biological diversity is classified in two ways: by ecological (functional) and evolutionary (phylogenetic) criteria. Ecologists and conservation biologists apply the term biodiversity most often at the species level (Colwell, 2012). In this chapter, plant biodiversity refers to the variety and variability of plants at species level. Pest management refers to methods and systems used in controlling pests in agricultural production and related activities (e.g. forestry). The concept could also include controlling of pests causing illness to human and livestock. This latter component is not treated in this chapter. Rather, it focuses on the variety of plants used to get hold of pests and their damages in agricultural production, at farm and post-harvest.

Crop pests and diseases represent important challenges for farmers, and particularly for smallholders in Africa. Two important migratory insect pests reported to cause more damage are the desert locust (*Schistocerca gregaria* Forskal) and the African armyworm (*Spodoptera exempta* Walker). They are responsible of important losses to yield. For instance, attacks by desert locust in the African Sahel have resulted in food insecurity and famine. Losses to cereals were estimated at 55 000 t in Sudan in 1954; 17 000 t in Senegal in 1957 and 167 000 t in Ethiopia in 1958 (CIRAD, n.d.). The same source estimated losses in 1974 at 368,000 tonnes of cereals for the whole African Sahel. Also grasshoppers cause between 70% to 90% losses in disastrous conditions and this is cyclic in the African Sahel (Abate *et al.*, 2000). About 342 million US dollars were spent between 1986 and 1993 to suppress the pest over about 30 million hectares of lands (CIRAD, n.d.). The damages by the desert locust threaten food supply and food security at a point that governments in most affected countries have created national centres to prevent locust attacks; for instance, the "Centre National de Lutte Contre le Criquet Pèlerin" in Mali. Several other diseases and unwanted weeds also challenge agricultural production in Africa.

Chemical or synthetic pesticides have been commonly used in fighting crop pests. These have been somehow successful in suppressing some of them. However, their impacts on the environment and human health have made them unsustainable. This is the case for several chemical products formerly used to control crop pest populations and diseases, for instance the DDT (Dichloro Diphenyl Trichloroethane), which have been banned worldwide for use in agriculture, under the Stockholm Convention on Persistent Organic Pollutants (POPs). However, there are controversial debates over this ban, especially on the use of DDT to control malaria. For instance, some governments in Africa are coming back to DDT to control malaria in the continent. The argument is that there is no alternative to the use of DDT to overcome malaria in their respective countries (Hecht, 2004). Meanwhile, there are scientific evidences on the dangers of chemical use on human's health (Pressinger and Sinclair, n.d.). For example, petroleum-based chemicals are responsible of several illnesses, including cancer, neurological dysfunctions, weakness of the immune system, infertility, asthma, and so on. An alternative solution to the intensive use of synthetic pesticides and their related environmental and toxicological problems has been the use of natural methods to deal with plant pests and diseases. Among these natural methods are the use of pesticide and repellence properties of some plants and their secondary metabolites as important sources of biopesticides (Cavoski *et al.*, 2011).

In consideration of the above, advocacy for the use of botanical products and biological methods in pest management has gained interest worldwide. This chapter presents a review of the African

*Novel Plant Bioresources: Applications in Food, Medicine and Cosmetics*, First Edition. Edited by Ameenah Gurib-Fakim.

plant biodiversity used in pest management and is structured around five key sections as follows:

1. History of humans' use of plant biodiversity in pest management
2. Methods and approaches in pest management
3. Research on plant utilization in pest management
4. Biodiversity of plants used in pest management
5. Benefits of the use of plants in crop pest management.

## 20.2 History of humans' use of plant biodiversity in pest management

### 20.2.1 Background and trends worldwide

Pest problems are probably as old as agriculture itself. Several indigenous and modern agricultural practices have evolved due to pest problems (OTA, 1979). Crop losses due to pests were more severe at that time in the humid tropics, just as they still are today. The pest problems were exacerbated by the introduction of new pests from other continents. For instance, several of the most important pests in the USA were introduced from other continents (OTA, 1979). In response to the outbreak of pest problems, humans have developed several methods to sustain productivity. Crop protection methods used for this end included the use of chemical pesticides, pest-resistant plants, cultural controls and biological controls, among others. However, the era of modern agriculture (1940s–1970s) emphasized more on sophisticated techniques and high inputs (chemical pesticides, high-yielding varieties, irrigation, etc.). This sophisticated agriculture is accompanied with high pesticide use, and this has recently raised concerns regarding risks to human health and environmental issues. Thus, since the end of the green revolution, and taking note of lessons of this era, there has been an advocacy for natural methods for pest control, which integrate the ecosystem as a whole. Nowadays, the trend worldwide is more oriented towards an approach of integrated pest management (IPM), with the development of several resistant varieties, multiplication and introduction of natural enemies, and more evolved from traditional practices, the use of plant extracts or cultural practices and mechanical methods.

### 20.2.2 Experiences from Africa

Traditional farming in Africa is characterized by mixed cropping, with two or more crops grown in the same field, thus enhancing diversity of natural enemies (Abate *et al.*, 2000). This is considered as a built-in process in smallholdings, and no specific activity is dedicated to pest management. It had been recognized that Africa's use of pesticides is among the lowest in the world. In many African regions, numbers of farmers still rely on traditional methods for pest management. This is mainly due to the limited financial strength of subsistence farmers, although extension programmes encourage use of pesticides (Abate *et al.*, 2000, Mihale *et al.*, 2009). Synthetic pesticides are mainly used on commercial crops (cotton, vegetables, coffee, cocoa, etc.). South Africa is one of the largest pesticides exporters in sub-Saharan Africa (Quinn *et al.*, 2011).

In late 20th century, pest management research focused on classical biological control and enhancement of host plants' resistance (Abate *et al.*, 2000). In the traditional agriculture, only about two dozen insect and mite species have an economic impact. The huge cortege is made up of introduced pests (Abate *et al.*, 2000). National programmes and international agricultural research centres have been carrying out IPM or IPM-related research since about four decades now.

## 20.3 Methods and approaches in pest management

Several approaches and methods have been developed in pest management. In Africa, pest management still relies on indigenous knowledge and practices and is based on experiences acquired by farmers. These traditional pest management practices include crop associations, planting and harvesting time, closed season, mechanical control, use of proper seed, use of herbal products and pest use for food and as income source (Abate *et al.*, 2000).

### 20.3.1 Diversity of traditional pest management practices

#### 20.3.1.1 On-farm pest management

***Intercropping***

Also called crop association, intercropping is the way whereby two or more species are grown on the same farm plot at the same time. This practice helps small farmers to meet their elementary needs. Different authors have reported that it better meets food security, promotes efficient land use, reduces weeding effort, and obviously reduces field pest infestation and increases natural enemy population (Abate, 1988; Kyamanywa and Ampofo, 1988). For instance, *Brassica* spp. infestation by the aphid *Brevicoryne brassicae* L. and the root fly *Delia brassicae* (Bohé) was significantly reduced, by over 60%, when the crop is associated with taxonomically unrelated plant species (Tukahirwa and Coaker, 1982). Single row intercropping is recommended so that the intercrop will better cover the ground at the infestation time for an efficient pest management. Furthermore, the ability of the association of sorghum (*Sorghum bicolor* [L.] Moench) and cowpea (*Vigna unguiculata* [L.] Walp.) to reduce stem borer (*Chilo partellus* (Swinhoe)) and thrips (*Megalurothrips sjostedti* Trybom) infestation respectively was showed by Ampong-Nyarko *et al.* (1994). Similarly, in Kenya, intercropping sorghum, maize (*Zea mays* L.) and cowpea reduced various stem borers (*Busseola fusca* Fuller, *Eldana saccharina* Wlk., and *Sesamia calamistis* Hmps.) on sorghum and thrips (*Megalurothrips sjostedti*, *Hydatothrips adolfifriederici* Karny) on cowpea (Dissemond and Hindorf, 1990). In Kenya, the positive effect of habitat diversity on the natural enemies of the African bollworm, *Heliothis armigera* (Hbn.) was reported by Abate (1991). He showed how tachinid parasitoids (*Tiphia* sp.) and a predatory wasp (*Tachytes sjostedti* Cameron) population could increase in strip-cropping haricot bean (*Phaseolus vulgaris* L.) with maize under weedy and weed-free conditions. In crop association, intercrops odours generally interfere and result in pest track muddle.

***Planting and harvesting time***

Cultural practices must be carried out as appropriate to avoid an eventual precocity or delay in crop planting or harvesting time, even though in some opinions early planting could perhaps be the most effective means to control some pests (Gebre-Amlak *et al.*, 1989). The importance of sowing at the appropriate time was emphasized by some authors. In a study carried out on bean insect pests in Africa, Abate and Ampofo (1996) reported that sowing bean at the appropriate time constitutes a major means for reducing bean stem maggots. In Ivory Coast, planting cassava between November and February helps to avoid the spread of cassava mosaic disease (Fargette *et al.*, 1994). Several other arguments in favour of planting and harvesting at appropriate times have been presented in Ambe (1993) and Adipala *et al.* (1998). However, fluctuation of pest population can result in planting time adjustment. Such strategies were reported on beans in Tanzania (Ampofo and Massomo, 1998) and on sweet potato in South Nyanza, Kenya (Smit and Matengo, 1995).

***Mechanical control, site selection and use of proper seed***

Selecting a suitable field site is crucial for infestation reduction. Thus, farmers, according to crops grown, can move over from lowland to upland or from upland to lowland. Similarity in the planting material is required when a better resistance to pests is desired. In Kenya, cuttings of high-altitude cassava were more resistant to African mosaic virus than lowlands ones were (Abate *et al.*, 2000). During land preparation, it is reported that farmers mechanically hand-pick important numbers of pests, including nocturnal beetles, larvae of the cotton leafworm (*Spodoptera littoralis* (Boisduval)), the orange dog (*Papilio demodocus* (Esper)) and the sweet potato butterfly (*Acraea* spp.) (Hillocks *et al.*, 1996).

***Pest exploitation for food and as income source***

Pests and often insects, although harmful, are also used as food and can constitute a relatively important source for income. This is the case for the edible grasshoppers, locusts and the cluster bug (*Agonoscelis pubescens* (Thumb.)), which are diversely exploited (Faure, 1944; Van Huis, 1996). This food and economic value will probably shape the management strategies developed by populations.

#### 20.3.1.2 Post-harvest pest management

Post-harvest pests have an important economic impact on products. Their impacts are categorized into three orders. There is direct loss due to consumption of products by pests, quality or value loss due to the presence of pests in the product and costs engaged to treat infested commodities (Campbell and Arthur, 2007). To deal with those pests, different approaches are used, including the use of chemical pesticides and natural products. Only methods employing plant products are presented here.

The use of herbal products is an old practice used both against field and storage pests. It is an eco-friendly way to control pests including use of different plant extracts, essential oils and some secondary metabolites. For instance, *Azadirachta indica* A.Juss (neem) cake extract, *Nerium oleander* L. leaf extract and *Trichoderma viride* Pers. had been reported to significantly reduce the incidence of rice brown spot by 66%, 52% and 45% respectively (Harish *et al.*, 2008). In an evaluation of efficacy of various botanical extracts for their repellence property, *Andrographis paniculata* L. and *Leucas aspera* L. leaf extracts and *A. indica* seed kernel extract were reported to be the most effective (Sathyaseelan and Bhaskaran, 2010). Koona *et al.* (2007) reported the ability of hexane extracts from *Tephrosia vogelii* Hook.f. to protect stored maize grain against the weevil *Sitophilus zeamais* (Motschulsky). In quantitative terms, the most important botanicals are by order pyrethrum, neem, rotenone and essential oils, typically used as insecticides (Cavoski *et al.*, 2011).

### 20.3.2 Classification of pest management systems

#### 20.3.2.1 Approach-based classification

Approach-based classification differentiates pest management systems into suppression, prevention and eradication (Willson, 1992).

***Prevention***

Prevention consists of any action undertaken to anticipate pest attacks from occurring. It consists of the use of proven methods to either prevent or reduce the probability of a significant pest problem from occurring. The prevention approach may include either chemical or non-chemical methods. In Florida, a preventive tactic consisting of scouting, ultraviolet reflective technologies, biological control, compatible insecticides, companion plants and fertility regulation has proven to be effective, economically and ecologically sound, and sustainable in the western flower thrips, *Frankliniella occidentalis* (Pergande) (Demirozer *et al.*, 2012).

***Suppression***

After the outbreak of a pest problem, all actions undertaken to manage it are termed suppressive. Post-emergence application of herbicides to reduce emerging weed populations is considered as suppression. The use of chemical methods is generally associated with suppression practices (Willson, 1992; Wright, 1995).

***Eradication***

Eradication is the approach used when a pest problem must be totally eliminated from a targeted area. For instance, coypu (*Myocastor coypus* Molina) accidentally introduced in Britain during the 1980s was successfully eradicated (Singleton *et al.*, 1999). Myers *et al.* (2000) pointed out the importance of eradication approaches in invasion of exotic species; even the attempt to remove or reduce the density of a species, even an exotic one, is a challenging undertaking.

#### 20.3.2.2 Method-based classification

Based on methods used, pest management systems are characterized as biological, chemical, cultural and mechanical, and legal (Willson, 1992). What makes the difference between one system and another is explained below.

***Biological control***

Biological control is the use of antagonist organisms (natural enemies) against others thought to be harmful. Three basic approaches are recognized: classical, natural and augmentation. However, it is worth mentioning that, intrinsically, biological control is focused primarily on predator–prey relationships between animal and microorganisms. Besides this, biological control does apply to plants as well, through exploitation of allelopathy phenomena between plant species.

*Classical biological control*. This is the importation of natural enemies from other countries and applies to pests which are not native to a given area; namely, a non-native pest is accidentally introduced in a new area. Classical biological control is used to determine (1) the pest's origin, (2) to locate beneficial organisms that naturally control the pest organism in its native area and (3) permanently establish them so that they will provide continuing pest control with little or no additional human intervention. An important success in classical biological pest management in Africa is the control of the cassava mealybug (*Phenacoccus manihoti* Matile-Ferrero) by *Epidinocarsis lopezi* De Santis (Herren and Neuenschwander, 1991). In southern Africa, Australian acacia (*Acacia* spp.), *Azolla filiculoides* Lam. and *Solanum elaeagnifolium* Cav. were subjected to classical biological control with weevils (*Melanterius maculatus* Lea, *Melanterius acaciae* Lea, etc.), wasps (*Trichilogaster* spp.) and flies (*Uromycladium tepperianum* (Sacc.) McAlpine and *Cylindrobasidium laeve* (Pers.:Fr.) Chamuris) as natural enemies (Dennill *et al.*, 1999). Moreover, cereal leaf beetle on oats and the alfalfa blotch leaf miner on alfalfa were dealt with by the release and establishment of beneficial parasitic wasps. Though relevant, that method is not directly conducted by the farmer or gardener owing to specific quarantine laws prohibiting private individuals or agencies from introducing non-native organisms (including natural enemies) without proper authorization.

*Augmentation*. This is the periodic mass rearing and releasing of beneficial organisms in order to implement the natural enemies' complex and achieve a reduction of the pest problem. Augmentation is mostly to protect greenhouse crops. Application of the augmentation approach to field crops is limited, although major efforts were underway in South Africa to control the Eldana borer on sugarcane via augmentative releases of parasites. This method is limited at the farm because it is the most costly and least sustainable form of biological control (Willson, 1992; Wright, 1995). In Mexico, augmentation release of *Diachasmimorpha longicaudata* (Ashmead) had been reported to be associated with an approximately 2.7-fold suppression of *Anastrepha* spp. populations in backyard orchards (Montoya *et al.*, 2010).

*Conservation of natural enemies*. Conservation of natural enemies is the most important concept in the practice of biological control. It means avoiding practices that harm natural enemies, implementing practices that benefit them and modifying practices to accommodate them. For instance, susceptibility of *Harmonia axyridis* (Pallas) to various pesticides depends on the developmental stage. Adults are often less susceptible than immature stages are. As for the *Amblyseius womersleyi* Schicha, *Phytoseiulus persimilis* Athias-Henriot and *Chrysopa pallens* Rambur, herbicide application weakly influences their activity (Ahn *et al.*, 2001).

#### *Chemical control*

Chemical control is the use of synthetic products to manage pests. The most common chemicals are insecticides, herbicides and fungicides. Chemical control constitutes an important way to manage pests in spite of its environmental cost. Indeed, pesticides not only destroy the target pest but also cause damage to other non-harmful or beneficial organisms.

#### *Cultural and mechanical control*

Prior to the advent of chemical pesticides, humanity relied primarily on cultural and mechanical methods of pest management. Cultural and mechanical control include crop rotation, tillage practices, barriers, hedge rows, traps, scarecrows and other forms of environmental modifications that influence the incidence of pest problems. For instance, adopting reduced spacing in a cassava field could contribute to reducing cassava mosaic disease spread (Thresh and Cooter, 2005). Abate (1988) showed the importance of trap-cropping in the integrated management of African bollworm, *H. armigera*, in Ethiopia. Respecting plant protection principles, including crop hygiene, disease-free planting material and systematic removal of unhealthy plants from within crops, was reported to contribute to pest management (Thresh and Cooter, 2005).

#### *Legal control*

Legal methods are regulatory actions and are often employed to prevent immigration of foreign pests or to prevent the dispersal of established pests. It is, for instance, fruit and other perishable products stopped for inspection and confiscated by customs to prevent entry of pests in a country. It also includes imported animals, which are often retained in quarantine for a period of time to allow inspection for pests and diseases. Legal actions are not always effective, but they enable intercepting several pests that could add to the pest problems already present in a given country (Wright, 1995).

## 20.4 Research on plant use in pest management

### 20.4.1 Research areas, concepts and organizations involved in Africa

Research on plant use in pest management is quite old. This must have been intensified during the green revolution era, where pest pressures increased due to monoculture. The trend has been a shift from research on synthetic pesticides against most damaging pests such as the desert locust and the malaria mosquito, to intensive research on more eco-friendly methods. Since the end of the green revolution era, with its weaknesses regarding impacts of synthetic pesticides on human health, research has been intensified towards alternatives that preserve the environment and human health. The development of pest resistance to some chemical compounds, together with health and environmental impacts of pesticides, constrained entomologists and plant protection scientists and ecologists to move to the concept of IPM. The IPM research seeks to define eco-friendly approaches to pest management that ensure a sustainable production in agricultural systems. This combines investigations on predator–prey relationships between the different pests and also integrates indigenous knowledge on pest management.

Pest management research in Africa is carried out by national and international agricultural research institutions and organizations. Among the leading international agricultural research centres there is the International Institute of Tropical Agriculture (IITA), which has been working on crop pest management for more than four decades now. IITA has developed different pest management measures by focusing on food crops like banana and plantain, maize, cassava, soybean, cowpea, yam and tree crops.

### 20.4.2 Gaps in knowledge

Owing to the harmful impacts of synthetic pesticides on human health and the environment, the trend today is oriented towards IPM approaches and emphasizing their sustainability. Although methods involving use of plants to handle pest infestations showed great comparative advantages, the main gap in knowledge is that there are no 100% efficient plant-based pesticides to handle a given pest of economic importance to African agriculture. Meanwhile, there is still lack of knowledge on the environmental impacts of these natural pesticides (Cavoski *et al.*, 2011) and quality assessement of biopesticides for human health (Mihale *et al.*, 2009).

## 20.5 Biodiversity of African plants used in pest management

As already mentioned, pests constitute important threats to agriculture as they lead to decreased yields and cause damage to agricultural products. Several methods are being used to overcome pest issues and their damage, as described above. From an ecological intensification perspective, exploiting biodiversity potentials is necessary for a sustainable pest management. Secoy and Smith (1983) presented the biodiversity of plants used worldwide to control agricultural and domestic pests, including human and livestock diseases vectors. This section focuses on the diversity of African plants used in crop pest management.

### 20.5.1 Biodiversity of plants used to control insect pests

Insects which cause at least 5% damage are considered pests. Minor pests are those which provoke between 5 and 10% damage, whereas major ones are responsible for above 10% crop loss (Navarajan, 2007). Insects destroy about 35% of crops all over the world (Shani, 2000), and for centuries their management was traditionally based on plants. Whatever insect type is considered, plants and/or plant products are proven to be effective to different degrees in their control.

#### 20.5.1.1 Storage pest control

The most common insect pests infesting stored grains and other crop products are *Tribolium castaneum* (Herbst.) and *Oryzaephilus surinamensis* (L.) (Al-Jabr, 2006).

In storage, plants are mostly used as botanicals for more efficiency (Abate *et al.*, 2000). Moreover, botanicals have low mammalian toxicity, are cost effective, mostly non-phytotoxic and are easily biodegradable (Isman, 2006). In this regard, spraying of *Annona squamosa* L., *Moringa oleifera* Lam. and *Eucalyptus globulus* Labill. leaf extracts at 2.0 g per 10 g of wheat grain can be recommended to small-scale farmers to control *T. castaneum* (Anita *et al.*, 2012). *T. castaneum* was also controlled by spraying *Punica granatum* L. and *Murraya koenigii* (L.) Spreng. leaf extracts (Gandhi *et al.*, 2010). Glycoalkaloids extracted from species of the *Solanum* genus inhibit the weight increase of *T. castaneum* (Weissenberg *et al.*, 1998). In Pakistan, propagation of *T. castaneum* and grain spoilage were significantly controlled by *Ricinus communis* L. and *A. indica*. Indeed, both crushed leaves and ground leaves of *R. communis* and *A. indica* reduce *T. castaneum* propagation by up to 78% and grain spoilage to 13% in storage (Tooba *et al.*, 2005). Rhizome of *Cyperus articulatus* L., a common plant in Nigeria, has more of an antifeedant property on *T. castaneum* than the light petroleum extract does (Abubakar *et al.*, 2000). Hexane extracts from *Ceriops tagal* (Perr.) C.B.Rob. stem and bud produce a robust feeding deterrent on *T. castaneum*, whereas capsaicin derived from *Capsicum* spp. constitutes a good repellence agent to *Sitophilus zeamais* (Ho and Ma, 1995; Du *et al.*, 2011). Powders of *Lantana camara* L. and *Tephrosia vogelii* used as natural pesticides considerably reduce maize infestation by *S. zeamais* without adverse effect on germination and, therefore, constitute a good alternative to Actellic Super 2% dust (Ogendo *et al.*, 2004). Some species, such as *Angelica dahurica* (Hoffm.) Benth. & Hook.f. ex Franch. & Sav., *Cnidium officinale* Makino, *Brassica juncea* (L.) Czern., *Pterocarpus indicus* Willd., *Allium sativum* L., *Illicium verum* Hook.f., *Eugenia caryophyllata* Thunb., *Lysimachia davurica* Ledeb., *Zanthoxylum schinifolium* Siebold & Zucc., *Kaempferia galanga* L., *Foeniculum vulgare* Mill., *Agastache rugosa* (Fisch. & C.A.Mey.) Kuntze and *Aquilaria agallocha* Roxb., exhibit high antifeedant activity on *Attagenus unicolor japonicus* Reitter (Soon-Il *et al.*, 2012). The traditional African plant products used to protect stored cowpea have been widely documented by Boeke (2002). Out of 33 plants collected in Benin and Tanzania, powder from ground leaves of three of them (*Nicotiana tabacum* L., *Tephrosia vogelii* and *Securidaca longepedunculata* Fresen.) were toxic, and five others (*Clausena anisata* (Willd.) Hook.f. ex Benth., *Dracaena arborea* (Willd.) Link, *T. vogelii*, *Momordica charantia* L. and *Blumea aurita* (L.f.) DC) showed repellence properties to *Callosobruchus maculatus* (F.). *C. macullatus* is well investigated owing to its destructive action, which starts in the fields and continues through into storage. In some plant species, efficacy tests on *C. maculatus* and *S. zeamais* proved *Vitex grandifolia* Gürke to be the most effective (Epidi *et al.*, 2008). In Nigeria, hexane extract from *Lantana camara*, *Monodora myristica* (Gaertn) Dunal and *Euphorbia lateriflora* Schum. and Thonn. against bean weevil *C. maculatus* caused high mortality to *C. maculatus* and maize weevil, *S. zeamais* (Ogunsina *et al.*, 2011). A survey carried out in the UK on sensitivity of three storage pests to *Securidaca longepedunculata* root extract found repellent and toxic effects against *S. zeamais* (Jayasekara *et al.*, 2005). *Caryedon serratus* (Olivier), the main threat to stored groundnut in Congo, is widely controlled by *Trephosia vogelii* and *Chenopodium ambrosioides* L. Powders of these two plants completely protect groundnut by shortening the pest's longevity and reducing its fecundity (Delobe and Malonga, 1987).

In Kenya, powders of *A. indica*, *Lantana camara* and *Tephrosia vogelii* showed strong repellent effect towards *Prostephanus truncatus* (Horn), which is a major problem in most areas that produce maize and cassava (Chebet *et al.*, 2013). Essential oils constitute other plant products used in storage pest control. The mixture of *A. indica* seeds with grains controls storage pests on pearl millet (Gahukar, 1988). In Tanzania, *Chenopodium opulifolium* Schrad. ex W.D.J.Koch&Ziz, *Ocimum suave* Willd., *Senna siamea* (Lam.) H.S.Irwin & Barneby and *Eucalyptus* spp. are used either in simple (two species) or complex (more than two species) mixtures for pest management. Globally, all those plants showed repellence and/or toxic activities against *Prostephanus truncatus*, *Sitophilus* spp., *Tribolium* spp., *Bruchus rufimanus* (Boheman) and *Rhyzopertha dominica* (Fabricius) (Mihale *et al.*, 2009).

#### 20.5.1.2 On-farm pest control

Leaves of *Cestrum parqui* L'Hérit. are reported to indirectly reduce *Schistocerca gregaria* (Forsk.) population by their toxicity on the imago (Barbouche *et al.*, 2001). In Uganda, maintaining *Tephrosia* plants in a farm environment protects potato tubers from mole rate damage, whereas juice of the species was used to control maize stem borer in southern Tanzania and northern Zambia (Wortmann *et al.*, 1998). Soaked maize grain in a solution of macerated leaves of *A. indica* before sowing is repellent for birds and termites (Zehrer, 1986). In the Sahel, smoke from burning the fruits of *Parkia biglobosa* (Jacq.) G.Don and leaves of *Terminalia* spp. repel blister beetles (Gahukar, 1988). The important substances contained in plants giving them the insecticidal properties are saponins (Chaieb, 2010). In Tanzania, spray of aqueous extracts of *Tamarindus indica* L. fruits mixed with *Nicotiana tabacum* leaves controlled cotton pests, whereas using *Aloe* spp. is known to indigenous pest management (Mihale *et al.*, 2009).

Saponins have various structures enabling their use in several contexts (beneficial or toxic activities). On the one hand, saponins interfere with the feeding behaviour of insects. For instance, saponins extracted from *Ilex opaca* Aiton reduce food uptake by *Oligonychus ilicis* (McGregor), a mite species, and two caterpillars, *Hyphantria cunea* Drury and *Malacosoma americanum* (F.) (Kreuger and Potter, 1994). It is also the case of *Balanites roxburghii* Planch., *Agave cantala* (Haw.) Roxb. ex Salm-Dyck and *Phaseolus vulgaris* on *Spilosoma obliqua* Walker (Jain and Tripathi, 1999) and the spirostanic saponin isolated from *Solanum laxum* Spreng. on *Schizaphis graminum* (Rondani) aphid (Soule *et al.*, 2000). Larvae weight losses were reported on *Manduca sexta* (L.) treated with glycoalkaloids extracted from *Solanum* species (Weissenberg *et al.*, 1998). On the other hand, saponins exhibit entomotoxicity. In this respect, crude saponins of *C. parqui* have variable toxicity on *Tribolium confusum* (Du Val). The saponins extracted from leaves and roots of *Medicago sativa* L. are toxic for *Leptinotarsa decemlineata* (Say) larvae (Szczepanik *et al.*, 2001) and caused prolongation of the larval and pupal stages, retarded growth, increased mortality, and reduced fecundity and fertility of *S. littoralis* (Adel *et al.*, 2000). Similarly, hexane extract from *Acorus calamus* L. exhibited antifeed activity against *Peridroma saucia* (Hübner) (Koul *et al.*, 1990). In Nigeria, leaf extract of *Tephrosia vogelii* and the root extract of *Petiveria alliacea* L. are both proven to act in the same way as decis against field cowpea pests such as *Ootheca mutabilis* (Sahlb.), *Maruca testulalis* (Geyer), *Zonocerus variegatus* (L.), *Riptortus dentipes* (Fabricius) and *Apion varium* (Wagner) (Adebayo *et al.*, 2007). Also from Nigeria, reports indicate that other dangerous pests were also sensitive to botanical mixtures (Oparaeke *et al.*, 2005). Indeed, synergetic action of *Xylopia aethiopica* (Dunal) A.Rich., *Allium sativum* L., *Capsicum* sp. and *Anacardium occidentale* L. strongly reduce *Megalurothrips sjostedti*, *Maruca vitrata* (Fab.) and *Clavigralla tomentosicollis* (Stal) on cowpea plants at farms. Spraying neem oil on rice plants reduced the leaf folder incidence (Mohan *et al.*, 1991), whereas applying neem cake (de-oiled) amendment on the soil at 150 kg/ha and neem oil spraying at 10-day intervals were reported to control the infestation by *Cnaphalocrocis medinalis* Guenee (Krishnaiah and Kalode, 1990). Similarly, rice seedling root soaked in neem extract minimized damage caused by *Sogatella furcifera* (Horvath) (Saxena, 1987), and various neem extracts containing azadirachtin induced reduction of weight and increased mortality of nymphs of *Nilaparvata lugens* (Stal) (Senthil Nathan *et al.*, 2007). Neem oil respectively applied at 0.5% and 1.5% doses reduced *H. armigera* infection on chickpea and *Lipaphis erysimi* (Kalt.) damage on soybean (Siddappaji *et al.*, 1986; Mani *et al.*, 1990). The same effects on those pests were also obtained with neem kernel extracts (38.7%) and leaf extracts and to aqueous leaf extract from *Catharanthus roseus* (L.) G.Don (Atwal and Pajni, 1964; Siddappaji *et al.*, 1986). The leaf extract of *A. squamosa* and 12% leaf extract of neem were reported to develop strong antifeeding activity against *Brevicornye brassicae* L. Losses induced by hairy caterpillar were reported to have been significantly minimized by *Euphorbia royleana* Boiss. or *Lantana camara* aqueous extract spraying (Sharma *et al.*, 1982). As for potato pests, the rhizome extract of *Acorus calamus* or the leaf extract of *Ageratum conyzoides* L. were reported to show up to 42% mortality to the third instar larvae of *Gnorimoschema operculella* Zell. (Pandey *et al.*, 1982). Moreover, *Epilachna vigintioctopuncata* (Fabr.), a dangerous pest affecting potato crops, is effectively controlled by rhizome extract of *A. calamus* (Chandel *et al.*, 1987).

Appendix 20.1 summarizes the biodiversity of African plants used to control insect pests, the plant parts used, the types of insects controlled and the type of control reported (on-farm or storage).

### 20.5.2 Biodiversity of plants used to treat crop diseases

Diseases also constitute an important threat to agricultural production. Their management using plants or plant-based approaches is proven to be effective. For instance, leaves, stem and flower powder of *Prosopis juliflora* (Sw.) DC. applied as amendment were proved to suppress root rot fungi on cowpea and mung bean (Ikram and Dawar, 2013).

In western Africa, plant use to control crop disease has been well documented. For instance, application of ground ginger (*Zingiber officinale* Roscoe) showed effective biocidal activity on *Fusarium moniliforme* (Sheldon), *Rhizopus stolonifer* (Vill) and *Geotrichum candidum* (Link) associated with rot on tomato (Chuku *et al.*, 2010). Among *Tridax procumbens* L., *Vernonia amygdalina* Delile, *Chromolaena odorata* (L.) R.M.King & H.Rob. and *Azadirachta indica* tested for their antifungal activity on post-harvest rot of tomato, *A. indica* was the most effective as a bio-protective agent on tomato fruits (Ijato *et al.*, 2011). In Nigeria, Enikuomehin (2005) demonstrated that spraying 7.5% extract of *Aspilia africana* (Pers.) C.D. Adams, *C. odorata*, *Musa paradisiaca* L. and *Tithonia diversifolia* (Hemsl.) A.Gray every 2 weeks from the third week after planting was comparable to Bentex T (20% Benlate + 20% thiram) application in suppressing *Cercospora* leaf spot disease in sesame (*Sesamum indicum* L.). Adesegun *et al.* (2013) studied *Sclerotium rolfsii* Sacc. (causal agent of southern blight disease) management in Nigeria and concluded that *Aframomum melegueta* K.Schum., *Ocimum gratissimum* L. and *Cymbopogon citratus* (DC.) Stapf extracts are recommendable, with a stress on *O. gratissimum*, which acts as funguforce (a synthetic fungicide). Tomato southern blight disease had also been reported in Benin but was treated with the sap of banana pseudo stem and banana column juice (Sikirou *et al.*, 2010, 2011). In Egypt, the use of *A. sativum* and *Datura stramonium* L. plant extracts in management of early blight disease on tomato was reported (Nashwa and Abo-Elyousr, 2012). Indeed, *A. sativum* and *D. stramonium* leaf extracts not only induced the highest reduction in mycelial growth of *Alternaria solani* Sorauer and the reduction of the induced disease, but also increased fruits yield. Amadioha (2003) reported that the mixture of alcohol and water with leaf extracts of *Piper nigrum* L. and *Ocimum sanctum* L. have considerable control on cowpea anthracnose disease (caused by *Colletotrichum lindemuthianum* (Sacc. & Magnus) Briosi & Cavara) and were potential substitutes for synthetic chemicals. *Colletotrichum destructivum* N.L. Horn, the agent of that disease, was also reported to be more sensitive to aqueous extracts of *Carica papaya* L. roots and seeds and of neem bark as concentration increased (Enyiukwu and Awurum, 2011). Cowpea root rot disease caused by *Pythium aphanidermatum* (Edson) Fitzp. was proved to be controlled by *Z. officinale* and *Aloe vera* (L.) Burm.f. plant extracts (Suleiman and Emua, 2009). *Myzus persicae* (Sulzer), very well known as a vector of potato virus Y disease, is well controlled by using pyrethrum and dried flower powder of *Chrysanthemum cinerariifolium* (Trevir.) Sch.Bip. leaf extract (Jacobson, 1975). *Cercospora* spot disease on avocado fruits is controlled with ethanol extracts of *O. gratissimum*, *Acalypha wilkesiana* Mull.Arg. and *Acalypha macrostachya* Jacq., which totally inhibited the causal agent *Cercospora purpurea* Cke. (Ogbo and Oyibo, 2008). Neem is also reported to be potentially effective in rice blast disease control in Nigeria. Neem leaf extracts (aqueous) and seed oil reduced radial growth of *Pyricularia oryzae* Cavara, and resulted in high reduction of blast development on rice (Amadioha, 2000). In this report, neem has almost the same action as carbendazim at 0.1%. Carré *et al.* (2006) reported that soaking banana fruits in *Artemisia camphorata* Vill. extract reduces anthracnose severity up to 67%. *Eucalyptus citriodora* Hook. is also useful in crop diseases control. The aqueous extract was reported to induce local resistance in cucumber plants against *Colletotrichum lagenarium* (Pass.) Ell. & Halst., responsible for anthracnose (Bonaldo *et al.*, 2007), whereas systemic resistance was induced by the aqueous leaf extract of *O. gratissimum* (Colpas *et al.*, 2009). Under *in vitro* assays, *E. citriodora* also inhibited mycelia growth of *Phytophthora* sp., *Sclerotium rolfsii* Sacc., *Rhizoctonia solani* Kuhn and *Alternaria alternata* (Fr.) Keissl. Mixing the extracts of *Bidens pilosa* L. and *Lippia alba* (Mill.) N.E.Br. ex Britton & P. Wilson had shown effective *in vitro* fungi-toxic activity against *Alternaria alternata* (Moreira *et al.*, 2008). Tohamy *et al.* (2002) reported in Egypt that *A. indica* (neem) and *Allium sativum* (garlic) and *Cinnamomum camphora* (L.) J.Presi (camphor) drastically reduced mildew disease, with an emphasis on the highest effect of garlic. In Germany, it has been demonstrated that the volatile phytoanticipin allicin contained in garlic (*Allium sativum*) extracts has applications in plant pathogen management. This secondary metabolite of *A. sativum* has the ability to control *Magnaporthe grisea* (T.T. Hebert) M.E. Barr infection in rice, and then reduces rice blast disease development, downy mildew of *Arabidopsis thaliana* (L.) Heynh. caused by *Hyaloperonospora parasitica* (Pers.) Constant., and *Phytophthora infestans* (Mont.) de Bary infesting potato tubers (Curtis *et al.*, 2004). Also, Kagale *et al.* (2004) demonstrated that a foliar application of *Datura metel* L. leaf extract significantly reduces the incidence of sheath blight and bacterial blight diseases on rice. *A. indica* and *Eucalyptus tereticornis* Sm. leaf extracts contribute to reduce pre- and post-emergence mortality and number of onion seedlings presenting symptoms of black mould disease (Gupta *et al.*, 2012). Van Tol *et al.* (2007) showed in a push–pull system that *Origanum majorana* L. has a strong repellence against the onion thrips, *Thrips tabaci* (Lindeman). *Xanthomonas campestris* (Pammel) Dowson, responsible of bacterial blight, common blight, fiscous blight and canker on cotton, tomato and rice paddy, was controlled by fresh leaf extracts of *Acacia arabica* (Lam.) Willd., *Prosopis juliflora* (Sw.) DC., *Achras zapota* L., *Enterolobium saman* (Jacq.) Prain, *Lawsonia inermis* L., *Oxalis corniculata* L., *Punica granatum* L. and *Viscum orientale* Willd. (Satish *et al.*, 1999).

Appendix 20.2 summarizes the biodiversity of African plants used to control plant diseases, the plant parts used, the types of diseases or pathogens controlled and the type of control reported (on-farm or storage).

### 20.5.3 Biodiversity of plants used to control farm weeds

Weeds represent a serious challenge to agricultural production and alone can reduce crop yield to about 34% (Oerke, 2006). The use of eco-friendly technology to manage weeds has become a necessity in view of the health and environmental hazards created by synthetic herbicides. In this respect, there is interest in botanicals and allelopathy phenomena occurring in plants for sustainable weed management (An *et al.*, 1998; Koul and Walia., 2009). In Japan, the potential of *Passiflora edulis* Sims to reduce some paddy rice weeds was reported (Khanh *et al.*, 2006). Indeed, *P. edulis* contained some substances, such as coumarin, lactones and fatty acids, which induce germination and growth suppression of *Echinochloa crus-galli* (L.) P.Beauv. and *Monochoria vaginalis* (Burm.f.) C.Presel ex Kunth. Alfalfa (*Medicago sativa* L.) pellets inhibited emergence of several weeds, such as abunome (*Doparium junceum* Hamilt.), false pimpernel (*Lindernia pyxidaria* L.), stemmed water wort (*Elatine triandra* Schk. var. *pedicellata* Krylov), *Eleocharis acicularis* (L.) Roem. & Schult and *Rotala indica* Koehne when applied at 1–3 t/ha and significantly reduced the number of *Echinochloa oryzicola* Vasinger plants (Xuan *et al.*, 2001). In India, *Tagetes minuta* L. leaf extract applied on rice field soil has been reported to exhibit herbicidal activity against *Echinochloa crus-galli* (L.) P.Beauv. and *Cyperus rotundus* L. and also to increase rice yields owing to its partial pesticidal properties against fungi, nematodes and insects (Batish *et al.*, 2007). The control ability of aqueous leaf extracts of *A. indica*, *Ficus benghalensis* L., *Melia azedarach* L., *Mangifera indica* L. and *Syzygium cumini* (L.) Skeels towards *Parthenium hysterophorus* L. (one of the world's worst weeds) was stressed by Shafique *et al.* (2005). The authors reported that, of all tested trees, *F. benghalensis* and *M. indica* extracts were the best inhibitors of *P. hysterophorus* seed germination, whereas *M. azedarach*, *S. cumini* and *F. benghalensis* had been found most effective in suppressing the radicle and plumule growth of *P. hysterophorus*. The ability of plant species to control this dangerous weed was confirmed by Javaid and Anjum (2006), who showed the herbicidal activity of *Dichanthium annulatum* Stapf, *Cenchrus pennisetiformis* Hochest and *Sorghum halepense* Pers roots and shoots aqueous extracts against *P. hysterophorus*. The most effective activity was obtained with *C. pennisetiformis* shoot extracts. Similarly, aqueous extracts of dried leaves of *Alstonia scholaris* (L.) R. Br., *A. indica* and *E. citriodora* Hook were proven to reduce by up to 43% the final germination of *Phalaris minor* Retz, an important weed of wheat fields (Javaid *et al.*, 2006). In Pakistan, herbicide use was reduced by 75% through an integration of sorghum (*Sorghum bicolor* L. Moench) plus sunflower (*Helianthus annuus* L.) extracts without adverse effects on wheat yield, in the management of the wild oat (*Avena fatua* L.) and canary grass (*P. minor*) (Mushtaq *et al.*, 2010). In Cambodia, a suitable use of phytotoxic rice crop residues was reported in barnyard grass (*Echinochloa crus-galli*), small umbrella sedge (*Cyperus difformis* L.) and water primrose (*Ludwigia octovalvis* (Jacq.) P.H.Raven) as a management strategy (Pheng *et al.*, 2010). Using root and stem aqueous extract of *Cirsium arvense* (L.) Scop. and *Ageratum conyzoides* constituted an ecological way to control *P. minor*, *Poa annua* L. and *Medicago polymorpha* L. in wheat (Akhtar *et al.*, 2001).

Appendix 20.3 summarizes the biodiversity of African plants used to control farm weeds, the plant parts used and the weed species controlled.

## 20.6 Benefits of the use of plants in crop pest management

### 20.6.1 Economic rewards of plant use in crop pest management

Africa is known to be the smallest user of chemical pesticides, and the main reason for this is that farming in the continent is characterized by smallholdings. Thus, the smallholders are financially limited in accessing those synthetic pesticides. The use of natural approaches developed by indigenous people, and especially botanicals in crop pest management, reduces their bills for pest management, and this is cost effective (Cavoski *et al.*, 2011). In the same way, this reduces expenditure for treating intoxications that often arise from the careless use of highly toxic pesticides. Moreover, the growth of the demand for organic food represents an opportunity and creates new markets for producers already experienced in the use of botanicals for pest control.

### 20.6.2 Health and environmental benefits

Among the severe criticisms of the green revolution are the health and environmental impacts of the heavy use of pesticide bulks. The use of plant-based products to manage pests reduces those health and environmental risks. The use of botanical pesticides in agricultural production offers an incomparable food quality by suppressing, if not significantly reducing, pesticide residues and risks for intoxication. Thus, much attention has been given to those methods as biopesticides, which are in most cases non-toxic to plants and are biodegradable (Cavoski *et al.*, 2011). These plant products generally have low toxicity to mammals. The overall agricultural biodiversity is also enhanced, by minimizing side effects on non-target useful insects and other micro-fauna, which play important functions in the agro ecosystems (e.g. pollinating, regulating).

### 20.6.3 Risks mitigation and food security

Use of eco-friendly pest management approaches and methods such botanicals and plants antagonisms offers incomparable advantages compared with synthetic pesticides. Botanicals and other plant-based methods are more resilient to environmental changes, as African small farmers have maintained agricultural ecosystems and related biodiversity through their practices. Thus, they are more sustainable and promote biodiversity conservation and enhancement. All these are concurrent contributors for better food security and poverty alleviation, as they are affordable for poor farmers.

## 20.7 Limits of the study

A major limit to this study is that it only covers crop pest management and does not treat human and animal disease vectors/agents. Another limit concerns the definition of what should be considered as 'African plant'. Several species have a wide distribution, with Africa as centre of origin or not, and occur also outside the African continent (e.g. Asia, Americas, Europe or Australia). Here, we have reported the use of plant species occurring in Africa and used in the African farming systems or that have been tested and confirmed by agricultural research centres, irrespective of their centres of origin. In addition, some species found in Africa but with no record of use in African pest management are employed in other continents to control crop pests. Those species are also reported here, with mention of places where their properties were studied.

## 20.8 Conclusion

Natural methods for pest control and organic farming are the least toxic possibilities. This chapter has presented the diversity of African plants and metabolites used in crop pest management. It has also showed that traditional pest management is an integral part of cropping systems, based on botanicals and natural processes. Biopesticides need to be promoted in order to develop sustainable methods for plant protection and to provide healthy and secure foods for the benefit of farmers, consumers and the environment. However, there is still need to evaluate their environmental sustainability.

## References

Abate, T. (1988) Experiments with trap crops against African bollworm, *Heliothis armigera*, in Ethiopia. *Entomologia Experimentalis et Applicata*, **48**, 135–140.

Abate, T. (1991) Intercropping and weeding: effects on some natural enemies of African bollworm, *Heliothis armigera* (Hbn.) (Lep., Noctuidae), in bean fields. *Journal of Applied Entomology*, **112**, 38–42.

Abate, T. and Ampofo, J.K.O. (1996) Insect pests of beans in Africa: their ecology and management. *Annual Review of Entomology*, **41**, 45–73.

Abate, T., Van Huis, A. and Ampofo, J.K.O. (2000) Pest management strategies in traditional agriculture: an African perspective. *Annual Review of Entomology*, **45**, 631–659.

Abubakar, M.S., Abdurahman, E.M. and Haruna, A.K. (2000) The repellant and antifeedant properties of *Cyperus articulatus* against *Tribolium casteneum* Hbst. *Phytotherapy Research*, **14**, 281–283.

Adebayo, O.T., Olaniran, O.A. and Akambi, W.I. (2007) Control of insects pest of cowpea in the fields with allelochems from *Tephrosia vogelli* and *Petiverai alliacea* in southern guinea savannah of Nigeria. *Agricultural Journal*, **2** (3), 365–369.

Adel, M.M., Sehnal, F. and Jurzysta, M. (2000) Effect of alfalfa saponins on the mouth *Spodoptera littoralis*. *Journal of Chemical Ecology*, **26**, 1065–1078.

Adesegun, E.A., Akintokun, A.K., Ajayi, E.O. *et al.* (2013) *Sclerotium rolfsii* management in tomato using *Aframomum melegueta*, *Ocimum gratissimum* and *Cymbopogon citratus*. *Archives of Phytopathology and Plant Protection*, **46** (6), 732–740.

Adipala, E., Kanyama-Phiri, G., Karanja, N. and Norman, D. (1998) Evaluation of the Forum Programme Supported by the Rockefeller Foundation, The Rockefeller Foundation, Malawi Office.

Ahn, Y., Kim Y. and Yoo, J. (2001) Toxicity of herbicide glufosinate-ammonium to predatory insects and mites of *Tetranychus urticae* (Acari: Tetranychidae) under laboratory conditions. *Journal of Economic Entomology*, **94**, 157–161.

Akhtar, N., Javaid, A. and Bajwa, R. (2001) Herbicidal activity of aqueous extracts of *Cirsium arvense* and *Ageratum conyzoides* against weeds of wheat. *Pakistan Journal of Biological Sciences*, **4** (11), 1364–1367.

Al-Jabr, A.M. (2006) Toxicity and repellency of seven plant essential oils to *Oryzaephilus surinamensis* (Coleoptera: Silvanidae) and *Tribolium castaneum* (Coleoptera: Tenebrioidae). *Scientific Journal of King Faisal University (Basic and Applied Sciences)*, **7** (1), 49–59.

Amadioha, A.C. (2000) Controlling rice blast in vitro and in vivo with extracts of *Azadirachta indica*. Crop Protection, **19**, 287–290.

Amadioha, A.C. (2003) Evaluation of some plant leaf extracts against *Colletotrichum lindemuthianum* in cowpea. *Acta Phytopathologica et Entomologica Hungarica*, **38** (3–4), 259–265.

Ambe, J.J. (1993) The effect of planting dates on three cassava diseases in Cameroon. *International Journal of Pest Management*, **39**, 309–311.

Ampofo, J.K.O. and Massomo, S.M. (1998) Some cultural strategies for management of bean stem maggots (Diptera: Agromyzidae) on beans in Tanzania. *African Crop Science Journal*, **6** (4), 351–356.

Ampong-Nyarko, K., Reddy, K.V.S., Nyang'or, R.A. and Saxena, K.N. (1994) Reduction of insect pest attack on sorghum and cowpea by intercropping. *Entomologia Experimentalis et Applicata*, **70**, 179–184.

An, M., Pratley, J.E and Haig, T. (1998) Allelopathy: from concept to reality, in *Agronomy, Growing a Greener Future? Proceedings of the 9th Australian Agronomy Conference* (eds D.L. Michalk and J.E. Pratley), http://www.regional.org.au/au/asa/1998/6/314an.htm#TopOfPage (accessed 16 November 2013).

Anita, S., Sujatha; P. and Prabhudas, P. (2012) Efficacy of pulverised leaves of *Annona squamosa* (L.), *Moringa oleifera* (Lam.) and *Eucalyptus globulus* (Labill.) against the stored grain pest, *Tribolium castaneum* (Herbst.) *Recent Research in Science and Technology*, **4** (2), 19–23.

Atwal, A.S. and Pajni, H.R. (1964) Preliminary studies on the insecticidal properties of drupes of *Melia azadirachta* against caterpillars of *Pieris brassicae* Linn. *Indian Journal of Entomology*, **26**, 221–227.

Barbouche, N., Hajjem, B., Lognay, G. and Ammar, M. (2001) Contribution à l'étude de l'activité biologique d'extraits de feuilles de *Cestrum parqui* L'Hérit. (Solanaceae) sur le criquet pèlerin *Schistocerca gregaria* (Forsk.). *Biotechnologie, Agronomie, Société et Environnement/Biotechnology, Agronomy, Society and Environment*, **5** (2), 85–90.

Batish, D.R., Arora, K., Singh; H.P. and Kohli, R.K.(2007) Potential utilization of dried powder of *Tagetes minuta* as a natural herbicide for managing rice weeds. *Crop Protection*, **26**, 566–571.

Boeke, S.J. 2002. Traditional African plant products to protect stored cowpeas against insect damage: the battle against the beetle. Thesis, Wageningen University.

Bonaldo, S.M., Schwan-Estrada, K.R.F., Stangarlin, J.R. *et al.* (2007) Contribution for the study of antifungal and phytoalexins elicitors in sorghum and soybean activities by eucalyptus (*Eucalyptus citriodora*). *Summa Phytopathologica*, **33**, 383–387.

Campbell, J.F. and Arthur F.H. (2007) Ecological implications for post harvest integrated pest management of grain and grain-based products, in *Ecologically Based Integrated Pest Management* (eds O. Koul and G.W. Cuperus), CAB International, pp. 406–431.

Carré, V., Stangarlin, J.R., Becker, A. *et al.* (2006) Postharvest control of *Colletotrichum musae* in banana (*Musa* sp.) by camphor (*Artemisia camphorata*) and chitosan. *Scientia Agraria Paranaensis*, **5**, 57–66.

Cavoski, I., Caboni, P. and Miano, T. (2011) Natural pesticides and future perspectives, in *Pesticides in the Modern World – Pesticides Use and Management* (ed. M. Stoytcheva), InTech, Rijeka, pp. 169–190.

Chaieb, I. (2010) Saponins as insecticides: a review. *Tunisian Journal of Plant Protection*, **5**, 39–50.

Chandel, B.S., Pandey, U.K. and Kumar, A. (1987) Insecticidal evaluation of some plant extracts against *Epilachna vigintioctopuncata* Fabr. (Coleoptera: Coccinellidae). *Indian Journal of Entomology*, **49** (2), 294–296.

Chebet, F., Deng, A.L., Ogendo, J.O. *et al.* (2013) Bioactivity of selected plant powders against *Prostephanus truncatus* (Coleoptera: Bostrichidae) in stored maize grains. *Plant Protection Science*, **49**, 34–43.

Chuku, E.C, Osakwe, J.A and Daddy-West, C. (2010) Fungal spoilage of tomato (*Lycopersicon esculentum* mill), using garlic and ginger. *Scientia Africana*, **9** (2), 42–50.

CIRAD (n.d.) *Une calamité qui coûte cher*, http://locust.cirad.fr/generalites/pdf/locust_6.pdf (accessed 16 November 2013).

Colpas, F.T., Schwan-Estrada, K.R.F., Stangarlin, J.R. *et al.* (2009) Induction of plant defense responses by *Ocimum gratissimum* L. (Lamiaceae) leaf extracts. *Summa Phytopathologica*, **35**, 191–195.

Colwell, R.K. (2012) Biodiversity: concepts, patterns, and measurement, in *The Princeton Guide to Ecology* (ed. S.A. Levin), Princeton University Press, Princeton, NJ, pp. 257–263, press.princeton .edu/chapters/s3_8879.pdf (accessed 18 March 2013).

Curtis, H., Noll, U., Störmann, J. and Slusarenko, A.J. (2004) Broad-spectrum activity of the volatile phytoanticipin allicin in extracts of garlic (*Allium sativum* L.) against plant pathogenic bacteria, fungi and oomycetes. *Physiological and Molecular Plant Pathology*, **65**, 79–89.

Delobe, A. and Malonga, P. (1987) Insecticidal properties of six plant materials against *Caryedon serratus* (ol.) (Coleoptera: Bruchidae). *Journal of Stored Product Research*, **23** (3), 173–176.

Demirozer, O., Tyler-Julian, K., Funderburk, J. *et al.* (2012) *Frankliniella occidentalis* (Pergande) integrated pest management programs for fruiting vegetables in Florida. *Pest Management Science*, **68**, 1537–1545.

Dennill, G.B., Donnelly, D., Stewart, K. and Impson, F.A.C. (1999) Insect agents used for the biological control of Australian *Acacia* species and *Paraserianthes lophanta* (Willd.) Nielsen (Fabacea) in South Africa, in *Biological Control of Weeds in South Africa (1990–1998)* (eds T. Olckers and M.P. Hill), African Entomology, Memoir No. 1, Entomological Society of Southern Africa, Pretoria, pp. 45–54.

Dissemond, A. and Hindorf, H. (1990) Influence of sorghum/maize /cowpea intercropping on the insect situation at Mbita/Kenya. *Journal of Applied Entomology*, **109**, 144–150.

Du, S.S., Wang, C.F., Li, J. *et al.* (2011) Antifeedant diterpenoids against *Tribolium castaneum* from the stems and twigs of *Ceriops tagal* (Rhizophoraceae). *Molecules*, **16** (7), 6060–6067.

Enikuomehin, O.A. (2005) *Cercospora* leaf spot disease management in sesame (*Sesamum indicum* L.) with plant extracts. *Journal of Tropical Agriculture*, **43** (1–2), 19–23.

Enyiukwu, D.N. and Awurum, A.N. (2011) Effects of phytochemicals from *Carica papaya* roots and seeds and *Piper guineense* seeds on germination of spores of *Colletotrichum destructivum*. *Continental Journal of Biological Sciences*, **4** (2), 55–59.

Epidi, T.T., Nwani, C.D and Udo, H.S. (2008) Efficacy of some plant species for the control of cowpea weevil (*Callosobruchus maculatus*) and maize weevil (*Sitophilus zeamais*). *International Journal of Agriculture and Biology*, **10**, 588–590.

Fargette, D., Jeger, M.J., Fauquet, C. and Fishpool, L.D.C. (1994) Analysis of temporal disease progress of African cassava mosaic virus. *Phytopathology*, **84**, 91–98.

Faure, J.C. (1944) Pentatomid bugs as human food. *Journal of the Entomological Society of South Africa*, **7**, 111–112.

Gahukar, R.T. (1988) Problems and perspectives of pest management in the Sahel: a case study of pearl millet. *Tropical Pest Management*, **34** (1), 35–38.

Gandhi, N. and Pillai, S. (2011) Control of *Rhyzopertha dominica* (Coleoptera: Bostrichidae) by pulverized leaves of *Punica granatum* (Lythraceae) and *Murraya koenigii* (Rutaceae). *International Journal of Agriculture and Biology*, **13** (4), 535–540.

Gandhi, N., Pillai, S. and Patel, P. (2010) Efficacy of pulverized *Punica granatum* (Lythraceae) and *Murraya koenigii* (Rutaceae) leaves against stored grain pest *Tribolium castaneum* (Coleoptera: Tenebrionidae). *International Journal of Agriculture and Biology*, **12**, 616–620.

Gebre-Amlak, A., Sigvald, R. and Pettersson, J. (1989) The relationship between sowing date, infestation and damage by the maize stalk borer, *Busseola fusca* (Noctuidae), on maize in Awassa, Ethiopia. *Tropical Pest Management*, **35**, 143–145.

Gupta, R., Khokhar, M.K. and Lal, R. (2012) Management of the black mould disease of onion. *Journal of Plant Pathology & Microbiology*, **3** (5), 1000133.

Gurjar, M.S., Ali, S., Akhtar, M. and Singh, K.S. (2012) Efficacy of plant extracts in plant disease management. *Agricultural Sciences*, **3** (3), 425–433.

Harish, S., Saravanakumar; D., Radjacommare, R. and Seetharaman, K. (2008) Use of plant extracts and biocontrol agents for the management of brown spot disease in rice. *BioControl*, **53**, 555–567.

Hecht, M.M. (2004) In Africa, DDT makes a comeback to save lives. *Executive Intelligence Review Science and Technology*, **31** (24), 66–71, http://www.larouchepub.com/other/2004/sci_techs/3124 ddt_africa.html (accessed 18 March 2013).

Herren, H.R. and Neuenschwander, P. (1991) Biological control of cassava pests in Africa. *Annual Review of Entomology*, **6**, 257–283.

Hillocks, R.J., Logan, J.W.M., Riches, C.R. *et al.* (1996) Soil pests in traditional farming systems in sub-Saharan Africa – a review. Part II. Management strategies. *International Journal of Pest Management*, **42** (4), 253–265.

Ho, S.H. and Ma, Y. (1995) Repellency of some plant extracts to the stored products beetles, *Tribolium castaneum* (Herbst) and *Sitophilus zeamais* Motsch, in *Proceedings of the Symposium on Pest Management for Stored Food and Feed*, Semeo Biotrop, Bogor, Indonesia, 5–7 September.

Ijato, J.Y., Adebiyi, A.O. and Ijadunola J.A. (2011) Antifungal effects of four tropical plant aqueous and ethanol extracts on post-harvest rot of tomato (*Lycopersicum esculentum*) in Ado–Ekiti, Nigeria. *New York Science Journal*, **4** (1), 64–68.

Ikram, N. and Dawar, S. (2013) Effect of *Prosopis juliflora* (Sw.) DC. in the control of root rot fungi of cowpea (*Vigna unguiculata* L.) and mung bean [*Vigna radiata* (L.) Wilczek]. *Pakistan Journal of Botany*, **45** (2), 649–654.

Isman, B. (2006) Botanical insecticides, deterrents, and repellents in modern agriculture and an increasingly regulated world. *Annual Review of Entomology*, **51**, 45–66.

Jacobson, M. (1975) *Insecticide from Plants. A Review of the Literature, 1954–1971*, Agricultural Handbook No. 461, United States Department of Agriculture, Washington, DC.

Jain, D.C. and Tripathi, A.K. (1999) Insect feeding-deterrent activity of some saponin glycosides. *Phytotherapy Research*, **5** (3), 139–141.

Javaid, A. and Anjum, T. (2006) Control of *Parthenium hysterophorus* L., by aqueous extracts of allelopathic grasses. *Pakistan Journal of Botany*, **38** (1), 139–145.

Javaid, A., Shafique, S. and Shafique, S. (2006) Herbicidal potential of aqueous leaf extract of allelopathic trees against *Phalarisminor*. *Pakistan Journal of Weed Science Research*, **12** (4), 339–346.

Jayasekara, T.K., Stevenson, P.C., Hall, D.R. and Belmain, S.R. (2005) Effect of volatile constituents from *Securidaca longepedunculata* on insect pests of stored grain. *Journal of Chemical Ecology*, **31** (2), 303–313.

Kagale, S., Marimuthu, T., Thayumanavan, B. *et al.* (2004) Antimicrobial activity and induction of systemic resistance in rice by leaf extract of *Datura metel* against *Rhizoctonia solani* and *Xanthomonas oryzae* pv. *oryzae*. *Physiological and Molecular Plant Pathology*, **65**, 91–100.

Khanh,T.D., Chung, I.M. Tawata, S. and Xuan, T.D. (2006) Weed suppression by *Passiflora edulis* and its potential allelochemicals. *Weed Research*, **46**, 296–303.

Koona, P., Malaa, D. and Koona, O.E.S. (2007) Hexane extracts from *Tephrosia vogelii* Hook. protect stored maize against the weevil *Sitophilus zeamais* Motschulsky (Coleoptera: Curculionidae). *Entomological Science*, **10**, 107–111.

Koul, O. and Walia, S. (2009) Comparing impacts of plant extracts and pure allelochemicals and implications for pest control. *CAB Reviews: Perspectives in Agriculture, Veterinary Science, Nutrition and Natural Resources*, **4**, (049), 1–30.

Koul, O., Smirle, M.J. and Isman, M.B. (1990) Asarones from *Acorus calamus* L. oil: their effect on feeding behavior and dietary utilization in *Peridroma saucia*. *Journal of Chemical Ecology*, **16** (6), 1911–1920.

Kreuger, B. and Potter, D.A. (1994) Changes in saponin and tannins in ripening holly fruits and effects of fruit consumption on non adapted insect herbivore. *American Midland Naturalist*, **132**, 183–191.

Krishnaiah, N.V. and Kalode, M.B. (1990) Efficacy of selected botanicals against rice insect pests under green house and field conditions. *Indian Journal of Plant Protection*, **18** (2), 197–205.

Kyamanywa, S. and Ampofo, J.K.O. (1988) Effect of cowpea/maize mixed cropping on the incident light at the cowpea canopy and flower thrips (Thysanoptera: Thripidae) population density. *Crop Protection*, **7** (3), 186–189.

Lee, B.H., Annis, P.C., Tumaalii, F. and Choi, W.S. (2004) Fumigant toxicity of essential oils from the Myrtaceae family and 1,8-cineole against 3 major stored-grain insects. *Journal of Stored Products Research*, **40** (5), 553–564.

Mani, A., Kumudanathan, K. and Jagadish, C.A. (1990) Relative efficacy of neem oil and endosulfan against insect pests of mustard. *Neem News Letter*, **7** (2), 129–131.

Mihale, M.J., Deng A.L., Selemani, H.O. *et al.* (2009) Use of indigenous knowledge in the management of field and storage pests around Lake Victoria basin in Tanzania. *African Journal of Environmental Science and Technology*, **3** (9), 251–259.

Mohan, K., Gopalan, M. and Balasubramanian, G. (1991) Studies on the effects of neem products and monocrotophos against major pests of rice and their safety to natural enemies. *Indian Journal of Plant Protection*, **19**, 23–30.

Montoya, P., Liedo, P., Benrey, B. *et al.* (2010) Biological control of *Anastrepha* spp. (Diptera: Tephritidae) in mango orchards through augmentative releases of *Diachasmimorpha longicaudata* (Ashmead) (Hymenoptera: Braconidae). *Biological Control*, **18** (3), 216–224.

Moreira C.G.A., Schwan-Estrada K.R.F., Bonaldo S.M. *et al.* (2008) Partial characterization of fractions from extracts of *Cymbopogon nardus* with elicitor activity of phytoalexins in sorghum and soy and effect on *Colletotrichum lagenarium*. *Summa Phytopathologica*, **34**, 332–337.

Mushtaq, M.N., Cheema, Z.A., Khaliq, A. and Naveed, M.R. (2010) A 75% reduction in herbicide use through integration with sorghum + sunflower extracts for weed management in wheat. *Journal of the Science of Food and Agriculture*, **90**, 1897–1904.

Myers, J.H., Simberloff, D., Kuris, A.M. and Carey, J.R. (2000) Eradication revisited: dealing with exotic species. *Trends in Ecology & Evolution*, **15** (8), 316–320.

Nashwa, S.M.A. and Abo-Elyousr, K.A.M. (2012) Evaluation of various plant extracts against the early blight disease of tomato plants under greenhouse and field conditions. *Plant Protection Science*, **48** (2), 74–79.

Navarajan, A.V. (2007) *Agriculture: entomology. Insect pests and their management*, Division of Entomology, Indian Agricultural Research Institute, New Delhi.

Oerke, E.-C. (2006) Crop losses to pests. *Journal of Agricultural Science*, **144** (01), 31–43.

Ogbo, E.M. and Oyibo, A.E. (2008) Effects of three plant extracts (*Ocimum gratissimum*, *Acalypha wilkesiana* and *Acalypha macrostachya*) on post harvest pathogen of *Persea americana*. *Jounal of Medicinal Plants Research*, **2** (11), 311–314.

Ogendo, J.O., Deng, A.L., Belmain, S.R. *et al.* (2004) Effect of insecticidal plant materials, *Lantana camara* L. and *Tephrosia vogelii* Hook, on the quality parameters of stored maize grains. *Journal of Food Technology in Africa*, **9** (1), 29–36.

Ogunsina, O.O., Oladimeji, M.O. and Lajide, L. (2011) Insecticidal action of hexane extracts of three plants against bean weevil, *Callosobruchus maculatus* (F.) and maize weevil, *Sitophilus zeamais* Motsch. *Journal of Ecology and the Natural Environment*, **3** (1), 23–28.

Oparaeke, A.M., Dike, M.C. and Amatobi C.I. (2005) Evaluation of botanical mixtures for insect pests management on cowpea plants. *Journal of Agriculture and Rural Development in the Tropics and Subtropics*, **106** (1), 41–48.

OTA (1979) *Pest Management Strategies in Crop Protection*, vol. 1, Office of Technology Assessment, OTA-F-98, US Government Printing Office, Washington, DC.

Pandey, U.K., Srivastava, A.K., Chandel, B.S. and Lekha, C. (1982) Response of some plant origin insecticides against potato tuber moth, *Gnorimoschema operculella* Zell. affecting solanaceous plants. *Zeitschrift für Angewandte Zoologie*, **69** (3), 267–270.

Pheng, S., Olofsdotter, M., Jahn, G. and Adkins, S. (2010) Use of phytotoxic rice crop residues for weed management. *Weed Biology and Management*, **10**, 176–184.

Pressinger, R. and Sinclair, W. (n.d.) Researching effects of chemicals and pesticides upon health, http://www.chem-tox.com/ (accessed 19 March 2013).

Quinn, L.P., De Vos, B.J., Fernandes-Whaley, M. *et al.* (2011). Pesticide use in South Africa: One of the largest importers of pesticides in Africa, in *Pesticides in the Modern World - Pesticides Use and Management* (ed. M. Stoytcheva), InTech, Rijeka, pp. 49–96.

Reddy, K.R.N., Reddy, C.S. and Muralidharan, K. (2009) Potential of botanicals and biocontrol agents on growth and aflatoxin production by *Aspergillus flavus* infecting rice grains. *Food Control*, **20** (2), 173–178.

Sathyaseelan, V. and Bhaskaran, V. (2010) Efficacy of some native botanical extracts on the repellencyproperty against the pink mealy bug, *Maconellicoccushirsutus* (Green) in mulberry crop. *Recent Research in Science and Technology*, **2** (10), 35–38.

Satish, S., Raveesha, K.A. and Janardhana, G.R. (1999) Antibacterial activity of plant extracts on phytopathogenic *Xanthosomas campestris* pathovars. *Letters in Applied Microbiology*, **28**, 145–147.

Saxena, R.C. (1987) Antifeedants in tropical pest management. *International Journal of Tropical Insect Science*, **8** (4–6), 731–736.

Secoy, D.M. and Smith, A.E. (1983) Use of plants in control of agricultural and domestic pests. *Economic Botany*, **37** (1), 28–57.

Senthil Nathan, S., Choi, M.Y., Paik, C.H. *et al.* (2007) The toxic effects of neem extract and azadiractin on the brown planthopper, *Nilaparvatha lugens* (Stal). *Chemosphere*, **67**, 80–88.

Shafique, S., Bajwa, R., Javaid, A. and Shafique, S. (2005) Biological control of parthenium IV: suppressive ability of aqueous leaf extracts of some allelopathic trees against germination and early seedling growth of *Arthenium hysterophorus* L. *Pakistan Journal of Weed Science Research*, **11** (1–2), 75–79.

Shani, A. (2000) Chemical communication agents (pheromones) in integrated pest management. *Drug Development Research*, **50** (3–4), 400–405.

Sharma, I.N.S., Singh, A.K and Singh, S.P. (1982) Allelopathic potential of some plant substances as anti-feedants against insect pests of jute, in *Proceedings of First National Symposium on Allelopathy in Agro-Ecosystems (Agriculture and Forestry)* (eds P. Tauro and S.S. Narwal), Indian Society of Allelopathy/ICAR, Hisar, India, pp. 157–176.

Siddappaji, C., Kumar, A.R.V. and Gangadhar, R. (1986) Evaluation of different insecticidal sprays against the chickpea borer, *Heliothis armigera* Hubn. *Pesticides*, **20**, 13–16.

Sikirou, R., Zannou, E., Gbèhounou, G. *et al.* (2010) Fungicide effect of banana column juice on tomato southern blight caused by *Sclerotium rolfsii*: technical and economic efficiency. *African Journal of Agricultural Research*, **5** (23), 3230–3238.

Sikirou, R., Assogba Komlan, F., Tosso F. and Agassounon T. (2011) Efficacité de la sève du pseudo-tronc du bananier contre le champignon *Sclerotium rolfsii*, un agent causal du flétrissement de la tomate. *Bulletin de la Recherche Agronomique du Bénin*, **70**, 47–59.

Singh, K. and Sharma, P.L. (1986) Studies on anti-feedant and repellant qualities of neem against aphid, *Brevicornye brassicae* Linn. on cauliflower and cabbage. *Research and Development Report*, **3** (1), 33–35.

Singleton, G.R., Hinds, L.A., Leirs, H. and Zhang, Z. (eds) (1999) *Ecologically-based Management of Rodent Pests*, ACIAR Monograph No. 59, Centre for International Agricultural Research, Canberra.

Smit, N.E.J.M. and Matengo, L.O. (1995) Farmers' cultural practices and their effects on pest control in sweet potato in South Nyanza, Kenya. *International Journal of Pest Management*, **41** (1), 2–7.

Soon-Il, K., Young-Joon, A. and Hyung-Wook, K. (2012) Toxicity of aromatic plants and their constituents against coleopteran stored products insect pests, in *New Perspectives in Plant Protection* (ed. A.R. Bandani), InTech, Rijeka.

Soule, S., Güntner, C., Vázquez, A. *et al.* (2000) An aphid repellent glycoside from *Solanum laxum*. *Phytochemistry*, **55**, 217–222.

Suleiman, M.N. and Emua, S.A. (2009) Efficacy of four plant extracts in the control of root rot disease of cowpea (*Vigna unguiculata* [L.] Walp). *African Journal of Biotechnology*, **8** (16), 3806–3808.

Szczepanik, M., Krystkowiak, K., Jurzysta, M. and Biały, Z. (2001) Biological activity of saponins from alfalfa tops and roots against Colorado potato beetle larvae. *Acta Agrobotanica*, **54**, 35–45.

Thresh, J.M. and Cooter, R.J. (2005) Strategies for controlling cassava mosaic virus disease in Africa. *Plant Physiology*, **54**, 587–614.

Tohamy, M.R.A., Aly, A.Z., Abd-El-Moity, T.H. *et al.* (2002) Evaluation of some plant species extracts in control of damping-off and mildew diseases of cucumber. *Egyptian Journal of Phytopathology*, **30** (2), 71–78.

Tooba, H., Usmani, N.F. and Tahir, A. (2005) Screening of plant leaves as grain protectants against *Tribolium castaneum* during storage. *Pakistan Journal of Botany*, **37** (1), 149–153.

Tukahirwa, E.M. and Coaker T.H. (1982) Effect of mixed cropping on some insect pests of brassicas: reduced *Brevicoryne brassicae* infestations and influences on epigeal predators and the disturbance of ovipositon behaviour in *Delia brassicae*. *Entomologia Experimentalis et Applicata*, **32**, 129–140.

Van Huis, A. (1996) The traditional use of arthropods in sub-Saharan Africa. *Proceedings of the Section Experimental and Applied Entomology of the Netherlands Entomological Society (NEV)*, **7**, 3–20.

Van Tol, R.W.H.M., James, D.E., Jan de Kogel, W. and Teulon D.A.J. (2007) Plant odours with potential for a push–pull strategy to control the onion thrips, *Thrips tabaci*. *Entomologia Experimentalis et Applicata*, **122**, 69–76.

Weissenberg, M., Levy, A., Svoboda, J.A. and Ishaaya, I. (1998) The effect of some *Solanum* steroidal alkaloids and glycoalkaloids on larvae of the red flour beetle, *Tribolium castaneum*, and the tobacco hornworm *Manduca sexta*. *Phytochemistry*, **47**, 203–209.

Willson, H.R. (1992) *Integrated pest management*, Field Crops Pest Management Circular #1, Ohio Pest Management and Survey Program, http://ohioline.osu.edu/icm-fact/fc-01.html (accessed 16 November 2013).

Wortmann, C.S., Fischler, M., Alifugani, F. and Iaizzi, C.K. (1998) Accomplishments of participatory research for systems improvement in Iganga District, Uganda 1993–97, in *Network on Bean Research in Africa*, Occasional Publications Series, No. 27, CIAT, Kampala, Uganda.

Wright, R.J. (1995) *Biological Control of Insect and Mite Pests*, University of Nebraska-Lincoln Extension, Institute of Agricultural and Natural Resources, http://www.ianrpubs.unl.edu/pages/publicationD.jsp?publicationId=520 (accessed 16 November 2013).

Xuan, T.D., Tsuzuki, E., Uematsu, H. and Terao, H. (2001) Weed control with alfalfa pellets in transplanting rice. *Weed Biology and Management*, **1**, 231–235.

Zehrer, W. (1986) Traditional agriculture and integrated pest management. *Low External Input for Sustainable Agriculture Newsletter*, **6** (Nov.), 4–6.

## Appendices

### Diversity of African plants used to control crop insect pests

| Family | Species | Plant part(s)/formulation | Application | Pest controlled | Source |
|---|---|---|---|---|---|
| Acanthaceae | *Acanthus montanus* (Nees) T. Anderson | Spiny leaf to keep rats from granaries | – | – | Secoy and Smith (1983) |
| Acanthaceae | *Duosperma* sp | Insect repellent | – | – | Secoy and Smith (1983) |
| Acoraceae | *Acorus calamus* L | Rhizome | Field | *Peridroma saucia* (Hübner) | Koul *et al.* (1990) |
| Acoraceae | *Acorus calamus* L. | Rhizome extract | Field | *Epilachna vigintioctopuncata* (Fabr) | Chandel *et al.* (1987) |
| Acoraceae | *Acorus calamus* L. | Rhizome extract | Field | *Gnorimoschema operculella* (Zell.) | Pandey *et al.* (1982) |
| Alliaceae | *Allium sativum* L. | Bulb | Field | *Megalurothrips sjostedti* (Trybom), *Maruca vitrata* (Fab.) and *Clavigralla tomentosicollis* (Stal.) | Oparaeke *et al.* (2005) |
| Amaranthaceae | *Chenopodium ambrosioides* L. | Plant powder | Storage | *Caryedon serratus* (OL.) | Delobe and Malonga (1987) |
| Amaranthaceae | *Pupalia lappacea* (L.) Juss. | Seed pods in rat poison | – | – | Secoy and Smith (1983) |
| Alliaceae | *Allium sativum* L. | Essential oil | Storage | *Aspergillus flavus* (Link) | Gurjar *et al.*, (2012), Reddy *et al.* (2009) |
| Anacardiaceae | *Anacardium occidentale* L. | Nutshell | Field | *Megalurothrips sjostedti* (Trybom), *Maruca vitrata* (Fab.) and *Clavigralla tomentosicollis* (Stal.) | Oparaeke *et al.* (2005) |
| Annonaceae | *Annona squamosa* L. | Leaf extract | Field | *Brevicornye brassicae* (L.) | Singh and Sharma (1986) |
| Annonaceae | *Monodora myristica* (Gaertn.) Dunal | Seed powder | Storage | *Callosobruchus maculatus* (F.) and *Sitophilus zeamais* (Motsch.) | Ogunsina *et al.* (2011) |
| Annonaceae | *Polyalthia suaveolens* Engl. & Diels | Decoction of leaves for ritual bath for crop protection against sorcery | – | – | Secoy and Smith (1983) |
| Annonaceae | *Xylopia aethiopica* (Dunal) A.Rich. | Fruits | Field | *Megalurothrips sjostedti* (Trybom), *Maruca vitrata* (Fab.) and *Clavigralla tomentosicollis* (Stal.) | Oparaeke *et al.* (2005) |
| Anonaceaea | *Annona squamosa* L. | Leaf powder | Storage | *Tribolium castaneum* (Herbst.) | Anita *et al.* (2012) |
| Apocynaceae | *Catharanthus roseus* (L.) G.Don | Aqueous leaf extract | Field | *Helicoverpa armigera* (Hubner) and *Lipaphis erysimi* (Kalt.) | Siddappaji *et al.* (1986) |
| Apocynaceae | *Nerium oleander* L. | Leaves against insect and rodents; oleandrin | – | – | Secoy and Smith (1983) |
| Apocynaceae | *Thevetia peruviana* (Pers.) K.Schum. | Insecticide | – | – | Secoy and Smith (1983) |
| Apocynaceae | *Voacanga obtusa* K. Schum | Fruit put in field to repel wild pigs | – | – | Secoy and Smith (1983) |
| Aquifoliaceae | *Ilex opaca* Aiton | Saponins extracted | Field | *Oligonychus ilicis* (McGregor), *Hyphantria cunea* Drury and *Malacosoma americanum* (F.) | Kreuger and Potter (1994) |
| Asparagaceae | *Agave cantala* (Haw.) Roxb. ex Salm-Dyck | Saponins extract of plant | Field | *Spilosoma obliqua* (Walker) | Jain and Tripathi (1999) |
| Asphodelaceae | *Aloe* spp. | Whole plant | Field | Lepidoptera and Coleoptera | Mihale *et al.* (2009) |
| Asteraceae | *Blumea aurita* (L.f.) DC | Leaf powder | Storage | *Callosobruchus maculatus* (F.) | Boeke (2002) |
| Asteraceae | *Tarchonanthus camphoratus* L. | Insect repellent | – | – | Secoy and Smith (1983) |

*(continued)*

| Family | Species | Plant part(s)/formulation | Application | Pest controlled | Source |
|---|---|---|---|---|---|
| Barringtoniaceae | *Barringtonia racemosa* (L.) Spreng. | Bark as insecticide, seeds as poison bait | – | – | Secoy and Smith (1983) |
| Brassicaceae | *Cochlearia armoracia* L. | Essential oil | Storage | *Callosobruchus chinensis* (L.), *Lasioderma serricorne* (F.) and *Sitophilus zeamais* (Motsch.) | Soon-Il *et al.* (2012) |
| Burseraceae | *Commiphora abyssinica* Engl. | Insect repellent and insecticide | – | – | Secoy and Smith (1983) |
| Burseraceae | *Commiphora africana* (A.Rich.) Engl. | Insect repellent and insecticide | – | – | Secoy and Smith (1983) |
| Burseraceae | *Commiphora myrrha* (Nees) Engl. | Insect repellent and insecticide | – | – | Secoy and Smith (1983) |
| Cesalpiniaceae | *Erythrophleum guineense* G. Don | Bark as rat poison, leaves in storage grain | – | – | Secoy and Smith (1983) |
| Chenopodiaceae | *Chenopodium opulifolium* Schrad. ex W.D.J.Koch&Ziz | whole plant | Storage and fields | *Prostephanus truncatus* (Hom), *Stophilus* sp., *Tribolium* spp., *Bruchus rufimanus* (Boheman) and *Rhyzopertha dominica* (Fabricius) | Mihale *et al.* (2009) |
| Chenopodiaceae | *Chenopodium ambrosioides* L. | Decoction of seed or leaves against insects | – | – | Secoy and Smith (1983) |
| Combretaceae | *Terminalia* sp. | Smoke from burning leaves | Field | Blister beetle [*Epicauta vittata* (Fabricius)] | Gahukar (1988) |
| Compositae | *Ageratum conyzoides* L. | Leaf extract | Field | *Gnorimoschema operculella* (Zell.) | Pandey *et al.* (1982) |
| Compositae | *Matricaria chamomilla* L. | Essential oil at 1% | Storage | *Oryzaephilus surinamensis* (L.) and *Tribolium castaneum* (Herbst.) | Al-Jabr *et al.* (2006) |
| Crassulaceae | *Bryophyllum* (=*Kalanchoe*) *pinnatum* (Lam.) Kruz | Wood in fetish for protection of fields | – | – | Secoy and Smith (1983) |
| Crassulaceae | *Kalanchoe crenata* (Andrews) Haw. | Wood in charm for protection of fields | – | – | Secoy and Smith (1983) |
| Crassulaceae | *Kalanchoe somaliensis* Baker | Sap as insect repellent | – | – | Secoy and Smith (1983) |
| Cucurbitaceae | *Luffa acutangula* Roxb. | Seed or plant as insecticide | – | – | Secoy and Smith (1983) |
| Cucurbitaceae | *Momordica charantia* L. | Leaf powder | Storage | *Callosobruchus maculatus* (F.) | Boeke (2002) |
| Cucurbitaceae | *Momordica foetida* Schumach. | Fruit pulp as insecticide and repellent | – | – | Secoy and Smith (1983) |
| Cucurbitaceae | *Momordica schimperiana* Naud. | Fruit as insecticide | – | – | Secoy and Smith (1983) |
| Cyperaceae | *Cyperus articulatus* L. | Rhizome | Storage | *Tribolium castaneum* (Herbst.) | Abubakar *et al.* (2000) |
| Dichapetalaceae | *Dichapetalum toxicarium* Baill. | Seeds as rat poison | – | – | Secoy and Smith (1983) |
| Dioscoreacea | *Dioscorea dumetorum* (Kunth) Pax | Bulb as poison bait for field-raiding monkey | – | – | Secoy and Smith (1983) |
| Ericaceae | *Agauria salicifolia* (Olivier) Hook. f. | Infusion as insecticide, especially for beetles | – | – | Secoy and Smith (1983) |
| Euphorbiaceae | *Antidesma membranaceum* Müll.Arg. | Fruit as charm for health of crops | – | – | Secoy and Smith (1983) |
| Euphorbiaceae | *Euphorbia lateriflora* Schum. and Thonn. | Stem powder | Storage | *Callosobruchus maculatus* (F.) and *Sitophilus zeamais* (Motsch.) | Ogunsina *et al.* (2011) |
| Euphorbiaceae | *Euphorbia royleana* Boiss. | Leaf extract | Field | *Spilosoma obliqua* (Walker) | Sharma *et al.* (1982) |
| Euphorbiaceae | *Jatropha curcas* L. | Leaves as general insecticide | – | – | Secoy and Smith (1983) |
| Euphorbiaceae | *Macaranga hypoleuca* (Rchb.f. & Zoll.) Müll.Arg. | Insect repellent | – | – | Secoy and Smith (1983) |

| Family | Species | Plant part(s)/formulation | Application | Pest controlled | Source |
|---|---|---|---|---|---|
| Euphorbiaceae | *Ricinus communis* L. | Crush leaves and ground leaves | Storage | *Tribolium castaneum* (Herbst.) | Tooba *et al.* (2005) |
| Euphorbiaceae | *Ricinus communis* L. | Seeds, oil insecticidal, poison bait for rats and coyotes, ricin | – | – | Secoy and Smith (1983) |
| Euphorbiaceae | *Spondianthus preussii* Engl. | Bark and seeds as rat bait | – | – | Secoy and Smith (1983) |
| Euphorbiaceae | *Toxicodendron* (=*Hyaenanche*) *capense* Thumb. | Seeds in carcasses to poison flock predators | – | – | Secoy and Smith (1983) |
| Leguminosae | *Daniellia oliveri* (Rolfe) Hutch. & Dalziel | Resin as termite repellent | – | – | Secoy and Smith (1983) |
| Leguminosae | *Daniellia thurifera* Benn. | Resin fumigant for flying insects; rotenone | – | – | Secoy and Smith (1983) |
| Leguminosae | *Derris uliginosa* (Willd.) Benth. | Insecticide, rotenone | – | – | Secoy and Smith (1983) |
| Leguminosae | *Dolichos pseudopachyrhizus* Harms | Root decoction as insecticide | – | – | Secoy and Smith (1983) |
| Leguminosae | *Indigofera* sp. | Root against insects | – | – | Secoy and Smith (1983) |
| Leguminosae | *Medicago sativa* L. | Leaves and roots saponin extract | Field | *Leptinotarsa decemlineata* (Say) | Szczepanik *et al.* (2001) |
| Leguminosae | *Medicago sativa* L. | Leaves and roots saponin extract | Field | *Spodoptera littoralis* (Boisd.) | Adel *et al.* (2000) |
| Leguminosae | *Mundulea sericea* (Willd.) A.Chev. | Bark as insecticide; rotenone | – | – | Secoy and Smith (1983) |
| Leguminosae | *Neorautanenia pseudopachyrrhiza* (Harms) Milne-Redh. | Root as insecticide | – | – | Secoy and Smith (1983) |
| Leguminosae | *Phaseolus vulgaris* L. | Saponins extract of plant | Field | *Spilosoma obliqua* (Walker) | Jain and Tripathi (1999) |
| Leguminosae | *Sesbania sesban* (L.) Merr. | Insecticide | – | – | Secoy and Smith (1983) |
| Leguminosae | *Swartzia madagascariensis* Desv. | Powdered fruit as repellent in stored grain | – | – | Secoy and Smith (1983) |
| Leguminosae | *Tephosia vogelii* Hook.f. | Leaf against fleas; rotenoids | – | – | Secoy and Smith (1983) |
| Leguminosae | *Tephrosia* spp. | Plants juice | Field | Blister beetle [*Epicauta vittata* (Fabricius)] | Wortmann *et al.* (1998) |
| Leguminosae | *Tephrosia vogelii* Hook.f. | Leaves extract | Field | *Ootheca mutabilis* (Sahlb.), *Maruca testulalis* (Geyer), *Zonocerus variegatus* (L.), *Riptortus dentipes* (Fabricius) and *Apion varium* (Wagner) | Adebayo *et al.* (2007) |
| Leguminosae | *Tephrosia vogelii* Hook.f. | Leaf powder | Storage | *Callosobruchus maculatus* (F.) | Boeke (2002) |
| Leguminosae | *Tephrosia vogelii* Hook.f. | Plant powder | Storage | *Caryedon serratus* (OL.) | Delobe and Malonga (1987) |
| Leguminosae | *Tephrosia vogelii* Hook.f. | Mixture of leaves and succulent stems (crude powder) | Storage | *Sitophilus zeamais* (Motsch.) | Ogendo *et al.* (2004) |
| Leguminosae | *Tephrosia vogelii* Hook.f. | Mixture of leaves and succulent stems (crude powder) | Storage | *Prostephanus truncatus* (Horn) | Chebet *et al.* (2013) |
| Hippocrateaceae | *Hippocratea velutina* Afzel. ex Spreng | Charm to protect yam crop from insects | – | – | Secoy and Smith (1983) |
| Hyperiaceae | *Psorospermum guineense* Hochr. | Bark and root in seed steep as bird poison | – | – | Secoy and Smith (1983) |
| Hypoxidaceae | *Hypoxis latifolia* Hook. | Bulb as poison bait for rodent | – | – | Secoy and Smith (1983) |

(*continued*)

| Family | Species | Plant part(s)/formulation | Application | Pest controlled | Source |
|---|---|---|---|---|---|
| Lamiaceae | *Hyptis spicigera* Lam. | Termite repellent in stored grain | – | – | Secoy and Smith (1983) |
| Lamiaceae | *Leonotis africana* (P.Beauv.) Briq. | Insect and rat repellent in stored grain | – | – | Secoy and Smith (1983) |
| Lamiaceae | *Leonotis nepetifolia* (L.) R.Br. | Insect and rodent repellent in stored grain | – | – | Secoy and Smith (1983) |
| Lamiaceae | *Ocimum sanctum* L. | Leaf, seed, fruit and crude extract | Storage | *Aspergillus flavus* (Link) | Gurjar *et al.*, (2012), Reddy *et al.*, (2009). |
| Lamiaceae | *Ocimum suave* Willd. | Leaves | Storage | *Prostephanus truncatus* (Hom), *Sitophilus* sp., *Tribolium* spp., *Bruchus rufimanus* (Boheman) and *Rhyzopertha dominica* (Fabricius) | Mihale *et al.* (2009) |
| Lamiaceae | *Origanum majorana L.* | Essential oil | Field | *Thrips tabaci* (Lindeman) | Van Tol *et al.* (2007) |
| Lamiaceae | *Origanum vulgare* L. | Essential oil | Storage | *Tribolium catsaneum* (Herbst.) | Soon-Il *et al.* (2012) |
| Lamiaceae | *Salvia repens* Burch. ex Benth. | Insect repellent | – | – | Secoy and Smith (1983) |
| Lamiaceae | *Salvia runcinata* L. | Insect fumigant | – | – | Secoy and Smith (1983) |
| Lamiaceae | *Salvia sisymbrifolia* Skan | Insect fumigant | – | – | Secoy and Smith (1983) |
| Lamiaceae | *Salvia stenophylla* Burch. ex Benth. | Insect fumigant | – | – | Secoy and Smith (1983) |
| Leguminosae | *Tephrosia* spp. | Plants juice | Field | Blister beetle [*Epicauta vittata* (Fabricius)] | Wortmann *et al.* (1998) |
| Leguminosae/ Caesalpinioideae | *Senna siamea* (Lam.) H.S.Irwin & Barneby | Root solution | Storage | *Prostephanus truncatus* (Hom), *Sitophilus* sp., *Tribolium* spp., *Bruchus rufimanus* (Boheman) and *Rhyzopertha dominica* (Fabricius) | Mihale *et al.* (2009) |
| Leguminosae | *Tamarindus indica* L. | Fruits | Field | *Agrotis* spp. | Mihale *et al.* (2009) |
| Leguminosae/ Mimosoideae | *Parkia biglobosa* (Jacq.) G.Don | Smoke from burning fruits | Field | Blister beetle [*Epicauta vittata* (Fabricius)] | Gahukar (1988) |
| Liliaceae | *Androcymbium leucanthum* Willd | Bird poison | – | – | Secoy and Smith (1983) |
| Liliaceae | *Baeometra collumerallis* Salisb. | Bird poison | – | – | Secoy and Smith (1983) |
| Liliaceae | *Dipidax ciliata* (L.f.) Baker | Bird poison | – | – | Secoy and Smith (1983) |
| Liliaceae | *Dipidax triquetra* (L.f.) Baker | Bird poison | – | – | Secoy and Smith (1983) |
| Liliaceae | *Dracaena arborea* (Willd) Link | Leaf powder | Storage | *Callosobruchus maculatus* (F.) | Boeke (2002) |
| Liliaceae | *Ornithoglossum glaucum* Salisb. | Bird poison | – | – | Secoy and Smith (1983) |
| Liliaceae | *Wurmbea capensis* Thunb. | Bird poison | – | – | Secoy and Smith (1983) |
| Lyrthraceae | *Punica granatum* L. | Leaf powder | Storage | *Tribolium castaneum* (Herbst.), *Rhyzopertha dominica* (Fabricius) | Gandhi *et al.* (2010), Gandhi and Pillai (2011) |
| Malvaceae | *Hibiscus abelmoschus* L. | Fruit as insecticide | – | – | Secoy and Smith (1983) |
| Meliaceae | *Azadirachta indica* A.Juss. | Seed (oil) | Field | *Cnaphalocrosis medinalis* (Guenee) | Mohan *et al.* (1991) |
| Meliaceae | *Azadirachta indica* A.Juss. | Seed (oil) | Field | *Cnaphalocrosis medinalis* (Guenee) | Krishnaiah and Kalode (1990) |
| Meliaceae | *Azadirachta indica* A.Juss. | Plants extracts | Field | *Sogatella furcifera* (Horváth) | Saxena (1987) |

| Family | Species | Plant part(s)/formulation | Application | Pest controlled | Source |
|---|---|---|---|---|---|
| Meliaceae | *Azadirachta indica* A.Juss. | Plants extracts | Field | *Nilaparvata lugens* (Stal) | Senthil Nathan *et al.* (2007) |
| Meliaceae | *Azadirachta indica* A.Juss. | Neem oil and kernel extracts | Field | *Helicoverpa armigera* (Hubner) | Siddappaji *et al.* (1986) |
| Meliaceae | *Azadirachta indica* A.Juss. | Neem oil | Field | *Lipaphis erysimi* (Kalt.) | Mani *et al.* (1990) |
| Meliaceae | *Azadirachta indica* A.Juss. | Kernel and leaves extracts | Field | *Helicoverpa armigera* (Hubner) and *Lipaphis erysimi* (Kalt.) | Atwal and Pajni (1964) |
| Meliaceae | *Azadirachta indica* A.Juss. | Leaves | Field | Birds and termites | Zehrer (1986) |
| Meliaceae | *Azadirachta indica* A.Juss. | Mixture of leaves, inflorescence and succulent stems (crude powder) | Storage | *Prostephanus truncatus* (Horn) | Chebet *et al.* (2013) |
| Meliaceae | *Azadirachta indica* A.Juss. | Crush leaves and ground leaves | Storage | *Tribolium castaneum* (Herbst.) | Tooba *et al.* (2005) |
| Meliaceae | *Azadirachta indica* A.Juss. | Leaf extract | Field | *Brevicornye brassicae* (L.) | Singh and Sharma (1986) |
| Mimosaceae | *Acacia concinna* (Willd.) DC. | Flowers and pot soap as insecticide | – | – | Secoy and Smith (1983) |
| Moraceae | *Ficus leprieurii* Miq. | Sap to destroy anthills | – | – | Secoy and Smith (1983) |
| Moringaceae | *Moringa oleifera* Lam. | Leaf powder | Storage | *Tribolium castaneum* (Herbst.) | Anita *et al.* (2012) |
| Musaceae | *Musa sapientum* L. | Fruit pulp on wall as insect repellent | – | – | Secoy and Smith (1983) |
| Myrtaceae | *Eucalyptus codonocarpa* Blakely & McKie | Essential oil | Storage | *Rhyzopertha dominica* (Fabricius) and *Sitophilus oryzae* (L.) | Lee *et al.* (2004) |
| Myrtaceae | *Eucalyptus globulus* Labill. | Leaf powder | Storage | *Tribolium castaneum* (Herbst.) | Anita *et al.* (2012) |
| Myrtaceae | *Eucalyptus globulus* Labill. | Leaf sap as insect repellent; eucalyptol | – | – | Secoy and Smith (1983) |
| Myrtaceae | *Eucalyptus* spp. | Air dried and ground leaves of *Eucalyptus* | Storage | *Prostephanus truncatus* (Hom), *Stophilus* sp, *Tribolium* spp, *Bruchus rufimanus* (Boheman) *and Rhyzopertha dominica* (Fabricius) | Mihale *et al.* (2009) |
| Myrtaceae | *Syzygium aromaticum* (L.) Merr. & L.M.Perry | Leaf, seed, fruit and crude extract | Storage | *Aspergillus flavus* (Link) | Gurjar *et al.*, (2012), Reddy *et al.*, (2009) |
| Papaveraceae | *Argemone mexicana L.* | Seed as insecticide and repellent | – | – | Secoy and Smith (1983) |
| Pedaliaceae | *Ceratotheca integribracteata* Engl. | Decoction as insecticide | – | – | Secoy and Smith (1983) |
| Phytolaccaceae | *Petiveria allicea* L. | Root extract | Field | *Ootheca mutabilis* (Schönherr), *Maruca testulalis* (Geyer), *Zonocerus variegatus* (L.), *Riptortus dentipes* (Fabricius) and *Apion varium* (Wagner) | Adebayo *et al.* (2007) |
| Pocaeae | *Andropogon marginatus* Steud. | In stored grain as rodent repellent | – | – | Secoy and Smith (1983) |
| Pocaeae | *Vetiveria zizanoides* (L.) Nash | Raceme use as repellent | – | – | Secoy and Smith (1983) |
| Polygalaceae | *Securidaca longepedunculata* Fresen | Leaf powder | Storage | *Callosobruchus maculatus* (F.) | Boeke (2002) |
| Polygalaceae | *Securidaca longepedunculata* Fresen | Root extract | Storage | *Sitophilus zeamais* (Motsch.) | Jayasekara *et al.* (2005) |
| Rhizophoraceae | *Ceriops tagal* (Perr.) C.B.Rob. | Stem and bud hexane extracts | Storage | *Tribolium castaneum* (Herbst.) | Du *et al.* (2011) |

*(continued)*

| Family | Species | Plant part(s)/formulation | Application | Pest controlled | Source |
|---|---|---|---|---|---|
| Rosaceae | *Prunus amygdalus* var. *amara* (DC.) Focke | Essential oil | Storage | *Tribolium castaneum* (Herbst.) | Al-Jabr (2006) |
| Rubiaceae | *Adina rubrostipulata* K.Schum | Insecticide | – | – | Secoy and Smith (1983) |
| Rubiaceae | *Xeromphis* (=*Randia*) *nilotica* Keay | Insecticide | – | – | Secoy and Smith (1983) |
| Rutaceae | *Clausena anisata* (Willd.) Hook.f. ex Benth. | Leaf powder | Storage | *Callosobruchus maculatus* (F.) | Boeke (2002) |
| Rutaceae | *Murraya koenigii* (L.) Spreng. | Leaf powder | Storage | *Tribolium castaneum* (Herbst.), *Rhyzopertha dominica* (Fabricius) | Gandhi *et al.* (2010), Gandhi and Pillai (2011) |
| Simaroubaceae | *Harrisonia abyssinica* Oliv. | Root in insecticide; harrisonin | – | – | Secoy and Smith (1983) |
| Simaroubaceae | *Mannia* (=*Quassia*) *gabonensis* (Pierre) Engl. | Seeds in poison bait for rodents | – | – | Secoy and Smith (1983) |
| Simaroubaceae | *Picraena* (=*Aeschrion*) *excelsa* Lind. | Wood extract as insecticide | – | – | Secoy and Smith (1983) |
| Simaroubaceae | *Quassia amara* L. | Flowers, wood extract as insecticide; quassia | – | – | Secoy and Smith (1983) |
| Solanaceae | *Capsicum annuum* L. | To protect stored kola nuts | – | – | Secoy and Smith (1983) |
| Solanaceae | *Capsicum frutescens* L. | In stored food against weevils, charms for crops | – | – | Secoy and Smith (1983) |
| Solanaceae | *Capsicum* sp. | Fruits | Field | *Megalurothrips sjostedti* (Trybom), *Maruca vitrata* (Fab.) and *Clavigralla tomentosicollis* (Stal.) | Oparaeke *et al.* (2005) |
| Solanaceae | *Capsicum* sp. | Capsaicin | Storage | *Sitophilus zeamais* (Motsch.) | Ho and Ma (1995) |
| Solanaceae | *Cestrum parqui* (Lam.) L'Hér. | Leaves extracts | Field | *Schistocerca gregaria* (Forsk.) | Barbouche *et al.* (2001) |
| Solanaceae | *Datura fastuosa* L. | Fruit as rat poison in stored grain | – | – | Secoy and Smith (1983) |
| Solanaceae | *Nicotiana tabacum* L. | Leaf powder | Storage | *Callosobruchus maculatus* (F.) | Boeke (2002) |
| Solanaceae | *Solanum laxum* Spreng. | Aerial parts extract | Field | *Schizaphis graminum* (Rondani) | Soule *et al.* (2000) |
| Solanaceae | *Solanum* sp. | Glycoalkaloids extracted | Field | *Manduca sexta* (L.) | Weissenberg *et al.* (1998) |
| Solanaceae | *Solanum* sp. | Glycoalkaloids | Storage | *Tribolium castaneum* (Herbst.) | Weissenberg *et al.* (1998) |
| Sterculiaceae | *Sterculia foetida* L. | Bark, leaves as repellent; known chemosterillant | – | – | Secoy and Smith (1983) |
| Verbenaceae | *Lantana camara* L. | Mixture of leaves, inflorescence and succulent stems (crude powder) | Storage | *Sitophilus zeamais* (Motsch.) | Ogendo *et al.* (2004) |
| Verbenaceae | *Lantana camara* L. | Leaf extract | Field | *Spilosoma obliqua* (Walker) | Sharma *et al.* (1982) |
| Verbenaceae | *Lantana camara* L. | Leaf powder | Storage | *Callosobruchus maculatus* (F.) and *Sitophilus zeamais* (Motsch.) | Ogunsina *et al.* (2011) |
| Verbenaceae | *Lantana camara* L. | Mixture of leaves, inflorescence and succulent stems (crude powder) | Storage | *Prostephanus truncatus* (Horn) | Chebet *et al.* (2013) |
| Verbenaceae | *Vitex grandifolia* Gürke | Leaf powder | Storage | *Callosobruchus maculatus* (F.) and *Sitophilus zeamais* (Motsch.) | Epidi *et al.*, (2008) |
| Vitidaceae | *Cissus quadrangularis* L. | Planted as termite repellent | – | – | Secoy and Smith (1983) |
| Zingiberaceae | *Curcuma longa* L. | Leaf, seed, fruit and crude extract | Storage | *Aspergillus flavus* (Link) | Gurjar *et al.* (2012), Reddy *et al.* (2009) |
| Zygophyllaceae | *Balanites roxburghii* Planch. | Saponins extract of plant | Field | *Spilosoma obliqua* (Walker) | Jain and Tripathi (1999) |

## Diversity of plants used in disease management

| Family | Species | Plants part/formulation | Application | Diseases/causal agents controlled | Source |
|---|---|---|---|---|---|
| Alliaceae | *Allium sativum* L. | Bulb extracts | Field | Downy mildew of *Arabidopsis thaliana* (L.) Heynh./ *Hyaloperonospora parasitica* (Pers.) Constant. | Curtis *et al.* (2004) |
| Alliaceae | *Allium sativum* L. | Bulb extracts | Field | Potato infection/*Phytophtora infestans* (Mont.) de Bary | Curtis *et al.* (2004) |
| Alliaceae | *Allium sativum* L. | Leaves extract | Field | Early blight disease/*Alternaria solani* Sorauer | Nashwa and Abo-Elyousr (2012) |
| Alliaceae | *Allium sativum* L. | Bulds tissues extract | Field | Mildew disease | Tohamy *et al.* (2002) |
| Alliaceae | *Allium sativum* L. | Bulb extracts | Field | Rice blast disease/*Magnaporthe grisea* (T.T. Hebert) M.E. Barr | Curtis *et al.* (2004) |
| Asparagaceae | *Aloe vera* (L.) Burm.f. | Fresh leaves and seeds | Field | Cowpea root rot disease/*Pythium aphanidermatum* | Suleiman and Emua (2009) |
| Caricaceae | *Carica papaya* L. | Root etxtract | Field | Cowpea antrachnose/ *Colletotrichum destructivum* N.L. Horn | Enyiukwu and Awurum (2011) |
| Compositae/ Asteraceae | *Artemisia camphorata* Vill. | Camphor extracts | Field | Banana anthracnose | Carré *et al.* (2006) |
| Compositae/ Asteraceae | *Aspilia africana* (Pers.) C. D. Adams | Fresh leaves extract | Field | Sesam cercospora leaf spot disease/*Cercospora sesami* (Zimm.) | Enikuomehin (2005) |
| Compositae/ Asteraceae | *Chromolaena odorata* (L.) R.M.King & H.Rob. | Fresh leaves extract | Field | Sesam cercospora leaf spot disease/*Cercospora sesami* (Zimm.) | Enikuomehin (2005) |
| Compositae/ Asteraceae | *Chrysanthemum cinerariifolium* (Trevir.) Sch.Bip. | Dried flower powder | Field | Potato Y disease virus/*Myzus persicae* | Jacobson (1975) |
| Compositae/ Asteraceae | *Tithonia diversifolia* (Hemsl.) A.Gray | Fresh leaves extract | Field | Sesam cercospora leaf spot disease/*Cercospora sesami* (Zimm.) | Enikuomehin (2005) |
| Compositae/ Asteraceae+ Verbanaceae | *Bidens pilosa* L. and *Lippia alba* (Mill.) N.E.Br. ex Britton & P.Wilson | Extracts mixture | Field | *Alternaria alternata* (Fr.) Keissl. | Moreira *et al.* (2008) |
| Euphorbiaceae | *Acalypha macrostachya* Jacq. | Fresh leaves | Field | *Cercospora* spot disease/ *Cercospora purpurea* Cke. | Ogbo and Oyibo (2008) |
| Euphorbiaceae | *Acalypha wilkesiana* Müll.Arg. | Fresh leaves | Field | *Cercospora* spot disease/ *Cercospora purpurea* Cke. | Ogbo and Oyibo (2008) |
| Lamiaceae | *Ocimum gratissimum* L. | Fresh leaves | Field | Southern blight disease/*Sclerotium rolfsii* Sacc. | Adesegun *et al.* (2013) |
| Lamiaceae | *Ocimum gratissimum* L. | Aqueous leaves extrcact | Field | Cucumber anthracnose/ *Colletotrichum lagenarium* (Pass.) Ell. & Halst. | Colpas *et al.* (2009) |
| Lamiaceae | *Ocimum gratissimum* L. | Fresh leaves | Field | *Cercospora* spot disease/ *Cercospora purpurea* Cke. | Ogbo and Oyibo (2008). |
| Lamiaceae | *Ocimum sanctum* L. | Leaves extract | Field | Anthracnose disease/ *Colletotrichum lindemuthianum* | Amadioha (2003) |
| Lauraceae | *Cinnamomum camphora* (L.) J.Presl | Leaves extract | Field | Mildew disease | Tohamy *et al.* (2002) |
| Leguminosae/ Mimosoideae | *Acacia arabica* (Lam.) Willd. | Leaves extract | Field | Bacterial blight, common blight, fiscous blight and canker /*Xanthomonas campestris* (Pammel) Dowson | Satish *et al.* (1999) |
| Leguminosae/ Mimosoideae | *Prosopis juliflora* (Sw.) DC. | Leaves, stem and flower powder | Field | Cowpea root rot fungi/*Fusarium* spp., *Rhizoctonia solani* Kuhn and *Macrophomina phaseolina* | Ikram and Dawar (2013) |
| Leguminosae/ Mimosoideae | *Prosopis juliflora* (Sw.) DC. | Leaves extract | Field | Bacterial blight, common blight, fiscous blight and canker/ *Xanthomonas campestris* (Pammel) Dowson | Satish *et al.* (1999) |

(*continued*)

| Family | Species | Plants part/formulation | Application | Diseases/causal agents controlled | Source |
|---|---|---|---|---|---|
| Lythraceae | *Lawsonia inermis* L. | Leaves extract | Field | Bacterial blight, common blight, fiscous blight and canker/*Xanthomonas campestris* (Pammel) Dowson | Satish *et al.* (1999) |
| Meliaceae | *Azadirachta indica* A. Juss | Leaves extract and seed oil extract | Field | Rice blat disease/*Pyricularia oryzae* Cavara | Amadioha (2000) |
| Meliaceae | *Azadirachta indica* A. Juss | Leaves extract | Field | Black mould disease | Gupta *et al.*, (2012) |
| Meliaceae | *Azadirachta indica* A. Juss | Leaves extract | Field | Mildew disease | Tohamy *et al.* (2002) |
| Meliaceae | *Azadirachta indica* A. Juss | Leaves extract | Storage | Tomato post-harvest rot / *Rhizopus stolonifer* [(Ehrenb. Ex. Fr) Lind.] and *Geotrichum candidum* (Link) | Ijato *et al.*, (2011) |
| Musaceae | *Musa paradisiaca* L. | Fresh leaves extract | Field | Sesame *Cercospora* leaf spot disease/*Cercospora sesami* (Zimm.) | Enikuomehin, (2005) |
| Musaceae | *Musa* spp. | Pseudo stem sap | Field | Southern blight disease/*Sclerotium rolfsii* Sacc. | Sikirou *et al.* (2010, 2011) |
| Myrtaceae | *Eucalyptus citriodora* Hook. | Leaves extract | Field | Cucumber anthracnose/*Colletotrichum lagenarium* (Pass.) Ell. & Halst. | Bonaldo *et al.* (2007) |
| Myrtaceae | *Eucalyptus citriodora* Hook. | Fresh leaves aqueous extrcact | Field | *Phytophthora* sp., *Sclerotium rolfsii* Sacc., *Rhizoctonia solani* Kuhn and *Alternaria alternata* (Fr.) Keissl. | Bonaldo *et al.* (2007) |
| Myrtaceae | *Eucalyptus tereticornis* Sm. | Leaves extract | Field | Black mould disease | Gupta *et al.* (2012) |
| Oxalidaceae | *Oxalis corniculata* L. | Leaves extract | Field | Bacterial blight, common blight, fiscous blight and canker/*Xanthomonas campestris* (Pammel) Dowson | Satish *et al.* (1999) |
| Piperaceae | *Piper nigrum* L. | Leaves extract | Field | Anthracnose disease/*Colletotrichum lindemuthianum* | Amadioha (2003) |
| Piperaceae | *Piper guineense* Schum & Thom | Seeds extract | Field | Cowpea antrachnose/*Colletotrichum destructivum* | Enyiukwu and Awurum (2011) |
| Poaceae | *Cymbopogon citratus* (DC.) Stapf | Fresh leaves | Field | Southern blight disease/*Sclerotium rolfsii* Sacc. | Adesegun *et al.* (2013) |
| Santalaceae | *Viscum orientale* Willd. | Leaves extract | Field | Bacterial blight, common blight, fiscous blight and canker/*Xanthomonas campestris* (Pammel) Dowson | Satish *et al.* (1999) |
| Sapotaceae | *Achras zapota* L. | Leaves extract | Field | Bacterial blight, common blight, fiscous blight and canker /*Xanthomonas campestris* (Pammel) Dowson | Satish *et al.* (1999) |
| Solanaceae | *Datura metel* L. | Leaves extract | Field | Sheath blight and bacterial blight diseases | Kagale *et al.* (2004) |
| Solanaceae | *Datura stramonium* L. | Leaves extract | Field | Early blight disease/*Alternaria solani* Sorauer | Nashwa and Abo-Elyousr (2012) |
| Zingiberaceae | *Aframomum melegueta* K.Schum. | Dried seeds | Field | Southern blight disease/*Sclerotium rolfsii* Sacc. | Adesegun *et al.* (2013) |
| Zingiberaceae | *Zingiber officinale* Roscoe | Fruits | Field | Tomato rot/*Fusarium moniliforme* (Sheld.s.str), *Rhizopus stolonifer* [(Ehrenb. Ex. Fr) Lind.] and *Geotrichum candidum* (Link) | Chuku *et al.* (2010) |
| Zingiberaceae | *Zingiber officinale* Roscoe | Fresh leaves and seeds | Field | Cowpea root rot disease/*Pythium aphanidermatum* | Suleiman and Emua (2009) |

## Diversity of plants used to control farm weed

| Family | Species | Plants part(s)/formulation | Application | Weeds controlled | Source |
|---|---|---|---|---|---|
| Anacardiaceae | *Mangifera indica* L. | Fresh leaves aqueous leaves extracts | Field | *Parthenium hysterophorus* L. | Shafique *et al.* (2005) |
| Apocynaceae | *Alstonia scholaris* (L.) R. Br. | Aqueous extracts of dry leaves | Field | *Phalaris minor* Retz | Javaid *et al.* (2006) |
| Compositae/ Asteraceae | *Ageratum conyzoides* L. | Root and stem aqueous extracts | Field | *Phalaris minor* Retz., *Poa annua* L. and *Medicago polymorphai* L. | Akhtar *et al.* (2001) |
| Compositae/ Asteraceae | *Cirsium arvense* (L.) Scop. | Root and stem aqueous extracts | Field | *Phalaris minor* Retz., *Poa annua* L. and *Medicago polymorphai* L. | Akhtar *et al.* (2001) |
| Compositae/ Asteraceae | *Tagetes minuta* L. | Dried leaves powder | Field | *Echinochloa crus-galli* (L.) P.Beauv. and *Cyperus rotundus* L. | Batish *et al.*, (2007) |
| Leguminosae | *Medicago sativa* L. | Leaves and stems | Field | *Doparium junceum* Hamilt., *Lindernia pyxidaria* L., *Elatine trianda* Schk., *Eleocharis acicularis* (L.) Roem. & Schult., *Rotala indica* Koehne and *Echinochloa oryzicola* Vasinger | Xuan *et al.* (2001) |
| Meliaceae | *Azadirachta indica* A. Juss. | Aqueous extracts of dry leaves | Field | *Phalaris minor* Retz | Javaid *et al.* (2006) |
| Meliaceae | *Azadirachta indica* A. Juss. | Fresh leaves aqueous leaves extracts | Field | *Parthenium hysterophorus* L. | Shafique *et al.* (2005) |
| Meliaceae | *Melia azedarach* L. | Fresh leaves aqueous leaves extracts | Field | *Parthenium hysterophorus* L. | Shafique *et al.* (2005) |
| Moraceae | *Ficus benghalensis* L. | Fresh leaves aqueous leaves extracts | Field | *Parthenium hysterophorus* L. | Shafique *et al.* (2005) |
| Myrtaceae | *Eucalyptus citriodora* Hook. | Aqueous extracts of dry leaves | Field | *Phalaris minor* Retz | Javaid *et al.* (2006) |
| Myrtaceae | *Syzygium cumini* (L.) Skeels | Fresh leaves aqueous leaves extracts | Field | *Parthenium hysterophorus* L. | Shafique *et al.* (2005) |
| Passifloraceae | *Passiflora edulis* Sims | Aqueous extract | Field | *Echinochloa crus-galli* (L.) P.Beauv. and *Monochoria vaginalis* (Burm.f.) C.Presl ex Kunth | Khanh *et al.* (2006) |
| Poaceae | *Cenchrus pennisetiformis* Steud. | Roots and shoots aqueous extracts | Field | *Parthenium hysterophorus* L. | Javaid and Anjum (2006) |
| Poaceae | *Dichanthium annulatum* (Forssk.) Stapf | Roots and shoots aqueous extracts | Field | *Parthenium hysterophorus* L. | Javaid and Anjum (2006) |
| Poaceae | *Oryza sativa* L. | Phytotoxic residues | Field | *Echinochloa crus-galli* (L.) P.Beauv., *Cyperus difformis* L. and *Ludwigia octovalvis* (Jacq.) P.H.Raven | Pheng *et al.* (2010) |
| Poaceae | *Sorghum halepense* (L.) Pers. | Roots and shoots aqueous extracts | Field | *Parthenium hysterophorus* L. | Javaid and Anjum (2006) |
| Poaceae+ Compositae | [*Sorghum bicolor* (L.) Moench] + [*Helianthus annuus* L.] | Herbage extract mixture | Field | *Avena fatua* L. *and Phalaris minor* Retz | Mushtaq *et al.* (2010) |

# 21 Commercialization of Ethnoveterinary Botanical Products

David R. Katerere

Department of Pharmaceutical Sciences, Faculty of Science, Tshwane University of Technology, Arcadia Campus, Pretoria, Republic of South Africa

## 21.1 Introduction

Traditional veterinary medical practice (more commonly called ethnoveterinary medicine) is thought to have evolved at the same time as humans and human medicine, especially at the point of domestication of animals. Ethnoveterinary medicine deals with people's knowledge, skills, methods, practices and beliefs about the care of their animals (McCorkle, 1986). In India and China, documentation of the use of ethnoveterinary medicine stretches over five millennia (Nair and Unnikrishnan, 2010; Teng *et al.*, 2010). Ethnoveterinary medicine continues to thrive in most post-colonial societies in Asia, Africa and Latin America because modern veterinary practice is limited in its geographical reach due to high cost and severe shortages of trained personnel (Tamboura and Okoli, 2010). Herbal medicines are the cornerstone of rational ethnoveterinary practices. Their use is largely supported by anecdotal evidence, and science is increasingly validating many of the claimed therapeutic indications.

At present, most ethnoveterinary botanicals are used in their crude form, especially in resource-poor settings and developing countries. In developed countries, on the other hand, formulated ethnoveterinary botanical products (or phytotherapies) have gained general acceptance. The interest in these products, especially for companion animals and pets, has mirrored the rise in the use of complementary and alternative medicines in humans in western Europe, North America and Japan. The general move towards natural therapies is mainly due to easy availability and accessibility of herbal products, particularly from India and China, popularization of alternative healing modalities by opinion makers (e.g. Steve Jobs and the singer Madonna), preference towards preventive medicine by an aging population, disaffection with conventional medicine (which is seen as offering no curative outcomes and has more adverse effects), and the user empowerment that comes through self-medication (Calixto, 2000; Mackenzie and Rakel, 2006; Weil, 2011).

The World Health Organization (WHO) defines phytotherapies as herbal drugs which contain complex mixtures of one or several plant species which may be standardized and formulated into pharmaceutical delivery forms, or used in their crude form (WHO, 1993). It is recognized that these products are based on traditional knowledge and folk use; they are generally more affordable than conventional pharmaceuticals, but their active principles are frequently unknown and their quality, safety and efficacy may not be assured (Calixto, 2000).

Owing to the burgeoning interest in the use of natural therapies, in particular herbal products, international trade has increased worldwide. The global demand for medicinal plants is projected to grow from US$14 billion in the next four decades to over US$5 trillion by the year 2050 (Aneesh *et al.*, 2009; Vargas-Murga *et al.*, 2011). This interest has made commercialization an attractive proposition to many pharmaceutical companies. However, in seeking to commercialize botanical products, whether for human or veterinary use, there are various considerations to be taken into account. These are discussed briefly below.

### 21.1.1 Safety, toxicity and efficacy

There is a general misconception that herbal products are 'safe because they are natural' (Naidoo and Seier, 2010; Nair and Unnikrishnan, 2010). As a consequence of the increasing number in herbal toxicity reports (Huxtable, 1990; Beuers *et al.*, 1991; Graham-Brown, 1992; Larrey 1997), warnings have been issued with regard to the quality and safety of products that have long enjoyed 'generally regarded as safe' (GRAS) status. Regulatory authorities in many countries and the media have issued warnings against some popular herbal products, and Chinese medicines in particular are increasing coming under scrutiny.

The toxicity of botanicals can be classified as intrinsic and extrinsic (Drew and Myers, 1997). The former is due to pharmacological interactions and is common with all xenobiotics and may be dose dependent or as a result of drug–herb or food–herb interactions. These adverse reactions may manifest as allergenic reactions, or organ toxicity such as cardiotoxicity and hepatoxicity. Nephrotoxicity is considered the most common manifestation of xenobiotic intoxication because the kidneys possess a large, highly vascularized endothelial surface area which results in high exposure to xenobiotics and thus increased risk of injury (Wojcikowski et al., 2004; Perazella, 2009; Jha, 2010). Commonly used products that have been shown to have a high risk of renal injury include *Ginkgo biloba*, liquorice (*Glycyrrhiza glabra*), alfalfa or noni juice (*Morinda citrifolia*) and willow bark (*Salix* spp.). These products should therefore be used with caution.

In animals, intrinsic toxicity may be species-specific, for example, cats are highly sensitive to various *Lilium* and *Hemerocallis* species (Fitzgerald, 2010).

Extrinsic toxicity, on the other hand, occurs due to lapses in the manufacturing process; for example, misidentification, adulteration

*Novel Plant Bioresources: Applications in Food, Medicine and Cosmetics*, First Edition. Edited by Ameenah Gurib-Fakim.

or inappropriate labelling. Extrinsic toxicity is probably more common in herbal preparations because many of them are not manufactured under current good manufacturing practice conditions and their distribution is largely unregulated. Manufacturing problems (e.g. misidentification of plant species, lack of standardization, microbial and mycotoxin contamination, substitution and deliberate adulteration of preparation) have been previously reported (Ernst, 2002; Thembo *et al.*, 2007; Arpadjan *et al.*, 2008). Plant products with known toxic components (e.g. pyrrolizidine alkaloids, aristolochic acid and berberine) or undeclared pharmaceuticals have, over the years, been found and withdrawn from commercial distribution (Ernst, 2002). However, they may continue to be used in less-regulated and informal markets.

Toxicity testing is an important consideration for herbal products. Naidoo and Seier (2010) suggested that single-dose acute, repeat-dose toxicity and reproductive toxicity testing should be conducted in the target animal species. Tolerance testing, in which the product is tested at the expected therapeutic dose and five and ten times the therapeutic dose, precludes the typical phase testing in humans. With production animals, special consideration should be paid to understand the elimination half-life of the product administered in order to determine withdrawal period after which it is safe to consume the eggs, milk or meat of the treated animals (Nair and Unnikrishnan, 2010). With herbal products, this is generally complicated because of the multiple constituents present, which are often not all known or chemically characterized.

In terms of efficacy, much of what is known about ethnoveterinary products is based on anecdotal evidence. In some cases, efficacy in human studies has been used to justify veterinary use; for example, the use of *Harpagophytum procumbens* to treat inflammatory disease. While some of the most commonly used herbals (e.g. *G. biloba* and *Hypericum perforatum*) have been subjected to controlled human clinical trials (Linde *et al.*, 1996; Bensoussan, 1998), there are few documented studies on veterinary botanicals.

### 21.1.2 Intellectual property rights and access and benefit-sharing regimes

There are now legal requirements which regulate the accessing and use of plant material and commercialization of indigenous knowledge at country level. The Convention for Biological Diversity (CBD), signed in 1992, recognized the principle that biological resources are the sole property of sovereign states and that these resources are tradable commodities (Kartal, 2007). In doing this, the CBD assigns ownership of biodiversity and the associated traditional knowledge (if any) to indigenous and local communities. There is, however, a general recognition that the current international legal frameworks may not be suited to the protection of indigenous knowledge. For example, it is difficult to protect indigenous knowledge using conventional intellectual property modalities, such as patents, copyrights and trademarks, because they may not show the 'novelty, originality and industrial utility' of other inventions (Kartal, 2007).

A *sui generis* (unique) protection system is currently evolving. The 'Bonn Guidelines of Access to Genetic Resources and Fair and Equitable Sharing of Benefits Arising Out of Their Utilization' were agreed upon as a cornerstone of such a protection regime (Nair and Unnikrishnan, 2010). They are meant to facilitate development of harmonized 'access and benefit-sharing' national laws and policies that should incorporate involvement of all relevant stakeholders, prior informed consent, condition for material transfer and sharing of benefits. The downside of these regulations is that they may create bureaucratic hurdles to basic research endeavours (WHO, 2008), raise unrealistic expectations of benefits leading to potential conflict, and also commodify cultural and traditional knowledge (Shiva, 2007).

### 21.1.3 Herbal medicinal products regulation

Establishing the quality of allopathic medicines is one of the most important requirements in pharmaceutical production to ensure the safety of the consumer. National medicines regulatory authorities use regulations, guidelines, protocols and pharmacopoeias to assess adherence to current good manufacturing practices, establish the quality, safety and efficacy of products, and grant marketing approvals. In contrast, herbal medicinal products (HMPs) are not strictly regulated as they may be largely traded in informal channels. However, the quality control of HMPs is important for various reasons, including:

- variation in the source and quality of the raw material (Figueiredo *et al.*, 2008);
- botanical substitution and misidentification (Barnes, 2003);
- adulteration (intentional or unintentional) and contamination with active synthetic drugs, pesticides and heavy metals (Keane *et al.*, 1999).
- intra-species chemotypic variation (Vermaak *et al.*, 2010; Pereira *et al.*, 2011).

Medicine regulation is aimed at protecting users from unsafe, poor quality and inefficacious medicinal products. However, because it has largely been concerned with conventional pharmaceuticals, which are single-compound agents manufactured in a highly regulated and systematically controlled environment, application of these regulations to herbal products (which may be based on cultural and magico-religious beliefs), made up of complex mixtures of several plant species with a cocktail of largely unknown actives, is sometimes viewed as inflexible.

Unlike with the pharmaceutical products, there are no harmonized regulations, and individual countries have adopted different regimes to deal with HMPs. In China, traditional Chinese medicine is practised alongside conventional medicine, circumventing the need for explicit controls; in the USA, most herbal products are classified under the Dietary Supplement Health and Education Act of 1994 as nutritional supplements/food (not drugs) and their regulation does not fall under the ambit of the Food and Drug Administration (Calixto, 2000). Some African countries are currently developing guidelines. However, there is the extra challenge of a huge informal trade in medicinal plant products as well as an entrenched culture of African traditional medicine where plants are not packaged but used extemporaneously.

Most other countries have regulations that are largely based on the German Commission E (Phytotherapy and herbal substances) of 1978, the Australian Therapeutic Good Act of 1990 and, to some extent, the WHO guidelines. The criteria they set for granting market approval takes into consideration traditional use, available scientific data (which includes chemical data, pharmacological, toxicological and clinical studies) and any patient records that may be available.

The lack of pharmacopoeial information about herbal constituents and the sheer number of potential species (between 30 000 and 70 000 by some estimates) (Affolter and Pengelly, 2007) and plant parts that can be used can also make designing and implementing guidelines complicated.

### 21.1.4 Standardization

Batch-to-batch variation of herbal products is common and occurs as a result of biological factors (plant species, plant parts used) environmental conditions (growing location and harvest season and conditions), storage conditions (light, temperature, oxygen tension and time) and agronomic conditions (the use of fertilizers, irrigation, soil type) (Huyghebaert *et al.*, 2011). Quality control of the whole value chain from sourcing the raw material to the preparation of a stability-tested formulated product is important. Protocols for quality control must encompass the source and traceability of raw material (wild-crafted or cultivated), time of collection, exclusion of microbial, heavy metal and pesticide residue contaminants, and the manufacturing and storage processes (Ernst and Dixon, 2006).

The WHO recommends chemical analysis as the best method for plant identification, authentication and detection of raw material contaminants, in addition to morphological characterization (Shinde *et al.*, 2009). Standardization can be used to, at a minimum, identify the marker compounds and determine the acceptance criteria (quantitative and qualitative). The major constituents that are easily identifiable are not necessarily the active ones, but they allow for some level of consistency which ensures that batches do not vary widely in composition and hopefully in biological activity.

Sensitive techniques that can profile the whole metabolome, including trace compounds, are ideal; however, the absence of commercial reference standards for all constituent molecules makes identification difficult and inconclusive (Ernst, 2004; Chan, 2009). The lack of reference standards is a huge barrier to the commercial development and standardization of HMPs. For example, of the more than 20 000 plant species known to grow in Africa, less than 100 have been documented according to pharmacopoeial requirements, which is a first step to enable commercial use (Brendler *et al.*, 2010).

Several techniques are currently employed for the chemical analysis of herbal products. The most common ones are conventional chromatographic techniques, such as gas and liquid chromatography that aim to identify and quantify chemical constituents within a mixture. The methods are highly selective, reproducible, accurate and have low detection limits (Breithaupt, 2004). Despite their widespread use, they have the disadvantages of requiring laborious and time-consuming sample clean-up, being expensive, prone to errors and not always reproducible between laboratories (Qiao and Van Kempen, 2004).

More elegant techniques, like near-infrared (NIR) and mid-infrared (MIR) vibrational spectroscopy and Raman spectroscopy, are increasingly being used. They have the advantages of being robust, efficient, non-destructive, inexpensive and require little or no sample preparation (Schulz *et al.*, 2004).

### 21.1.5 Supply-chain management

For commercialization, the availability of the plant material in large enough quantities to ensure constant year-round supply is critical. Wild harvesting generally does not assure this and can inevitably lead to loss of habitats and negative environmental consequences for the target species, as well as for other organisms in its ecological niche. At present, of the 50 000 medicinal plants in commercial use, over two-thirds are collected from the wild (Canter *et al.*, 2005).

Wild-harvested plants show much greater variability in physical quality (e.g. plant health, leaf size) and chemical composition, which poses a quality problem (Affolter and Pengelly, 2007). Domestication and cultivation are therefore the chosen methods for countering the problems of wild-crafting and conserving biodiversity (Sen *et al.*, 2011). In some cases, this is combined with genetic modification to increase yield, reduce growing season or enhance biological activity (Canter *et al.*, 2005). However, domestication and/or cultivation may also pose a problem, in that the bioactive compounds may not to be produced in sufficient quantities because of environmental changes. In addition, the desired plant species might be more prone to disease once they are in monoculture.

## 21.2 Therapeutic areas for ethnoveterinary applications

In veterinary medicine and animal science, there are certain therapeutic areas that are underserved by the current portfolio of pharmaceuticals because of, among other reasons, growing antibiotic resistance to current drugs and emerging concerns about ecotoxicity and residues in food. In bioprospecting for therapeutic leads from ethnoveterinary plants, it would be advisable to focus on these indications for potential maximum commercial and clinical impact. These are internal and topical parasitic diseases, inflammatory conditions and growth promotion. These are discussed in more detail below.

### 21.2.1 Herbs for treating helminthosis

Helminthosis is infection with a wide range of intestinal worms. The most common are protozoal and flagellate species, roundworm (*Ascaris strongyloides*) and hookworm (*Ancylostoma caninum* and *Necator americanus*) (Wynn and Fougère, 2007).

Helminths in ruminants are of major significance in a wide range of agroclimatic zones, and lead to high mortalities, reduced food conversion efficiency and other losses of production (Tembely *et al.*, 1996). Various factors determine the extent and effects of helminthic infections; for example, livestock species and breed, age and health status of host, the parasite species and worm population involved. Co-morbidities, weather conditions and husbandry practices are also important considerations.

In the past, extracts of betel nut palm (*Areca catechu*), male fern (*Dryopteris fillix-mas*), wormwood (*Artemisia* spp.) and pomegranate (*Punica granatum*) were the mainstay of anthelmintic therapy in veterinary medicine (Wynn and Fougère, 2007). The anthelmintic activities of extracts have been shown to be species specific and may depend on how different species metabolize certain chemicals; for example, condensed tannins were shown to be more active against abomasal worms in goats but not in sheep (Paolini *et al.*, 2003; Athanasiadou *et al.*, 2007).

Whole plant of *Artemisia brevofolia* reduced abomasal nematode (*Haemonchus contortus*) egg counts by 62% (Iqbal *et al.*, 2004). Cassava (*Manihot esculenta*) hay showed efficacy in reducing faecal egg counts and worm population in sheep (Sokerya, 2003).

*Peltophorum africanum* is widely used for treating intestinal worms, and acetone extracts of various parts have shown some activity against *Haemonchus contortus* and *Trichostrongylus colubriformis* with activity at concentrations less than 0.2 mg/mL (Bizimenyera, 2007).

In India, seeds of various fruits and culinary herbs are widely used in multi-herb mixtures. Pastes are made from papaya (*Carica papaya*) seeds alone or in combination of *Embelia ribes* seeds, *Kaempferia galangal* rhizomes, *Allium sativum* bulbs and *Vitex negundo* leaves. Citrus and sesame seeds are also widely used (Nair and Unnikrishnan, 2010). Papaya seeds are also widely used in South America and the West Indies for de-worming animals (Quinlan *et al.*, 2002). Activity has been shown against *Caenorhabditis elegans*, which has been attributed to the presence of cysteine protein enzymes and possibly benzyl isothiocynate (El Moussaoui *et al.*, 2001).

Papaya latex has been shown to reduce worm count in pigs infected with *Ascaris suum* when administered in high doses (Satrija *et al.*, 1994). Mild diarrhoea was a common side effect, which is beneficial because it may expel the worms as well (Wynn and Fougère, 2007).

Waller *et al.* (2001) have extensively reviewed the use of culinary and dietary herbs to control intestinal worms. There is now good evidence to support the use of carrot and various brassicas, chicory, caraway, thyme and garlic against nematode infections. Allicin, diallyl trisulfide and ajoene, constituents of garlic, have been reported to have broad-spectrum activity against protozoal and flagellate species (Wynn and Fougère, 2007; Williams and Lamprecht, 2008). Boiling or cooking garlic has been shown to inactivate the bioactive allyl sulfides, which are highly volatile (Cavagnaro and Galmarini, 2012). Allicin is formed through the action of allinase on alliin, which occurs on crushing fresh garlic. Crushed garlic was shown to be active against *Capillaria* species in carp, but whole bulbs demonstrated inferior activity in donkeys (Wynn and Fougère, 2007). This is probably as a result of the formation of allicin from alliin (diallyl disulfide) during the crushing or mincing process. However, animal feeds high in garlic may cause anaemia.

By far the most interesting herbs with potential for use in this therapeutic area are *Hydrastis canadensis* (goldenseal) and *Inula helenium* (elecampane). They are widely used and readily available, which means that, from a safety and commercial availability perspective, the barriers to market entry may be low. *In vitro* and *in vivo* studies on tapeworm (*Echinococcus granulosus*) have shown a larval mortality of 70% when treated with low concentrations of hydrastine (Chen *et al.*, 1991; Wynn and Fougère, 2007). Alantolactone isolated from elecampane has shown good anthelmintic activity (Wynn and Fougère, 2007). However, there has been no follow up on these compounds and extracts into veterinary clinical testing and product development.

### 21.2.2 Herbs for treating ectoparasitosis

Ectoparasites are parasites which infest and live on or in the skin; for example, lice, scabies, cutaneous larval migrans and sandflea (Heukelbach and Feldmeier 2004). They include blood-sucking flies, ticks, fleas and lice and may cause blood-borne diseases; for example, babesiosis and zoonotic diseases like psittacosis. Tick-borne diseases are a major cause of animal and human morbidity and mortality (Gitau *et al.*, 1999; Jongegan and Uilenberg, 2004). Treatment involves the use of ovicidal or larvacidal compounds that interrupt the early stages of the parasite's life cycle, or anthropocidal or repellent compounds (Blagburn and Dryden, 2009). Conventional treatments largely involve the use of insecticides (e.g. pyrethroids or anthelmentics like ivermectin) which are used in an off-label manner in these circumstances. However, widespread resistance has been reported, mainly because of the irrational use of the products (i.e. overuse of the same products or similar ones, underdosing or incomplete treatment). As a result of growing resistance and the environmental toxicity of conventional treatments, various natural remedies are now being investigated, many of which have a long history of ethnoveterinary use (Heukelbach and Feldmeier, 2004; George *et al.*, 2008).

*Eupatorium adenophorum* extracts were tested against *Sarcoptes scabiei* in New Zealand rabbits (Nong *et al.*, 2013). The extracts showed efficacy similar to the ivermectin positive controls over a 2–4-week period at high concentrations. The active compounds isolated from *E. adenophorum* are 2-acetoxy-3,4,6,11-tetrahydroca dinen-7-one and the sesquiterpenoid 8aα-hydroxy-1-isopropyl-4,7-dimethyl-1,2,3,4,6,8a-hexahydro-naphthalene-2,6-dione (He *et al.*, 2006). Clinical application in other animal species needs to be done to confirm efficacy of the extracts and the isolated compound.

Nchu *et al.* (2005) showed that acetone extracts of *Allium porum* had a high repellency effect on *Hyalomma marginatum rufipes*, while *Allium sativum* dichloromethane extracts were highly toxic to the ticks. *Lippia javanica* and *Tagetes minuta*, both widely used in villages in southern Africa as acaricides, were found to have activity against *H. marginatum rufipes* in the same study.

Essential oils from *Melaleuca alternifolia* (tea tree), *Cedrus deodara*, *Pongamia glabra*, *Jatropha curcas*, *Cinnamomum zeylanicum* and *Azadirachta indica* (neem) have also shown good acaricidal activity in various studies (Nong *et al.*, 2013). Neem is widely used for ectoparasites and has been shown in various studies to have broad-spectrum insecticidal activity owing to the presence of azadirachtins and salannin (Wynn and Chalmers, 2002). Neem oil is formulated into pet products and the oil has been shown to suppress fecundity, induce sterility, modulate oviposition and induce changes in biological fitness against various ectoparasites (Mulla and Su, 1999). Tea tree oil has acaricidal properties (Iori *et al.*, 2005) and has been formulated together with neem. Aromatic plants (e.g. lemon, lavender and *Mentha* species) have been used with good repellency against fleas and other insects (Lans *et al.*, 2006). However, essential oils pose challenges with formulating pharmaceutically stable products.

### 21.2.3 Herbs for treating inflammation and pain

Pain is the most common symptom of injuries and inflammatory diseases. At the biochemical level, various mediators are involved, the most important ones being the transcription factor, nuclear factor-kappa B (NF-κB), inducible nitric oxide synthase (iNOS – which inhibits production of nitric oxide (NO)), prostaglandin E2 ($PGE_2$), cyclooxygenase-2 (COX-2 – which stimulates prostaglandin production in inflamed tissue), tumour necrosis factor (TNF-α) and interleukin (IL-1β) (Kassim *et al.*, 2010a; Wang *et al.*, 2013).

Chronic pain and musculoskeletal inflammation have both performance and economic consequences on livestock and companion animals. This has become an increasingly important therapeutic area, especially for dogs and horses. Inflammation may be short term, or as a result of age or physical activity, which may result in lameness and arthritis. Treatment involves the use of analgesic or anti-inflammatory medication, while longer term alternatives are being employed (Wynn and Fougère, 2007).

Central and South American Indians historically used *Capsicum annum* (cayenne), which today is a registered topical rubefacient for the relief of rheumatoid arthritis and osteoarthritis (Nair and Unnikrishnan, 2010). Capsaicinoids are known to be the bioactive compounds.

*Curcuma longa* (turmeric) and *Zingiber officinale* (ginger) are used for rheumatism in traditional Chinese medicine and ayurveda. It is thought that the curcuminoid and gingerol compounds inhibit pro-inflammatory prostaglandins (Ali *et al.*, 2008; Vitetta *et al.*, 2008). Curcuminoids also inhibit platelet aggregation and stabilize lysosomal membranes. At high concentrations they stimulate adrenal glands to release cortisone (dos Santos Jaques *et al.*, 2011). A clinical trial with dogs and two different species of turmeric (i.e. *Curcuma domestica* and *Curcuma xanthorrhiza*) showed activity in osteoarthritis significantly better than the control based on veterinary evaluations (Innes *et al.*, 2003).

Devil's claw (*H. procumbens*), flax seed (*Linum usitatissimum*), yucca (*Yucca schidegera*) and arnica (*Arnica montana*) are by far the herbs of choice in the treatment of joint and arthritic pains in animals. All of these have well-established efficacy and many are formulated into various commercially available products in humans. *H. procumbens*, which has a long history of use in Namibia, south-western South Africa and Botswana, has been shown to suppress $PGE_2$ synthesis and NO production by inhibiting COX-2 and iNOS mRNA expression (Mncwangi *et al.*, 2012). It has shown clinical efficacy when used to treat lower back pain in human subjects in controlled trials.

Extracts and compounds from *L. usitatissimum* (flax seed) have been shown to interfere with cyclo-oxygenase enzymes, the former being more selective for COX-2 (Pearson *et al.*, 1999; Williams and Lamprecht, 2008).

*Y. schidegera* and arnica extracts have been shown to inhibit NF-κB resulting in the suppression of iNOS, and consequently of the inflammatory mediator NO (Klaas *et al.*, 2002; Cheeke *et al.*, 2006). Yuccaol C isolated from *Y. schidegera* was found to be the active compound. However, more work (e.g. randomized clinical trials in target species) needs to be done on these plant species in order to commercialize them on a broader scale.

*Salix alba* is also widely used to treat pain, arthritis and lameness in horses and ruminants. It is the source of the glycoside salicilin and other anti-nociceptive phenolic glycosides.

*Urtica dioica* (stinging nettle) has a long history of use in treating inflammation and joint disease. This has now been elucidated to be exerted via modulation of various cytokines, including IL-1β and NF-κB (Teucher *et al.*, 1996). Schulze-Tanzil *et al.*, 2002 have reported that there is reduced joint cartilage degradation because nettle extracts suppress the activity of extracellular matrix metalloproteinase. The phenolic acid caffeoyl malic acid has been shown to inhibit cyclo-oxygenase and 5-lipoxygenase cytokines (Obertreis *et al.*, 1996). Toxicity at higher doses manifested as ataxia, distress and muscle weakness in horses (Bathe, 1994).

*G. glabra* (liquorice) is a popular source of flavour in confectionaries and, therefore, easily commercializable. It is rich in triterpenoids, of which most biological activity is attributed to the glycyrrhizin, which has been shown to inhibit prostaglandins and leukotrienes (Wynn and Fougère, 2007). It would appear that many of the other species that have shown good anti-inflammatory effects possess pentacyclic triterpenoids, either free or glycosylated or saponins; for example, *Bupleurum falcatum* (which contains saikosaponins), *Centella asiatica* (asiaticoside and madecassoside), *Boswellia serrata* (boswellic acids), *Dioscorea villosa* (diosgenin), *H. procumbens* (harproside) (Wynn and Fougére, 2007; Mncwangi *et al.*, 2012). The anti-inflammatory effects of triterpenoids are well known (Ríos *et al.*, 2000) and have been shown in various *in vivo* and *in vitro* experiments. The mechanism of action appears to involve modulation of various pro-inflammatory processes/pathways, including 5-lipoxygenase, elastase, phospholipase $A_2$, histamine, collagenase and interleukin release, corticosteroid metabolism and complement and protein-kinase activities (Huguet *et al.*, 2000). Structure–activity relationships have been shown to be important; for example, ursane-type triterpenoids were better at inhibiting 5-lipoxygenase activity than oleanane derivatives were, which showed better effects on corticoid metabolism.

Phenolics are also important anti-inflammatory compounds, of which flavonoids are well studied. For example, hesperetin and naringin induce haem-oxygenase-1, an antioxidant enzyme that triggers an anti-inflammatory response (Goven *et al.*, 2010) and also inhibits lipopolysaccharide-induced NO production. Other flavonoids, like genistein, kaempferol, quercetin, daidzein, caffeic acid, chrysin and ellagic acid, interfere with production of iNOS, in some cases by down-regulation of NF-κB (Kassim *et al.*, 2010b). In general, phenolic compounds are said to act by inhibiting molecular vasodilators, such as NO, and also inhibiting $PGE_2$ (Kassim *et al.*, 2010b). The presence of phenolics may therefore explain the rationale for the use of *Zingiber* spp. (which contain gingerol), *Yucca* (yuccaol), *Salix* (salicins), *Curcumin* (curcuminoids) and carnosol-containing herbs; for example, *Rosmarinus officinalis* and *Salvia* spp. (Kamatou *et al.*, 2008).

Alkaloids are the quintessential pain killers, but plants known to contain these compounds are not very popular in ethnoveterinary practice for this indication. Such plants are more widely used for anti-parasitic remedies, for example, *Vernonia* spp. for anthelmintic (Nair and Unnikrishnan, 2010) or *Zanthoxylum chalybeum* for theileriosis and trypanosomosis (Nair and Unnikrishnan, 2010).

### 21.2.4 Phytogenic growth promoters

Antibiotic feed additives (AFAs) are widely used in livestock production. Administering sub-therapeutic concentrations of antibiotics to production animals reduces the incidence and severity of subclinical infections and improves luminal absorption of nutrients, thus enhancing growth (Bach Knudsen, 2001; Alexander *et al.*, 2008). However, the use of AFAs has raised concerns about the potential for residual exposure and cross-resistance in the human population and potential toxicities to the host animals (Salem *et al.*. 2011). In

2006, a total ban on AFA use was enforced in the European Union (EU) region, and other territories are predicted to follow. As a result of this ban, there is an increasing interest in finding alternative growth promoters (AGPs); for example, organic acids, probiotics and phytobiotics (also more commonly known as phytogenic feed additives) (Huyghebaert *et al.*, 2011).

Plants, especially spices and so-called culinary herbs, have been shown to be beneficial to gut health, exerting laxative and spasmolytic effects and stimulating digestive secretions, all of which enhance nutrient absorption (Windisch *et al.*, 2008). They are also said to possess antioxidant, antimicrobial and immunomodulating properties. The Lamiaceae (mint) species (e.g. rosemary, thyme and oregano) is an important family for deriving phytogenic feed additives. This is because the herbs have GRAS status and they are cultivated on a commercial basis and so can be easily used for product development. The active components are known to be the essential oils.

Essential oils have been shown to stimulate endogenous digestive enzymes, enhance production of digestive secretions, stimulate blood circulation, and exert antioxidant and/or antimicrobial properties, all resulting in growth promotion (Brenes and Roura, 2010; Huyghebaert *et al.*, 2011). In one study, Arbor Acres broilers fed essential oils had longer duodenum villi, more tender breast muscles and juicier thigh muscles compared with the control and antibiotic groups (Hong *et al.*, 2012). Feed/gain ration was also improved and survival rate was comparable to the antibiotic control. In another study on swine, gastric emptying was reduced when the animals were fed essential oils and capsaicin (Manzanilla *et al.*, 2004).

Saponin-containing plants (e.g. *Y. schidegera*) are also important feed additives as the saponins are said to reduce intestinal ammonia formation, whose emission is known to cause stress in animals in the housing environment (Windisch *et al.*, 2008). *Yucca* has the advantage of also possessing anti-inflammatory properties, as discussed earlier.

Other phytogenic additives have also shown potential for use as AGPs. A mixture of *Houttuynia cordata* and *Taraxacum officinale* extract powder fed to finishing pigs resulted in increased growth performance, dry matter and nitrogen digestibility, white blood cell concentration and improved antioxidant status (as measured by thiobarbituric-acid-reacting substances) (Yan *et al.*, 2011a).

Fermentation of plant products has also been shown to give different effects. For example, fermented garlic powder showed improved growth performance and better meat marbling and firmness score in pigs when compared with controls (Yan *et al.*, 2011b). *Aspergillus niger*-fermented *G. biloba*-leaves supplementation in broilers resulted in improved physiological outcomes, namely an increase in luminal protease and amylase activities, higher ($P < 0.05$) mucosal alkaline phosphatase activities in the duodenum and jejunum of treated groups and increased levels of sodium glucose co-transporter 1 (Zhang *et al.*, 2012). The overall effect was improved feed efficiency and intestinal function. The effects of fermented *G. biloba* leaves was superior to that of non-fermented leaves, possibly because of more bioactive metabolites and increased bioavailability of other metabolites resulting from the fermentation process. Fermentation is increasingly being seen as an ideal way of resolving overproduction of fruits and vegetables by preserving them for later silage use.

A whole range of phytochemicals, including polysaccharides, organic acids, alkaloids and essential oil constituents – for example, *p*-cymene, coriander (linalol), oregano (carvacrol), rosemary (cineol), sage (cineol) and thyme (thymol), clove (eugenol), mustard (allylisothiocyanate), cinnamon (cinnamaldehyde) and garlic (allicin) – have been shown to promote growth (Huyghebaert *et al.*, 2011; Salem *et al.*, 2011). The modes of action are poorly understood, but may be by improving the immune function, enhancing gut functions (e.g. enzyme activity, microbial eubiosis), improving nutrient digestibility and protein metabolism, decreasing ruminal methane production or reducing nutritional stress, such as bloat (Liu *et al.*, 2011).

While there is great potential in the use of phytobiotics, safety to the animal, the user (from occupational hazards), the consumer of the products and the environment needs to be established. Regulatory authorities, particularly in the EU, also have stringent criteria for approval of such use, which includes the identity and traceability of the entire formulated product and its claimed efficacy.

## 21.3 Conclusion

From the foregoing, it is evident that ethnoveterinary medicines have the potential to be commercialized. However, navigating the road to the market is perilous. It requires that all issues concerning pharmacological development (safety and efficacy) and pharmaceutical, agronomic (quality) and legal matters should be addressed. This chapter has outlined these and also identified potentially profitable therapeutic areas where the need for botanicals exists.

## Acknowledgements

Dr M. Sandasi is thanked for useful discussions and peer review. This chapter is dedicated to the late Professor Gordon Chavhunduka, the father of traditional medicine in southern Africa and an inspiration to many of us.

## References

Affolter, J.M. and Pengelly, A. (2007) Conserving medicinal plant biodiversity, in *Veterinary Herbal Medicine* (eds S. Wynn and B. Fougère), Mosby, St. Louis, pp. 257–263.

Alexander, T.W., Yanke, L.J., Topp, E. *et al.* (2008) Effect of subtherapeutic administration of antibiotics on the prevalence of antibiotic-resistant *Escherichia coli* bacteria in feedlot cattle. *Applied and Environmental Microbiology*, **74** (14), 4405–4416.

Ali, B.H., Blunden, G., Tanira, M.O. and Nemmar, A. (2008) Some phytochemical, pharmacological and toxicological properties of ginger (*Zingiber officinale* Roscoe): a review of recent research. *Food and Chemical Toxicology*, **46** (2), 409–420.

Aneesh, T.P., Hisham, M., Sonal Sekhar, M. *et al.* (2009) International market scenario of traditional Indian herbal drugs – India declining.... *International Journal of Green Pharmacy*, **3**, 184–190.

Arpadjan, S., Çelik, G., Taşkesen, S. and Güçer, Ş. (2008) Arsenic, cadmium and lead in medicinal herbs and their fractionation. *Food and Chemical Toxicology*, **46** (8), 2871–2875.

Athanasiadou, S., Githiori, J. and Kyriazakis, I. (2007) Medicinal plants for helminth parasite control: facts and fiction. *Animal*, **1** (9), 1392–1400.

Bach Knudsen E.K. (2001) Development of antibiotic resistance and options to replace antimicrobials in animal diets. *Proceedings of the Nutrition Society*, **60** (03), 291–299.

Barnes, J. (2003) Quality, efficacy and safety of complementary medicines: fashions, facts and the future. Part I. Regulation and quality. *British Journal of Clinical Pharmacology*, **55** (3), 226–233.

Bathe, A.P. (1994) An unusual manifestation of nettle rash in three horses. *Veterinary Record*, **134** (1), 11–12.

Bensoussan, A., Talley, N.J., Hing, M. *et al.* (1998) Treatment of irritable bowel syndrome with Chinese herbal medicine: a randomized controlled trial. *JAMA*, **280** (18), 1585–1589.

Beuers, U., Spengler, U. and Pape, G.R. (1991) Hepatitis after chronic abuse of senna. *Lancet*, **337** (8737), 372–373.

Bizimenyera, E. S. (2007) The potential role of antibacterial, antioxidant and antiparasitic activity of *Peltophorum africanum* Sond. (Fabaceae) extracts in ethnoveterinary medicine. PhD thesis, University of Pretoria.

Blagburn, B. L. and M. W. Dryden (2009) Biology, Treatment, and Control of Flea and Tick Infestations. *Veterinary Clinics of North America: Small Animal Practice* **39**(6): 1173-1200.

Breithaupt, D.E. (2004) Simultaneous HPLC determination of carotenoids used as food coloring additives: applicability of accelerated solvent extraction. *Food Chemistry*, **86** (3), 449–456.

Brendler, T., Eloff, J., Gurib-Fakim, A. and Phillips, L.D. (eds) (2010) *African Herbal Pharmacopeia*, AAMPS Publishing, Mauritius.

Brenes, A. and Roura E. (2010) Essential oils in poultry nutrition: main effects and modes of action. *Animal Feed Science and Technology*, **158** (1–2): 1–14.

Calixto, J.B. (2000) Efficacy, safety, quality control, marketing and regulatory guidelines for herbal medicines (phytotherapeutic agents). *Brazilian Journal of Medical and Biological Research*, **33**, 179–189.

Canter, P.H., Thomas, H. and Ernst, E. (2005) Bringing medicinal plants into cultivation: opportunities and challenges for biotechnology. *Trends in Biotechnology*, **23** (4), 180–185.

Cavagnaro, P.F. and Galmarini, C.R. (2012) Effect of processing and cooking conditions on onion (*Allium cepa* L.) induced antiplatelet activity and thiosulfinate content. *Journal of Agricultural and Food Chemistry*, **60** (35), 8731–8737.

Chan, T.Y. (2009) Aconite poisoning. *Clinical Toxicology*, **47** (4), 279–285.

Cheeke, P.R., Piacente, S. and Oleszek. W. (2006) Anti-inflammatory and anti-arthritic effects of *Yucca schidigera*: a review. *Journal of Inflammation*, **3** (1), 1–7.

Chen, Q.M., Ye, Y.C., Chai, F.L. and Jin, K.P. (1991) Protoscolicidal effect of some chemical agents and drugs against *Echinococcus granulosus*. *Zhongguo Ji Sheng Chong Xue Yu Ji Sheng Chong Bing Za Zhi [Chinese Journal of Parasitology and Parasitic Diseases]* **9** (2), 137–139 (in Chinese).

Dos Santos Jaques, J.A., Peres Rezer, J.F., Betsch Rezel, J. *et al.* (2011) Lung and blood lymphocytes NTPDase and acetylcholinesterase activity in cigarette smoke-exposed rats treated with curcumin. *Biomedicine & Preventive Nutrition*, **1** (2), 109–115.

Drew, A. and Myers, S. (1997) Safety issues in herbal medicine: implications for the health professions. *Medical Journal of Australia*, **166** (10), 538–541.

El Moussaoui, A., Nijs, M., Paul, C. *et al.* (2001) Revisiting the enzymes stored in the laticifers of *Carica papaya* in the context of their possible participation in the plant defence mechanism. *Cell and Molecular Life Sciences*, **58**, 556–570.

Ernst, E. (2002) Toxic heavy metals and undeclared drugs in Asian herbal medicines. *Trends in Pharmacological Sciences*, **23** (3), 136–139.

Ernst, E. (2004) Risks of herbal medicinal products. *Pharmacoepidemiology and Drug Safety*, **13** (11), 767–771.

Ernst, E. and Dixon, A. (2006) Alternative medicines in Europe, in *Regulating Pharmaceuticals in Europe: Striving for Efficiency, Equity and Quality* (eds E. Mossialos, M. Mrazek and T. Walley), Open University Press, Maidenhead, pp. 307–322.

Figueiredo, A.C., Barroso, J.G., Pedro, L.G. and Scheffer, J.J.C. (2008) Factors affecting secondary metabolite production in plants: volatile components and essential oils. *Flavour and Fragrance Journal*, **23** (4), 213–226.

Fitzgerald, K.T. (2010) Lily toxicity in the cat. *Topics in Companion Animal Medicine*, **25** (4), 213–217.

George, D.R., Guy, J.H., Arkle, S. *et al.* (2008) Use of plant-derived products to control arthropods of veterinary importance: a review. *Annals of the New York Academy of Sciences*, **1149** (1), 23–26.

Gitau, G.K., Perry, B.D. and McDermott, J.J. (1999) The incidence, calf morbidity and mortality due to *Theileria parva* infections in smallholder dairy farms in Murang'a District, Kenya. *Preventive Veterinary Medicine*, **39** (1), 65–79.

Goven, D., Boutten, A., Leçon-Malas, V. *et al.* (2010) Induction of heme oxygenase-1, biliverdin reductase and H-ferritin in lung macrophage in smokers with primary spontaneous pneumothorax: role of HIF-1α. *PLoS ONE*, **5** (5), e10886.

Graham-Brown, R. (1992) Toxicity of Chinese herbal remedies. *Lancet*, **340** (8820), 673–674.

He, L., Yang, J., Cao, A.-C. *et al.* (2006) A new sequiterpenoid from *Eupatorium adenophorum* Spreng. *Chinese Journal of Chemistry*, **24** (10), 1375–1377.

Heukelbach, J. and H. Feldmeier (2004) Ectoparasites – the underestimated realm. *Lancet*, **363** (9412), 889–891.

Hong, J.-C., Steiner, T., Aufy, A. and Lien, T.-F. (2012) Effects of supplemental essential oil on growth performance, lipid metabolites and immunity, intestinal characteristics, microbiota and carcass traits in broilers. *Livestock Science*, **144** (3), 253–262.

Huguet, A., del Carmen Recio, M., Máñez, S. *et al.* (2000) Effect of triterpenoids on the inflammation induced by protein kinase C activators, neuronally acting irritants and other agents. *European Journal of Pharmacology*, **410** (1), 69–81.

Huxtable, R.J. (1990) The harmful potential of herbal and other plant products. *Drug Safety*, **1**, 126–136.

Huyghebaert, G., Ducatelle, R. and Van Immerseel, F. (2011) An update on alternatives to antimicrobial growth promoters for broilers. *Veterinary Journal*, **187** (2), 182–188.

Innes, J.F., Fuller, C.J., Grover, E.R. *et al.* (2003) Randomised, double-blind, placebo-controlled parallel group study of P54FP for the treatment of dogs with osteoarthritis. *Veterinary Record*, **152** (15), 457–460.

Iori, A., Grazioli, D., Gentile, E. *et al.* (2005) Acaricidal properties of the essential oil of *Melaleuca alternifolia* Cheel (tea tree oil) against nymphs of *Ixodes ricinus*. *Veterinary Parasitology*, **129** (1–2), 173–176.

Iqbal, Z., Lateef, M., Ashraf, M. and Jabbar, A. (2004) Anthelmintic activity of *Artemisia brevifolia* in sheep. *Journal of Ethnopharmacology*, **93** (2–3), 265–268.

Jha, V. (2010) Herbal medicines and chronic kidney disease. *Nephrology*, **15**, 10–17.

Jongejan, F. and Uilenberg, G. (2004) The global importance of ticks. *Parasitology*, **129** (Suppl S1), S3–S14.

Kamatou, G.P., Makunga, N.P., Ramogola, W.P. and Viljoen, A.M. (2008) South African *Salvia* species: a review of biological activities and phytochemistry. *Journal of Ethnopharmacology*, **119** (3), 664–672.

Kartal, M. (2007) Intellectual property protection in the natural product drug discovery, traditional herbal medicine and herbal medicinal products. *Phytotherapy Research* **21**(2): 113-119.

Kassim, M., Achoui, M., Mansor, M. and Yusoff, K.M. (2010a) The inhibitory effects of Gelam honey and its extracts on nitric oxide and prostaglandin E2 in inflammatory tissues. *Fitoterapia*, **81** (8), 1196–1201.

Kassim, M., Achoui, M., Mustafa, M.R. *et al.* (2010b) Ellagic acid, phenolic acids, and flavonoids in Malaysian honey extracts demonstrate *in vitro* anti-inflammatory activity. *Nutrition Research*, **30** (9), 650–659.

Keane, F.M., Munn, S.E., du Vivier, A.W.P. *et al.* (1999) Analysis of Chinese herbal creams prescribed for dermatological conditions. *BMJ*, **318** (7183), 563–564.

Klaas, C.A., Wagner, G., Laufer, S. *et al.* (2002) Studies on the anti-inflammatory activity of phytopharmaceuticals prepared from *Arnica* flowers. *Planta Medica*, **68** (5), 385–391.

Lans, C., Turner, N., Brauer, G. *et al.* (2006) Ethnoveterinary medicines used for horses in Trinidad and in British Columbia, Canada. *Journal of Ethnobiology and Ethnomedicine*, **2** (1), 31.

Larrey, D. (1997) Hepatotoxicity of herbal remedies. *Journal of Hepatology*, **1**, 47–51.

Linde, K., Ramirez, G., Mulrow, C.D. *et al.* (1996) St John's wort for depression – an overview and meta-analysis of randomised clinical trials. *BMJ*, **313** (7052), 253–258.

Liu, H.-w., Tong, J.-m. and Zhou, D.-w. (2011) Utilization of Chinese herbal feed additives in animal production. *Agricultural Sciences in China*, **10** (8), 1262–1272.

Mackenzie, E. and Rakel, B. (2006) *Complementary and Alternative Medicine for Older Adults: A Guide to Holistic Approaches to Healthy Aging*, Springer.

Manzanilla, E.G., Perez, J.F., Martin, M. *et al.* (2004) Effect of plant extracts and formic acid on the intestinal equilibrium of early-weaned pigs. *Journal of Animal Science*, **82** (11), 3210–3218.

McCorkle, C.M. (1986) An introduction to ethnoveterinary research and development. *Journal of Ethnobiology and Ethnomedicine*, **6** (1), 129–149.

Mncwangi, N., Chen, W., Vermaak, I. *et al.* (2012) Devil's claw – a review of the ethnobotany, phytochemistry and biological activity of *Harpagophytum procumbens*. *Journal of Ethnopharmacology*, **143** (3), 755–771.

Mulla, M.S. and Su, T. (1999) Activity and biological effects of neem products against arthropods of medical and veterinary importance. *Journal of the American Mosquito Control Association*, **15** (2), 133–152.

Naidoo, V. and Seier, J. (2010) Preclinical safety testing of herbal remedies, in *Ethnoveterinary Botanical Medicine: Herbal Medicines for Animal Health* (eds D.R. Katere and D. Luseba), CRC Press, Boca Raton, FL.

Nair, M. and Unnikrishnan, P. (eds) (2010) Revitalizing ethnoveterinary medical traditions: a perspective from India, in *Ethnoveterinary Botanical Medicine: Herbal Medicines for Animal Health* (eds D.R. Katere and D. Luseba), CRC Press, Florida, FL.

Nchu, F., Magano, S.R. and Eloff, J.N. (2005) *In vitro* investigation of the toxic effects of extracts of *Allium sativum* bulbs on adults of *Hyalomma marginatum rufipes* and *Rhipicephalus pulchellus*. *Journal of the South African Veterinary Association*, **76** (2), 99–103.

Nong, X., Ren, Y.-J., Wang, J.H. *et al.* (2013) Clinical efficacy of botanical extracts from *Eupatorium adenophorum* against the *Sarcoptes scabiei* (Sarcoptidae: Sarcoptes) in rabbits. *Veterinary Parasitology*, **195** (1–2), 157–164.

Obertreis, B., Giller, K., Teucher, T. *et al.* (1996). Anti-inflammatory effect of *Urtica dioica* folia extract in comparison to caffeic malic acid. *Arzneimittelforschung*, **46** (1), 52–56 (in German).

Paolini, V., Bergeaud, J.P., Grisez, C. *et al.* (2003) Effects of condensed tannins on goats experimentally infected with *Haemonchus contortus*. *Veterinary Parasitology*, **113** (3–4), 253–261.

Pearson, W., McKee, S. and Clarke, A.F. (1999) Effect of a proprietary herbal product on equine joint disease. *Journal of Nutraceuticals, Functional & Medical Foods*, **2** (2), 31–46.

Perazella, M.A. (2009) Renal vulnerability to drug toxicity. *Clinical Journal of the American Society of Nephrology*, **4** (7), 1275–1283.

Pereira, C.A., Pereira, L.L.S. and Corrêa, A.D. (2011). High performance liquid chromatography (HPLC) of *Hoodia gordonii* commercial powder. *Journal of Medicinal Plants Research*, **5** (24), 5766-5772.

Qiao, Y. and van Kempen, T.A. (2004) Technical note: comparison of Raman, mid, and near infrared spectroscopy for predicting the amino acid content in animal meals. *Journal of Animal Science*, **82** (9), 2596–2600.

Quinlan, M.B., Quinlan, R.J. and Nolan, J.M. (2002) Ethnophysiology and herbal treatments of intestinal worms in Dominica, West Indies. *Journal of Ethnopharmacology*, **80** (1), 75–83.

Ríos, J.L., Recio, M.C., Máñez, S. and Giner, R.M. (2000) *Natural triterpenoids as anti-inflammatory agents*. Studies in Natural Products Chemistry, Volume 22, Bioactive Natural Products (Part C) (ed. Atta-ur-Rahman ), Elsevier, pp. 93–143.

Salem, A.Z.M., Olivares, M., Lopez, S. *et al.* (2011) Effect of natural extracts of *Salix babylonica* and *Leucaena leucocephala* on nutrient digestibility and growth performance of lambs. *Animal Feed Science and Technology*, **170** (1–2), 27–34.

Satrija, F., Nansen, P., Bjørn, H. *et al.* (1994) Effect of papaya latex against *Ascaris suum* in naturally infected pigs. *Journal of Helminthology*, **68** (4), 343–346.

Schulz, H., Baranska, M., Belz, H.-H. *et al.* (2004) Chemotaxonomic characterisation of essential oil plants by vibrational spectroscopy measurements. *Vibrational Spectroscopy*, **35** (1–2), 81–86.

Schulze-Tanzil, G., de Souza, P., Behnke, B. *et al.* (2002) Effects of the antirheumatic remedy hox alpha – a new stinging nettle leaf extract – on matrix metalloproteinases in human chondrocytes *in vitro*. *Histology and Histopathology*, **17** (2), 477–485.

Sen, S., Chakraborty, R. and De, B. (2011) Challenges and opportunities in the advancement of herbal medicine: India's position and role in a global context. *Journal of Herbal Medicine*, **1** (3–4), 67–75.

Shinde, V.M., Dhalwal, K., Potdar, M. and Mahadik, K.R. (2009) Application of quality control principles to herbal drugs. *International Journal of Phytomedicine*, **1**, 4–8.

Shiva, V. (2007) Bioprospecting as sophisticated biopiracy. *Signs*, **32** (2), 307–313.

Sokerya, S. (2003) Effect of grass or cassava foliage on growth and nematode parasite infestation in goats fed low or high protein diets in confinement. MSc Thesis, Swedish University of Agricultural Sciences, Uppsala.

Tamboura, H. and Okoli, I. (2010) Ethnoveterinary medicine and sustainable livestock management in West Africa, in *Ethnoveterinary Botanical Medicine: Herbal Medicines for Animal Health* (eds D.R. Katere and D. Luseba), CRC Press, Boca Raton, FL.

Tembely, S., Lahlou-Kassi, A., Sovani, S. *et al.* (1996) Small ruminant management practices and control of helminthosis under traditional production systems in the cool Ethiopian highlands, in *Small Ruminant Research and Development in Africa. Proceedings of the Third Biennial Conference of the African Small Ruminant Research Network, Kampala, Uganda, 5-9 December*, International Livestock Research Institute, Nairobi, Kenya.

Teng, L., Shaw, D. and Barnes, J. (2010) Traditional Chinese veterinary medicine, in *Ethnoveterinary Botanical Medicine: Herbal Medicines for Animal Health* (eds D.R. Katere and D. Luseba), CRC Press, Boca Raton, FL.

Teucher, T., Obertreis, B., Ruttkowski, T. and Schmitz, H. (1996) Cytokine secretion in whole blood of healthy subjects following oral administration of *Urtica dioica* L. plant extract. *Arzneimittelforschung*, **46** (9), 906–910 (in German).

Thembo, K.M., Katerere, D.R., Stockenström, S. *et al.* (2007) Mycological and mycotoxin contamination of traditional herbal medicines. *South African Journal of Botany*, **73** (2), 316–317.

Vargas-Murga, L., Garcia-Alvarez A., Roman-Viñas, B. *et al.* (2011) Plant food supplement (PFS) market structure in EC Member States, methods and techniques for the assessment of individual PFS intake. *Food and Function*, **2**, 731–739.

Vermaak, I., Hamman, J.H. and Viljoen, A.M. (2010) High performance thin layer chromatography as a method to authenticate *Hoodia gordonii* raw material and products. *South African Journal of Botany*, **76** (1), 119–124.

Vitetta, L., Cicuttini, F. and Sali, A. (2008) Alternative therapies for musculoskeletal conditions. *Best Practice & Research Clinical Rheumatology*, **22** (3), 499–522.

Waller, P., Bernes, G., Thamsborg, S.M. *et al.* (2001) Plants as de-worming agents of livestock in the Nordic countries: historical perspective, popular beliefs and prospects for the future. *Acta Veterinaria Scandinavica*, **42** (1), 31–44.

Wang, Q., Kuang, H., Su, Y. *et al.* (2013) Naturally derived anti-inflammatory compounds from Chinese medicinal plants. *Journal of Ethnopharmacology*, **146** (1), 9–39.

Weil, A. (2011) Steve Jobs and alternative cancer care, http://edition.cnn.com/2011/10/27/opinion/weil-steve-jobs/ (accessed 17 November 2013).

WHO (1993) *Research Guidelines for Evaluating the Safety and Efficacy of Herbal Medicines*, World Health Organization, Geneva.

WHO (2008) *Traditional medicine*, WHO Fact Sheet No. 134, World Health Organization, Geneva, www.who.int/mediacentre/factsheets (accessed 20 November 2013).

Williams, C.A. and Lamprecht, E.D. (2008) Some commonly fed herbs and other functional foods in equine nutrition: a review. *Veterinary Journal*, **178** (1), 21–31.

Windisch, W., Schedle, K., Plitzner, C. and Kroismayr, A. (2008) Use of phytogenic products as feed additives for swine and poultry. *Journal of Animal Science*, **86** (14 Suppl), E140–E148.

Wojcikowski, K., Johnson, D.W. and Gobé, G. (2004) Medicinal herbal extracts – renal friend or foe? Part one: the toxicities of medicinal herbs. *Nephrology*, **9**: 313–318.

Wynn, S.G. and Chalmers, S. (2002) Alternative therapies for pruritic skin disorders. *Clinical Techniques in Small Animal Practice*, **17** (1), 37–40.

Wynn, S.G. and Fougère, B.J. (2007) Veterinary herbal medicine: a systems-based approach, in *Veterinary Herbal Medicine* (eds S. Wynn and B. Fougère), St. Louis, Mosby, pp. 291–409.

Yan, L., Meng, Q.W. and Kim, I.H. (2011a) The effects of dietary *Houttuynia cordata* and *Taraxacum officinale* extract powder on growth performance, nutrient digestibility, blood characteristics and meat quality in finishing pigs. *Livestock Science*, **141** (2–3), 188–193.

Yan, L., Meng, Q.W., Ao, X. *et al.* (2011b) Effects of fermented garlic powder supplementation on growth performance, blood characteristics and meat quality in finishing pigs fed low-nutrient-density diets. *Livestock Science*, **137** (1–3), 255–259.

Zhang, X., Cao, F., Sun, Z. *et al.* (2012) Effect of feeding *Aspergillus niger*-fermented *Ginkgo biloba*-leaves on growth, small intestinal structure and function of broiler chicks. *Livestock Science*, **147** (1–3), 170–180.

# 22 Plants Used for Pest Management in Malawi

Cecilia Maliwichi-Nyirenda[1], Lucy Lynn Maliwichi[2] and John F. Kamanula[3]

[1]Indigenous Knowledge Centre, Blantyre, Malawi
[2]University of Venda, Thohoyando, Republic of South Africa
[3]Chemistry Department, Mzuzu University, Luwinga, Mzuzu, Malawi

## 22.1 Introduction

Food security is the principal need for most African farmers. However, many farmers are unable to keep their produce to the next season, mostly due to insect pest infestations (Masamba *et al.*, 2003). Post-harvest handling and agro processing of produce are important components of food security in the food chain in both developed and developing countries. In most cases, agro processing refers to modern processing methods such as canning.

One of the main causes of food insufficiency in East Africa is the high prevalence of storage pests in grains, dry legume seeds and tubers. This often leads to food insecurity among smallholder and subsistence farmers. The main storage pests, apart from rodents, are beetles and moths. Some pests, such as grain borers, weevils and Angoumois grain moths, are able to feed on whole, healthy grains, and they are considered primary pests. Secondary pests, such as flour beetles, can attack only broken grain, moist (and thus soft) grain damaged by primary pests or processed products such as flour. The main beetle pests of storage are bruchids (e.g. cowpea seed beetles and bean bruchid), grain borers (e.g. the larger grain borer, *Prostephanus truncatus* and the lesser grain borer, *Rhizopertha dominica*), weevils (e.g. grain weevils), flour beetles and dried fruit beetles (ARC·LNR, 1997).

In Africa and most developing countries where smallholder subsistence farmers produce food to feed their families, the households often face challenges resulting from a high prevalence of pests affecting farm produce. In most cultures, various traditional methods are used to protect mostly stored grains, vegetables and fruits against insect damage. Grain can be stored in a variety of containers, such as clay pots, grass baskets, grain wells, bags, grain huts and ferrumbus.

In southern Africa, especially Malawi and Zambia, vegetables are valuable as a relish, providing dietary vitamins and minerals in the largely maize-based diet of rural households. However, pest damage severely limits vegetable production in many parts of southern Africa (Sibanda *et al.*, 2000; Kuntashula *et al.*, 2006; Obopile *et al.*, 2008; Grzywacz *et al.*, 2010).

Since ancient times, humans have been using different strategies to control field and storage insect pests. These have comprised indigenous techniques of crop processing and storage that have been practiced way before the arrival of chemical pest management. Some of the techniques have scientific backing. For example, in the Uttarakhand hills (India), people store green gram/mung bean with a mixture of immature turmeric, dried leaves of walnut and mustard oil. This technique has been proved to have the capability of reducing the chances of disease development and also acts as a physical barrier between stored grain insects pests and grains (Chandola *et al.*, 2011).

Traditionally, most farmers have relied on traditional grain storage techniques. These techniques include the use of wood ash, a mixture of fine dry clay or fine sand (this limits the locomotion of stored cereal insects and bruchids and causes death through desiccation and also prevents immigration of pest insects), use of toxic plants/admixture of plant parts, use of vegetable oils, use of flower/fruit/seed powders/aqueous extracts, storing cereals as unthrashed grains, admixture of green plant parts/powders made from dried green parts, and the use of bark and root powders. It is a common practice to store cereals unthrashed in order to give some protection against insects. Husks, pods and outer coverings offer some protection against insect attack provided they are undamaged. Some farmers in East Africa use chillies in grains. Other farmers store unthreshed pods as a strategy to minimize grain damage by bruchids (Minja *et al.*, 1999).

A growing body of literature suggests that many farming communities possess traditional knowledge of pests that affect their crops and alternative approaches to their control (Altieri, 1993; Price and Björnsen Gurung, 2006; Sileshi *et al.*, 2008; Chhetry and Belbahri, 2009). In a study that was conducted in Malawi and Zambia, over 70% of the respondents were aware of pesticidal plants and reported using them. While over 20 different plant species were mentioned by respondents, *Tephrosia vogelii* accounted for 61% and 53% of the pesticidal species known to respondents in Malawi and Zambia respectively (Nyirenda *et al*, 2011a). The use of pesticidal plants constituted a smaller portion of the pest control options in both crucifers and tomato. This implies the need for more research and promotion of the use of pesticidal plants among smallholder farmers.

African agriculture is largely traditional, characterized by a large number of smallholdings of no more than 1 ha per household. Farmers often select well-adapted, stable crop varieties, and cropping systems are such that two or more crops are grown in the same field at the same time. These diverse traditional systems enhance natural enemy abundance and generally keep pest numbers at low levels (Abate *et al.*, 2000).

*Novel Plant Bioresources: Applications in Food, Medicine and Cosmetics*, First Edition. Edited by Ameenah Gurib-Fakim.

The traditional methods used for crop storage in Malawi include construction of a wooden rack/platform in the kitchen on which dry maize on cobs is stored. The smoke produced from a food preparation fireplace located under the rack protects the grains from insect damage. Ash is particularly useful for protecting small amounts of seed for replanting or small quantities of grains for consumption. Other traditional methods include the use of mineral substances such as wood ash, which is added to cereals and legume products stored in a container/granary or to fresh sweet potatoes stored in a pit.

### 22.1.1 Use of wood ash

Wood ash, either alone or as a mixture, is an efficient method of pest control. However, ashes may have an affect on the taste of the treated product. The success of this method depends on the amount and types of ashes being added. Ash is particularly useful for protecting small amounts of seed for replanting or small quantities of grains for consumption. The use of wood ash for storage of food grains is also practised in Uttarakhand, India. Ash collected by burning of fuel wood is dusted on the heap of food grains and mixed well by hands and then food grains are stored. Scientists considered it as a rational practice since, by mixing ash with the grain, ash forms a protective coating around the grains and works as physical barrier between insect pests and grains, thereby reducing their activity. Use of ash from burnt cow dung as an insecticide against *Tribolium casteneum*, *Sitophilous granaries* (L.) and *Cryptolestes ferrugineus* larvae has also been reported. Ash kills insects by desiccation or by filling the intergranular spaces, restricting insect movement and emergence (Chandola *et al.*, 2011).

Fruit wood is a major source of energy; consequently, wood ash is plentiful. It is used to wash and clean utensils as well as clothes in many areas of Himachal Pradesh. It is a common practice to sprinkle wood ash on vegetable crops (especially those growing in a kitchen garden) and to spread it around plants to ward off pests and to enhance nutrient status of the soil. To achieve this, a thick layer of ash is spread on the soil around plants and it is also sprinkled on foliage to protect it against a variety of pests. This is because it is a source of phosphorus for plants and it also acts as a physical poison, usually causing abrasion of epicuticular waxes and thus exposing pests to death through desiccation. It also interferes in the chemical signals emanating from the host plants, thus obstructing the initial host location by pests. The treated foliage further becomes unpalatable for foliage feeders like cutworms, caterpillars, grasshoppers, and so on. But since ash provides only temporary protection against pests, insecticides that have a quick knock-down effect have replaced the use of wood ash today (Verma, 1998).

### 22.1.2 Other traditional methods

Man has developed other traditional methods to curb pests. The methods include construction of a wooden rack/platform in the kitchen on which dry maize on cobs is stored. The smoke produced from a food preparation fireplace located under the rack protects the grains from insect damage. Some farmers in East Africa use smoke from cooking fire to preserve seeds for planting (Minja *et al.*, 1999).

Sun is also used to prolong the shelf life of products. For example, in Malawi, most indigenous vegetables are sun-dried (locally called *Mfutso*). Households use the direct sun drying method where vegetables are spread on mats or raised racks in the sun. Most of these indigenous vegetables grow during the rainy season but are not available in the dry season. The vegetables are normally dried and stored in gourds, pots or storage bags made from leaves (locally known as *chikwatu*). The dried vegetables are also stored in plastic bags. All these methods of storage assist in ensuring proper preservation of the products for use in times of scarcity. Consequently, the vegetables are available throughout the year rather than just the 3 months in which fresh vegetables grow.

## 22.2 Merits and demerits of pest management systems in Malawi

### 22.2.1 Pest management systems in Malawi

As previously mentioned, farmers use different methods in managing pests. In Malawi, the situation is the same. Farmers employ cultural practices, synthetic chemical insecticides, biological control, botanical pesticides and integrated pest management (IPM) to protect field crops and stored products from pests (Gwinner *et al.*, 1996). These methods have advantages and disadvantages.

#### 22.2.1.1 Cultural practices

Cultural practices usually involve modifying the relationship between a pest population and its environment. Cultural practices include crop rotation, intercropping, sanitation, and proper planting and harvesting time.

***Crop rotation***

In monoculture, farmers grow one type of crop year after year in the same field. In this way, insect populations have enough time to become established and cause massive damage to the crop. However, by rotating the field to a different crop, pests that do not adapt to a different host plant can be starved and their population reduced because there is a discontinuity in the pest's food supply. In addition, crop rotation gives improvement in soil quality and better distribution of nutrients in the soil profile, as deep-rooted crops bring up nutrients from below.

***Intercropping (mixed cropping)***

Intercropping or mixed cropping refers to the growing of more than one crop in the same field and at the same time. This could be maize–sweet potato, maize–beans or maize–pumpkin. Intercropping increases the diversity of the farming system, and consequently reduces the incidence of pests and diseases in the field. However, intercropped crops may compete for water, light and nutrients from the soil, resulting in low yields (van Wolfswinkel, 2007). In addition, the leaves or roots of the other crop may repel or attract the pests of the other crop, which could be an advantage or disadvantage. Therefore, before embarking on a large-scale intercropping, it is advisable to start with small-scale intercropping and then scale up based on the results.

***Sanitation***

Like in many African countries, farmers in Malawi know that removing crop debris from maize or cotton fields after harvesting can eliminate the populations of cotton or maize pests. This has a huge advantage, since no chemicals are used and the cost is assessed in terms of labour only.

***Proper planting and harvesting time***

Farmers who harvest crops at the right time have a high chance of reducing losses due to insects, as fewer insects would be carried together with the commodity into storage (Baidoo *et al.*, 2010).

### 22.2.1.2 Use of synthetic pesticides

Like many other countries in the world, farmers in Malawi use synthetic pesticides to manage field and storage pests. Some of the pesticides that are used in Malawi include insecticides, fungicides, herbicides, fumigants, nematocides, acaricides and rodenticides (Malawi Government, 2012). Synthetic pesticides are usually effective, with well-known knock-down effect (e.g. Actellic Super dust/Actellic Super E.C., DDT), whilst some are good fumigants (e.g. phosphine). Although synthetic pesticides are effective, research has documented many negative impacts associated with them. Many synthetic pesticides cause harm to non-target pests and reduce their populations, which in turn results in changes in the biodiversity of an area and affects the natural biological balance. In addition, frequent use of the same chemical pesticides results in insects developing resistance to the pesticides, making them ineffective (Isman, 2006; Stevenson *et al.*, 2012). Chemical pesticides also deplete organic soil nutrients.

Because of their environmental and health hazards, such as contaminating soil and ground water aquifers and toxic effects to the user and commodity consumers, some chemical insecticides (e.g. DDT) were banned in developed and some developing countries (National Pesticide Information Center, 2000). The other notable problem of synthetic insecticides is that their costs are usually high to poor farmers (Sighamony *et al.*, 1990; Stevenson *et al.*, 2012), thereby restricting their distributions in rural areas (Stevenson *et al.*, 2012). In addition, synthetic pesticides are poorly labelled, unavailable and often contaminated. They are also a health hazard to mankind owing to frequent incorrect use (ADAPPT, n.d.).

### 22.2.1.3 Biological control

Biological control involves the use of natural enemies of crop pests, often called beneficials, which include parasites, predators and insect pathogens (Malawi Government, 2012). These natural enemies are reared and released to feed or attack the pests in order to reduce their numbers. In Malawi, an economic insect pest of stored maize and dried cassava chips, the larger grain borer, *P. truncatus* (Horn) (Coleoptera: Bostrichidae), was first identified in the northern Malawi in 1991 (Singano *et al.*, 2008). To suppress its population in the country, the Department of Agricultural Research Services released a natural enemy (predator), *Teretrius nigrescens* Lewis (TN), to feed on it (Paliani and Muwalo, 2001). Both adult and larvae of *T. nigrescens* eat the immature stages of *P. truncatus* (Hodges, 1998). Although *T. nigrescens* was released in Malawi and other parts of Africa, the larger grain borer problem has not been entirely solved (Gwinner *et al.*, 1996) partly because the predator has no or only a very low impact on other storage pests such as *Sitophilus* spp. and *Triborium* spp. (Gwinner *et al.*, 1996). Despite this, biological control methods have the advantages that they reduce environmental and health risks, are easily applied and do integrate well with other methods.

### 22.2.1.4 Botanical pesticides

Since antiquity, African farmers have used botanical pesticides for the protection of field crops and stored products from insect pests. Like the rest of the world, farmers in Malawi are knowledgeable about pesticidal plants as being effective, environmentally friendly, less toxic and cost effective compared with synthetic pesticides (Kamanula *et al.*, 2011; Nyirenda *et al.*, 2011a). Plant-based pesticides are now preferred because synthetic chemical insecticides give health risks to consumers and they are also associated with residue problems. Thus, plants have the potential of providing alternatives because of the rich source of bioactive molecules they contain (Rajashekar *et al.*, 2012).

Since plant extracts contain different constituents with broad-spectrum bioactivity, the issue of insect resistance of botanical pesticides is minimal. In addition, most small-scale farmers harvest and produce pesticidal plants themselves, so the problem of adulteration in this case is overcome (Stevenson *et al.*, 2012). Although botanical pesticides technologies are familiar to farmers, their priority in agricultural policy is low; as a result, few farmers use pesticidal plants (Kamanula *et al.*, 2011). The other problem could be that the registration process of pesticidal plants is equated with that of synthetic pesticides (Stevenson *et al.*, 2012).

### 22.2.1.5 Integrated pest management

IPM involves use of available, suitable and compatible methods, which include resistant varieties, cultural methods, biological control and safe pesticides use to maintain pests below levels that cause economic damage and loss (FAO, 1968; Malawi Government, 2012). Because IPM combines all possible techniques, it is believed to be cost effective, environmental friendly and pose less health risks.

## 22.3 Plant species used in pest management

The problems of using synthetic pesticides for the control of insect pests have well been highlighted in this chapter. Farmers have knowledge of plants that have the effect of controlling pests. Traditionally, people use different types of plants against pests that attack stored products. In many countries, flowers, roots, leaves and fruits of plants are known to have insect-pest-repellent properties. Notable examples are neem leaves, derries, lippia, citronella and pyrethrum.

In Africa, there are reports in Kenya that the larger grain borer can be effectively repelled by storing cassava or grains with a fairly large amount of dried lantana or eucalyptus leaves (personal communication, field officer of Meru herbs). Some farmers in East Africa use wood ash in grain stores for food or seed for planting, or chillies or smoke from cooking fires to preserve seeds for planting. Other farmers store unthreshed pods as a strategy to minimize grain damage by bruchids (Minja *et al.*, 1999).

Many farmers in southern Africa know that pesticidal plants are effective alternatives to synthetic pesticides (Kamanula *et al.*, 2011; Nyirenda *et al.*, 2011a), and their promotion would bring a big impact on the ability of farmers to manage insect pests. Plant materials offer poor farmers a low-cost and effective but environmentally benign alternative to expensive pesticides and provide development programmes with a valuable tool for enhancing the livelihoods of the poorest farmers. In South Africa and Zimbabwe, a poisonous tree commonly called the tamboti tree (*Spirostachys africana*) is used as pesticide. The tree is toxic and poisonous. Bark is used in some parts of the Limpopo province (South Africa), while sap from the bark serves as a pesticide in granaries in Zimbabwe. The tree bark is placed in a bag containing maize grains to protect grain damage from storage insects, such as the larger and lesser grain borers (personal communication, L.L. Maliwichi, 2013). Hill farmers of Uttarakhand also use leaves of walnut (*Juglans regia* Linn.) and timur (*Zanthoxylum alatum* Roxb.) trees for storing grains and pulses. For storing paddy seed, it is dried for 2 days just after harvesting; then, before loading it in the storage structure, leaves of walnut (*Juglans regia* Linn.) and timur (*Z. alatum* Roxb.) trees are mixed with the rice. In Gujarat, farmers incorporate leaves of nafatia (*Ipomoea fistulosa*) in the sorghum and rice grains which are kept for consumption. It is believed that the bitter smell of these leaves protects stored grains against any pest (Chandola *et al.*, 2011).

Malawi is endowed with many plants that have potential to control field and storage pests. People use different methods of application. For example, essential oils from *Eucalyptus citriodora*, *Eucalyptus. camaldulensis*, *Cymbopogon nardus*, *Cymbopogon citratus*, *Tagetes minuta* and *Lippia javanica* were shown to be toxic to *Sitophilus zeamais* and *P. truncatus*, the primary insect pests of stored maize, under laboratory conditions (Masamba *et al.*, 2003; Kamanula, 2005; Nyirenda *et al.*, 2011b). Some plant species known to be toxic to insect pests are listed in Table 22.1.

In our previous study that was conducted in Nchenachena and Champhira extension planning areas of Rumphi and Mzimba districts (northern Malawi) respectively, dry powders of *Securidaca longepedunculata* (locally called *Muuluka*) root bark and *Azadirachta indica* (local name: nimu) seed kernel, admixed with maize grain at 2% (w/w), reduced maize grain damage from 90% in the untreated to 16% in treated maize grain (Kamanula *et al.*, 2013). The study further showed that over 75% of untreated beans were damaged, mainly by *Acanthoscelides obtectus*, compared with only 3% in beans treated with *A. indica* seed kernel, which compared favourably with the commercial standard Actellic Super dust. Farmer surveys carried out in Malawi and Zambia showed that all farmers were knowledgeable about pesticidal plants as alternatives to synthetic pesticides; however, only a few used them (Kamanula *et al.*, 2011).

All in all, several plant species are known to be used as pesticides. This chapter has documented 42 genera and 43 species (Table 22.2).

**Table 22.1** Plants with pesticidal properties in Malawi

| Plant species | Organ | Insect species | Action |
|---|---|---|---|
| *Azadirachta indica* | Seed kernel | *S. zeamais, P. truncatus* | Contact |
| *Cymbopogon citratus* | Entire plant | *S. zeamais, P. truncatus* | Contact, fumigant |
| *Cymbopogon nardus* | Entire plant | *S. Zeamais, P. truncatus* | Contact, fumigant |
| *Dicoma sessiliflora* | Root | *P. truncatus, S. oryzae* | Contact |
| *Lecaniodiscus fraxinifolius* | Root | *P. truncatus, S. oryzae* | Contact |
| *Lippia javanica* | extract | *S. zeamais, P. truncatus* | Contact, fumigant, repellent |
| *Lippia javanica* | essential oil | *S. zeamais, P. truncatus* | Contact, fumigant, repellent |
| *Neorautanenia mitis* | Tuber | *P. truncatus, S. oryzae* | Contact |
| *Securidaca longepedunculata* | Root bark | *P. truncatus, S. zeamais* | Contact, fumigant, repellent |
| *Securidaca longepedunculata* | Root bark | *R. dominica, C. maculatus* | Contact, fumigant, repellent |
| *Tephrosa vogelii* | Leaf | *C. mculatus* | Contact |
| *Tephrosa vogelii* | Leaf | *S. zeamais* | Contact, repellent |

*Sources*: Belmain *et al.* (2012), Chimbe and Galley (1996), Jayasekara *et al.* (2005), Kamanula *et al.* (2013), Kamanula (2005), Masamba *et al.* (2003) and Ogendo *et al.* (2003).

**Table 22.2** Plant species used for pest management in Malawi

| Serial no. | Plant species used as pesticides | English/local name | Plant part used | Preparation and application method | Pest |
|---|---|---|---|---|---|
| 1 | *Acorus calamus* | Sweet flag/calamus | Roots and leaves | Dry, grind into powder and apply | Numerous pests |
| 2 | *Allium sativum* | Garlic | Cloves | Extract essential oils and apply | Beetles |
| 3 | *Aloe castanea* | Cat's tail aloe | Leaves | Burn and apply ashes to grain | Weevils |
| 4 | *Aloe* spp. | | Whole plant | Dry and grind into powder. Apply 50 g of powder to 1 kg of stored product | Storage pests |

**Table 22.2** (*Continued*)

| | | | | | |
|---|---|---|---|---|---|
| 5 | *Annona senegalensis* | Wild soursop/wild custard apple | Leaves | Add to product in layers; or dry, grind into powder and mix with stored product | Bruchids and sorghum and millet pests |
| 6 | *Arachis hypogaea* | Groundnut | Nut | Distil oil and apply. Add 5 mL oil per 1 kg of grain/seed | Bruchids |
| 7 | *Arachis hypogaea* | Groundnut | Nut | Mix with grain legumes at 5 mL/kg | Bruchids (eggs) |
| 8 | *Artemisia annua* | Sweet Annie/sweet wormwood | Leaves | Distil oil and apply | *Tribolium castaneum* and *Callosobruchus maculates* |
| 9 | *Azadirachta indica* | Neem | Leaves | Use while dry or grind into powder and mix with grain or apply in layers | Stored product beetles |
| 10 | *Azadirachta indica* | Neem | Leaves, fruits and seeds | Grind dried leaves or seeds or fruits into powder and mix with grains | *Sitophilus oryzae, Tribolium cataneum, Rhyzopertha dominica* and *Callosobruchus chinensis* |
| 11 | *Azadirachta indica* | Neem | Leaves, crushed seeds and their extracts and oils | Apply 5 mL oil per 1 kg of grain/seed | Larger grain borer |
| 12 | *Azadirachta indica* | Neem | Leaves | Use while dry or grind into powder and either apply in layers, sprinkle or mix with grain legumes and cereals at 2–3 mL per 1 kg grain | Stored-product beetles and pulse beetles |
| 13 | *Capsicum annuum.* | Chilli pepper | Pods | Either grind into powder and mix with beans or mix while uncrushed. Apply 50 g of product per 1 kg of stored product | Storage pests |
| 14 | *Capsicum annuum* | Red pepper | Fruits | Dry, grind into powder and mix with stored product | Storage pests |
| 15 | *Capsicum annuum* | Red pepper | Fruits | Dry and use it to smoke the grain | Storage pests |
| 16 | *Carthamus tinctorius* | Safflower | Seed | Extract oil and apply 5 mL oil per 1 kg of grain/seed | Bruchids |
| 17 | *Chrysanthemum cinerariifolium* | Pyrethrum | Flower heads | Pick on hot days. Dry under shade and grind into powder. Add 50 g of powder to 1 kg of stored product | Storage pests |
| 18 | *Chrysanthemum cinerariifolium* | Pyrethrum | Flower heads | Dry and grind into powder. Apply powder to product and storage structures | Storage pests |
| 19 | *Chrysanthemum cinerariifolium* | Pyrethrum | Flower heads | Sprinkle on commodity | Storage pests |
| 20 | *Cocos nucifera* | Coconut | Seed | Extract oil and apply to mung bean (*Vigna radiata*) | Chinese bruchid/ oriental cowpea bruchid/southern cowpea weevil (*Callosobruchus chinensis*) |
| 21 | *Cocos nucifera* | Coconut | Seed | Mix with grain legumes at 5-10 mL/kg | *Rhizopertha dominica* (the lesser grain borer) and pulse beetles |
| 22 | *Crotalaria juncea* | Sunnhemp | Seeds | Mix with stored grains (50 g of seed per 1 kg of grains) | Storage pests |
| 23 | *Cuminum cyminum* | Cumin | Seeds | Distil oil and apply | *Tribolium confisum* and *Ephesia kuehniella* |
| 24 | *Curcuma longa* | Turmeric | Root | Grind into powder and mix with grain | Pests attacking food grains, e.g. rice and wheat |

(*continued*)

**Table 22.2** (*Continued*)

| Serial no. | Plant species used as pesticides | English/local name | Plant part used | Preparation and application method | Pest |
|---|---|---|---|---|---|
| 25 | *Datura stramonium* | Thorn apple | Leaves and stems | Dry and mix with produce. Apply 50 g of product per 1 kg of stored product | Storage pests |
| 26 | *Decalepis hamiitonii* | Swallow root | Root | Grind into powder and apply to grain | Stored grain pests |
| 27 | *Derris* spp. | Common derris | All parts | Grind to powder and dust/spray on stored product. Apply 50 g of pesticide per 1 kg of stored product | Storage pests |
| 28 | *Eucalyptus* spp. | Bluegum | Leaves | Layered or mixed with stored produce. Apply 50 g of product per 1 kg of stored product | *Prostephanus truncatus* (the larger grain borer) |
| 29 | *Eucalyptus* spp. | Bluegum | Leaves | Burn into ashes and mix with grain | Pests living outside grain |
| 30 | *Eucalyptus* spp. | Bluegum | Leaves | Distil oil and apply | *Tribolium confisum* and *Ephesia kuehniella* |
| 31 | *Glycine max* | Soybean | Bean | Add 5 mL oil per 1 kg of grain/seed | Cowpea bruchids (*Callosobruchus* spp.) |
| 32 | *Gossypium* spp. | Cotton | Seed | Add 5 mL oil per 1 kg of grain/seed | Storage pests |
| 33 | *Hyptis spicigera* | Black beni-seed/black sesame | Leaves | Add to product in layers or just mix | Bruchids, groundnut seed beetles (*Caryedon serratus*) and termites |
| 34 | *Khaya senegalensis* | African mahogany | Bark | Grind into powder and mix with grains | *Bruchus maculatus* in cowpeas |
| 35 | *Khaya senegalensis* | African mahogany | Whole tree | Burn into ashes and mix with grain | Pests living outside grain |
| 36 | *Lantana* spp. | Lantana | Leaves | Crushed and placed among seeds. Apply 50 g of product per 1 kg of stored product | Cowpea bruchids (*Callosobruchus* spp.) |
| 37 | *Lantana camara* | Lantana | Leaves | Grind and apply powder in layers | Bruchids |
| 38 | *Lantana camara* | Lantana | Leaves | Grind and apply powder to grain | *Sitophilus zeamais* (pest on stored maize) |
| 39 | *Lippia javanica* | Fever tea/lemon bush | Root | Grind and apply powder to grain | Beetles |
| 40 | *Melia azedarach* | Syringa | Leaves, ripe seeds | Use while dry or grind into powder and mix with stored grain. Apply 50 g of product per 1 kg of stored product | Storage pests |
| 41 | *Melia* spp. | Persian lilac | Leaves | Use while dry or grind into powder and either mix to grain or apply in layers | Stored-product beetles |
| 42 | *Mentha spicata* | Spearmint | Whole plant | Grind into powder and apply 50 g of powder per 1 kg of stored product | Stored-product beetles |
| 43 | *Mentha* spp. | Mint | Leaves | Mix leaves with grain | Insects on grain |
| 44 | *Sinapis alba* | Mustard | Seeds | Extract oil and mix 5 mL oil per 1 kg of grain/seed | Stored-grain insects pests and grains |
| 45 | *Ocimum canum* | Hoary basil | Leaves | Use entire leaves or grind into powder and apply in layers | Beetles on legumes |
| 46 | *Ocimum suave* | Tree basil | Leaves and cloves | Mix leaves and cloves. Add mixture to grain | Stored grain |
| 47 | *Origanum vulgare* | Oregano | Leaves | Distil oil and apply | *Tribolium confisum* and *Ephesia kuehniella* |
| 48 | *Pimpinella anisum* | Anise | Seeds | Distil oil and apply | *Tribolium confisum* and *Ephesia kuehniella* |
| 49 | *Piper nigrum* | Black pepper | Fruits | Sprinkle on commodity | Stored-product pests |

Table 22.2 (*Continued*)

| | | | | | |
|---|---|---|---|---|---|
| 50 | *Piper nigrum* | Black pepper | Pods | Dry, grind into powder and mix with product | Stored-product pests |
| 51 | *Raphia* spp. | Palm | Fruits | Mix with grain legumes at 5 mL/kg | Pulse beetles |
| 52 | *Ricinus communis* | Castor bean | Seeds | Add 5 mL oil per 1 kg of grain/seed | Pulse beetles (bruchids) |
| 53 | *Rosmarinus officinalis* | Rosemary | Sprigs | Distil oil and apply | *Tribolium confisum* and *Ephesia kuehniella* |
| 54 | *Securidaca longepedunculata* | Violet tree/fibre tree/Rhodesian violet | Roots | Dry and grind into powder. Mix with maize and cowpea | *Sitophilus zeamais* and *Callosobruchus maculatus* |
| 55 | *Sesamum indicum* | Sesame | Seeds | Mix with grain legumes at 5 mL/kg | Pulse beetles |
| 56 | *Tagetes lucida* | Mexican marigold | Entire plant | Either dry and mix with product in layers, or grind into powder and mix with product. Can also put crushed plants at the base of bins where grains are stored | Storage pests |
| 57 | *Tephrosia vogelii* | Fish-poison bean | Leaves | Grind and apply powder to grain | *Sitophilus zeamais* |
| 58 | *Veratrum album* | Common white hellebore | Root | Dry, grind into powder and apply to stored product | Rodenticide |
| 59 | *Vitellaria paradoxa* | Shea | Nuts | Melt and apply to cereals and grain legumes at 5 mL/kg | Beetles |
| 60 | *Zea mays* | Maize | Leaves | Dry, grind into powder and mix with grain | Insects on grain |

*Sources*: Gwinner *et al.* (1990), Williamson (1975), B. Chidzaro (personal communication, 17 February 2013), Orwa *et al.* (2009), Crop Protection Compendium (2013), Burkill (1985) and *Science News* (2008).

## References


Abate, T., van Huis, A. and Ampofo, J.K. (2000) Pest management strategies in traditional agriculture: an African perspective. *Annual Review of Entomology*, **45** (1), 631–659.

ADAPPT (n.d.) Information flyer, African Dryland Alliance for Pesticidal Plant Technologies, http://www.nri.org/projects/adappt/docs/adapptflyer.pdf (accessed 21 May 2013).

Altieri, M.A. (1993) Ethnoscience and biodiversity: key elements in the design of sustainable pest management systems for small farmers in developing countries. *Agriculture, Ecosystems & Environment*, **46**, 257–272.

ARC·LNR (1997) *How to store grain*, National Department of Agriculture and the Agricultural Research Council, www.daff.gov.za/docs/Infopaks/grain.htm (accessed 21 November 2013).

Baidoo, P.K., Mochiah, M.B. and Owusu-Akya, M. (2010) The effect of time of harvest on the damage caused by the cowpea weevil *Callosobruchus maculates* (Fab.) (Coleoptera: Bruchidae). *Journal of Stored Products and Postharvest Research*, **1** (3), 24–28.

Belmain, S.R., Amoah, B.A., Nyirenda, S.P. *et al.* (2012) Highly variable insect control efficacy of *Tephrosia vogelii* chemotypes. *Journal of Agricultural and Food Chemistry*, **60** (40), 10055–10063.

Burkill, H.M. (1985) *The useful plants of west tropical Africa*, vol. 3, http://plants.jstor.org/upwta/3_22 (accessed 21 May 2013).

Chandola, M., Rathore, S., and Kumar, B. (2011) Indigenous pest management practices prevalent among hill farmers of Uttarakhand. *Indian Journal of Traditional Knowledge*, **10** (2), 311–315.

Chhetry, G.K.N. and Belbahri, L. (2009) Indigenous pest and disease management practices in traditional farming systems in north east India: a review. *Journal of Plant Breeding and Crop Science*, **1**, 028–038.

Chimbe, C.M. and Galley, D.J. (1996) Evaluation of material from plants of medicinal importance in Malawi as protectants of stored grain against insects. *Crop Protection*, **15** (3), 289–294.

Crop Protection Compendium (2013) http://www.cabi.org/cpc/ (accessed 17 December 2013).

FAO (1968) Report of the first session of the FAO Panel of Experts on integrated pest control, Rome, Italy, 18–22 Sep 1967, Report No. PL-1967/M/7, The Food and Agriculture Organization of the United Nations, Rome.

Grzywacz, D., Rossbach, A., Rauf, A. *et al.* (2010) Current control methods for diamond back moth and other brassica insect pests and the prospects for improved management with lepidopteran-resistant Bt vegetable brassicas in Asia and Africa. *Crop Protection*, **29**, 68–79.

Gwinner, J., Harnisch, R. and Mück, O. (1990) *Manual of the Prevention of Post-Harvest Grain Losses*, 1st edn, Post-Harvest Project, Deutsche Gesellschat für Technische Zusammenarbeit (GTZ), Eschborn.

Gwinner, J., Harnisch, R. and Mück, O. (1996) *Manual of the Prevention of Post-Harvest Grain Losses*, 2nd edn, Post-Harvest Project, Deutsche Gesellschat für Technische Zusammenarbeit (GTZ), Eschborn.

Hodges, R. (1998) *Larger Grain Borer*, 3rd edition, Technical Leaflet (Group for Assistance on Systems Relating to Grain after Harvest), No. 1, Technical Centre for Agricultural and Rural Cooperation, Wageningen.

Isman, M.B. (2006) Botanical insecticides, deterrents, and repellents in modern agriculture and an increasingly regulated world. *Annual Review of Entomology*, **51**,45–66.

Jayasekara, T.K., Stevenson, P.C., Hall, D.R. and Belmain, S.R. (2005) Effect of volatile constituents from *Securidaca longepedunculata* on insect pests of stored grain. *Journal of Chemical Ecology*, **31** (2), 303–313.

Kamanula, J., Sileshi, G.W., Belmain, S.R. *et al.* (2011) Farmers' insect pest management practices and pesticidal plant use in the protection of stored maize and beans in southern Africa. *International Journal of Pest Management*, **57** (1), 41–49.

Kamanula, J.F., Sileshi, G.W., Belmain, S.R. *et al.* (2013) Field evaluation of pesticidal plants for management of storage insect pests of dried maize and beans in northern Malawi, presented at the *First International Conference on Pesticidal Plants: Harnessing Pesticidal Plant Technologies for Improved Livelihoods*, Icipe, Nairobi, Kenya, 21–24 January, http://projects.nri.org/adappt/docs/ICPPabstracts&timetable.pdf (accessed 21 November 2013).

Kamanula, J.F.M. (2005) Essential oils from *Cymbopogon nardus*, *Cymbopogon citratus*, *Eucalyptus camaldulensis* and *Eucalyptus citriodora* grown in Malawi: extraction, characterization and pesticidal activity against *Prostephanus truncatus* (larger grain borer) in maize. MSc thesis, University of Malawi.

Kuntashula, E., Sileshi, G., Mafongoya, P.L. and Banda, J. (2006) Farmer participatory evaluation of the potential for organic vegetable production in the wetlands of Zambia. *Outlook on Agriculture*, **35**, 299–305.

Malawi Government (2012) Pest management plan, Ministry of Agriculture, Irrigation and Water Development, Agricultural Sector Wide Approach-Support Project, draft report.

Masamba, W.R.L., Kamanula, J.F.M., Henry, E.M.T. and Nyirenda, G.K.C. (2003) Extraction and analysis of lemongrass (*Cymbopogon citratus*) oil: an essential oil with potential to control the larger grain borer (*Prostephanus truncatus*) in stored products in Malawi. *Malawi Journal of Agricultural Sciences*, **2** (1), 56–64.

Minja *et al.* (1999) Pest control at farm level, http://www.nzdl.org/gsdlmod.

National Pesticide Information Center (2000) DDT: Technical Fact Sheet, Oregon State University, Corvallis, OR, http://npic.orst.edu/factsheets/ddttech.pdf (accessed 21 November 2013).

Nyirenda, S.P., Sileshi, G.W., Belmain, S.R. *et al.* (2011a) Farmers's ethno-ecological knowledge of vegetable pests and pesticidal plant use in Malawi and Zambia. *African Journal of Agricultural Research*, **6** (6), 1525–1537.

Nyirenda, S., Belmain, S., Stevenson, P. *et al.* (2011b) Optimized pest management with *Tephrosia* on legume cropping systems in Malawi and Tanzania, Botanicals Project Annual Report 2011 for the MacNight Foundation Collaborative Crop Research Program.

Obopile, M., Munthali, D.C. and Matilo, B. (2008) Farmers' knowledge, perceptions and management of vegetable pests and diseases in Botswana. *Crop Protection*, **27**, 1220–1224.

Ogendo, J.O., Belmain, S.R., Deng, A.L. and Walker, D.J. (2003) Comparison of toxic and repellent effects of *Lantana camara* L. with *Tephrosia vogelii* Hook and synthetic pesticide against *Sitophilus zeamais* Motschulsky (Coleoptera: Curculionidae) in stored maize grain. *International Journal of Tropical Insect Science*, **23**, 127–135.

Orwa, C., Mutua, A., Kindt, R. *et al.* (2009) Agroforestree Database: a tree reference and selection guide, version 4.0, World Agroforestry Centre, Kenya.

Paliani, A.L.P. and Muwalo, E.S. (2001) Evaluation of insecticides for the protection of stored grain insect pests in Malawi, in *Agricultural Technologies for Sustainable Development in Malawi. Proceedings of the First Annual Scientific Conference held at the Malawi Institute of Management (MIM), Lilongwe, Malawi, 6–10 November, 2000*, Department of Agricultural Research and Technical Services, Ministry of Agriculture and Irrigation, Lilongwe, Malawi, pp. 162–172.

Price, L.L. and Björnsen Gurung, A.B. (2006) Describing and measuring ethnoentomological knowledge of rice pests: tradition and change among Asian farmers. *Environment, Development and Sustainability*, **8**, 507–517.

Rajashekar, Y., Nandagopal, B. and Shivanandappa, T. (2012) *Botanicals as Grain Protectants*, National Bureau of Agriculturally Important Insects, Bangalore.

Science News (2008) *Study*: Oregano oil works as an insecticide. United Press International, Inc. URL http://www.upi.com/Science_News/2008/05/27/Study-Oregano-oil-works-as-an-insecticide/UPI-74571211916484/ (accessed 23 May 2013).

Sibanda, T., Dobson, H.M., Cooper, J.F. *et al.* (2000) Pest management challenges for smallholder vegetable farmers in Zimbabwe. *Crop Protection*, **19**, 807–815.

Sighamony, S., Anees, I., Chandrakala, T.S. and Kaiser, J. (1990) Indigenous plant products as grain protectants against *Sitophilus oryzae* (L) and *Rhyzopertha dominica* (F). *Journal of Stored Products Research*, **22**, 21–23.

Sileshi, G., Kuntashula, E., Matakala, P. and Nkunika, P.O. (2008) Farmers' perceptions of tree mortality, pests and pest management practices in agroforestry in Malawi, Mozambique and Zambia. *Agroforestry Systems*, **72**, 87–101.

Singano, C.D., Nkhata, B.T. and Mhango, V. (2008) National annual report on larger grain borer monitoring and *Teretrius nigrescens* rearing and release in Malawi, in *Plant Protection Progress Report for the 2007/2008 Season, presented at the Department of Agricultural Research Services Planning and Review Meeting, Andrews Hotel, Mangochi, 14–20 September*, Ministry of Agriculture and Food Security, Lilongwe, Malawi, pp. 1–8.

Stevenson, P.C., Nyirenda, S.P., Mvumi, B. *et al.* (2012) Pesticidal plants: a viable alternative insect pest management approach for resource-poor farming in Africa, in *Biopesticides in Environment and Food Security: Issues and Strategies* (eds K. Opender, G.S. Dhaliwal, S. Khokhar and R. Singh), Scientific Publishers (India), pp. 212–238.

Van Wolfswinkel, M. (2007) *Intercropping of Annual Foodcrops*, AgroBrief No. 4, AGROMISA Foundation, Wageningen, Netherlands, http://www.agrifoodgateway.com/articles/intercropping-annual-foodcrops (accessed 21 November 2013).

Verma, L.R. (1998) *Indigenous technology knowledge for watershed management in upper north-west Himalayas of India* (eds A. Negi and P.N. Sharma), PWMTA Field Document No. 15, Chapter 6 – Plant protection practices, www.fao.org/docrep/x5672e/x5672e07.htm#chapter%206%20%20%20plant%20protection%20practices (accessed 21 November 2013).

Williamson, J. (1975) *Useful Plants of Malawi*, Mbadzi Book Trust, Lilongwe.

## Part Three

# Food (Spices, Fruit and Vegetables, etc.)

# 23 Aromatic Plants: Use and Nutraceutical Properties

Lucia Guidi and Marco Landi
Department of Agriculture, Food and Environment, University of Pisa, Pisa, Italy

## 23.1 Introduction

### 23.1.1 Use of aromatic plants: from prehistoric days to the present

Evidence demonstrating the use of aromatic plants dates back to prehistoric times. Archaeologists have found bodies covered with dried aromatic plants and seeds in eastern Europe and Iraq, and the existence of tribal rituals where the smoke from the burning of dried aromatic leaves or woods has been confirmed by cave paintings.

In the ancient worlds, the use of aromatic plants was highly popular:

- *The Sumerians* (4000 BC) used aromatic plants such as fennel, galbanum and pine and they were the first population that had written formulas for plant remedies.
- *The Egyptians* (3000 BC) were experts on the use of aromatic plants to produce perfumes, cosmetics and medicine. Body, mind and soul practice with aromatic plants provided the uses of aniseed, cinnamon, cardamom, cumin, ferula, frankincense, mastic, laurel, mint, myrrh and terebenth, among several other plant species. Complex rites that utilized aromatic oils were involved in the mummification processes to preserve the body of the dead during the journey to the afterlife.
- *The Chineses* (2800 BC) have long been known to have been among the first to have discovered medicinal plants properties. Aniseeds, cinnamon and ginger were among some of the species that they were routinely using as remedies.
- *The Indians* (1500 BC) created the basis for the ayurvedic medicine, a Hindu system of folk-traditional medicine native to India as a form of alternative medicine. In the *Ayurveda* (Brahma's sacred book on plants), names of aromatic plants were mentioned as a part of the secret for eternal youth.
- *The Greeks* (1000 BC) made a strong contribution on the use of aromatic plants as medicine. Hippocrates is remembered today as the 'Father of Medicine', and he is also remembered for saving Athens from the plague by ordering the inhabitants of the city to burn large urns of aromatic plants throughout the streets. Pline the Elder, in his *Natural History*, mentioned 22 different types of essential oils concocted with marjoram, cypress, cystus, iris, nard, myrtle, laurel, lemon and rose. A century later, Galen established certain rules concerning the preparation of remedies obtained from plant extracts and oils. Aristotle was the first physician to handle scented vapours and perfumes in a scientific manner, and he also proposed a method of extraction to obtain essential oils. Dioscorides classified and described many plants and remedies, and he is considered as being the 'Father of Pharmacy'.
- *The Ancient Romans* (1st century BC–5th century BC) were also very hungry for herbs and fragrance, and they widely used incense and other aromatic plant parts. Nero, the notorious Roman Emperor, burned incredible amounts of incense at the funeral of his wife and had every house in his city oozing out fragrant fumigations. To attest to the importance that aromatic plants had gained until this historical age, one should remember that two among the three gifts that the Magi gave to Jesus when he was born were derived from aromatic plants: frankincense (burned to honour God) and myrrh (to honour humankind). The Bible also quoted Jesus as having instructed people to make special healing unguents with aromatic substances in order to cure various ills and requested the same when he was about to restore the life of Lazarus.
- *Medieval*. One has to wait until nearer to the Middle Ages to see any further progress in medical advances in western Europe, as it had regressed into the obscurity of 'the Dark Ages', where the written word practically disappeared. This does not mean that there was no knowledge on matter but the write information, including those related to the use of plants, were the privilege of few people such as wizards, alchemists and in particular monks. However, folk medicine survived through oral transmission down to the generations.
- *Renaissance and the Modern age*. After the dark period represented by the Medieval age, a strong impulse on the desire of knowledge occurred. Soldiers and knights returning from the Crusade areas, where culture and medicine flourished, brought back many treasures of knowledge that have helped western Europe move forward on many fronts. In the next centuries, western Europe probably becomes the epicentre of worldwide culture, and botanical and chemical advances were also made in improving the use of plants as medicine. The capacity to obtain even more purified plant extracts and essential oils increased the degree of success in the cure of serious diseases (for the time), such as fevers, wound infections, diarrhoea, and so on. The demand for perfume (used a lot to cover up unpleasant bodily odors!) and beauty and youth ointments dramatically

*Novel Plant Bioresources: Applications in Food, Medicine and Cosmetics*, First Edition. Edited by Ameenah Gurib-Fakim.

increased to cater for an emergent middle class and the nobility. The economical relevance of spice and oils (from Mediterranean species or imported from eastern Europe or Asia) made the apothecaries and the medical practitioners the depositaries of medical and drugs knowledge. Thus, they were considered a group of important people until the end of the 1800s.

- *Aromatic plants of the twenty-first century*. From the early 1900s, with increased medical knowledge, the use of oils and drugs as medicine was increasingly being entrusted to chemists and physicians. However, the use of dried plants, extracts and oils was not completely abandoned, and the folk medicine continues to be an alternative to the scientific medicine. One also had to consider that the progress just described above took place only in industrialized countries, while natural and folk medicine in less advanced areas of the world remained the only alternative for a long time. In fact, today, in many parts of the globe, and in particular in undeveloped countries, medicinal plants represent the main cure for human diseases. The turn of this century also marked the birth of aromatherapy, as we know with the work of Gattefosse, a perfumer chemist who inherited the business from his family. His research on essential oils took a boost when he accidentally burnt his hand in his laboratory and quickly immersed it in a tank of essential oil: lavender! He was surprised to see that it immediately soothed the pain and that its wound healed more rapidly than expected. He established some of the therapeutic properties of the essential oils through his research, and more particularly noted that many of them had antiseptic and antibiotic properties. Aromatherapy was made more widely known when Valnet, a doctor, scientist and professor at 'La Faculté de la Sorbonne' in Paris, began to use essential oils to treat his patients, contributing to the expansion of the aromatherapy profession. Essential oils cannot be considered only for their chemical quality; as well as their obvious healing properties on skin and body systems, they also have a subtle and beneficial influence on the nervous system, on the mind and on hormone balance. Many explanations have been put forward as to how these aromatic plants work, but so far the mystery remains. The rapid expansion of alternative and complementary medicine is a strong indication of the necessity to have a more wholesome approach to illness. This should include taking into account the root cause of health problems and any psychological factors, rather than operating on a strictly scientific basis, which often turns out to be a limitation of the classical Western medicine.

During the last decades, owing to the low cost associated with the extraction of essential oils from aromatic species, a lot of interesting applications, in addition to the herbal and medical uses, have taken place. Food preserving, antimicrobials, insecticides and antifungals, as well as additives for perfumes and beauty creams, are only some of the applications of aromatic plants and their essential oils. The latter are extracted from a number of aromatic crops through steam distillation, with each plant species often having a distinct temperature and duration for the distillation process. The steam distillation of any aromatic crop results in four products: (i) essential oil, (ii) steam-distilled (residual) plant material, (iii) residual water and (iv) hydrolat. The essential oil is the targeted valuable product, but the residual plant material may be used as forage for animals, depending on the plant species.

### 23.1.2 Modern use of aromatic plants: constituents and biological properties

Aromatic plants have a unique aroma and flavour that are derived from compounds known as phytochemicals or secondary metabolites (Fisher, 1992). The phytochemicals are antimicrobial substances present in aromatic plants; whilst they are beneficial to the plants, they also repel harmful organisms. They act as photoprotectants and respond to environmental changes. Numerous classes of phytochemicals, including isoflavones, anthocyanins, flavonoids and terpenes, are found in plant species.

The use of aromatic plants is on the increase in the world, and according to the World Health Organization (WHO) nearly 80% of the planet's population, especially in developing countries, still depends on plants for their healthcare (Gurib-Fakim, 2006). As reported by Lubbe and Verpoorte (2011), plants, primarily used for their medicinal or aromatic properties in pharmacy and perfumery are defined as medicinal and aromatic plants, respectively, in the EU. Aromatic species can also be used for cosmetic purposes (Aburjai and Natsheh, 2003), in the production of dyes and colourants (Gilbert and Cook, 2001), crop protection products (Tripathi *et al.*, 2009), alcoholic beverages (Tonutti and Liddle, 2010), food additives (Priya *et al.*, 2011), food preservatives (Ait-Ouazzou *et al.*, 2011) and substances for animal nutrition (Christaki *et al.*, 2012). These applications of aromatic plants and the subsequent increasing demand for these species have transformed them in to industrial crops. They are known to be natural, eco-friendly and generally recognized as safe products (Lubbe and Verpoorte, 2011). Clearly, the essential oils, which are responsible for the flavour and fragrance of aromatic plants, are taking centre stage from an industrial point of view.

Essential oils are odorous, volatile, hydrophobic and highly concentrated compounds that are obtained from various parts of plants, including flowers, buds, seeds, leaves, twigs, bark, wood, fruits and roots, and are stored in secretory cells, cavities, canals, epidermic cells or glandular trichomes (Bakkali *et al.*, 2008).

Generally, two or three major components at fairly high concentrations (20–70%) tend to characterize a plant, with other components present in trace amounts. These major components are responsible for the biological properties of the essential oils, albeit that in some cases minor compounds also contribute to the biological activity. The chemical components belong to distinct biosynthetic pathways: terpenes and aromatic compounds (Bakkali *et al.*, 2008). The main group is composed of terpenes and terpenoids and the rest are aromatic and aliphatic constituents, all characterized by low molecular weight.

Terpenes are derived biosynthetically from units of isoprene, which has the molecular formula $C_5H_8$. All terpenes are multiples of the previous empirical formula $(C_5H_8)_n$, where $n$ is the number of linked isoprene units. The biosynthesis of terpenes consists of synthesis of the isopentenyl diphosphate (IPP) precursor, following by repetitive additions of IPPs to form the prenyldiphosphate, the precursor of the various classes of terpenes. Subsequently, modifications of the allylic prenyldiphosphate by terpene-specific synthetases to form the terpene skeleton occur. Finally, secondary enzymatic modifications of the skeleton to attribute functional properties to the different terpenes take place. The main terpenes are the monoterpenes (C10) and sesquiterpenes (C15), but hemiterpenes (C5), diterpenes (C20), triterpenes (C30) and tetraterpenes

(C40) also exist. Monoterpenes are the most representative in essential oils (90%) and contain several functional groups: alcohols, aldehydes, ketone, esters, peroxides and phenols (Bakkali *et al.*, 2008).

Aromatic compounds derived from phenylpropane occur less frequently than the terpenes. The biosynthetic pathways concerning terpenes and phenylpropanic derivatives generally are separate in plants, but may coexist in some, with one major pathway taking over (e.g. cinnamon oil with cinnamaldehyde as major constituent and eugenol as minor constituent; also, clove oil, fennel, etc.). Aromatic compounds comprise aldehyde (cinnamaldehyde), alcohol (cinnamic alcohol), phenols (chavicol, eugenol), methoxy derivatives (anethole, elemicine, estragole and methyleugenols) and methylene dioxy compounds (apiole, myristicine, and safrole) (Bakkali *et al.*, 2008). About 3000 plants are used for the production of essential oils, and 300 of these are commonly traded on the global market (CBI, 2009). Recently, an amount of 104 000 tons of the most important 20 essential oils were produced in the world (CBI, 2010).

Aromatic species are widely distributed in the world, and many originate from the Mediterranean area, either wild or cultivated (Carrubba and Scalenghe, 2012; Christaki *et al.*, 2012). Aromatic species from the Mediterranean represent an incredible source of essential oils and other biologically active moieties.

The mode of action of essential oil is related to antimicrobial and antioxidant activity. The antimicrobial activity is attributable to the hydrophobicity of the essential oil that disturbs and rends more permeable the bacterial cell structure (Solorzano-Santos and Miranda-Novales, 2012). In addition, some aromatic plant extracts have been reported to stimulate growth of some bacteria and have a prebiotic-type effect (Greathead, 2003) and for this reason they are utilized in ruminants to manipulate rumen metabolism (Benchaar *et al.*, 2009). Obviously, the essential oil of aromatic plants also represents a good source of natural antioxidant attributable principally to the presence of phenolic compounds. These compounds occur as glycosides, even though the antioxidant properties are attributable to aglycon and mainly to catechol (Rice-Evans *et al.*, 1993; Sakakibara *et al.*, 2003). It is clear that the antioxidant capacity of these compounds contributes to the delay of many oxidative-stress-related diseases, such as cardiovascular diseases, cancer, diabetes and Alzheimer's (Shetty, 1997; Miron *et al.*, 2010).

## 23.2 Mediterranean aromatic plants

The term Mediterranean is used to describe the climate synonymous with the region surrounding the Mediterranean Sea, although it is not exclusive to this region. Mediterranean climates are found in California, Australia, New Zealand, England and South Africa, in addition to the 'true' Mediterranean regions of Spain, France, Italy, Greece, Cyprus, Malta and Turkey. The Mediterranean climate is characterized by a long, hot summer with little rainfall and a mild, rainy winter. In these conditions plants are subjected to high solar irradiances and, concomitantly, with low water availability. Consequently, Mediterranean plant species have evolved the ability to synthesize a wide range of secondary metabolites involved in the tolerance to these constrained environmental factors (Galli, 2012).

Many plant species are present in the Mediterranean area, and many of these are medicinal and aromatic plants (Carrubba and Scalenghe, 2012). Mediterranean herbs and flowers spring up underneath the trees and shrubs and are commonly referred to as macquis. Macquis plants include rosemary, thyme and lavender, autochthonous species of Mediterranean areas. It is also possible to find Spanish broom, juniper and gorse amongst the macquis. Mediterranean macquis is known as chaparral in California, heathlands in Australia and fynbos in South Africa. Depending on the elevation levels and associated growing conditions, Mediterranean plants may include native cushion plants (e.g. *Anthyllis*), aromatic herbs and both native and non-native plants. The latter include the *Allium* species, introduced into the Mediterranean region, and now believed to be 'native' by many, owing to their longevity in the area.

In this review, the most important aromatic Mediterranean species are characterized for the constituents of essential oil and other relevant chemical compounds; a summarized review of the biological activity connected to the principal uses is also reported.

### 23.2.1 *Ocimum* spp.

Basil (*Ocimum* spp.), a member of the Lamiaceae family, is an annual herb widely distributed in the world and probably the major essential oil crop commercially cultivated in many countries (Makri and Kintzios, 2007). The most important species of genus *Ocimum* are *sanctum* L., *gratissimum* L., *viride* Willd., *basilicum* L., *americanum* L., *canum* Sims., *kilimandscharicum* Guerc and *suave* Willd. Within each species there are an extremely elevated number of cultivars, with different leaf size and colour (green to dark purple), flower colour (white, red, lavender and purple), growth characteristics (shape, height and flowering time) and aroma (Morales *et al.*, 1993). Among these *Ocimum* species, the most important and characterized is certainly *O. basilicum*. Its importance is evidenced by the wide range of selected cultivars and ecotypes belonging to this species (Makri and Kintzios, 2007).

The specific aroma and flavour of this herb are attributable to the presence of essential oils, which determined its use as condiments (Klimankova *et al.*, 2008). Fresh or dried, basil is widely used in the Mediterranean kitchen, such as tomato products, vegetables, salads, pizza, meat, soups and marine foods (Marotti *et al.*, 1996). However, in addition to the culinary use, this aromatic species is also attributed a folk medicinal value and it is accepted officially in a number of countries (Sajjadi, 2006; Zeggwagh *et al.*, 2007).

The major aroma constituents of *Ocimum* spp. are 3,7-dimethyl-1, 6-octadien-3-ol (linalool; 3.94 mg $g^{-1}$ dry weight), 1-methoxy-4-(2-propenyl) benzene (estragole; 2.03 mg $g^{-1}$ dry weight), methyl cinnamate (1.28 mg $g^{-1}$ dry weight), 4-allyl-2-methoxyphenol (eugenol; 0.896 mg $g^{-1}$ dry weight) and 1,8-cineole (0.288 mg $g^{-1}$ dry weight) (Lee *et al.*, 2005). The quality and quantity of essential oil produced by plants depends on leaf developmental stages (Fisher *et al.*, 2011), but also on various environmental factors, such as seasonal variation (Hussain *et al.*, 2008), method of harvest (Kothari *et al.*, 2004), climate (Pamplona-Roger, 2004), salt stress (Bernstein *et al.*, 2010) and soil type (Pamplona-Roger, 2004). Even nitrogen fertilization and the use of ethephon and gibberellins (Ono *et al.*, 2011) can influence the quali-quantitative yield of essential oil in basil (Nurzyńska-Wierdak, 2012).

Banchio *et al.* (2009) reported biochemical evidence that the plant growth-promoting soil bacterium *Bacillus subtilis* GB03 releases volatile chemicals that elevate fresh weight essential oil accumulation and emissions along with plant size. The two major essential oil components from sweet basil, *R*-terpineol and eugenol, increased approximately 2- and 10-fold respectively in plants exposed only to the elicitor GB03 or with root inoculation as compared with water controls. Similarly, the effect of arbuscular mycorrhizal fungi on root colonization, growth, essential oil content and composition and nutrient acquisition of *O. basilicum* was investigated (Rasouli-Sadaghiani *et al.*, 2010). The authors concluded that *Glomus fasciculatum* was more effective than other species, which may indicate effective symbiotic potential of this strain with basil roots.

As previously reported, essential oil from basil shows a wide range of biological and pharmacological activities. The different species of genus *Ocimum* show various medicinal potentialities, as reported by Bashin (2012). Table 23.1 summarizes the medical and agro-industrial applications of different basil species.

Basil also contains high concentrations of phenolic compounds that would explain its potent antioxidant capacity (Javanmardi *et al.*, 2002; Lee and Scagel, 2009). The most prevalent basil phenolic is rosmarinic acid, but a high concentration of chicoric acid was also found (Javanmardi *et al.*, 2002; Lee and Scagel, 2009). However, the high antioxidant capacity of basil has been attributed primarily to rosmarinic acid (Petersen and Simmonds, 2003). Polyphenols having catechol (1,2-dihydroxybenzene) nuclei are strong antioxidants, *in vitro*. They react by rapidly reducing the reactive oxygen species, bind transition metal ions as inert complexes and regenerate the potent chain-breaking antioxidant α-tocopherol. Rosmarinic acid has antiviral, antimicrobial and anti-inflammatory pharmacological properties (Javanmardi *et al.*, 2002). Bora *et al.* (2011a) concluded that *O. basilicum* possesses neuroprotective effects, as is evident from the reduction of cerebral infarct size and lipid peroxidation, restoration of glutathione content and attenuation of motor dysfunctions.

The *O. basilicum* samples indicated also the presence of one or two major flavone aglycones, which were identified as salvigenin and nevadensin, and up to 10 minor ones: cirsileol, cirsilineol, eupatorin, apigenin, acacetin, genkwanin, apigenin 7,4′-dimethyl ether, cirsimaritin, ladenein and gardenin B (Grayer *et al.*, 1996). Even though relatively little information is available on the medical effects of purified flavonoid moieties from basil, Gajula *et al.* (2009) reported a chemopreventive role of *Ocimum* spp. extracts from leaves, while Yesiloglu and Sit (2012) attributed an antioxidant role to flavonoid compounds.

An interesting application of basil is the use of the plant extract at room temperature for the green synthesis of noble metal nanoparticles (NPs) (Ahmad *et al.*, 2010), such as silver NPs. It appears important to study the formation of NPs in *Ocimum* root and stem extracts for the understanding of the mechanism of biological synthesis and for exploring new green ways to prefabricate nanomaterials. Along this line, Ahmad *et al.* (2010) recently reported the synthesis and antibacterial activity of silver NPs from the sun-dried stems and roots of basil plant. Biosynthesis of stable and spherical-shaped NPs of average size ~10 nm has been described using different parts of the basil plant. Phenolic and flavonoid compounds play an important role in the formation of nanosilver.

### 23.2.2 *Salvia officinalis* L.

The genus *Salvia* is another important and widespread member of the Lamiaceae family. It is a perennial, evergreen sub-shrub, with woody stems, greyish leaves, and blue to purplish flowers. It is native to the Mediterranean region, though it has been naturalized in many areas across the world. It has a long history of medicinal and culinary use, and in modern times as an ornamental garden plant. The genus *Salvia* counted about 900 species and the species *Salvia officinalis* is used as a culinary herb and spice or for medicinal purposes against inflammations and infections of the mouth and throat. In fact, *Salvia* species have been used in folk medicine for wound healing and in alleviating stomach, liver and rheumatic pains and for treating the common cold in the form of infusion and decoction in various parts of the world (Bayrak and Akgul, 1987).

Until the beginning of 2011, a total of 773 compounds have been isolated from 134 *Salvia* species (Wu *et al.*, 2012), subdivided into seven groups: sesquiterpenoids, diterpenoids, sesterterpenoids, triterpenoids, steroids, polyphenols and others. However, the main secondary metabolites are terpenoids (more than 80%) and flavonoids, which are also responsible for the medicinal properties of the *Salvia* plant (Pan *et al.*, 2010; Ramos *et al.*, 2010; Wu *et al.*, 2012). Wu *et al.* (2012) carefully reported single compounds found in *Salvia* species belonging to the different groups. Among terpenoids, abietane and clerodane are the most representative compounds, whereas sesquiterpenoids and triterpenoids are rare in *Salvia* spp.

Taking into consideration the numerous species available in genus *Salvia* and the numerous reports on the medicinal properties of these species, in this review we refer principally to *S. officinalis* typical of the Mediterranean basin. This species is largely cultivated for culinary, medicinal and ornamental purposes. The curative properties of sage have long been known; it is used as an antihydrotic, spasmolytic, antiseptic and anti-inflammatory and in the treatment of mental and nervous conditions (Baricevic and Bartol, 2000). On the other hand, it has been demonstrated that sage essential oil can improve the memory, showing promise in the treatment of Alzheimer's disease (Perry *et al.*, 2003). Finally, sage also has a potential in treating cancer, as it shows strong antitumorigenic activities (Russo *et al.*, 2013). The name of this genus identifies perfectly its uses: name *Salvia* means 'to cure' and species name '*officinalis*' means medicinal. Many studies reported the composition of essential oils from sage, and all of them demonstrated that it is extremely variable (Mirjalili *et al.*, 2006; Ben-Farhat *et al.*, 2009). ISO 9909 for medicinal uses regulates the amounts of the following constituents in essential oil of sage: *cis*-thujone (18.0–43.0%), camphor (4.5–24.5%), 1,8-cineole (5.5–13.0%), *trans*-thujone (3.0–8.5%), α-humulene (≤12.0%), α-pinene (1.0–6.5%), camphene (1.5–7.0%), limonene (0.5–3.0%), bornyl acetate (≤2.5%) and linalool+linalylacetate (≤1.0%) (Santos-Gomes and Fernandes-Ferreira, 2001; Bernotienė *et al.*, 2007). These results are also recently confirmed by Russo *et al.* (2013). These authors used gas chromatography (GC) and GC–mass spectrometry (GC–MS) to analyse the sage essential oils, extracted by hydrodistillation, and found that the main components were α-thujone, camphor, borneol, γ-murolene and sclareol for all the samples. Similar results have been reported by Baranauskiene *et al.* (2011). Ben-Farhat *et al.* (2009) reported that the analysis of some

**Table 23.1** Principal species of the genus *Ocimum* and their medicinal properties and other application in the agricultural and food industries

| Species | Medicinal potentialities | Other uses |
|---|---|---|
| *O. sanctum* L. | • Antitubercolosis (Farivar *et al.*, 2006)<br>• Antiseptic and antibiotic (Mishra and Mishra, 2011)<br>• Anticancerous (Rahman *et al.*, 2011)<br>• Anticoagulant (Singh *et al.*, 2012)<br>• Antiulcer (Kath and Gupta, 2012)<br>• Antidiabetic (Muralikrishnan *et al.*, 2012)<br>• Antidepressant (Chatterjee *et al.*, 2011)<br>• Hepatoprotective (Lahon and Das, 2011)<br>• Antiviral effect (Yucharoen *et al.*, 2011)<br>• Lipid-lowering effect (Suanarunsawat *et al.*, 2009)<br>• Antifungal and antiflatoxigenic (Amber *et al.*, 2010; Khan *et al.*, 2010; Kumar *et al.*, 2010)<br>• Antibacterial (Mahmood *et al.*, 2008)<br>For general overview, see Pattanayak *et al.* (2010), Rahman *et al.* (2011), Mahajan *et al.* (2012) and Singh *et al.* (2012) | • Insecticides: larvicidal (Anees, 2008); aphidicidal (Satpathi and Ghatak, 1990)<br>• Fungicide (Sinha and Saxena, 1989) |
| *O. gratissimum* L. | • Brain diseases (Bora *et al.*, 2011b)<br>• Cardioprotective (Lee M.-J. *et al.*, 2011)<br>• Hepatoprotective disease of liver (Chiu *et al.*, 2012)<br>• Spleen disease (Shehu *et al.*, 2012)<br>• Laxative (Effraim *et al.*, 2001)<br>• Antidiabetic (Mohammed *et al.*, 2007)<br>• Antimicrobial (Matasyoh *et al.*, 2007; Braga *et al.*, 2011)<br>• Antimutagenic (Obaseiki-Ebor *et al.*, 1993)<br>For general overview, see Prabhu *et al.* (2009) | • Germicidal (Pessoa *et al.*, 2002; Holets *et al.*, 2003)<br>• Nematocidal (Chatterje *et al.*,1982).<br>• Toothpastes and mouthwashes additive (Ahonkhai *et al.*, 2009)<br>• Fungicide (Dambolena *et al.*, 2010)<br>• Ovicidal (Pessoa *et al.*, 2002) |
| *O. viride* Willd. | • Antibacterial (Isu, 2005)<br>• Anticancerous (Sharma *et al.*, 2010) | • Insects repellant (Owusu, 2000) |
| *O. basilicum* L. | • Antidiabetic (El-Beshbishy and Bahashwan, 2012)<br>• Respiratory diseases and antituberculosis (Siddiqui *et al.*, 2012; Singh *et al.*, 2012)<br>• Antimutagenic (Berić *et al.*, 2008)<br>• Brain diseases (Bora *et al.*, 2011a)<br>• Heart diseases (Amrani *et al.*, 2009)<br>• Antihypertensive (Umar *et al.*, 2010)<br>• Antiherpes (Yucharoen *et al.*, 2011)<br>• Hepatoprotective (Meera *et al.*, 2009)<br>• Antiulcer (Singh, 1999)<br>• Chemomodulator (Dasgupta *et al.*, 2004)<br>• Immunomodulator (Neelam and Nilofer, 2010)<br>For general overview, see Bilal *et al.* (2012) | • Insecticidal (Kéita *et al.*, 2001)<br>• Food preservant (Lachowicz *et al.*, 1998)<br>• Fungistatic (Reuveni *et al.*, 1984)<br>• Toothpastes and mouthwashes additive (Ahonkhai *et al.* 2009)<br>• Antifungal (Dambolena *et al.*, 2010) |
| *O. americanum* L. | • Antiherpes (Yucharoen *et al.*, 2011)<br>• Antibacterial (Carović-Stanko *et al.*, 2010)<br>• Antidiabetic (Bihari *et al.*, 2011) | • Insecticides (Ogendo *et al.*, 2008)<br>• Pharmaceutical tablets formulation (Sheth *et al.*, 2010)<br>• Mouthpaste antimicrobial additive (Thaweboon and Thaweboon, 2009) |
| *O. canum* Sims. | • Antimicrobial (Bassole *et al.*, 2005)<br>• Hepatoprotective (Behera *et al.*, 2012)<br>• Antiacetylcholinesterase (Kiendrebeogo *et al.*, 2011)<br>• Antiplasmodial (Inbaneson *et al.*, 2012) | • Antifungal (Singh P. *et al.*, 2011)<br>• Larvicidal (Tchoumbougnang *et al.*, 2009)<br>• Antibacterial (Prakash and Karmegam, 2012) |

(*continued*)

**Table 23.1** (*Continued*)

| Species | Medicinal potentialities | Other uses |
|---|---|---|
| *O. kilimandscharicum* Guerk | • Antibacterial (Shinde *et al.*, 2010; Carović-Stanko *et al.*, 2010)<br>• Wound healing (Paschapur *et al.*, 2009)<br>• Analgesics/antirheumatics (Mwangi, 2011)<br>For general overview, see Vaidyaratnam (1980) and Kashyap *et al.* (2011) | • Insecticidal (Kweka *et al.*, 2009) |
| *O. suave* Willd. | • Antiinflammatory (Masresha *et al.*, 2012)<br>• Wound healing (Hassan *et al.*, 2011)<br>• Antidiabetic (Umar *et al.*, 2012)<br>• Antiulcer (Tan *et al.*, 2002)<br>• Antipyretic (Makonen *et al.*, 2003) | • Insecticidal (Kweka *et al.*, 2009)<br>• Ovicidal (Ojianwuna and Umoru, 2010) |

representative polyphenolic compounds revealed that rosmarinic acid, carnosol and carnosic acid were the prevalent compounds of *S. officinalis* methanolic extracts and their profile changed in relation to the geographical distribution.

It is a well-known fact that the biosynthesis of the sage volatile oils is influenced by various environmental factors, namely the soil mineral fertilization (Piccaglia and Marotti, 1993), light intensity (Li *et al.*, 1996) and inoculation with *Glomus intraradices*, which resulted in improved essential oil yield and quality (Geneva *et al.*, 2010). In addition, influences in essential oil yield have been found in relation to the climate conditions (Máthé *et al.*, 1992) and the soil salinity level. Interestingly, Ben-Taarit *et al.* (2009) reported that salt stress strongly affected the sage essential oil compounds. An increase in sage essential oil yield and the main essential oil constituents camphor, α-thujone and 1,8-cineole has also been reported by Bettaieb *et al.* (2009) following a moderate water stress. Russo *et al.* (2013) reported that the percentages of these compounds varied depending on environmental factors, such as altitude, water availability and pedo-climatic conditions. Phosphorus nutrition can also improve essential oil in sage (Nell *et al.*, 2009).

Sage has a wide range of medicinal effects: antidiabetic (Baddar *et al.*, 2011), antihydrotic (Eidi and Eidi, 2009), antinociceptive (Alves-Rodrigues *et al.*, 2012) and spasmolytic (Khan *et al.*, 2011b). Sage has analgesic effects that are attributed to linalyl acetate, linalool, eucalyptol, and β-caryophyllene (Qu *et al.*, 2012). In addition, Generalic *et al.* (2012) reported antioxidant and antibacterial activity of sage and found the presence of phenolic compounds responsible for antibacterial activities against Gram-positive (e.g. *Bacillus cereus* and *Staphylococcus aureus*) and Gram-negative bacteria (e.g. *Salmonella infantis* and *Escherichia coli*). Bauer *et al.* (2012) reported on anti-inflammatory activity in *S. officinalis* and this validated the common uses of this plant in traditional practices for antiseptic and anti-inflammatory activity.

*S. officinalis* also possesses an anticarcinogenic effect, as reported by Ramos *et al.* (2012) and Deeb *et al.* (2011). Also, Sertel *et al.* (2011) suggested for the first time the ability of *S. officinalis* essential oil to inhibit human head and neck squamous cell carcinoma cell growth.

Many other medical properties are attributed to *S. officinalis*: treatment of mental and nervous conditions (Baricevic and Bartol, 2000) and against cardiovascular diseases (Kianbakht *et al.*, 2011). Antimutagenic action (Vuković-Gačić *et al.*, 2006), antimicrobial action (Miguel *et al.*, 2011), antiangiogenic (Keshavarz *et al.*, 2005) and antiviral (Loizzo *et al.*, 2008) properties have also been reported.

Sage is also traditionally used in food preparation (Piccaglia, 1998) and as flavouring agents in perfumery and cosmetics (Delamare *et al.*, 2007). Çalikoglu *et al.* (2009) studied the influence of various sage applications (leaves, deodorized leaves and essential oil) on the oxidative stability of sunflower oil during accelerated storage. The most appealing result was that the residue can be used as a natural antioxidant.

The use of sage as food or medicine is therefore regulated for the neurotoxic effect of thujone (Lachenmeier and Uebelacker, 2010). In 1988, the European food law deregulated the use of thujone-containing plants in food. However, the previous limit for foodstuffs containing preparations based on sage ($25\,mg\,kg^{-1}$) and the general limit for foodstuffs ($0.5\,mg\,kg^{-1}$) (EEC, 1988) were removed from the new flavouring regulation. Only artemisia-containing beverages (i.e. mainly absinthe) remain regulated (EU-EC, 2008). Nevertheless, thujone as such (i.e. in chemically pure form) is not allowed in foods; it may only be indirectly introduced into foods through the use of thujone-containing plants (EU-EC, 2008).

### 23.2.3 *Rosmarinus officinalis* L.

*Rosmarinus officinalis*, commonly known as rosemary, is a woody, perennial herb with fragrant, evergreen, needle-like leaves and white, pink, purple or blue flowers. It is typically native to the Mediterranean region and belonging to the family of Lamiaceae. Rosemary is used as a decorative plant in gardens and has many culinary and medical uses. The leaves are used to flavour various foods, such as stuffing and roast meats, but it is also utilized as a natural antioxidant for food conservation (Ho *et al.*, 2000; Moreno *et al.*, 2012).

In this species, the main active constituents are camphene, camphor, cineol, borneol, resin, bitters matter, rosmarinic acid and flavonoids (Begum *et al.*, 2013). The major constituents of rosemary essential oil are camphor (5–21%), 1,8-cineole (15–55%), α-pinene

(9–26%), borneol (2–5%), camphene (2–12%), β-pinene (2–9%) and limonene (1–5%). Phenolic compounds are represented by flavonoids with a methylated aglycon (e.g. genkwanin) and by phenolic acids (>3%), particularly rosmarinic, chlorogenic and caffeic acids. Tricyclic diterpenes are also present (i.e. rosmaridiphenol, carnosol, carnosic acid and rosmanol), and diterpenes, including secohinokio (Salido *et al.*, 2003; Cantrell *et al.*, 2005; EDQM, 2004). Some authors have reported the distribution of six compounds with three different polyphenol skeletons in *R. officinalis*: phenolic diterpenes (carnosic acid, carnosol, and 12-*O*-methylcarnosic acid), caffeoyl derivatives (rosmarinic acid) and flavones (isoscutellarein 7-*O*-glucoside and genkwanin), each showing a characteristic behaviour and distribution during the vegetative cycle (del Baño *et al.*, 2003).

Essential oil derived from wild rosemary plants shows high variations in the chemical composition according to geographical area origin (Angioni *et al.*, 2004; Jamshidi *et al.*, 2009). This factor is extremely important in rosemary. Rosemary essential oil can be classified into three chemotypes from a chemical point of view: cineoliferum (high content of 1,8-cineol); camphoriferum (camphor >20%); and verbenoniferum (verbenone >15%) (Napoli *et al.*, 2010b). The differences in essential oil composition in different chemotypes of rosemary also induced a different biological activity (Jordán *et al.*, 2013). Obviously, other factors influence the composition of essential oil in rosemary, such as environmental and agronomic conditions (Moghtader and Afzali, 2009), season (Celiktas *et al.*, 2007), developmental stage (Yosr *et al.*, 2013), harvesting time (Singh and Guleria, 2013), biofertilization (Abdullah *et al.*, 2012), water stress (Singh and Ramesh, 2000), storage (Turek and Stintzing, 2012) and the method of extraction (Okoh *et al.*, 2010). In another study, Kassem *et al.* (2009) showed that increasing salinity levels reduced all vegetative growth and influenced chemical composition and oil production.

Rosemary species contains bioactive ingredients that provide a complementary value besides the nutritional one, to be applied in the food industry (Moreno *et al.*, 2012). These authors focused on the most significant rosemary biological properties, reviewing the free-radical scavenging and antibacterial actions of non-volatile constituents and essential oils. Antibacterial activity of the main rosemary bioactives in combination with antibiotics, as well as a possible antibacterial mechanism of action, is proposed, among others. The numerous phytochemicals contained in rosemary plants constitute a potential source of natural compounds. Within the different secondary metabolites, about 90% of the antioxidant activity is attributed mainly to the high content of non-volatile components as carnosic acid and carnosol (phenolic diterpenes) and rosmarinic acid (Bradley, 2006). Taking into consideration that rosemary is the only spice commercially available for use as an antioxidant in Europe and the USA, there is no doubt regarding the economic relevance of this species. In addition, rosemary contains both lipophilic (monoterpenes and diterpenes) and hydrophilic molecules (caffeic acid and rosmarinic acid) that can act as an antioxidant either in lipophilic parts of the body or in aqueous fluids.

Clearly, this species possesses other beneficial effects, like antimicrobial, antiviral, anti-inflammatory, anticarcinogenic activities, and is also known to be an effective chemopreventive agent, as deeply reviewed by Begum *et al.* (2013). Of note is that Bozin *et al.* (2007) found the most important antibacterial activity of rosemary essential oil was expressed against *E. coli*, *Salmonella typhi*, *Salmonella enteritidis* and *Shigella sonei*. Sharabani *et al.* (2006) suggested the possibility to adopt a novel protocol for chemoprevention or differentiation therapy of myeloid leukaemia by using carnosic acid that reduced the negative effects of chemotherapeutic vitamin D3.

Beside the above-mentioned effects on human diseases, rosemary essential oil was demonstrated to be effective against all the larval stages of *Aedes aegypti* (Waliwitiya *et al.*, 2009) and against *Sitophilus granarius* (Zoubiri and Baaliouamer, 2011), suggesting a potential application of this species as a bio-pest controller.

### 23.2.4 *Laurus nobilis* L.

Laurel is a perennial pyramid-shaped tree or a large shrub with aromatic, evergreen leaves and shiny grey bark of the Lauraceae, a family that comprises 32 genera and about 2000–2500 species. The leaves are alternate, lanceolate, coriaceous and shining with wavy margins. Clusters of small yellow flowers are produced on the female plants in spring, followed by shiny purplish black berries (Longo and Vasapollo, 2005).

Laurel is native of the Mediterranean region and Asia Minor; currently, it is cultivated and collected from the wild in many Mediterranean countries (Turkey, Greece, Spain, Portugal and Morocco) and Mexico. It is utilized mainly in Europe and in the USA as an ornamental plant (Barla *et al.*, 2007), but as with other aromatic species, the global demand of bay leaves (which exceeds 3000 tons per year) represents another relevant market segment (UNCTAD, 2006).

The commercial value of leaves derives from their richness in essential oils. Thus, the chemical composition of the volatile fraction, as well as the composition and bioactivities extracts, have been studied extensively (Pino *et al.*, 1993; Caredda *et al.*, 2002; Dadalioglu and Evrendilek, 2004; Conforti *et al.*, 2006). The oil yields on a dry weight basis ranged from 1.4% to 2.6% (Özcan and Chalchat, 2005), while the main components of the essential oil were cineol (44.12%), eugenol (15.16%), sabinene (6.20%), 4-terpineol (3.60%), α-pinene (2.74%), methyleugenol (2.48%), α-terpineol (2.19%) and β-pinene (2.05%) (Riaz *et al.*, 1989). The cineol has always been reported as the major constituent of essential oils, ranging from 30% up to 56% of the total volatiles compounds (Özcan and Chalchat, 2005; Sangun *et al.*, 2007; Di Leo Lira *et al.*, 2009). However, both quantitative essential oils yield and variations in the percentage of each moiety have been found as a consequence of seasonal variation (Müller-Riebau *et al.*, 1997; Di Leo Lira *et al.*, 2009), area of cultivation (Özcan and Chalchat, 2005; Sangun *et al.*, 2007) and drying or extraction method (Diaz-Maroto *et al.*, 2002). Conversely, Verdian-rizi (2008) showed that the time of harvesting of this species did not have a major effect on chemical composition of the essential oil but that it affected the essential oil content, and the flowering stage was the best time for harvesting to maximize the essential oils yield.

Intensive researches have also been conducted on this species in order to characterize other interesting compounds, such as flavones (apigenin and luteolin) and flavonols (kaempferol, myricetin, and quercitin) (Sÿkerget *et al.*, 2005), sesquiterpenoid lactones, isoquinoline alkaloids (Bruneton, 1999) in leaves and anthocyanin and tannins from berry (Longo and Vasapollo, 2005).

The leaves have been used in folk medicine to treat epilepsy (Zargari, 1990). The essential oil obtained from the leaves of this species has been used for relieving haemorrhoid and rheumatic pains (Zargari, 1990). Pharmacological studies have demonstrated the anaesthetic, hypothermic, muscle relaxant and anticonvulsant activity of eugenol and methyleugenol (Dallmeier and Carlini, 1981; Sayyah *et al.*, 2001) and also an anti-stress effect of eugenol (Sen *et al.*, 1992). Furthermore, some analogs of α-pinene prevent audiogenic seizures in susceptible rats (Consroe and Fish, 1981). Essential oils have been used with positive results in cancer treatment (Sylvestre *et al.*, 2006; Loizzo *et al.*, 2007), while Derwich *et al.* (2009) found an inhibitory effect of oils on the growth of *Staphylococcus* spp. and *Klebsiella pneumoniae*. Remarkably, Sayyah *et al.* (2003) found that the essential oils effects were comparable to that exhibited by the administering of analgesics and non-steroid anti-inflammatory drugs morphine and piroxicam.

Besides the potential medical applications, laurel ground leaves or crude extracts are widely utilized in the food industry simply for seasoning of soups, meat, fish and beverages (Bauer and Garbe, 1997). However, as well as this common use, it is also used in the food industry owing to its essential oils, which have been shown to possess antibacterial (Milhau *et al.*, 1997; Hammer *et al.*, 1999), antifungal, antiviral, insecticidal and antioxidant properties (reviewed by Burt (2002)). The antimicrobial (Oussalah *et al.*, 2006) and the insecticidal properties (Saim and Meloan, 1986) have been successfully exploited for food preservation against *Pseudomonas putida* isolated from meat and *Triboleum castaneum* in wheat. In addition to essential oils, laurel vapours demonstrated toxic effects on reproduction of *Acanthoscelides obtectus* (Say), one of the most destructive pests of *Phaseolus vulgare* L., suggesting a potential application of these volatiles as a bio-control agent (Papachristos and Stamopoulos, 2002). Similar application as a bio-controller, but against fungi, was demonstrated by Soylu *et al.* (2006), who found a complete growth inhibition of *Phytophthora infestans* by laurel essential oils even at very low concentrations (0.4–2.0 $\mu g\, mL^{-1}$ air concentrations), suggesting a higher antifungal activity. Utilization of oils as food preservatives is also due to the antioxidant capacity (Faid *et al.*, 1995; Politeo *et al.*, 2007).

### 23.2.5 *Thymus vulgaris* L.

*Thymus vulgaris*, or common thyme, is a low-growing herbaceous plant, sometimes becoming lignified. It is native to southern Europe, where it is often cultivated as a culinary herb. It typically grows as a sub-shrub, between 15 and 20 cm tall.

The genus *Thymus* (Labiateae) consists of about 215 species (Nickavar *et al.*, 2005) and it has been reported that the Mediterranean region can be described as the centre of the genus (Stahl-Biskup and Saez, 2002). It is one of the species most utilized in folk medicine, popular for stimulating action in all organism functions (Bruneton, 1993) and also for the antimicrobial activity of the essential oil (Dob *et al.*, 2006). *Thymus* species are commonly used as herbal tea, flavouring agents (condiment and spice) and medicinal plants (Stahl-Biskup and Saez, 2002).

In the essential oil of *T. vulgaris*, 27 components have been identified, accounting for 97.2% of the oil (Soković *et al.*, 2009). The main components were thymol (48.9%) and *p*-cymene (19.0%), but other volatile constituents were obtained from the aerial parts of the plant, such as geranial, linalool, γ-terpineol, carvacrol and *trans*-thujan-4-ol/terpinen-4-ol (Piccaglia *et al.*, 1993).

However, the composition of essential oil in *Thymus* depends on the method of extraction (Grosso *et al.*, 2010) and much more in relation to the geographical origins. In the Mediterranean area, there are many ecotypes of thyme, which differ in the composition of essential oils (Corticchiato *et al.*, 1998). Clearly, also the stage of development of the plant (Abu-Darwish *et al.*, 2012), the harvesting time (Senatore, 1996) and the field environment conditions can influence the composition of essential oil. In fact, organic fertilization increased the percentage of oxygenated components, especially thymol (51%), and shifted the major components of thyme to the high margin specified by European pharmacopoeias (Edris *et al.*, 2009; Hendawy *et al.*, 2010).

Extracts from thyme have been used in traditional medicine against several respiratory diseases, like asthma and bronchitis (Ram *et al.*, 2011), but also for the treatment of other pathologies in relation to its properties, such as antiseptic, antispasmodic, antitussive, antimicrobial, antifungal, antioxidative and antiviral (Soliman and Badeaa, 2002; Stahl-Biskup and Saez, 2002). On the other hand, thyme has strong bactericide properties against Gram-positive and Gram-negative bacteria and acts also as a bronchospasmolytic (Stahl-Biskup and Saez, 2002). Thyme oil reduces NO production in murine macrophages (Vigo *et al.*, 2004). The most important components of essential oil of thymus are thymol and carvacrol, which exhibit multiple biological activities, including anti-inflammatory (Fachini-Queiroz *et al.*, 2012), immunomodulating (Khan *et al.*, 2012), antioxidant (Lagouri *et al.*, 2011), antibacterial (Karami-Osboo *et al.*, 2010) and antifungal activity (Samara de Lira Mota *et al.*, 2012). Carvacrol also possesses antimutagenic and anticarcinogenic effects (Kaliora and Kountouri, 2012).

Lee *et al.* (2010) reported that carvacrol, β-citronellol and borneol extract from *Thymus* species could be useful as a new preventive agent against damage caused by arthropod pests. Previously, Pavela *et al.* (2009) tested oil samples from four chemotypes of *T. vulgaris* for mosquitocidal activity and found that thymol oil type was the more efficient. On the other hand, Szczepanik *et al.* (2012) showed that thymol and carvacrol from *Thymus* were more active against *Alphitobius diaperinus* larvae than thyme oil. Thus, these two pure components (thymol and carvacrol) can effectively control this pest.

An interesting application reported by Razzaghi-Abyaneh *et al.* (2009) showed that the essential oils of *Thymus* are a potential candidate to protect foods and feeds from toxigenic fungus growth and subsequent aflatoxin contamination. El-Nekeety *et al.* (2011) used essential oil of *Thymus* in male rats against oxidative stress induced by aflatoxin and found that the essential oil had a potential antioxidant activity and a protective effect against aflatoxin; the protection was dose dependent.

Interestingly, Navarrete *et al.* (2010) used *Thymus* essential oil in aquaculture as an alternative to antibiotics.

### 23.2.6 *Origanum majorana* L.

Sweet marjoram, belonging to the Lamiaceae family, is indigenous to Cyprus and southern Turkey, and was known to the Greeks and Romans as a symbol of happiness. This species has great economic and industrial importance (Werker *et al.*, 1993). The fresh leaves and flowering tops of sweet marjoram retain their aromas and are widely used to flavour foodstuffs. The leaf oil is used as a spice and condiment (Novak *et al.*, 2000). The volatile aromatic compounds are used in the food industry as flavouring in foods and beverages. The leaf oils are appreciated in perfumery for their spicy herbaceous notes (Vera and Chane-Ming, 1999).

Sweet marjoram also contains phenolic acid (*trans*-2-hydroxycinnamic, rosmarinic, vanillic, chlorogenic, gallic and cinnamic acids), flavonoids (diosmetin amentoflavone, apigenin, quercetin, luteolin, coumarin and rutin), terpenoids (thymol and carvacrol), tannins, hydroquinones, phenolic glycosides (arbutin, methyl arbutin, vitexin, orientin and thymonin) and triterpenoids (ursolic acid, carnosic acid and oleanolic acid) (El-Ashmawy *et al.*, 2005; Sellami *et al.*, 2009).

Essential oils yield ranged from 0.04 to 0.09%; the highest was obtained by Sellami *et al.* (2009) during the full-flowering stage. Analysis of the essential oils by GC and GC–MS revealed that the main compounds were represented by oxygenated monoterpenes (64–71%), followed by monoterpene (22–30%) and sesquiterpene (1.5–4%). The main components were terpinen-4-ol (29–33%), *cis*-sabinene hydrate (20–29%), *trans*-sabinene hydrate (3.5–12%), γ-terpinene (2–8%), bornyl acetate (1.52–2.94%) and linalool (1.1–1.4%) (Sellami *et al.*, 2009). On the other hand, phenolic contents varied from 2.7 to 6.8 mg $g^{-1}$ of dry weight, and the highest concentration was found in samples collected during the later vegetative stage (Komaitis *et al.*, 1992; Sellami *et al.*, 2009).

In addition, Banchio *et al.* (2008) observed that plants inoculated with *Pseudomonas fluorescens* and *Bradyrhizobium* spp. showed significant increases in growth parameters and essential oils yield.

Traditionally, the species has been used as a folk remedy against asthma, indigestion, headache and rheumatism (Jun *et al.*, 2001), cramps, depression, dizziness, gastrointestinal disorders, migraine, nervous headaches, paroxysmal coughs and as a diuretic (El-Ashmawy *et al.*, 2005). Recent research, however, showed more scientific evidence that essential oils extract as well as single isolated compounds from the oils may be usefully utilized to counteract serious human diseases. The ability to reduce the diffusion of human breast cancer was recently pointed out by Al-Dhaheri *et al.* (2012); apoptotic and anti-proliferative activity of *O. majorana* extracts on human leukaemic cell line was found by Abdel-Massiha *et al.* (2010), while Yazdanparast and Shahriyary (2007) observed that essential oils of this species can reduce thrombosis by reducing platelet hyperactivity and, consequently, platelet adhesion to the vessel. Carnosic and ursolic acid were considered to be responsible for the reduction of atherosclerosis by Yu and Tzeng (2009) and cardiac alterations by Yu *et al.* (2010), while the extract from leaves can prove to be beneficial in the concomitant treatment of gastric ulcers and diabetes (Pimple *et al.*, 2012).

Essential oils have shown positive effects as fungicides or insecticides, in pharmaceutical and industrial products (Deans and Svoboda, 1990). Among many essential oils, which may be usefully used as antimicrobial agents, marjoram oils showed broad-spectrum antimicrobial properties against food-borne bacteria and mycotoxigenic fungi. Therefore, it may have the greatest potential for use in industrial applications as an alternative for synthetic preservatives routinely utilized in food preservation (Baratta *et al.*, 1998; Daferera *et al.*, 2000; Ezzeddine *et al.*, 2001; Mahmoud *et al.*, 2007). Sweet marjoram has strong antioxidant activity, mainly because of its high content of phenolic acids and flavonoids, which are useful in health supplements and food preservation (Vagi *et al.*, 2005). Another interesting result achieved on the application of sweet marjoram essential oils was recently pointed out by Qari (2008), who revealed antimutagenic properties of this species against sodium-azide-induced chromosomal aberrations in *Vicia faba* L. root meristematic cells.

### 23.2.7 *Allium sativum* L.

Garlic, belonging to Amaryllidaceae, is closely related to shallot, leek, chive and rakkyo. It is a bulb with long, narrow and flat-like grass leaves. The bulb is composed of numerous bulblets (cloves) grouped together between the membraneous scales and enclosed within a whitish skin. The flowers are placed at the end of a stalk rising direct from the bulb and are whitish, grouped together in a globular head, or umbel, with an enclosing kind of leaf or spathae, and among them there are small bulbils.

This species is native to central Asia and has long been largely used in the Mediterranean region, as well as in Asia, Africa and continental Europe. It was known to Ancient Egyptians, and it has been used for both culinary and medicinal purposes.

The oldest reports of health-promoting properties of garlic date back to the 16th century BC, when, in the so-called 'Ebers Papyrus' from Egypt, over 20 ailments were supposed to be efficiently cured by garlic (Block, 1985).

The most important compounds occurring in the intact garlic bulbs are sulfur derivates, especially alliin (*S*-allylcysteine sulfoxide). The whole bulbs contain also γ-glutamyl-*S*-allylcysteine, *S*-methylcysteine sulfoxide (methiin), *trans*-1-propenylcysteine sulfoxide and *S*-2-carboxypropylglutathione and *S*-allylcysteine, though at much smaller amounts (Amagase, 2006). Damaging of a garlic bulb by crushing, grinding or cutting induces the release of the vacuolar enzyme alliinase (alliin: lyase EC. 4.4.1.4), which in few seconds transforms alliin into allicin, which is easily transformed into oil-soluble polysulfides, mostly diallyl disulfide (DADS), but also into diallyl sulfide (DAS), diallyl trisulfide (DATS) and diallyl tetrasulfide.

Garlic bulb has been extensively used in folk medicine for centuries to counteract parasitic fungal, bacterial and viral infections. Recently, following the chemical characterization of its sulfur compounds, it has been confirmed that they are the main active antimicrobial agents (Rose *et al.*, 2005). However, some proteins, saponins and phenolic compounds can also contribute to this activity (Griffiths *et al.*, 2002). Garlic extract resulted in effective protection against parasitic protozoa such as *Opalina ranarum*, *Opalina dimidicita*, *Balantidium entozoon*, *Entamoeba histolytica*, *Tripanosoma brucei*, *Leishmania*, *Leptomonas* and *Crithidia* (Reuter *et al.*, 1996). Owing to the occurrence of unpleasant side effects and increasing resistance to the synthetic pharmaceuticals recommended for the treatment of giardiasis, interesting natural alternatives have been explored, and Lun *et al.* (1994) found that garlic is effective against *Giardia lamblia* and *Giardia intestinalis*. As an applicative result, in China, an allicin breakdown product (more stable than the extremely volatile allicin) is commercially available as a preparation, called Dasuansu, prescribed for the treatment of giardiasis and infections induced by *E. histolytica* and *Trichomonas vaginalis* (Lun *et al.*, 1994). Similar to the antibacterial activity, the main active antifungal agents from garlic extracts are the breakdown products of allicin, including DATS, DADS, DAS and ajoene, which have a greater antifungal effect than allicin (Tansey and Appleton, 1975). The active compounds of garlic destroy fungal cells, decreasing the oxygen uptake, reducing cellular growth, inhibiting the synthesis of lipids, proteins and nucleic acids, changing the lipid profile of the cell membrane and inhibiting the synthesis of the fungal cell wall (Gupta and Porter, 2001). Finally, it is necessary to consider that certain steroid saponins, such as eruboside-B isolated from the garlic bulb, also exhibits antifungal activity against *Candida albicans* (Matsuura *et al.*, 1988). Garlic has been proven to inhibit the growth of Gram-positive, Gram-negative and acid-fast bacteria, as well as toxin production (Du *et al.*, 2008a, 2008b, 2009). It also inhibited the beneficial intestinal microflora, but it is more effective against potentially harmful enterobacteria, probably due to a greater sensitivity of enterobacteria to allicin (Miron *et al.*, 2000). Recently, it has been reported that onion and garlic extracts exert bactericidal effects towards *Streptococcus mutans*, *Streptococcus sobrinus*, *Porphyromonas gingivalis* and *Prevotella intermedia*, considered as the main bacteria responsible of dental caries and adult periodontitis (Bakri and Douglas, 2005).

The first report on the evidence of garlic as a possible anticancer agent dates back to the 1950s when Weisberger and Pensky (1958) demonstrated *in vitro* and *in vivo* that thiosulfinate extracts from garlic inhibited tumour cell growth. During the last decades, much scientific evidence has confirmed that and it has deepened the knowledge on matter demonstrating that garlic intake diminishes the risk of sarcoma and carcinoma in various tissues and organs, such as stomach, colon, oesophagus, prostate, bladder, liver, lungs, mammae, skin and brain (for an overview, see Corzo-Martínez *et al.* 2007). Those effects appear to be mediated by various mechanisms, which, however, are not fully understood. Allicin and its derivative compounds also appear as the main active substances responsible for the hypolipidaemic and hypocholesterolaemic effects of garlic, as much in humans as in experimentation animals (Liu and Yeh, 2000, 2002; Gupta and Porter, 2001; Singh and Porter, 2006). The bioactive constituents from garlic, such as methiin and *S*-allyl cysteine sulfoxide, exert their antidiabetic action (Urios *et al.*, 2007) by stimulating insulin production and secretion by the pancreas, interfering with dietary glucose absorption (Srinivasan, 2005). Moreover, anti-inflammatory (Dkhil *et al.*, 2011), antihypertensive (Al-Qattan *et al.*, 1999; Harauma and Moriguchi, 2006), antiplatelet and antithrombotic (Allison *et al.*, 2012), antihyperhomocysteinaemia (Weiss *et al.*, 2006), antiasthmatic (Breu and Dorsch, 1994) and immunomodulatory properties (Butt *et al.*, 2009) for garlic extract, or isolated compounds from this species, have been demonstrated.

The biological activity of garlic was also pointed out against diseases not directly correlated with human health, but which were otherwise economically important to reduce plant and food diseases. Indeed, *Aspergillus niger* (Irkin and Korukluoglu, 2007; Matan *et al.*, 2012) and *Botrytis cinerea* mycelia growth (Cheng and Shao, 2011) were significantly decreased by garlic oil. The effect of garlic-supplemented diets in *Epinephelus coioides* showed positive results on resistance against *Streptococcus iniae*, a fish pathogen, which naturally occurs in marine and estuarine environments deleterious for aquaculture (Guo *et al.*, 2012a). Owing to the great antimicrobial activity that garlic possesses, it could also be used as a natural preservative, to control food spoilage by microbial growth (Pszczola, 2002).

The adverse effects of garlic in humans can be most common, less frequent and rare (Tattelman, 2005). The most common side effects produced by intake of small amounts of garlic are bad breath and body odour. However, several studies have demonstrated that consumption of excessive amounts of these vegetables, especially when the stomach is empty, can cause other less frequent undesirable effects, such as gastrointestinal upsets (burning sensation and diarrhoea), flatulence and changes in the intestinal flora (Ackermann *et al.*, 2001). Within the group of rare effects can be included dermatological problems in susceptible individuals (allergic dermatitis, burns and blisters), produced by the direct application on skin of fresh or crushed garlic (Friedman *et al.*, 2006), and asthmatic effects in workers exposed to garlic powder in garlic growing or industrial processing (Lybarger *et al.*, 1982).

### 23.2.8 *Mentha* spp.

*Mentha* is a member of the tribe Mentheae in the subfamily Nepetoideae. The tribe groups about 65 genera even though the relationships within it remain obscure. However, the most common and popular mints for cultivation are peppermint (*Mentha* × *piperita*), spearmint (*Mentha spicata*) and, recently, apple mint (*Mentha suaveolens*).

Peppermint is a hybrid mint, a cross between watermint (*Mentha hirsuta* Huds.) and spearmint. The plant, indigenous to Europe, is now widespread in cultivation throughout all regions of the world. It is occasionally find in the wild with its parent species. It is a herbaceous rhizomatous perennial plant growing to 30–90 cm tall, with smooth stems, square in cross-section.

Spearmint is native to much of Europe and southwest Asia, though its exact natural range is uncertain due to extensive early cultivation. Even spearmint is a herbaceous rhizomatous perennial plant growing 30–100 cm tall, with variably hairless to hairy stems and foliage, and a wide-spreading fleshy underground rhizome. The stem is square shaped, a trademark of the Lamiaceae family.

However, the genus *Mentha* includes 25–30 species cultivated in the world (America, Europe, China, Brazil, India, etc.) with a production of 16 000 tons of mint oil (Dorman *et al.*, 2003; Khanuja, 2007).

Species of the genus *Mentha* have been reported to contain a wide range of components, including cinnamic acids (Triantaphyllou *et al.*, 2001), aglycon, glycoside or acylated flavonoids (Fialova *et al.*, 2008), and steroidal glycosides (Ali *et al.*, 2002), although the main active component of *Mentha* is essential oil, which is responsible of its various properties. However, it has been reported (Chauhan *et al.*, 2009) that different species of *Mentha* show high polymorphism and vary in their essential oil content and composition. In relation to this, different species of *Mentha*, subjected to varying geographical conditions and on the basis of biosynthetic pathway followed in the different species of *Mentha*, could be divided into menthol-rich, carvone-rich and pulegone/piperitone-rich essential oils (Kumar *et al.*, 2011).

Among the menthol-rich species, the most important mint species is peppermint (Gul, 1994). Menthol is a compound widely utilized for various medical purposes, such as to relieve skin irritation, sunburn, sore throat, fever, muscle aches and in nasal congestion, while menthone is used in perfumery and as a flavour agent. Generally, essential oil of peppermint has 30–55% of menthol, while the composition of menthone is between 14 and 32% (ESCOP, 1997).

Carvone is a monocyclic monoterpene ketone, which exists both as *R* and *S* enantiomers in natural products. It has strong antiseptic properties and it is used as mosquito repellent and in the food industry as a flavouring agent. Carvone is the main component of essential oil of *M. spicata* (Hussain *et al.*, 2010).

Pulegone, a monoterpene, is a naturally occurring organic compound and is commonly found in essential oil of *Mentha pulegium* L. and *Mentha microphylla* K. Koch, among *Mentha* species. This compound is used as flavouring agents, in perfumery and in aromatherapy.

Definitively, *M. piperita* essential oil contains α-terpinene (19.7%), isomenthone (10.3%), *trans*-carveol (14.5%), pipertitinone oxide (19.3%), β-caryophyllene (7.6%) and menthol (3.6%) as the major compounds (Yadegarinia *et al.*, 2006). Major components in cultivated spearmint oil are instead *S*-carvone (76.6%), limonene (9.6%), and 1,8-cineole (1.9%) (Maffei *et al.*, 1986; Kokkini *et al.*, 1995; Chauhan *et al.*, 2009).

The main components menthol and menthone showed significant variation with different ecologies. In fact, it has been shown that menthol contents were higher in temperate locations, while menthone contents were lower (Telci *et al.*, 2011). Even UV treatment can influence the essential oil composition (Karousou *et al.*, 1998) as well as the harvest time (Zheljazkov *et al.*, 2010). However, Behn *et al.* (2010) reported that the absence of UV-B radiation led to reduced menthol and increased menthone contents and thereby induced a substantial decrease in oil quality. Fletcher *et al.* (2010) reported that cold stress, impacted over a 6-week period had no effect on rosmarinic acid production of four spearmint and two peppermint clonal lines, whereas the soil type played an essential role in the accumulation of rosmarinic acid in the leaf tissue, probably due to retention of moisture, which favours rosmarinic acid production.

Menthol is used as a raw material in toothpaste, toothpowder, chewing tobacco, confectionary, mouth fresheners, analgesic balms, cough drops, perfumes, chewing gums, candies and the tobacco industry, which constitutes about 40% of the total oil consumption followed, by pharmaceutical and confectionary industries. The fresh or dried leaves of *Mentha* spp. are the culinary source of mint and are used in breath fresheners, drinks, antiseptic mouth rinses, toothpaste, chewing gum, mint chocolate teas, beverages, jellies, syrups, candies, ice creams and also used as a necessary ingredient in Touareg tea, a popular tea in the northern African and Arab countries.

Peppermint oil is one of the most popular and widely used essential oils, with use in various medical conditions, such as to relieve skin irritation, sunburn, sore throat, fever, muscle aches and in nasal congestion, and also in perfumery and as a flavouring agent (Kumar *et al.*, 2011).

Besides their medicinal properties, oils of peppermint and its components are also reported for their antibacterial, antifungal, and insecticidal properties (Kumar *et al.*, 2011).

The insecticidal activity of *Mentha* oil has been tested and established against various insects/pests, even if many studies are carried out against adult insects (repellency and adulticidal activity) and very few studies have targeted assessment of *Mentha* efficacy for change in reproductive behaviour and other stages (larvae/pupae) of the insect life cycle (Kumar *et al.*, 2012). For a detailed review, see Kumar *et al.* (2011).

Spearmint oils are used in the food confectioneries and pharmaceuticals and this species is also known to be endowed with a variety of biological and pharmacological properties due to the high content of secondary metabolites (Choudhury *et al.*, 2006). In fact, it has been reported that spearmint possesses antiallergic (Yamamura *et al.*, 1998), antioxidant (Kanatt *et al.*, 2007), antiplatelet (Tognolini *et al.*, 2006), antiproliferative (Manosroi *et al.*, 2006) and chemopreventive (Saleem *et al.*, 2000) properties. Arumugam and Ramesh (2009) indicated that the aqueous fraction of *M. spicata* mediates antigenotoxic effects by the modulation of lipid peroxidation and antioxidant enzymes.

### 23.2.9 *Origanum vulgare* L.

As mostly *Origanum* species, oregano is a perennial herb native to the dry, rocky calcareous soils in the mountainous areas of the southern Europe, south-west Asia and Mediterranean countries. The species reaches a height of 0.8 to 1 m and has a pubescent stem, ovate and dark green leaves, and white or purple flowers. Oregano is used as a flavouring in salads, meat, sauces and soups. Before the introduction of hops, oregano was used to flavour ale and beer. The essential oils used extensively in place of the plant material are found in food products, liqueurs, beverages and cosmetics. The reported life zone for *Origanum vulgare* is 5–28 °C with annual precipitation of 0.4–2.7 m and soil pH of 4.5–8.7. This demonstrates a high degree of adaptability to environmental conditions, such as most of the other Mediterranean aromatic species (Kintzios, 2002).

Analyses of the GC–MS spectrum performed by de Barros *et al.* (2009) identified 16 compounds making 100% of the oil total mass. Carvacrol (57.7 g per 100 g) was the most prevalent compound, followed for *p*-cymene (10.9 g per 100 g), γ-terpinene (7.2 g per 100 g), terpinen-4-ol (6.7 g per 100 g) and thymol (3.8 g per 100 g). Other compounds, such as α-pinene (1.9 g per 100 g), α-terpinene (2.0 g per 100 g), linalool (2.6 g per 100 g) and *trans*-β-caryophyllene (2.1 g per 100 g), were found in minor percentages. However, one has to consider that many factors have a significant effect on the yield and the essential oils profile, such as the season of collection and drying of samples (Jerkovicä *et al.*, 2001), plant portion (Verma *et al.*, 2010), plant stage and ontogeny (Verma *et al.*, 2012), distillation time of the leaf material (Zheljazkov *et al.*, 2012), temperature, geographical distribution and ecotypes (Vokou *et al.*, 1993).

As for the results from research targeting human diseases, remarkable are those pointed out by Guvenalp *et al.* (2010), who investigated the antitumour properties of oregano by using the MCF-7 breast cancer cell. A positive activity against colon cancer was also shown by Savini *et al.* (2009). The antiurolithic effect is another notable property of oregano extract, as reported by Khan *et al.* (2011a). In this study, it was shown to have an inhibitory effect on the slope of nucleation and aggregation of calcium oxalate monohydrate crystals. For the oregano extract, potential antiviral activity was also reported against influenza virus in host animals (Mancini *et al.*, 2009) and against yellow fever virus (Meneses *et al.*, 2009). The anti-hyperglycaemic activity of oregano has long been recognized by folk medicine (Eddouks *et al.*, 2002); in view of this information, Lemhadri *et al.* (2004) studied the effects of an aqueous extract on diabetic rats and found that it may be effectively used against the diabetes mellitus.

Recent research demonstrates that oregano essential oils possess interesting antimicrobial and antifungal activities against spoilage and pathogenic food-related microorganisms such as *Aeromonas hydrophilla*, *Enterobacter aerogenes*, *K. pneumoniae*, *Salmonella choleraesius*, *Serratia* spp., *Shigella* spp., *Yersinia enterocolitica* (Souza *et al.*, 2006), *Bacillus* spp., *E. coli*, *S. typhi*, *Streptococcus* spp., *Proteus* spp., *Pseudomonas* spp. (Suoza *et al.*, 2006; De Martino *et al.*, 2009), *Helicobacter pulori* (Chun *et al.*, 2005), *S. aureus* (de Barros *et al.*, 2009; Suoza *et al.*, 2009; Oliveira *et al.*, 2010; Kazemi *et al.*, 2012), *Listeria monocytogenes* (Suoza *et al.*, 2006; de Azerêdo *et al.*, 2012), *Candida* spp. (Suoza *et al.*, 2007; Cleff *et al.*, 2010), *Acremonium strictum*, *Aspergillus* spp., *Cladosporium sphaerospermum* and *Mucor racemosus* (Chavez-López *et al.*, 2012). Although the above-mentioned assays with the essential oils showed promising antimicrobial properties, when applied to food matrices, the amounts required to substantially inhibit the bacterial growth were often higher than would be organoleptically acceptable (Naveena *et al.*, 2006). For this reason, a compromise between bio-control and taste is necessary, and sometimes mixing essential oils from more than one species can represent a valid solution (de Azarêdo *et al.*, 2012). In addition to their action as a bio-control agent, oregano extract and essential oils were positively utilized in a small quantity as an antioxidant additive in food or in order to avoid heat or light oxidation of several compounds (Boroski *et al.*, 2012).

Insecticidal activity was also reported, and it seems particularly related to the toxicity exhibited by phenols against the targets (termites: Pandey *et al.*, 2012).

### 23.2.10 *Foeniculum vulgare* Mill.

*Foeniculum vulgare* is a member of the family Apiaceae (formerly the Umbelliferae). It is a hardy, perennial, umbelliferous herb, with yellow flowers and feathery leaves. It is indigenous to the shores of the Mediterranean but has become widely naturalized in many parts of the world, especially on dry soils near the sea coast and on riverbanks. It is an aromatic and flavourful species with culinary and medicinal uses and is one of the primary ingredients of absinthe, an alcoholic beverage. In addition, Florence fennel, or finocchio, is a selection with a swollen, bulb-like stem base that is used as a vegetable. Essential oil of fennel is also used as flavouring agents in food products such as bread, pickles, pastries and cheese and as a constituent of cosmetic and pharmaceutical products (Piccaglia and Marotti, 2001).

The major component is *trans*-anethole (34.8–82.0%); the other principal compounds in oils are fenchone (1.6–22.8%), estragole (2.4–17.0%), limonene (0.8–16.5%), and *cis*-anethole (0.1–8.6%) (Raal *et al.*, 2012). The most intense odour of fennel fruits is attributable to *trans*-anisole, estragole, fenchone and 1-octen-3-ol (Diaz-Maroto *et al.*, 2005). Napoli *et al.* (2010a) identified 78 compounds from fennel fruits by GC-flame ionization detection–MS, representing more than 98% of the oils. They can be divided into monoterpene hydrocarbons, oxygenated monoterpenes and phenylpropanoids. The main compound, estragole, ranged from 34 to 89%. The essential oil of bitter fennel fruits was characterized by a relatively high concentration of α-pentene and fenchone, and a low concentration of *trans*-anethole and estragole, unlike sweet fennel oils (Akgül and Bayrak, 1988).

Fennel is used as a food flavourant, in alcoholic beverages and in mouth-care products. The mericarps (commonly referred as seed) are the most aromatic and widely used part of the plant, but all the aerial parts contain the typical oleoresin.

Various studies carried out on fennel from different geographic origins have indicated that the estragole content in mericarps varies from 2 to 86% and that of *trans*-anethole from 0 to 89% (Pank *et al.*, 2003). More recently, Gross *et al.* (2009) reported that the levels of estragole and *trans*-anethole varied during plant development, being maximal in flowers and developing mericarps even though the ratio between estragole and *trans*-anethole remained constant throughout development. Similarly, Telci *et al.* (2009) reported an increase in the content of *trans*-anethole during maturation stages; these authors also found that some components, particularly monoterpenes, α-pinene, β-myrcene, limonene, and α-terpinene, varied significantly during maturation stages. Fertilization influenced the composition and yield of fennel essential oil, too. Abdallah *et al.* (1978) reported that a high dose of nitrogen fertilization gave a higher number of compounds and increased oil percentage. In contrast, Kandil *et al.* (2002) and Chatzopoulou *et al.* (2006) found that application of nitrogen had no marked effect on seed yield, yield components and seed essential oil content of fennel. In addition, growth and essential oil yield of fennel were influenced by irrigation, nutrient levels and crop geometry (Mehta *et al.*, 2012).

From a medicinal properties point of view, fennel is widely used as an antinociceptive (Mikaili *et al.*, 2010), antimicrobial (Anwara *et al.*, 2009), antiacetylcholinesterase (Mata *et al.*, 2007) and clearly antioxidant. However, Barros *et al.* (2009), in a study on systematic evaluation of the antioxidant potential of different parts of fennel, demonstrated that the shoots seemed to have the highest radical-scavenging activity and lipid peroxidation inhibition, which was in agreement with the highest content in phenolics found in that part.

Fennel is an antibacterial agent (Kaur and Arora, 2009). In particular, Çetin *et al.* (2010) reported anethole, the main component of Florence fennel essential oil, was responsible for the antimicrobial activity and that the essential oils as well as the hexane extract can be used as a food preservative. However, Gulfraz *et al.* (2008) reported that fennel fruit essential oil showed more inhibitory activity against *B. cereus*, *Bacillus megaterium*, *Bacillus pumilus*, *B. subtilis*, *E. coli*, *K. pneumoniae*, *Micrococcus lutus*, *Pseudomonas pupida*, *Pseudomonas syringae*, and *C. albicans* than the effect of methanolic and ethanolic fennel seed extracts.

Kim *et al.* (2012) found that fennel extract inhibited the osteoclast differentiation and bone resorptive activity of mature osteoclasts, representing in this way a possible and effective therapeutic approach to the treatment of postmenopausal bone loss.

Essential oil of *F. vulgare* also showed hepatoprotective activity (Ozbek *et al.*, 2003), owing to its immunomodulatory effect of NFK-B activity that plays a critical role in apoptosis and immunomodulation (Kaileh *et al.*, 2007). Finally, the chemopreventive potential of fennel against carcinogenesis has also been reported (Singh and Kale, 2008). It has also been demonstrated that extract of fennel seed showed anticarcinogenic (Mohamad *et al.*, 2011) and antimutagenic activity (Tripathi *et al.*, 2013).

Recently, Ghanem *et al.* (2012) found that the hepatoprotective antioxidant activity demonstrated by fennel can be attributed to the presence of phenolic compounds, in particular 3,4-dihydroxy-phenethylalcohol-6-*O*-caffeoyl-β-D-glucopyranoside and 3′,8′-binaringenin, isolated from the fennel wild herb for the first time. About 22 dietary sources along with fennel have been shown to possess good aldose reductase inhibitory action and thus is useful as an antidiabetic (Saraswat *et al.*, 2010). Fennel, increasing elasticity of connective tissues, acts as an antiaging agent too (Arslan *et al.*, 1989).

Guo *et al.* (2012b) reported the use of fennel extract on pine wood nematode and Zahid *et al.* (2012) demonstrated that extract of fennel inhibited the growth of three soil-borne fungi, namely *Macrophomina phaseoli*, *Rhizocotina solani* and *Fusarium moniliforme*.

An interesting and novel application was reported by Lahhit *et al.* (2011): this species was tested as a corrosion inhibitor of carbon steel in HCl solution and the authors reported that its essential oil provided a good inhibition of corrosion. Extract of fennel has also been used for allelopathic properties (Colvin and Gleissman, 2011).

### 23.2.11 *Juniperus communis* L.

*Juniperus communis*, namely common juniper, is a species in the genus *Juniperus*, in the family Cupressaceae. It is perhaps the most widely distributed plant throughout the cool temperate Northern Hemisphere from the Arctic south in mountains to around 30°N latitude in North America, Europe and Asia. This species is a shrub or small coniferous evergreen tree, very variable and often a low-spreading shrub, but occasionally reaching 10 m tall. Common juniper has needle-like leaves in whorls of three; the leaves are green, with a single white stomatal band on the inner surface. The seed cones are berry-like, green ripening in 18 months to purple–black with a blue waxy coating.

Its astringent blue–black seed cones, commonly known as 'juniper berries', are too bitter to eat raw and are usually sold dried and used to flavour meats, sauces and stuffings. They are generally crushed before use to release their flavour. Since juniper berries have a strong taste, they should be used sparingly. They are generally used to enhance meat with a strong flavour, such as game, including game birds, or tongue. The cones are used to flavour certain beers and gin, the latter name being derived from juniper.

The essential oil from common juniper is recognized by the *European Pharmacopoeia* (EDQM, 2007). Analysis of needles and berries of common juniper revealed a total of 87 compounds, representing over 95% of the oil (Orav *et al.*, 2012). The major compounds in the needle oil were monoterpenes α-pinene (33.3–45.6%), sabinene (0.2–15.4%), limonene (2.8–4.6%) and sesquiterpenes β-caryophyllene (0.8–10.3%), α-humulene (0.8–6.2%) and germacrene D (3.0–7.8%). The juniper berry oil was rich in α-pinene (53.6–62.3%), β-myrcene (6.5–6.9%) and germacrene D (4.5–6.1%). The main oxygenated terpenoids found in the needle oil were germacrene D-4-ol (0.4–4.0%) and α-cadinol (to 2.7%). The oil from fresh needles also contained high amounts of (*E*)-2-hexenal (3.7–11.7%).

The medicinal properties of *J. communis* mostly depend on monoterpene α-pinene (Leite *et al.*, 2007), and its content in plants varies greatly, reaching 80.4% of the total essential oil content (Filipowicz *et al.*, 2009). However, the enantiomerization characteristic of α-pinene, the (1*R*)-(+) and (1*S*)-(−) enantiomers, induced a quite different biological activity (Filipowicz *et al.*, 2003). The reasons for the differences in the enantiomeric composition of terpenes in the essential oils of juniper are still unknown. Recently, Ložienė and Labokas (2012) reported that the amount of α-pinene and its enantiomeric composition did not correlate significantly with the illumination and soil characteristics of habitats, either in leaves or in unripe cones.

Orav *et al.* (2010) reported that the oil obtained by supercritical carbon dioxide extraction from juniper berries contained more sesquiterpenes and higher boiling-point compounds than that obtained by micro-distillation and extraction. The oil yield ranged from 0.7 to 2.1%; and the content of α-pinene of juniper-berry essential oil was 47.9%, and that of juniper needle oil was 36.4%.

According to Lesjak *et al.* (2010), *J. communis* species could be regarded as a promising source of bioactive natural compounds, which can be used both as food supplement and as a remedy. In fact, juniper oil has been used as a carminative, diuretic and as a steam inhalant in the management of bronchitis. It has antimicrobial activity (Marino *et al.*, 2010) that has been ascribed to isocupressic acid, communic acid and deoxypodophyllotoxin found in the aerial parts of *J. communis* (Carpenter *et al.*, 2012). In addition, the presence of antimycobacterial terpenoids (longifolene, totarol and *trans*-communic acid) in *J. communis* aerial parts and roots justifies, to some extent, the ethnomedicinal use of this species as a traditional anti-tubercolosis remedy (Gordien *et al.*, 2009). This species demonstrated also anti-inflammatory and antinociceptive activities (Akkol *et al.*, 2009). *J. communis* oil was also administered to rats within a diet rich in cholesterol and obtained positive results in the heart tissue of rats (Gümral *et al.*, 2013). *In vitro* acetylcholinesterase and butyrylcholinesterase inhibitory and antioxidant activities of the aqueous and ethanol extracts of the leaves, ripe fruits and unripe fruits of common juniper have been found (Orhan *et al.*, 2011).

More interesting, Kusari *et al.* (2009) isolated, identified and characterized an endophytic fungus from *J. communis* as a novel producer of deoxypodophyllotoxin and its *in vitro* antimicrobial assay. Although the current accumulation of deoxypodophyllotoxin (an anticancer pro-drug) by the endophyte is not very high, it could be scaled up to provide adequate production to satisfy new drug development and clinical needs.

### 23.2.12 *Petroselinum crispum* Hoffm.

Parsley, also known as *Petroselinum hortense* or *Apium petroselinum* L. (Apiaceae), is a bright green, hairless, biennial, herbaceous plant in temperate climates, or an annual herb in subtropical and tropical areas. Where it grows as a biennial, in the first year, it forms a rosette of tripinnate leaves 10–25 cm long with numerous 1–3 cm leaflets, and a taproot used as a food store over the winter. In the second year, it grows a flowering stem to 75 cm tall with sparser leaves and flat-topped 3–10 cm diameter umbels with numerous 2 mm diameter yellow to yellowish-green flowers. The seeds are ovoid, 2–3 mm long, with prominent style remnants at the apex. This species is native to the central Mediterranean region (southern Italy, Algeria and Tunisia), is naturalized elsewhere in Europe, and is widely cultivated as a herb, a spice and a vegetable.

The culinary use of the plant dates back to the Roman times, but Albertus Magnus in the 13th century was seemingly already aware of the pharmacological activities of parsley, which he considered '*plus medicina quam cibus*' (more a medicine than a food) (Tschirch,

1912). So, it has long been used in folk-traditional medicine for its diuretic, vermifuge, emmenagogue, laxative properties (Marczal *et al.* 1997; López *et al.*, 1999) and largely utilized to induce abortions (Leporatti *et al.*, 1996). The latter property had already been reported by Hippocrates. Parsley extract promotes menstruation by stimulating contractions of the uterus (Tyler, 1994). These negative effects of parsley are attributable to apiol and particularly myristicin found in its essential oil.

The composition of parsley oil given down below would explain, to some extent, the benefits of this essential oil. The leaf oil (≥96%), as identified by Vokk *et al.* (2011), has 34 components, of which five are found in highest quantity: myristicin (30.7–42.7%), β-phellandrene (21.8–35.9%), *p*-1,3,8-menthatriene (5.4–10.0%), β-myrcene (4.5–8.7%) and apiol (1.8–0.1%).

The same authors demonstrated significant seasonal changes in the major constituents of parsley oils. The level of myristicin and *p*-1,3,8-menthatriene was higher in summer parsley (42.7% versus 30.7%, while in winter parsley they were at 10.0% versus 5.4% respectively). Moreover, summer parsley oil exhibited stronger antimicrobial activity against *E. coli*, *Staphylococcus albus*, *Bacillus mesentericus* and *Aspergillus flavus*. This would further confirm the findings of other authors who had suggested that myristicin could be one of the major compounds that could be responsible for this activity (Manderfeld *et al.*, 1997; Tolba *et al.*, 2008; Osman *et al.*, 2009). The level of activity of apiol can range from 3% to 0.1% (sometimes at trace level and not detectable), and this would depend on the parsley genotype and agronomical practices (Petropoulos *et al.*, 2008b). Relative concentrations of the principal aroma constituents of the foliar essential oil also appeared to be affected by NaCl or $CaCl_2$, even though in a way that differed between cultivars. Generally, salinity increased oil yield, while among different cultivars the differences in oil profile were sometimes not predictable. Conversely, oil yield from parsley roots was very low and appeared unaffected by salinity (Petropoulos *et al.*, 2009). Nonetheless, ammonium nitrate addition maximizes the oil yield to about 150 mg $kg^{-1}$. More intense nitrogen supply resulted only in biomass accumulation, not in the concentration of foliar oil (Petropoulos *et al.*, 2008a). Fertilization of *P. crispum* with a high concentration of nickel (50 mg $kg^{-1}$) represents another strategy to increase oil yield (Atta-Aly, 1999).

Modern medicine has also demonstrated antidiabetic (Bolkent *et al.*, 2004; Ozsoy-Sacan *et al.*, 2006), diuretic (Marczal *et al.*, 1997; Wright *et al.*, 2007; de Campos *et al.*, 2009), antimicrobial (Al-Kareemi, 2012; Teixeira *et al.*, 2013), antiurolithiasic and hepatocurative (Saeidi *et al.*, 2002, 2012), laxative (Kerydiyyeh *et al.*, 2001), antihypertensive, anticoagulant, antihyperlipidaemic, membrane protective (Fejes *et al.*, 2000) and antioxidant effects (Nielsen *et al.*, 1999; Romeilah *et al.*, 2010). The latter authors also showed antiviral activity against herpes simplex virus 1. Recently, evaluation of 16 different genotypes of *P. crispum* showed that the contents of some volatile compounds highly and significantly correlated either with the resistance against *Septoria petroselini* (e.g. hexanal and α-copaene) or to its susceptibility (e.g. *p*-menthenol) (Ulrich *et al.*, 2011). Abbas (2010) found that dietary parsley resulted in significant improvement in live body weight, feed efficiency and feed intake in broiler chickens, while Al-Daraji *et al.* (2012) illustrated the influence of parsley in haematological traits of geese. This improvement in haematological traits may be explained by the presence of iron, β-carotene and vitamin C, which are all very useful for good health as reported by Duke (2009) and Ragab *et al.* (2010).

In conclusion, parsley improves the digestion of proteins and fats and promotes intestinal absorption, liver assimilation and storage due to its high enzyme content (Bahnas *et al.*, 2009).

### 23.2.13 *Cuminum cyminum* L.

Cumin, a member of the Apiaceae family, is an annual herb, growing to a height of about 25–30 cm. It has white or pink flowers, in compound umbels form, and the seeds come as paired or separate carpels, between 3 and 6 mm long.

Cumin originated from Egypt, Turkistan, and the eastern Mediterranean, and is now widely cultivated in Iran, China, India, Morocco, south Russia, Japan, Indonesia, Algeria and Turkey (Tunçtürk and Murat, 2006). Leaf shape and colour, a reduced size and surface cover of plant parts depict the various adaptations of the plant to its environment, especially to drought conditions (Kizil *et al.*, 2003).

Today, cumin is the second most popular spice in the world after black pepper (Daniel and Maria, 2000). The natural sweetness and distinctive aroma of this herb, both ground or as whole seeds, season meat and fish dishes. The spice is a familiar taste in Tex–Mex dishes and is extensively used in the Indian and Chinese cuisines. Curiously, cumin was also used heavily in ancient Roman cuisine (Hajlaoui *et al.*, 2010). Farmers have observed beneficial effects on lactating animals when they were fed with cumin straw (Bettaieb *et al.*, 2010).

In fact, this herb is used in traditional and veterinary medicine as a stimulant, carminative and astringent and as a remedy against indigestion, flatulence, diarrhoea (Norman, 1990), toothache, dyspepsia, epilepsy and jaundice (Nostro *et al.*, 2005). It is also used because of its diuretic, emmanogogic and antispasmodic properties (Janahmadi *et al.*, 2006).

The essential oil is a complex mixture of terpenic hydrocarbons, alcohols, aldehydes, ketones, epoxides, phenols, acids and esters. Twenty-one components were identified in seeds of *Cuminum cyminum*, including cuminlaldehyde (39%), γ-terpinene (15.2%), *O*-cymene (11.8%), β-pinene (11.1%), 2-caren-10-al (7.9%), *trans*-carveol (4.5%) and myrtenal (3.5%) as major compounds (Hajlaoui *et al.*, 2010). Essential oil yields were <0.05% in roots, 0.1% in stem and leaves, and 1.7% in flowers. Major components of the oils were bornyl acetate (23%), α-terpinene (34%) and γ-terpinene (51%) in roots, stems and leaves, and flowers respectively. In all *C. cyminum* organs, total phenolics content ranged from 11.8 to 19.2 mg of gallic acid equivalents per gram of dry weight. The major phenolic compound in the roots was quercetin (26%), whereas in the stems and leaves *p*-coumaric, rosmarinic, *trans*-2-dihydrocinnamic acids and resorcinol were predominant. In the flowers, vanillic acid was the main compound (51%) (Bettaieb *et al.*, 2010).

Water stress (Rebey *et al.*, 2012) and manure (Ahmadian *et al.*, 2011) significantly affected essential oil yield, antioxidant ability and the profile of its constituents. Three irrigation times with manure treatment caused the highest amount of cuminaldehyde and *p*-cymene and the lowest of β-pinene, γ-terpinene and α-pinene. Since cuminaldehyde has been described as the main molecule involved in cumin essential oil biological activity, Chen *et al.* (2012) found the optimum conditions maximizing yield and purities of this compound occurred at distilling temperature of 30 °C and under a pressure of 19 Pa (cuminaldehyde averaged 55.75% of total oil content).

The essential oil is also active against human diseases. Gagandeep *et al.* (2003) found the cancer chemopreventive potentials of cumin seed against forestomach and uterine cervix tumorigenesis. The authors hypothesized the ability to modulate carcinogen metabolism for cumin extract. Mekawey *et al.* (2009) attributed the antitumour properties against six types of tumour cell lines to 1-(2-ethyl, 6-heptyl) phenol, a biologically active compound extracted from *C. cyminum*. Recently, Lee (2005) pointed out that cuminaldehyde can inhibit α-glucosidase, the enzyme that catalyses the final step in the digestive process of carbohydrates. As an inhibitor, cuminaldehyde can retard the dietary uptake of carbohydrates and suppress post-prandial hyperglycaemia; thus, it could be useful for treating diabetic and/or obese patients. Moreover, anti-inflammatory (Shivakumar *et al.*, 2010) and antinociceptive activity of cumin fruit essential oil (Sayyah *et al.*, 2002) were demonstrated in rats.

The essential oil or its components appear promising also for possible use as bactericides to control bacterial plant diseases or, additionally, to treat seeds against phytopathogenic bacteria. Iacobellis *et al.* (2005) observed antibacterial activity against a large collection of both Gram-positive and Gram-negative species. The oil effectiveness was particularly high against the genera *Clavibacter*, *Curtobacterium*, *Rhodococcus*, *Erwinia*, *Xanthomonas*, *Ralstonia* and *Agrobacterium*, which are responsible for a high number of crop or cultivated mushroom diseases worldwide. In this work, at certain concentrations of oil, the inhibition found on the growth of many microorganisms was comparable to that induced by the use of rifampicin. In general, a lower activity was observed against bacteria belonging to the genus *Pseudomonas*. Antimicrobial properties were confirmed by Jirovetz *et al.* (2005), Viuda-Martos *et al.* (2008), Oroojalian *et al.* (2010) and Hajlaoui *et al.* (2010) against some other bacteria commonly used in the food industry or related to food spoilage.

Remarkably, when used as fumigants, cumin essential oils reduced oviposition, egg hatching rate, pupal formation and emergence of adults of $F_1$ progeny of *Callosobruchus chinensis*, a serious pest of stored grain and growpea and beans (Chaubey, 2008).

### 23.2.14 *Nasturtium officinalis* R.Br.

Watercress is a fast-growing, aquatic or semi-aquatic, perennial plant native to Europe and Asia and one of the oldest known leaf vegetables consumed by humans. It is a member of Brassicaceae, botanically related to garden cress, mustard and radish, all noteworthy for a peppery, tangy flavour. The hollow stems allow watercress to float, and the leaves are pinnately compound. Small, white and green flowers are produced in clusters.

Watercress has been grown in many locations worldwide, such as in many European countries (France, UK) and in the USA especially by using large-scale hydroponic-system cultivation. Curiously, the village of Méréville, 60 km outside of Paris, yields 10% of France's total watercress production and the town of Alresford, near Winchester, holds a Watercress Festival that brings in more than 15 000 visitors every year.

Like other cruciferous vegetables, watercress is highly nutritious and, in addition, it has been empirically used for a long time in the treatment of several diseases, including diabetes, anaemia, eczema, and kidney and liver disorders (Casanova and Carballo, 2011). Remarkably, Zaki *et al.* (2011) pointed out the benefits of adding *Nasturtium officinals* extract to poultry ration in modulating the undesirable effects of raw soya bean meals in terms of nutritional values in old male chicks.

Analyses of the essential oils of the leaves, stems and flower resulted in the identification of 9, 8 and 15 compounds, representing 97%, 100% and 94.7% of the oils respectively. The main compounds in leaves were myristicin (57.6%), α-terpinolene (8.9%) and limonene (6.7%). Caryophyllene oxide (37.2%), *p*-cymene-8-ol (17.6%), α-terpinolene (15.2%) and limonene (11.8%) represented the main molecules in the stems, whereas limonene (43.6%), α-terpinolene (19.7%), *p*-cymene-8-ol (7.6%) and caryophyllene oxide (6.7%) were the major constituents in the oil of flowers (Amiri, 2012).

Abu-Zinadah (2008) demonstrated the benefits of watercress oil on healing the partial-thickness burn wounds in rabbit. The author hypothesized that the main reason of this effect is the antimicrobial activities of phenethyl isothiocyanate (PEITC) and, in addition, the free-radical scavenging ability attributable to these compounds.

However, just like for other member of the Brassicaceae family, research on watercress has focused on the other compounds normally synthesized by this plant family rather than on its essential oil. Watercress is known to be one of the richest sources of glucosinolates, which are hydrolysed by the enzyme myrosinase to produce PEITC. Interest in watercress is rising, since the demonstration that PEITC may have a protective effect against cancer (Abu-Zinadah, 2008). Isothiocyanates are released upon chewing or maceration of watercress and other cruciferous vegetables in which they occur as thioglycoside conjugates, usually known as glucosinolates (Tookey *et al.*, 1980). The PEITC level within watercress tissues can increase linearly with the age of plant within 40 days (Palaniswamy *et al.*, 2003) and it is also affected by the photoperiod (Palaniswamy *et al.*, 1996). The PEITC was found to inhibit phase I enzymes, which are responsible for the activation of many carcinogens in animals, and induced phase II enzymes, which are associated with enhanced excretion of carcinogens (Rose *et al.*, 2000; Hecht and Shiny, 2007). Consumption of average portions of cruciferous vegetables can result in the release of tens of milligrams of isothiocyanates (Chung *et al.*, 1992). Moreover, Casanova and co-workers (Casanova and Carballo, 2011; Casanova *et al.*, 2012) showed the protective effect of watercress extract against oxidative damage to DNA. This potential antimutagenic effect may be related to the antioxidant capacity of the extract, since watercress leaves contain a high concentration of glucosinolates, carotenoids, chlorophylls and polyphenols (Gill *et al.*, 2007). It is likely that the observed effects occurred as the result of a combination of these bioactive substances, acting on different targets and perhaps via different mechanisms.

### 23.2.15 *Lavandula officinalis* L.

The genus *Lavandula* (Lamiaceae) includes annual or short-lived herbaceous perennial plants, suffrutescent perennials, subshrubs or small shrubs. Leaf shape is diverse across the genus. They are simple in some commonly cultivated species; in others they are pinnately toothed, or pinnate, sometimes multiple pinnate and dissected. In most species the leaves are covered in fine hair or indumenta, which normally contain the essential oils. Flowers are borne in whorls, held on spikes rising above the foliage, the spikes being branched in some species. Some species produce coloured bracts at the apices. The flowers may be blue, violet or lilac in the wild species, occasionally blackish purple or yellowish. The calyx is tubular. The corolla is also tubular, usually with five lobes (the upper lip often cleft, and the lower lip has two clefts) (Upson and Andrews, 2004).

Lavenders fall into four main categories: *Lavandula latifolia*, a Mediterranean grass-like lavender; *Lavandula angustifolia*, a stockier plant with a fuller flower, commonly known as English lavender (formerly known as *Lavandula vera* or *Lavandula officinalis*); *Lavandula stoechas*, which has butterfly-like bracts on top of the flowers and is sometimes known as French lavender; and *Lavandula* × *intermedia*, which is a sterile cross between *L. latifolia* and *L. angustifolia*.

The various lavenders have similar ethnobotanical properties and major chemical constituents; however, there are some differences in the reported therapeutic uses for different species. For example, most lavenders are believed to have carminative actions, but *L. stoechas* is traditionally used for headache, *L. latifolia* as an abortifacient and *L. officinalis* as a diuretic. The essential oil of *L. officinalis* is also used in aromatherapy as a relaxant, carminative and sedative agent (Cavanagh and Wilkinson, 2002), as an antiseptic agent in swabbing of wounds, for burns and insect bites and in veterinary practice to kill lice and other animal parasites (Evandri *et al.*, 2005). The most commonly and widely cultivated is *L. officinalis* and could explain why there is so much data on this particular species. For these reasons, the properties and the uses described below refer to *L. officinalis* even when it is commonly known as lavender.

Phytochemical studies revealed that mono- and sesquiterpenes, flavonoids (luteolin), triterpenoids (ursolic acid) and coumarins (umbelliferone) were the main components of the aerial parts and flowers of the plant. The leaves yielded 1.1% of a pale-yellowish essential oil; 21 components were characterized, representing 99.1% of the total oil (Hajhashemi *et al.*, 2003). The primary components of this oil are the monoterpenoids linalool, linalyl acetate, 1,8-cineole, β-ocimene, terpinen-4-ol and camphor. Sesquiterpenoids, such as caryophyllene and nerolidol, and other terpenoid compounds, such as perillyl alcohol, are also present in trace quantities (Woronuk *et al.*, 2010). Many of the identified compounds in the oil, are also present in the aerial parts and flower essential oils of *L. angustifolia*, have been reported before (Renaud *et al.*, 2001). Each of these constituents can vary significantly in the oils derived from different cultivars. The relative level of each component is the main determinant of the market value, application and aroma (Cavanagh and Wilkinson, 2002; Evandri *et al.*, 2005). In addition, climatic, geographical and seasonal differences, as well as cultural practices used, can have an influence on the oil yields and composition (for revision, see Harborne and Williams (2002) and Lis-Balchin (2002)). Thus, it is possible to manage lavender as an industrial crop in order to achieve a relatively constant oil yield and composition.

The oil is reported to have anti-inflammatory, analgesic properties (Hajhashemi *et al.*, 2003) and antimutagenic properties (Evandri *et al.*, 2005). Nonetheless, lavender is still best known for its capacity to enhance sleep (Hudson, 1996; Field *et al.*, 2008), to act as an anxiolytic (Tasev *et al.*, 1969; Tisserand 1988; Kritsidima *et al.*, 2009) and sedative (Huang *et al.*, 2008). The anxiolytic effect has been linked to the activity of linalool, and Hoferl *et al.* (2006) demonstrated that linalool fragrances alone reversed the psychological parameters produced by stress. Linalool, on the other hand, has potential allergenic properties if it is exposed to air, as auto-oxidation processes take place spontaneously (Sköld *et al.*, 2004). Other medical benefits expressed by *L. officinalis* essential oils and attributable to linalool are anti-inflammatory, analgesic (Abe *et al.*, 2003), antinociceptive (Peana *et al.*, 2003, 2006) and anti-anxiety properties (Hoferl *et al.*, 2006). Decrease of tumour formation (Wattenberg, 1991) was instead attributed to the presence of 1,8-cineole in lavender oil, while apoptosis of cancer cells (Moteki *et al.*, 2002) was attributed to other constituents, such as terpinen-4-ol, nerolidol and perillyl alcohol (Calcabrini *et al.*, 2004; Loutrari *et al.*, 2004) . Woronuk *et al.* (2011) give a complete overview of the pharmaceutical properties of lavender.

Over and above its multiple benefits on human health, the oil has been positively assayed for its antimicrobial activity (Zuzarte *et al.*, 2009), and positive results were obtained against *Aspergillus nidulans*, *Trichophyton mentagrophytes*, *Leptosphaeria maculans* and *Sclerotinia sclerotiorum* (Moon *et al.*, 2007).

Novel uses for lavender essential oil have recently been proposed. Halambek *et al.* (2010) demonstrated that *L. officinalis* oil provides a good protection to Al-3-Mg alloy against pitting corrosion in sodium chloride solution, while Easton and Jacob (2009) found that essential oil from lavender, owing to its richness in hydrocarbon components, may be usefully utilized to produce environmentally friendly organic polymer thin film.

### 23.2.16 *Matricaria recutita* L.

Chamomile (*Matricaria recutita*, or also known as *Marticaria chamomilla* and *Chamomilla recutita*) belongs to the Asteraceae family and is also commonly known as German chamomile, English chamomile, camomilla, and Flos chamomile. Chamomile has a branched, erect and smooth stem, which grows to a height of 15–65 cm. The long and slim leaves are bipinnate or tripinnate. The flowers are borne in paniculate flower heads (capitula). The white ray florets are furnished with a ligule, while the disc florets are yellow. The hollow receptacle is swollen and lacks scales. The flowers bloom in early to mid summer, and have a strong, aromatic smell. Although numerous varieties of chamomile exist, the two most popular are Roman chamomile (*Anthemis nobilis*) and German chamomile (*M. recutita*), both belonging to the Asteraceae. German chamomile, considered the more potent of the two, has received more scientific evaluation and is more widely cultivated than Roman chamomile.

This annual herbaceous species, indigenous to Europe and western Asia, is now naturalized in Australia, Britain and the USA. Today, it is widely cultivated in Germany, Hungary, Russia and other southern and eastern European countries for the flower heads. Infusions and essential oils from fresh or dried flower heads have aromatic, flavouring and colouring properties. Both are used in a number of commercial products, including soaps, detergents, perfumes, lotions, ointments, hair products, baked goods, confections, alcoholic beverages and herbal teas (McKay and Blumberg, 2006).

Chamomile is believed to possess anti-inflammatory, vulnerary, deodorant, bacteriostatic, antimicrobial, anticatarrhal, carminative, sedative, antiseptic and spasmolytic properties (Newall *et al.*, 1996; Blumenthal, 1998). Its extract is also used with positive results to counteract anxiety, hysteria, nightmares, insomnia and other sleep problems, convulsions, and even delirium tremens (Martens, 1995) and to treat gastrointestinal disturbances, including flatulence, indigestion, diarrhoea, anorexia, motion sickness, nausea and vomiting. (Nemecz, 1998). For these reasons, it has been empirically used for long time. Chamomile is utilized both internally and externally to treat an extensive list of conditions, such as wounds, ulcers, eczema, gout, skin irritations, neuralgia, sciatica, rheumatic pain, haemorrhoids, mastitis and leg ulcers (Newall *et al.*, 1996).

More than 120 constituents have been identified in the flower extract (Mann and Staba, 1986), and the yield of essential oil from the flowers ranges from 0.4 to 2%. Eleven bioactive phenolic compounds (coumarins: herniarin, umbelliferone; phenylpropanoids: chlorogenic acid, caffeic acid; flavones: apigenin, apigenin-7-*O*-glucoside, luteolin, luteolin-7-*O*-glucoside; flavonols: quercetin, rutin; flavanone: naringenin) are found in chamomile extract (Gupta *et al.*, 2010). The main components of the flower oil include the terpenoids α-bisabolol and its oxides (≤78%) and azulenes, including chamazulene (1–15%). The latter constituent is responsible for the light blue colour of the essential oil. Farnesene (12–28%), spathulenol and spiroethers, including the *cis*/*trans*-en-in-dicycloethers (8–20%), are also present in the volatile oil (Stanev *et al.*, 1996; Pino *et al.*, 2002).

Molecules that are highly relevant medically include chamazulene, α-bisabolol, bisabololoxides, bisabolonoxide A, *trans*-beta-farnesene, α-farnesene, spathulenol and the *cis*/*trans*-en-in-dicycloethers. Flavonoids, coumarins, mucilages, and mono- and oligosaccharides also have pharmacological effects (Máday *et al.*, 1999). Chamomile is also rich in apigenin ($0.8\,\mu g\,g^{-1}$ dry weight), a molecule that modulates cell cycle progression, enzyme activities and intercellular communication (McKay and Blumberg, 2006).

Chamomile adapts well to a multitude of climates, soils and even environmental stresses, including, amongst others, drought and salinity (Pirzard *et al.*, 2006; Salamon 2007; Razmjoo *et al.*, 2008), which are known to affect yield and profile of chamomile essential oil.

The antibacterial and antiviral profiles of chamomile are well documented (Aggag and Yousef, 1972; Nogueira *et al.*, 2008). McKay and Blumberg (2006) have shown growth inhibition of 25 different Gram-positive and Gram-negative bacteria and 20 strains of *L. monocytogenes* with chamomile oil, albeit with lower effectiveness compared with essential oil from other species. Among chamomile essential oil components, α-bisabolol had the strongest activity against Gram-positive and Gram-negative bacteria. Chamazulene also had strong antimicrobial activity (McKay and Blumberg, 2006), while spiroethers had weak activity against Gram-positive bacteria and they were inactive against Gram-negative bacteria (Kedzia, 1991). German chamomile esters and lactones showed activity against *Mycobacterium tuberculosis* and *Mycobacterium avium* (Lu *et al.*, 1998). Chamazulene, α-bisabolol, flavonoids and umbelliferone displayed antifungal properties against *T. mentagrophytes*, *Trichophyton rubrum*, and *C. albicans* (Szalontai, 1976, 1977; Ahmed *et al.*, 1994). An ethanolic extract of German chamomile inhibited the growth of herpes and poliovirus (Koch *et al.*, 2008).

Animal model studies showed anti-inflammatory and antiallergenic (Kobayashi *et al.*, 2003), antigenotoxic (Hernandez-Ceruelos *et al.*, 2002), antiplatelet (Pierre *et al.*, 2005), hypocholesterolaemic (Al-Jubouri *et al.*, 1990), uterotonic (Shipochliev, 1981), antiulcer (Khayyal *et al.*, 2001), antispasmodic (Achterrath-Tuckermann *et al.*, 1980) and hepatic modulation effects (Babenko and Shakhova, 2006). Positive results were achieved by Wei *et al.* (1990) in the treatment of skin tumorigenesis in mice by topical application of apigenin.

Furthermore, human studies have shown that chamomile cream was as effective as 0.25% hydrocortisone against dermatitis or eczema and more effective than a glucocorticoid and non-steroidal anti-inflammatory agents commonly used for therapy (Aertgeerts *et al.*, 1985). The extract also exhibited wound-healing properties (Nayak *et al.*, 2007; Jarrahi, 2008; Martins *et al.*, 2009).

### 23.2.17 *Artemisia absinthium* L.

Absinthe (also known as absinthe wormwood, wormwood, common wormwood, green ginger or grand wormwood) is a herbaceous, perennial plant also from the Asteraceae, and growing up to 0.8–1.2 m. It is grooved, branched and silvery-green. The leaves are greenish-grey above and white below, covered with trichomes and bearing minute oil-producing glands. Basal leaves are longer than the cauline ones (those on the stem). Its flowers are pale yellow, tubular and clustered in spherical bent-down heads (capitula), which are in turn clustered in leafy and branched panicles. Flowering is from early summer to early autumn and the fruit is a small achene. It is a species native to temperate regions of Eurasia and northern Africa, but today is also widely cultivated outside its native area, as are other members of *Artemisia*.

In Greece and Morocco, for example, this species is used to season and colour many traditional dishes. Nonetheless, this plant is perhaps best known for forming the basis of the spirit 'absinthe'. It is also used for the flavouring of other spirits and wines, including bitters, vermouth and pelinkovac. Curiously, in the Middle Ages, it was used to spice up meat, and during 18th century England wormwood was also used as an alternative to hops in beer. From the ethnopharmacological point of view, *Artemisia absinthium* has been used for its antihelmintic, stomachic, antibacterial, antifeedant, antifertility, antipyretic, cytostatic, antitumour and antimalarial action (Tucakov, 1973; Nin *et al.*, 1995).

The dry leaves and stems contain 0.3–1.3% essential oil, and the major components include absinthin, anabsin, anabsinthin, artabsin and matricin. The essential oil also contains thujone and thujyl alcohol and other terpene-derivatives, which are neurotoxic (Omer *et al.*, 2007).

Analysis of the chemical composition of *A. absinthium* oils extracted from plants grown in the USA showed β-thujone (17.5–42.3%) and *cis*-sabinyl acetate (15.1–53.4%) as the main components (Lawrence, 1992). Recently, the chemical composition of absinthe oils from Turkey was also investigated. The results showed chamazulene (17.8%), nuciferol butanoate (8.2%), nuciferol propionate (5.1%) and caryophyllene oxide (4.3%) as the main components (Kordali *et al.*, 2005a, 2005b). On the other hand, the main constituents found in the essential oil isolated from the aerial parts of *A. absinthium* collected at Mokra (Serbia) were α-thujone (19.8%), *cis*-α-epoxyocimene (10.7%), *trans*-sabinyl acetate (8.8%), sabinene (8.1%) and linalyl 3-methylbutanoate (7.5%). The oil was dominated by monoterpenes (84.6%), most of all belonging to the thujane-type compounds (40.6%). As just reported, there is a high variability, not only in quantity of the major compounds, but also in the corresponding class. Thus, it is necessary to investigate wild wormwood populations for the essential oil composition prior to their application in food and beverage production. According to Council Directive 88/388/ (EEC, 1988), on the approximation of the laws of the Member States relating to flavourings for use in foodstuffs and source materials for their production, the addition of thujone-containing plants has been allowed once again in the European Union. For this reason, there has been a surge in the industrial consumption of *A. absinthium*. This was indeed restricted in the last century as the result of absinthe prohibition (Patocka and Plucar, 2003). There are, nonetheless, restrictions on the maximum acceptable levels of thujones in final products for consumption.

Thujones are recognized as the most active constituents of absinthe oil affecting the growth rate of a large number of bacterial strains (Juteau *et al.*, 2003). For this reason essential oils of *A. absinthium* were also positively tested as anti-leishmanial compounds against two *Leishmania* strains (*L. aethiopica* and *L. donovani*). The latter are vector-borne parasitic diseases infecting humans and other mammals (Tariku *et al.*, 2011). Absinthe extracts and essential oils were also assessed for other biological activities, including antitrypanosomal (Nibret and Wink, 2010), antiplasmodial (Zafar *et al.*, 1990), analgesic and anti-inflammatory (Ahmad *et al.*, 1992), antipyretic (Khattak *et al.*, 1985), antidepressant and antioxidant (Mahmoudi *et al.*, 2009), and antifungal effects (Bora and Sharma, 2011). A weak insecticidal activity was observed by Pavela (2008) against housefly, whereby absinthe extract was used both topically and as a fumigant. Kordali *et al.* (2006) reported on its ability to control *Sitophilus granaries* population.

Furthermore, Kojima *et al.* (1984) filed a patent in which they described a substance isolated from *A. absinthium* and other plants belonging to the *Artemisia* family which could induce the production of interferon and would be non-toxic to animal cells. Omer *et al.* (2007) also reported on absinthe extract, which could be used to treat patients suffering Crohn's disease and thus act as a substitute to steroids.

### 23.2.18 *Satureja hortensis* L.

Summer savory reaches between 0.3 and 0.6 m and has slender, bronze–green leaves; lilac tubular flowers bloom from July to September. Leaves, flowers and stem are frequently used as a tea or additive in commercial spice mixtures for many foods to offer aroma and flavour. This species has been also used as folk remedies to treat various ailments, such as cramps, muscle pains, nausea, indigestion, diarrhoea and infectious diseases (Madsen *et al.*, 1996; Deans and Svoboda, 1989; Hajhashemi *et al.*, 2000).

The efficacy of *Satureja hortensis* extract is attributable to its essential oil profile. For the latter, if one excludes the volatiles, phenols are present and manifest a slight insecticidal activity (Pavela *et al.*, 2008). Güllüce *et al.* (2003) reported a total of 22 constituents, which represent 99.9% of the essential oil, in which thymol (29.0%), carvacrol (26.5%), γ-terpinene (22.6%) and *p*-cymene (9.3%) were the main components. On the other hand, Saharkhiz *et al.* (2011) found the amount of essential oils to be 2.3 wt%, 2.5 wt%, 2.0 wt% and 1.8 wt% at floral budding, full flowering, immature fruit and ripened fruit stages respectively. The major compound present at all developmental stages was γ-terpinene. It is replaced by carvacrol (46.4%) at the ripened fruit stage. Therefore, one has to remember that with different cultivation techniques, water management, nutrient administration and, in addition, extraction techniques (Pavela *et al.*, 2008) sometimes give quite different results in oil yield and profile. The factors mentioned have to be taken into consideration, especially as carvacrol appears as the main molecule responsible for the biological activity expressed by *S. hortensis* extract (Baser, 2008). Baser *et al.* (2002) have demonstrated that moderate drought stress increased the amount of carvacrol, while γ-terpinene content decreased under moderate and severe water stress. Conversely, NaCl-imposed stress reduced the carvacrol amount (Najafia *et al.*, 2010).

The essential oil and non-polar sub-fraction of the methanol extract of *S. hortensis* can be used as an antimicrobial agent in pharmaceutical and natural therapies of infectious diseases in humans, management of plant diseases or additionally as environmentally safer alternatives to protect the spoilage of food products from pathogenic and saprophytic fungi. Indeed, Güllüce *et al.* (2003) confirmed growth inhibition of *B. subtilis*, *Enterococcus faecalis*, *Pseudomonas aeruginosa*, *S. enteritidis*, and *Streptococcus pyogenes* by summer savory essential oil, while other researches pointed out its effectiveness against *C. albicans* (Naeini *et al.*, 2009) and *Bacillus* spp. (Sahin *et al.*, 2003). However, the latter authors underlined that the type of extraction used can influence the effectiveness of the extracts. Methanol extraction enhanced its effectiveness against a broader spectrum of bacteria belonging to genus *Bacillus*.

Razzaghi-Abyaneh *et al.* (2008) found that carvacrol and thymol purified from *S. hortensis* inhibit aflatoxins production by *Aspergillus parasiticus*, while Dikbas *et al.* (2008) confirmed the biological activity of essential oils against *A. flavus*, another aflatoxigenic fungus. Similarly, Boyraz and Özcan (2006) demonstrated the biological efficiency of summer savory extract against *Alternaria mali* and *B. cinerea*. For this reason, essential oils can find application as a potential source of eco-friendly fungicide for crop production.

Other properties emanating from folk medicine include antispasmodic, anti-diarrhoeal, gastrointestinal protection (Hajhashemi *et al.*, 2000), antinociceptive and anti-inflammatory activities (Hajhashemi *et al.*, 2002). Summer savory extracts also helped suppress abdominal fat accumulation and prevented obesity, fatty liver, hyperlipidaemia and diabetes (Park, 2011).

### 23.2.19 *Ruta graveolens* L.

Rue, or herb of grace, belonging to the Rutaceae family, is a perennial herbaceous or half-shrubby plant, reaching 0.5–0.8 m in height. It is characterized by a strong, heavy and unpleasant smell. Fruits are dry, hard rounded, four- or five-lobed at the top, greyish-brown and rough. Seeds are ovoid, rounded on the back, flattish in front, angular, testa blackish, rough (Parray *et al.*, 2012). Rue is native to the Balkan Peninsula and south-east Europe. Today it is widely cultivated all over the world as an ornamental plant, especially because of its bluish leaves and yellow flowers, and also for its tolerance of hot season and dry soil conditions. It is also cultivated as a condiment, medicinal herb and, to a lesser extent, as an insect repellent.

Rue, as well as other *Ruta* species, contains different classes of secondary metabolites, including coumarins (Srivastava *et al.*, 1998), flavonoids (rutin and quercitine), furanocoumarins (psoralens, xanthotoxin, bergapten) (Milesi *et al.*, 2001) and acridone or furoquinolone alkaloids (arborinine, evoxanthine, graveoline, gravelinine, rutamine, rutamarine) (Oliva *et al.*, 2003; Borbala *et al.*, 2007). The acridone alkaloids were detected in all organs, particularly in endodermal and vascular tissues (Parray *et al.*, 2012). The essential oil of *Ruta graveolens* contains two main constituents: undecan-2-one (47%) and nonan-2-one (19%), which together total more than the 60% of the total oil yield (De Feo *et al.*, 2002).

Although infusions, extractions or decoctions have several positive attributes, these products may still be toxic due to the effects of some alkaloids, which are known to be responsible for blocking the calcium and potassium channels in nerve membranes (EMEA, 1999). The essential oil may be a local irritant at a certain concentration owing to the presence of psoralen and bergaptene; thus, in Europe, the limit of furocoumarins is not allowed to exceed 1 mg $kg^{-1}$ in sun protection and bronzing products (EMEA, 1999). Rue is also an active irritant whether applied externally or taken internally; sometimes it produces painful vomiting, always great prostration, and confusion of mind, cloudy vision, feebleness and slowness of pulse, coldness of extremities and allergic reactions (Dymock *et al.*, 2005).

Ethnobotanical lore indicates that this species has potent antimicrobial activity, and for this reason many researchers have been addressed on this matter so as to apply a more scientific approach to empirical evidence. Methanol, petroleum ether, ethyl acetate and water–methanol extracts of rue have been proposed as topical pharmaceutical fungicides against *Microsporum canis*, *T. mentagrophytes* and *Trichophyton violaceum*, which cause dermatitis (Ali-Shtayeh and Abu Ghdeib, 1999), but also against *Micrococcus pyogenes* and *E. coli* and other Gram-negative human pathogens (Ivanova *et al.*, 2005). In the traditional system of medicine, the plant is also used as an anti-inflammatory (Raghav *et al.*, 2006), stimulant (Dymock *et al.*, 2005), emmenagogue and rubefacient (Trease and Evans, 2002), anthelmintic (Graves, 1996), antispasmodic and antiepileptic (Trease and Evans, 2002), vermifuge (Lindley, 2001) and resolvent (Graves, 1996; Nadkarni, 2005), although some of its main uses remain connected to its abortifacient (Nadkarni, 2005), anticonceptive and antifertility activities (Al-Mahmoud *et al.*, 2003). Flavonoids are believed to be the main compounds responsible for the positive effects of rue. Rutin and quercetin, in particular, have been reported to possess anti-inflammatory (Afanas'eva *et al.*, 2001), anti-arthritis (Guardia *et al.*, 2001), antithrombotic (López-Revuelta *et al.*, 2006) and anticholesterol effects (Wu *et al.*, 2003). On the other hand, furanoalkaloids present in the plant confer toxicity to this species, but they can be advantageously used to counteract tumour diffusion and propagation. The total extract is known to decrease solid tumours developing from Dalton's lymphoma ascites and Ehrlich ascites carcinoma (Preethi *et al.*, 2006), while arborinine and furanoacridones isolated from leaves extract of rue were found to be good at counteracting the propagation and reproduction of murine cells due to their apoptotic and citotoxic activity (Rethy *et al.*, 2007). The compounds were applied *in vitro* to three types of human cancer cells: breast adenocarcinoma, cervix adenocarcinoma and skin epidermoid carcinoma. Arborine displayed the highest antiproliferative effects. Recently, the capacity of furanoacridones to induce apoptosis in a murine lymphoma cell line was demonstrated also by Borbala *et al.* (2007, 2008); thus, this species appears to be a wide source of anticancer compounds which requires investigation. *R. graveolens* folk and medical uses were thoroughly reviewed by Parray *et al.* (2012).

Besides these potential applications directly involved counteracting human diseases, other agronomical applications of rue extract were proposed. *In vitro* fungistatic activity of rue extract against six species of fungus has been evaluated (Oliva *et al.*, 1999, 2003), and hydroalcoholic extract tested on the leaves of tobacco against *Erysipe orontii* showed biological activity on conidia and provided a good pest control similar to that obtained with the commercial fungicide Bupirimate. The essential oil of *R. graveolens* also demonstrated a potential application as an allelopathic agent. De Feo *et al.* (2002) found inhibition of both germination and radical growth in radish, while the elongation of roots of cucumber, maize and garden pea was inhibited to different degrees by 8-methoxypsoralen (Kupidlowska *et al.*, 1994). Hale *et al.* (2004) reported the reduction of biomass production in some other species due to furanocoumarins treatment. Meepagala *et al.* (2005) also showed an algicidal activity against *Oscillatoria perornata* and *Selenastrum capricornutum* of some compounds extracted from *R. graveolens* roots comparable to the use of chemically synthesized molecules. Coumarins from rue were also found suitable for the control of Coleoptera pests of stored products (Moreira *et al.*, 2007).

### 23.2.20 *Pimpinella anisum*

*Pimpinella anisum*, belonging to the family of Apiaceae, has been known since medieval times, when it was best known as a drug. *P. anisum* is native to the eastern Mediterranean region and south-west Asia, and variations in the essential oil composition of this species obtained from different geographical areas were reported (Orav *et al.*, 2008). The aniseed is of economic importance as a flavouring agent in the food and perfumery industries (Baser, 2002; Tabanca *et al.*, 2003). In folk medicine, aniseed has been used for the treatment of nausea, abdominal colic, insomnia and epilepsy (Said *et al.*, 1996; Kreydiyyeh *et al.*, 2003). The phytotherapeutic applications of aniseed are based on its digestive, carminative, diuretic, antiseptic and expectorant actions (Said *et al.*, 1996).

Aniseed contains 1.5–6.0 mass% of a volatile oil consisting primarily of *trans*-anethole and also as much as 8–11 mass% of lipids rich in fatty acids, such as palmitic and oleic acids, as well as approximately 4 mass% of carbohydrates and 18 mass% of protein (Besharati-Seidani *et al.*, 2005). Other studies have demonstrated the presence of eugenol *trans*-anethole, methylchavicol, anisaldehyde, estragole, coumarins, scopoletin, umbelliferone, estrols, terpene hydrocarbons, polyenes and polyacetylenes as the major compounds of the essential oil of aniseed (Gülçın *et al.*, 2003). Anisaldehyde is an acaricidal component used against house dust mites, *Dermatophagoides farinae* and *Dermatophagoides pteronyssinus* (Lee, 2004). It is also used in healing peptic ulcers (Al-Mofleh *et al.*, 2007), and it has antispasmodic (Tirapelli *et al.*, 2007), anticonvulsant effects (Pourgholami *et al.*, 1999) and antifungal activity (Kosalec *et al.*, 2005).

The presence of other compounds has also been reported: (*E*)-methyleugenol, α-cuparene, α-himachalene, β-bisabolene, *p*-anisaldehyde and *cis*-anethole (Özcan and Chalchat, 2006). In addition to the major components reported, another study aimed at determinining the composition of essential oil of *P. anisum* fruits obtained from different geographical areas of Europe reported on the presence of compounds like *trans*-pseudoisoeugenyl 2-methylbutyrate, *p*-anisaldehyde and methylchavicol (Orav *et al.*, 2008).

Differences in the oil composition have also been reported when the plant is exposed to different environmental conditions (Figueiredo *et al.*, 2008). Those conditions that affect plant growth and the production of secondary metabolites of anise are: sowing date, plant density, water supply and harvesting time studied under field or greenhouse conditions. For example, Zehtab-Salmasi *et al.* (2001) reported that aniseed harvested at the waxy stage had higher oil content than those harvested at earlier or later stages, whereas Ullah and Honermeier (2013) reported that anise is sensitive to sowing date. A higher concentration of essential oil was synthesized in delayed sowing date compared with earlier sown anise plants. Conversely, the essential oil concentration of anise showed no remarkable variation regarding plant density (Ullah and Honermeier, 2013). Even water deficit decreased the oil yield of anise (Zehtab-Salmasi *et al.*, 2001).

As mentioned above, aniseed has many pharmacological properties, such as antibacterial (Akhtar *et al.*, 2008), antifungal (Gülçın *et al.*, 2003), antiviral (Lee J.B. *et al.*, 2011), analgesic (Tas, 2009), antiulcer (Al-Mofleh *et al.*, 2007) and antidiabetic properties (Rajeshwari *et al.*, 2011), and it also impacts morphine dependence (Sahraei *et al.*, 2002). Recently, the pharmacological properties of constituents of *P. anisum* were reviewed by Shojaii and Fard (2012).

As far as other industrial application uses are concerned, Aminifard and Mohammadi (2012) reported that anise essential oils increased the shelf life and inhibited the grey mould growth on tomato fruits. Aniseed essential oil also has an antiviral effect against potato virus, tobacco mosaic virus and tobacco ring spot virus (Shukla, 1989) and can be used as a food preservant (Singh G. *et al.*, 2011).

Another important agro-industrial application of aniseed is its use in flavouring alcoholic beverages. In fact, pastis is a famous anise-flavoured beverage, symbolic of hot summer days in the south of France. The compound responsible for its characteristic aniseed flavour is anethole, extracted from aniseed, star-aniseed (*Illicium verum*) and fennel (*Foeniculum vulgare*).

### 23.2.21 *Myrtus communis* L.

Myrtle is an evergreen, sclerophyll shrub, which grows naturally in the Mediterranean area. Myrtle is well known as a medicinal plant, but it is also utilized in the food and cosmetic industries. Finally, liquors prepared from myrtle berries became popular especially in Sardinia (Italy), while its leaves have been used as a hop substitute in beer.

This species is a very aromatic plant because of the high essential oil content in its leaves, flower and fruit glands. A total of 27 components were detected in essential oil of myrtle, accounting for 90.6–98.7% of the total essential oil composition even though there was a strong chemical variability depending on the origin of the samples (Tuberoso *et al.*, 2006). The major compounds in the essential oils from leaves and berries respectively were α-pinene (30.0% and 28.5%), 1,8-cineole (28.8% and 15.3%), and limonene (17.5% and 24.1%). The fruits of this species are mostly composed of volatile oils, sugars and organic acids, such as citric and malic acids (Martín *et al.*, 1999). Fruits are also rich in vitamins, and for this reason are utilized as food integrators, whereas essential oil obtained from leaves is important in perfumery (Baytop, 1999). The analysis of essential oil from floral buds revealed that they had a significant antioxidant effect attributable to the high hydrocarbon monoterpenes and oxygenated monoterpenes content (Snoussi *et al.*, 2011). Finally, phenolic acids, as gallic and ellagic acids, flavonoids (Romani *et al.*, 1999), fatty acids (Cakir , 2004), tannins (Diaz and Abeger, 1986) and anthocyanin pigments (Martín *et al.*, 1990) have been found.

Myrtle oils can be separated into two groups, depending on the level of myrtenyl acetate present (Bradesi *et al.*, 1997). The analyses of myrtle essential oil from seven Mediterranean regions (Chalchat *et al.* (1998) have shown that myrtenyl acetate is present in some samples, but absent in some others. Pereira *et al.* (2009) also reported that Portuguese myrtle belongs to that group of myrtles which are characterized by the presence of myrtenyl acetate as one of the major components.

Myrtle berry also contains small amounts of myricetin derivates (myricetin-3-*O*-galactoside and myricetin-3-*O*-rhamnoside), while anthocyanins (malvidin-3-*O*-glucoside, petunidin-3-*O*-glucoside) are present in large amounts (Viuda-Martos *et al.*, 2011).

The essential oil obtained from this species has been widely investigated and it has been reported that its composition is quite variable. In fact, the chemical composition of 27 oil samples of myrtle isolated from leaves collected in three locations in north-eastern Algeria was similar to that reported for myrtle oils from Corsica, Sardinia and Tunisia, but differed from that of Moroccan and Spanish myrtle oils (Bouzabata *et al.*, 2010).

Numerous investigations have confirmed that the chemical composition of myrtle essential oil depends greatly not only on the geographical area, but harvesting time also influenced the fruit essential oil yield in myrtle ranging between 0.003% and 0.01%. It has also been showed a remarkable increase at 60 days after flowering to reach a maximum of 0.11% (Wannes *et al.*, 2009). The extraction method also influenced the essential oil composition (Ghasemi *et al.*, 2010; Berka-Zougali *et al.*, 2012).

*Myrtus communis* was moderately active against Gram-positive and Gram-negative bacteria (Akin *et al.*, 2010). Zanetti *et al.* (2010) reported that the essential oil manifested activity towards *Mycobacterium tuberculosis*. Although the individual compounds (except α-pinene) manifested higher minimal inhibitory concentrations (MICs), it would appear in this case that a synergistic effect of all the molecules together would account for its activity. Myrtle has also been shown to manifest antimutagenic activity (Mimica-Dukić *et al.*, 2010). In addition, it has been reported that the aqueous and ethanolic extracts of the aerial parts of *M. communis* showed antinociceptive effects which are mediated by opioid receptors (Hosseinzadeh *et al.*, 2011). Another study indicated a significant antihyperglycaemic effect of *M. communis* leaves and supported its traditional usage in the treatment of diabetes mellitus (Malekpour *et al.*, 2012). Al-Anbori *et al.* (2008) found the leaf ethanolic extract to have an MIC value of 106.6 $\mu g\, mL^{-1}$ against dental pathogens in adults. The effect was attributed to flavonoids. Tumen *et al.* (2012) also demonstrated for the first time *in vitro* neuroprotective effects of myrtle.

Myrtle essential oil was also insecticidal against the Culicidae mosquito *Aedes albopictus* (Conti *et al.*, 2010), and Oka *et al.* (2012) suggested that the leaf powder and paste extract of *M. communis* were potential nematicides against root-knot nematodes. In addition, it has been reported that crude water extract and a flavonoid fraction of *M. communis* possessed molluscicidal activity against the aquatic snail *Biomphalaria glabrata* vector of schistosomiasis (Deruaz *et al.*, 1993).

The methanol, ethanol and ethyl acetate extracts of myrtle leaves and berries were also active against a range of food-borne pathogens and food spoilage bacteria (Amensour *et al.*, 2010).

## 23.3 Concluding remarks

Plants are sessile organisms and consequently show a great plasticity of their genome, which reaches the maximal expression in Mediterranean climate areas. In these regions, plant species (both native and alien imported long ago) have evolved in the harsh and extremely variable environmental conditions, depending on seasonal or even daily variations to which plants are subjected. Clearly, in these constrained conditions, the susceptibility to biotic stress is enhanced, too. In addition to morphanatomical adaptations, a large pool of compounds, especially secondary metabolites, represent a necessary 'investment' for plants in order to acquire better protection against a large number of both biotic and abiotic stresses.

Humans have empirically discovered Mediterranean aromatic plants since pre-historic times and, over time, the knowledge on the properties of these plant species has been improved. In the last decades, the cumulative knowledge, especially in industrialized countries, of folk-popular uses of Mediterranean aromatic species have been supported (and partially substituted) by more medical and scientific approaches.

In spite of the potential of these plants as highlighted in this chapter, a lot is still unknown on the properties of a wide range of other aromatic plants, and thus greater attention needs to be focused on their use. Indeed, several properties (such as antimicrobial or bactericidal) are attributable either to the toxicity of one molecule, a series of compounds or to the total pool deriving from the essential oil or extracts of various plant parts. Thus, for some compounds (e.g. eugenol or thujone) the border line between beneficial and toxic effects might be extremely narrow and only a thorough investigation would allow us to take full advantage of the potential of aromatic plants.

## References


Abbas, R.J. (2010) Effect of using fenugreek, parsley and sweet basil seeds as feed additives on the performance of broiler chickens. *International Journal of Poultry Science*, **9**, 278–282.

Abdallah, N., El-Gengaihi, S. and Sedrak, E. (1978) The effect of fertilizer treatments on yield of seed and volatile oil of fennel (*Foeniculum vulgare* Mill.). *Pharmazie*, **33**, 607–608.

Abdel-Massiha, R.M., Faresa, R., Bazzia, S. *et al.* (2010) The apoptotic and anti-proliferative activity of *Origanum majorana* extracts on human leukemic cell line. *Leukemia Research*, **34**, 1052–1056.

Abdullah, A.T., Hanafy, M.S., El-Ghawwas, E.O. and Ali, Z.H. (2012) Effect of compost and some biofertilizers on growth, yield, essential oil productivity and chemical composition of *Rosmarinus officinalis*, L. plants. *Journal of Horticultural Science & Ornamental Plants*, **4**, 201–214.

Abe, S., Maruyama, N., Hayama, K. *et al.* (2003) Suppression of tumor necrosis factor-alpha-induced neutrophil adherence responses by essential oils. *Mediators of Inflammation*, **12**, 323–328.

Abu-Darwish, S.M., Alu'datt, M.H., Al-Tawaha, A.R. *et al.* (2012) Seasonal variation in essential oil yield and composition from *Thymus vulgaris* L. during different growth stages in the south of Jordan. *Natural Product Research: Formerly Natural Product Letters*, **26**, 1310–1317.

Aburjai, T. and Natsheh, F.M. (2003) Plants used in cosmetics. *Phytotherapy Research*, **17**, 987–1000.

Abu-Zinadah, O.A. (2008) Effects of watercress oil on the thermal and chemical burn injuries in rabbits. *Medical Science*, **15**, 3–17.

Achterrath-Tuckermann, U., Kunde, R., Flaskamp, E. *et al.* (1980) Pharmacological investigations with compounds of chamomile; V. Investigations on the spasmolytic effect of compounds of chamomile and Kamillosan on the isolated guinea pig ileum. *Planta Medica*, **39**, 38–50.

Ackermann, R.T., Mulrow, C.D., Ramirez, G. *et al.* (2001) Garlic shows promise for improving some cardiovascular risk factors. *Archives of Internal Medicine*, **161**, 813–824.

Aertgeerts, P., Albring, M., Klaschka, F. *et al.* (1985) Comparative testing of Kamillosan cream and steroidal (0.25% hydrocortisone, 0.75% fluocortin butyl ester) and nonsteroidal (5% bufexamac) dermatologic agents in maintenance therapy of eczematous diseases. *Zeitschrift für Hautkrankheiten*, **60**, 270–277.

Afanas'eva, I.B., Ostrakhovitch, E.A., Mikhal'chik, E.V. *et al.* (2001) Enhancement of antioxidant and anti-inflammatory activities of bioflavonoid rutin by complexation with transition metals. *Biochemical Pharmacology*, **61**, 677–684.

Aggag, M.E. and Yousef, R.T. (1972) Study of antimicrobial activity of chamomile oil. *Planta Medica*, **22**, 140–144.

Ahmad, F., Khan, R.A. and Rasheed, S. (1992) Study of analgesic and anti-inflammatory activity from plant extracts of *Lactuca scariola* and *Artemisia absinthium*. *Journal of Islamic Academy of Sciences*, **5**, 111–114.

Ahmad, N., Sharma, S., Alam, M.K. *et al.* (2010) Rapid synthesis of silver nanoparticles using dried medicinal plant of basil. *Colloids and Surfaces B: Biointerfaces*, **81**, 81–86.

Ahmadian, A., Tavassoli, A. and Amiri, E. (2011) The interaction effect of water stress and manure on yield components, essential oil and chemical compositions of cumin (*Cuminum cyminum*). *African Journal of Agricultural Research*, **6**, 2309–2315.

Ahmed, F.H., El Badri, A.A., Ibrahim, M.M.K. *et al.* (1994) Comparative studies of antifungal potentialities for some natural plant oils against different fungi isolated from poultry. *Grasas y Aceites*, **45**, 260–264.

Ahonkhai, I., Ayinde, B., Edogun, O. and Uhuwmangho, M. (2009) Antimicrobial activities of the volatile oils of *Ocimum bacilicum* l. and *Ocimum gratissimum* L. (Lamiaceae) against some aerobic dental isolates. *Pakistan Journal of Pharmacological Science*, **22**, 405–409.

Ait-Ouazzou, A., Cherrat, L., Espina, L. *et al.* (2011) The antimicrobial activity of hydrophobic essential oil constituents acting alone or in combined processes of food preservation. *Innovative Food Science & Emerging Technologies*, **12**, 320–329.

Akgül, A. and Bayrak, A. (1988) Comparative volatile oil composition of various parts from Turkish bitter fennel (*F. vulgare* var. *vulgare*). *Food Chemistry*, **30**, 319–323.

Akhtar, A., Deshmukh, A.A., Bhonsle, A.V. *et al.* (2008) *In vitro* antibacterial activity of *Pimpinella anisum* fruit extracts against some pathogenic bacteria. *Veterinary World*, **1**, 272–274

Akin, M., Aktumsek, A. and Nostro, A. (2010) Antibacterial activity and composition of the essential oils of *Eucalyptus camaldulensis* Dehn. and *Myrtus communis* L. growing in northern Cyprus. *African Journal of Biotechnology*, **9**, 531–535,

Akkol, E.K., Güvenç A. and Yesilada, E. (2009) A comparative study on the antinociceptive and anti-inflammatory activities of five *Juniperus* taxa. *Journal of Ethnopharmacology*, **125**, 330–336.

Al-Anbori, D.K.A., Al-Nimer, M.S.M. and Al-Weheb, A.M. (2008) Antibacterial activity of ethanolic extract of *Myrtus communis*. L leaves against salivary *Mutans streptococci*. *Saudi Dental Journal*, **20**, 82–87.

Al-Daraji, H.J., Al-Mashadani, H.A., Al-Hassani, A.S. *et al.* (2012) The influence of parsley (*Petroselinum crispum*) as feed additive on hematological traits of local Iraqi geese. *Advanced Nutritional Research*, **1**, 1–5.

Al-Dhaheri, Y., Eid, A., Arafat, K. *et al.* (2012) *Origanum majorana* extract induces apoptosis and suppresses migration and invasion of MDA-MB-231 human breast cancer cell line through inactivation of the NFκB pathway. *European Journal of Cancer*, **48** (Suppl 5) S233–S234.

Ali, M.S., Saleem, M., Ahmad, W. *et al.* (2002) A chlorinated monoterpene ketone, acylated β-sitosterol glycosides and a flavanone glycoside from *Mentha longifolia* (Lamiaceae). *Phytochemistry*, **59**, 889–895.

Ali-Shtayeh, M.S. and Abu Ghdeib, S.I. (1999) Antifungal activity of plant extracts against dermatophytes. *Mycoses*, **42**, 665–672.

Al-Jubouri, H.H.F., Al-Jalil, B.H., Farid, I. *et al.* (1990) The effect of chamomile on hyperlipidemias in rats. *Journal of Faculty of Medicine University of Baghdad*, **32**, 5–11.

Al-Kareemi, K.K. (2012) Inhibitory effect of parsley (*Petroselinum crispum*) juice against some urinary pathogens *in vitro*. *The Iraqi Postgraduate Medical Journal*, **11**, 336–342.

Allison, G.L, Lowe, G.M. and Rahman, K. (2012) Aged garlic extract inhibits platelet activation by increasing intracellular cAMP and reducing the interaction of GPIIb/IIIa receptor with fibrinogen. *Life Sciences*, **91**, 25–26.

Al-Mahmoud, M.S., Elbetieha, A., Al-Muhur, R.A. (2003) Anticonceptive and antifertility activities of various *Ruta graveolens* extracts in female rats. *Acta Pharmaceutica Turcica*, **45**, 203–213.

Al-Mofleh, I., Alhaider, A., Mossa, J.S. *et al.* (2007) Aqueous suspension of anise *Pimpinella anisum* protects rats against chemically induced gastric ulcers. *World Journal of Gastroenterology*, **13**, 1112–1118.

Al-Qattan, K.K., Alnaqeeb, M.A. and Ali, M.J. (1999) The antihypertensive effect of garlic (*Allium sativum*) in the rat two-kidney–one-clip Goldblatt model. *Ethnopharmacology*, **66**, 217–222.

Alves-Rodrigues, M.R., Kanazawa, L.K., das Neves, T.L.M. *et al.* (2012) Antinociceptive and anti-inflammatory potential of extract and isolated compounds from the leaves of *Salvia officinalis* in mice. *Journal of Ethnopharmacology*, **139**, 519–526.

Amagase, H. (2006) Clarifying the real bioactive constituents of garlic. *Journal of Nutrition*, **136**, 716–725.

Amber, K.,Aijaz, A., Immaculata, X. *et al.* (2010) Anticandidal effect of *Ocimum sanctum* essential oil and its synergy with fluconazole and ketoconazole. *Phytomedicine*, **17**, 921–925.

Amensour, M., Bouhdid, S., Fernández-López, J. *et al.* (2010) Antibacterial activity of extracts of *Myrtus communis* against food-borne pathogenic and spoilage bacteria. *International Journal of Food Properties*, **13**, 1215–1224.

Aminifard, M.H. and Mohammadi, S. (2012) Effect of essential oils on postharvest decay and quality factors of tomato *in vitro* and *in vivo* conditions. *Archives of Phytopathology and Plant Protection*, **45**, 1280–1285.

Amiri, H. (2012) Volatile constituents and antioxidant activity of flowers, stems and leaves of *Nasturtium officinale* R. Br. *Natural Product Research: Formerly Natural Product Letters*, **26**, 109–115.

Amrani S., Harnafi H., Gadi D., Mekhfi H., Legssyer A., Aziz M., Martin-Nizard F., Bosca L. (2009) Vasorelaxant and anti-platelet aggregation effects of aqueous *Ocimum basilicum* extract. *Journal of Ethnopharmacology*, **125**, 157–162.

Anees, A.M. (2008) Larvicidal activity of *Ocimum sanctum* Linn. (Labiatae) against *Aedes aegypti* (L.) and *Culex quinquefasciatus* (Say). *Parasitology Research*, **103**, 1451–1453.

Angioni, A., Barra, A., Cereti, E. *et al.* (2004) Chemical composition, plant genetic differences, antimicrobial and antifungal activity investigation of the essential oil of *Rosmarinus officinalis* L. *Journal of Agricultural and Food Chemistry*, **52**, 3530–3535.

Anwara, F., Alia, M., Hussaina, A.I. and Shahida, M. (2009) Antioxidant and antimicrobial activities of essential oil and extracts of fennel (*Foeniculum vulgare* Mill.) seeds from Pakistan. *Flavour and Fragrance Journal*, **24**, 170–176.

Arslan, N., Bayrak, A. and Akgul, A. (1989) The yield and component of essential oil in fennel of different origin (*Foeniculum vulgare* Mill) grown in Ankara conditions. *Herba Hungarica*, **28**, 27–31.

Arumugam, P. and Ramesh, A. (2009) Antigenotoxic and antioxidant potential of aqueous fraction of ethanol extract of *Mentha spicata* (L.) against 4-nitroquinoline-1-oxide-induced chromosome damage in mice. *Drug and Chemical Toxicology*, **32**, 411–416.

Atta-Aly, M.A. (1999) Effect of nickel addition on the yield and quality of parsley leaves, *Scientia Horticulturae*, **82**, 9–24.

Babenko, N.A. and Shakhova, E.G. (2006) Effects of *Chamomilla recutita* flavonoids on age-related liver sphingolipid turnover in rats. *Experimental Gerontology*, **41**, 32–39.

Baddar, N.W., Aburjai, T.A., Taha, M.O. and Disi, A.M. (2011) Thujone corrects cholesterol and triglyceride profiles in diabetic rats. *Natural Product Research: Formerly Natural Product Letters*, **25**, 1180–1184.

Bahnas, M.S., Ragab, M.S., Ascer, N.E.A. and Eman, R.M.S. (2009) Effect of using parsley or its by product with or without enzyme supplementation on performance of growing Japanese quail. *Egyptian Poultry Science*, **29**, 241–262.

Bakkali, F., Averbeck, S., Averbeck, D. and Idaomar, M. (2008) Biological effects of essential oils – a review. *Food and Chemical Toxicology*, **46**, 446–475.

Bakri, I.M. and Douglas, C.W.I. (2005) Inhibitory effect of garlic extract on oral bacteria. *Archives of Oral Biology*, **50**, 645–651.

Banchio, E., Bogino, P.C., Zygadlo, J. and Giordano, W. (2008) Plant growth promoting rhizobacteria improve growth and essential oils yield in *Origanum majorana* L. *Biochemical Systematics and Ecology*, **36**, 766–771.

Banchio, E., Xie, X., Zhang, H. and Parè, P.W. (2009) Soil bacteria elevate essential oil accumulation and emissions in sweet basil. *Journal of Agricultural and Food Chemistry*, **57**, 653–657.

Baranauskiene, R., Dambrauskiene, E., Venskutonis, P.R. and Viskelis, P. (2011) Influence of harvesting time on the yield and chemical composition of sage (*Salvia officinalis* L.), in *Proceedings of the 6th Baltic Conference on Food Science and Technology "Innovations for Food Science and Production" FOODBALT-2011*, 5–6 May, Jelgava, Latvia, pp. 103–109.

Baratta, M.T., Dorman, H.J.D., Deans, S.G. *et al.* (1998) Antimicrobial and antioxidant properties of some commercial essential oils. *Flavour and Fragrance Journal*, **13**, 235–244.

Baricevic, D. and Bartol, T. (2000) The biological/pharmacological activity of the *Salvia* genus, in *Sage: The Genus Salvia* (ed. S.E. Kintzios), Harwood Academic Publishers, Amsterdam, pp. 143–184.

Barla, A., Topcu, G., Oksuz, S. *et al.* (2007) Identification of cytotoxic sesquiterpenes from *Laurus nobilis* L. *Food Chemistry*, **104**, 1478–1484.

Barros, L., Heleno, S.A., Carvalho, A.M. and Ferreira, I.C.F.R. (2009) Systematic evaluation of the antioxidant potential of different parts of *Foeniculum vulgare* Mill. from Portugal. *Food and Chemical Toxicology*, **47**, 2458–2464.

Baser, K.H.C. (2002) Recent advances on the Umbelliferae essential oils of Turkey, in *Natural Products Chemistry at the Turn of the Century, Proceedings of the 8th International Symposium on Natural Product Chemistry, 18–22 January 2000* (eds Atta-ur-Rahman, M.I. Choudhary and K.M. Khan), Print Arts, Karachi, pp. 271–289.

Baser, K.H.C. (2008) Biological and pharmacological activities of carvacrol and carvacrol bearing essential oils. *Current Pharmaceutical Design*, **14**, 3106–3119.

Bashin, M. (2012) *Ocimum* – taxonomy, medicinal potentialities and economic valu of essential oil. *Journal of Biosphere*, **1**, 48–50.

Bassole, I.H.N., Nebie, R., Savadogo, A. *et al.* (2005) Composition and antimicrobial activities of the leaf and flower essential oils of *Lippia chevalieri* and *Ocimum canum* from Burkina Faso. *African Journal of Biotechnology*, **4**, 1156–1160.

Bauer, J., Kuehnl, S., Rollinger, J.M. *et al.* (2012) Koeberle, carnosol and carsonic acids from *Salvia officinalis* inhibit microsomal prostaglandin $E_2$ synthase-1. *Journal of Biosphere*, **342**, 169–176.

Bauer, K. and Garbe, D. (1997) *Common Fragrance and Flavor Materials, Preparation, Properties and Uses*, 3rd edn, Wiley–VCH GmbH, Weinheim.

Bayrak, A. and Akgul, A. (1987) Composition of essential oil from Turkish *Salvia* species. *Phytochemistry*, **26**, 846–847.

Baytop, T. (1999) *Therapy with Medicinal Plants in Turkey (Past and Present)*, Nobel Tıp Kitapevleri Press, Istanbul.

Begum, A., Sandhya, S., Ali, S.S. *et al.* (2013) An in-depth review on the medicinal flora *Rosmarinus officinalis* (Lamiaceae). *ACAT Scientiarum Polonorum Technologia Alimentaria*, **12**, 61–73.

Behera, S., Panigrahi, R., Ramani, Y. *et al.* (2012) Evaluation of antioxidant activity of *Ocimum canum* hydro-alcoholic leaf extract in the prevention of hepatic ischaemia. *Pharmacology*, **3**, WMC003584.

Behn, H., Albert, A., Marx, F. *et al.* (2010) Ultraviolet-B and photosynthetically active radiation interactively affect yield and pattern of monoterpenes in leaves of peppermint (*Mentha × piperita* L.). *Journal of Agricultural and Food Chemistry*, **58**, 7361–7367.

Benchaar, C., Hristov, A.N. and Greathead, H. (2009) Essential oils as feed additives in ruminant nutrition, in *Phytogenics in Animal Nutrition* (ed. T. Steiner), Nottingham University Press, Nottingham, pp. 111–146.

Ben-Farhat, M., Jordán, M.J., Chaouech-Hamada, R. *et al.* (2009) Variations in essential oil, phenolic compounds, and antioxidant activity of Tunisian cultivated *Salvia officinalis* L. *Journal of Agricultural and Food Chemistry*, **57**, 10349–10356.

Ben-Taarit, M., Msaada, K., Hosni, K. *et al.* (2009) Plant growth, essential oil yield and composition of sage (*Salvia officinalis* L.) fruits cultivated under salt stress conditions. *Industrial Crops and Products*, **30**, 333–337.

Berić, T., Nikolić, B., Stanojević, J. *et al.* (2008) Protective effect of basil (*Ocimum basilicum* L.) against oxidative DNA damage and mutagenesis. *Food and Chemical Toxicology*, **46**, 724–732.

Berka-Zougali, B., Ferhat, M.-A., Hassani, A. *et al.* (2012) Comparative study of essential oils extracted from Algerian *Myrtus communis* L. leaves using microwaves and hydrodistillation. *Internation Journal of Molecular Science*, **13**, 4673–4695.

Bernotienė, G., Nivinskienė, O., Butkienė, R. and Mockutė, D. (2007) Essential oil composition variability in sage (*Salvia officinalis* L.). *Chemija*, **18**, 38–43.

Bernstein, N., Kravchik, M. and Dudai, N. (2010) Salinity-induced changes in essential oil, pigments and salts accumulation in sweet basil (*Ocimum basilicum*) in relation to alterations of morphological development. *Annals of Applied Biology*, **156**, 167–177.

Besharati-Seidani, A., Jabbari, A. and Yamini, Y. (2005) Headspace solvent microextraction: a very rapid method for identification of volatile components of Iranian *Pimpinella anisum* seed. *Analytica Chimica Acta*, **530**, 155–161.

Bettaieb, I., Zakhama, N., Aidi Wannes, W. *et al.* (2009) Water deficit effects on *Salvia officinalis* fatty acids and essential oils composition. *Scientia Horticulturae*, **120**, 271–275.

Bettaieb, I., Bourgou, S., Wannes, A.W. *et al.* (2010) Essential oils, phenolics, and antioxidant activities of different parts of cumin (*Cuminum cyminum* L.). *Journal of Agricultural and Food Chemistry*, **58**, 10410–10418.

Bihari, C.G., Manaswini, B., Keshari, P.S. and Kumar, T.S. (2011) Phytochemical investigation & evaluation for antidiabetic activity of leafy extracts of various *Ocimum* (Tulsi) species by alloxan induced diabetic model. *Journal of Pharmacy Research*, **4**, 28–29.

Bilal, A., Jahan, N., Ahmed, A. *et al.* (2012) Phytochemical and pharmacological studies on *Ocimum basilicum* Linn – a review. *International Journal of Current Research and Review – Healthcare*, **04**, 73–83.

Block, E. (1985) The chemistry of garlic and onions. *Scientific American*, **252**, 114–119.

Blumenthal, M. (1998) *The Complete German Commission Monographs: Therapeutic Guide to Herbal Medicines*, American Botanical Council, Austin, TX.

Bolkent, S., Yanardag, R., Ozsoy-Sacan, O. and Karabulut-Bulan, O. (2004) Effects of parsley (*Petroselinum crispum*) on the liver of diabetic rats: a morphological and biochemical study. *Phytotherapy Research*, **18**, 996–999.

Bora, K.S. and Sharma, A. (2011) Evaluation of antioxidant and free-radical scavenging potential of *Artemisia absinthium*. *Pharmaceutical Biology*, **49**, 1216–1223.

Bora, K.S., Arora, S. and Shri, R. (2011a) Role of *Ocimum basilicum* L. in prevention of ischemia and reperfusion-induced cerebral damage, and motor dysfunctions in mice brain. *Journal of Ethnopharmacology*, **137**, 1360–1367.

Bora, K.S., Shri, R. and Monga, J. (2011b) Cerebroprotective effect of *Ocimum gratissimum* against focal ischemia and reperfusion-induced cerebral injury. *Pharmaceutical Biology*, **49**, 175–181.

Borbala, R., Istvan, Z., Renata, M. *et al.* (2007) Investigation of cytotoxic activity on human cancer cell lines of arborinine and furanoacridones isolated from *Ruta graveolens*. *Planta Medica*, **73**, 41–48.

Borbala, R., Hohmann, J., Minorics, R. *et al.* (2008) Antitumour properties of acridone alkaloids on a murine lymphoma cell line. *Anticancer Research*, **28**, 2737–2744.

Boroski, M., Giroux, H.J., Sabik, H. *et al.* (2012) Use of oregano extract and oregano essential oil as antioxidants in functional dairy beverage formulations. *LWT – Food Science and Technology*, **47**, 167–174.

Bouzabata, A., Boussaha, F., Casanova, J. and Tomi, F. (2010) Composition and chemical variability of leaf oil of *Myrtus communis* from north-eastern Algeria. *Natural Product Communications*, **5**, 1659–1662.

Boyraz, N. and Özcan, M. (2006) Inhibition of phytopathogenic fungi by essential oil, hydrosol, ground material and extract of summer savory (*Satureja hortensis* L.) growing wild in Turkey. *International Journal of Food Microbiology*, **107**, 238–242.

Bozin, B., Mimica-Dukic, N., Samojlik, I. and Jovin, E. (2007) Antimicrobial and antioxidant properties of rosemary and sage (*Rosmarinus officinalis* L. and *Salvia officinalis* L., Lamiaceae) essential oils. *Journal of Agricultural and Food Chemistry*, **55**, 7879–7885.

Bradesi, P., Tomi, F., Casanova, J. *et al.* (1997) Chemical composition of myrtle leaf essential oil from Corsica (France). *Journal of Essential Oil Research*, **9**, 283–288.

Bradley, P. (2006) *British Herbal Compendium: A Handbook of Scientific Information on Widely Used Plant Drugs*, British Herbal Medicine Association, Bournemouth.

Braga, L.K.A., de Macedo, A.K.C., Cunha, A.A. *et al.* (2011) Potentiation of *in vitro* antibiotic activity by *Ocimum gratissimum* L. *African Journal of Pharmacy and Pharmacology*, **5**, 2145–2149.

Breu, W. and Dorsch, W. (1994) *Allium cepa* L. (onion), chemistry, analysis and pharmacology, in *Economic and Medicinal Plant Research* (eds H. Wagner and N.R. Farnsworth), Academic Press, London, pp. 115–147.

Bruneton, J. (1993) *Pharmacognosie: Phytochimie, Plantes Medicinales*, 2nd edn, Tecnique et Documentation, Lavoisier, Paris.

Bruneton, J. (1999) *Pharmacognosy, Phytochemistry, Medicinal Plants*, 2nd edn, Lavoisier Publishing, Paris.

Burt, S. (2000) Essential oils: their antibacterial properties and potential applications in foods: a review. *International Journal of Food Microbiology*, **94**, 223–253.

Butt, M.S., Sultan, M.T., Butt, M.S. and Iqbal, J. (2009) Garlic: nature's protection against physiological threats. *Critical Reviews in Food Science and Nutrition*, **49**, 538–551.

Cakir, A. (2004) Essential oil and fatty acid composition of the fruits of *Hippophae rhamnoides* L. and *Myrtus communis* L. from Turkey. *Biochemical Systematics and Ecology*, **32**, 809–816.

Calcabrini, A., Stringaro, A., Toccacieli, L. *et al.* (2004) Terpinen-4-ol, the main component of *Melaleuca alternifolia* (tea tree) oil inhibits the *in vitro* growth of human melanoma cells. *Journal of Investigation in Dermatology*, **122**, 349–360.

Çalikoglu, E., Kiralan, M., Bayrak, A. (2009) Effect of direct applications of sage (*Salvia officinalis* L.) leaves on oxidative stability of sunflower oil during accelerated storage. *Journal of Food Quality*, **32**, 566–576.

Cantrell, C.L., Richheimer, S.L., Nicholas, G.M. *et al.* (2005) Seco-hinokiol, a new abietane diterpenoid from *Rosmarinus officinalis*. *Journal of Natural Products*, **68**, 98–100.

Caredda, A., Marongiu, B., Porcedda, S. and Soro, C. (2002) Supercritical carbon dioxide extraction and characterization of *Laurus nobilis* essential oil. *Journal of Agricultural and Food Chemistry*, **50**, 1492–1496.

Carović-Stanko, K., Orlić, S., Politeo, O. *et al.* (2010) Composition and antibacterial activities of essential oils of seven *Ocimum* taxa. *Food Chemistry*, **119**, 196–201.

Carpenter, C.D., O'Neill, T., Picot, N. *et al.* (2012) Anti-mycobacterial natural products from the Canadian medicinal plant *Juniperus communis*. *Journal of Ethnopharmacology*, **143**, 695–700.

Carrubba, A. and Scalenghe, R. (2012) The scent of *Mare Nostrum*: medicinal and aromatic plants in Mediterranean soils. *Journal of the Science of Food and Agriculture*, **92**, 1150–1170.

Casanova, N.A. and Carballo, M.A. (2011) Antigenotoxic activity of watercress extract in an *in vitro* mammalian system using comet assay. *Phytotherapy Research*, **25**, 1743–1746.

Casanova, N.A., Ariagno, J.I., Lopez Nigro, M.M. *et al.* (2012) *In vivo* antigenotoxic activity of watercress juice (*Nasturtium officinale*) against induced DNA damage. *Journal of Applied Toxicology*, **33** (9), 880–885.

Cavanagh, H.M.A. and Wilkinson, J.M. (2002) Biological activities of lavender essential oil. *Phytotherapy Research*, **16**, 301–308.

CBI (2009) *Natural ingredients for cosmetics: The EU market for essential oils for cosmetics*, CBI, The Hague, http://www.cbi.eu/system/files/marketintel/2009_-_Essential_oils_for_cosmetics1.pdf.

CBI (2010) *The natural colours, flavours and thickeners market in the EU*, CBI, The Hague, http://www.cbi.eu/system/files/marketintel/2010_-_Natural_colours_flavours_and_thickeners_-_EU1.pdf.

Celiktas, O.Y., Kocabas, E.E.H., Bedir, E. *et al.* (2007) Antimicrobial activities of methanol extracts and essential oils of *Rosmarinus officinalis*, depending on location and seasonal variations. *Food Chemistry*, **100**, 553–559.

Çetin, B., Özer, H., Cakir, A. *et al.* (2010) Antimicrobial activities of essential oil and hexane extract of florence fennel [*Foeniculum vulgare* var. *azoricum* (Mill.) Thell.] against foodborne microorganisms. *Journal of Medicinal Food*, **13**, 196–204.

Chalchat, J., Garry, R., Michet, A. (1998) Essential oils of myrtle of the Mediterrenean litoral. *Journal of Essential Oil Research*, **10**, 613–617.

Chatterje, A., Sukul, N.C., Laskal, S. and Ghoshmajumdar, S. (1982) Nematicidal principles from two species of Lamiaceae. *Journal of Nematology*, **14**, 118–120.

Chatterjee, M., Verma, P., Maurya, R. and Palit, G. (2011) Evaluation of ethanol leaf extract of *Ocimum sanctum* in experimental models of anxiety and depression. *Pharmaceutical Biology*, **49**, 477–483.

Chatzopoulou, P., Koutsos, T.V., Katsiotis, S.T. (2006) Study of nitrogen fertilization rate of fennel cultivars for essential oil yield and composition. *Journal of Vegetable Science*, **12**, 85–93.

Chaubey, M.K. (2008) Fumigant toxicity of essential oils from some common spices against pulse beetle, *Callosobruchus chinensis* (Coleoptera: Bruchidae). *Journal of Oleo Science*, **57**, 171–179.

Chauhan, R.S., Kaula, M.K., Shahia, A.K. *et al.* (2009) Chemical composition of essential oils in *Mentha spicata* L. accession [IIIM(J)26] from north-west Himalayan region, India. *Industrial Crops and Products*, **29**, 654–656.

Chaves-López, C., Martin-Sánchez, A.M., Fuentes-Zaragoza, E. *et al.* (2012) Role of oregano (*Origanum vulgare*) essential oil as a surface fungus inhibitor on fermented sausages: evaluation of its effect on microbial and physicochemical characteristics. *Journal of Food Protection*, **1**, 104–111.

Chen, Q., Hu, X., Wang, Y. *et al.* (2012) Enrichment of cuminaldehyde and *p*-mentha-1,4-dien-7-al in cumin (*Cuminum cyminum* L.) oil by molecular distillation. *Separation and Purification Technology*, **98**, 136–144.

Cheng, S. and Shao, X. (2011) In vivo antifungal activities of the tea tree oil vapor against *Botrytis cinerea*, in *New Technology of Agricultural Engineering (ICAE), International Conference*, 27–29 May, pp. 949–951.

Chiu, C.C., Huang, C.-Y., Chen, T.-Y. *et al.* (2012) Beneficial effects of *Ocimum gratissimum* aqueous extract on rats with $CCl_4$-induced acute liver injury. *Evidence-Based Complementary and Alternative Medicine*, **2012**, article ID 736752, www.ncbi.nlm.nih.gov/pmc/articles/PMC3390165/.

Choudhury, R.P., Kumar, A. and Garg, A.N. (2006) Analysis of Indian mint (*Mentha spicata*) of essential, trace and toxic elements and its antioxidant behaviour. *Journal of Pharmaceutical and Biomedical Analysis*, **41**, 825–832.

Christaki, E., Bonos, E., Giannenas, I. and Florou-Paneri, P. (2012) Aromatic plants as a source of bioactive compounds. *Agriculture*, **2**, 228–243.

Chun, S.K., Vatterm, D.A., Lin, Y.T. and Shetty, K. (2005) Phenolic antioxidants from clonal oregano (*Origanum vulgare*) with antimicrobial activity against *Helicobacter pylori*. *Process Biochemistry*, **40**, 809–816.

Chung, F.L., Morse, M.A., Eklind, K.I. and Lewis, J. (1992) Quantitation of human uptake of the anticarcinogen phenethyl isothiocyanate after a watercress meal. *Cancer Epidemiology Biomarkers and Prevention*, **1**, 383–388.

Cleff, M.B., Meinerz, A.R., Xavier, M. *et al.* (2010) *In vitro* activity of *Origanum vulgare* essential oil against *Candida* species. *Brazilian Journal of Microbiology*, **41** (1) Jan/Mar.

Colvin, W.I. and Gliessman, S.R. (2011) Effects of fennel (*Foeniculum vulgare* L.) interference on germination of introduced and native plant species. *Allelopathy Journal*, **28**, 41–51.

Conforti, F., Statti, G., Uzunov, D. and Menichini, F. (2006) Comparative chemical composition and antioxidant activities of wild and cultivated *Laurus nobilis* L. leaves and *Foeniculum vulgare* subsp. *piperitum* (Ucria) Coutinho seeds. *Biological and Pharmaceutical Bulletin*, **29**, 2056–2064.

Consroe, P. and Fish, B.S. (1981) Rabbit behavior model of marijuana psychoactivity in humans. *Medical Hypotheses*, **7**, 1079–1090.

Conti, B., Canale, A., Bertoli, A. *et al.* (2010) Essential oil composition and larvicidal activity of six Mediterranean aromatic plants against the mosquito *Aedes albopictus* (Diptera: Culicidae). *Parasitology Research*, **107**, 1455–1461.

Corticchiato, M., Tomi, F., Bernardin, A.F. and Casanova, J. (1998) Composition and infraspecific variability of essential oil from *Thymus herba barona* Lois. *Biochemical Systematics and Ecology*, **26**, 915–932.

Corzo-Martínez, M., Corzo, N. and Villamiel, M. (2007) Biological properties of onions and garlic. *Trends in Food Science & Technology*, **18**, 609–625.

Dadalioglu, I. and Evrendilek, G. (2004) Chemical compositions and antibacterial effects of essential oils of Turkish oregano (*Origanum minutiflorum*), bay (*Laurus nobilis*), Spanish lavender (*Lavandula stoechas* L.) and fennel (*Foeniculum vulgare*) on common food-borne pathogens. *Journal of Agricultural and Food Chemistry*, **52**, 8255–8260.

Daferer, D.J., Ziogas, B.N. and Polissiou, M.G. (2000) GC–MS analysis of essential oils from some Greek aromatic plants and their fungitoxicity on *Penicillium digitatum*. *Journal of Agricultural and Food Chemistry*, **48**, 2576–2581.

Dallmeier, K. and Carlini, E.A. (1981) Anesthetic, hypothermic, myorelaxant and anticonvulsant effects of synthetic eugenol derivatives and natural analogues. *Pharmacology*, **22**, 113–127.

Dambolena, J.S., Zunino, M.P., López, A.G. *et al.* (2010) Essential oils composition of *Ocimum basilicum* L. and *Ocimum gratissimum* L. from Kenya and their inhibitory effects on growth and fumonisin production by *Fusarium verticillioides*. *Innovative Food Science & Emerging Technologies*, **11**, 410–414.

Daniel, Z. and Maria, H. (2000) *Domestication of Plants in the Old World*, 3rd edn, Oxford University Press, Oxford.

Dasgupta, T., Rao, A.R. and Yadava, P.K. (2004) Chemomodulatory efficacy of basil leaf (*Ocimum basilicum*) on drug metabolizing and antioxidant enzymes, and on carcinogen-induced skin and forestomach papillomagenesis. *Phytomedicine*, **11**, 139–151.

De Azerêdo, G.A., de Figueiredo, R.C.B.Q, de Souza, E.L. and Stamford, T.L.M. (2012) Changes in *Listeria monocytogenes* induced by *Origanum vulgare* L. and *Rosmarinus officinalis* L. essential oils alone and combined at subinhibitory amounts. *Journal of Food Safety*, **32**, 226–235.

De Barros, J.C., da Conceição, M., Neto, N.J.G. *et al.* (2009) Interference of *Origanum vulgare* L. essential oil on the growth and some physiological characteristics of *Staphylococcus aureus* strains isolated from foods. *LWT - Food Science and Technology*, **42**, 1139–1143.

De Campos, K.E., Coelho Balbi, A.P. and de Freitas Alves, M.J. (2009) Diuretic and hipotensive activity of aqueous extract of parsley seeds (*Petroselinum sativum* Hoffm.) in rats. *Brazilian Journal of Pharmacognosy*, **19**, 41–45.

De Feo, V., De Simone, F. and Senatore, F. (2002) Potential allelochemicals from the essential oil of *Ruta graveolens*. *Phytochemistry*, **61**, 573–578.

De Martino, L., de Feo, V., Formisano, C. *et al.* (2009) Chemical composition and antimicrobial activity of the essential oils from three chemotypes of *Origanum vulgare* L. ssp. *hirtum* (Link) Ietswaart growing wild in Campania (southern Italy). *Molecules*, **14**, 2735–2746.

Deans, S.G. and Svoboda, K.P. (1989) Antibacterial activity of summer savory (*Satureja hortensis* L.) essential oil and its constituents. *Journal of Horticultural Science*, **64**, 205–210.

Deans, S.G. and Svoboda, K.P. (1990) The antimicrobial proprieties of marjoram (*Origanum majorana* L.) volatile oil. *Flavour and Fragrance Journal*, **5**, 187–190.

Deeb, S.J., El-Baba, C.O., Hassan, S.B. *et al.* (2011) Sage components enhance cell death through nuclear factor kappa-B signaling. *Frontiers in Bioscience*, **3**, 410–420.

Del Baño, M.J., Lorente, J., Castillo, J. *et al.* (2003) Phenolic diterpenes, flavones, and rosmarinic acid distribution during the development of leaves, flowers, stems, and roots of *Rosmarinus officinalis*. Antioxidant activity. *Journal of Agricultural and Food Chemistry*, **51**, 4247–4253.

Delamare, A.P.L., Moschen-Pistorello, I.T., Artico, L. *et al.* (2007) Antibacterial activity of the essential oils of *Salvia officinalis* L. and *Salvia triloba* L. cultivated in South Brazil. *Food Chemistry*, **100**, 603–608.

Deruaz, D., Reynaud, J. and Raynaud, J. (1993) Evaluation of the molluscicidal properties of *Myrtus communis* L. (*Myrtaceae*). *Phytotherapy Research*, **7**, 428–430.

Derwich, E., Benziane, Z. and Boukir, A. (2009) Chemical composition and antibacterial activity of leaves essential oil of *Laurus nobilis* from Morocco. *Australian Journal of Basic and Applied Sciences*, **3**, 3818–3824.

Diaz, A.M. and Abeger, A. (1986) Study of the polyphenolic compounds present in alcoholic extracts of *Myrtus communis* L. seeds. *Anales de la Real Academia Nacional de Farmacia*, **52**, 541–546.

Diaz-Maroto, M.C., Díaz-Maroto Hidalgo, I.J., Sánchez-Palomo, E. and Pérez-Coello, M.S. (2005) Volatile components and key odorants of fennel (*F. vulgare* Mill.) and thyme (*Thymus vulgaris* L.) oil extracts obtained by simultaneous distillation-extraction and supercritical fluid extraction. *Journal of Agricultural and Food Chemistry*, **53**, 5385–5389.

Diaz-Maroto, M.C., Rez-Coello, M.S.P. and Cabezudo, M.D. (2002) Effect of drying method on the volatiles in bay leaf (*Laurus nobilis* L.). *Journal of Agricultural and Food Chemistry*, **50**, 4520–4524.

Dikbas, N., Kotan, R., Dadasoglu, F. and Sahin, F. (2008) Control of *Aspergillus flavus* with essential oil and methanol extract of *Satureja hortensis*. *International Journal of Food Microbiology*, **124**, 179–182.

Di Leo Lira, P., Retta, D., Tkacik, E. *et al.* (2009) Essential oil and by-products of distillation of bay leaves (*Laurus nobilis* L.) from Argentina. *Industrial Crops and Products*, **30**, 259–264

Dkhil, M.A., Abdel-Baki, A.S., Wunderlich, F. *et al.* (2011) Anticoccidial and antiinflammatory activity of garlic in murine *Eimeria papillata* infections. *Veterinary Parasitology*, **175**, 66–72.

Dob, T., Dahmane, D., Benabdelkader, T. and Chelghoum, C. (2006) Studies on the essential oil composition and antimicrobial activity of *Thymus algeriensis* Boiss. *International Journal of Aromatherapy*, **16**, 95–100.

Dorman, D.H.J., Koşar, M., Kahlos, K. *et al.* (2003) Antioxidant properties and composition of aqueous extracts from *Mentha* species, hybrids, varieties, and cultivars. *Journal of Agricultural and Food Chemistry*, **51**, 4563–4569.

Du, W.-X., Olsen, C.W., Avena-Bustillos, R.J. *et al.* (2008a) Antibacterial activity against *E. coli* O157:H7, physical properties, and storage stability of novel carvacrol-containing edible tomato films. *Journal of Food Science*, **73**, 378–83.

Du, W.-X., Olsen, C.W., Avena-Bustillos, R.J. *et al.* (2008b) Storage stability and antimicrobial activity against *Escherichia coli* O157:H7 of carvacrol in edible apple films prepared by two different casting methods. *Journal of Agricultural and Food Chemistry*, **56**, 3082–3088.

Du, W.-X., Olsen, C.W., Avena-Bustillos, R.J. *et al.* (2009) Antibacterial effects of allspice, garlic, and oregano essential oils in tomato films determined by overlay and vapor-phase methods. *Journal of Food Science*, **74**, 390–397.

Duke, J.A., Godwin, M.J.B. and Ottesen, A.R. (2009) *Medicinal Plants of Latin America*, Taylor & Francis, London.

Dymock, W., Warden, C.J.H. and Hooper, D. (2005) *Pharmacographia Indica: A History of the Principal Drugs*, Shastri Book, New Delhi.

Easton, C.D. and Jacob, M.V. (2009) Optical characterisation of radio frequency plasma polymerised *Lavandula angustifolia* essential oil thin films. *Thin Solid Films*, **517**, 4402–4407.

Eddouks, M., Maghrani, M., Lemhadri, A. *et al.* (2002) Ethnopharmacological survey of medicinal plants used for the treatment of diabetes mellitus, hypertension and cardiac diseases in the south-east region of Morocco (Tafilalet). *Journal of Ethnopharmacology*, **82**, 97–103.

EDQM (2004) *European Pharmacopoeia*, 5th edn, European Directorate for the Quality of Medicines & HealthCare, Strasbourg.

EDQM (2007) *European Pharmacopoeia*, 6th edn, European Directorate for the Quality of Medicines & HealthCare, Strasbourg.

Edris, A.E., Shalaby, A.S. and Fadel, H.M. (2009) Effect of organic agriculture practices on the volatile flavor components of some essential oil plants growing in Egypt: III. *Thymus vulgaris* L. essential oil. *Journal of Essential Oil-Bearing Plants*, **12**, 319–326.

EEC (1988) European Council. Council Directive No. 88/388 on the approximation of the laws of the Member States relating to flavorings for use in foodstuff and to source materials for their production. *Official Journal of European Community*, **L 184**, 61–66.

Effraim, K., Jacks, T.W. and Sodipo, O.A. (2001) Histopathological studies on the toxicity of *Ocimum gratissimum* leave extract on some organs of rabbit. *African Journal of Biomedical Research*, **6**, 21–25.

Eidi, A. and Eidi, M. (2009) Antidiabetic effects of sage (*Salvia officinalis* L.) leaves in normal and streptozotocin-induced diabetic rats. *Diabetes & Metabolic Syndrome: Clinical Research & Reviews*, **3**, 40–44.

El-Ashmawy, I.M., El-Nahas, A.F. and Salama, O.M. (2005) Protective effect of volatile oils, alcoholic and aqueous extracts of *Origanum majorana* on lead acetate toxicity in mice. *Basic Clinical Pharmacology Toxicology*, **97**, 238–243.

El-Beshbishy, H.A. and Bahashwan, S.A. (2012) Hypoglycemic effect of basil (*Ocimum basilicum*) aqueous extract is mediated through inhibition of α-glucosidase and α-amylase activities. An *in vitro* study. *Toxicology and Industrial Health*, **28**, 42–50.

El-Nekeety, A.A., Mohamed, S.R., Hathout, A.S. *et al.* (2011) Antioxidant properties of *Thymus vulgaris* oil against aflatoxin-induce oxidative stress in male rats. *Toxicon*, **57**, 984–991.

EMEA (1999) Committee for Veterinary Medicinal Products – *Ruta graveolens* summary report. EMEA/MRL/542/98-final, European Agency for the Evaluation of Medicinal Products, http://www.ema.europa.eu/docs/en_GB/document_library/Maximum_Residue_Limits_-_Report/2009/11/WC500015840.pdf (accessed 21 November 2013).

ESCOP (1997) *Menthae piperitae aetheroleum (Peppermint Oil)*, ESCOP Monographs, European Scientific Cooperative on Phytotherapy, Exeter.

EU-EC (2008) Regulation (EC) No. 1334/2008 of the European Parliament and of the Council of 16 December 2008 on flavourings and certain food ingredients with flavouring properties for use in and on foods and amending Council Regulation (EEC) No. 1601/91, Regulations (EC) No. 2232/96 and (EC) No. 110/2008 and Directive 2000/13/EC. *Official Journal of European Commission*, **L 354**, 34–50.

Evandri, M.L., Battinelli, L., Daniele, C. *et al.* (2005) The antimutagenic activity of *Lavandula angustifolia* (lavender) essential oil in the bacterial reverse mutation assay. *Food and Chemical Toxicology*, **43**, 1381–1387.

Ezzeddine, N.B., Abdelkefi, M.M., Ben-Aissa, R. and Chaabouni, M.M. (2001) Antibacterial screening of *Origanum majorana* L. oil from Tunisia. *Journal of Essential Oil Research*, **13**, 295–297.

Fachini-Queiroz, F.C., Kummer, R., Estevão-Silva, C.F. *et al.* (2012) Effects of thymol and carvacrol, constituents of *Thymus vulgaris* L. essential oil, on the inflammatory response. *Evidence-Based Complementary and Alternative Medicine*, **2012**, 657026.

Faid, M., Bakhy, K., Anchad, M. *et al.* (1995) Physicochemical and microbiological characterizations and preservation with sorbic acid and cinnamon. *Journal of Food Products*, **58**, 547–550.

Farivar, T.N., Fard, A.H.M., Zahedani, S.S. *et al.* (2006) Anti-tubercolosis effect of *Ocimum sanctum* extracts in *in vitro* and macrophage culture. *Journal of Medical Science*, **6**, 348–351.

Fejes, S., Blázovics, A., Lemberkovics, E. *et al.* (2000) Free radical scavenging and membrane protective effects of methanol extracts from *Anthriscus cerefolium* L. (Hoffm.) and *Petroselinum crispum* (Mill.) nym. ex A.W. Hill. *Phytotherapy Research*, **14**, 362–365.

Fialová, S., Tekel'ová, D., Mrlianová, M. and Grančai, D. (2008) The determination of phenolics compounds and antioxidant activity of mints and balms cultivated in Slovakia. *Acta Facultatis Pharmaceuticae Universitatis Comenianae*, **LV**, 96–102.

Field, T., Field, T., Cullen, C. *et al.* (2008) Lavender bath oil reduces stress and crying and enhances sleep in very young infants. *Early Human Development*, **84**, 399–401.

Figueiredo, A.C., Barroso, J.G., Pedro, L.G. and Scheffer, J.J.C. (2008) Factors affecting secondary metabolite production in plants: volatile components and essential oils. *Flavour and Fragrance Journal*, **23**, 213–226.

Filipowicz, N., Kamiński, M., Kurlenda, J. *et al.* (2003) Antibacterial and antifungal activity of juniper berry oil and its selected components. *Phytotherapy Research*, **17**, 227–231.

Filipowicz, N., Madanecki, P., Gołębiowski, M. *et al.* (2009) HS-SPME/GC analysis reveals the population variability of terpene contents in *Juniperus communis* needles. *Chemical Biodiversity*, **6**, 2290–2301.

Fisher, C. (1992) Phenolic compounds in spices, in *Phenolic Compounds in Food and their Effects on Health I* (eds C.-T. Ho, C.Y. Lee and M.-T. Huang), ACS Symposium Series, vol. 506, ACS Publications, Washington, DC, pp. 118–129.

Fisher, R., Nitzan, N., Chaimovitsh, D. *et al.* (2011) Variation in essential oil composition within individual leaves of sweet basil (*Ocimum basilicum* L.) is more affected by leaf position than by leaf age. *Journal of Agricultural and Food Chemistry*, **59**, 4913–4922.

Fletcher, R.S., Slimmon, T., Kott, L.S. (2010) Environmental factors affecting the accumulation of rosmarinic acid in spearmint (*Mentha spicata* L.) and peppermint (*Mentha piperita* L.). *Open Agriculture Journal*, **4**, 10–16.

Friedman, T., Shalom, A. and Westreich, M. (2006) Self-inflicted garlic burns: our experience and literature review. *International Journal of Dermatology*, **45**, 1161–1163.

Gagandeep, D.S., Dhanalakshmi, S., Mendiz, E. *et al.* (2003) Chemopreventive effects of *Cuminum cyminum* in chemically induced forestomach and uterine cervix tumors in murine model systems. *Nutrition and Cancer*, **47**, 171–180.

Gajula, D., Verghese, M., Boateng, J. *et al.* (2009) Determination of total phenolics, flavonoids and antioxidant and chemopreventive potential of basil (*Ocimum basilicum* L. and *Ocimum tenuiflorum* L.). *International Journal of Cancer Research (USA)*, **5**, 130–143.

Galli, C. (2012) Bioactive components in Mediterranean diets. *Nutrafoods*, **11**, 11–17.

Generalic, I., Skroza, D., Surjak, J. *et al.* (2012) Seasonal variations of phenolic compounds and biological properties in sage (*Salvia officinalis* L.). *Chemical Biodiversity*, **9**, 441–457.

Geneva, M.P., Stancheva, I.V., Boychinova, M.M. *et al.* (2010) Effects of foliar fertilization and arbuscular mycorrhizal colonization on *Salvia officinalis* L. growth, antioxidant capacity, and essential oil composition. *Journal of the Science of Food and Agriculture*, **90**, 696–702.

Ghanem, M.T.M., Radwan, H.M.A., Mahdy, E.-S.M. *et al.* (2012) Phenolic compounds from *Foeniculum vulgare* (subsp. *piperitum*) (Apiaceae) herb and evaluation of hepatoprotective antioxidant activity. *Pharmacognosy Research*, **4**, 104–108.

Ghasemi, E., Raofie, F. and Mashkouri-Najafi, N. (2010) Application of response surface methodology and central composite design for the optimisation of supercritical fluid extraction of essential oils from *Myrtus communis* L. leaves. *Food Chemistry*, **126**, 1449–1453.

Gilbert, K.G. and Cooke, D.T. (2001) Dyes from plants: past usage, present understanding and potential. *Plant Growth Regulation*, **34**, 57–69.

Gill, C.I.R., Haldar, S., Boyd, L.A. *et al.* (2007) Watercress supplementation in diet reduces lymphocyte DNA damage and alters blood antioxidant status in healthy adults. *American Journal of Clinical Nutrition*, **85**, 504–510.

Gordien, A.Y., Gray, A.I., Franzblau, S.G. and Seidel, V. (2009) Antimycobacterial terpenoids from *Juniperus communis* L. (Cuppressaceae). *Journal of Ethnopharmacology*, **126**, 500–505.

Graves, G. (1996) *Medicinal Plants: An Illustration Guide to more than 180 Herbal Plants*, Bracken Books, London.

Grayer, R.J., Bryan, S.E., Veitch, N.C. *et al.* (1996) External flavones in sweet basil, *Ocimum basilicum*, and related taxa. *Phytochemistry*, **43**, 1041–1047.

Greathead, H. (2003) Plants and plant extracts for improving animal productivity. *Proceedings of the Nutrional Society*, **62**, 279–290.

Griffiths, G., Trueman, L., Crowther, T. *et al.* (2002) Onions a global benefit to health. *Phytotherapy Research*, **16**, 603–615.

Gross, M., Lewinsohn, E., Tadmor, Y. *et al.* (2009) The inheritance of volatile phenylpropenes in bitter fennel (*Foeniculum vulgare* Mill. var. *vulgare*, Apiaceae) chemotypes and their distribution within the plant. *Biochemical Systematics and Ecology*, **37**, 308–316.

Grosso, C., Figueiredo, A.C., Burillo, J. *et al.* (2010) Composition and antioxidant activity of *Thymus vulgaris* volatiles: comparison between supercritical fluid extraction and hydrodistillation. *Journal of Separation Science*, **33**, 2211–2218.

Guardia, T., Rotelli, A.E., Juarez, A.O. and Pelzer, L.E. (2001) Anti-inflammatory properties of plant flavonoids. Effects of rutin, quercetin and hesperidin on adjuvant arthritis in rat. *Il Farmaco*, **56**, 683–687.

Gul, P. (1994) Seasonal variation of oil and menthol content in *Mentha arvensis* Linn. *Pakistan Journal of Forestry*, **44**, 16–20.

Gülçın, İ., Oktay, M., Kireçcı, E. and Küfrevıoğlu, Ö.İ. (2003) Screening of antioxidant and antimicrobial activities of anise (*Pimpinella anisum* L.) seed extracts. *Food Chemistry*, **83**, 371–382.

Gulfraz, M., Mehmood, S., Minhas, N. *et al.* (2008) Composition and antimicrobial properties of essential oil of *Foeniculum vulgare*. *African Journal of Biotechnology*, **7**, 4364–4368.

Güllüce, M., Sökmen, M., Daferera, D. *et al.* (2003) *In vitro* antibacterial, antifungal, and antioxidant activities of the essential oil and methanol extracts of herbal parts and callus cultures of *Satureja hortensis* L. *Journal of Agricultural and Food Chemistry*, **51**, 3958–3965.

Gümral, N., Kumbul, D.D., Aylak, F. *et al.* (2013) *Juniperus communis* Linn oil decreases oxidative stress and increases antioxidant enzymes in the heart of rats administered a diet rich in cholesterol. *Toxicology and Industrial Health*, in press. doi: 10.1177/0748233712469995.

Guo, J.J., Kuo, C.M., Chuang, Y.C. *et al.* (2012a) The effects of garlic-supplemented diets on antibacterial activity against *Streptococcus iniae* and on growth in orange-spotted grouper, *Epinephelus coioides*. *Aquaculture*, **364-365**, 33–38.

Guo, Y., Qin, L., Qiao, C. *et al.* (2012b) Inhibition effects of fennel extract (*Foeniculum vulgare* Mill) on pine wood nematode and the bacteria it carries. *African Journal of Microbiology Research*, **6**, 1837–1843.

Gupta, N. and Porter, T.D. (2001) Garlic and garlic-derived compounds inhibit human squalene monooxygenase. *Journal of Nutrition*, **131**, 1662–1667.

Gupta, V., Mittal, P., Bansal, P. *et al.* (2010) Pharmacological potential of *Matricaria recutita* – a review. *International Journal of Pharmaceutical Sciences and Drug Research*, **2**, 12–16.

Gurib-Fakim, A. (2006) Medicinal plants: traditions of yesterday and drugs of tomorrow. *Molecular. Aspects of Medicine*, **27**, 1–93.

Guvenalp, Z., Turan, M., Sümer, Z. *et al.* (2010) Study on antitumoral activity of some chemical compounds isolated from *Origanum vulgare* ssp. *Planta Medica*, **76**, 271.

Hajhashemi, V., Sadraei, H., Ghannadi, A.R. and Mohseni, M. (2000) Antispasmodic and anti-diarrhoeal effect of *Satureja hortensis* L. essential oil. *Journal of Ethnopharmacology*, **71**, 187–192.

Hajhashemi, V., Ghannadi, A. and Pezeshkian, S.K. (2002) Antinociceptive and anti-inflammatory effects of *Satureja hortensis* L. extract and essential oils. *Journal of Ethnopharmacology*, **82**, 83–87.

Hajhashemi, V., Ghannadi A. and Sharif B. (2003) Anti-inflammatory and analgesic properties of the leaf extracts and essential oil of *Lavandula angustifolia* Mill. *Journal of Ethnopharmacology*, **89**, 67–71.

Hajlaoui, H., Mighri, H., Noumi, E. *et al.* (2010) Chemical composition and biological activities of Tunisian *Cuminum cyminum* L. essential oil: a high effectiveness against *Vibrio* spp. strains. *Food and Chemical Toxicology*, **48**, 2186–2192.

Halambek, J., Berković, K. and Vorkapić-Furač, J. (2010) The influence of *Lavandula angustifolia* L. oil on corrosion of Al-3-Mg alloy. *Corrosion Science*, **52**, 3978–3983.

Hale, I., Kumudini, M., Meepagala, A.O. *et al.* (2004) Phytotoxins from the leaves of *Ruta graveolens* amber. *Journal of Agricultural and Food Chemistry*, **52**, 3345–3349.

Hammer, K.A., Carson, C.F. and Riley, T.V. (1999) Antimicrobial activity of essential oils and other plant extracts. *Journal of Applied Microbiology*, **86**, 985–990.

Harauma, A. and Moriguchi, T. (2006) Aged garlic extract improves blood pressure in spontaneously hypertensive rats more safely than raw garlic. *Journal of Nutrition*, **136**, 769–773.

Harborne, J.B. and Williams, C.A. (2002) Phytochemistry of the genus *Lavandula*, in *Lavender* (ed. M. Lis-Balchin), Taylor & Francis, New York, NY, pp. 86–99.

Hassan, K.A., Deogratius, O., Nyafuono, J.F. *et al.* (2011) Wound healing potential of the ethanolic extracts of *Bidens pilosa* and *Ocimum suave*. *African Journal of Pharmacy and Pharmacology*, **5**, 132–136.

Hecht, S.S. and Shiny, F.L. (2007) Effects of watercress consumption on metabolism. *Cancer Epidemiology Biomarkers and Prevention*, **428**, 877–884.

Hendawy, S.F., Azza, A, El-Din, E. *et al.* (2010) Productivity and oil quality of *Thymus vulgaris* L. under organic fertilization conditions. *Ozean Journal of Applied Sciences*, **3**, 203–216.

Hernandez-Ceruelos, A., Madrigal-Bujaidar, E. and de la Cruz, C. (2002) Inhibitory effect of chamomile essential oil on the sister chromatid exchanges induced by daunorubicin and methyl methanesulfonate in mouse bone marrow. *Toxicology Letters*, **135**, 103–110.

Ho, C.-T., Wang, M., Wei, G.-J. *et al.* (2000) Chemistry and antioxidative factors in rosemary and sage. *Bio-Factors*, **13**, 161–166.

Hoferl, M., Krist, S. and Buchbauer, G. (2006) Chirality influences the effects of linalool on physiological parameters of stress. *Planta Medica*, **72**, 1188–1192.

Holets, F.B., Ueda-Nakamura, T., Filho, B.P.D. *et al.* (2003) Effect of essential oil of *Ocimum gratissimum* on the trypanosomatid *Herpetomonas samuelpessoai*. *Acta Protozoologica*, **42**, 269–276.

Hosseinzadeh, H., Khoshdel, M. and Ghorbani, M. (2011) Antinociceptive, anti-inflammatory effects and acute toxicity of aqueous and ethanolic extracts of *Myrtus communis* L. aerial parts in mice. *Journal of Acupuncture and Meridian Studies*, **4**, 242–247

Huang, L., Abuhamdah, S., Howes, M.J.R. *et al.* (2008) Pharmacological profile of essential oils derived from *Lavandula angustifolia* and *Melissa officinalis* with anti-agitation properties: focus on ligand-gated channels. *Journal of Pharmacy and Pharmacology*, **11**, 1515–1522.

Hudson, R. (1996) The value of lavender for rest and activity in the elderly patient. *Complementary Therapies in Medicine*, **4**, 52–57.

Hussain, A.I., Anwar, F., Hussain Sherazi, S.T. and Przybylski, R. (2008) Chemical composition, antioxidant and antimicrobial activities of basil (*Ocimum basilicum*) essential oils depends on seasonal variations. *Food Chemistry*, **108**, 986–995.

Hussain, A.I., Anwar, F., Shahid, M. *et al.* (2010) Chemical composition, and antioxidant and antimicrobial activities of essential oil of spearmint (*Mentha spicata* L.) from Pakistan. *Journal of Essential Oil Research*, **22**, 78–84.

Iacobellis, N.S., Lo Cantore, P., Capasso, F. and Senatore, F. (2005) Antibacterial activity of *Cuminum cyminum* L. and *Carum carvi* L. essential oils. *Journal of Agricultural and Food Chemistry*, **53**, 57–61.

Inbaneson, S.J., Ravikumar, S. ans Suganthi, P. (2012) *In vitro* antiplasmodial effect of ethanolic extracts of traditional medicinal plant *Ocimum* species against *Plasmodium falciparum*. *Asian Pacific Journal of Tropical Medicine*, **5**, 103–106.

Irkin, R. and Korukluoglu, M. (2007) Control of *Aspergillus niger* with garlic, onion and leek extracts. *African Journal of Biotechnology*, **6**, 384–387.

Isu, N.R. (2005) Antibacterial effect of *Aframomum meleguata*, *Xylopia aethiopicum* and *Ocimum viride*. *Nigerian Journal of Natural Products and Medicine*, **9**, 22–25.

Ivanova, A., Mikhova, B., Najdenski, H. *et al.* (2005) Antimicrobial and cytotoxic activity of *Ruta graveolens*. *Fitoterapia*, **76**, 344–347.

Jamshidi, R., Afzali, Z. and Afzali, D. (2009) Chemical composition of hydrodistillation essential oil of rosemary in different origins in Iran and comparison with other countries. *American-Eurasian Journal of Agriculture and Environment Science*, **5**, 78–81.

Janahmadi, M., Niazi, F., Danyali, S., Kamalinejad, M. (2006) Effects of the fruit essential oil of *Cuminum cyminum* Linn. (*Apiaceae*) on pentylenetetrazol induced epileptiform activity in F1 neurones of *Helixaspersa*. *Journal of Ethnopharmacology*, **104**, 278–282.

Jarrahi, M. (2008) An experimental study of the effects of *Matricaria chamomilla* extract on cutaneous burn wound healing in albino rats. *Natural Product Research: Formerly Natural Product Letters*, **22**, 422–427.

Javanmardi, J., Khalighi, A., Kashi, A. *et al.* (2002) Chemical characterization of basil (*Ocimum basilicum* L.) found in local accessions and used in traditional medicines in Iran. *Journal of Agricultural and Food Chemistry*, **50**, 5878–5883.

Jerkovică, I., Mastelică, J., Milos, M. (2001) The impact of both the season of collection and drying on the volatile constituents of *Origanum vulgare* L. ssp. *hirtum* grown wild in Croatia. *International Journal of Food Science and Technology*, **36**, 649–654.

Jirovetz, L., Buchbauer, G., Stoyanova, A.S. *et al.* (2005) Composition, quality control and antimicrobial activity of the essential oil of cumin (*Cuminum cyminum* L.) seeds from Bulgaria that had been stored for up to 36 years. *International Journal of Food Science and Technology*, **40**, 305–310.

Jordán, M.J., Lax, V., Rota, M.C. *et al.* (2013) Effect of bioclimatic area on the essential oil composition and antibacterial activity of *Rosmarinus officinalis* L. *Food Control*, **30**, 463–468.

Jun, W.J., Han, B.K., Yu, K.W. *et al.* (2001) Antioxidant effects of *Origanum majorana* L. on superoxide anion radicals. *Food Chemistry*, **75**, 439–444.

Juteau, F., Jerkovic, I., Masotti, V. *et al.* (2003) Composition and antimicrobial activity of the essential oil of *Artemisia absinthium* from Croatia and France. *Planta Medica*, **69**, 158–161.

Kaileh, M., Berghe, W.V., Boone, E. *et al.* (2007) Screening of indigenous Palestinian medicinal plants for potential anti-inflammatory and cytotoxic activity. *Journal of Ethnopharmacology*, **113**, 510–516.

Kaliora, A.C. and Kountouri, A.M. (2012) Chemopreventive activity of Mediterranean medicinal plants, in *Cancer Prevention – From Mechanisms to Translational Benefits*, (ed. A.G. Georgakilas), InTech, Rijeka, http://www.intechopen.com/books/cancer-prevention-from-mechanisms-totranslational-benefits/chemopreventive-activity-of-mediterranean-medicinal-plants (accessed 21 November 2013)).

Kanatt, S.R., Chander, R. and Sharma, A. (2007) Antioxidant potential of mint (*Mentha spicata* L.) in radiation-processed lamb meat. *Food Chemistry*, **100**, 451–458.

Kandil, M., Salah, A., Omer, E.S.E. *et al.* (2002) Fruit and essential oil yield of fennel (*Foeniculum vulgare* Mill.) grown with fertilizer sources for organic farming in Egypt. *Landbauforschung Volkenrode*, **52**, 135–139.

Karami-Osboo, R., Khodaverdi, M., Ali-Akbari, F. (2010) Antibacterial effect of effective compounds of *Satureja hortensis* and *Thymus vulgaris* essential oils against *Erwinia amylovora*. *Journal of Agricultural Science and Technology*, **12**, 35–45.

Karousou, R., Grammatikopoulos, G., Lanaras, T. *et al.* (1998) Effects of enhanced UV-B radiation on *Mentha spicata* essential oils. *Phytochemistry*, **49**, 2273–2277.

Kashyap, C.P., Ranjeet, K., Vikrant, A. and Vipin, K. (2011) Therapeutic potency of *Ocimum kilimandscharicum* Guerke – a review. *Global Journal of Pharmacology*, **5**, 191–200.

Kassem, A.H., Mehasen, H.R.A. and Mousa, F.R. (2009) Effect of salinity and compost levels on growth, chemical composition and oil production of rosemary (*Rosmarinus officinalis* L.) plant. *Egyptian Journal of Horticulture*, **36**, 219–230.

Kath, R.K. and Gupta, R.K. (2012) Evaluation of antiulcer activity of *Ocimum sanctum* and its mechanism in animal models. *Internet Journal of Pharmacology*, **10** (1), http://ispub.com/IJPHARM/10/1/14343.

Kaur, G.J. and Arora, D.S. (2009) Antibacterial and phytochemical screening of *Anethum graveolens*, *Foeniculum vulgare* and *Trachyspermum ammi*. *BMC Complementary and Alternative Medicine*, **9**, 30.

Kazemi, M., Rostami, H. and Ameri, A. (2012) The study of composition and antimicrobial properties of essential oils of *Origanum vulgare* and *Rosmarinus officinalis* on human phatogens. *Current Research in Bacteriology*, **5**, 1–12.

Kedzia, B. (1991) Antimicroorganisms activity of oil. Chamomillae and its components. *Herba Polonica*, **37**, 29–38.

Kéita, S.M., Vincent, C., Schmit, J.-P. *et al.* (2001) Efficacy of essential oil of *Ocimum basilicum* L. and *O. gratissimum* L. applied as an insecticidal fumigant and powder to control *Callosobruchus maculatus* (Fab.) [Coleoptera: Bruchidae]. *Journal of Stored Products Research*, **37**, 339–349.

Kerydiyyeh, S.I., Uste, J., Kaouk, I. and Al-Sadi R. (2001) The mechanism underlying the laxative properties of parsley extract. *Phytomedicine*, **8**, 382–388.

Keshavarz, M., Mostafaie, A., Mansouri, K. *et al.* (2010) *In vitro* and *ex vivo* antiangiogenic activity of *salvia officinalis*. *Phytotherapy Research*, **24**, 1526–1531.

Khan, A., Ahmad, A., Manzoor, N. and Khan, L.A. (2010) Antifungal activities of *Ocimum sanctum* essential oil and its lead molecules. *Natural Products Community*, **5**, 345–349.

Khan, A., Bashir, S., Khan, S.R. and Gilani, A.H. (2011a) Antiurolithic activity of *Origanum vulgare* is mediated through multiple pathways. *BMC Complementary and Alternative Medicine*, **11**, 96.

Khan, A., Rehman, N., AlKharfy, K.M. and Gilani, A.-H. (2011b) Antidiarrheal and antispasmodic activities of *Salvia officinalis* are mediated through activation of $K^+$ channels. *Bangladesh Journal of Pharmacology*, **6** (2), http://www.banglajol.info/index.php/BJP/article/view/9156 (accessed 21 November 2013).

Khan, R.U., Naz, S., Nikousefat, Z. *et al.* (2012) *Thymus vulgaris*: alternative to antibiotics in poultry feed. *World's Poultry Science Journal*, **68**, 401–408.

Khanuja, S.P.S. (2007) Employ contract farming to boost area under cultivation for essential oil bearing crops. *Chemical Weekly*, **53** (19), 180–184.

Khattak, S.G., Gilani, S.N. and Ikram, M. (1985) Antipyretic studies on some indigenous Pakistani medicinal plants. *Journal of Ethnopharmacology*, **14**, 45–51.

Khayyal, M.T., el-Ghazaly, M.A., Kenawy, S.A. *et al.* (2001) Antiulcerogenic effect of some gastrointestinally acting plant extracts and their combination. *Arzneimittelforschung*, **51**, 545–53.

Kianbakht, S., Abasi, B., Perham, M. and Hashem Dabaghian, F. (2011) Antihyperlipidemic effects of *Salvia officinalis* L. leaf extract in patients with hyperlipidemia: a randomized double-blind placebo-controlled clinical trial. *Phytotherapy Research*, **25**, 1849–1853.

Kiendrebeogo, M., Coulibaly, A.Y., Nebie, R.C.H. *et al.* (2011) Antiacetylcholinesterase and antioxidant activity of essential oils from six medicinal plants from Burkina Faso. *Revista Brasileira de Farmacognosia*, **21**, 63–69.

Kim, T.H., Kim, H.J., Lee, S.H. and Kim, S.Y. (2012) Potent inhibitory effect of *Foeniculum vulgare* Miller extract on osteoclast differentiation and ovariectomy-induced bone loss. *International Journal of Molecular Medicine*, **29**, 1053–1059.

Kintzios, S.E. (2002) *Oregano: The Genera Origanum and Lippia*, Taylor & Francis, London.

Kizil, S., Arslan, N. and Ipek, A. (2003) A research on adaptation of different cumin (*Cuminum cyminum* L.) lines in Diyarbakır ecological conditions. *Journal of Agricultural Science*, **9**, 340–343.

Klimankova, E., Holadova, K., Hajslova, J. *et al.* (2008) Aroma profiles of five basil (*Ocimum basilicum* L.) cultivars grown under conventional and organic conditions. *Food Chemistry*, **107**, 464–472.

Kobayashi, Y., Nakano, Y., Inayama, K. *et al.* (2003) Dietary intake of the flower extracts of German chamomile (*Matricaria recutita* L.) inhibited compound 48/80-induced itch–scratch responses in mice. *Phytomedicine*, **10**, 657–664.

Koch, C., Reichling, J., Schneele, J. and Schnitzler, P. (2008) Inhibitory effect of essential oils against *Herpes simplex* virus type 2. *Phytomedicine*, **15**, 71–78.

Kojima, Y., Konno, S., Tamamura, S. *et al.* (1984) Interferon inducer a process for producing the same and pharmaceutical composition containing the same. United States Patent, Application no. US 06/282,468.

Kokkini, S., Karousou, R. and Lanaras, T. (1995) Essential oils of spearmint (carvone-rich) plants from the island of Crete (Greece). *Biochemical Systematic and Ecology*, **23**, 423–430.

Komaitis, M.E., Ifanti-Papatragianni, N. and Melissari-Panagiotou, E. (1992) Composition of the essential oils of marjoram (*Origanum majorana* L.). *Food Chemistry*, **45**, 117–118.

Kordali, S., Cakir, A., Mavi, A. *et al.* (2005a) Screening of chemical composition and antifungal and antioxidant activities of the essential oils from three Turkish *Artemisia* species. *Journal of Agricultural and Food Chemistry*, **53**, 1408–1416.

Kordali, S., Kotan, R., Mavi, A. *et al.* (2005b) Determination of the chemical composition and antioxidant activity of the essential oil of *Artemisia dracunculus* and of the antifungal and antibacterial activities of Turkish *Artemisia absinthium*, *A. dracunculus*, *Artemisia santonicum*, and *Artemisia spicigera* essential oils. *Journal of Agricultural and Food Chemistry*, **53**, 9452–9458.

Kordali, S., Aslan, I., Calmasur, O. and Cakir, A. (2006) Toxicity of essential oils isolated from three *Artemisia* species and some of their major components to granary weevils, *Sitophilus granarius* (L) (Coleoptera: Curculionidae). *Industrial Crops Production*, **23**, 162–170.

Kosalec, I., Pepeljnjak, S. and Kustrak, D. (2005) Antifungal activity of fluid extract and essential oil from anise fruits (*Pimpinella anisum* L., Apiaceae). *Acta Pharmacologica*, **55**, 377–385.

Kothari, S.K., Bhattacharya, A.K. and Ramesh, S. (2004) Essential oil yield and quality of methyl eugenol rich *Ocimum tenuiflorum* L.f. (syn. *O. sanctum* L.) grown in south India as influenced by method of harvest. *Journal of Chromatography A*, **1054**, 67–72.

Kreydiyyeh, S.I., Usta, J., Knio, K. *et al.* (2003) Aniseed oil increases glucose absorption and reduces urine output in the rat. *Life Science*, **74**, 663–673.

Kritsidima, M., Newton, T. and Asimakopoulou, K. (2009) The effects of lavender scent on dental patient anxiety levels: a cluster randomised-controlled trial. *Community Dentistry and Oral Epidemiology*, **38**, 83–87.

Kumar, A., Shukla, R., Singh, P. and Dubey, N.K. (2010) Chemical composition, antifungal and antiaflatoxigenic activities of *Ocimum sanctum* L. essential oil and its safety assessment as plant based antimicrobial. *Food and Chemical Toxicology*, **48**, 539–543.

Kumar, P., Mishra, S., Malik, A., Satya, S. (2011) Insecticidal properties of *Mentha* species: a review. *Industrial Crops and Products*, **34**, 802–817.

Kumar, P., Mishra, S., Malik, A. and Satya, S. (2012) Efficacy of *Mentha×piperita* and *Mentha citrata* essential oils against housefly, *Musca domestica* L. *Industrial Crops and Products*, **39**, 106–112.

Kupidlowska, E., Kowalec, M., Sulkowski, G. and Zobel, A.M. (1994) The effect of coumarins on root elongation and ultrastructure of meristematic cell protoplast. *Annals of Botany*, **73**, 525–530.

Kusari, S., Lamshöft, M. and Spiteller, M. (2009) *Aspergillus fumigatus* Fresenius, an endophytic fungus from *Juniperus communis*

L. Horstmann as a novel source of the anticancer pro-drug deoxypodophyllotoxin. *Journal of Applied Microbiology*, **107**, 1019–1030.

Kweka, E.J., Nkya, H.M., Lyaruu, L. *et al.* (2009) Efficacy of *Ocimum kilimandscharicum* plant extracts after four years of storage against *Anopheles gambiae*. *Journal of Cell and Animal Biology*, **3**, 171–174.

Lachenmeier, D.W. and Uebelacker, M. (2010) Risk assessment of thujone in foods and medicines containing sage and wormwood – evidence for a need of regulatory changes? *Regulatory Toxicology and Pharmacology*, **58**, 437–443.

Lachowicz, K.J., Jones, G.P., Briggs, D.R. *et al.* (1998) The synergistic preservative effects of the essential oils of sweet basil (*Ocimum basilicum* L.) against acid-tolerant food microflora. *Letters in Applied Microbiology*, **26**, 209–214.

Lagouri, V., Guldas, M. and Gurbuz, O. (2011) *In vitro* antioxidant/free radical scavenging and antibacterial properties of endemic oregano and thyme extracts from Greece. *Food Science and Biotechnology*, **20**, 1487–1493.

Lahhit, N., Bouyanzer, A., Desjobert, J.-M. *et al.* (2011) Fennel (*Foeniculum vulgare*) essential oil as green corrosion inhibitor of carbon steel in hydrochloric acid solution. *Portugaliae Electrochimica Acta*, **29**, 127–138.

Lahon, K. and Das, S. (2011) Hepatoprotective activity of *Ocimum sanctum* alcoholic leaf extract against paracetamol-induced liver damage in Albino rats. *Pharmacognosy Research*, **3**, 13–18.

Lawrence, B.M. (1992) Progress in essential oils. *Perfumer and Flavorist*, **17**, 39–42.

Lee, C.-H., Lee, S.G. and Lee, H.S. (2010) Acaricidal effects of *Thymus vulgaris* leaf-derived materials and monoterpene alcohols against *Dermatophagoides* spp. *Journal of the Korean Society for Applied Biological Chemistry*, **53**, 170–174.

Lee, H.S. (2004) *p*-Anisaldehyde: acaricidal component of *Pimpinella anisum* seed oil against the house dust mites *Dermatophagoides farina* and *Dermatophagoides pteronyssinus*. *Planta Medica*, **70**, 279–281.

Lee, H.S. (2005) Cuminaldehyde: aldose reductase and α-glucosidase inhibitor derived from *Cuminum cyminum* L. seeds. *Journal of Agricultural and Food Chemistry*, **53**, 2446–2450.

Lee, J. and Scagel, C.F. (2009) Chicoric acid found in basil (*Ocimum basilicum* L.) leaves. *Food Chemistry*, **115**, 650–656.

Lee, J.B., Yamagishi, C., Hayashi, K. and Hayashi T. (2011a) Antiviral and immunostimulating effects of lignin–carbohydrateprotein complexes from *Pimpinella anisum*. *Bioscience, Biotechnology and Biochemistry*, **75**, 459–465.

Lee, M.-J., Chen, H.-M., Tzang, B.-S. *et al.* (2011b) *Ocimum gratissimum* aqueous extract protects H9c2 myocardiac cells from $H_2O_2$-induced cell apoptosis through Akt signaling. *Evidence-Based Complementary and Alternative Medicine*, **2011**, article ID 578060, http://www.ncbi.nlm.nih.gov/pmc/articles/PMC2952314/.

Lee, S.J., Umano, K., Shibamoto, T. and Lee, K.G. (2005) Identification of volatile components in basil (*Ocimum basilicum* L.) and thyme leaves (*Thymus vulgaris* L.) and their antioxidant properties. *Food Chemistry*, **91**, 131–137.

Leite, A.M., Lima, E.O., Souza, E.L. *et al.* (2007) Inhibitory effect of β-pinene, α-pinene and eugenol on the growth of potential infections endocarditis causing Gram-positive bacteria. *Brazilian Journal of Pharmacological Science*, **43**, 121–125.

Lemhadri, A., Zeggwagh, N.-A., Maghrani, M. *et al.* (2004) Anti-hyperglycaemic activity of the aqueous extract of *Origanum vulgare* growing wild in Tafilalet region. *Journal of Ethnopharmacology*, **92**, 251–256.

Leporatti, M.L., Guarrera, P.M. and De Giacomo, M. (1996) Wild and ornamental toxic plants in Latium region (central Italy). *Fitoterapia*, **67**, 513–534.

Lesjak, M., Beara, I., Orcic, D. *et al.* (2010) Anti-infammatory and free radical scavenging activity of *Juniperus communis* L. 1753 var. *communis* leaves and cones extract. *Planta Medica*, **76**, P331.

Li, Y., Craker, L.E. and Potter, T. (1996) Effect of light level on essential oil production of sage (*Salvia officinalis*) and thyme (*Thymus vulgaris*). *Acta Horticulturae*, **426**, 419–426.

Lindley, J. (2001) *Flora Medica*, Ajay Book Service, New Delhi.

Lis-Balchin, M. (2002) Chemical composition of essential oils from different species, hybrids and cultivars of *Lavandula*, in *Lavender* (M. Lis-Balchin), Taylor & Francis, New York, NY, pp. 251–262.

Liu, L. and Yeh, Y.Y. (2000) Inhibition of cholesterol biosynthesis by organosulfur compounds derived from garlic. *Lipids*, **35**, 197–203.

Liu, L. and Yeh, Y.Y. (2002) *S*-Alk(en)yl cysteines of garlic inhibit cholesterol synthesis by deactivating HMG-CoA reductase in cultured rat hepatocytes. *Journal of Nutrition*, **132**, 1129–1134.

Loizzo, M.R., Tundis, R., Menichini, F. *et al.* (2007) Cytotoxic activity of essential oils from Labiatae and Lauraceae families against *in vitro* human tumor models. *Anticancer Research*, **27**, 3293–3300.

Loizzo, M.R., Saab, A.M., Tundis, R. *et al.* (2008) Phytochemical analysis and *in vitro* antiviral activities of the essential oils of seven Lebanon species. *Chemistry and Biodiversity*, **5**, 461–470.

Longo, L. and Vasapollo, G. (2005) Anthocyanins from bay (*Laurus nobilis* L.) berries. *Journal of Agricultural and Food Chemistry*, **53**, 8063–8067.

López, M.G., Sánchez-Mendoza, I.R. and Ochoa-Alejo, N. (1999) Comparative study of volatile components and fatty acids of plants and in vitro cultures of parsley (*Petroselinum crispum* (Mill) Nym ex Hill). *Journal of Agricultural and Food Chemistry*, **47**, 3292–3296.

López-Revuelta, A., Sánchez-Gallego, J.I., Hernández-Hernández, A. *et al.* (2006) Membrane cholesterol contents influence the protective effects of quercetin and rutin in erythrocytes damaged by oxidative stress. *Chemico-Biological Interactions*, **161**, 79–91.

Loutrari, H., Hatziapostolou, M., Skouridou, V. *et al.* (2004) Perillyl alcohol is an angiogenesis inhibitor. *Journal of Pharmacological Experimental Therapy*, **311**, 568–575.

Ložienė, K. and Labokas, J. (2012) Effects of abiotic environmental conditions on amount and enantiomeric composition of α-pinene in *Juniperus communis* L. *Biochemical Systematics and Ecology*, **44**, 36–43.

Lu, T., Cantrell, C.L., Robbs, S.L. *et al.* (1998) Antimycobacterial matricaria esters and lactones from *Astereae* species. *Planta Medica*, **64**, 665–667.

Lubbe, A. and Verpoorte, R. (2011) Cultivation of medicinal and aromatic plants for specialty industrial materials. *Industrial Crops and Products*, **34**, 785–801.

Lun, Z.R., Burri, C., Menzinger, M., Kaminsky, R. (1994) Antiparasitic activity of diallyl trisulfide (Dasuansu) on human and animal pathogenic protozoa (*Trypanosoma* sp., *Entamoeba histolytica* and *Giardia lamblia*) *in vitro*. *Annales de la Societe Belge de Medecine Tropicale*, **74**, 51–59.

Lybarger, J.A., Gallagher, J.S., Pulver, D.W. *et al.* (1982) Occupational asthma induced by inhalation and ingestion of garlic. *Journal of Allergy and Clinical Immunology*, **69**, 448–454.

Máday, E., Szöke, E., Muskáth, Z. and Lemberkovics, E. (1999) A study of the production of essential oils in chamomile hairy root cultures. *European Journal of Drug Metabolism Pharmacokinetics*, **24**, 303–308.

Madsen, H.L., Andersen, L., Christiansen, L. *et al.* (1996) Antioxidative activity of summer savory (*Satureja hortensis* L.) and rosemary (*Rosmarinus officinalis* L.) in minced, cooked pork meat. *Zeitschrift fur Lebensmittel Untersuchung und Forschung*, **203**, 333–338.

Maffei, M., Codignola, A. and Fieschi, M. (1986) Essential oil from *Mentha spicata* (spearmint) cultivated in Italy. *Flavour and Fragrance Journal*, **1**, 105–109.

Mahajan, N., Rawal, S., Verma, M. *et al.* (2012) A phytopharmacological overview on *Ocimum* species with special emphasis on *Ocimum sanctum*. *Biomedicine & Preventive Nutrition*, **3** (2), 185–192.

Mahmood, K., Yaqoob, U. and Bajwa, R. (2008) Antibacterial activity of essential oil of *Ocimum sanctum* L. Mycopathology, **6**, 63–65

Mahmoud, B.S.M., Kawai, Y., Yamazaki, K. *et al.* (2007) Effect of treatment with electrolyzed NaCl solutions and essential oils compounds on the proximate composition, amino acid and fatty acid composition of carp fillets. *Food Chemistry*, **101**, 1492–1498.

Mahmoudi, M, Ebrahimzadeh, M.A., Ansaroudi, F. *et al.* (2009) Antidepressant and antioxidant activities of *Artemisia absinthium* L. at flowering stage. *African Journal of Biotechnology*, **8**, 7170–7175.

Makonnen, E., Debella, A., Zerihun, L. *et al.* (2003) Antipyretic properties of the aqueous and ethanol extracts of the leaves of *Ocimum suave* and *Ocimum lamiifolium* in mice. *Journal of Ethnopharmacology*, **88**, 85–91.

Makri, O. and Kintzios, S. (2007) *Ocimum* sp. (basil): botany, cultivation, pharmaceutical properties, and biotechnology. *Journal of Herbs, Spices & Medicinal Plants*, **13**, 123–150.

Malekpour, A., Dehghani, S., Zahedi, S. and Eskandari, F. (2012) Effects of the hydro-ethanol extract of *Myrtus communis* L. on blood glucose level and histopathological changes in alloxan-induced diabetic rats. *Middle-East Journal of Scientific Research*, **12**, 517–522.

Mancini, D.A.P., Mendoça, R.M.Z., Pinto, J.R. *et al.* (2009) Evaluation of compounds from oregano (*Origanum vulgare*) that inactivate the influenza virus in host animals. *Ciências Exatas e da Terra, Agrárias e Engenharias*, **15** (1), 45–52.

Manderfeld, M.M., Schafer, H.W., Davidson, P.M. and Zottola, E.A. (1997) Isolation and identification of antimicrobial furanocoumarins from parsley. *Journal of Food Protection*, **60**, 72–77.

Mann, C. and Staba, EJ. (1986) The chemistry, pharmacology, and commercial formulations of chamomile, in *Herbs, Spices, and Medicinal Plants: Recent Advances in Botany, Horticulture, and Pharmacology* (eds L.E. Craker and J.E. Simon), Oryx Press, Phoenix, AZ, pp. 235–280.

Manosroi, J., Dhumtanom, P. and Manosroi, A. (2006) Anti-proliferative activity of essential oil extracted from Thai medicinal plants on KB and P388 cell lines. *Cancer Letters*, **235**, 14–120.

Marczal, G., Balogh, M. and Verzar-Petri, G. (1997) Phenol-ether components of diuretic effect in parsley. *Acta Agronomy of Budapest*, **26**, 7–13.

Marino, A., Bellinghieri, V., Nostro, A. *et al.* (2010) *In vitro* effect of branch extracts of *Juniperus* species from Turkey on *Staphylococcus aureus* biofilm. *FEMS Immunology & Medical Microbiology*, **59**, 470–476.

Marotti, M., Piccaglia, R., Giovanelli, E. (1996) Differences in essential oil composition of basil (*Ocimum basilicum* L.) Italian cultivars related to morphological characteristics. *Journal of Agricultural and Food Chemistry*, **14**, 3926–3929.

Martens, D. (1995) Chamomile: the herb and the remedy. *Journal of the Chiropractic Academy of Homeopathy*, **6**, 15–18.

Martín, T., Villaescusa, L., De Sotto, M. *et al.* (1990) Determination of anthocyanin pigments in *Myrtus communis* berries. *Fitoterapia*, **61**, 85.

Martín, T., Rubio, B., Villaescua, L. *et al.* (1999) Polyphenolic compounds from pericarps of *Myrtus communis*. *Pharmaceutical Biology*, **37**, 28–31.

Martins, M.D., Marques, M.M., Bussadori, S.K. *et al.* (2009) Comparative analysis between *Chamomilla recutita* and corticosteroids on wound healing. An *in vitro* and *in vivo* study. *Phytotherapy Research*, **23**, 274–278.

Masresha, B., Makonnen, E. and Debella, A. (2012) *In vivo* anti-inflammatory activities of *Ocimum suave* in mice. *Journal of Ethnopharmacology*, **142**, 201–205.

Mata, A.T., Proença, C., Ferreira, A.R. *et al.* (2007) Antioxidant and antiacetylcholinesterase activities of five plants used as Portuguese food spices. *Food Chemistry*, **103**, 778–786.

Matan, N., Matan, N. and Ketsa, S. (2012) Effect of heat curing on antifungal activities of anise oil and garlic oil against *Aspergillus niger* on rubberwood. *International Biodeterioration and Biodegradation*, **75**, 150–157.

Matasyoh, L.G., Matasyoh, J.C., Wachira, F.N. *et al.* (2007) Chemical composition and antimicrobial activity of the essential oil of *Ocimum gratissimum* L. growing in eastern Kenya. *African Journal of Biotechnology*, **6**, 760–765.

Máthé Jr, I., Oláh, L., Máthé, A. *et al.* (1992) Changes in the essential oil production of *Salvia officinalis* under climatic conditions of the temperature belt. *Planta Medica*, **58**, A680.

Matsuura, H., Ushiroguchi, T., Itakura, Y. *et al.* (1988) A furostanol glycoside from garlic bulbs of *Allium sativum*. *Chemical and Pharmaceutical Bulletin*, **36**, 3659–3663.

McKay, D.L. and Blumberg, J.B. (2006) A review of bioactivity and potential health benefits of chamomile tea (*Matricaria recutita* L.). *Phytotherapy Research*, **20**, 519–530.

Meepagala, K.M., Schrader, K.K., Wedge, D.E. and Duke, O.S. (2005) Algicidal and antifungal compounds from the roots of *Ruta graveolens* and synthesis of their analogs. *Phytochemistry*, **66**, 2689–2695.

Meera, R., Devi, P., Kameswari, B. *et al.* (2009) Antioxidant and hepatoprotective activities of *Ocimum basilicum* Linn. and *Trigonella foenum-graecum* Linn. against $H_2O_2$ and $CCl_4$ induced hepatotoxicity in goat liver. *Indian Journal of Experimental Biology*, **47**, 584–590.

Mehta, R.S., Anwer, M.M. and Aishwath, O.P. (2012) Growth and yield of fennel (*Foeniculum vulgare* Mill.) as influenced by irrigation, nutrient levels and crop geometry. *Journal of Spices and Aromatic Crops*, **20**, 77–80.

Mekawey, A.A.I., Mokhtar, M.M. and Farrag, R.M. (2009) Antitumor and antibacterial activities of [1-(2-ethyl, 6-heptyl) phenol] from *Cuminum cyminum* seeds. *Journal of Applied Sciences Research*, **5**, 1881–1888.

Meneses, R., Ocazionez, R.E., Martinez, J.R. and Stashenko, E.E. (2009) Inhibitory effect of essential oils obtained from plants grown in Colombia on yellow fever virus replication *in vitro*. *Annals of Clinical Microbiology and Antimicrobials*, **8** (1), article 8.

Miguel, G., Cruz, C., Faleiro, M.L. *et al.* (2011) *Salvia officinalis* L. essential oils: effect of hydrodistillation time on the chemical composition, antioxidant and antimicrobial activities. *Natural Product Research: Formerly Natural Product Letters*, **25**, 526–541.

Mikaili, P., Nezhady, M.A.M., Shayegh, J. and Asghari, M.H. (2010) Study of antinociceptive effect of *Thymus vulgaris* and *Foeniculum vulgare* essential oil in mouse. *International Journal of Academic Research*, **2**, 374–376.

Milesi, S., Massot, B., Gontier, E. *et al.* (2001) *Ruta graveolens* L.: a promising species for the production of furanocoumarins. *Plant Science*, **161**, 189–199.

Milhau, G., Valentin, A., Benoit, F. *et al.* (1997) *In vitro* antimicrobial activity of eight essential oils. *Journal of Essential Oil Research*, **9**, 329–333.

Mimica-Dukić, N., Bugarin, D., Grbović, S. *et al.* (2010) Essential oil of *Myrtus communis* L. as a potential antioxidant and antimutagenic agents. *Molecules*, **15**, 2759–2770.

Mirjalili, M.H., Salehi, P., Sonboli, A. and Vala, M.M. (2006) Essential oil variation of *Salvia officinalis* aerial parts during its phenological cycle. *Chemistry of Natural Compounds*, **42**, 19–23.

Miron, T., Rabinkov, A., Mirelman, D. *et al.* (2000) The mode of action of allicin: its ready permeability through phospholipid membranes may contribute to its biological activity. *Biochimica et Biophysica Acta*, **1463**, 20–30.

Miron, T.L., Gazi, I. and Plaza del Moral, M. (2010) Romanian aromatic plants as source of antioxidants. *Innovative Romanian Food Biotechnology*, **6**, 18–24.

Mishra, P. and Mishra, S. (2011) Study of antibacterial activity of *Ocimum sanctum* extract against gram positive and gram negative bacteria. *American Journal of Food Technology*, **6**, 336–341.

Moghtader, M. and Afzali, D. (2009) Study of the antimicrobial proprieties of the oil of rosemary. *American–Eurasian Journal of Agriculture and Environment Science*, **5**, 393–397.

Mohamad, R.H., El-Bastawesy, A.M., Abdel-Monem, M.G. *et al.* (2011) Antioxidant and anticarcinogenic effects of methanolic extract and volatile oil of fennel seeds (*Foeniculum vulgare*). *Journal of Medicinal Food*, **14**, 986–1001.

Mohammed, A., Tanko, Y., Okasha, M.A. *et al.* (2007) Effects of aqueous leaves extract of *Ocimum gratissimum* on blood glucose levels of streptozotocin induced diabetic wistar rats. *African Journal of Biotechnology*, **6**, 2087–2090.

Moon, T., Cavanagh, H.M.A. and Wilkinson, J.M. (2007) Antifungal activity of Australian Grown *Lavandula* spp. essential oils against *Aspergillus nidulans*, *Trichophyton mentagrophytes*, *Leptosphaeria maculans* and *Sclerotinia sclerotiorum*. *Journal of Essential Oil Research*, **19**, 171–175.

Morales, M.R., Charles, D.J., Simon, J.E. (1993) New aromatic lemon basil germplasm, in *New Crops* (eds J. Janick and J.E. Simon), John Wiley & Sons, Inc., New York, NY, pp. 632–635.

Moreira, M.D., Picanco, M.C., de Almeida Barbosa, L.C. *et al.* (2007) Plant compounds insecticide activity against Coleoptera pests of stored products. *Pesquisa Agropecuaria Basileira*, **42**, 7.

Moreno, S., Sana, A.M.O., Gaya, M. (2012) Rosemary compounds as nutraceutical health products, food additive, in *Food Additive* (ed. Y. El-Samragy), InTech, Rijeka, http://www.intechopen.com/books/foodadditive/rosemary-compounds-as-nutraceutical-health-products (accessed 21 November 2013).

Moteki, H., Hibasami, H., Yamada, Y. *et al.* (2002) Specific induction of apoptosis by 1,8-cineole in two human leukemia cell lines, but not a in human stomach cancer cell line. *Oncology Reports*, **9**, 757–760.

Müller-Riebau, F.J., Berger, B.M., Yegen, O. and Cakir, C. (1997) Seasonal variations in the chemical compositions of essential oils of selected aromatic plants growing wild in Turkey. *Journal of Agricultural and Food Chemistry*, **45**, 4821–4825.

Muralikrishnan, G., Pillai, P.K. and Shakeel, F. (2012) Protective effects of *Ocimum sanctum* on lipid peroxidation and antioxidant status in streptozocin-induced diabetic rats. *Natural Product Research: Formerly Natural Product Letters*, **26**, 474–478.

Mwangi, P.W. (2011) The neuropharmacological evaluation of *Ocimum kenyense, Ocimum kilimandscharicum* and *Ocimum masaiense* species found in Kenya. PhD thesis, University of Nairobi, Kenya.

Nadkarni, K.M. (2005) *Indian Plants and Drugs*, 5th edn, Srishti Book Distributers, New Delhi.

Naeini, A., Khosravi, A.R, Chitsaz, M. *et al.* (2009) Anti-*Candida albicans* activity of some Iranian plants used in traditional medicine. *Journal of Mycologia Medica*, **19**, 168–172.

Najafia, F., Khavari-Nejada, R.A. and Siah Alia, M. (2010) The effects of salt stress on certain physiological parameters in summer savory (*Satureja hortensis* L.) plants. *Journal of Stress Physiology and Biochemistry*, **6**, 13–21.

Napoli, E.M., Curcuruto, G. and Ruberto, G. (2010a) Screening the essential oil composition of wild Sicilian fennel. *Biochemical Systematic and Ecology*, **38**, 213–223.

Napoli, E.M., Curcuruto, G. and Ruberto, G. (2010b) Screening of the essential oil composition of wild Sicilian rosemary. *Biochemical Systematics and Ecology*, **38**, 659–670.

Navarrete, P., Toledo, I., Mardones, P. *et al.* (2010) Effect of *Thymus vulgaris* essential oil on intestinal bacterial microbiota of rainbow trout, *Oncorhynchus mykiss* (Walbaum) and bacterial isolates. *Aquaculture Research*, **41**, e667–e678.

Naveena, B.M., Muthukumar, M., Sen, A.R. *et al.* (2006) Improvement of shelf-life of buffalo meat using lactic acid, cove oil and vitamin C during retail display. *Meat Science*, **74**, 409–415.

Nayak, B.S., Raju, S.S. and Rao, A.V. (2007) Wound healing activity of *Matricaria recutita* L. extract. *Journal of Wound Care*, **16**, 298–302.

Neelam, L.D. and Nilofer, S.N. (2010) Preliminary immunomodulatory activity of aqueous and ethanolic leaves extracts of *Ocimum basilicum* Linn in mice. *International Journal of PharmTech Research*, **2**, 1342–1349.

Nell, M., Vötsch, M., Vierheilig, H. *et al.* (2009) Effect of phosphorus uptake on growth and secondary metabolites of garden sage (*Salvia officinalis* L.). *Journal of the Science of Food and Agriculture*, **89**, 1090–1096.

Nemecz, G. (1998) Chamomile. *U.S. Pharmacist*, **23**, 115–116.

Newall, C.A., Anderson, L.A. and Phillipson, J.D. (1996) *Herbal Medicines: A Guide for Health-Care Professionals*, Pharmaceutical Press, London.

Nibret, E. and Wink, M. (2010) Volatile components of four Ethiopian *Artemisia* species extracts and their *in vitro* antitrypanosomal and cytotoxic activities. *Phytomedicine*, **17**, 369–374.

Nickavar, B., Mojab, F. and Dolat-Abadi, R. (2005) Analysis of the essential oils of two *Thymus* species from Iran. *Food Chemistry*, **90**, 609–611.

Nielsen, S.E., Young, J.F., Daneshvar, B. *et al.* (1999) Effect of parsley (*Petroselinum crispum*) intake on urinary apigenin excretion, blood antioxidant enzymes and biomarkers for oxidative stress in human subjects. *British Journal of Nutrition*, **81**, 447–455.

Nin, S., Afraioli, P. and Bosetto, M. (1995) Quantitative determination of some essential oil components of selected *Artemisia absinthium* plants. *Journal of Essential Oil Research*, **7**, 271–277.

Nogueira, J.C., Diniz Mde, F. and Lima, E.O. (2008) *In vitro* antimicrobial activity of plants in acute otitis externa. *Brazilian Journal of Otorhinolaryngology*, **74**, 118–124.

Norman, J. (1990) *The Complete Book on Spices*, Dorling Kindersley, London.

Nostro, A., Cellini, L., Di Bartolomeo, S. *et al.* (2005) Antibacterial effect of plant extracts against *Helicobacter pylori*. *Phytotherapy Research*, **19**, 198–202.

Novak, J., Christina, B., Langbehn, J. *et al.* (2000) Ratios of *cis*- and *trans*-sabinene hydrate in *Origanum majorana* L. and *Origanum midrophyllum* (Bentham) vogel. *Biochemical Systematics and Ecology*, **28**, 697–704.

Nurzyńska-Wierdak, R. (2012) Sweet basil essential oil composition: relationship between cultivar, foliar feeding with nitrogen and oil content. *Journal of Essential Oil Research*, **24**, 217–227.

Obaseiki-Ebor, E.E., Odukoya, K., Telikepalli, H. *et al.* (1993) Antimutagenic activity of extracts of leaves of four common edible vegetable plants in Nigeria (West Africa). *Mutation Research*, **2**, 109–117.

Ogendo, J.O., Deng, A.L., Kostyukovsky, M. *et al.* (2008) Bioactivity of *Ocimum americanum* L. essential oils against four insect pests of stored food commodities. *African Journal of Traditional, Complementary and Alternative Medicines (AJTCAM)*, Abstract of the World Congress on Medicinal and Aromatic plants, Cape Town, November.

Ojianwuna, C.C. and Umoru, P.A. (2010) Effects of *Cymbopogon citratus* (lemon grass) and *Ocimum suave* (wild basil) applied as mixed and individual powders on the eggs laid and emergence of adult *Callosobruchus maculatus* (cowpea bruchids). *African Journal of Agricultural Research*, **5**, 2837–2840.

Oka, Y., Ben-Daniel, B. and Cohen, Y. (2012) Nematicidal activity of the leaf powder and extracts of *Myrtus communis* against the root-knot nematode *Meloidogyne javanica*. *Plant Pathology*, **61**, 1012–1020.

Okoh, O.O., Sadimenko, A.P. and Afolayan, A.J. (2010) Comparative evaluation of the antibacterial activities of the essential oils of *Rosmarinus officinalis* L. obtained by hydrodistillation and solvent free microwave extraction methods. *Food Chemistry*, **120**, 308–312.

Oliva, A., Lahoz, E., Contillo, R. and Aliotta, G. (1999) Fungistatic activity of *Ruta graveolens* extract and its allelochemicals. *Journal of Chemical Ecology*, **25**, 519–526.

Oliva, A., Meepagala, K.M., Wedge, D.E. *et al.* (2003) Natural fungicides from *Ruta graveolens* L. leaves, including a new quinolone alkaloid. *Journal of Agricultural Food Chemistry*, **51**, 890–896.

Oliveira, C.E.V., Stamford, T.L.M., Neto, N.J.G. and Souza, E. L. (2010) Inhibition of *Staphylococcus aureus* in broth and meat broth using synergies of phenolics and organic acids. *International Journal of Food Microbiology*, **137**, 308–311.

Omer, B., Krebs, S., Omer, H. and Noor, T.O. (2007) Steroid-sparing effect of wormwood (*Artemisia absinthium*) in Crohn's disease: a double-blind placebo-controlled study. *Phytomedicine*, **14**, 87–95.

Ono, E.O., Rodrigues, J. D., Barreiro, A.P. and Marques, M.O.M. (2011) Yield and composition of the essential oil of basil on plant growth regulators application. *Journal of Essential Oil Research*, **23**, 17–22.

Orav, A., Raal, A., Arak, E. (2008) Essential oil composition of *Pimpinella anisum* L. fruits from various European countries. *Natural Product Research: Formerly Natural Product Letters*, **22**, 227–232.

Orav, A., Koel, M., Kailas, T. and Müürisepp, M. (2010) Comparative analysis of the composition of essential oils and supercritical carbon dioxide extracts from the berries and needles of Estonian juniper (*Juniperus communis* L.). *Procedia Chemistry*, **2**, 161–167.

Orav, A., Kailas, T. and Müürisepp, M. (2012) Chemical investigation of the essential oil from berries and needles of common juniper (*Juniperus communis* L.) growing wild in Estonia. *Natural Product Research: Formerly Natural Product Letters*, **24**, 1789–1799.

Orhan, N., Orhan, I.E. and Ergun, F. (2011) Insights into cholinesterase inhibitory and antioxidant activities of five *Juniperus* species. *Food and Chemical Toxicology*, **49**, 2305–2312.

Oroojalian, F., Kasra-Kermanshahi, R., Azizi, M. and Bassami, M.R. (2010) Phytochemical composition of the essential oils from three Apiaceae species and their antibacterial effects on food-borne pathogens. *Food Chemistry*, **120**, 765–770.

Osman, Y.A.H., Yaseen, E.M. and Farag, M.M. (2009) Antimicrobial effect of some essential oils mixtures. *Journal of Applied Sciences Research*, **5**, 1265–1276.

Oussalah, M., Caillet, S., Saucier, L., Lacroix, M. (2006) Antimicrobial effects of selected plant essential oils on the growth of a *Pseudomonas putida* strain isolated from meat. *Meat Science*, **73**, 236–244.

Owusu, E.O. (2000) Effect of some Ghanaian plant components on control of two stored-product insect pests of cereals. *Journal of Stored Products Research*, **37**, 85–91.

Ozbek, H., Ugras, S., Dulger, H. *et al.* (2003) Hepatoprotective effect of *Foeniculum vulgare* essential oil. *Fitoterapia*, **74**, 317–319.

Özcan, M.M. and Chalchat, J.C. (2005) Effect of different locations on the chemical composition of essential oils of laurel (*Laurus nobilis* L.) leaves growing wild in Turkey. *Journal of Medicinal Food*, **8**, 408–411.

Özcan, M.M. and Chalchat, J.C. (2006) Chemical composition and antifungal effect of anise (*Pimpinella anisum* L.) fruit oil at ripening stage. *Annals of Microbiology*, **56**, 353–358.

Ozsoy-Sacan, O., Yanardag, R., Orak, H. *et al.* (2006) Effects of parsley (*Petroselinum crispum*) extract versus glibornuride on the liver of streptozotocin-induced diabetic rats. *Journal of Ethnopharmacology*, **104**, 175–181.

Palaniswamy, U.R., McAvoy, R.J. and Bible, B.B. (1996) The effect of light intensity before harvest on phenethyl isothiocyanate concentration in watercress (*Nasturtium officinale* R.Br.) depends on photoperiod. *HortScience*, **31**, 578–585.

Palaniswamy, U.R., McAvoy, R.J., Bible, B.B. and Stuart, J.D. (2003) Ontogenic variations of ascorbic acid and phenethyl isothiocyanate concentrations in watercress (*Nasturtium officinale* R.Br.) leaves. *Journal of Agricultural and Food Chemistry*, **51**, 5504–5509.

Pamplona-Roger, G.D. (2004) *Encyclopedia of Medicinal Plants*, Marpa Artes Graficas, Alfajarín, Spain.

Pan, Z.-H., Wang, Y.-Y., Li, M.-M. *et al.* (2010) Terpenoids from *Savia trijuga*. *Journal of Natural Products*, **73**, 1146–1150.

Pandey, A., Chattopadhyay, P., Banerjee, S. *et al.* (2012) Antitermitic activity of plant essential oils and their major constituents against termite *Odontotermes assamensis* Holmgren (Isoptera: Termitidae) of north east India. *International Biodeterioration and Biodegradation*, **75**, 63–67.

Pank, F., Schneider, E. and Krüger, H. (2003) Possibilities and limitations of estragole content reduction of fennel (*Foeniculum vulgare* Mill.) and its preparations. *Journal of Medicinal and Spice Plants*, **4**, 164–172.

Papachristos, D.P. and Stamopoulos, D.C. (2002) Repellent, toxic and reproduction inhibitory effects of essential oil vapours on *Acanthoscelides obtectus* (Say) (Coleoptera: Bruchidae). *Journal of Stored Products Research*, **38**, 117–128.

Park, T.S. (2011) Use of *Thymus capitatus* extract, *Satureja hortensis* extract, or carvacrol for treating metabolic diseases. United States Patent 2011/0281956 A1, http://www.google.it/patents?hl=it&lr=&vid=USPATAPP13190020&id=glr9AQAAEBAJ&oi=fnd&dq=satureja+hortensis+anticancer+essential+oils&printsec=abstract#v=onepage&q=satureja%20hortensis%20anticancer%20essential%20oils&f=false (accessed 21 November 2013).

Parray, S.A., Bhat, J.U., Ahmad, G. *et al.* (2012) *Ruta graveolens*: from traditional system of medicine to modern pharmacology: an overview. *American Journal of Pharmaceutical Technologies Research*, **2**, 239–252.

Paschapur, M.S., Patil, M.B., Kumar, R. and Patil, R.S. (2009) Evaluation of aqueous extract of leaves of *Ocimum kilimandscharicum* on wound healing activity in albino wistar rats. *International Journal of PharmTech Research*, **1**, 544–550.

Patocka, J. and Plucar, B. (2003) Pharmacology and toxicology of absinth. *Journal of Applied Biomedicine*, **1**, 199–205.

Pattanayak, P., Behera, P., Das, D. and Panda, S.K. (2010) *Ocimum sanctum* Linn. A reservoir plant for therapeutic applications: an overview. *Pharmacognosy Reviews*, **4**, 95–105.

Pavela, R. (2008) Insecticidal properties of several essential oils on the house fly (*Musca domestica* L.). *Phytotherapy Research*, **22**, 274–278.

Pavela, R., Sajfrtova, M., Sovova, H. and Bernet, M. (2008) The insecticidal activity of *Satureja hortensis* extracts obtained by supercritical fluid extraction and traditional extraction tecniques. *Applied Entomology and Zoology*, **43**, 377–382.

Pavela, R., Vrchotová, N. and Tříska, J. (2009) Mosquitocidal activities of thyme oils (*Thymus vulgaris* L.) against *Culex quinquefasciatus* (Diptera: Culicidae). *Parasitology Research*, **105**, 1365–1370.

Peana, A.T., D'Aquila, P.S., Chessa, M.L. *et al.* (2003) (−)-Linalool produces antinociception in two experimental models of pain. *European Journal of Pharmacology*, **403**, 37–41.

Peana, A.T., Rubattu, P., Piga, G.G. *et al.* (2006) Involvement of adenosine A1 and A2A receptors in (−)-linalool-induced antinociception. *Life Science*, **78**, 2471–2474.

Pereira, P.C., Cebola, M.-I. and Bernardo-Gil, M.G. (2009) Evolution of the yields and composition of essential oil from Portuguese myrtle (*Myrtus communis* L.) through the vegetative cycle. *Molecules*, **14**, 3094–3105.

Perry, N.S.L., Bollen, C., Perry, E.K. and Ballard, C. (2003) *Salvia* for dementia therapy: review of pharmacological activity and pilot tolerability clinical trial. *Pharmacology, Biochemistry and Behavior*, **75**, 651–659.

Pessoa, L.M., Morais, S.M., Bevilaqua, C.M.L. and Luciano, J.H.S. (2002) Anthelmintic activity of essential oil of *Ocimum gratissimum* Linn and eugenol against *Haemoachus contortus*. *Veterinary Parasitology*, **109**, 59–63.

Petersen, M. and Simmonds, M.S.J. (2003) Rosmarinic acid. *Phytochemistry*, **62**, 121–125.

Petropoulos, S.A., Olympios, C.M. and Passam, H.C. (2008a) The effect of nitrogen fertilization on plant growth and the nitrate content of leaves and roots of parsley in the Mediterranean region. *Scientia Horticulturae*, **118**, 255–259.

Petropoulos, S.A., Daferera, D, Polissiou, M.G. and Passam, H.C. (2008b) The effect of water deficit stress on the growth, yield and composition of essential oils of parsley. *Scientia Horticulturae*, **115**, 393–397.

Petropoulos, S.A, Daferera, D., Polissiou, M.G. and Passam, H.C. (2009) The effect of salinity on the growth, yield and essential oils of turnip-rooted and leaf parsley cultivated within the Mediterranean region. *Journal of the Science of Food and Agriculture*, **89**, 1534–1542.

Piccaglia, R. (1998) Aromatic plants: a world of flavouring compounds. *Agro Food Industry Hi-Tech*, **9**, 12–15.

Piccaglia, R. and Marotti, M. (1993) Characterization of several aromatic plants grown in northern Italy. *Flavour and Fragrance Journal*, **8**, 115–122.

Piccaglia, R. and Marotti, M. (2001) Characterization of some Italian types of wild fennel (*Foeniculum vulgare* Mill.). *Journal of Agricultural and Food Chemistry*, **49**, 239–244.

Piccaglia, R., Marotti, M., Giovanelli, E. *et al.* (1993) Antibacterial and antioxidant properties of Mediterranean aromatic plants. *Industrial Crops and Products*, **2**, 47–50.

Pierre, S, Crosbie, L. and Duttaroy, AK. (2005) Inhibitory effect of aqueous extracts of some herbs on human platelet aggregation *in vitro*. *Platelets*, **16**, 469–473.

Pimple, B.P., Kadam, P.V. and Patil, M.J. (2012) Ulcer healing properties of different extracts of *Origanum majorana* in streptozotocin–nicotinamide induced diabetic rats. *Asian Pacific Journal of Tropical Disease*, **2**, 312–318.

Pino, J., Borges, P. and Roncal, E. (1993) The chemical composition of bay leaf oil from various origins. *Die Nahrung*, **37**, 592–595.

Pino, J.A., Bayat, F., Marbot, R. and Aguero, J. (2002) Essential oil of chamomile *Chamomilla recutita* (L.) Rausch. from Iran. *Journal of Essential Oil Research*, **14**, 407–408.

Pirzard, A., Alyari, H., Shakiba, M.R. *et al.* (2006) Essential oil content and composition of German chamomile (*Matricaria chamomilla* L.) at different irrigation regimes. *Journal of Agronomy*, **5**, 451–455.

Politeo, O., Juki, M. and Milo, M. (2007) Chemical composition and antioxidant activity of free volatile aglycones from laurel (*Laurus nobilis* L.) compared to its essential oil. *Croatica Chemica Acta*, **80**, 121–126.

Pourgholami M.H., Majzoob S., Javadi M. *et al.* (1999) The seeds essential oil of *Pimpinella anisum* exerts anticonvulsant effects in mice. *Journal of Ethnopharmacology*, **66**, 211–215.

Prabhu, K.S., Lobo, R., Shirwaikar, A.A. and Shirwaikar, A. (2009) *Ocimum gratissimum*: a review of its chemical, pharmacological and ethnomedicinal properties. *Open Complementary Medicine Journal*, **1**, 1–15.

Prakash, M. and Karmegam, N. (2012) *In-vitro* antibacterial activity of certain plant extracts against plant disease causing bacteria isolated from citrus plant. *International Journal of Current Microbiology and Applied Science*, **1**, 1–11.

Preethi, K.C., Kuttan, G. and Kuttan, R. (2006) Anti-tumour activity of *Ruta graveolens* extract. *Asian Pacific Journal of Cancer Prevention*, **7**, 439–443.

Priya, K., Gupta, V.R.M. and Srikanth, K. (2011) Natural sweeteners: a complete review. *Journal of Pharmacy Research*, **4**, 2034–2039.

Pszczola, D.E. (2002) Antimicrobials: setting up additional hurdles to ensure food safety. *Food and Technology*, **56**, 99–107.

Qari, S.H. (2008) *In vitro* evaluation of the anti-mutagenic effect of *Origanum majorana* extract on the meristemetic root cells of *Vicia faba*. *Journal of Taibah University for Science*, **1**, 6–11.

Qu, M.C., Hsu, T.F., Lai, A.C. *et al.* (2012) Pain relief assessment by aromatic essential oil massage on outpatients with primary dysmenorrheal: a randomized, double blind clinical trial. *Obstetric and Gynecology Research*, **38**, 817–822.

Raal, A., Orav, A. and Arak, E. (2012) Essential oil composition of *Foeniculum vulgare* Mill. fruits from pharmacies in different countries. *Natural Product Research: Formerly Natural Product Letters*, **26**, 1173–1178.

Ragab, M.S., Magda, R.A. and Farhat, G.S. (2010) Effect of molukhyia or parsley feeding on carcass characteristic, glutathione peroxidase enzyme activity and meat quality of two broiler strains. *Egyptian Poultry Science*, **30**, 353–389.

Raghav, S.K, Gupta, B., Agrawal, C. *et al.* (2006) Anti-inflammatory effect of *Ruta graveolens* L. in murine macrophage cells. *Journal of Ethnopharmacology*, **104**, 234–239.

Rahman, S., Islam, R., Kamruzzaman, M. *et al.* (2011) *Ocimum sanctum* L.: a review of phytochemical and pharmacological profile. *American Journal of Drug Discovery and Development*, http://docsdrive.com/pdfs/academicjournals/ajdd/0000/29842-29842.pdf.

Rajeshwari, C.U., Abirami, M. and Andallu, B. (2011) *In vitro* and *in vivo* antioxidant potential of aniseeds (*Pimpinella anisum*). *Asian Journal of Experimental Biological Sciences*, **2**, 80–89.

Ram, A., Balachandar, S., Vijayananth, P. and Singh, V.P. (2011) Medicinal plants useful for treating chronic obstructive pulmonary disease (COPD): current status and future perspectives. *Fitoterapia*, **82**, 141–151.

Ramos, A., Azqueta, A., Pereira-Wilson, C. and Collins, A.R. (2010) Polyphenolic compounds from *Salvia* species protect cellular DNA from oxidation and stimulate DNA repair in cultured human cells. *Journal of Agricultural and Food Chemistry*, **58**, 7465–7471.

Ramos, A.A., Pedro, D., Collins, A.R. and Pereira-Wilson, C. (2012) Protection by *Salvia* extracts against oxidative and alkylation damage to DNA in human HCT 15 and CO 115 cells. *Toxicological and Environmental Health*, **75**, 765–775.

Rasouli-Sadaghiani, M.H., Hassani, A., Barin, M. *et al.* (2010) Effects of arbuscular mycorrhizal (AM) fungi on growth, essential oil production and nutrients uptake in basil. *Journal of Medicinal Plants Research*, **4**, 2222–2228.

Razmjoo, K., Heydarizadeh, P. and Sabzalian, M.R. (2008) Effect of salinity and drought stresses on growth parameters and essential oil content of *Matricaria chamomila*. *International Journal of Agriculture and Biology*, **10**, 451–454.

Razzaghi-Abyaneh, M., Shams-Ghahfarokhi, M., Yoshinari, T. *et al.* (2008) Inhibitory effects of *Satureja hortensis* L. essential oil on growth and aflatoxin production by *Aspergillus parasiticus*. *International Journal of Food Microbiology*, **123**, 228–233.

Razzaghi-Abyaneh, M., Shams-Ghahfarokhi, M., Rezaee, M.-B. *et al.* (2009) Chemical composition and antiaflatoxigenic activity of *Carum carvi* L., *Thymus vulgaris* and *Citrus aurantifolia* essential oils. *Food Control*, **20**, 1018–1024.

Rebey, I.B., Jabri-Karoui, I., Hamrouni-Sellami, I. *et al.* (2012) Effect of drought on the biochemical composition and antioxidant activities of cumin (*Cuminum cyminum* L.) seeds. *Industrial Crops and Products*, **36**, 238–245.

Renaud, E.N.C., Charles, D.J. and Simon, J.E. (2001) Essential oil quantity and composition from 10 cultivars of organically grown lavender and lavandin. *Journal of Essential Oil Research*, **13**, 269–273.

Rethy, B., Zupko, I., Minorics, R. *et al.* (2007) Investigation of cytotoxic activity on human cancer cell lines of arborinine and furanoacridones isolated from *Ruta graveolens*. *Planta Medica*, **73**, 41–48.

Reuter, H.D., Koch, H.P. and Lawson, L.D. (1996) Therapeutic effects and applications of garlic and its preparations, in *Garlic, the Science and Therapeutic Application of Allium sativum L. and Related Species* (eds H.P. Koch and L.D. ), Williams and Wilkins, Baltimore, MD, pp. 135–213.

Reuveni, R., Fleischer, A. and Putievsky, E. (1984) Fungistatic activity of essential oils from *Ocimum basilicum* chemotypes. *Journal of Phytopathology*, **110**, 20–22.

Riaz, M., Ashraf, C. and Chaudhary, F. (1989) Studies of the essential oil of the Pakistani *Laurus nobilis* Linn in different seasons. *Pakistan Journal Scientific and Industrial Research*, **32**, 33– 35.

Rice-Evans, C.A., Miller, N.J., Bolwell, P.G. *et al.* (1993) The relative antioxidant activities of plant derived polyphenolic flavonoids. *Free Radical Research*, **22**, 375–383.

Romani, A., Pinelli, P., Mulinacci, N. *et al.* (1999) Identification and quantification of polyphenols in leaves of *Myrtus communis*. *Chromatographia*, **49**, 17–20.

Romeilah, R.M., Fayed, S.A. and Mahmoud, G.I. (2010) Chemical compositions, antiviral and antioxidant activities of seven essential oils. *Journal of Applied Sciences Research*, **6**, 50–62.

Rose, P., Faulkner, K., Williamson, G. and Mithen, R. (2000) 7-Methylsulfinylheptyl and 8-methylsulfinyloctyl isothiocyanates from watercress are potent inducers of phase II enzymes. *Carcinogenesis*, **21**, 1983–1988.

Rose, P., Whiteman, M., Moore, P.K. and Zhu, Y.Z. (2005) Bioactive salk (en)yl cysteine sulfoxide metabolites in the genus *Allium*: the chemistry of potential therapeutic agents. *Natural Product Reports*, **22**, 351–368.

Russo, A., Formisano, C., Rigano, D. *et al.* (2013) Chemical composition and anticancer activity of essential oils of Mediterranean sage (*Salvia officinalis* L.) grown in different environmental conditions. *Food and Chemical Toxicology*, **55**, 42–47.

Saeidi, J., Bozorgi, H., Zendehdel, A. and Mehrzad, J. (2002) Therapeutic effects of aqueous extracts of *Petroselinum sativum* on ethylene glycol-induced kidney calculi in rats. *Urology Journal*, **9**, 361–364.

Saeidi, J., Bozorgi, H., Zendehdel, A. and Mehrzad, J. (2012) The antiurolithiasic and hepatocurative activities of aqueous extracts of *Petroselinum sativum* on ethylene glycol-induced kidney calculi in rats. *Scientific Research and Essays*, **7**, 1577–1583.

Saharkhiz, M.J., Zomorodian, K., Rezaei, M.R. *et al.* (2011) Influence of growth phase on the essential oil composition and antimicrobial activities of *Satureja hortensis*. *Natural Product Community*, **6**, 1173–1178.

Sahin, F., Karaman, I., Güllüce, M. *et al.* (2003) Evaluation of antimicrobial activities of *Satureja hortensis* L. *Journal of Ethnopharmacology*, **87**, 61–65.

Sahraei, H., Ghoshooni, H., Hossein Salimi, S. *et al.* (2002) The effects of fruit essential oil of the *Pimpinella anisum* on acquisition and expression of morphine induced conditioned place preference in mice. *Journal of Ethnopharmacology*, **80**, 43–47.

Said, H.M., Saeed, A., D'Silva, L.A. *et al.* (1996) *Medicinal Herbal: A Textbook for Medical Students and Doctors*, Hamdard Foundation Pakistan, Pakistan, pp. 1–82.

Saim, N. and Meloan, E.C. (1986) Compounds from leaves of bay (*Laurus nobilis* L.) as repellents for *Tribolium castaneum* (Herbst) when added to wheat flour. *Journal of Stored Products Research*, **22**, 141–144.

Sajjadi, S.E. (2006) Analysis of the essential oils of two cultivated basil (*Ocimum basilicum* L.) from Iran. *DARU Journal of Pharmaceutical Sciences*, **14**, 128–130.

Sakakibara, H., Hona, Y., Nagawaga, S. *et al.* (2003) Simultaneous determination of all polyphenols in vegetables, fruits and teas. *Journal of Agricultural and Food Chemistry*, **51**, 571–581.

Salamon, I. (2007) Effect of the internal and external factors on yield and qualitative–quantitative characteristics of chamomile essential oil. *Acta Horticulturae*, **749**, 45–65.

Saleem, M., Alam, A. and Sultana, S. (2000) Attenuation of benzoyl peroxide-mediated cutaneous oxidative stress and hyperproliferative response by the prophylactic treatment of mice with spearmint (*Mentha spicata*). *Food and Chemical Toxicology*, **38**, 939–948.

Salido, S., Altarejos, J., Nogueras, M. *et al.* (2003) Chemical composition and seasonal variations of rosemary oil from southern Spain. *Journal of Essential Oil Research*, **15**, 10–14.

Samara de Lira Mota, K., de Oliveira Pereira, F., Araújo de Oliveira, W. *et al.* (2012) Antifungal activity of *Thymus vulgaris* L. essential oil and its constituent phytochemicals against *Rhizopus oryzae*: interaction with ergosterol. *Molecules*, **17**, 14418–14433.

Sangun, M.K., Aydin, E., Timur, M. *et al.* (2007) Comparison of chemical composition of the essential oil of *Laurus nobilis* L. leaves and fruits from different regions of Hatay, Turkey. *Journal of Environmental Biology*, **28**, 731–733.

Santos-Gomes, P.C. and Fernandes-Ferreira, M. (2001) Organ- and season-dependent variation in the essential oil composition of *Salvia officinalis* L. cultivated at two different sites. *Journal of Agricultural and Food Chemistry*, **49**, 2908–2916.

Saraswat, M., Muthenna, P., Suryanarayana, P. *et al.* (2010) Dietary sources of aldose reductase inhibitors: prospects for alleviating diabetic complications. *Asia Pacific Journal of Clinical Nutrition*, **17**, 558–565.

Satpathi, C.R. and Ghatak, S.S. (1990) Evaluation of the efficacy of some indigenous plant extracts against *Henosepilachna vigintioctopunctata* (Coccinellidae: Coleoptera), a pest of brinjal. *Environmental Ecology*, **8**, 1287–1289.

Savini, I., Arnone, R., Catani, M.V., Avigliano, L. (2009) *Origanum vulgare* induces apoptosis in human colon cancer CaCo2 cells. *Nutrition and Cancer*, **61**, 381–389.

Sayyah, M., Saroukhani, G., Peirovi, A. and Kamalinejad, M. (2001) Analgesic and anti-inflammatory activity of the leaf essential oil of *Laurus nobilis* L. *Phytotheraphy Research*, **17**, 733–736.

Sayyah, M., Peirovi, A. and Kamalinejad, M. (2002) Anti-nociceptive effect of the fruit essential oil of *Cuminum cyminum* L. in rat. *Iranian Biomedical Journal*, **6**, 141–145.

Sayyah, M., Javad-Pour, M. and Ghazi-Khansari, M. (2003) The bacterial endotoxin lipopolysaccharide enhances seizure susceptibility in mice: involvement of proinflammatory factors: nitric oxide and prostaglandins. *Neuroscience*, **122**, 1073–1080.

Sellami, I.H., Maamourib, E., Chaheda, T. *et al.* (2009) Effect of growth stage on the content and composition of the essential oil and phenolic fraction of sweet marjoram (*Origanum majorana* L.). *Industrial Crops and Products*, **30**, 395–402

Sen, P., Maiti, P.C. and Puri, S. (1992) Mechanism of anti-stress activity of *Ocimum sanctum* L., eugenol and *Tinospora malabaria* in experimental animals. *Indian Journal of Experimental Biology*, **30**, 592–596.

Senatore, F. (1996) Influence of harvesting time on yield and composition of the essential oil of a thyme (*Thymus pulegioides* L.) growing wild in Campania (southern Italy). *Journal of Agricultural and Food Chemistry*, **44**, 1327–1332.

Sertel, S., Eichhorn, T., Plinkert, P.K. and Efferth, T. (2011) Anticancer activity of *Salvia officinalis* essential oil against HNSCC cell line (UMSCC1). *HNO*, **59**, 1203–1208 (in German).

Sharabani, H., Izumchenko, E., Wang, Q. *et al.* (2006) Cooperative antitumor effects of vitamin D3 derivatives and rosemary preparations in a mouse model of myeloid leukemia. *International Journal of Cancer*, **118**, 3012–3021.

Sharma, M., Agrawal, S.K., Sharma, P.R. *et al.* (2010) Cytotoxic and apoptotic activity of essential oil from *Ocimum viride* towards COLO 205 cells. *Food and Chemical Toxicology*, **48**, 336–344.

Shehu, A., Umana, U.E., Timbuak, J.A. *et al.* (2012) Effect of aqueous and ethanolic extracts of *Ocimum gratissimum* on the histology of the spleen, hematological indices of wistar rats and its antimicrobial properties. *Asian Journal of Medical Sciences*, **4**, 130–133.

Sheth, N.S., Shah, N.V. and Shah, N.C. (2010) Extraction of mucilage from *Ocimum americanum* Linn & its role as disintegrant in tablets formulation. *Journal of Global Pharma Technology*, **2**, 25–31.

Shetty, K.(1997) Biotechnology to harness the benefits of dietary phenolics: focus on lamiaceae. *Asia Pacific Journal of Clinical Nutrition*, **6**, 162–171.

Shinde, K., Shinde, V., Mahadik, K. and Gibbons, S. (2010) Phytochemical and antibacterial studies on *Ocimum kilimandscharicum*. *Planta Medica*, **76**, P412.

Shipochliev, T. (1981) Uterotonic action of extracts from a group of medicinal plants. *Veterinary Medicin Nauki*, **18**, 94–98.

Shivakumar, S.I., Shahapurkar, A.A., Kalmath, K.V. and Shivakumar, B. (2010) Antiinflammatory activity of fruits of *Cuminum cyminum* Linn. *Der Pharmacia Lettre*, **2**, 22–24.

Shojaii, A. and Fard, M.A. (2012) Review of pharmacological properties and chemical constituents of *Pimpinella anisum*. *ISRN Pharmaceutics*, **2012**, article ID 510795, http://www.ncbi.nlm.nih.gov/pmc/articles/PMC3405664/pdf/ISRN.PHARMACEUTICS2012-510795.pdf (accessed 21 November 2013).

Shukla, H.S. (1989) Antiviral properties of essential oils of *Foeniculum vulgare* and *Pimpinella anisum* L. *Agronomie*, **9**, 277–279.

Siddiqui, B.S., Bhatti, H.A., Begum, S. and Perwaiz, S. (2012) Evaluation of the antimycobacterium activity of the constituents from *Ocimum basilicum* against *Mycobacterium tuberculosis*. *Journal of Ethnopharmacology*, **144**, 220–222.

Singh, B. and Kale, R-K. (2008) Chemomodulatory action of *Foeniculum vulgare* (Fennel) on skin and forestomach papillomagenesis, enzymes associated with xenobiotic metabolism and antioxidant status in murine model system. *Food and Chemical Toxicology*, **46**, 3842–3850.

Singh, D.K. and Porter, T.D. (2006) Inhibition of sterol 4 alpha-methyl oxidase is the principal mechanism by which garlic decreases cholesterol synthesis. *Journal of Nutrition*, **136**, 759–764.

Singh, E., Sharma, S., Dwivedi, J. and Sharma, S. (2012) Diversified potentials of *Ocimum sanctum* Linn (Tulsi): An exhaustive survey. *Journal of Natural Products and Plant Resource*, **2**, 3–48.

Singh, G., Kapoor, I.P.S. and Singh, P. (2011a) Effect of volatile oil and oleoresin of anise on the shelf life of yogurt. *Journal of Food Processing and Preservation*, **35**, 778–783.

Singh, M. and Guleria, N. (2013) Influence of harvesting stage and inorganic and organic fertilizers on yield and oil composition of rosemary (*Rosmarinus officinalis* L.) in a semi-arid tropical climate. *Industrial Crops and Products*, **42**, 37–40.

Singh, M. and Ramesh S. (2000) Effect of irrigation and nitrogen on herbage, oil yield and water-use efficiency in rosemary grown under semi-arid tropical conditions. *Journal of Medicinal Aromatic Plant Science*, **22**, 659–662.

Singh, P., Pandey, A.K., Sonker, N. and Tripathi, N.N. (2011b) Preservation of *Buchnania Lanzan* Spreng. seeds by *Ocimum canum* Sims. essential oil. *Annals of Plant Protection Sciences*, **19**, 407–410.

Singh, S. (1999) Evaluation of gastric anti-ulcer activity of fixed oil of *Ocimum basilicum* Linn. and its possible mechanism of action. *Indian Journal of Experimental Biology*, **37**, 253–257.

Sinha, P. and Saxena, S.K. (1989) Effect of treating tomatoes with leaf extracts of certain plants on the development of fruit rot caused by *Aspergillus niger* in the presence of *Drosophila busckii*. *Journal of Phytological Research*, **2**, 97–102.

Sköld, M., Börje, A., Harambasic, E. and Karlberg, A.-T. (2004) Contact allergens formed on air exposure of linalool. Identification and quantification of primary and secondary oxidation products and the effect on skin sensitization. *Chemical Research in Toxicology*, **17**, 1697–1705.

Snoussi, A., Chaabouni, M.M., Bouzouita, N. and Kachouri, F. (2011) Chemical composition and antioxidant activity of *Myrtus communis* L. floral buds essential oil. *Journal of Essential Oil Research*, **23**, 10–14.

Soković, M-D., Vukojević, J., Marin, P.D. *et al.* (2009) Chemical composition of essential oils of *Thymus* and *Mentha* species and their antifungal activities. *Molecules*, **14**, 238–249.

Soliman, K.M. and Badeaa R.I. (2002) Effect of oil extracted from some medicinal plants on different mycotoxigenic fungi. *Food and Chemical Toxicology*, **40**, 1669–1675.

Solorzano-Santos, F. and Miranda-Novales, M.G. (2012) Essential oils from aromatic herbs as antimicrobial agents. *Current Opinion in Biotechnology*, **23**, 136–141.

Souza, E.L., Stamford, T.L.M. and Lima, E.O. (2006) Sensitivity of spoiling and pathogen-food related bacteria to *Origanum vulgare* L. (*Lamiaceae*) ssential oil. *Brazilian Journal of Microbiology*, **37**, 527–532.

Souza, E.L., Stamford, T.L.M., Lima, E.O. and Trajano, V.N. (2007) Effectiveness of *Origanum vulgare* L. essential oil to inhibit the growth of food spoiling yeasts. *Food Control*, **18**, 409–413.

Souza, E.L., Barros, J.C.B., Conceição, M.L. *et al.* (2009) Combined application of *Origanum vulgare* L. essential oil and acetic acid for controlling the growth of *Staphylococcus aureus* in foods. *Brazilian Journal of Microbiology*, **40**, 387–393.

Soylu, E.M., Soylu, S. and Kurt, S. (2006) Antimicrobial activities of the essential oils of various plants against tomato late blight disease agent *Phytophthora infestans*. *Turkey Mycopathologia*, **161**, 119–128.

Srinivasan, K. (2005) Plant foods in the management of diabetes mellitus: spices as beneficial antidiabetic food adjuncts. *International Journal of Food Sciences and Nutrition*, **56**, 399–414.

Srivastava, S.D., Srivastava, S.K. and Halwe, K. (1998) New coumarins and limonoids of *Ruta graveolens*. *Fitoterapia*, **69**, 7–12.

Stahl-Biskup, E., and Saez, F. (2002) *Thyme*, Taylor & Francis, London.

Stanev, S., Zheljazkov, V. and Janculoff, Y. (1996) Variation of chemical compounds in the essential oil from some native forms of chamomile (*Chamomilla recutita* L.). *Beiträge zur Züchtungsforschung*, **2**, 214–217.

Suanarunsawat, T., Ayutthaya, W.D.N., Songsak, T. *et al.* (2009) Antioxidant activity and lipid-lowering effect of essential oils extracted from *Ocimum sanctum* L. leaves in rats fed with a high cholesterol diet. *Journal of Clinical Biochemistry and Nutrition*, **46**, 52–59.

Sÿkerget, M., Kotnik, P., Hadolin, M. *et al.* (2005) Phenols, proanthocyanidins, flavones and flavonols in some plant materials and their antioxidant activities. *Food Chemistry*, **89**, 191–198.

Sylvestre, M., Pichette, A., Longtin, A. *et al.* (2006) Essential oil analysis and anticancer activity of leaf essential oil of *Croton flavens* L. from Guadeloupe. *Journal of Ethnopharmacology*, **103**, 99–102.

Szalontai, M., Verzar-Petri, G. and Florian, E. (1976) Data on the antifungal effect of the biologically active components of *Matricaria chamomilla* L. *Acta Pharmaceutica Hungarica*, **46**, 232–247.

Szalontai, M., Petri-Verzar, G. and Florian, E. (1977) Contribution to the study of antimycotic effect of biologically active components of *Matricaria chamomilla* L. *Parfuemerie und Kosmetik*, **58**, 121–127.

Szczepanik, M., Zawitowska, B. and Szumny, A. (2012) Insecticidal activities of *Thymus vulgaris* essential oil and its components (thymol and carvacrol) against larvae of lesser mealworm, *Alphitobius diaperinus* Panzer (Coleoptera: Tenebrionidae). *Allelopathy Journal*, **30**, 129–142.

Tabanca, N., Bedir, E., Kirimer, N. *et al.* (2003) Antimicrobial compounds from *Pimpinella* species growing in Turkey. *Planta Medica*, **69**, 933–938.

Tan, P.V., Nyasse, B., Dimo, T. and Mezui, C. (2002) Gastric cytoprotective anti-ulcer effects of the leaf methanol extract of *Ocimum suave* (Lamiaceae) in rats. *Journal of Ethnopharmacology*, **82**, 69–74.

Tansey, M.R. and Appleton, J.A. (1975) Inhibition of fungal growth by garlic extract. *Mycologia*, **67**, 409–413.

Tariku, Y., Hymete, A., Hailu, A. and Rohloff, J. (2011) *In vitro* evaluation of antileishmanial activity and toxicity of essential oils of *Artemisia absinthium* and *Echinops kebericho*. *Chemical Biodiversity*, **8**, 614–623.

Tas, A. (2009) Analgesic effect of *Pimpinella anisum* L. essential oil extract in mice. *Indian Veterinary Journal*, **86**, 145–147.

Tasev, T., Toléva, P. and Balabanova, V. (1969) Neurophysical effect of Bulgarian essential oils from rose, lavender and geranium. *Folia Medica*, **11**, 307–317.

Tattelman, E.M.D. (2005) Health effects of garlic. *American Family Physician*, **72**, 103–106.

Tchoumbougnang, F., Dongmo, P.M.J., Sameza, M.L. *et al.* (2009) Larvicidal activity against *Anopheles gambiae* Giles and chemical composition of essential oils from four plants cultivated in Cameroon. *Biotechnologie, Agronomie, Société et Environnement*, **13**, 77–84.

Teixeiraa, B., Marquesa, A., Ramosa, C. *et al.* (2013) Chemical composition and antibacterial and antioxidant properties of commercial essential oils. *Industrial Crops and Products*, **43**, 587–595.

Telci, I., Demirtas, I., Sahin, A. (2009) Variation in plant properties and essential oil composition of sweet fennel (*Foeniculum vulgare* Mill.) fruits during stages of maturity. *Industrial Crops and Products*, **30**, 126–130.

Telci, I., Kacar, O., Bayram, E. *et al.* (2011) The effect of ecological conditions on yield and quality traits of selected peppermint (*Mentha piperita* L.) clones. *Industrial Crops and Products*, **34**, 1193–1197.

Thaweboon, S. and Thaweboon, B. (2009) *In vitro* antimicrobial activity of *Ocimum americanum* L. essential oil against oral microorganisms. *Southeast Asian Journal of Tropical Medicine and Public Health*, **40**, 1025–1033.

Tirapelli, C.R., De Andrade, C.R., Cassano, A.O. *et al.* (2007) Antispasmodic and relaxant effects of the hydroalcoholic extract of *Pimpinella anisum* L. (Apiaceae) on rat anococcygeus smooth muscle. *Journal of Ethnopharmacology*, **110**, 23–29.

Tisserand, R. (1988) *The Essential Oil Safety Data Manual*, Tisserand Aromatherapy Institute, Brighton.

Tognolini, M., Barocelli, E., Ballabeni, V. *et al.* (2006) Comparative screening of plant essential oils: phenylpropanoid moiety as basic core for antiplatelet activity. *Life Sciences*, **78**, 1419–1432.

Tolba, K.H., Youssef, S.M. and Abd-Elwahab, E.S. (2008) Physicochemical properties and its application of dill (*Anethum graveolens*) and parsley (*Petroselinum sativum*) and their essential oils. *Arab Universities Journal of Agricultural Sciences*, **16**, 379–387.

Tonutti, I. and Liddle, P. (2010) Aromatic plants in alcoholic beverages. A review. *Flavour and Fragrance Journal*, **25**, 341–350.

Tookey, H.L., Van Etten, C.H., Daxenbichler, M.E. (1980) Glucosinolates, in *Toxic Constituents of Plant Stuffs* (ed. I.E. Liener), Academic Press, New York, NY, pp. 103–142.

Trease, T.E. and Evans, W.C. (2002) *Pharmacognosy*, 15th edn, Elsevier India Private Limited, New Delhi.

Triantaphyllou, K., Blekas, G. and Boskou, D. (2001) Antioxidative properties of water extracts obtained from herbs of the species Lamiaceae. *International Journal of Food Science and Nutrition*, **52**, 313–317.

Tripathi, A.K., Upadhyay, S., Bhuiyan, M. and Bhattacharya, P.R. (2009) A review on prospects of essential oils as biopesticide in insect-pest management. *Journal of Pharmacognosy and Phytotherapy*, **1**, 52–63.

Tripathi, P., Tripathi, R., Patel, R.K. and Pancholi, S.S. (2013) Investigation of antimutagenic potential of *Foeniculum vulgare* essential oil on cyclophosphamide induced genotoxicity and oxidative stress in mice. *Drug and Chemical Toxicology*, **36**, 35–41.

Tschirch, A. (1912) *Handbuch der Pharmacognosie*, Tauchnitz, Leipzig.

Tuberoso, C.I.G., Barra, A., Angioni, A. *et al.* (2006) Chemical composition of volatiles in Sardinian myrtle (*Myrtus communis* L.) alcoholic extracts and essential oils. *Journal of Agricultural and Food Chemistry*, **54**, 1420–1426.

Tucakov, J. (1973) Izdavacko preduzece rad: beograd Jugoslavija. Lecenje biljem. *Fitoterapija*, **402**, 535–537.

Tumen, I., Senol, F.S. and Orhan, I.E. (2012) Inhibitory potential of the leaves and berries of *Myrtus communis* L. (myrtle) against enzymes linked to neurodegenerative diseases and their antioxidant actions. *International Journal of Food Sciences and Nutrition*, **63**, 387–392.

Tunçtürk, R. and Murat, T. (2006) Effects of different phosphorus levels on the yield and quality components of cumin (*Cuminum cyminum* L.). *Research Journal of Agriculture and Biological Science*, **2**, 336–340.

Turek, C. and Stintzing, F.C. (2012) Impact of different storage conditions on the quality of selected essential oils. *Food Research International*, **46**, 341–353.

Tyler, V. (1994) *Herbs of Choice: the Therapeutic Use of Phytomedicinals*, Pharmaceutical Product Press, Binghamton, NY.

Ullah, H. and Honermeier, B. (2013) Fruit yield, essential oil concentration and composition of three anise cultivars (*Pimpinella anisum* L.) in relation to sowing date, sowing rate and locations. *Industrial Crops and Products*, **42**, 489–499.

Ulrich, D., Bruchmüller, T., Krüger, H. and Marthe, F. (2011) Sensory characteristics and volatile profiles of parsley (*Petroselinum crispum* [Mill.] Nym.) in correlation to resistance properties against septoria blight (*Septoria petroselini*). *Journal of Agricultural and Food Chemistry*, **59**, 10651–10656.

Umar, A., Imam, G., Yimin, W. *et al.* (2010) Antihypertensive effects of *Ocimum basilicum* L. (OBL) on blood pressure in renovascular hypertensive rats. *Hypertension Research*, **33**, 727–730.

Umar, I.A., Mohammed, A., Dawud, F.A. *et al.* (2012) Anti-diabetic action of the aqueous extract of *Ocimum suave* in alloxan-induced diabetic rats. *African Journal of Biotechnology*, **11**, 9243–9247.

UNCTAD (2006) *World Market in the Spices Trade 2000–2004*, UNCTAD, Geneva.

Upson, T. and Andrews, S. (2004) *The Genus Lavandula*, Royal Botanic Gardens, Kew, London.

Urios, P., Grigorova-Borsos, A.M. and Sternberg, M. (2007) Flavonoids inhibit the formation of the cross-linking AGE pentosidine in collagen incubated with glucose, according to their structure. *European Journal of Nutrition* **46**, 139–146.

Vagi, E., Simándi, B., Suhajda, A. and Hethelyi, E. (2005) Essential oils composition and antimicrobial activity of *Origanum majorana* L. extracts obtained with ethyl alcohol and supercritical carbon dioxide. *Food Research International*, **38**, 51–57.

Vaidyaratnam, P.S. (1980) *Ocimum kilimandscharicum* Guerke, in *Varier's Arya Vaidya Sala – Indian Medicinal Plants: A Compendium of 500 Species*, vol. 4, Orient Longman, Chennai.

Vera, R.R. and Chane-Ming, J. (1999) Chemical composition of the essential oil of marjoram (*Origanum majorana* L.) from Reunion Island. *Food Chemistry*, **66**, 143–145.

Verdian-rizi, M. (2008) Variation in the essential oil composition of *Laurus nobilis L.* of different growth stages cultivated in Iran. *Journal of Basic and Applied Sciences*, **5**, 33–36.

Verma, R.S., Chauhan A., Verma, R.K. and Yadav A.K. (2010) Volatile terpenoid composition of *Origanum vulgare* L. derived from top, middle and lower portions of the plant cultivated in Uttarakhand. *Journal of Essential Oil-Bearing Plants*, **13**, 692–698.

Verma, R.S., Padalia, R.C. and Chauhan, A. (2012) Volatile constituents of *Origanum vulgare* L., thymol chemotype: variability in north India during plant ontogeny. *Natural Product Research*, **26**, 1358–1362.

Vigo, E., Cepeda, A., Gualillo, O. and Perez-Fernandez, R. (2004) *In-vitro* anti-inflammatory effect of *Eucalyptus globulus* and *Thymus vulgaris*: nitric oxide inhibition in J77RA.1 murine macrophages. *Journal of Pharmacy and Pharmacology*, **56**, 257–263.

Viuda-Martos, M., Ruiz-Navajas, Y., Fernández-López, J. and Pérez-Álvarez, J.A. (2008) Antibacterial activity of different essential oils obtained from spices widely used in Mediterranean diet. *International Journal of Food Science and Technology*, **43**, 526–531.

Viuda-Martos, M., Sendra, E., Pérez-Alvarez, J.A. *et al.* (2011) Identification of flavonoid content and chemical composition of the essential oils of Moroccan herbs: myrtle (*Myrtus communis* L.), rockrose (*Cistus ladanifer* L.) and Montpellier cistus (*Cistus monspeliensis* L.). *Journal of Essential Oil Research*, **23**, 1–9.

Vokk, R., Lõugas, T., Mets, K. and Kravets, M. (2011) Dill (*Anethum graveolens* L.) and parsley (*Petroselinum crispum* (Mill.) Fuss) from Estonia: seasonal differences in essential oil composition. *Agronomy Research*, **9**, 515–520.

Vokou, D., Kokkini, S. and Bessiere, J.M. (1993) Geographic variation of Greek oregano (*Origanum vulgare* spp. *hirtum*) essential oils. *Biochemical Systematic and Ecology*, **21**, 287–295.

Vuković-Gačić, B., Nikčević, S., Berić-Bjedov, T. *et al.* (2006) Antimutagenic effect of essential oil of sage (*Salvia officinalis* L.) and its monoterpenes against UV-induced mutations in *Escherichia coli* and *Saccharomyces cerevisiae*. *Food Chemistry and Toxicology*, **44**, 1730–1738.

Waliwitiya, R., Kennedy, C.J. and Lowenberger, C.A. (2009) Larvicidal and oviposition-altering activity of monoterpenoids, *trans*-anithole and rosemary oil to the yellow fever mosquito *Aedes aegypti* (Diptera: Culicidae). *Pest Management Science*, **65**, 241–248.

Wannes, W.A., Mhamdi, B. and Marzouk, B. (2009) Variations in essential oil and fatty acid composition during *Myrtus communis* var. *italica* fruit maturation. *Food Chemistry*, **112**, 621–626.

Wattenberg, I.V. (1991) Inhibition of azoxymethane-induced neoplasia of the large bowel by 3-hydroxy-3,7,11-trimethyl-1,6,10-dodecatriene (nerolidol). *Carcinogenesis*, **12**, 151–152.

Wei, H., Tye, L., Bresnick, E. and Birt, D.F. (1990) Inhibitory effect of apigenin, a plant flavonoid, on epidermal ornithine decarboxylase and skin tumor promotion in mice. *Cancer Research*, **50**, 499–502.

Weisberger, A.S. and Pensky J. (1958) Tumor inhibition by a sulfhydrylblocking agent related to an active principle of garlic (*Allium sativum*). *Cancer Research*, **18**, 1301–1308.

Weiss, N., Ide, N., Abahji, T. *et al.* (2006) Aged garlic extract improves homocysteine-induced endothelial dysfunction in macro- and microcirculation. *Journal of Nutrition*, **136**, 750–754.

Werker, E. (1983) Functional of essential oil secreting glandular hairs in aromatic plants of Lamiacea – a review. *Flavour and Fragrance Journal*, **8**, 249–255.

Woronuk, G., Demissie, Z., Rheault, M. and Mahmoud, S. (2011) Biosynthesis and therapeutic properties of *Lavandula* essential oil constituents. *Planta Medica*, **77**, 7–15.

Wright, C.I, Van-Buren, L., Kroner, C.I. and Koning, M.M.G. (2007) Herbal medicines as diuretics: a review of the scientific evidence. *Journal of Ethnopharmacology*, **114**, 1–31.

Wu, T.S., Shi, L.S., Wang, J.J. *et al.* (2003) Cytotoxic and antiplatelet aggregation principles of *Ruta graveolens*. Journal of the Chinese Chemical Society of Taipei, **50**, 171–178.

Wu, Y.-B., Ni, Z.-Y., Shi, Q.-W. *et al.* (2012) Constituents from *Salvia* species and their biological activities. *Chemical Reviews*, **112**, 5967–6026.

Yadegarinia, D., Gachkar, L., Bagher Rezaei, M. *et al.* (2006) Biochemical activities of Iranian *Mentha piperita* L. and *Myrtus communis* L. essential oils. *Phytochemistry*, **67**, 1249–1255.

Yamamura, S., Ozawa, K., Ohatani, K. *et al.* (1998) Antihistaminic flavones and aliphatic glycosides from *Mentha spicata*. *Phytochemistry*, **48**, 131–136.

Yazdanparast, R. and Shahriyary, L. (2008) Comparative effects of *Artemisia dracunculus*, *Satureja hortensis* and *Origanum majorana* on inhibition of blood platelet adhesion, aggregation and secretion. *Vascular Pharmacology*, **48**, 32–37.

Yesiloglu, Y. and Sit, L. (2012) Antioxidant properties of various solvent extracts from purple basil. *Spectrochimica Acta Part A: Molecular and Biomolecular Spectroscopy*, **95**, 100–106.

Yosr, Z., Hnia, C., Rim, T. and Mohamed, B. (2013) Changes in essential oil composition and phenolic fraction in *Rosmarinus officinalis* L. var. *typicus* Batt. organs during growth and incidence on the antioxidant activity. *Industrial Crops and Products*, **43**, 412–419.

Yu, Y.M. and Tzeng, Y.W. (2009) Carnosic acid and ursolic acid (*Origanum majorana* L.) prevent atherosclerosis caused by obesity: suppression of leptin-induced proliferation in vascular smooth muscle cells. *Atherosclerosis Supplements*, **10** (2), e1472.

Yu, Y.M., Tzeng, Y.W., Chang, W.C. (2010) *Origanum majorana* L. prevents cardiovascular disease caused by obesity: suppression of leptin-induced proliferation in vascular smooth muscle cells. *Atherosclerosis Supplements*, **11** (2), 137.

Yucharoen, R., Anuchapreeda, S. and Tragoolpua, Y. (2011) Anti-herpes simplex virus activity of extracts from the culinary herbs *Ocimum sanctum* L., *Ocimum basilicum* L. and *Ocimum americanum* L. *African Journal of Biotechnology*, **10**, 860–866.

Zafar, M.M., Hamdard, M.E. and Hameed, A. (1990) Screening of *Artemisia absinthium* for antimalarial effects on *Plasmodium berghei* in mice: a preliminary report. *Journal of Ethnopharmacology*, **30**, 223–226.

Zahid, N.Y., Abbasi, N.A., Hafiz, I.A. *et al.* (2012) Antifungal activity of local fennel (*Foeniculum vulgare* Mill) extract to growth responses of some soil diseases. *African Journal of Microbiology Research*, **6**, 46–51.

Zaki, A.A., Elbarawy, A.M. and Darwish, A.S. (2011) Biochemical studies on the effect of *Nasturtium officinalis* plant extract in chickens fed raw soya bean meals. *Australian Journal of Basic and Applied Sciences*, **5**, 755–761.

Zanetti, S., Cannas, S., Molicotti, P. *et al.* (2010) Evaluation of the antimicrobial properties of the essential oil of *Myrtus communis* L. against clinical strains of *Mycobacterium* spp. *Interdisciplinary Perspectives on Infectious Diseases*, **2010**, article ID 931530 www.hindawi.com/journals/ipid/2010/931530/abs/ (accessed 21 November 2013)).

Zargari, A. (1990) *Medicinal Plants*, vol. IV, Tehran University Press, pp. 325–328.

Zeggwagh, N.A., Sulpice, T. and Eddouks, M. (2007) Antihyperglycaemic and hypolipidemic effects of *Ocimum basilicum* aqueous extract in diabetic rats. *American Journal of Pharmacology and Toxicology*, **2**, 123–129.

Zehtab-Salmasi, S., Javanshir, A., Omidbaigi, R. *et al.* (2001) Effects of water supply and sowing date on performance and essential oil production of anise (*Pimpinella anisum* L.). *Acta Agronomica Hungarica*, **49**, 75–81.

Zheljazkov, V.D., Cantrell, C.L., Astatkie, T. and Ebelhar, M.W. (2010) Peppermint productivity and oil composition as a function of nitrogen, growth stage, and harvest time. *Agronomy Journal*, **102**, 124–128.

Zheljazkov, V.D., Astatkie, T., Schlegel, V. (2012) Distillation time changes oregano essential oil yields and composition but not the antioxidant or antimicrobial activities. *HortScience*, **47**, 777–784.

Zoubiri, S. and Baaliouamer, A. (2011) Chemical composition and insecticidal properties of some aromatic herbs essential oils from Algeria. *Food Chemistry*, **126**, 179–182.

Zuzarte, M., Gonçalves, M.J., Cavaleiro, C. *et al.* (2009) Chemical composition and antifungal activity of the essential oils of *Lavandula pedunculata* (Miller) Cav. *Chemistry and Biodiversity*, **6**, 1283–1292.

# 24 'Let Your Food Be Your Medicine': Exotic Fruits and Vegetables as Therapeutic Components for Obesity and Other Metabolic Syndromes

Mohamad Fawzi Mahomoodally

Department of Health Sciences, Faculty of Science, University of Mauritius, Réduit, Mauritius

## 24.1 Introduction

'Let food be thy medicine and medicine be thy food' by Hippocrates, the father of medicines (431 BC), has become a golden reference statement throughout history and has stood the test of time (Cardenas, 2013). Indeed, with the increasing evidence of chronic non-communicable metabolic diseases such as hypertension, cardiovascular problems, obesity and diabetes, there has been a high interest in how foods, particularly fruits, herbs and spices, can exert positive health effects and well-being beyond basic nutritional needs (Mohamed, 2013). Such foods have become known as functional foods.

Although there is no globally accepted standard definition of functional foods, a generally accepted one is: foods that, by virtue of physiologically active food components, provide health benefits beyond basic nutrition (Crowe and Francis, 2013). Functional foods are not pills, capsules or any form of dietary supplement (Mohamed, 2013), but their functionality is attributed to the fact that they are taken as whole foods and are part of the normal food pattern (Crowe and Francis, 2013; Mohamed, 2013). The term 'functional food' itself was first used in Japan, in the 1980s, for food products fortified with special constituents that possess advantageous physiological effects (Siro *et al.*, 2008). Recently, the Academy of Nutrition and Dietetics has clarified its position in relation to the definition of functional food (Crowe and Francis, 2013). Although it was recognized that all foods provide some level of physiological function, the term functional foods is defined as whole foods along with fortified, enriched or enhanced foods that have a potentially beneficial effect on health when consumed as part of a varied diet on a regular basis at effective levels based on significant standards of evidence (Crowe and Francis, 2013). It is also agreed that functional foods may improve the general conditions of the body (e.g. pre- and probiotics), decrease the risk of some diseases (e.g. cholesterol-lowering products), and could even be used for curing some illnesses. It was recognized that there is a demand for these products, as different demographical and epidemiological studies revealed that the medical service of the aging population is rather expensive (Menrad, 2003).

The recent surge in non-communicable diseases, such as cardiovascular diseases, obesity, diabetes mellitus and cancer, and neurodegenerative diseases, such as Alzheimer's disease, has contributed to the shift from conventional medicines to functional foods. The active ingredients found in functional foods are secondary metabolites, such as phenolic compounds, genistein in soy beans or flavonoids in red wine, microorganisms such as lactobacillus behaving as probiotics, fibre and oligosaccharides, omega-3 and 6 fatty acids and essential oils (Croteau *et al.*, 2000).

## 24.2 Obesity, diabetes and metabolic syndromes

Obesity and related metabolic syndromes are major health problems that have reached epidemic proportions worldwide. The impact is such that the World Health Organization (WHO) has coined a new term, globesity, to describe this expanding situation (WHO, 2013). Indeed, obesity and overweight are the fifth leading risk for global deaths, with at least 2.8 million adults, aged 20 years and older, dying each year as a result of excess weight (WHO, 2013). Most overweight or obese persons experience increased morbidity and mortality rates due to excess body fat. Contrary to popular belief, obesity and overweight are not diseases of the rich. In fact, 65% of the world's population lives in countries where overweight and obesity kill more people than underweight, and these include both developed and developing countries (WHO, 2013).

It is now established that being overweight or obese is a major risk factor for non-communicable metabolic diseases, such as cardiovascular diseases, diabetes, musculoskeletal disorders, psychological disorders, complicated pregnancies and certain types of cancer (Dedeli *et al.*, 2011). These health problems create a substantial burden for a nation and need to be addressed. Nevertheless, overweight, obesity and related metabolic syndromes are preventable diseases (Dedeli *et al.*, 2011; WHO, 2013). Indeed, the fundamental cause of excess body weight is the imbalance between energy taken in and energy used (NIH, 1998; WHO, 2003). This is due to the fact that globally there has been an increased intake of energy-dense foods and a decrease in physical activity (WHO, 2013).

*Novel Plant Bioresources: Applications in Food, Medicine and Cosmetics*, First Edition. Edited by Ameenah Gurib-Fakim.

An estimated 41 million people in the USA are postulated to have metabolic syndrome characterized by elevated blood sugar, hyperlipidaemia, hypertension and obesity. The National Cholesterol Education Program has established the current criteria for the metabolic syndrome as three or more of the following: (1) waist circumference >102 cm for men or >88 cm for women; (2) a fasting total serum triacylglycerol concentration of 1.7 mmol/L; (3) a fasting HDL-cholesterol concentration <1.0 mmol/L for men or <1.2 mmol/L for women; (4) a fasting glucose >6.1 mmol/L; or blood pressure >130 mmHg (systolic) and/or 85 mmHg (diastolic) and/or the use of antihypertensive medication.

One metabolic disease that is reaching pandemic proportion is diabetes. There are two types of diabetes: types 1 and 2. Type 1, often called insulin-dependent diabetes mellitus (IDDM), is caused by autoimmune destruction of pancreatic β-cells, leading to a near total deficiency in insulin secretion. IDDM patients require lifelong insulin for the control of blood sugar levels (Mahomoodally, 2012). About 90% of all diabetics are considered to have non-insulin-dependent diabetes mellitus (NIDDM, or tDM). Type 2 diabetes is usually caused by a combination of pancreatic β-cell failure and insulin resistance in target tissues like liver, muscle and fat (Mahomoodally, 2012). The relative proportions of IDDM to NIDDM vary from 15:85 for western populations to 5:95 in developing countries (Bilous and Donelly, 2010).

Type 2 diabetes has increased dramatically in the past two decades, with 1.6 million cases diagnosed each year in the USA (Ahmad and Crandall, 2010). This is closely linked to the emergence of obesity in developed and developing countries. Diabetes prevalence is highest among the elderly and in certain ethnic groups, especially African Americans, Hispanic Americans and Native Americans. People with diabetes have a two- to fourfold increased risk of developing cardiovascular disease, peripheral vascular disease and stroke. These complications account for 65% of mortality from diabetes and, as of 2006, have made diabetes the seventh leading cause of death in the USA (Ahmad and Crandall, 2010). According to a WHO report, type 2 diabetes is characterized by either resistance to the blood glucose-regulating hormone insulin or its relative deficiency (WHO, 2006). Insulin is responsible for glucogenesis following the removal of glucose from the blood by cells of the liver and muscles.

## 24.3 Medicinal food plants against metabolic diseases

Medicinal herbs and food plants, unlike pharmacological drugs, commonly have several chemicals working together catalytically and synergistically to produce a combined effect that surpasses the total activity of the individual constituents (Bagghi, 2001). The combined action of these substances increases the activity of the main, medicinal constituent by speeding up or slowing down its assimilation in the body. Secondary substances from plant origins might increase the stability of the active compound(s) or phytochemicals, minimize the rate of undesired side effects and have an additive, potentiating or antagonistic effect (Mahomoodally, 2012). One metabolic disease where medicinal plants have been comprehensively used is diabetes mellitus, and to a lesser extent obesity.

Various plants and natural products have been assessed and validated for their potential anti-obesity effect both *in vitro* and *in vivo*. There is proof that functional food and their products can exert anti-obesity effects through various mechanisms, such as anti-lipase, an anti-adipogenesis effect or suppressing appetite. Some researchers also suggest that a synergy between different plants gives a better effect in terms of anti-obesity compared with the single plants. Several mechanisms have been proposed for the hypoglycemic and anti-obesity effect of phytochemicals: manipulation of glucose transporters and β-cell regeneration, enhancing insulin-releasing activity, the inhibition of key lipid and carbohydrate hydrolysing and metabolizing enzymes, disruption of adipogenesis and modulation of its factors, or appetite suppression are some of the plethora of targeted approaches to probe the anti-obesity potential of medicinal food plants (Mahomoodally, 2012).

### 24.3.1 Vegetables and spices

#### 24.3.1.1 Fenugreek (*Trigonella foenum-graecum*, Apiaceae)

Seeds of fenugreek have been used as a medicinal agent in folk medicine in the treatment and/or management of diabetes, cellulitis, tuberculosis, hypercholesterolaemia, inflammation and gastrointestinal disturbances. A significant drop in blood glucose with no change in insulin levels in type 2 diabetic subjects was observed when they were given 15 g fenugreek per day (Madar and Arad, 1989; Sharma *et al.*, 1996). Sharma (1986) reported an improvement in plasma glucose response with a reduction in insulin levels when 25 g fenugreek seeds were administered to diabetics for 21 days. Moreover, diabetic symptoms, such as polyuria, polydipsia and polyphagia, were controlled and decreases in urinary glucose and serum cholesterol levels were noted.

Type 1 diabetic patients fed on 25 g fenugreek seed daily led to improved plasma glucose profile, glycosuria and reduced insulin requirement after 8 weeks of treatment. In a long-term trial, 100 g defatted fenugreek seeds fed to either type 1 diabetics (Sharma *et al.*, 1990) or type 2 diabetics (Sharma and Raghuram, 1990) for 10 days reduced fasting blood glucose levels and urinary glucose and improved glucose tolerance as well as led to a significant reduction in serum cholesterol levels. The hypoglycaemic effect of fenugreek seeds was also predicted in a human trial involving 60 type 2 diabetic subjects given a daily administration of 25 g fenugreek over a period of 24 weeks (Sharma *et al.*, 1990). Not only was fasting blood glucose lowered, but an improvement in glucose tolerance, reduced sugar excretion and diminished glycated haemoglobin were also observed.

In a placebo-controlled study, where type 2 diabetic patients received a daily dose of 1 g alcoholic extracts of fenugreek seeds per day for a period of 2 months, oral glucose tolerance test, lipid levels, fasting C-peptide, glycosylated haemoglobin and insulin resistance were similar to control initially (Gupta *et al.*, 2001). No difference was observed with respect to fasting blood glucose levels or the 2 h post-glucose tolerance test at the end of the 2 months treatment. However, a decrease in insulin secretion and an improved sensitivity to insulin were observed after the fenugreek treatment. Decreased serum triglycerides and increased HDL-cholesterol were also observed. Hence, it was derived that fenugreek seeds improve glycaemic control and decrease insulin resistance in mild type 2 diabetic patients with attendant favourable effect on hyper-triglyceridaemia.

In order to understand if the beneficial effect on glucose tolerance in diabetic patients can be attributed to the effect of fenugreek on the absorption or metabolism of glucose, a study in a crossover design was carried out on type 2 diabetics given 25 g fenugreek seeds for 15 days (Raghuram *et al.*, 1994). An intravenous glucose tolerance test at the end of the study showed that fenugreek reduced the area under the plasma glucose curve and increased the metabolic clearance rate and erythrocyte insulin receptors, suggesting that fenugreek improves peripheral glucose utilization, contributing to an improved glucose tolerance. It became apparent that fenugreek may exert its hypoglycaemic effect by acting at the insulin receptor and at the gastrointestinal level.

In essence, fenugreek seed was confirmed in its ability to diminish hyperglycaemia in individuals with diabetes. Fasting blood glucose, urinary sugar excretion and serum lipid levels in diabetics are significantly reduced, while clinical symptoms are improved. Antidiabetic properties of sub-fractions of fenugreek seeds have also been studied. The antidiabetic effect of fenugreek was found to be highest with whole seeds, followed by the gum isolate, extracted seeds and cooked seeds. The beneficial effects of fenugreek seeds are due to the fibre present in them. Chemical constituents of fenugreek include saponins, many of which are glycosides of diosgenin. The seeds also contain the alkaloids trigonelline, gentianine and carpaine compounds. Other components of the seeds include several C-glycosides. The seeds contain up to 50% mucilaginous fiber. Other seed constituents include 4-hydroxyisoleucine, an amino acid, and fenugreekine. Thus, fenugreek seeds included in the daily diet in amounts of 25–50 g can be an effective supportive therapy in the management of diabetes (Pasupuleti and Anderson, 2008; Mahomoodally, 2013).

### 24.3.1.2 Turmeric (*Curcuma longa*, Zingiberaceae)

Turmeric is a famous spice used globally and commonly known as the 'golden spice' as well as the 'spice of life'. It has been used in Asian countries, particularly India, as a medicinal food plant, and held sacred from time immemorial. Turmeric has strong associations with the socio-cultural life of the people of the Indian subcontinent. The rhizome of turmeric has been used as a medicine, spice and colouring agent for more than two millennia. Turmeric is extensively used in India, the Middle East and the Far East in food preparations. In addition, turmeric is traditionally used in women's cosmetics for its characteristic fragrance and colour, due to its 'cosmeceutical' and 'anti-aging' benefits (Ravindran *et al.*, 2007).

The beneficial effects of curcuma on blood sugar levels in diabetic rabbits were reported nearly half a century ago. A decrease in blood sugar levels has also been reported in diabetic rats given oral curcumin (0.08 g/kg body weight) or turmeric (1 g/kg), and this correlated with the decrease in oxidative stress (Arun and Nalini, 2002; Ravindran *et al.*, 2007). This effect was linked with a decrease in the enzyme sorbitol dehydrogenase, which converts sorbitol to fructose. In these studies, curcumin was more effective than turmeric, which contains only 2–5% curcumin. Using genetically modified diabetic KK-A mice, Kuroda *et al.* (2005) also showed that turmeric (0.2–1 g per 100 g diet for 4 weeks) has hypoglycaemic effects. These results are in agreement with those reported by Nishiyama *et al.* (2005). The C57BL/Ks-db/db diabetic mouse is considered to be a good model for type 2 diabetes. It displays many features of human disease, including hyperphagia, hyperglycaemia, insulin resistance and progressive obesity. Seo *et al.* (2008) used this model to examine the effect of curcumin on insulin resistance, glucose homeostasis, and oxidative damage in C57BL/Ksdb/db diabetic mice and their age-matched lean non-diabetic db/+ mice. They were fed curcumin (0.02% wt/wt) for 6 weeks. Curcumin lowered the blood glucose and glycosylated haemoglobin levels and suppressed body weight loss in db/db mice. Curcumin also increased the plasma insulin levels and increased hepatic glucokinase activity. However, it lowered the activities of glucose-6-phosphatase and phosphoenolpyruvate carboxykinase and the hepatic activities of fatty acid synthase, β-oxidation, 3-hydroxy-3-methylglutaryl coenzyme reductase, acyl-CoA and cholesterol acyltransferase. In addition, curcumin lowered plasma free fatty acid, cholesterol and triglyceride levels and increased hepatic glycogen and skeletal muscle lipoprotein lipase. Moreover, polyphenol antioxidant activities (superoxide dismutase (SOD), catalase and glutathione peroxidase) in erythrocytes and liver led to a reduction in lipid peroxidation. These changes were not observed in control db/+ mice. Overall, Seo's group concluded that curcumin has the potential to lower blood glucose levels.

Chronic hyperglycaemia associated with diabetes can lead to secondary complications, including diabetic cataracts, as characterized by cloudiness or opacification of the crystalline eye lens, leading to blindness (Josifova *et al.*, 2008; Madsen-Bouterse and Kowluru, 2008; Crawford *et al.*, 2009). Turmeric (0.5%) and curcumin (0.01%) were found to delay the progression and maturation of cataract in rats with (streptozotocin) STZ-induced diabetes (Suryanarayana *et al.*, 2005). Interestingly, turmeric was found to be more effective than curcumin. This is consistent with another study in which attenuation of galactose-induced cataracts by curcumin in rats was reported (Pandya *et al.*, 2000). The mechanism of galactose-induced cataract, however, differs from that of STZ-induced diabetic cataract.

Diabetes is a major risk factor for neuropsychiatric deficits such as stroke, cerebrovascular diseases, diabetic encephalopathy, depression and anxiety. Diabetic encephalopathy is characterized by impaired cognitive functions due to direct neuronal damage caused by intracellular glucose. Chronic treatment with curcumin (60 mg/kg; p.o.) significantly attenuated cognitive deficit, cholinergic dysfunctions, oxidative stress and inflammation in diabetic rats (Sharma *et al.*, 2006; Kuhad and Chopra, 2007). Curcumin was also found to enhance wound healing in diabetes-impaired healing in diabetic rats (Sidhu *et al.*, 1999).

Since the report in 1972 that curcumin could lower blood glucose levels in human diabetic subjects (Srinivasan, 1972), several pilot studies have been done in human subjects with curcumin to examine its effect on obesity-related parameters. One of the first studies showed that curcumin lowered blood sugar levels in diabetic patients (Srinivasan, 1972). Another study examined the effect of curcumin in humans on the levels of HDL- and LDL-cholesterol. Administration of 10 mg curcumin per day for 30 days to eight human subjects increased HDL-cholesterol, decreased LDL-cholesterol, and increased apolipoprotein A (APO A) but decreased APO B and APO A/B. The same group reported another study with curcumin in human subjects with atherosclerosis (Ramirez-Bosca *et al.*, 2000). In this study, 10 mg curcumin was administered twice a day for

15 days to 16 men and 14 women. Curcumin significantly lowered the levels of plasma fibrinogen in both men and women.

Curcumin, along with its mono- and bisdemethoxy derivatives, collectively called curcuminoids, constitutes the major orange–yellow colouring matter and the biologically active constituents of *Curcuma longa*. Besides curcuminoids, minor amounts of oils and resins are also present in curcumin. These include α-turmerone, β-turmerone, curlon and zingiberene. Turmeric contains a wide variety of phytochemicals, including curcumin, demethoxycurcumin, bisdemethoxycurcumin, zingiberene, curcumenol, curcumol, eugenol, tetrahydrocurcumin, triethylcurcumin, turmerin, turmerones and turmeronols. The main curcuminoid is curcumin, which is responsible for the yellow colour of turmeric and the therapeutic effect of curcuma. Curcumin has antioxidant and anti-inflammatory properties. Many studies have shown that curcumin is not toxic to humans, but rather it is as powerful an antioxidant as vitamins C, E and β-carotene. Overall, turmeric consists of 2–6% of curcuminoid, 3–7% volatile (essential) oil, 2–7% fibre, 3–7% mineral matter, 6–8% protein, 5–10% fat, 6–13% moisture and 60–70% carbohydrate (Sharma *et al.*, 2006; Mahomoodally, 2013).

#### 24.3.1.3 Cinnamon (*Cinnamomum verum, Cinnamomum cassia, Cinnamomum zeylanicum*, Lauraceae)

Cinnamon is among the earliest known spice used by humankind, with frequent references available in both pre-biblical and post-biblical writings. It is a famous spice obtained from the bark of several trees from the genus *Cinnamomum* and is used in a panoply of recipes. Additionally, it is traditionally used for metabolic diseases (mainly diabetes) in China, Korea and Russia; and it has been reported to reduce fasting plasma concentrations (Yu and Lyons, 2005; Ziegenfuss *et al.*, 2006; Hlebowicz *et al.*, 2007). Following the observations that cinnamon potentiates insulin action *in vitro*, Khan *et al.* (2003) conducted a human study involving 60 people with type 2 diabetes mellitus, and after 40 days all three levels of cinnamon tested were found to reduce the mean fasting serum glucose (18–29%), triglycerides (23–30%), total cholesterol (12–26%) and LDL-cholesterol (7–27%).

In a separate study (Ziegenfuss *et al.*, 2006) involving 22 subjects with the metabolic syndrome, subjects were divided into two groups and given either 500 mg/day of a commercially available aqueous extract of cinnamon or a placebo for 12 weeks. Subjects in the group receiving the capsules containing the aqueous extract of cinnamon displayed decreases in fasting blood glucose (−8.4%), systolic blood pressure (−3.8%) and increases in lean mass (+1.1%) compared with the placebo group. There were also significant decreases in body fat (−0.7%) in the cinnamon group. Recent intervention studies on cinnamon supplementation on glycaemic control in healthy and metabolic disorders are summarized in Table 24.1.

Oxidative stress, which is increased in obesity, plays an important role in the development of diabetes and cardiovascular diseases in obese people (Yu and Lyons, 2005). Hyperglycaemia causes the auto-oxidation of glucose, glycation of proteins and the activation of polyol metabolism (Robertson, 2004). These changes accelerate the generation of reactive oxygen species (ROS) and increase oxidative modifications of lipids and proteins (Osawa and Kato, 2005). A significant positive correlation between plasma glucose levels and plasma malondialdehyde (MDA), a measure of lipid peroxidation, was also found. In similar studies, cinnamon improved antioxidant status of people with the metabolic syndrome. The improvement of impaired fasting glycaemia due to cinnamon was correlated with the antioxidant effects of cinnamon supplementation assessed by plasma MDA, sulfhydryl groups and plasma antioxidant status evaluated using ferric reducing antioxidant power (FRAP). A significant positive correlation between plasma glucose levels and plasma MDA confirms a previous study showing that plasma glucose levels play a role in determining oxidative status (Hayden and Tyagi, 2004).

In subjects with metabolic syndrome, plasma MDA levels were reduced by the aqueous extract of cinnamon, indicating decreased lipid peroxidation, while plasma sulfhydryl groups were increased, indicating a protection of antioxidant sulfhydryl groups against oxidation (Roussel *et al.*, 2009). The oxidation of lipids is thought to play a crucial role in the generation of atherosclerotic lesions in obese patients (Davi and Falco, 2005). For patients suffering from obesity, cardiovascular diseases and diabetes, it is well known that decreasing lipid peroxidation is an important health challenge to avoid oxidative damage of the arterial walls and oxidative complications (Couillard *et al.*, 2005). Functional consequences of SH group losses include protein misfolding, catalytic inactivation and decreased antioxidative capacity (Balcerczyk and Bartosz, 2003). In the group receiving cinnamon, plasma SH groups were increased after 12 weeks of supplementation, suggesting that cinnamon acts in protecting both lipids and proteins against oxidation. In parallel, the FRAP, which is a measure of the total antioxidant capacity of plasma, was increased, thereby providing a contributory factor to the protective effects of cinnamon supplementation (Roussel *et al.*, 2009). In contrast, cinnamon intake did not alter the activities of renal antioxidant enzymes, while plasma antioxidant capacity is enhanced following red wine consumption (Rodrigo and Bosco, 2006). However, in rats fed a high-fat diet with spices rich in cinnamon, antioxidant enzyme activities were enhanced (Dhuley, 1999).

The antioxidant effects of cinnamon may favourably impact hormone signalling via adipokine cross-talk and may help explain the effects of cinnamon on body composition (Ziegenfuss *et al.*, 2006). For instance, Chrysohoou *et al.* (2006) recently reported an inverse relationship between body fat and antioxidant capacity, even after controlling for smoking, physical activity patterns, dietary habits, blood pressure, glucose levels and lipid concentrations. It is possible, therefore, that the observed improvements in fasting blood glucose and plasma antioxidant status are responsible, at least in part, for cinnamon's beneficial effects on body composition and features of the metabolic syndrome (Ziegenfuss *et al.*, 2006). Hence, it has been postulated that compounds found in cinnamon not only improve the function of insulin, but also function as antioxidants and may be anti-inflammatory. In conclusion, human studies involving subjects with type 2 diabetes mellitus show beneficial effects of whole cinnamon and aqueous extracts of cinnamon on glucose, insulin, lipids and antioxidant status.

#### 24.3.1.4 Aloe (*Aloe vera*, Asphodelaceae)

*Aloe vera* is one of nature's most revered therapeutic healing herbs. The beneficial effects of this tropical succulent have spurred recurrent legends about its properties that have persisted from the fourth century BC throughout world history (Reynolds, 2004). It is used in the folk medicine of the Arabian Peninsula for the management

**Table 24.1** Recent intervention studies on cinnamon supplementation on glycaemic control in healthy and metabolic disorders (adapted from Rudkowska, 2009)

| Study design | Study diet | Results |
|---|---|---|
| Khan *et al.* (2003)<br>60 people with type 2 diabetes in controlled study of parallel design | Randomly divided into six groups: groups 1, 2 and 3 respectively consumed 1 g, 3 g and 6 g of cinnamon (*Cinnamomum cassia*) daily, and groups 4, 5 and 6 were given placebo capsules corresponding to the number of capsules for the three levels of cinnamon for 40 days | ↓ FBG in all three levels of cinnamon<br>↓ TG, LDL-C and TC levels in all three levels of cinnamon<br>↔ HDL-C |
| Ziegenfuss *et al.* (2006)<br>22 subjects with prediabetes and the metabolic syndrome in study of parallel design | Randomly assigned to supplement their diet with either Cinnulin$^{PF}$ (500 mg/day, which is equivalent of 10 g of whole cinnamon powder) or a placebo for 12 weeks | ↓ Body fat and ↑ lean body mass<br>↓ FBG<br>↔ HbA1c levels<br>↓ Systolic blood pressure<br>↔ Blood lipids |
| Vanschoonbeek *et al.* (2006)<br>25 postmenopausal patients with type 2 diabetes in study of parallel design | Supplemented with either cinnamon (*Cinnamomum cassia*, 1.5 g/day) or a placebo for 6 weeks | ↔ FBG, fasting insulin or HbA1c levels<br>↔ Whole-body insulin resistance/sensitivity<br>↔ Blood lipids |
| Hlebowicz *et al.* (2007)<br>14 healthy subjects using a crossover trial in one session | Randomly ingested of 300 g rice pudding or 300 g rice pudding and 6 g cinnamon | ↓ Postprandial glucose response<br>Delayed gastric emptying<br>↔ Satiety |
| Wang *et al.* (2007)<br>15 women with polycystic ovary syndrome in study of parallel design | Randomized to daily oral cinnamon extract (1 g/day) or placebo for 8 weeks | ↓ FBG<br>↔ Fasting insulin and insulin index<br>↓ Insulin resistance |
| Blevins *et al.* (2007)<br>58 subjects with type 2 diabetes in study of parallel design | Randomized to receive either cinnamon (*Cinnamomum cassia*, 1 g/day) or placebo (wheat flour) for 3 months | ↔ Body weight<br>↔ FBG or HbA1c levels<br>↔ Blood lipids |
| Tang *et al.* (2008)<br>11 young and healthy subjects participated in randomly assigned cross-over study | Randomized to be supplemented doses of cinnamon (3 g/day) and turmeric (2.8 g/day) for 4 weeks each | ↔ FBG<br>↔ TC and TG |

FBG: fasting blood glucose; HDL-C: HDL-cholesterol; LDL-C: LDL-cholesterol; TC: total cholesterol; TG: triglycerides. ↑: increase; ↓: decrease; ↔: no change.

of diabetes. There have been controversial reports on the hypoglycaemic activity of *Aloe* species, probably due to differences in the parts of the plant used or to the model of diabetes studied. The dried sap of the plant has shown significant hypoglycaemic effect in clinical trials (Pasupuleti and Anderson, 2008).

Extracts of aloe gum are reported to effectively increase glucose tolerance in both normal and diabetic rats (Al-Awadi and Gumaa, 1987). *A. vera* leaf pulp extract showed hypoglycaemic activity in type 1 and type 2 diabetic rats, the effectiveness being higher for type 2 (Okyar *et al.*, 2001). Daily oral administration of *A. vera* gel extract (300 mg/kg) to STZ-diabetic rats for 21 days resulted in a significant reduction in fasting blood glucose and improvement in plasma insulin, which is accompanied by reduction in plasma and tissue lipids (Rajasekaran *et al.*, 2006).

The acute and chronic effects of the exudate of *Aloe barbadensis* leaves (500 g/kg p.o.) and its bitter principle (5 mg/kg i.p.) were studied on plasma glucose levels of alloxan-diabetic mice (Ajabnoor, 1990). The hypoglycaemic effect of a single oral dose of bitter principle was significant over a period of 24 h, with maximum hypoglycaemia observed at 8 h. Chronic administration of aloes or the bitter principle resulted in significant reduction in plasma glucose level on the fifth day. The hypoglycaemic effect of aloes may be mediated through stimulated synthesis and/or release of insulin from the β-cells. Both *A. vera* and *Aloe gibberellin* (2–100 mg/kg) inhibited inflammation in a dose-dependent manner and improved wound healing in STZ-diabetic mice (Chitra *et al.*, 1998).

*Aloe* species have been reported to be rich in anthraquinones, saccharides, enzymes, vitamins, minerals and cholesterol, amongst others. The principal constituents present in the leaves are barbaloin, chrysophanol glycoside and the aglycone aloe-emodin. The phytosterols present in the *A. vera* gel have been reported to possess anti-hyperglycaemic activity which may have a long-term blood glucose level control effect, and thus would be useful for treatment of IDDM and NIDDM. Another study reports the presence of inorganic trace elements (vanadium, zinc, copper, manganese and traces of chromium) in aloe leaf gel possessing hypoglycaemic property, probably through improvement in impaired glucose tolerance, thereby indirectly contributing to management of diabetes. Phytochemicals from *Aloe* have been found to stimulate glucose uptake and insulin release (Mahomoodally, 2013).

#### 24.3.1.5 Bitter melon (*Mormordica charantia*, Cucurbitaceae)

Bitter melon (*Mormordica charantia*) of the Cucurbitaceae family is commonly consumed as a vegetable in Asian countries. The fruit, leaves, seeds and roots of *M. charantia*, all of which taste bitter, are used in the Indian systems of medicine and in the folk medicine as an antidiabetic and for other ailments (Pasupuleti and Anderson, 2008). Several human studies have demonstrated significant blood sugar control and overall blood sugar lowering effects of bitter melon. The first clinical report on the antidiabetic potential of bitter melon was by Lakholia (1956), who observed complete disappearance of all symptoms of diabetes (polyuria, polydipsia and polyphagia) in 2 months as a result of consuming fresh juice (up to 1 oz/day) with attendant reduction in urinary sugar after 15 days in spite of discontinuing insulin injections (Reynolds, 2004).

Fresh bitter melon juice significantly reduced blood glucose and improved the glucose tolerance in type 2 diabetic patients who underwent a glucose tolerance test (Welihinda *et al.*, 1986). Fried bitter melon fruits consumed as a daily supplement to the diet (250 g) significantly improved glucose tolerance in diabetic subjects without any increase in serum insulin (Leatherdale *et al.*, 1981). Aqueous extract of bitter melon was hypoglycaemic in diabetic patients at the end of a 3-week trial (Srivastava *et al.*, 1993). Adaptogenic properties were indicated by the delay in appearance of cataract and relief in neurological symptoms even before the hypoglycaemia occurred. Type 2 diabetic patients who consumed cooked/fried bitter melon had reduced plasma glucose at the end of 1 h (William *et al.*, 1993).

Homogenized pulp of *M. charantia* fruit given to 100 moderate type 2 diabetic patients caused a significant reduction in postprandial blood glucose (Ahmad *et al.*, 1999). Contradictory to these findings, absence of any beneficial influence of bitter melon (given as fresh juice or dry powder for 6–12 weeks) on blood glucose levels or glucose tolerance in diabetic subjects is also reported (Patel *et al.*, 1968). This contradiction is attributable to the differences in methodology (different forms of test materials) and study design (with regard to number of subjects, their age group and duration of disease) employed by the investigators (Pasupuleti and Anderson, 2008).

### 24.3.2 Exotic fruits

#### 24.3.2.1 Pomegranate (*Punica garanatum*, Punicaceae)

The pomegranate is an ancient fruit native to regions from the Himalayas in northern India to Iran. In recent years, however, cultivation of pomegranate has disseminated throughout many dry regions of the world, including parts of the USA. The fruit itself has a distinctive rich red, leathery skin and can grow up to 13 cm. Inside the fruit are many seeds covered by little amounts of red juice, each separated by a membranous white pericarp. A 100 g edible portion of pomegranate contains water (77.93 g), protein (1.67 g), lipids (1.17 g), ash (0.53 g), carbohydrates (18.7 g), fibre (4 g) and sugars (13.67 g).

Pomegranate has been heavily used in various ancient medicines, such as ayurveda, for treatment of diabetes (Devalaraja *et al.*, 2011). As a result of its rapidly increasing production and consumption throughout the world, a considerable amount of recent research has explored the potential of pomegranate to fight obesity and diabetes. A study in CD-1 male mice showed that consumption of pomegranate seed oil reduced weight gain, improved key markers that lead to the development of type 2 diabetes and improved insulin sensitivity – all suggesting a diminished development to type 2 diabetes (McFarlin *et al.*, 2009).

Experiments in 3T3-L1 pre-adipocytes demonstrated that punicic acid, a conjugated linolenic acid found in pomegranate, activates peroxisome proliferator-activated receptor-gamma (PPAR-γ), an important target of insulin action and energy metabolism. Punicic acid also augmented PPAR-γ downstream gene expression, and demonstrates an impaired ability to alleviate the pathogenesis of diabetes in PPAR-γ knockdown immune cells (Devalaraja *et al.*, 2011). This suggests that pomegranate may be a natural complement to the synthetic thiazolidinediones (antidiabetic drugs and PPAR-γ agonists), and alleviate pathogenesis of obesity and diabetes through a PPAR-γ-mediated mechanism. In the Otsuka Long-Evans Tokushima fatty rat model, pomegranate seed oil rich in punicic acid significantly decreased hepatic triacylglycerol contents and levels of monounsaturated fatty acid (MUFA), which can be preventive for development of hepatic steatosis (Devalaraja *et al.*, 2011).

In obese Zucker rats, a model of metabolic syndrome, supplementation with pomegranate fruit extract and juice markedly decreased expression of vascular inflammatory markers such as thrombospondin and TGF-β (Devalaraja *et al.*, 2011). Also, a 4-week clinical trial administering 400 mg of pomegranate seed oil twice a day in hyperlipidaemic subjects significantly improved the lipid profile as shown by a decreased triglyceride:HDL-cholesterol ratio (Mirmiran *et al.*, 2010). Furthermore, pomegranate has shown dramatic antioxidant potential. It has been shown that pomegranate fruit extract, which is rich in polyphenolic antioxidants, represses the expression of oxidation-sensitive genes at the site of stress. More recently, studies show that pomegranate juice extract may prevent high blood pressure induced by angiotensin II in diabetic rats by ameliorating oxidative stress and inhibiting angiotensin-converting enzyme activity (Devalaraja *et al.*, 2011). Research is also heading towards development of a synergistic combination of various extracts. Recently, Fenercioglu *et al.* (2010) demonstrated antagonizing effects on oxidative stress and lipid peroxidation of a supplement containing pomegranate extract, green tea extract and ascorbic acid in type 2 diabetic patients.

All the above-mentioned results clearly suggest that pomegranate can play an important role in the prevention of diabetes, obesity and their associated metabolic complications. On the basis of these findings, it can be suggested that pomegranate can be considered as a rational complementary therapeutic agent to ameliorate obesity, diabetes and the resultant metabolic syndrome.

#### 24.3.2.2 Avocado (*Persea americana*, Lauraceae)

Avocado is a tropical fruit distributed mainly in Mexico, South America and Central America. A 100 g edible portion contains water (73.23 g), protein (2 g), lipids (14.66 g), ash (1.58 g), carbohydrates (8.53 g), fibres (6.70 g) and sugars (0.66 g). Although a panoply of claims have been made about the unhealthy effects of avocados, key research debunks the myth that avocados are fattening and must be avoided in any sort of energy-restricted diet. Surprising data from a clinical trial came out that the feeding of 55 subjects with 200 g/day of avocado for 6 weeks with an energy-restricted diet dramatically decreased demonstrated body weight, body mass index and percentage body fat, suggesting that consumption of avocado plays a role in regulating energy homeostasis and maintains body weight balance (Devalaraja *et al.*, 2011). Indeed, the MUFAs highly present in avocado might be responsible agents showing an improved lipid profile and adequately maintained glycaemic control in patients with type diabetes (Devalaraja *et al.*, 2011). In another study, MUFAs decreased serum total cholesterol levels by 17%, LDL-cholesterol by 22%, triglycerides by 11%, and increased HDL-cholesterol by 11% in mild hypercholesterolaemic patients (López Ledesma *et al.*, 1996).

Furthermore, the antioxidant and anti-inflammatory capacities of avocado are well established, suggesting that incorporation of avocado into the diet might ameliorate the pathogenesis of metabolic syndrome (Kim *et al.*, 2000). Persenone A, a compound isolated from the avocado fruit, nearly wholly suppressed protein expression of iNOS and COX-2 in RAW 264.6 mouse macrophage cells (Kim *et al.*, 2000). Overall, results from studies exploring the role of avocado in various models can be well correlated with the preventive effects of avocado in obesity and diabetes. However, it is critical to note that an optimal quantity of avocado must be consumed for the observed antidiabetic and anti-obese effects, which is true of any natural components or food function.

#### 24.3.2.3 Persimmon (*Diospyros digyna*, Ebenaceae)

The persimmon fruit is native to Japan and China. The published literature demonstrates a potent antidiabetic and anti-obesity capacity of the persimmon fruit (Matsumoto *et al.*, 2006; Lee *et al.*, 2007; Dewanjee *et al.*, 2009). Proanthocyanidin is the major component isolated from persimmon peel and has been demonstrated to play a role in obesity and diabetes. Administration of proanthocyanidin from the peel of persimmon in STZ-induced diabetic rats decreased the elevation of lipid peroxidation, suppressed generation of ROS, and decreased serum glucose, glycosylated haemoglobin (HbA1c), serum urea nitrogen, urinary protein and renal advanced glycation end-products under diabetic conditions. This clearly suggests an overall protective effect against oxidative stress-related inflammatory processes and diabetes (Lee *et al.*, 2007).

In addition, administration of persimmon exhibited dose-dependent hypoglycaemic and hypolipidaemic activity in STZ-induced diabetic rats. In addition, the activity of SOD, catalase and glutathione content was increased in the liver and kidney, which is highly correlated with preventing oxidative-stress-mediated diabetic complications in this severe diabetic model (Dewanjee *et al.*, 2009). In the diet-induced obesity mouse model, feeding of persimmon significantly attenuated the elevation in plasma lipids (total cholesterol, triglyceride, LDL cholesterol) (Matsumoto *et al.*, 2006). In male db/db mice (leptin receptor deficient mice), administration of proanthocyanidins reduced hyperglycemia through reduction of serum glucose and glycosylated protein levels, and hyperlipidemia (reduction of triglyceride, total cholesterol and non-esterified fatty acids) (Devalaraja *et al.*, 2011). Polymers from proanthocyanidins of persimmon exhibited a strong inhibitory effect on α-amylase, while oligomers exerted a stronger protective activity against α-glucosidase activity and AGE formation, suggesting that oligomers may have more potential as anti-diabetic agents (Lee *et al.*, 2007). Proanthocyanidins from persimmon also attenuated the increased oxidative stress in db/db mice by suppressing lipid peroxidation, ROS, protein expression of iNOS and COX-2, and increasing the reduced glutathione/oxidized ratio (Devalaraja *et al.*, 2011).

Another study in Wistar albino diabetic rats also suggests persimmon is rendered a hypoglycaemic effect from its antioxidant defence mechanisms (Dewanjee *et al.*, 2009). Tannins, also isolated from persimmon pulp, have been demonstrated to have hydroxyl radical scavenging antioxidant capacities (Devalaraja *et al.*, 2011). Overall, the studies suggest that polymerization of proanthocyanidin has an important effect on its ability to combat the obese and diabetic phenotype, and that its oligomers are the more potent and promising compounds.

#### 24.3.2.4 Guava (*Psidium guajava*, Myrtaceae)

According to ancient Chinese medicine, guava is highly useful in the treatment of diabetes and other chronic metabolic diseases. However, results regarding the antidiabetic potential of guava fruit are ambiguous, clearly suggesting the need for further characterization for its hypoglycaemic effects (Cheng and Yang, 1983; Roman-Ramos *et al.*, 1995). In alloxan-treated diabetic mice, intra-peritoneal administration of 1 g/kg of guava juice dramatically reduced the blood glucose levels (Cheng and Yang, 1983). However, another study reported contradictory findings, whereby a non-significant effect of guava treatment on glucose homeostasis in rabbits was found (Roman-Ramos *et al.*, 1995). But here this might be because of the difference in doses and duration of treatment. In STZ-induced diabetic rats, oral administration of guava fruit peel extract actually induced a hyperglycaemic effect, suggesting that guava fruit peel should be peeled before eating in diabetic patients (Rai *et al.*, 2007). In contrast, another study found that oral administration of guava fruit peel extract demonstrated significant hypoglycaemic and hypolipidaemic effects in the same model (Rai *et al.*, 2010).

Hence, it is highly evident that the literature is inconclusive regarding the hypoglycaemic and hypolipidaemic effects of guava; larger studies in diverse diabetic and obesity models are critical to solidly characterize the role of guava in glucose and energy homeostasis. Moreover, chemical composition analysis of guava constituents will give further opportunity to develop a unique antidiabetic and anti-obese component (Devalaraja *et al.*, 2011).

#### 24.3.2.5 Grapefruit (*Citrus paradise*, Rutaceae)

Grapefruit has been traditionally used as an antiseptic, aperitif, cardiotonic, detoxicant, hypocholesterolaemic, sedative, antidiabetic, dysuric, and against high cholesterol. Recently, the scientific literature has confirmed that grapefruit reduces atherosclerotic plaque formation, inhibits breast cancer cell proliferation and mammary cell tumorigenesis, and is antigenotoxic, anti-inflammatory, antibacterial and anti-obesogenic (Alvarez-Gonzalez *et al.*, 2011). Grapefruit-seed extract has been reported to be successful in combating a variety of common infectious agents. In another *in vivo* study, the effect of grapefruit and grapefruit juice was tested on body weight loss, caloric restriction, energy balance and cardio-mebatolic risk in free-living obese adults. It was found that subjects experienced 7.1% weight loss overall, with significant decreases in percentage body, trunk, android and gynoid fat, as well as waist circumferences. The amount and direction of change in serum HDL-cholesterol levels among subjects were significant (Kalidhar and Meera, 2008; Silver *et al.*, 2011).

### 24.3.3 Nuts

Nuts, including peanuts (*Arachis hypogaea* L., Fabaceae) and almonds (*Prunus amygdalus* Linn., Roseaceae), are well documented as having the potential to improve the blood lipid profile in humans (Rudkowska and Jones, 2007). Several epidemiological studies have revealed that people who consume nuts regularly are less likely to suffer from cardiovascular disease (Kelly and Sabate, 2006; Kris-Etherton *et al.*, 2008). More recently, interest has developed in the potential value of including nuts in the diets of individuals with diabetes. Data from the Nurses Health Study indicate that there are benefits to consuming higher nut and peanut butter consumption to lower risk of type 2 diabetes in women (Jiang *et al.*, 2002). In this study, the relative risk of developing diabetes was reduced by 27% in women who ate nuts five or more times per week compared with those who rarely or never ate nuts.

In general, nuts contain 70–80% fat, and most fatty acids in nuts are unsaturated – being either polyunsaturated or monounsaturated, which may be beneficial for glucose and insulin homeostasis. Several studies have shown that a high amount of monounsaturated and polyunsaturated fat versus saturated fat improves glucose homeostasis (Due *et al.*, 2008; Brehm *et al.*, 2009). The mechanisms by which specific types of dietary fat affect insulin sensitivity are not well understood. It has been shown that the fatty acid composition of phospholipids in skeletal muscle cell membranes is directly related to insulin sensitivity in humans (Vessby, 2000). Therefore, changes in dietary fatty acids might result in compositional changes of membrane phospholipids, perhaps in turn modulating insulin action and sensitivity in skeletal muscle. In addition, other components of nuts, such as fibre and magnesium, may decrease insulin demand and resistance. Nuts are a rich source of vitamins, minerals, antioxidants and plant protein, which could also be beneficial (Kris-Etherton *et al.*, 2008). In this respect, nuts contain many bioactive components, which may have potential advantages for glycaemic control.

Only a few long-term (Lovejoy *et al.*, 2002; Tapsell *et al.*, 2004) and short-term (Jenkins *et al.*, 2006, 2008; Josse *et al.*, 2007) clinical studies have investigated the effects of nuts and peanut consumption on glycaemic control in type 2 diabetic patients, and these are summarized in Table 24.2. Two long-term, randomized controlled trials (Lovejoy *et al.*, 2002; Tapsell *et al.*, 2004) confirmed the beneficial effects of nuts on blood lipids in non-diabetic and type 2 diabetic patients, but did not report improvement in glycaemia. This particular study by Lovejoy *et al.* (2002) demonstrated that an almond-enriched diet for 1 month did not alter insulin sensitivity in healthy adults, or glycaemia in patients with diabetes. Similarly, Tapsell *et al.* (2004) reported that a moderate-fat diet with 30 g of walnuts did not affect long-term glycaemic control, as indicted by $HbA1_C$ levels, in healthy subjects. In sum, obvious long-term effects of nuts and peanuts have yet to be recognized. However, acute postprandial studies suggest that nuts should be combined with the carbohydrate portion of the meal to reduce postprandial glycaemia. A lower glycaemic response is often thought to equate to a lower insulin demand, better long-term blood glucose control and a reduction in blood lipids. Josse *et al.* (2007) demonstrated that almonds may reduce the glycaemic impact of carbohydrate foods with which they are eaten. Additionally, another study concluded that reductions in 24 h insulin secretion appear to be a further metabolic advantage of nuts, which the investigators suggest may in the longer term help to explain the association of nut consumption with reduced cardiovascular disease risk (Jenkins *et al.*, 2008). Furthermore, Jenkins *et al.* (2006) also suggested that almonds are likely to lower this risk by decreasing the glycaemic response and by providing antioxidants, since inflammation and oxidation may be causative in the initial pathogenesis of insulin resistance, loss of β-cell response to glucose, and development of both micro- and macrovascular complications in type 2 diabetes.

Recently, it has been reported that diets rich in almond were found to increase plasma and red blood cell levels of α-tocopherol and increase α-tocopherol-to-cholesterol ratios in healthy adults (Jambazian *et al.*, 2005). *P. amygdalus* can also reduce the brain cholinesterase activity in rat, and thus proves to be a useful memory-restorative agent, and it would be worthwhile to explore the potential of this plant in the management of Alzheimer's disease (Dwivedi *et al.*, 2012). *P. amygdalus* seeds have significant hypoglycaemic effect (Abraham and Adithan, 2000), antioxidant effects and can decrease oxidized LDL levels, thus decreasing the incidence of cardiovascular diseases (Jenkins *et al.*, 2006). Overall, acute feeding studies have established the ability of nuts, when eaten with carbohydrate, to depress postprandial glycaemia, but long-term benefits of consuming nuts have not yet been established.

In conclusion, more long-term randomized controlled clinical trials should evaluate the effect of nuts on glycaemic control. However, nuts may have favourable effects on the postprandial response in both people with normal blood glucose and those with type 2 diabetes. Thus, nuts can be incorporated into a healthy diet to lower the glycaemic index of a meal or snack for individuals with diabetes or those at risk for diabetes, on condition that overall caloric intake is regulated to maintain a healthy body weight (Rudkowska, 2009).

**Table 24.2** Recent intervention studies on nuts and peanuts on glycaemic control in healthy and type 2 diabetes populations (adapted from Rudkowska (2009))

| Study design | Study diet | Results |
|---|---|---|
| Lovejoy *et al.* (2002)<br>Study 1<br>20 free-living healthy volunteers with normoglycaemia | Received 100 g almonds per day for 4 weeks | ↑ Body weight<br>↔ FBG and fasting insulin<br>↔ Insulin sensitivity<br>↓ TC and LDL-C |
| Lovejoy *et al.* (2002)<br>Study 2<br>30 subjects with type 2 diabetes in a randomized cross-over study | Randomized to one of four diets: (1) high fat, high almond (37% total fat, 10% from almonds); (2) low fat, high almond (25% total fat, 10% from almonds); (3) high-fat control (37% total fat, 10% from olive or canola oil); (4) low-fat control (25% total fat, 10% olive or canola oil) for 4 weeks each | ↔ FBG, fasting insulin or HbA1c levels<br>↔ 2 h glucose and 2 h insulin<br>↓ TC with the high-fat, high-almond diet<br>↓ HDL-C with the almond diets |
| Tapsell *et al.* (2004)<br>58 men and women in a parallel randomized controlled study | Randomized to one of three diets with 30% energy as fat: (1) low fat; (2) modified low fat; (3) modified low fat inclusive of 30 g of walnuts per day for 6 months | ↔ Body weight, percentage body fat or total antioxidant capacity<br>↔ HbA1c levels<br>↑ HDL-C-to-TC ratio and HDL-C in walnut diet versus control<br>↓ LDL-C in walnut diet versus control |
| Jenkins *et al.* (2006)<br>15 healthy subjects in five sessions with minimum 1 week washout | Consumed two bread control meals and three test meals: (1) almonds and bread; (2) parboiled rice; (3) instant mashed potatoes, balanced in carbohydrate, fat and protein | ↓ Blood glucose and insulin for the rice and almond meals versus for the potato meal<br>↔ Total antioxidant capacity<br>↑ Serum protein thiol concentration following the almond meal indicating ↓ oxidative protein damage |
| Josse *et al.* (2007)<br>Nine healthy volunteers in five sessions with minimum 1 week washout | Randomly consumed two white bread control meals and three test meals containing each 50 g of available carbohydrate from white bread alone or with 30, 60 or 90 g of almonds | ↓ Glycaemic index of the composite meal in a dose-dependent manner for the 30 g, 60 g and 90 g doses of almonds to white bread<br>↓ Glycaemic index and glucose peak height in 90 g almond meal versus white bread control meal |
| Jenkins *et al.* (2008)<br>27 hyperlipidaemic men and postmenopausal women in a randomized cross-over study | Randomized to consume one of three isoenergetic supplements: (1) full-dose almonds (73 ± 3 g/day); (2) half-dose almonds plus half-dose muffins; (3) full-dose muffins for 1 month each | ↔ FBG, insulin, or insulin resistance<br>↓ 24 h urinary C-peptide output (marker of 24 h insulin secretion) on the half- and full-dose almonds by comparison with control |

FBG: fasting blood glucose; HDL-C: HDL-cholesterol; LDL-C: LDL-cholesterol; TC: total cholesterol. ↑: increase; ↓: decrease; ↔: no change.

## 24.4 Conclusion

As drugs failed to give desirable and satisfactory outcomes, more and more interest is generated in the use of medicinal food plants to alleviate chronic metabolic diseases. Clinical studies investigating the effects of specific medicinal food plants as potential pharmacophores in the management/treatment of metabolic diseases have yielded promising results, including normalization of blood sugar level, prevention and delaying the digestion and absorption of glucose, amongst others.

As successful preventive and therapeutic strategies for these life-threatening health ailments often come with adverse side effects, nutritional elements are widely used in many countries as preventive therapies to prevent or manage metabolic syndrome. Fruits, vegetables, nuts and seeds, as well as herbs and spices, are important dietary components and contain various bioactive constituents. The bioactive components in these functional foods have become popular and considered as complementary or alternative therapeutic agents to help manage and/or treat chronic diseases. Alone or together, a combination of these health benefits rendered by the reported functional foods is very important to prevent the development of complex pathophysiology of chronic metabolic diseases.

However, the bioactive constituents, therapeutic effects, drug–herb interactions and adverse effects certain functional foods, particularly the exotic fruits and vegetables, certain nuts and seeds, as well as herbs and spices reviewed in this chapter, are clearly understudied. Further large-scale and defined basic and clinical studies are critical to determine optimal dietary regimens to achieve the desired beneficial health effects. In addition, components of these functional foods are clearly attractive targets for the scientific community to develop novel functional ingredients/foods compounds for future treatment/prevention of these life-threatening diseases (Devalaraja *et al.*, 2011).

## References


Abraham, B.K. and Adithan, C. (2000) Review of endocrine pharmacology. *Indian Journal of Pharmacology*, **32**, 67–80.

Ahmad, L.A. and Crandall, J.P. (2010) Type 2 diabetes prevention: a review. *Clinical Diabetes*, **28** (2), 53–59.

Ahmad, N., Hassan, M.R., Haider, H. and Bennoor, K.S. (1999) Effect of *Momordica charantia* (Karella) extracts on fasting and postprandial serum glucose levels in NIDDM patients. *Bangladesh Medical Research Council Bulletin*, **25**, 11–13.

Ajabnoor, M.A. (1990) Effect of aloes on blood glucose levels in normal and alloxan diabetic mice. *Journal of Ethnopharmacology*, **28**, 215–220.

Al-Awadi, F.M. and Gumaa, K.A. (1987) Studies on the activity of individual plants of an antidiabetic plant mixture. *Acta Diabetologica Latina*, **24**, 37–41.

Alvarez-Gonzalez, I., Mojica, R., Madrigal-Bujaidar, E. *et al.* (2011) The antigenotoxic effects of grapefruit juice on the damage induced by benzo(*a*)pyrene and evaluation of its interaction with hepatic and intestinal cytochrome P450 (Cyp)1a1. *Food and Chemical Toxicology*, **49**, 807–811.

Arun, N. and Nalini, N. (2002) Efficacy of turmeric on blood sugar and polyol pathway in diabetic albino rats. *Plant Foods for Human Nutrition*, **57**, 41–52.

Bagghi, A.K. (2001) Alternative medicine – old wine in a new bottle. *Journal of the Indian Medical Association*, **98**, 332–333.

Balcerczyk, A. and Bartosz, G. (2003) Thiols are main determinants of total antioxidant capacity of cellular homogenates. *Free Radical Research*, **37**, 537–541.

Bilous, R. and Donnelly, R. (2010) *Handbook of Diabetes*, 4th edn, Blackwell, Oxford.

Blevins, S.M., Leyva, M.J., Brown, J. *et al.* (2007) Effect of cinnamonon glucose and lipid levels in non-insulin-dependent type 2 diabetes. *Diabetes Care*, **30**, 2236–2237.

Brehm, B.J., Lattin, B.L. Summer, S.S. *et al.* (2009) One-year comparison of a high monounsaturated fat diet with a high-carbohydrate diet in type 2 diabetes. *Diabetes Care* **32** (2), 215–220.

Cardenas, D. (2013) Let not thy food be confused with thy medicine: the Hippocratic misquotation. *e-SPEN Journal*, **8**, e260–e262.

Cheng, J.T. and Yang, R.S. (1983) Hypoglycemic effect of guava juice in mice and human subjects. *American Journal of Chinese Medicine*, **11**, 74–76.

Chitra, P., Sajithlal, G.B. and Chandrahasan, G. (1998) Influence of *Aloe vera* on the healing of dermal wounds in diabetic rats. *Journal of Ethnopharmacology*, **59**, 195–201.

Chrysohoou, C., Panagiotakos, D.B., Pitsavos, C. *et al.* (2006) The implication of obesity on total antioxidant capacity in apparently healthy men and women: the ATTICA study. *Nutrition, Metabolism & Cardiovascular Diseases*, **17**, 590–597.

Couillard, C., Ruel, G., Archer, W.R. *et al.* (2005) Circulating levels of oxidative stress markers and endothelial adhesion molecules in men with abdominal obesity. *Journal of Clinical Endocrinology & Metabolism*, **90**, 6454–6459.

Crawford, T.N., Alfaro, D.V., Kerrison, J.B. and Jablon, E.P. (2009) Diabetic retinopathy and angiogenesis. *Current Diabetes Reviews*, **5**, 8–13.

Croteau, R., Kutchan, T.M., and Lewis, N.G. (2000) Natural products (secondary metabolites), in *Biochemistry and Molecular Biology of Plants* (eds B. Buchanan, W. Gruissem and R. Jones), American Society of Plant Physiologists, Rockville, MD, pp. 1250–1318.

Davi, G. and Falco, A. (2005) Oxidant stress, inflammation and atherosclerosis. *Lupus*, **14**, 760–764.

Dedeli, O., Fadiloglu, C. and Kalimeler, A. (2011) Development and evaluation of the health belief model scale in obesity. *TAF Preventive Medical Bulletin*, **10** (5), 533–542.

Devalaraja, S., Jain, S. and Yadav, H. (2011) Exotic fruits as therapeutic complements for diabetes, obesity and metabolic syndrome. *Food Research International*, **44**, 1856–1865.

Dewanjee, S., Das, A. K., Sahu, R. and Gangopadhyay, M. (2009) Antidiabetic activity of *Diospyros peregrina* fruit: effect on hyperglycemia, hyperlipidemia and augmented oxidative stress in experimental type 2 diabetes. *Food and Chemical Toxicology*, **47**, 2679–2685.

Dhuley, J.N. (1999) Anti-oxidant effects of cinnamon (*Cinnamomum vernum*) bark and greater cardamom (*Amomum subulatum*) seeds in rats fed high fat diet. *Indian Journal of Experimental Biology*, **37**, 238–242.

Due, A., Larsen, T.M., Mu, H. *et al.* (2008) Comparison of ad libitum diets for weight-loss maintenance, risk of cardiovascular disease, and diabetes: a 6-month randomized, controlled trial. *American Journal of Clinical Nutrition*, **88**, 1232–1241.

Dwivedi, P., Singh, R., Malik, M.T. and Jawaid, T. (2012) A traditional approach to herbal nootropic agents: an overview. *International Journal of Pharmaceutical Sciences and Research*, **3** (3), 630–636.

Fenercioglu, A.K., Saler, T., Genc, E. *et al.* (2010) The effects of polyphenol-containing antioxidants on oxidative stress and lipid peroxidation in type 2 diabetes mellitus without complications. *Journal of Endocrinological Investigation*, **33**, 118–124.

Gupta, A., Gupta, R. and Lal, B. (2001) Effects of *Trigonellafoenum-graecum* (fenugreek) seeds on glycaemic control and insulin resistance in type 2 diabetes mellitus: a double blind placebo controlled study. *Journal of the Association of Physicians of India*, **49**, 1055–1056.

Hayden, M.R. and Tyagi, S.C. (2004) Neural redox stress and remodelling in metabolic syndrome, type 2 diabetes mellitus, and diabetic neuropathy. *Medical Science Monitor*, **10**, RA291–RA307.

Hlebowicz, J., Darwiche, G., Bjorgell, O. and Almer, L.O. (2007) Effect of cinnamon on postprandial blood glucose, gastric emptying, and satiety in healthy subjects. *American Journal of Clinical Nutrition*, **85**, 1552–1556.

Jambazian, P.R., Haddad, E., Rajaram, S. *et al.* (2005) Almonds in the diet simultaneously improve plasma α-tocopherol concentrations and reduce plasma lipids. *Journal of the American Dietetic Association*, **105**, 449–454.

Jenkins, D.J., Kendall, C.W., and Marchie, A. (2008) Effect of almonds on insulin secretion and insulin resistance in non diabetic hyperlipidemic subjects: a randomized controlled crossover trial. *Metabolism*, **57**, 882–887.

Jenkins, D.J.A., Kendall, C.W.C., Josse, A.R. *et al.* (2006) Almonds decrease postprandial glycemia, insulinemia, and oxidative damage in healthy individuals. *Journal of Nutrition*, **136**, 2987–2992.

Jiang, R., Manson, J.E., Stampfer, M.J. *et al.* (2002) Nut and peanut butter consumption and risk of type 2 diabetes in women. *JAMA*, **288**, 2554–2560.

Josifova, T., Schneider, U., Henrich, P.B. and Schrader, W. (2008) Eye disorders in diabetes: potential drug targets. *Infectious Disorders – Drug Targets*, **8**, 70–75.

Josse, A.R., Kendall, C.W., Augustin, L.S. *et al.* (2007) Almonds and postprandial glycemia – a dose–response study. *Metabolism*, **56**, 400–404.

Kalidhar, S.B., and Meera, S. (2008) A new coumarin from *Citrus paradisi* Macf. *Indian Journal of Pharmaceutical Sciences*, **70**, 517–519.

Kelly, J.R.J.H. and Sabate, J. (2006) Nuts and coronary heart disease: an epidemiological perspective. *British Journal of Nutrition*, **96** (2), S61–S67.

Khan, A., Safdar, M., Ali Khan, M.M. *et al.* (2003) Cinnamon improves glucose and lipids of people with type 2 diabetes. *Diabetes Care*, **26**, 3215–3218.

Kim, O.K., Murakami, A., Nakamura, Y. *et al.* (2000) Novel nitric oxide and superoxide generation inhibitors, persenone A and B, from avocado fruit. *Journal of Agricultural and Food Chemistry*, **48**, 1557–1563.

Kris-Etherton, P.M., Hu, F.B., Ros, E. and Sabate, J. (2008) The role of tree nuts and peanuts in the prevention of coronary heart disease: multiple potential mechanisms. *Journal of Nutrition*, **138**, S1746–S1751.

Kuhad, A. and Chopra, K. (2007) Curcumin attenuates diabetic encephalopathy in rats: behavioural and biochemical evidences. *European Journal of Pharmacology*, **576**, 34–42.

Kuroda, M., Mimaki, Y., Nishiyama, T. *et al.* (2005) Hypoglycemic effects of turmeric (*Curcuma longa* L. rhizomes) on genetically diabetic KK-Ay mice. *Biological & Pharmaceutical Bulletin*, **28**, 937–939.

Lakholia, A.N. (1956) The use of bitter gourd in diabetes mellitus. *Antiseptic*, **53**, 608–610.

Leatherdale, B.A., Panesar, R.K., Singh, G. *et al.* (1981) Improvement in glucose tolerance due to *Momordica charantia* (Karela). *British Medical Journal*, **282**, 1823–1824.

Lee, Y.A., Cho, E.J., Tanaka, T. and Yokozawa, T. (2007) Inhibitory activities of proanthocyanidins from persimmon against oxidative stress and digestive enzymes related to diabetes. *Journal of Nutritional Science and Vitaminology (Tokyo)*, **53**, 287–292.

Lopez Ledesma, R., Frati Murari, A.C., Hernandez Dominguez, B.C. *et al.* (1996) Monounsaturated fatty acid (avocado) rich diet for mild hypercholesterolemia. *Archives of Medical Research*, **27**, 519–523.

Lovejoy, J.C., Most, M.M., Lefevre, M. *et al.* (2002) Effect of diets enriched in almonds on insulin action and serum lipids in adults with normal glucose tolerance or type 2 diabetes. *American Journal of Clinical Nutrition*, **76**, 1000–1006.

Madar, Z. and Arad, J. (1989) Effect of extracted fenugreek on post-prandial glucose levels in human diabetic subjects. *Nutrition Research*, **9**, 691–692.

Madsen-Bouterse, S.A. and Kowluru, R.A. (2008) Oxidative stress and diabetic retinopathy: pathophysiological mechanisms and treatment perspectives. *Reviews in Endocrine and Metabolic Disorders*, **9**, 315–327.

Mahomoodally, M.F. (2012) Advances to evaluate anti-diabetic medicinal plants, in *Medicinal Plants: Biodiversity and Drugs* (eds M. Rai, L. Rastrell, M. Marinoff et al.), CRC Press, pp. 526–569.

Mahomoodally, M.F. (2013) Phytochemicals: occurrence in nature, health effects and antioxidant properties, in *Phytochemicals as Potential Pharmacophores to Target Diabetes* (ed. H.-X. Kuang), Nova Publishers, Hauppauge, NY.

Matsumoto, K., Watanabe, Y., Ohya, M.A. and Yokoyama, S. (2006) Young persimmon fruits prevent the rise in plasma lipids in a diet-induced murine obesity model. *Biological & Pharmaceutical Bulletin*, **29**, 2532–2535.

McFarlin, B.K., Strohacker, K.A. and Kueht, M.L. (2009) Pomegranate seed oil consumption during a period of high-fat feeding reduces weight gain and reduces type 2 diabetes risk in CD-1 mice. *British Journal of Nutrition*, **102**, 54–59.

Menrad, K. (2003) Market and marketing of functional food in Europe. *Journal of Food Engineering*, **56**, 181–188.

Mirmiran, P., Fazeli, M. R., Asghari, G. *et al.* (2010) Effect of pomegranate seed oil on hyperlipidaemic subjects: a double-blind placebo controlled clinical trial. *British Journal of Nutrition*, **104**, 402–406.

Mohamed, S. (2013) Functional foods against metabolic syndrome (obesity, diabetes, hypertension and dyslipidemia) and cardiovasular disease. *Trends in Food Science & Technology*. doi: 10.1016/j.tifs.2013.11.001

NIH (1998) Clinical guidelines on the identification, evaluation and treatment of overweight and obesity in adults. *The evidence report*. NIH Publication No. 98-4083. National Institutes of Health, US Department of Health and Human Services, www.nhlbi.nih.gov/guidelines/obesity/ob_gdlns.pdf (accessed 12 January 2013).

Nishiyama, T., Mae, T., Kishida, H. *et al.* (2005) Curcuminoids and sesquiterpenoids in turmeric (*Curcuma longa* L.) suppress an increase in blood glucose level in type 2diabetic KK-Ay mice. *Journal of Agricultural and Food Chemistry*, **53**, 959–963.

Okyar, A. Can, A., Akev, N. *et al.* (2001) Effect of *Aloe vera* leaves on blood glucose level in type I and type II diabetic rat models. *Phytotherapy Research*, **15**, 157–161.

Osawa, T. and Kato, Y. (2005) Protective role of antioxidative food factors in oxidative stress caused by hyperglycaemia. *Annals of the New York Academy of Sciences*, **1043**, 440–451.

Pandya, U., Saini, M.K., Jin, G.F. *et al.* (2000) Dietary curcumin prevents ocular toxicity of naphthalene in rats. *Toxicology Letters*, **115**, 195–204.

Pasupuleti, V.K. and Anderson, J.W. (2008) *Nutraceuticals, Glycemic Health & Type 2 Diabetes*, Wiley–Blackwell, Hoboken, NJ.

Patel, J.C., Dhirawani, M.K. and Doshi, J.C. (1968) Karela in the treatment of diabetes mellitus. *Indian Journal of Medical Sciences*, **22**, 30–32.

Raghuram, T.C., Sharma, R.D., Sivakumar, B. and Sahay, B.K. (1994) Effect of fenugreek seeds on intra-venous glucose disposition in non-insulin dependent diabetic patients. *Phytotherapy Research*, **8**, 83–86.

Rai, P.K., Singh, S.K., Kesari, A.N. and Watal, G. (2007) Glycaemic evaluation of *Psidium guajava* in rats. *Indian Journal of Medical Research*, **126**, 224–227.

Rai, P.K., Mehta, S. and Watal, G. (2010) Hypolipidaemic & hepatoprotective effects of *Psidium guajava* raw fruit peel in experimental diabetes. *Indian Journal of Medical Research*, **131**, 820–824.

Rajasekaran, S., Ravi, K., Sivagnanam, K. and Subramanian, S. (2006) Beneficial effects of *Aloe vera* leaf gel extract on lipid profile status in rats with streptozotocin diabetes. *Clinical and Experimental Pharmacology and Physiology*, **33**, 232–237.

Ramirez-Bosca, A., Soler, A., Carrion, M.A., *et al.* (2000) An hydroalcoholic extract of *Curcuma longa* lowers the apo B/apo A ratio. *Implications for atherogenesis prevention. Mechanisms of Ageing and Development*, **119**, 41–47.

Ravindran, P.N., Nirmal Babu, K. and Sivaraman, K. (2007) *Turmeric: Genus Cucurma*, Medicinal and Aromatic Plants – Industrial Profiles, CRC Press.

Reynolds, T. (2004) *Aloes: The Genus Aloe*, Medicinal and Aromatic Plants – Industrial Profiles, CRC Press.

Robertson, R.P. (2004) Chronic oxidative stress as a central mechanism for glucose toxicity in pancreatic islet beta cells in diabetes. *Journal of Biological Chemistry*, **279**, 42351–42354.

Rodrigo, R. and Bosco, C. (2006) Oxidative stress and protective effects of polyphenols: comparative studies in human and rodent kidney. A review. *Comparative Biochemistry and Physiology Part C: Toxicology & Pharmacology*, **142**, 317–327.

Roman-Ramos, R., Flores-Saenz, J.L. and Alarcon-Aguilar, F.J. (1995) Anti-hyperglycemic effect of some edible plants. *Journal of Ethnopharmacology*, **48**, 25–32.

Roussel, A.M., Hininger, I., Benaraba, R. *et al.* (2009) Antioxidant effects of a cinnamon extract in people with impaired fasting glucose that are overweight or obese. *Journal of the American College of Nutrition*, **28** (1), 16–21.

Rudkowska, I. (2009) Functional foods for health: focus on diabetes. *Maturitas*, **62**, 263–269.

Rudkowska, I. and Jones, P.J. (2007) Functional foods for the prevention and treatment of cardiovascular diseases: cholesterol and beyond. *Expert Review of Cardiovascular Therapy*, **5**, 477–490.

Seo, K.I., Choi, M.S., Jung, U.J. *et al.* (2008) Effect of curcumin supplementation on blood glucose, plasma insulin, and glucose homeostasis related enzyme activities in diabetic db/db mice. *Molecular Nutrition & Food Research*, **52**, 995–1004.

Sharma, R.D. (1986) Effect of fenugreek seeds and leaves on blood glucose and serum insulin responses in human subjects. *Nutrition Research*, **6**, 1353–1364.

Sharma, R.D. and Raghuram, T.C. (1990) Hypoglycaemic effect of fenugreek seeds in non-insulin dependent diabetic subjects. *Nutrition Research*, **10**, 731–739.

Sharma, R.D., Raghuram, T.C. and Sudhakar Rao, N. (1990) Effect of fenugreek seeds on blood glucose and serum lipids in type I diabetes. *European Journal of Clinical Nutrition*, **44**, 301–306.

Sharma, R.D., Sarkar, A., Hazra, D.K. *et al.* (1996) Use of fenugreek seed powder in the management of non-insulin-dependent diabetes mellitus. *Nutrition Research*, **116**, 1331–1339.

Sharma, S., Kulkarni, S.K. and Chopra, K. (2006) Curcumin, the active principle of turmeric (*Curcuma longa*), ameliorates diabetic nephropathy in rats. *Clinical and Experimental Pharmacology and Physiology*, **33**, 940–945.

Sidhu, G.S., Mani, H. and Gaddipati, J.P. (1999) Curcumin enhances wound healing in streptozotocin induced diabetic rats and genetically diabetic mice. *Wound Repair*, **7**, 362–374.

Silver, H.J., Dietrich, M.S. and Niswender, K.D. (2011) Effects of grapefruit, grapefruit juice and water preloads on energy balance, weight loss, body composition, and cardiometabolic risk in free-living obese adults. *Nutrition and Metabolism*, **8**, 1–11.

Siro, I., Kapolna, E., Kapolna, B. and Lugasi, A. (2008) Functional food. Product development, marketing and consumer acceptance – a review. *Appetite*, **51**, 456–467.

Srinivasan, M. (1972) Effect of curcumin on blood sugar as seen in a diabetic subject. *Indian Journal of Medical Sciences*, **26**, 269–270.

Srivastava, Y., Bhatt, H.V., Verma, Y. *et al.* (1993) Antidiabetic and adaptogenic properties of *Momordica charantia* extract: AN experimental and clinical evaluation. *Phytoetherapy Research*, **7**, 285–289.

Suryanarayana, P., Saraswat, M., Mrudula, T. *et al.* (2005) Curcumin and turmeric delay streptozotocin-induced diabetic cataract in rats. *Investigative Ophthalmology & Visual Science*, **46**, 2092–2099.

Tang, M., Larson-Meyer, D.E. and Liebman, M. (2008) Effect of cinnamon and turmeric on urinary oxalate excretion, plasma lipids, and plasma glucose in healthy subjects. *American Journal of Clinical Nutrition*, **87**, 1262–1267.

Tapsell, L.C., Gillen, L.J., and Patch, C.S. (2004) Including walnuts in a low-fat/modified-fat diet improves HDL cholesterol-to-total cholesterol ratios in patients with type2 diabetes. *Diabetes Care*, **27**, 2777–2783.

Vanschoonbeek, K., Thomassen, B.J., Senden, J.M. *et al.* (2006) Cinnamon supplementation does not improve glycemic control in postmenopausal type 2 diabetes patients. *Journal of Nutrition*, **136**, 977–980.

Vessby, B. (2000) Dietary fat and insulin action in humans. *British Journal of Nutrition*, **83** (1), S91–S96.

Wang, J.G., Anderson, R.A., and Graham III, G.M. (2007) The effect of cinnamon extract on insulin resistance parameters in polycystic ovary syndrome: a pilot study. *Fertil Steril*, **88**, 240–243.

Welihinda, J., Karunanayake, E.H., Sheriff, M.H. and Jayasinghe, K.S. (1986) Effect of *Momordica charantia* on the glucose tolerance in maturity onset diabetes. *Journal of Ethnopharmacology*, **17**, 277–282.

WHO (2003). *Diet, Nutrition and the Prevention of Chronic Diseases*, Technical Report Series 916, Joint WHO/FAO Expert Consultation, Geneva, http://whqlibdoc.who.int/trs/who_trs_916.pdf (accessed 2 December 2012).

WHO (2006) *The World Health Report 2006 – Working Togther for Health*, http:www.who.int/whr/2006/en/ (accessed 1 December 2012).

WHO (2013) *Overweight and obesity*, Fact sheet No. 311, http://www.who.int/mediacentre/factsheets/fs311/en/ (accessed 9 December 2013).

William, F., Lakshminarayanan, S. and Hariprasad, C. (1993) Effect of some Indian vegetables on glucose and insulin response in diabetic subjects. *International Journal of Food Sciences and Nutrition*, **44**, 191–196.

Yu, Y. and Lyons, T.G. (2005) A lethal tetrad in diabetes: hyperglycaemia, dyslipidaemia, oxidative stress and endothelial dysfunction. *American Journal of the Medical Sciences*, **330**, 227–232.

Ziegenfuss, T.N., Hofheins, J.E., Mendel, R.W. *et al.* (2006) Effects of a water-soluble cinnamon extract on body composition and features of the metabolic syndrome in pre-diabetic men and women. *Journal of the International Society of Sports Nutrition*, **3**, 45–53.

# 25 Strategic Repositioning African Indigenous Vegetables and Fruits with Nutrition, Economic and Climate Change Resilience Potential

M.O. Abukutsa-Onyango

Jomo Kenyatta University of Agriculture and Technology, Kenya; Department of Horticulture, Nairobi, Kenya

## 25.1 Introduction

### 25.1.1 Background

In 400 BC, Hippocrates, the father of modern medicine said, 'Let food be thy medicine' (Gala *et al.*, 2007; Price, 2012). This statement should hold true today since Nature has provided a great diversity of amazing plants that can be used as food to provide mankind with good nutrition, health and quality life. Unfortunately, the world largely depends on a very limited number of selected crops to provide energy, while many vegetables and fruits remain largely underutilized (Williams and Haq, 2002; Shackleton *et al.*, 2009). The world derives 95% of its food energy from about 30 crop species, whilst over 7000 species are underutilized due to neglect and are held in low esteem (Williams and Haq, 2002). Part of this neglect has resulted in some of the highly nutritious African indigenous vegetables and fruits becoming threatened with extinction (Franzel *et al.*, 2008; Abukutsa-Onyango, 2010, 2012). The Millenium Development Goals (MDGs) call for the world to reduce by half the proportion of people who suffer from hunger before the year 2015 (IPGRI, 2004). The goals that directly or indirectly relate to food security, nutrition, health, economic development and climate change cannot be adequately achieved in sub-Saharan Africa without interventions that recognize the hidden potential in indigenous vegetables and fruits (IPGRI, 2004; Abukutsa-Onyango, 2010).

### 25.1.2 Food security, nutrition and diet-related diseases

Food security is a situation that exists when all people at all times have physical, social and economic access to sufficient, safe and nutritious food that meets their dietary needs and food preferences for an active and healthy life (FAO, 2001; Burke and Lobell, 2010). Food insecurity, hunger and malnutrition are issues of greatest concern in sub-Saharan Africa, where the proportion of undernourished population is the highest in the world (Juma, 2011). Over 50% of children, under 5 years in developing countries are malnourished, and many are stunted. The most serious malnutrition problems are as a result of inadequate consumption of micronutrients, which are mainly provided for by vegetables and fruits (Smith and Eyzaguirre, 2007). Inadequate consumption of vegetables and fruits and micronutrient malnutrition results in impaired intellectual development, anaemia, blindness and mortality in children (Smith and Eyzaguirre, 2007; Ganry, 2009). It is also associated with increased risk of obesity, cardiovascular diseases, diabetes and certain cancers (FAO/WHO, 2004). Of the people living in sub-Saharan Africa, 30% are obese, and this pandemic is increasing at 40% every 10 years. Some 44% of the diabetes burden, 23% of the heart disease burden and 41% of the cancer burden are attributed to obesity. If obesity is not controlled, obesity-related cardiovascular complications will be the leading cause of death in sub-Saharan Africa by the year 2020 (Dalal *et al.*, 2011; Staggers-Hakim, 2012).

Research shows that physical activity has many proven benefits, but may not be key to curbing the obesity epidemic. Key causes of obesity have been reported to be the consumption of foods rich in fats and extracted sugars coupled with consuming inadequate amounts of vegetables and fruits (Dalal *et al.*, 2011). For instance, the recommended consumption of vegetables and fruits by World Health Organization (WHO) is 73 kg/person/year (FAO/WHO, 2004). In Africa, the average regional vegetable and fruit consumption is estimated to be about 30 kg/person/year, which is less than half the WHO-recommended amount. The inadequate consumption of vegetables and fruits has contributed immensely to the upsurge of diet-related ailments, like diabetes, cancer, cardiovascular diseases and blindness in sub-Saharan Africa (FAO/WHO, 2004). Vegetables and fruits are important components of a healthy diet, and if consumed daily in sufficient amounts they could help prevent major non-communicable diseases like cardiovascular diseases, diabetes and certain cancers and nutrient deficiencies (Gockowski *et al.*, 2003; FAO/WHO, 2004).

### 25.1.3 Poverty and economic development

Poverty is defined as a situation where one cannot afford goods and services such as food, clothing, housing, health care and education (Juma, 2011). Half of the world's population live on less than \$2 a day,

*Novel Plant Bioresources: Applications in Food, Medicine and Cosmetics*, First Edition. Edited by Ameenah Gurib-Fakim.

and about 1 billion live on less than $1 a day, leading to their inability to acquire basic necessities (Burke and Lobell, 2010). The situation is particularly pathetic in sub-Saharan Africa, where nearly half of the population lives on less than $1 a day (Juma, 2011). Despite very good economic growth rates over the past decade, the continent continues to face serious challenges meeting the MDGs of poverty and hunger (Juma, 2011).

Agricultural development is linked to overall economic development in African countries, and the sector provides the basis for subsistence for the majority of the people (Juma, 2011). The horticulture sector has grown in the last decade to become the most vibrant industry, and a major foreign exchange earner, employer and source of food in a number of African countries (HCDA, 2011). Horticultural biodiversity has not been strategically repositioned in the horticulture sector to play the rightful role in the development of the economies of African countries. Strategic repositioning will ensure that the rural populace in Africa can make use of the diversity of vegetables and fruits to improve their livelihood and achieve sustainable development (IPGRI, 2004).

### 25.1.4 Climate change and food security

Climate change has been defined as a change in the statistical properties of the climate system when considered over long periods of time (Burke and Lobell, 2010). The term sometimes is used to refer specifically to climate change caused by human activity, as opposed to changes in climate that may have resulted as part of Earth's natural processes. Global climate has been changing, and the last decade of the 20th century and the beginning of 21st century have been the warmest period in the entire global instrumental period (Burke and Lobell, 2010).

Climate change has serious effects on agricultural production and food security; therefore, appropriate mitigation and adaptation strategies must be put in place to avoid disaster. Mitigation is a reduction in greenhouse gas emissions, which eventually leads to a reduction in climate change (Burke and Lobell, 2010). Adaptation, on the other hand, refers to changes made in the system impacted by climate change. Adaptation can include both changes that either reduce negative outcomes or enhance positive outcomes. Many possible adaptations involve direct changes to agricultural systems, such as changing when and where crops are grown (Burke and Lobell, 2010).

Climate change can also shape the decisions farmers make about what crops to grow, and this could potentially alter planting decisions, and therefore micronutrient availability (Rosenzweig and Binswanger, 1993). Farmers might seek to diversify the location or diversity of crops or crop varieties with different sensitivities to climate (Burke and Lobell, 2010).

The prevalence of food insecurity and malnutrition in sub-Saharan Africa has been attributed to the focus on a few high-yielding cultivars of a few staple crops (Akinnifesi *et al.*, 2008). As larger tracts of land were brought under agricultural production and human settlement, the diversity of indigenous fruit trees and vegetable species in farming systems decreased, leading to reduced dietary diversity and micronutrient intake (Welch, 2002). The resultant disruption of natural ecosystems has led to environmental challenges, most notably global warming and climate change. The most affected population segments are the resource-poor rural communities who derive their livelihoods from subsistence farming and the urban poor who tend to overconsume high-energy foods, with low intake of micronutrient-rich foods (Ruel *et al.*, 2004). Dietary diversification is important in providing quality nutrition.

### 25.1.5 Agricultural biodiversity

Agricultural biodiversity is the variety and variability of animals, plants and microorganisms that are used directly or indirectly for food and agriculture (FAO, 2001). It comprises the diversity of genetic resources and species used for food, feed, fibre, fuel and pharmaceuticals. It also includes the diversity of non-harvested species that support production and those in the wider environment that support agro-ecosystems. There is a great need to exploit this agricultural biodiversity in food security, nutrition and economic development (IPGRI, 2004). It is critical to human survival, but it is currently undervalued. This calls for strategically repositioning them so that they can contribute effectively to the challenges facing populations in sub-Saharan Africa. Agricultural biodiversity contributes to dietary diversity, diet intake and morbidity of children (Ekesa *et al.*, 2010). Many of the African indigenous vegetables and fruits have an inbuilt ability to withstand stress and could have higher climate change resilience than exotic ones; for example, baobab thrive during periods of water deficit (Franzel *et al.*, 2008) and can withstand drought.

## 25.2 African indigenous vegetables and fruits

### 25.2.1 Background

African indigenous vegetables and fruits are those that are native to Africa and whose primary or secondary center of origin is believed to be Africa (Maundu *et al.*, 1999; Schippers, 2000). Indigenous vegetables and fruits have originated and evolved over time. They may include some introduced tropical species that have become part of the culture of the African people. Africans have traditionally made use of edible leaves and fruits of plant species growing wild as weeds. These plants were well known to the rural people and were either collected in the wild or planted in home gardens as intercrops with staples. For some indigenous fruits and vegetables, wild collection is still practised in many parts of Africa, especially southern Africa. Indigenous fruit trees grew wild and were not domesticated. In Kenya, for example, there are over 400 indigenous fruit tree plants (Maundu *et al.*, 1999; Franzel *et al.*, 2008).

The introduction of exotic vegetables and fruits to the African continent had a negative impact on the domestication, cultivation and consumption of indigenous vegetables and fruits (Ganry, 2009; Abukutsa-Onyango, 2010). During colonial times, a deliberate suppression of the indigenous vegetables and fruits was done and a lot of effort was made to promote the exotic vegetables and fruits, such as cabbage and oranges. The net effect of such suppression flowed into the post-independence era, where the governments perpetuated the agricultural policies developed by the colonial rulers (Abukutsa-Onyango, 2010). It is only recently that a significant interest toward Africa's indigenous vegetables and fruits has been developed (Williams and Haq, 2002). Even in the area of crop research, indigenous vegetables and fruits have only won some recognition recently at the international, regional and national institutions (Williams and Haq, 2002).

The use of indigenous vegetables and fruits was part of the cultural heritage, playing a significant role in customs and traditions and in maintaining equity within the family structure, since the appearance on the family table depends largely on the activities of women (Mnzava, 1997; Franzel *et al.*, 2008). African indigenous vegetables and fruits have been grown and utilized traditionally by many African communities. They possess several advantages and potentials that need to be fully exploited (Saka *et al.*, 2008; Schippers, 2000).

## 25.2.2 Value and potential of African indigenous vegetables and fruits

### 25.2.2.1 Nutritive value and health benefits

African indigenous vegetables and fruit trees are essential because they are sources of vital micronutrients. They provide minerals and vitamins to diets that are dominated by cereals (Mnzava, 1997; Franzel *et al.*, 2008). In most cases the mineral and vitamin contents are equivalent to or higher than those found in popular exotic vegetables and fruits, like cabbage and oranges. Table 25.1 shows that, on average, 100 g of fresh vegetable contain levels of calcium, iron and vitamins that would provide 100% of the daily requirement (Abukutsa-Onyango, 2003).

African indigenous fruits can be considered as 'super fruits', as they contain both high levels of nutrients and antioxidants that help fight certain diseases. Marula contains 180 mg per 100 g fresh weight of vitamin C compared with orange, which contains 50–75mg per 100 g fresh weight. Baobab fruit, found in most parts of Africa, is a valued source of vitamins and minerals. It contains three times more vitamin C than oranges and 50% more calcium than spinach (Shackleton, 2004).

African oil bean seeds contain 44% protein with 20 essential amino acids, which makes it an ideal substitute for meat. Findings in a study conducted in Nigeria suggest that patients who regularly eat African oil bean products had reduced risks of cancer (Tchoundjeu *et al.*, 2008). White star apple contains more vitamin C than guava or orange. It is an excellent source of calcium, potassium, phosphorus, magnesium, flavonoids and phytochemicals. African mango has been scientifically credited with weight loss properties (Franzel *et al.*, 2008).

**Table 25.1** Nutrient content of 100 g fresh weight edible portion of some indigenous and exotic leafy vegetables

| | Ca (mg) | Fe (mg) | Vit. A (mg) | Vit. C (mg) |
|---|---|---|---|---|
| **Indigenous vegetables** | | | | |
| *Amaranthus spp* | 480 | 10 | 10.7 | 135 |
| *Cleome gynandra* | 262 | 19 | 8.7 | 144 |
| *Solanum villosum* | 442 | 12 | 8.8 | 131 |
| *Vigna unguiculata* | 152 | 39 | 5.7 | 8.7 |
| *Curcubita moschata* | 40 | 2.1 | 3.9 | 170 |
| *Corchorus olitorius* | 360 | 7.7 | 6.4 | 187 |
| **Exotic vegetables** | | | | |
| *Brassica oleracea* | | | | |
| var. *acephala* | 187 | 32 | 7.3 | 93 |
| var. *capitata* | 44 | – | 1.2 | 33 |
| *Lactuca sativa* | 62 | 2.2 | 0.04 | 18 |
| *Spinacia oleracea* | 93 | 32 | 5.1 | 28 |
| RDA | 1.4 | 20 | 12 | 45 |

*Sources*: Maundu *et al.* (1999) and Abukutsa-Onyango (2003).

Vegetables and fruits have also been known to contain substantial amounts of chemicals normally classified as antioxidants. These are phytochemicals that have been credited with disease-fighting properties and help the body reduce the risk of many diet-related diseases, like cancer and heart diseases (FAO/WHO, 2004; Gunasekera *et al.*, 2009).

African indigenous vegetables and fruits have medicinal properties. The bitter species have been known to heal stomach-related ailments (Olembo *et al.*, 1995). Most such vegetables and fruits have been reported to have medicinal properties (Kokwaro, 1993; Olembo *et al.*, 1995); for instance, spider plant has been reported to aid constipation and facilitate birth, while African nightshades have been reported to cure stomachache (Schippers, 2000).

Many of the indigenous vegetables have, nonetheless, been reported to contain anti-nutrient factors. Oxalates found in Amaranthaceae and Solanaceae vegetables may bind calcium and render it unavailable; alkaloids found in the bitter types of *Crotalaria* and *Solanum* species may cause stomachache if eaten in excess (Schippers, 2000). Spider plants contain phenolic compounds, which bind proteins and thereby reduce the nutritional value of the vegetable. The smell of spider plant, caused by an acrid volatile oil, has a high phenolic content and glucosinolates, interfering with iodine metabolism, as occurs in *Brassica carinata* (Schippers, 2000). These factors need to be investigated further, as some of the phenols can be antioxidants. Nevertheless, African indigenous vegetables and fruits have the potential to be the source of revolutionary remedies (Chikamai *et al*, 2004; Abukutsa-Onyango, 2010).

### 25.2.2.2 Economic potential

Indigenous vegetables and fruits play a vital role in the livelihood of over 60% of rural communities in sub-Saharan Africa (Schippers, 2000). Socio-economic surveys on traditional vegetables conducted in various parts of Africa, particularly in Central, West and East Africa, revealed that indigenous vegetables are important commodities in household food security (Schippers, 2000; Abukutsa-Onyango, 2002).

Tamarind and marula are some of the fruits that farmers from several parts of Africa are harvesting and processing to make jams and wines for sale. A participatory fruit tree domestication programme with International Centre for Research in Agroforestry (ICRAF) has revealed that they are of strategic importance to economic empowerment, poverty alleviation and sustainable development. African plum has become an important staple food in Central and West Africa and is becoming a cash crop with an annual trade value of US$ 1.5 million. Domestication of indigenous fruits is pertinent to ensuring the contribution of these fruits to income generation to the rural communities (Jordaan *et al.*, 2008).

For these crops to contribute to economic development, it is important to develop niche and sustainable strategic markets for the products, possibly linking farmers to appropriate domestic and export markets.

#### 25.2.2.3 Agronomic advantages and climate change resilience potential

Indigenous vegetables and fruits are well adapted to harsh climatic conditions and disease infestation and are easier to grow in comparison with their exotic counterparts (Abukutsa-Onyango *et al.*, 2010). African indigenous vegetables and fruits produce a lot of seed under tropical conditions, unlike the exotic vegetables, and also have a short growth period, with most of them being ready for harvesting within 3–4 weeks, and they respond very well to organic fertilizers (Abukutsa-Onyango, 2009a).

Most indigenous vegetables and fruits have an inbuilt ability to withstand and tolerate some biotic and abiotic stresses (Mwai *et al.*, 2004). They can also flourish under sustainable and environmental friendly cropping conditions, like intercropping and use of organics (Abukutsa-Onyango, 2003; Oseko *et al.*, 2005). Furthermore, because most of them have not been intensively selected, they have wide genetic bases, which will be important in sourcing for new genotypes and/or genes for adaptation to climate change (Franzel *et al.*, 2008).

### 25.2.3 Multipurpose uses and genetic diversity

Although leaves and fruits are the most obvious products of indigenous vegetables and fruit trees, they have many other uses as well. Trees grown in the homestead provide important shade for crops. Leaves may be used for fodder or compost. Leaves, fruits, stem and roots may also be used for medicinal purposes. Bark is normally used for fibre and the timber for furniture, house building poles, fences and other constructions (Kokwaro, 1993; Olembo *et al.*, 1995). This, therefore, makes African indigenous fruit trees a very economically efficient undertaking.

Given that most indigenous vegetables and fruit trees are not domesticated, these species provide an important source of genetic diversity, which is important for preserving characteristics that are well adapted to local conditions.

### 25.2.4 Constraints to optimal production and utilization

#### 25.2.4.1 Neglect and stigmatization

Changed food habits in favour of introduced temperate vegetables and Asian and South American fruits lowered the demand of indigenous vegetables and fruits, due to the fact that the former fetched higher prices in local markets. Indigenous vegetables and fruits were considered out of fashion, a poor man's food or child food that could only be used as a last resort. They enjoyed less social prestige, being associated with the low-income group.

They were considered as some kind of non-serious indulgence, and their nutritional benefits were dismissed; therefore, they not given the due attention they deserved and were regarded inferior (Williams and Haq, 2002). As the poor sought to imitate the eating habits of the affluent and were exposed to more fashionable exotic species, the indigenous species became neglected. The neglect and stigmatization have been perpetuated by stakeholders, including policy makers, agricultural training institutions, traders, researchers, consumers and the traders (Mnzava, 1997). Having been branded and denoted by the agriculturalists and researchers as weeds, the tendency was to eradicate them (Maundu *et al.*, 1999; Schippers, 2000).

#### 25.2.4.2 Lack of quality seed

Lack of quality seed has been a major hinderance to sustainable production and utilization of indigenous vegetables and fruits (Abukutsa-Onyango, 2007b; Franzel *et al.*, 2008). Seed production has for a long time virtually remained in the hands of farmers, although seed sales in markets were common. Normally, indigenous vegetables are grown as a subsistence crop, and most farmers save their own seed from season to season and sell surplus to other growers. The quality of such seeds is poor in terms of purity, viability and seed dormancy issues (Abukutsa-Onyango, 2007b). There is a need for production and supply of quality seed to increase yields and quantities produced to meet the unsatisfied market demands of priority indigenous vegetables and fruits, especially in urban centres.

#### 25.2.4.3 Lack of agronomic and utilization technical packages

Indigenous vegetables have often been grown as intercrops with other vegetables or staples; however, there has been hardly any technical information on optimal production and appropriate cropping systems (Oseko, 2007). There has been a lack of agronomic and preparation packages, and access to technical information has been very limited; therefore, extension workers have limited knowledge to advise indigenous vegetable and fruit tree growers. Information on the propagation, cultivation and nutritional contents of indigenous fruits is not normally widely known because of the strong oral tradition in Africa, where recipes have, in many cases, been lost over generations (Musotsi *et al.*, 2005). Developing technical information for the priority species would be a useful strategy for repositioning indigenous vegetables and fruits (Abukutsa-Onyango, 2009a).

## 25.3 Strategic repositioning of indigenous vegetables and fruits in the horticulture

### 25.3.1 Background

Repositioning involves changing the identity of the product relative to the identity of competing products, in the collective minds of the target market. Strategic repositioning is seeking to exploit opportunities or deliberately influence any audience of significance without controlling them. One key to successful strategic repositioning is recognizing that success involves innovative change and requires advocacy, capacity building and marketing for a technology or commodity. This should, as much as possible, involve as many players as possible in the value chain. In repositioning African indigenous vegetables in the horticultural sector, the following strategies are vital: advocacy and promotion; capacity building; sustainable seed supply system conservation; marketing provision of technical information (Abukutsa-Onyango, 2010).

## 25.3.2 Repositioning strategies

### 25.3.2.1 Research

***Multidisciplinary and participatory research***

Multidisciplinary, multi-institutional and participatory research is more effective in the promotion of new commodities like African indigenous vegetables and fruits. A participatory approach ensures success and sustainability as it involves players along the value chain. Fifteen national, regional and international research projects were undertaken between 1992 and 2009 on various aspects of African indigenous vegetables at Jomo Kenyatta University of Agriculture and Technology and Maseno University in Kenya (Abukutsa-Onyango, 2010). One of the most effective ways of achieving scaling up of indigenous vegetable and fruit trees cultivation is to involve farmers in the entire process of participatory selection, propagation, nursery and tree establishment and management of superior planting materials and seed processing (Franzel *et al.*, 2008).

***Identification of priority indigenous vegetables and fruits and domestication***

Priority setting in agricultural research has received great attention, especially for the non-domesticated species (Franzel *et al.*, 2008). Priority setting seeks to bring about consensus among various stakeholders. The designed procedure must ensure simplicity, transparency and participation (Franzel *et al.*, 2008,). Priority setting by various organizations in combination with a series of household, baseline and market surveys has identified priority indigenous vegetables and fruits with great potential (Abukutsa-Onyango, 2007a; Franzel *et al.*, 2008).

Some of the indigenous vegetables identified that are utilized and consumed throughout the continent are African nightshade (*Solanum scabrum*), spider plant (*Cleome gynandra*), jute mallow (*Corchorus olitorius*), vegetable cowpeas (*Vigna unguiculata*), slender leaf (*Crotalaria ochroleuca*), African kale (*Brassica carinata*), vine spinach (*Basella alba*) and West African okra (*Abelmoschus esculentus*) (Schippers, 2000; Abukutsa-Onyango, 2003, 2010).

With regard to African indigenous fruits, the species identified include, amongst others, baobab fruit (*Adansonia digitata*), African oil bean (*Pentaclethra macrophylla*), kola nut (*Cola* sp.), prekese (*Tetrapleura tetraptera*), white star apple (*Chrysophyllum albidum*), African mango (*Irvingia gabonensis*), desert date (*Balanites aegyptiaca*), ebony (*Diospyros* sp.), marula (*Sclerocarya birrea*), tamarind (*Vitex payos*), African fan palm (*Borassus aethiopum*), yeheb (*Cordeauxia edulis*), African plum (*Dacryodes edulis*), shea butter (*Vitellaria paradoxa*) and chocolate berry (*Vitex payos*) (Shackleton and Shackleton, 2005; Franzel *et al.*, 2008).

Domesticating fruit trees means taking them out of the wild and planting them in the garden. This will ensure access to fruits, which will be free and available throughout the year. Most indigenous fruit trees generally grow wild. The fruits are harvested and eaten at home, sold at the market or processed into jams and juices (Franzel *et al.*, 2008). If domesticated and planted on farms, scientists envisage a situation where indigenous fruit trees could save millions of lives by providing for children and adults who are vulnerable to malnutrition. Planting trees will also help rehabilitate degraded environments and reduce soil erosion or desertification. Despite the numerous advantages, many indigenous fruit trees are neglected.

***Germplasm collection and evaluation***

In a study carried out in western Kenya, collected seeds of selected priority African indigenous vegetables were subjected to laboratory analysis and field evaluation. Selection was based on the weight, percentage germination and moisture content. Seeds of selected accessions of spider plant, African nightshade, jute mallow, slender leaf, vegetable cowpea and African kale were multiplied, evaluated and packaged for a test run of seed distribution in Kenya (Abukutsa-Onyango, 2007b). In another study in the East African region, the germplasm of vegetable cowpea, pumpkin, vegetable amaranths and African nightshades were collected and evaluated (Abukutsa-Onyango *et al.*, 2005). Selection of promising lines is in progress and should contribute to development of new varieties.

***Agronomic, physiological and nutritional studies***

While most African indigenous vegetables have reportedly been growing mainly in home gardens, within the homestead (Mnzava, 1997), there are some hardy vegetables like cowpea and slender leaf that could be grown in main cropping lands as intercrops (Abukutsa-Onyango, 2007a). The vegetables were cultivated at a subsistence level, in home gardens and intercrop systems with organic manures, where the use of chemical fertilizers and chemicals was almost non-existent (Abukutsa-Onyango, 2007a). Home gardens, where most of indigenous vegetables and fruits were grown, play a significant role in food security and nutrition in western Kenya according to Musotsi *et al.* (2008). Topping, the removal of the apical stem, could be used to delay flowering in those indigenous vegetables. Those that flower early, like spider plant, tend to have increased plant productivity.

Nutritional evaluation of priority African indigenous vegetables and fruits indicated that they are super foods with regard to minerals, vitamins and antioxidants (Saka *et al.*, 2008; Abukutsa-Onyango *et al.*, 2010).

Cooking and processing has been shown to affect nutrient contents of vegetables, yet many African communities boil vegetables over prolonged periods (Habwe *et al.*, 2008). In light of this, traditional recipes from western Kenya were collected, standardized and evaluated (Musotsi *et al.*, 2005) to ensure minimal loss of nutrients during cooking. The findings from the indigenous knowledge survey on recipes indicated that the use of pots was preferred and recipes were based on a mixture of different vegetable species (Musotsi *et al.*, 2005). Thirty high-iron recipes for African leafy vegetables were developed and evaluated, and it was observed that recipes prepared with lye (traditional salt) were significantly acceptable by testers and had high iron content (Habwe *et al.*, 2009). Recipes of 'The recipes' have been documented and could be adopted by industry, companies, hotels, hospitals and institutions and individuals, especially the younger generation. Many consumers had expressed concern that traditional methods of preparation were time consuming and tedious; this was especially so for the younger generation (Musotsi *et al.*, 2005). Several standardized, consumer-friendly nutritious recipes have been documented for African leafy vegetables for distribution and dissemination (Woomer and Bwari, 2001; Woomer, 2002; Habwe *et al.*, 2009).

### 25.3.2.2 Awareness creation, advocacy and promotion

Advocacy and promotion are powerful tools in repositioning a commodity. This strategy has been used since 1999 to reposition African indigenous vegetables and fruits in the horticulture sector (Abukutsa-Onyango, 2010). The main target groups included farmers, students, researchers, policy makers and consumers. Methodologies used must be diverse and user friendly and encourage participation. These have included orature, song, dance and narratives, demonstration plots, the print media, leaflets, posters and newspapers, lectures, seminars and workshops, exhibitions and shows, radio and TV, documentaries and cooking competitions (Abukutsa-Onyango, 2010).

### 25.3.2.3 Capacity building

University students from agricultural faculties normally form the major human resource in research and extensions services. It is important to build capacity of the human resource that will undertake research on indigenous vegetables and those who will effectively disseminate information on indigenous vegetables (Abukutsa-Onyango, 2009b). Restructuring undergraduate and postgraduate programmes has been undertaken to include indigenous vegetables and fruits in many universities in Kenya and other African countries. National and international students have undertaken research on African indigenous vegetables and fruits (Herbst, 2007).

The World Agroforestry Centre offers training programmes for communities who want to set up their own nurseries, cultivating and domesticating indigenous fruit trees. There are also programmes that train farmers on value addition and processing so that communities can make the most of indigenous fruits both nutritionally and economically. The success of indigenous fruit tree domestication depends on the extent to which the establishment of nurseries by smallholder farmers is accomplished. However, this could be strengthened through training to ensure success and sustainability (Oduol *et al.*, 2008).

To successfully reposition African indigenous vegetables, farmers are very important as they are the producers; so, apart from creating awareness, capacity building and training them are vital. In this endeavour, the study conducted in Kenya between 1992 and 2012 on African indigenous vegetables has trained over 100 farmers on seed production and it is envisaged that these farmers can act as catalysts to the promotion and repositioning of indigenous vegetables at the grassroots (Abukutsa-Onyango, 2009b).

### 25.3.2.4 Conservation

African indigenous vegetables and fruits have been neglected, and some are threatened with extinction. Planned and deliberate conservation efforts as a repositioning strategy need to be put in place to ensure that the genetic diversity is maintained. This could be accomplished through *in-situ* or *ex-situ* conservation. One of the *ex-situ* conservation strategies is the development of botanic gardens to conserve the endangered species. According to Botha *et al.* (2000), 10 gardens were founded in southern Africa between 1960 and 1995. The Maseno University botanic garden in Kenya, which was founded in 2001, has conserved 235 indigenous plants, and 20% are indigenous vegetables and fruits (Abukutsa-Onyango and Onyango, 2005; Omondi, 2013). Some of the indigenous vegetables and fruits conserved at Maseno University botanic garden include *Amaranthus blitum*, *Colocasia esculentum*, *Ipomea batatas*, *Crotalaria brevidens*, *Moringa oleifera*, *Solanum scabrum* and *Psidium guajava* (Omondi, 2013).

### 25.3.2.5 Development of dissemination materials

Diversity of effective dissemination materials would enhance the repositioning of indigenous vegetables and fruits with regard to all players along the value chain. In this endeavour, designing a variety of appropriate and user-friendly methods would be instrumental in promoting sustainable production and utilization of indigenous vegetables and fruits. Some of the methods tested in repositioning underutilized vegetables include leaflets, documentaries and policy briefs (Abukutsa-Onyango, 2010). Development of appropriate resource packs, where listeners are provided with practical information on indigenous fruits and vegetables about the best ways to cultivate, harvest, process and store vegetables and fruits, would maximize their economic potential.

### 25.3.2.6 Development of sustainable quality seed-supply systems

Farmers have consistently cited lack of quality planting materials and seed as a major constraint to diversifying and expanding indigenous fruit tree production (Carandang *et al.*, 2006). Timely provision of quality planting material available to farmers is key to effective scaling up. Farmers can be organized to produce high-quality seed, seedlings and vegetative propagule, as evidenced in a small-scale nursery in West Africa (Akinnifesi *et al.*, 2006) and indigenous vegetable seed production in Kenya (Abukutsa-Onyango, 2010). Three pathways in which improved germplasm can be made available to farmers include: distribution of seed originating from national research centres, community-based NGOs and the private sector; dissemination of selected seeds originating from farmers and farmer groups; and direct diffusion through informal farmer to farmer exchange. Selections made from the germplasm collected for indigenous vegetables have been multiplied, evaluated, bulked, packaged and distributed to farmers (Abukutsa-Onyango, 2010).

### 25.3.2.7 Marketing, commercialization and pricing

Markets in most tropical countries seem to favour the well-known exotic vegetables and fruits, while the indigenous species are restricted to home consumption (Schippers, 2000). In most parts of Africa, marketing of indigenous fruits and vegetables is done by women at village level (Abukutsa-Onyango and Onyango, 2005). At a commercial level, however, it is mostly done by men. Minimal amounts of indigenous vegetables and fruits are exported, and marketing systems and linkages are normally poorly developed. However, African indigenous fruits and vegetables can offer a very good source of income, especially if they are processed (Saka *et al.*, 2008). Farmers should produce what consumers want. Long-distance markets would require proper packaging.

Markets surveys have indicated that the demand for indigenous vegetables is not fully met in Kenyan urban and peri-urban markets (Shackleton *et al.*, 2009). There are potential urban, national, regional and international markets. There is a need for product development research, and private-sector involvement is strongly advocated. ICRAF has facilitated the process of market development through effective partnership and strategic alliances with enterprise development partners and the private sector (Saka *et al.*, 2002). In Malawi, Tanzania and Zambia, cottage processing of indigenous fruits into juice, wine, yogurts and other products has been done (Saka *et al.*, 2002).

#### 25.3.2.8 Development of appropriate policies

Although there is a growing international and national interest in the promotion of African indigenous vegetables and the domestication of indigenous fruit trees, this theoretical support still needs to be confirmed by policy action (Djombo, 2004). The multifaceted nature of the support required for domestication of indigenous fruit trees demands a coherent and consistent policy approach. Policies can broadly be grouped into several categories, including legislative and regulatory issues, marketing issues, and extension and research issues (Ndoye *et al.* 2004). The management and sustainable use of indigenous vegetables and fruits should be governed by an institutional framework and policies at local, national and international levels (Oduol *et al.*, 2008).

The existing policies have several inadequacies, and gaps exist that need to be addressed to ensure conservation and sustainable use of indigenous vegetables and fruits (Chishakwe, 2008).

## 25.4 Concluding remarks

African indigenous vegetables and fruits have been used by Africans for a long period of time and have played a very important role in nutrition, food security and quality of life. They have several values and advantages that need to be exploited for improving the livelihood of the people in sub-Saharan Africa. To fully exploit their potential and reposition them, appropriate strategies need to be selected and implemented.

## References


Abukutsa-Onyango, M.O. (2002) Market survey on African indigenous vegetables in western Kenya, in *Proceedings of the Second Horticulture Seminar on Sustainable Horticultural Production in the Tropics, 6th–9th August 2002. Jomo Kenyatta University of Agriculture and Technology (JKUAT), Juja – Kenya*, (eds J.M.T. Wesonga, G.K. Losenge, K. *et al.*), Department of Horticulture, Jomo Kenyatta University of Agriculture and Technology (JKUAT)/Institute of Vegetable and Fruit Science and Institute of Plant Diseases & Plant Protection, University of Hannover/German Academic Exchange Service (DAAD), Nairobi/Hannover/Nairobi, pp 39–46 (ISBN: 9966-923-27-6).

Abukutsa-Onyango, M.O. (2003) Unexploited potential of indigenous African vegetables in western Kenya. *Maseno Journal of Education Arts and Science*, **4** (1), 103–122.

Abukutsa-Onyango, M.O. (2007a) The diversity of cultivated African leafy vegetables in three communities in western Kenya. *African Journal of Food, Agriculture, Nutrition and Development*, 7 (3), 1–15, http://www.ajfand.net/Volume7/No3/Abukutsa-IPGRI1-3.pdf

Abukutsa-Onyango, M.O. (2007b) Seed production and support systems for African leafy vegetables in three communities in western Kenya. *African Journal of Food, Agriculture, Nutrition and Development*, **7** (3), 1–16, http://www.ajfand.net/Volume7/No3/Abukutsa-IPGRI1-2.pdf.

Abukutsa-Onyango, M.O. (2009a) Development and promotion of technologies for sustainable production and utilization of indigenous vegetables for nutrition security and wealth creation in Kenya, in *Agricultural Innovations for Sustainable Development. Contributions from the Finalists of the African Women Professionals in Science Competition* (eds K. Urama, J. Francis, M. Momanyi *et al.*), Volume 2, Issue 1, African Technology Policy Studies Network, Nairobi, Kenya, pp. 100–108.

Abukutsa-Onyango, M.O. (2009b) The role of universities in promoting underutilized crops: a case of African indigenous vegetables at Maseno University. *Acta Horticulturae*, **806**, 155–162.

Abukutsa-Onyango, M.O. (2010) African indigenous vegetables in Kenya: strategic repositioning in the horticultural sector. Inaugural Lecture, Jomo Kenyatta University of Agriculture and Technology, Nairobi, Kenya, 30 April.

Abukutsa-Onyango, M.O. (2012) Good agricultural practices for under-utilized vegetables in sub-Saharan Africa: current status and future perspectives, in *Proceedings of a Technical Consultation Workshop Held in Arusha, Tanzania, 7–8 December 2009* (eds R. Nono-Womdin, C. Ojiewo, M. Abang and M.O. Oluoch).

Abukutsa-Onyango, M.O. and Onyango, J.C. (2005). Conservation and seed production of African leafy vegetables at Maseno University botanic garden, Kenya. *African Crop Science Conference Proceedings*, 7, 1201–1204, http://www.acss.ws/Upload/XML/Research/726.pdf.

Abukutsa-Onyango, M.O., Kavagi, P., Amoke, P. and Habwe, F.O. (2010) Iron and protein content of priority African indigenous vegetables in the Lake Victoria basin. *Journal of Agricultural Science and Technology*, **2** (1), 67–69.

Akinnifesi, F.K., Kwesiga, F., Mhango, J. *et al.* (2006) Towards the development of miombo fruit trees as commercial tree crops in southern Africa. *Forests, Trees and Livelihoods*, **16**, 103–121.

Akinnifesi, F.K., Chirwa, P.W., Ajayi, O.C. *et al.* (2008) Contributions to agro-forestry research to livelihood of smallholder farmers in southern Africa: 1. Taking stock of the adaptation, adoption and impact of fertilizer tree options. *Agricultural Journal*, **3**, 58–75.

Botha, D.J., Willis, C.K. and Winter, J.H.S. (2000) *Southern African Botanical Gardens Needs Assessment*, Southern African Botanical Diversity Network Report 11, SABONET, South Africa (ISBN 1-919795-54-5).

Burke, M. and Lobell, D. (2010) Climate effects on food security: an overview, in *Climate Change and Food Security: Adapting Agriculture to a Warmer World* (eds D. Lobell and M. Burke), Advances in Global Change Research, vol. 37, Springer, pp. 13–30.

Carandang, W.M., Tolentino, E.L. Jr and Roshetko, J.M. (2006) Smallholder tree nursery operations in Southern Philippines – supporting mechanisms for timber tree domestication. *Forests, Trees and Livelihoods*, **16** (1), 71–83

Chikamai, B., Eyong-Matig, O. and Mbogga, M. (eds) (2004) *Review and appraisal on the status of indigenous fruits in eastern Africa*. A report prepared for IPGRI–SAFORGEN in the framework of AFREA/FORNESSA, IPGRI, Nairobi.

Chishakwe, N.E. (2008) The role of policy in the conservation and extended use of underutilized plant species: across national policy analysis. Global Facilitation Unit for Underutilized Species, Bioversity International, Rome.

Dalal, S., Beunza, J.J. and Volmink, J. (2011) Non-communicable diseases in sub-Saharan Africa: what we know. *International Journal of Epidemiology*, **40**, 885–901.

Djombo, H. (2004) Foreword, in *Forest Products, Livelihoods and Conservation. Case Studies of Non-Timber Forest Product Systems. Volume 2 – Africa* (eds T. Sunderland and O. Ndoye), CIFOR, Indonesia, pp. x–xi.

Ekesa B, Walingo M. and Abukutsa-Onyango, M. (2010) *Agricultural Biodiversity, Diet Intake and Morbidity in Western Kenya: Effect of Agricultural Biodiversity on Diet Intake and Morbidity of Preschoolers in Matungu Division, Western Kenya*, Lambert Academic Press.

FAO (2001) *The State of Food Insecurity in the World*, Food and Agriculture Organization of the United Nations, Rome.

FAO/WHO (2004) Fruit and vegetables for health. Report of a joint FAO/WHO Workshop, 1–3 September, Kobe, Japan.

Franzel, S., Akinnifesi, F.K. and Ham, C. (2008) Setting priorities among indigenous fruit tree species in Africa: examples from southern, eastern and western Africa regions, in *Indigenous Fruit Trees in the Tropics: Domestication, Utilization and Commercialization* (eds F.K. Akinnifesi, R.R.B. Leaky, O.C. Ajayi *et al.*), CABI Publishing, Wallingford, pp. 1–27.

Gala, D.R., Gala, D. and Gala, S. (2007) *Nature Cure for Common Diseases*, Navneet Publications, Mumbai.

Ganry, J. (2009) Current status of fruits and vegetables production and consumption in Francophone African countries – potential impact of health. *Acta Horticulturae*, **841**, 249–256.

Gockowski, J., Mbazo'o, J., Mbah, G. and Moulende, T.F. (2003) African traditional – leafy vegetables and the urban and peri-urban poor. *Food Policy*, **28**, 221–235.

Gunasekera, R.S., Cokenour, J., Arrambide, D. and Somasundaram, S.G. (2009) Bioactive molecules from fruits and vegetables significantly potentiate traditional chemotherapy. *Acta Horticlturae*, **841**, 55–60.

Habwe, F.O., Walingo, K.M. and Abukutsa-Onyango, M.O. (2008) Food processing and preparation technologies for sustainable utilization of African indigenous vegetables for nutrition security and wealth creation in Kenya, in *Using Food Science and Technology to Improve Nutrition and Promote National Development* (G.L. Robertson and J.R. Lupien), International Union of Food Science & Technology, Oakville, Ontario.

Habwe, F.O., Walingo, M.K., Abukutsa-Onyango, M.O. and Oluoch, M.O. (2009). Iron content of the formulated East African indigenous vegetable recipes. *African Journal of Food Science*, **3** (12): 393–397.

HCDA (2011) Horticulture Validated Report. Horticultural Crops Development Authority, Ministry of Agriculture, USAID.

Herbst, M. (2007) Maseno University botanical garden and activities on AIVs and conservation of endangered tropical plants. Student internship report, University of Hannover, Germany.

IPGRI (2004) *Meeting the Millenium Development Goals with Agricultural Biodiversity*, International Plant Genetic Resources Institute, Rome, http://www.underutilized-species.org/documents/Publications/Meeting%20the%20MDGs.pdf.

Jordaan, D.P.S., Akinnifesi, F.K., Ham, C. and Ajayi, O.C. (2008) The feasibility of small scale indigenous fruit processing enterprises in Southern Africa, in *Indigenous Fruit Trees in the Tropics* (eds F.K. Akinnifesi, R.R.B. Leakey, O.C. Ajayi *et al.*), CAB International, Wallingford, pp. 273–287.

Juma, C. (2011) *The New Harvest, Agricultural Innovation in Africa*, Oxford University Press.

Kokwaro, J.O. (1993) *Medicinal Plants of East Africa*, 2nd edn, Kenya Literature Bureau, Nairobi.

Maundu, P.M., Ngugi, G.W. and Kabuye, C.H.S. (1999) *Traditional* Food Plants of Kenya, KENRIK National Museums of Kenya, Nairobi.

Mnzava, N.A. (1997) Vegetable crop diversification and the place of traditional species, in *Traditional African Vegetables. Promoting the Conservation and Use of Underutilized and Neglected Crops. 16. Proceedings of the IPGRI International Workshop on Genetic Resources of Traditional Vegetables in Africa. Conservation and Use, 29–31 August 1995, ICRAF-HQ, Nairobi, Kenya* (ed. L. Guarino), Institute of Plant Genetic and Crop Plant Research/International Plant Genetic Resources Institute, Gatersleben/Rome, pp. 1–15.

Musotsi, A.A., Sigot, A. and Abukutsa-Onyango, M.O. (2005) African Indigenous vegetables recipe documentation and their role in food security, in *Proceedings of the Third Horticulture Workshop on Sustainable Horticultural Production in the Tropics, 26th–29th November 2003. Maseno University (MSU), Maseno-Kenya* (eds M.O. Abukutsa-Onyango, A.N. Muriithi, V.E. Anjichi *et al.*), Department of Horticulture, Maseno University/Horticultural Association of Kenya/Department of Horticulture, Jomo Kenyatta University of Agriculture and Technology (JKUAT)/Institute of Vegetable and Fruit Science and Institute of Plant Diseases & Plant Protection, University of Hannover/German Academic Exchange Service (DAAD), Maseno/Nairobi/Nairobi/Hannover/Nairobi, pp. 105–111 (ISBN: 9966-758-11-9).

Musotsi, A.A., Sigot, A.J. and Abukutsa-Onyango, M.O. (2008) The role of home gardening in household food security in Butere Division of western Kenya. *African Journal of Food Agriculture, Nutrition and Development (AJFAND)*, **8** (4), 375–390.

Mwai, G.N., Onyango, J.C. and M.O.A. Abukutsa-Onyango, M.O. (2004) Effect of salinity on growth and leaf yield of spiderplant (*Cleome gynandra* L.). *African Journal of Food, Agriculture, Nutrition and Development (AJFAND)*, **4** (2).

Ndoye, O., Awono, A., Schreckenberg, K. and Leakey, R. (2004) Commercialising indigenous fruit for poverty alleviation. A policy briefing note for governments in the African Humid Tropics Region, ODI/CIFOR, London/Cameroon.

Oduol, P.A., Ajayi, O.C., Matakala, P. and Akinnifesi F.K. (2008) The role of institutional arrangements and policy on the conservation, utilization and commercialization of indigenous fruits in southern Africa, in *Indigenous Fruit Trees in the Tropics: Domestication,*

*Utilization and Commercialization* (eds F.K. Akinnifesi, R.R.B. Leakey, O.C. Ajayi *et al.*), CABI Publishing, Wallingford, pp. 310–319.

Olembo, N.K., Fedha, S.S. and Ngaira, E.S. (1995) *Medicinal and Agricultural plants of Ikolomani Division, Kakamega District*, Development Partners, Kakamega, Kenya.

Omondi, S. (2013) Annotated checklist of the vascular plants conservation, phytochemical analysis, educational and economic values of the university botanic garden, Maseno. MSc thesis, Maseno University, Kenya.

Oseko J.K. (2007) Performance of African Indigenous vegetables under sole and intercropping systems: a case of African kale (*Brassica carinata*). MSc thesis, Maseno University, Kenya.

Oseko, J.K., Abukutsa-Onyango, M.O. and Onyango, J.C. (2006) Performance of African kale (*Brassica carinata*) to intercropping with other indigenous vegetables, in *Proceedings of the Fifth Workshop on Sustainable Horticultural Production in the Tropics, 23rd–26th November 2005. Agricultural Resource Center, Egerton University, Kenya* (eds D.K. Isutsa, J.M. Wesonga, M.O. Abukutsa-Onyango *et al.*), Horticultural Association of Kenya/Department of Horticulture, Egerton University/Department of Horticulture, Jomo Kenyatta University of Agriculture and Technology/Institute of Biological Production Systems and Institute of Plant Diseases and Plant Protection, Leibniz Universität Hannover/German Academic Exchange Service (DAAD), Nairobi/Egerton/Nairobi/Hannover/Nairobi, pp. 235–250 (ISBN: 9966-923-64-0).

Price, W.A. (2012) *Nutrition and Physical Degeneration*, Price-Pottenger Nutrition Foundation, Lemon Grove, CA.

Rosenzweig, M. and Binswanger, H. (1993) Wealth, weather risk and the composition and profitability of agricultural investments. *Economic Journal*, **103** (416), 56–78.

Ruel, M., Graham, J., Murphy, S. and Allen, L. (2004) Validating simple indicators of dietary diversity and animal source food intake that accurately reflect nutrient adequacy in developing countries. Report submitted to GL-CRSP.

Saka, J.D.K., Mwendo-Phiri, E. and Akinnifesi, F.K. (2002) Community processing and nutritive value of some miombo indigenous fruits in central and southern Malawi, in *Proceedings of the 14th Southern Africa Regional Review and Planning Workshop, 3–7 September 2001, Harare, Zimbambwe* (eds F. Kwesiga, E. Ayuk and A. Agumya), ICRAF, Regional office, Harare, pp. 165–169.

Saka, J.D.K., Kadzere, I., Ndabikunze, B.K. *et al.* (2008) Product development: nutritional value, processing and utilization of indigenous fruits from the Miombo ecosystem, in *Indigenous Fruit Trees in the Tropics: Domestication, Utilization and Commercialization* (eds F.K. Akinnifesi, R.R.B. Leakey, O.C. Ajayi *et al.*), CABI Publishing, Wallingford, pp. 288–309

Schippers, R.R. (2000) *African Indigenous Vegetables: An Overview of the Cultivated Species*, Natural Resources Institute/ACP-EU Technical Centre for Agricultural and Rural Cooperation, Chatham.

Shackleton, C.M. (2004) Use and selection of *Sclerocarya birrea* (marula) in the Bushbackridge lowveld, South Africa, in *Proceedings of Regional Agroforestry Conference on Agroforestry Impacts on Livelihoods in Southern Africa: Putting Research into Practice* (eds M.R. Rao and F.R. Kwesiga), World Agroforestry Centre (ICRAF) Nairobi, pp. 77–92.

Shackleton S.E. and Shackleton C.M. (2005) The contribution of marula (*Sclerocarya birrea*) fruit and fruit products to rural livelihoods in the Bushbuckridge district, South Africa: balancing domestic needs and commercialisation. *Forests, Trees and Livelihoods*, **15**, 3–24.

Shackleton, C.M., Pasquini, M.W. and Drescher, A.W. (eds) (2009) *African Indigenous Vegetables in Urban Agriculture*, Earthscan, London.

Smith, I.F. and Eyzaguirre, P. (2007) African leafy vegetables: their role in the World Health Organization's global fruit and vegetable initiative. *African Journal of Food, Agriculture, Nutrition and Development*, **7** (3), http://www.ajfand.net/Volume7/No3/Smith-IPGRI1-1.pdf (accessed 9 December 2013).

Staggers-Hakim, R. (2012) The impact of globalization on the prevalence of obesity: trends and health implications in sub-Saharan Africa, in *Global Awareness Society International 21st Annual Conference*, New York, May.

Tchoundjeu, Z., Atangana, A., Asaah, E. *et al.* (2008) Domestication, utilization and marketing of indigenous fruit trees in West and Central Africa, in *Indigenous Fruit Trees in the Tropics* (eds F.K. Akinnifesi, R.R.B. Leakey, O.C. Ajayi *et al.*), CAB International, Wallingford, pp. 171–185.

Welch, R.M. (2002) Breeding strategies for bio-fortified staple plants foods to reduce micro-nutrient malnutrition globally. *Journal of Nutrition*, **132**, 495–499.

Williams, J.T. and Haq, N. (2002) *Global Research on Underutilized Crops: An Assessment of Current Activities and Proposals for Enhanced Cooperation*, International Centre for Underutilised Crops, Southampton.

Woomer P.L. (2002) *The Traditional Green Vegetable Cookbook*, FORMAT, Nairobi.

Woomer P.L. and Bwari, L. (2001) *The Traditional Green Vegetable Cookbook*, FORMAT, Nairobi.

# 26 Hepatoprotective, Antiulcerogenic, Cytotoxic and Antioxidant Activities of *Musa acuminata* Peel and Pulp

Fatimah Corazon Abdullah, Lida Rahimi, Zainul Amiruddin Zakaria and Abdul Latif Ibrahim

Natural Products Discovery Laboratory, Institute of Bio-IT Selangor, Universiti Selangor, Malaysia

## 26.1 Introduction

Since ancient times, plants have been commonly used as a source of medicines and health supplements. Despite the availability of synthetic drugs, natural remedies, particularly plant-based products, remain the preferred source of treatment for various diseases worldwide as they are deemed as safe and effective.

Banana of the genus *Musa* are ubiquitously found in Southeast Asia, including Malaysia, as part of the world's primary centre of origin and centre of diversity for this plant. In many tropical and subtropical countries, bananas are popular not just as a nutritious fruit, but also as a traditional source of cure for several ailments, such as gastrointestinal problems, liver disorders, swelling and tumours (Masran, 2011; Rahimi, 2012).

The Natural Products Discovery Laboratory at the Institute of Bio-IT Selangor, Universiti Selangor in Malaysia, has carried out research to obtain scientific evidence on the many uses of banana in traditional medicine. This chapter summarizes the findings from the investigations of the hepatoprotective, antiulcerogenic, antioxidant and cytotoxic activities of *Musa acuminata* cv Kapas, a popular banana cultivated variety (cultivar) commonly eaten raw or cooked in Malaysia, but its pharmacological activities have not been studied.

### 26.1.1 *Description of* Musa acuminata *Colla. cv Kapas*

The fruit of *M. acuminata* Colla. cv Kapas, known in Malaysia as *pisang kapas*, is usually smaller in size than other local bananas, being only 8–12 cm in length and 3–4 cm in diameter. The peel is thin and green but turns yellow when it ripens. The pulp is firm, soft and pale, light yellow in colour, which could be the basis of its local name 'kapas', which means cotton. The tree can grow up to 2–9 m in height and vegetatively grows from the rhizome. The stem part of the plant consists of the pseudostem that forms from the overlapping and tightly packed leaf sheath. Each leaf is 150–400 cm long and 70–100 cm wide (Valmayor *et al.*, 2000).

Unripe fruits of *M. acuminata* Colla. cv Kapas (Figure 26.1) were collected from its natural habitat in a small village in Johol, Negeri Sembilan, Malaysia, and the identification was authenticated by Dr Shamsul Khamis, a botanist from the Institute of Bioscience, Universiti Putra Malaysia. Methanolic extracts of the peel and the pulp were obtained using the method described by Hashim (2011).

### 26.1.2 Phytochemical screening

The phytochemical screening of the methanolic extracts of the peel and pulp of *M. acuminata* Colla. cv. Kapas (Said, 2011; Rahimi, 2012) resulted in the detection of saponins, flavonoids and triterpenes. The presence of tannins and steroids were only detected in the peel extract, while alkaloids were not detected in either the peel or pulp extracts.

### 26.1.3 Acute toxicity test

Rahimi (2012) carried out the acute oral toxicity test on three male and three female Sprague Dawley rats, using 3000 mg/kg b.w. of the methanolic extract of *M. acuminata* peel (MMAPL) and 3000 mg/kg b.w. of the methanolic extract of *M. acuminata* pulp (MMAPP). The rats were observed during the first 30 min after dosing with the extracts, periodically during the first 24 h and daily thereafter for a total of 14 days. The rats did not show behavioural changes, clinical abnormalities, gross lesions, mortality or any other toxic effects during the 2 weeks of observation.

#### 26.1.3.1 Body weight change and liver index

A week after the administration of 3000 mg/kg of peel and pulp extracts, the percentage change of body weight in male and female groups increased. However, only the male rats had an increase in body weight shortly before sacrificing. The female group treated with MMAPP showed no further change in body weight before sacrificing, while the female group given MMAPL showed a decrease in body weight change. This indicates that the female body could be more sensitive than the males (OECD, 2006). For the liver index, all groups showed no significant difference (Table 26.1).

*Novel Plant Bioresources: Applications in Food, Medicine and Cosmetics*, First Edition. Edited by Ameenah Gurib-Fakim.

**Figure 26.1** Unripe fruit of cultivar Kapas

### 26.1.3.2 Analysis of biochemical parameters

Blood samples were obtained from the rats and processed for the analysis of biochemical markers for liver function. Using commercial kits (Roche Diagnostic, USA), serum levels of enzymes, namely alanine transaminase (ALT), aspartate transaminase (AST) and alkaline phosphatase (ALP), as well as total bilirubin (Tbil), total protein (TP) and reduced glutathione (GSH), were analysed. The male and female rats that were given 3000 mg/kg of MMAPL and MMAPP showed almost the same serum level for the enzymes AST, ALT and ALP. For the Tbil, TP and GSH serum levels, the males and females treated with MMAPP showed no significant difference. However, the female group treated with MMAPL showed an increase in Tbil and a decrease in TP and GSH levels compared with the other groups (Table 26.2).

### 26.1.3.3 Histopathological assessment of the liver

An examination of the liver sections from the treated rats revealed normal histoarchitecture. The hepatocyte nucleus and cytoplasm, bile canaliculi, central vein and hepatic sinusoids appeared normal (Figure 26.2). Since there was no necrosis in any of the liver sections, the morphological changes were scored zero.

Since 3000 mg/kg was not toxic to the experimental rats, doses of 100, 200 and 400 mg/kg for MMAPL and MMAPP were used for the studies of the hepatoprotective and antiulcerogenic effects of the banana extracts.

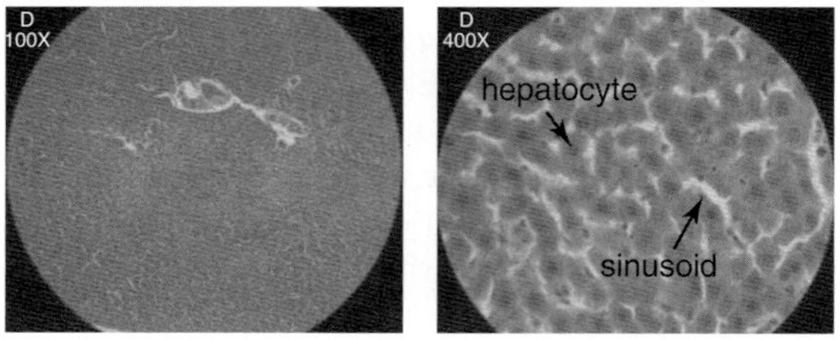


**Figure 26.2** Photomicrographs of haematoxylin-and-eosin (H&E)-stained liver sections of male rats treated with MMAPP. Liver sections from other rat groups showed a similar histoarchitecture. *See plate section for colour detail*

**Table 26.1** Body weight (b.w.) changes and liver index

| Group | B.w change 1 week after administration of the extract (%) | B.w. change shortly before sacrificing the rats (%) | Liver weight/100 g b.w. |
|---|---|---|---|
| Female rats + MMAPL | $8.93 \pm 4.49^b$ | $1.04 \pm 1.04^c$ | $3.91 \pm 0.11^a$ |
| Male rats + MMAPL | $25.19 \pm 0.19^a$ | $44.51 \pm 8.61^a$ | $3.33 \pm 0.14^a$ |
| Female rats + MMAPP | $20.03 \pm 6.67^a$ | $20.04 \pm 6.67^b$ | $3.19 \pm 0.07^a$ |
| Male rats + MMAPP | $15.83 \pm 4.63^a$ | $25.58 \pm 1.68^a$ | $3.27 \pm 0.21^a$ |

Value is mean ± SEM. Number of rats in each group is $n = 3$. Means with different superscripts differ significantly at $p < 0.05$.

**Table 26.2** Effect of MMAPL and MMAPP on the serum level of enzymes, total protein, total bilirubin and reduced glutathione in the female and male rats

| Serum | Female rat + peel extract | Male rat + peel extract | Female rat + pulp extract | Male rat + pulp extract |
|---|---|---|---|---|
| AST (U/I) | $1.0996 \pm 0.60078^a$ | $1.0253 \pm 1.85272^a$ | $1.0834 \pm 2.60154^a$ | $1.0478 \pm 3.12106^a$ |
| ALT (U/I) | $1.3127 \pm 5.22374^a$ | $1.2199 \pm 7.38625^a$ | $1.3687 \pm 10.38827^a$ | $1.3106 \pm 6.04773^a$ |
| ALP (U/I) | $3.1819 \pm 7.69014^a$ | $3.4728 \pm 7.53844^a$ | $3.5495 \pm 8.29521^a$ | $3.4664 \pm 11.15344^a$ |
| Tbil (mg/dL) | $0.5400 \pm 0.12288^b$ | $0.2433 \pm 0.03383^a$ | $0.3167 \pm 0.01453^a$ | $0.2433 \pm 0.02848^a$ |
| TP (mg/dL) | $3.0700 \pm 0.07000^b$ | $3.8367 \pm 0.78231^a$ | $4.0700 \pm 0.47089^a$ | $3.6600 \pm 0.33005^a$ |
| GSH (μg/mg of protein) | $4.8200 \pm 0.46004^b$ | $5.6633 \pm 0.33168^a$ | $5.7233 \pm 0.54743^a$ | $5.2200 \pm 0.47149^a$ |

Value is mean ± SEM. Number of rats in each group is $n = 3$. Means with different superscripts differ significantly at $p < 0.05$.

## 26.2 Hepatoprotective activity

### 26.2.1 Introduction

The liver is a vital organ that serves as a major site of carbohydrate, protein and lipid metabolism, detoxification, secretion of bile and storage of vitamins (Nirmala *et al.*, 2012). Liver diseases still remain as a worldwide health problem. Jaundice and hepatitis are major liver disorders that account for high death rates (Amat *et al.*, 2010). Liver injury can be induced either by toxic chemicals (e.g. carbon tetrachloride ($CCl_4$) and alcohol) or drugs (e.g. paracetamol (PCM), a widely used analgesic, and isoniazid) (Krishna *et al.*, 2009; Malar and Bai, 2009). Drug-induced liver injury is a major health concern and hepatotoxicity are reported to be much higher in developing countries than in advanced countries with a similar dose schedule (Raj *et al.*, 2009).

Although conventional or synthetic drugs are available for the treatment of liver diseases, they are still inadequate, and sometimes can cause adverse side effects. Silymarin (SMN), a mixture of flavonolignans from the fruits of *Silybum marianum*, is the standard drug for treating hepatoxicity, but cases of nausea, epigastric discomfort, arthralgia, pruritus, headache and urticaria among patients had been reported (Yam *et al.*, 2010; Saller *et al.*, 2011). Therefore, there is a need to find alternative solutions from natural sources that can provide a better effect than existing drugs.

Traditionally, a tonic from banana stem has been used to relieve liver problems (Nirmala *et al.*, 2012). Hence, we carried out studies to investigate the hepatoprotective activity of MMAPL and MMAPP in $CCl_4$- and PCM-induced hepatotoxicity models in rats.

### 26.2.2 Pretreatment and induction of hepatoxicity

In our laboratory, Sprague Dawley healthy adult male rats weighing 180–200 g were used in the experiments following the methods described by Masran (2011) and Ali (2011). The rats (102) were divided into 17 groups containing six rats each. MMAPL and MMAPP were orally administered for 7 days to the test groups at doses of 100, 200 and 400 mg/kg. For PCM-induced hepatotoxicity, PCM was orally administered with a dose of 3 g/kg b.w. (p.o.); for the $CCl_4$-induced hepatotoxicity, 1 ml/kg b.w. of $CCl_4$ was intraperitoneally administered to the rats on the eighth day. For the control groups, the amount of 50 mg/kg b.w. of standard drug of SMN in two positive control groups and normal saline in two negative control groups were orally administered for seven consecutive days. The normal control group only received ordinary feeding and did not receive MMAPL, MMAPP or hepatotoxins.

### 26.2.3 Evaluation of hepatoprotective activity

Various parameters were considered to assess the hepatoprotective effect of MMAPL and MMAPP. These included body weight change, liver weight, biochemical markers and histopathological studies as performed by Rahimi (2012), Masran (2011) and Ali (2011). Results were analysed statistically by one-way analysis of variance followed by a *post hoc* Tukey test using SPSS 11.5 for Windows software. Data are expressed as the mean plus/minus the standard error of the mean (SEM). A $p$ value <0.05 is regarded as statistically significant.

### 26.2.4 Body weight change and liver weight

After induction of liver damage, rats suffered from pain, causing them to lose appetite and consequently lose body weight, as was observed in the negative control groups. However, pretreatment with MMAPL and MMAPP significantly reduced body weight loss compared with the negative control group in both PCM (Masran, 2011) and $CCl_4$ (Ali, 2011) experimental groups (Table 26.3).

An increase in liver weight is expected after inducing hepatotoxicity due to increased lipid accumulation, hepatocellular death and inflammation (Masran, 2011). Hence, the highest liver weight was observed in the negative groups, and a significant reduction was observed in the groups treated with MMAPL and MMAPP and SMN.

For the groups treated with the hepatotoxin $CCl_4$, the hepatoprotective effect of MMAPL and MMAPP is comparable to SMN. However, with PCM as hepatotoxin, the hepatoprotective effect of MMAPL is comparable to SMN but the effect of MMAPP is significantly milder (Table 26.3).

### 26.2.5 Biochemical parameters

Liver function can be assessed using biological parameters such as the enzymes ALT, AST and ALP, and GSH, Tbil and TP.

**Table 26.3** Percentage change in body weight and liver weight after pretreatment with MMAPL and MMAPP followed by hepatotoxin

| Prescription | Change in body weight (%) | | Liver weight (g/100 g b.w.) | |
|---|---|---|---|---|
| | PCM | $CCl_4$ | PCM | $CCl_4$ |
| Normal (control) | $12.24 \pm 0.28^a$ | $12.24 \pm 0.28^a$ | $3.07 \pm 0.26^a$ | $3.07 \pm 0.26^a$ |
| SMN + hepatotoxin (positive control) | $12.40 \pm 0.14^a$ | $12.73 \pm 0.30^b$ | $3.25 \pm 0.13^a$ | $3.25 \pm 3.25^a$ |
| Normal saline + hepatotoxin (negative control) | $22.75 \pm 1.05^b$ | $21.59 \pm 0.74^c$ | $7.16 \pm 0.41^b$ | $4.86 \pm 5.53^b$ |
| 100 mg/kg peel + hepatotoxin | $13.54 \pm 0.67^a$ | $14.54 \pm 0.65^b$ | $3.44 \pm 0.11^a$ | $3.97 \pm 3.97^a$ |
| 200 mg/kg peel + hepatotoxin | $13.61 \pm 1.02^a$ | $15.44 \pm 0.39^b$ | $3.78 \pm 0.32^a$ | $3.48 \pm 3.48^a$ |
| 400 mg/kg peel + hepatotoxin | $13.54 \pm 2.15^a$ | $13.54 \pm 0.88^b$ | $3.45 \pm 0.12^a$ | $3.24 \pm 3.24^a$ |
| 100 mg/kg pulp + hepatotoxin | $17.66 \pm 1.97^c$ | $13.83 \pm 0.71^b$ | $5.89 \pm 0.31^c$ | $4.65 \pm 3.77^a$ |
| 200 mg/kg pulp + hepatotoxin | $17.33 \pm 1.72^c$ | $14.17 \pm 0.66^b$ | $5.12 \pm 0.35^c$ | $3.66 \pm 3.66^a$ |
| 400 mg/kg pulp + hepatotoxin | $17.45 \pm 2.14^c$ | $13.13 \pm 0.63^b$ | $5.61 \pm 0.16^c$ | $3.53 \pm 3.53^a$ |

Value is in mean ± SEM. Number of animals in each group is $n = 6$. Means with different superscripts differ significantly $p < 0.05$.

**Table 26.4** Effect on serum enzyme levels after pre-treatment with MMAPL and MMAPP followed by PCM or $CCl_4$

| Group | Serum enzyme (U/l) | | | | | |
|---|---|---|---|---|---|---|
| | PCM as inducer | | | $CCl_4$ as inducer | | |
| | ALT | AST | ALP | ALT | AST | ALP |
| Normal (control) | $121.48 \pm 5.50^a$ | $44.02 \pm 1.94^a$ | $220.97 \pm 5.27^a$ | $121.48 \pm 5.50^a$ | $44.02 \pm 1.94^a$ | $220.97 \pm 5.27^a$ |
| SMN (positive control) | $127.83 \pm 9.17^a$ | $49.58 \pm 3.88^a$ | $241.03 \pm 23.83^a$ | $148.17 \pm 4.40^a$ | $47.91 \pm 1.83^a$ | $246.30 \pm 3.83^a$ |
| Normal saline (negative control) | $243.00 \pm 13.29^b$ | $250.78 \pm 13.22^b$ | $512.55 \pm 13.53^b$ | $232.10 \pm 9.15^b$ | $159.45 \pm 5.83^b$ | $378.75 \pm 18.94^b$ |
| 100 mg/kg MMAPL | $125.83 \pm 7.26^a$ | $63.26 \pm 6.23^a$ | $228.35 \pm 8.55^a$ | $133.83 \pm 6.37^a$ | $59.47 \pm 2.97^a$ | $233.57 \pm 6.19^a$ |
| 200 mg/kg MMAPL | $129.87 \pm 5.94^a$ | $66.38 \pm 11.06^a$ | $248.78 \pm 12.28^a$ | $126.53 \pm 7.61^a$ | $60.15 \pm 4.39^a$ | $246.97 \pm 11.56^a$ |
| 400 mg/kg MMAPL | $137.17 \pm 8.03^a$ | $60.15 \pm 9.03^a$ | $240.02 \pm 14.35^a$ | $150.78 \pm 4.57^a$ | $63.71 \pm 3.63^a$ | $243.05 \pm 9.21^a$ |
| 100 mg/kg MMAPP | $176.38 \pm 6.24^c$ | $122.00 \pm 9.58^c$ | $370.43 \pm 17.45^c$ | $137.71 \pm 4.61^a$ | $59.70 \pm 5.35^a$ | $255.67 \pm 5.52^a$ |
| 200 mg/kg MMAPP | $178.93 \pm 7.67^c$ | $137.11 \pm 13.63^c$ | $384.08 \pm 32.56^c$ | $145.60 \pm 6.19^a$ | $54.08 \pm 5.65^a$ | $257.00 \pm 5.34^a$ |
| 400 mg/kg MMAPP | $180.18 \pm 4.70^c$ | $140.06 \pm 9.42^c$ | $363.10 \pm 15.68^c$ | $143.51 \pm 2.75^a$ | $58.45 \pm 4.53^a$ | $260.07 \pm 3.71^a$ |

Value is mean ± SEM. Number of animals in each group is $n = 6$. Means with different superscripts differ significantly at $p < 0.05$.

Table 26.4 shows that PCM and $CCl_4$ overdosage caused a high increase in the levels of ALT, AST and ALP. Owing to liver injury, the transport function of the hepatocytes gets disturbed, resulting in the leakage of the plasma membrane, endoplasmic reticulum, mitochondria and Golgi apparatus, thereby causing increased enzyme levels in the serum (Dikshit *et al.*, 2011).

Supplementation with 100, 200 and 400 mg/kg of MMAPL and MMAPP showed a hepatoprotective effect comparable to the positive control in the $CCl_4$-induced rats. However, for the PCM-induced groups, the liver-protecting effect of the MMAPP was significantly lower than the MMAPL. These results agree with the observations in the body weight change and liver weight.

The effect of pretreatment with MMAPL and MMPP on serum levels of Tbil, TP and GSH due to hepatotoxicity with PCM and $CCl_4$ is summarized in Table 26.5. A significant elevation in Tbil and a significant reduction in TP and GSH serum levels were observed in the negative control group compared with the normal control group.

For both the PCM- and $CCl_4$-induced groups, there was an apparent decrease in Tbil level with increasing dose of the peel and pulp extracts. In the PCM-induced group pretreated with 100, 200 and 400 mg/kg doses of MMAPL and 200 mg/kg dose of MMAPP, the Tbil serum levels were reduced to normal level, which was significantly lower ($p < 0.05$) than the negative control group. In the $CCl_4$-induced group, a similar pattern was observed, although the MMAPL doses gave slightly better reduction than the 100 and 200 mg/kg doses of MMAPP. The most significant reduction of Tbil serum level ($p < 0.05$) was obtained in the rats pretreated with 400 mg/kg of MMAPP in both hepatotoxin groups when compared with the other doses of peel and pulp extracts, and with the normal control and positive control.

Tbil is a useful indicator of liver damage, where bilirubin is a byproduct of the breakdown of red blood cells in response to cellular peroxidative changes in the liver. Jaundice or icterus can result in humans due to a high level of Tbil (Chatterjea and Shinde, 2005).

In addition, TP serum levels increased in a dose-dependent manner in the groups pretreated with peel and pulp extracts compared with the negative control. The most significant elevations were

**Table 26.5** Effect of MMAPL and MMAPP followed by PCM or $CCl_4$ on serum levels of Tbil, TP and GSH

| Group | PCM-induced hepatotoxicity | | | $CCl_4$-induced hepatotoxicity | | |
|---|---|---|---|---|---|---|
| | Tbil (mg/dL) | TP (mg/dL) | GSH (μg/mg of protein) | Tbil (mg/dL) | TP (mg/dL) | GSH (μg/mg of protein) |
| Normal (control) | $0.35 \pm 0.08^a$ | $4.58 \pm 0.47^a$ | $6.67 \pm 0.35^a$ | $0.35 \pm 0.08^a$ | $4.58 \pm 0.47^a$ | $6.67 \pm 0.35^a$ |
| SMN (positive control) | $0.39 \pm 0.03^a$ | $4.42 \pm 0.37^a$ | $5.56 \pm 0.39^{b,c}$ | $0.45 \pm 0.02^b$ | $4.58 \pm 0.48^a$ | $5.41 \pm 0.40^c$ |
| Normal saline (negative control) | $0.83 \pm 0.06^d$ | $1.38 \pm 0.17^d$ | $3.60 \pm 0.23^d$ | $0.84 \pm 0.58^d$ | $1.44 \pm 0.16^d$ | $4.06 \pm 0.30^d$ |
| 100 mg/kg MMAPL | $0.42 \pm 0.05^a$ | $2.60 \pm 0.20^c$ | $5.22 \pm 0.31^c$ | $0.39 \pm 0.06^a$ | $2.38 \pm 0.19^c$ | $5.06 \pm 0.27^c$ |
| 200 mg/kg MMAPL | $0.33 \pm 0.05^a$ | $2.94 \pm 0.30^c$ | $6.03 \pm 0.19^b$ | $0.29 \pm 0.04^a$ | $2.97 \pm 0.27^c$ | $6.14 \pm 0.25^b$ |
| 400 mg/kg MMAPL | $0.30 \pm 0.04^a$ | $4.09 \pm 0.55^b$ | $6.82 \pm 0.25^a$ | $0.30 \pm 0.05^a$ | $4.62 \pm 0.49^a$ | $6.54 \pm 0.25^a$ |
| 100 mg/kg MMAPP | $0.53 \pm 0.03^c$ | $2.72 \pm 0.44^c$ | $5.31 \pm 0.27^c$ | $0.58 \pm 0.15^c$ | $2.99 \pm 0.51^c$ | $5.44 \pm 0.26^c$ |
| 200 mg/kg MMAPP | $0.41 \pm 0.04^a$ | $3.93 \pm 0.28^b$ | $5.88 \pm 0.32^b$ | $0.46 \pm 0.05^b$ | $3.79 \pm 0.21^b$ | $5.88 \pm 0.32^b$ |
| 400 mg/kg MMAPP | $0.23 \pm 0.03^b$ | $3.63 \pm 0.45^b$ | $6.57 \pm 0.41^a$ | $0.22 \pm 0.27^e$ | $3.80 \pm 0.35^b$ | $6.91 \pm 0.19^a$ |

Value is in mean ± SEM. Number of rats in each group is $n = 6$. Means with different superscripts differ significantly at $p < 0.05$.

with 200 and 400 mg/kg doses of MMAPP and 400 mg/kg doses of MMAPL in the PCM- and $CCl_4$-induced rats. However, the TP serum levels were lower ($p < 0.05$) in these extract-treated groups in comparison with the normal and positive control groups, except the $CCl_4$-induced rats pretreated with 400 mg/kg doses of MMAPL, which had a TP level comparable to the normal control.

The liver synthesizes amino acids and most of the blood plasma proteins. Many plasma proteins and immunoglobulins are synthesized in the rough endoplasmic reticulum in the hepatocytes and later discharged into the plasma. Hepatic necrosis due to increased *N*-acetyl-*p*-benzoquinone-imine (NAPQI) leads to a decline in TP level in the blood serum (Saladin, 2005).

With regard to the GSH serum level, similar results were obtained from the PCM- and $CCl_4$-induced groups, where MMAPL and MMAPP increased the GSH levels in a dose-dependent manner. The GSH serum levels with 400 mg/kg dose of peel and pulp extracts for both hepatotoxins, and with 200 mg/kg dose of peel and pulp extracts for $CCl_4$-induced groups, were comparable to the normal control group. For the $CCl_4$-induced groups, treatment with 100 mg/kg dose of MMAPL and MMAPP have increased to the range of the positive control, while the 100 and 200 mg/kg doses of MMAPL and MMAPP in PCM-treated rats have similar GSH levels to the positive control group.

GSH is a non-enzymatic antioxidant widely distributed in liver cells that functions mainly as a free-radical scavenger. $CCl_4$ forms the trichloromethyl free radical $CCl_3^{\cdot}$ that subsequently interacts with polyunsaturated fatty acids and initiates lipid peroxidation (Dikshit *et al.*, 2011). Hence, an increase in the amount of free radicals in the negative control leads to a drastic fall in GSH serum level, but SMN and the MMAPL and MMAPP treatments helped elevate GSH level. On the other hand, PCM forms the toxic metabolite NAPQI, which reacts with GSH to yield a nontoxic 3-GS-yl-paracetamol. Depletion of GSH causes the remaining quinine to bind to DNA and cellular proteins to produce protein adducts that result in hepatocyte dysfunction and death (Nirmala *et al.*, 2012).

### 26.2.6 Gross morphology of the liver

Normal healthy liver was observed to be smaller than damaged liver, and the colour is dark brick-red with a smooth surface. In the PCM-damaged liver, the colour becomes pale red and many red spots indicating inflammation were observed. In the $CCl_4$-damaged liver, an unevenly distributed pale brownish colour is observed, indicating lipid peroxidation. However, pretreatment with MMAPL and MMAPP in both models showed an improvement in the liver colour (Figure 26.3).

### 26.2.7 Histopathological assessment of the liver

Liver sections of the normal control group were examined to serve as a basis of comparison for the experimental groups. Normal hepatic cells, each containing a purplish small round nucleus, sometimes with nucleolus, surrounded by the pink cytoplasm, were clearly visible. The bile canaliculi between hepatocytes, the central vein surrounded by hepatic cord of cells and hepatic sinusoids are also clearly seen (Figure 26.4).

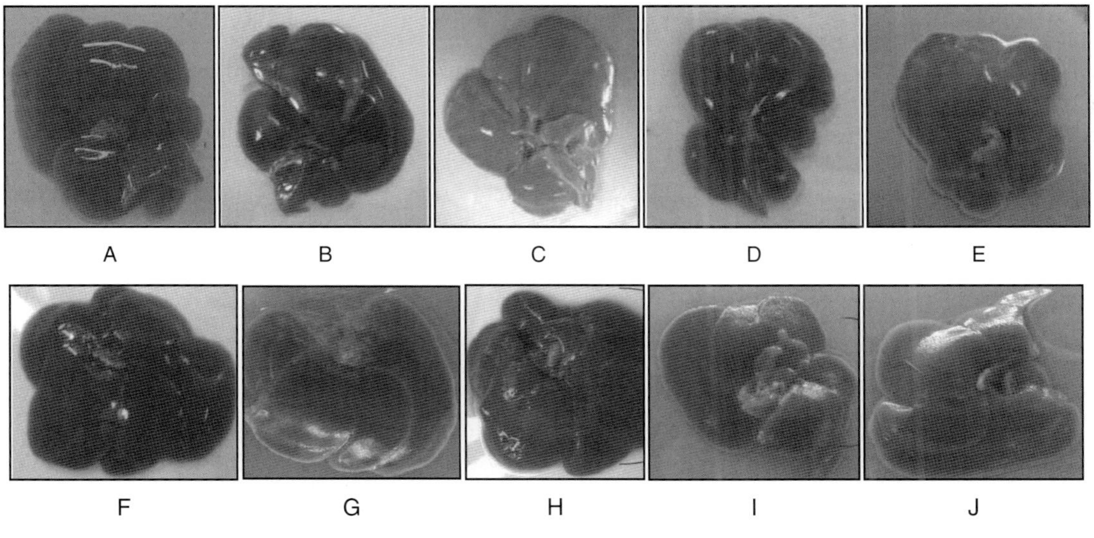


**Figure 26.3** Colour change observed in the livers isolated from the control and treated groups. Pale brownish-red colour reflects toxicity level, hepatic leakage and lipid accumulation. Red spots indicate hepatocellular necrosis and inflammation. (a) Dark brick-red of normal liver (control); (b) dark brick-red of SMN + PCM-induced liver (positive control); (c) pale brownish-red of saline + PCM-induced liver (negative control); (d) dark brick-red of 400 mg/kg MMAPL + PCM-induced liver; (e) pale brownish-red of liver 400 mg/kg MMAPP + PCM-induced liver; (f) dark brick-red of normal liver (control); (G) bright red of SMN + $CCl_4$-induced liver (positive control); (h) brown of saline + $CCl_4$-induced liver (negative control); (i) bright red of 400 mg/kg MMAPL + $CCl_4$-induced liver; (j) bright red of 400 mg/kg MMAPP + $CCl_4$-induced liver. *See plate section for colour detail*

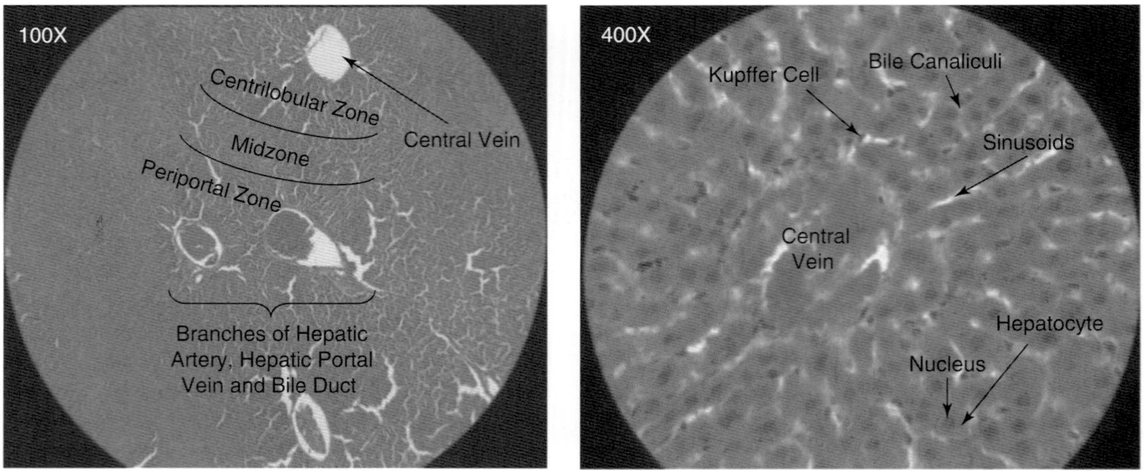


**Figure 26.4** Photomicrographs under light microscope of H&E-stained section from normal liver group. *See plate section for colour detail*

#### 26.2.7.1 Effect of methanolic extracts of *Musa acuminata* peel and pulp on the histoarchitecture of the liver due to hepatotoxicity with paracetamol

In the negative control group of PCM-intoxicated rats, severe hepatic necrosis was observed in the liver sections as shown by the presence of abnormal sinusoids, invisible bile canaliculi, inflammation and severe hepatic macrovesicular steatosis in all zones of lobules. In contrast, the histoarchitecture of the liver sections in the positive control group showed more than 90% normal hepatic cells, normal sinusoids and bile canaliculi, and the absence of hepatic macrovesicular steatosis (Figure 26.5). In rats pretreated with 100 and 200 mg/kg doses of MMAPL, liver sections were similar to the negative control. However, the liver sections from rats pretreated with 400 mg/kg of MMAPL (Figure 26.6) presented normal hepatocytes, sinusoids and bile canaliculi at the periportal zone and midzone, but some inflammation at the centrilobular zone. Rat groups pretreated with 200 and 400 mg/kg doses of MMAPP (Figure 26.7) showed the absence of inflammation and hepatic macrovesicular steatosis and almost all of the hepatocytes were of normal shape. Sinusoids and bile canaliculi in the centrilobular zone, midzone and periportal zone also appeared normal.

#### 26.2.7.2 Effect of the peel and pulp of *Musa acuminata* extracts on the histoarchitecture of the liver due to hepatotoxicity with carbon tetrachloride

In $CCl_4$ hepatotoxicity, the liver sections in the negative control group showed abnormality in sinusoids and bile canaliculi with evidence of extensive hepatic necrosis, hepatic macrovesicular steatosis and inflammation. As expected, the positive control group showed normal sinusoids and bile canaliculi with normality in all hepatic cells and no evidence of hepatic macrovesicular steatosis and inflammation. Liver sections pretreated with 100 mg/kg dose of MMAPL and 200 mg/kg of MMAPP, showed almost the same histoarchitecture as the negative control group. Likewise, in 100 mg/kg of MMAPP, severe macrovesicular steatosis, inflammation, abnormal sinusoids were observed with invisible bile canaliculi. Only some normal hepatocytes were visible at the periportal zone. However, the histoarchitecture of the liver sections which were pretreated with 200 and 400 mg/kg of MMAPL presented abnormal sinusoids, severe hepatic macrovesicular steatosis and inflammation, especially at the centrilobular zone. Some normal hepatocytes at the periportal zone (400 mg/kg dose MMAPL), normal nucleus of hepatocytes in all zones and normal bile canaliculi (200 mg/kg dose MMAPP) are observed. Finally, pretreatment

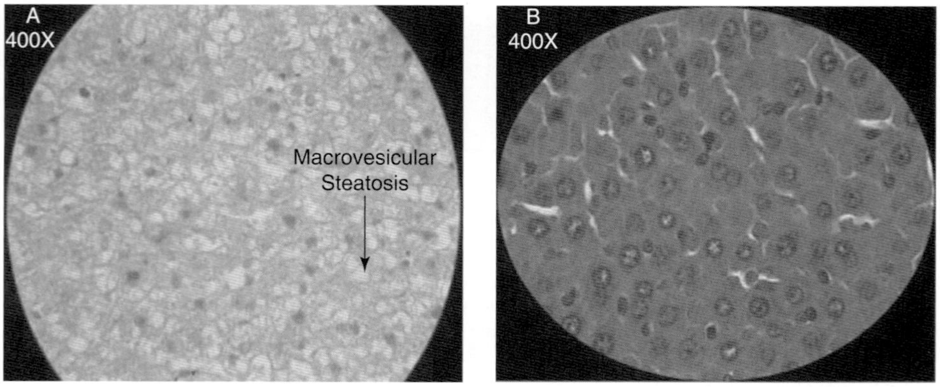


**Figure 26.5** (a) Negative control group; (b) positive control group. *See plate section for colour detail*

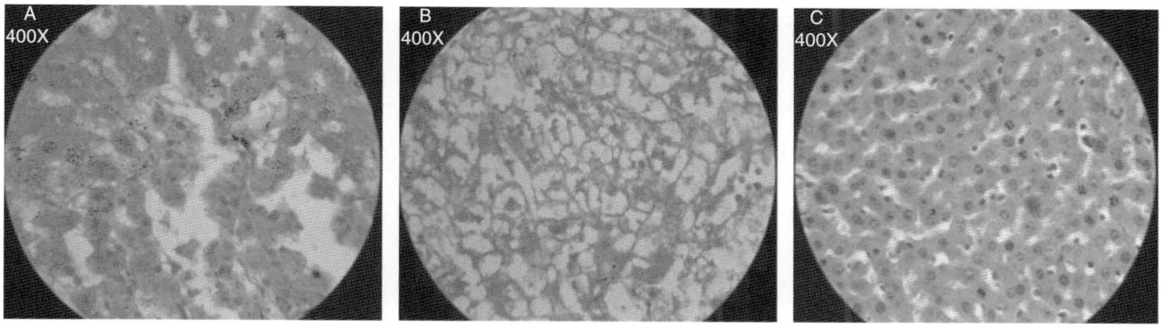


**Figure 26.6** Pretreated with (a) 100 mg/kg, (b) 200 mg/kg and (c) 400 mg/kg MMAPL. *See plate section for colour detail*

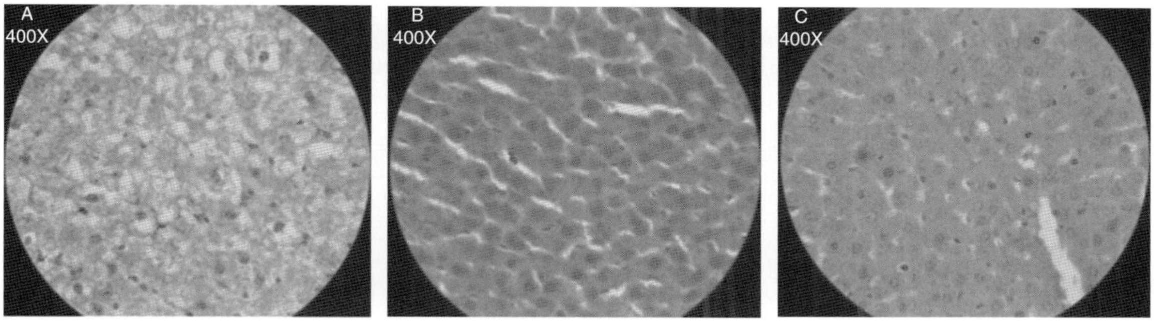


**Figure 26.7** Pretreated with (a) 100 mg/kg, (b) 200 mg/kg and (c) 400 mg/kg MMAPP. *See plate section for colour detail*

with 400 mg/kg of MMAPP presented normal hepatocytes in all zones of lobule, mild macrovesicular steatosis, no inflammation, normal bile canaliculi and sinusoids approaching normal shape. Based on the morphological changes, the liver sections were graded for hepatic necrosis due to hepatotoxicity with PCM and $CCl_4$ (Table 26.6).

**Table 26.6** Necrosis score

| Group | Necrosis score | |
|---|---|---|
| | PCM-induced | $CCl_4$-induced |
| Normal control group | $0.00 \pm 0.00^a$ | $0.00 \pm 0.00^a$ |
| Normal saline + hepatotoxin (negative control) | $4.00 \pm 0.00^c$ | $3.50 \pm 0.22^c$ |
| SMN + hepatotoxin (positive control) | $0.50 \pm 0.29^a$ | $0.50 \pm 0.22^a$ |
| 100 mg/kg peel extract + hepatotoxin | $2.00 \pm 1.15^b$ | $3.20 \pm 0.58^c$ |
| 200 mg/kg peel extract + hepatotoxin | $2.00 \pm 0.41^b$ | $3.00 \pm 0.71^c$ |
| 400 mg/kg peel extract + hepatotoxin | $0.00 \pm 0.00^a$ | $1.50 \pm 1.50^b$ |
| 100 mg/kg pulp extract + hepatotoxin | $3.25 \pm 0.48^c$ | $2.50 \pm 0.50^c$ |
| 200 mg/kg pulp extract + hepatotoxin | $2.00 \pm 0.82^b$ | $2.50 \pm 0.34^c$ |
| 400 mg/kg pulp extract + hepatotoxin | $0.75 \pm 0.75^a$ | $2.20 \pm 0.37^b$ |

Value is mean ± SEM. Number of animals each group is $n = 6$. Means with different superscripts differ significantly at $p < 0.05$.

The observations in all the parameters studied, namely body weight change and liver weight, biochemical parameters, gross morphology and histopathological studies, support each other. Damage induced by the hepatotoxins were alleviated by MMAPP and MMAPL, and at certain doses, especially 400 mg/kg MMAPP, showed hepatoprotection similar to the commercial drug SMN.

Saponins, flavonoids and triterpenes present in both MMAPP and MMAPL could be responsible for the antihepatotoxic activity observed in our studies, which could mainly be due to their antioxidant and anti-inflammatory activities (Rahman, 2008). Rao *et al.* (2012) have previously proven that saponins in the methanolic extract of *Musa paradisiaca* L. var. Bontha pulp possess hepatoprotective action. Saponins from ivy plant also have high antioxidant activity (Gülçin *et al.*, 2004).

Flavonoids also act as antioxidants, protecting the liver from free radicals and increasing the synthesis of glutathione (Hemat, 2004; Rahman, 2008). Tannins in MMAPL also exhibit radical scavenging property (Ogbonna *et al.*, 2011).

## 26.3 Antiulcerogenic activity

Many previous studies had already been done to investigate the antiulcerogenic activity of banana extracts (Yahaya, 2011). Nevertheless, our laboratory still carried out a study to assess the antiulcerogenic effect of *M. acuminata* Colla. cv Kapas as no previous study had yet been done on this cultivated variety.

The gastroprotective effect of the MMAPL and MMAPP was evaluated in rats using two models, namely, ethanol-induced ulcer (Abdul Aziz, 2011) and indomethacin-induced gastric lesion (Hashim, 2011).

### 26.3.1 Ethanol-induced gastric damage

Eight groups, each consisting of six healthy Sprague Dawley male rats, were orally pretreated with saline solution (ulcer control group), omeprazole 30 mg/kg (reference group), and 100, 200 and 400 mg/kg of MMAPL and MMAPP an hour before oral administration of absolute alcohol to generate mucosal injury. The animals were sacrificed after 4 h, the stomach dissected along the greater curvature and the areas of striated haemorrhagic lesions in the gastric walls were scored to calculate mean ulcer index per group and percentage of inhibition of ulceration.

The ulcer control group exhibited severe mucosal injury, whereas the groups pretreated with extracts showed significantly reduced mucosal damage as indicated by the ulcer index and percentage of ulcer inhibition (Table 26.7). The peel extract inhibited gastric mucosal injury more effectively than the pulp extract did. Interestingly, the different concentrations of MMAPL gave a gastroprotective effect comparable to omeprazole, a standard drug for curing gastric ulcer. It appears that 200 mg/kg of MMAPL gave the highest percentage inhibition of ulceration (95.6%). However, a further increase of dosage to 400 mg/kg of MMAPL and MMAPP did not lead to a significant increase in gastroprotective effect.

Histological sections of the stomach from the different rat groups were examined and gastric injury was scored using the method described by Al-Qarawi *et al.* (2005) and Abdul Hisam (2013). The severity of histopathological changes were expressed according to an arbitrary scale (between − and +++). Results from the histopathological examination of stomach sections showed that the different concentrations of MMAPL and MMAPP significantly reduced haemorrhage, oedema and congestion in a non-dose-dependent manner (Figure 26.8 and Table 26.8). This supports the observations of the gross morphology of the dissected stomach.

**Table 26.7** The effect of MMAPL, MMAPP and standard drug omeprazole on induced gastric mucosal injury in rats

| Pretreatment | Dose (mg/kg) | Ethanol-induced ulcer | | Indomethacin-induced ulcer | |
|---|---|---|---|---|---|
| | | Ulcer index | Inhibition of ulceration (%) | Ulcer index | Inhibition of ulceration (%) |
| Normal saline (negative control) | – | $37.67 \pm 4.40^{a}$ | | $47.50 \pm 2.28^{a}$ | |
| Omeprazole (positive control) | 30 | $4.33 \pm 0.63^{b}$ | 88.51 | $4.00 \pm 0.42^{b}$ | 91.34 |
| MMAPL | 100 | $3.17 \pm 1.74^{b}$ | 91.58 | $9.00 \pm 0.76^{b}$ | 80.51 |
| MMAPL | 200 | $1.67 \pm 0.54^{c}$ | 95.57 | $4.00 \pm 0.42^{b}$ | 91.34 |
| MMAPL | 400 | $4.33 \pm 0.68^{b}$ | 88.51 | $6.83 \pm 0.83^{b}$ | 85.20 |
| MMAPP | 100 | $21.67 \pm 2.10^{b}$ | 42.47 | $6.17 \pm 0.54^{b}$ | 86.64 |
| MMAPP | 200 | $12.17 \pm 1.30^{b}$ | 67.69 | $6.33 \pm 0.67^{b}$ | 86.28 |
| MMAPP | 400 | $13.00 \pm 1.42^{b}$ | 65.49 | $8.67 \pm 0.63^{b}$ | 81.22 |

Value is mean ± SEM. Number of animals each group is $n = 6$. Means with different superscripts differ significantly at $p < 0.05$.

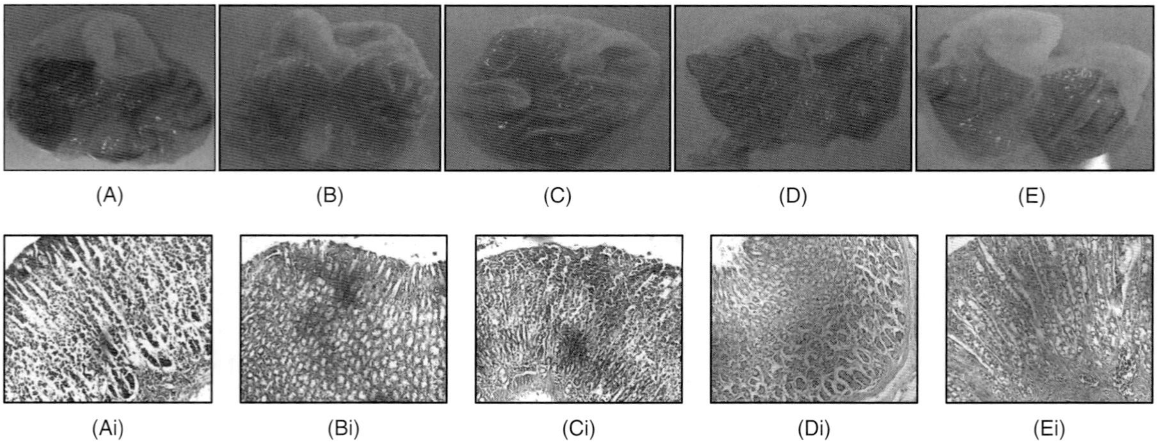


**Figure 26.8** Histological evaluation of antiulcer activity of MMAPL and MMAPP against ethanol-induced gastric lesions in rats. (a) Stomach of an ulcer control rat; (b) stomach of a rat pretreated with omeprazole; (c) stomach of a rat treated with 200 mg/kg MMAPL; (d) stomach of a rat treated with 400 mg/kg MMAPL; (e) stomach of a rat pretreated with 400 mg/kg MMAPP. The respective histopathological sections (ai)–(ei) are also shown: (ai) stomach of the negative control animal showing moderate effect on mucosal epithelium, severe haemorrhage, moderate oedema and congestion, and mild leucocytes infiltration and cellular debris; (bi) stomach of the positive control animal showing mild effect on mucosal epithelium, mild haemorrhage, moderate oedema and mild congestion; (ci) stomach of the animal that received 200 mg/kg of MMAPL showing mild effect on mucosal epithelium, moderate haemorrhage and mild oedema, leucocytes infiltration and congestion; (di) Stomach of the animal that received 400 mg/kg of MMAPL showing mild effect on mucosal epithelium, mild haemorrhage, oedema, congestion and leucocytes infiltration; (ei) stomach of the animal that received 400 mg/kg of MMAPP showing mild effect on mucosal epithelium, moderate haemorrhage and oedema

**Table 26.8** Histopathological evaluation of MMAPL and MMAPP on ethanol-induced gastric ulcer in rats

| Pretreatment | Mucosal epithelium | Glands | Haemorrhage | Oedema | Congestion | Leucocytes infiltration | Cellular debris |
|---|---|---|---|---|---|---|---|
| Saline (negative control) | ++ | – | +++ | ++ | ++ | + | + |
| Omeprazole (positive control) | + | – | + | ++ | + | – | – |
| 100 mg/kg MMAPL | + | – | - | + | – | – | – |
| 200 mg/kg MMAPL | + | – | ++ | + | + | + | – |
| 400 mg/kg MMAPL | + | – | + | + | + | + | – |
| 100 mg/kg MMAPP | + | – | + | – | – | + | – |
| 200 mg/kg MMAPP | ++ | – | + | + | + | + | – |
| 400 mg/kg MMAPP | + | – | ++ | ++ | – | – | – |

The severity of histopathological changes is expressed according to an arbitrary scale (between – and +++) adopted from Al-Qarawi *et al.* (2005) and Abdul Hisam (2013). –, normal; +, mild ulcer effect; ++, moderate ulcer effect; +++, severe ulcer effect.

### 26.3.2 Indomethacin-induced gastric lesion

For this model, induction of ulcer by indomethacin was done 1 h after administering the extracts, and 4 h after treatment the rats were sacrificed and the stomachs were removed for evaluation of the gastric lesion. As in the ethanol-induced ulcer, the results indicate that 200 mg/kg of MMAPL showed the most effective dose, where the percentage of inhibition is 91.43%. However, there is no significant difference in the antiulcerogenic effect among the different doses of MMAPL and MMAPP, all of which have values similar to omeprazole (Table 26.7). Results based on the histopathological analysis of the stomach sections are not included as they show very similar data to the ethanol-induced groups.

### 26.3.3 Mechanisms of antiulcerogenic activity

Ethanol and indomethacin induce ulcers via different, but related, mechanisms that weaken the integrity of the gastric mucosa. Absolute ethanol is highly corrosive to the gastric mucosa. The ethanol-induced gastric lesion has been proven to involve a combination of direct toxic action of ethanol, superficial necrosis, a decrease in gastric wall mucus and reduction in the secretion of bicarbonate (Okoli *et al.*, 2009). Moreover, ethanol has also been reported to increase release of histamine and stimulation of leukotriene C4, which act on gastric microvasculature and result in destruction of the mucosa and submucosa (Abdul Aziz, 2011). Ethanol also reduces endogenous glutathione and prostaglandin levels, as well as increases influx of $Ca^{2+}$ and oxygen free-radical synthesis. This, in turn, leads to lipid peroxidation, causing changes in cell membrane permeability and, consequently, cell damage (Okoli *et al.*, 2009). Release of reactive oxygen-derived species (ROS), such as hydroperoxy and superoxide anion free radicals, resulting from the metabolism of ethanol leads to acute and chronic ulceration in the gastric mucosa. Based on several studies, ROS and lipid peroxidation initially attack and damage many biological molecules (e.g. prostaglandins) followed by a continuing chain of reactions that will lead to cell injury and ultimately cell death (Londonkar and Poddar, 2009). The ability of MMAPL and MMAPP to reduce ethanol-induced gastric ulcer in the present study suggests the involvement of the cytoprotective effect associated with the antioxidant and free-radical scavenging effects of the extracts.

Indomethacin is a non-steroidal anti-inflammatory drug commonly used to relieve fever, pain, stiffness and swelling. It induces ulcer by acting as a nonselective inhibitor of cyclooxygenase 1 and 2, which are enzymes that contribute in the synthesis of prostaglandins from arachidonic acid through the cyclooxygenase pathway (Okokon and Nwafor, 2009; Okoli *et al.*, 2009). Prostaglandins play a protective role in the stomach by maintaining the mucosal blood flow, regulating mucosal turnover and repair, and promoting the secretion of mucus and bicarbonate within the gastric compartment (Hiruma-Lima *et al.*, 2006; Okoli *et al.*, 2009). Suppression of prostaglandin synthesis by indomethacin causes instability of the mucous lining, thus making the stomach susceptible to mucosal injury and gastroduodenal ulceration. Hence, the ability of MMAPP and MMAPL to inhibit ulcer formation induced by indomethacin suggests that the extract may have cytoprotective activity, possibly by increasing synthesis of prostaglandins.

The antiulcerogenic effect of both the MMAPL and MMAPP could be due to the flavonoids, triterpenes and saponins present in them as detected through phytochemical screening. These compounds exhibit several biological effects, including anti-inflammatory and antioxidant activities (Vincken *et al.*, 2006; Rahman, 2008; Rao *et al.*, 2012), which could be associated with their gastroprotective effect. A flavonoid isolated from plantain banana *Musa sapientum* L. var *paradisiaca*, leucocyanidin, was found to protect the gastric mucosa from aspirin-induced erosions by significantly increasing mucus thickness (Lewis and Shaw, 2001).

Tannins and steroids were also present in the MMAPL. These compounds have also been reported to have anti-inflammatory, antioxidant and antiulcer activities (Oliveira *et al.*, 2005; Prabha *et al.*, 2011).

## 26.4 Cytotoxic activity

Concoctions from different banana plant parts, such as the rhizome, flowers, bract and fruits, have been traditionally used to cure tumours. Several scientific studies have also been done to explore the anticancer and antitumour potential of various *Musa* species, such as *M. sapientum, M. paradisiaca* and *Musa acuminata* (Abdullah *et al.*, 2011; Dass, 2012).

Research was performed in our laboratory by Dass (2012) to evaluate the cytotoxic activity of the Malaysian wild indigenous banana, *Musa acuminata malaccensis*, locally known in Malaysia as *pisang hutan*, which means 'forest banana'. This plant can grow 3–4 m high with a slender pseudostem, and the fruit bunch uniquely

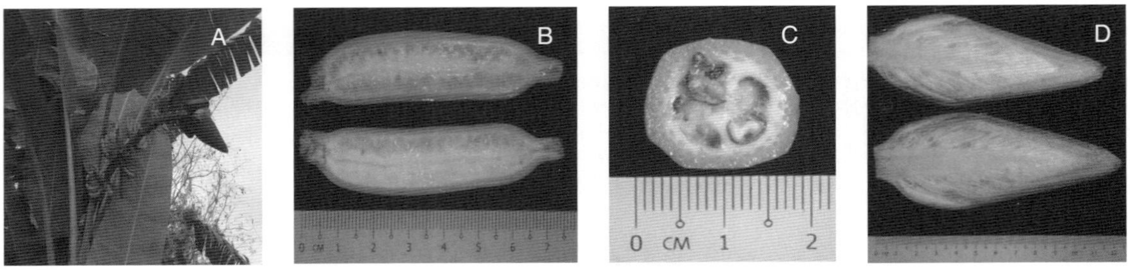


**Figure 26.9** (a) Fruit bunch; (b,c) highly seeded fruit without substantial pulp; (d) longitudinally cut heart (bracts and flowers)

emerges upward and the long empty rachis terminates with the downward-pointing male bud. The fruit is edible but is not commonly consumed by humans because it contains more seeds than pulp (Figure 26.9).

Unripe fruit and heart samples of *M. acuminata malaccensis* were collected from the plant's natural habitat in Shah Alam, Selangor, Malaysia, and authenticated by Dr Shamsul Khamis, a botanist from the Institute of Bioscience, Universiti Putra Malaysia. Methanolic fruit extract of *M. acuminata malaccensis* (FEMAM) and heart extract (HEMAM) were obtained and tested against the cancer cell lines, hepatic carcinoma cell line (HepG2), breast adenocarcinoma cell line (MCF-7) and fetal hepatic normal cells (WRL-68). The cytotoxicity of both extracts (at decreasing concentrations of 100 µg/mL , 50, 25, 12.5 and 6.25 µg/mL) was evaluated using an MTT-based cytotoxic assay, and the effectiveness of the extracts was determined based on the $IC_{50}$ value obtained from the plotted concentration-dependent curve.

Table 26.9 gives the $IC_{50}$ values for FEMAM and HEMAM against MCF-7, HepG2 and WRL-68 for 24, 48 and 72 h treatment durations. FEMAM successfully obtained an $IC_{50}$ value (47.74 µg/mL) at 72 h treatment duration against HepG2 cells. However, FEMAM exhibited only a weak cytotoxic effect against MCF-7, as there was cell inhibition but it did not reach an $IC_{50}$ value. HEMAM caused 50% cell inhibition on MCF-7 cells at 96.36 µg/mL after 72 h of treatment but it did not exhibit cytotoxic activity against HepG2 cells except in the 72 h treatment, where mild cytotoxicity was observed. Interestingly, both FEMAM and HEMAM exhibited similar non-cytotoxic effects against the fetal hepatic normal cell line WRL-68, and the percentage of cell viability increased as the treatment time increased.

The observed cytotoxic effect of HEMAM against MCF-7 agrees with the findings of Jenshi *et al.* (2011), which proved that *M. acuminata* bracts exhibited strong anticancer activities towards MCF-7 breast adenocarcinoma. Crude extracts from the rhizome of *M. acuminata* also indicated anticancer activity (Adinarayana and Ajay, 2011). Natural Musaceae plant extracts, including banana extracts, are shown to inhibit proteasome activity and induce apoptosis selectively in human tumours (Aslamuzzaman *et al.*, 2003).

The cytotoxic activity of HEMAM and FEMAM against the cancer cell lines in this study could be attributed to the flavonoids, saponins, triterpenes and tannins present in the extracts. These compounds are associated with strong antioxidant activity and anti-inflammatory activity which are closely linked to anticancer effects (Dass, 2012; Rahimi, 2012).

## 26.5 Antioxidant activity

Plants in general are known to be rich in antioxidants that serve as effective free-radical scavengers, and thus prevent or cure various illnesses. Evaluation of the antioxidant activity MMAPL and MMAPP, as well as the methanolic extracts of *Musa balbisiana* cv Nipah heart (MMBHT), which consists of flowers and bracts, and *M. balbisiana* cv Nipah pulp (MMBPP) was done at the Biology Laboratory, Natural Product Division, Forest Research Institute Malaysia. Four assays were used: a superoxide radical scavenging system, which evaluates the scavenging activity of the extract on superoxide free-radical anions; the 1,1-diphenyl-2-picrylhydrazyl (DPPH) radical scavenging assay, which assesses the reducing activity of the sample that determines its antioxidant potential; total phenolic content (TPC) analysis, which indicates the amount of polyphenols in the extract tested equivalent to gallic acid standard;

**Table 26.9** $IC_{50}$ values of FEMAM and HEMAM against MCF-7, HepG2 and WRL-68 for three treatment periods

| Cell line | FEMAM | | | HEMAM | | |
|---|---|---|---|---|---|---|
| | 24 h | 48 h | 72 h | 24 h | 48 h | 72 h |
| MCF-7 | * | * | * | * | * | + (96.36 µg/mL) |
| HepG2 | – | * | + (47.74 µg/mL) | – | – | * |
| WRL-68 | * | * | – | * | * | – |

+: 50% inhibition of cell ($IC_{50}$); –: no inhibition of cells; *: some inhibition but it did not reach $IC_{50}$.

**Table 26.10** Antioxidant profile for a sample concentration of 200 μg/mL based on superoxide scavenging and DPPH scavenging capacity

| Extract | Radical scavenging (%) | |
|---|---|---|
| | DPPH[a] | Superoxide[b] |
| MMPHT | 11.7 ± 0.4 L | 73.60 ± 0.55 H |
| MMPPP | 0.58 ± 0.0 L | 60.90 ± 0.25 M |
| MMAPP | 2.85 ± 0.3 L | 55.70 ± 2.20 M |

Data expressed as percentage of radical scavenging activity, is mean value of triplicate wells in duplicate experiments; SEM <5%.
[a]Standard: ascorbic acid 200 μg/mL.
[b]Standard: superoxide dismutase $6 \times 10^{-3}$ U/mL.

**Table 26.11** TPC of MMBHT, MMBPP and MMAPP for a sample concentration of 200 μg/mL against gallic acid (GAE) standard curve

| Extract | TPC mg/100 g GAE |
|---|---|
| MMBHT | 32.18 ± 0.75 |
| MMBPP | 33.29 ± 0.53 |
| MMAPP | 61.43 ± 1.12 |

Data are mean values of triplicate wells in duplicate experiments; SEM <5%.

and oxygen radical absorbance capacity (ORAC), which measures more accurately the antioxidant capacity of samples.

Table 26.10 shows the antioxidant profile for MMBHT, MMBPP and MMAPP based on their DPPH scavenging and superoxide scavenging activity. MMBHT has a high superoxide scavenging capacity, whereas MMBPP and MMAPP only have moderate superoxide scavenging capacity compared with the standard superoxide dismutase. However, all these samples exhibited a low DPPH radical scavenging capacity compared with the standard ascorbic acid.

TPC analysis (Table 26.11) shows that MMAPP has almost twice the TPC of MMBHT and MMBPP.

MMBHT has the highest ORAC value, which is more than twice the ORAC value of MMAPL and five times more than that of MMAPP (Table 26.12). In our study of the antiulcerogenic effect of the methanolic extracts *M. acuminata* Colla. cv Kapas against ethanol and indomethacin-induced gastric lesion, the MMAPL

**Table 26.12** ORAC of MMPHT, MMAPL and MMAPP

| Extract | ORAC value (μmol TE/100 g) | SEM |
|---|---|---|
| MMBHT | 19 330.8 | 2 420.0 |
| MMAPL | 8 845.6 | 347.7 |
| MMAPP | 3 487.0 | 35.0 |

TE: Trolox equivalent. Data are mean values of triplicate wells in duplicate experiments, SEM <30%.

showed a better gastroprotective effect than MMAPP. Likewise, in the PCM-induced hepatotoxicity, MMAPL also alleviated liver damage more significantly than MMAPP. These observations could be due to the higher antioxidant activity (as reflected by ORAC value) of the peel compared with the pulp.

## 26.6 Conclusion

Results of the investigations indicate that the methanolic extracts of unripe *M. acuminata* Colla. cv Kapas peel and pulp do possess significant hepatoprotective and gastroprotective effects as demonstrated in the animal model. These ameliorative effects could be related to the saponins, flavonoids and triterpenes in the peel and pulp extracts, and to the tannins present in the peel extract.

Methanolic extracts of the heart (consisting of bracts and flowers) and fruit (including peel and pulp) of *M. acuminanata malaccensis* also have moderate cytotoxic activity against cancer lines, but exhibit non-cytotoxic effect on the normal cell line.

Methanolic extract of the heart of *M. balbisiana* cv Nipah was shown to have a high superoxide-radical scavenging capacity and a high oxygen-radical absorbance capacity. Further investigations need to be carried out to explore the pharmacological activities of the three *Musa* species used in our studies.

## References


Abdullah, F.A., Vun, Y.L., Mohamad Wali, H.S. *et al.* (2011) *Peeling the Scientific Facts of Banana*, 1st edition, Shah Alam, Universiti Selangor.

Abdul Aziz, M.A.A. (2011) Anti-ulcerogenic activity of peel and pulp methanolic extracts of *Musa acuminata* against ethanol-induced ulcer *in vivo*. Bachelor's thesis, Universiti Selangor, Bestari Jaya Campus, Kuala Selangor, Selangor, Malaysia.

Abdul Hisam, E.E. (2013) Antiulcer activity of *Bauhinia purpurea* leaf extracts. Master's thesis, Universiti Teknologi MARA, Puncak Alam Campus, Bandar Puncak Alam, Selangor, Malaysia.

Adinarayana, K.P.S. and Ajay, P.B. (2011) Anti-oxidant activity and cytotoxicity of ethanolic extracts from rhizome of *Musa acuminata*. *Natural Science*, **3**, 291–294.

Ali, M.F. (2011) Hepatoprotective potential of peel and pulp methanolic extracts of *Musa acuminata* against carbon tetrachloride induced hepatotoxicity. Bachelor's thesis, Universiti Selangor, Bestari Jaya Campus, Kuala Selangor, Selangor, Malaysia.

Al-Qarawi, A.A., Abdel-Rahman, H., Ali, B.H. *et al.* (2005) The ameliorative effect of dates (*Phoenix dactylifera* L.) on ethanol-induced gastric ulcer in rats. *Journal of Ethnopharmacology*, **98**, 313–317.

Amat, N., Upura, H. and Blazekoví, B. (2010) *In vivo* hepatoprotective activity of the aqueous extract of *Artemisia absinthium L.* against chemically and immunologically induced liver injuries in mice. *Journal of Ethnopharmacology*, **131**, 478–484.

Aslamuzzaman, K., Daniel, A.U., Deborah, J.K. *et al.* (2003) A natural Musaceae plant extract inhibits proteasome activity and induces apoptosis selectively in human tumor and transformed, but not normal and non-transformed cells. *International Journal of Molecular Medicine*, **12**, 879–887.

Chatterjea, M.M. and Shinde, R. (2005) *Textbook of Medical Biochemistry*, 6th edn, Jaypee Brothers Medical Publishers, New Delhi.

Dass, S.A. (2012) Evaluation of the cytotoxic effect of wild banana (*Musa acuminata malaccensis*) against selected cell lines. Bachelor's thesis, Universiti Selangor, Bestari Jaya Campus, Kuala Selangor, Selangor, Malaysia.

Dikshit, P., Tyagi, M.K., Shukla, K. *et al.* (2011) Hepatoprotective effect of stem of *Musa sapientum* Linn in rats intoxicated with carbon tetrachloride. *Annals of Hepatology*, **10** (3), 333–339.

Gülçin I.O., Küfrevioglu, I., Oktayb, M. and Büyükokuroglu, M.E. (2004) Antioxidant, antimicrobial, antiulcer and analgesic activities of nettle (Urtica dioica L.). *Journal of Ethnopharmacology*, **90** (2004), 205–215.

Hashim, M.H. (2011) The effect of peel and pulp methanolic extracts of *Musa acuminata* on indomethacin-induced gastric mucosal injury in rats. Bachelor's thesis, Universiti Selangor, Bestari Jaya Campus, Kuala Selangor, Selangor, Malaysia.

Hemat, R.A.S. (2004) *Orthomolecularism: Principles and Practice*. Urotext, Bloomington, IN .

Hiruma-Lima, C.A., Calvo, T.R., Rodrigues, C.M. *et al.* (2006) Antiulcerogenic activity of *Alchornea castaneafolia*: effects on somatostatin, gastrin and prostaglandin. *Journal of Ethnopharmacology*, **104**, 215–224.

Krishna, K.L., Mruthunjaya, K. and Patel, J.A. (2009) Antioxidant and hepatoprotective activity of leaf extract of *Justicia gendarussa* burm. *International Journal of Biological Chemistry*, **3** (3), 99–110.

Lewis, D.A. and Shaw, G.P. (2001) A natural flavonoid and synthetic analogues protect the gastric mucosa from aspirin-induced erosions. *Journal of Nutritional Biochemistry*, **12** (2), 95–100.

Londonkar, R.L. and Poddar, P.V. (2009) Studies on activity of various extracts of *Mentha arvensis* Linn against drug induced gastric ulcer in mammals. *World Journal of Gastrointestinal Oncology*, **1**, 82–88.

Malar, H.L.V. and Bai, S.M.M. (2009) Hepatoprotective activity of *Phyllanthus emblica* against paracetamol induced hepatic damage in Wister Albino rats. *African Journal of Basic and Applied Science*, **1** (2), 21–25.

Masran, S.H. (2011) Evaluation of the hepatoprotective activity of methanol extract of peel and pulp of Malaysian banana variety 'pisang kapas' (*Musa acuminata)* on the paracetamol-induced liver damage in rats. Bachelor's thesis, Universiti Selangor, Bestari Jaya Campus, Kuala Selangor, Selangor, Malaysia.

Nirmala, M., Girija, K., Lakshman, K. and Divya, T. (2012) Hepatoprotective activity of *Musa paradisiaca* on experimental animal models. *Asian Pacific Journal of Tropical Biomedicine*, **2**, 11–15.

OECD (2006) *Guidelines for the Testing of Chemicals (OECD) Guideline 425. Acute Oral Toxicity: Up-and-Down Procedure*, adopted 23 March 2006, Organization for Economic Cooperation and Development, Paris.

Ogbonna, N., Matthew, M.C. and Okechukwu, D. (2011) Inhibition of copper corrosion by acid extracts of *Gnetum africana and Musa acuminata* peel. *International Journal of Multidisciplinary Sciences and Engineering*, **2** (5), 5–9.

Okokon, J.E. and Nwafor, P.A. (2009) Antiulcer and anticonvulsant activity of *Croton zambesicus. Pakistan Journal of Pharmaceutical Science*, **22** (4), 384–390.

Okoli, C.O., Ezike, A.C., Akah, P.A. *et al.* (2009). Studies on wound healing and antiulcer activities of extract of aerial parts of *Phyllanthus niruri* L. (Euphorbiaceae). *American Journal Pharmacological Toxicology*, .**4**, 118–126.

Oliveira, L., Freire, C.S.R., Silvestre, A.J.D. *et al.* (2005) Steryl glucosides from banana plant *Musa acuminata* Colla var Cavendish. *Industrial Crops and Products*, **22** (3), 187–192.

Prabha, P., Karpagam, T., Varalakshmi, B. and Packiavathy, S.A. (2011) Indigenous anti-ulcer activity of *Musa sapientum* on peptic ulcer. *Pharmacognosy Research*, **3** (4), 232–238.

Rahimi, L. (2012) Hepatoprotective activity of *Musa acuminata* peel and pulp methanolic extracts against carbon tetrachloride and paracetamol induced liver damage in rats. Master's thesis, Universiti Selangor, Bestari Jaya Campus, Kuala Selangor, Selangor, Malaysia.

Rahman, A.U. (2008) *Studies in Natural Products Chemistry*, Vol. **34**, Elsevier, Oxford.

Raj, D.S., Venilla, J.J., Aiyavu, C. and Panneerselvam, K. (2009) The hepatoprotective effect of alcoholic extract of *Annona squamosa* leaves on experimentally induced liver injury in Swiss albino mice. *International Journal of Integrative Biology*, **5** (3), 182–186.

Rao, M.N., Prasad, S.H.K.R and Jyothirmayi, N. (2012) Efficacy of ripened and unripened fruit extracts of *Musa* x *paradisiaca* l. (Bontha cultivar) against human pathogens. *International Journal of Pharmacy and Pharmaceutical Sciences*, **4** (1), 455–460.

Said, M.R. (2011) Phytochemical screening and brine shrimp toxicity of the methanolic extracts of selected Malaysian banana varieties (*Musa* spp). Bachelor's thesis, Universiti Selangor, Bestari Jaya Campus, Kuala Selangor, Selangor, Malaysia.

Saladin, K.S. (2005) *Human Anatomy*, McGraw-Hill, New York, NY.

Saller, R., Meier, R. and Brignoli, R. (2011) The use of silymarin in the treatment of liver diseases. *Drugs*, **61**, 2035–2063.

Valmayor, R.V., Jamaluddin, S.H., Silayao, B. *et al.* (2000). *Banana cultivar names and synonyms in Southeast Asia*, International Network for the Improvement of Banana and Plantain – Asia and The Pacific Office, Laguna, Philippines, http://kukr.lib.ku.ac.th/Fulltext_kukr/KU0222075c.pdf (accessed 5 December 2013).

Vincken, J.P., Heng, L., Groot, A.D. and Gruppen, H. (2006) Saponins, classification and occurence in the plant kingdom. *Journal of Phytochemistry*, **68**, 275–297.

Yahaya, M.A.F. (2011) Gastroprotective effect of peel and pulp methanolic extracts of *Musa acuminata* on experimental gastric lesions in rats. Bachelor's thesis, Universiti Selangor, Bestari Jaya Campus, Kuala Selangor, Selangor, Malaysia.

Yam, M.F., Ang, L.F., Lim, C.P. *et al.* (2010) Antioxidant and hepatoprotective effects of *Murdannia bracteata* methanol extract. *Journal of Acupuncture and Meridian Studies*, **3**, 197–202.

# 27 Plant Bioresources and their Nutrigenomic Implications on Health

Maznah Ismail[1,2] and Mustapha Umar Imam[1]

[1]Laboratory of Molecular Biomedicine, Institute of Bioscience, Universiti Putra Malaysia, Serdang, Selangor, Malaysia
[2]Department of Nutrition and Dietetics, Faculty of Medicine and Health Sciences, Universiti Putra Malaysia, Serdang, Selangor, Malaysia

## 27.1 Introduction

Diseases have always been a part of human existence, and so has the search for remedies, not only to restore health, but also to improve the quality of life. For many years, people have realized that dietary components can promote health and alleviate diseases, as observed by Hippocrates' 'Let food be thy medicine and medicine be thy food'. In recent years, this realization has been supported by evidence-based studies that show a direct link between diet and the development of chronic disease (WHO/FAO, 2002; Carrera-Bastos *et al.*, 2011). Interestingly, chronic diseases are the leading cause of mortality throughout the world, despite advances in technology (WHO, 2008). Chronic diseases are the leading causes of death in developed countries, and, even in developing countries, an epidemiologic transition appears to reflect a changing pattern of morbidity and mortality from predominantly infectious aetiologies to chronic diseases, making them very prominent causes of morbidity and mortality (WHO, 2008). Given this dismal picture and the high cost of managing these diseases, preventive strategies have been a priority, and genuine advances in managing chronic diseases have been reported, though on a limited scale.

The emerging patterns of diet–disease interactions have led to more in-depth studies of dietary components, especially in the form of plant materials and their health benefits (Briskin, 2000). In addition, the need to reduce the cost of managing chronic diseases, especially in developing countries, has given more impetus to research in this field to provide chronic disease remedies from natural sources, which are abundantly available in many of these countries (Hoareau and DaSilva, 1999; Gurib-Fakim, 2006). However, this approach may have disadvantages, given that plants play a crucial role in maintaining a balanced ecosystem. Unlike the simplistic view, which suggests that only a thin line exists between health and disease, a more encompassing approach to health management must be taken to cover the multiple 'grey areas' that exist between the two states of wellbeing. These 'grey areas' must consider anything that has an impact on disease and/or health. This consideration has brought forth concepts that now try to minimize the adverse impact of human activities on the environment, relating to the processes of planting, harvesting, processing or developing any agricultural product with health benefits. As such, advances in technology have ensured that adverse environmental effects due to the utilization of plants for their health benefits are minimized.

Similarly, important advances have been recorded in the area of nutritional sciences. Nutrition research previously focused on the deficiency states of major food macromolecules, but its focus has changed over the last decade due to an increasing awareness of the interaction of food and its constituents with macromolecules at different molecular levels (genome, epigenome, proteome and metabolome). This is largely what has given birth to the field of nutrigenomics, which focuses on the interactions at the diet–genome interface (Gillies, 2003; Kaput and Rodriguez, 2004; Mutch *et al.*, 2005; Afman and Muller, 2006). It is through these interactions that plant and food bioactive compounds are believed to influence the molecular and biochemical processes in biological systems (Muller and Kersten, 2003). Additionally, the ever-increasing availability of molecular biology tools is making it possible to study these interactions with unprecedented precision. Specifically, in nutrigenomics, the particular focus is on understanding the molecular relationships between nutrition and the response of genes and how such changes can affect human health (Afman and Muller, 2006; Subbiah, 2007). Advanced molecular biology tools and the mapping of the entire human genome make these studies possible within a considerably shorter amount of time than previously anticipated through the use of high-throughput genomic tools that generate data within a very short time.

The basis for interactions at the diet–genome interface that underlie nutrigenomic studies is that dietary components act as signals that convey information about the diet and its components to cells or subcellular components in biological systems. These dietary components are then detected by sensor systems in the cell at different levels (genome, epigenome, proteome and metabolome) to change gene and protein expression and metabolite production in accordance with the level of nutrients sensed by the cell. Dietary signatures are thought to underlie the health benefits of plant bioactive compounds, commonly referred to as nutraceuticals (Kauwell, 2005; Afman and Muller, 2006; Subbiah, 2007) . The word 'nutraceutical' was coined in 1989 by Dr Stephen DeFelice (Hardy, 2000; Kalra, 2003) and refers to foods or parts of foods that provide medical or health benefits, including the prevention and/or treatment of disease (Hardy, 2000).

*Novel Plant Bioresources: Applications in Food, Medicine and Cosmetics*, First Edition. Edited by Ameenah Gurib-Fakim.

## 27.2 Plant bioresources for health uses: beyond traditional uses

Plants have been traditionally used to manage communicable and non-communicable diseases, especially in Africa and Asia. The realization that medicinal plants provide a cheaper means of disease management has generated interest in their use globally (Hoareau and DaSilva, 1999; Gurib-Fakim, 2006). Many of these plants have been used solely based on a traditional hunch, and studies are now providing evidence of their efficacy; what used to be herbal practice is now appearing in mainstream science (Martin *et al.*, 2011). In fact, in developing countries, there appears to be an increasing use of medicinal plants, even among the urban population, largely because of their popularity (Andel and Carvalheiro, 2013). The desire to improve health and prevent diseases continues to drive the search for newer therapies. Interest in the medicinal properties of natural products has grown considerably due to concerns of the side effects and other adverse effects resulting from synthetic compounds (Hoareau and DaSilva, 1999). This interest has yielded a better understanding of the role of plant bioactive compounds in the promotion of health and disease prevention, and hence their functional potential is currently the subject of close scrutiny (Gurib-Fakim, 2006; Martin *et al.*, 2011). This need-driven shift led to increased investments by stakeholders attempting to investigate the potential of medicinal plants. However, increased medicinal plant research and the development of nutraceuticals from these plants will most likely lead to overcultivation without proper replacement. The consequences are that many of these plants may become extinct, leaving a void in the ecosystem that needs to be critically considered in view of its environmental implications (Ramakrishappa, 2002).

The appeal for nutraceuticals over drugs is increasing for a number of reasons that include their natural origin, lower side effects and less stringent regulation (Hoareau and DaSilva, 1999; Briskin, 2000). In the US, for example, nutraceuticals are given the same scrutiny as supplements, making them more likely to get approved than regular drugs (Briskin 2000). This fact, however, creates a problem, because many nutraceuticals may pass through the loose regulatory process without meeting ideal standards. It may be argued that drugs may have better efficacy than nutraceuticals owing to their targeted action, but the convenience of taking natural products and the relative lack of side effects could increase the efficacy of nutraceuticals in terms of consumer acceptance. All drugs are known to be poisons, even in small doses, and though the benefits outweigh the risks for those used in therapy, in the absence of side effects, individuals are more likely to accept and consume nutraceuticals on a continuous basis than drugs to avoid any unwanted side effects. Continuous consumption and the supply of bioactive compounds could improve efficacy long term with better health improvements and disease prevention than regular drugs (Korver, 1998). Additionally, some compounds (drugs and nutraceuticals alike) are known to accumulate in some tissues in the body; and when they do, this can increase the toxicity of drugs. In the case of nutraceuticals, accumulation would be expected to improve the delivery of their beneficial bioactive compounds over a longer period of time. Often times, nutraceuticals, especially when present in their bioactive-rich fractions, are known to possess multiple mechanisms of action or have multiple targets (Briskin, 2000), which may not be common among drugs. The multiple mechanisms may also result in better efficacy when compared with drugs. The concept of bioactive-rich fractions, as discussed later in the chapter, is also gaining attention.

It is imperative to consider the cost, appeal and convenience to the target population and their effectiveness when developing nutraceuticals. Cost may not seem at first an important issue, but in the long run, because a nutraceutical for use in chronic disease prevention or treatment (such as in type 2 diabetes) will need to be taken continuously, the financial burden may reduce compliance. An estimated 382 million people have type 2 diabetes in 2013, and the number is projected to almost double by 2035 (IDF, 2013). This increase will mostly be observed in the low- and middle-income countries of Asia and Africa, where poverty is taking its toll on individuals. Individuals from these developing countries are expected to be the target population for nutraceuticals for diabetes; therefore, these agents must be inexpensive for consumers to afford them on a continuous basis. Additionally, the effectiveness of any such nutraceutical must be proven to be superior to other available products or their usage will be low. Local herbs and traditional concoctions for managing the disease must also be taken into consideration to convince individuals from developing countries to use a new and different product from what is already in use. If this product also fits easily into a daily routine, it is likely to be accepted better and faster. This product could be in the form of food or a part of food taken regularly by individuals. In this regard, mass public enlightenment may be necessary to educate people on the need to use nutraceuticals as natural products with fewer side effects than drugs using evidence.

## 27.3 Bioactivity of plant bioresources: nutrigenomic implications

The properties of many plants now receive global attention for their potential in preventing chronic diseases and, thereby, reducing the global burden of such diseases. Table 27.1 shows examples of plants that have been extensively studied around the world. Interest in plant bioresources as potential sources of remedies for chronic diseases has now become a global phenomenon across all continents. This rising trend is driven primarily by the realization that chronic diseases are the leading causes of morbidity and mortality, what humans ingest plays a major role in the pathogenesis of chronic diseases, and dietary modifications can prevent chronic diseases (WHO/FAO, 2002). With more advances, plants are being aggressively studied to determine which of their bioactive compounds, if any, may be effective in the prevention and treatment of chronic diseases and even in the amelioration of communicable and even non-communicable diseases. Furthermore, most studies on plant bioactive compounds not only report their bioactivity, but also which compounds are responsible for their bioactivity (Table 27.1) and what mechanisms are involved. Many traditional Chinese herbs, for example, have been shown to reduce inflammatory changes in the kidney and prevent progression to chronic kidney disease through the regulation of the p38 mitogen-activated protein kinase signalling pathway (Zhao *et al.*, 2012a). In addition, the underlying nutrigenomic mechanisms involved in the functional effects of many other bioactive compounds are continuously being elucidated, as shown in Table 27.2. Considering the mechanism of action of plant bioresources, Kim (2012) argued that a better understanding of

**Table 27.1** Representative studies of the bioactivity of plant bioresources showing the plant part used, active ingredients responsible and study location

| Plant | Bioactivity[a] | Plant part[b] | Active ingredients responsible | Study location[c] | Reference |
|---|---|---|---|---|---|
| *Agaricus brasiliensis* | Antioxidant and antimicrobial activities | Whole mushroom | Linoleic and palmitic acids | Brazil | Mazzuti *et al.* (2012) |
| *Anacardium occidentale*/*Averrhoa bilimbi*/*Lagerstroemia speciosa*/ *Parkia speciosa Phyllanthus amarus*/*Pithecellobium jiringa* | Antidiabetic activity | Whole plant/leaf and stem/leaf and stem/ leaf/seed and pericarp/seed | Pentacyclic triterpenoids, oleanolic acid, ursolic acid and lupeol | UK/USA | Ali *et al.* (2006) |
| *Andrographis echioides* | Anti-inflammatory activity | Whole plant | Flavonoids | Taiwan | Shen *et al.* (2013) |
| *Apios americana* | Antioxidative activity | Nut | Isoflavone: genistein-7-*O*-gentiobioside | Japan | Takashima *et al.* (2013) |
| *Azadirachta indica* | Cytotoxic and anti-leshmanial activities | Leaf and fruit | Not reported | Brazil | Carneiro *et al.* (2012) |
| *Berberis thunbergii* var. *atropurpurea* | Antioxidant and anti-inflammatory activities | Roots | Not reported | USA | Zhang *et al.* (2013) |
| *Centaurea jacea* | Antiproliferative | Aerial part | Flavonoids, sesquiterpenes | Hungary | Forgo *et al.* (2012) |
| *Cocos nucifera* | Antidiabetic activity | Coconut water | Not reported | India | Preetha *et al.* (2012) |
| *Citrus aurantium* L. var. *amara* | Antioxidant, anti-inflammatory, antivirus and antitumour activities | Flower | 5-Hydroxy-6,7,3′,4′-tetramethoxy flavone and limonexic acid | China | Zhao *et al.* (2012b) |
| *Combretum zeyheri* | Antifungal activity | Leaf | Triterpenoids, ursolic acid, oleanolic acid, maslinic acid, 2α,3β-dihydroxyurs-12-en-28-oic acid, 6β-hydroxymaslinic acid and terminolic acid | India/Tanzania | Runyoro *et al.* (2013) |
| *Curcuma longa* | Antidiabetic activity | Rhizome | Bisdemethoxycurcumin | India | Ponnusamy *et al.* (2012) |
| *Cyrtomium fortumei* | Antitumour activity | Rhizome | Pimpinellin | China | Yang *et al.* (2013) |
| *Evodia officinalis* | Anticholinesterase activity | Fruit | Rutaecarpine, evodiamine, limonin and dehydroevodiamine | South Korea | Lee *et al.* (2012) |
| *Ficus deltoidea* | Bioactivity against alpha-glucosidase | Leaf | Vitexin and isovitexin | Malaysia | Choo *et al.* (2012) |
| *Matricaria chamomilla* | Antioxidant and antitumour activities | Flower head and leafy stem | Phenolic compounds | Portugal/Spain | Guimarães *et al.* (2013a) |
| *Gynandropsis gynandra* | Bioactivity against spider mite | Leaf | Isothiocyanates, aldehydes, esters, alcohols and terpenes | Denmark | Nyalala *et al.* (2013) |
| *Holarrhena antidysenterica* | Antidiabetic activity | Seed | 5-Caffeoylquinic acid (chlorogenic acid) | India | Kumar *et al.* (2013) |
| *Ipomoea hederacea* | Antibacterial and antifungal activities | Seed | Not reported | India | Singh and Singh (2012) |
| *Juniperus oxycedrus* | Antidiabetic activity | Leaf | Palmitic, linoleic and linolenic acids | Turkey | Orhan *et al.* (2012) |
| *Kaempferia galanga* | Anti-inflammatory activity | Rhizome | Ethyl-*p*-methoxycinnamate | Malaysia | Umar *et al.* (2012) |
| *Lagenaria breviflora* | Anti-inflammatory and analgesic activities | Seeds, pulp, bark or whole fruit | Not reported | Nigeria | Oridupa and Saba, (2012) |
| *Litchi chinensis* | Antioxidant activity | Fruit | 2-(2-Hydroxyl-5-(methoxycarbonyl) phenoxy) benzoic acid, kaempferol, isolariciresinol, stigmasterol, butylated hydroxytoluene, 3,4-dihydroxyl benzoate, methyl shikimate and ethyl shikimate | China/USA | Jiang *et al.* (2013) |

*(continued)*

**Table 27.1** (*Continued*)

| | | | | | |
|---|---|---|---|---|---|
| *Maclura amboinenesis* | Antimetastatic activity | Root | Prenylated xanthones: macluraxanthone and gerontoxanthone-I | Thailand | Siripong *et al.* (2012) |
| *Morinda citrifolia* | Anti-angiogenic activity | Fruit and leaf | Scopoletin | Malaysia | Beh *et al.* (2012) |
| *Narcissus* spp. | Anticholinesterase activity | Leaf/bulb | Sanguinine and galanthamine | Spain/Bulgaria | Torras-Claveria *et al.* (2013) |
| *Nigella sativa* | Antibacterial and antifungal activities | Seed | Not reported | Italy | Piras *et al.* (2013) |
| *Nigella sativa* | Bioactivity against chemical-induced colitis | Seed | Thymoquinone | China | Lei *et al.* (2012) |
| *Oroxylum indicum* | Antioxidant, antimicrobial, cytotoxic and apoptotic activities | Bark | Flavonoids and phenolics | India | Moirangthem *et al.* (2013) |
| *Oryza sativa* | Antihyperglycaemic and hypocholesterolaemic activities | Seed | Phenolics, gamma amino butyric acid, oryzanol and acylated steryl glycoside | Malaysia | Imam and Ismail (2013), Imam *et al.* (2013a,2013b) |
| *Parthenium hysterophorus* | Antifungal and antibacterial activities | Leaf | Not reported | India | Malarkodi and Manoharan (2013a,2013b) |
| *Pittosporum* spp. | Antimicrobial activity | Leaf and fruit | Phenolics, saponins, flavonoids, tanins, triterpenoids, | Australia | Sadgrove and Jones, (2012) |
| *Rauwolfia tetraphylla* | Antipsychotic activity | Leaf | Alkaloids | India | Gupta *et al.* (2012) |
| *Chamaemelum nobile* | Antioxidant and antitumour activities | Flower head and leafy stem | Phenolic compounds | Portugal/Spain | Guimarães *et al.* (2013b) |
| *Salicornia herbacea* | Antiadipogenic activity | Whole plant | Flavonoids | South Korea | Kong *et al.* (2012) |
| *Solanum nigrum* | Hypocholesterolaemic activity | Seed | Stigmast-5,22-dien-3β-ol | India | Sharma *et al.* (2012) |
| *Thymus linearis* and *Thymus serpyllum* | Antibacterial, antioxidant, antimalarial and antiproliferative activities | Whole plant | Monoterpene hydrocarbons, oxygenated monoterpenes, sesquiterpene hydrocarbons and oxygenated sesquiterpene | Pakistan, UK, Germany, Malaysia | Hussain *et al.* (2013) |
| *Vigna vexillata* | Anti-inflammatory | Seed | Daidzein, abscisic acid, and quercetin | Taiwan | Leu *et al.* (2012) |
| *Vitis vinifera* | Antioxidant activity | Leaf | Phenolics: *trans*-caffeoyltartaric and *trans*-coumaroyltartaric acids, myricetin-3-*O*-glucoside, quercetin-3-*O*-glucoside, quercetin-3-*O*-galactoside and kaempferol-3-*O*-glucoside | Portugal | Fernandes *et al.* (2013) |
| *Waldheimia glabra* | Antioxidant activity | Whole plant | Phenolics | Italy | Giorgi *et al.* (2013) |
| *Quercus resinosa*, *Quercus laeta*, *Quercus grisea*, *Quercus obtusata* | Antioxidant, antimicrobial, anti-topoisomerase and anti-proliferative activities | Leaf | Not reported | Mexico/Spain | Sánchez-Burgos *et al.* (2013) |
| *Zingiber officinale* | Antiplatelet and vasorelaxing activities | Rhizome | 6-Gingerol and 6-shogaol | Taiwan | Liao *et al.* (2012) |

[a] Different plant parts may contain the biologically active ingredients in the plants.
[b] Bioactivity studies cover communicable and non-communicable diseases.
[c] The global distribution of the study locations points to the growing interest in the use of plant bioresources for medicinal purposes, not just in biodiverse countries but all over the world.

**Table 27.2** Nutrigenomic mechanisms underlying the functional effects of some plant bioactive compounds

| Compound/plant source | Bioactivity studied | Nutrigenomic mechanisms involved | Reference |
|---|---|---|---|
| 4-Hydroxy-17-methylincisterol/*Agaricus blazei* | Anti-inflammatory activity | Through suppression of the production of cytokine via downregulation of signal transduction pathways involving NF-AT and NF-κB | Tsai *et al.* (2013) |
| α-Tocopherol/synthetic from Sigma (St. Louis, MO) | Antioxidative and antifibrotic activities | Through downregulation of the expression of TGF-β1 and collagen-α1 (I) | Jiang *et al.* (2011) |
| α-Tocopherol/synthetic from Sigma (St. Louis, MO) | Anti-inflammatory activity | Through inhibition of the production of nitric oxide, prostaglandin E and proinflammatory cytokine as well as blocking the expression of nitric oxide synthase, cyclooxygenase-2 and NF-κB | Ng and Ko (2012) |
| ASG, GABA, oryzanol and phenolics/GBR | Antidiabetic activity | Through upregulation of PPARγ, APO A1 and LDLR genes, and downregulation of *Pck* and *Fbp* genes | Imam and Ismail (2013), Imam *et al.* (2013a,2013b) |
| Brazilein/*Caesalpinia sappan* L. | Antiadipogenic and apoptotic activities | Through suppression of PPARγ and activation of caspase-3 | Liang *et al.* (2013) |
| Gamma-oryzanol/rice bran | Regulation of antioxidant and oxidative-stress-related genes | Through upregulation of antioxidant genes and downregulation of oxidative-stress-related genes | Ismail *et al.* (2010) |
| TQ/*Nigella sativa* | Regulation of cholesterol metabolism genes | Through upregulation of APO A1 and LDLR genes and downregulation of APO B100 and HMGCR genes | Al-Naqeep and Ismail (2009), Al-Naqeep *et al.* (2009a,2009b) |
| TQ/*Nigella sativa* | Regulation of antioxidant genes | Through upregulation of antioxidant genes | Ismail *et al.* (2010) |
| Triptolide/*Tripterygium wilfordii* Hook f. | Chrodroprotective and anti-inflammatory activities | Through suppression of osteoclastic activity in inflamed joints by reducing the expression of the receptor activator of NF-κB and its ligand and increasing the expression of OPG | Liu *et al.* (2013) |
| Vanillin/synthetic from Sigma (St. Louis, MO) | Regulation of cholesterol metabolism genes | Through upregulation of LDLR gene and downregulation of HMGCR gene | Al-Naqueep *et al.* (2010) |

APO A1, apolipoprotein A1; APO B100, apolipoprotein B100; ASG, acylated steryl glycoside; GABA, gamma amino butyric acid; GBR, germinated brown rice; TQ, thymoquinone; HMGCR, 3-hydroxy-3-methylglutaryl-coenzyme A reductase;LDLR, low density lipoprotein receptor; NF-κB, nuclear factor kappa B; NF-AT, nuclear factor of activated T cells; OPG, osteoprotegerin; PPARγ, peroxisome proliferator-activated receptor γ; TGF-β1, transforming growth factor beta 1.

the pharmacokinetics and pharmacodynamics of natural products could help improve the quality of life of patients with colonic and other diseases, especially considering the colon's central importance in nutrition (Kim, 2012). Additionally, a detailed understanding of the mechanisms involved in a compound's bioactivity is important because side effects may potentially occur with plant bioresources, contrary to traditional beliefs. More reliable analytical control methods may be needed to ensure that the composition, bioactivity and even quality of extracts of plant bioresources, especially in their finished products, are of the reported/claimed standards, as argued by Lavecchia *et al.* (2012). In addition, unwanted effects could be ruled out as complications due to the consumption of plant bioresources if the mechanisms of action are known extensively. In this regard, clinical trials are essential because preclinical findings are often not corroborated by clinical data. It is hoped that once the bioactivity of plant bioresources and their underlying mechanisms of action are established, this knowledge can be related through public health enlightenment campaigns and policies that will enable individuals to benefit the most from the food/nutraceutical choices they make.

In the past, plants were used traditionally for medicinal purposes. In recent years, however, studies have revealed many bioactive compounds in plants that are responsible for their medicinal properties. A better understanding of how bioactive compounds elicit their effects is now fuelling the desire to maximize the potential of bioactive compounds from plants. This motivation has led to the development of rich fractions in which a lead compound is usually isolated with smaller amounts of other bioactive compounds that will potentiate its effects through synergy (Briskin, 2000). The concept of bioactive-rich fractions is somewhat similar to that of food synergy (Jacob and Tapsell, 2007), in which it is claimed that the food, not its nutrients, is the fundamental unit of nutrition. Its proponents claim that through the synergy of the bioactive compounds in food, better nutritive value is derived than the individual nutrients would provide. In bioactive-rich fractions, it is believed that a bioactive compound in the presence of other bioactive compounds with similar effects, even in small amounts, will produce a significantly higher effect. The advantages are that the benefits from the various compounds are derived from a single source and a better functional effect is attained.

Several studies, which are summarized in Table 27.3, have reported the health-promoting and disease-preventing effects of many bioactive compounds and compared such effects with those of their corresponding bioactive-rich fractions. Most of these studies, however, are at the preclinical stages of study. Nevertheless, there seems

**Table 27.3** Summary of studies comparing single bioactive compounds and their rich fractions

| Compound/rich fraction | Method of preparation of rich fraction | Study design | Summary of findings | Reference |
|---|---|---|---|---|
| α-Tocopherol/TRF | TRF supplied by Carotech (Malaysia) | Antioxidative and antifibrotic effects of α-tocopherol (800 mg/kg per day) and TRF (800 mg/kg per day) were evaluated on Wistar rats following 4 weeks of intervention | TRF was better than α-tocopherol at lowering inflammation, pancreatic fibrosis and expression of TGF-β1 and collagen-α1 (I).<br>Both lower oxidative stress but with no significant difference for MDA level, and TRF significantly lowers hydroxyproline levels compared with α-tocopherol | Jiang *et al.* (2011) |
| α-Tocopherol and α-tocopheryl acetate/TRF | TRF supplied by Carotech (Malaysia) | Effects of TRF, α-tocopherol and α-tocopheryl acetate in the prevention of chronic inflammatory diseases using mouse peritoneal macrophage cells were studied | TRF showed most potent anti-inflammatory activity by inhibiting the production of nitric oxide, prostaglandin E and proinflammatory cytokine as well as blocking the expression of nitric oxide synthase, cyclooxygenase-2 and NF-κB | Ng and Ko (2012) |
| α-Tocopherol/TRF | PalmVitee from Lanelabs (FL, USA) | Protective effect of 8-week diet supplemented with α-tocopherol or TRF (300 mg/kg food) on aspirin-induced gastric lesions in Sprague-Dawley rats was evaluated | Both TRF and α-tocopherol demonstrated equal potency (no significant difference) in preventing aspirin-induced gastric lesions through reduction of the lipid peroxidation process | Nafeeza *et al.* (2002) |
| α-Tocopherol/TRF | Supplied by the Malaysian Palm Oil Board processing unit | Effects of α-tocopherol and TRF on lipid peroxidation of LDL and HUVEC were evaluated. LDL samples were obtained from blood collected from subjects involved in human dietary feeding trials (on 15% enriched and 5% enriched PUFA diets for 6 weeks) | At 50 μM, TRF was better than α-tocopherol at preventing LDL oxidation. In HUVEC, 25 μM TRF treatment decreased TBARS by 73% compared with 50% by the same concentration of α-tocopherol. Arachidonic-acid-induced oxidative stress (conjugated dienes) was reduced to a similar level by 10 μM TRF or 25 μM α-tocopherol | Mutalib *et al.* (2003) |
| α-tocopherol/TRF | TRF supplied by Carotech (Malaysia) | Effects of α-tocopherol (25 mg/kg per day) and TRF (25 and 50 mg/kg per day) were evaluated on oxidative stress and exercise endurance in swimming rats for 28 days | At 25 mg/kg of TRF, rats showed longer swimming time, higher levels of liver antioxidant enzymes (SOD, catalase, GPx) and liver glycogen, higher levels of muscle glycogen and SOD, and lower levels of blood lactate and oxidative stress markers (plasma and liver TBARS, liver and muscle protein carbonyls) compared with the same concentration of α-tocopherol. Parameters were even better for 50 mg/kg of TRF | Lee S.-P. *et al.* (2009) |
| Fisetin/flavonol-rich fraction | Extracted with refluxing water, and then partitioned using *n*-hexane | Effects of fisetin and flavonol-rich fraction were evaluated on inflammation and angiogenesis in RA FLS cells and *in vivo* models | Flavonol-rich fraction was better than fisetin at reducing RA FLS proliferation, cytokine/chemokine levels (TNF-α, IL-6, IL-8, MCP-1) and angiogenesis factors (VEGF). In mice, it was better at reducing acetic-acid-induced vascular permeability, leukocyte migration and disease progression in collagen-induced arthritis | Lee J.D. *et al.* (2009) |

**Table 27.3** (*Continued*)

| Compound/rich fraction | Method of preparation of rich fraction | Study design | Summary of findings | Reference |
|---|---|---|---|---|
| Gamma-oryzanol/ ORF | Supercritical fluid extraction | Regulation of gene expression related to antioxidant and oxidative stress in stressed Sprague Dawley rats by gamma-oryzanol and ORF | ORF demonstrated greater antioxidant activity than gamma-oryzanol through regulation of antioxidants and oxidative stress related genes | Ismail *et al.* (2010) |
| TQ/TQRF | Supercritical fluid extraction | Effects of 2 ppm TQ and 80 ppm TQRF (containing 2 ppm TQ) on expression of APO A1 and APO B100 genes in HEPG2 cells were evaluated | TQRF at 0.05% upregulated APO A1 gene and downregulated APO B100 gene significantly more than TQ | Al-Naqeep and Ismail (2010) |
| TQ/TQRF | Supercritical fluid extraction | Effects of 2 ppm TQ and 80 ppm TQRF (containing 2 ppm TQ) on expression of LDLR and HMGCR genes in HEPG2 cells were evaluated | TQRF at 0.05% upregulated LDLR gene and downregulated HMGCR gene significantly more than TQ | Al-Naqeep *et al.* (2009a) |
| TQ/TQRF | Supercritical fluid extraction | Hypocholesterolaemic effect of TQ (20, 50 and 100 mg/kg body weight) and TQRF (0.5, 1.0 and 1.5 g/kg body weight) in high-fat-diet-fed Sprague Dawley rats following 8-week intervention | Both TQ and TQRF showed equal hypocholesterolaemic effect (no significant difference) in a dose-dependent manner | Al-Naqeep *et al.* (2009b) |
| TQ/TQRF | Supercritical fluid extraction | Effects of TQ (20, 50 and 100 mg/kg body weight) and TQRF (0.5, 1.0 and 1.5 g/kg body weight) were evaluated on antioxidant capacity of hypercholesterolaemic rats after 8 weeks of feeding | Within the range of doses tested, TQRF was better than TQ at reducing LDL-C and significantly improved liver hydroxyl radical scavenging activity and upregulation of antioxidant genes | Ismail *et al.* (2010) |
| TQ/TQRF | Supercritical fluid extraction | Cytotoxicity of TQ and TQRF on colon cancer (HT29), lymphoblastic leukaemia (CEMSS) and promyelocytic leukaemia (HL60) cell lines were evaluated | TQRF was more cytotoxic on all cells ($IC_{50}$ values for HT29, CEMSS and HL60 cells were 400, 350 and 250 μg/mL) when compared to TQ ($IC_{50}$ values of 8, 5 and 3 μg/mL, respectively) | Norsharina *et al.* (2011) |
| Vanillin/VRF | Supercritical fluid extraction | Antioxidant capacities of vanillin and VRF were evaluated, and their effects (100 ppm of vanillin, and 100 and 200 ppm of VRF) on LDLR and HMGCR genes were determined | VRF showed better antioxidant activities than vanillin using BCB and DPPH assays. At 100 ppm, VRF upregulated LDLR gene and downregulated HMGCR gene better than vanillin. Higher concentration (20 ppm) of VRF was more effective at regulating genes | Al-Naqeep *et al.* (2010) |

Bioactive-rich fractions appear to be more active than single compounds likely due to synergistic effects of the multiple compounds in the bioactive-rich fractions. APO A1, apolipoprotein A1; APO B100, apolipoprotein B100; BCB, beta-carotene bleaching; BMI, body mass index; DPPH, 2,2-diphenyl-1-picrylhydrazyl; GPx, glutathione peroxidase; HEPG2, hepatoma cells; HMGCR, 3-hydroxy-3-methylglutaryl-coenzyme A reductase; HUVEC, human umbilical vein endothelial cells; IL-6, interleukin 6; IL-8, interleukin 8; LDL-C, low-density lipoprotein cholesterol; LDLR, low-density lipoprotein receptor; MCP-1, monocyte chemotactic protein 1; MDA, malondialdehyde; NF-κB, nuclear factor kappa B; ORF, gamma-oryzanol-rich fraction; PUFA, polyunsaturated fatty acid; RA FLS, rheumatoid arthritis fibroblast like synovial; SOD, superoxide dismutase; TBARs, thiobarbituric acid reactive species; TGF-β1, transforming growth factor beta 1; TNF-α, tumour necrosis factor alpha; TQ, thymoquinone;, TQRF, thymoquinone-rich fraction; VEGF, vascular endothelial growth factor; VRF, vanillin-rich fraction.

to be an informal consensus among these studies that bioactive-rich fractions produce better functional effects than single compounds do, through multiple regulatory steps in addition to synergy as reflected by the different targets of gene regulation for different compounds in a rich fraction. Hence, we hypothesize that the isolation and use of bioactive-rich fractions from plant sources instead of single bioactive compounds may be a way of maximizing the health benefits from such plants and should be encouraged among researchers working in ethno-, phyto-, complementary and alternative medicine fields. Additionally, this concept of rich fractions ought to be extended to more extracts of plant bioresources because only a limited number of rich-fraction compounds have currently been reported in the literature. However, possible challenges in the preparation of rich fractions include their standardization and mass production for commercial usage.

## 27.4 Potential implications of the rising trend in the use of plant bioresources for remedies

The plants that served as sources of nutraceuticals prior to the recent developments were not intended for such purposes. As such, most of them are not available in commercial quantities sufficient to be cultivated and used for disease prevention, without causing hazards to humans and other life forms. Interest in the study and use of plant bioresources is largely fuelled by the biodiversity present in any country. In view of the promising findings from the plant bioresources present in Brazil, Bolzani *et al.* (2012) argue that the biodiversity of Brazil offers a great opportunity for the use of plant bioresources for medicinal purposes, while Aremu *et al.* (2012) also argue that the lack of adequate antihelminthic agents and the abundance of promising medicinal plants in South Africa is enough reason to focus attention on the search for a treatment from those plants for the poorer South African populations afflicted by diseases. These opinions may also be valid for many countries enriched with biodiversity. However, we are faced with a dilemma: whether to harness the protective properties of these plants or maintain the balance of the ecosystem by leaving them unperturbed. Plants have roles to play in maintaining a balanced ecosystem. Cultivating them for use as nutraceuticals will mean disturbing the natural balance that exists in an already fragile ecosystem and risking the increased effects of global warming. However, the world cannot afford to allow people to die from chronic diseases when there are natural, God-given and affordable remedies that will effectively reduce morbidity and mortality from such diseases. Herein lies the major problem of fighting chronic diseases while at the same time exposing the Earth to the detrimental effects of adverse climate change. This issue will therefore need to be explored pragmatically to strike a balance that will not endanger the existence of any life form on Earth.

For many years, there has been a global environmental threat largely caused by human activities that produce greenhouse gases. These gases are known to affect the environment adversely and are therefore a hazard not only to humans but also to other life forms. Human industrial activities, such as mining, driving cars that produce excessive amounts of gas and other activities increase the Earth's mean temperature. The increased temperatures have been suggested to affect the Earth and its inhabitants adversely through global warming. In particular, the physical environment and especially plants receive a significant part of this impact. Environmental temperature stressors can affect the secondary metabolites and other compounds produced by plants (Kirakosyan *et al.*, 2003; Zobayed *et al.*, 2005). For instance, *Rhodiola rosea*, which is a medicinal plant used in Europe, Asia, and North America for the treatment of a wide variety of ailments, including fatigue, depression and infections, is now believed to be extinct due to climate change (Brown *et al.*, 2002; Cavalieri, 2009). Cavaliere (2009) also describes in detail how climate change has influenced medicinal and aromatic plants. He maintains that climate change has caused a tremendous effect on the life cycles and distribution of the world's vegetation, including wild medicinal and aromatic plants.

The world is already facing problems from climate change and global warming. Deforestation has also increased, and commercial harvesting of available plant sources for their nutraceutical properties will only increase environmental hazards, potentially causing more problems than solutions (Culas, 2009). However, as more studies unravel the potential of plants in the management of chronic diseases, the natural temptation will be to harvest and harness every available source within reach. This will tilt the balance in the ecosystem toward higher amounts of atmospheric carbon dioxide and lower oxygen due to reduced plant populations. The continued massive use of medicinal plants and a concurrent lack of or disregard for legislation to control it may soon lead us to a realization that we had taken for granted the importance of plants in maintaining a balanced ecosystem. The temptation to harvest plants proven to have medicinal value may create a void that will push global warming and its effects further.

## 27.5 Conclusions

Chronic diseases cause a considerable number of deaths annually, and nutraceuticals from plants are showing great promise in managing these diseases. A balance should be maintained by ensuring that plants with medicinal potential can be produced in commercial quantities before they are harvested and harnessed for nutraceutical purposes. Despite modest attempts by regional bodies and some countries, not enough is being done to pre-empt the likely environmental effects of mass cultivation of medicinal plants. Global policies are needed for practices that will ensure commercial production of these plants prior to being used for medicinal purposes. After all, the environmental effects of human activities that promote climate change are felt globally, even if they are predominantly carried out in certain regions of the world. Efficient agricultural practices must be established to ensure that this does not happen. Given the current level of technological advancements in agriculture, the commercial production of any food material or plant should not pose difficulties. Another way to prevent this problem will be to produce synthetic forms of the bioactive compounds found in plants for commercial purposes. However, this approach also has disadvantages because there are risks of losing some benefits or experiencing harm from synthetic compounds. Not only should the mass production of the plants be a priority, standardization of the extracts from plants and their rich fractions should also be taken seriously to maximize the benefits from any plants. Preclinical studies dominate the studies on extracts from plants and their rich-fractions; therefore, there is a need for more clinical studies on their efficacy in an attempt to harness the benefits of not only one compound, but also of the multiple compounds present in plants. Interestingly, green technologies, such as supercritical carbon dioxide fluid extraction, can now be used to maximize the extraction of some bioactive-rich fractions and minimize the environmental hazards associated with extraction of plant bioresources. Advanced molecular biology techniques make it possible to produce high-throughput data on the effects of plant bioactive compounds in biological systems. In the future, therefore, good collaborative research is expected between agriculturists, biotechnologists, pharmacologists, molecular biologists, pharmacists and other health care professionals to fully benefit from plant bioresources.

## Acknowledgements

We thank Universiti Putra Malaysia (UPM), PadiBeras Nasional Berhad (BERNAS Sdn. Bhd.), Malaysian Ministry of Science, Technology and Innovation (MOSTI), and the Ministry of Higher Education (MOHE) for sponsoring our various research projects on germinated brown rice and *Nigella sativa*. We also thank the staff of the Institute of Bioscience (IBS) and the animal house at the Faculty of Medicine and Health Sciences for their support in carrying out the studies reported here and during preparation of the current chapter.

## References

Afman, L. and Muller, M. (2006) Nutrigenomics: from molecular nutrition to prevention of disease. *Journal of the American Dietetic Association*, **106** (4), 569–576.

Ali, H., Houghton, P. and Soumyanath, A. (2006) α-Amylase inhibitory activity of some Malaysian plants used to treat diabetes; with particular reference to *Phyllanthus amarus*. *Journal of Ethnopharmacology*, **107** (3), 449–455.

Al-Naqeep, G. and Ismail, M. (2009) Regulation of apolipoprotein A-1 and apolipoprotein B100 genes by thymoquinone rich fraction and thymoquinone in HepG2 cells. *Journal of Food Lipids*, **16** (2), 245–258.

Al-Naqeep, G., Ismail, M. and Allaudin, Z. (2009a) Regulation of low-density lipoprotein receptor and 3-hydroxy-3-methylglutaryl coenzyme A reductase gene expression by thymoquinone-rich fraction and thymoquinone in HepG2 cells. *Journal of Nutrigenetics and Nutrigenomics*, **2**, 163–172.

Al-Naqeep, G., Ismail, M. and Yazan, L.S. (2009b) Effects of thymoquinone rich fraction and thymoquinone on plasma lipoprotein levels and hepatic low density lipoprotein receptor and 3-hydroxy-3-methylglutaryl coenzyme A reductase genes expression. *Journal of Functional Foods*, **1** (3), 298–303.

Al-Naqeep, G., Ismail, M., Bagalkotkar, G. and Adamu, H.A. (2010) Vanillin rich fraction regulates LDLR and HMGCR gene expression in HepG2 cells. *Food Research International*, **43** (10), 2437–2443.

Andel, T.V. and Carvalheiro, L.G. (2013) Why urban citizens in developing countries use traditional medicines: the case of Suriname. *Evidence-Based Complementary and Alternative Medicine*, **2013**, ID 687197.

Aremu, A., Finnie, J. and Van Staden, J. (2012) Potential of South African medicinal plants used as anthelmintics – their efficacy, safety concerns and reappraisal of current screening methods. *South African Journal of Botany*, **82**, 134–150.

Beh, H.-K., Seow, L.-J., Asmawi, M. *et al.* (2012) Anti-angiogenic activity of *Morinda citrifolia* extracts and its chemical constituents. *Natural Product Research*, **26** (16), 1492–1497.

Bolzani, V., Valli, M., Pivatto, M. and Viegas Jr, C. (2012) Natural products from Brazilian biodiversity as a source of new models for medicinal chemistry. *Pure and Applied Chemistry*, **84** (9), 1837–1846.

Briskin, D.P. (2000) Medicinal plants and phytomedicines. linking plant biochemistry and physiology to human health. *Plant Physiology*, **124** (2), 507–514.

Brown, R.P. Gerbarg, P.L. and Ramazanov, Z. (2002) *Rhodiola rosea*: a phytomedicinal overview. *HerbalGram*, **56**, 40–52.

Carneiro, S., Carvalho, F., Santana, L. *et al.* (2012) The cytotoxic and antileishmanial activity of extracts and fractions of leaves and fruits of *Azadirachta indica* (A Juss.). *Biological Research*, **45** (2), 111–116.

Carrera-Bastos, P., Fontes-Villalba, M., O'Keefe, J.H. *et al.* (2011) The Western diet and lifestyle and diseases of civilization. *Research Reports in Clinical Cardiology*, **2011** (2), 15–35.

Cavaliere, C. (2009) The effects of climate change on medicinal and aromatic plants, *HerbalGram*, **81**, 44–57.

Choo, C., Sulong, N., Man, F. and Wong, T. (2012) Itexin and isovitexin from the leaves of *Ficus deltoidea* with *in-vivo* α-glucosidase inhibition. *Journal of Ethnopharmacology*, **142** (3), 776–781.

Culas, R.J. (2009) The causes and consequences of tropical deforestation: a review. *International Journal of Climate Change: Impacts and Responses*, **1** (2), 127–139.

Fernandes, F., Ramalhosa, E., Pires, P. *et al.* (2013) *Vitis vinifera* leaves towards bioactivity. *Industrial Crops and Products*, **43** (1), 434–440.

Forgo, P., Zupkó, I., Molnár, J. *et al.* (2012) Bioactivity-guided isolation of antiproliferative compounds from *Centaurea jacea* L. *Fitoterapia*, **83** (5), 921–925.

Gillies, P.J. (2003) Nutrigenomics: the Rubicon of molecular nutrition. *Journal of the American Dietetic Association*, **103** (12), 50–55.

Giorgi, A., Panseri, S., Mattara, M. *et al.* (2013) Secondary metabolites and antioxidant capacities of *Waldheimia glabra* (Decne.) Regel from Nepal. *Journal of the Science of Food and Agriculture*, **93** (5), 1026–1034.

Guimarães, R., Barros, L., Dueñas, M. *et al.* (2013a) Infusion and decoction of wild German chamomile: bioactivity and characterization of organic acids and phenolic compounds. *Food Chemistry*, **136** (2), 947–954.

Guimarães, R., Barros, L., Dueñas, M. *et al.* (2013b) Nutrients, phytochemicals and bioactivity of wild Roman chamomile: a comparison between the herb and its preparations. *Food Chemistry*, **136** (2), 718–725.

Gupta, S., Khanna, V., Maurya, A. *et al.* (2012) Bioactivity guided isolation of antipsychotic constituents from the leaves of *Rauwolfia tetraphylla* L. *Fitoterapia*, **83** (6), 1092–1099.

Gurib-Fakim, A. (2006) Medicinal plants; traditions of yesterday and drugs for tomorrow. *Molecular Aspects of Medicine*, **27**, 1–93.

Hardy, G. (2000) Nutraceuticals and Functional Foods, *Nutrition*, **16**, 688–697.

Hoareau, L. and DaSilva, E.J. (1999) Medicinal plants: a re-emerging health aid. *Electronic Journal of Biotechnology*, **2** (2).

Hussain, A., Anwar, F., Chatha, S. *et al.* (2013) Chemical composition and bioactivity studies of the essential oils from two *Thymus* species from the Pakistani flora. *LWT – Food Science and Technology*, **50** (1), 185–192.

IDF (2013) *IDF Diabetes Atlas*, 6th edn, International Diabetes Federation, www.idf.org/diabetesatlas (accessed 1 December 2013).

Imam, M.U. and Ismail, M. (2013) Nutrigenomic effects of germinated brown rice and its bioactives on hepatic gluconeogenic genes in type 2 diabetic rats and HepG2 cells. *Molecular Nutrition and Food Research*, **57** (3), 401–411.

Imam, M.U., Ismail, M., Omar, A.R. and Ithnin, H. (2013a) The hypocholesterolemic effect of germinated brown rice involves the upregulation of the apolipoprotein A1 and low-density lipoprotein receptor genes. *Journal of Diabetes Research*, **2013**, ID 134694.

Imam, M.U., Ismail, M., Ithnin, H. *et al.* (2013b) Effects of germinated brown rice and its bioactive compounds on the expression of the peroxisome proliferator-activated receptor gamma gene. *Nutrients* , **5**, 468–477.

Ismail, M., Al-Naqeep, G. and Chan, K.W. (2010a) *Nigella sativa* thymoquinone-rich fraction greatly improves plasma antioxidant capacity and expression of antioxidant genes in hypercholesterolemic rats. *Free Radical Biology and Medicine*, **48** (5), 664–672.

Ismail, M., Al-Naqeep, G., Mamat, W.A. and Ahmad, Z. (2010b) Gamma-oryzanol rich fraction regulates the expression of antioxidant and oxidative stress related genes in stressed rats liver. *Nutrition and Metabolism*, **7**, 23.

Jacob Jr, D.R. and Tapsell, L.C. (2007) Food, not nutrients, is the fundamental unit in nutrition. *Nutrition Reviews*, **65** (10), 439–450.

Jiang, F., Liao, Z., Hu, L.H. *et al.* (2011) Comparison of antioxidative and antifibrotic effects of α-tocopherol with those of tocotrienol-rich fraction in a rat model of chronic pancreatitis. *Pancreas*, **40** (7), 1091–1096.

Jiang, G., Lin, S., Wen, L. *et al.* (2013) Identification of a novel phenolic compound in litchi (*Litchi chinensis* Sonn.) pericarp and bioactivity evaluation. *Food Chemistry*, **136** (2), 563–568.

Kalra, E. (2003) Nutraceutical – definition and introduction. *AAPS Pharmaceutical Science*, **5**, article 25.

Kaput, J. and Rodriguez, R.L. (2004) Nutritional genomics: the next frontier in the postgenomic era. *Physiological Genomics*, **16**, 166–177.

Kauwell, G.P.A. (2005) Emerging concepts in nutrigenomics: a preview of what is to come. *Nutrition in Clinical Practice*, **20** (1), 75–87.

Kim, H. (2012) Natural products to improve quality of life targeting for colon drug delivery. *Current Drug Delivery*, **9** (2), 132–147.

Kirakosyan, A., Seymour, E., Kaufman, P.B. *et al.* (2003) Antioxidant capacity of polyphenolic extracts from leaves of *Crataegus laevigata* and *Crataegus monogyna* (hawthorn) subjected to drought and cold stress. *Journal of Agricultural and Food Chemistry*, **51**, 3973–3976.

Kong, C.-S., Lee, J., Kim, Y. *et al.* (2012) Evaluation on anti-adipogenic activity of flavonoid glucopyranosides from *Salicornia herbacea*. *Process Biochemistry*, **47** (7), 1073–1078.

Korver, O. (1998) Functional foods: the food industry and functional foods: some European perspectives, in *Functional Foods for Disease Prevention: II. Medicinal Plants and Other Foods* (eds T. Shibamoto, J. Terao and T. Osawa), American Chemical Society, Washington, DC, pp. 22–25.

Kumar, D., Datta, S., Roy, S. *et a;*. (2013) Bioactivity-guided isolation and quantification of anti-diabetic principle *in vitro* from *Holarrhena antidysenterica* L. (Wall). *Journal of Herbs, Spices and Medicinal Plants*, **19** (1), 54–65.

Lavecchia, T., Rea, G., Antonacci, A. and Giardi, M.T. (2012) Healthy and adverse effects of plant-derived functional metabolites: the need of revealing their content and bioactivity in a complex food matrix. *Critical Reviews in Food Science and Nutrition*, **53** (2), 198–213.

Lee, J., Cha, M.-R., Choi, C. *et al.* (2012) Cholinesterase inhibitors isolated from the fruits extract of *Evodia officinalis*. *Korean Journal of Pharmacognosy*, **43** (2), 122–126.

Lee, J.-D., Huh, J.-E., Jeon, G. *et al.* (2009a) Flavonol-rich RVHxR from *Rhus verniciflua* Stokes and its major compound fisetin inhibits inflammation-related cytokines and angiogenic factor in rheumatoid arthritic fibroblast-like synovial cells and *in vivo* models. *International Immunopharmacology*, **9** (3), 268–276.

Lee, S.-P., Mar, G.-Y. and Ng,L.-T. (2009b) Effects of tocotrienol-rich fraction on exercise endurance capacity and oxidative stress in forced swimming rats. *European Journal of Applied Physiology*, **107** (5), 587–595.

Lei, X., Liu, M., Yang, Z. *et al.* (2012) Thymoquinone prevents and ameliorates dextran sulfate sodium-induced colitis in mice. *Digestive Diseases and Sciences*, **57** (9), 2296–2303.

Leu, Y.-L., Hwang, T.-L., Kuo, P.-C. *et al.* (2012) Constituents from *Vigna vexillata* and their anti-inflammatory activity. *International Journal of Molecular Sciences*, **13** (8), 9754–9768.

Liang, C.-H., Chan, L.-P., Chou, T.-H. *et al.* (2013) Brazilein from *Caesalpinia sappan* L. antioxidant inhibits adipocyte differentiation and induces apoptosis through caspase-3 activity and anthelmintic activities against *Hymenolepis nana* and *Anisakis simplex*. *Evidence-Based Complementary and Alternative Medicine*, **2013**, ID 8648923.

Liao, Y.-R., Leu, Y.-L., Chan, Y.-Y. *et al.* (2012) Anti-platelet aggregation and vasorelaxing effects of the constituents of the rhizomes of *Zingiber officinale*. *Molecules*, **17** (8), 8928–8937.

Liu, C., Zhang, Y., Kong, X. *et al.* (2013) Triptolide prevents bone destruction in the collagen-induced arthritis model of rheumatoid arthritis by targeting RANKL/RANK/OPG signal pathway. *Evidence-Based Complementary and Alternative Medicine*, **2013**, ID 626038.

Malarkodi, E. and Manoharan, A. (2013a) Study on antibacterial activity of *Parthenium hysterophorus* L. *Journal of Chemical and Pharmaceutical Research*, **5** (1), 134–136.

Malarkodi, E. and Manoharan, A. (2013b) Antifungal activity of parthenium hysterophorus L. *Journal of Chemical and Pharmaceutical Research*, **5** (1), 137–139.

Martin, C., Butelli, E., Petroni, K. and Tonelli, C. (2011) How Can research on plants contribute to promoting human health? *Plant Cell*, **23**, 1685–1699.

Mazzutti, S., Ferreira, S., Riehl, C. *et al.* (2012) Supercritical fluid extraction of *Agaricus brasiliensis*: antioxidant and antimicrobial activities. *Journal of Supercritical Fluids*, **70**, 48–56.

Moirangthem, D., Talukdar, N., Bora, U. *et al.* (2013) Differential effects of *Oroxylum indicum* bark extracts: antioxidant, antimicrobial, cytotoxic and apoptotic study. *Cytotechnology*, **65** (1), 83–95.

Muller, M. and Kersten, S. (2003) Nutrigenomics: goals and strategies. *Nature Reviews Genetics*, **4**, 315–322.

Mutalib, M.S.A., Khaza'ai, H. and Wahle, K.W. (2003) Palm-tocotrienol rich fraction (TRF) is a more effective inhibitor of LDL oxidation and endothelial cell lipid peroxidation than α-tocopherol *in vitro*. *Food Research International*, **36** (5), 405–413.

Mutch, D.M., Wahli, W. and Williamson, G. (2005) Nutrigenomics and nutrigenetics: the emerging faces of nutrition. *FASEB Journal*, **19** (12), 1602–1616.

Nafeeza, M.I., Fauzee, A.M., Kamsiah, J. and Gapor, M.T. (2002) Comparative effects of a tocotrienol-rich fraction and tocopherol in aspirin-induced gastric lesions in rats. *Asia Pacific Journal of Clinical Nutrition*, **11** (4), 309–313.

Ng, L.T. and Ko, H.J. (2012) Comparative effects of tocotrienol-rich fraction, α-tocopherol and α-tocopheryl acetate on inflammatory mediators and nuclear factor kappa B expression in mouse peritoneal macrophages. *Food Chemistry*, **134** (2), 920–925.

Norsharina, I., Maznah, I., Aied, A.-A. and Ghanya, A.-N. (2011) Thymoquinone rich fraction from *Nigella sativa* and thymoquinone are cytotoxic towards colon and leukemic carcinoma cell lines. *Journal of Medicinal Plant Research*, **5** (15), 3359–3366.

Nyalala, S., Petersen, M. and Grout, B. (2013) Volatile compounds from leaves of the African spider plant (*Gynandropsis gynandra*) with bioactivity against spider mite (*Tetranychus urticae*), *Annals of Applied Biology*, **162** (3), 290–298.

Orhan, N., Aslan, M., Demirci, B. and Ergun, F. (2012) A bioactivity guided study on the antidiabetic activity of *Juniperus oxycedrus* subsp. *oxycedrus* L. leaves, *Journal of Ethnopharmacology*, **140** (2), 409–415.

Oridupa, O. and Saba, A. (2012) Relative anti-inflammatory and analgesic activities of the whole fruit, fruit bark, pulp and seed of *Lagenaria breviflora* Roberty. *Journal of Pharmacology and Toxicology*, **7** (6), 288–297.

Piras, A., Rosa, A., Marongiu, B. *et al.* (2013) Chemical composition and *in vitro* bioactivity of the volatile and fixed oils of *Nigella sativa* L. extracted by supercritical carbon dioxide. *Industrial Crops and Products*, **46**, 317–323.

Ponnusamy, S., Zinjarde, S., Bhargava, S. *et al.* (2012) Discovering bisdemethoxycurcumin from *Curcuma longa* rhizome as a potent small molecule inhibitor of human pancreatic α-amylase, a target for type-2 diabetes. *Food Chemistry*, **135** (4), 2638–2642.

Preetha, P., Devi, V. and Rajamohan, T. (2012) Hypoglycemic and antioxidant potential of coconut water in experimental diabetes. *Food and Function*, **3** (7), 753–757.

Ramakrishappa, K. (2002) Impact of cultivation and gathering of medicinal plants on biodiversity: case studies from India. Biodiversity and the ecosystem approach in agriculture, forestry and fisheries, in *Biodiversity and the Ecosystem Approach in Agriculture, Forestry and Fisheries*, Food and Agriculture Organization, Rome.

Runyoro, D., Srivastava, S., Darokar, M. *et al.* (2013) Anticandidiasis agents from a Tanzanian plant, *Combretum zeyheri*. *Medicinal Chemistry Research*, **22** (3), 1258–1262.

Sadgrove, N. and Jones, G. (2013) Chemical and biological characterisation of solvent extracts and essential oils from leaves and fruit of two Australian species of *Pittosporum* (Pittosporaceae) used in aboriginal medicinal practice. *Journal of Ethnopharmacology*, **145** (3), 813–821.

Sánchez-Burgos, J., Ramírez-Mares, M., Larrosa, M. *et al.* (2013) Antioxidant, antimicrobial, antitopoisomerase and gastroprotective effect of herbal infusions from four *Quercus* species. *Industrial Crops and Products*, **42** (1), 57–62.

Sharma, B., Iyer, D. and Patil, U. (2012) Bioactivity guided fractionation in experimentally induced hyperlipidemia in rats and characterization of phytoconstituent from *Solanum nigrum*. *Journal of Herbs, Spices and Medicinal Plants*, **18** (3), 257–267.

Shen, D.-Y., Juang, S.-H., Kuo, P.-C. *et al.* (2013) Chemical constituents from *Andrographis echioides* and their anti-inflammatory activity. *International Journal of Molecular Sciences*, **14** (1), 496–514.

Singh, B. and Singh, S. (2012) Chemical investigation of seed of *Ipomoea hederacea* and its biological activity. *Journal of Chemical and Pharmaceutical Research*, **4** (2), 1441–1448.

Siripong, P., Rassamee, K., Piyaviriyakul, S. *et al.* (2012) Anti-metastatic effects on B16F10 melanoma cells of extracts and two prenylated xanthones isolated from *Maclura amboinensis* Bl. roots. *Asian Pacific Journal of Cancer Prevention*, **13** (7), 3519–3528.

Subbiah, M. (2007) Nutrigenetics and nutraceuticals: the next wave riding on personalized medicine. *Translational Research*, **149**, 55–61.

Takashima, M., Nara, K., Niki, E. *et al.* (2013) Evaluation of biological activities of a groundnut (*Apios americana* Medik) extract containing a novel isoflavone. *Food Chemistry*, **138** (1), 298–305.

Torras-Claveria, L., Berkov, S., Codina, C. *et al.* Daffodils as potential crops of galanthamine. Assessment of more than 100 ornamental varieties for their alkaloid content and acetylcholinesterase inhibitory activity. *Industrial Crops and Products*, **43** (1), 237–244.

Tsai, W.-J., Yang, S.-C., Huang, Y.-L. *et al.* 4-Hydroxy-17-methylincisterol from *Agaricus blazei* decreased cytokine production and cell proliferation in human peripheral blood mononuclear cells via inhibition of NF-AT and NF-κB activation. *Evidence-Based Complementary and Alternative Medicine*, **2013**, ID 435916.

Umar, M., Asmawi, M., Sadikun, A. *et al.* (2012) Bioactivity-guided isolation of ethyl-*p*-methoxycinnamate, an anti-inflammatory constituent, from *Kaempferia galanga* L. extracts. *Molecules*, **17** (7), 8720–8734.

WHO (2008) *The global burden of disease: 2004 update*, World Health Organization, http://www.who.int/healthinfo/global_burden_disease/GBD_report_2004update_full.pdf (accessed 13 February 2013).

WHO/FAO (2002) *Diet, nutrition and the prevention of chronic diseases*, WHO Technical Report Series 916, Joint WHO/FAO Expert Consultation on Diet, Nutrition and the Prevention of Chronic Diseases, World Health Organization, Geneva.

Yang, S., Liu, M., Liang, N. *et al.* (2013) Discovery and antitumor activities of constituents from *Cyrtomium fortumei* (J.) Smith rhizomes. *Chemistry Central Journal*, **7** (1), article 24.

Zhang, C.-R., Schutzki, R. and Nair, M. (2013) Antioxidant and anti-inflammatory compounds in the popular landscape plant *Berberis thunbergii* var. *atropurpurea*. *Natural Product Communications*, **8** (2), 165–168.

Zhao, Q., Wan, Y., Wang, C. *et al.* (2012a) Regulatory mechanism of p38MAPK signaling pathway on renal tissue inflammation in chronic kidney disease and interventional effect of traditional Chinese medicine. *Zhongguo Zhongyao Zazhi*, **37** (12), 1700–1704.

Zhao, H.-Y., Yang, L., Wei, J. *et al.* (2012b) Bioactivity evaluations of ingredients extracted from the flowers of *Citrus aurantium* L. var. *amara* Engl. *Food Chemistry*, **135** (4), 2175–2181.

Zobayed, S.M.A., Afreen, F. and Kozai, T. (2005) Temperature stress can alter the photosynthetic efficiency and secondary metabolite concentrations in St. Johns wort. *Plant Physiology and Biochemistry*, **43**, 977–984.

# 28 Safety of Botanical Ingredients in Personal Healthcare: Focus on Africa

R. Vihotogbé, C.N.A. Sossa-Vihotogbé and G.E. Achigan-Dako

Faculty of Agronomic Sciences, University of Abomey-Calavi (FSA/UAC, Benin), Cotonou, Benin

## 28.1 Introduction

Biological diversity with the integration of its components ensures its own maintenance and remains the most valuable heritage on Earth, since it plays key and direct roles in ecology, culture, agriculture, nutrition and healthcare (Jose, 2012). The value of biodiversity depends most likely on the way its components are directly used by humans. Plant species are a vital component of the biodiversity, since they determine agrobiodiversty and hold biological codes in which human health is rooted (Wezel and Bender, 2003; Bertrand, 2009; Jose, 2012). In particular, herbs as well as shrubs and trees are an integral part of traditional human nutrition (via foods/beverages, spices) and health systems (Alhusainy *et al.*, 2012). Therefore, it is quite impossible for millions of people to avoid the daily utilization of any type of botanical extract and, especially, 75–80% of the world population uses raw botanical extracts for healthcare (Cunningham, 1993; Bandanarayake, 2003; Parekh *et al.*, 2005; Surekha *et al.*, 2011; Ainooson *et al.*, 2012). Ethnobotanical data suggest that the diversity of valuable medicinal plants depends on plant and disease ecology and people's traditional knowledge, since this diversity is rooted in human–environment interactions (Begossi, 1996; Teklehaymanot and Giday, 2007). Therefore, the local and large-scale commercialization of medicinal plants over the world (Cunningham, 1993; Laird, 2005) might be seen as an information system that broadens people's knowledge, which increases diversity and use for a given community. This learning system is increasing the potential users of botanical extracts and, consequently, their economic asset worldwide (Walker, 2004), since trade barriers that limit the cross-continental commercialization of other commodities do not operate efficiently (Lambert *et al.*, 2005). In the USA, for example, almost 79% of drugs are made of natural botanical extracts, and tropical worlds are increasingly explored for growing pharmaceutical industries valuing botanical materials (Roberson, 2008).

There are at least three main ways in which biochemical properties of botanical extracts can be valued in healthcare: foods/beverages /spices, cosmetics and drugs (Saxena *et al.*, 2007; Srianta *et al.*, 2009; Downie, 2010; Antignac *et al.*, 2011; Sossa-Vihotogbé *et al.*, 2012). The increasing use of natural botanical materials in healthcare is also rooted in the fact that they may supply, apart from specific compounds targeting a range of diseases, a conjunction of additional valuable nutritive elements, such as vitamins, fatty acids, minerals and antioxidants (Saxena *et al.*, 2007; Gruenwald, 2009; Andarwulan, 2012). Because of the specific biochemical properties of taxonomically different plant species, extracts of different species may be active against different parasites. It is also acknowledged that many species have common medical properties and that many parts (leaf, bark, flower, root, etc.) of the same plant species are often involved in the treatment of a unique or a wide range of infections – see Okolo *et al.* (1995), Marr *et al.* (2004), Shinde *et al.* (2009), Vihotogbé *et al.* (2008), Igbe *et al.* (2009), Sossa-Vihotogbé *et al.* (2012) and Ainooson *et al.* (2012). For example, Vihotogbé *et al.* (2013, 2014) reported 23 different types of use involving the root, bark, fruit and leaves for for bush mango tree species in Benin and Togo (West Africa). Mostly in the African traditional situation, the combination of many species is also desired to have a rapid and efficient treatment of infections (Burnie *et al.*, 1997, Bandeira *et al.*, 2001, Ignacimuthu *et al.*, 2006). This implies that cocktails containing numerous biochemical compounds are often consumed as vegetables or spices or even applied on hair or skin with limited desired and targeted effects. Therefore, it is clear that there is no clear limit between the use of plant species as foods or drugs in healthcare (Walker, 2004). Moreover, adverse effects of botanical materials occur, and plant ecology and physiology impact the concentration of biochemical compounds (Walker, 2004). In the context of their growing potential users with their widespread distribution due to their increasing market, it is important to think of the safety of botanical extracts, especially in areas where these remain the main alternative in healthcare.

We hope this chapter will present a concise state of knowledge of these important issues and will give challenges for future research in the sense of improving world, and particularly the African population's, healthcare.

## 28.2 Safety in healthcare via food consumption

Foods provide vital nutritional elements to the human organism (proteins, carbohydrates and minerals), allowing energy supply and growth, while vitamins (e.g. carotene, ascorbic acid and riboflavin), antioxidants and other important substances are also required to ensure healthier eating, which is highly recommended in today's rapid disease-spreading world (Akubugwo *et al.*, 2007; Afolayan and Jimoh, 2009). This indicates that the medical role is inherent

*Novel Plant Bioresources: Applications in Food, Medicine and Cosmetics*, First Edition. Edited by Ameenah Gurib-Fakim.

to food in any balanced nutritional system – see Marr *et al.* (2004). Therefore, the consumption of fruits and leafy vegetables meets this principle of 'eating while treating diseases', since these botanical materials are good sources of vitamins, antioxidants, proteins, carbohydrates and fibre, preventing complex diseases from occurring (Lesley and Brown, 2004; Sossa-Vihotogbé *et al.*, 2012). In tropical Africa, a great diversity of leafy vegetables and fruits, mostly harvested from natural ecosystems as well as from traditional agroforestry systems, is part of food systems contributing to the daily nutritional equilibrium of local and urban communities (Malaisse and Parent 1985; FAO/WHO, 1988; Dansi *et al.*, 2008; Maundu *et al.*, 2009; Achigan-Dako *et al.*, 2010). Herbs (*Justicia tenella*, *Acmella uliginosa*, *Talinum triangulare*, *Sesamum radiatum*, *Cleome* spp., *Heliotropium indicum*, *Solanum nigrum*, *Gonphrena globosa*, *Bidens pilosa*, *Amaranthus* spp., etc.), as well as higher trees (*Adansonia digitata*, *Vitex doniana*, *Triplochiton scleroxylon*, *Moringa oleifera*, *Tamarindus indica*, *Acacia* spp., *Balanites aegyptiaca*, *Grewia tenax*, *Ziziphus spina-christi*, *Hyphaene thebaica*, *Borassus aethiopum*, etc.), provide African communities with nutrient-rich vegetables and/or fruits (Olson; 1991; Ndoye *et al.*, 1997; Eyog Matig *et al.*, 2002; Codjia *et al.*, 2003; Sossa-Vihotogbé *et al.*, 2012). A large number of these fruits are demonstrated to have medical properties, not only due to their high antioxidants content, but also because of specific biochemicals useful in the treatment of many complex diseases, like malaria, diabetes and cancers (see Table 28.1). An extensive review by Chadare *et al.* (2008) indicated the power of many food materials from the baobab tree with effective medicinal properties. Many researches (Adefegha and Oboh, 2005; Aliero and Afolayan, 2006; Sossa-Vihotogbé *et al.*, 2012) demonstrated the high antimicrobial and antioxidant effects, and thus medicinal properties, of African traditional leafy vegetable used on daily basis by local communities. Fruits of *Adansonia digitata*, *Irvingia gabonensis*, *Acacia senegal*, *Acacia seyal*, *Balanites aegyptiaca*, *Grewia tenax*, *Ziziphus spina-christi*, *Hyphaene thebaica*, *Borassus aethiopum*, *Tamarindus indica*, *Cissus quadrangularis*, *Gnetum* spp., *Ricinodendron heudelotii*, *Psidium guajava*, *Lannea microcarpa*, *Detarium microcarpum*, *Acacia macrostachya*, *Bombax costatum* and many other African fruit tree species with their important nutritional potentials are also demonstrated to have medicinal properties against many diseases (Eyog Matig *et al.*, 2002; Clark and Sunderland, 2004). Eyog Matig *et al.* (2002) and Clark and Sunderland (2004) presented very good summaries of the diversity of the highly valuable food plant species with clearly local or widely valued medical properties.

In general, extensive data exist on botanical and morphological characterization as well as on the integration of these important resources in traditional agroforestry systems, and their economic values are widely documented. However, their full biochemical characterization leading to differential medical properties and ecological, morphological and genetic factors in which these differences are rooted, is profoundly lacking for tropical African taxa. For example, Okolo *et al.* (1995) and Egras *et al.* (2011) presented in-depth laboratory studies clearly demonstrating the analgesic property and weight-loss-inducing capacity of *Irvingia gabonensis*, which is the top priority and multipurpose food tree species in humid lowland sub-Saharan Africa. Indeed, some great efforts have recently been made on the determination of the nutritional capacity of these resources, and many studies have reported the main challenges of evaluating and handling anti-nutritional factors (like tannins, cooking methods, etc.) that complex the nutrients elements available in leafy vegetables (Eyzaguirre *et al.*, 2006; Ejoh *et al.*, 2009). However, in-depth research on botanical extracts of traditional food materials and on these anti-nutritional factors are crucially needed, and Sossa-Vihotogbé *et al.* (2012) suggested a thorough evaluation of the impacts of food processing methods of local leafy vegetable dishes. This is extremely important when promoting new food resources. For example, death cases after the consumption of African bush mango seed-based sauces was mentioned in Benin, while this diet is recommended for obesity treatment (Egras *et al.*, 2011). In particular, the first author of this review does not support the bitter bush mangoes' seeds-based diets, while those of the sweet fruits give him no digestive trouble.

## 28.3 Medicinal plants in healthcare

### 28.3.1 Advantages of extracts from medicinal plants in healthcare and practice in Africa

Plant species with medical power range from the smaller herbs to higher tree taxa (Lis-Balchin, 1999), and it is often believed in African cultures that there is no plant species without specific beneficial compounds. Bioactive compounds with acknowledged healing power are extremely diverse (Table 28.1), and their concentration in the plant body depends on factors such as the species' taxonomic position, the organ used, geographic origin and soil type (Walker, 2004).

Plant species provide low-cost drugs mostly in developing countries where people basically rely on them for healthcare (de Boer *et al.*, 2004). In a global situation of large-scale emergence of new and complex diseases (HIV/AIDS, malaria, diabetes, heart disease, cancer and psychiatric disorders), botanical extracts, being a cocktail of numerous biochemically active compounds, limit the causative group of pathogens' resistance capabilities and are demonstrated to be a powerful medical strategy where drugs with unique synthesized molecules rapidly fail (Raskin *et al.*, 2002). This means that pathogens as well as plants species' medical codes captured by human cells in disease treatment are dynamic. A good example based on the bitter melon (*Momordica charantia*, Cucurbitaceae) was reported by Aruna and Nirmala (2012); this supplies to the human organism of a large array of elements, ranging from simple minerals to a huge diversity of antioxidants and specific anticancer cell-death-inducing biochemicals. Therefore, in this specific situation, because many causative factors are involved in the occurrence of diabetes, a higher medical efficiency of botanical extracts from numerous plant species, supposedly inducing less side effects after application, is demonstrated (Parekh *et al.*, 2005; Aruna and Nirmala, 2012). Apart from the specific biochemical compounds of each species, the important role of the widely common substances like antioxidants needs to be emphasized. The question is: how important are these antioxidants? The explanation in the case of drugs lies in the fact that those complex diseases are always followed by other symptoms (fatigues, vomiting, low food intake, fever, anaemia, etc.) and, therefore, induced by a consortium of pathogens. Each of these pathogens may be sensitive to at least one substance in the cocktail represented by a specific botanical extract. This simultaneously reinforces the immunity of a patient while targeting the specific health disorder to be repaired. Parekh *et al.* (2005) argued that those secondary products (alkaloids, antioxidants) are often common to the entire plant body and may play a catalytic role for enhancing the effectiveness of the main desired biochemically unique substance and by mitigating its side effects.

All of these arguments tend to link the natural character of botanical extracts and the fact that they belong to an old traditional use system, to their absolute safety (Setty and Sigal, 2005; Cassileth *et al.*, 2009). Therefore, most of the studies sampled in this review lack any caution that might be required in the use of extracts from wild plant species (see Table 28.1). Despite the growing scientific investigations on medicinal plants, little is still done on their biochemical properties, given the diversity of use and the increasing diversity of plants valuable in human healthcare (Roberson, 2008; Schmelzer *et al.*, 2010). This is an important issue to be addressed, mostly in the low scientific capability and less-educated communities who remain the most important and traditional users of botanical extracts for healthcare.

Nowadays, biochemical evaluations are revealing the potential toxicity of many plant species whose extracts are highly used for today's simple as well as complex and expensive treatment of diseases around the world. Kayne (2010) emphasized the global lack of information across the different components of health systems based on the traditional botanicals, raising the need for assessing the balance between desired substances and other poisonous compounds associated with botanical extracts. This supposes that poisons are often inevitably associated with beneficial substances, and this characterizes medicinal plants. Therefore, the beneficial substance acts as a second level of poisonous element in botanical ingredients when overdose situations occur. Moreover, Fennell *et al.* (2004) pointed out a major source of many toxicological risks as being ignorance of the valuable taxa as well as their use. These two basic issues refer to the low educational levels of the professional as well as the passive users of these resources who should be trained accordingly – see Heron and Yarnell (2011).

This lack of training results in four major risks: (i) the misidentification of the exact plant species efficient for a specific disease treatment; (ii) the misidentification of the organ to be valued in the species; (iii) the specific processing methods required to optimize the extraction of the needed cocktail to be valued; and (iv) determining the required doses (Parekh *et al.*, 2005; Setty and Sigal, 2005; Kayne, 2010). The misidentification might be related both to the species' morphological and biochemical characteristics and is source of many intoxication issues, since they determine both species and organ to which the bioactive component is associated (Marr *et al.*, 2004; Antignac *et al.*, 2011). The biochemical characterization appears to be the most important challenge because the healthier consumption of botanical extracts is rooted in their full standardization (Ribnicky *et al.*, 2008). This suggest the full identification of biochemical components in the extract derived from a well-known material, their relative concentrations, their specific and overall functions in the human organism and the doses to be recommended (Ribnicky *et al.*, 2008; Alhusainy *et al.*, 2012).

Despite this call for caution, standardization remains difficult to reach at this stage of research on botanical extracts for many reasons, among which are that plant ecology (soil and climate) and origin (domesticated or wild) are demonstrated to be significant sources of the variation in biochemical concentrations (Donald and Grollman, 2003; Ribnicky *et al.*, 2008). The large amount of substances contained in the botanical extracts renders any standardization very difficult to achieve (Kroes and Walker, 2004). These authors proposed a series for an attempt to overcome the standardization issue based on plant taxonomical position, origin, conservation/storage system, transformation and dosage plus safety assessment.

### 28.3.2 Potential dangers from botanical extracts with emphasis on Africa

Lis-Balchin (1999) presented detailed specific information related to the dangers of botanical oils in the fields of aromatherapy. This study demonstrated that any plant species is potentially toxic to humans and ranges from raw poison for different organs (intestine, heart, sliver, skin, kidneys, etc.), to allergy, abortion, blood and death in the case of high dose consumption. In a rheumatology treatment system, Setty and Sigal (2005) revealed that each plant species valued for the purpose presented important side effects on the human organism. Even though Aruna and Nirmala (2012) lacked data related to the exact side effects following the application of the numerous plant species used in diabetes treatment, these authors nevertheless have the merit for calling for caution and predicting the urgent need of assessment of those undesirable effects. This caution might particularly be seriously taken when botanical extracts are applied to sensitive or highly vulnerable individuals, like babies, young children and nursing mothers (Lis-Balchin, 1999).

In Africa, the power of African plant species as efficient natural drugs is widely documented (Eyog-Matig *et al.*, 2002; de Boer *et al.*, 2004; Okigbo and Mmeka, 2006). However, botanical extracts from numerous plant species are used locally, while there are limited data on the plant body part used as well as on the exact disease for which they are valorized – see Eyog-Matig *et al.* (2002). De Boer *et al.* (2004) reported that different degrees of toxicity are always associated with doses of most of the botanical extracts used for healthcare in Africa, even though these compounds are often multi-disease treatment substances (Igbe *et al.*, 2009). Thus, many tropical species produce multipurpose botanical extracts, so that a large array of complex diseases involving taxonomically different pathogens may be sensitive to botanical products from few taxa (Lambert *et al.* 2005, Lulekal *et al.* 2008). However, in-depth toxicological assessment for the priority species is still lacking. This is a priority issue in Africa, where the medicinal plant commercialization system is mostly managed by less- or non-educated people. Many challenges still exist in order to strengthen and gain from African medicinal plant research systems. First, the number of plant species to be associated with obtaining a bioactive compound depends on the ethnobotanical knowledge of the seller, who in most situations is not educated regarding the specificity of the user's organism and their medical history and plant ecology linkage to morphological and biochemical compounds' variations (Figure 28.1). Therefore, because there is no feedback of the users towards sellers and that the users often consult many sellers for the same disease, there is no evaluation of the efficiency of specific African plant species against a specific disease. Second, in many situations the processing systems of medicinal plants depend on the user, so that the same species or a combination of different parts of the same species or a combination of many different species could be processed in different

**Table 28.1** Health function of different botanical materials

| Species | Scientific name | English name | Family | Chemical identified | Health function | Reference |
|---|---|---|---|---|---|---|
| Leafy vegetables | *Acca sellowiana* | Feijoa | Myrtaceae | Polyphenolics compound | Antioxidant activity | Beyhan *et al.* (2010) |
| | *Acmella uliginosa* (L.) Jansen | Toothache plant | Asteraceae | Polyphenolics compound | Antioxidant activity | Sossa-Vihotogbé *et al.* (2013a, b) |
| | *Amaranthus gangeticus* | Amaranth | Amaranthaceae | Polyphenolic compound | Antioxidant activity | Saxena *et al.* (2007) |
| | *Aphanes arvensis* (*Alchemilla arvensis*, *Aphanes occidentalis*) | Parsley piert | Rosaceae | | Kidney and bladder calculi, strangury and oedema of renal and hepatic origin, antioxidant activity | Ebadi (2006) |
| | *Asteracantha longifolia* Nees | Long leaves barleria | Acanthaceae | Polyphenolics compounds | Antioxidant activity | Dasgupta and De (2007) |
| | *Bacopa monnieri* (Linn.) Pennell | – | Scrophulariaceae | Polyphenolics compounds | Antioxidant activity | Dasgupta and De (2007) |
| | *Bauhinia racemosa Lam.* | Bidi leaf tree | Caesalpiniaceae | Polyphenolics compounds | Antioxidant activity | Dasgupta and De (2007) |
| | *Caudatus cosmos* | King's salad | Asteraceae | Polyphenolics compounds (quercetin, kaempferol) | Antioxidant activity, antimicrobial activity | Huda-Faujan *et al.* (2009), Andarwulan *et al.* (2010), Rasdi *et al.* (2010) |
| | *Centella asiatica* | | Apiceae | Asiaticoside | Vulnerary | Farnsworth *et al.* (1985), Dasgupta and De (2007), Gupta and Prakash (2009), Huda-Faujan *et al.* (2009), Andarwulan *et al.* (2010) |
| | | | | Polyphenolics compounds (quercetin, kaempferol, myricetin), ascorbic acid, carotenoids | Antioxidant activity | |
| | *Ceratotheca sesamoïdes* Endl | False sesame | Pedaliaceae | Polyphenolics compound | Antioxidant activity | Sossa-Vihotogbé *et al.* (2013a, b) |
| | *Chenopodium album* Linn. | Lamb's quarters | Chenopodiaceae | Polyphenolics compounds | Antioxidant activity | Dasgupta and De (2007) |
| | *Brassica nigra* (L.) Koch | | Brassicaceae | Allyl isothiocyanate | Rubefacient | Farnsworth *et al.* (1985) |
| | *Erechtites hieracifolia* | American burnweed | Asteraceae | Ascorbic acid, polyphenolic compound | Antioxidant activity | Srianta *et al.* (2012) |
| | | | | | Against abdomen bloating, increase appetite, improving immunity | |
| | *Ipomoea reptans* (Linn.) Poir. | Water spinach | Convolvulaceae | Polyphenolics compounds | Antioxidant activity | Dasgupta and De (2007) |
| | *Justicia tenella* (Nees) T. Anderson | – | Acanthaceae | Polyphenolics compound | Antioxidant activity | Sossa-Vihotogbé *et al.* (2013a, b) |
| | *Moringa oleifera lam* | Drumstick tree | Moringaceae | Polyphenolics compounds, vitamins A, C and E | Antioxidant activity | Yang *et al.* (2006), Dasgupta and De (2007) |
| | *Murraya koenigii* | Curry tree | Rutaceae | Polyphenolics compound, ascorbic acid, carotenoids | Antioxidant activity | Gupta and Prakash (2009), Huda-Faujan *et al.* (2009) |
| | *Nothopanax scutellarius* (Burm.f.) Merr | – | Araliacea | Polyphenolics compounds (quercetin, kaempferol) | Antioxidant activity | Andarwulan *et al.* (2010) |
| | *Ocimum americanum* L. | Hoary basil | Lamiaceae | Polyphenolics compounds (quercetin, kaempferol, luteolin) | Antioxidant activity | Andarwulan *et al.* (2010) |
| | *Ocimum basilicum* | Common basilic | Lamiaceae | Polyphenolic compound | Antioxidant activity | Cioroi (2009) |

| | | | | | | |
|---|---|---|---|---|---|---|
| | *Portulaca oleracea* | Purslane | Portulacaceae | Polyphenolics compounds (quercetin) | Antioxidant activity | Andarwulan *et al.* (2010) |
| | *Sauropus androgynus* (L) Merr | Sweet leaf | Phillanthaceae | Polyphenolics compounds (quercetin, kaempferol) | Antioxidant activity | Andarwulan *et al.* (2010) |
| | *Sesamum radiatum* L. | Black beniseed | Pedaliaceae | Polyphenolics compound | Antioxidant activity | Sossa-Vihotogbé *et al.* (2013a, b) |
| | *Spinacia oleracea* | Spinach | Amaranthaceae | Polyphenolic compound | Antioxidant activity | Saxena *et al.* (2007) |
| | *Talinum triangulare* (Jacq.) Willd. | Waterleaf | Portulacaceae | Polyphenolics compounds (quercetin, kaempferol) | Antioxidant activity | Andarwulan *et al.* (2010) |
| | *Trigonella foenum graecum* | Fenugreek | Fabaceae | Polyphenolics compound, ascorbic acid, carotenoids | Antioxidant activity | Dasgupta and De (2007), Gupta and Prakash (2009), Huda-Faujan *et al.* (2009) |
| | *Vernonia colorata* (Willd.) Drake | Bitter leaf | Asteraceae | Vernodalin, sesquiterpene lactones | Antibacterial | Rabe *et al.* (2002) |
| Fruits | *Acca sellowiana* | Feijoa | Myrtaceae | Polyphenolics compound, vitamins | Antioxidant activity, antimicrobial (bacteria and fungi) activity | Basile *et al.* (1997), Beyhan *et al.* (2010) |
| | *Ananas comosus* (L.) Merrill | Ananas | Bromeliaceae | Bromelain | Anti-inflammatory; proteolytic agent | Farnsworth *et al.* (1985) |
| | *Aronia melanocarpa* (Michx) Elliot | Black chokeberry | Rosaceae | Polyphenolic compound (anthocyanins) | Antioxidant activity | Valcheva-Kuzmanova *et al.* (2007) |
| | *Capsicum annuum* L. | Sweet bell pepper | Solanaceae | Carotenoid, polyphenolic compound, ascorbic acid | Antioxidant activity | Hallmann and Rembialkowska (2012) |
| | *Carica papaya* L. | Pawpaw | Caricaceae | Chymopapain, papain | Proteolytic; mucolytic | Farnsworth *et al.* (1985) |
| | *Cucurbita moschata* | Pumpkin | Cucurbitaceae | Carotenoid content, polyphenolic compound | Antioxidant activity | Azizah *et al.* (2009) |
| | *Oenanthe javanica* | Chinese celery | Apiaceae | Polyphenolics compounds | Antioxidant activity | Huda-Faujan *et al.* (2009) |
| | *Origanum vulgare* L. | Oregano | Lamiaceae | Polyphenolic compound | Antioxidant activity | Cioroi (2009) |
| | *Paederia foetida* Linn. | Chinese fever vine | Rubiaceae | Ascorbic acid, polyphenolic compound | Antioxidant activity | Srianta *et al.* (2012), Dasgupta and De (2007) |
| | | | | | Against abdomen bloating, increase appetite, improving immunity | |
| | *Pilea melastomoides* (Poir.) Bl. | – | Urticaceae | Polyphenolics compounds (quercetin, kaempferol, luteolin) | Antioxidant activity | Andarwulan *et al.* (2010) |
| | *Pluchea indica* Less. | Indian camphorweed | Asteraceae | Polyphenolics compounds (quercetin, kaempferol, myricetin) | Antioxidant activity | Andarwulan *et al.* (2010) |
| | *Polygonum minus* | Tearthumb | Polygonaceae | Polyphenolics compounds | Antioxidant activity | Andarwulan *et al.* (2010) |
| | *Polyscias pinnata* | Dinner plate aralia | Araliaceae | Polyphenolics compounds (quercetin, kaempferol) | Antioxidant activity | Andarwulan *et al.* (2010) |
| | *Portulaca oleracea* | Purslane | Portulacaceae | Polyphenolics compounds (quercetin) | Antioxidant activity | Andarwulan *et al.* (2010) |
| | *Salvia officials* L. | Sage | Lamiaceae | Polyphenolic compound | Antioxidant activity | Cioroi (2009) |
| | *Warburgia salutaris* (Bertol.f.) Chiov. | Pepper-bark tree | Canellaceae | | Sores and inflammation | Rabe and van Staden (2000) |
| | *Kigelia africana* (Lam.) Benth. | | Bignoniaceae | Palmitic acid | Antibacterial | Grace *et al.* (2002) |

*(continued)*

**Table 28.1** (*Continued*)

| Species | Scientific name | English name | Family | Chemical identified | Health function | Reference |
|---|---|---|---|---|---|---|
| | *Hunteria umbellata* K Schum | – | Apocynaceae | Saponins, flavonoids, alkaloids and steroidal compounds. | Antipyretic, analgesic | Igbe *et al.* (2009) |
| | *Laurus nobilis* L. | Laurel berries | Lauraceae | Polyphenolic compound | Antioxidant activity | Bozan and Karakaplan (2007) |
| | *Lycopersicon esculentum* | Tomato | Solanaceae | Polyphenolic compounds | Antioxidant activity | Saxena *et al.* (2007) |
| | *Psidium guajava* L. | Guava | Myrtaceae | Carotenoid, ascorbic acid, phenolic compounds | Antioxidant activity | Mercadante *et al.* (1999), Setiawan *et al.* (2001), Miean and Mohamed (2001), Thaipong *et al.* (2006) |
| | *Syzygium cumini* | Black plum | Myrtaceae | Gallic acid, tannins, anthocyanins | Antioxidant activity, antidiabetic, anticancer | Benherlal and Arumughan (2007) |
| Roots and tubers | *Allium cepa* | Onion | Amarillydaceae | Polyphenolic compounds | Antioxidant activity | Saxena *et al.* (2007) |
| | *Daucus carota* | Carrot | Apiaceae | Polyphenolic compounds | Antioxidant activity | Saxena *et al.* (2007) |
| Algua | *Parmelia caperata* (L.) Ach. | Lichens | Parmeliaceae | Polyphenolic compounds, flavonoid content | Antioxidant activity (anticancer), antimicrobial activity (bacteria and fungi) | Kosanic *et al.* (2012) |
| | *Parmelia saxatilis* (L.) Ach. | Lichens | Parmeliaceae | Polyphenolic compounds, flavonoid content | Antioxidant activity (anticancer), antimicrobial activity (bacteria and fungi) | Gulluce *et al.* (2006), Kosanic *et al.* (2012) |
| | *Parmelia sulcata* (Taylor) | Lichens | Parmeliaceae | Polyphenolic compounds, flavonoid content | Antioxidant activity (anticancer), antimicrobial activity (bacteria and fungi) | Kosanic *et al.* (2012) |
| | *Platismatia glauca* (L.) W.L. club. and C.F. culb | Lichens | Parmeliaceae | Polyphenolic compounds | Antioxidant activity, antimicrobial activity (bacteria, fungi and yeasts) | Gulluce *et al.* (2006) |
| | *Ramalina pollinaria* (Wesstr) Arch. | Lichens | Ramalinaceae | Polyphenolic compounds | Antioxidant activity, antimicrobial activity (bacteria, fungi and yeasts) | Gulluce *et al.* (2006) |
| | *Ramalina polymorpha* (Liljeblad) Arch. | Lichens | Ramalinaceae | Polyphenolic compounds | Antioxidant activity, antimicrobial activity (bacteria, fungi and yeasts) | Gulluce *et al.* (2006) |
| | *Umbilicaria nylanderiana* (Zahlbr) | Lichens | Umbilicariaceae | Polyphenolic compounds | Antioxidant activity, antimicrobial activity (bacteria, fungi and yeasts) | Gulluce *et al.* (2006) |
| Medicinal plants | *Anacardium occidentale* | – | Anacardiaceae | – | Antitusive and treat mouth infection | Eyog Matig *et al.* (2002) |
| | *Adonis vernalis* L. | Green false hellbore | Ranunculaceae | Adoniside | Cardiotonic | Farnsworth *et al.* (1985) |
| | *Aesculus hippocastanum* L. | Horse- chestnut | Sapindaceae | Aescin | Anti-inflammatory | Farnsworth *et al.* (1985) |
| | *Agrimonia eupatoria* L. | – | Rosaceae | Agrimophol | Anthelmintic | Farnsworth *et al.* (1985) |
| | *Ammi majus* L. | Laceflower | Apiaceae | Xanthotoxin | Leukoderma; vitiligo | Farnsworth *et al.* (1985) |
| | *Ammi visnaga* (L.) Lamk. | Toothpickweed | Apiaceae | Khellin | Bronchodilator | Farnsworth *et al.* (1985) |
| | *Amplelocissus grantii* (Baker) Planch. | – | Vitacea | Polyphenolics compounds | Antimicrobial and antioxidant activities | Zongo *et al.* (2010) |
| | *Anamirta cocculus* (L.) W.&A. | Levant nut | Menispermaceae | Picrotoxin | Analeptic | Farnsworth *et al.* (1985) |
| | *Andrographis paniculata* Nees | King of bitter | Acanthaceae | Andrographolide, neoandrographolide Polyphenolics compounds | Bacillary dysentery, antioxidant activity | Farnsworth *et al.* (1985), Arash *et al.* (2010) |

| | | | | | |
|---|---|---|---|---|---|
| *Anisodus tanguticus* (Maxim.) Pascher | – | Solanaceae | Anisodamine | Anticholinergic | Farnsworth *et al.* (1985) |
| *Ardisia japonica* Bl. | – | Myrsinaceae | Bergenin | Antitussive | Farnsworth *et al.* (1985) |
| *Areca catechu* L. | Betel tree | Arecaceae | Arecoline | Anthelmintic | Farnsworth *et al.* (1985) |
| *Artemisia maritima* L. | Old woman | Asteracea | Santonin | Ascaricide | Farnsworth *et al.* (1985) |
| *Atropa belladonna* L. | Belladonna | Solanaceae | Atropine | Anticholinergic | Farnsworth *et al.* (1985) |
| *Azadirachta Indica* A. Juss | Neem | Meliaceae | Polyphenolic compounds | Antioxidant activity | Ghimeray *et al.* (2009) |
| *Berberis vulgaris* L. | Barberry | Berberidaceae | Berberine | Bacillary dysentery | Farnsworth *et al.* (1985) |
| *Calpurnia aurea* (Ait.) Benth. | – | Fabaceae | Alkaloids (calpurmenine and 13-2′-pyrrolylcarboxyl calpurmenine), total polyphenolic compound (flavonoids, proanthocyanidins, flavonols) | Antibacterial and antioxidant activities | Asres *et al.* (1986), Adedapo *et al.* (2008) |
| *Camellia sinensis* (L.) Kuntze | Tea plant | Theaceae | Caffeine, theophylline | Stimulant, diuretic; bronchodilator | Farnsworth *et al.* (1985) |
| *Cassia* spp. | | Ceasalpiniaceae | Danthron, sennosides A & B | Laxative | Farnsworth *et al.* (1985) |
| *Cephaelis ipecacuanha* (Brotero) A. Richard | – | Rubiuaceae | Emetine | Amoebicide; emetic | Farnsworth *et al.* (1985) |
| *Chondodendron tomentosum* R. & P. | | Menispermaceae | Tubocurarine | Skeletal muscle relaxant | Farnsworth *et al.* (1985) |
| *Cichorium pumilum* | Wild endive | Asteraceae | Polyphenolic compounds | Antioxidant activity | Azaizeh *et al.* (2005) |
| *Cinchona ledgeriana* Moens ex.Trimen | – | Rubiaceae | Quinine | Antimalarial | Farnsworth *et al.* (1985) |
| *Colchicum autumnale* L. | Naked lady | Colchicaceae | Colchicine | Antitumor agent; antigout | Farnsworth *et al.* (1985) |
| *Convallaria majalis* L. | – | Liliaceae | Convallotoxin | Cardiotonic | Farnsworth *et al.* (1985) |
| *Corydalis ambigua* (Pallas) Cham. & Schltal. | – | Papaveraceae | Tetrahydropalmatine | Analgesic; sedative | Farnsworth *et al.* (1985) |
| *Crotolaria sessiliflora* L. | Purpleflower crotalaria | Fabaceae | Monocrotaline | Antitumour agent | Farnsworth *et al.* (1985) |
| *Croton lobatus* | – | Euphorbiaceae | Tiliroside (kaempferol), vitexin | Antimalarial activity | Lagnika (2005) |
| *Cynara scolymus* L. | Artichoke | Asteraceae | Cynarin | Choleretic | Farnsworth *et al.* (1985) |
| *Daphne genkwa* Seib. & Zucc. | – | Thymelaeceaea | Yuanhuacine | Abortifacient | Farnsworth *et al.* (1985) |
| *Datura metel* L. | Devil's trumpet | Solanaceae | Scopolamine | Sedative | Farnsworth *et al.* (1985) |
| *Digenea simplex* (Wulf.) Agardh | – | Rhodomelaceae | Kainic acid | Ascaricide | Farnsworth *et al.* (1985) |
| *Digitalis lanata* Ehrh. | Wooly foxglove | Plantaginaceae | Acetyldigoxin, deslanoside, digoxin, lanatosides A, B, C | Cardiotonic | Farnsworth *et al.* (1985) |
| *Digitalis purpurea* L. | Lady's glove, common foxglove | Plantaginaceae | Digitoxin, digitalin, gitalin | Cardiotonic | Farnsworth *et al.* (1985) |
| *Enhydra fluctuans* Lour. | | Compositeae | Polyphenolics compounds | Antioxidant activity | Dasgupta and De (2007) |

(*continued*)

**Table 28.1** (*Continued*)

| Species | Scientific name | English name | Family | Chemical identified | Health function | Reference |
|---|---|---|---|---|---|---|
| | *Ephedra sinica* Stapf. | – | Ephedraceae | Pseudoephedrine, ephedrine | Sympathomimetic | Farnsworth *et al.* (1985) |
| | *Eryngium creticum* | – | Apiacea | Polyphenolic compounds | Antioxidant activity | Azaizeh *et al.* (2005) |
| | *Erythroxylum coca* Lamk | | Erythroxylaceae | Cocaine | Local anaesthetic | Farnsworth *et al.* (1985) |
| | *Etlingera elatior* (Jack) R. M. Sm | Torch ginger | Zingiberaceae | Polyphenolics compounds (quercetin) | Antioxidant activity | Andarwulan *et al.* (2010) |
| | *Fraxinus rhynchophylla* Hance | – | Oleaceae | Aesculetin | Antidysentery | Farnsworth *et al.* (1985) |
| | *Glycyrrhiza glabra* L. | Reglisse | Fabaceae | Glycyrrhizin | Sweetener | Farnsworth *et al.* (1985) |
| | *Gomphrena celosioides* | – | Amaranthaceae | – | Antimalarial activity | Lagnika (2005) |
| | *Gossypium* spp. | Cotton plant | Malvaceae | Gossypol | Male contraceptive | Farnsworth *et al.* (1985) |
| | *Helmsleya amabilis* Diels | – | Cucurbitaceae | Hemsleyadin | Bacillary dysentery | Farnsworth *et al.* (1985) |
| | *Hydrangea macrophylla* (Thunb.) DC | French hydrangea | Hydrangeaceae | Phyllodulcin | Sweetener | Farnsworth *et al.* (1985) |
| | *Hydrastis canadensis* L. | Orangeroot | Ranunculaceae | Hydrastine | Haemostatic; astringent | Farnsworth *et al.* (1985) |
| | *Hyoscamus niger* L. | Stinking nightshade | Solanaceae | Hyoscamine | Anticholinergic | Farnsworth *et al.* (1985) |
| | *Hypericum monogynum* | – | Clusiaceae | Polyphenolics compounds | Antioxidant activity | Raziq *et al.* (2011) |
| | *Hypericum perforatum* | – | Clusiaceae | Polyphenolics compounds | Antioxidant activity | Raziq *et al.* (2011) |
| | *Jasminum nervosum* Lour. | – | Oleaceae | Polyphenolic compounds such as gallic acid and lutin | Antioxidant activity | Huo *et al.* (2011) |
| | *Lobelia inflata* L. | Indian tobacco | Campalunaceae | Lobeline | Smoking deterrent; respiratory stimulant | Farnsworth *et al.* (1985) |
| | *Lonchocarpus nicou* (Aubl.) DC. | Barbasco | Fabaceae | Rotenone | Piscicide | |
| | *Micromeria croatica (Pers.) Schott,* | – | Lamiaceae | Acid phenolics, tannins, flavonoids | Heart disorders, headache, colds, wounds and skin infections, antimicrobial | Vladimir-Knežević *et al.* (2011) |
| | *M. thymifolia* (Scop.) Fritsch and *M. juliana* (L.) Bentham ex Reichenb | | | | | |
| | *Momordica charantia* | Bitter gourd/bitter melon | Cucurbitaceae | Polyphenolic compounds | Antioxidant activity/diabetes | Saxena *et al.* (2007), Aruna and Nirmala (2012) |
| | *Nyctanthes arbortristis* Linn. | Night flowering jasmine | Oleaceae | Polyphenolics compounds | Antioxidant activity | Dasgupta and De (2007) |
| | *Papaver somniferum* L. | Opium poppy | Papaveraceae | Codeine, morphine, noscapine | Analgesic; antitussive | Farnsworth *et al.* (1985) |
| | *Pausinystalia yohimbe* (K.Schum.) Pierre | – | Rubiaceae | Yohimbine | Aphrodisiac | Farnsworth *et al.* (1985) |
| | *Pelargonium reniforme* Curtis | Kidney-leaved perlagonium | Geraniaceae | Total polyphenolic compounds (flavonoids, proanthocyanidins, flavonols) | Liver diseases, antibacterial and antioxidant activity | Adewusi and Afolayan (2009) |
| | *Physostigma venenosum* Balf. | – | Fabaceae | Physostigmine | Cholinesterase inhibitor | Farnsworth *et al.* (1985) |
| | *Pilocarpus jaborandi* Holmes | – | Rutaceae | Pilocarpine | Parasympathomimetic | Farnsworth *et al.* (1985) |

| | | | | | | |
|---|---|---|---|---|---|---|
| | *Piper methysicum* Forst. f. | Kava | Piperaceae | Kawain | Tranquilizer | Farnsworth *et al.* (1985) |
| | *Pistacia palaestina* | Terebinth | Anacardiaceae | Polyphenolic compounds | Antioxidant activity | Azaizeh *et al.* (2005) |
| | *Podophyllum peltatum* L. | Umbrella plant | Berberidaceae | Podophyllotoxin, teniposide, etoposide | Condylomata acuminata, antitumour agent | Farnsworth *et al.* (1985) |
| | *Potentilla fragaroides* L. | - | Rosaceae | (+)-Catechin | Haemostatic | Farnsworth *et al.* (1985) |
| | *Quisqualis indica* L. | Chinese honeysuckle | Combretaceae | Quisqualic acid | Anthelmintic | Farnsworth *et al.* (1985) |
| | *Ranunculus repens* L | Creeping buttercup | Ranunculaceae | Flavonoids | Antihyperglycaemic, antidiabetic, antioxidant | Kebieche (2009) |
| | *Rauvolfia canescens* L. | Black snake root | Apocynaceae | Deserpidine | Antihypertensive; tranqulizer | Farnsworth *et al.* (1985) |
| | *Rauvolfia serpentina* (L.) Benth ex. Kurz | Snakeroot | Apocynaceae | Ajmalicine, reserpine rescinnamine | Circulatory disorders, antihypertensive; tranquilizer | Farnsworth *et al.* (1985) |
| | *Rhododendron molle* G. Don | - | Ericaceae | Rhomitoxin | Antihypertensive | Farnsworth *et al.* (1985) |
| | *Rorippa indica* (L.) Hochr. | - | Brassicaceae | Rorifone | Antitussive | Farnsworth *et al.* (1985) |
| | *Salix alba* L. | - | Slicaceae | Salicin | Analgesic | Farnsworth *et al.* (1985) |
| | *Satureja hortensis* L. | Summer savory | Lamiaceae | Polyphenolic compounds | Antioxidant activity | Farnsworth *et al.* (1985) |
| | *Silybum marianum* (L.) Gaertn. | Milk thistle | Asteraceae | Silymarin | Antihepatotoxic | Farnsworth *et al.* (1985) |
| | *Simarouba glauca* DC. | Paradise tree | Simaroubaceae | Glaucaroubin | Amoebicide | Farnsworth *et al.* (1985) |
| | *Stephania sinica* Diels | - | Menispermaceae | Rotundine | Analgesic; sedative | Farnsworth *et al.* (1985) |
| | *Stevia rebaudiana* Bertoni | Stevia | Asteraceae | Stevioside | Sweetener | Farnsworth *et al.* (1985) |
| | *Strophanthus gratus* Baill. | Spider tresses | Apocynaceae | Ouabain | Cardiotonic | Farnsworth *et al.* (1985) |
| | *Strychnos nux-vomica* L. | Nux vomica | Loganaceae | Strychnine CNS | Stimulant | Farnsworth *et al.* (1985) |
| | *Tagetes erectus L.* | African marigold | Asteraceae | - | Antinociceptive and anti-inflammatory effects | Shinde *et al.* (2009) |
| | *Teucrium polium* | Felty germander | Lamiaceae | Polyphenolic compounds | Antioxidant activity | Azaizeh *et al.* (2005) |
| | *Thalia geniculata* | - | Marantaceae | Geranylfarnesol | Antimalarial activity | Lagnika (2005) |
| | *Theobroma cacao* L. | Cacao tree | Malvaceae | Theobromine | Diuretic; bronchodilator | Farnsworth *et al.* (1985) |
| | *Thymus vulgaris* L. | | | Trichosanthin | Abortifacient | Farnsworth *et al.* (1985) |
| | *Urginea maritima* (L.) Baker | Sea onion | Asparagaceae | Scillarin A | Cardiotonic | Farnsworth *et al.* (1985) |
| | *Valeriana officinalis* L. | - | Valerianaceae | Valepotriates | Sedative | Farnsworth *et al.* (1985) |
| | *Veratrum album* L. | White veratrum | Liliaceae | Protoveratrines A & B | Antihypertensive | Farnsworth *et al.* (1985) |
| | *Vinca minor* L. | Dwarf periwinkle | Apocynaceae | Vincamine | Cerebral stimulant | Farnsworth *et al.* (1985) |
| | *Vitex diversifolia* | - | Verbenaceae | - | Skin disease, toothhache | Eyog Matig *et al.* (2002) |
| Spices | *Allium sativum* | Garlic | Amaryllidaceae | Polyphenolic compounds | Antioxidant activity | Saxena *et al.* (2007) |
| | *Capsicum chinense* | Chili pepper | Solanaceae | Polyphenolic compounds | Antioxidant activity | Núñez-Ramírez *et al.* (2011) |
| | *Cuminum cyminum* L. | Cumin | Apiaceae | Flavonoid content | Antibaterial activity, antioxidant activity | Athamena *et al.* (2010) |
| | *Curcuma longa* L. | | Zingiberacea | Curcumin | Antiparasitic, antispasmodic, anti-inflammatory and gastrointestinal effects; inhibits carcinogenesis and cancer growth | Farnsworth *et al.* (1985), Araújo and Leon (2001) |
| | *Piper nigrum* | Black pepper | Piperaceae | Polyphenolic compounds | Antioxidant activity | Saxena *et al.* (2007) |
| | *Zingiber officinale* (L.) Rosc. | Ginger | Zingiberaceae | Polyphenolic compounds (6-gingerol and its derivatives) | Antioxidant activity | Saxena *et al.* (2007), Stoilova *et al.* (2007) |

**Figure 28.1** Partial overview of the use of botanical extract for healthcare in Benin. (a) Dried bark and roots in local market in Cotonou. (b) Woman acting as doctor selecting botanical material for an indicated disease in local market of Cotonou. (c) Uncontrolled dose service for water extracted compound of *Lippia multiflora*, *Occimum graticimum*, *Momordica charantia* and *Citrus limon*. (d) Uncontrolled dose service for alcohol extracted compound of barks, roots, leaves and flowers of unknown plant species. (e) *Momordica charantia* and *Dialium guineensis* for regular household consumption in Cotonou. (f) *Occimum gratissimun* for regular household consumption as medicine or vegetable in Cotonou. *Pictures*: R. Vihotogbé

ways with no concrete dose prescription (Figure 28.1). Lastly, there is no judiciary system that medicinal plant sellers or traditional practitioners will fear in order to improve their professionalism as 'traditional doctors' in the use of traditional botanical extracts. This judiciary system does not exist owing to the lack of scientific data that might act as proof of overdose or inadequacy with specific human organisms or diseases, or even extraction method. A simple example in Benin is that small businesses of alcohol-based botanical extracts are increasingly flourishing in Cotonou and in its related areas. It is locally recognized that people who consume those extracts (even of scientifically recognized medicinal species) on high frequency basis, often waste away and quickly die. In this specific situation, three main questions rise up: (i) is the types of alcohol used adequate for human health? (ii) what is the impact of those botanical components on human health? and (iii) what are the human organ ism's biological particularities to tolerate or not these unknown components? Consequently, nobody often knows what exactly kills or saves.

## References


Achigan-Dako, E.G., Pasquini, M.W., Assogba Komlan, F. *et al.* (2010) Traditional Vegetables in Benin. *Institut National des Recherches Agricoles du Bénin*. Imprimeries du CENAP, Cotonou

Adedapo, A.A., Jimoh, F.O., Koduru, S. *et al.* (2008) Antibacterial and antioxidant properties of the methanol extractsof the leaves and stems of *Calpurnia aurea*. *BMC Complementary and Alternative Medicine*, **8**, 35–42.

Adefegha, S.A. and Oboh, G. (2005) Cooking enhances the antioxidant properties of some tropical green leafy vegetables. *African Journal of Biotechnology*, **10** (4), 632–639.

Adewusi, E.A. and Afolayan, A.J. (2009) Antibacterial, antifungal and antioxidant activity of the roots and leaves of *Pelargonium reniforme* Curtis (Geraniaceae). *African Journal of Biotechnology*, **8** (22), 6425–6433.

Afolayan, A.J. and Jimoh, F.O. (2009) Nutritional quality of some wild leafy vegetables in South Africa. *International Journal of Food Science and Nutrition*, **60** (5), 424–431.

Ainooson, G.K., Owusu, G., Woode, E. *et al.* (2012) *Trichilia monadelpha* bark extracts inhibit carrageenan-induced foot-oedema in the 7-day old chick and the oedema associated with adjuvant-induced arthritis in rats. *African Journal of Traditional, Complementary, and Alternative Medicine*, **9** (1), 8–16.

Akubugwo, I.E., Obasi, N.A., Chinyere, G.C. and Ugbogu, A.E. (2007) Nutritional and chemical value of *Amaranthus hybridus* L. leaves from Afikpo, Nigeria. *African Journal of Biotechnology*, **6** (24), 2833–2839.

Alhusainy, W., van den Berg, J.P.L., Paini, A. *et al.* (2012) Matrix modulation of the bioactivation of estragole by constituents of different alkenylbenzene-containing herbs and spices and physiologically based biokinetic modeling of possible *in-vivo* effects. *Toxicological Sciences*, **129**, 174–187.

Aliero, A.A. and Afolayan, A.J. (2006) Antimicrobial activity of *Solanum tomentosum*. *African Journal of Biotechnology*, **5** (4), 369–372.

Andarwulan, N., Batari, R., Sandrasari, D.A., *et al.* (2010) Flavonoid content and antioxidant activity of vegetables from Indonesia. *Food Chemistry*, **121**, 1231–1235.

Antignac, E., Nohynek, G.J., Re, T. *et al.* (2011) Safety of botanical ingredients in personal care products/cosmetics. *Food and Chemical Toxicology*, **49**, 324–341.

Arash, R., Koshy, P. and Sekaran, M. (2010) Antioxidant potential and content of phenolic compounds in ethanolic extracts of selected parts of *Andrographis paniculata*. *Journal of Medicinal Plants Research*, **4** (3), 197–202.

Araújo, C.C. and Leon, L.L. (2001) Biological activities of *Curcuma longa* L. *Memórias do Instituto Oswaldo Cruz*, **96**, 723–728.

Aruna, D.R. and Nirmala, A.R.R. (2012) Herbal solution for the treatment of diabetes mellitus. *Journal of Pharmaceutical and Scientific Innovation*, **1**, 27–31.

Asres, K., Gibbons, W.A., Phillipson, J.D. and Mascgani, P. (1986) Alkaloids of Ethiopian *Calpurnia aurea* subsp. *aurea*. *Phytochemistry*, **25**, 1443–1447.

Athamena, S., Chalghemi, I., Kassah-Laouar, A. *et al.* (2010) Activité antioxydante et antimicrobienne d'extraits de *Cuminum cyminum* L. *Lebanese Science Journal*, **11** (1), 69–81.

Azaizeh, H., Lyubuncic, P., Potrtnaya, I. *et al.* (2005) Fertilization induced changes in growth parameters and antioxidant activity of medicinal plants used in traditional Arab medicine. *Evidence-Based Complementary and Alternative Medicine*, **2** (4), 549–556.

Azizah, A.H., Wee, K.C., Azizah, O. and Azizah, M. (2009) Effect of boiling and stir frying on total phenolics, carotenoids and radical scavenging activity of pumpkin (*Cucurbita moschato*). *International Food Research Journal*, **16**, 45–51.

Bandaranayake, W.M. (2003) Quality control, screening, toxicity, and regulation of herbal drugs, in *Modern Phytomedicine – Turning Medicinal Plants into Drugs* (eds I. Ahmad, F. Aqil and M. Owais), Wiley-VCH Verlag, Weinheim, pp. 25–57.

Bandeira, S.O., Gaspar, F. and Pagula, F.P. (2001) African ethnobotany and healthcare: emphasis on Mozambique. *Pharmaceutical Biology*, **39** (1), 70–73.

Basile, A., Vuotto, M.L., Violante, U. *et al.* (1997) Antibacterial activity in *Actinidia chinensis*, *Feijoa sellowiana* and *Aberia caffra*. *International Journal of Antimicrobial Agents*, **8**, 199–203.

Begossi, A. (1996) Use of ecological methods in ethnobotany: diversity indices. *Economic Botany*, **50**, 280–289.

Benherlal, P.S. and Arumughan, C. (2007) Chemical composition and *in vitro* antioxidant studies on *Syzygium cumini* fruit. *Journal of the Science of Food and Agriculture*, **87**, 2560–2569.

Bertrand, Y.A. (2009) Home. http://www.youtube.com/watch?v=jqxENMKaeCU (accessed 1 December 2013).

Beyhan, O., Elmastas, M. and Gedikli, F. (2010) Total phenolic compounds and antioxidant capacity of leaf, dry fruit and fresh fruit of feijoa (*Acca sellowiana*, Myrtaceae). *Journal of Medicinal Plants Research*, **4**, 1065–1072.

Bozan, B. and Karakaplan, U. (2007) Antioxidants from laurel (*Laurus nobolis* L.) berries: influence of extraction procedure on yield and antioxidant activity of extracts. *Acta Alimentaria*, **36**, 321–328.

Burnie, J.P., Naderi-Nasab, M., Loudon, K.W. and Matthews, R.C. (1997) An epidemiological study of blood culture isolates of coagulase-negative staphylococci demonstrating hospital-acquired infection. *Journal of Clinical Microbiology*, **35** (7), 1746–1750.

Cassileth, B.R., Heitzer, M. and Wesa, K. (2009) The public health impact of herbs and nutritional supplements. *Pharmaceutical Biology*, **47**, 761–767.

Chadare, F.J., Linnemann, A.R., Hounhouigan, J.D. *et al.* (2008) Baobab food products: a review on their composition and nutritional value. *Food Sciences and Nutrition*, **49** (3), 254–274.

Cioroi, M. (2009) Study on total polyphenols and reducing power of aqueous extracts from selected Lamiaceae species. *Journal of Agroalimentary Processes and Technologies*, **15**, 521–524.

Clark, L.E. and Sunderland, T.C.H. (eds) (2004) The Key Non-Timber Forest Products of Central Africa: State of the Knowledge. Technical Paper No. 122. SD Publication Series, USAID.

Codjia, J.T.C., Assogbadjo, A.E. and Ekué, M.R.M. (2003) Diversité et valorisation au niveau des ressources forestières alimentaires végétales du Bénin. *Cahiers Agriculture*, **12**, 321–331.

Cunningham, A.B. (1993) African medicinal plants: setting priorities at the interface between conservation and primary health care, People and Plants Working Paper 1, UNESCO, Paris.

Dansi, A., Adjatin, A., Vodounhè, R. *et al.* (2008) *Biodiversité des Légumes-Feuilles Traditionnels Consommés au Bénin*. Bibliothèque Nationale, Porto-Novo, Bénin.

Dasgupta, N. and De, B. (2007) Antioxidant activity of some leafy vegetables of India: a comparative study. *Food Chemistry*, **101**, 471–474.

De Boer, H.J., Kool, A., Broberg, A. *et al.* (2004) Anti-fungal and anti-bacterial activity of some herbal remedies from Tanzania. *Journal of Ethnopharmacology*, **96**, 461–469.

Donald, M.M. and Grollman, A.P. (2003) Botanical medicines – the need for new regulations. *New England Journal of Medicine*, **347** (25), 2073–2076.

Downie, B.J. (2010) Inside popular botanical ingredients in pediatric skincare products. Feverfew, colloidal oatmeal, and sunflower seed oil may provide benefits for children with eczematous skin conditions. *Pediatric Skin Care*, (July–August) 19–23.

Ebadi, M. (2006) *Pharmacodynamic Basis of Herbal Medicine*, CRC, Boca Raton, FL.

Egras A.M., Hamilton, W.R., Lenz, T.L. and Monaghan, M.S. (2011) An evidence-based review of fat modifying supplemental weight loss products. *Journal of Obesity*, **2011**, article ID 297315.

Ejoh, R.A., Djuikwo, V.N., Gouado, I. and Mbofung, C.M. (2009) Effect of different postharvest treatments on antinutritional factors in some commonly consumed leafy vegetables in Cameroon. *Journal of Food Processing and Preservation*, **33**, 161–174.

Eyog Matig, O., Gaoué, O.G. and Dossou, B. (2002) *Programme de Ressources Génétiques Forestières en Afrique au Sud du Sahara*, Première réunion, Réseau 'Espèces Ligneuses Alimentaires', IPGRI, CNSF, Ouagadougou, Burkina Faso.

Eyzaguirre, R.Z., Nienaltowska, K., de Jong, L.E.Q. *et al.* (2006) Effect of food processing of pearl millet (*Pennisetum glaucum*) IKMP-5 on the level of phenolics, phytate, iron and zinc. *Journal of the Science of Food and Agriculture*, **86**, 1391–1398.

FAO/WHO (1988) *Requirements of vitamin A, iron, folate and vitamin B12*. Report of a Joint FAO/WHO Expert Consultation. FAO Food and Nutrition Series No. 23, Food and Agricultural Organization of the United Nations, Rome.

Farnsworth, N.R., Akerele, O., Bingel, A.S. *et al.* (1985) Medicinal plants in therapy. *Bulletin of the World Health Organization*, **63**, 965–981.

Fennell, C.W., Lindsey, K.L., McGaw, L.J. *et al.* (2004) Assessing African medicinal plants for efficacy and safety: pharmacological screening and toxicology. *Journal of Ethnopharmacology*, **94**, 205–217.

Ghimeray, A.K., Jin, C.-W., Ghimire, B.K. and Cho, D.H. (2009) Antioxidant activity and quantitative estimation of Azadirachtin and nimbin in *Azadirachta indica* A. Juss. *African Journal of Biotechnology*, **8**, 3084–3091.

Grace, O.M., Light, M.E., Lindsey, K.L. *et al.* (2002) Antibacterial activity and isolation of active compounds from fruit of the traditional African medicinal tree *Kigelia africana*. *South African Journal of Botany*, **68**, 220–222.

Gulluce, M., Aslan, A., Sokmen, M. *et al.* (2006) Screening the antioxidant and antimicrobial properties of the lichens *Parmelia saxatilis*, *Plastismatia glauca*, *Ramalina pollinaria*, *Ramalina polymorpha* and *Umblilicardia nylanderiana*. *Phytomedecine*, **13**, 515–521.

Gruenwald, J. (2009) Novel botanical ingredients for beverages. *Clinics in Dermatology*, **27**, 210–216.

Gupta, S. and Prakash, J. (2009) Studies on Indian green leafy vegetables for their antioxidant activity. *Plant Foods and Human Nutrition*, **64**, 39–45.

Hallmann, E. and Rembialkowska, E. (2012) Characterization and antioxidant compounds in sweet bell pepper (*Capsicum annuum* L.) under organic and conventional growing systems. *Journal of the Science of Food and Agriculture*, **92**, 2409–2415.

Heron, S. and Yarnell, E. (2001) The safety of low-dose *Larrea tridentate* (DC) Coville (creosote bush or chaparral): a retrospective clinical study. *The Journal of Alternative and Complementary Medicine*, **7** (2), 175–185.

Huda-Faujan, N., Noriham, A., Norrakiah, A.S. and Babji, A.S. (2009) Antioxidant activity of plants methanolic extracts containing phenolic compounds. *African Journal of Biotechnology*, **8**, 484–489.

Huo, L, Lu, R., Li, P. *et al.* (2011) Antioxidant activity, total phenolic, and total flavonoid of extracts from the stems of *Jasminum nervosum* Lour. *Grasas y Aceites*, **62** (2), 149–154.

Igbe, I., Ozolua, R.I., Okpo, S.O. and Obasuyi, O. (2009) Antipyretic and analgesic effects of the aqueous extract of the fruit pulp of *Hunteria umbellata* K Schum (Apocynaceae). *Tropical Journal of Pharmaceutical Research*, **8** (4), 331–336.

Ignacimuthu, S., Ayyanar, M. and Sivaraman, S.K. (2006) Ethnobotanical investigations among tribes in Madurai District of Tamil Nadu (India). *Journal of Ethnobiology and Ethnomedicine*, **2** (25), 1–7.

Jose, S. (2012) Agroforestry for conserving and enhancing biodiversity. *Agroforestry Systems*, **85**, 1–8.

Kayne, S.B. (2010) *Traditional Medicine: A Global Perspective*, Pharmaceutical Press, London.

Kebieche, M. (2009) Activité biochimique des extraits flavonoïdiques de la plante *Ranunculus repens* L: effet sur le diabète expérimental et l'hépatotoxicité induite par l'Epirubicine. PhD thesis, Université Mentouri Constantine.

Kosanic, M.M., Rankovic, B.R. and Stanojkovic, T.P. (2012) Antioxidant, antimicrobial and anticancer activities of three *Parmelia* species. *Journal of the Science of Food and Agriculture*, **92**, 1909–1916.

Kroes, R. and Walker, R. (2004) Safety issues of botanicals and botanical preparations in functional foods. *Toxicology*, **198**, 213–220.

Lagnika, L. (2005) Etude phytochimique et activité biologiques des substances naturelles isolées de plantes béninoises. Thèse de doctorat, Université Louis Pasteur de Strasbourg/Université d'Abomey-calavi, Bénin.

Laird, S.A. (2005) Medical plants in international trades: conservation and equity issues. *UNESCO-EOLSS, Ethnopharmacology*, **2**, 7.

Lambert, J.D.H., Ryden, P.A. and Esikuri, E.E. (2005) Capitalization on the bioeconomic value of multipurpose medical plants for the rehabilitation of dry lands in sub-Saharan Africa. Global Environment Facility Program/The World Bank, Washington DC.

Lesley, L. and Brown, N. (2004) Bush mango (*Irvingia gabonensis* and *I. wombolu*), in *Key Non–Timber Forest Products of Central Africa: State of the Knowledge* (eds L.E. Clark and T.C.H. Sunderland), SD Publication Series, USAID, pp. 15–35.

Lis-Balchin, M. (1999) Possible health and safety problems in the use of novel plant essential oils and extracts in aromatherapy. *Journal of the Royal Society for the Promotion of Health*, **119** (4), 240–243.

Lulekal, E., Kelbessa, E., Bekele, T. and Yinege, H. (2008) An ethnobotanical study of medicinal plants in Mana Angetu District, southeastern Ethiopia. *Journal of Ethnobiology and Ethnomedicine*, **4**, 10.

Malaisse, F. and Parent, G. (1985) Edible wild vegetable products in the Zambian wood-land area: a nutritional and ecological approach. *Ecology of Food and Nutrition*, **18**, 43–82.

Marr, K.L., Mei, X.Y. and Bhattarai, N.K. (2004) Allozyme, morphological and nutritional analysis bearing on the domestication of *Momordica charantia* L. (Cucurbitaceae). *Economic Botany*, **58** (3), 435–455.

Maundu P., Achigan-Dako E. and Morimoto Y. (2009) Bioversity of African vegetables, in *African Indigenous Vegetables in Urban Agriculture* (eds C.M. Shackleton, M.W. Pasquini and A.W. Drescher), Earthscan, London, pp. 65–104.

Mercadante, A.Z., Steck, A. and Pfander, H. (1999) Carotenoids from guava (*Psidium guajava* L): isolation and structure elucidation. *Journal of Agricultural and Food Chemistry*, **4**, 145–151.

Miean, K.H. and Mohamed, S. (2001) Flavonoid (myricetin, quercetin, kaempferol, luteolin, and apigenin) content of edible tropical plants. *Journal of Agricultural and Food Chemistry*, **9**, 3106–3112.

Ndoye, O., Ruiz–Perez, M. and Eyebe A. (1997) The market of non-timber forest products in the humid forest zone of Cameroon. Rural Development Forestry Network, network paper 22c, 20, ODI, London.

Núñez-Ramírez, F., González-Mendoza, D., Grimaldo-Juárez, O. and Díaz, L.C. (2011) Nitrogen fertilization effect on antioxidants compounds in fruits of habanero chili pepper (*Capsicum chinense*). *International Journal of Agricultural Biology*, **13**, 827–830.

Okigbo, R.N. and Mmeka, E.C. (2006) An appraisal of phytomedicine in Africa. *KMITL Science and Technology Journal*, **6** (2), 83–94.

Okolo, C.O., Johnson, P.B., Abdurahman, E.M. *et al.* (1995) Analgesic effect of *Irvingia gabonensis* stem bark extract. *Journal of Ethnopharmacology*, **45** (2), 125–129.

Olson, G. (1991) Socio-economic importance of non-timber forest products in the South Pacific. Focus on Vanuatu. *Unasylva*, **42** (65), 24–30

Parekh, J., Jadeja, D. and Chanda, S. (2005) Efficacy of aqueous and methanol extracts of some medicinal plants for potential antibacterial activity. *Turkish Journal of Biology*, **29**, 203–210.

Rabe, T. and van Staden, J. (2000) Isolation of an antibacterial sesquiterpenoid from *Warburgia salutaris*. *Journal of Ethnopharmacology*, **73**, 171–174.

Rabe, T., Mulholland, D. and van Staden, J. (2002) Isolation and identification of antibacterial compounds from *Vernonia colorata* leaves. *Journal of Ethnopharmacology*, **80**, 91–94.

Rasdi, N.H.M., Samah, O.A., Sule, A. and Ahmed, Q.U. (2010) Antimicrobial studies of *Cosmos caudatus* Kunth. (Compositae). *Journal of Medicinal Plants Research*, **4** (8), 669–673.

Raskin, I., Ribnicky, D.M., Komarnytsky, S. *et al.* (2002) Plants and human health in the twenty-first century. *Trends in Biotechnology*, **20**, 522–531.

Raziq, N., Muhammad, N., Chishti, K.A. *et al.* (2011) Correlation of the antioxidant capacity with the phenolic contents of *Hypericum monogynum* and *Hypericum perforatum*. *African Journal of Pharmacy and Pharmacology*, **5**, 1872–1876.

Ribnicky, D.M., Poulev, A., Schmidt, B. *et al.* (2008) Evaluation of botanicals for improving human health. *American Journal of Clinical Nutrition*, **87**, 1–4.

Roberson, E. (2008) Medicinal plants at risk – nature's pharmacy, our treasure chest: why we must conserve our natural heritage, Native Plant Conservation Campaign Report, Center for Biological Diversity, Tucson, AZ.

Saxena, R., Venkaiah, K., Anitha, P. *et al.* (2007) Antioxidant activity of commonly consumed plant foods of India: contribution of their phenolic content. *International Journal of Food Sciences and Nutrition*, **58** (4), 250–260.

Schmelzer G.H., Achigan-Dako, E.G. and Bosch. C.H. (2010) Medicinal Plants of Tropical Africa. Conclusions and Recommendations based on PROTA 11(1): 'Medicinal Plants', PROTA Foundation, Nairobi, Kenya.

Setiawan, B., Sulaeman, A., Giraud, D.W. and Driskell, J.A. (2001) Carotenoid content of selected Indonesian fruits. *Journal of Food Composition and Analysis*, **14**, 169–176.

Setty, A.R. and Sigal, L.H. (2005) Herbal medications commonly used in the practice of rheumatology: mechanisms of action, efficacy, and side effects. *Seminars in Arthritis and Rheumatism*, **34**, 773–784.

Shinde, N.V., Kanase, K.G., Shilimkar, V.C. *et al.* (2009) Antinociceptive and anti-inflammatory effects of solvent extracts of *Tagetes erectus* Linn (Asteraceae). *Tropical Journal of Pharmaceutical Research*, **8** (4), 325–329.

Sossa-Vihotogbé, C.N.A., Akissoe, N.H., Anihouvi, V.B. *et al.* (2012) Endogenous knowledge of four leafy vegetables used by rural populations in Benin. *Ecology of Food and Nutrition*, **51** (1), 22–39.

Sossa-Vihotogbé, C.N.A., Akissoe, N.H., Anihouvi, B.V. *et al.* (2013a) Effect of organic and mineral fertilization on nutritive value of three leafy vegetables harvested at different periods.

*International Journal of Biological and Chemical Sciences*, **7** (1), 271–286.

Sossa-Vihotogbé, C.N.A., Akissoe, N.H., Anihouvi, V.B. *et al.* (2013b) Effect of fertilization and harvesting time on antioxidant activity of three leafy vegetables commonly used in Benin. *African Journal of Food Agriculture Nutrition and Development*, **13** (5), 1–20.

Srianta, I., Arisasmita,, J.H., Patria, H.D. and Epriliati, I. (2012) Ethnobotany, nutritional composition and DPPH radical scavenging of leafy vegetables of wild *Paederia foetida* and *Erechtites hieracifolia*. *International Food Research Journal*, **19** (1), 245–250.

Stoilova, I., Krastanov, A., Stoyanova, A. *et al.* (2007) Antioxidant activity of a ginger extract (*Zingiber officinale*). *Food Chemistry*, **102**, 764–770.

Surekha, D., Rajan, V.S.T., Kumar, N.A. *et al.* (2011) A review on role of quality control and quality assurance system in regulation of herbal drugs. *International Journal of Review in Life Sciences*, **1** (3), 97–105.

Teklehaymanot, T. and Giday, M. (2007) Ethnobotanical study of medicinal plants used by people in Zegie Peninsula. *Northwestern Ethiopia Journal of Ethnobiology and Ethnomedicine*, **3**, 12–26.

Thaipong. K., Boonprakob, U., Crosby, K. *et al.* (2006) Comparison of ABTS, DPPH, FRAP, and ORAC assays for estimating antioxidant activity from guava fruit extracts. *Journal of Food Composition and Analysis*, **19**, 669–675.

Valcheva-Kuzmanova, S., Gadjeva, V., Ivanova, D. and Belcheva, A (2007) Antioxidant activity of *Aronia melanocarpa* fruit juice *in vitro*. *Acta Alimentaria*, **36**, 425–428.

Vihotogbé, R., Houessou, L.G., Ponette, Q. *et al.* (2008) Ethnobotany and endogenous conservation of *Irvingia gabonensis* (Aubry-Lecomte) Baill. in traditional agroforestry systems in Benin. *Indilinga*, **6** (2), 196–209.

Vihotogbé, R., van Den Berg, R.G. and Sosef, M.S.M. (2013) Morphological characterization of African bush mango trees (*Irvingia* species) in the Dahomey Gap (West Africa). *Genetic Resources and Crop Evolution*, **60**, 1597–1614.

Vihotogbé, R., Glèlè Kakaï, R., Bongers, F. *et al.* (2014) Impact of traditional uses and potential economic values on food tree conservation status: a case study of African bush mango trees (Irvingiaceae) in the Dahomey Gap (West Africa). *Plant Ecology and Evolution*, in press.

Vladimir-Knežević S, Blažeković B, Štefan MB, *et al.* (2011) Antioxidant activities and polyphenolic contents of three selected *Micromeria* species from Croatia. *Molecules*, **16**, 1454–1470.

Walker R (2004) Criteria for risk assessment of botanical food supplements. *Toxicology Letters*, **149**, 187–195.

Wezel, A. and Bender, S. (2003) Plant species diversity of home gardens of Cuba and its significance for household food supply. *Agroforestry Systems*, **57**, 39–49.

Yang, R.-Y., Chang, L.-C., Hsu, J.-C. *et al.* (2006) Propriétés nutritionnelles et fonctionnelles des feuilles de Moringa – du germoplasme, à la plante, à l'aliment et à la santé, in *Moringa et Autres Végétaux à Fort Potentiel Nutritionnel: Stratégies, Normes et Marchés pour un Meilleur Impact sur la Nutrition en Afrique, Accra, Ghana, 16–18 novembre*, pp. 16–18.

Zongo, C., Savadogo, A., Ouattara, L. *et al.* (2010) Polyphenols contents, antioxidants activities of *Amplelocissus grantii* (Baker) Planch. (Vitacea): a medicinal plant from Burkina-Faso. *International Journal of Pharmacology*, **6** (6), 880–887.

**Part Four**

# Cosmetics (Including Dyes, Aromas)

# 29 Aromatic and Medicinal Plants in North Africa: Opportunities, Constraints and Prospects

Mohamed Ghanmi[1], Abderrahman Aafi[1], Badr Satrani[1], Mohamed Aberchane[1], Abderrahim Khia[2] and Salah Eddine Bakkali Yakhlef[1]

[1]Forest Research Center, High Commission for Water, Forests and Combat Desertification, Agdal, Rabat, Morocco
[2]Laboratory of Biotechnology, Environment and Quality, Faculty of Sciences, University Ibn Tofail, Kénitra Morocco

## 29.1 Introduction

To date, the plant kingdom has offered mankind the essential resources for nutrition, hygiene and health. It seems that human beings have known and used aromatic and medicinal plants (AMPs) since antiquity, and all major ancient civilizations (Chinese, Greek, Roman, Muslim, etc.) have used them for their medicinal properties, fragrance and ritual uses. Demand for AMPs and their essential oils (EOs) is currently increasing in both developed and developing countries for various reasons. For some, it is because natural products are known to have little or no side effects; for others, it is for their ease of access and cost, which is still affordable (Ghanmi *et al.*, 2011).

According to the World Health Organization (WHO), 80% of the world's population relies on traditional medicines to meet the needs of primary health care. More than 20 000 to 25 000 plants are used in the human pharmacopoeia, and more than 50% of pharmaceutical products available on the market have natural origin (Hamilton, 2003). Indeed, the WHO statistics indicate an average increase of 10% per year of global market for herbal plants industries, reaching about €45 billion, whence €10 billion relate to medicinal plants.

AMPs are used whole or in part only, or they can be distilled to extract EO. They are used in the pharmaceutical, cosmetics, perfumery and food industries, among others . However, the evaluation of herbal properties as an antioxidant and antimicrobial remains a very interesting and useful task, especially for plants that are either commonly known, but also for those which are lesser known both to medicine and traditional practice. These plants represent a new source of active compounds.

It is a fact that secondary metabolites are the subject of research; in particular, the search for new natural constituents, such as phenolic compounds, saponins and EO. Many secondary metabolites are also important for our food (taste, colour), while others (among them the alkaloids, anthocyanins, flavonoids, quinines, lignans, steroids and terpenoids) have commercial applications in the pharmaceutical and biomedical domains, as well as drugs, dyes, flavours, perfumes and insecticides (Teixeira Da Silva, 2004).

The African continent is endowed with the richest biodiversity in the world, with a large number of plants used as herbs, natural foods and for therapeutic purposes. This is due, in large part, to the vast geography of forest areas around 216 634 000 ha. More than 5000 different natural substances have been identified, and many of them have proven useful in traditional medicine for the prophylaxis and treatment of disease. However, the North African countries known as the Maghreb countries, comprising Morocco, Algeria, Tunisia and Libya, are part of the Mediterranean basin. They have a variety of landscapes with a rich biodiversity in terms of flora and fauna and they contain many types of coastal, island, mountainous, desert oases and wetlands ecosystems (Farombi, 2003).

## 29.2 Aromatic and medicinal plants in North Africa: a snapshot on the countries of the Maghreb (Morocco, Algeria and Tunisia)

The AMPs find application in many areas, such as medicine and the pharmaceutical, perfumery, cosmetics and food industries. They are used either in their natural form as a condiment or in traditional medicine. Natural substances derived from plants have multiple interests in food, cosmetics and pharmaceuticals industries (Bahorun *et al.*, 1997). The past few years have witnessed a growing renewed interest in using AMPs both in the developed and developing countries. One of the main reasons is the herbs heal without much in the way of adverse side effects and they also provide a good source for the search of natural substances of choice. Since 1850, medicinal plants have provided the pharmaceutical industry with highly effective medicines. The discovery of new products has so far taken place through the search for actives from medicinal plants. The development of extraction methods has yielded products that have shown high activity towards many diseases (Lanhers *et al.*, 1990; Rolland *et al.*, 1991).

Aromatic plants are rich in EOs and they are among the many secondary metabolites the plant produces. They are found throughout the plant kingdom, but they are particularly abundant in some families: Coniferae, Rutaceae, Umbelliferae, Myrtaceae, Labiatae. All

*Novel Plant Bioresources: Applications in Food, Medicine and Cosmetics*, First Edition. Edited by Ameenah Gurib-Fakim.

the organs can contain EOs, especially the flowering tops (lavender, mints, etc.), but they are also found in roots or rhizomes (vetiver, ginger), bark (cannelles), wood (camphor), fruits (peppers) and seeds (Paris and Moyse, 1967; Arctander, 1969).

Their exact function in the life processes of the plant is still unknown (Rai and Rai, 2003). The volatile oils are produced in the cytoplasm of the secretory cells (Sens, 1979). They are then transported in the plant during the growth period to various other parts. Thus, they are found in wood, fruits, bark, seeds and roots. However, the quantity of oil produced is highly variable, depending on the species. They can also be found in vegetative and reproductive organs (Muchalal and Crouzet, 1985).

EOs are mixtures of aromatic components, extracted by steam distillation or solvent extraction (Smallfield *et al.*, 2001). The distillation process implies driving the volatile constituents of the plant material by steam. The plant material is in direct contact with water. Thus, in the hydrodistillation, the plant material is in direct contact with the boiling water, while in steam-hydrodistillation the plant material is supported by a grid or a perforated plate placed at a suitable distance from the bottom of the still (Pellerin, 2000). These processes get rid of unwanted molecules and isolate some of the active principles and noble elements of the crude oils, thus increasing their power, finesse and solubility. The content of EO in plants is generally low and ranges in the order of 1–5%, with the exceptions of clove (14–19%), mace (10–13%), nutmeg (8–9%) and cardamom (4–10%). The EO yield is expressed in millilitres of distillate per 100 g of dry matter.

The best quantitative and qualitative means to identify EO is based on the gas chromatographic profile, which gives the exact chemical composition. Gas chromatography (GC) is an analysis method based on separation that applies to gaseous compounds or those liquids that may be vaporized by heating without decomposition. GC is the usual technique in the analysis of EO. It allows the separation of complex volatile components and a quantitative analysis of the results from a reduced injection volume (Arpino *et al.*, 1995). Coupling GC and mass spectrometry in electron-impact mode is the most common technique used in analysis of EO. It leads to knowing, in most cases, the molecular weight of a compound and the structural information linked to a molecule from its fragmentation (Constantin, 1996).

In spite of their potential, EOs must meet analytical standards imposed by importing or exporting countries, and established by national and international experts' commissions. These standards describe accurately the physico-chemical and chromatographic characteristics that the quality of the EO must have. The control of EO is performed by different tests, such as miscibility with ethanol and some physical measurements, including the refractive index, rotation optical power and relative density. The colour and smell are also considered as important parameters. Control is exercised against falsification of EOs based on geographical origin, botanical species, the producer organ (leaves, flowers, fruit, bark, etc.) and organoleptic characteristics (colour, odour, density, refractive index) (AFNOR, 2000).

EOs have applications in the cosmetics industry, perfumery and aromatherapy. They are increasingly being added to a large number of products, such as food additives (spices and herbs) and even certain pharmaceutical preparations, many perfumes and detergents. They have important applications in medicine in helping relieve pain, and they have general physiological efficiency. According to many researchers, the EO of plants can reduce competition of other plant species (allelopathy) by chemical inhibition of seed germination and protection against microbial infection by fungicidal and bactericidal properties and against herbivores by taste and adverse effects on the nervous system.

Some authors believe that the plant uses oil to repel or attract insects to promote pollination. Others consider EO as an energy source, facilitating chemical reactions and retaining moisture in arid and desert climates.

Aromatic fragrant liquid substances are more or less coloured, volatile, hydrophobic in nature, and completely soluble in alcohol, ether and vegetable and minerals oils. Their density is generally less than that of water, they have a high refractive index and most EOs deviate polarized light. Unlike vegetable oils, EOs do not contain fat and are sensitive to decomposition under the effect of heat. EOs contain thousands of components. They are very effective, but also dangerous. Some aromatic compounds can be harmful, allergenic, teratogenic or carcinogenic. EO toxicity and mutagenic or carcinogenic properties of their compounds is still poorly understood. However, the neurotoxicity of AMPs with thujone from *Artemesia absinthum* and pinochaphene from *Hyssopus officinalis* is fairly well known. In fact, these ketone substances induce mental and sensory disorders. Other monoterpenes that are toxic at high doses are camphor, menthol, cineole, anethole, and so on.

Indeed, the toxicity of AMPs depends on the nature of the compounds they contain and also the dose. For example, the EOs of different varieties of oregano showed strong cytotoxicity on human cancer cells (Sivropoulou *et al.*, 1996). Among the various other chemical constituents present are esters, which are known for their calming and relaxing effect on the body; aldehydes, which possess calming and sedative properties; ketones, which help to stimulate cell regeneration and promote tissue formation; terpenes, which can inhibit the accumulation of toxins in the body and can help the liver and kidneys to get rid of the accumulation of toxins and acids such as benzoic acid and alcohols that have antimicrobial properties. In addition, sesquiterpenol molecules have anti-inflammatory properties and can help the functioning of the immune system. And phenol compounds are among the most beneficial of all aromatic groups; they contain high levels of oxygenating molecules and possess powerful antioxidant properties.

Aromatherapy is another branch of herbal medicine that uses EOs to treat a number of human diseases and disorders. Currently, a very clear return to EOs for disinfection and treatment of infectious diseases has been reported. This return is stimulated by the danger of the use of certain antibiotics. The EOs have longer lasting effect and no resistance from the microbes, unlike antibiotics (Valnet *et al.*, 1978). The efficiency of oils is due to the fact that they contain more than 20 different active molecules, while in a synthetic drug we can only evaluate the interactions of not more than three molecules. Therefore, the EOs are increasingly being quoted for their comprehensive action and their 'broad spectrum' of activity on the entire physiology (Franchomme, 1981).

A large number of EOs and their components are used in the development of the majority of perfumes and toilet products. Owing to their antiseptic activity, these substances serve to preserve these cosmetics and ensure a pleasant scent. 1,8-Cineole, commercially known as eucalyptol, is the main constituent of the majority of EO of *Eucalyptus*. It is used in the manufacture of medicines, soaps, and toothpaste and deodorant lotions (Farah *et al.*, 2002).

In addition to these uses, more than 20 EO compounds with bacteriostatic effects were assembled in a harmonious formulation from the olfactory point of view (Beylier-Maurel, 1976). Also, a high activity of several compounds of EO on the skin's microflora has been demonstrated, and hence their use in therapeutic and cosmetic formulations (Lahlou, 2004).

The use of EOs in the processes of disinfection is quite common (Mallea *et al.*, 1979). They are endowed with a bactericidal and fungistatic power and still have no toxicity to human beings independently of the dose used (Mallea *et al.*, 1979).

Herbs and their extracts are used in food preservation. Studies have demonstrated the antifungal efficiency of some spices and herbs as well as some EO and aqueous extracts resulting from the infusion of plants (Beraoud, 1990). Other studies have shown the potential of using some EOs, such as rosemary and thyme, in combating cryptogamic floras that grow on food (Pellecuer *et al.*, 1976).

The synergy between the antiseptic compounds of the EO, on the one hand, and the salt, sugar and fatty acids, on the other hand, has been proven. This action gives a very strong activity to the mixture (Kurita and Koike, 1982).

The antimicrobial potential of EOs has been evaluated against microbial strains such as *Bacillus subtilis*, *Escherichia coli*, *Staphylococcus aureus* and *Micrococcus luteus*, and fungi such as *Aspergillus niger*, *Penicillium expansum* and *Penicillium digitatum* for their high frequencies to contaminate food and their pathogenicity. Also, other brown and white rot mushrooms of wood (*Gloeophyllum trabeum*, *Poria placenta*, *Coniophora puteana*, *Coriolus versicolor*) were tested. These fungi cause considerable damage to wood and wood products. The bacterial strains tested are normally from the American Type Culture Collection. They are maintained by subculture on nutrient agar favourable to their growth. The fungi used belong to the collection of the fungus culture collection of the Mycology Laboratory of Forestry Research Center (Morocco) (Ghanmi *et al.*, 2011).

The study of antibacterial and antifungal activity of EOs was performed according to the dispersion technique in agar at 0.2%. This technique has the advantage of promoting the incorporation of the compound in the solid or liquid solution. It maximizes the contact germ/compound intake while excluding a foreign agent input (Tantaoui-Elaraki *et al.*, 1993; Satrani *et al.*, 2006). In addition to the antimicrobial test, a method for insecticides testing was used for EO of *Mentha rotundifolia*.

## 29.3 Aromatic and medicinal plants in North Africa: overview and prospects

In North Africa, forest cover ranges from 9% in Morocco to less than 1% in Libya. The main tree species are based on the following genera: *Cedrus*, *Pinus*, *Abies*, *Juniperus*, *Quercus*, *Argania*, *Tetraclinis*, *Cupressus*, *Populus*, and so on. Most of the forest areas have already been cleared and have been replaced by matorrals, maquis and woodlands.

In fact, North Africa has one of the oldest and richest traditions associated with the use of AMPs. Medicinal plants are very important to the local population, especially in rural areas, where they are the only source of available medicines. Even in urban areas, where modern medicine prices are rising, people are turning to traditional medicinal plants.

The ecological and climate heterogeneities prevailing in the northern Africa's countries ensure a rich phytogenetic bank of AMPs. This diversity is particularly highlighted when comparing the richness of AMPs between these Mediterranean countries, including Morocco (4200), Algeria (3150), Tunisia (2200), Libya (1800), Egypt (2100), Mauritania (1100), Lebanon (2100), Jordan (2500), Syria (2100) and Portugal (3100). In France and Spain, the number of species is respectively 4500 and 4900. Finally, Turkey is the country in the region that has the largest number of species (9000) (UICN-Med, 2008).

In terms of richness, Morocco is a leader (first rank) in the area of AMPs in North Africa and second to Turkey in the Mediterranean basin. Actually, Morocco remains an example to other countries in North Africa, especially Algeria and Tunisia, for the promotion and protection of AMPs (Satrani, 2010; Aafi, 2011; Ghanmi, 2011).

## 29.4 Aromatic and medicinal plants in Morocco: opportunities, constraints and prospects

Morocco, through its geographical location, offers a full range of bioclimates favouring rich and varied flora with a high rate of endemism. AMPs occupy a niche ecological and economic position in Morocco. These plant species are very often endemic, and this caters for the revival of the forests in arid, semi-arid and humid regions. Of the 4200 species and subspecies, 537 are endemic and 1625 are rare or endangered (Benabid, 2000; Aafi *et al.*, 2002, 2005). While 800 species of AMPs are potentially exploitable, only 10 have so far passed the tests: rosemary, thyme, chamomile, sage, oregano, pennyroyal, laurel, lichen, capers and pyrethrum. Thus, further value-addition of this impressive heritage is suffering from lack of specific knowledge, as well as the potential of phytomass of AMPs, the chemical and biological characteristics of their extracts, the optimal time for collection of the raw material and for the extraction (Satrani, 2010; Aafi, 2011; Ghanmi, 2011).

In the northern African region, Morocco represents a model to emulate by countries in both Africa and those from the Mediterranean basin in terms of the development and conservation of AMPs. Nonetheless, there is a danger that the exploitation of these plant species will be done in an irrational manner, and this would jeopardize the survival and sustainability of some of the species. The average exploitation techniques are not suitable for all AMPs because they do not take into account the specific circumstances of each species, operating conditions (level and intensity of cuts, plant and harvest time, etc.) (Ghanmi *et al.*, 2011).

Considering the requirement of the international market, the quality of AMPs products is subject to international standards. Product quality and consistency of production rates fluctuate significantly, which usually results in a loss of market. These problems stem from the lack of the professional organization and the lack of promotion of the culture of the main species. Some of the cultivated species are on the verge of disappearing. The size and technical supervision of the processing units are often very weak. The lack of the industrial integration is another handicap for the industry, because the oils are exported unprocessed (Ghanmi *et al.*, 2011).

AMPs are multipurpose species. They contribute to the enrichment of pasture and honey, which has exceptional quality. They are used in traditional medicine and are occasionally transformed before entering the trade. Cultivated or spontaneous, AMPs provide alternative income for local communities. This sector provides annually an amounts of about 33 000 tons and generate 500 000 working days, the equivalent of 25 million Morroccan dirham (MAD). The majority of AMPs in Morocco, including wormwood and rosemary, are exported as dried plants or as aromatic extracts (EO). Currently, the Moroccan processing units usually have new processing technologies according to international standards (HCEFLCD, 2008).

There are two important routes that AMPs can take before they reach the final market. The first is made up of large exporting companies who have staff members throughout the year, controlling the quality of the product before packaging them. These companies are often integrated downstream, by treating a portion of the collection themselves or by organizing the collection of the product. They can also work with intermediaries who deliver the dried product. The second is mainly composed of traders and exporters, as well as gatherers. Exporters receive an order and then start looking for a collector or wholesaler who has the product available.

Indeed, the importance of the application recorded in the 1980s by some domestic or foreign processing industries encouraged the intensification and exploitation of spontaneous AMPs, which makes Morocco one of the largest traditional suppliers in the global market of AMPs and their extracts. This activity uses green or dried biomass, fruits, flowers or EOs for the herbalism, medicine, cosmetics, perfumery and food industries.

A large number of species are used in EO production or for other aromatic extracts primarily intended for the perfumery and cosmetic industry, as well as for the preparation of hygiene products and formulation of flavours.

According to the revenues from the export of around 526 million MAD and average annual revenue in export of about 5.28 million MAD, the sector of AMPs plays a great socio-economic role in Morocco (Office des Change, 2006). In addition to their use as fodder resources with a positive impact on the quality of the meat and honey, AMPs have played an important role because of their therapeutic properties, especially when medicinal plants have been quoted as providing for over 80% of the world's population for their primary health care (OMS, 2003).

World production of EOs was 28 000 tons per year in 2003. Of particular importance are citrus (orange, lemon) and mint (mint and eucalyptus), which are among the most commonly traded species. The food, beverages and the cosmetics industries are the biggest consumers of EOs. The global market was estimated between €10 billion and €15 billion in 2004. The export figures for EOs are not readily available owing to a lack of production data recorded in the export markets.

In Morocco, the products market derived from these plants has experienced a growth in tonnage. In addition to EOs, this activity generates an export of about 100 tons of dry matter. Currently, the processing and valorization activities of AMPs allow an export of nearly 1000 tons of EOs and various extracts and about 400 tons of dried matter for a total value of 300 million MAD (MDSFS, 2007).

At the socio-economic level, the collection of AMPs is meant to diversify agricultural production and it remains an income-generating activity for the local population. This activity provides about 500 000 working days for an income of 25 million MAD and generates a booster revenue for rural and local communities (HCEFLCD, 2008).

Economic valorization of AMPs is characterized by an important development at the international level, especially with globalization and free trade. Currently, the AMPs appear among the sectors with high and great potential. Plants from organic farming are very popular in the international market (Bellakhdar, 1997). As the international market is very demanding in terms of product quality, the AMPs are governed by international quality standards.

Morocco exports more than €26 million in the AMPs sector and at least €14 million of EO, with a great production potential (1 million hectares of *Rosmarinus officinalis* producing 60 tons of EO).

Degradation of natural resources has become a major concern in African countries. The African region is an important treasure for AMPs and EO localized mainly in fragile ecosystems. These plants have great medicinal and commercial value, but are now threatened due to the absence of a conservation strategy and because of uncontrolled and abusive trade practices. For example, in Morocco, the main causes are the use of natural resources for heating, construction or related activities and inappropriate harvesting techniques, which is a major impediment to regeneration. A sustainable management of these resources can help to conserve the biodiversity of plants and maintain critical resources that support living standards for local populations over time.

Morocco's EO and aromatic extracts exportations are continuously fluctuating from one year to another, but in 2008 they reached 725 tons (248 million MAD). Among the EOs and EA exported by Morocco, the main and important are from rosemary, citrus, pennyroyal, resinoids, extracted oleoresins, aromatic waters and EO solutions. Spices and herbs represent an important value among food products exported by Morocco. During the 2008–2009 period, Morocco exported 40 000 tons of spices and herbs with a value of 862 million MAD.

In this class of compounds, the most important products are paprika, coriander seeds and saffron. The export value of these products was €13 million per year between 2003 and 2008. The main destinations are Europe (61%), Asia (19.7%) and America (14.7%). Five products are the main ones exported in this segment of AMPs: cultivated verbena and mint, wild collected rosemary, thyme and oregano. Spices values are around €17 million.

Partnership opportunities in the AMPs are quite diversified and should aim at improving the livelihoods of the local population, encouraging operators to transform the raw material at the local level and have the best added value and help improve knowledge, especially in terms of mode of propagation and culture, and characterization and crop improvement.

The local population is the most disadvantaged and derives least benefit from the commercialization of AMPs. To be more involved in the management and exploitation of AMPs, a certain amount of the harvest could be offered for sale and be granted annually to distillers' and harvesters' cooperatives. This would generate additional revenue for users, encourage them to participate directly towards the preservation and development of the sector, better use of the resource at local level and help add value of the product. To ensure the success for this organization and to enable the beneficiaries to cope with any changes in the sector, coaching, education, training cooperators, monitoring the granting of financial aid and research opportunities are needed.

A large part of AMPs is exported as raw material and yet the value-addition should have been created locally. Partnership with farmers and industrialists should focus on processing at the local level, in addition to their contribution to the development of the production areas of the resource. Opportunities for development and valorization of AMPs remain understudied. There is a need to involve research institutions in the experimental activities and to provide technical assistance and training in order to effectively guide the production of AMPs. Rural communities and non-governmental organizations must partner in the process of development of the AMPs sector.

The socio-economic development of the sector requires the implementation of a comprehensive vision focused on the sustainable management of the resource and the professionalization of the sector. All research and study in this area should take into account the scientific, economic, social and environmental aspects.

## 29.5 Development of the aromatic and medicinal plants sector in Morocco: the strategy adopted

Recognizing the value of these natural resources at the local development, the High Commissariat for Water, Forests and Combating Desertification, which manages the forest sector in Morocco, has developed a national strategy for development of AMPs in consultation with all actors involved in this sector. This strategy is based on the consolidation and development of specific knowledge to Morocco's AMPs (inventory, mapping, technical, research), the valorization of the offer (domestication, intensification, standardization and marketing), the organization of the sector (regulation and partnership) and sustainable development of the sector (stakeholder faming and resource conservation) (HCEFLCD, 2008).

Actually, other countries of North Africa are trying to learn from the Moroccan experience to develop their own strategy in this area.

## 29.6 Research conducted in the field of aromatic and medicinal plants: achievements and prospects

The unit of AMPs at the Forest Research Center (FRC) has carried out many studies, especially on rosemary (*Rosmarinus officinalis* and *Rosmarinus tournefortii*), myrtle (*Myrtus communis*), oregano (*Origanum elongatum* and *Origanum compactum*), lavender (*Lavandula dentata* and *Lavandula stoechas*), sagebrush (*Artemisia herba-alba*, *Artemisia mesatlantica* and *Artemisia campestris*), thyme (*Thymus capitatus*, *Thymus riatarum*, *Thymus ciliatus* var. *munbyanus*, *Thymus bleicherianus* and *Thymus ciliatus*), pennyroyal (*Mentha pulegium*), mint (*Mentha rotundifolia*), satureja (*Satureja calamintha* and *Satureja alpina*), Khalla (*Ammi visnaga*), wild chamomile (*Cladanthus mixtus*), yew (*Taxus baccata*), junipers (*Juniperus thurifera*, *Juniperus phoenicea* and *Juniperus oxycedrus*), *Tetraclinis articulata*, Atlas cedar (*Cedrus atlantica*), *Eucalyptus*, and so on (Ghanmi *et al.*, 2011)

In this context, an overview of the scientific work relating to rosemary and Atlas cedar is produced (Ghanmi *et al.*, 2011).

Many of the AMPs are endemic, and this adds an edge to the country, as it can earn capital through the value-addition of these unique crops. The domestication and valorization of the rare and endemic species can only help preserve the phytogenetic richness that the country may lose under the effect of climate change and whose impact is already being felt and has been observed since the 1980s (Ghanmi *et al.*, 2011).

### 29.6.1 Rosemary

The genus of *Rosmarinus* of the Lamiaceae family includes two plant species: *R. officinalis* and *R. tournefortii*. Rosemary is a shrub growing in the wild in semi-arid and sub-humid bioclimate areas, with an average height of 60 cm, but can reach 1.50 m in some specific cases. This evergreen shrub has leaves with a leathery texture. They are long and thin with slightly rolled edges and a dark green upper surface and white undersurface. The flowers have a blue hue with purple spots on top of the branches.

Rosemary grows in almost all countries of the Mediterranean basin. It is found in North Africa (Morocco, Algeria, Tunisia and Libya) and from Mediterranean Europe (France, Spain, Portugal, Greece, Turkey and Italy) to Asia Minor through the island Cyprus. Rosemary is also grown in the Middle East (Egypt, Lebanon and Palestine).

In Morocco, this species grows on the mountains of Feggous, Taouarirt, Tamokrant, channels Gueb-er-Bou Rahal, Bou Iblane, Ouarirt south east of Taza and Oued Atchane's valley near Boulmane. In the High Atlas, it is common in the eastern part of Titahouine and Tizi-n-Telghemt. Rosemary is encountered in the highlands (Figuig). It is also abundant in the region of Midelt and Debdou. It is present in clear stands and in matorrals of low and medium altitudes. It is found from the arid to the subhumid climates in the thermo-Mediterranean and meso-Mediterranean parts of the country. It is indifferent to the chemical qualities of the soil but thrives best in calcareous and well-drained soils (Benabid, 2000).

Rosemary is known as a medicinal and aromatic plant which has multiple uses ranging from simple use of traditional medicine to many industrial uses (pharmacies, food, cosmetics, etc.). This shrub is well appreciated by beekeepers, producing flowers throughout the year that bees visit willingly. It allows them to make a very perfumed honey. Owing to its antioxidant properties, rosemary is used in the manufacturing of meat products industry. Its EO is used in the manufacture of perfumes, shampoos and insecticides. Rosemary stimulates the functioning of the gallbladder. Also, it calms digestive spasms by its spasmolytic action on the intestines and stomach. It is also prescribed in traditional medicine as leaves or buds infusions. It is used against headache and dyspepsia. The leaves are applied against swelling and joint rheumatism. Also, its diuretic properties are little used in popular medicine.

Rosemary herb is the second most exported AMP in Morocco. It represents 19% of all the exported spices. Quantities exported and revenues made rose during the period spanning between 2001 and 2006, from 0.6 tons (574 666 MAD) to 13 tons (12 305 333 MAD), which enabled Morocco to occupy the second rank after Spain in the international market.

A very large amount of rosemary is used as culinary herbs and dried leaves, which are exported to France and the USA. At the same time, Morocco imports some rosemary products, which are directly used in the formulation of medicines, perfumes and cosmetics (Fechtal, 2000).

Several sources of rosemary (*R. officinalis*) were studied from the chemical and biological points of view at the FRC. These studies have shown that rosemary EO yield (*R. officinalis*) is variable according to the origin of the product and the harvesting period. Depending on the origin of the product, rosemary provides a variable yield of 1.3% for Aknoul (Taza), 2.2% for the Oriental region (El Ayat, Debdou and Rchida) and 2.5% for Maghraoua (Taza). Other earlier studies on this species in the same regions showed similar or slightly lower results: El Ayat (1.6%), Aknoul (0.9%), Rchida (1.3%) and 0.9% for Errachidia (southern Morocco). Based on the harvest period, the maximum yield recorded corresponded to the flowering period. The chemical quality of *R. officinalis* EO of six regions is characterized by the presence of 1,8-cineole as a main component, with values ranging from 41.6 to 50.8%. In addition to this compound, we also have the presence of α- and β-pinene and camphor. These EOs preserve their chemical profile with a slight variation in the content that depends on the origin and the harvesting period.

For *R. tournefortii* collected in Tafoughalet area, a low yield of EO is obtained compared with *R. officinalis*. It varies slightly from 0.5% to 0.7% between April and October respectively. The chemical composition of the EO is dominated by α-pinene, camphene, β-pinene, 1,8-cineole and camphor, with a slightly variation in contents that depends on the harvest period.

Fractional distillation of rosemary EO in a pilot unit allowed the aggregation of different fractions into three groups. The first group is rich in α- and β pinene with an overall percentage of about 41% and represents 23% of all oils processed. The second group constitutes 56% of rectified oils, providing an overall rate of about 90% of 1,8-cineole. It can be classified among EO with 80–90% of the active principle as recommended by ISO R 869 (AFNOR, 1992) and the last fraction was found to be rich in camphor.

The results of the antibacterial activity of *R. officinalis* EO showed that all strains of bacteria and fungi tested had almost the same sensitivity towards these oils. These microorganisms were somewhat resistant because they were inhibited at a concentration of 1/100 v/v.

In Debdou forest, *R. officinalis* was characterized by moderately dense stands (1580 seedlings/ha). The total phytomass is estimated at 1400 kg of green matter per hectare, green leaf phytomass is around 692 kg/ha and dry matter leaf phytomass is about 499 kg/ha. Rosemary stands of Rchida forest are at 1500 plants/ha. The total green phytomass is estimated at 900 kg/ha, green leaf phytomass at 250 kg/ha and dry matter leaf phytomass at 200 kg/ha.

In the Tafoughalt forest, *R. tournefortii* was characterized by stands with a density of around 1800 seedlings/ha. The total green phytomass was evaluated to be 810 kg/ha, green leaf phytomass was around 324 kg/ha and the dry matter leaf phytomass was around 260 kg/ha.

Rosemary has a very low rate of seed germination; the easiest method for its propagation is by cuttings. This form of vegetative propagation consists of cultivating a fragment of plant organ in favourable environmental conditions. The cutting can eventually turn into a whole organism able to live independently and with the fundamental characteristics of the individual strain. As natural rosemary is a difficult rooting species, direct planting of stem fragments presents significant failures, and hence the interest in cuttings breeding in the nursery in favourable culture conditions before planting in the field.

In the case of rosemary, herbaceous cuttings (terminal shoots of stems) give a better rooting rate, unlike hardwood cuttings (pieces of stem base). This type of cutting gives a success rate of over 80%. Cuttings of rosemary are successful in winter, but beyond March the rate of rooting decreases. Raising cuttings of rosemary can be done in the open air in polythene bags filled with soil. Before planting the cuttings, the soil should be slightly moist. The cuttings should be of a size of 10–15 cm in length.

During the first days of cutting, daily watering is essential to prevent drying of the cuttings. The frequency of watering depends on climatic conditions: either one to two waterings per day if the conditions are severe (winds, high temperatures).

Using a shade during the first 2–3 weeks of cuttings is desirable. Once the cuttings begin to show the first shoots, progressive sunshine can give strong individuals. The residence time of the plants in the nursery is around 3 months.

Rosemary, once rooted, is strong enough. It is a plant that likes sunny places and where the soil is well drained. From the first season, it grows to about 30 cm in diameter and 60 cm in height. Planting is done earlier in spring. Success rates achieved over 80% in the Oriental region of Morocco (rate calculated after the passage of the summer cap) and about 100% in the region of Rabat. Watering, during the summer, in the first year of planting, improves the success rate of rosemary plants.

For reforestation, the planting method consists of digging a ditch of about 20 cm wide and 20 cm deep along the contours. The application of fertilizers must take into account several factors, in particular the type of soil and its physico-chemical characteristics, its disability, type of climate and species.

In the case of rosemary, mineral fertilization improves biomass production and EO significantly. This improvement ranges from 40 to 64% for dry biomass production and from 15 to 20% EO yield. From the chemical composition point of view, there is no difference between rosemary that has been fertilized or left unfertilized. The content of major constituents (α- and β-pinene, 1,8-cineole and camphor) remains the same regardless of fertilization treatment applied.

Dry biomass production of rosemary is around 1 ton/ha in the first year of cultivation. In the case of fertilizer, the production can reach up to 2 tons/ha per year.

Irrigation improves the AMPs' production biomass; however, their EO yield remains relatively low compared with that obtained from the non-irrigated rosemary. Nursery trials have shown that drying of crops (no irrigation) for 1 month before the harvest improves EO yield by 10%.

Also, this treatment allowed a significant improvement in the content of 1,8-cineole and a slight decrease in that of α-pinene. Regarding the other components, the chemical composition remains stable.

Populations of rosemary from the Oriental region had been transplanted to the FRC nursery in Rabat provide an EO content slightly lower than samples from the field, averaging 1.68%. The chemical quality of EO of these areas is characterized by a high content of 1,8-cineole (active principle prized by processing industries). Transplanted populations have retained their EO chemical quality with a slight decrease of α- and β-pinene.

The cultivation of rosemary populations of this region is encouraging. Its transplantation to another environment does not affect the quality of the EO.

Preliminary results of the experimental device installed in Debdou's forest (Oriental region), which aimed to evaluate different methods and cutting periods on the regeneration and biomass production of rosemary, indicate that better recovery rates of tufts of rosemary are obtained with the cutting method which normally implies cutting half of each tuft to a height of 10 cm and leaving a minimum of green biomass (cutting method 2). Regarding the exploitation period, the recovery tufts of rosemary are difficult if the exploitation is performed during the summer season regardless of the cutting method used; then, it is better to wait for the fall.

With respect to technological advances, non-woody parts of species can be valorized (branches, twigs, fruits and needles, sawdust and fruits, buds, buds, etc.). This would provide additional revenue to the common exploitation of these species (production of timber, firewood, pulp, etc.).

Indeed, these activities enable sustainable development at local and regional levels by the revenue they generate and the jobs they would create. Products used in the extraction of EO can be harvested during the fulfillment of the silvicultural operations (artificial pruning, thinning, etc.) provided for in the management plans. Until today, the real value of these products had not been estimated as per its real value. It is now recommended that they be taken into consideration during the inventory operation and assessment of the potential in the review of the current management. Moreover, EO, provided from sawdust, branches, fruits and leaves, gives an added value and contributes to the valorization of a number of forest trees.

### 29.6.2 Atlas cedar

*Cedrus atlantica* is a native and endemic tree to Morocco and Algeria. It can reach 40–50 m in height, with a diameter of 2 m. It differs from other species by its distinctive character, such as the port, the dimension of the cone and the appearance of leaves and twigs.

The Atlas cedar is a noble tree of North Africa and of Morocco in particular, where the conditions are favourable. It is a treasure from the ecological, economic, social and cultural points of view. It has a pyramidal aspect when it is still young and becomes tabular later. Its branches are usually long and erect; its leaves are isolated (green needles) and about 1–2 cm in length. Fruits are in the form of brown ellipsoid cones of 5–8 cm in length.

The cedar wood is soft and light, with a penetrating and persistent aromatic odour. It gives a noble and durable timber. It is used in carpentry, cabinetmaking, peeling, slicing and paper pulp mixed with pinewood. It is also suitable for the manufacture of pencils. The Moroccan cedar covers an area of 133 000 ha. It is the main source of timber production. Its contribution to national production is around 80%.

The cedar's EO is antiseptic, healing, antifungal, detoxifying and relaxing. It is said that this oil is characterized by a warm aroma, woody and balsamic. This EO is used in perfumery and cosmetics, and is indicated against dermatitis and skin inflammation. Because of its richness with the himachalenes molecules, it is an essential component in perfumery and cosmetics.

Economically, the cedar forests generate an important economic activity by their products quality: pasture (production of 116 million units fodder), raw materials for saw industries, sources of income for the local population (353 400 000 MAD equivalent work around 6.5 million man-days) and multiple easements that provide high-value products such as timber, edible mushrooms, tar, herbs, lichens, and so on. The cedar EO and its tar, which are rich in sesquiterpenes and sesquiterpene alcohols, are a raw material for the pharmaceutical and cosmetic industries. The himachalenes form part of a medicinal product sold in the market. EO yield is variable and depends on the origin. It is around 3.8 mL per 100 g of dry matter. The cedar forests of the Middle Atlas and High Atlas Oriental yield plants that are richest in these oils. β-Himachalene is the main constituent of these oils. It is a precursor in the preparation of the products used in the perfume industry, mainly epoxides, which can raise the olfactory power.

Chromatographic analysis of three samples of EO studied showed that there are differences between the chemical compositions of these oils. EO from the unattacked cedar wood and wood attacked by *Trametes pini* showed substantially identical chemical compositions, with himachalenes and *E*-α-atlantone as major constituents.

These compositions differ from the EO composition of the wood infected by *Ungulina officinalis*. The latter is characterized by low levels of himachalenes and higher levels of other constituents. In this case, the overall rate of himachalenes (α+β+γ) is low and is around 15.85%, compared with an EO of healthy wood of 51.57% and with that of wood attacked by *Trametes pini* of 55.90%. The percentages of major constituents in the non-attacked wood and in wood attacked by *Trametes pini* and *Ungulina officinalis* are respectively 10.87%, 13.42% and 5.97% for α-himachalene, 6.89%, 8.32% and 4.20% for γ-himachalene, and 33.81%, 34.16% and 5.68% for β-himachalene. *E*-α-Atlantone, which appears with 11.24% and 10.35% respectively in the non-attacked and attacked by *Trametes pini*, is almost a trace (0.14%) in EO of wood infected by *Ungulina officinalis* (Aberchane *et al*., 2001, 2003, 2004).

The bioactivity study of different samples of EO shows that oils obtained from non-attacked wood and wood infected by *Trametes pini* do not possess any *in vitro* bacteriostatic or fungistatic power at the concentrations used. This is in marked contrast to the EO of cedar wood attacked by *Ungulina officinalis*, which is reactive from a concentration of 2 μL/mL. These EOs possess an inhibitory activity towards bacteria but have low antifungal power.

## 29.7 Medicinal and aromatic plants in Algeria

According to its geographical location, Algeria overlaps between two floral empires: Holarctis and Paleotropis. This position gives it a very diverse flora, with species belonging to different geographical features. Algeria has an impressive biodiversity of both floral and faunal origin. This biodiversity is represented by 3139 species of plants described in the new flora of Algeria and southern desert regions and from where 289 species are very rare, 675 rare and 168 endemic (Quezel and Santa, 1963; INRAA, 2006).

In Algeria, there are many plants that are commonly used in the food industry (condiments, preservatives and flavourings), as medicines, perfumes and cosmetics: rose, rosemary, sage, lavender, geranium, laurier, thyme, nasturtium, castor, jasmine, verbena, linden, acacias, and so on. Algeria also has a culture of extracting natural products from plants grown under organic conditions. Algeria has more than 600 species of AMPs, and the Hoggar region contributes more than 300 species, out of which 25% are used in traditional medicinal. Many of them are in a precarious state due to the impacts of frequent and excessive droughts. The impact of desertification, unsuitable land-exploitation for crops, deforestation and overgrazing are all being felt. Herbs can be classified as a renewable natural resource. This resource has undergone irreversible damage with the result that, over the past decade, dozens of AMPs have been lost.

## 29.8 Medicinal and aromatic plants in Tunisia

Tunisia has many favourable parameters (climate, soil and hydrology) that enable the country to engage in the intensive development of AMP crops from its rich floral heritage.

There are several AMPs growing spontaneously in the vast forest area of around 409 000 ha located mainly in the northern parts of the country. An area of 100 000–250 000 ha is known to be under rosemary, thyme, myrtle, mint and sagebrush cultivation. The Tunisian flora is particularly rich and is estimated to contain over 2150 species, reflecting the great diversity of ecological systems in the country. In Tunisia, the use of plants in traditional medicine goes back a long time (APIA, 2003).

The interest in AMPs in Tunisia lies not only in their various uses and economic interests, but also in their social and cultural values. Indeed, the local know-how about the use of these plants is widespread. Thus, it is reported that 'among the nomads of southern Tunisia, there is no woman or old man who does not know, by experience, how to diagnose a common disease or ailment and remedy by using plants'. Moreover, the ecological importance of AMPs, and especially wild plants, is not likely to be neglected because these plants are the best guarantee of water and soil conservation and combating desertification (APIA, 2003).

Taking into account the climatic, environmental, cultural and economic factors, some species are becoming increasingly rare and their recovery is justified. Various studies have been conducted, in particular the national programme to establish an inventory of AMPs. The plants listed are identified and stored in herbaria.

Currently, about 149 medicinal plants are commonly used in human therapy, particularly in rural areas. To this list, other plants traditionally used as medicines whose medicinal properties have been highlighted recently may be mentioned: acacias (*Acacia tortilis* and *Acacia horrida*), *Aloe vera* and *Sylibum marianum*. Some species of iris and eucalyptus have been introduced and are well adapted to Tunisian conditions (APIA, 2003).

It has been reported that there are around 48 herbs which are most commonly used in Tunisia. Areas reserved for AMP crops vary from one year to another and from one species to another. Adopted crops techniques, climate conditions, import and export would explain the extension or reduction of land allocated to them annually (APIA, 2003).

Marketed species are available under several forms: raw state, packaged, EO and concrete. Tunisia is a major producer and exporter of fresh plants, EO and concrete. The Tunisian EO is well known in the international market, especially in Europe and the USA. The main EOs made in Tunisia are neroli, rosemary, sagebrush, marjoram and citrus.

The specificity of Tunisia for certain plants, such as bitter orange, geranium, myrtle, rosemary, rose and jasmine, helped it to maintain its ranking as being an important exporter.

Tunisia also exports large quantities of rosemary and occupies second place after Morocco.

Tunisian neroli ranks first in terms of quality and quantity. This product is highly sought after by major manufacturers of cosmetics and perfumes brands. The existing distillation units were 24 in 2004, located in Sfax, Nabeul, Ben Arous, Gafsa, Kasserine, Kef, Siliana and Zaghouan Tunis. The average national production of major AMPs and their derivatives is of the order of 8500 tons, of which nearly one-quarter are exported mainly to Europe. The value of this production is estimated at 14.2 million Tunisian dinars.

## 29.9 Molecular techniques as tools for conservation and valorization of aromatic and medicinal plants

The use of AMPs for research purposes is not new. However, the interest in the commercial exploitation and legal protection of those plants is quite recent.

In fact, AMPs are a valuable resource for research and development. They are used, worldwide, as sources of new molecules for the synthesis of chemopreventive and anticancer drugs (Balunas and Kinghorn, 2005) and as medicinal, cosmetic, antibacterial, antiparasitic (Bakkali *et al.*, 2008), antimicrobial, antioxidant and anti-inflammatory agents (Cowan, 1999; Svoboda and Hampson, 1999). AMPs can appear in different settings. They be found in nature (wild AMPs) or developed through plant breeding (cultivated AMPs).

In the Maghreb countries (Morocco, Algeria and Tunisia) there is an increasing interest in studying the composition and biochemical properties of EOs from different native AMPs, and their potential uses.

*A. herba-alba* is of big economic interest. Its EO content varies from 1 to 1.5% of dry material. Its oil is destined for the beauty care industry and also for the perfume trade.

Morocco and Tunisia both share the international market to cater for this oil, but Morocco supplies nearly 90% of the world market. More so, the cultivation of *A. herba-alba* is recommended for the restoration of degraded ecosystems. In view of its medicinal value, mainly as natural antioxidant plant, both *in situ* and *ex situ* conservation strategies are needed through forestry policies and legislation to protect this species.

Understanding the genetic variation within and between populations is essential for the establishment of effective and efficient conservation for plants (Shafie *et al.*, 2009). Earlier approaches that made selection of specific traits easier were based on the evaluation of morphological traits, isozymes, and storage proteins like glutenins, gliadins, hordeins, and so on.

However, DNA markers seem to be the best candidates for efficient evaluation and selection of plant material. Unlike protein markers, DNA markers target single genes and are not affected by the environment like morphological markers.

The research in this AMPs species must be accompanied by: (i) the development of cultivable lines in plant breeding programmes assisted by molecular markers from wild materials to ensure the proper agronomic exploitation and the conservation of genetic resources (Gómez-Galera *et al.*, 2007; Kumar and Gupta, 2008); (ii) the application of molecular and genomic tools appropriate for handling the modulation in the production of these oils and plant secondary metabolites (Gómez-Galera *et al.*, 2007). Also, the current focus on chemotype-driven fingerprinting and related techniques requires integration with genotype-driven molecular techniques so that an optimal characterization of botanical materials is possible.

Various DNA-based technologies are useful in genotyping and for the quick evaluation of DNA polymorphism. DNA-based techniques have been widely used for the authentication of plant species of medicinal importance. This is especially useful in the case of those that are frequently substituted or adulterated with other species or varieties that are morphologically and/or phytochemically indistinguishable. Estimation of genetic variation is increasingly being based upon information at the DNA level by various molecular markers, such as randomly amplified polymorphic DNA (RAPD), amplified fragment length polymorphism, restriction fragment length polymorphism, simple sequence repeat (SSR), or microsatellite, and so on.

RAPD and inter-SSR (ISSR) markers are widely used in Morocco and Tunisia to assess the genetic variation of the AMPs. These are polymerase chain reaction (PCR)-based methods. PCR-based markers involve *in vitro* amplification of particular DNA sequences or loci, with the help of specific or arbitrary oligonucleotide primers and a thermostable DNA polymerase enzyme.

However, the isolation of DNA in AMPs species is a difficult, tedious and time-consuming procedure (Khanuja *et al.*, 1999; Michiels *et al.*, 2003; Padmalatha and Prasad, 2006). Basically, the problem is the formation of complexes between DNA and plant secondary metabolites, such as polyphenols and polysaccharides. Among these metabolites are tannins, flavonoids, quinones, terpenes and alkaloids, which not only make this group of plants attractive, but also make the procedure of isolation difficult, because they degrade the extracted DNA and directly or indirectly inhibit the action of enzymes used in downstream molecular applications such as PCR amplification (Porebski *et al.*, 1997; Khanuja *et al.*, 1999; Michiels *et al.*, 2003; Padmalatha and Prasad, 2006; Ivanova *et al.*, 2008).

Various protocols for DNA extraction have been successfully applied to many plant species (Doyle and Doyle, 1987; Ziegenhagen and Scholz, 1993), which were further modified to provide DNA suitable for several kinds of analyses (Wang and Taylor, 1993; Ziegenhagen and Scholz, 1998).

In this chapter, we present examples of molecular characterization of the main AMPs encountered in Morocco, Algeria and Tunisia.

In Morocco, Ourid *et al.* (2011) published the first report on the development of a protocol for high-purity genomic DNA isolation from the *A. herba-alba* schrub leaves and optimization of conditions for RAPD-PCR analysis. Two DNA extraction protocols were specifically developed: QIAgen DNA Kit and a protocol developed by Ouenzar and collaborators. The Ouenzar and collaborators' protocol gave a higher yield but was more time consuming than the QIAgen Kit. However, both techniques gave DNA of good quality that is amenable for genetic diversity studies within this species. Additionally, RAPD profiling from the isolated DNA was optimized to produce scorable and clear amplicons (Table 29.1, Figure 29.1).

Moreover, a similar study of genetic variation among 216 accessions of *A. herba-alba* was conducted in semi-arid regions of Tunisia, using ISSR markers to assess the polymorphism at the species level (Haouari and Ferchichi, 2008).

**Table 29.1** Summary of the RAPD-PCR parameters tested and the optimum conditions selected

| PCR parameter | Range tested | Optimum conditions |
|---|---|---|
| DNA concentration (ng) | 5, 10, 20, 30, 40, 50, 70, 100, 200 | 5 ng |
| Magnesium chloride (mM) | 1.5, 2.5, 3, 3.5, 4, 4.5, 5 | 1.5 mM |
| Primer concentration (μM) | 0.1, 0.3, 0.5, 0.7 | 0.3 μM |
| Taq polymerase (units) | 0.2, 0.4, 0.6, 0.8, 1, 1.2, 1.4, 1.6, 1.8, 2 | 0.4 unit |
| dNTPs concentration (mM) | 0.05, 0.1, 0.15, 0.2, 0.25 | 0.1 mM |
| Annealing temperature (°C) | 33, 37 and 40 | 37 °C |

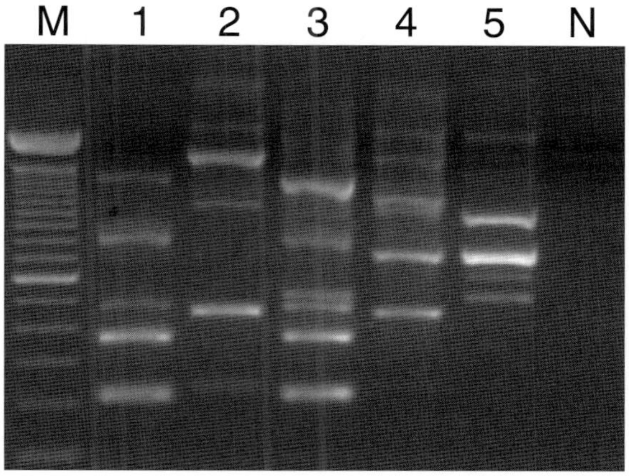


**Figure 29.1** A representative agarose gel fractionation of RAPD amplification products from *A. herba-alba* genomic DNA (5 ng) derived from a single *Artemisia* shrub using the Operon primers: OPA-01 (1), OPA-02 (2), OPA-03 (3), OPA-04 (4), OPA-05 (5). Amplification products were fractionated in a 1.8% agarose gel. M: 100 bp DNA ladder. N: PCR negative control (no template DNA)

A total of 60 polymorphic loci were scored using four primers, revealing a high level of genetic polymorphism for this species. Genetic distances vary in a large interval, indicating a high level of intra-specific heterogeneity. This heterogeneity reflects the ecologic differences between accessions as shown by the non-metric multidimensional scaling (NMDS) distribution in three groups as a function of the bioclimatic stage (Figure 29.2).

Correlation analysis revealed no direct relation between morphological traits, geographic distance and genetic distance. Correlogram analysis showed a patchy distribution of the genetic variability of *A. herba-alba* accessions, revealing the contribution of local ecological and geographic conditions on variability (Figure 29.3).

In Algeria, the genus *Teucrium* EOs have been the subject of several investigations and their diverse chemical compositions were recently reviewed by Awadh *et al.* (2008). It is one of the richest sources of neoclerodane diterpenes: more than 220 diterpenes have been described, many of which are of interest because of their

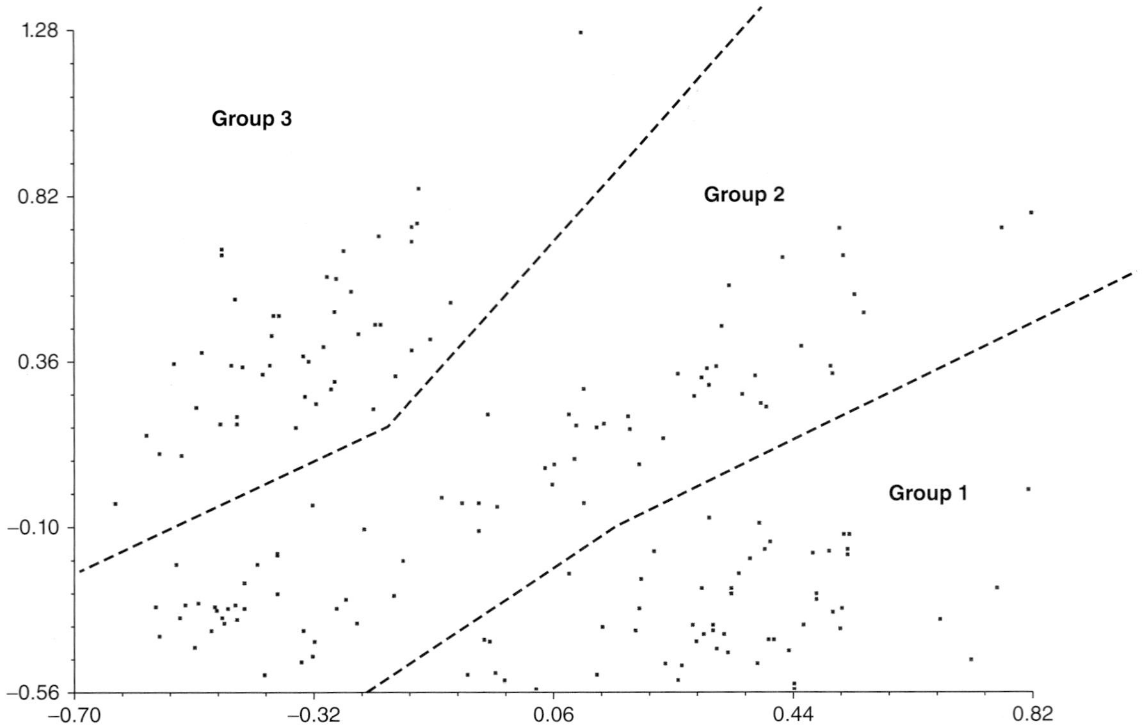


**Figure 29.2** NMDS distribution of the genetic diversity generated by ISSR markers

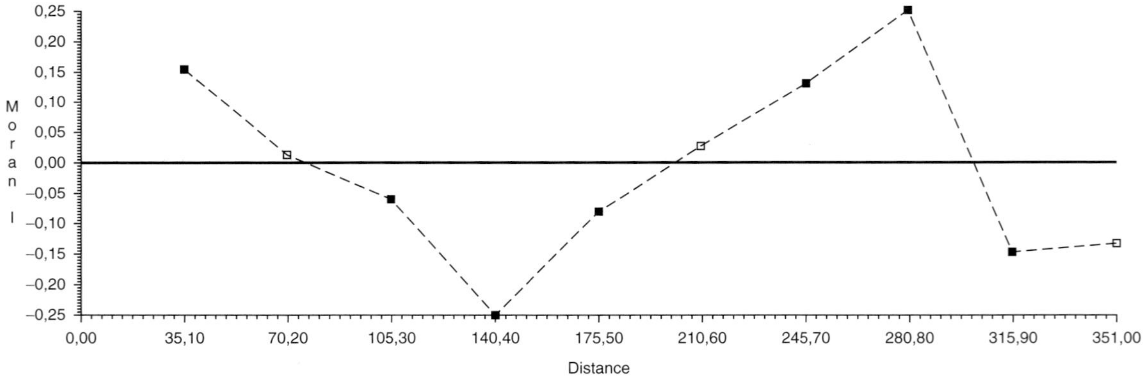


**Figure 29.3** Moran's *I* correlogram showing the correlation between classes of geographic distances and genetic distances (no significant value; significant value)

insect-repellent and medicinal properties (Coll and Tandrón, 2005; Menichini *et al.*, 2009).

Genetic sequences of different species of *Teucrium* were performed using three molecular markers: two chloroplast markers (TrnL-F and RPL32-TRNL) and a nucleic marker (ITS) (Djabou, 2012). The objectives of this study were to correlate DNA with the chemical compositions of EOs of the different sub-species of the same species, especially for *Teucrium scorodonia*, *Teucrium flavum* and *Teucrium polium*. Comparative studies have been established between the sequences of each *Teucrium*. The author showed that the production of secondary metabolites can be influenced by environmental factors, but is most often linked to the genetic diversity of plants.

In conclusion, molecular biology offers promising tools for the full exploitation of the biosynthetic potential of AMPs. In the Maghreb countries, the molecular-biological techniques should be used in programmes for conservation and valorization of AMPs.

Another challenge for future research on AMPs is the elucidation of the pleiotropic modes of actions of single ingredients of plants and even more of complex herbal mixtures. Recently emerging techniques of systems biology, such as genomics, transciptomics, and metabolomics, as well as their evaluation by bioinformatic tools, represent a new area in AMPs research. This can be an effective approach for comprehensive evaluation of the effectiveness and qualities of AMPs.

## 29.10 Sector of aromatic and medicinal plants in North Africa: prospects

The AMPs sector is poised for major changes for several reasons:

- The desire of countries to establish a mode of rational management of AMPs.
- The use of the 'natural' label on a global scale.
- Awareness of local populations and the AMPs being used.
- The socio-economic environment of the country being very favourable to the transfer of research results to the development of different industries.
- The lack of research in the sector supported in aromatic extracts remains a major handicap for these mutations.
- The mutation from an unprocessed provider sector to a real industry.
- Promoting quality through support of learning and the creation of new products for new markets. Promoting quality also requires traceability to establish a system of trust with the consumer. The procedure is to train the people who produce the raw material and develop the new technologies needed.

## References


Aafi, A. (2011) Evaluation de la Biodiversité des Écosystèmes Naturels du Maroc, 'Etude de la Diversité Floristique de L'écosystème de Chêne-liège de la Forêt de la Mamora', Editions Universitaires Européennes, Sarrebruck, Germany (ISBN-13: 978-613-1-59627-8).

Aafi, A., Taleb, M.S. and Fechtal, M. (2002) *Espèces remarquables de la flore du Maroc*, CNRF/MCEF/Maroc.

Aafi, A., Achhal, A.K., Benabid, A. and Rouchdi, M. (2005) Richesse et diversité floristique de l'écosystème de chêne-liège de la forêt de la Mamora, *Acta Botanica Malacitana*, **30**, 127–138.

Aberchane, M., Fechtal, M., Chaouch, A. and Bouayoune, T. (2001) Influence de la durée et de la technique d'extraction sur le rendement et la qualité des huiles essentielles du cèdre de l'Atlas (*Cedrus atlantica* M.). *Annales de la Recherche Forestière au Maroc*, **34**, 110–118.

Aberchane, M., Satrani, B., Fechtal, M. and Chaouch, A. (2003) Effet de l'infection du bois de Cèdre de l'Atlas par *Trametes pini* et *Ungulina officinalis* sur la composition chimique et l'activité antibactérienne et antifongique des huiles essentielles. *Acta Botanica Gallica*, **150** (2), 223–229.

Aberchane, M., Fechtal, M. and Chaouch, A. (2004) Analysis of Moroccan Atlas cedarwood oil (*Cedrus atlantica* Manetti). *Journal of Essential Oil Research*, **16**, 542–547.

AFNOR (1992) *Recueil des Normes Françaises sur les Huiles Essentielles*, 2nd edn, Association Française de Normalisation, Paris.

AFNOR (2000) Huiles essentielles. Echantillonnage et Méthodes d'Analyse *(Tome 1)* – Monographies Relatives aux Huiles Essentielles *(Tome 2)*, Association Française de Normalisation, Paris.

APIA (2003) AGRO-SERVICES, La culture des plantes aromatiques et médicinales en Tunisie, Document de synthèse, http://www.apia.com.tn/ar/images/stories/pdf/synt2.pdf.

Arctander, S. (1969) *Perfume and Flavor Chemicals, Volume II, Aroma Chemicals*, Arctander, Montclair, NJ.

Arpino, P., Prévôt, A., Serpinet, J. *et al.* (1995) *Manuel Pratique de Chromatographie en Phase Gazeuse*, Masson, Paris.

Awadh, A., Wurster, M., Arnold, N. *et al.* (2008) Chemical composition of the essential oil of *Teucrium yemense* Deflers. *Records of Natural Products*, **2**, 25–32.

Bahorun, T., Gressier, D., Trotin, F. *et al.* (1997) Oxygen species scavenging activity of phenolic extracts from hawthorn fresh plant organs and pharmaceutical preparations. *Arznieiminittelforschung/Drug Research*, **46** (11), 1086–1089.

Bakkali, F., Averbeck, S., Averbeck, D. and Idaomar, M. (2008) Biological effects of essential oils – a review. *Food and Chemical Toxicology*, **46** (2), 446–475.

Balunas, M.J. and Kinghorn, A.D. (2005) Drug discovery from medicinal plants. *Life Sciences*, **78**, 431–441.

Bellakhdar, J. (1997) *La Pharmacopée Marocaine Traditionnelle*, Ibis Press, Paris, pp. 307, 400–402, 760.

Benabid, A. (2000) *Flore et Écosystèmes du Maroc, Evaluation et Préservation de la Biodiversité*, Ibis Press, Paris.

Beraoud, L. (1990) Effet de certains épices et plantes aromatiques et de leurs extraits sur la croissance et l'aflatoxinogénèse d'*Aspergillus parasiticus* NRRL 2999. Thèse de doctorat de troisième cycle, Faculté des Sciences Rabat, Maroc.

Beylier-Maurel, M.F. (1976) Activité bactériostatique des matières premières de parfumerie. *Rivista Italiana EPPOS*, **58**, 283–286.

Coll, J. and Tandrón, Y.A. (2005) Isolation and structure elucidation of three neoclerodane diterpenes from *Teucrium fruticans* L. (Labiatae). *Phytochemistry*, **66**, 2298–2303.

Constantin, E. (1996) *Spectrométrie de Masse*, Lavoisier Technique et Documentation, Paris, pp. 1–14.

Cowan, M.M. (1999) Plant products as antimicrobial agents. *Clinical Microbiology Reviews*, **12** (4), 564–582.

Djabou, N. (2012) Caractérisation et variabilité des plantes à parfums aromatiques et médicinales de Corse et de l'ouest Algérien. Thèse de Doctorat, Université Pascal Paoli.

Doyle, J.J. and Doyle J.L. (1987) A rapid DNA isolation procedure for small quantities of fresh leaf tissue. *Phytochemical Bulletin*, **19**, 11–15.

Farah, A., Fechtal, M. and Chaouch, A. (2002) Effet de l'hybridation interspécifique sur la teneur et la composition chimique des huiles essentielles d'eucalyptus cultivés au Maroc. *Biotechnology, Agronomy, Society and Environment*, **6** (3), 163–169.

Farombi, E.O. (2003) African indigenous plants with chemotherapeutic potentials and biotechnological approach to the production of bioactive prophylactic agents. *African Journal of Biotechnology*, **2** (12), 662–671.

Fechtal, M. (2000) Production et marché des huiles essentielles au Maroc, in *Conférence plénière au 2eme colloque Andalou – Marocain sur la chimie des produits naturels*, 11 et 12 Avril, Faculté des sciences et Technique de Mohammedia.

Franchomme, P. (1981) L'aromatologie à visée anti-infectieuse. *Phytomédecine*, **1-2**, 25–47.

Ghanmi, M. (2011) *Valorisation de la Résine des Pins du Maroc*, Éditions Universitaires Européennes (ISBN: 978-613-1-59501-1).

Ghanmi, M., Satrani, B., Aberchane, M. *et al.* (2011) *Plantes Aromatiques et Médicinales du Maroc: Les Milles et une Vertus*, Collection Maroc Nature, Éditions Centre de Recherche Forestière (ISBN: 978-9981-824-28-7).

Gómez-Galera, S., Pelacho, A.M., Gené, A. *et al.* (2007) The genetic manipulation of medicinal and aromatic plants. *Plant Cell Reports*, **26**, 1689–1715.

Hamilton, A. (2003) *Medicinal plants and conservation: issues and approaches*, International Plants Conservation Unit, WWF-UK, Godalming.

Haouari, M. and Ferchichi, A. (2008) Study of genetic polymorphism of *Artemisia herba-alba* from Tunisia using ISSR markers. *African Journal of Biotechnology*, **7** (1), 44–50.

HCEFLCD (2008) Stratégie nationale de développement du secteur des plantes aromatiques et médicinales au Maroc, Rapport Final, Haut Commissariat aux Eaux et Forets et à la Lutte Contre le Désertification, http://www.eauxetforets.gov.ma/admin/telechargement/fr/PAM_strategie_nationale.pdf.

INRAA (2006) *Rapport national sur l'état des ressources phytogénétiques pour l'alimentation et l'agriculture*, Deuxième rapport national sur l'état des ressources phytogénétiques.

Ivanova, N.V., Fazekas, A.J. and Hebert, P.D.N. (2008) Semiautomated, membrane-based protocol for DNA isolation from plants. *Plant Molecular Biology Reporter*, **26**, 186–198.

Khanuja, S.P.S., Shasany, A.K., Darokar, M.P. and Kumar, S. (1999) Rapid isolation of DNA from dry and fresh samples of plants producing large amounts of secondary metabolites and essential oils. *Plant Molecular Biology Reporter*, **17**, 1–7.

Kumar, J. and Gupta, P.K. (2008) Molecular approaches for improvement of medicinal and aromatic plants. *Plant Biotechnology Reports*, **2**, 93–112.

Kurita, N. and Koike, S. (1982) Synergetic antimicrobial effect of sodium chloride and essential oils components. *Agricultural and Biological Chemistry*, **46**, 159–165.

Lahlou, M. (2004) Methods to study the phytochemistry and bioactivity of essential oils. *Phytotherapy Research*, **18** (6), 435–448.

Lanhers, M.C., Fleurentin, J., Cabalion, P. *et al.* (1990) Behavioral effects of *Euphorbia hirta* L.: sedative and anxiolytic properties. *Journal of Ethnopharmacology*, **29**, 189–198.

Mallea, M., Soler, M., Anfossa, F. and Charpin, J. (1979) Antifungal activity of aromatic essential oils. *Pathologie Biologie*, **27**, 597–602.

MDSFS (2007) Contribution à la mise en œuvre de l'INDH, activités génératrices de revenue, module 2, Ministère du Développement Social, de la Famille et de la Solidarité.

Menichini, F., Conforti, F., Rigano, D. *et al.* (2009) Phytochemical composition, anti-inflammatory and antitumour activities of four *Teucrium* essential oils from Greece. *Food Chemistry*, **115**, 679–686.

Michiels, A., Van den Ende, W., Tucker, M. *et al.* (2003) Extraction of high-quality genomic DNA from latex-containing plants. *Analytical Biochemistry*, **315**, 85–89.

Muchalal, M. and Crouzet, J. (1985) Volatile components of clove essential oil (*Eugenia caryophyllus* Spreng): neutral fraction. *Agricultural and Biological Chemistry*, **49**, 1853–1859.

Office des Changes (2006) Annuaire de commerce extérieur, Importations et exportation par produits et pays, Maroc, Office des Changes, Ministère du Commerce et de l'Industrie.

OMS (2003) Directives OMS sur les bonnes pratiques agricoles et les bonnes pratiques de récolte (BPAR) relatives aux plantes médicinales, Organisation Mondiale de la Santé, Geneva.

Ourid, I., Ghanmi, M., El Ghadraoui, L. *et al.* (2011) Genomic DNA extraction protocols from a Moroccan medicinal and aromatic plant *Artemisia herba-alba* Asso for RAPD-PCR studies. *International Journal for Biotechnology and Molecular Biology Research*, **2** (8), 131–137.

Padmalatha, K. and Prasad, M.N.V. (2006) Optimization of DNA isolation and PCR protocol for RAPD analysis of selected medicinal and aromatic plants of conservation concern from peninsular India. *African Journal of Biotechnology*, **5** (3), 230–234.

Paris, R.R. and Moyse, H. (1967) *Précis de Matière Médicale*, Tome I, II, III, Masson et Cie, Editeur, Paris.

Pellecuer, J., Allegrini, S. and De Buochberg, M.S. (1976) Huiles essentielles bactéricides et fongicides. *Revue de l'Institut Pasteur de Lyon*, **9**, 135–159.

Pellerin, P. (2000) Technique d'extraction des huiles essentielles, Actes de la journée de réflexion sur les plants aromatiques et médicinales. *Annales de la Recherche Forestière au Maroc*, **33**, 74–103.

Porebski, S., Bailey, L.G. and Baum, B.R. (1997) Modification of a CTAB DNA extraction protocol for plants containing high polysaccharide and polyphenol components. *Plant Molecular Biology Reporter*, **15**, 8–15.

Quezel, P. and Santa, S. (1963) *Nouvelle Flore de l'Algérie et des Régions Désertiques Méridionales*, Tome II, CNRS, Paris.

Rai, A.K. and Rai, V. (2003) Effect of NaCl on growth, nitrate uptake and reduction and nitrogenase activity of *Azolla pinnata–Anabaene azollae*. *Plant Science*, **164**, 61–69.

Rolland, A., Fleurentin, J., Lanhers, M.C. *et al.* (1991) Behavioural effects of the American traditional plant *Eschscholzia californica*: sedative and anxiolytic properties. *Planta Medica*, **57** (3), 212–217.

Satrani, B. (2010) *Valorisation de Plantes Aromatiques et Médicinales du Maroc*, Éditions Universitaires Européennes (ISBN: 978-613-1-51855-3).

Satrani, B., Aberchane, M., Farah, A. *et al.* (2006) Composition chimique et activité antimicrobienne des huiles essentielles extraites par hydrodistillation fractionnée du bois de *Cedrus atlantica* Manetti. *Acta Botanica Gallica*, **153** (1), 97–104.

Sens, O.G. (1979) Les huiles essentielles, généralités et définition, in *Traité de Phytotherapy et d'Aromathérapie*, Tome I, Masson, Paris, pp. 135–177.

Shafie, M.S.B., Hasan, S.M.Z. and Shah, R.M. (2009) Study of genetic variability of wormwood capillary (*Artemisia capillaris*) using inter simple sequence repeat (ISSR) in Pahang region, Malaysia. *Plant Omics Journal*, **2**, 127–134.

Sivropoulou, A., Papanikolaou, E., Nikolaou, C. *et al.* (1996) Antimicrobial and cytotoxic activities of *Origanum* essential oils. *Journal of Food Chemistry*, **44**, 1202–1205.

Smallfield, B.M., van Klink, J.W. Perry, N.B. and Dodds, K.G. (2001) Coriander spice oil: effects of fruit crushing and distillation time on yield and composition. *Journal of Agricultural and Food Chemistry*, **49**, 118–123.

Svoboda, K.P. and Hampson, J.B. (1999) Bioactivity of essential oils of selected temperate aromatic plants: antibacterial, antioxidant, antiinflammatory and other related pharmacological activities, in *Special Chemicals for the 21st Century*, Ademe, Paris, pp. 43–49.

Tantaoui-Elaraki, A., Lattaoui, N. and Errifi, A. (1993) Composition and antimicrobial activity of the essential oil of *Thymus broussonetti*, *T. zygis* and *T. satureïoides*. *Journal of Essential Oil Research*, **5**, 45–53.

Teixeira Da Silva, J.-A. (2004) Mining the essential oils of the Anthemideae. *African Journal of Biotechnology*, **3** (12), 706–720.

UICN-Med (2008) *Programme Sous-Régional Afrique du Nord 2008–2012*, Centre de coopération pour la Méditerranée de l'UICN, http://cmsdata.iucn.org/downloads/prog_uicn_afrique_nord_fr.pdf.

Valnet, J., Duraffourd, Ch. and Lapraz, J.-Cl. (1978) *Une Médecine Nouvelle: Phytothérapie et Aromathérapie – Comment Guérir les Maladies Infectieuses par les Plantes*, Presses de la Renaissance, Paris (ISBN 2-85616-121-9).

Wang, Y. and Taylor, D.E. (1993) A rapid CTAB DNA isolation technique useful for RAPD fingerprinting and other PCR applications. *Biotechniques*, **14**, 748–750.

Ziegenhagen, B. and Scholz, F. (1993) A procedure for mini-preparation of genomic DNA from needles of silver fir (*Abies alba* Mill.). *Plant Molecular Biology Reporter*, **11**, 117–121.

Ziegenhagen, B. and Scholz, F. (1998) Methods for difficult plant species/tissues, in *Molecular Tools for Screening Biodiversity: Plants and Animals* (eds A. Karp, P.G. Issaac and D.S. Ingram), Chapman and Hall, London, pp. 32–35.

# 30 Development of Natural Cosmeceuticals: Harnessing Asia's Biodiversity

Azila Abdul-Aziz[1], Mariani Abdul Hamid[2], Norhayati Mohammad Noor[2], Harisun Yaakob[2], Rosnani Hasham[2] and Mohamad Roji Sarmidi[2]

[1]Institute of Bioproduct Development, Universiti Teknologi Malaysia Kuala Lumpur, Jalan Semarak, Kuala Lumpur, Malaysia
[2]Institute of Bioproduct Development, Universiti Teknologi Malaysia, Skudai, Johor, Malaysia

## 30.1 Introduction

It is human nature to desire beauty. As such, the use of cosmetics and personal care products can be traced back to ancient times, where ingredients such as milk, honey, lemon juice, clay, mud and even arsenic have been used in the name of beauty. In the present day, the cosmetics and personal care industry is a vital industry that contributes significantly to the world GDP. Cosmeceuticals represent a hybrid between cosmetics and drugs and they stand for products that contain active ingredients capable of altering structure or function (Elsner and Maibach, 2000; Millikan, 2001).

The active ingredients for cosmeceuticals can be naturally derived or chemically synthesized. Ideally, the active ingredients must be safe, efficacious, novel, stable, inexpensive to manufacture and can be metabolized within skin (Dooley, 1997). Nowadays, consumers are more discerning about the products that they use. Most consumers consider natural products to be safer than their synthetic counterparts, and hence the upsurge in the demand for natural cosmeceuticals.

Worldwide, it has been loosely estimated that more than 250 000 species of plants exist, thus providing an abundant source of phytochemicals (Angerhofer *et al.*, 2008). Plants have been the source for medicinal and cosmetic uses in Asia for ages, with the plant sources being highly documented in ayurveda and traditional Chinese medicine. In this chapter, five plants that are available within the context of Asia's biodiversity are reviewed in terms of their potential use in natural cosmeceuticals.

## 30.2 Mangosteen: a 'fruity' depigmenting agent

Mangosteen, *Garcinia mangostana*, is a tropical evergreen tree. The mangosteen fruit, triangular segments of white flesh surrounded by a purple rind when ripe, is consumed fresh and can be processed into jams, preserves and candies. It has also been traditionally used for nutritional and medicinal purposes for many centuries to alleviate fever, diarrhoea, headaches and heal wounds. Some species have been shown to exhibit significant antimicrobial and pharmacological activities (Valdir *et al.*, 2000).

The mangosteen tree and its fruit are rich in a variety of oxygenated and prenylated xanthones (Mahabusakaram *et al.*, 1987; Suksamrarn *et al.*, 2002; Nilar, 2002). These highly biologically active xanthones demonstrate anti-inflammatory properties through the inhibition of cyclooxygenase (COX) enzymes, and have cardiovascular protective effects (Jiang *et al.*, 2004). These xanthones have also demonstrated antibacterial (Iinuma *et al.*, 1996), antifungal (Gopalakrishnan *et al.*, 1997), antitumour-promotion (Suksamrarn *et al.*, 2002) and cytotoxic characteristics (Katsumoto *et al.*, 2003).

The fruit hulls of *G. mangostana* have been shown to contain 8-desoxygartanin, α-, β-, and γ-mangostins (Chairungsrilerd *et al.*, 1996; Huang *et al.*, 2001) and other xanthones. Studies have demonstrated that prenylated xanthones can be used in the treatment of tuberculosis. Of the xanthones tested, α- and β-mangostins, and also garcinone B, had the strongest inhibitory effects against mycobacterium, which is known to cause tuberculosis (Sunit *et al.*, 2003).

Mangosteen pulp and pericarp also contain many antioxidants, which may have antitumour activity (Moongkarndi *et al.*, 2004). α-Mangostin has been reported to exhibit strong antioxidant activity (Yokota *et al.*, 1998). These findings suggest that α-mangostin might have anti-melanogenic activity by prohibiting the process of tyrosinase oxidation.

Development of novel depigmenting phytochemicals from natural resources has become a new popular trend. These depigmenting materials could be applied to melanogenic disorders such as melasma and also for skin-lightening cosmetic products (Surace *et al.*, 2000; Lupo, 2001). Melanogenesis is regulated by the activity of tyrosinase, a rate-limiting enzyme in melanin biosynthesis (Schallreuter *et al.*, 2003). Since melanin is derived from the precursor DOPA quinone that is formed by tyrosinase oxidation of L-tyrosine, tyrosinase plays an important role in melanin synthesis. Decreased tyrosinase activity in melanocytes could be achieved either by direct inhibition of tyrosinase itself or suppression of tyrosinase gene expression, leading to decreased amounts of the protein level in cells.

Decreased melanin biosynthesis for both secreted and intracellular melanin content (data not shown) have been shown with

*Novel Plant Bioresources: Applications in Food, Medicine and Cosmetics*, First Edition. Edited by Ameenah Gurib-Fakim.

the treatment of α-mangostin isolated from mangosteen's leaves on B16F1 melanoma cell (Abdul Hamid, 2012). To elucidate the mechanism of melanogenesis inhibited by α-mangostin, B16F1 melanoma cells were treated with different concentrations of the compound (0.156, 0.313, 0.625, 1.25, 2.5 and 5 μg/mL). Each percentage value of tyrosinase activity in the treated cells was calculated with respect to the control group. The same amount of cell lysate calibrated with respect to protein concentration was applied to the oxidation reaction with L-dopa. Upon treatment with tyrosinase, activity in B16F1 cells was increased by twofold, as shown in Figure 30.1a. Treatment with 800 μM of kojic acid and 2 mM of arbutin did not reduce the tyrosinase activity. However, treatment with α-mangostin yielded statistically significant reduction of the enzyme activity in a dose-dependent manner, suggesting that, unlike kojic acid and arbutin, which directly inhibit tyrosinase activity, α-mangostin has the ability to both directly and indirectly inhibit tyrosinase activity (data not shown) as well as suppress tyrosinase gene expression. Tyrosinase activity decreased by 60% at the highest concentration of α-mangostin (5 μg/mL).

In order to clarify further the mechanism of tyrosinase deactivation by α-mangostin, the levels of tyrosinase expression in B16F1 melanoma cells were examined by tyrosinase zymography. Figure 30.1b shows that that α-MSH treatment resulted in a dark black band compared with that of untreated control. Tyrosinase activity was suppressed to the level of α-MSH untreated control at a concentration of 5 μg/mL α-mangostin and virtually no tyrosinase activity was exhibited at this concentration. This result proved that tyrosinase zymography seemed to be well correlated with the measurement of intracellular tyrosinase activity, suggesting that α-mangostin can inhibit melanin biosynthesis by inhibiting tyrosinase activity and suppressing tyrosinase gene expression, making it a potent depigmenting agent.

## 30.3 *Ficus deltoidea*: the 'golden' treasure from nature

*Ficus deltoidea* or commonly known by Malaysians as *Mas cotek* (literally small gold) or *serapat angin*, is one of the popular herbs among Malaysian women (Figure 30.2). *F. deltoidea* is part of the Moraceae family. The genus *Ficus* comprises about 750 species, and most of them grow in the tropics (Grison-Pigé *et al.*, 2002). The plant is commonly used for medicinal applications and is widely used as a post-partum herbal drink to help strengthen the uterus

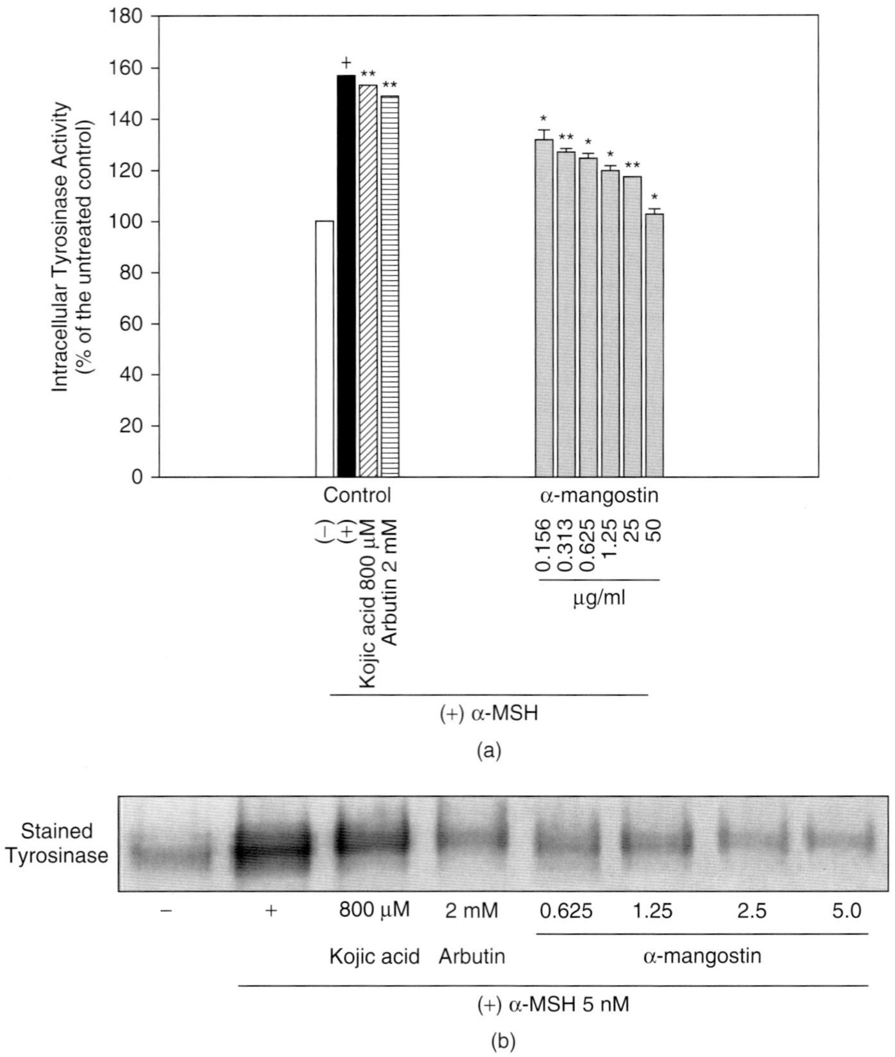


**Figure 30.1** Intracellular tyrosinase activity of B16F1 melanoma was inactivated by α-mangostin. (a) α-Mangostin effectively diminished α-melanocyte-stimulating hormone (α-MSH)-induced intracellular tyrosinase activity. A 60% decrease in intracellular tyrosinase activity was observed. (b) Decrease in tyrosinase activity by α-mangostin treatment was confirmed through the result of tyrosinase zymography. Results are the average of three independent experiments plus/minus the standard deviation. $^{*}P < 0.01$ compared with (+) α-MSH control. *Source*: Abdul Hamid (2012)

Figure 30.2 (a) *Ficus deltoidea* subsp. *deltoidea* Jack taken from a Malaysian source. (b) Dried leaves of the plant

(Sulaiman *et al.*, 2008). In Malaysia, *F. deltoidea* is sold in the form of dried or powdered leaves, stems, roots or fruits, as an ingredient in bottled drinks or mixed with other beverages such as coffee (Musa, 2006).

Among others, *F. deltoidea* contains flavonoid, coumarin, alkaloid, steroid, triterpene and salicylic acid (Ong *et al.*, 2009). The plant possesses stronger antioxidant activity than Trolox and vitamin C (Oh *et al.*, 2011).

Pigmentation is one of the defence mechanisms (melanogenesis process) against oxidative stress or ultraviolet (UV) irradiation, but abnormal hyperpigmentation in human skin may pose a serious aesthetic problem (Wu *et al.*, 2011). Antioxidants or inhibitors of tyrosinase have been purported to be able to suppress melanogenesis in the epidermal layer of the skin (Hanamura *et. al.*, 2008). Owing to its strong antioxidant activity, *F. deltoidea* has been studied by a research group from Dongguk University, South Korea, and the Institute of Bioproduct Development, Universiti Teknologi Malaysia, Malaysia, as a potential skin-lightening agent.

α-MSH-induced melanin synthesis in B16F1 melanoma cells was reduced dose dependently when the cells were treated with *F. deltoidea* extract at concentrations between 0.0125% (w/v) to 0.1% (w/v), with complete inhibition observed at the concentration of 0.1% (Oh *et al.*, 2011). The treatment of *F. deltoidea* extract at the concentration of 0.1%, presented more effective reduction of mushroom tyrosinase activity than 800 μM kojic acid or 2 mM arbutin. The group also investigated the effect of *F. deltoidea* extract on intracellular tyrosinase activity. They discovered that intracellular tyrosinase activity of B16F1 melanoma cells was inactivated by *F. deltoidea* extract and that this was due to the suppression of gene expression in B16F1 melanoma cells.

It seems that, similar to α-mangostin, *F. deltoidea* extract manages to inhibit melanin synthesis in melanocytes by direct inhibition of tyrosinase activity and also suppression of the genes that encode tyrosinase (Oh *et al.*, 2011). Further pre-clinical studies may need to be done in order to determine the safety of *F. deltoidea* extract, followed by a clinical study on human volunteers to test the efficacy of the extract.

## 30.4 *Labisia pumila*: Malaysia's queen of herbs

*Labisia pumila*, which is also known as Kacip Fatimah (Figure 30.3), is a popular herb amongst Malaysian women (Burkill, 1935; Choi *et al.*, 2010). *L. pumila* belongs to the Myrsinaceae family. Other local Malaysian names for the herb are Akar Fatimah, Selusoh Fatimah, Tadah Matahari, Rumput Siti Fatimah, Bunga Belangkas Hutan and Pokok Pinggang (Jamal *et al.*, 2003). There are three varieties of *L. pumila*: *L. pumila* var. *alata*, *L. pumila* var. *pumila* and *L. pumila* var. *lanceolata* (Arifin, 2005). Each variety has its respective uses, and local healers tend to use var. *alata* and var. *pumila* traditionally (Ibrahim and Jaafar, 2011).

This herb's extract is prepared by boiling the roots, leaves or the whole plant with water and the extract is taken orally (Zainon *et al.*, 1999). The herb has a very long and intimate history with Malaysian women as it is commonly used by them to treat post-partum conditions (Ayida *et al.*, 2007), to assist contraction of the birth channel (Jamal *et al.*, 2003), to shrink the uterus, to improve the menstrual cycle and for weight loss (Arifin, 2005), It was also reported that *L. pumila* can be used for delaying fertility and to regain body strength; some other folkloric uses include treatment of flatulence, dysentery, dysmenorrhoea, gonorrhoea and 'sickness in the bones'.

*L. pumila* contains favonols, flavonols, flavonoids, phenolics, saponins, benzoquinoid and others (Chua *et al.*, 2012). Its reported biological activities range from antioxidant, to antibacterial to antifungal (Chua *et al.*, 2012), making it a likely candidate to be considered for a cosmeceutical formulation. A research group from Dongguk University, South Korea, and the Institute of Bioproduct Development, Universiti Teknologi Malaysia, Malaysia, studied the potential anti-photo-aging effects of *L. pumila*.

(a) (b)

**Figure 30.3** (a) *Labisia pumila* var. *alata* and (b) dried *Labisia pumila* var. *alata*

*L. pumila* plant extract has been shown to have the ability to protect human skin fibroblast and keratinocytes from continuous attack of reactive oxygen species (ROS) generated by critical UVB exposure, making it a potential photo-protective agent against extrinsic aging factor (Choi *et al.*, 2010). In the study, the plant extract has been shown to possess higher free-radical scavenging activity than ascorbic acid. This plant extract also has the ability to stimulate the growth of human skin fibroblast cells *in vitro* after exposure to UV radiation. Apart from that, *L. pumila* could also up regulate the synthesis of collagen in human dermal fibroblast cells in a dose-dependent manner, markedly reduce the production of the tumour necrotic factor (TNF)-α, a pro-inflammatory cytokine, and markedly inhibit the expression of COX-2, an enzyme that is involved in the production of prostaglandins that mediate pain and support the inflammatory process. Interestingly, *L. pumila* extract also possesses the ability to downregulate matrix metalloproteinase (MMP)-1, the enzyme responsible for the degradation of dermal collagen, and attenuate MMP-9, a gelatinase that is also responsible for the degradation of collagen type IV. This makes *L. pumila* a unique photo-protective agent as it can simultaneously promote collagen synthesis and reduce collagen degradation. However, the toxicological profile of this herb has not yet been fully elucidated.

**Figure 30.4** *Andrographis paniculata* plant

## 30.5 *Andrographis paniculata*: a 'bitter' therapy for the skin

*Andrographis paniculata* from the family of Acanthaceae is a popular medicinal herb in tropical Asian countries. The plant is called 'Hempedu bumi' in Malaysia, which literally means 'bile of the earth', as it is the 'king of bitter' among herbal medicines. The leaves of *A. paniculata* are dark green with a smooth oval surface measuring 5–10 cm × 1.2–2.5 cm (Figure 30.4). Most commonly, the aerial parts are used for medicinal purposes. Traditionally, *A. paniculata* is used for the treatment of hypertension, diabetes, obesity, and insect and snake bites (Bhatnagar *et al.*, 1961). In primary modern medicine, the plant has been used orally to prevent and treat the common cold, influenza, pharyngotonsilitis, allergies and sinusitis (Khory and Katrak, 1985). Its major compounds are diterpenoids, flavonoids and polyphenols. Andrographolide is the major diterpenoid found in *A. paniculata*, making up of about 4% of the dried whole plant (Koteswara Rao *et al.*, 2004).

The phytochemical profiling of *A. paniculata* suggests that this plant can be utilized in a diverse range of beneficial cosmeceutical applications. The cosmeceutical potential of this plant can be realized in terms of its antioxidant, anti-inflammatory, anti-allergy, immuno-stimulant and wound-healing properties.

Oxygen-derived free radicals are responsible for skin damage (aging) at cellular and tissue levels. Natural antioxidant could act as a defence against oxidation and further cellular function impairment. Methanolic extract of *A. paniculata* was found to inhibit the formation of ROS *in vitro* (Sheeja *et al.*, 2006). Pretreatment with andrographolide also significantly reduced accumulation of phorbol-12-myristate-13-acetate-induced formation of ROS and *N*-formyl-methionyl-leucyl-phenylalanine-induced adhesion *in vivo* (Shen *et al.*, 2000). Furthermore, andrographolide had been found to inhibit the production of neutrophils, macrophage migration and the production of TNF-α and interleukin (IL)-12 (Shen *et al.*, 2000). *A. paniculata* also showed significant alleviation of pro-inflammatory (NO, IL-1β and IL-6), inflammatory (PGE2 and TXB2) and allergic (LTB4) mediators (Shen *et al.*, 2002). Since chronic oxidative damage and inflammation is a major culprit in visible skin aging, these findings provide a rationale for the application of *A. paniculata* in cosmeceutical products such as anti-aging treatment.

*A. paniculata* or its constituents have also been reported to possess antibacterial, antimicrobial and antifungal activities (George and Pandalai, 1949; Nakanishi, 1965; Tsuchiya, *et al.*, 1996; Shukla *et al.*, 1999; Singha *et al.*, 2003; Limsong *et al.*, 2004). These activities have been reported to play an important role in the promotion of wound healing. Ethanol extract of *A. paniculata* had been found to inhibit the growth of *Escherichia coli* and *Staphylococcus aureus* (George and Pandalai, 1949), whereas methanolic extracts of *A. paniculata* inhibited the growth of *Proteus vulgaris in vitro* (Nakanishi, 1965). Limsong *et al.* (2004) discovered that ethanol extract of *A. paniculata* inhibited the adherence of *Streptococcus mutans* ATCC 25175 and *S. mutans* TPF-1. Aqueous extract of *A. paniculata* was reported to have antifungal activity (Singha *et al.*, 2003; Limsong *et al.*, 2004). Andrographolide, neoandrographolide and 14-deoxy-11,12-didehydroandrographolide have also been reported to be viricidal against herpes simplex virus 1 (HSV-1) (Wiart *et al.*, 2005). Based on these properties of *A. paniculata* extracts, Al-Bayaty *et al* (2010) treated wounds using the plant extracts and they found that treatment with the extracts significantly enhanced the rate of wound healing *in vivo*.

Thus, under experimental conditions of antioxidant, anti-inflammatory and wound healing, it can be suggested that *A. paniculata* or its constituents can be utilized as a potential functional cosmetic ingredient.

## 30.6 *Centella asiatica*: herbs' jack of all trades

*Centella asiatica* is a perennial creeping plant belonging to the Apiaceae family that is used in cuisine or drinks in Asian countries. The plant is also traditionally used to treat a variety of illnesses (Bown, 1995). *C. asiatica* can be found in swampy areas of tropical and subtropical countries such as Malaysia, India, Sri Lanka, Japan, Madagascar and South Africa. The plant has small heart-shaped to almost circular green leaves arising from long petioles with pink to reddish flowers.

The main chemical groups of the plants are essential oil, flavone derivatives, sesquiterpenes, triterpene steroids, triterpene acids and triterpene acid sugar esters (Kartnig and Hoffmann-Bohm, 1992). The active constituents of this plant that are of therapeutic interests mostly come from the triterpene acids and their sugar esters groups, with the most important being asiatic acid, madecassic acid, asiaticoside, asiaticoside A and asiaticoside B (Brinkhaus *et al.*, 2000). *C. asiatica* has been found to exhibit the following therapeutic properties: facilitation of wound healing (Sunilkumar *et al.*, 1998), strengthening of weakened veins (Allegra, 1981), inhibition of gastric lesions (Cheng and Koo, 2000), a sedative agent (Ramaswamy *et al.*, 1970), an antidepressant agent (Chen *et al.*, 2003), an anti-epileptic agent (Ganachari *et al.*, 2004), a cognitive-enhancing agent (Veerendra Kumar and Gupta, 2003), an antinociceptive agent (Somchit *et al.*, 2004), an anti-inflammatory agent (Nhiem *et al.*, 2011) and a radio-protective agent (Sharma and Sharma, 2002).

Apart from the previously described pharmacological properties of *C. asiatica*, the plant has also been used in skincare products as the plant has been reported to possess chemical constituents that can positively affect skin health. Tenni *et al.* (1988) studied the effect of total triterpenoid fraction from *C. asiatica* on the biosynthesis of collagen, fibronectin and proteoglycans using human skin fibroblast culture. They concluded that 25 μg/mL of the triterpenoid fraction significantly increased collagen and fibronectin. The effect of the mixture of three terpenes from *C. asiatica* (asiatic acid (30%, w/w), madecassic acid (30%, w/w) and asiaticoside (40%, w/w)) on human foreskin fibroblast monolayer culture have also been studied (Maquart *et al.*, 1990). Asiatic acid was reported to be the component responsible for collagen synthesis stimulation. All three terpenes increased free proline. More recently, Lee *et al.* (2006) investigated the effect of asiaticoside on type I collagen synthesis. They postulated that the mechanism of induction of type I collagen synthesis by asiaticoside is through the activation of TβRI kinase-independent Smad pathway. The research studies on the collagen synthesis ability of *C. asiatica* prompted Haftek *et al.* (2008) to perform a clinical study on the effect of 0.1% madecassoside and 5% vitamin C on photo-aging. They obtained promising results where two-thirds of the subjects showed functional and structural remodelling of chronically sun-damaged skin after 6 months of treatment. *C. asiatica* can also be potentially used for cellulite reduction (Rossi and Vergnanini, 2000). Reduction of cellulite may be due to the action of the plant extracts on connective tissue and microcirculation. Sampson *et al.* (2001) suggested that *C. asiatica* could be used in the treatment of psoriasis. The anti-psoriatic effects may be due the inhibition of the proliferation of keratinocyte by triterpenoid glycosides. Aqueous extract of *C. asiatica* has also been shown to show some photo-protective effect against UVA. Water extract of the plant, particularly the ones that were nano-encapsulated, reduced matrix metalloproteinase (MMP)-1 in a dose-dependent manner, where the largest drop of MMP1 in UVA-irradiated human skin fibroblasts was from 136% to 78% (Kwon *et al.*, 2011).

## 30.7 Future trends

One of the most critical factors in the development of plant-based cosmeceutical products is the issue of consumer acceptance. The consumers in this product segment are generally health and wellness-conscious, better educated, well travelled, more demanding and endowed with a higher purchasing power. The products that they are looking for should be of higher quality with good value for money and should be safe, natural, efficacious and ethical.

The extensive research carried out on these five plants indicates a tremendous potential for them to be used as bioactive ingredients in novel cosmeceutical products. The cosmetic industry should capitalize on the uniqueness of the skin protective and regenerative bioactivities of phytochemicals from these native plants, especially for regional Asian market.

The Asian region comprises a diverse ethnic group with different skin physiology and chemistry. Malaysian skins are known to have a higher content of melanin and as a result Malaysians have darker complexion than the rest of the ethnic groups, such as the Koreans and the Vietnamese (Hasham, 2013). A darker pigmented skin is correlated with a more compact stratum corneum and intercellular cohesion. In the recent study on skin barrier function using confocal reflectance microscopy, Hasham (2013) discovered that Malaysian skins are generally endowed with a thicker epidermis, followed by Koreans and Vietnamese. However, in terms of skin barrier function expressed as transepidermal water loss, Malaysians showed the lowest skin barrier function as compared with Koreans and Vietnamese. Malaysians have higher hydration and skin sebum content, followed by Koreans and Vietnamese.

The information on the epidermal thickness, skin barrier function, hydration and sebum content are crucial in the formulation of a more efficacious cosmeceutical product. There is clearly a need to develop and formulate cosmeceutical products based on ethnic group and region to optimize the product effectiveness. A targeted research and development effort is required to address the specific needs and requirements of a particular ethnic group. The ideal cosmeceutical product is, of course, a personalized product based on individual health status and skin physiology. Custom-blended cosmeceutics should follow the traditional custom-blended cosmetics sooner or later.

To ensure the authenticity, consistency and efficacy of cosmeceutical products, standardized purified extracts of the plants are required by the industries. Appropriate chemical markers and biomarkers should be developed and made available (Malaysian Monograph Committee, 1999). Isovitexin and vitexin are the possible chemical markers that can be used to standardize *F. deltoidea* whole extract (Choo *et al.*, 2012), while madecassic acid, asiaticoside and asiatic acid can be used for *C. asiatica* (Maquart *et al.*, 1990). In the case of *A. paniculata*, andrographolide can be used as the appropriate chemical marker (Koteswara Rao *et al.*, 2004). Concerted research effort is needed to establish the cosmeceutical biomarkers for these plants to support the product development and commercialization activities.

In addition to the issue of safety and efficacy in cosmeceutical products, there is the emerging concern on the Halal issue especially by Muslim consumers. The awareness of Halal cosmetics amongst the Muslim consumers provides an opportunity for the industry to develop new Halal product ranges. Halal cosmetics require the process and the components in the formulation to conform to the Halal standard. There is no major issue with the plant- and marine-based cosmetic ingredients as long as they are extracted using appropriate solvents.

In light of the recent developments in research and commercialization activities on these five unique plants, more cosmeceutical products are expected to hit the market in the near future. The synergistic effects of the five unique plants on skin protection and regeneration have not been fully explored. For example, the synergistic effects of *L. pumila* and *F. deltoidea* on skin protection from UV radiation and collagen synthesis are being investigated by the authors. The outcome of the research on the synergistic effect is interesting. In addition, the emergent synergistic effects of topical and oral applications of a specific combination of the five plants need to be investigated.

## References


Abdul Hamid, M. (2012) Study on Malaysian tropical plants that modulate melanogenesis in B16F1 melanoma cells. PhD Thesis, Dongguk University, South Korea.

Al-Bayaty, F.H., Abdulla, M.A., Abu Hassan, M.A. and Mohd Ali, H. (2010) Effect of *Andrographis paniculata* leaf extract on wound healing in rats. *Natural Product Research*, **26** (5), 423–429.

Allegra, C. (1981) Comparative capillaroscopic study of certain bioflavonoids and total triterpenic fractions of *Centella asiatica* in venous insufficiency. *Clinical Therapeutics*, **99**, 507–513.

Angerhoofer, C.K., Maes, D. and Giacomoni, P.U. (2008) The use of natural compounds and botanicals in the development of anti-aging skin care products, in *Skin Aging Handbook: An Integrated Approach to Biochemistry and Product Development* (ed. N. Dayan), William Andrew Inc., New York, NY, pp. 205–263.

Arifin, N. (2005) *Penyembuhan semula jadi dengan herba*, PTS Publications and Distributors Sdn Bhd, Kuala Lumpur.

Ayida, A.W., Wan Nazaimoon, W.M., Farihah, H.S. and Azian, A.L. (2007) Effect of variectomy, *Labisia pumila* var. *alata* treatment and estrogen replacement therapy on the morphology of adipose tissue in ovariectomized Sprague Dawley rats. *Journal of Medical and Biological Sciences*, **1**, 1–7.

Bhatnagar, S.S., Santapau, H., Desa, J.D. *et al.* (1961) Biological activity of Indian medicinal plants. I. Antibacterial, antitubercular and antifungal action. *Indian Journal of Medical Research*, **49**, 799–813.

Bown, D. (1995) *Encyclopaedia of Herbs and their Uses*, Dorling Kindersley, London, pp. 361–365.

Brinkhaus, B., Linder, M., Schuppan, D. and Hahn, E.G. (2000) Chemical, pharmacological and clinical profile of the East Asian medical plant *Centella asiatica*. *Phytomedicine*, **7** (5), 427–448.

Burkill, I.H. (1935) *A Dictionary of Economic Products of the Malay Peninsula* (ed Ministry of Agriculture (Malaysia)), Crown Agents for the Colonies, London.

Chairungsrilerd, N., Takeuchi, K., Ohizumi, Y. *et al.* (1996) Xanthones from the green fruit hulls of *Garcinia mangostana*. *Phytochemistry*, **43**, 1099–1102.

Chen, Y., Han, T., Qin, L. *et al.* (2003) Effect of total triterpenes from *Centella asiatica* on the depression behaviour and concentration of amino acid in forced swimming mice. *Zhong Yao Cai*, **26**, 870–873.

Cheng, C.L. and Koo, M.W. (2000) Effects of *Centella asiatica* on ethanol induced gastric mucosal lesions in rats. *Life Sciences*, **67**, 2647–2653.

Choi, H., Kim, D., Kim, J.W. *et al.* (2010) *Labisia pumila* extract protects skin cells from photoaging caused by UVB irradiation. *Journal of Bioscience and Bioengineering*, **109** (3), 291–296.

Choo, C.Y., Sulong, N.Y., Man, F. and Wong, T.W. (2012) Vitexin and isovitexin from the leaves of *Ficus deltoidea* with *in-vivo* α-glucosidase inhibition. *Journal of Ethnopharmacology*, **142**, 776–781.

Chua, L.S., Lee, S.Y., Abdullah, N. and Sarmidi, M.R. (2012) Review on *Labisia pumila* (Kacip Fatimah): bioactive phytochemicals and skin collagen synthesis promoting herb. *Fitoterapia*, **83** (8), 1322–1335.

Dooley, T.P. (1997) Is there room for a moderate regulatory oversight?, in *Drug Discovery Approaches for Developing Cosmeceuticals: Advanced Skincare and Cosmetics Products* (ed. W. Hori), IBC Library Series, Drug & Market Development Publications, Southborough, MA.

Elsner, P. and Maibach, H.I. (2000) *Cosmeceuticals: Drugs vs Cosmetics*, Marcel Dekker, New York, NY.

Ganachari, M.S., Babu, V. and Katare, S. (2004) Neuropharmacology of an extract derived from *Centella asiatica*. *Pharmaceutical Biology*, **42**, 246–252.

George, M. and Pandalai, K.M. (1949) Investigations on plant antibiotics. Part IV. Further search for antibiotic substances in Indian medicinal plants. *Indian Journal of Medical Research*, **37**, 169–181.

Gopalakrishnan, G., Banumathi, B. and Suresh, G. (1997) A review of mangosteen (*Garcinia mangostana*) Linn. *Journal of Natural Products*, **60**, 519–524.

Grison-Pigé, L., Hossaert-McKey, M., Greeff, J.M. and Bessière, J.-M. (2002) Fig volatile compounds – a first comparative study. *Phytochemistry*, **61** (1), 61–71.

Haftek, M., Mac-Mary, S., Le Bitoux, M.A. *et al.* (2008) Clinical, biometric and structural evaluation of the long-term effects of a topical treatment with ascorbic acid and madecassoside in photoaged human skin. *Experimental Dermatology*, **17** (11), 946–952.

Hanamura, T., Uchida, E. and Aoki, H. (2008) Skin-lightening effect of a polyphenol extract from Acerola (*Malpighia emarginata* DC.) fruit on UV-induced pigmentation. *Bioscience, Biotechnology, and Biochemistry*, **72** (12), 3211–3218.

Hasham, R. (2013) Study on comparison of ethnic variations of skin barrier properties and other skin conditions. PhD Thesis, Dongguk University, South Korea.

Huang Y.L., Chen C.C., Chen Y.J. *et al.* (2001) Three xanthones and a benzophenone from *Garcinia mangostana*. *Journal of Natural Products*, **64** (7), 903–906.

Ibrahim, M.H. and Jaafar, H.E. (2011) Photosynthetic capacity, photochemical efficiency and chlorophyll content of three varieties of *Labisia pumila* Benth. exposed to open field and greenhouse growing conditions. *Acta Physiologiae Plantarum*, **33**, 2179–2185.

Iinuma M., Tosa H., Tanaka T. *et al.* (1996) Antibacterial activity of xanthones from guttiferaeous plants against methicillin-resistant *Staphyloococcus aureus*. *Journal of Pharmacy and Pharmacology*, **48**, 861–865.

Jamal, J.A., Houghton, P.J., Milligan, S.R. and Jantan, I. (2003) The oestrogenis and cytotoxic effects of the extracts of *Labisia pumila* var. alata and *Labisia pumila* var. *pumila in vitro*. *Sains Kesihatan*, **1**, 53–60.

Jiang, D.J., Dai, Z. and Li, Y.J. (2004) 2-Methylxanthen-9-one. *Cardiovascular Drug Reviews*, **22**, 91–102.

Kartnig, T. and Hoffmann-Bohm, K. (1992) Centella, in Hager's Handbuch der Pharmazeutischen *Praxis,* Band 4 (eds R. Hansel, K. Keller, H. Rimpler and G. Schneider), Springer Verlag, Berlin.

Katsumoto, K., Akao, Y., Kobayashi, E. *et al.* (2003) Induction of apoptosis by xanthones from mangosteen in human leukemia cell lines. *Journal of Natural Products*, **66**, 1124–1127.

Khory, R.N. and Katrak, N.N. (1985) *Materia Medica of India and their Therapeutics*, Neeraj Publishing House, Delhi.

Koteswara Rao, Y., Vimalamma, G., Rao, C.V. and Tzeng, Y.M. (2004) Flavonoids and andrographolides from *Andrographis paniculata*. *Phytochemistry*, **65**, 2317–2321.

Kwon, M.C., Choi, W.Y., Seo, Y.C. *et al.* (2012) Enhancement of the skin-protective activities of *Centella asiatica* L. Urban by a nano-encapsulation process. *Journal of Biotechnology*, **157** (1), 100–106.

Lee, J., Jung, E., Kim, Y. *et al.* (2006) Asiaticoside induces human collagen I synthesis through TGFbeta receptor I kinase (TbetaRI kinase)-independent Smad signaling. *Planta Medica*, **72** (4), 324–328.

Limsong, J., Benjavongkulchai, E. and Kuvatanasuchati, J. (2004) Inhibitory effect of some herbal extracts on adherence of *Streptococcus mutans*. *Journal of Ethnopharmacology*, **92**, 281–289.

Lupo, M.P. (2001) Antioxidants and vitamins in cosmetics. *Clinics in Dermatology*, **19**, 467–473.

Mahabusakaram, W., Iriyachitra, P. and Taylor, W.C.J. (1987) Chemical constituents of *Garcinia mangostana*. *Journal of Natural Products*, **50**, 474–478.

Malaysian Monograph Committee (1999) *Malaysian Herbal Monograph*, Forest Research Institute Malaysia, Kuala Lumpur.

Maquart, F.X., Bellon, G., Gillery, P. *et al.* (1990) Stimulation of collagen synthesis in fibroblast cultures by a triterpene extracted from *Centella asiatica*. *Connective Tissue Research*, **24** (2), 107–120.

Millikan, L.E. (2001) Cosmetology, cosmetics, cosmeceuticals: definitions and regulations. *Clinics in Dermatology*, **19**, 371–374.

Moongkarndi, P., Kosem, N., Kaslungka, S. *et al.* (2004) Antiproliferation, antioxidation and induction of apoptosis by *Garcinia mangostana* (mangosteen) on SKBR3 human breast cancer cell line. *Journal of Ethnopharmacology*, **90**, 161–166.

Musa, Y. (2006) Evaluation of growth performance and yield potential of selected emas cotek (*Ficus deltoidea*) accessions on bris soils. *Journal of Tropical Agriculture and Food Science*, **34** (2), 229–235.

Nakanishi, K. (1965) Phytochemical survey of Malaysian plants: preliminary chemical and pharmacological screening. *Chemical and Pharmaceutical Bulletin*, **13**, 882–890.

Nhiem, N.X., Tai, B.H., Quang, T.H. *et al.* (2011) A new ursane-type triterpenoid glycoside from *Centella asiatica* leaves modulates the production of nitric oxide and secretion of TNF-α in activated RAW 264.7 cells. *Bioorganic & Medicinal Chemistry Letters*, **21** (6), 1777–1781.

Nilar, L.J.H. (2002) Xanthones from the heartwood of *Garcinia mangostana*. *Phytochemistry*, **60**, 541–548.

Oh, M.J., Abdul Hamid, M., Ngadiran, S. *et al.* (2011) *Ficus deltoidea* (Mas cotek) extract exerted anti-melanogenic activity by preventing tyrosinase activity *in vitro* and by suppressing tyrosinase gene expression in B16F1 melanoma cells. *Archives of Dermatological Researchs*, **303**, 161–170.

Ong, C.Y., Ling, S.K., Ali, R.M. *et al.* (2009) Systematic analysis of *in vitro* photo-cytotoxic activity in extracts from terrestrial plants in Peninsula Malaysia for photodynamic therapy. *Journal of Photochemistry and Photobiology B: Biology*, **96** (3), 216–222.

Ramaswamy, A.S., Pariyaswami, S.M. and Basu, N. (1970) Pharmacological studies on *Centella asiatica*. Linn. *Indian Journal of Medical Research*, **4**, 160–164.

Rossi, A.B.R. and Vergnanini, A.L. (2000) Cellulite: a review. *European Academy of Dermatology and Venereology*, **14**, 251–262.

Sampson, J.H., Raman, A., Karlsen, G. *et al.* (2001) *In vitro* keratinocyte antiproliferant effect of *Centella asiatica* extract and triterpenoid saponins. *Phytomedicine*, **8** (3), 230–235.

Schallreuter K.U., Kothari, S., Hasse, S. *et al.* (2003) *In situ* and *in vitro* evidence for DCoH/HNF-1 alpha transcription of tyrosinase in human skin melanocytes. *Biochemical and Biophysical Research Communications*, **301**, 610–616.

Sharmam, J. and Sharmam, R. (2002) Radioprotection of Swiss albino mouse by *Centella asiatica* extract. *Phytotherapy Research*, **16**, 785–786.

Sheeja, K., Shihab, P.K. and Kuttan, G. (2006) Antioxidant and anti-inflammatory activities of the plant *Andrographis paniculata* Nees. *Immunopharmacology and Immunotoxicology*, **28**, 129–140.

Shen, Y.C., Chen, C.F. and Chiou, W.F. (2000) Suppression of rat neutrophil reactive oxygen species production and adhesion by the diterpenoid lactone andrographolide. *Planta Medica*, **66**, 314–317.

Shen, Y.C., Chen, C.F. and Chiou, W.F. (2002) Andrographolide prevents oxygen radical production by human neutrophils: possible mechanism(s) involved in its anti-inflammatory effect. *British Journal of Pharmacology*, **135**, 399–406.

Shukla, A., Rasik, A.M. and Dhawan, B.N. (1999) Asiaticoside-induced elevation of antioxidant levels in healing wounds. *Phytotherapy Research*, **13**, 50–54.

Singha, P.K., Roy, S. and Dey, S. (2003) Antimicrobial activity of *Andrographis paniculata*. *Fitoterapia*, **74**, 692–694.

Somchit, M.N., Sulaiman, M.R., Zuraini, A. *et al.* (2004) Antinociceptive and antiinflammatory effects of *Centella asiatica*. *Indian Journal of Pharmacology*, **36**, 377–380.

Suksamrarn, S., Suwannapoch, N., Ratananukul, P. *et al.* (2002) Xanthones from the green fruit hulls of *Garcinia mangostana*. *Journal of Natural Products*, **65**, 761–763.

Sulaiman, M.R., Hussain, M.K., Zakaria, Z.A. *et al.* (2008) Evaluation of the antinociceptive activity of *Ficus deltoidea* aqueous extract. *Fitoterapia*, **79** (7–8), 557–561.

Sunilkumar, Parameshwaraiah S., and Shivakumar H.G. (1998) Evaluation of topical formulations of aqueous extract of *Centella asiatica* on open wounds in rats. *Indian Journal of Experimental Biology*, **36**, 569–572.

Sunit, S., Narisara, S., Wong, P. *et al.* (2003) Antimycobacterial activity of prenylated xanthones from the fruits of *Garcinia mangostana*. *Indian Chemical and Pharmaceutical Bulletin*, **51**, 857–859.

Surace, E.M., Angeletti, B., Ballabio, A. and Marigo, V. (2000) Expression pattern of the ocular albinism type 1 (Oa1) gene in the murine retinal pigment epithelium. *Investigative Ophthalmology & Visual Science*, **41** (13), 4333–4337.

Tenni, R., Zanaboni, G., De Agostini, M.P. *et al.* (1988) Effect of the triterpenoid fraction of *Centella asiatica* on macromolecules of the connective matrix in human skin fibroblast cultures. *Italian Journal of Biochemistry*, **37** (2), 69–77.

Tsuchiya, H., Sato, M., Miyazaki, T. *et al.* (1996) Comparative study on the antibacterial activity of phytochemical flavanones against methicillin-resistant *Staphylococcus aureus*. *Journal of Ethnopharmacology*, **50**, 27–34.

Valdir, P., Nagem, T.J. and Oliveira, F.F.D. (2000) Tetraoxygenated naturally occurring xanthones. *Phytochemistry*. **55**, 683–710.

Veerendra Kumar, M.H. and Gupta, Y.K. (2003) Effect of *Centella asiatica* on cognition and oxidative stress in an intracerebroventricular streptozotocin model of Alzheimer's disease in rats. *Clinical and Experimental Pharmacology and Physiology*, **30**, 336–342.

Wiart, C., Kumar, K., Yusof, M.Y. *et al.* (2005) Antiviral properties of ent-labdene diterpenes of *Andrographis paniculata* Nees, inhibitors of herpes simplex virus type 1. *Phytotherapy Research*, **19**, 1069–1070.

Wu, L.C., Lin, Y.Y., Yang, S.Y. *et al.* (2011) Antimelanogenic effect of c-phycocyanin through modulation of tyrosinase expression by upregulation of ERK and downregulation of p38 MAPK signaling pathways. *Journal of Biomedical Science*, **18**, 74.

Yokota, T., Nishio, H., Kubota, Y. and Mizoguchi, M. (1998) The inhibitory effect of glabridin from licorice extracts on melanogenesis and inflammation. *Pigment Cell Research*, **11**, 355–361.

Zainon, A.S., Musaadah, M., Ismail, M. and Wan-Fadhilah, W.Z.A. (1999) Ethobotany of medicinal plants at Pos Lanai, Lipis, Pahang, in *Interdisciplinary Approaches in Natural Products Research* (eds K. Mawardi, A. Zhari, M. Suparjo and S. Chong), Department of Chemistry, University Putra Malaysia, Serdang, Malaysia, pp. 35–42.

# 31 Unique Bioresources from Ethiopia for Food, Medicine and Cosmetics

E. Dagne

African Laboratory for Natural Products (ALNAP), Department of Chemistry, Addis Ababa University, Addis Ababa, Ethiopia

## 31.1 Introduction

Ethiopia's diverse topography provides habitats for approximately 7000 higher plant species, where approximately 10% are endemic. The rich traditional knowledge of the people has, over the centuries, led to the application of a wide array of plants used for food and as medicines. Many of these plant medicines are today readily available at local markets, where they are customarily sold side by side with spices and other food items.

At the principal market in Addis Ababa, the most conspicuous natural products include stem bark of the plant with anti-tumour properties locally known as *bissana* or *Croton macrostachyus*, beans of *buna* or *Coffea arabica*, roots of the analgesic *dengetegna* or *Taverniera abyssinica*, berries of the molluscicide *endod* or *Phytolacca dodecandra*, rhizomes of the fumigant *kebericho* or *Echinops kebericho*, leaves of the stimulant plant khat or *Catha edulis*, fermented food from stem of *enset or Ensete ventricosum*, flowers of the anthelmintic *koso* or *Hagenia abyssinica*, *kerbe* or myrrh gum-resin or *Commiphora myrrha*, leaves of *sheferaw* or *Moringa stenopetala*, seeds of *tequr azmud* or black cumin or *Nigella sativa*, and so on. Add to these *zebad* or civet, a unique animal product used since time immemorial as additive to perfumes.

## 31.2 *Boswellia* species (Burseraceae), *etan* (Amharic)

The frankincense tree or shrub belongs to the genus *Boswellia* (Figure 31.1), and like myrrh is a member of the Burseraceae family. There are a dozen or so species in Africa, Arabia and India. Ethiopia has probably the largest number of *Boswellia* species and are known by the following scientific names: *Boswellia papyrifera*, *Boswellia rivae*, *Boswellia neglecta*, *Boswellia pirottae*, *Boswellia ogadensis* and *Boswellia microphylla*. Most of the incense traded in Ethiopia is obtained from the first three species, collectively known locally as *etan* and in trade as Tigray, Ogaden and Borana types of frankincense respectively. Somalia has two species: *Boswellia sacra* (also found in Arabia) and *Boswellia frereana*, which occurs only in Somalia. Only one species, *Boswellia dalzielii*, is known from West Africa, while *Boswellia serrata* is the species that occurs in Asia and used in India as a traditional remedy in ayurvedic medicine for the treatment of inflammatory diseases, rheumatoid arthritis, bronchial asthma, and so on (Siddiqui *et al.*, 2011).

Owing to its pleasant and refreshing aroma, frankincense is burnt in homes, in places of worship and during special occasions, such as weddings. The incense is believed to purify the atmosphere. Local people chew the fresh resins as mouth cleanser and also to suppress the feeling of thirst. Modern uses include flavouring in foods, making of adhesives, chewing gums, perfumes, and so on. Frankincense, when steam distilled, yields a pleasant essential oil used in aromatherapy and perfume manufacture. It is also an important additive in the formulation of skin care products for use against acne, wrinkles and other problems. Frankincense oil combined with myrrh is sometimes recommended for the treatment of throat and larynx infections, mouth thrush, stomach and liver ailments, and so on (de Rapper *et al.*, 2012).

Ethiopia is one of the largest exporters of frankincense in the world, most of it going to Catholic and Orthodox churches, where burning of incense is an essential religious ritual. There is also huge local consumption, in particular due to the common practice of burning incense at coffee and other ceremonies.

However, the ecological zones where these resources occur are very vulnerable to climate change and other human interference, which include increasing population pressures and conversion of woodlands to agricultural lands. Unregulated grazing by cattle is also another problem that hinders natural growth and regeneration. Hence, measures should be taken to protect the fragile eco zones where these valuable resources are found.

## 31.3 *Catha edulis* (Celastraceae), khat

This is an erect, evergreen, glabrous indigenous African shrub or tree 2–25 m high, also occurring in the Arabian Peninsula and Afghanistan (Figure 31.2).

Khat is now widely cultivated, and many types are recognized mainly based on the locality and ecology where it grows. The tender leaves of khat are customarily ingested by chewing. Khat use is highly significant for the social and cultural life of the people, where it is used to welcome and entertain guests, both on happy occasions such as weddings and sad ceremonies such as after burial gatherings. Many believe that chewing khat leaves helps one to get relief from fatigue, to increase alertness and reduce the sensations of hunger. On the negative side, excessive khat chewing causes cardiovascular,

*Novel Plant Bioresources: Applications in Food, Medicine and Cosmetics*, First Edition. Edited by Ameenah Gurib-Fakim.

Figure 31.1 Frankincense market stand in Addis Ababa

Figure 31.2 Bundle of khat leaves ready for sale in a market

(a)

(b)

Figure 31.3 (a) Coffee leaf and (b) typical coffee drink

respiratory, gastrointestinal and central nervous system disorders, for example (Al-Motarreb *et al.*, 2010).

The stimulant effect is mainly caused by two simple alkaloids, known as cathine (**1**) and cathinone (**2**), which have amphetamine-like pharmacological effects.

## 31.4 *Coffea arabica* (Rubiaceae), *buna* (Amharic)

Ethiopians, probably in ancient times, used to eat coffee, by adding the beans to milk or butter and which, by infusion, extracted the bean's power. They also knew how to brew fresh or dried coffee leaves to make herbal tea drink called *amertasa* (Figure 31.3). The gently roasted leaves when brewed give the *quti* drink and the coffee husk would give *hoja*. Coffee found its way from Ethiopia to Yemen, where the art of roasting the beans probably started, and from where coffee radiated to different parts of the world, impacting the world economy and culture in an unprecedented manner. Nevertheless, Ethiopia still remains as the primary centre of genetic diversity of coffee plants, making this country an ideal location for coffee research, particularly in the search for low caffeine-containing varieties. Coffee drink is the most popular beverage in the world, rivaled only by tea. Not only do both have the same stimulant substance (caffeine) in common, they also share a similar fascinating history. In many ways, coffee has played a significant role in spurring modern civilization, as it continues to do the same now.

There is an elaborate ritual and ceremony for the preparation and drinking of coffee in Ethiopia. The key to the unique taste and flavour of coffee is roasting, which produces over 400 compounds resulting from different chemical transformations of the few compounds present in the green bean. Interestingly, caffeine is present both in the green and roasted beans. But there are many other secondary metabolites, such as the antioxidant polyphenols, mainly the chlorogenic acids (CGAs), which make up over 10% of the green coffee bean by weight. The CGAs are one of the main sources of a wide array of compounds that are produced during the roasting process. Coffee can be roasted to three varying degrees: light, medium and dark. Both extremes of roasting are not desired. There is increasing evidence in the scientific literature on the beneficial effects of coffee, including protection against gout through lowering of uric acid, improving memory and some protection against type-2 diabetes (Davies, 2011).

## 31.5 *Commiphora myrrha* (Burseraceae), *kerbe* (Amharic)

In the biblical story of Christmas, the three wise men carried gold, frankincense and myrrh. Myrrh is a thorny shrub or a tree reaching up to 4 m in height. The tree is known in science as *Commiphora myrrha* and is native to the countries of the Horn of Africa and Arabia. This plant is known in these countries by different names.

In Somali, the tree is known as *dhidin* and the gum-resin molmol (Amharic: *kerbe*). In English, myrrh is used as the common name for the tree as well the gum-resin. The bark is silvery and easily peels off in small pieces. Strongly aromatic gum-resin (Figure 31.4), brown in colour, exudes out from natural cracks on the trunk of the tree with a characteristic deep balsamic aroma, not particularly pleasant, but not repulsive either. This exudate has a unique bitter taste that serves as a quick aid to distinguish it from many other related resins. The term gum-resin indicates a product that is partly soluble in water, like a gum, and is also soluble in hydrocarbon solvents, as resins usually are.

Best quality myrrh gum-resin is obtained when the exudates come out of the trunk naturally without tapping. Tapping of myrrh trees is not common in the Ogaden and Bale, two regions that produce most of the myrrh of Ethiopia. But in the Wajir and Mandera regions of Kenya tapping is common and it results in light brown gum-resin of inferior quality. Ethiopia is one of the major suppliers of myrrh in the world. The principal buyers are countries in the Middle East, Europe and China. Myrrh is an important ingredient in Chinese traditional medicine.

(a)

(b)

**Figure 31.4** (a) Myrrh gum-resin and (b) its brown essential oil

Myrrh has been used since ancient times as a medicine and for ceremonial and religious purposes. There is also ample evidence that indicates myrrh was used by the Pharaonic Egyptians for embalming. Owing to its wound healing properties, the British Herbal Pharmacopoeia recommends myrrh tincture in mouthwashes to treat ulcers and inflammation of the gums. The tincture may be prepared by infusing one part gum-resin in five parts alcoholic drink, such as brandy. Then 5–10 drops of the tincture may be added to a glass of water to make an instant mouthwash.

Somali culture has a rich tradition of using myrrh. Its decoction, prepared by boiling about 5 g in a litre or so of water, is used to treat stomach complaints; when burnt in houses and in the bush its smoke chases away snakes; when mixed with powdered charcoal it yields ink used for writing on parchments.

Modern uses include flavouring foods, drinks and confectionary items, and as an additive to products for personal use, such as perfumes, deodorants, shampoos, bath lotions, toilet soaps, toothpastes, mouthwashes, air fresheners, and so on. In commerce, myrrh gum-resin is often processed to give value-added products. It can be extracted by organic solvents such as hexane, which after removal of solvent gives a viscous mass, known in commerce as 'resinoid'. If the solvent used for extraction is alcohol, one obtains an 'absolute'.

One of the most popular value-added products of myrrh is its essential oil, which is obtained by steam or hydro-distillation. Myrrh oil (Figure 31.4) is viscous and dark-brown in colour. It is a popular ingredient in the preparation of skin care products and is also employed as a fixative for perfumes.

## 31.6 *Croton macrostachyus* (Euphorbiaceae), *bissana* (Amharic)

This plant is among the most commonly used medicinal plants of Ethiopia. It is an evergreen tree (Figure 31.5) widespread in Africa from Guinea to Ethiopia, and also found in Angola and Mozambique. In Ethiopian traditional medicine, the root bark is used against tapeworm infestation, and the fruits are used to treat chronic skin diseases locally known as *chirt* and *quaqucha*, eye disease, ringworm, constipation, to induce abortion and as a purgative. Ethanol extracts of the seeds was shown to contain crotepoxide (**3**), which has a significant tumour-inhibitory activity (Kupchan *et al.*, 1968).

## 31.7 *Echinops kebericho* (Asteraceae), *kebericho* (Amharic)

This is a perennial herb, known only from Ethiopia, usually found at altitudes 2300–2600 m in the wild, in particular in the Tigray, Gojam and Shewa regions. *Kebericho's* large rhizomes are considered as one of the most important medicinal plants of Ethiopia and are among the commonly traded items in markets. *Kebericho* root is customarily burnt by placing it on red-hot charcoal, resulting in a characteristic smoke that is used to fumigate homes in order to get

**Figure 31.5** The *bissana* tree (*Croton macrostachyus*)

rid of insects and other disease-causing pests. Women also smoke their body parts with it on account of its healing properties. Two compounds have been identified as the principal components of the roots, namely dehydrocostus lactone (**4**) and costunolide (**5**). It has been shown that these bioactive compounds are present in the smoke, giving some credence to the traditional practice of using the smoke of this plant for fumigation and other purposes (Abegaz *et al.*, 1991).

## 31.8 *Ensete ventricosum* (Musaceae), *enset* (Amharic)

*Enset*, commonly known as wild banana, is a perennial root crop widely cultivated for food in southern Ethiopia, but also found distributed in central and eastern Africa.

However, this plant has been domesticated only in Ethiopia, where it is a staple food for over 10 million people. The plant parts that are used are mainly the massive underground stems and leaf sheaths. The corm and scraped leaf bases are chopped and allowed to ferment for about 20 days. The resultant starch-rich product is called *kocho* (Amh), which is mainly used to make flat *enset* bread served with the meat dish *kitfo* – a spiced beef mince. The refined starch-rich, flour-like product of *enset* locally known as *bulla* is used to make porridge, and in textile industries the extracted starch is used to make adhesives and other products. The gorgeous *enset* leaves (Figure 31.6) are commonly used for wrapping fresh khat leaves, for covering dough when baking bread and for food serving. The stem fibre is processed to make rope. Bioassay-guided chemical investigation led to the isolation of phenylphenalenones (**6**, **7**), which possess anti-tumour and antibacterial, nematicidal and anti-fungal activities (Holscher *et al.*, 1998).

## 31.9 *Eragrostis tef* (Poaceae), *tef* (Amharic)

*Tef* is a small grass of up to 80 cm in height bearing relatively narrow leaves and exceedingly small seeds (Figure 31.7a). *Tef* is

**Figure 31.6** A young *enset* plant

(a)

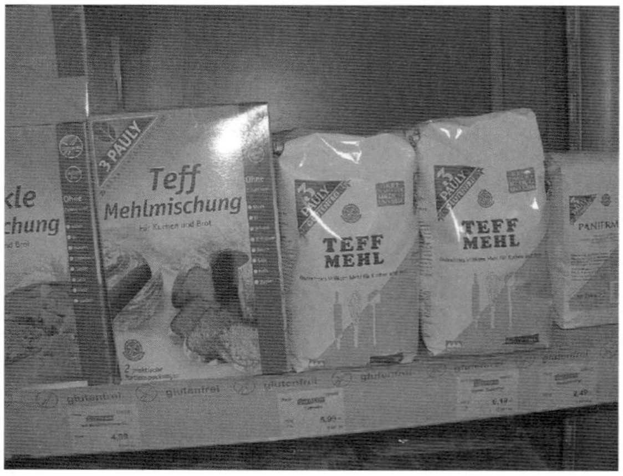


(b)

**Figure 31.7** (a) *Tef* plant and (b) its powder in a supermarket in Leipzig, Germany

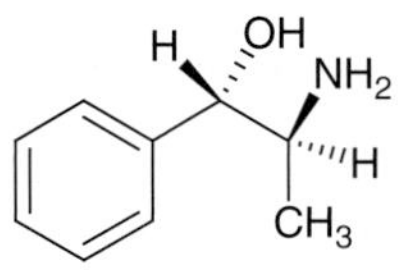

Cathine (1)

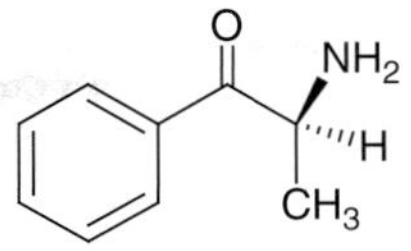

Cathinone (2)

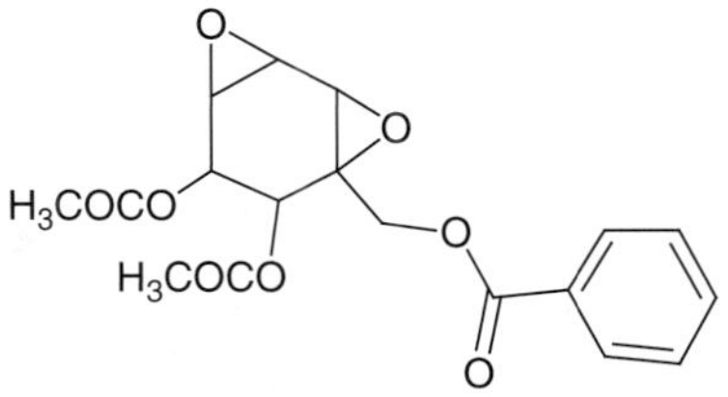

Crotepoxide (3)

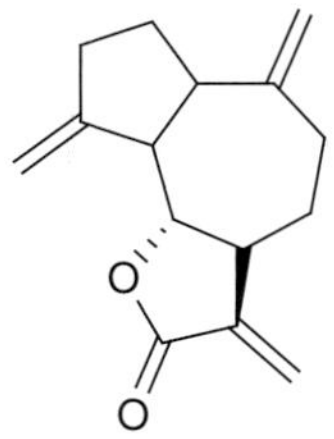

Dehydrocostus lactone (4)

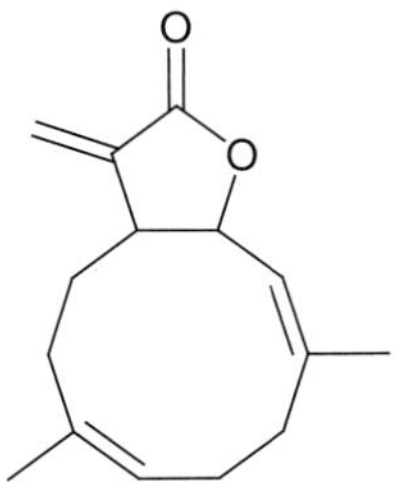

Costunolide (5)

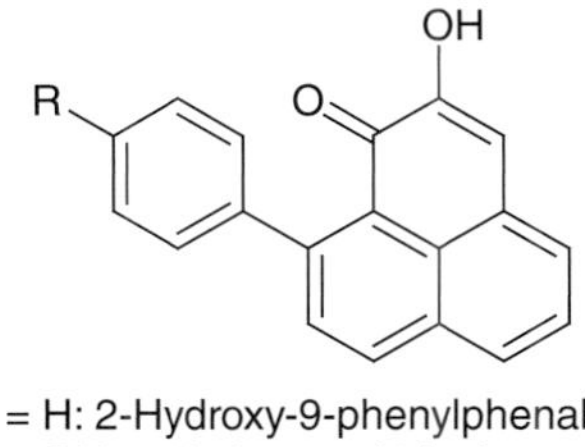

R = H: 2-Hydroxy-9-phenylphenalenone (6)
R = OH: 2-Hydroxy-9-(4-hydroxyphenyl)-phenalen-one(7)

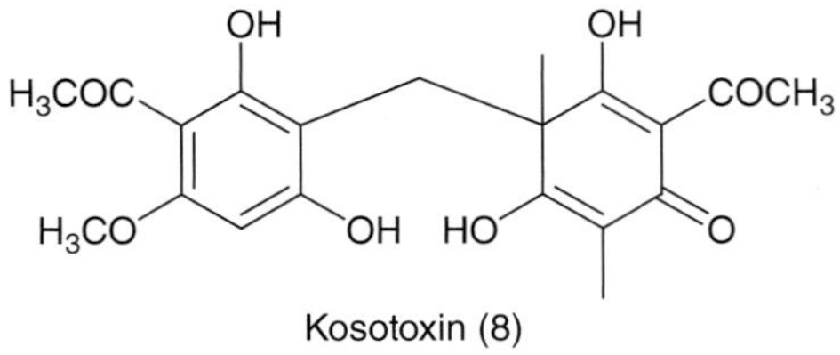

Kosotoxin (8)

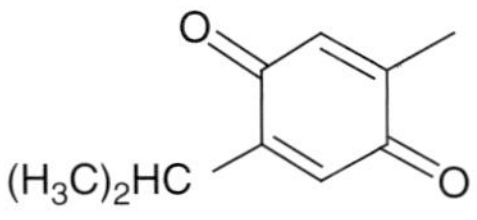

Thymoquinone (9)

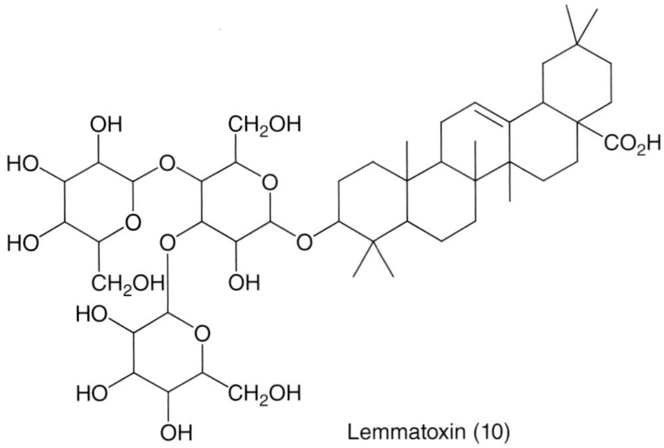

Lemmatoxin (10)

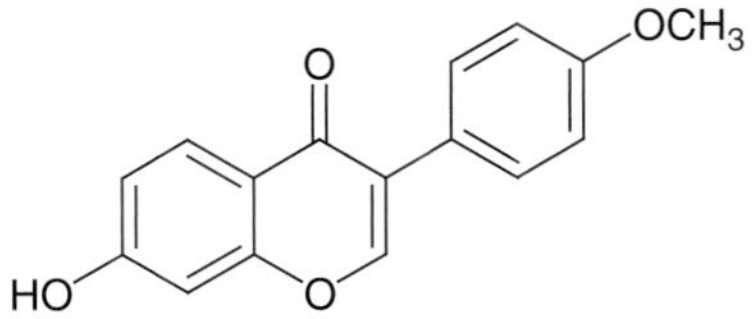

Formononetin (11)

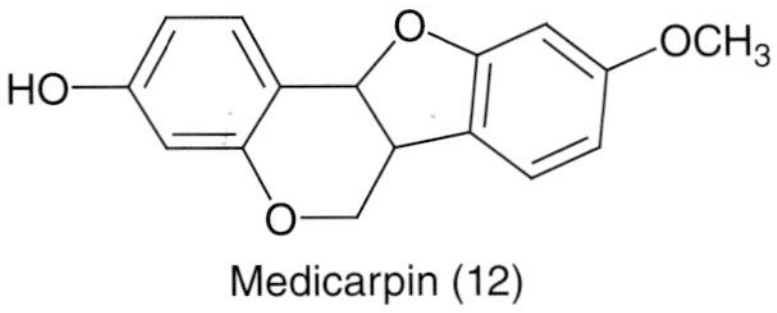

Medicarpin (12)

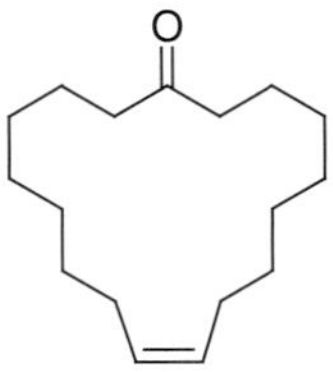

Civetone (13)

indigenous to the Ethiopian highlands and appears to have been cultivated here since antiquity. It has also been introduced to a small extent into other tropical countries (mainly within Africa). It is also grown on a small scale in the USA, due to demand for the crop by migrant Ethiopian communities. *Tef is* mainly used to make *enjera*, a special Ethiopian bread, which is made by fermenting *tef* flour (Figure 31.7b) for 2–3 days and baking it on a special clay pan known as a *mettad*. A flat bread, *qitta*, can also be prepared by baking unfermented *tef* flour. *Tef* contains 9% protein, 2% fat and 74% carbohydrate. It is considered a health food, because it is rich in lysine and unlike wheat it is gluten free, making it desirable for persons who are allergic to wheat.

**Figure 31.8** A street vendor in Addis Ababa, Ethiopia, carrying flowers of *H. abyssinica*

## 31.10 *Hagenia abyssinica* (Rosaceae), *koso* (Amharic)

This plant, the sole member of the genus *Hagenia* (Rosaceae), is commonly known as the African red wood. It is a dioecious tree that grows to a height of about 20 m, and a characteristic tree of the Afro Montane Forests of Central and East Africa. In Ethiopia, the dried female flowers (Figure 31.8) are traditionally used as an anthelmintic, especially to expel tapeworm from humans, which is common in Ethiopia due to the regular traditional consumption of raw beef. Historically, it was one of the most famous African plants that were included in the European pharmacopoeia. Traditionally, the dried powdered female *koso* flowers are steeped in water and the whole decoction is taken by a patient to get rid of the worm *Taenia saginata* from the intestine. Children infested with the parasite resist taking the drug owing to its bitter taste and are sometimes punished by their parents in order to force them to swallow it.

The main nonpolar constituents of the flowers are protokosin and kosotoxin (**8**) (Lounasma *et al.*, 1974). Recently, we investigated for the first time the polar constituents of the female flowers using the non-parasitic nematode *Caenorhabditis elegans* for the anthelmintic assay and found out that the polar compounds, which are mainly flavonoids, are not responsible for the anthelimintic activities, confirming earlier assumptions that the activity is due to the nonpolar kosins, such as kosotoxin (Thomsen *et al.*, 2012).

## 31.11 *Moringa stenopetala* (Moringaceae), *sheferaw* (Amharic)

The most well-known species is *Moringa oleifera* (Figure 31.9a), a plant originally from India, now cultivated in many parts of the world, including tropical Africa. It is known by at least three common English names: 'drumstick tree', owing to its unique long pods; 'horseradish tree', because its roots taste like horseradish; and 'ben oil tree', after the name of the oil obtained from the seeds also used in cosmetics. There is evidence in the annals of scientific literature of the many positive health application of *M. oleifera*. It is rich in vitamins and amino acids, is a cardiac and circulatory stimulant, has anti-tumour, antiepileptic, anti-inflammatory, anti-ulcer, anti-spasmodic, diuretic, anti-hypertensive, cholesterol lowering, anti-diabetic, hepato-protective, anti-bacterial and anti-fungal properties, and is a rich source of vitamins and minerals (Gilani *et al.*, 1994).

(a)

(b)

**Figure 31.9** (a) *M. oleifera* and (b) *M. stenopetala*

*M. oleifera* has recently been introduced in tropical Africa, including in the lowlands of the Ogaden region of Ethiopia. Present in Ethiopia (and also in Kenya) is another unique and distinct species known as *Moringa stenopetala* (Figure 31.9b), which is known by at least three other common names: the cabbage tree (Eng), and locally as *aleko* and *sheferaw*. In south Ethiopia, the local people cook the leaves and eat it as cabbage, particularly when other foods are scarce.

The leaves can also be dried and crushed into a powder and used as a base for soups and sauces. The seeds and root suspensions are as effective as aluminium sulfate in removing dirt, suspended matter and even bacteria from polluted water samples. To treat 20 L of water, about 2 g of crushed seeds are required.

Although there are more scientific papers on the species *M. oleifera*, some evidence is also emerging that shows the leaves of its sister species *M. stenopetala*, have blood glucose and high blood pressure lowering properties (Makonnen *et al.*, 1997). *Aleko* leaves are also used to allay stomach pain and to expel any retained placenta in women. More research work needs to be done on this species in order to find out to what extent its positive attributes match with those of its Indian sister species *M. oleifera*.

## 31.12 *Nigella sativa* (Ranunculaceae), *tequr azmud* (Amharic)

*Nigella sativa* is an annual herb cultivated in Mediterranean countries and in eastern Africa for its black seeds, also known as black cumin. Ethiopia is also among the top producing countries of black seed, locally called *tequr azmud*. It is used as a spice and medicine and to flavour bread. The crushed seeds are mixed with butter, wrapped in a piece of cloth and sniffed to relieve headache. Other local uses include the treatment of stomachache and to induce abortion.

When the seeds are pressed one obtains the popular medicinal black seed oil with a yield of up to 40%. Biological activities of black cumin corroborated by scientific studies include analgesic and anti-inflammatory, anti-fungal, anti-asthmatic, anti-allergic, anti-convulsant, blood glucose lowering, anti-viral and anti-oxidant. The most important biological active compound of black seed, as well as of its oil, is thymoquinone (**9**).

Thymoquinone is a labile compound that is gradually converted to other biologically less active compounds, in particular when exposed to light. It is possible to monitor its level in pressed oil using UV–Vis spectral measurements due to the fact that thymoquinone has maximum absorption at 253 nm (Dagne *et al.*, 2011), which gradually decreases as the amount of thymoquinone decreases.

## 31.13 *Phytolacca dodecandra* (Phytolaccaceae), *endod* (Amharic)

Also known as the soapberry plant (Eng), this is a scrambling, much-branched shrub with stems up to 8 m long, growing from a woody base (Figure 31.10). *Endod* is commonly found in degraded vegetation along river banks and around villages at altitudes between 1500 and 3000 m. In Ethiopia, it is the most widely used soapberry plant for washing clothes. The main molluscicidal saponin, lemmatoxin (**10**), is named after Dr Aklilu Lemma, an Ethiopian scientist who first discovered the molluscicidal effects of the water extract of *endod*.

Lemmatoxin and the related *endod* compounds are derivatives of oleanolic acid substituted in the 3-position with a branched trisaccharide containing two glucose and galactose units. The water extract of *endod* is lethal to snails, which makes it an ideal agent for curbing the spread of the debilitating disease bilharzia. The extract also reduces the population of leeches in ponds and streams. Other

**Figure 31.10** Fruits of *endod*

uses of *endod* include as an abortifacient, against ascariasis, eczema, infertility and liver disease (Parkhurst *et al.*, 1974).

## 31.14 *Sorghum bicolor* (Poaceae), *mashla* (Amharic)

*Mashla*, *zengada* (Amh) or sorghum (Eng) is a cultivated annual grass species growing up to 4 m high (Figure 31.11). *S. bicolor* is the fourth important world cereal after wheat, rice and maize. Sorghum originated in north-eastern Africa, including Ethiopia, where it was domesticated at least 5000 years ago and from where it spread to India and China. Sorghum is now cultivated in many parts of the world. In Ethiopia, it is also used to make the thin flat bread known as *enjera* and for brewing the local beverage *tella*. Like *tef* (*Eragrostis tef*), sorghum is also gluten-free and can be used as a wheat substitute for persons sensitive to gluten. Stems and leaves are used for animal fodder. Sorghum contains 60–75% carbohydrates and 8–13% proteins (Van Wyk, 2005). A potent anti-fungal substance was reported from sorghum seedlings (Suzuki *et al.*, 1998).

**Figure 31.11** Stalk and seeds of *Sorghum bicolor*

## 31.15 *Taverniera abyssinica* (Leguminosae), *dengetegna* (Amharic)

This shrub grows up to 2 m high, with a long and deep-seated tap root. It is so far known only from Ethiopia in the dry bush land areas of the Tigray and Shewa regions. The name of the plant *dengetegna* means medicine for sudden illness, and the root bundles (Figure 31.12) are sold in markets for people to chew and swallow the juice to treat headache, stomachache and fever.

We have studied crude extracts as well as purified substances from the roots and observed significant anti-spasmodic, anti-pyretic and analgesic activities, clearly showing that there are sound pharmacological bases for the ethno-medical use of the roots of this plant to treat stomachache, fever and general pain. Chemical work on this plant by our research group led to the isolation and characterization of the bioactive isoflavonoid formononetin (**11**) and the pterocarpan medicarpin (**12**) (Noamesi *et al.*, 1990).

**Figure 31.13** Picture showing containers for civet made from cattle horn, and used since ancient times in Ethiopia for the export of civet to faraway lands

## 31.16 *Civettictis civetta*: source of civet *zebad* (Amharic)

Civet is among the very limited commercially important natural cosmetics obtained from an animal. It is produced as a natural secretion from the odoriferous glands of the African civet cat (*Civettictis civetta*, known in Ethiopia as *teregn*). Civet is one of the most expensive natural perfumes, which has been supplied to the world mainly by Ethiopia since ancient times. Legend has it that among the prized gifts presented to King Solomon during the Queen of Sheba's historic visit were frankincense, myrrh, spices, ivory and civet. Ethiopia produces 90% of the world's civet, most of which is intended for export. In olden days civet was packed and exported in special containers made from horn of cattle (Figure 31.13). Determining the quality of civet has always been a great challenge for chemists. The main constituent and important marker compound of civet is civetone (**13**), a 17-membered macrocyclic ketone. Also present are a wide range of minor saturated and unsaturated cyclic ketones and alcohols.

One of the most serious challenges in civet trade is the determination of the quality of the product. We have found that the percentage of civetone in civet product can be established by the method of quantitative $^1$H nuclear magnetic resonance. In this method, to a known amount of civet dissolved in $CDCl_3$, a known amount of caffeine is added. Using the molar ratio of the olefinic proton of civet with one of the methyl signals of caffeine, it is possible to deduce the amount of civetone in a civet product.

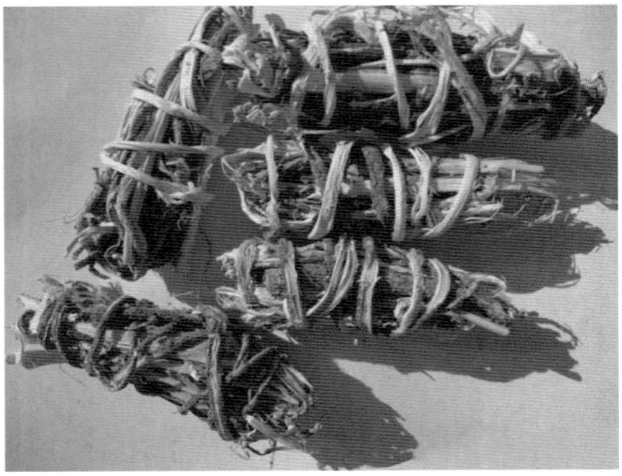

**Figure 31.12** Bundles of roots of *T. abyssinica* as sold in market

## 31.17 Conclusion

The above brief accounts on the selected plants of Ethiopia show that this country hosts several unique and interesting plants. More botanical and chemical information on numerous plants of Ethiopia and other African countries can be readily accessed in a database developed by our research group, which is readily available on the internet at www.alnapnetwork.com.

## References


Abegaz, B.M., Tadesse, M. and Majinda, R. (1991) Distrbution of sesquiterpene lactones and polyacetylenic thiophenes in *Echinops*. *Biochemical Systematics and Ecology*, **19**, 323–328.

Al-Motarreb, A., Al-Habori, M. and Broadley, K.J. (2010) Khat chewing, cardiovascular diseases and other internal medical problems: the current situation and directions for future research. *Journal of Ethnopharmacology*, **132**, 540–548.

Dagne, E., Belete, Y., Kassaye, H. and Melaku, Y. (2011) Application of UV–VIS spectroscopy to evaluate quality of medicinal and edible oils, in *Proceedings of 14th NAPRECA Symposium*, pp. 48–51.

Davies, E. (2011) Chemistry in every cup. *Chemistry World*, **8** (5), 36–39.

De Rapper, S., van Vuuren, S.F., Kamatou, G.P.P. *et al.* (2012) The additive and synergistic antimicrobial effects of select frankincense and myrrh oils – a combination from the Pharaonic pharmacopoeia. *Letters in Applied Microbiology*, **54**, 352–358.

Gilani, A.H., Aftab, K., Suria, A. *et al.* (1994) Pharmacological studies on hypotensive and spasmolytic activities of pure compounds from *Moringa oleifera*. *Phytotherapy Research*, **8**, 87–91.

Holscher, D. and Schneider, B. (1998) Phenylphenalenones from *Ensete ventricosum*. *Phytochemistry*, **49**, 2155–2157.

Kupchan, S.M., Davies, A.P., Barboutis, S.J. *et al.* (1968) Crotepoxide, a novel cyclohexane diepoxide tumor inhibitor from *Croton macrostachys*. *Journal of the American Chemical Society*, **90**, 2982–2983.

Lounasma, M., Widen, C.J. and Huhtikangas, A. (1974) Phloroglucinol derivatives of *Hagenia abyssinica*. II. The structure determination of kosotoxin and protokosin. *Acta Chemica Scandinavica B*, **28**, 1200–1208.

Makonnen, E., Hunde, A. and Damecha, G. (1997) Hypoglycemic effect of *Moringa stenopetala* aqueous extract in rabbits. *Phytotherapy Research*, **11**, 147–148.

Noamesi, B.K., Bogale, M. and Dagne, E. (1990) Intestinal smooth muscle spasmolytic actions of the aqueous extract of the roots of *Taverniera abyssinica*. *Journal of Ethnopharmacology*, **30**, 107–113.

Parkhurst, R.M., Thomas, D.W., Skinner, W.A. and Cary, L.W. (1974) Molluscicidal saponins of *Phytolacca dodecandra*: lemmatoxin. *Canadian Journal of Chemistry*, **52**, 702–705.

Siddiqui, M.Z. (2011) *Boswellia serrata*, a potential antiinflammatory agent: an overview. *Indian Journal of Pharmaceutical Sciences*, **77** (3), 255–261.

Suzuki, Y., Kono, Y., Inoue, T. and Sakurai, A.A. (1998) Potent antifungal benzoquinone in etiolated sorghum seedlings and its metabolites. *Phytochemistry*, **47** (6), 997–1001.

Thomsen, H., Reider, K., Franke, K. *et al.* (2012) Characterization of constituents and anthelmintic properties of *Hagenia abyssinica*. *Scientia Pharmaceutica*, **80**, 433–446.

Van Wyk, B.E. (2005) Food plants of the world: *Pimpinella* to *Tropaeolum*, in *Food Plants of the World* (ed. R. Ferreira), Briza Publications, Pretoria, pp. 296–370.

# 32 Aromatic Plants from Reunion Island (France)

Anne Bialecki and Jacqueline Smadja

Laboratoire de Chimie des Substances Naturelles et des Sciences des Aliments, Faculté des Sciences et Technologies, Université de la Réunion, La Réunion, France

## 32.1 Introduction

Research work on aromatic plants began in 1974 within two distinct chemistry laboratories of Reunion Island University: the Organic Chemistry Laboratory and the Agrochemical Laboratory.

In the first one, the work was devoted to the survey and study of several species of aromatic plants from Reunion Island and included the famous 'Geranium Bourbon' and the 'Vetiver Bourbon'. The identification of their chemical composition and their valorization were done either by hemisynthesis or through the study of climatic influence on their composition. This research work performed over several years has yielded several PhD theses as well as many academic publications (The numerous PhD theses and academic papers are proof of the relevance of these works (Conan *et al.*, 1976, Conan, 1977a,1977b; Pieribattesti *et al.*, 1979, 1980a,1980b, 1981, 1984; Pieribattesti, 1982; Smadja *et al.*, 1986, 1988, 1989, 1990a,1990b; Smadja, 1987, 1990, 1991, 1994; Smadja and Gaydou, 1991; Viano *et al.*, 1991a,1991b; Bombarda, 1994; Bombarda *et al.*, 1996; Vernin *et al.*, 1998). From 1989 to 1992, the research team of the Organic Chemistry Laboratory also collaborated with the 'Société Chatel' and the Nestlé Research Centre of Vevey. This collaboration led to the identication of the aromatic components of an endemic orchid of Reunion Island, *Jumellea fragrans*, commonly named Faham, widely used for flavouring rums. This chemical composition elucidation needed sophisticated analytical means as bidimensional chromatography (Shum-Cheong-Sing, 1992; Shum-Cheong-Sing *et al.*, 1992).

With respect to the Agrochemical Laboratory, from 1984 to 1998 the research on aromatic plants was focused on several species growing in Reunion: *Callistemon citrinus*, *Anethum graveolens*, *Origanum majorana*, *Salvia officinalis*, *Psidium cattleianum*, *Psidium guajava* and some species of th Zingiberaceae family. The main purpose was to optimize the extraction process of essential oil and to ensure the preservation of the fresh fruits (Chane-Ming *et al.*, 1998; Chane-Ming, 1999; Paniandy, 1999; Vera and Chane-Ming, 1998, 1999; Vera *et al.*, 1999; Paniandy *et al.*, 2000).

In 1998, the two chemistry laboratories merged to form the current Chemistry Laboratory of Natural Substances and Food Sciences (LCSNSA). In this new laboratory, research on essential oils and other volatile compounds is still actively going ahead. Thus, after a brief presentation of the severe economic situation in the sector of fragrant and aromatic plants in Reunion, this chapter will provide an overview of all the work done in the LCSNSA from the year 1998. Some examples will be described, to highlight the research topics devoted to plant volatiles and developed by the laboratory. Most of the results presented in this paper were obtained from a close collaboration between the researchers, botanists and biologists who have been actively working in environmental protection and biodiversity conservation. The reference section will guide the reader in getting a bigger idea of what has been realized in our laboratory.

## 32.2 Aromatic plant production: economic data

The perfume and aromatic plant sector has weakened over the past 30 years. The production of *Geranium*, which is now estimated to be around 2 tons per year, used to be over 9 tons in 2000. The production of vetiver is almost non-existent and has regressed from 35 tons in 1970, to a mere 50 kg or less! Minor quantities of various essential oils, such as pink pepper, cryptomeria, camphor wood and lemongrass, are also produced, quite often anecdotally.

### 32.2.1 Vetiver Bourbon

The production of Vetiver Bourbon essential oil (*Chrysopogon zizanioides* (L.) Roberty, formerly *Vetiveria zizanioides* (L.) (Poaceae)), traditionally from the highlands of Reunion Island, has almost disappeared since 1998. Only 1 ha of culture has persisted. This production began to increase slightly in 2000: from 250 kg in 2000, it was close to 400 kg in 2001 with 25 growers. It may have reached 800 kg in 2002, but the tropical cyclone Dina in January 2002 damaged all the fields. During the subsequent following 3 years, no production was carried out. Yet the Oil of the Vetiver Bourbon is priced at around 300 €/kg, whilst the Oil of Vetiver from Haiti ranged from €75 to €120.

### 32.2.2 Geranium Bourbon

The situation for the Bourbon Geranium is less dramatic than that of vetiver. In 2000, the production of the Oil of Geranium Bourbon or Geranium rosat extracted from *Pelargonium graveolens* L. (Geraniaceae) was estimated to be around 9 tons emanating from the cooperative of 450 growers over a cultivated area of 500 ha. The production decreased significantly to 3.5 tons in 2002 because of the tropical cyclone Dina, from which it has never recovered. It got reduced further to 2 tons in 2011 with 120 growers for a cultivated

*Novel Plant Bioresources: Applications in Food, Medicine and Cosmetics*, First Edition. Edited by Ameenah Gurib-Fakim.

area of 110 ha. The weather during the recent years has played havoc with the production, and there is a general lack of motivation among the smallholders, especially as there are other more lucrative productions. The essential oil of Bourbon Geranium at 140 €/kg has to compete with that coming from China priced at 70–160 €/kg and that from Egypt (35–100 €/kg).

### 32.2.3 Pink pepper

The annual production of *Schinus terebinthifolius* (Anacardiaceae), commonly known as wild pepper or pink pepper in Reunion, is unstable and varies from 40 to 120 tons (9 €/kg). The market is strictly for export and the oil is produced under supercritical (SC) $CO_2$ extraction. Hydrodistilled pink pepper essential oil does not sell because of its organoleptic properties. The one produced from the SC $CO_2$ extraction remains a reference and is two to three times more expensive. At the present time, the Coopérative Agricole des Huiles Essentielles de Bourbon (CAHEB) has, nevertheless, a stock of 100 kg of this essential oil (210 €/kg).

### 32.2.4 Lemongrass

From 2011, because of its repellent activity on mosquitoes, lemongrass essential oil has been produced for local domestic consumption and the tourist market (120 €/kg). Two species of lemongrass are cultivated in Reunion. The first is *Cymbopogon citratus* (DC.) Stapf. (Poaceae), also called verbena in India or citronnella in Madagascar. The major components of this essential oil are citral, geranial, nerol, neral, limonene, linalool, β-caryophyllene and myrcene. The second species is *Cymbopogon nardus* (L.), known as citronnella in Ceylan. Its essential oil is mainly composed of citronellal, geraniol and limonene. The production of the essential oil of *C. citratus* is about 100 kg/year, while that of *C. nardus* is around 50 kg/year.

### 32.2.5 Cryptomeria

Essential oil of *Cryptomeria japonica* var. *japonica* (Cupressaceae) or Japanese cedar has been produced for the past 4 years. It is extracted from felled trees or prunings of the Forest National Office. The current stock of essential oil is about 50 kg/year with a sale price of 120 €/kg. Tourists are the largest consumers of this essential oil.

### 32.2.6 Camphor wood

The essential oil of *Cinnamomum camphora* (Lauraceae) or camphor, whose production varies between 50 and 100 kg/year, is exclusively destined for the tourist market (120 €/kg).

### 32.2.7 Other essential oils

Some essential oils, such as ginger, mango ginger, curcuma, citronnella geranium are also produced, albeit in small yields of a few kilograms per year and also destined for the tourist market.

## 32.3 Extraction techniques used in Reunion Island

See Gauvin (2006), Smadja (2009), Tandon (2008), Ormeno *et al.* (2011), Chemat *et al.* (2003, 2004) and Visinoni *et al.* (2006).

### 32.3.1 Conventional techniques

#### 32.3.1.1 Direct steam distillation

***Principle***

In direct steam distillation, the plant material is distilled with steam generated outside the tank, in a steam generator or boiler. In water and steam distillation, the plant material is supported on a perforated grid above the steam inlet; the steam is at atmospheric pressure, and hence its maximum temperature is 100 °C, whereas steam in a modern pressure boiler operating at, for example, 50 psi pressure, will have a temperature correspondingly higher. Moreover, there is no limitation to the steam generation when an external boiler is used as a source of steam.

***Users and plants concerned***

Steam distillation under pressure is the technique chosen by CAHEB to extract Vetiver Bourbon essential oil from the dried roots. Indeed, at atmospheric pressure, it is too hard to obtain this essential oil from roots.

#### 32.3.1.2 Hydrodistillation

***Principle***

Hydrodistillation is the simplest and oldest process available for obtaining essential oils from plants. Hydrodistillation differs from steam distillation mainly in that the plant material is almost entirely covered with water in the still, which is placed on a furnace. An important factor to consider in water distillation is that the water present in the tank must always be enough to last throughout the distillation process, otherwise the plant material may overheat and char. In this method, water is made to boil and the essential oil is carried over to the condenser with the steam that is formed. Water-distilled oil is slightly darker in colour and has much stronger still notes than oils produced by other methods.

***Users and plants concerned***

Field distillation at atmospheric pressure is mostly used by small-scale producers. The latter use the hydrodistillaton technique for extracting Geranium Bourbon essential oil from fresh flowering tops and leaves.

#### 32.3.1.3 Distillation with cohobation

***Principle***

Cohobation is a technique that can be used for hydrodistillation or for water and steam distillation. It uses the process of returning the distillate water to the still after the oil has been separated from it so that it can be reboiled. This method has been developed for oils that have partial solubility in water. Indeed, although most of the essential oils have finite solubility in water, some oils, like those of rose, lavender and geranium, have comparatively higher solubility. In such extractions, the loss with the outgoing water of distillation can become alarmingly high. This problem can be solved by returning the condensate water from the separator back to the still; this is known as cohobation. It is evident that this cannot be done with steam distillation as the water level in the still will keep building up to continuous steam injection.

*Users and plants concerned*

Distillation at atmospheric pressure with cohobation is used at CAHEB. Essential oils of pink pepper, gingember, cryptomeria and lemongrass are extracted in this way. This technique is also the method of choice at the LCSNSA. The hydrodistillation with cohobation is then carried out thanks to a modified Clevenger apparatus for all the plants studied.

### 32.3.2 Innovative techniques

#### 32.3.2.1 Solvent-free microwave extraction

*Principle*

Microwaves are increasingly being used as the heat source to assist the extraction of volatile compounds. The microwave extraction methods developed include solvent-free microwave extraction (SFME), microwave-accelerated Dean–Stark, microwave steam diffusion and microwave-integrated Soxhlet. The SFME method in particular is the most commonly used. This method involves placing the fresh plant material in the microwave reactor without addition of water or any other solvent. The internal heating of the water within the sample distends its cells and leads to rupture of the glands and oleiferous receptacles. This process thus frees essential oil, which is evaporated by the *in-situ* water of the plant material. A cooling system outside the microwave oven continuously condenses the vapours, which are collected on specific glassware. The excess of water is refluxed back to the extraction vessel in order to restore the *in-situ* water to the sample.

*Users and plants concerned*

The microwave extraction of essential oil at atmospheric pressure is a technique developed by LCSNSA 10 years ago. By this technique, several essential oils from aromatic herbs and seeds, such as basil, mint, thyme and cardamom, were extracted and studied.

#### 32.3.2.2 Solid phase micro-extraction

*Principle*

An important advance in static headspace analysis was the development of solid phase micro- extraction (SPME), which is a true 'green revolution' in sample preparation techniques. Cheap, fast, simple and solvent free, this method enables the collection of volatiles at detection limits in the parts per billion by volume region. SPME is based on ad/absorption and desorption of volatiles from an inert fibre coated with different types of ad/absorbents. The fibre is attached within the needle of a modified syringe and volatiles can be sampled by inserting the needle through a septum of a headspace collection container and pushing the plunger to expose the fibre. Following equilibration between the fibre and the volatile sample (a few minutes to half an hour), the fibre is retracted into the needle and can be transferred to a gas chromatograph injector for direct thermal desorption. SPME fibres can be reused approximately 100 times. By carefully selecting the polarity and thickness of the fibre coating, compounds of different polarity and volatility ranging from high-boiling point or semi-volatile to volatile compounds can be sampled. SPME does not provide directly quantitative information, since the amount adsorbed depends on the fibre-coating affinity for the compound in addition to its concentration in the headspace. Thus, the results obtained may misrepresent some volatiles and overrepresent others. However, quantification of volatiles by SPME may be possible by the application of internal or external calibration.

*Users and plants concerned*

SPME is a fast and easy method used in LCSNSA for trapping volatiles from fragrant parts of plants (leaves, flowers, fruits, etc.).

## 32.4 Analysis of essential oils and plant headspace in the Chemistry Laboratory of Natural Substances and Food Sciences

See Gauvin (2006) and Smadja (2009).

Exhaustive analysis of essential oils and headspace combines qualitative and quantitative gas chromatographic methods using two compulsory types of columns: polar (e.g. Supelcowax) and nonpolar (e.g. SPB-5).

Gas chromatography equipped with a flame ionization detector (GC-FID) is used for quantitative analyses. With such a detector, organic compounds are ionized in a hydrogen–air flame, producing a signal proportional to the mass flow of carbon. They are commonly used for quantitative analyses because of their wide linear dynamic range, their very stable response and their high sensitivity with detection limits of the order of picograms and nanograms per compound.

Qualitative analyses are performed by gas chromatography combined with mass spectrometry (GC–MS). This is the most popular type of detector for routine plant volatile GC analysis. In the mass spectrometer, compounds exiting the GC column are ionized by electron impact and the resulting positively charged molecules and molecule fragments are selected according to their mass-to-charge ($m/z$) ratio by entering a quadrupole mass filter. Total ion chromatograms are obtained, which provide information on the retention time of each compound and its mass spectrum consisting of a characteristic ion fragmentation pattern.

For identification of compounds in GC–MS analysis, suggestions are obtained from popular computerized mass spectral libraries, such as Wiley (McLafferty, 2009), NIST (2011) or MassFinder (Koenig *et al.*, 2008) and other databases not computerized developed by Adams (2001) or Joulain and König (1998). However, because MS characteristics are not always sufficiently unique, GC retention indices should be used as a second dimension for data comparison. The retention index (RI) or Kovats index (KI) is a relative value determined by comparing the retention time of the compounds with retention times of two standard compounds. Several homologous series of organic compounds can be used as standards, but *n*-alkanes have been used exclusively. The formula for RI calculation depends on temperature conditions:

for isothermal conditions:

$$\mathrm{RI}(x) = 100\frac{\log(t_{R(x)}) - \log(t_{R(z)})}{\log(t_{R(z+1)}) - \log(t_{R(z)})} + 100z$$

for linear temperature program conditions:

$$\mathrm{RI}(x) = 100\frac{t_{R(x)} - t_{R(z)}}{t_{R(z+1)} - t_{R(z)}} + 100z$$

where RI($x$) is the retention index of compound $x$, $z$ is the number of carbon atoms in alkane $z$, $t_{R(x)}$ is the retention time of compound $x$, $t_{R(z)}$ is the retention time of alkane $z$, $t_{R(z+1)}$ is the retention time of alkane $z + 1$.

To use the retention index system, a measure of gas chromatograms of all $n$-alkanes covering the desired retention time range needs to be done; for example, to investigate monoterpenic and sesquiterpenic compounds, a range of $C_9$–$C_{26}$ is needed, so all terpene peaks will elute between known alkanes. In order to measure the complete alkane pattern, with only a single injection, mixtures of $n$-alkanes are usually prepared from pure chemicals at 5% concentration in pentane. The measurement of the alkane pattern obviously has to be carried out under exactly the same experimental conditions as the future analyses. Few compilations of retention indices are currently available (Davies, 1990; Kondjoyan and Berdague, 1996; Joulain and Konig 1998; Adams, 2001; Koenig *et al.*, 2008).

Thus, by combining MS data and retention indices, a complex mixture of volatile compounds from plants is systematically and rapidly screened for compounds that are already known.

Particularly demanding is the identification of unknown (new) compounds that cannot be identified by computer matching with MS libraries and RI. In most cases, the isolation of this constituent and its investigation by one- (1D) and two-dimensional (2D) nuclear magnetic resonance (NMR) techniques is inevitable. For the isolation, liquid chromatographic techniques such as high-performance liquid chromatography and thin-layer chromatography are usually applied.

## 32.5 Identification of volatile compounds at the Chemistry Laboratory of Natural Substances and Food Sciences

### 32.5.1 Essential oils

#### 32.5.1.1 Development of new extractive methods

In the LCSNSA, studies on extraction of essential oil using the microwave heating technique started in 1999. Several attempts were conducted to develop the process. Thus, the SFME, combining of microwave heating and dry distillation, was performed at atmospheric pressure without any added solvent or water. In this method, isolation and concentration of volatile compounds were performed in a single stage. SFME has also been compared with a conventional technique, hydrodistillation, for the extraction of essential oil from three aromatic herbs belonging to the Lamiaceae family: basil (*Ocimum basilicum* L.), garden mint (*Mentha crispa* L.) and thyme (*Thymus vulgaris* L.). The essential oils extracted by SFME for 30 min were quantitatively (yield) and qualitatively (aromatic profile) similar to those obtained by conventional hydrodistillation for 4.5 h. The SFME method yields an essential oil with higher amounts of more valuable oxygenated compounds, and allows substantial savings of costs, in terms of time, energy and plant material.

Another example of study was the extraction of cardamom essential oil (*Elletaria cardamomum* L., Zingiberaceae) through SFME. A multivariate study based on a central composite design (CCD) was used to evaluate the influence of three major variables affecting the performance of the SFME of cardamom seed. The yield and the composition of the essential oils from the dry cardamom seeds obtained by SFME were determined and compared with those obtained by the traditional hydrodistillation. Statistical treatment of the results provided by the CCD revealed that the selected parameters of extraction time, irradiation power and moisture content of the seeds were significant. The essential oils were analysed by GC–MS. Essential oils provided by SFME are dominated by the oxygenated fraction, which is the more valuable and composed of highly odoriferous aromatic compounds. Cardamom seeds treated by SFME and hydrodistillation were observed by scanning electronic microscopy. Micrographs provide more evidence of the destruction of cardamom seeds treated by SFME, in contrast to conventional hydrodistillation.

To sum up, SFME is a green technology and appears as a good alternative for the extraction of essential oils from aromatic plants. All of these studies permitted the holding of three patents and many academic papers (Lucchesi *et al.*, 2004a–c, 2007; Lucchesi, 2005).

#### 32.5.1.2 Comparison of five flavouring *Citrus*

Investigations on some *Citrus* species (Rutaceae) from Reunion Island (Figure 32.1) were also carried out between 2001 and 2005. The Petitgrain essential oils of *Citrus latifolia* Tanaka (two varieties, Lime Tahiti and Lime de Perse) and *Citrus limon* (L.) Burm.f. (three varieties, Meyer, Eureka, Doux) have been analysed by GC–MS. Some major components found on the Supelcowax column are β-pinene, sabinene, limonene, citronellal, linalool, neral, geranial and neryl acetate, in different amounts according to variety. The chemical composition of each variety is responsible for the typical olfactory notes. The two varieties of lime (Lime Tahiti, Lime de Perse) belong to the limonene/citral chemotype. A lemon variety (citron Eureka) has this lime chemotype plus β-pinene, while the two other lemon varieties (citron Meyer, citron Doux) have a limonene/citronellal/linalool chemotype. However, citron Doux has, in addition, an appreciable amount of sabinene. The key component of lime is citral, which lends the leaf oil its fruity lemony-fresh character. Limonene, the key component of lemon, is responsible for typical *Citrus* and pine notes. Moreover, the presence of particular compounds (β-pinene, sabinene) in some lemon varieties (citron Eureka, citron Doux) is significant for their odour and taste profile. In both lime and lemon varieties, esters such as geranyl acetate and neryl acetate contribute to the fruity and floral flavour (Rondeau *et al.*, 2003; Smadja *et al.*, 2005).

#### 32.5.1.3 Comparison of two scented geraniums

At the request of the CAHEB, a project was carried out by the LCSNSA in 2003 with a view to comparing the essential oils emanating from two different cultivars of geraniums. The essential oils

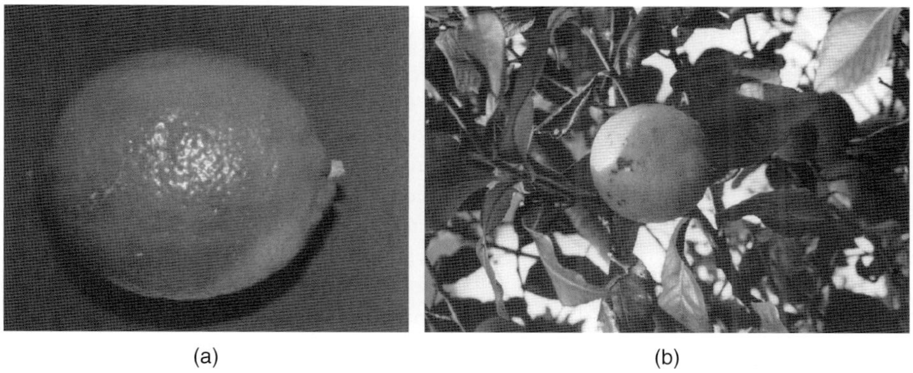

**Figure 32.1** (a) *Citrus latifolia* Tanaka var. Persion lime and (b) *Citrus limon* (L.) Burm.f. var. Eureka. *Photo*: Philippe Rondeau. Reproduced with permission

were analysed by GC–MS. The specific odour attributes of each single constituent were correlated with the olfactory impression of the two essential oils. The scent of the essential oil extracted from the well-known rose-scented geranium cultivar, *Pelargonium* cv. Rosé (Figure 32.2), is based on the rosy, fruity, minty and faintly citrusy aromas of the main components, citronellol (21.9%), geraniol (18.3%), linalool (16.0%), citronellyl formate (11.6%) and isomenthone (7.6%), accompanied by minor compounds of olfactory significance, including geranyl formate (4.0%), nerol (1.7%), geranyl tiglate (1.6%) and neryl propanoate (1.5%) in particular. The volatile constituents of the essential oil obtained from the second geranium cultivar, *Pelargonium* sp. (Figure 32.2), were dominated by monoterpene hydrocarbons, of which the most abundant is *p*-cymene (35.8%). This oil had a very unusual, although very pleasant, 'citrusy-peppery-spicy' and herbaceous scent, somewhat pungent, and rather reminiscent of thyme (Gauvin *et al.*, 2004a).

#### 32.5.1.4 Characterization of *Ayapana triplinervis* essential oil

The study of this plant has been the theme of one PhD thesis. The research work was performed at the LCSNSA between 2006 and 2008. Three specimens of *Ayapana triplinervis* R.M. King & H. Rob (Asteraceae), commonly known as Ayapana (Figure 32.3), were collected from two different locations (flowering: samples 1 and 3; vegetative: sample 2). Their leaf oil composition was investigated with a view to correlating the composition with the development stage of the plant and its geographic location. Analysis by GC-FID and GC–MS allowed the identification of 39 constituents accounting for 97.1–98% of the oil composition.

The three essential oil samples, all obtained by hydrodistillation, showed a high percentage of the aromatic compound thymohydroquinone dimethyl ether (89.9–92.8%). All other minor components remained more or less unchanged both qualitatively and quantitatively with respect to the stage of growth. On the contrary, variations were observed with geographic distribution. The geographical variation of the chemical composition of the volatile oil of *A. triplinervis* from several sites in the world was also briefly discussed (Gauvin-Bialecki and Marodon, 2008).

#### 32.5.1.5 Characterization of *Toddalia asiatica* essential oil

The analysis of the essential oil of *Toddalia asiatica* (Rutaceae) (Figure 32.4) has been investigated. The aim of that study was to characterize *T. asiatica*, growing wild in Reunion Island, through the composition of its essential oil in order to establish whether there was homogeneity in the composition or whether there was chemical variability among specimens from different geographical regions of the world. Thus, upon comparison of the results obtained for plant material from Reunion Island with those reported for the same species from India, profound differences were observed in the composition of essential oils: monoterpenes such as linalool (12.4%), α-borneol (9.4%), *p*-cymene (8.8%), α-terpineol (8.6%) are reported as the characteristic components of Indian specimen, while the two aldehydes prenal (24.6%) and 2-hydroxy-3-methyl-but-2-enal

**Figure 32.2** (a) *Pelargonium* cv. Rosé and (b) *Pelargonium* sp. *Photos*: LCSNSA

**Figure 32.3** *Ayapana triplinervis. Photo*: Claude Marodon. Reproduced with permission

**Figure 32.4** *Toddalia asiatica. Photo*: LCSNSA

(22.8%) are the most important volatiles of the Reunion Island specimen. Moreover, it is well known that aldehydes usually play a significant role in odour composition. So, these two major constituents could be putatively considered as key odorants of *T. asiatica* essential oil principally characterized by choking, pungent, powerful, aldehydic and fruity notes (Smadja *et al.*, 2010a).

#### 32.5.1.6 Chemotaxonomic studies

Among the many ongoing scientific projects, the phytochemical study of the species of *Psiadia* (Asteraceae) and *Melicope* (Rutaceae) has also been envisaged by the LCSNSA. This was done with a view to establishing their chemotaxonomic profiles and has been performed over a period of 10 years. For all these species, the chemical compositions of the leaf essential oil obtained by hydrodistillation were examined. The chromatographic analyses by GC-FID and GC–MS allowed the quantification and identification of their volatile constituents. However, some of them could not be identified by computer matching with MS libraries (laboratory-made and commercial) and their retention index. So, their complete structure was elucidated by means of mono- and bidimensional NMR spectroscopic techniques after isolation and purification by repeated column chromatography. In this context, several new molecules were identified.

***Psiadia* genus** Nine endemic species of the genus *Psiadia* from both Reunion Island and Mauritius have been studied. The chemical profiles of the essential oils were established. The essential oils of four wild species of *Psiadia* endemic to Reunion Island – *Psiadia anchusifolia, Psiadia argentea, Psiadia boivinii, Psiadia salaziana* (Figure 32.5) – were obtained by hydrodistillation and subjected to detailed GC–MS analysis in order to determine possible similarities among them and also the differences in their chemical compositions depending on the stage of the life cycle (vegetative and flowering). A total of 57 compounds were identified, including the two unusual components 7,7-dimethyl-2-methylenebicyclo[3.3.1]heptan-6-ol acetate (**1**) and 6,6,8,9-tetramethyltricyclo[3.3.3.0]-undec-7-en-2-ol (**2**), which, until now, have been found only in species of the genus *Psiadia*. For *P. anchusifolia, P. argentea* and *P. salaziana*, the qualitative and quantitative composition of the oils appeared to be constant in the different stages of their development. In contrast, considerable differences were found to exist in the composition of essential oils of *P. boivinii*. The occurrence of these unusual terpenoids in these four species supposes that the latter have the same biosynthetic pathway and it may also suggest that the two new molecules could be used as a chemotaxonomic tool for the characterization of some *Psiadia* species (Gurib-Fakim *et al.*, 1995, 2000; Gauvin *et al.*, 2004b; Govinden-Soulange *et al.*, 2004; Gauvin and Smadja, 2005; Gauvin-Bialecki and Smadja, 2007).

***Melicope* genus** To evaluate the interpopulation variability of volatile compounds in *Melicope obscura* (Figure 32.6), four samples representing four populations were collected all over the distribution area of the species on Reunion Island (Indian Ocean). The samples were extracted by hydrodistillation, and analysed using GC-FID and GC–MS techniques. The study revealed that, in the four essential oils obtained, oxygenated sesquiterpenes were one of the major chemical classes (9.2–35.2%), mainly consisting of a new compound, (+)-6-ethenyl-2-hydroxy-6,10-dimethylundeca-2,9-dien-4-one (**3**), called melicopenol (8.6–30.1%). The compound was isolated by column chromatography and identified by spectral analyses including 1D- and 2D-NMR (Smadja *et al.*, 2010b).

The three new compounds isolated from *Psiadia* and *Melicope* species are presented in Figure 32.7.

### 32.5.2 Headspace

The studies performed in headspace relate to two thematic areas: the study of some plants–pollinators interactions and the research of odoriferous molecules. The first thematic area has been conducted at the request of the researchers of the UMR C_53 PVBMT of Reunion University.

#### 32.5.2.1 Plants–pollinators interactions

Environment protection and biodiversity conservation requires a better knowledge of inter-specific interactions between plants and their pollinators. Thus, in order to better understand the reproductive biology of the tropical orchids on Reunion Island, biologists and chemists from the University of Reunion investigated the pollination syndromes – that is, floral morphology, breeding system,

**Figure 32.5** The four *Psiadia species* studied: (a) *Psiadia anchusifolia*; (b) *Psiadia argentea*; (c) *Psiadia boivinii*; (d) *Psiadia anchusifolia*. *Photos*: LCSNSA

**Figure 32.6** *Melicope obscura*. *Photo*: LCSNSA

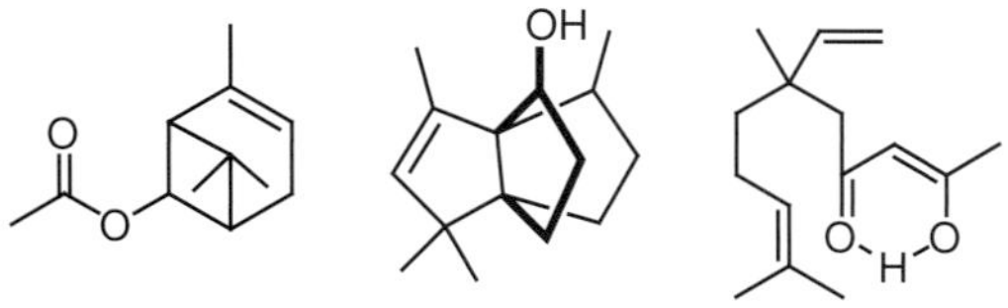

1 2 3

**Figure 32.7** New compounds isolated from *Psiadia* and *Melicope* species

**Figure 32.8** *Jumellea stenophylla*. *Photo*: Claire Micheneau. Reproduced with permission

pollinator diversity, floral scent profile and fruity success – of orchid species of the genera *Angraecum*, *Jumellea* and *Bulbophyllum*. Floral scents were studied using headspace SPME combined with GC–MS analyses.

### *Auto-pollination in an endemic orchid: Jumellea stenophylla*

Since Darwin, long-spurred angraecoid orchids have been known for their fascinating evolutionary relationship with long-tongued hawkmoths (Sphingidae) on Madagascar. The reproductive biology of the long-spurred endemic *Jumellea stenophylla* (Figure 32.8) on Reunion was studied. Despite the species exhibiting flowers with the typical sphingophilous pollination syndrome (i.e. spur length averaged 137.9 mm, mean nectar volume was 6.1 μL and nectar concentration was 10.7% sugar in sucrose equivalent), it does not require pollinators to develop fruits. Compared with other hawkmoth-pollinated orchids, flower longevity is very short, lasting less than 5 days, and the species do not emit the characteristic strong and sweet scent at dusk. Fruit set ranged from 66.7 to 83.9% when pollinators were excluded, and 56–77.5% under natural conditions. Auto-pollination is a consequence of structural modifications. On Reunion, such a breeding system is not rare within long-spurred species, and seems linked to the absence of specific pollinators during island colonization and species establishment.

*J. stenophylla* emits fragrance on a nocturnal rhythm. Thus, the volatile compounds extraction was performed at night. In these conditions, *J. stenophylla* fragrance is largely dominated by indole (83.3–86.6%), along with lower levels of hydrocarbons, aldehydes, ketones, mono- and sesquiterpene hydrocarbons and aromatic compounds, including benzaldehyde (3.4–3.9%), methylbenzoate (0.7–0.9%) and benzylbenzoate (0.4–0.9%) (Micheneau *et al.*, 2008).

### *Orthoptera, a new order of pollinator*

The background and aims of pollinator-mediated selection and evolution of floral traits have long fascinated evolutionary ecologists. No other plant family shows as wide a range of pollinator-linked floral forms as Orchidaceae. In spite of the large size of this model family and a long history of orchid pollination biology, the identity and specificity of most orchid pollinators remains inadequately studied, especially in the tropics, where the family has undergone extensive diversification. *Angraecum* (Vandeae, Epidendroideae), a large genus of tropical Old World orchids renowned for their floral morphology specialized for hawkmoth pollination, has been a model system since the time of Darwin.

The pollination biology of *Angraecum cadetii* (Figure 32.9), an endemic species of the islands of Mauritius and Reunion (Mascarene Islands, Indian Ocean) displaying atypical flowers for the genus (white and medium-size, but short-spurred), was investigated. Natural pollinators were observed by means of hard-disk camcorders. Pollinator-linked floral traits, namely spur length, nectar volume and concentration and scent production, were also investigated. Pollinator efficiency (pollen removal and deposition) and reproductive success (fruit set) were quantified in natural field conditions weekly during the 2003, 2004 and 2005 flowering seasons (January to March).

*A. cadetii* is self-compatible but requires a pollinator to reach to the fruit stage. Only one pollinator species was observed, an undescribed species of raspy cricket (Gryllacrididae, Orthoptera). These crickets, which are nocturnal foragers, reach the flowers by climbing up leaves of the orchid or jumping across from neighbouring plants and probed the most 'fresh-looking' flowers on each plant. Visits to flowers were relatively long (if compared with the behaviour of birds or hawkmoths), averaging 16.5 s with a maximum of 41.0 s. At the study site of La Plaine des Palmistes (Pandanus forest), 46.5% of flowers had pollen removed and 27.5% had pollen deposited on stigmas. The proportion of flowers that set fruit ranged from 11.9% to 43.4%, depending on the sites sampled across the island. Although orthopterans are well known for being herbivores, this represents the first clearly supported case of orthopteran-mediated pollination in flowering plants.

The fragrance of *A. cadetii* is largely dominated by monoterpenes, of which (*E*)-β-ocimene (52.6%) and geraniol (18.3%) are the two dominant compounds. These major volatiles are accompanied by a lower content of aldehydes, esters and other mono- and sesquiterpenes (Micheneau *et al.*, 2010; Hugel *et al.*, 2010).

**Figure 32.9** *Angraecum cadetii. Photo*: Claire Micheneau. Reproduced with permission

### *Sapromyiophily in the orchid Bulbophyllum variegatum*

Orchid species are well known for their highly specialized pollinator interactions. To better understand the reproductive biology of the tropical epiphytic orchid *Bulbophyllum variegatum* (Figure 32.10) on Reunion, investigations were conducted on the floral morphology, breeding system, pollinator diversity, floral scent profile and fruiting success of about 30 individuals in three natural populations during two consecutive flowering seasons. Controlled hand-pollination experiments in two populations showed that the species is self-compatible, but requires pollinator service to achieve reproduction. Videotape pollinator observations conducted during

**Figure 32.10** *Bulbophyllum variegatum. Photo*: LCSNSA

two flowering seasons for 56 h revealed that *B. variegatum* is pollinated by a single species of fly from the Platystomatidae. This fly seems to be attracted by the unpleasant scent produced by the flowers, and does not receive any reward after achieving pollination. In addition, no egg-laying behaviour was observed. *B. variegatum* thus exhibits atypical sapromyiophilous pollination syndrome, which constitutes the first proven case of sapromyiophily within the genus *Bulbophyllum* on the Mascarene Archipelago. Hand pollinations further showed that fruit set was not significantly higher for flowers that received outcross pollen than for those that were self-crossed (53% and 44% respectively). Fruit sets under natural conditions were significantly different among populations, ranging from 0.5% to 24.3%. This low fruit production is most likely due to infrequent pollinator visits, particularly in disturbed forests where the pollinator has never been observed.

This species displays large reddish flowers that emit a very unpleasant urine-like scent, suggesting a probable case of sapromyiophilous pollination. To verify this hypothesis, fieldwork and floral measurements were conducted on Reunion in order to determine the pollination system of *B. variegatum*, and to quantify its reproductive success across its natural range. The study was carried out during two consecutive flowering seasons (2006 and 2007) at three different localities: (i) Mare Longue (hereafter ML), considered to be the best-preserved lowland wet forests of Reunion, (ii) Bassin Boeuf (hereafter BB) and (iii) Eden (hereafter ED), which are situated in more disturbed and fragmented areas that have been severely altered by the invasion of alien plants. The composition of the identified headspace of the three *B. variegatum* populations revealed 51 compounds, accounting for approximately 66.8% (sample 1, ED), 85.3% (sample 2, ML) and 47.3% (sample 3, BB) of the total composition. Nineteen major compounds were detected, such as nonan-2-one, nonan-2-ol, nonanal, decanal, indole, toluene, *p*-cresol and *p*-cresyl acetate (Humeau *et al.*, 2011).

#### 32.5.2.2 Fragrant volatiles compounds

***From liverwort***

Three populations of the epiphyllous liverwort *Drepanolejeunea madagascariensis* (Figure 32.11) collected in the cloud forests of Reunion Island (Mascarene Archipelago) were investigated for their volatile compounds, because of the pleasant, sweet, warm, woody-spicy and herbaceous fragrance, slightly reminiscent of dill. By applying the headspace SPME technique coupled to GC–MS analysis, 34 compounds were detected in total, with *p*-menth-1-en-9-ol (28.8–43.5%), limonene (10.5–14.7%), β-phellandrene (8.8–11.6%) and the so-called dill ether (8.5–16.6%) as the main components. The presence of 1-epi-α-pinguisene confirms the possible use of pinguisane-type sesquiterpenoids as a characteristic chemical marker for the order Jungermanniales (Gauvin-Bialecki *et al.*, 2010).

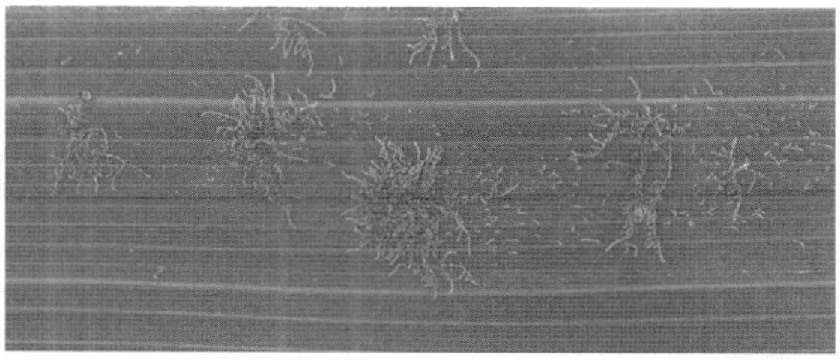

**Figure 32.11** *Drepanolejeunea madagascariensis. Photo*: LCSNSA

***From flowers***

For over 10 years, in order to discover new fragrances, the investigation of the headspace of around 100 flowers *in vivo*, from wild or ornamental, endemic or indigenous plants of Reunion was studied by SPME. The apparatus for trapping the flower headspace is presented in Figure 32.12. More than 60 flowers were selected and the results were described in an ongoing book being published for perfumers.

## 32.6 Conclusion

The objectives of the work on aromatic plants are numerous: research of odoriferous plants, plants–pollinators interactions, environment protection and biodiversity conservation. Investigations by LCSNSA will continue because of the richness of the island.

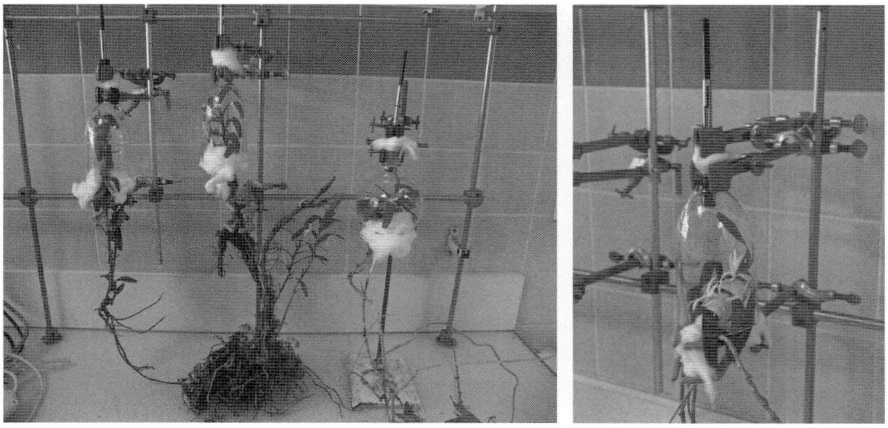

(a) (b)

**Figure 32.12** Apparatus for trapping flower headspace by SPME. *Photos*: LCSNSA

The latter is indeed one of the biodiversity hotspots of the world[1] and a trans-disciplinary study of this exceptional biodiversity is fully relevant. All the research topics of LCSNSA underline the unreserved commitment of the laboratory to valorize the particular fauna (specifically marine invertebrates) and flora (medicinal and aromatic plants) of Reunion Island and more broadly that of the Indian Ocean area. That type of zone offers an untapped reservoir for the identification of new molecules for a multitude of applications, such as pharmaceutical, nutritional, perfumery and cosmetics uses.

## Acknowledgements

The LCSNSA thanks the Regional Council of La Réunion and the Environment-Biodiversity-Health Federation of Reunion University (FEBS-UR) for financial support. We are grateful to the 'Amis des plantes et de la Nature (APN)', the 'Office National des forêts (ONF)', the 'Parc National de La Réunion (PNR)' and Dominique Strasberg and Jacques Fournel of the 'UMR C_53 PVBMT' for collection and identification of the plants. We wish to express in particular our gratefulness to Claude Marodon, Hermann Thomas, Max Felicite, Jean-Maurice Tamon and Antoinette Chane Ki for their help in plant material collection. Thanks to Laurent Janci from CAHEB for all the information relating to the production of essential oils in Reunion. Special thanks to Dominique Strasberg, Thierry Pailler, Laurence Humeau, Claire Micheneau and Claudine Ah-Peng and to all the members of the Chemistry Department of Mauritius University with whom we have worked very cooperatively for many years.

## References

Adams, R.P. (2001) *Identification of Essential Oil Components by Gas Chromatography/Quadrupole Mass Spectroscopy*, Allured Publishing, Carol Stream, IL.

Bombarda I. (1994) Hémisynthèse d'alcools sesquiterpéniques: application à la valorisation de la fraction hydrocarbure de l'huile essentielle de Vétyver Bourbon. Doctorat en Sciences, Université de Droit, d'Ecomonie et des Sciences d'Aix–Marseille.

Bombarda, I., Gaydou, E.M., Smadja, J. and Faure, R. (1996) Sesquiterpene epoxides and alcohols derived from hydrocarbons of Vetiver essential oil. *Journal of Agricultural and Food Chemistry*, **44** (1), 217–222.

Chane-Ming, J. (1999) Contribution à l'étude des Zingibéracées à l'Ile de La Réunion. Optimisation du mode d'extraction sur la composition des extraits aromatiques. Thèse de Doctorat en Sciences, Université de La Réunion.

Chane-Ming, J., Vera, R.R. and Fraisse, D.J. (1998) Chemical composition of essential oil of *Callistemon citrinus* (Curtis) skeel from Reunion. *Journal of Essential Oil Research*, **10**, 429–431.

Chemat, F., Lucchesi, M. and Smadja, J. (2003) Brevet sur un procédé d'extraction par Chauffage micro-ondes. Solventless microwave extraction of natural plants. European Patent, EP 1 439 218 A1.

[1] Pitons, Cirques and Remparts of Reunion Island are listed in UNESCO World Heritage.

Chemat, F., Lucchesi, M. and Smadja, J. (2004) Brevet sur un procédé d'extraction par chauffage micro-ondes. Solventless microwave extraction of volatile natural substance. *Brevet Américain*, 10/751,988.

Conan, J.Y. (1977a) Essai de définition d'un label bourbon pour quelques huiles essentielles de La Réunion. *Rivista Italiana EPPOS*, **10**, 544.

Conan, J.Y. (1977b) Essais d'amélioration de l'huile essentielle dite de Géranium Bourbon. *Parfums, Cosmétiques et Savons de France*, **15**, 55.

Conan, J.Y., Darant, R. and Michellon, R. (1976) Sur les Huiles essentielles dites de Géranium Bourbon. *Rivista Italiana EPPOS*, **7**, 556.

Davies, N.W. (1990) Gas chromatographic retention indices of monoterpenes and sesquiterpenes on methyl silicone and carbowax 20M phases. *Journal of Chromatography*, **503**, 1–24.

Gauvin, A. (2006) Hyphenated methods. 8-7-01 Analysis and characterisation of essential oils using GC–MS, in *Encyclopedia of Mass Spectrometry*, Volume 8 (eds M. Gross and R. Caprioli), Elsevier, Oxford (ISBN 978–0080438474).

Gauvin, A. and Smadja, J. (2005) Essential oil composition of four *Psiadia* species from Reunion Island: a chemotaxonomic study. *Biochemical Systematics and Ecology*, **33**, 705–714.

Gauvin, A., Lecomte, H. and Smadja, J. (2004a) Comparative investigations of the essential oils of two scented geranium (*Pelargonium* species) cultivars grown on Reunion Island. *Flavour and Fragrance Journal*, **19**, 455–460.

Gauvin, A., Susperregui, A., Barth, P. *et al.* (2004b) An acetylated monoterpene and a sesquiterpene alcohol from *Psiadia anchusifolia*. *Phytochemistry*, **65**, 897–901.

Gauvin-Bialecki, A. and Marodon, C. (2008) Essential oil of *Ayapana triplinervis* from Reunion Island: a good natural source of thymohydroquinone dimethyl ether. *Biochemical Systematics and Ecology*, **36**, 853–858.

Gauvin-Bialecki, A. and Smadja, J. (2007) Analyse des huiles essentielles de quatre espèces du genre *Psiadia* de l'Île de La Réunion: étude chimiotaxonomique. *Journal of Nature*, **19**, 3–17.

Gauvin-Bialecki, A., Ah-Peng, C., Smadja, J. and Strasberg, D. (2010) Fragrant volatile compounds in the liverwort *Drepanolejeunea madagascariensis* (Steph.) Grolle: approach by HS-SPME technique. *Chemistry and Biodiversity*, **7**, 639–643.

Govinden-Soulange, J., Magan, N., Gurib-Fakim, A. *et al.* (2004) Chemical composition and *in vitro* antimicrobial activities of the essential oils from endemic *Psiadia* species growing in Mauritius. *Biological and Pharmaceutical Bulletin*, **27** (11), 1814–1818.

Gurib-Fakim, A., Bourrel, C., Kodja, H. and Govinden, J. (1995) Chemical composition of the essential oils of *Psiadia lithospermifolia* (Lam.) Cordem. and *P. viscosa* (Lam) A.J. Scott of the Astereceae family. *Journal of Essential Oil Research*, **7**, 533–535.

Gurib-Fakim, A., Gauvin, A., Smadja, J. *et al.* (2000) Composition of the essential oil of the endemic *Psiadia arguta* Pers. (Voigt) from Mauritius. *Journal of Essential Oil Research*, **12**, 556–558.

Hugel, S., Micheneau, C., Fournel, J. *et al.* (2010) *Glomeremus* species from the Mascarene Islands (Orthoptera, Gryllacrididae) with the description of the pollinator of an endemic orchid from the island of Réunion. *Zootaxa*, **2545**, 58–68.

Humeau, L., Micheneau, C., Jacquemin, H. *et al.* (2011) Sapromyiophily in the native orchid, *Bulbophyllum variegatum*, on Réunion

(Mascarene Archipelago, Indian Ocean). *Journal of Tropical Ecology*, **27**, 591–599.

Joulain, D. and Konig, W.A (1998) *The Atlas of Spectral Data of Sesquiterpene Hydrocarbons*, EB-Verlag, Hamburg.

Koenig, W.A., Joulain, D. and Hochmuth, D.H. (2008) Terpenoids and related constituents of essential oils. *MassFinder 4* (ed. D.H. Hochmuth), Convenient and Rapid Analysis of GCMS, Hochmuth Scientific Consulting, Hamburg, Germany.

Kondjoyan, N. and Berdague, J.L. (1996) *A Compilation of Relative Retention Indices for the Analysis of Aromatic Compounds*, 1st edn, Laboratoire Flaveur, Clermont-Ferrand.

Lucchesi, M.E. (2005) Extraction sans solvant assistée par micro-ondes conception et application à l'extraction des huiles essentielles. Thèse de Doctorat en Sciences, Université de la Réunion.

Lucchesi, M.E., Chemat, F. and Smadja J. (2004a) Solvent free microwave extraction of essential oils from aromatic herbs: comparison with conventional hydro-distillation. *Journal of Chromatography A*, **1043**, 323–327.

Lucchesi, M.E., Chemat, F. and Smadja, J. (2004b) An original solvent free microwave extraction of essential oils from spices. *Flavour and Fragrance Journal*, **19**, 134–138.

Lucchesi, M.E., Chemat, F. and Smadja, J. (2004c) Solvent-free microwave extraction: an innovative tool for rapid extraction M.E. of essential oil from aromatic herbs and spices. *Journal of Microwave Power and Electromagnetic Energy*, **39**, 1–5.

Lucchesi, M.E., Chemat, F., Smadja, J. *et al.* (2007) Solvent free microwave extraction of *Elletaria cardamomum*: a multivariate study of a new technique for the extraction of essential oil. *Food Engineering*, **79**, (3), 1079–1086.

McLafferty, F.W. (2009) Wiley Registry of Mass Spectral Data, 9th edn.

Micheneau, C., Fournel, J., Gauvin-Bialecki, A. and Pailler, T. (2008) Auto-pollination in a long-spurred endemic orchid (*Jumellea stenophylla*) on Reunion Island (Mascarene Archipelago, Indian Ocean). *Plant Systematics and Evolution*, **272**, 11–22.

Micheneau, C., Fournel, J., Warren, B.H. *et al.* (2010). Orthoptera, a new order of pollinator. *Annals of Botany*, **105**, 355–364.

NIST (2011) *Mass Spectral Database for Windows* (data version: NIST11), National Institute of Standards and Technology.

Ormeno, E., Goldstein, A. and Niinemets, U. (2011) Extracting and trapping biogenic volatile organic compounds stored in plant species. *Trends in Analytical Chemistry*, **30**, 978–989.

Paniandy, J.C. (1999) Etude d'un fruit tropical: le goyavier-fraise (*Psidium cattleianum* Sabine). Caractérisation physico-chimique, biochimique et conservation en frais en vue de sa valorisation. Thèse de Doctorat en Sciences, Université de La Réunion.

Paniandy, J.C., Chane-Ming, J. and Pieribattesti, J.C. (2000) Chemical composition of the essential oil and headspace solid phase micro-extraction of the guava fruit (*Psidium guajava* L.). *Journal of Essential Oil Research*, **12**, 153–158.

Pieribattesti, J.C. (1982) Contribution à l'étude de quelques huiles essentielles de La Réunion. Thèse de Doctorat d'Etat, Université de Droit, d'Economie et des Sciences d'Aix Marseille.

Pieribattesti, J.C., Conan, J.Y. and Guerrere, M. (1979) Etude de certaines anomalies constatées dans l'huile essentielle de Géranium Bourbon. *Rivista Italiana EPPOS*, **10**, 119.

Pieribattesti, J.C., Conan, J.Y., Buill, P. *et al.* (1980a) Terpenoïdes et esters polyacétyléniques de l'huile essentielle d'*Erigeron naudini*. *Phytochemistry*, **20** (3), 507–508.

Pieribattesti, J.C., Conan, J.Y. and Vincent, E.J. (1980b) *L'huile essentielle de* Schinus terebenthifolius *de La Réunion*. Rivista Italiana EPPOS.

Pieribattesti, J.C., Conan, J.Y., Vincent, E.J. *et al.* (1981) Etude chimique des Baies Roses de Bourbon. *Annales des Falsifications, de l'Expertise Chimique et Toxicologique (Janvier)*.

Pieribattesti, J.C., Conan, J.Y., Guerrere, M. and Payet, C.A. (1984) Etude statistique de l'huile essentielle de Vétyver Bourbon en vue de la normalisation. *Annales des Falsifications, de l'Expertise Chimique et Toxicologique*, **77**, 825.

Rondeau, P., Smadja, J. and Shum-Cheong-Sing, A. (2003) Constituants volatils de trois espèces de *Citrus* de La Réunion. *Journal of Nature*, **15** (1), 89–95.

Shum Cheong Sing, A. (1992) Etude des composants aromatiques de *Jumella Fragrans* – Recherche des critères de qualité des rhums de La Réunion. Thèse de Doctorat en Sciences, Institut National Polytechnique de Toulouse.

Shum-Cheong-Sing, A., Smadja, J., Brevard, H. *et al.* (1992) Volatile constituents of Faham (*Jumellea fragrans* (Thou.) Schltr.). *Journal of Agricultural and Food Chemistry*, **40** (4), 642–646.

Smadja, J. (1987) Influence climatique et régionale sur la qualité de l'huile essentielle de Vétyver Bourbon. Thèse de Doctorat d'Etat, Université des Sciences et Techniques du Languedoc.

Smadja, J. (1990) L'huile essentielle de Vétyver. *Journal of Nature*, **2**, 15–31.

Smadja, J. (1991) Mise au point sur la composition chimique de l'huile essentielle de Vétyver. *Journal of Nature*, **3**, 3–17.

Smadja, J. (1994) Mise au point bibliographique sur l'huile essentielle de Vétyver. *Rivista Italiana EPPOS*, **14**, 15–36.

Smadja, J. (2009) Chemical composition and localization of essential oils, in *Essential Oils and Aromas: Green Extractions and Applications* (ed. F. Chemat), Har Krishan Bhalla and Sons, Dehradun, India, pp. 122–146 (ISBN: 978-81-905771-4-4).

Smadja, J. and Gaydou, E.M. (1991) Recherche de caractéres les plus discriminants par analyse factorielle discriminante. *Analusis*, **19**, 39–40.

Smadja, J., Gaydou, E., Lamaty, G. and Conan, J.Y (1986) Variation des constantes physico-chimiques de l'huile essentielle de Vétyver Bourbon. Influences régionale et climatique. *Parfums, Cosmétiques, Arômes*, **69**, 69–73.

Smadja, J., Gaydou, E., Lamaty, G. and Conan, J.Y. (1988) Essai d'identification des constituants de l'huile essentielle de Vétyver Bourbon. *Parfums, Cosmétiques, Arômes*, **84**, 61–66.

Smadja, J., Gaydou, E.M., Lamaty, G. and Conan, J.Y. (1989) Influence régionale et annuelle de l'huile essentielle de Vétyver Bourbon sur les constantes physicochimiques. *Analusis*, **17**, 319–325.

Smadja, J., Gaydou, E.M., Lamaty, G. and Conan, J.Y (1990a) Etude des facteurs de variation de la composition de l'huile essentielle de Vétyver Bourbon par analyse factorielle discriminante. *Analusis*, **18**, 343–351.

Smadja, J., Gaydou, E.M., Peyron, L. *et al.* (1990b) Etude comparative d'huiles essentielles de Vétyver d'origine géographique diverse. *Rivista Italiana EPPOS*, **1**, 7–21.

Smadja, J., Rondeau, P. and Shum-Cheong-Sing, A. (2005) Volatile constituents of five *Citrus* Petitgrain essential oils from Reunion. *Flavour and Fragrance Journal*, **20** (4), 399–402.

Smadja, J., Laurencon, L., Thomas, H. and Bialecki, A. (2010a) Medicinal plant potentialities from Reunion Island. The particular case of *Toddalia asiatica* (Rutaceae), in *VIII International Ethnobotany Symposium Proceedings*, Lisbon, pp. 625–635.

Smadja, J., Strasberg, D., Legoff, G. and Gauvin-Bialecki, A. (2010b) Volatiles compounds from *Melicope obscura*. *Chemistry and Biodiversity*, **7**, 467–475.

Tandon, S. (2008) Distillation technology for essential oils, in *Extraction Technologies for Medicinal and Aromatic Plants* (eds S.S. Handa, S.P.S. Khanuja, G. Longo and D.D. Rakesh), ICS UNIDO, Trieste, pp. 115–143.

Vera, R.R. and Chane-Ming, J. (1998). Chemical composition of essential oil of dill (*Anethum graveolens* L.) growing in Reunion Island. *Journal of Essential Oil Research*, **10**, 539–542.

Vera, R.R. and Chane-Ming, J. (1999). Chemical composition of essential oil of majoram (*Origanum majorana* L.) from Reunion Island. *Food Chemistry*, **66**, 143–145.

Vera, R., Chane-Ming, J. and Fraisse, D.J. (1999) Chemical composition of the essential oil of sage (*Salvia officinalis* L.) from Reunion Island. *Journal of Essential Oil Research*, **11**, 399–402.

Vernin, G., Vernin, C., Pieribattesti, J.C. and Roque, C. (1998) Analysis of the volatile compounds of *Psidium cattleianum* Sabine from Reunion Island. *Journal of Essential Oil Research*, **10**, 353–362.

Viano, J., Smadja, J., Conan, J.Y. and Gaydou, E.M. (1991a) Ultrastructure des racines de *Vetiveria zizanioides* Stapf (Gramineae). *Bulletin du Museum National d'Histoire Naturelle, Paris, 4me Série, Section B, Adansonia*, **13**, 61–69.

Viano, J., Smadja, J. and Gaydou, E.M. (1991b) Sécrétion des huiles essentielles dans les racines de *Vetiveria zizanioides* (L.) Stapf. *Revue de Cytologie Végétale et de Botanique*, **14**, 65–70.

Visinoni, F., Chemat, F. and Smadja, J. (2006) Brevet sur un procédé d'entraînemant à la vapeur par chauffage micro-ondes. Brevet Allemand, 2006, EP 1 629 725 A1. Assistée par Micro-ondes – conception et application à l'extraction des huiles essentielles, Université de La Réunion.

# 33 Anti-Parasitic Activity of Essential Oils and their Active Constituents against *Plasmodium*, *Trypanosoma* and *Leishmania*

Joanne Bero[1], Salomé Kpoviessi[1,2,3] and Joëlle Quetin-Leclercq[1]

[1]Pharmacognosy Research Group, Louvain Drug Research Institute, Université catholique de Louvain, Bruxelles, Belgium
[2]Laboratory of Physic and Synthesis Organic Chemistry (LaCOPS), University of Abomey-Calavi (UAC), Faculty of Sciences and Technics (FAST), Cotonou Benin
[3]Laboratory of Pharmacognosy and Essential Oils (LAPHE), University of Abomey-Calavi (UAC), Faculty of Health Sciences (FSS), Faculty of Sciences and Technics (FAST), Cotonou Benin

## 33.1 Introduction

Infectious diseases, such as leishmaniasis, malaria, sleeping sickness and Chagas disease, which particularly affect the developing countries, have often been relegated by the pharmaceutical industries. All these diseases are caused by protozoan parasites, and some forms are lethal for humans. Another common characteristic of these diseases is the absence of an efficient treatment that would not cause toxicity, resistance or other side-effects. As an example, the number of actual cases of sleeping sickness is currently estimated at 30 000 (Drugs for Neglected Diseases Initiatives, 2011). Resistance, toxicity and variable efficacy between strains or species of most of the drugs used, as well as for some of them the need for a long course parenteral administration, show how the search for new antiparasitic compounds is needed, particularly from plants used in traditional medicine, as a source of new leads with new mechanisms of action.

Aromatic plants have been used since ancient times for many medicinal properties (Bakkali *et al.*, 2008). Several essential oils (EOs) extracted from them are known to possess antimicrobial properties and may also be used as alternatives or adjuvants to current antiparasitic therapies. The emergence of parasites resistant to current chemotherapies highlights the importance of plant EOs as potential novel antiparasitic agents (Anthony *et al.*, 2005).

Activities of EOs and their respective constituents were assessed on different strains of several parasites (*Plasmodium falciparum* resistant (W2, FcB1, K1) or not to chloroquine (3D7, FCR-3, F32, D6, D10), *Leishmania mexicana*, *Leishmania amazonensis*, *Leishmania braziliensis*, *Leishmania aethiopica*, *Leishmania donovani*, *Leishmania panamensis*, *Leishmania chagasi*, *Leishmania guyanensis*, *Leishmania infantum*, *Leishmania tropica*, *Leishmania major*, *Trypanosoma cruzi* or *Trypanosoma brucei brucei*) to find effective compounds against these various diseases. We considered that those having an $IC_{50} \leq 2$ μg/mL (or 2 μM for pure compounds) may have some interest for further development and are cited as having good activity, those with an $IC_{50}$ between 2 and 20 μg/mL (or micromolar for pure compounds) are considered as having moderate activity, while those with lower activity were less interesting. In this review, we only detail the EOs and pure compounds having activities below 20 μg/mL or nl/mL (1 nl is a little less than 1 μg depending on the density of the EO), or 20 μM for pure compounds (corresponding for example to 3 μg/mL of thymol). Activity criteria (Figure 33.1) have been proposed by the Tropical Diseases Research of the World Health Organization to consider a pure compound as a hit antiprotozoal compound (Nwaka and Hudson, 2006).

In 2005, another publication in *Nature Reviews* expanded the criteria: a pure compound is considered as a hit if the $IC_{50}$ is less than 1 μg/mL and a selectivity of at least 10 compared with a cell line (Pink *et al.*, 2005). For more complex extracts, a higher $IC_{50}$ may be considered as interesting because the active compound(s) is(are) diluted with other substances.

- *Plasmodium falciparum* (K1) $IC_{50}$ <0.2 μg per ml, SI*>100
- *Trypanosoma brucei rhodesiense* (STIB 900) $IC_{50}$ <0.2 μg per ml, SI*>100
- *Trypanosoma cruzi* (Tulahuen) $IC_{50}$ <1.0 μg per ml, SI*>50
- *Leishmania donovani* (L82)

Axenic amastigotes $IC_{50}$ <0.5 μg per ml, SI*>20
Amastigotes in macrophage $IC_{50}$ <1 μg per ml, SI*>20

SI* = $IC_{50}$ L-6/$IC_{50}$ parasite

**Figure 33.1** Activity criteria for hit compounds

## 33.2 Essential oils

### 33.2.1 Annonaceae

The EO from leaves of *Annona foetida* Mart. showed significant antileishmanial activity with $IC_{50}$ of 16.2 ± 1.92 μg/mL, 9.9 ± 1.22 μg/mL and 4.1 ± 0.2 μg/mL respectively against *L. amazonensis*,

*Novel Plant Bioresources: Applications in Food, Medicine and Cosmetics*, First Edition. Edited by Ameenah Gurib-Fakim.

*L. braziliensis* and *L. guyanensis*. A preliminary cytotoxicity assay for this oil was carried out on hamster and mice peritoneal macrophages (Balb/c) and considered to be not cytotoxic to macrophages (Costa *et al.*, 2009).

The EO from leaves of *Xylopia discreta* Sprague & Hutch. was evaluated on J774 and on U937 cells infected with *L. panamensis* and presented $IC_{50}$ of 6.35 ± 4.00 μg/mL and 6.25 ± 1.1 μg/mL respectively. The cytotoxicities were evaluated on J774 ($IC_{50}$ = 698.45 ± 98 μg/mL) and on U937 ($IC_{50}$ = 160.0 ± 8.7 μg/mL) (Lopez *et al.*, 2009).

Four EOs extracted from the Cameroonian plants *Xylopia phloiodora* Mildbr., *Pachypodanthium confine* Engl. & Diels, *Xylopia aethiopica* A.Rich. and *Hexalobus crispiflorus* A.Rich. were evaluated with respect to their antiplasmodial activity against the W2 strain of *P. falciparum*. These oils were active with $IC_{50}$ of 17.9, 16.6, 17.8 and 2.0 μg/mL (Boyom *et al.*, 2003).

EOs extracted from the stem barks and leaves of *Cleistopholis patens* Engl. & Diels and *Uvariastrum pierreanum* Engl. & Diels were evaluated *in vitro* for antiplasmodial activity against W2 strain. The oils showed $IC_{50}$ values of 9.19 ± 0.08 μg/mL and 15.19 ± 0.13 μg/mL respectively for the stem bark and leaf oils of *Cleistopholis patens* and 6.08 ± 0.01 μg/mL and 13.96 ± 0.05 μg/mL respectively for those from *Uvariastrum pierreanum* (Boyom *et al.*, 2011).

### 33.2.2 Apiaceae

The effect of EO from *Coriandrum sativum* L. was evaluated on *L. chagasi* promastigotes and amastigotes and the toxicity was also examined on the murine monocyte cell line RAW 264.7. The EO was effective against amastigotes ($IC_{50}$ of 1.51 μg/mL) but not on promastigotes ($IC_{50}$ of 181.00 μg/mL), and a viability of 49.9% was shown at a concentration of 100 μg/mL on murine monocyte cells (RAW 264.7) (Rondon *et al.*, 2012).

The hydrodistilled oils from the aerial parts of *Bupleurum montanum* Coss. & Durieu and *Bupleurum plantagineum* Desf., which are endemic to North Africa, were evaluated against *P. falciparum* (FcB1-Colombia) and exhibited $IC_{50}$ of 0.083 μg/mL and 1.39 μg/mL respectively (Laouer *et al.*, 2009).

The EO of *Carum carvi* L. showed antiplasmodial activity *in vitro* with $IC_{50}$ = 1 μg/mL against *P. falciparum*. This oil was also studied for its *in vivo* antimalarial activity against *Plasmodium berghei* on mice. The effective dose was 50 μl oil by percutaneous administration as indicated by a significant reduction in parasitaemia on day 8 (8.3% of remaining parasitaemia) (Fujisaki *et al.*, 2012).

### 33.2.3 Asteraceae

The EO of *Achillea millefolium* L. demonstrated significant dose-dependent antileishmanial activity against promastigotes of *L. amazonensis* with $IC_{50}$ of 7.8 ± 0.3 μg/mL. The oil had a similar effect on the cultures of intracellular amastigotes, with an $IC_{50}$ of 6.5 μg/mL. The cytotoxic effects on J774G8 macrophage cells were assessed and the $IC_{50}$ was 72.0 ± 2.0 μg/mL (Santos *et al.*, 2010).

EOs of *Artemisia absinthium* L. and *Echinops kebericho* Mesfin were screened for antileishmanial activity and toxicity *in vitro*. Both oils showed activity against promastigote with MICs of 0.0765 μl/mL and 0.0097 μl/mL respectively for *E. kebericho* on *L. donovani* and *L. aethiopica* and 0.1565 μl/mL for *Artemisia absinthium* on the two strains. The $IC_{50}$ values for *Artemisia absinthium* on axenic amastigote forms of both *Leishmania* species were 42.00 ± 9.80 nl/mL and 7.94 ± 5.58 nl/mL respectively and for *Echinops kebericho* 0.5 ± 0.7 nl/mL and 0.24 ± 0.010 nl/mL respectively. The cytotoxicity on the human monocytic leukaemia (THP-1) cell line was 152.04 ± 8.40 nl/mL for *Artemisia absinthium* and 0.40 ± 0.01 nl/mL for *Echinops kebericho* (Tariku *et al.*, 2011).

The EO of another sample of *Artemisia absinthium* was evaluated in another study and showed an $IC_{50}$ of 14.4 ± 0.36 μg/mL against promastigotes of *L. donovani* (Sariego *et al.*, 2008).

The EO of *Artemisia abyssinica* Sch. Bip. from Ethiopia was screened for leishmanicidal activity and toxicity *in vitro*. The oil showed an effect on promastigotes (MIC = 312.5 nl/mL and 76.5 nl/mL respectively on *L. donovani* and *L. aethiopica*) and amastigotes ($IC_{50}$ = 131.00 ± 7.48 nl/mL and 12.44 ± 2.78 nl/mL respectively on *L. donovani* and *L. aethiopica*). Cytotoxicity on THP-1 cells was $IC_{50}$ = 350 ± 5.00 nl/mL (Tariku *et al.*, 2010b).

*Artemisia gorgonum* Webb EO from Cape Verde was found to exhibit antiplasmodial activity against FcB1 with $IC_{50}$ = 5.2 ± 0.7 μg/mL (Ortet *et al.*, 2010).

The EO of *Artemesia herba-alba* Asso was found to exhibit antileishmanial activity. At a concentration of 2 μg/mL the cultures of promastigotes of *L. major* and *L. tropica* were not growing from the third day, and at concentrations of 4 and 10 μg/mL the antileishmanial effect was direct and total (Hatimi *et al.*, 2001).

The EO of *Matricaria chamomilla* L. showed no activity *in vitro* against axenic amastigotes of *L. braziliensis* and presented $IC_{50}$ of 2.87 ± 0.08 and 10.30 ± 3.37 μg/mL, respectively against intracellular amastigotes of *L. braziliensis* and *L. panamensis*. The EO also showed potential toxicity to mammalian cells of the promonocytic human cell line U937 with $IC_{50}$ of 30.21 ± 1.55 μg/mL (Rios *et al.*, 2008).

The EO from *Vanillosmopsis arborea* (Gardner) Baker was assessed for its *in vitro* activity against *P. falciparum* ($IC_{50}$ = 7.00 ± 3.50 μg/mL) and for its *in vivo* activity in mice infected with *P. berghei* by subcutaneous route (inhibition of 33% and 47% respectively for 50 mg/kg per day and 100 mg/kg per day). The acute toxicity was assessed in healthy mice ($LD_{50}$ = 7 g/kg) and *in vitro* cytotoxicity was determined against HeLa cells and mice macrophages (both $IC_{50}$ > 500 μg/mL) (Mota *et al.*, 2012).

The *in vitro* antimalarial activity and the cytotoxicitiy (lymph nodes of Balb/c mice) of EOs of *Ambrosia tenuifolia* Spreng and *Ambrosia scabra* Hook. & Arn. were evaluated. They showed significant antiplasmodial activity, with $IC_{50}$ of 3.0 μg/mL (selectivity index, SI = 59.0) and 2.8 μg/mL (SI = 18.6) respectively against F32 and $IC_{50}$ of 7.0 (SI = 25.3) and 3.0 (SI = 17.3) μg/mL respectively against W2 (Sulsen *et al.*, 2008).

Antimalarial and cytotoxicity profiles were determined for the EO of *Helichrysum cymosum* D.Don. $IC_{50}$ values were 1.25 ± 0.77 μg/mL against *P. falciparum* FCR-3 strain and 17.47 ± 3.10 μg/mL against transformed human kidney epithelial cells (Van Vuuren *et al.*, 2006).

### 33.2.4 Burseraceae

The EO of *Boswellia carteri* Birdw. showed antiplasmodial activity *in vitro* with $IC_{50}$ = 10 μg/mL against *P. falciparum* (Fujisaki *et al.*, 2012).

### 33.2.5 Caesalpinaceae

Eight different varieties of Brazilian copaiba oils were screened for antileishmanial activity *in vitro* on promastigotes of *L. amazonensis*. Copaiba oils showed variable levels of activity with $IC_{50}$ of 5.0 ± 0.8, 14.0 ± 0.9 µg/mL, 18.0 ± 0.0 µg/mL, 11.0 ± 0.4 µg/mL, 20.0 ± 0.8 µg/mL, 20.0 ± 0.4 µg/mL, 10.0 ± 0.8 µg/mL and 20.0 ± 0.9 µg/mL respectively for *Copaifera reticulata* Ducke, *Copaifera martii* Hayne, *Copaifera cearensis* Huber ex Ducke, *Copaifera paupera* (Herzog) Dwyer, *Copaifera langsdorffii* Desf., *Copaifera officinalis* L., *Copaifera multijuga* Hayne and *Copaifera lucens* Dwyer. The most active EO (*Copaifera reticulata*) showed $IC_{50}$ of 15 µg/mL and 20 µg/mL on axenic amastigote and intracellular amastigote forms respectively and low cytotoxicity against J774G8 macrophages (SI = 8.0 and 2.5 respectively for promastigotes and amastigotes) (Santos *et al.*, 2008).

The effect of the oleoresin from *Copaifera reticulata* was also evaluated on *L. chagasi* promastigotes and amastigotes and the toxicity was also examined on the murine monocyte cell line RAW 264.7. It was effective against both promastigotes ($IC_{50}$ of 7.88 µg/mL) and amastigotes ($IC_{50}$ of 0.52 µg/mL) and a viability of 78.45% was shown at a concentration of 100 µg/mL on murine monocyte cells (RAW 264.7) (Rondon *et al.*, 2012).

### 33.2.6 Chenopodiaceae

Only one plant EO was studied from this family: *Chenopodium ambrosioides* L.

On *L. amazonensis*, it showed potent inhibitory action against promastigote and amastigote forms, with $IC_{50}$ of 3.7 ± 0.6 µg/mL and 4.6 ± 0.7 µg/mL respectively. It also showed a moderate toxicity on macrophages from Balb/c mice ($IC_{50}$ = 58.2 ± 0.05 µg/mL). An optimal dose of 30 mg/kg per day was effective (0% of mortality) when administered during 15 days by intraperitoneal route to Balb/c mice infected experimentally (Monzote *et al.*, 2006). Moreover, three distinct EOs from green material, dried sample at room temperature or fermented in water for 3 days were evaluated on promastigotes ($IC_{50}$ = 2.9 ± 1.1 µg/mL, 20.7 ± 1.2 µg/mL and 12.0 ± 1.9 µg/mL respectively) and amastigotes ($IC_{50}$ = 4.7 ± 0.8 µg/mL, 12.4 ± 2.6 µg/mL and 6.0 ± 1.9 µg/mL respectively) of *L. amazonensis*. The selectivity indices on macrophages of normal Balb/c mice versus amastigotes were 71, 21 and 36 respectively (Monzote *et al.*, 2011). The EO from the dried sample further showed synergistic activity with pentamidine against promastigotes of *L. amazonensis*. However, no synergy was found for combinations with meglumine antimoniate or amphotericin B (Monzote *et al.*, 2007c). Balb/c mice infected with *L. amazonensis* received two cycles of treatment by different routes (intraperitoneal, oral or intralesional routes). The intraperitoneal administration at a dose of 30 mg/kg prevented lesion development and decreased the parasite burden. Oral administration retarded the infection in the experimental model compared with untreated mice, although it was less effective than the intraperitoneal route. The administration by intralesional route did not show any activity. No resistance was detected in *L. amazonensis* isolates obtained from treated mice (Monzote *et al.*, 2007b). A dose of 150 mg/kg was the most effective against mice infected with *L. amazonensis* administrated by oral route and no macroscopic toxic effects were observed. The 50% effective concentration was 51.4 mg/kg (Monzote *et al.*, 2009). Its *in vitro* antileishmanial effect against another subspecies of *Leishmania*, *L. donovani*, was investigated and showed $IC_{50}$ of 4.45 µg/mL and 5.1 µg/mL against promastigotes and amastigotes respectively. The EO caused an irreversible inhibition of the growth of promastigotes after a treatment with 100 µg/mL or 10 µg/mL for 1 h or 24 h respectively (Monzote *et al.*, 2007a).

### 33.2.7 Cochlospermaceae

EOs from the tubercles of *Cochlospermum tinctorium* Perr. and *Cochlospermum planchonii* Hook.f. ex Planch. were tested for their *in vitro* antiplasmodial activity against FcB1-Columbia and a Nigerian strain. These oils presented an $IC_{50}$ of 21 ± 3 µg/mL and 15 ± 1 µg/mL respectively for *Cochlospermum planchonii* and $IC_{50}$ of 8 ± 1.5 µg/mL and 5 ± 0 µg/mL, respectively, for *Cochlospermum tinctorium* (Benoit-Vical *et al.*, 2001).

### 33.2.8 Cupressaceae

Aomori Hiba, the EO of *Thujopsis dolabrata* Hort. ex Carrière var. *hondae*, showed antiplasmodial activity *in vitro* with $IC_{50}$ = 1 µg/mL against *P. falciparum* (Fujisaki *et al.*, 2012).

### 33.2.9 Cyperaceae

The EO from *Cyperus kyllingia* Endl. showed significant activities, with $IC_{50}$ values of 7.52 µg/mL and 7.72 µg/mL respectively against *P. falciparum* (K1) and NCI-H187 (small cell lung cancer) (Khamsan *et al.*, 2011).

### 33.2.10 Euphorbiaceae

The EO of *Croton cajucara* Benth. was tested on promastigotes and amastigotes of *L. amazonensis* and showed $IC_{50}$ of 0.0083 µg/mL and 0.022 µg/mL respectively. No cytotoxic effects on mammalian cells (mouse macrophages) were observed at the concentrations used (unaffected at 0.015 µg/mL) (Rosa *et al.*, 2003).

The volatile oil from berries of *Croton macrostachyus* Hochst. ex Del. was tested for its *in vitro* antileishmanial activity and its cytotoxicity. The oil was effective against *L. donovani* and *L. aethiopica* promastigotes (MIC = 0.08 µl/mL and 0.16 µl/mL respectively) and axenic amastigote stages ($IC_{50}$ = 20.00 ± 2.10 nl/mL and 6.66 ± 0.99 nl/mL respectively). The $IC_{50}$ value for the oil in terms of its toxicity on the human monocytic leukemia (THP-1) cell line was evaluated to be 10.00 ± 4.00 nl/mL (Tariku *et al.*, 2010a).

The EO from *Croton zehntneri* Pax & K. Hoffm. was assessed for its *in vitro* activity against *P. falciparum* ($IC_{50}$ = 10.50 ± 2.80 µg/mL) and for its *in vivo* activity in mice infected with *P. berghei* by oral route (inhibition of 28% and 53.50% respectively for 100 mg/kg per day and 500 mg/kg per day). The acute toxicity was assessed in healthy mice ($LD_{50}$ = 3 g/kg), and *in vitro* cytotoxicity was determined against HeLa cells and mice macrophages (both $IC_{50}$ > 500 µg/mL) (Mota *et al.*, 2012).

### 33.2.11 Lamiaceae

The EO of leaves of *Rosmarinus officinalis* L. exhibited antileishmanial activity against promastigotes of *L. braziliensis* with $IC_{50}$ of 17.4 ± 0.43 μg/mL (Arévalo *et al.*, 2009).

The EO of *Satureja punctata* R. Br. ssp. *punctata* from Ethiopia was screened for leishmanicidal activity and toxicity *in vitro*. The oil was active on the promastigotes (MIC = 156.5 nl/mL and 312.5 nl/mL respectively on *L. donovani* and *L. aethiopica*) and amastigotes ($IC_{50}$ = 8.70 ± 0.96 nl/mL and 4.06 ± 0.88 nl/mL respectively on *L. donovani* and *L. aethiopica*). Cytotoxicity on THP-1 cells was $IC_{50}$ = 0.013 ± 0.002 nl/mL (Tariku *et al.*, 2010b).

*Satureja thymbra* L. was collected in three vegetative periods (before, during and after flowering) and was tested on D10 and W2 strains of *P. falciparum in vitro*. The EO was effective with an inhibitory activity independent from the time of collection (before flowering: $IC_{50}$ = 26.1 ± 10 μg/mL and 11.3 ± 6.5 μg/mL respectively against the two strains; during flowering: $IC_{50}$ = 17.5 ± 6.6 μg/mL and 9.0 ± 3.3 μg/mL respectively against the two strains; after flowering: $IC_{50}$ = 21.3 ± 4.2 μg/mL and 9.7 ± 2.8 μg/mL respectively against the two strains) (Dell'Agli *et al.*, 2012).

EOs of *Satureja macrantha* C.A. Mey., *Dracocephalum kotschyi* Boiss. and *Nepeta cataria* L. showed effect on epimastigotes of *T. cruzi* (MIC = 12.5 μg/mL, 6.2 μg/mL and 6.2 μg/mL respectively) (Saeidnia *et al.*, 2007a, 2007b, 2008).

EO of *Salvia albicaulis* Benth. and *Salvia dolomitica* Codd were evaluated against FCR-3 strain of *P. falciparum* and they exhibited antiplasmodial activity with $IC_{50}$ values of 6.4 ± 2.0 μg/mL and 4.8 ± 0.7 μg/mL respectively. The toxicity profiles of the two oils were evaluated against human kidney epithelium cells ($IC_{50}$ = 2.6 ± 0.7 μg/mL and 7.7 ± 0.3 μg/mL) (Kamatou *et al.*, 2007). EOs from aerial parts of *Salvia recognita* Fisch. & C.A.Mey. were evaluated for their antimalarial activity against *P. falciparum* ($IC_{50}$ = 17 μg/mL and 12 μg/mL respectively on D6 and W2) (Tabanca *et al.*, 2006). Three closely related South African species (*Salvia stenophylla* Burch. ex Benth., *Salvia repens* Benth. and *Salvia runcinata* L.f.) were investigated for their antimalarial activity on an undefined strain and presented $IC_{50}$ = 4.38 ± 1.07 μg/mL, 1.23 ± 0.31 μg/mL and 1.68 ± 0.26 μg/mL respectively (Kamatou *et al.*, 2005).

The EO of *Ocimum gratissimum* L. was evaluated for its *in vivo* antimalarial activity on mice infected with *P. berghei*. The oil showed significant activity with 55.0%, 75.2% and 77.8% of suppression of parasitemia by oral route at 200 mg/kg per day, 300 mg/kg per day and 500 mg/kg per day respectively (Tchoumbougnang *et al.*, 2005). This oil also inhibited trypomastigote forms of *T. cruzi* (Dm28c) ($IC_{50}$ = 11.5 ± 0.3 μg/mL). The cytotoxicity was also assessed against macrophages from Balb/c mice, and $IC_{50}$ was evaluated at 180.4 ± 1.0 μg/mL (Borges *et al.*, 2012).

The EO from aerial parts of an Ethiopian medicinal plant, *Leonotis ocymifolia* (Burm.f.) Iwarsson, was investigated for its trypanocidal and cytotoxic activities. It showed activity and cytotoxicity with $IC_{50}$ of 15.41 μg/mL and 81.88 μg/mL respectively against bloodstream forms of *T. brucei brucei* (TC221) and HL-60 (Nibret and Wink, 2010).

*Melissa officinalis* L., *Menthae piperitae* L., *Rosmarinus officinalis* L. and *Thymus vulgaris* L. EOs showed antitrypanosomal activity with $IC_{50}$ = 3.9 μg/mL, 4.2 μg/mL, 19.1 μg/mL and 0.4 μg/mL respectively against *T. brucei* and cytotoxicity with $IC_{50}$ = 25.5 μg/mL, 200.2 μg/mL, 91.0 μg/mL and 24.6 μg/mL, respectively, against HL-60. Moreover, *Melissa officinalis* oil showed antileishmanial activity with $IC_{50}$ of 7.0 μg/mL against promastigotes of *L. major* (Mikus *et al.*, 2000).

*Origanum heracleoticum* L., *Perilla frutescens* Britton and *Thymus vulgaris* L. EOs showed antiplasmodial activity *in vitro* with $IC_{50}$ = 10 μg/mL against *P. falciparum* (Fujisaki *et al.*, 2012).

The EO of *Thymus hirtus* Link sp. *algeriensis* has antileishmanial activity *in vitro* against *L. major* and *L. infantum* with $IC_{50}$ = 0.43 μg/mL and 0.25 μg/mL respectively. The cytotoxicity was evaluated on murine macrophages RAW264.7 and showed $IC_{50}$ = 0.68 μg/mL (Ahmed *et al.*, 2011).

### 33.2.12 Lauraceae

The leaf EO of *Licaria canella* (Meisn.) Kosterm. was evaluated for its antileishmanial activities and showed $IC_{50}$ of 19.0 ± 0.9 μg/mL against *L. amazonensis* and cytotoxity of only 6.2% at a concentration corresponding to this $IC_{50}$ on mice Balb/c macrophages (Silva *et al.*, 2009).

### 33.2.13 Loganiaceae

The EO of *Strychnos spinosa* Lam. was active on *T. brucei brucei* bloodstream forms with $IC_{50}$ of 13.5 μg/mL and selectivity index of 4.4 (Hoet *et al.*, 2006).

### 33.2.14 Meliaceae

The EO from seeds of *Carapa guianensis* Aubl. (Andiroba oil) was evaluated for its *in vitro* antiplasmodial activity against W2 (inhibition of 100% at 8.2 μg/mL) and Dd2 (inhibition of 71% at 8.2 μg/mL; $IC_{50}$ = 9.4 μg/mL) strains of *P. falciparum*. Andiroba oil in acute toxicity test with a fixed dose was not toxic by oral route in mice (during 14 days) ($LD_{50} \gg 2000$ mg/kg) (Miranda Junior *et al.*, 2012).

### 33.2.15 Moringaceae

The EO from seeds of the Ethiopian medicinal plant *Moringa stenopetala* (Baker f.) Cufod. was investigated for its trypanocidal and cytotoxic activities. It showed activity and cytotoxicity with $IC_{50}$ of 5.03 μg/mL and 11.63 μg/mL respectively against bloodstream forms of *T. brucei brucei* (TC221) and HL-60 (Nibret and Wink, 2010).

### 33.2.16 Myristicaceae

Nutmeg oil is the EO of *Myristica fragrans* Houtt., and it showed antiplasmodial activity *in vitro* with $IC_{50}$ = 1 μg/mL against *P. falciparum*(Fujisaki *et al.*, 2012).

### 33.2.17 Myrtaceae

*Myrtus communis* L. was collected in three vegetative periods (before, during and after flowering) and the three EOs were tested on D10 and W2 strains of *P. falciparum in vitro*. They were effective with an inhibitory activity independent from the time of collection ($IC_{50}$ of *M. communis* after flowering was 21.8 ± 2.2 μg/mL and

18.3 ± 12.5 μg/mL respectively against the two strains) (Dell'Agli *et al.*, 2012).

EOs of *Leptospermum scoparium* Forst., *Melaleuca alternifolia* Cheel, *Syzygium aromaticum* (L.) Merr. & L.M.Perry and *Kunzea ericoides* (A.Rich.) Joy Thomps. showed antitrypanosomal activity with $IC_{50}$ = 16.9 μg/mL, 0.5 μg/mL, 1.9 μg/mL and 13.6 μg/mL respectively against *T. brucei* and cytotoxicity with $IC_{50}$ = 248.5 μg/mL, 43.0 μg/mL, 34.4 μg/mL and 15.5 μg/mL respectively against HL-60 (Mikus *et al.*, 2000).

### 33.2.18 Pinaceae

The EO of *Pinus caribaea* Morelet from Cuba was evaluated against promastigotes of *L. donovani* and showed $IC_{50}$ = 5.2 ± 0.42 μg/mL (Sariego *et al.*, 2008).

### 33.2.19 Piperaceae

The EO of *Piper betle* L. (landrace Bangla Mahoba) showed a dose-dependent inhibitory effect on promastigotes and intracellular transgenic amastigotes of *L. donovani.* The $IC_{50}$ values for promastigotes and intracellular amastigotes were 11.2 ± 1.23 μg/mL and 9.31 ± 0.53 μg/mL respectively. The oil was found to be devoid of any cytotoxic effect towards macrophages (J774) (Misra *et al.*, 2009).

*Piper duckei* C.DC. EO inhibited the growth in *L. guyanensis* promastigotes, with $IC_{50}$ of 15.2 μg/mL. The non-parasitized peritoneal macrophages were preserved at the concentrations of the EO which were toxic to the parasite (Moura do Carmo *et al.*, 2012).

The EO from *Piper auritum* Kunth was active against the promastigotes of *L. donovani* with $IC_{50}$ = 12.8 ± 2.8 μg/mL with SI = 8 against peritoneal macrophages from Balb/c mice. This *Piper*-oil inhibited the growing of intracellular amastigotes of *L. donovani* with $IC_{50}$ = 22.3 ± 1.8 μg/mL (Monzote *et al.*, 2010). The EO from another sample was evaluated in another study and showed an $IC_{50}$ = 12.9 ± 0.26 μg/mL against promastigotes of *L. donovani* (Sariego *et al.*, 2008).

The EO of *Piper aduncum* L. from Cuba was evaluated against promastigotes of *L. donovani* and showed $IC_{50}$ = 7.7 ± 0.18 μg/mL (Sariego *et al.*, 2008).

### 33.2.20 Poaceae

The EO of *Cymbopogon citratus* Stapf showed $IC_{50}$ values of 1.7 ± 0.1 μg/mL and 3.2 ± 0.03 μg/mL respectively after 72 h of incubation on promastigotes and amastigotes forms of *L. amazonensis.* The cytotoxicity was also evaluated on J774G8 (IC50 = 25.0 ± 0.1 μg/mL) (Santin *et al.*, 2009). The $IC_{50}$/24 h of the oil against trypomastigotes of *T. cruzi* (strain Y) was 15.5 μg/mL, and no cytotoxicity was observed on mouse peritoneal macrophages with concentrations up to 31.2 μg/mL (Santoro *et al.*, 2007). This EO was also evaluated for its *in vivo* antimalarial activity on mice infected with *P. berghei.* It showed significant activity with 62.1%, 81.7% and 86.6% of suppression of parasitaemia by oral route at 200 mg/kg per day, 300 mg/kg per day and 500 mg/kg per day respectively (Tchoumbougnang *et al.*, 2005).

The EO of *Cymbopogon martini* Stapf var. *motia* showed antiplasmodial activity *in vitro* with $IC_{50}$ = 10 μg/mL against *P. falciparum* (Fujisaki *et al.*, 2012).

### 33.2.21 Rubiaceae

The EO of leaves of *Keetia leucantha* (K.Krause) Bridson (syn. *Plectronia leucantha* Krause) showed an antitrypanosomal activity with $IC_{50}$ of 20.9 ± 12.6 μg/mL on Tbb bloodstream forms and >100 μg/mL on Tbb procyclic forms. An $IC_{50}$ of 39.7 ± 7.2 μg/mL was obtained for the antiplasmodial activity on *P. falciparum* (3D7) (Bero *et al.*, 2013).

### 33.2.22 Rutaceae

EOs from the leaves and fruits of *Toddalia asiatica* Lam. showed antiplasmodial activity with $IC_{50}$ values of 7.0 ± 0.6 μg/mL and 9.0 ± 0.7 μg/mL respectively against *P. falciparum* W2 strain and of 19.5 ± 1.0 μg/mL and 14.1 ± 1.8 μg/mL respectively against D6 strain (Nyahanga *et al.*, 2010).

Monarda is the EO of *Citrus bergamia* Risso and showed antiplasmodial activity *in vitro* with $IC_{50}$ = 1 μg/mL against *P. falciparum* (Fujisaki *et al.*, 2012).

The EO of *Ruta chalepensis* L. has antileishmanial activity *in vitro* against *L. major* and *L. infantum* with $IC_{50}$ = 8.0 μg/mL and 1.13 μg/mL respectively. The cytotoxicity was evaluated on murine macrophages RAW264.7 and showed $IC_{50}$ = 1.52 μg/mL (Ahmed *et al.*, 2011).

### 33.2.23 Santalaceae

The EO of *Santalum album* L. showed antiplasmodial activity *in vitro* with $IC_{50}$=1 μg/mL against *P. falciparum.* This oil was also studied for its *in vivo* antimalarial activity against *P. berghei* on mice: 200 μl oil by subcutaneous administration significantly reduced parasitaemia on days 6, 7 and 8 (10.2%, 21.1% and 22.7% of parasitaemia respectively) (Fujisaki *et al.*, 2012).

### 33.2.24 Verbenaceae

Some EOs obtained from *Lippia* species grown in Colombia were tested on different parasites. Antitrypanosomal activity was observed for eight oils from *Lippia alba* (Mill.) N.E.Br., *Lippia citrodora* (Paláu) Kunth and *Lippia origanoides* Kunth from different places in Colombia on epimastigotes of *T. cruzi,* with $IC_{50}$ in the range of 5.5–19.7 μg/mL and two oils from *Lippia alba* (Mill.) N.E.Br. were active on amastigotes forms with $IC_{50}$ values of 12.2 and 17.47 μg/mL. Tests on *L. chagasi* promastigotes of five oils from *Lippia alba, Lippia citrodora* and *Lippia origanoides* from different places in Colombia showed interesting activities (the $IC_{50}$ values ranged from 4.4 to 18.9 μg/mL) (Escobar *et al.*, 2010).

The effect of the EO of *Lippia sidoides* Cham. was evaluated on *L. chagasi* promastigotes and amastigotes and the toxicity was also examined on the murine monocyte cell line RAW 264.7. The EO was effective against both promastigotes ($IC_{50}$ = 19.76 μg/mL) and amastigotes ($IC_{50}$ = 5.07 μg/mL) and viability of 57.8% was shown at a concentration of 100 μg/mL on murine monocyte cells (Rondon *et al.*, 2012). The EO was also assessed for its *in vitro* activity against *P. falciparum* ($IC_{50}$ = 15.20 ± 3.30 μg/mL) and for its *in vivo* efficacy in mice infected with *P. berghei* by subcutaneous route (inhibition of 55% at 100 mg/kg per day). The acute toxicity was assessed in healthy mice ($IC_{50}$ = 7 g/kg) and *in vitro* cytotoxicity was determined against HeLa cells ($IC_{50}$ = 340 ± 34 μg/mL) and mice macrophages

($IC_{50} = 415 \pm 30$ µg/mL) (Mota *et al.*, 2012). This oil also inhibited trypomastigote forms of *T. cruzi* (Dm28c) ($IC_{50} = 10.3 \pm 0.5$ µg/mL). Its cytotoxicity was also assessed against macrophages from Balb/c mice, and $IC_{50}$ was evaluated at 192.7 ± 0.3 µg/mL (Borges *et al.*, 2012).

The volatile oil of *Lippia javanica* Spreng was tested for antiplasmodial activity against *P. falciparum* (D10) and showed $IC_{50}$ of 8 µg/mL (Manenzhe *et al.*, 2004).

The EO of *Lippia multiflora* Moldenke showed $IC_{50}$ of 1/21 000 (2–4 µg/mL) against FcB1 and F32 strains of *P. falciparum* (Valentin *et al.*, 1995).

## 33.3 Compounds isolated from essential oils

The structures of the pure compounds found to have the most antiparasitic effect are shown in Table 33.1. The most antiparasitic compounds isolated from EOs ($IC_{50} \leq 20$ µM) were as follows:

### 33.3.1 Antiplasmodial activity

- α-Pinene, a hydrocarbon isolated from the aerial part of *Daucus crinitus* Desf. (Apiaceae) (Lanfranchi *et al.*, 2010), but also present in different other EOs, showing an antiplasmodial activity with $IC_{50}$ of 1.2 µM on W2 (Van Zyl *et al.*, 2006; Lanfranchi *et al.*, 2010).
- The ester linalylacetate, isolated from the leaves and fruits of *T. asiatica* Lam. (Rutaceae) (Nyahanga *et al.*, 2010) with $IC_{50}$ of 1.4 µM on W2 (Nyahanga *et al.*, 2010; Van Zyl *et al.*, 2006).
- The sesquiterpenic alcohols (*E,Z*)-(±)-nerolidol, isolated from the aerial parts of *Baccharis dracunculifolia* DC. (Asteraceae), with $IC_{50}$ of 0.9 µM on W2 (Parreira *et al.*, 2010).
- The ketone (−)-pulegone, isolated from the aerial parts of *Mentha longifolia* L. (Lamiaceae) (Oyedeji and Afolayan, 2006) with $IC_{50}$ of 1.1 µM on W2 (Oyedeji and Afolayan, 2006; Van Zyl *et al.*, 2006).
- Isochavicol and thymol, two phenols (one phenylpropanoid and one monoterpene), isolated respectively from the aerial parts of *Daucus crinitus* Desf. (Apiaceae) and *Satureja thymbra* L. (Lamiaceae) with $IC_{50}$ of 14.16 µM and 18.64 µM on W2 respectively (Lanfranchi *et al.*, 2010; Dell'Agli *et al.*, 2012). Thymol (**7**) was also tested on D10 strain and was less active ($IC_{50}$ 41.27 µM) (Dell'Agli *et al.*, 2012).
- Three ketones, thymoquinone, hinokitiol and (−)-*R*-carvone, isolated from the wood of *Thujopsis dolabrata* Hort. (Cupressaceae) and the aerial parts of *Monarda fistulosa* Hook. (Lamiaceae) and *Mentha spicata* L. (Lamiaceae) respectively, with $IC_{50}$ of 1.22 µM, 6.09 µM and 6.67 µM respectively on FCR-3/Gambia, which is sensitive to chloroquine and artemisinin (Fujisaki *et al.*, 2012).

### 33.3.2 Antileishmanial activity

The monoterpenic alcohol (±)-linalool isolated from the leaves of *C. cajucara* Benth. (Leguminosae) showed antileishmanial activity with $IC_{50}$ of 0.03 µM and 0.1 µM respectively on promastigotes and amastigotes of *L. amazonensis* (Raimundo strain, MHOM/BR/76/Ma-5) (Rosa *et al.*, 2003). (+)-δ-Cadinene, a sesquiterpenic hydrocarbon isolated from aerial parts of *Ocimum basilicum* L. (Lamiaceae), showed antileishmanial activity with $IC_{50}$ of 19.57 µM against *L. donovani* promastigote (Zheljazkov *et al.*, 2008).

### 33.3.3 Antitrypanosomal activity

(*E*)-Nerolidol (**3**) and (−)-terpinen-4-ol, two monoterpenic alcohols isolated from the leaves of *Strychnos spinosa* Lam. (Loganiaceae) and from the seeds of *Moringa stenopetala* (Baker f.) Cufod. (Moringaceae), showed antitrypanosomal activity with $IC_{50}$ of 7.64 µM and 0.13 µM respectively on bloodstream forms of *T. brucei brucei* (Hoet et al., 2006). The aromatic sulfur-containing compound benzyl isothiocyanate, cinnamaldehyde (a phenylpropanoid) and the sesquiterpenic hydrocarbon (−)-α-cedrene were isolated from the seeds of *Moringa stenopetala* (Baker f.) Cufod. (Moringaceae) and showed antitrypanosomal activity with $IC_{50}$ of 8.5 µM, 22.19 µM and 19.92 µM respectively on bloodstream forms of *T. brucei brucei* (Nibret and Wink, 2010). Alloaromadendrene, a sesquiterpenic hydrocarbon isolated from *Sclerocarya birrea* (A. Rich.) Hochst (Kpoviessi *et al.*, 2011), showed antitrypanosomal activity with $IC_{50}$ of 9.30 µM on bloodstream forms of *T. brucei brucei* (Nibret and Wink, 2010).

## 33.4 Discussion and conclusion

Several plants and EOs are used in traditional medicine in many countries for the treatment of parasitic diseases. This review shows that some of them may be considered, as well as some of their constituents, as having good ($IC_{50} < 2$ µg/mL or µM) or moderate activities ($IC_{50} < 21$ µg/mL or µM). Our literature review indicates that 123 EOs possessed interesting antiparasitic activities *in vitro* on at least one parasite. Most of them were tested for their antileishmanial activity (70), about half as many for their antiplasmodial effect (36) and 26 on *Trypanosoma* (Figure 33.2). Some of these oils were also tested for their cytotoxicity, but only a few of them for their *in vivo* antiparasitic effect. This is the case for EOs from *Vanillosmopsis arborea* (Gardner) Baker (Asteraceae), *Croton zehntneri* Pax & K. Hoffm. (Euphorbiaceae), *Ocimum gratissimum* L. (Lamiaceae), *Cymbopogon citratus* Stapf (Poaceae) and *Lippia sidoides* Cham. (Verbenaceae), which were evaluated for their *in vivo* activity in mice infected with *P. berghei*; by subcutaneous route for *Vanillosmopsis arborea*, with inhibitions of 33% and 47% parasites at 50 mg/kg per day and 100 mg/kg per day, respectively; by oral route for *Croton zehntneri*, with inhibitions of 28% and 53.50% at 100 mg/kg per day and 500 mg/kg per day respectively; by oral route for *Ocimum gratissimum*, with inhibitions of 55.0%, 75.2% and 77.8% at 200 mg/kg per day, 300 mg/kg per day and 500 mg/kg per day respectively; by oral route for *Cymbopogon citratus*, with inhibitions of 62.1%, 81.7% and 86.6 % at 200 mg/kg per day, 300 mg/kg per day and 500 mg/kg per day respectively; and by subcutaneous route for *Lippia sidoides*, with an inhibition of 55% at 100 mg/kg per day (Tchoumbougnang *et al.*, 2005; Mota *et al.*, 2012). EOs of *Carum carvi* L. (Apiaceae) and *Santalum album* L. (Santalaceae) showed *in vivo* antimalarial activity against *P. berghei* on mice with an effective dose of 50 µl and 200 µl respectively by subcutaneous administration with a significant reduction in parasitaemia on day 8 (8.3% and 22.7% of parasitaemia respectively) (Fujisaki *et al.*, 2012).

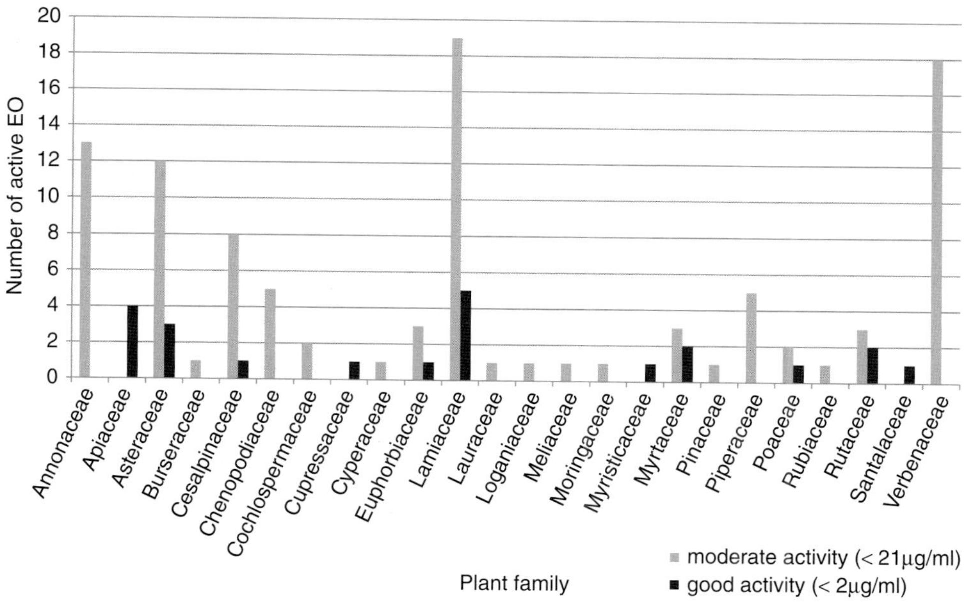


**Figure 33.2** Number of active EOs *in vitro* (on at least a parasite), classified according to their plant families

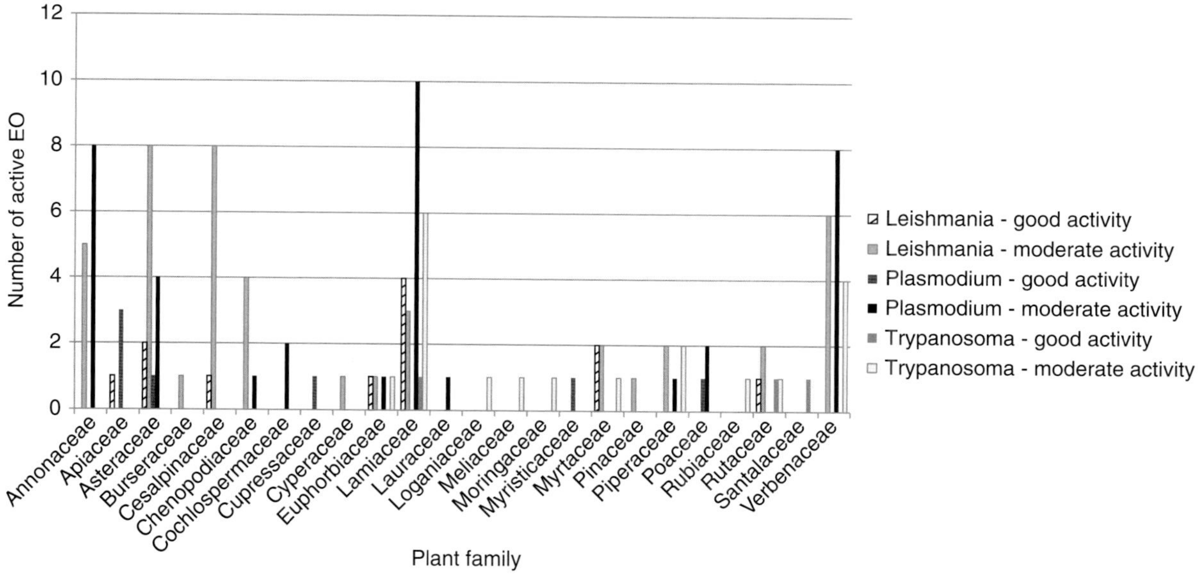


**Figure 33.3** Number of active EOs *in vitro* with good ($IC_{50}$ < 2 µg/mL) or moderate activity ($IC_{50} \leq 20$ µg/mL) on *Leishmania*, *Trypanosoma* and *Plasmodium*, classified according to their plant families

Among them, only a few exhibited what can be considered as a good activity *in vitro* ($IC_{50} \leq 2$ µg/mL) (Figure 33.3). When considering families from which these EOs were extracted, we observed that they belong mainly to Apiaceae, Asteraceae, Cesalpinaceae, Cupressaceae, Euphorbiaceae, Lamiaceae, Myristicaceae, Myrtaceae, Poaceae, Rutaceae and Santalaceae.

There are different sources of antileishmanial oils (12) with good activities (Apiaceae, Asteraceae, Caesalpinaceae, Euphorbiaceae, Lamiaceae, Myrtaceae and Rutaceae), while interesting antiplasmodial oils (7) were mainly found in Apiaceae, Asteraceae, Cupressaceae, Myristicaceae and Poaceae and antitrypanosomal oils (3) in Lamiaceae, Rutaceae and Santalaceae (Figure 33.3). However, it is possible that highly active compounds are isolated from oils with only moderate activity.

The mode of action of these oils was usually not analysed, but two studies evaluated oils against a specific enzyme of the parasite:

- The leaf EO of two species of *Ocotea* from Monteverde, Costa Rica – *Ocotea meziana* C.K.Allen and *Ocotea whitei* Woodson – were tested for their enzyme inhibitory activities against cruzain, a cysteine protease from *T. cruzi* that is key to replication and differentiation of the parasite and has been identified as an important biochemical target for treatment of this parasitic infection. The $IC_{50}$ values are 14.9 ± 0.9 µg/mL and 15.8 ± 0.2 µg/mL, respectively (Setzer *et al.*, 2006).

- The nerolidol-rich EO from *Piper claussenianum* (Miq.) C. DC. (Piperaceae) was assayed on arginase activity of *L. amazonensis*. The results showed an enzyme inhibition of 62.2% at the ICSO. The EO showed no toxic effect to fibroblasts or to macrophages cell lines in any concentration tested (ranging from 40 to 0.56 mg/mL) (Marques *et al.*, 2011).

Different compounds (mono- and sesquiterpenes, phenolics, phenylpropanoids) isolated from active antiparasitic EOs were separately tested against *Plasmodium*, *Trypanosoma* and *Leishmania* (Table 33.1). Some of them were hydrocarbons, but others were alcohols, phenols, ethers, epoxides, aldehydes, ketones, esters or isothiocyanate. Our review showed that only 50 volatile pure compounds were separately tested against at least one of the three parasites, with 23 tested against *Plasmodium*, 36 against *Trypanosoma* and 20 on *Leishmania*. Among them, five compounds – α-pinene, linalylacetate, (*E*,*Z*)-(±)-nerolidol, (−)-pulegone and thymoquinone – showed a good antiplasmodial activity ($IC_{50}$ < 2 μM). The most active were the sesquiterpenic alcohols (*E*,*Z*)-(±)-nerolidol, with $IC_{50}$ = 0.9 μM (Van Zyl *et al.*, 2006). Only one compound, the monoterpenic alcohol (±)-linalool, showed a good antileishmanial effect, and (−)-terpinen-4-ol showed a good antitrypanosomal activity. The other compounds were moderately active or less interesting. This low number of compounds tested compared with EOs can be explained by a problem of availability. More often, the activities of the pure compounds

**Table 33.1** Pure compounds shown to possess some antiparasitic activities

| Compounds | Chemical class | | Chemical subclass | Activity ($IC_{50}$) (μM) | | | Reference |
|---|---|---|---|---|---|---|---|
| Name | Structure* | | | *Plasmodium* | *Trypanosoma* | *Leishmania* | |
| (−)-terpinen-4-ol | 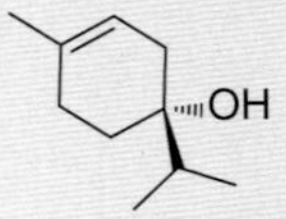 | monoterpenes | alcohols | – | 0.13 | 2177.63 | (Mikus *et al.*, 2000) |
| (±)-linalool |  | monoterpenes | alochols | 254.4 | 254.91 | 0.03<br>0.1 | (Van Zyl *et al.*, 2006)<br>(Nibret and Wink, 2010)<br>(Rosa *et al.*, 2003) |
| (−)-myrtenal | 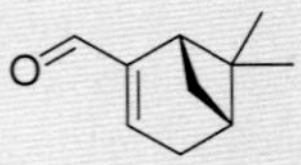 | monoterpenes | aldehydes | – | 114.77 | – | (Nibret and Wink, 2010) |
| (±)-citronellal | 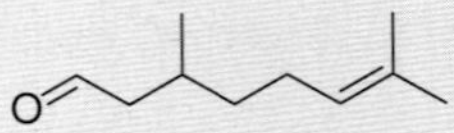 | monoterpenes | aldehydes | 698.5 | 86.52 | – | (Van Zyl *et al.*, 2006)<br>(Nibret and Wink, 2010) |
| citral | 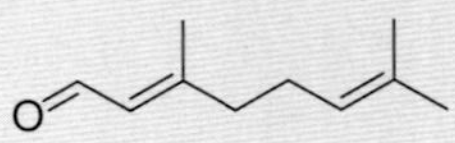 | monoterpenes | aldehydes | – | – | 52.55 | (Machado *et al.*, 2012) |
| limonene epoxide | 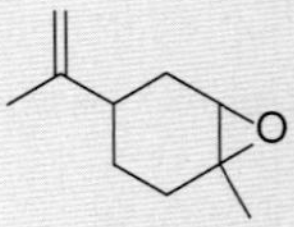 | monoterpenes | epoxides | – | 148.32 | – | (Nibret and Wink, 2010) |
| isobornyl acetate | 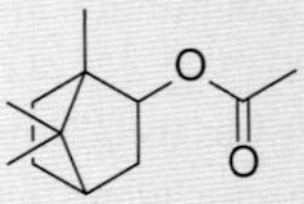 | monoterpenes | esters | – | 203.22 | – | (Nibret and Wink, 2010) |
| linalylacetate | 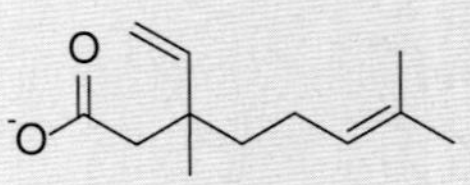 | monoterpenes | esters | 1.4 | – | – | (Van Zyl *et al.*, 2006) |
| (−)-rose oxide | 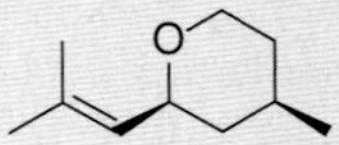 | monoterpenes | ethers | – | 600.45 | – | (Nibret and Wink, 2010) |

**Table 33.1** (*Continued*)

| Compounds | Chemical class | | Chemical subclass | Activity ($IC_{50}$) (μM) | | | Reference |
|---|---|---|---|---|---|---|---|
| Name | Structure* | | | *Plasmodium* | *Trypanosoma* | *Leishmania* | |
| 1,8-cineole | 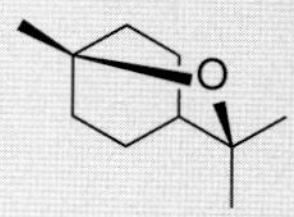 | monoterpenes | ethers | 70.2 | 539.06 | – | (Van Zyl *et al.*, 2006) (Nibret and Wink, 2010) |
| γ-terpinene | 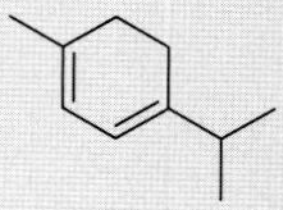 | monoterpenes | hydrocarbons | – | 1006.7 | 1065.11 | (Escobar *et al.*, 2010) |
| myrcene | 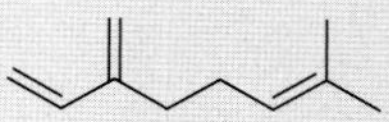 | monoterpenes | hydrocarbons | – | – | 1203.85 | (Machado *et al.*, 2012) |
| R(+)-limonene | 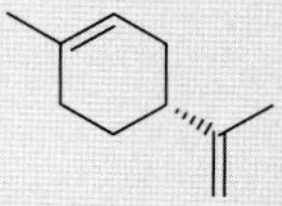 | monoterpenes | hydrocarbons | 533.5 | 260.96 | 261.0 | (Escobar *et al.*, 2010) |
| sabinene | 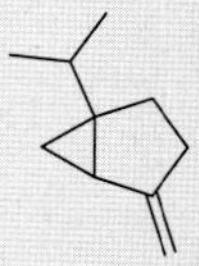 | monoterpenes | hydrocarbons | – | 129.93 | 929.31 | (Mikus *et al.*, 2000) |
| α-phellandrene | 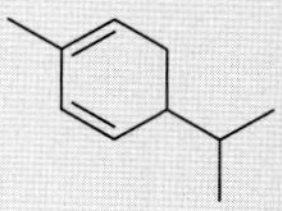 | monoterpenes | hydrocarbons | – | 67.53 | 240.77 | (Mikus *et al.*, 2000) |
| α-pinene | 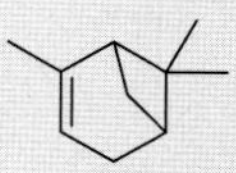 | monoterpenes | hydrocarbons | 1.2 | 30.12 | 406.23 | (Van Zyl *et al.*, 2006) (Mikus *et al.*, 2000) |
| β-pinene | 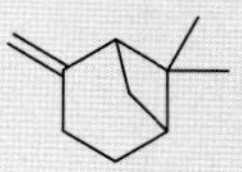 | monoterpenes | hydrocarbons | 318.5 | 402.56 | 1469.92 | (Van Zyl *et al.*, 2006) (Mikus *et al.*, 2000) |
| (−)-pulegone | 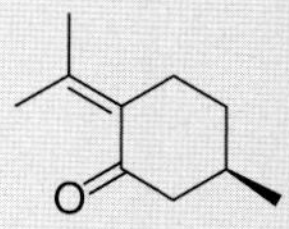 | monoterpenes | ketones | 1.1 | – | – | (Van Zyl *et al.*, 2006) |
| (−)-*R*-carvone | 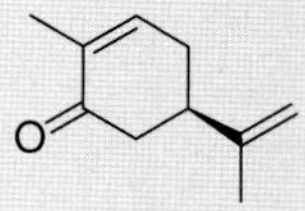 | monoterpenes | ketones | 6.67 | 86.14 | 1192.25 | (Fujisaki *et al.*, 2012) (Nibret and Wink, 2010) (Escobar *et al.*, 2010) |
| (−)-verbenone | 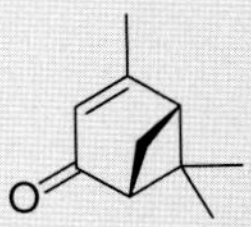 | monoterpenes | ketones | – | 181.88 | – | (Nibret and Wink, 2010) |

(*continued*)

**Table 33.1** (*Continued*)

| Compounds Name | Chemical class Structure* | | Chemical subclass | Activity ($IC_{50}$) (μM) *Plasmodium* | *Trypanosoma* | *Leishmania* | Reference |
|---|---|---|---|---|---|---|---|
| (+)-*S*-carvone | 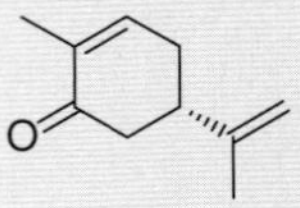 | monoterpenes | ketones | 66.57 | – | 1997.1 | (Escobar *et al.*, 2010) (Van Zyl *et al.*, 2006) |
| camphor | 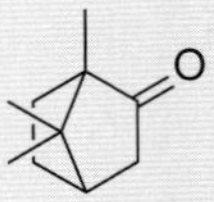 | monoterpenes | ketones | – | 245.62 | – | (Nibret and Wink, 2010) |
| hinokitiol | 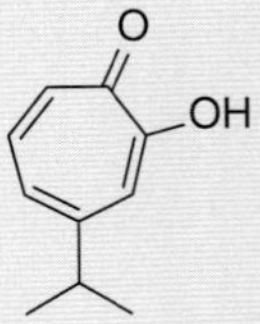 | monoterpenes | ketones | 6.09 | – | – | (Fujisaki *et al.*, 2012) |
| piperitone | 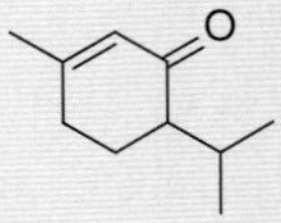 | monoterpenes | ketones | – | 270.12 | – | (Nibret and Wink, 2010) |
| thujone | 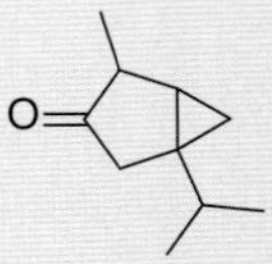 | monoterpenes | ketones | 528.6 | 254.81 | – | (Van Zyl *et al.*, 2006) (Nibret and Wink, 2010) |
| thymoquinone | 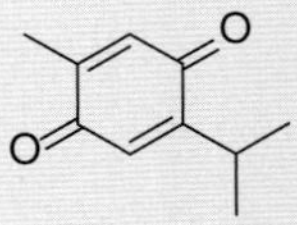 | monoterpenes | ketones | 1.22 | – | – | (Fujisaki *et al.*, 2012) |
| carvacrol | 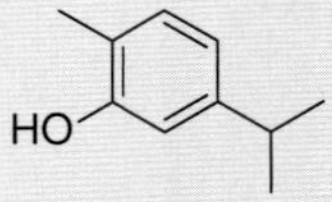 | monoterpenes | phenols | 1067.9 | 74.89 | 186.39 | (Escobar *et al.*, 2010) |
| thymol | 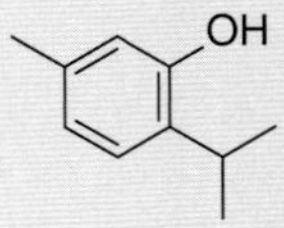 | monoterpenes | phenols | 18.64 41.27 | 152.18 | 434.03 | (Dell'Agli *et al.*, 2012) (Escobar *et al.*, 2010) (Mota *et al.*, 2012) |
| epoxystyrene | 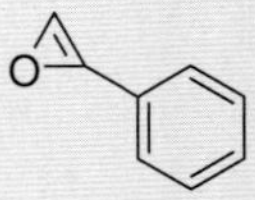 | phenolics | epoxides | – | 31.29 | – | (Nibret and Wink, 2010) |
| benzyl isothiocyanate | 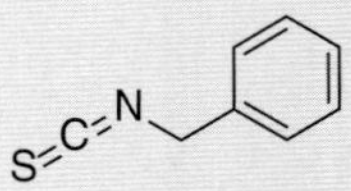 | phenolics | sulphur containing component | – | 8.04 | – | (Nibret and Wink, 2010) |

**Table 33.1** (*Continued*)

| Compounds | | Chemical class | Chemical subclass | Activity ($IC_{50}$) (μM) | | | Reference |
|---|---|---|---|---|---|---|---|
| Name | Structure* | | | *Plasmodium* | *Trypanosoma* | *Leishmania* | |
| cinnamaldehyde | 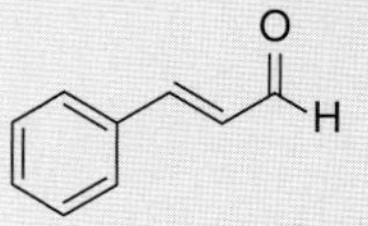 | phenylpropanoids | aldehydes | – | 22.19 | – | (Nibret and Wink, 2010) |
| ethyl cinnamate | 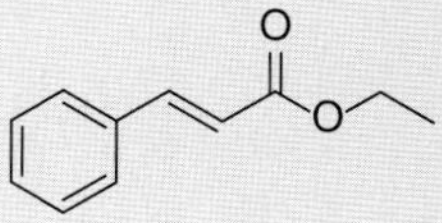 | phenylpropanoids | esters | – | 160.32 | – | (Nibret and Wink, 2010) |
| isochavicol isobutyrate | 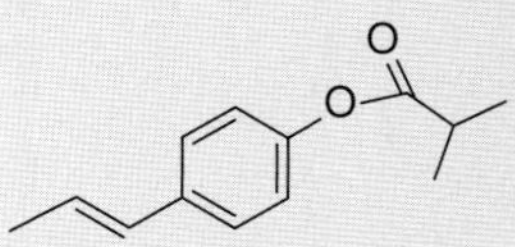 | phenylpropanoids | esters | 68.54 | – | – | (Lanfranchi *et al.*, 2010) |
| isochavicol propionate | 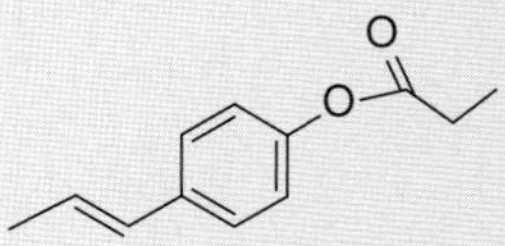 | phenylpropanoids | esters | 69.91 | – | – | (Lanfranchi *et al.*, 2010) |
| estragole | 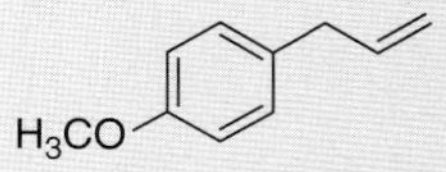 | phenylpropanoids | ethers | 186.23<br>228.07 | 216.46 | – | (Van Zyl *et al.*, 2006) (Nibret and Wink, 2010) |
| safrol | 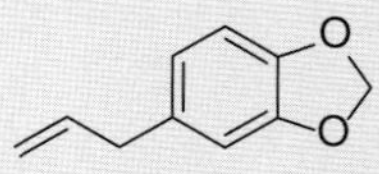 | phenylpropanoids | ethers | – | 113.45 | – | (Nibret and Wink, 2010) |
| α-asarone | 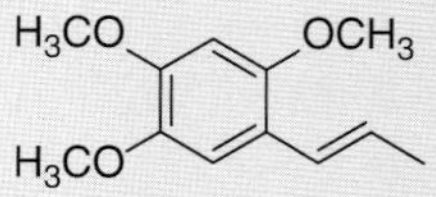 | phenylpropanoids | ethers | – | 96.95 | – | (Nibret and Wink, 2010) |
| eugenol | 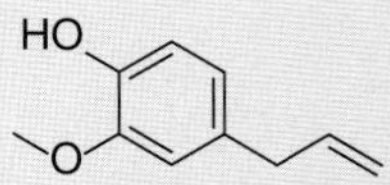 | phenylpropanoids | phenols | 753.7 | 226.55 | – | (Nibret and Wink, 2010) |
| isochavicol | 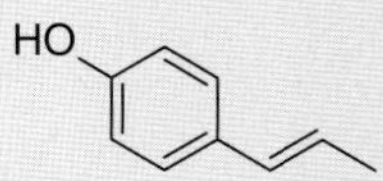 | phenylpropanoids | phenols | 14.16 | – | – | (Lanfranchi *et al.*, 2010) |
| (*E*)-nerolidol | 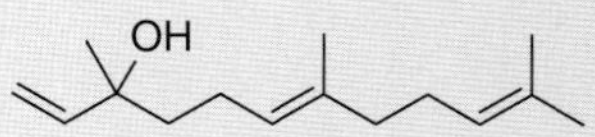 | sesquiterpenes | alcohols | 0.9 | 7.64 | – | (Van Zyl *et al.*, 2006) (Hoet *et al.*, 2006) |

(*continued*)

**Table 33.1** (*Continued*)

| Compounds | Chemical class | | Chemical subclass | Activity ($IC_{50}$) (μM) | | | Reference |
|---|---|---|---|---|---|---|---|
| Name | Structure* | | | *Plasmodium* | *Trypanosoma* | *Leishmania* | |
| (*Z*)-nerolidol | 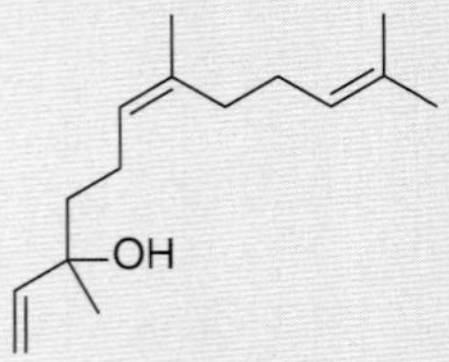 | sesquiterpenes | alcohols | 0.9 | 70.96 | – | (Van Zyl *et al.*, 2006) (Nibret and Wink, 2010) |
| α-bisabolol | 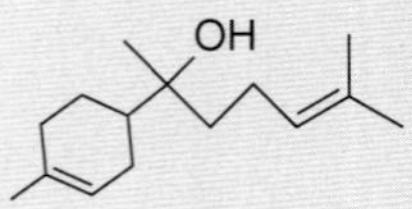 | sesquiterpenes | alcohols | 307.3 | – | 49.42 | (Mota *et al.*, 2012) (Morales-Yuste *et al.*, 2010) |
| caryophyllene oxide | 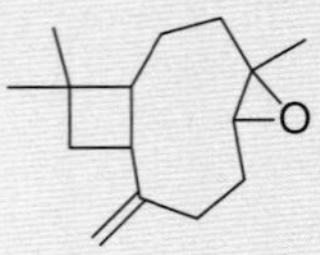 | sesquiterpenes | epoxides | – | 80.33 | – | (Nibret and Wink, 2010) |
| (−)-α-cedrene | 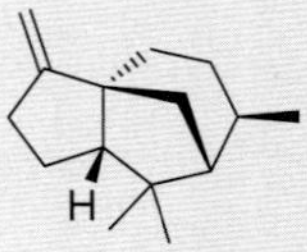 | sesquiterpenes | hydrocarbons | – | 19.92 | – | (Nibret and Wink, 2010) |
| (+)-δ-cadinene | 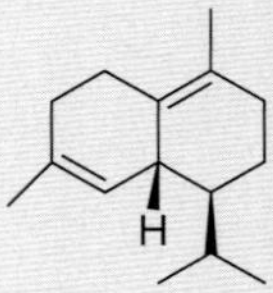 | sesquiterpenes | hydrocarbons | – | – | 19.57 | (Zheljazkov *et al.*, 2008) |
| alloaromadendrene | 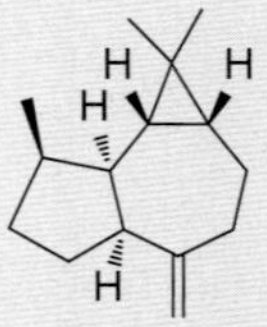 | sesquiterpenes | hydrocarbons | – | 9.30 | 451.68 | (Mikus *et al.*, 2000) |
| aromadendrene | 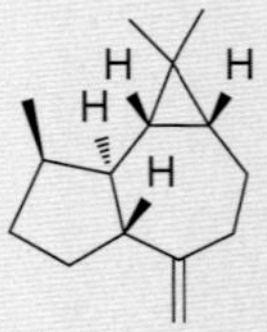 | sesquiterpenes | hydrocarbons | – | 92.00 | 312.21 | (Mikus *et al.*, 2000) |
| α-humulene | 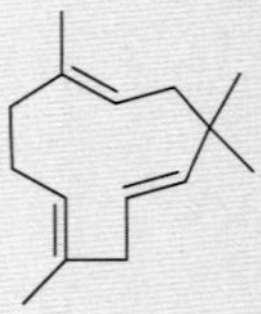 | sesquiterpenes | hydrocarbons | – | – | 92.98 | (Zheljazkov *et al.*, 2008) |
| β-caryophyllene | 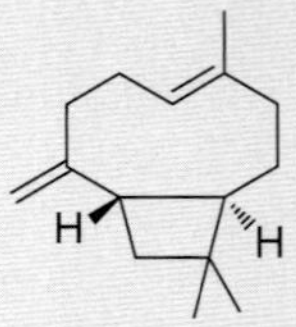 | sesquiterpenes | hydrocarbons | – | 67.43 | 96 | (Moura do Carmo *et al.*, (2012) |

tested separately could not explain the strong activity of their corresponding EOs, but synergic effects are possible.

This review shows that EOs could be considered as a source of new antiparasitic extracts/compounds. However, EOs are known to have variable compositions depending on their growing conditions, stages and periods of harvesting, and methods of EO extraction, although in some cases no variability was observed for different oils of the same species, so results on only one sample must be repeated on other samples from the same species of the same chemotype.

Considering pure compounds present in these oils, most of them were not tested alone or could not explain the high activity of the EO. Further research is needed to test more pure compounds, evaluating synergistic activities between them (as can be found in EO), but also with synthetic antiparasitic drugs.

Only a limited number of EOs or compounds were tested *in vivo* and for their toxicity. This is very important to do before assessing or using these compounds in therapy. Furthermore, further research is needed to analyse their mode(s) of action.

## References


Ahmed, S.B., Sghaier, R.M., Guesmi, F. *et al.* (2011) Evaluation of antileishmanial, cytotoxic and antioxidant activities of essential oils extracted from plants issued from the leishmaniasis-endemic region of Sned (Tunisia). *Natural Product Research*, **25**, 1195–1201.

Anthony, J.P., Fyfe, L. and Smith, H. (2005) Plant active components – a resource for antiparasitic agents? *Trends in Parasitology*, **21**, 462–468.

Arévalo, Y., Robledo, S., Muñoz, D.L. *et al.* (2009) *In vitro* evaluation of essential oils activity from Colombian plants on *Leishmania brasiliensis*. *Revista Colombiana de Ciencias Químico - Farmacéuticas*, **38**, 131–141 (in Spanish).

Bakkali, F., Averbeck, S., Averbeck, D. and Idaomar, M. (2008) Biological effects of essential oils – a review. *Food and Chemical Toxicology*, **46**, 446–475.

Benoit-Vical, F., Valentin, A. and Mallie, M. (2001) Antiplasmodial activity of *Cochlospermum planchonii* and *C. tinctorium* tubercle essential oils. *Journal of Essential Oil Research*, **13**, 65–67.

Bero, J., Beaufay, C., Hannaert, V. *et al.* (2013) Antitrypanosomal compounds from the essential oil and extracts of *Keetia leucantha* leaves with inhibitor activity on *Trypanosoma brucei* glyceraldehyde-3-phosphate dehydrogenase. *Phytomedicine*, **20**, 270–274.

Borges, A.R., Aires, J.R., Higino, T.M. *et al.* (2012) Trypanocidal and cytotoxic activities of essential oils from medicinal plants of northeast of Brazil. *Experimental Parasitology*, **132**, 123–128.

Boyom, F.F., Ngouana, V., Zollo, P.H. *et al.* (2003) Composition and anti-plasmodial activities of essential oils from some Cameroonian medicinal plants. *Phytochemistry*, **64**, 1269–1275.

Boyom, F.F., Ngouana, V., Kemgne, E.A. *et al.* (2011) Antiplasmodial volatile extracts from *Cleistopholis patens* Engler & Diels and *Uvariastrum pierreanum* Engl. (Engl. & Diels) (Annonaceae) growing in Cameroon. *Parasitology Research*, **108**, 1211–1217.

Costa, E.V., Pinheiro, M.L., de Andrade Silva, J.R. *et al.* (2009) Antimicrobial and antileishmanial activity of essential oil from the leaves of *Annona foetida* (Annonaceae). *Quimica Nova*, **32**, 78–81.

Dell'Agli, M., Sanna, C., Rubiolo, P. *et al.* (2012) Anti-plasmodial and insecticidal activities of the essential oils of aromatic plants growing in the Mediterranean area. *Malararia Journal*, **11**, 219.

Drugs for Neglected Diseases Initiatives (2011) Human African Trypanosomiasis, www.dndi.org/diseases-projects/diseases/hat.html (accessed 28 November 2013).

Escobar, P., Milena, L.S., Herrera, L.V. *et al.* (2010) Chemical composition and antiprotozoal activities of Colombian *Lippia* spp essential oils and their major components. *Memórias do Instituto Oswaldo Cruz*, **105**, 184–190.

Fujisaki, R., Kamei, K., Yamamura, M. *et al.* (2012) *In vitro* and *in vivo* anti-plasmodial activity of essential oils, including hinokitiol. *Southeast Asian Journal of Tropical Medicine and Public Health*, **43**, 270–279.

Hatimi, S., Boudouma, M., Bichichi, M. *et al.* (2001) *In vitro* evaluation of antileishmania activity of *Artemisia herba alba* Asso. *Bulletin de la Société de Pathologie Exotique*, **94**, 29–31 (in French).

Hoet, S., Stevigny, C., Herent, M.F. and Quetin-Leclercq, J. (2006) Antitrypanosomal compounds from the leaf essential oil of *Strychnos spinosa*. *Planta Medica*, **72**, 480–482.

Kamatou, G.P., Viljoen, A.M., Gono-Bwalya, A.B. *et al.* (2005) The *in vitro* pharmacological activities and a chemical investigation of three South African *Salvia* species. *Journal of Ethnopharmacology*, **102**, 382–390.

Kamatou, G.P.P., Viljoen, A.M., Figueiredo, A.C. *et al.* (2007) Trichomes, essential oil composition and biological activities of *Salvia albicaulis* Benth. and *S. dolomitica* Codd, two species from the Cape region of South Africa. *South African Journal of Botany*, **73**, 102–108.

Khamsan, S., Liawruangrath, B., Liawruangrath, S. *et al.* (2011) Antimalarial, anticancer, antimicrobial activities and chemical constituents of essential oil from the aerial parts of *Cyperus kyllingia* Endl. *Records of Natural Products*, **5**, 324–327.

Kpoviessi, S., Gbaguidi, F., Kossouoh, C. *et al.* (2011) Chemical composition and seasonal variation of essential oil of *Sclerocarya birrea* (A. Rich.) Hochst subsp birrea leaves from Benin. *Journal of Medicinal Plants Research*, **5**, 4640–4646.

Lanfranchi, D.A., Laouer, H., El Kolli, M. *et al.* (2010) Bioactive phenylpropanoids from *Daucus crinitus* Desf. from Algeria. *Journal of Agricultural and Food Chemistry*, **58**, 2174–2179.

Laouer, H., Hireche-Adjal, Y., Prado, S. *et al.* (2009) Chemical composition and antimicrobial activity of essential oil of *Bupleurum montanum* and *B. plantagineum*. *Natural Product Communications*, **4**, 1605–1610.

Lopez, R., Cuca, L.E., Delgado, G. (2009) Antileishmanial and immunomodulatory activity of *Xylopia discreta*. *Parasite Immunology*, **31**, 623–630.

Machado, M., Pires, P., Dinis, A.M. *et al.* (2012) Monoterpenic aldehydes as potential anti-*Leishmania* agents: activity of *Cymbopogon citratus* and citral on *L. infantum*, *L. tropica* and *L. major*. *Experimental Parasitology*, **130**, 223–231.

Manenzhe, N.J., Potgieter, N. and van Ree, T. (2004) Composition and antimicrobial activities of volatile components of *Lippia javanica*. *Phytochemistry*, **65**, 2333–2336.

Marques, A.M., Barreto, A.L., Curvelo, J.A. *et al.* (2011) Antileishmanial activity of nerolidol-rich essential oil from *Piper claussenianum*. *Brazilian Journal of Pharmacognosy*, **21**, 908–914.

Mikus, J., Harkenthal, M., Steverding, D. and Reichling, J. (2000) *In vitro* effect of essential oils and isolated mono- and sesquiterpenes on *Leishmania major* and *Trypanosoma brucei*. *Planta Medica*, **66**, 366–368.

Miranda Junior, R.N., Dolabela, M.F., da Silva, M.N. *et al.* (2012) Antiplasmodial activity of the andiroba (*Carapa guianensis* Aubl., Meliaceae) oil and its limonoid-rich fraction. *Journal of Ethnopharmacology*, **142**, 679–683.

Misra, P., Kumar, A., Khare, P. *et al.* (2009) Pro-apoptotic effect of the landrace Bangla Mahoba of *Piper betle* on *Leishmania donovani* may be due to the high content of eugenol. *Journal of Medical Microbiology*, **58**, 1058–1066.

Monzote, L., Montalvo, A.M., Almanonni, S. *et al.* (2006) Activity of the essential oil from *Chenopodium ambrosioides* grown in Cuba against *Leishmania amazonensis*. *Chemotherapy*, **52**, 130–136.

Monzote, L., Garcia, M., Montalvo, A.M. *et al.* (2007a) *In vitro* activity of an essential oil against *Leishmania donovani*. *Phytotherapy Research*, **21**, 1055–1058.

Monzote, L., Montalvo, A.M., Scull, R. *et al.* (2007b) Activity, toxicity and analysis of resistance of essential oil from *Chenopodium ambrosioides* after intraperitoneal, oral and intralesional administration in BALB/c mice infected with *Leishmania amazonensis*: a preliminary study. *Biomedicine & Pharmacotherapy*, **61**, 148–153.

Monzote, L., Montalvo, A.M., Scull, R. *et al.* (2007c) Combined effect of the essential oil from *Chenopodium ambrosioides* and antileishmanial drugs on promastigotes of *Leishmania amazonensis*. *Revista do Instituto de Medicina Tropical de São Paulo*, **49**, 257–260.

Monzote, L., Garcia, M., Montalvo, A.M. *et al.* (2009) Effect of oral treatment with the essential oil from *Chenopodium ambrosioides* against cutaneous leishmaniasis in BALB/c mice, caused by *Leishmania amazonensis*. *Forschende Komplementärmedizin*, **16**, 334–338.

Monzote, L., Garcia, M., Montalvo, A.M. *et al.* (2010) Chemistry, cytotoxicity and antileishmanial activity of the essential oil from *Piper auritum*. *Memórias do Instituto Oswaldo Cruz*, **105**, 168–173.

Monzote, L., Nance, M.R., Garcia, M. *et al.* (2011) Comparative chemical, cytotoxicity and antileishmanial properties of essential oils from *Chenopodium ambrosioides*. *Natural Product Communications*, **6**, 281–286.

Morales-Yuste, M., Morillas-Marquez, F., Martin-Sanchez, J. *et al.* (2010) Activity of (−)α-bisabolol against *Leishmania infantum* promastigotes. *Phytomedicine*, **17**, 279–281.

Mota, M.L., Lobo, L.T., Costa, J.M. *et al.* (2012) *In vitro* and *in vivo* antimalarial activity of essential oils and chemical components from three medicinal plants found in northeastern Brazil. *Planta Medica*, **78**, 658–664.

Moura do Carmo, D.F., Amaral, A.C., Machado, G.M. *et al.* (2012) Chemical and biological analyses of the essential oils and main constituents of *Piper* species. *Molecules*, **17**, 1819–1829.

Nibret, E. and Wink, M. (2010) Trypanocidal and antileukaemic effects of the essential oils of *Hagenia abyssinica*, *Leonotis ocymifolia*, *Moringa stenopetala*, and their main individual constituents. *Phytomedicine*, **17**, 911–920.

Nwaka, S. and Hudson, A. (2006) Innovative lead discovery strategies for tropical diseases. *Nature Reviews Drug Discovery*, **5**, 941–955.

Nyahanga, T., Jondiko, J.I., Manguro, L.O.A. and Orwa, J.A. (2010) Chemical composition, antiplasmodial, larvicidal and antimicrobial activities of essential oils of *Toddalia asiatica* L. leaves and fruits. *International Journal of Essential Oil Therapeutics*, **4**, 54–58.

Ortet, R., Thomas, O.P., Regalado, E.L. *et al.* (2010) Composition and biological properties of the volatile oil of *Artemisia gorgonum* Webb. *Chemistry & Biodiversity*, **7**, 1325–1332.

Oyedeji, A.O. and Afolayan, A.J. (2006) Chemical composition and antibacterial activity of the essential oil isolated from South African *Mentha longifolia* (L.) L. subsp. *capensis* (Thunb.) Briq. *Journal of Essential Oil Research*, **18**, 57–59.

Parreira, N.A., Magalhaes, L.G., Morais, D.R. *et al.* (2010) Antiprotozoal, schistosomicidal, and antimicrobial activities of the essential oil from the leaves of *Baccharis dracunculifolia*. *Chemistry & Biodiversity*, **7**, 993–1001.

Pink, R., Hudson, A., Mouries, M.A. and Bendig, M. (2005) Opportunities and challenges in antiparasitic drug discovery. *Nature Reviews Drug Discovery*, **4**, 727–740.

Rios, Y.K., Otero, A.C., Muñoz, D.L. *et al.* (2008) *In vitro* leishmanicidal and cytotoxic activities of *Matricaria chamomilla* essential oil. *Revista Colombiana de Ciencias Químico – Farmacéuticas*, **37**, 200–211 (in Spanish).

Rondon, F.C., Bevilaqua, C.M., Accioly, M.P. *et al.* (2012) *In vitro* efficacy of *Coriandrum sativum*, *Lippia sidoides* and *Copaifera reticulata* against *Leishmania chagasi*. *Revista Brasileira de Parasitologia Veterinaria*, **21**, 185–191.

Rosa, M.S., Mendonca-Filho, R.R., Bizzo, H.R. *et al.* (2003) Antileishmanial activity of a linalool-rich essential oil from *Croton cajucara*. *Antimicrobial Agents and Chemotherapy*, **47**, 1895–1901.

Saeidnia, S., Gohari, A.R. and Hadjiakhoondi, A. (2007a) Trypanocidal activity of the essential oil of *Satureja macrantha* and its main volatile components. *International Journal of Essential Oil Therapeutics*, **1**, 184–186.

Saeidnia, S., Gohari, A.R., Hadjiakhoondi, A. and Shafiee, A. (2007b) Bioactive compounds of the volatile oil of *Dracocephalum kotschyi*. *Zeitschrift für Naturforschung C: Biosciences*, **62**, 793–796.

Saeidnia, S., Gohari, A.R. and Hadjiakhoondi, A. (2008) Trypanocidal activity of oil of the young leaves of *Nepeta cataria* L. obtained by solvent extraction. *Journal of Medicinal Plants*, **7**, 54–57.

Santin, M.R., dos Santos, A.O., Nakamura, C.V. *et al.* (2009) *In vitro* activity of the essential oil of *Cymbopogon citratus* and its major component (citral) on *Leishmania amazonensis*. *Parasitology Research*, **105**, 1489–1496.

Santoro, G.F., Cardoso, M.G., Guimaraes, L.G. *et al.* (2007) Antiproliferative effect of the essential oil of *Cymbopogon citratus* (DC) Stapf (lemongrass) on intracellular amastigotes, bloodstream trypomastigotes and culture epimastigotes of *Trypanosoma cruzi* (Protozoa: Kinetoplastida). *Parasitology*, **134**, 1649–1656.

Santos, A.O., Ueda-Nakamura, T., Dias Filho, B.P. *et al.* (2008) Effect of Brazilian copaiba oils on *Leishmania amazonensis*. *Journal of Ethnopharmacology*, **120**, 204–208.

Santos, A.O., Santin, A.C., Yamaguchi, M.U. *et al.* (2010) Antileishmanial activity of an essential oil from the leaves and flowers of

*Achillea millefolium*. *Annals of Tropical Medicine and Parasitology*, **104**, 475–483.

Sariego, I., Monzote, L., Scull, R. *et al.* (2008) Activity of essential oils from cuban plants against *Leishmania donovani* and *Trichomonas vaginalis*. *International Journal of Essential Oil Therapeutics*, **2**, 172–174.

Setzer, W.N., Takaku, S., Stokes, S.L. and Penton, A.F. (2006) Inhibition of cruzain by *Ocotea* leaf essential oils from Monteverde, Costa Rica. *Pharmacologyonline*, **3**, 785–793.

Silva, J.R., do Carmo, D.F.M., Reis, E.M. *et al.* (2009) Chemical and biological evaluation of essential oils with economic value from Lauraceae species. *Journal of the Brazilian Chemical Society*, **20**, 1071–1076.

Sulsen, V., Cazorla, S.I., Frank, F.M. *et al.* (2008) *In vitro* antiprotozoal activity and chemical composition of *Ambrosia tenuifolia* and *A. scabra* essential oils. *Natural Product Communications*, **3**, 557–562.

Tabanca, N., Demirci, B., Baser, K.H.C. *et al.* (2006) Chemical composition and antifungal activity of *Salvia macrochlamys* and *Salvia recognita* essential oils. *Journal of Agricultural and Food Chemistry*, **54**, 6593–6597.

Tariku, Y., Hymete, A., Hailu, A. and Rohloff, J. (2010a) Constituents, antileishmanial activity and toxicity profile of volatile oil from berries of *Croton macrostachyus*. *Natural Product Communications*, **5**, 975–980.

Tariku, Y., Hymete, A., Hailu, A. and Rohloff, J. (2010b) Essential-oil composition, antileishmanial, and toxicity study of *Artemisia abyssinica* and *Satureja punctata* ssp. *punctata* from Ethiopia. *Chemistry & Biodiversity*, **7**, 1009–1018.

Tariku, Y., Hymete, A., Hailu, A. and Rohloff, J. (2011) *In vitro* evaluation of antileishmanial activity and toxicity of essential oils of *Artemisia absinthium* and *Echinops kebericho*. *Chemistry & Biodiversity*, **8**, 614–623.

Tchoumbougnang, F., Zollo, P.H., Dagne, E. and Mekonnen, Y. (2005) *In vivo* antimalarial activity of essential oils from *Cymbopogon citratus* and *Ocimum gratissimum* on mice infected with *Plasmodium berghei*. *Planta Medica*, **71**, 20–23.

Valentin, A., Pelissier, Y., Benoit, F. *et al.* (1995) Composition and antimalarial activity *in vitro* of volatile components of *Lippia multiflora*. *Phytochemistry*, **40**, 1439–1442.

Van Vuuren, S.F., Viljoen, A.M., Van Zyl, R.L. *et al.* (2006) The antimicrobial, antimalarial and toxicity profiles of helihumulone, leaf essential oil and extracts of *Helichrysum cymosum* (L.) D. Don subsp. *cymosum*. *South African Journal of Botany*, **72**, 287–290.

Van Zyl, R.L., Seatlholo, S.T. and Van Vuuren, S.F. (2006) The biological activities of 20 nature identical essential oil constituents. *Journal of Essential Oil Research*, **18**, 129–133.

Zheljazkov, V.D., Cantrell, C.L., Tekwani, B. and Khan, S.I. (2008) Content, composition, and bioactivity of the essential oils of three basil genotypes as a function of harvesting. *Journal of Agricultural and Food Chemistry*, **56**, 380–385.

# 34 Metabolomic Analysis of a Commercially Important Aromatic Plant from the Indian Ocean: *Vanilla planifolia*

Tony L. Palama

Unité Mixte de Recherche – Peuplement Végétaux et Bioagresseurs en Milieu Tropical, Université de La Réunion, Saint-Denis, La Réunion, France

## 34.1 Introduction

'Vanilla' is a well-known flavouring agent commonly used in ice creams, cola drinks, cosmetics and tobacco. Its natural source is the fruit of an orchid: *Vanilla planifolia* G. Jackson. Major growing countries are regularly hit by tropical storms, causing damage to crops and resulting in fluctuations in the price. Thus, and because of the large consumption of vanilla-flavoured products, vanillin (the major flavour compound) is also made along other routes, such as (bio) conversion of related natural products or via synthesis. Less than 1% of the approximately 12 000 tons of vanillin used in the flavour market is derived naturally from vanilla plants (Walton *et al.*, 2003). While vanillin extracted from pods cost up to $4000 per kilogram, the so-called natural vanillin derived from microbial production has a price of about $1000 per kilogram and synthetic vanillin costs about $11–15 per kilogram (Schrader *et al.*, 2004; Serra *et al.*, 2005). However, the flavour of vanillin itself is nothing comparable to the 200 compounds present in cured vanilla pods. Indeed, mature *V. planifolia* pods are odourless and have to follow a 'curing process' in order to release the pleasant aroma. This curing process transforms the green pods to the well-known black-coloured pods with a high amount of vanillin. Most of the studies conducted on vanilla have focused on the composition of cured pods and potential improvement of the curing process. Despite its importance for the vanilla flavour, the chemical mechanisms occurring during this process are still not completely understood and studies following this objective have to be done.

Besides the improvement of the curing and extraction processes, it would also be useful to work upstream and increase the quality of the green pods. Nonetheless, up to now, the biochemistry of the plant and the vanillin biosynthesis pathway still remains uncertain. Several studies have been conducted and reviewed, but authors still do not agreed on one pathway (Zenk, 1965; Ranadive, *et al.*, 1983; Kanisawa, *et al.*, 1994; Havkin-Frenkel, *et al.*, 1999; Dixon, 2010). Most of these studies used potential precursors on vanilla cultured cells or pod sections and followed its bioconversion to vanillin. So as to better understand the production of flavour compounds, Palama *et al.* (2009, 2010a,2010b, 2011, 2012) conducted metabolomic studies in vanilla green pods. In parallel with the terms 'genomics', 'transcriptomics' and 'proteomics', 'metabolomics' is defined as both the qualitative and quantitative analysis of all primary and secondary metabolites synthesized by a biological system (Fiehn, 2002; Verpoorte *et al.*, 2007). In the past few years, plant metabolomics has rapidly become a valuable technology used to gain a global picture of the molecular organization in multicellular organisms (Hall, 2006). Metabolomics information can also assist in a deeper understanding of plant metabolic networks and response to genetic changes; but it also provides insights into the fundamental nature of plant phenotypes in relation to development, physiology and environment (Ward *et al.*, 2007). Thus, the metabolomic approach has been developed to *V. planifolia* in order to monitor primary and secondary metabolites and then to improve the knowledge on the biochemistry and the physiology of the plant.

## 34.2 Vanilla description

*V. planifolia* is indigenous to Central America, and especially from the region of Mexico, where it was called 'tlilxochitl'. This name comes from the words 'tlilli' and 'xochitl', which mean 'black' and 'pod' respectively (Bouriquet, 1954; Purseglove *et al.*, 1981). When the Spaniards discovered America, they called it 'Vanilla', which comes from the Spanish word 'vaina' meaning 'pod'. The latin word *planifolia* described the flatness ('plani') of the leaves ('folia').

### 34.2.1 Distribution

The genus *Vanilla* belongs to the largest family of flowering plants, the Orchidaceae. Among the 800 genera and the 25 000 species within the family, the genus *Vanilla* contains by itself more than 100 species (Bory *et al.*, 2008). *V. planifolia* Andrews and *Vanilla* × *tahitensis* J.W. Moore are the only cultivated species. *Vanilla* plants are epiphytic, hemi-epiphytic or terrestrial orchids that are distributed between latitudes 27°N and 27°S on all continents except Australia (Bouriquet, 1954; Soto Arenas, 2003).

*Novel Plant Bioresources: Applications in Food, Medicine and Cosmetics*, First Edition. Edited by Ameenah Gurib-Fakim.

Domestication of *V. planifolia* was first initiated in Mexico. The Totonac people, accredited as the first producers of vanilla, used it to flavour chocolate beverages (Bruman, 1948; Lubinsky *et al.*, 2008). Subsequently, *V. planifolia* has been spread with both the desire and demand for its cultivation worldwide. Unfortunately, the lack of pollinators in the areas of introduction prevented natural reproduction and pod production. Nevertheless, a 12-year-old French-owned slave by the name of Edmond Albius is known to have discovered in 1841, in the Island of La Réunion, a practical method to hand-pollinate the plant, allowing global cultivation of vanilla. Nowadays, vanilla is currently grown in various tropical countries, such as Madagascar, Uganda, Papua New Guinea, Indonesia, India and Mexico, and islands like Comoros, Mayotte, Tahiti and La Réunion (Bory *et al.*, 2008; Lubinsky *et al.*, 2008; Gleason-Allured, 2009).

### 34.2.2 Biology

Vanilla is an epiphytic orchid (Figure 34.1). Aerial roots help the plants to climb on trees or various supports, but also help absorb water from humidity in the air. Ideal soil for vanilla is light, rich in humus and porous, thus allowing the roots to spread without excessive moisture (Bouriquet, 1954). Established from vine cuttings in the crops, vanilla plants grow better at a temperature between 20 and 30°C with around 2000 mm of rainfall per year (Bouriquet, 1954). Higher temperatures or insufficient shading are fatal to the plant. Nevertheless, excessive humidity and shading can induce susceptibility of the plant to diseases (mildew and root rot) (Bouriquet, 1954). A water stress of about 50 days is necessary to induce flowering; and depending on the region of the world, the flowering period will vary (Fouché and Jouve, 1999).

The first flowers generally appear 3 years after planting. Each vanilla plant can bear 10–12 or more inflorescences if plants are vigorous enough (Figure 34.2). But for each inflorescence, only a few flowers open in a day and close up at midday. After fecundation, ovary develops into a fruit (usually called a pod or bean), and 5–6 weeks later the fruit reaches full-size development. To ensure good fruit size, a maximum of 10 fruits should be kept on the same inflorescence. Generally, *V. planifolia* pods are mature at 8–9 months after pollination, when their colours turn from green to yellow (Figure 34.3). Although 15 vanilla species can bear aromatic fruits, *V. planifolia* is the most widely grown species because of its vanillin content.

### 34.2.3 The curing process

Mature green pods of *V. planifolia* are odourless and develop their aroma and flavour after a 'curing' process. A number of procedures have been developed for curing vanilla, but they are all characterized by four phases to obtain a commercially viable product (Balls and Arana, 1941; Odoux, 2000; Dignum *et al.*, 2002; Havkin-Frenkel

**Figure 34.1** *V. planifolia* vine cultivated in a shade house in La Réunion. Photograph by T.L. Palama

**Figure 34.2** Flower of *V. planifolia*. Photograph by T.L. Palama. *See plate section for colour detail*

**Figure 34.3** *V. planifolia* green pods. Photograph by T.L. Palama

*et al.*, 2004; Pérez-Silva, 2006). These four phases are called 'killing', 'sweating', 'drying' and 'conditioning'. *V. planifolia* fruits are dehiscent capsules that normally open at maturity. To keep the pods closed, and thus maintain the aroma compound inside, a 'killing' step is necessary to stop the vegetative development. At the same time, killing methods allow cell structure disruption, and thus various enzymes can come into contact with their substrates. The killing process can be carried out by hot water scalding, sun/oven heating, scarification or freezing. The commonly used methods are sun, oven and hot water killing (Havkin-Frenkel and Dorn, 1997). The 'sweating' step allows the moisture to escape rapidly to reach a level that will avoid microbial spoilage during the subsequent operations. This step is also crucial because it lets indigenous enzymes take effect to develop the characteristic vanilla aroma and flavour. During the 'drying' step, the decrease in the moisture content reduces undesirable enzyme activities and biochemical changes (Figure 34.4). Then, for the 'conditioning' step, vanilla pods are stored in closed boxes for several months to refine their flavour and develop new compounds by oxidation. The full curing process usually lasts 6 months, but it can go for 1 year or more.

### 34.2.4 Phytochemistry

Only a few studies have reported the composition of vanilla green pods. Vanillin, 4-hydroxy-3-methoxybenzaldehyde, is the major flavour compound of vanilla cured pods and was isolated for the first time by Gobley (1858). Vanillin is found under its glucosylated form (glucovanillin) in green pods (Figure 34.5). Indeed, phenolic compounds can be toxic for the plant. Later, during the curing process, vanillin is released after an enzymatic hydrolysis. Since other diverse reactions (oxidation, hydroxylation, etc.) are taking place during the curing process, the focus here will be on the phytochemistry of the vanilla plant and pod prior to curing. Most of the previous studies have focused on the glycosidic content of the green pods, as this is supposed to be the precursor of the flavour compounds present in the cured pods.

Prat and Subitte (1969) reported a high amount of *p*-hydroxy benzyl alcohol in extracts of green pods. They suggested that this compound may be an important intermediate in aroma development. Tokoro *et al.* (1990) and Kanisawa (1993; Kanisawa *et al.*, 1994) isolated the most abundant glucosides from green pods and identified them as glucovanillin, *p*-hydroxybenzyl alcohol glucoside, bis[4-(β-D-glucopyranosyloxy)-benzyl]-2-isopropyltartrate (glucoside A) and bis[4-(β-D-glucopyranosyloxy)-benzyl]-2-(2-butyl)-tartrate (glucoside B). These glucosides A and B were reported for the first time in vanilla and have a structure very similar to the loroglossins found in other Orchidaceae plants (Cota *et al.*, 2008; Huang *et al.*, 2004). Various other minor glucosides were identified after enzymatic hydrolysis: glucosides of vanillic acid, vanillyl alcohol, *p*-vinyl-guaiacol, acetovanillon, caffeic acid, ferulic acid, methyl-3,4-dihydroxycinnamic acid, 2-methoxy-4-cresol, homovanillyl alcohol, 3,4-dihydroxybenzoic acid, ethyl-4-hydroxy-3-methoxyphenylacetate, *p*-cresol, *p*-vinylphenol, methylsalicylate, *p*-hydroxybenzylethyl ether, *p*-hydroxybenzaldehyde, *p*-hydroxy cinnamic acid, cinnamic alcohol, cinnamic acid, phenethyl alcohol, 3-phenylpropanol (Kanisawa *et al.*, 1994). Negishi and Ozawa (1996) synthesized 12 different glucosides but were able to identify only glucovanillin and *p*-hydroxybenzaldehyde from green pod extracts using high-performance liquid chromatography (HPLC) analysis. The other glucosides were probably not formed yet in the 6-month-old pods studied experimentally. Leong *et al.* (1989) identified the glucosides of vanillin, *p*-hydroxybenzaldehyde, vanillic acid and *p*-hydroxybenzoic acid in green pod extract by HPLC (Figure 34.5).

Several hydrocarbons, *p*-demethylsterols, triterpene alcohols and long-chain aliphatic β-diketones were detected in vanilla green pods (Ramaroson-Raonizafinimanana *et al.*, 1997, 1998, 2000). Long-chain γ-pyranones were also isolated from crushed vanilla pods and were said to come from the epicuticular wax (Ramaroson-Raonizafinimanana *et al.*, 1999). After dissection of vanilla pods, Odoux and Brillouet (2009) argued that these long-chain γ-pyranones were more likely found in trichomes. Odoux and Brillouet (2009) also demonstrated that trichomes produce a mucilage that is made from a mixture of a glucomannan and a pectic polysaccharide.

In other parts of the plant, only a few studies have been conducted. Despite a smaller amount than in green pods, Tokoro *et al.* (1990) reported the presence of glucosides A and B in leaves and stems of *V. planifolia*. Sun *et al.* (2001) reported the presence of *p*-ethoxymethylphenol, *p*-butoxymethylphenol, vanillin, *p*-hydroxy-2-methoxycinnamaldehyde and 3,4-dihydroxyphenyl acetic acid in the ethanol extract of vanilla leaves and stems. They observed that, among the compounds detected, *p*-ethoxymethyl phenol was the most abundant. Interestingly, *p*-butoxymethylphenol showed a strong toxicity to mosquito larvae (Sun *et al.*, 2001).

No previous studies have reported the organic acid content of vanilla green pods. However, Fitelson and Bowden (1968) identified malic acid as a major organic acid of vanilla extract, and the presence of citric acid, isocitric acid and succinic acid was suspected. Additionally, Martin *et al.* (1977) used the total and free amino acid contents to determine the authenticity of vanilla extract.

Figure 34.4 *V. planifolia* black pods drying in the sun. Photograph by T.L. Palama

## 34.3 Vanilla metabolomics

The metabolome of vanilla has been investigated to have a qualitative and quantitative analysis of all primary and secondary

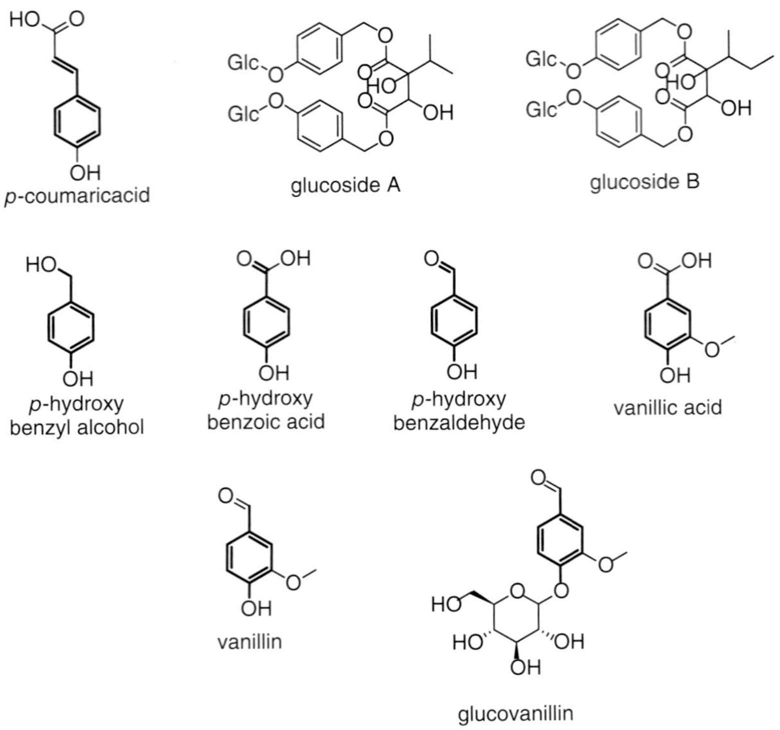


**Figure 34.5** Phenolics compounds usually detected in *V. planifolia*

metabolites. Conventional metabolic analysis covers a too limited number of metabolites to show a broad view of the biochemical status. Chemical analysis techniques to be applied to metabolite profiling should be rapid, reproducible and stable over time, while requiring only simple sample preparation. Since it potentially meets the requirements, nuclear magnetic resonance (NMR) spectroscopy has been used on vanilla. The huge amount of data provided is generally analysed using multivariate data analysis (MVDA). This combination allows one to get a good visualization of the dataset and has already been applied to the metabolic profiling of various plant species (Ward *et al.*, 2003; Choi *et al.*, 2004a–c; Kim, *et al.*, 2005; Abdel-Farid *et al.*, 2007; Choi and Yoon, 2007), but also to wine (Brescia *et al.*, 2002), coffee (Charlton *et al.*, 2002), fruit juice (Vogels *et al.*, 1996) and beer (Duarte *et al.*, 2002). In vanilla, NMR-based metabolomic profiles of *V. planifolia* green pods, leaves and organogenic calli were characterized, but we will discuss here only the results on green pods and leaves.

### 34.3.1 Vanilla green pods

Vanilla pods were first used to follow the metabolic profile during pod development (Palama *et al.*, 2009). Pods from 3–8 months after pollination were analysed by $^1$H-NMR. A good separation was obtained from the principal component analysis (PCA) score plot (Figure 34.6). Indeed, the authors were able to identify primary and secondary metabolites responsible for this separation. MVDA shows that glucose and organic acids contents decrease during maturation, while sucrose and phenolics (vanillin, *p*-hydroxybenzaldehyde and their glucosides) increase with time. Even if no standard pure compounds were used, the identification of two glucosides was made possible by NMR. Signals corresponding to glucosides A and B (Figure 34.5) were detected in the metabolic profile and structures were confirmed using two-dimensional NMR experiments. These two compounds were previously detected in vanilla (Tokoro *et al.*, 1990) and their concentration increased from 3 to 5 months and then decreased from 5 to 8 months after pollination. Thus, glucosides A and B were supposed to be precursors of vanillin and glucovanillin.

NMR spectroscopy was also used to analyse vanilla mature pods (8 months after pollination) from different accessions of *V. planifolia* (Palama *et al.*, 2011). It has been shown that a combination of NMR and MVDA could be used to discriminate green pods based on their metabolic profile. Moreover, this combination was useful to select accessions giving pods with higher characteristics (glucovanillin content, pod length and weight, etc.). Then, those accessions could be considered as 'elite'. Furthermore, the metabolic profiles obtained in this study contrasted from the previous ones. It has been shown that vanillin was either in both free and glucosylated forms or in glucosylated form only. Nevertheless, total vanillin content remains unaffected. The glycosylation mechanism is clearly affected. Since pods were coming from two shade houses (but the same accession and age), it would be interesting to investigate the impact of light intensity, pluviometry and relative humidity on the metabolic profile of vanilla mature pods.

### 34.3.2 Vanilla leaves

The impact of different parameters on the metabolic profile of *V. planifolia* leaves was also investigated by $^1$H-NMR (Palama *et al.*, 2010a). It was observed that environment (day time and period of the year) can affect the content of primary and secondary metabolites. Indirectly, it might be concluded that light and temperature have an impact on the metabolic profile of vanilla leaves.

The glucosides A and B were also detected in vanilla leaves samples (Palama *et al.*, 2010a). From the metabolomic results obtained from the different studies, it appeared that the glucosides A and B could be

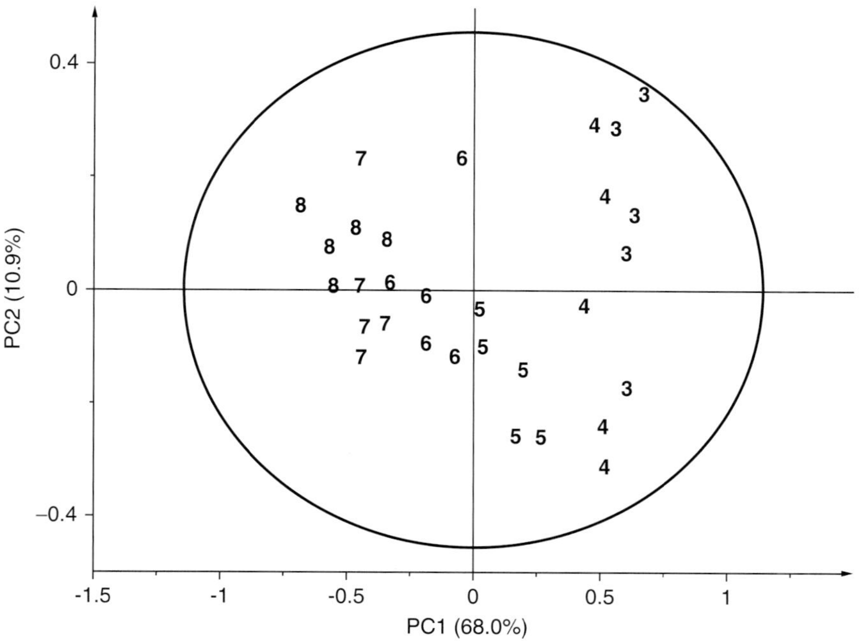


**Figure 34.6** Score plot (PC1 versus PC2) of PCA results obtained from $^1$H-NMR spectra of five accessions of *V. planifolia* pods of (3) 3 months old, (4) 4 months old, (5) 5 months old, (6) 6 months old, (7) 7 months old, and (8) 8 months old. Reprinted with permission from Palama *et al.* (2009) Copyright 2009 American Chemical Society

precursors of glucovanillin and vanillin. They have been detected in large amounts in developing green pods until 4–5 months after pollination. In addition, these glucosides have also been identified in the leaves of field plants, but also in young plants (i.e. *in vitro* plants) (Palama *et al.*, 2012). These results highlight their omnipresence in *Vanilla* plants. These glucosides could be locally produced in the pods and then converted to glucovanillin or they could be synthesized in the leaves and subsequently transported to the pods via the stem to follow the same conversions. Further studies have to focus on the synthesis and potential transport of glucosides A and B in the plant. These results will give significant insight into the detailed elucidation of the vanillin biosynthesis pathway.

Analysis of leaves from accessions belonging to four different species of *Vanilla* showed that the metabolomic approach combining $^1$H-NMR spectroscopy with MVDA allows good separation between accessions from different species and even between hybrid and parents based on their metabolic profile (Palama *et al.*, 2012). It has been observed that CR18 (*Vanilla pompona*) has a metabolic profile clearly different from the other related species. It has been observed that glucose and phenolic levels are higher in younger *Vanilla* leaves than in older leaves. Phenolic compounds might have a protective role against pests, particularly in the younger leaves, whereas when the leaves are older, the surface of the leaves might become stronger and thus a barrier against insect attack. Then, phenolic compounds might no longer be required in the older leaves. It has been observed that glucose and phenolic compounds levels of CR18 (*V. pompona*) old leaves are higher than in other species (Palama *et al.*, 2012). These phenolic compounds could explain the fact that CR18 (*V. pompona*) shows fewer symptoms upon *Cymbidium* mosaic virus (CymMV) infection.

In spite of no clear effects of CymMV observed on the leaves collected in the shade house, interesting results were obtained from materials issued from *in vitro* propagation. Indeed, it has been observed that virus infection is correlated with an increase in sugars content in leaf and stem samples, whereas phenolic compounds content decreased after virus infection. Nevertheless, one accession of *V. pompona* (CR18) showed that phenolics content could be used to investigate the resistance against pests and diseases. In addition to the aromatic profile, prospection of *Vanilla* diversity has to be investigated in order to look at compounds awarding resistance and that can be used in plant breeding.

## 34.4 Other future prospects

Besides the outlooks that have been mentioned previously, and based on the previous studies done on the characterization of the vanilla metabolome, further experiments can now be conducted. Preliminary results are shown here to serve as examples for future investigations on vanilla: the use of NMR for quality control of pods and for chemotaxonomy of vanilla pods production are just direct examples following on from the previous results.

### 34.4.1 Quality control

In La Réunion, vanilla production has spread in the east coast of the island generally from 0 to 600 m above sea level. These conditions are optimal for the efficient growth of the vanilla vines. Nevertheless, it has yet to be demonstrated that optimal growth is directly correlated with the quality of its aroma. In the green pods, both vanillin and glucovanillin can be monitored; hence, an estimation of the vanillin potential in the final product can be made.

NMR spectroscopy has also been used to monitor the vanillin content (free or glucosylated). A collection of 10 mature vanilla pod samples (yellow at the blossom end) was conducted among 18 vanilla growers. The vanillin content was determined for each pod by NMR, and the results are presented in Figure 34.7 as the total vanillin (vanillin and glucovanillin) content in percentage of dry matter. Pods collected from the 18 growers have similar quality even though the vanillin content ranges from 2.5 to 7.5%. Odoux *et al.*

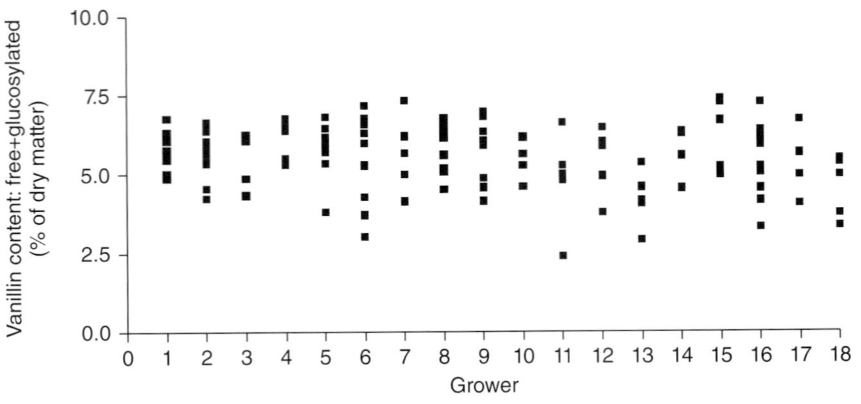


**Figure 34.7** Vanillin content (% of dry matter) of different growers in La Réunion in 8-month-old green pods obtained by NMR quantification

(2000) reported 2.8% of vanillin after curing from 5% vanillin green pods. Indeed, the curing process is characterized by a loss of vanillin potential. But still, it can be observed that vanilla green pods quality is good, since most of pods have more than 5% of vanillin content. Results also show that vanillin content variation is different between growers. Further experiments will have to confirm these findings and look for possible explanations.

### 34.4.2 Chemotaxonomy

*Vanilla* is the most diverse genus of the Vanilloideae subfamily (Cameron, 2011). The first systematics of the genus classified the species based on vegetative features and flower morphology (Rolfe, 1896; Portères, 1954). Indeed, for example, a section Aphyllae was created for the leafless species of the genus. Recently, since the rise of molecular biology, the classification systems for *Vanilla* have been revised (Bouetard *et al.*, 2010; Cameron, 2011).

$^1$H-NMR spectroscopy is a powerful tool that can rapidly provide a metabolic profile of a sample. This profile can be used as a fingerprint to discriminate samples. As performed on *Ilex* species, NMR-based metabolomics can be used to classify different species samples (Choi *et al.*, 2005; Kim *et al.*, 2010). The same methodology has been applied to *Vanilla* for the extraction of plant material, NMR measurements and data analysis. A total of 147 stem samples from different species were collected in the Biological Resource Center VATEL, La Réunion. Stem samples give the opportunity to work on both leafy and leafless species and have the advantage of being available all year long. A dendrogram of the hierarchical clustering analysis (HCA) results was obtained (Figure 34.8). This dendrogram shows similarities in the metabolic profile of the samples. Thus, we can observe a first cluster containing the species *V. crenulata*, and *V. imperialis* with the aphyllous species *V. humblotii* and *V. madagascariensis*. The second cluster contains *V. leuprieurii* samples. Stem samples of *V. pompona* and *V. grandiflora* are found in a third cluster. The fourth cluster regroups samples of *V. planifolia*, *V.* × *tahitensis*, *V. bahiana*, hybrids and self-pollination (Figure 34.8). These results are very similar to those obtained from molecular phylogeny using chloroplast genes (Bouetard *et al.*, 2010). Indeed, the authors were able to obtain three major groups. Comparison of the different results shows that cluster 1 in Figure 34.8 corresponds to Old World and Caribbean species, while the other clusters represent American fragrant species (Bouetard *et al.*, 2010). Even if this is a preliminary study, NMR spectroscopy shows here its potential for rapid metabolic fingerprinting of *Vanilla* species, which could be a first step for the chemotaxonomy of the genus.

## 34.5 Conclusions

Metabolomics has shown its potential in the analysis of vanilla. More interesting results will come from the correlation with other 'omics' data, such as proteomics and transcriptomics. Indeed, studies are in progress to correlate metabolite levels with gene expression. This will be very helpful to unravel the vanillin biosynthetic pathway, although clues have already been obtained. Furthermore, metabolomics will help in the comprehensive understanding of the mechanisms taking place in the vanilla plant. A first overview has been done, but more has to follow. From now, NMR has been

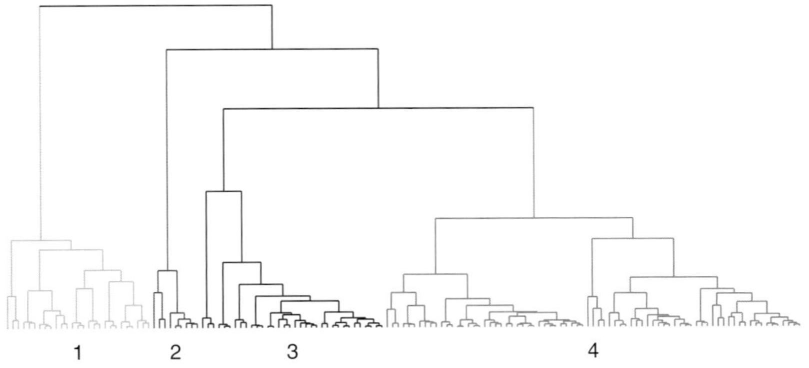


**Figure 34.8** Dendrogram of HCA results based on 47 PCs obtained from partial least squares discriminant analysis: (1) *Vanilla crenulata*, *Vanilla imperialis*, *Vanilla humblotii* and *Vanilla madagascariensis*; (2) *Vanilla leuprieurii*; (3) *V. pompona* and *Vanilla grandiflora*; (4) *V. planifolia*, *V.* × *tahitensis*, *Vanilla bahiana* and hybrids

used to analyse the metabolome of vanilla, but a combination with chromatography and mass spectrometry techniques will allow a more sensitive detection of metabolites and help to correlate observations with changes in contents.

## References


Abdel-Farid, I.B., Kim, H.K., Choi, Y.H. and Verpoorte, R. (2007) Metabolic characterization of *Brassica rapa* leaves by NMR spectroscopy. *Journal of Agricultural and Food Chemistry*, **55** (19), 7936–7943.

Balls, A.K. and Arana, F.E. (1941) The curing of vanilla. *Industrial and Engineering Chemistry*, **33** (8), 1073–1075.

Bory, S., Grisoni, M., Duval, M.-F. and Besse, P. (2008) Biodiversity and preservation of vanilla: present state of knowledge. *Genetic Resources and Crop Evolution*, **55**, 551–571.

Bouetard, A., Lefeuvre, P., Gigant, R. *et al.* (2010) Evidence of transoceanic dispersion of the genus Vanilla based on plastid DNA phylogenetic analysis. *Molecular Phylogenetics and Evolution*, **55** (2), 621–630.

Bouriquet, G. (ed.) (1954) Le Vanillier et la Vanille dans le Monde, *Encyclopédie Biologique, XLVI*, Paul Lechevalier, Paris.

Brescia, M.A., Caldarola, V., De Giglio, A. *et al.* (2002) Characterization of the geographical origin of Italian red wines based on traditional and nuclear magnetic resonance spectrometric determinations. *Analytica Chimica Acta*, **458** (1), 177–186.

Bruman, H. (1948) The culture history of Mexican vanilla. *Hispanic American Historical Review*, **28**, 360–376.

Cameron, K.M. (2011) Vanilla phylogeny and classification, in *Handbook of Vanilla Science and Technology* (eds D. Havkin-Frenkel and F.C. Belanger), Blackwell Publishing Ltd, pp. 243–255.

Charlton, A.J., Farrington, W.H.H. and Brereton, P. (2002) Application of $^1$H NMR and multivariate statistics for screening complex mixtures: quality control and authenticity of instant coffee. *Journal of Agricultural and Food Chemistry*, **50** (11), 3098–3103.

Choi, H.-K., Choi, Y.H., Verberne, M. *et al.* (2004a) Metabolic fingerprinting of wild type and transgenic tobacco plants by $^1$H NMR and multivariate analysis technique. *Phytochemistry*, **65**, 857–864.

Choi, H.-K. and Yoon, J.–H. (2007) Metabolomic profiling of *Vitis vinifera* cell suspension culture elicited with silver nitrate by $^1$H NMR spectrometry and principal components analysis. *Process Biochemistry*, **42** (2), 271–274.

Choi, Y.H., Kim, H.K., Hazekamp, A. *et al.* (2004b) Metabolic differentiation of *Cannabis sativa* cultivars using $^1$H-NMR spectroscopy and principal component analysis. *Journal of Natural Products*, **67**, 953–957.

Choi, Y.H., Sertic, S., Kim, H.K. *et al.* (2005) Classification of *Ilex* species based on metabolic fingerprinting using nuclear magnetic resonance and multivariate data analysis. *Journal of Agricultural and Food Chemistry*, **53**, 1237–1245.

Choi, Y.H., Tapias, E.C., Kim, H.K. *et al.* (2004c) Metabolic discrimination of *Catharanthus roseus* leaves infected by phytoplasma using $^1$H-NMR spectroscopy and multivariate data analysis. *Plant Physiology*, **135**, 2398–2410.

Cota, B.B., Magalhaes, A., Pimenta, A.M.C. *et al.* (2008) Chemical constituents of *Habenaria petalodes* Lindl. (Orchidaceae). *Journal of the Brazilian Chemical Society*, **19** (6), 1098–1104.

Dignum, M.J.W., Kerler, J. and Verpoorte, R. (2002) Vanilla curing under laboratory conditions. *Food Chemistry*, **79**, 165–171.

Dixon, R.A. (2010) Vanillin biosynthesis – not as simple as it seems?, in *Handbook of Vanilla Science and Technology* (eds D. Havkin-Frenkel and F.C. Belanger), Wiley–Blackwell, pp. 292–298.

Duarte, I., Barros, A., Belton, P.S. *et al.* (2002) High-resolution nuclear magnetic resonance spectroscopy and multivariate analysis for the characterization of beer. *Journal of Agricultural and Food Chemistry*, **50** (9), 2475–2481.

Fiehn, O. (2002) Metabolomics – the link between genotypes and phenotypes. *Plant Molecular Biology*, **48**, 155–171.

Fitelson, J. and Bowden, G.L. (1968) Determination of organic acids in vanilla extracts by GLC separation of trimethylsilylated derivatives. *Journal of the Association of Official Agricultural Chemists*, **51**, 1224–1231.

Fouché, J.G. and Jouve, L. (1999) *Vanilla planifolia*: history, botany and culture in Réunion island. *Agronomie*, **19**, 689–703.

Gleason-Allured, J. (2009) The state of vanilla: challenges and opportunities. *Perfumer & Flavorist*, **34**, 19–22.

Gobley, M. (1858) *Recherches sur le Principe Odorant de la Vanille*, Thunot et Co., Paris.

Hall, R.D. (2006) Plant metabolomics: from holistic hope, to hype, to hot topic. *New Phytologist*, **169** (3), 453–468.

Havkin-Frenkel, D. and Dorn, R. (1997) Vanilla, in *Spices, Flavor Chemistry and Antioxidant Properties* (eds S.J. Risch and C.-T. Ho), American Chemical Society, Washington, DC, pp. 29–40.

Havkin-Frenkel, D., French, J.C., Graft, N. *et al.* (2004) Interrelation of curing and botany in vanilla (*Vanilla planifolia*) bean. *Acta Horticulturae*, **629**, 93–102.

Havkin-Frenkel, D., Podstolski, A., Witkowska, E. *et al.* (1999) Vanillin biosynthetic pathways: an overview, in *Plant Cell and Tissue Culture for the Production of Food Ingredients* (eds T.J. Fu, G. Singh and W.R. Curtis), Kluwer Academic/Plenum, New York, NY, pp. 35–43.

Huang, S.Y., Li, G.Q., Shi, J.G. *et al.* (2004) Chemical constituents of the rhizomes of *Coeloglossum viride* var. *bracteatum*. *Journal of Asian Natural Products Research*, **6** (1), 49–61.

Kanisawa, T. (1993) Flavor development in vanilla beans. *Kouryou*, **180**, 113–123.

Kanisawa, T., Tokoro, K. and Kawahara, S. (1994) Flavor development in the beans of *Vanilla planifolia*, in *Olfaction and Taste XI* (eds K. Kurihara, N. Suzuki and H. Ogawa), Springer Japan, Tokyo, pp. 268–270.

Kim, H.K., Choi, Y.H., Erkelens, C. *et al.* (2005) Metabolic fingerprinting of *Ephedra* species using $^1$H-NMR spectroscopy and principal component analysis. *Chemical & Pharmaceutical Bulletin*, **53** (1), 105–109.

Kim, H.K., Saifullah , Khan, S. *et al.* (2010) Metabolic classification of South American *Ilex* species by NMR-based metabolomics. *Phytochemistry*, **71** (7), 773–784.

Leong, G., Uzio, R. and Derbesy, M. (1989) Synthesis, identification and determination of glucosides present in green vanilla beans (*Vanilla fragrans* Andrews). *Flavour and Fragrance Journal*, **4**, 163–167.

Lubinsky, P., Bory, S., Hernández Hernández, J. *et al.* (2008) Origins and dispersal of cultivated vanilla (*Vanilla planifolia* Jacks. [Orchidaceae]). *Economic Botany*, **62** (2), 127–138.

Martin, G.E., Ethridge, M.W. and Kaiser, F.E. (1977) Determining the authenticity of vanilla extracts. *Journal of Food Science*, **42** (6), 1580–1583.

Negishi, O. and Ozawa, T. (1996) Determination of hydroxycinnamic acids, hydroxybenzoic acids, hydroxybenzaldehydes, hydroxybenzyl alcohols and their glucosides by high-performance liquid chromatography. *Journal of Chromatography A*, **756**, 129–136.

Odoux, E. (2000) Changes in vanillin and glucovanillin concentrations during the various stages of the process traditionally used for curing *Vanilla fragrans* beans in Réunion. *Fruits*, **55**, 119–125.

Odoux, E. and Brillouet, J.M. (2009) Anatomy, histochemistry and biochemistry of glucovanillin, oleoresin and mucilage accumulation sites in green mature vanilla pod (*Vanilla planifolia*; Orchidaceae): a comprehensive and critical reexamination. *Fruits*, **64** (4), 221–241.

Odoux, E., Languille, S., Leblé, J.-P. and Lombard, A. (2000) The effect of the nature of the soil amendments on the main aromatic compounds in *Vanilla fragrans* beans. *Fruits*, **55**, 63–71.

Palama, T.L., Fock, I., Choi, Y.H. *et al.* (2010a) Biological variation of *Vanilla planifolia* leaf metabolome. *Phytochemistry*, **71**, 567–573.

Palama, T.L., Grisoni, M., Fock-Bastide, I. *et al.* (2012) Metabolome of *Vanilla planifolia* (Orchidaceae) and related species under *Cymbidium* mosaic virus (CymMV) infection. *Plant Physiology and Biochemistry*, **60**, 25–34.

Palama, T.L., Khatib, A., Choi, Y.H. *et al.* (2011) Metabolic characterization of green pods from *Vanilla planifolia* accessions grown in La Réunion. *Environmental and Experimental Botany*, **72**, 258–265.

Palama, T.L., Khatib, A., Choi, Y.H. *et al.* (2009) Metabolic changes in different developmental stages of *Vanilla planifolia* pods. *Journal of Agricultural and Food Chemistry*, **57** (17), 7651–7658.

Palama, T.L., Menard, P., Fock, I. *et al.* (2010b) Shoot differentiation from protocorm callus cultures of *Vanilla planifolia* (Orchidaceae): proteomic and metabolic responses at early stage. *BMC Plant Biology*, **10** (1), 82–100.

Pérez-Silva, A. (2006) Contribution à l'étude de la génèse des composés d'arôme au cours du procédé mexicain de transformation de la vanille (*Vanilla planifolia* G. Jackson). Thèse de doctorat, Université de Montpellier II, Montpellier.

Portères, R. (1954) La genre *Vanilla* et ses espèces, in *Le Vanillier et la Vanille dans le Monde* (ed. G. Bouriquet), Encyclopédie Biologique XLVI, Paul Lechevalier, Paris, pp. 94–290.

Prat, Y. and Subitte, J. (1969) Contribution à l'étude des constituants aromatiques de la vanille – Detection de l'alcool *p*-hydroxybenzylique par chromatographie sur couche minche. *Annales des Falsifications et de l'Expertise Chimique*, **62**, 225–230.

Purseglove, J.W., Brown, E.G., Green, C.I. and Robbins, S.R.J. (1981) Vanilla, in *Spices*, vol. 2, Longman, New York, NY.

Ramaroson-Raonizafinimanana, B., Gaydou, E.M. and Bombarda, I. (1997) Hydrocarbons from three vanilla bean species: *V. fragrans*, *V. madagascariensis*, and *V. tahitensis*. *Journal of Agricultural and Food Chemistry*, **45**, 2542–2545.

Ramaroson-Raonizafinimanana, B., Gaydou, E.M. and Bombarda, I. (1998) 4-Demethylsterols and triterpene alcohols from two *Vanilla* bean species: *Vanilla fragrans* and *V. tahitensis*. *Journal of the American Oil Chemists Society*, **75** (1), 51–55.

Ramaroson-Raonizafinimanana, B., Gaydou, E.M. and Bombarda, I. (1999) Long-chain gamma-pyrones in epicuticular wax of two *Vanilla* bean species: *V. fragrans* and *V. tahitensis*. *Journal of Agricultural and Food Chemistry*, **47** (8), 3202–3205.

Ramaroson-Raonizafinimanana, B., Gaydou, E.M. and Bombarda, I. (2000) Long-chain aliphatic β-diketones from epicuticular wax of *Vanilla* bean species. Synthesis of nervonoylacetone. *Journal of Agricultural and Food Chemistry*, **48** (10), 4739–4743.

Ranadive, A.S., Szkutnica, K., Guerrera, J.G. and Frenkel, C. (1983) Vanilla biosynthesis in vanilla beans, in *Proccedings of the IX International Congress of Essential Oils*, Singapore, pp. 147–154.

Rolfe, R.A. (1896) A revision of the genus *Vanilla*. *Journal of the Linnaean Society*, **32**, 439–478.

Schrader, J., Etschmann, M.M.W., Sell, D. *et al.* (2004) Applied biocatalysis for the synthesis of natural flavour compounds – current industrial processes and future prospects. *Biotechnology Letters*, **26**, 463–472.

Serra, S., Fuganti, C., and Brenna, E. (2005) Biocatalytic preparation of natural flavours and fragrances. *Trends in Biotechnology*, **23** (4), 193–198.

Soto Arenas, M.A. (2003) Vanilla, in *Genera Orchidacearum: Orchidoideae (Part 2), Vanilloideae* (eds A.M. Pridgeon, P.J. Cribb, M.W. Chase and F.N. Rasmussen), vol. 3, Oxford University Press, New York, NY, pp. 321–334.

Sun, R., Sacalis, J.N., Chin, C.K. and Still, C.C. (2001) Bioactive aromatic compounds from leaves and stems of *Vanilla fragrans*. *Journal of Agricultural and Food Chemistry*, **49** (11), 5161–5164.

Tokoro, K., Kawahara, S., Amano, A. *et al.* (1990) Glucosides in vanilla beans and changes of their contents during maturation, in *Flavour Science and Technology* (eds Y. Bessiere and A.F. Thomas), John Wiley & Sons, Ltd, Chichester, pp. 73–76.

Verpoorte, R., Choi, Y.H. and Kim, H.K. (2007) NMR-based metabolomics at work in phytochemistry. *Phytochemistry Reviews*, **6**, 3–14.

Vogels, J.T.W.E., Terwel, L., Tas, A.C. *et al.* (1996) Detection of adulteration in orange juices by a new screening method using proton NMR spectroscopy in combination with pattern recognition techniques. *Journal of Agricultural and Food Chemistry*, **44** (1), 175–180.

Walton, N.J., Mayer, M.J. and Narbad, A. (2003) Molecules of interest – vanillin. *Phytochemistry*, **63**, 505–515.

Ward, J.L., Baker, J.M. and Beale, M.H. (2007) Recent applications of NMR spectroscopy in plant metabolomics. *FEBS Journal*, **274** (5), 1126–1131.

Ward, J.L., Harris, C., Lewis, J. and Beale, M.H. (2003) Assessment of $^{1}$H NMR spectroscopy and multivariate analysis as a technique for metabolite fingerprinting of *Arabidopsis thaliana*. *Phytochemistry*, **62**, 949–957.

Zenk, M.H. (1965) Biosynthese von Vanillin in *Vanilla planifolia* Andr. *Zeitschrift für Pflanzenphysiologie*, **53**, 404–414.

# 35 Natural Dyes for Photonics Applications

M. Maaza[1,2]

UNESCO–UNISA Africa Chair in Nanosciences/Nanotechnology, College of Graduate Studies, University of South Africa (UNISA), Muckleneuk Ridge, Pretoria, South Africa
Nanosciences African Network (NANOAFNET), iThemba LABS–National Research Foundation, Somerset West, Western Cape Province, South Africa

## 35.1 Introduction

Natural dyes are chemical cycles extracted from plants, invertebrates or minerals. From the animal world, one could distinguish those derived from cochineal insect (red), cow urine (Indian yellow), lac insect (red, violet), murex snail (purple) and octopus or cuttlefish (sepia brown). From the plant world, the major ones are extracted from catechu or cutch tree (brown), Cambodge tree resin (dark mustard yellow), Himalayan rubhada root (yellow), indigofera plant (blue), kamala tree (orange–yellow, golden yellow), madder root (red, pink, orange) and pomegranate peel (yellow), among others. In 1856, William Henry Perkin discovered the first synthetic dye 'Mauve', which is a basic dye (aniline), while searching for a cure for malaria, indirectly pioneering a new industry of synthetic dyes. This was followed by the design of the second synthetic dye, alizarin (synthetic madder), by Graebe and Liebermann and W.H. Perkin in 1868. The discovery of man-made synthetic dyes in the mid 19th century triggered a long decline in the large-scale market for natural dyes. Synthetic dyes, which could be produced in large quantities, quickly superseded natural dyes for commercial textile production, and unlike natural dyes they were suitable for the synthetic fibres industry.

While the usage of natural dyes was/is blooming in the food industry as a colorant and in modern medicine, there is also an obvious growing interest from the photonics community in such natural chemical systems due to their $\pi$–$\pi^*$/conjugated electronic structures. In nonlinear photonics, it was demonstrated that extracts from *Hibiscus sabdariffa*, commonly known as Roselle, carminic acid of the cochineal scale and saffron exhibit exceptional nonlinear optical (NLO) properties of a prime importance in optics. Yet, as early as the birth stage of lasers, coumarin ($C_9H_6O_2$), which is found naturally in high concentration in the tonka bean (*Dipteryx odorata*), was used in dye lasers. As shown in Figure 35.1, coumarin dye is still the basic active medium for many tunable dye laser sources (Duarte and Hillman, 1990; Schäfer, 1990). The second sector where natural dyes are attracting a growing interest is the dye solar cells (DSCs) field. Indeed, recent efforts in DSCs are focusing on increasing the solar photo-conversion while lowering the manufacturing costs by using ruthenium (Ru)/osmium (Os) free organic dyes. Owing to their intermolecular $\pi$–$\pi^*$ transitions, natural dyes exhibit very large absorption coefficients in the visible and near-infrared spectral range, conjugated with a high light-harvesting efficiency, and therefore potential photosensitizing applications in DSCs.

This chapter reports on specific nano-photonics applications of selected natural dyes in the field of nonlinear optics, third-order nonlinear susceptibility $\chi^{(3)}$ and optical limiting, as well as green photosensitizing media in DCSs technology.

## 35.2 Nonlinear optical properties of natural dyes: $\chi^{(3)}$ and optical limiting applications

The NLO properties of natural dyes-based compounds is becoming a subject of numerous investigations in the very recent years owing to their unique potential photonic applications. Generally, these compounds contain donor and acceptor groups positioned at either ends of a conjugated system of an appropriate length. As in the case of synthetic cycles, it is established that the extent of this conjugation and the size of the $\pi$-delocalized bonds are the major factors controlling the large third-order nonlinearities. Similar to the so-called poled organic materials, which are asymmetric organic molecules with a large molecular hyperpolarizability caused by their electron delocalization along the conjugated backbone (Sekkat *et al.*, 1992), specific natural dyes have been found to exhibit very large NLO characteristics with nonlinear coefficients $10^4$–$10^6$ times higher than that of the standard $CS_2$. This natural dye family includes chlorophyll, red carmine, Chinese tea, betanin, *Hibiscus sabdariffa*, as well as, recently, functionalized DNA if one has to push the border limit of biophotonics to a certain extent (Zhang *et al.*, 1989; Uma Maheswari *et al.*, 1994; Henari and Al-Saie, 2006; Alarcon-Aguilar *et al.*, 2007; Thankappan *et al.*, 2012; Kouissa *et al.*, 2013).

*H. sabdariffa* L. (Rosella), which is considered in this section as a typical example in NLO, is an annual shrub that grows in tropical and subtropical regions, including West Africa. The seedpods are enclosed in red fleshy calyces, which are commonly used for making food and tea as well as for several bio- and phytochemicals applications in relation to its antioxidant properties (Samoc *et al.*, 2006; Rekha and Ramalingam, 2009). The calyx of *H. sabdariffa* L. (Rosella) is originally of a deep red colour (Figure 35.2). This red colouring component is very soluble in

*Novel Plant Bioresources: Applications in Food, Medicine and Cosmetics*, First Edition. Edited by Ameenah Gurib-Fakim.

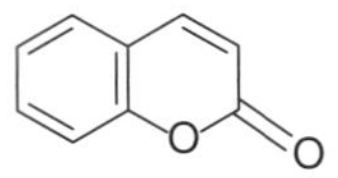

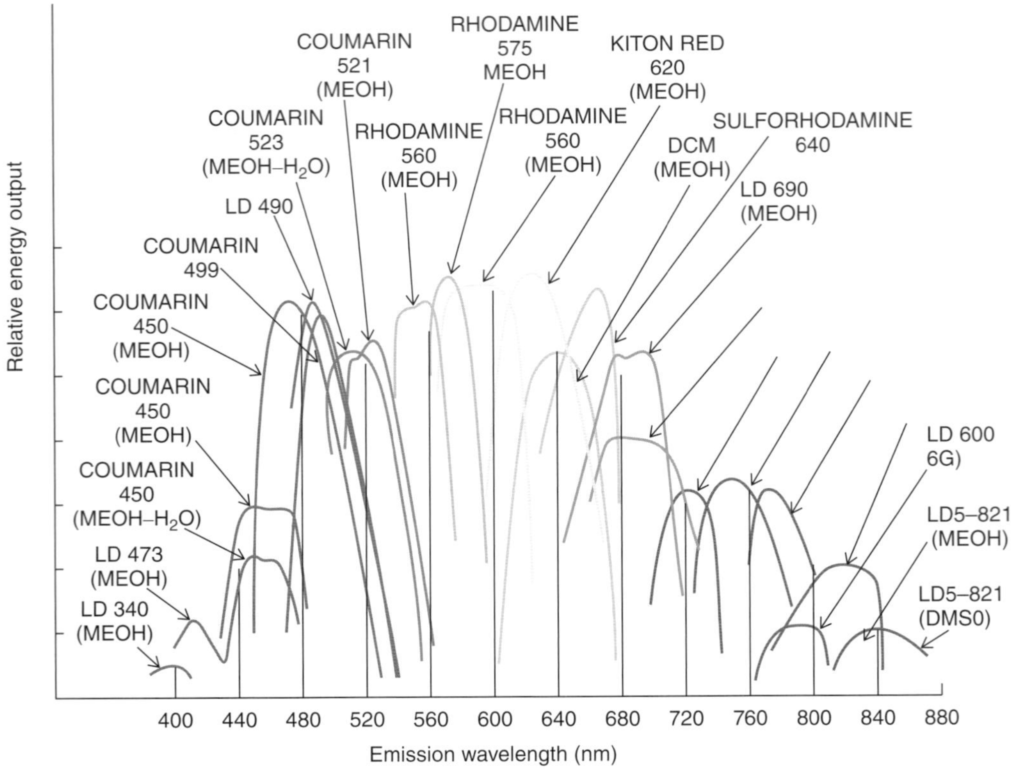


**Figure 35.1** Dye spectral emission characteristics of some dye laser sources including coumarin

$H_2O$ and changes to fine red by HCl and by alkali to bluish green, indicating the presence of anthocyanin cycles. The two major anthocyanin cycles are delphinidin 3-xylosylglucoside (also called delphinidin 3-sambubioside or hibiscin) and cyanidin 3-xylosylglucoside (also called cyanidin 3-sambubioside or gossypicyanin). As illustrated in Figure 35.2, *H. sabdariffa's* anthocyanin is glycosylated in the 3-position with the chemical structure given as delphinidin-3-*O*-(2-*O*-D-xylopyranosyl)-D-glucopyranoside (R = OH, **1**), cyanidin 3-*O*-(2-*O*-D-xylopyranosyl)-D-glucopyranoside (R = H, **2**) (Frank, 2005; Morsten, 2006).

The NLO characteristics of the *H. sabdariffa* roselle dye are due to the high population of delocalized π-electrons within their anthocyanin skeleton leading to a large third-harmonic generation and two-photon absorption phenomena. This includes the appealing intensity-dependent refractive index and optical limiting. From the NLO viewpoint, Henari and Al-Saie (2006) reported the first experimental observation of self-action phenomena, such as self-phase modulation and beam fanning in an *H. sabdariffa* roselle solution for a fixed concentration. This natural dye has been found to be an attractive novel type of organic nonlinear media, with a large negative nonlinear refractive index of $n_2 \approx 1.0 \times 10^{-4}$ esu, likely from a thermal origin. Relative to Henari and Al-Saie's (2006) pioneering work on the NLO properties of *H. sabdariffa's* roselle dye under argon laser radiations ($\lambda = 514$ nm), the recent experimental investigations by Diallo *et al.* (2013), reported on a complementary *z*-scan as well as additional optical limiting properties of such natural dye solutions within a wide concentration range with a doubled Q-switched Nd:YAG laser source in the nanosecond regime.

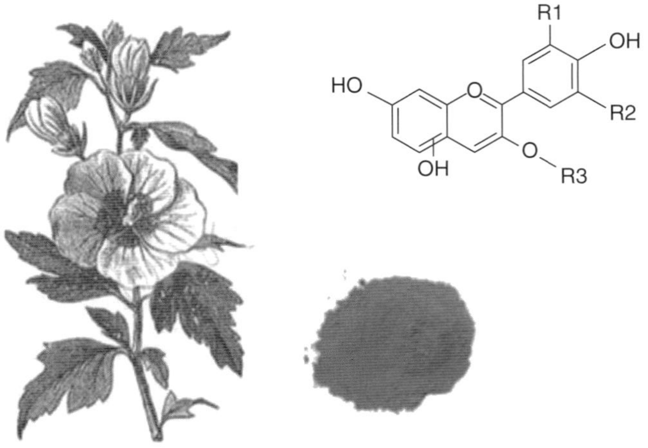


**Figure 35.2** Structure of *Hibiscus sabdariffa* anthocyanins: delphinidin 3-*O*-(2-*O*-D-xylopyranosyl)-D-glucopyranoside (R = OH, **1**), cyanidin 3-*O*-(2-*O*-D-xylopyranosyl)-D-glucopyranoside (R = H, **2**). *See plate section for colour detail*

More precisely, the coefficient of the intensity-dependent refractive index $n_2$ ($n = n_0 + n_2 I$), the third-order nonlinear susceptibility $\chi^{(3)}$ ($\chi^{(3)} = n_2 n_0{}^2/12\pi^2$) and the optical limiting thresholds are presented in a comprehensive way within this section.

As mentioned previously, *H. sabdariffa* roselle calyces contain high amounts of anthocyanins, especially delphinidin 3-sambubioside and cyanidin 3-sambubioside, and this up to 2.5 g/100 g dry weight (Herrera-Arellano *et al.*, 2004; Cisse *et al.*, 2009, 2011). Based on the work of Cisse *et al.* (2009, 2011) related to the kinetics and modelling of the aqueous extraction of the anthocyanins from the calyces, fresh calyces were collected and immersed in $H_2O$ solutions for an effective ultrasound surface cleaning. The original collected calyces material were dried under sunny conditions at ambient temperature and then ground with an electric mill of ~1 mm in mesh. Five extracts were prepared in a fixed volume of deionized $H_2O$ for 48 h each (0.5 L, 25 °C). The initial mass of the milled calyce powders was 0.5, 2, 4, 6 and 8 g. The liquid extract samples were filtered twice to eliminate residual solids, if any. In the following part, the samples were designed as $S_{1/2}$, $S_2$, $S_4$, $S_6$ and $S_8$ where the subscript $i$ in $S_i$ is the mass used during the extraction phase.

Figure 35.3 shows the room-temperature optical UV–VIS–NIR spectra of the five liquid solutions through a quartz cell of 10 mm thickness. Excluding the slight absorption band centred around 962 nm, which is due to $H_2O$ vibrational modes, two major regions can be distinguished: $200 \leq \lambda \leq 500$ nm and $\lambda \geq 500$ nm. While the optical transmission of the concentrated samples is low in the first spectral region of $200 \leq \lambda \leq 500$ nm, the less concentrated one (i.e. $S_{1/2}$) does exhibit a large absorption band (~347 to ~513 nm), which is in fact a convolution of at least three peaks. These latter peaks can be assigned to $\pi$–$\pi^*$ and $\sigma$–$\pi^*$ transitions. While the optical transmission is nearly 90% for all samples above 500 nm, there is a redshift in the cut-off wavelength as the concentration increases. Consequentially, and considering such optical transmission profiles,

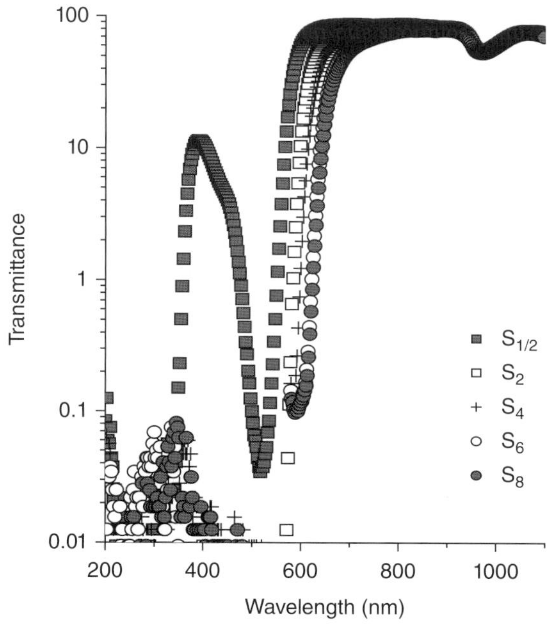


**Figure 35.3** UV–VIS–NIR optical transmittance of the various *H. sabdariffa* roselle dye extract solutions contained in 10 mm quartz cell

*H. sabdariffa* solutions are, *a priori*, attractive materials for optical limiting applications, possibly for a wide variety of laser sources whose emission wavelengths lie in the visible region. The linear refractive indices of these liquid solution samples were measured using an Abbe type refractometer, while their UV–VIS–NIR spectra were recorded using a CECIL scanning spectrophotometer (model 2021). The NLO properties, in addition to the optical limiting, were investigated by *z*-scan technique, which has the advantages of simplicity and high sensitivity, as well as enabling simultaneous measurement of the magnitude and sign of the nonlinear refractive index and the nonlinear absorption coefficient of the samples (Herrera-Arellano *et al.*, 2004; Cisse *et al.*, 2011). Basically, the method consists of translating a sample through the focus of a Gaussian beam and monitoring the changes in the far-field intensity pattern (Figure 35.4). When the intensity of the incident laser beam is sufficient to induce nonlinearity in the sample, it either converges or diverges the beam, depending on the nature of that nonlinearity. By moving the sample through the focus, the intensity-dependent absorption is measured as a change of transmittance through the sample (open aperture). The nonlinear refraction is determined by the intensity variation at the plane of a finite aperture placed in front of the detector (closed aperture), as the sample itself acts as a thin lens with varying focal length as it moves through the focal plane. A standard Q-switched Nd:YAG laser with a pulse width of 8 ns at 532 nm and a pulse repetition rate of about 10 Hz was used as an excitation source. The laser beam, with a Gaussian intensity profile, was focused by using a lens of focal length ~50 mm into the sample solution contained in an approximately 1 mm standard quartz cuvette. The resulting beam spot at the focus was about 30 μm. The corresponding Rayleigh length, derived from the formula $z_R = \pi w_0^2/\lambda$, was found to be ~1.33 mm. Consequentially, and as the sample thickness of 1 mm is less than the Rayleigh length, it can be treated as a thin medium, hence ensuring the validity of the *z*-scan investigations. The *z*-scan of the various *H. sabdariffa* samples was performed with a laser power of 5 mW, which results in an average fluence of 7.1 GW/m$^2$. The optical limiting measurements were performed keeping the sample at the focal point and varying the input energy while recording the output energy

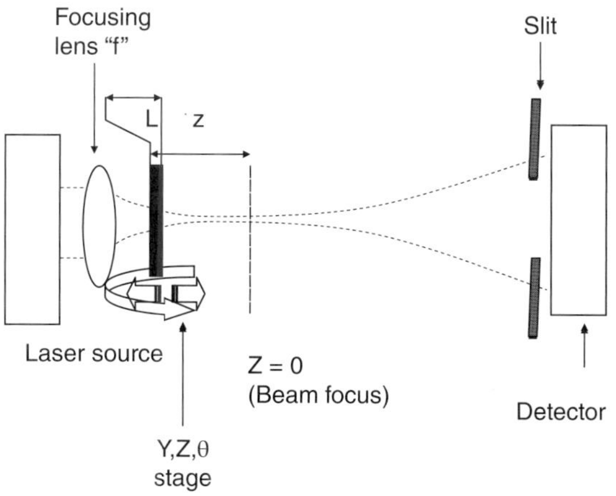


**Figure 35.4** The *z*-scan arrangement for the closed-aperture measurements of the various *H. sabdariffa* roselle dye extract solutions contained in 1 mm quartz cell

without placing the aperture in front of the detector. In order to avoid a cumulative thermal effect, the data were collected in a single-shot mode.

Figure 35.5 shows the closed-aperture *z*-scan measurements carried out at room temperature on the single beam set-up (described in Figure 35.4) for the determination of the NLO, the coefficient of the intensity-dependent refractive index $n_2$ and the third-order nonlinear susceptibility $\chi^{(3)}$ as well as the optical limiting characteristics. In the *z*-scan investigations (Herrera-Arellano *et al.*, 2004; Cisse *et al.*, 2011), each *H. sabdariffa* solution sample in an approximately 1 mm thickness quartz cell is scanned along the optical axis (*z*-direction: propagation direction) of the focused Gaussian laser beam, around its focus. The high intensity of the electromagnetic field in this region induces on the sample a nonlinear lensing of variable focal length. This causes a refractive divergence or convergence of the laser beam, depending on the sample position relative to the beam focus and the sign of the nonlinearity. The resulting variations in the beam radius within the closed aperture, and therefore the normalized transmission of the irradiance in the far field, are able to give adequate information for the calculation of the refractive index nonlinearities. More precisely, the normalized transmission of the irradiance in the far field is given approximately by $\Delta T(z,\phi) = 1 - \Delta\phi_0 x/[(x^2+1)(x^2+9)]$, with $x = z/z_R$ and $\Delta\phi_0$ the on-axis phase change at the laser beam focus (Sheik-Bahae *et al.*, 1989, 1990). The parameter $\Delta\phi_0(t)$ is the on-axis phase shift induced to the beam by the refractive nonlinearities of the material, and hence it is related to the refractive index change at the focus $\Delta n_0(t) = n_2$ by the relation $n_2 = \lambda|\Delta\phi_0|/(2\pi L_{\text{eff}} I_0)$, where $I_0$ is the intensity of the laser beam at focus ($z = 0$), $\lambda$ is the laser source wavelength, $L_{\text{eff}}$ is the effective thickness of the sample ($L_{\text{eff}} = [1 - \exp(-\alpha L)]/\alpha$), $L$ is the sample thickness length and $\alpha$ is the linear absorption coefficient. Owing to the mathematical formulation of the normalized transmission $\Delta T(z,\phi)$, which presents an asymmetry versus $z$, $n_2$ is directly related to the difference between the normalized peak and valley transmittances, $\Delta T_{p-v} = T_p - T_v$. Such a quantity is given by $\Delta T_{p-v} = 0.406(1-S)^{0.25}|\Delta\phi_0|$ (Matthews *et al.*, 2007). Figure 35.4 shows such characteristic asymmetric profiles of the various *H. sabdariffa* liquid solutions for the Q-switched Nd:YAG laser radiations (532 nm). It can be observed that the difference peak-to-valley transmission $\Delta T_{p-v}$ as well as the corresponding peaks' width increased with the dye concentration in the sample solutions. The dashed line superposed on the $S_8$ profile is a typical simulation curve allowing the determination of $n_2$ and $\chi^{(3)}$ of the various *H. sabdariffa* liquid solutions, and hence their concentration evolution. Table 35.1 summarizes the corresponding values and Figure 35.6 shows their variation with the concentration. Accordingly, one could then generalize the remark by Henari and Al-Saie (2006) on the large value of $n_2$ of the *H. sabdariffa* roselle dye solutions that they investigated for a fixed concentration (The concentration by weight of their sample solution was $2.02 \times 10^{-2}$.) The current $n_2$ values are all larger by one order of magnitude than the value reported for the Chinese tea and thermotropic liquid crystals (Zhang *et al.*, 1989; Gömeze *et al.*, 2003), and much larger indeed than the reported values for photorefractive crystals, liquid crystals and fullerenes (Prasad and Williams, 1991; Li *et al.*, 2005). Both the $n_2$ and $\chi^{(3)}$ values obtained are equivalent to those of Rekha and Ramalingam (2009) and Balaji *et al.* (2011) obtained on carmine ($C_{22}H_{20}O_{13}$) and safranin O ($C_{20}H_{19}N_4Cl$) natural dye solutions with a similar, yet a CW Nd-YAG laser source emitting at 532 nm. As illustrated in Figure 35.6, there is an increasing trend for the values of $n_2$ and $\chi^{(3)}$ as the concentration increases. This should be attributed to the fact that the dye molecules' population increases with the concentration, and therefore more thermally agitated particles, resulting in an enhanced effect. The negative sign of the closed *z*-scan profiles is indicative of the dominance of the thermo-optic effect. Moreover, as a similar behaviour is observed in continuous mode (CW Nd:YAG), the observed optical nonlinearity of the various *H. sabdarifa* dye solutions is likely to be of a thermal origin indeed, arising from the temperature dependence of the refractive index. The phase of the propagating beam would be distorted due to the presence of this thermal lens. Based on Figure 35.5, one could deduce that the peak–valley difference in terms of *z*-position for all samples investigated is ≥14 mm; that is, nearly 10 times the Rayleigh range, hence confirming the presence of the thermal component. It is well established indeed that a peak-to-valley separation larger than $1.7z_R$ is an indication of a Kerr type of nonlinearity (Matthews *et al.*, 2007). As was concluded in the above-mentioned natural dyes of carmine and safranin O, the organic dyes in solvent will have a tendency to form non-fluorescent dimers, optimizing the conversion of the absorption energy into heat (Rashidian *et al.*, 2009). With increasing dye concentration there is an increase in dimer concentration. Such dimers are expected to absorb light, eventually inducing a rise in the temperature of the medium. If the observed nonlinearity in the various *H. sabdariffa*–$H_2O$ solutions is of a Kerr type in the nanosecond regime, it would certainly be of an additional nature in the picosecond and sub-picosecond regimes. As pointed out by Zhang *et al.* (1989) and He *et al.* (1991) in their studies of Chinese tea dye solutions with both He–Ne and CW-mode locked Nd:YAG (532 nm, 70 ps) lasers, the large observed nonlinearity was attributed to the pull of higher refractive index molecules into

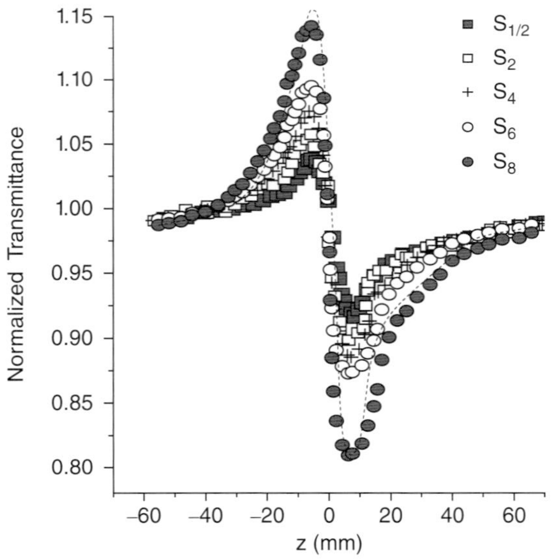


**Figure 35.5** Closed-aperture *z*-scan measurements of the various *H. sabdariffa* roselle dye extract solutions contained in 1 mm quartz cell. The Rayleigh length is about 1.33 mm

Table 35.1 Major derived nonlinear parameters of the z-scan and optical limiting studies. *Source*: Diallo *et al.* (2010). Reproduced with permission of Elsevier

| Sample | Concentration (g/L) | $\Delta T_{p-v}$ | $n_2$ ($10^{-4}$ esu) | $\chi^{(3)}$ ($10^{-6}$ esu) | ⟨Saturated output power⟩ (mW) | ⟨Threshold input power⟩ (mW) |
|---|---|---|---|---|---|---|
| $S_{1/2}$ | 0.5 | 0.12 | 1.31 | 05.77 | 2.67 | 31.69 |
| $S_2$ | 2 | 0.15 | 1.63 | 07.18 | 2.70 | 22.86 |
| $S_4$ | 4 | 0.19 | 2.07 | 09.12 | 2.14 | 20.38 |
| $S_6$ | 6 | 0.23 | 2.50 | 11.16 | 1.86 | 18.96 |
| $S_8$ | 8 | 0.34 | 3.70 | 16.52 | 1.66 | 14.60 |

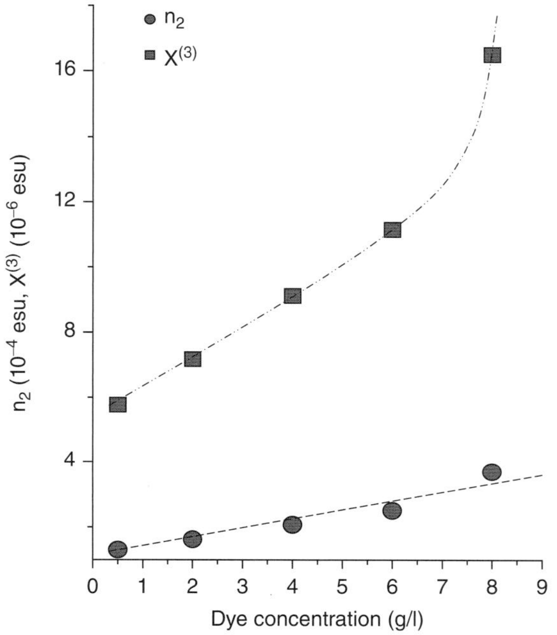


Figure 35.6 Variation of the nonlinear refractive index $n_2$ and the third-order NLO susceptibility $\chi^{(3)}$ of the *H. sabdariffa* roselle dye extract solutions; Q-switched Nd:YAG radiation at 532 nm with a pulse width of 8 ns and pulse repetition rate of about 10 Hz

regions of high optical density by light pressure as well as to their reorientation and redistribution in the tea dye solution.

During the z-scan investigations and high peak intensities closer to the focus ($z^{\pm} \approx 0$ mm), one could observe a diffraction-type pattern with concentric ring structures which are likely to be caused by self-phase modulation, as in the case of Henari and Al-Saie (2006). Hence, for the optical limiting studies which were conducted on the five *H. sabdariffa* solutions, the 1 mm quartz cell containing the dye solution was placed away from the focus. Figure 35.7 shows the corresponding optical limiting profiles within the range of 0–40 mW for the laser beam input power. One could distinguish two typical regions; in the first one (i.e. linear regime) the output power varies linearly with the input power according to the Lambert–Beer law, while for the second region the output power goes through a plateau. The value of this latter is very low relative to the input power and seems not to exceed 3.0 mW for an input as high as 40 mW. The values of the threshold input power for the various dye solutions have been estimated as 31.69 mW, 22.86 mW, 20.38 mW, 18.96 mW and 14.60 mW for $S_{1/2}$, $S_2$, $S_4$, $S_6$ and $S_8$ respectively (Table 35.1). Consequentially, the optical limiting effect is enhanced and the transmittance decreases with increasing concentration as more molecules per unit volume participate in the laser-dye cycles interaction during nonlinear absorption. The trend of $n_2$ and $\chi^{(3)}$ with the concentration in Figure 35.6 clearly indicates that the optical limiting behaviour of the *H. sabdariffa* roselle originates from reverse saturable absorption (Henari and Al-Saie, 1989; Rekha and Ramalingam, 2009; Diallo *et al.*, 2010).

As already mentioned, these interesting NLO properties are not proper to *H. sabdariffa* dye only, but are a characteristic of a significant class of other natural dyes too, such as carmine and safranin O, as well as chlorophyll. Rekha and Ramalingam (2009) have investigated the NLO characteristics of both Carmine solutions and carmine–polymer nanocomposite thin films. Carmine, also known as crimson lake, cochineal or natural red 4, is a natural

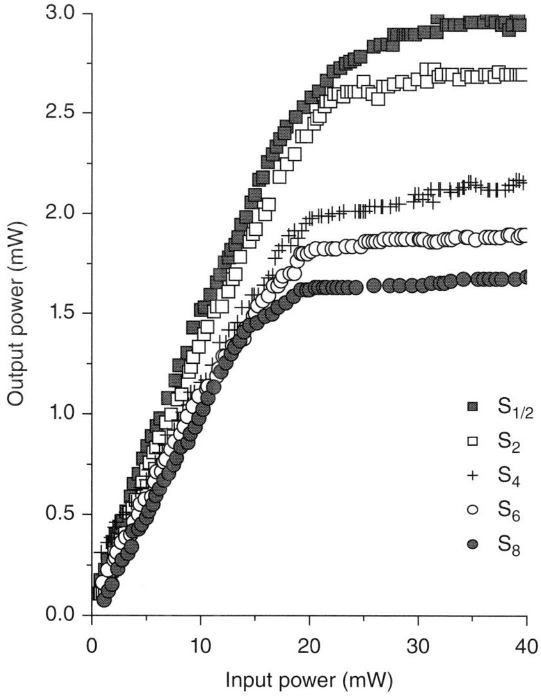


Figure 35.7 Optical limiting profiles of the various *H. sabdariffa* roselle dye extract solutions contained in 1 mm quartz cell. To avoid the observed diffraction caused by self-phase modulation, the quartz cell containing the dye solution was placed away from the focus

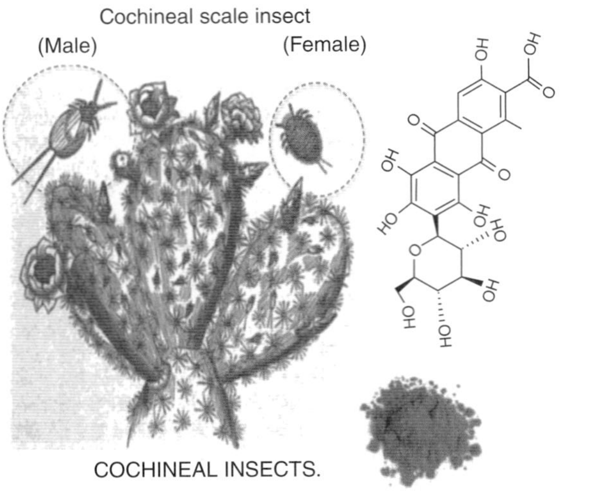


**Figure 35.8** Schematic representation of the cochineal scale insect and the plant on which it lives (genus *Opuntia*), molecular structure of red carmine ($C_{22}H_{20}O_{13}$) and typical powder sample of the orange–red carmine that can be extracted from the cochineal scale insect

pigment of a bright-red colour obtained from carminic acid, which is produced by some insects, such as the cochineal scale and the Polish cochineal. The cochineal is a scale insect in the suborder Sternorrhyncha. A primarily sessile parasite native to tropical and subtropical South America and Mexico, this insect lives on cacti in the genus *Opuntia*, feeding on plant moisture and nutrients (Figure 35.8). The insect produces carminic acid that deters predation by other insects. The carminic acid, which is typically 17–24% of the dried insect's weight, can be extracted from the body and eggs and then mixed with aluminium or calcium salts to make carmine dye. To prepare carmine, the powdered scale insect bodies are boiled in ammonia or a sodium carbonate solution. The insoluble matter is removed by filtering, and alum is added to the clear salt solution of carminic acid to precipitate the red aluminium salt. Figure 35.8 also shows the chemical structure of carmine, $C_{22}H_{20}O_{13}$, and a typical corresponding powdered sample. More precisely, Rekha and Ramalingam (2009) have conducted the NLO studies of carmine dye in methanol using a CW Nd:YAG laser at 532 nm as the source of excitation, both in solution and solid film mixed with polymethylmethacrylate (PMMA. The optical response was characterized by measuring the intensity-dependent refractive index $n_2$ of the medium using the $z$-scan technique of Figure 35.4 both in open- and closed-aperture configurations (Rekha and Ramalingam, 2009). Therefore, both the real $\text{Re}\chi^{(3)}$ and imaginary $\text{Im}\chi^{(3)}$ parts of the third-order susceptibility were deduced for each concentration ($\text{Re}\chi^{(3)}$ (esu) $= 10^{-4}\epsilon_0 c^2 n_0^2 n_2/\pi$ (cm²/W), where $\epsilon_0$ is the vacuum permittivity, $c$ is the celerity of light in vacuum and $n_2$ is the nonlinear refractive index; $\text{Im}\chi^{(3)}$ (esu) $= 10^{-2}\epsilon_0 c^2 n_0^2\,\breve{}fi/4\pi^2$ (cm/W), where β is a nonlinear absorption coefficient) and hence the absolute value of $\chi^{(3)}$ ($|\chi^{(3)}| = [(\text{Re}\chi^{(3)})^2 + (\text{Im}\chi^{(3)})^2]^{0.5}$]). Table 35.2 summarizes the major nonlinear parameters of the various carmine solutions and thin-film samples investigated; namely $n_2$, β and $\chi^{(3)}$.

As reported in Table 35.2, the carmine dye exhibits a negative (defocusing) nonlinearity and large nonlinear refractive index $n_2$ of the order of $10^{-7}$ cm²/W. Both $n_2$ and β increase with the carmine concentration, a trend that could be correlated to an enhanced thermal agitation with the concentration increase in view of the continuous laser pump source used in this case. If so, the optical nonlinearity of the carmine dye is likely to be of thermal origin arising from the temperature dependence of the refractive index $n(T)$. Moreover, as one can notice, the carmine–PMMA thin-film's NLO characteristics are relatively larger than the corresponding one in the solution form (0.05 mM concentration). This may be caused by a difference in heat dissipation; heat dissipation would be faster in liquids than in a solid form. As pointed out correctly by Rekha and Ramalingam (2009), while the NLO properties of the carmine dye originate from both refraction and absorption thermo-optical phenomena, $\chi^{(3)}$ is larger than those of standard NLO materials such as chalcone and its derivatives, as well as organic dyes such as croconium (Tsigaridas *et al.*, 2003; Ravindra *et al.*, 2007). As the NLO response was dominated by nonlinear refraction, which leads to a strong optical limiting response, the result reveals that carmine dye is indeed a promising material for optical limiting applications too.

Complementing the *H. sabdariffa* and carmine dye NLO characteristics are the recent studies conducted by Balaji *et al.* (2011) on safranin O. Safranin, an orange–red natural compound extracted from saffron, presents a conjugated molecular structure ($C_{20}H_{19}ClN_4$), as shown in Figure 35.9. While it is mostly used as a biomarker and staining protocols in histology and cytology, its significant optical absorbance centred at about 500 nm makes it a suitable candidate for NLO applications. Table 35.3 reports the NLO characteristics of safranin O dye in methanol-based solutions and polymer thin films at various concentrations for an incident probe diode-pumped Nd:YAG laser beam (λ = 532 nm) with an incident intensity of 3.5 kW/cm² in a $z$-scan configuration. As for the *H. sabdariffa* and carmine dyes, and in view of the evolution of the NLO parameters with the concentration with a continuous pump laser source, the observed large nonlinearities are likely to be of

**Table 35.2** Nonlinear parameters of carmine dye in methanol and as polymer-based nano-composites. *Source*: Rekha and Ramalingum (2009)

| Sample form | Concentration (mM) | $\Delta T_{p-v}$ | $n_2$ ($10^{-4}$ esu) | β ($10^{-3}$ cm/W) | $\Delta n = n_2 I_0$ ($10^{-4}$ cm/W) | $\chi^{(3)}$ ($10^{-6}$ esu) |
|---|---|---|---|---|---|---|
| Solution | 0.02 | 0.61 | 0.22 | 0.36 | 1.48 | 0.97 |
| Solution | 0.03 | 0.88 | 0.32 | 0.43 | 2.16 | 1.41 |
| Solution | 0.04 | 1.00 | 0.37 | 0.73 | 2.48 | 1.62 |
| Solution | 0.05 | 1.18 | 0.43 | 0.86 | 2.88 | 1.92 |
| Polymer film | 0.05 | 1.22 | 0.46 | 0.95 | 3.12 | 2.03 |

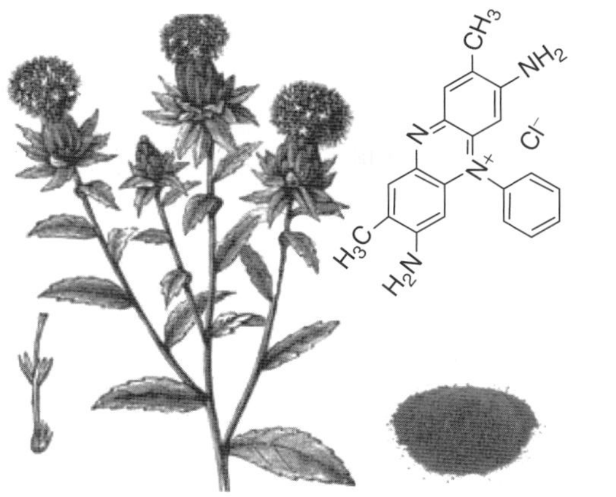


**Figure 35.9** Schematic representation of the saffron plant, molecular structure of safranin O ($C_{20}H_{19}ClN_4$) and typical powder sample of the orange–red safranin O

thermal origin arising from the thermal dependence of the refractive index $n(T)$. The peculiar enhancement in the polymer–safranin O composite was argued by Balaji *et al.* (2011) to be likely due to Anderson localization of photons (Sharma *et al.*, 2011).

## 35.3 Linear optical properties of natural dyes: Grätzel dye solar cells

The Sun is arguably the most abundant source of energy on Earth. Yet, harnessing solar energy in an efficient, low-cost way remains a challenge. Photoelectrochemical dye-sensitized solar cells, also known as DSCs or Grätzel cells, have been proposed as a cost-effective, easy-to-manufacture alternative to conventional solar cells (Calogero *et al.*, 2009, 2012). DSC technology mimics the redox functions of the light-dependent reactions of photosynthesis. The electron acceptors nicotinamide adenine dinucleotide phosphate ($NADP^+$) and $CO_2$ are replaced by a few micrometres thin film of mesoporous nanocrystalline titania 'nano-$TiO_2$'. The electron donor $H_2O$ and the oxidation product $O_2$ are replaced by a $I^-/I_3^-$ redox couple. To absorb incident light and produce excited electrons, a photoactive sensitizer that can adsorb onto the $TiO_2$ surface via its -OH and =O groups is required. Currently, efficient Ru-based (sometimes Os-based) dyes, known as RuL3 (or N3), RuL2 and RuL2(NCS)2, are the most used owing to their high spectral incident photon-to-electron efficiency (IPCE) values (O'Regan and Grätzel, 1991; Kuciauskas *et al.*, 2001; Nazeeruddin *et al.*, 2001; Argazzi *et al.*, 2004; Altobello *et al.*, 2005; Calogero *et al.*, 2009; Zhou *et al.*, 2011; Calogero *et al.*, 2012). However, the severe toxicity and the elevated total cost of the sensitizing dye due to the rare aspect of ruthenium mining deposits are now of concern.

The first highly efficient nanocrystalline DSC was pioneered by Grätzel in the early 1990s, with efficiency reaching 10% on laboratory prototypes (O'Regan and Grätzel, 1991). A typical DSC, as shown in Figure 35.10, consists of two transparent electrodes, with the first one covered with nano-scaled semiconductor particles (generally nano-$TiO_2$) each coated with dye molecules. This photoactive-dye molecules–nano $TiO_2$ porous layer is interfaced with the two electrodes via a redox electrolyte, generally an iodine-based solution. Once the solar radiation is transmitted through one of the transparent electrodes, it is absorbed by the dye molecules coating the nano-$TiO_2$, creating an excitonic $e^-$–hole pair (Grätzel, 2003). The electrons created then flow through the nano-scaled semiconductor particles to reach the second transparent electrode and hence generate an electric current. From a technological viewpoint, a DSC is assembled placing in succession the first transparent photoanode, an electrolyte solution containing the redox system and the counter-electrode. More accurately, the photoanode consists of a porous film of $TiO_2$ nano-particles deposited onto a transparent conductive-oxide glass support, sensitized with artificial dye molecules. The electrolyte system, placed between the two electrodes, is usually based on the iodide/iodine redox couple ($I$, $I_3^-$). The main photochemical processes of a DSC are summarized as follows and described in Figure 35.11 (Calogero *et al.*, 2009; Zhou *et al.*, 2011):

$$\text{Dye} + h\nu \rightarrow \text{Dye}^* \tag{35.1}$$

$$\text{Dye}^* + \text{TiO}_2 \rightarrow \text{Dye}^+ + e^-_{CB}(\text{TiO}_2) \tag{35.2}$$

$$\text{Dye}^* \rightarrow \text{Dye} \tag{35.3}$$

$$2\text{Dye}^+ + 3\text{I}^- \rightarrow 2\text{Dye} + \text{I}_3^- \tag{35.4}$$

$$\text{Dye}^+ + e^-_{CB}(\text{TiO}_2) \rightarrow \text{Dye} + \text{TiO}_2 \tag{35.5}$$

$$\text{I}_3^- + 2e^-(\text{catalyst}) \rightarrow 3\text{I}^- \tag{35.6}$$

$$\text{I}_3^- + 2e^-(\text{TiO}_2) \rightarrow 3\text{I}^- + \text{TiO}_2 \tag{35.7}$$

The dye initially at a lower energy state ($S^0/S^+$), upon absorption of a photon $h\nu$, goes to an electronically excited state Dye$^*$ ($S^*$) following Equation 35.1, which energetically lies above the conduction band edge of the $TiO_2$ semiconductor nanoparticles (Equation 35.2) and

**Table 35.3** Nonlinear parameters of safranin O dye in methanol and as polymer-based nano-composites. *Source*: Balaji *et al.* (2011)

| Sample form | Concentration (mM) | $\Delta T_{p-v}$ | $n_2$ ($10^{-4}$ esu) | β ($10^{-3}$ cm/W) | $\Delta n = n_2 I_0$ ($10^{-4}$ cm/W) | $\chi^{(3)}$ ($10^{-6}$ esu) |
|---|---|---|---|---|---|---|
| Solution | 0.02 | 0.59 | −0.40 | −0.36 | −1.33 | 1.80 |
| Solution | 0.03 | 0.99 | −0.78 | −0.40 | −2.60 | 3.51 |
| Solution | 0.04 | 1.47 | −1.16 | −0.43 | −3.87 | 5.18 |
| Solution | 0.05 | 1.84 | −1.45 | −0.47 | −4.86 | 6.54 |
| Polymer film | 0.05 | 2.18 | −1.72 | −0.52 | −5.76 | 7.76 |

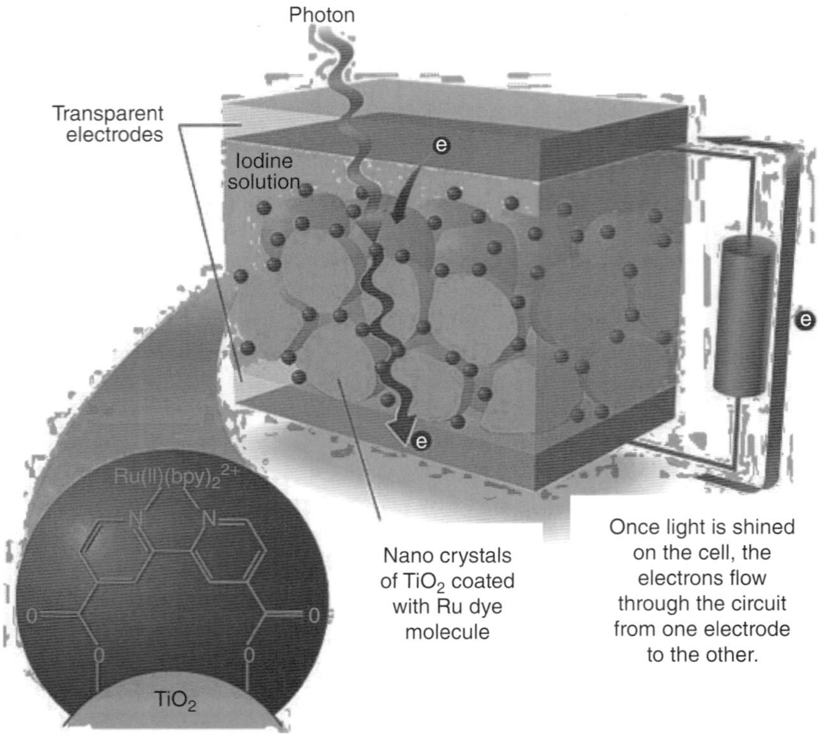


**Figure 35.10** Principle of electron generation in a Grätzel DSC via excitation of dye molecules (small spheres) coating nano-$TiO_2$ mesoporous film

injects an electron into the $TiO_2$ conduction band. The deactivation reaction governed by Equation 35.3, is a relaxation of the excited states, which occurs in competition with the electron injection into the nano-$TiO_2$. The electron dynamics of Equations 35.1–35.5 are governed by ultrafast physical–chemical phenomena and temporal regimes as reported in Figure 35.12. The collection efficiency of the photo-injected electrons at the anode back contact is affected by two major recombination processes, which are described in Equations 35.5 and 35.7; that is, (i) back electrons transfer and (ii) the $TiO_2$ conduction band electron capture by the oxidized redox couple respectively. These two processes are in competition with the oxidation of iodide (Equation 35.4) and are among the major causes of reducing the current production of the cell. In the external circuit, the injected electrons give a current flow while providing for the reduction of iodine at the counter electrode (Equation 35.6).

An efficient sensitizing dye should absorb light over a broad range from the visible to the near-infrared, and the energy of its electronic excited state should lie energetically above the conduction band edge of the $TiO_2$ (Figure 35.11). Thermal and photochemical stability are other additional required characteristics of the dyes used in DSCs. Generally, transition metal coordination compounds such as Ru and Os polypyridil complexes (Kuciauskas *et al.*, 2001; Nazeeruddin *et al.*, 2001; Grätzel, 2003; Argazzi *et al.*, 2004; Altobello *et al.*, 2005; Calogero and Di Marco, 2008; Zhou *et al.*, 2011) and synthetic organic dyes (Campbell *et al.*, 2007; Yum *et al.*, 2007), are used as effective sensitizers in DSCs. Over the last two decades, Ru complexes equipped with appropriate ligands and anchoring groups have been the most widely used choice of charge transfer sensitizers (Grätzel, 2003; Polo *et al.*, 2004).

In addition to their elevated cost, the synthesis routes for metal complexes are based on multiphase procedures involving complexe purification methodologies (Nazeeruddin *et al.*, 2001; Grätzel, 2003). In parallel with Ru and Os complexes, there are other organic dyes that the DSCs community is exploring. Their conversion efficiency within the DSC devices has reached values close to Ru/Os-based cells: 9.5% for the indoline dye D205 developed by

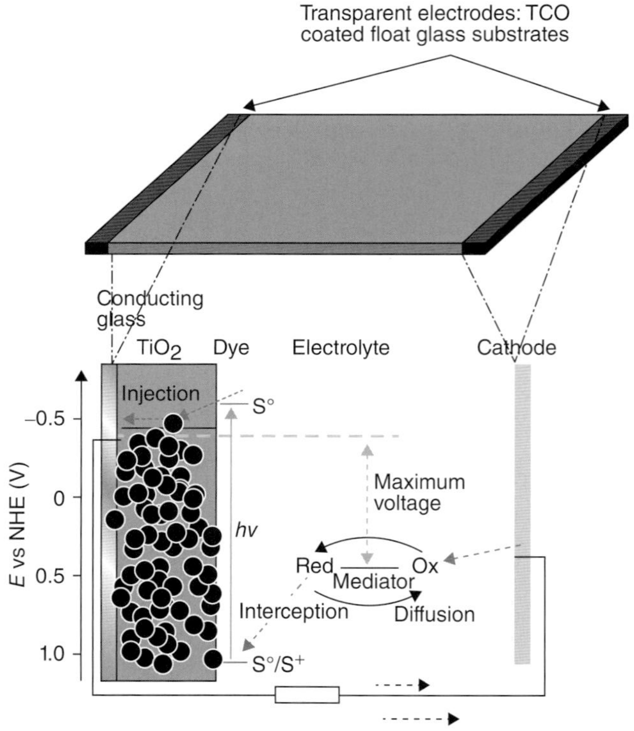


**Figure 35.11** Schematic representation of the various physical–chemical reactions taking place in a standard DSC unit

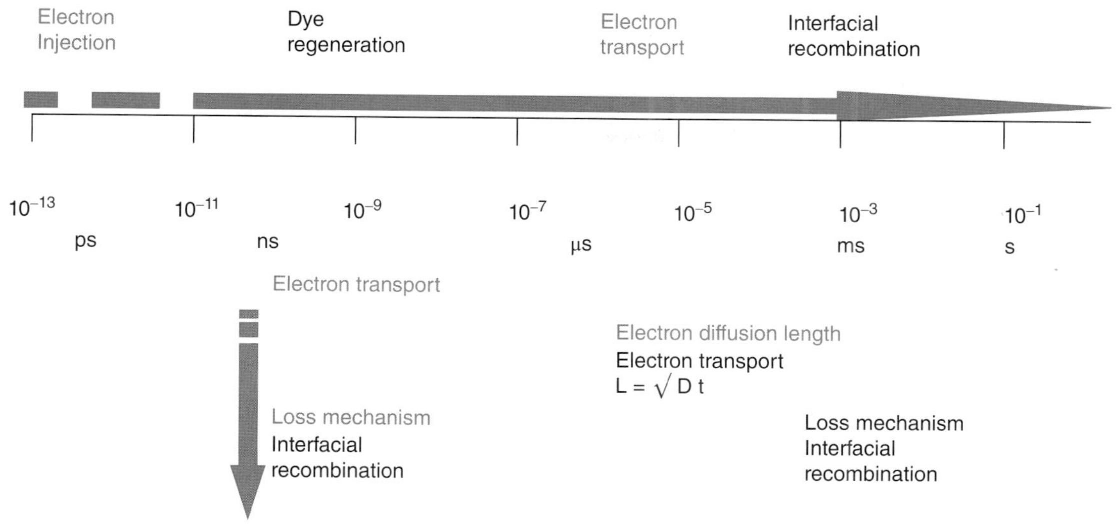


**Figure 35.12** Time regime of the dynamic competition between the transport of the photo-generated electrons and their interfacial recombination

Uchida and co-workers (Ito *et al.*, 2008) and 9.8% for the dye C217 developed by Zhang *et al.* (2008), which are promising if compared to the 11.1% obtained with standard DSCs based on Ru complexes (Nazeeruddin *et al.*, 2005). Other Ru-free sensitizers, based on the so-called push–pull substituted porphyrins, with high solar-to-electric power conversion efficiency have been reported (Bessho *et al.*, 2010).

While the standard Ru dye is the critical component in the DSC device in relation to the efficiency of conversion, which has reached currently 12.3% (Yella *et al.*, 2011), its weight in the final cost is elevated. As per estimated by Calogero and Di Marco (2008) and Zhou *et al.* (2011), a DSC with an Ru-based sensitizer has a dye uptake $C$ of approximately $10^{-7}$ mol/cm$^2$ (layer thickness ~10 μm), which corresponds to 1 g of dye per square metre. The price of the dye per watt peak ($W_p$) can be deduced as $P_{total} = CM_{dye}P_{mass}/(I_{AM1.5}\eta)$, where $P_{total}$ is the price per watt peak, $C$ is the surface coverage described above, $M_{dye}$ (g/mol) is the molecular mass of the dye, $P_{mass}$ is the price per mass, $I_{AM1.5}$ (W/m$^2$) is the light intensity of air mass (AM) 1.5, and $\eta$ is the efficiency of the DSC. For Ru-based compounds, $P_{mass}$ is currently estimated to be of the order €300/g, which could decrease with large-scale production. The cost of the dye for a DSC-based panel of 3 $kW_p$ will be about €2400 (Calogero and Di Marco, 2008). The same applies for organic dyes: the production costs and the $P_{total}$ are estimated similar to Ru dyes. Independently from cost-related issues (Smestad *et al.*, 1994; Smestad, 1998; Campbell *et al.*, 2007), indirect environmental implications, due to the chemical by-products, also have to be considered. Unlike synthetic dye-sensitizers, natural ones are easy to extract, cost effective, non-toxic and environmentally friendly while fully biodegradable. Among natural pigments, three main families of compounds have been exploited as sensitizer in DSCs: chlorophylls (Kay and Grätzel, 1993; Chang *et al.*, 2010), anthocyanins (Cherepy *et al.*, 1997; Polo and Murakami Iha, 2006) and betalains (Calogero *et al.*, 2010) or their mixtures (Zakeeruddin *et al.*, 2003; Kumara *et al.*, 2006; Chang *et al.*, 2010). In this context, the possibility to obtain interesting photo-conversion efficiencies, using natural dyes, was demonstrated by various teams (Yamazaki *et al.*, 2007; Calogero and Di Marco, 2008; Chang *et al.*, 2010; Furukawa *et al.*, 2009; Gomez-Ortiz *et al.*, 2010; Zhou *et al.*, 2011; Calogero *et al.*, 2012).

The recent sound results were reported by Calogero and Di Marco (2008) and Zhou *et al.* (2010). They have focused their attention on anthocyanins and betalains. Anthocyanins are a very large group of red–blue plant pigments that naturally occurs in all higher plants. They are responsible for the coloration of flowers and fruits (Timberlake and Bridle, 1975). Blackberries, grapes, blueberries, aubergine, red cabbage, red radicchio, and so on are the major sources of anthocyanins, whose production in nature is estimated to be about $10^9$ t/year (Calogero and Di Marco, 2008; Zhou *et al.*, 2011). Over 500 different types of anthocyanins have been isolated from plants and all are derivatives of the single basic core structure of 2-phenylbenzopyrylium (flavylium) ion. More precisely, betalains are a class of red and yellow derived pigments that are found in plants of the order Caryophyllales, comprising a modest number of nearly 55 structures. The most promising source of betalains is the prickly pear Cactaceae fruit, which originates from Mexico and spread widely throughout Latin America, South Africa and the Mediterranean area. Like the anthocyanins, the betalains exhibit a significant light absorption in the visible region and are antioxidant compounds (Timberlake and Henry, 1988; Takeoka and Dao, 2002; Stintzing and Carle 2007; Azeredo, 2009); conversely to the anthocyanins, which present functional groups (AOH), betalains have the requisite functional groups (ACOOH) to bind better to the $TiO_2$ nanoparticles in standard DSCs (Hao *et al.*, 2006; Quin and Clark, 2007), than the functional groups (AOH) present in the former. Indeed, the interaction between $TiO_2$ meso-porous film and carboxylic functions should bring about a stronger electronic coupling and rapid forward and reverse electron transfer reactions.

The efficiency of a DSC is characterized by several parameters, among which are the energy conversion efficiency $\eta$ (%), the short-circuit photocurrent $I_{SC}$ (mA/cm$^2$), the open-circuit voltage $V_{OC}$ (V), the fill factor FF and the IPCE. Generally, the smaller the $V_{OC}$ is and the larger are $I_{SC}$, $\eta$ and IPCE, the better is the performance of the DSCs. Table 35.4 reports the photochemical parameters of various DSC devices based on several natural dye extracts.

**Table 35.4** Photoelectrochemical parameters of various DSCs sensitized with natural dyes

| Compound | Dye | $J_{SC}$ (mA/cm$^2$) | $V_{OC}$ (V) | FF (%) | η (%) | Reference |
|---|---|---|---|---|---|---|
| Sicilian prickly pear | | 7.85 | 0.267 | 62.0 | 1.87 | Calogero and Di Marco (2008), Calogero et al. (2012) |
| Blackberry | | 5.85 | 0.320 | 57.0 | 1.07 | Calogero and Di Marco (2008), Calogero et al. (2012) |
| Red Sicilian orange extract | | 5.13 | 0.329 | 59.0 | 1.01 | Calogero and Di Marco (2008), Calogero et al. (2012) |
| Red mulberry | | 4.45 | 0.340 | 64.0 | 0.99 | Calogero and Di Marco (2008), Calogero et al. (2012) |
| Radicchio | | 5.05 | 0.322 | 55.0 | 0.90 | Calogero and Di Marco (2008), Calogero et al. (2012) |
| Red Sicilian orange | | 4.98 | 0.325 | 48.0 | 0.78 | Calogero and Di Marco (2008), Calogero et al. (2012) |
| Eggplant | | 3.48 | 0.346 | 53.0 | 0.64 | Calogero and Di Marco (2008), Calogero et al. (2012) |
| Giacchè grapes | | 3.06 | 0.333 | 56.0 | 0.57 | Calogero and Di Marco (2008), Calogero et al. (2012) |
| Begonia | | 0.63 | 0.537 | 72.2 | 0.24 | Zhou *et al.* (2011) |
| Tangerine peel | | 0.74 | 0.592 | 63.1 | 0.28 | Zhou *et al.* (2011) |
| Rhododendron | | 1.61 | 0.585 | 60.9 | 0.57 | Zhou *et al.* (2011) |
| Fructus lycii | | 0.53 | 0.689 | 46.6 | 0.17 | Zhou *et al.* (2011) |
| Marigold | | 0.51 | 0.542 | 83.1 | 0.23 | Zhou *et al.* (2011) |
| Perilla | | 1.36 | 0.522 | 69.2 | 0.50 | Zhou *et al.* (2011) |
| Herba artemisiae scopariae | | 1.03 | 0.484 | 68.2 | 0.34 | Zhou *et al.* (2011) |
| China loropetal | | 0.84 | 0.518 | 62.6 | 0.27 | Zhou *et al.* (2011) |
| Yellow rose | | 0.74 | 0.609 | 57.1 | 0.26 | Zhou *et al.* (2011) |
| Flowery knotweed | | 0.60 | 0.554 | 62.7 | 0.21 | Zhou *et al.* (2011) |
| Bauhinia tree | | 0.96 | 0.572 | 66.0 | 0.36 | Zhou *et al.* (2011) |
| Petunia | | 0.85 | 0.616 | 60.5 | 0.32 | Zhou *et al.* (2011) |
| Lithospermum | | 0.14 | 0.337 | 58.5 | 0.03 | Zhou *et al.* (2011) |
| Violet | | 1.02 | 0.498 | 64.5 | 0.33 | Zhou *et al.* (2011) |
| Chinese rosa | | 0.90 | 0.483 | 61.9 | 0.27 | Zhou *et al.* (2011) |
| Mangosteen pericarp | | 2.69 | 0.686 | 63.3 | 1.17 | Zhou *et al.* (2011) |
| Lily | | 0.51 | 0.498 | 66.7 | 0.17 | Zhou *et al.* (2011) |
| Coffee | | 0.85 | 0.559 | 68.7 | 0.33 | Zhou *et al.* (2011) |
| Broadleaf holly leaf | | 1.19 | 0.607 | 65.4 | 0.47 | Zhou *et al.* (2011) |
| N-719 | | 13.74 | 0.773 | 57.6 | 6.11 | Zhou *et al.* (2011) |
| Achiote seeds | Bixin | 1.1 | 0.57 | 59 | 0.37 | Gomez-Ortiz *et al.* (2010) |
| | Norbixin | 0.38 | 0.53 | 64 | 0.13 | Gomez-Ortiz *et al.* (2010) |
| Gardenia fruit | Crocetin | 2.84 | 0.43 | 46 | 0.56 | Yamazaki *et al.* (2007) |
| | Crocin | 0.45 | 0.58 | 60 | 0.16 | Yamazaki *et al.* (2007) |
| Pomegranate | Chlorophyll II | 2.05 | 0.560 | 52 | 0.597 | Chang *et al.* (2010) |
| Pomegranate+mulberry | | 2.8 | 0.53 | 49 | 0.722 | Chang *et al.* (2010) |
| Red cabbage | | 0.68 | 0.52 | 70 | 0.50 | Furukawa *et al.* (2009) |
| Curcumin | | 0.53 | 0.53 | 72 | 0.41 | Furukawa *et al.* (2009) |
| Red cabbage+curcumin | | 0.81 | 0.53 | 69 | 0.60 | Furukawa *et al.* (2009) |

As indicated in Table 35.5 and reported in by Calogero *et al.* (2012), the most proven effective natural dye in DSC technology so far is the prickly pear's extract dye coating a meso-porous nano-$TiO_2$ of ~13.3 μm in thickness. It has the highest IPCE of ~69%, close to the Ru synthetic dye complexes; the IPCE corresponds to the number of electrons, measured as photocurrent in the external circuit, divided by the monochromatic photon flux that strikes the cell. The corresponding *J–V* curve of the prickly pear dye-based DSC is shown in Figure 35.13. The $V_{OC}$ and short-circuit current density $J_{SC}$ values are the highest relative to other prominent natural dyes: 0.389 V and 8.82 mA/cm$^2$ respectively. Likewise, it was found that among the best additional selected natural dyes (mulberry, blackberry, Sicilian red orange juice), the IPCE spectra of the Sicilian prickly pear dye (from the blue region of the spectrum up to 630 nm) clearly showed the highest value on the ordinate axis when compared with the other dyes.

There are additional requirements upon the usage of natural dyes in DSC technology, among which one could mention the following:

**Table 35.5** IPCE maximum values, measured at $\lambda_{max}$ for some selected dyes adsorbed on $TiO_2$ multilayer structures (13.3 μm thick, including a scattering layer) (Calogero and Di Marco, 2008). *Source:* Zhou *et al.* (2011). Reproduced with permission of Elsevier

| Dye | Electrolyte | $\lambda_{max}$ (nm) | IPCE (%) |
|---|---|---|---|
| N719 | A6979 | 550 | 89 |
| N719 | A6141 | 550 | 86 |
| **Sicilian prickly pear** | **A6979** | **450** | **69** |
| Blackberry | G1 | 550 | 46 |
| Blackberry | A6979 | 550 | 44 |
| Red Sicilian orange concentrated juice | A6979 | 550 | 42 |
| Red Sicilian orange concentrated juice | G1 | 550 | 40 |

(i) effective anchoring of the dye molecules onto the surface of the nano-$TiO_2$, (ii) dye purification and (iii) its time stability. In terms of dye anchoring onto the surface of the nano-$TiO_2$, its role through effective functional groups in the DSCs is exemplified by the studies conducted by Yamazaki *et al.* (2007) on *Gardenia* fruit dye, which is used in Asian countries as a natural food colourant. *Gardenia* yellow dye derives from two carotenoids: crocetin and crocin. These two carotenoids have the same frame, which are different in structure: with or without free carboxylic groups (Figure 35.14a). The observed large broadening of the absorbance spectrum of crocetin onto nano-$TiO_2$ films implies a strong interaction between crocetin and the nano-$TiO_2$ surface with the delocalized $\pi^*$ state of the crocetin injecting effectively into the $TiO_2$ conduction band, similar to the case of standard Ru complexes. This study supports that terminal carboxylic groups of crocetin can achieve a formation of covalent bonds with metallic sites of $TiO_2$ (Figure 35.14b). Figure 35.14c confirms indeed the importance of such a surface anchoring of the natural dye onto the nano-$TiO_2$ and its effect on the final DSC efficiency. Figure 35.14c reports typical *J–V* curves of DSCs sensitized with crocetin and crocin. $J_{SC}$, $V_{OC}$ and η for the optimized DSCs are 2.84 mA/cm², 0.43 V and 0.56% respectively for crocetin and 0.45 mA/cm², 0.58 V and 0.16% respectively for crocin. The effectiveness of the anchoring in this case specifically is clearly shown via the $J_{SC}$ value, which is about 2.84 mA/cm² for the DSCs sensitized with crocetin; that is, more than six times higher than the one with crocin. In terms of the importance of natural dyes' purifity before any DSC technology usage, its role

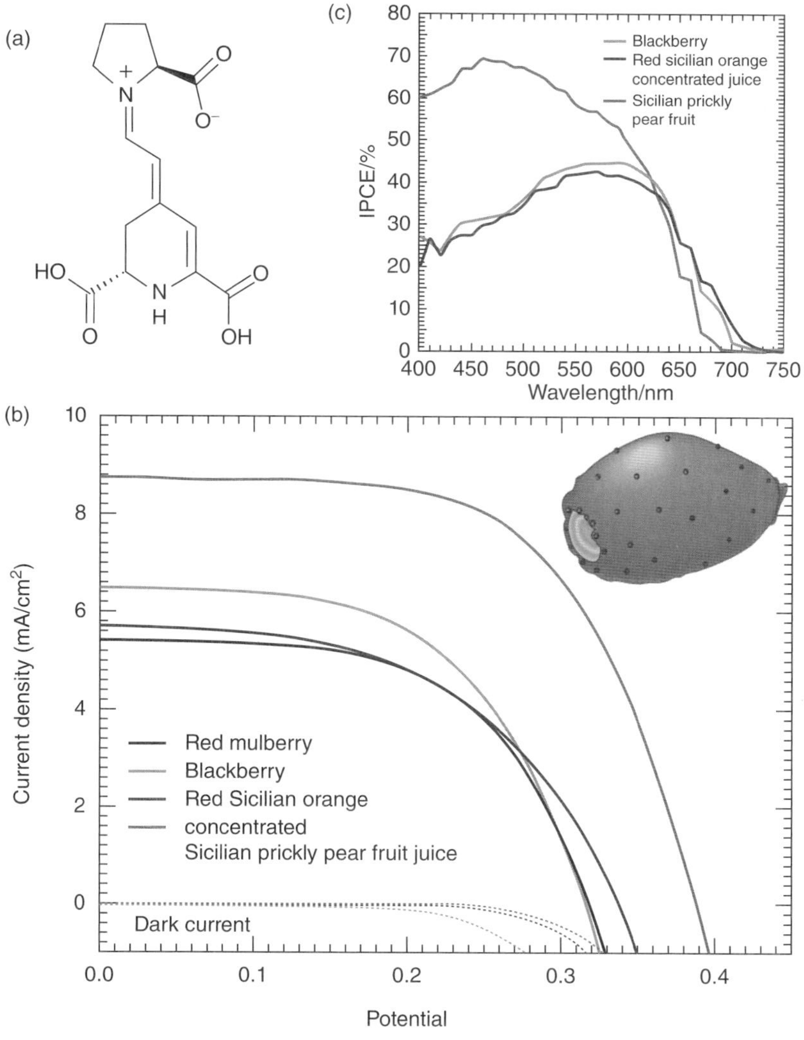


**Figure 35.13** (a) Molecular structures of typical Sicilian prickly pear dye. (b) Current density–voltage characteristics curves of DSCs sensitized with red mulberry natural pigments, red Sicilian orange concentrated juice, blackberry and Sicilian prickly pear. (c) IPCE spectra of DSCs using A6979, nanosized Pt/FTO as counter electrode and $TiO_2$ film sensitized with blackberry extract, red Sicilian orange concentrated juice and Sicilian prickly pear (Calogero and Di Marco, 2008). *Source:* Zhou *et al.* (2011). Reproduced with permission of Elsevier

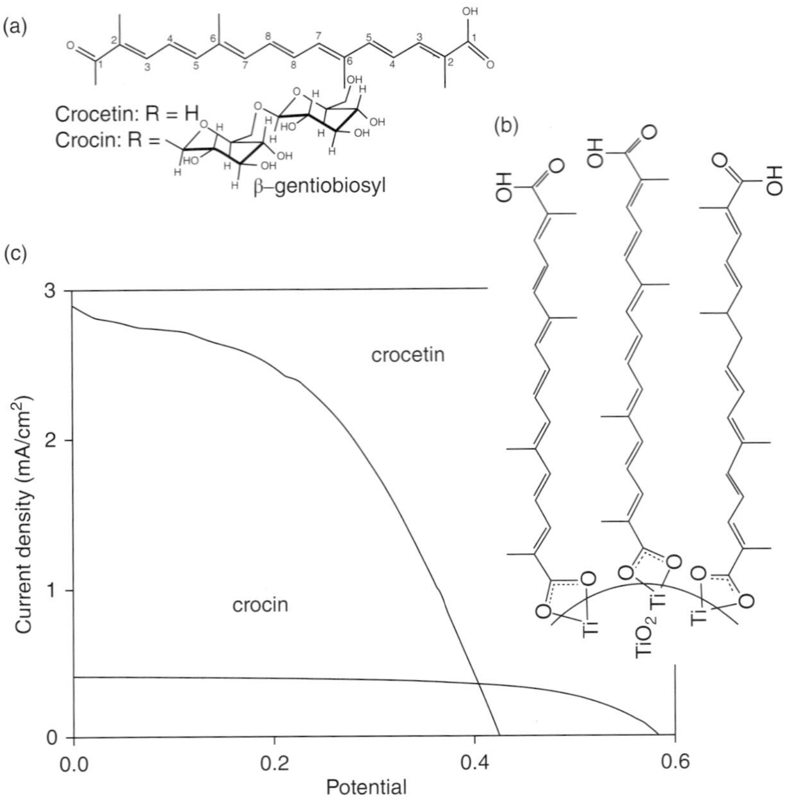


Figure 35.14 (a) Molecular structures of crocetin and crocin, (b) possible mechanism of the binding between crocetin and $TiO_2$ surface and (c) *J–V* characteristics of DSCs sensitized with crocetin and crocin. *Source*: Yamazaki *et al.* (2007). Reproduced with permission of Elsevier

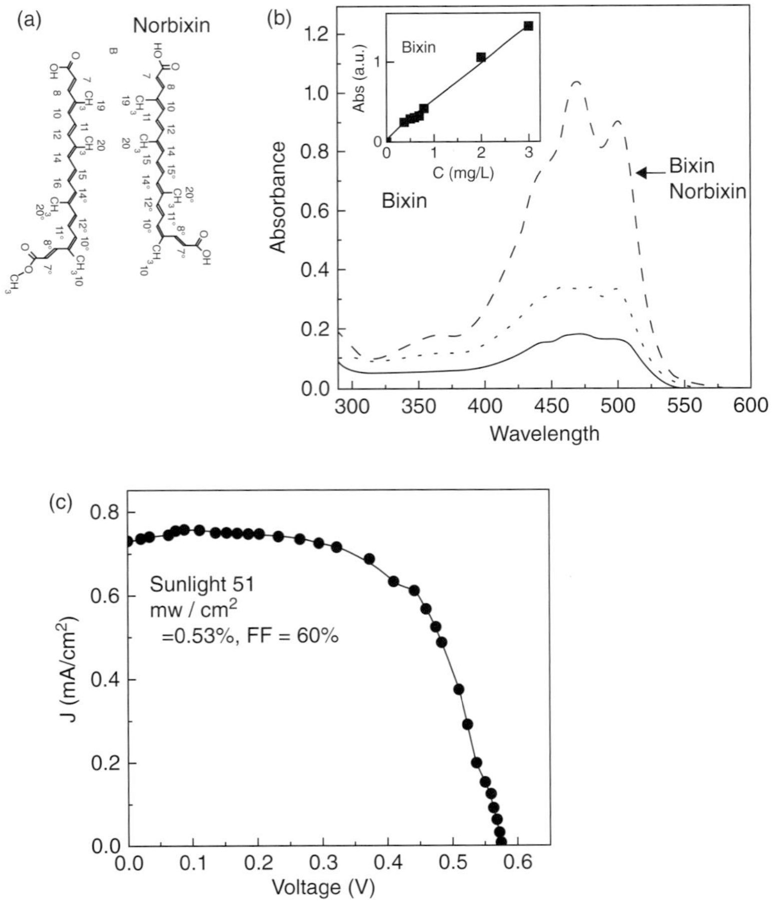


Figure 35.15 (a) Molecular structures of *cis*-norbixin and *cis*-bixin, (b) their optical absorbance profiles in solution and (c) the *J–V* characteristic of a *cis*-bixin-sensitized mesoporous nano-$TiO_2$ DSC cell. *Source*: Gomez-Ortiz *et al.* (2010). Reproduced with permission of Elsevier

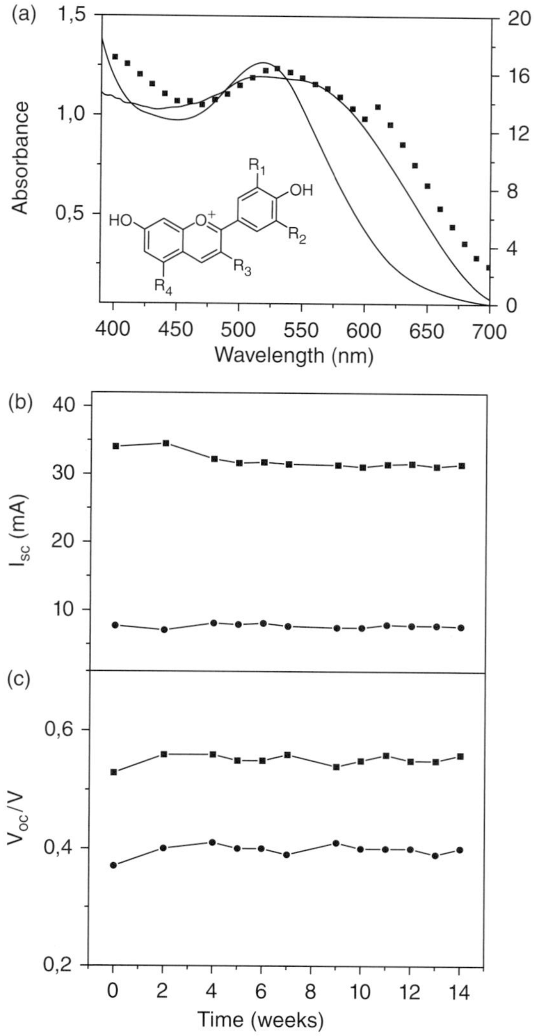


**Figure 35.16** (a) Optical absorbance of aqueous pure mulberry dye (continuous line), optical absorbance of $TiO_2$ film sensitized by mulberry dye (dashed line) and its IPCE (filled squares). (b) Variation of $I_{SC}$ versus time for DSCs sensitized by mulberry (filled circles) and N3 (filled squares); AM 1.5, 100 mW/cm$^2$. *Source*: Patrocinio *et al.* (2009). Reproduced with permission of Elsevier

via various chromatography phases is illustrated by the studies conducted by Gomez-Ortiz *et al.* (2010) on achiote seeds. Achiote, *Bixa orellana*, is a tropical shrub from the American continent that bears a red fruit containing about 50 red seeds. These later contain a dark red dye used mainly in the textile, cosmetics and food industries. More accurately, the seeds contain a high concentration of carotenoids, composed of up to ~80% of carotenoid *cis*-bixin – that is, 9-*cis*-bixin ($C_{25}H_{30}O_4$) – with the remaining ~20% including *trans*-bixin and norbixin ($C_{24}H_{28}O_4$). Yet while both bixin and norbixin have a quasi-similar conjugated skeleton (Figure 35.15a), they exhibit completely different optical absorbances in the visible range (Figure 35.15b), and hence a different efficiency once used in DSCs. Indeed, their respective efficiencies in mesoporous nano-$TiO_2$ DSCs are about 0.37% and 0.13% (Table 35.4, Figure 35.15c). The time stability of natural dye-based DCSs is an imperative parameter. It depends mainly on the rate of electron injection efficiency at the natural dye–nano-$TiO_2$ interface and its constancy with time. Hence, the thermal and time stability of the dye itself is paramount relative to synthetic Ru-based dyes. The investigations of Patrocinio *et al.* (2009) report the results in this regard. The efficiency and stability of DSC devices based on natural sensitizers such as mulberry (*Morus alba* Lam), blueberry (*Vaccinium myrtillus* Lam) and jaboticaba skin (*Myrtus cauliflora* Mart) were studied. It was observed that mulberry-extract-based DSCs, which exhibit relatively both large optical absorbance and IPCE over a spectral range of about 400–600 nm (Figure 35.16a), presented the highest $P_{max}$ value of about 1.6 mW/cm$^2$, with $J_{SC} \approx 6.14$ mA/cm$^2$ and $V_{OC} \approx 0.49$ V. Subsequent to such a phase, the photoelectrochemical parameters of a 16 cm$^2$ active area device sensitized by mulberry-dye-based DSC were measured for 14 weeks of continuous evaluation. Figure 35.16b and c shows the variation of $I_{SC}$ and $V_{OC}$ values as a function of time for the mulberry-sensitized DSC compared with the Ru-based N3 one. The N3 dye was chosen as a standard for comparison owing to the well-known stability of solar cells with Ru polypyridyl complexes as sensitizers. Despite the lower photocurrent of the mulberry-sensitized DSC in comparison with that sensitized by N3, the system exhibited a similar stability during the period investigated. In fact, it was demonstrated that devices sensitized by mulberry can remain operative for about 36 weeks (Patrocinio *et al.*, 2009) and hence sustain the potential commercial feasibility of natural extracts-based DSCs.

## 35.4 Conclusion

This chapter reported on the potentialities of natural dyes in the field of nano-photonics in general and in nonlinear optics and photochemical solar conversion specifically. Via three natural dyes, *H. sabdariffa* (Rosella), carmine and safranin O extracts, it was demonstrated that such extracts exhibit significant values of $n_2$ and $\chi^{(3)}$ that are larger by one order of magnitude at least than the values reported for thermotropic liquid crystals, photorefractive crystals, liquid crystals and fullerenes. In the DSC field, following the investigation of several natural dyes, it is established that extract from Sicilian prickly pear exhibits an IPCE close to that of Ru complex N719. The additional time stability of some natural dyes enhances their potential applications towards industrial requirements.

## Acknowledgements

This research programme is conducted within the framework of the UNESCO UNISA Africa Chair in Nanosciences & Nanotechnology. It was generously supported by grants from the National Research Foundation of South Africa, the Abdus Salam ICTP via the Nanosciences African Network (NANOAFNET), the Organization of Women in Science for the Developing World (OWSDW), the Academy of Science for the Developing World (TWAS), the French Embassy in Pretoria, South Africa, and the African Laser Centre, to whom we are grateful. I express my thanks to Professor G. Calogero, Professor E. Yamazaki, Professor N.M. Gomez-Ortiz, Professor A.O.T. Patrocinio and their colleagues, as well as to fellows Dr P. Mthunzi, Dr G. Mhlongo, Mr S. Zongo and Mr A. Diallo.

## References


Alarcon-Aguilar, F.J. Zamilpa, A., Perez-Garcia, D. *et al.* (2007) *J. Ethnopharmacol.*, **114**, 66–71.

Altobello, S., Argazzi, R., Caramori, S. *et al.* (2005) *J. Am. Chem. Soc.*, **127**, 15342.

Argazzi, R., Larramona, G., Contado, C., and Bignozzi, C.A. (2004) *J. Photochem. Photobiol. A: Chem.*, **164**, 15.

Azeredo, H.M.C. (2009) *Int. J. Food Sci. Technol.*, **44**, 2365.

Balaji, G., Rekha, R.K. and Ramalingam, A. (2011) *Acta Polon. A*, **119** (3), 359.

Bessho, T., Yeh, C.Y., Diau, E.W.G. *et al.* (2010) *Angew. Chem. Int. Edit.*, **49**, 6646.

Calogero, G. and Di Marco, G. (2008) *Sol. Energy Mater. Sol. Cells*, **92**, 1341.

Calogero, G., Di Marco, G., Caramori, S. *et al.* (2009) *Energy Environ. Sci.*, **2**, 1162.

Calogero, G., Di Marco, G., Cazzanti, S. *et al.* (2010) *Int. J. Mol. Sci.*, **11**, 254.

Calogero, G., Yum, J.–H., Sinopoli, A. *et al.* (2012) *Sol. Energy*, **86**, 1563–1575.

Campbell, W.M., Jolley, K.W., Wagner, P. *et al.* (2007) *J. Phys. Chem. C*, **111**, 11760.

Chang, H. and Lo, Y.-J. (2010) *Sol. Energy*, **84**, 1833–1837.

Chang, H., Wu, H.M., Chen, T.L. *et al.* (2010) *J. Alloys Compd.*, **495**, 606.

Cherepy, N.J., Gratzel, M. and Zhang, J.Z. (1997) *J. Phys. Chem. B*, **101**, 9342.

Cisse, M. Vaillant, F. and Dornier, M. (2009) *J. Agric. Food Chem.*, **57** (14), 6285–6291.

Cisse, M., Vaillant, F., Soro, D. *et al.* (2011) *J. Food Eng.*, **106**, 20–27.

Diallo, A. Zongo, S., Mthunzi, P. *et al.* (2013) *Optics Commun.*, submitted.

Duarte, F.J. and Hillman, L.W. (eds) (1990) *Dye Laser Principles*, Academic Press, New York, NY.

Frank, T. (2005) *J. Clin. Pharmacol.*, **45**, 203.

Furukawa, S., Iino, H., Iwamoto, T. *et al.* (2009) *Thin Solid Films*, **518**, 526–529.

Gömeze, S.L., Cuppo, F.L.S. and Figueiredo Neto, A.M. (2003) *Braz. J. Phys.*, **33**, 813.

Gomez-Ortiz, N.M., Vasquez-Maldonado, I.A., Perez-Espadas, A.R. *et al.* (2010) *Sol. Energy Mater. Sol. Cells*, **94**, 40–44.

Grätzel, M. (2003) *J. Photochem. Photobiol. C: Photochem. Rev.*, **4**, 145–153.

Hao, S., Wu, J, Huang, Y. and Lin, J. (2006) *Sol. Energy*, **80**, 209–214.

He, K.X., Abdeldayem, H., Venkateswaralu, P. and George, M.C. (1991) *Opt. Commun.*, **81**, 101.

Henari, F.Z. and Al-Saie, A. (2006) *Nonlinear Quant. Opt. Laser Phys.*, **16** (12), 1664–1667.

Herrera-Arellano, A., Flores-Romero, S. and Tortoriello, J. (2004) *Phytomedicine*, **11**, 375–38.

Ito, S., Miura, H., Uchida, S. *et al.* (2008) *Chem. Commun.*, 5194.

Kay, A. and Grätzel, M. (1993) *J. Phys. Chem.*, **97**, 6272.

Kouissa, B., Bouchouit, K., Abed, S. *et al.* (2013) *Opt. Commun.*, **293**, 75–79.

Kuciauskas, D., Freund, M.S., Gray, H.B. *et al.* (2001) *J. Phys. Chem. B*, **105**, 392.

Kumara, G.R.A., Kaneko, S., Okuya, M. *et al.* (2006) *Sol. Energy Mater. Sol. Cells*, **90**, 1220.

Li, Z., Jin, Z.–H., Kasatani, K. *et al.* (2005) *Jpn. J. Appl. Phys.*, **44**, 4956–4961.

Mathews, S.J., Chaitanya Kumar, S., Giribabu, L. and Venugopal Rao, S. (2007) *Mater. Lett.*, **61**, 4426–4431.

Mortensen, A. (2006) *Pure Appl. Chem.*, **78** (8), 1477–1491.

Nazeeruddin, M.K., Pechy, P., Liska, P. *et al.* (2001) *J. Am. Chem. Soc.*, **123**, 1613.

Nazeeruddin, M.K., De Angelis, F., Fantacci, S. *et al.* (2005) *J. Am. Chem. Soc.*, **127**, 16835.

O'Regan, B. and Grätzel, M., (1991) *Nature*, **353**, 737.

Patrocinio, A.O.T., Mizoguchi, S.K., Paterno, L.G. *et al.* (2009) *Synth. Met.*, **159**, 2342–2344.

Polo, A.S., Itokazu, M.K. and Murakami Iha, N.Y. (2004) *Coord. Chem. Rev.*, **248**, 1343.

Polo, A.S. and Murakami Iha, N.Y. (2006) *Sol. Energy Mater. Sol. Cells*, **90**, 1936.

Prasad, P.N. and Williams, D.J. (1991) *Introduction to Nonlinear Optical Effects in Molecules and Polymers*, John Wiley & Sons, Inc., New York, NY.

Quin, C. and Clark, A.E. (2007) *Chem. Phys. Lett.*, **438**, 26.

Rashidian, M., Dorranian, D., Ahmadi Darani, S. *et al.* (2009) *Optik*, **120**, 1000–1006.

Ravindra, H.J., Johkiran, A., Chandrasekaran, K. *et al.* (2007) *Appl. Phys. B. Lasers Opt.*, **88**, 105–110.

Rekha, R.K. and Ramalingam, A. (2009) *Indian J. Sci. Technol.*, **2** (8), 27–31.

Samoc, M. Samoc, A. and Grote, J.G. (2006) *Chem. Phys. Lett.*, **431**, 132–134.

Schäfer, F.P. (ed.) (1990) *Dye Lasers*, 3rd edn, Springer-Verlag, Berlin.

Sekkat, Z. and Dumont, M. (1992) *Appl. Phys. B*, **45**, 486–489.

Sharma, S., Mohan, D. and Ghoshal, S.K. (2011) *Opt. Commun.*, **281**, 2923–2928.

Sheik-Bahae, M., Said, A.A. and Van Stryland, E.W. (1989) *Opt. Lett.*, **14**, 955–957.

Sheik-Bahae, M., Said, A.A., Wei, T. *et al.* (1990) *IEEE J. Quant. Elect.*, **26**, 760–769.

Smestad, G.P. (1998) *Sol. Energy Mater. Sol. Cells*, **55**, 157.

Smestad, G., Bignozzi, C.A. and Argazzi, R. (1994) *Sol. Energy Mater. Sol. Cells*, **32**, 259.

Stintzing, F.C. and Carle, R. (2007) *Trends Food Sci. Technol.*, **18**, 514.

Takeoka, G. and Dao, L. (2002) Anthocyanins, in *Methods of Analysis for Functional Foods and Nutraceuticals* (ed. W.J. Hurst), CRC Press, Boca Raton, FL, pp. 219–241.

Thankappan, A., Thomas, S. and Nampoori, V.P.N. (2012) *J. Appl. Phys.*, **112**, 123104.

Timberlake, C.F. and Bridle, P. (1975) The anthocyanins, in *The Flavonoids* (eds J.B. Harborne, T.J. Mabry and H. Mabry), Chapman and Hall, London, pp. 214–266.

Timberlake, C.F. and Henry, B.S. (1988) *Prog. Clin. Biol. Res.*, **280**, 107–121.

Tsigaridas, G., Fakis, M., Polyzos, I. *et al.* (2003) *Appl. Phys. B Lasers Opt.*, **76**, 83–86.

Uma Maheswari, R., Kadono, H. and Jaaskelainen, T. (1994) *Opt. Laser Technol.*, **26** (2), 136.

Yamazaki, E., Murayama, M., Nishikawa, N. *et al.* (2007) *Sol. Energy*, **81**, 512–516.
Yella, A., Lee, H.-W., Tsao, H.N. *et al.* (2011) *Science*, **334**, 629.
Yum, J.H., Walter, P., Huber, S. *et al.* (2007) *J. Am. Chem. Soc.*, **129**, 10320.
Zakeeruddin, S.M., Comte, P., Charvet, R. *et al.* (2003) *J. Phys. Chem. B*, **107**, 14336.
Zhang, D., Lanier, S.M., Downing, J.A. *et al.* (2008) *J. Photochem. Photobiol. A: Chem.*, **195**, 72.
Zhang, H.J., Dai, J.H., Wang, P.Y. and Wu, L.A. (1989) *Opt. Lett.*, **14**, 695.
Zhou, H., Wu, L., Gao, Y. and Ma, T. (2011) *J. Photochem. Photobiol. A: Chem.*, **219**, 188–194.

# 36 The Host Innate Immune Response to *Propionibacterium acnes* and the Potential of Natural Products as Cosmeceutical Agents

Marco Nuno de Canha, Smeetha Singh and Namrita Lall
University of Pretoria, Department of Plant Science, Lynwood, Pretoria, South Africa

## 36.1 The skin and its function

The skin is considered to be the largest organ in the human body. It accounts for 16% of total body weight. This organ is particularly complex, requiring the use of specialized cell structures in order to perform its functions optimally. The skin consists of two main layers: the epidermis (a superficial layer) and the dermis (the deepest layer of the organ) (Venus *et al.*, 2010).

The skin is able to act as a protective feature through various morphological and physiological traits.

The epidermis consists mainly of keratinocyte cells. It is also home to the stratum corneum, also known as the horny layer (Figure 36.1). This is a structure that is formed through the lattice-like arrangement of keratinocyte cells. It is responsible for the skin barrier function, which reduces the time taken for foreign matter to absorb through the skin and also acts as a waterproofing structure. Keratinocyte cells are also responsible for the production of keratin and cytokines that are secreted in response to injury (Venus *et al.*, 2010).

Melanocyte cells within the dermis are responsible for the production of melanin. This pigment is transported to the stratum corneum keratinocyte cells where it acts as a natural absorbent of ultraviolet (UV) radiation (Venus *et al.*, 2010).

The skin is the first line of defence against invading pathogens. It is home to various physiologically advanced cells that play various roles as defence mechanisms. The production of antimicrobial peptides is a common method in which cells of the epidermis can combat Gram-positive bacteria, Gram-negative bacteria, fungi and some viruses. There are two common families of compounds, known as the defensins and cathelicidins, responsible for the antimicrobial activity. The cathelicidin family can also act as signals to a number of inflammatory cells. Langerhans cells are classified as antigen-presenting cells. They function in the lymph nodes, where they present antigen-MHC class II complexes to T cells. These T cells are responsible for the host immunity (Venus *et al.*, 2010).

## 36.2 The impact of skin disorders with focus on acne

Problems associated with the skin make up 15–20% of consults at family practices. The field of psychodermatology is one that is dedicated to the evaluation of the relationship between skin diseases and their psychological impact. Patients suffering from imperfections of the face, scalp, hands and genital areas are prone to psychological distress related to the importance of these bodily parts with regard to image. Acne, atopic dermatitis and psoriasis are the most common skin disorders that are observed by many primary care physicians. They are also the most studied in terms of their effects on psychosocial health. Some common problems, such as blemishes, have been known to cause distress. However, people with skin disorders that may result in body dismorphia, such as acne and psoriasis, have been known to cause depression, which is a large risk for suicides (Barankin and DeKoven, 2002).

The physical symptoms of acne may include soreness, itching and even pain. These symptoms all have an influence on quality of life. Because acne affects the more visible areas of the skin and because it peaks during teenage years, where self-esteem and confidence building are important, it has a large impact on psychological health as well. Acne patients may suffer from depression, suicidal ideation, anxiety, psychosomatic symptoms, embarrassment, shame and social inhibition. Studies have shown that these symptoms all improve with effective treatment of the disorder. In another study, the presence of acne was linked to unemployment of young men and women (Williams *et al.*, 2012).

## 36.3 *Propionibacterium acnes*: is it the culprit?

*Propionibacterium acnes* has been identified as a key factor in acne for over a 100 years. Its traces go back as early as the year 1896. More recent studies, where keratinocyte cysts injected with *P. acnes* ruptured, has linked the bacterium to inflammatory processes

*Novel Plant Bioresources: Applications in Food, Medicine and Cosmetics*, First Edition. Edited by Ameenah Gurib-Fakim.

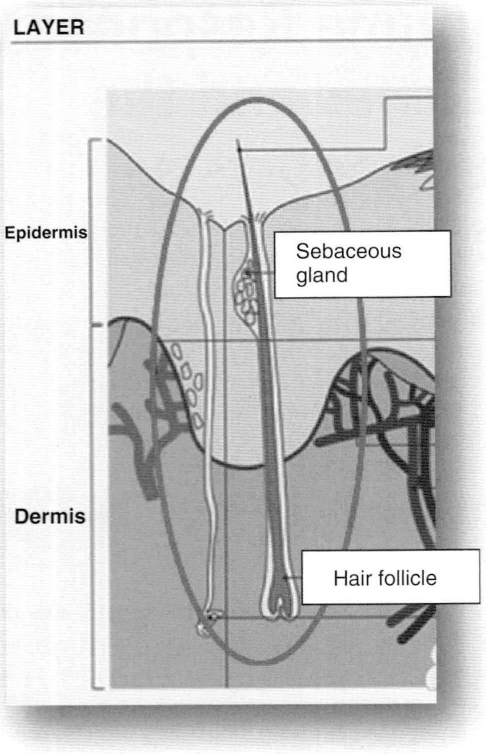


**Figure 36.1** Pilosebaceous unit consisting of the hair follicle and sebaceous glands which provide the skin with moisture and waterproofing and are affected by *P. acnes* colonisation. *Source*: Venus *et al.* (2010). Reproduced with permission of Elsevier

associated with the disorder (Dessinioti and Katsambas, 2010). There are, however, suspicions that some other microorganisms are involved, such as *Staphylococcus epidermidis*.

*P. acnes* are anaerobic, Gram-positive, rod-shaped bacteria (Figure 36.2) (Leyden *et al.*, 2009). The bacteria are known to synthesize propionic acid, catalases, indoles and nitrates. *P. acnes* are resident bacteria of the human skin and grow optimally under the surface of the skin, but predominantly in areas abundant in

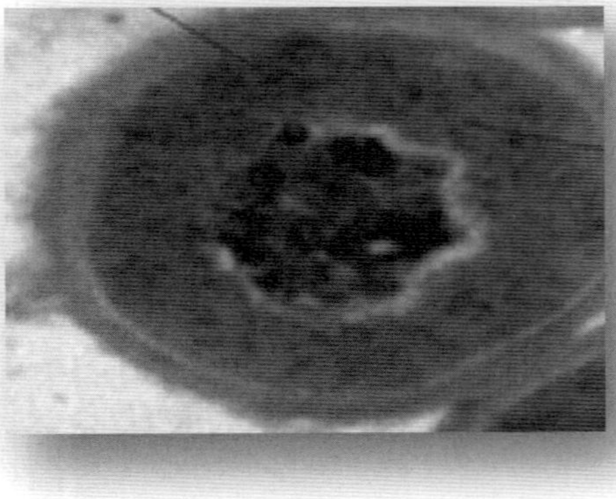

**Figure 36.2** Transmission electron micrograph of *P. acnes*. *Source*: Sharma and Lall (2013). Reproduced with permission of Elsevier

sebaceous glands. This can be indicated by the colonization density levels of *P. acnes*, which are higher in the face and scalp in normal skin. *P. acnes* colonization has also been shown to be age related, where the bacterial recovery levels in pre-pubertal children are very low but are denser in children that have reached puberty, and this has been attributed to maturation of sebaceous gland function (Leyden, 2001). Another interesting fact to note is the density of *P. acnes* levels and the production of sebum, which increase in direct proportions (Bojar and Holland., 2004). This indicates the use of sebum by the bacteria as a source of enrichment for growth and development (Leyden, 2001). This evidence suggests that the *P. acnes* bacterium is involved in the progression of the disease.

## 36.4 Acne vulgaris (acne)

Acne vulgaris (AV) is a chronic inflammatory disease that affects the pilosebaceous unit. The pilosebaceous unit is classified as the hair follicles in the skin that are associated with an oil (sebaceous) gland. AV is said to be a culmination of various events that include the production of excess sebum, altered keratinocyte morphology, inflammation and the colonization of the hair follicles by *P. acnes* (Williams *et al.*, 2012).

The disorder affects approximately 80% of people aged between 11 and 30 years old, with 60% of these cases requiring some form of treatment. Owing to differences in the onset of puberty between different genders, females tend to develop the disorder at an earlier stage than males. The disorder is generally observed in females between the ages of 14 and 17 and in males between the ages of 15 and 19 (Truter, 2009).

The disorder is characterized by non-inflammatory and inflammatory lesions (Figure 36.3).

Non-inflammatory lesions are known as comedones. There are two types of comedones: open and closed. Open comedones are commonly known as blackheads (Figure 36.3a) and closed comedones as whiteheads. These lesions are the result of altered morphology of keratinocyte cells in a process known as keratinization (Shaw and Kennedy, 2007).

The two types of inflammatory lesions are called pastules and papules (Figure 36.3b). These occur as a result of the host immune response to comedone lesions that have been colonized by the *P. acnes* bacteria. More severe lesion, known as cysts or nodules, are deep pustules that commonly cause scarring (Shaw and Kennedy, 2007).

Acne can be rated on a scale that varies in severity. The different levels of severity range from mild acne, moderate acne to severe acne. Each of these levels is characterized by the absence and presence of acne-associated lesions and generally requires different approaches with regard to types and number of treatment methods (Williams *et al.*, 2012).

Patients with mild acne may observe redness of the skin in certain areas, blackheads or a mixture of both (Williams *et al.*, 2012) but less inflammatory lesions (Layton, 2005). When the acne severity is graded as moderate, the appearance of papules and pustules is clear and are usually present in larger numbers when compared with mild acne. Severe acne, on the other hand, has a larger number of lesions that range from inflammatory, non-inflammatory to even nodules or cysts. Severe acne is also prone to scarring (Layton, 2005). These scars can be either atrophic or hypertrophic and generally result

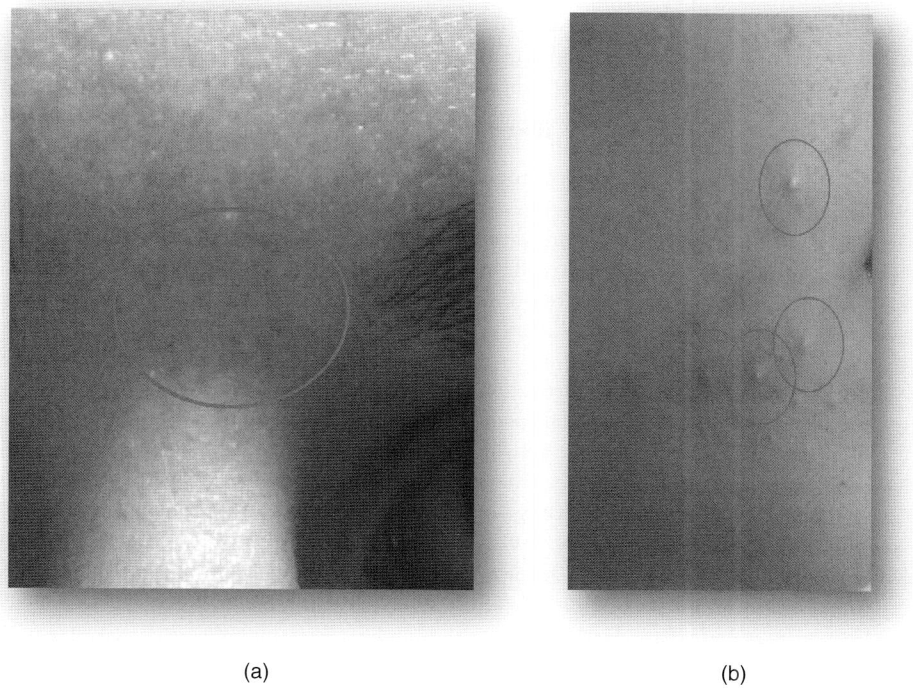


**Figure 36.3** (a) Blackheads on the forehead and (b) inflammatory lesions on the face. *Source*: Photographs were taken by author M.N. de Canha. *See plate section for colour detail*

in changes in the skin located at affected areas. Hypertrophic scars are those that are slightly elevated, and atrophic scars are generally sunken areas where structural features of the skin may have been affected. There may also be the formation of keloid scars, which are most commonly observed on the chest, upper back and jaw line. These scars can be difficult to deal with because of their tendency to grow out of the areas of wounding and become a mass of tumorous or benign tissue (Shaw and Kennedy, 2007).

## 36.5 The activation of innate and adaptive immune system

The innate immune system is a non-specific immunity that protects the body from any type of invader. It is the body's immediate defence against toxins and pathogens. This type of immunity acts as the first line of defence once pathogenic invaders penetrate the epithelial surfaces of the body. Once the pathogen has entered the body it

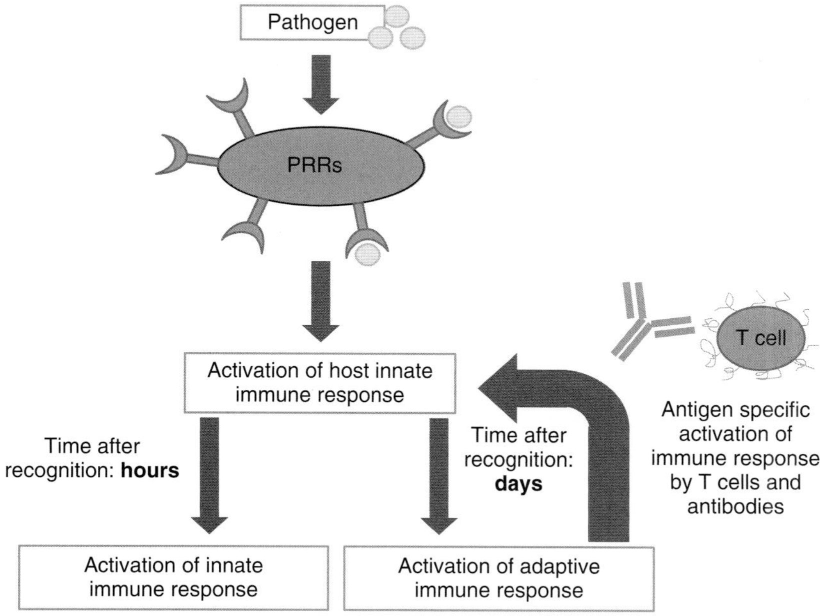


**Figure 36.4** Recognition of pathogens through PRRs and the activation of the host immune response. *Source*: Adapted from Medzhitov (2007)

is immediately exposed to phagocytes. These cells contain surface receptors that bind to bacterial constituents on bacterial cells. Through this recognition the bacterial cells become engulfed, which results in secretion of other biologically active molecules. These activated phagocytes secrete cytokines that attract neutrophils and monocytes from the bloodstream. Secreted cytokines are involved in the inflammatory response, most likely due to activation of complement. Complement is defined as plasma proteins which result in proteolytic reactions located at microbial cell surfaces, coating these surfaces, so they are phagocytized. This chain of reactions results in secretion of small peptides involved in the inflammatory response as well as the attraction of leukocytes to sites of infection (Janeway *et al.*, 2001).

The innate immune response is also responsible for the activation of the adaptive immunity in humans. Activation is achieved through various mechanisms. Increased flow of lymph containing antigen and antigen-cell complexes is carried to lymphoid tissues where immune cells are activated, complement-coated microbial cells and invaded cells activate lymphocytes, macrophages activate T cells and dentritic cells present antigen molecules to T cells (Janeway *et al.*, 2001).

The activation of the adaptive immune system is initiated by the ingestion of pathogens by immature dentritic cells in affected tissues. These cells have a long lifespan and patrol the peripheral blood in many tissues. These dentritic cells ultimately migrate back to the lymph nodes, where they interact with naive lymphocytes. Dentritic cells can be activated or, if not, induce tolerance to the antigen which they carry. These dentritic cells detect common bacterial cell surface structures, such as proteoglycans, through the expression of cell surface receptors. Once they are activated they become antigen-presenting cells that take the antigen to T cells. These T cells in the adaptive immune system are pathogen specific. Cytotoxic T cells are responsible for killing infected cells. T helper cells are responsible for presentation of pathogenic antigen to active cytotoxic T cells but have no cytoxocic activity of their own. B cells of the adaptive immune system are also activated when they encounter pathogens. They then become either plasma cells or memory cells. Plasma B cells secrete antibodies that either inactivate pathogens directly or make them accessible to other immune cells. Memory B cells have a longer lifespan than plasma B cells and remain in the body for detection of the same pathogen; thereafter the response is more rapid (Janeway *et al.*, 2001).

## 36.6 The host immune response to infection by *Propionibacterium acnes*

The host pilosebaceous unit is constantly exposed to *P. acnes* due to favourable environmental and chemical composition available for optimal growth. AV is characterized by four main events, and each of these has a distinct immune response (Burkhart *et al.*, 1999).

The first of these factors, mentioned earlier, is the hypercornification of keratinocyte cells. This process occurs due to altered lipid composition in sebum. Linoleic acid, squalene and other free fatty acids have been identified as enhancers of comedone formation. Various studies have implicated lipids in the proliferation and differentiation of keratinocyte cells, where *N*-acetylated sphingolipids, such as ceramides, promote keratinocyte differentiation and sphingosine and sphingophosphorylcholine promote keratinocyte proliferation (Wakita *et al.*, 1998). Cytokines have also been implicated in comedone formation where IL-1α enhanced comedogenesis. This cytokine has been found in large quantities in non-inflammatory lesions (Koreck *et al.*, 2003).

One of the most important factors in the pathogenesis of acne is the host inflammatory response. The hypothesis is that inflammation associated with AV is largely due to the immunological response to the presence of *P. acnes*. The role of *P. acnes* in inflammation is said to be due to several factors, which include the production of lytic enzymes and lipases that are responsible for the disruption of the follicular epithelium in the pilosebaceous unit and the inflammatory response. The bacteria are able to secrete chemoattractant molecules for attracting immune cells such as neutrophils, monocytes and lymphocytes (Grange *et al.*, 2010). *P. acnes* may also be involved in the amplification of immune cell function through activation of complement, secreting potent C5a (Koreck *et al.*, 2003). Neutrophils are responsible for destruction of the bacteria through lysis, but lysis of the bacteria produces hydrolases that are involved in further degradation of epithelial structures that can also enhance inflammation (Grange *et al.*, 2010).

It has been hypothesized that *P. acnes* activates the immune system through a set of receptors known as pattern recognition receptors (PRRs) (Figure 36.4). Examples of these are Toll-like receptors (TLRs). These receptor molecules can differentiate between Gram-positive and Gram-negative bacterial ligands. Gram-positive bacterial ligands are said to activate immune monocyte cells through recognition by TLR-2 or TLR-4 receptors (Kim *et al.*, 2002). TLR-2 is expressed on the cell surface of macrophage cells which are found around the pilosebaceous follicles within acne lesions, and the secretion of IL-12 and IL-8 is said to be triggered through the TLR-2 activation pathway (Koreck *et al.*, 2003). IL-8 is responsible for attracting neutrophils to the pilosebaceous unit where they secrete lysosomal enzymes. These enzymes ultimately lead to bursting of the follicular epithelium and continued inflammation (Webster *et al.*, 1980). IL-12 enhances the Th-1 immune response; the overproduction of this cytokine is said to lead to tissue injury and autoimmune and inflammatory disorders (Kim *et al.*, 2002). In keratinocytes, both of these receptors are responsible for the detection of peptidoglycan (by TLR-2) and lipopolysaccharide (by TLR-4) (Koreck *et al.*, 2003). *In vivo* experiments have shown that the expression TLR-2 and TLR-4 in keratinocyte cells increased after 1 h of incubation with bacterial fractions. These experiments also showed an increase in the expression and secretion of metalloproteinase (MMP)-9, which is associated with inflammatory processes. These MMPs are endopeptidases produced by a variety of cells and are responsible for the degradation of components of the extracellular matrix. MMPs also result in the rupture of the pilosebaceous follicle, which could account for their link to inflammation (Jugeau *et al.*, 2005).

Infections by *P. acnes* bacteria trigger antimicrobial peptide and cytokine release in human keratinocytes. These cells secrete β-defensin-2 upon infection with the bacteria. In the stationary phase of growth, *P. acnes* is able to stimulate IL-1α, tumour necrosis factor (TNF)-α and granulocyte-macrophage colony-stimulating factor (GM-CSF) in keratinocyte monolayers. TNF-α and GM-CSF are said to be involved in the maturation and transportation of Langerhans cells from the skin and presentation of antigens to T cells in the lymph nodes. Some components of the *P. acnes* cell envelope, such as GroEL (heat shock protien), DnaK and lipoglycans, are able to enhance the production of proinflammatory cytokines. These components are also said to be ligands to the TLRs mentioned previously (Dessinioti and Katsambas, 2010).

Some strains of *P. acnes* have been shown to decrease the efficacy of antibiotic treatments through acquisition of resistance. Patients suffering from severe acne can show abnormal immune responses. These responses range from high levels of complement-fixing antibodies (Ashbee *et al.*, 1997), enhanced lymphocyte stimulation in nodulocystic acne (Puhvel *et al.*, 1977) and delayed skin reactivity to *P. acnes* in proportion to levels of inflammation (Kersey *et al.*, 1980).

## 36.7 Conventional treatments available for acne vulgaris

There are various forms of conventional treatments available for AV. These include topical treatments, oral treatments (Layton, 2005), laser therapy and hormonal therapies (Williams *et al.*, 2012). The treatment regimens of acne are dependent on the level of severity. Treatments are chosen according to the number and type of visible lesions (Shaw and Kennedy, 2007).

Topical treatments can be used alone or in combination. They are usually successful for the treatment of mild acne (Layton, 2005) where there are more comedones than inflammatory lesions. Topical applications for acne include retinoids, benzoyl peroxide, azelaic acid and some antibiotics. Retinoids are derivatives of vitamin A. They are mainly applied for the treatment of comedones. The most common retinoids are tretinoin and isotretinoin. The use of these treatments comes with some side-effects, which includes dryness, redness and irritation of the skin. These products are recommended together with the use of a moisturizer that must be non-comedogenic. Benzoyl peroxide is a chemical ingredient in many non-prescription products. The active constituent can be present in percentages ranging from 2.5 to 10%. This active is used to reduce the number and prevalence of comedones. However, it is also known to be a good antimicrobial and anti-inflammatory agent. This ingredient causes dryness and irritation, but the percentage can be altered in order to treat more-susceptible patients. Another problem with this product is its bleaching potential of hair and clothing, but this is a minor problem. Azelaic acid has a weaker effect as an antibacterial, and treatment for comedones but can be used in place of harsher products. This product can cause hypopigmentation, which may be advantageous in dark skins where post-inflammatory hyper-pigmentation may be observed. Topical antibiotics that are commonly found are erythromycin and clindamycin. They are used to target inflammatory lesions, but the problem with these treatments is the increased resistance to these antibiotics. For this reason, the combination of topical treatments is recommended (Williams *et al.*, 2012). The use of tetracycline has been known to leave a UV-fluorescent residue on the skin (Layton, 2005).

Sometimes the treatment of AV will require combination therapies using topical agents, as the combined effect may be a more effective. Examples of these include benzoyl peroxide–antibiotic treatments or retinoid–antibiotic treatments. A major advantage to this approach is also the decrease in emergence of antibiotic-resistant strains (Williams *et al.*, 2012).

Various oral treatments are available for the treatment of AV. Tetracycline and its derivative oxytetracycline are common first-line antibiotic treatments for patients older than 12 years old. The use of tetracyclines is recommended twice daily, without meals. If treatment is not successful they can be substituted with minocycline or doxycycline, other derivatives. Minocycline does not induce *P. acnes* resistance; however, one of the major side-effects is the appearance of a bluish pigment in the skin. All tetracyclines have been associated with photosensitivity, and patients should be aware of increased sun exposure. In children, tetracyclines can cause dental staining. Erythromycin is the preferred substitute for tetracyclines, but erythromycin-resistant strains and gastro-intestinal side-effects associated with its use make it less favourable. Another problem with the sole use of antibiotic treatments is that it may take up to 6 months before improvement in the skin is observed (Williams *et al.*, 2012).

The benchmark oral treatment known as Roaccutane (isotretinoin) has proven to be very effective in treatment of AV. The mechanisms of action include the reduction of sebum, anti-inflammatory activity and interference with differentiation of follicular epithelial cells. Although efficacy is important, there are some setbacks to using this treatment. It is relatively expensive and can only be used under the supervision of a certified dermatologist. The treatment period may take 4–5 months, with dose and duration being dependent on patient to patient responses. A major concern is that the side-effects associated with the treatment are common. Dryness of skin and mucous membranes are common in all patients. In the first week of treatment, the appearance of lesions may seem to worsen. Hepatitis and hyper-lipidaemia are a major concern, but this is why medical supervision is required. Oral isotretinoin is also a potent teratogen, and fertile females need to take contraceptives concurrently with the treatment and for an additional month after the termination of treatment (Williams *et al.*, 2012). Other side-effects that have been reported include fissured tongue (Figure 36.5a), hypo-pigmentation in type 4 skins (Figure 36.5b) and formation of keloid scars (Lowenstein and Lowenstein, 2011).

Hormonal treatments are usually in the form of oral contraceptives. These drugs contain oestrogen and progestogen. Oestrogen targets sebaceous gland activity and progestogen decreases the formation of androgens originating from both the ovaries and the adrenal glands (Williams *et al.*, 2012). Contraceptives containing only progestogen, however, can worsen acne in women due to the stimulation of the adrenal gland activity (Layton, 2005).

Complementary and alternative medications for treatment of acne are widespread. The use of *Aloe vera*, pyridoxine, fruit-derived acids, Japanese herbal preparations and ayurvedic herbal preparations has been recorded for their use against the disorder. The problem with these recorded studies is that they indicate the biological potential of these preparations but fail to show any form of quality or conclusive results (Williams *et al.*, 2012). A study performed in Iran in 2007 on 60 patients with moderate–mild acne showed a decrease in lesion number and little adverse effects when compared with placebo controls. This study suggests that more large-scale clinical trials on complementary medications are worth looking into in the future (Enshaieh *et al*, 2007).

With the number of treatments available for the disorder, it makes it difficult for dermatologists to provide a set treatment. Another major problem is that a prescribed treatment may differ in efficacy between patients (Williams *et al.*, 2012).

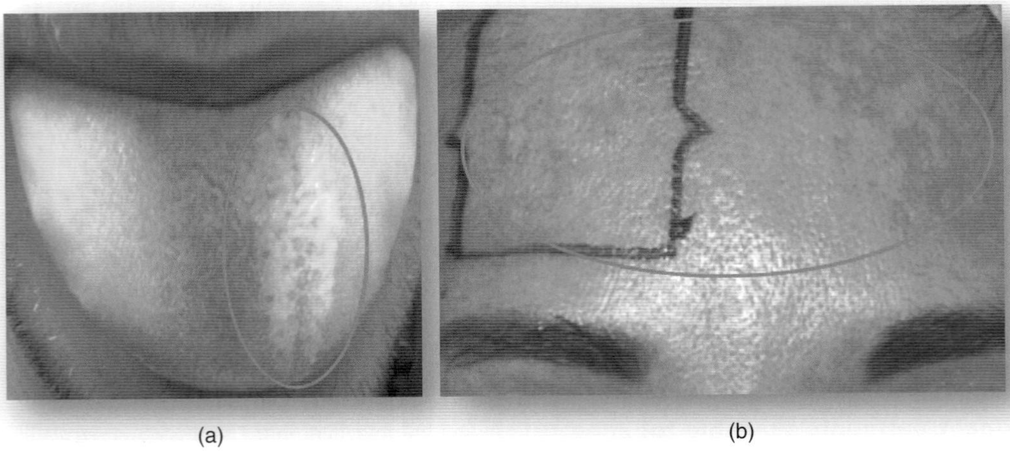


**Figure 36.5** (a) Fissured tongue and (b) hypopigmentation of the forehead. *Source*: Lowenstein and Lowenstein (2011). Reproduced with permission of Elsevier. *See plate section for colour detail*

## 36.8 Potential of natural products to treat acne vulgaris

The emergence of many anti-biotic resistant *P. acnes* strains has become a worldwide problem with regard to treatment approach. For this reason alternatives to conventional drugs are being studied and developed in order to combat the disorder. Natural products are extensively studied for their biological activity (Chomnawang *et al.*, 2007).

In a study on Thai medicinal plants, the anti-inflammatory and anti-oxidant activities of 19 ethanol plant extracts were investigated. The selection of these plants was based on their traditional use. The best anti-oxidant was *Garcinia mangostana* with an $IC_{50}$ value of 6.13 μg/mL. This plant was also able to decrease the production of TNF-α, a pro-inflammatory cytokine when PBMC (Peripheral Blood Monocyte Cells) were treated with heat-killed *P. acnes* cultures. The significance of these results is that through the inhibition of reactive oxygen species there is a decrease in the production of inflammatory mediators and their most detrimental effect, scarring of the skin. *G. mangostana* fruit rinds have been known to treat skin infections, wounds and diarrhoea. Phytochemistry of the plant revealed a xanthone compound α-mangostin responsible for antimicrobial activity against *P. acnes* and *Staphylococcus epidermidis* which have both been implicated in AV pathogenicity (Chomnawang *et al.*, 2007). This natural resource shows great potential as it may in fact reduce colony number in pilosebaceous follicles and reduce inflammation in acne lesions. It is important to remember; however that these results are *in vitro* and more research *in vivo* would benefit the results of the study. The use of these natural products could help combat the antibiotic-resistant *P. acnes* strains; however biomarkers from natural products are available in small quantities but may be advantageous where synthesis is possible (Cowan, 1999).

A study done using controlled Phase II clinical trials on oral Ayurvedic formulations concurrently with or without polyherbal dermatological formulations was tested at Pune. Various plant extracts were incorporated in different proportions and formulated. In this study 53 patients were selected between ages 14 and 28. The patients were selected on the basis of acne lesions with a minimum of 10 inflammatory and 5 non-inflammatory being required. The study followed Good Clinical Practices (GCP) guidlines and certain parameters were selected in order to determine results. The patients were split into groups and given the following treatments (Lalla *et al.*, 2001):

- Group I: oral tablets + topical aqueous gel
- Group II: oral tablets + topical cream formulation
- Group III: oral tablets + placebo topical formulation
- Group IV: placebo tablets + placebo topical formulation.

A single physician was responsible for the recording of results to reduce variability. The patients reported back to the clinic every week for a course of 4 weeks. Any side-effects were also recorded. The results showed that treatment with oral tablets and the topical cream formulation was better than the oral tablets and gel combination. This was attributed to the oil present in the water based cream which helped to increase penetration. The results also showed a larger decrease in inflammatory lesions and little effect on other acne lesions. Group II had 57.89% with good to excellent improvement of acne lesions. The drug therapies were also well tolerated over the treatment period. Slight itching (in 2 patients) and increase in gastric motility (in 2 patients) were recorded as side-effects. However, these side-effects decreased over the course of the treatment. The study suggests that a combination therapy is more beneficial in terms of efficacy (Lalla *et al.*, 2001).

The increased interest in the use of herbal preparations for treatment means that the potential for profits is high (Figure 36.6). However, only with good quality results proving efficacy *in vitro* and *in vivo* can this be achieved. Products on the market may use the limited regulatory rules on herbal products to make claims that may not be scientifically proven which affect their integrity. If herbal preparations are well researched and reports are conclusive their manufacture and marketing will be made easier.

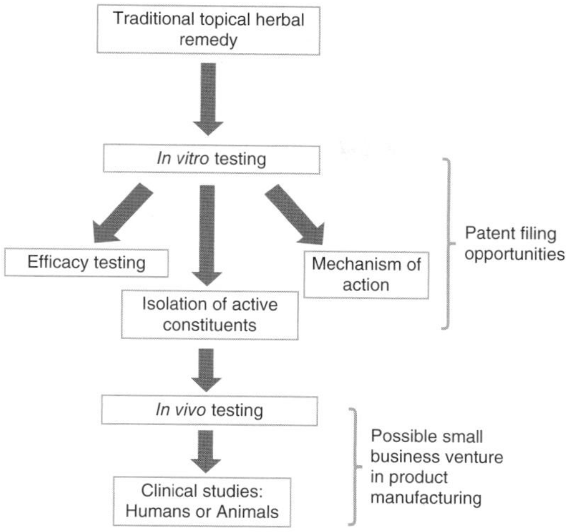


**Figure 36.6** The potential benefits for *in vitro* and *in vivo* research. *Source*: Adapted from Krajewski *et al.* (2010). Reproduced with permission of Elsevier

## 36.9 The importance of the emergence of plant life on Earth

Plants are believed to have made their appearance on land during the Palaeozoic era, which began 542 million years ago (Rafferty, 2011), whereas humankind only came into existence approximately 3–5 million years ago during the Cenozoic era (Kim and Weave, 1994). This significant difference in time favours plants with the advantage of having almost 500 million additional years to have evolved appropriately for survival.

By taking the time factor into consideration, as well as the fact that plants are sessile and respond to the environment accordingly (Huey *et al.*, 2002), it is most obvious to question how their survival was possible. A major part of the evolution of plants was the synthesis of secondary compounds by these organisms (Pichersky and Gang, 2000). The role of many of these compounds was to combat damage from herbivores or as a general defence mechanism against insects (Cotton, 1996).

Many scientists once believed that these secondary compounds were waste products (Whittaker and Feeny, 1971; Taiz and Zieger, 2006), and even now, as these compounds are studied, their role in the plant may not always be clear. The discoveries that have been made up to this point, however, indicate that secondary plant compounds are responsible for the beneficial and medicinal value of plants to humans (Cotton, 1996).

There are numerous commercially available plant-derived medicines, cosmetics and nutritional supplements on the market today (Talalay and Talalay, 2001; Gershwin *et al.*, 2010; Gellenback, 2012). The potential to discover novel and more efficient medicinal, cosmeceutical and nutritional compounds from unexplored plant species may seem infinite owing to the fact that only a minute fraction of plant life on Earth has been explored (Wheeler, 1995). There may even be a plant for every ailment a human being could experience.

With regard to cosmetics, many cultures of people have used plants for hundreds of years as the remedy to treat inflammation, acne and even pigmentation (Yokota *et al.*, 1998; Messele 2004; Kapoor and Saraf, 2011). Today, in addition to naturally derived actives, a large proportion of the actives of cosmetic products are synthetically produced (Pieroni *et al.*, 2004; Farris, 2010). There are many advantages and disadvantages to using either naturally derived or synthetically produced actives in cosmetics.

However, considering that a significant portion of people are turning to naturally derived products (Swisher, 1988; Artuso, 2002; Govindsamy *et al.*, 2011) as well as products that are friendlier to the environment (Bhate and Lawler, 1997), it is worth considering the pathway from plant extract to cosmetic product. This pathway from plant to product has not been explored extensively probably due to the uncertainty surrounding the regulatory processes (Ma *et al.*, 2005).

## 36.10 A proposed stepwise approach from plant extract to cosmeceutical product

An overview of the pathway is outlined in Figure 36.7. Each step is now discussed in detail.

### 36.10.1 Identification of a plant with desired activity

A number of plants can be screened for specific activity, such as anti-acne (Kubo *et al.*, 1994), anti-inflammatory or anti-ageing. For example, *Garcinia mangostana* was identified to have an MIC of 0.039 mg/mL when tested against *P. acnes* as well as *S. epidermidis* (Chomnawang *et al.*, 2005).

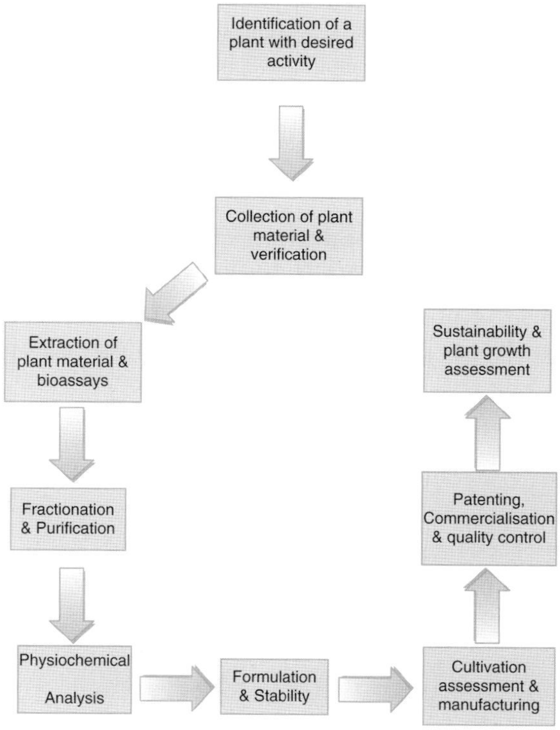


**Figure 36.7** A diagram to represent a summary of the steps that may be required to develop a product from a plant

(a) (b)

**Figure 36.8** The instruments and ground plant material used by a traditional healer

In this case, a hypothetical product from this plant will be relevant to the market as the activity was targeted. One could select plants based on their ethnobotanical usage: the traditional use of a plant. This method is recommended as it has high success rates (Cox, 1994). Ethnobotany explores the interaction between local people and specific plants with the aim of expanding and gaining a complete understanding of traditional knowledge of plants. An ethnobotanical survey can be carried out to select plants for further screening (Martin, 2004). A survey like this may involve meeting and questioning traditional healers (Figure 36.8) and local people who are familiar with medicinal plants.

### 36.10.2 Extensive research on plant, verification and collection of sufficient plant material

A literature review should be carried out to compile all data on the plant; that is, existing products or patents, published data on activity, traditional usage: cosmetic or other, distribution, diversity, taxonomy and seasonal data. It would also be advisable to consider whether the part of the plant used will be easily available to produce an extract. A voucher specimen number of the plant should be prepared at a herbarium. Sufficient plant material should be collected for the subsequent steps.

### 36.10.3 Extract preparation of plant material and appropriate bioassays

An extract of an active plant can then be prepared in a suitable manner to carry out all relevant bioassays. To elaborate with an example; a plant has been identified on an ethnobotanical basis to be used for the treatment of an inflammatory skin disorder such as acne vulgaris or to treat a similar symptom associated with the disorder. The plant is then screened using an antibacterial assay on *P. acnes* (the pathogen associated with the disease). If the extract is found to be a successful inhibitor of *P. acnes* growth *in vitro*, further steps can be carried out to characterize the plant extract that was screened.

Further steps may include the testing of the mutagenicity of the plant extract. This would potentially be the next step or assay to be tested. A mutagen is considered a substance that is capable of altering one's DNA. The Ames test can be used as an *in vitro* test where it is merely an initial screen in the testing of mutagenic potential; this test is used quite frequently. Additional studies on animals must be carried out to confirm the safety of the extract. This test investigates the effect of an extract or compound on specific bacterial strains and monitors the number of revertant colonies that arise from the interaction between the extract and the bacteria. The test uses colonies of *Salmonella typhimurium* with mutations in a number of genes of the histidine operon. The TA98 and TA100 strains (Figure 36.9) are used commonly and detect the presence of frame-shift mutations or base-pair substitutions respectively. These mutations are areas that are highly prone to changes in DNA by mutagenic products. The strains of bacteria are grown on minimal media with traces of histidine, where the colonies that are visible have reverted to histidine independence in the presence of a mutagen. Based on the numbers of revertant colonies with reference to a known mutagen and other control plates, one can determine the mutagenic factor of an extract (Mortelmans and Zeiger, 2000).

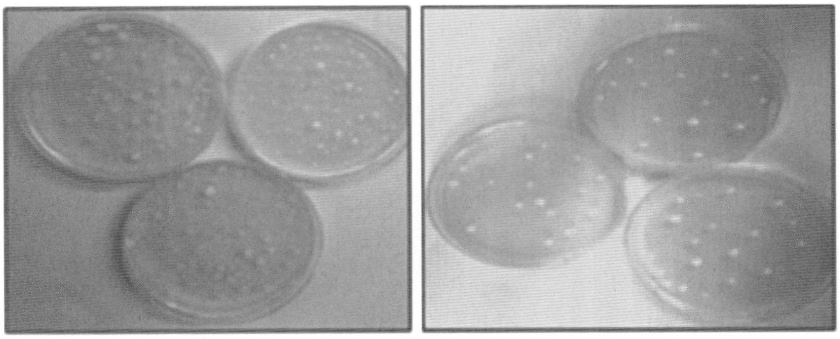

(a) (b)

**Figure 36.9** An image of revertant colonies for the bacterial strain *S. typhimurium* TA98 and TA100

### 36.10.4 Extraction of large amounts of plant material for the purpose of fractionation and purification to identify the active compounds

This step entails the collection of large amounts of plant material in order to identify the pure compounds in the plant that are responsible for the activity that was previously observed. The crude extract that showed activity as well as the active pure compounds can then be tested for cytotoxicity on cell lines that are relevant to the dosage form and the area in which the dosage form will be used in order to determine the efficacy of the extract (Lee and Houghton, 2005). Additional bio-assays can be done on the pure compounds in order to select the appropriate bio-markers for the extract. These assays are usually similar to the assays performed during the screening process of the crude extract. These active bio-markers are required for the standardization of an extract and are used as a tool in quality control (Wojdyłoa *et al.*, 2007).

### 36.10.5 Physiochemical analysis of the extract or actives

A compilation of data regarding all the physiochemical properties of the extract should be prepared. This is carried out in order to characterize the extract as extensively as possible. An extract that will be used in a cream/lotion should have the following physiochemical data: heavy metal count, pH value, microbial count, preservative challenge status. The reason for obtaining such data is to enable one to monitor changes that may affect activity while the extract or formulation is undergoing a stability trial (Soni *et al.*, 2010; Ogbunugafor *et al.*, 2011). A brief section on heavy metals will be outlined so as to elaborate on the importance of characterizing an extract.

Heavy metals are found everywhere in nature, and humans can come into contact with them through exposure with plants as heavy metals may be abundant in soil. The accumulation of such metals in the body may have fatal effects; therefore, it is important to identify these heavy metals in a plant extract that may be used in a cosmetic product. The most common heavy metals to be analysed for are lead, mercury and arsenic (Table 36.1).

### 36.10.6 Formulation and stability

A formulation can be prepared, ideally similar to the intended final product, including the extract in the allowed amount where activity is still present. A formulation could be in the form of a cream gel (Figure 36.10), body lotion, cream or balm. Stability studies can then be carried out on the formulation to identify any changes in physiochemical characteristics, appearance and, more importantly, activity (Table 36.2). A stability trial can be carried out over 3 or 6 months depending on the product that is prepared (Table 36.3).

**Table 36.1** The common heavy metals and some implications that may arise when they accumulate in the human body (Peralta-Vidia *et al.*, 2009)

| Metal | Implications |
|---|---|
| Lead | Sterility, neurological dysfunction and anaemia |
| Mercury | Acrodynia, rash, stomatitis and gingivitis |
| Arsenic | Acute renal failure, neuropathy, abdominal pain and constipation |

*Source: Reproduced with permission of Elsevier.*

**Table 36.2** The different types of stability testing (International Conference on Harmonisation, 2011)

| Type | Description | Indications/purpose |
|---|---|---|
| Shelf life/ long-term studies | Three product batches should be used for determining shelf-life claims | Determination of the expiration period of a product in the final packaging and under a defined set of conditions |
| Stress testing (transport simulation) | The samples are exposed to a set of predetermined changes in environmental conditions, such as temperature, light exposure and humidity. This is carried out over a period of time with the aim of creating a worst-case scenario for the product | Identification of the most suitable transporting and handling conditions for a product |
| In-use claims | The product will undergo conditions typical to that exposed by a customer who has opened the packaging and begun using the product | Period of time that the product remains stable after being placed into use |

### 36.10.7 Cultivation viability assessment, sustainability and manufacturing plan

From step 2, one should have a clear idea of the ease of growing the plant. In this step, one should explore the viability of producing the plant on a commercial scale. This can be determined by carrying out appropriate field trials. Figure 36.11 illustrates a field trial carried out in a glasshouse at an agricultural research facility. The trial aims to optimize rooting of cutting in order to increase the number of plants over a shorter period of time to maintain supply to companies when the plant is commercialized.

The sustainability of a large-scale production of the plant should be considered. Based on positive results from the cultivation and sustainability trial, a manufacturing plan can then be devised.

### 36.10.8 Clinical trials

The formulation of the extract can then be submitted for appropriate clinical trials in order to evaluate the safety and efficacy of a potential

**Figure 36.10** An image of the cream gel prepared to undergo storage under different temperatures for the purpose of stability testing

(a) (b)

**Figure 36.11** A field trial carried out to optimize rooting of a plant and the counting of roots at the end of the trial. *Source:* Courtesy of Future Cosmetics cc, 2013

product. These trials expose individuals to the final product within controlled parameters of application and instruction so that the effects can be monitored uniformly.

There are a number of clinical trials that a plant-based cosmeceutical can be submitted for:

- even-skin tone
- de-pigmentation
- anti-wrinkle
- anti-acne
- collagen enhancing
- hydration, and so on.

Perhaps one of the more important clinical trials that are highly recommended for products that are applied to the skin is the irritany test (Figure 36.12). This clinical trial determines whether an active product causes irritation to the skin, and monitors several factors from inflammation to redness and the degree of severity of these parameters. In this test, the active or formulation containing the active is placed on a particular area of the skin for a specified period of time. The area of application is usually a very sensitive area, like the skin on the underside of the arm. After the specified period, the area of application is observed and data is recorded using a scale of severity. The test will include various controls, like known irritants and negative controls in which to compare the test sample (potential products) using statistical analysis to determine the significance of the data.

### 36.10.9 Patenting, commercialization, advertising and quality control

The next step is to patent the extract; to protect the company that is producing the product (Moreira *et al.*, 2006). The product can then

**Table 36.3** The different categories of stability studies (International Conference on Harmonisation, 2011)

| Type of study | Storage conditions | Minimum time period of test (months) |
|---|---|---|
| Accelerated | 40 ± 2 °C/75 ± 5% relative humidity (RH) | 3 |
| | 50–60°C ± 2°C/Ambient RH ± 5% RH | |
| | 25 ± 2 °C/60 ± 5% RH or 30 ± 2 °C/65 ± 5% RH | 6 |
| Intermediate | 30 ± 2 °C/65 ± 5% RH | 6 |
| Long term | 5 ± 3 °C 12 months | 12 |

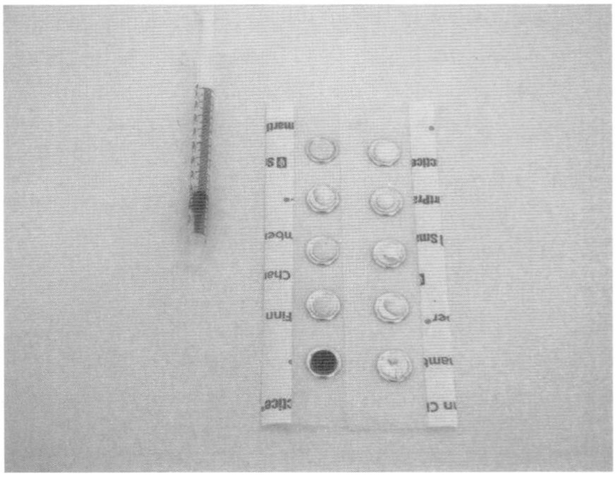

**Figure 36.12** The chambers that contain the sample to be applied to the skin for irritancy testing. *Photo*: courtesy of Future Cosmetics Cc

be included in the appropriate formulation, commercialized and also advertised. Quality control is a tool that checks each batch of product and makes sure that it is as similar to the standardized product as possible. This will test the manufacturing process and is a task that is performed regularly.

With the advertising, one should focus on the specific target market. A study indicated that different age groups of women respond to different colours and information in magazines and pamphlets. A mature age group preferred details about the product being advertised, as well as less eccentric colours, compared with younger age groups (Bertrand and Larcher, 2009).

### 36.10.10 Continuous assessment of plant growth and sustainability

The growth and harvesting of the plant resource need to be examined to ensure the continuous production, sustainability and viability that will continue over the period in which there is a high demand from consumers (Peters, 1996).

## References


Ashbee, H.R., Muir, S.R., Cunliffe, W.J. and Ingham, E. (1997) IgG subclasses specific to *Staphylococcus epidermidis* and *Propionibacterium acnes* in patients with acne vulgaris, *British Journal of Dermatology*, **136** (5), 730–733.

Artuso, A. (2002) Bioprospecting, benefit sharing, and biotechnological capacity building. *World Development*, **30**, 1355–1368.

Barankin, B. and DeKoven, J. (2002) Psychosocial effect of common skin diseases. *Canadian Family Physician Medecin de Famille Canadien*, **48**, 712–716.

Bertrand, S. and Larcher, L. (2009) Halmstad University, http://hh.divaportal.org/smash/get/diva2:239877/FULLTEXT01 (accessed 13 March 2013).

Bhate, S. and Lawler, K. (1997) Environmentally friendly products: factors that influence their adoption. *Technovation*, **17**, 457–465.

Bojar, R.A. and Holland, K.T. (2004) Acne and *Propionibacterium acnes*. *Clinics in Dermatology*, **22** (5), 375–379.

Burkhart, C.G., Burkhart, C.N. and Lehmann, P.F. (1999) Acne: a review of immunologic and microbiologic factors, *Postgraduate Medical Journal*, **75**, 328–331.

Chomnawang, M.T., Surassmo, S., Nukoolkarn, V.S. and Gritsanapan, W. (2005) Antimicrobial effects of Thai medicinal plants against acne-inducing bacteria. *Journal of Ethnopharmacology*, **101**, 330–333.

Chomnawang, M.T., Surassmo, S., Nukoolkarn, V.S. and Gritsanapan, W. (2007) Effect of *Garcinia mangostana* on inflammation caused by *Propionibacterium acnes*, *Fitoterapia*, **78**, 401–408.

Cotton, C.M. (1996) *Ethnobotany: Principles and Applications*, John Wiley & Sons, Ltd, Chichester.

Cowan, M.M. (1999) Plant products as antimicrobial agents. *Clinical Microbiology Reviews*, **12**, 564–582.

Cox, P.A. (1994) The ethnobotanical approach to drug discovery: strengths and limitations, in *Ciba Foundation Symposium 185 - Ethnobotany and the Search for New Drugs* (eds D.J. Chadwick and J. Marsh), John Wiley & Sons, Ltd, Chichester, pp. 25–41.

Dessinioti, C. and Katsambas, A.D. (2010) The role of *Propionibacterium acnes* in acne pathogenesis: facts and controversies. *Clinics in Dermatology*, **28**, 2–7.

Enshaieh, S., Jooya, A., Siadat, A.H. and Iraji, F. (2007) The efficacy of 5% topical tea tree oil gel in mild to moderate acne vulgaris: a randomized, double-blind placebo-controlled study. *Indian Journal of Dermatology, Venereology and Leprology*, **73**, 22–25.

Farris, K.F. (2010) Natural ingredients and their applications in dermatology. *Practical Dermatology*, (February), 51–54.

Gellenback, K.W. (2012) Utilization of algal materials for nutraceutical and cosmeceutical applications - what do manufacturers need to know? *Journal of Applied Psychology*, **24**, 309–313.

Gershwin, M.E., Borchers, A.T., Keen, C.L. *et al.* (2010) Public safety and dietary supplementation. *Annals of the New York Academy of Sciences*, **1190**, 104–117.

Govindasamy, R., Jehle, P., El Abidine, G.Z. *et al.* (2011) Aromatic and medicinal plants of Tunisia: a market study. *Journal of Food Products Marketing*, **17**, 470–486.

Grange, P.A., Weill, B., Dupin, N. and Batteux, F. (2010) Does inflammatory acne result from imbalance in the keratinocyte innate immune response? *Microbes and Infection/Institut Pasteur*, **12**, 1085–1090.

Huey, R.B., Carlson, M., Crozier, L. *et al.* (2011) Topical herbal therapies an alternative and complementary choice to combat acne. *Research Journal of Medicinal Plant*, **5**, 650–669.

International Conference on Harmonisation (2011) ICH Guidelines, http://www.ich.org/products/guidelines.html (accessed 12 October 2011).

Janeway, C.A. Jr, Traver, P., Walport, M. and Schlomchik, M. (2001) Basic concepts in immunology, in *Immunobiology: The Immune System in Health and Disease*, 5th edn, Garland Science, New York, NY.

Jugeau, S., Tenaud, I., Knol, A.C. *et al.* (2005) Induction of toll-like receptors by *Propionibacterium acnes*. *British Journal of Dermatology*, **153**, 1105–1113.

Kapoor, S. and Saraf, S. (2011) Topical herbal therapies an alternative and complementary choice to combat acne. *Research Journal of Medicinal Plant*, **5**, 650–669.

Kersey, P., Sussman, M. and Dahl, M. (1980) Delayed skin test reactivity to *Propionibacterium acnes* correlates with severity of inflammation in acne vulgaris. *British Journal of Dermatology*, **103** (6), 651–655.

Kim, J., Ochoa, M.T., Krutzik, S.R. *et al.* (2002) Activation of toll-like receptor 2 in acne triggers inflammatory cytokine responses. *Journal of Immunology (Baltimore, Md.: 1950)*, **169**, 1535–1541.

Kim, K.C. and Weave, R.D. (1994) *Biodiversity and Landscapes: A Paradox of Humanity*, Cambridge University Press, Melbourne.

Koreck, A., Pivarcsi, A., Dobozy, A. and Kemeny, L. (2003) The role of innate immunity in the pathogenesis of acne. *Dermatology (Basel, Switzerland)*, **206**, 96–105.

Krajewski, A., Garg, M. and Chandawarkar, R.Y. (2010) Topical herbal remedies: research opportunities for plastic surgeons. *Journal of Plastic, Reconstructive & Aesthetic Surgery*, **63**, 896–905.

Kubo, I., Muro, H. and Kubo, A. (1994) Naturally occurring anti-acne agents. *Journal of Natural Products*, **57**, 9–17.

Lalla, J.K., Nandedkar, S.Y., Paranjape, M.H. and Talreja, N.B. (2001) Clinical trials of ayurvedic formulations in the treatment of acne vulgaris. *Journal of Ethnopharmacology*, **78**, 99–102.

Layton, A.M. (2005) Acne vulgaris and similar eruptions. *Medicine*, **33**, 44–48.

Lee, C.C. and Houghton, P. (2005) Cytotoxicity of plants from Malaysia and Thailand used traditionally to treat cancer. *Journal of Ethnopharmacology*, **100**, 237–243.

Leyden, J.J. (2001) The evolving role of *Propionibacterium acnes* in acne. *Seminars in Cutaneous Medicine and Surgery*, **20**, 139–143.

Leyden, J.J., Del Rosso, J.Q. and Webster, G.F. (2009) Clinical considerations in the treatment of acne vulgaris and other inflammatory skin disorders: a status report. *Dermatologic Clinics*, **27**, 1–15.

Lowenstein, E.B. and Lowenstein, E.J. (2011) Isotretinoin systemic therapy and the shadow cast upon dermatology's downtrodden hero. *Clinics in Dermatology*, **29**, 652–661.

Ma, J.K.-C., Chikwamba, R., Sparrow, P. *et al.* (2005) Plant-derived pharmaceuticals – the road forward. *Trends in Plant Science*, **10**, 580–585.

Martin, G.J. (2004) *Ethnobotany: A Methods Manual*, Earthscan, London.

Medzhitov, R. (2007) Recognition of microorganisms and activation of the immune response. *Nature*, **449** (7164), 819–826.

Messele, B. (2004) Studies on extracts of some medicinal plants traditionally used for dermatological disorders in Ethiopia. MSc thesis, Addis Ababa University, http://etd.aau.edu.et/dspace/bitstream/123456789/3014/1/8208476959170070928180908245852695840 (accessed 8 March 2013).

Moreira, A.C., Müller, A.C.A., Pereira, N. and Antunes, A.M.D. (2006) Pharmaceutical patents on plant derived materials in Brazil: policy, law and statistics. *World Patent Information*, **28**, 34–42.

Mortelmans, K. and Zeiger, E. (2000) The Ames *Salmonella*/microsome mutagenicity assay. *Mutation Research*, **455**, 29–60.

Muhammad, B.A. and Awaisu, A. (2008) The Need for Enhancement of Research, Development, and Commercialization of Natural Medicinal Products in Nigeria: Lessons from the Malaysian Experience. *Africa Journal of Traditional, Complementary and Alternative medicines*, **5** (2), 120–130.

Pan, C., Chen, J., Lin, T. and Lin, C. (2009) *In vitro* activities of three synthetic peptides derived from epinecidin-1 and an anti-lipopolysaccharide factor against *Propionibacterium acnes*, *Candida albicans*, and *Trichomonas vaginalis*. *Peptides*, **30**, 1058–1068.

Peralta-Videa, J.R., Lopez, M.L., Narayan, M. *et al.* (2009) The biochemistry of environmental heavy metal uptake by plants: implications for the food chain. *International Journal of Biochemistry & Cell Biology*, **41**, 1665–1677.

Peters, C.M. (1996) Observations on the sustainable exploitation of non-timber tropical forest products, in *Current Issues in Non-Timber Forest Products Research* (eds M.R. Perez and J.E.M. Arnold), Centre for International Forestry Research, Indonesia, Chapter 2.

Pichersky, E. and Gang, D.R. (2000) Genetics and biochemistry of secondary metabolites in plants: an evolutionary perspective. *Trends in Plant Science*, **5**, 439–445.

Pieronia, A., Quavec, C.L., Villanellid, M.L. *et al.* (2004) Ethnopharmacognostic survey on the natural ingredients used in folk cosmetics, cosmeceuticals and remedies for healing skin diseases in the inland Marches, central-eastern Italy. *Journal of Ethnopharmacology*, **91**, 331–344.

Puhvel, S.M., Amirian, D., Weintraub, J. and Reisner, R.M. (1977) Lymphocyte transformation in subjects with nodulo-cystic acne. *British Journal of Dermatology*, **97** (2), 205–211.

Ogbunugafor, H.A., Eneh, F.U., Ozumba, A.N. *et al.* (2011) Physiochemical and antioxidant properties of *Moringa oleifera* seed oil. *Pakistan Journal of Nutrition*, **10**, 409–414.

Rafferty, J.P. (ed.) (2011) *The Paleozoic Era: Diversification of Plant and Animal Life*, Britannica Educational Publishing, New York, NY.

Sharma, R., Lall, N. (2013) Acne, a review on epidemiology, pathogenesis and treatment options. In: *Acne, etiology, treatment options and social effects*. Nova Science Publishers, New York. pp. 1–14.

Shaw, L. and Kennedy, C. (2007) The treatment of acne. *Paediatrics and Child Health*, **17**, 385–389.

Soni, H.K., Ribadiya, N.C., Bhatt, B.S. and Sheth, N.R. (2010) Evaluation of herbal formulation (capsule) containing ashwagandha as a single herb with their nutritional value determination. *International Journal of Applied Biology and Pharmaceutical Technology*, **1**, 960–966.

Swisher, H.E. (1988) Avocado oil from food use to skin care. *Journal of the American Oil Chemists' Society*, **65**, 1704–1706.

Taiz, L. and Zeiger, E. (2006) *Plant Physiology*, 4th edn, Sinauer Associates, Inc., Sunderland, MA.

Talalay, P. and Talalay, P. (2001) The importance of using scientific principles in the development of medicinal agents from plants. *Academic Medicine*, **76**, 238–247.

Truter, I. (2009) Acne vulgaris. *South African Pharmaceutical Journal*, **76**, 12–19.

Venus, M, Waterman, J. and McNab, I. (2010) Basic physiology of the skin. *Surgery (Oxford)*, **28**, 469–472.

Wakita, H., Matsushita, K., Nishimura, K. *et al.* (1998) Sphingosylphosphorylcholine stimulates proliferation and upregulates cell surface-associated plasminogen activator activity in cultured human keratinocytes. *Journal of Investigative Dermatology*, **110**, 253–258.

Webster, G.F., Leyden, J.J., Tsai, C.C. *et al.* (1980) Polymorphonuclear leukocyte lysosomal release in response to *Propionibacterium acnes in vitro* and its enhancement by sera from inflammatory acne patients. *Journal of Investigative Dermatology*, **74**, 398–401.

Wheeler, Q.D. (1995) Systematics, the scientific basis for inventories of biodiversity. *Biodiversity and Conservation*, **4**, 476–489.

Whittaker, R.H. and Feeny, P.P. (1971) Allelochemics: chemical interactions between species. *Science*, **171**, 757–770.

Williams, H.C., Dellavalle, R.P. and Garner, S. Acne vulgaris. *Lancet*, **379**, 361–372.

Wojdyłoa, A., Oszmiański, J. and Czemerys, R. (2007) Antioxidant activity and phenolic compounds in 32 selected herbs. *Food Chemistry*, **105**, 940–949.

Yokota, T., Nishio, H., Kubota, Y. and Mizoguchi, M. (1998) The inhibitory effect of glabridin from licorice extracts on melanogenesis and inflammation. *Pigment Cell Research*, **11**, 355–361.

# 37 New Natural Aromatic Products: Search, Evaluation and the Development Issues

Murray Hunter

School of Business Innovation & Technoentrepreneurship, University Malaysia Perlis, Malaysia

## 37.1 Introduction

The search for, identification and development of new natural aromatic materials is fraught with difficulties. Finding new materials is an extremely time-consuming and expensive process. Identifying the potential usefulness as an aromatic material is complex. Finally, developing the material as a commercial product has many issues involved, thus creating barriers. These costs and complexities are depressing the general outlook for the future growth of natural aromatic compounds usage in industry.

This chapter will define the family of natural aromatic materials, look at the search process for new materials, suggest a set of criteria to assist in evaluating commercial potential and, finally, look at some of the issues involved in commercial development.

## 37.2 The family of natural aromatic extracts

Natural aromatic materials can be extracted from the roots, rhizomes, wood, bark, leaves, stems, fruits, flowers and seeds, from a wide variety of plant, shrub and tree species. Different parts of the same plant may contain compounds that differ in their chemical composition, and may or may not require different methods to extract these compounds effectively. Primarily, the extraction method used determines whether the aromatic extract is called an essential oil, concrete, absolute, tincture, pomade, oleoresin or balsam. These methods are summarized in Figure 37.1.

Essential oils result from extracting volatile compounds, mainly terpenes and oxygenated compounds, through one of the methods of distillation. The constituents of essential oils are synthesized in the plant through the mevalonic and shikimic acid pathways as secondary metabolites, stored in glandular trichomes, oil cells or ducts in plant tissue. The specific method of distillation used will depend upon the characteristics of the aromatic materials present in the plant, the structure of the plant material and the economics of extraction. Three basic types of distillation exist: hydro-distillation, water and steam distillation, and steam distillation.

Other specialized extraction methods, like vacuum distillation, high-pressure distillation and destructive distillation, are used in certain cases, where specific conditions are required. For example, low temperatures are sometimes needed to prevent highly volatile materials from being destroyed or lost during distillation, so vacuum distillation would be used. Where greater heat and pressure are needed to extract high boiling point volatile materials from the plant, high-pressure distillation can be used. Where the plant material itself must be destroyed to remove volatiles, destructive distillation can be used.

Other oils are produced from citrus fruits through a process called cold expression. Orange, lime or other citrus peels are pressed until a fine spray of aromatic oil is released. This is specifically called an *expressed oil*. Cold expressed oils can be distilled under vacuum to remove the terpene hydrocarbons from the rest of the oxygenated components to produce what is called a *terpeneless oil*. The other resulting product is known as a *terpene oil*, *terpenes* or *terpene tails*.

Where aromatic materials are susceptible to chemical changes from heat processes and this is undesirable, extraction can be undertaken through washing the plant material with a solvent in a method called solvent extraction. Petroleum ether, acetone, hexane or ethyl acetate can be used to wash out specific plant materials, usually volatiles, pigments and waxes, into the solvent. Once the solvent has been evaporated off through vacuum distillation, the remaining material is called a *concrete*, or in some cases a *resinoid*. The volatile materials can be further extracted from the concrete through washing the material with alcohol and then removing the alcohol through vacuum distillation. The remaining material is called an *absolute*.

Enfleurage is an old method of extracting aromatic materials from plants, which is being revived in aromatherapy. Enfleurage involves laying plant material on specially prepared fat and lard to absorb the volatile materials. After a period of time the odorous material is extracted from the fat by heating. The material is then filtered from the solid particles, leaving a *pomade*. When the ethanol is distilled off through vacuum distillation, the resulting product is called an *absolute*.

The use of liquid carbon dioxide as a solvent under pressure to remove volatile plant substances is one of the newer methods of extraction. Products produced from this process resemble nature very closely as there is no heat involved. Carbon dioxide is an inert material, which does not promote any undesirable chemical reactions. Natural materials extracted in this way are called *$CO_2$ extracts*.

*Novel Plant Bioresources: Applications in Food, Medicine and Cosmetics*, First Edition. Edited by Ameenah Gurib-Fakim.

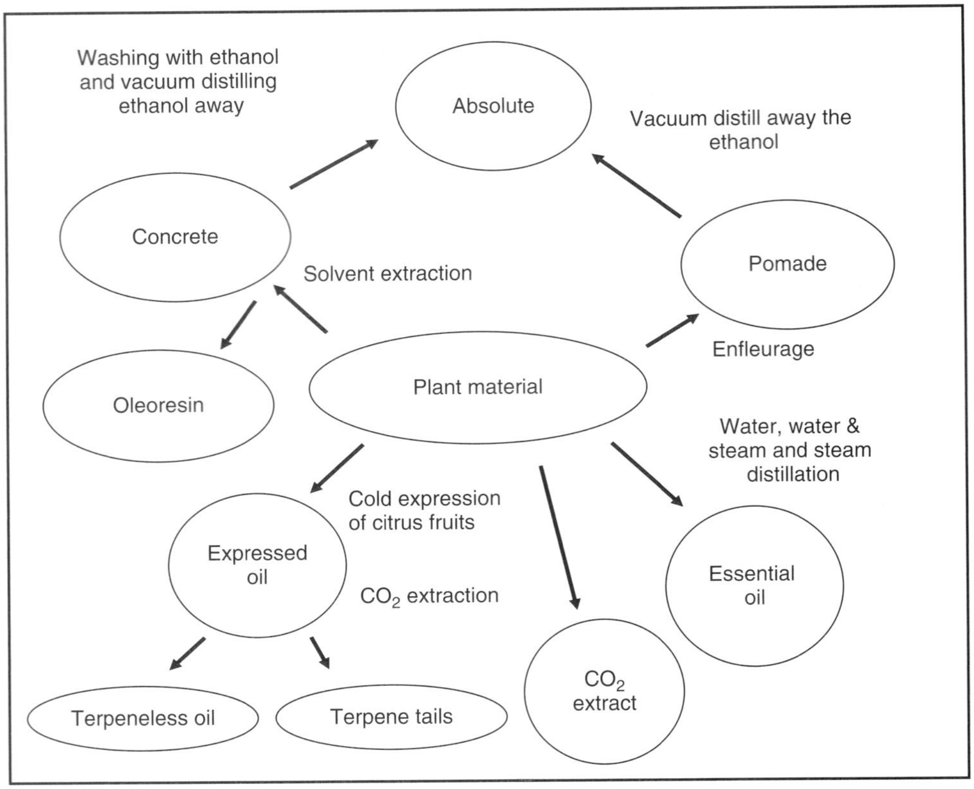


**Figure 37.1** The various methods of natural material extraction and products

Very low and non-volatile aromatic materials that are present in woods are extracted through a more complex form of solvent extraction, which yields an *oleoresin*.

## 37.3 The search and screening process

Most search projects for new natural aromatic materials commence with some form of preliminary screening process, either formal or informal. The aim of this process is to develop a short list of probable candidates for closer evaluation. Only flora that has some realistic commercial potential should be included on the short list. This process involves simultaneous biological–environment and commercial development potential evaluation. Adequate time should be allocated for the screening process.

Whether the screening process is restricted to desktop evaluation or involves some preliminary field work will depend upon available time, information and resources available. Without a proper screening process there is risk that any future field research will be *ad hoc* and lead to an unmarketable material due to technical, resource and/or market reasons. Still today, many projects commence without any formal technical or market studies about crop feasibility (Gallagher, 1993).

The evaluation of essential oil market potential is only part of the total screening process. The screening process must also take into account the limits imposed by the physical environment (climate, rainfall, soil, sunlight hours, soil type and pH, water holding capacity, drainage, topography, etc.) and production economies. All these factors must be considered simultaneously. Lack of appreciation of the interdependence between production and market development has led to many failures by new producers (Green and Hone, 1992).

The process of screening aromatic plants for development potential is complex due to the multiplicity of issues – biological screening and development potential evaluation. All factors must be found to be compatible and potentially meet regulatory requirements, as the product must be able to achieve registration to be sold legally. Given that over 50 potential aromatic plants may be screened for potential, some system of handling the information in a manageable manner must be devised. The more ordered and systematic the system to handle the evaluation, the easier and more meaningful will be the process.

The first part of the screening process involves the creation of a list of crop possibilities that are both physically and market feasible. This infers converging knowledge in two areas: agronomic and market. This can be approached by generating a list of crops that can be commercially produced in the selected geographical area in question. A comparison between potential crops and the physical characteristics of the selected potential geographical area can be checked off using similar criteria listed in Table 37.1.

Direct comparisons between the physical conditions and potential plant suitability will not always reveal all potential crops. Some crops cultivated in subtropical climates may grow better in tropical climates, with more consistent weather and higher rainfall. There may be no data to support these possibilities, and the exercise will rely upon experience, hunches and educated guesses. This was the

**Table 37.1** Comparison factors between potential cultivation site and potential crop

| Actual conditions | Range of possible growing conditions for potential crop |
|---|---|
| • General climate<br>• Range of microclimates<br>• Topography that influences microclimates<br>• Rainfall range (access to irrigation)<br>• Temperature ranges<br>• Daylight hours<br>• Soil types<br>• Soil characteristics (pH, humus profile, soil layers, etc.) | • Preferred climate(s)<br>• Preferred microclimates<br>• Preferred topography<br>• Preferred rainfall levels<br>• Preferred temperature ranges<br>• Preferred daylight hours<br>• Preferred soil types<br>• Preferred soil characteristics |

case with Australian tea tree, originally cultivated in central and southern New South Wales, Australia. Trials in northern Queensland, Australia, and Malaysia found that not only was the crop adaptable to tropical climates, but it also produced more biomass, leading to higher yields with a comparable oil quality to the original area of cultivation (Hunter, 1996).

An alternative way to look for potential crops is to examine existing crops and plants already cultivated along similar latitudes, in the Southern and Northern Hemispheres. This way, possibilities can be looked at by region. This is also appropriate for specific geographic areas where the micro-climate may be similar to the climate at a different latitude; that is, *at higher altitudes in tropical areas where temperatures are lower than the lowlands.* This is another good rough method of quickly creating a list of potential crops to examine. Figure 37.2 shows the 44° latitudes, north and south of the Equator. One can see that many essential oil crops currently cultivated in Tasmania and New Zealand are similar to those cultivated in central Europe.

## 37.4 Sources of potential plant opportunity identification

An active search for potential crops is the next step of the process. This can be achieved through field work (bio-prospecting) and/or a desktop search utilizing existing published data.

### 37.4.1 Bio-prospecting

The South-East Asia-Pacific region is abundant in flora, where the majority of species are yet to be examined for their economic potential. In Malaysia alone, there are over 15 000 species of plants, of which only around 1200 have been investigated (Latiff *et al.*, 1984). On a global scale, of the 350 000 known plant species, less than 0.5% of them have been chemically investigated (Rao and Arora, 2004). Even with the flora that has already been investigated, much of the data recorded and reported lacks any economic evaluation. Many books containing flora references and catalogues may be out of print, which will require some time and effort in locating copies.

There are many approaches to bio-prospecting. Forest and jungle areas can be systematically and blanket searched, where all aromatic plants are investigated that the team comes across. This approach will be very time and resource consuming and may only lead to very long-term results. Alternatively, plants with specific characteristics and properties identified in ethno-botanical literature are searched for in specific geographical locations in a bio-prospecting programme.

Decisions have to be made about the scope of screening. Screening for multiple characteristics is very complex and will increase time and resources required. Potential screening protocols depend on the project focus. Some of these would include:

- the chemical constituents of the plant oil (leaves, flowers, roots, bark, fruits, seeds, etc.);
- antimicrobial properties;
- anti-inflammatory properties;
- skin whitening properties;
- UV absorbing properties;
- anti-age properties;
- flavour and fragrance application; and
- aromatherapy application.

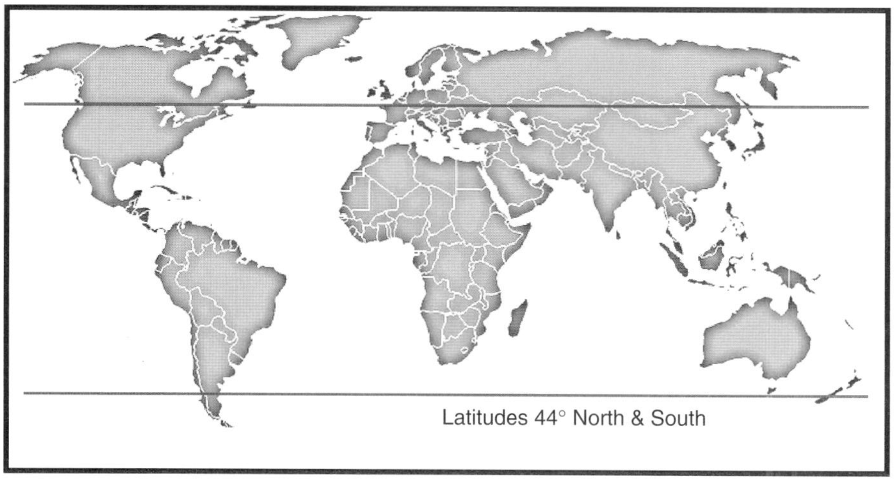


**Figure 37.2** The 44° latitude lines north and south of the Equator

### 37.4.2 Desktop studies

Potential new crops can be discovered through desktop studies. This requires a literature search through books, periodicals, journals and other monographs pertaining to essential oil crops. Studies would include studies of plants cultivated in similar latitudes. The aspects listed in Table 37.2 would be used to screen each potential possibility.

Through bio-prospecting and a desktop study, a list of potential crops can be developed. The next step is to eliminate plants of little potential until a short list is obtained.

### 37.4.3 Screening for development potential

The major question in screening for development potential is, 'What value does the potential natural aromatic material have for any industry?' This question can be understood better through evaluating each possibility with a series of sub-questions, answered with existing reported data on each material. These questions would include, for example:

- What are the chemical constituents?
- Are these chemical constituents similar to other natural materials?
- Is the odour profile of any value?
- Is this a similar odour profile to other materials?

These questions and corresponding answers should provide some clue as to whether the essential oil is of interest and has any potential application in industry.

If there are some potential applications, the next set of questions would relate to potential yields and probable cost of development. If the reviewed material is similar to something already existing commercially, then the questions will relate to relative yields, growth periods and percentage of valuable constituents in the oils. If the material is novel and contains potentially valuable chemical constituents or interesting odour profile, what would be the potential yields? At this stage these estimates should be rough or very 'ball park' based on what published data are available or through observation and sample distillations of the plant, extended out on a worksheet. A simple template can be used for manipulating very rough labour, operational and yield estimates. A sample template for this calculation is shown in Table 37.2.

After undertaking the above exercise, some ideas on potential and critical issues should be understood and answered in some sort of preliminary manner. These issues are:

- Is there a match between chemical constituents and possible market uses?
- What is the application potential of the essential oil?
- What could be possible yields and cost of production?
- What would be the cost of development?

## 37.5 The characteristics and classification of natural aromatic materials

Natural aromatic materials have many of the characteristics of commodities, but are by no means a homogeneous group of products. Trade statistics indicate the market size for many individual aromatic materials, but they may not be complete and thus of limited use in evaluating essential oil potential. To evaluate the commercial potential of new essential oils, a market orientation (i.e., potential uses, characteristics, product strengths and weaknesses) is required, rather than a commodity view of the essential-oil trade. Some basic classification system is necessary to use as a tool for evaluating development potential and to indicate the types of business strategies required. A basic classification system is outlined in Table 37.3.

**Group 1.** The olfactory profiles of aromatic materials in this group are dominated by a single or a few major constituents. The production of many essential oils in this group relies on high capital investment and mechanization as the key to competitive advantage. The mint industry in the USA is a good example of this (Borst, 1988). Other essential oil crops in this group are produced by smallholders in developing countries. The smallholder production sector has only survived because the cultivation of certain essential oils is not profitable in industrialized countries, full mechanization is not possible and cultivation of some plant species is only possible a few months each year (Verlet, 1993a). The citronella industry in Indonesia is also a good example (Harris, 1987). Prices in this group of essential oils are kept relatively low due to the availability of substitutes (other essential oils and synthetic aroma chemicals) and competition between producers of the same crop in developing countries.

**Group 2.** Most aromatic materials in this group are produced by smallholders in developing countries along the tropical latitudes. Some are produced on plantations, with low-cost labour supply as a source of competitive advantage, but labour shortages due to rural/urban migration are beginning to erode many producers' competitive positions (Ministry of Agriculture, 1991). Fluctuations in prices occur when there is a mismatch of supply and demand on the world market. Generally, in times of supply shortages, higher prices will encourage existing producers to increase production and marginal growers will be attracted back into production. High prices and shortages of essential oils in this group may not necessarily indicate potential opportunities to develop these materials in new geographical locations, as existing producer countries can easily increase production levels (Robbins, 1982). There are generally very low barriers to entry and exit to producers within this group. There are also few cost-effective substitutes available to the flavour and fragrance industry in this group. This lack of close substitutes enables producers to gain slightly higher marginal returns for these oils compared with those in the first group (Hunter, 1992). However, with rapid technology advances, substitutes for these materials are becoming available.

**Group 3.** Natural aromatic materials in this group are primarily used in perfumery. These materials are valued for their depth, quality, complexity and richness in olfactory profile. Many of these materials are still traditionally cultivated (Konur and Robertet, 1983). However, as production costs are rising in developed countries, the production of these materials is being encouraged in developing countries (Anon., 1993), often under the sponsorship and close working relationships with major European companies (McKay, 1990). Generally, these materials are becoming too expensive to use in functional perfumery, where reconstitutions and specialties have mostly replaced them (Fenn, 1988). Herbaceous oils are limited to minor use in flavour compounds (Robbins and Greenhalgh, 1979). Changing market trends have increased the application of some herbaceous oils in fragrance applications, but aggregate demand is not increasing above natural market growth.

Table 37.2 Worksheet for rough calculation of financial viability at initial screening stage

1. **Costs of crop domestication**
   - Can they potential crop be domesticated into field production easily?
   - If not, will biomass be wild-collected?
   - What method would be most suitable for propagation: from seed, cuttings, tissue culture, other?
   - Does nursery propagation of the potential crop require any other special care?
   - What staffing will be required?
   - What would be the approximate costs of achieving the above?
2. **Field preparation and infrastructure**
   - What overall infrastructure will be needed (i.e. nursery, road access, fencing, outbuildings, farming equipment, etc.)?
   - What land preparation is needed (land levelling and contouring, drainage, etc.)?
   - Does the crop require large amounts of water to thrive during growth?
   - Is there adequate water available through rainfall to satisfy this?
   - Will irrigation be required? If so, what method? Will dams and catchment areas have to be constructed to ensure a plentiful water supply?
   - What will be the approximate costs of this?
   - Are there any other potential costs?
3. **Planting and maintenance**
   - Approximately how long will the crop take from field planting to harvest maturity?
   - How will the potential crop be planted (manually/automated)? What will be the costs involved?
   - What would the approximate planting density be?
   - Will nutrients have to be applied? If so, how regularly? How much? What method will be used to apply them? What will be the approximate costs of this?
   - How often are replantings required? After each harvest, after a number of seasons, after how many years, what are the costs involved to prepare for each replanting?
4. **Harvesting, extraction and post-extraction**
   - Is harvest timing crucial (i.e. a time of day, a very short window in a particular month, etc.)?
   - What are the costs involved in achieving this harvest window?
   - What method of harvest will be utilized (manual, semi-mechanized, fully mechanized)?
   - What would be the approximate costs of building the harvest equipment?
   - What method of extraction will be required (hydro-distillation, steam distillation, destructive distillation, vacuum distillation, solvent extraction, other)?
   - What power sources will be utilized? What are their costs?
   - How will spent biomass be dealt with? Does it have any economic value or can it be used back in the farming process?
   - Is the technology understood for the above processes? If not, what will be the costs of acquiring it?
   - What will be the fabrication costs to build the above?
   - What regulations (i.e. EPA) are relevant to the processes? And how much will development and compliance cost?
   - Will specialist staff be required?
   - What would the approximate cost of energy to oil yield?
5. **Estimated (guessed) project size and yields**
   - How many hectares do you anticipate to cultivate?
   - How many years will it take to achieve this?
   - What (based on literature and other knowledge) would be the approximate biomass per hectare achievable (minimum and maximum estimates)?
   - Does the biomass have to be wilted, stored or otherwise processed before extraction?
   - What would be the yield as a percentage of biomass after extraction?
6. **Estimated financial viability**
   - 1. Research costs =
   - 2. Costs of crop domestication =
   - 3. Field preparation and infrastructure costs =
   - 4. Propagation, planting and maintenance costs =
   - 5. Harvesting, extraction and post-extraction costs =
   - Total capital costs (1 + 3) =
   - Total operational costs (2 + 4 + 5) =
   - Total amount of oil yielded =
   - Total oil value =
   - Value − total operational costs =
   - Return/total capital costs × 100 = Return on investment

*Notes*: This is a rough estimate based on the whole project cost estimates. Alternatively, this can be calculated on a hectare basis. Allowances must be made for varying yields on some crops (especially tree crops).

**Table 37.3** A basic natural aromatic material classification scheme

| Group | Characteristics | Examples | Uses |
|---|---|---|---|
| 1. Flavour/odour profile due to one or few major constituents | • Usually high volume/low- to medium-value products.<br>• Level of chemical constituents very important in trade<br>• Synthetic aroma chemicals usually good substitutes | • Peppermint oil<br>• Lemongrass oil<br>• Some citrus oils<br>• Eucalyptus oils<br>• Clove oils | • Perfume and flavour compounds<br>• Flavours where natural status is desired<br>• Some citrus oils used for cleaning solvents<br>• Isolation of natural aroma chemicals, e.g. eugenol from clove oil |
| 2. Flavour/odour profile due to one or a few major constituents and essential oil cannot be easily reconstituted | • Usually medium- to high-volume, medium-priced oils<br>• Olfactory and flavour characteristics more important in purchase decisions<br>• Difficult to reconstitute | • Vetiver oil<br>• Sandalwood oil<br>• Patchouli oil | • Perfumery and flavour compounds (both functional and fine perfumery) |
| 3. Character from main constituents, but richness and complexity from minor constituents | • Usually low-volume/high-priced oils<br>• Olfactory characteristics important in purchase decisions and pricing<br>• In most cases these oils can be reconstituted efficiently | • Rose oil<br>• Jasmine absolute<br>• Many herb oils | • Fine perfumery<br>• Reconstitutions used for functional perfumes<br>• Limited flavour use<br>• Majority of herb oils used for flavours but beginning to be used for fragrances |
| 4. None of the main constituents contribute decisively to the desired odour/flavour profile | • Usually low-volume/high-priced oils<br>• Olfactory characteristics most important in purchase decisions<br>• In most cases reconstitutions can be produced | • Mimosa absolute | • Fine perfumery (usually too expensive for functional products) |

**Group 4**. Natural aromatic materials in this group, like materials in Group 3, are valued for their olfactory profile and are mainly traditionally produced, although some new ventures are utilizing new technologies to enable competitive production. The availability of good reconstitutions of these materials has limited their use to fine perfumery applications.

Aromatic materials in Groups 1 and 2 tend to be traded like pseudo-commodities, but strong product differentiation, based on olfactory profiles, is also an important purchasing criteria. Other industrial uses for many of these oils add to traded volumes. Essential oils in Groups 3 and 4 are traded more like specialty and fine chemical products, where olfactory quality plays a major role in price determination. Demand tends to be inelastic in the short term, as manufacturers are locked into using selected essential oils in the formulation of existing products. Past price trends and a poor supply history for some individual oils has encouraged the use of alternative materials in new products.

The above classification system can be used to generally identify natural aromatic material use and trade characteristics. This establishes a structure from where further specific market analysis can be undertaken.

## 37.6 Evaluating the characteristic strengths and weaknesses of natural aromatic materials

The market potential of a new natural aromatic materials in the flavor and fragrance industry depends upon potential flavor and fragrance applications. The scope of potential uses and applications directly corresponds to the characteristic strengths and weaknesses of the essential oil. After the objective identification of characteristic strengths and weaknesses, estimates of the essential oil's perceived value to potential end users can be made. The list in Table 37.4 provides criteria to evaluate new essential oils.

## 37.7 The development issues

The essential oil industry faces a number of major issues and challenges, including climate change and the food and economic crises. Other important issues the industry is facing include wild plant collection, the bargaining power of producers, the regulatory system, the rural crisis, the synthesis of natural aromatic molecules and patents, research and sustainability. These issues are discussed briefly in the following sections.

### 37.7.1 Global warming, and the food and financial crises

Three major issues, climate change and the food and economic crises form a great part of the underlying general environment that the natural aromatic material industry must exist and operate within. Issues generated by climate change and the food and economic crises are extremely complex and subject to intense debate and analysis, both scientifically and politically. Figure 37.3 depicts some of the underlying issues and challenges to the industry from climate change and the food and economic crises.

**Table 37.4** Characteristic strength and weakness potential for natural aromatic materials

| Characteristic | Comments |
|---|---|
| The novelty of a new aromatic material | The major factor determining the novelty is the perceived uniqueness of the material's organoleptic profile. Thus, the degree of novelty is limited by the closeness of potential substitutes. The concept of novelty extends to essential oils that are more cost-effective sources of natural aroma chemicals. New natural sources of aroma chemicals would also fit into this criterion of novelty. |
| The potential uses and applications of a new aromatic material | Without perfumers and flavourists perceiving applications potential, a new material will remain in the realm of curiosity. Time, effort and imagination on the part of perfumers and flavourists is required to discover useful applications for new essential oils. It is under this criteria that most new essential oils will struggle to find acceptance as a new aromatic material. |
| The closeness of any substitutes | It is difficult to find materials that cannot be duplicated by reconstitutions. New essential oils with close substitutes are of little value to the flavour and fragrance industry, unless they can offer a significant cost or stability advantage. The only exception is when a new material is a source of a natural aroma material. |
| The stability of a new aromatic material | One of the major problems associated with aromatic materials is stability in end products. Many processed food products undergo harsh cooking procedures during manufacture. Cosmetic bases often contain free fatty acids, even after neutralization. Aromatic materials that contain high levels of terpenes tend to polymerize on exposure to light and air, discolour end products or are not stable in alkaline or acidic media. Synthetic aroma chemicals and specialties are generally more stable than natural materials and used more extensively in functional perfumery. |
| The cost price/performance ratio | The cost price/performance ratio is important to the application potential of a new material. If a new essential oil does not offer a perceptible odour/flavour at a low concentration, then its value to the flavour and fragrance industry is greatly diminished unless it is very cheap. Poor performance under this criterion will negate the potential of most new materials for application in functional perfumery. |
| The toxicity | The cost of proving a new material is safe to use in flavours and fragrances is a major obstacle to the development of new aromatic materials. The industry has an impeccable reputation for self-regulation, and added EU regulations increase the cost of preparing dossiers on new materials even more. In markets outside the EU, most international flavour and fragrance houses would not consider using a new essential oil unless it meets International Fragrance Association (IFRA) safety and toxicity recommendations and is included on the Generally Recognized as Safe list. |
| The general consistency of quality and supply | Natural materials will vary in quality according to geographic origin, type of soil, level of nutrients in the soil, climate and weather, rainfall, time of harvest, season, method of extraction, altitude and the incidence of pests and diseases. Likewise, there are risks with continual supply of natural materials because of adverse weather conditions, changes in climate, floods and other natural disasters, wars, political upheavals and the inexperience of new producers. Launching new consumer products requires large investments on the part of the end-product manufacturer. Flavour and fragrance houses do not want to be placed in a position of being unable to supply a manufacturer with a flavour or fragrance compound because of the unavailability of a raw material. |
| The prevailing market/product trends | Market and product trends slowly evolve. Changes in market trends are the result of complex forces, including technology (which makes new trends possible), advertising, and cultural influences upon consumer tastes and preferences. A particular essential oil may become more or less important to the flavour and fragrance industry, depending upon these trends. |
| The current level of technology | New technology advances influence the value of existing aromatic materials to the flavour and fragrance industry. The development of new essential oil reconstitutions is aimed at eliminating some of the potential toxicity and solubility problems of existing materials. Reconstitutions are generally more stable and cheaper than their more expensive natural counterparts. As better and more cost-effective reconstitutions are developed in the future, the use of some essential oils will decline. Since the advent of more sophisticated analytical techniques, like gas chromatography–mass spectrometry, headspace analysis, electronic noses, aroma chemical and specialty product manufacturers have been better able to isolate powerful aromatic molecules from essential oils and synthesize these compounds. The discovery of new aroma chemicals in essential oils due to increased equipment sensitivity is more likely to lead to synthesis rather than cultivation. |

Population growth requires a corresponding increase in the production of goods and services to meet growing consumption demands (needs and wants). Rapid population growth is increasing consumption, leading to higher rates of production, which increases carbon dioxide emissions. The growing carbon dioxide traps infra-red radiation in the atmosphere, rather than allowing it to escape back into outer space. This creates a condition where the atmospheric temperature will slowly rise. The increase in atmospheric temperature creates a number of secondary effects, including altering weather and tidal patterns, which in turn cause floods, droughts, heat waves and cold snaps. These secondary effects give rise to changing pest and disease cycles, faster loss of top soils due to adverse weather and the increase of arid land area. Continued warming is melting the land-based ice caps, which cause rising sea

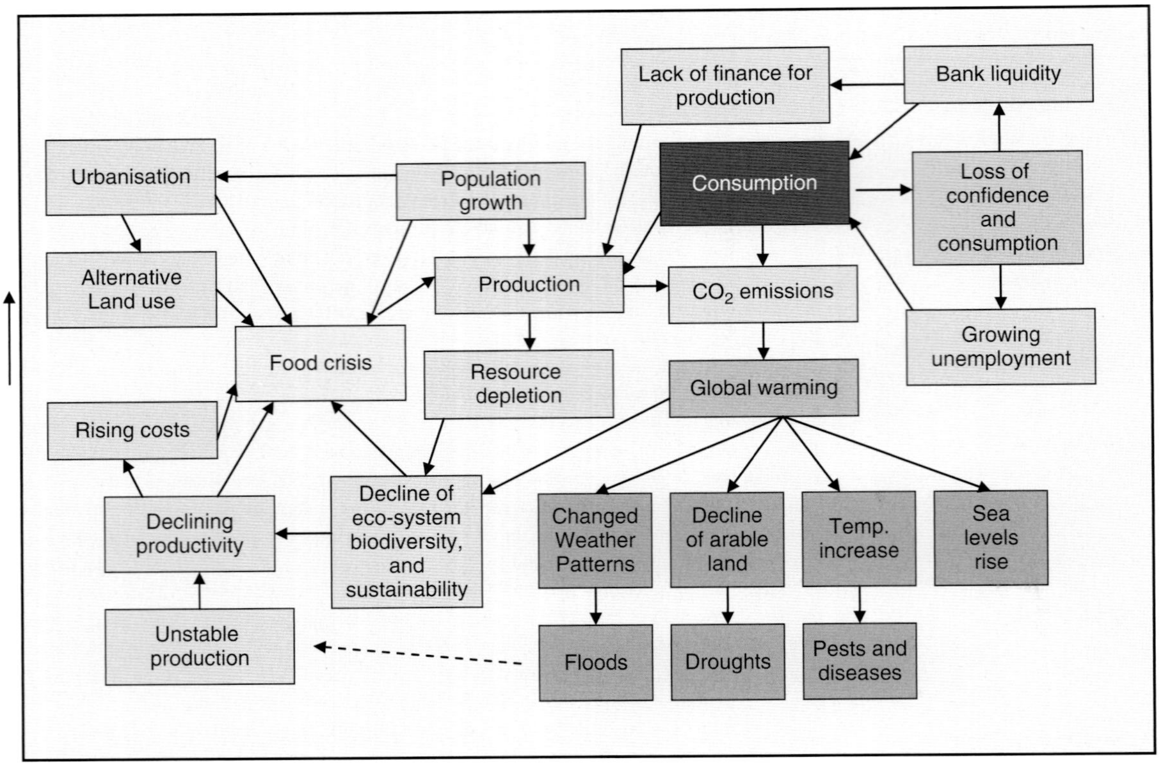


**Figure 37.3** Some of the underlying challenges on the natural aromatic material industry from climate change, the food and economic crisis

levels. The results of these environmental activities are an increase in the incidence of crop failures.

Mankind's unsustainable production practices at our current state of technology are leading to the depletion of the Earth's natural resources to the point where some may become extremely scarce. This is also leading to the decline of biodiversity. This causes the cost of production to rise due to the need to use more inputs per unit of output. There are also increasing levels of outright crop failures. Food costs are further heightened through the rising costs of land due to greater urbanization and industrialization, the shortage of labour and increasing operational costs.

In 2008, US housing prices rose to the point where they reached a peak and began to collapse. Prices started to decrease to the point where housing mortgage equity was not supported by equity in the house values. When many borrowers defaulted on their repayments and banks could not recover their loans through mortgage auctions, a liquidity crisis began. Banks cut lending to manufacturers and retailers, which need liquidity to operate. This led to a loss of confidence, decline in consumption and corresponding decline in production and worker lay-offs. The loss of confidence, increasing unemployment and corresponding credit squeeze started a downward spiral in demand, production and consumption leading into deep recession.

Through the interrelated world economy, the effects of the US liquidity and confidence crisis spread to Europe and Japan. European banks have lent money to banks in the USA and southern Europe where there is risk of major loan defaults due to these economies' shrinking economic activity. Decreasing consumption in developed countries is leading to lower product orders from Asian manufacturers. The crisis spread to the Asian region, where there were risks that unemployment would rise.

There are concerns due to the wide extent of this global economic crisis that the issues of climate change and food scarcity will be put into the background by governments at a time where many believe they cannot be ignored.

Global warming and the food and economic crises are already affecting the production of natural aromatic materials in a number of ways. Global warming increases the incidence of risk in general agricultural production. As regions become warmer, droughts set in and seasons change, some areas will become unsuitable for the production of certain essential oils, while other areas will become suitable. Agronomic practices will have to take account of the technical issues brought about from the effects of global warming. The global economic crisis will challenge the stewardship of business. The marketing of aromatic materials will take place in markets that are declining, rather than growing. The funding of natural product research and production will also be expected to decline because of the financial crisis.

Some of the basic issues underlining the food crisis are also relevant to natural product production. Natural products will have to compete for land against other potential land uses. This means that natural products must make a higher return per unit of land than food crops for farmers to be enticed to cultivate them. This will mean higher prices for natural aromatic materials in years to come.

### 37.7.2 Market turbulence

Like other commodities, natural aromatic materials are subjected to price fluctuations. This is the result of mismatches in supply and demand, particularly on the supply side. As natural aromatic materials are basically agricultural products, they are affected by adverse weather, especially heat, droughts, natural disasters, pest and diseases, labour shortages and competition for land use. The attraction of other crops also causes supply problems (e.g. sugarcane production in Brazil).

The situation for a number of natural products over the last few years has proved no exception, where a number of oils have suffered supply problems. There have been shortages of aniseed oil (weather), bergamot oil (weather), coriander oil (seeds used for spice as prices more attractive), dill and fennel oils (excessively warm weather), geranium oil (farmers switched to growing food crops), lavender and lavendin oils (excessively warm weather), lemon oil (weather), *Litsea cubeba* oil (shortage of labour), and rose oil (shortage of labour during harvests).

Sometimes price fluctuations and supply shortages are enough to force customers to switch to substitutes. This has happened to a number of essential oils, which became priced out of the market or the shortage forced cessation of product use. Tea tree oil was discontinued as a product line in some European retail chains in 2005, when there were chronic shortages of supply (Bolt, 2000). Patchouli oil was replaced by many users in fragrance formulations with synthetic aroma chemicals that are cheaper and stable in price. Shortages of *Litsea cubeba* oil encouraged the use of synthetic citral. The rising price of sandalwood will encourage the use of synthetic alternatives, which may leave only a small remaining market for fine fragrance use, if confidence is lost in supply by the majority of users. There has also been a drastic decline of essential oils that are reliant on wild collection. Natural product price fluctuations can threaten the long-term usage of an essential oil and force substitution to a low-cost mix of stable synthetic substitutes.

Other factors that cause market turbulence include the rise in the price of petroleum, which increases the costs of distillation. Some oils, like ylang-ylang, require over 20 h of distillation, and clove bud requires 48–50 h. Thus, diesel and bunker fuel is not a minor cost in the production of the oils. Steel prices are increasing, making the costs of drums go up, China is tightening environmental regulations and urbanization is creating shortages of rural labour. Adulteration of essential oils is still of great concern. There are still a number of cases of adulteration of essential oils, which cause concern and loss of integrity in the marketplace.

### 37.7.3 Wild collection and threatened plant species

There are between 300 and 400 natural aromatic materials, including essential oils, concretes, absolutes, resins, oleoresins and balsams, available commercially in the market. Almost 50% of these materials are extracted from trees that are hundreds of years old or rare plants that cannot be easily domesticated (Lawrence, 1993). These are collected from the forests, jungles and bush lands around the world. This raises the issue of sustainability. Over the last few decades a number of flora species have become threatened with potential extinction, due to over harvesting.

The pricing mechanism heavily influences what happens in wild plant collection. When prices are high, two major things can happen.

The initial response is often a frenzy of collection by various collectors, which results in harvests well above demand levels. This reckless manner usually occurs because of lack of law enforcement in remote areas that fails to preserve stocks of the species for future harvesting.

Second, if the plant can be domesticated, entrepreneurs will commence planting of the crop. For example, supply shortages and continued high prices have encouraged a large number of people to cultivate both *Aquilaria* and *Santalum* species. This will most likely lead to a massive oversupply situation in the future.

The Convention on International Trade in Endangered Species of Wild Fauna and Flora (CITES) is an intergovernmental agreement ratified by 189 nations under the International Union for the Conservation of Natural Resources (IUCN), which compiles a list of threatened species that should not be traded. Table 37.5 shows a list of potentially endangered species relevant to the aromatic trade. Listing plant species as endangered is one issue; enforcement is another, as many of these endangered species are still traded in significant quantities (Burfield, 2008).

The harvesting of wild plants often involves travelling over difficult terrain to collect material and process it into an essential oil or aromatic extract. The supply of wild plant material is unstable and subject to both natural interventions and regulation. Wild collection in the majority of cases is difficult to regulate (www.cropwatch.org) and ensure sustainability.

### 37.7.4 Bargaining power of producers

Approximately 55% of the world's natural aromatic materials come from developing countries (Verlet, 1993b), particularly those countries along the tropical belt. The vast majority of producers in these regions are subsistence farmers who suffer from poor yields, poor quality, unfavourable economics and markets outside of their control. The power of producers to influence price in the market does not exist, particularly where producers of one geographic region also compete against producers from other geographical regions. To survive, subsistence farmers must skillfully decide which crops to cultivate, so their income is maximized. However, when large numbers of farmers receive the same information, aggregate overresponses usually occur through dramatic increases or decreases in production. This creates large price fluctuations and variances in income. Where crops are perennial, producers have no choice but to continue in the market and take what price is offered, unless they decide to get out of production completely.

The longer the supply chain is, the less influence the producer will have over it. Traders act on behalf of a large group of fragmented producers from remote locations, thus exercising great power over producers. The supply chain can be manipulated by hoarding stocks and waiting until prices rise, allowing the sell off of stocks at large profits. How widespread this practice is unknown. There are few organizations that represent the true interests of the producer; most represent the interests of traders.

The ethics of purchasing from primary producers in the supply chain is now an important issue among many companies. Some large companies have set up community projects in developing countries with ethical supply chains. A number of non-government international agencies have put in large efforts to shorten the supply chains so that sellers can deal directly with buyers or with intermediaries on agreed buyer–seller charters through programmes like *Fair-trade*. These moves are aimed at bringing equity into buyer–seller relationships.

**Table 37.5** A list of potentially endangered aromatic flora species

| Common and botanical name | Location | Status | Use and comments |
|---|---|---|---|
| Chinese rice flower, *Aglaia odorata* (Lour.) | China, Cambodia, Laos, Myanmar | IUCN red list of threatened species in 2007 | A flower oil and absolute used in fine perfumery |
| West Indian sandalwood, *Amyris balsamifera* L. | Florida, Belize, Costa Rica, Haiti, Honduras, Nicaragua, Cuba, Jamaica, Puerto Rio, Venezuela, Colombia, Ecuador | Predicted to disappear because of overexploitation (Joulain, 1994) | Limited use in soap and deodorant perfumery |
| Aniseed myrtle, *Backhousia anistata* | Australia, in north-east NSW | Species has a very small geographic range of 100 km (Briggs and Leigh, 1995) | Very minor quantities sold for aromatherapy use. Some small cultivation is going on |
| Agarwood, various *Aquilaria* spp. | Over the forests and jungles of South-East Asia | Most species listed on CITES Appendix II.[a] *A. banaensae, A. beccariana, A. crassna, A. cumingiana, A. hirta, A. malaccensis, A. microcarpa, A. rostrata,* and *A. sinensis*, on IUCN red list of threatened species in 2007 | Sold as an incense material and essential oil for fine perfumery. Large plantations being developed in Vietnam, Thailand, Laos, Malaysia and Indonesia. Human infected wood said not to be as good quality as naturally infected wood |
| Wormwood, various *Artemisia* species | Europe | Considered rare | Essential oil of *A. gracilis* used as a flavour in a local European liqueur *genipi* |
| Asafoetida, *Ferula assa-foetida* L. | Iran | Stocks deteriorating due to unsustainable exploitation | Used as a material in savoury flavours |
| Boldo, *Peumus boldus* Molina | South America | Local authorities restricting exploitation | Leaves produce an essential oil used in flavours (FDA approved). |
| Buchu leaf, *Agathosma betulina* | South Africa | Vulnerable stocks in the wild, due to harvesting for oil and tea, diminishing habitat (Hoegler, 2000) | Essential oil used in flavouring for blackcurrant flavours. Also used in fine perfumery. Now a cultivated crop |
| Holy weed, *Bursera glabrifolia* | Mexico | Overharvesting threatening existence of species (Peter *et al.*, 2003). | Now an essential oil of minor importance |
| Calamus, *Acorus calamus* L. | Widely distributed throughout Europe and India, also in South-East Asia | Becoming rare in India (CIMAP, 1997), but variety of opinion as to whether should be restricted due to wide distribution | Essential oil from rhizome. Employed as a traditional incense oil. Also used as a flavour ingredient, although restricted[b] use by EU food regulations (EEC, 1991) |
| Candeia plant, *Eremanthus erthropappus* (DC) MacLeish, *Vanillosmopsis arborea* (Aguiar) Ducke | Found in the Atlantic Brazilian rainforest | The tree is becoming rare due to exploitation | The wood is a source of (−)-α-bisabolol (Lopes *et al.*, 1991) |
| *Canarium zeylanicum* (Retz.) Blume | Sri Lanka | IUCN red list of threatened species 2007 | Used as an oleoresin for incense |
| Cedarwood Atlas, *Cedrus atlantica* (Endl.) Manetti ex Carr | At 1500–2500 m altitude in Atlas Mountains, Morocco | IUCN red list of threatened species 2007. Production has declined substantially | Essential oil produced from waste wood. An absolute is also produced |
| Cedarwood Himalayan, *Cedrus deodora* | Above 2000 m on the slopes of the Himalayan Mountains from Afghanistan to Pakistan. | IUCN red list of threatened species in 2007 | Crude and rectified oils from sawdust and wood shavings, an absolute and a fir needle oil available. Production has decreased dramatically to almost nothing. Also used in aromatherapy, perfumery and a product extracted through destructive distillation as a veterinary medicine |

Table 37.5 (*Continued*)

| Common and botanical name | Location | Status | Use and comments |
|---|---|---|---|
| Cedarwood Kenyan, *Juniperus procera* Hochst ex Endl. | 1000–3000 m range along the mountains of Kenya, and Ethiopia. Also in other African countries and Saudi Arabia, Somalia | IUCN red list of threatened species in 2002 | A now almost unattainable essential oil. Species suffered from insect infestations and dieback (Ciesla, 2002) |
| Cigar box wood, *Cedrela odorata* L. | South and Central America | Illegal export in many countries but trade continuing | A wood essential oil |
| *Cinnamomum* spp. *Cinnamomum camphora* L. var. *linaloolifera*, *Cinnamomum camphora* Sieb var. *glavescens* Hayata (ho wood) | China and India | IUCN red list, Chinese authorities ban harvesting | Rectified crude oil produces ho-wood oil, which is used as a source of linalool, acetylated oil produced through esterification with acetic anhydride used to reconstitute some essential oils and extend lavender oil. Also used a natural linalyl acetate |
| *Cinnamomum parthenoxylon* (Jack) Meisn. | Vietnam and China | IUCN red list of threatened species in 2007, Chinese authorities ban harvesting | As a source of safrole (>95%), which is used to produce piperonyl butoxide |
| Tejpat oil, *Cinnamomum tamala*, | Northern India | Overexploitation of species | Essential oil from steam distillation of the leaves |
| Opoponax, *Commiphora* spp. | Ethiopia, Eritrea, Kenya, Somalia, also species in India | IUCN red list of threatened species in 2007 | A source of opoponax, resin exudes from the bark naturally, but can be collected through making incisions. Also has numerous medicinal uses |
| *Copaiba* spp. *C. reticulate*, *C. guianensis*, *C. multijuga* and *C. officinalis* | Rainforests in Brazil | Threatened by logging (Plowden, 2003) | Used as a fixative in perfumery |
| Costus, *Saussurea lappa* C.B. Clarke | India and China | Critically endangered and listed as a negative export by the Ministry of Commerce, India. Some cultivation occurring (Guha and Pramanaik, 2005) | Essential oil of roots and plant, concrete and resinoid from roots. Material banned by IFRA for use in cosmetics. Now used for aromatherapy |
| Elemi, *Canarium luzonicum* | Indonesia, Philippines, Pacific Islands and Papua New Guinea | IUCN red list of threatened species in 2007 | Essential oil distilled from the pathological resin of the tree. Also $CO_2$ extracts produced, probably in Europe. Used in perfumery |
| Canadian fir balsam, *Abies balsamea* (L.) Mill. | Canada | IUCN red list of threatened species in 2007 | Essential oil distilled from the needles and twigs of the tree. Used in perfumery and as a varnish |
| Silver fir, *Abies alba* (L.) Mill. | Northern Europe | IUCN red list of threatened species in 2007 | Oil distilled from cones, needles and twigs and used for pine-type fragrances |
| Galbanum, *Ferula gummosa* Boiss | Pakistan, Turkmenistan, Iran, and India | Overexploitation | Gum exported to Europe, oil distilled from gum for perfumery use |
| *Gaultheria fragrantissma* Wall. (wintergreen) | India, Himalayas, China, Java | Overexploitation | Medicinal and perfumery use |
| *Gentiana* spp. *Gentiana lutera* L. | Central Europe, also France, Spain, Turkey, Albania and Romania | Overexploitation, wild harvesting banned in Spain | An absolute developed from the roots for flavouring |
| Ginger lily, *Hedychium coronarium* Koenig | Himalayas, throughout South-East Asia, India, China and Hawaii | Overcollection in India | Used as an essential oil in perfumery |

(*continued*)

**Table 37.5** (*Continued*)

| Common and botanical name | Location | Status | Use and comments |
|---|---|---|---|
| *Gonystylus* spp. | South-East Asia | IUCN red list of threatened species in 2007 | These trees produce a lower quality of gaharu than the *Aquilaria* spp. |
| Gurjun, *Dipterpcarpus* spp. | A number of species throughout South-East Asia | Logging has dwindled tree populations extensively. IUCN red list of threatened species in 2007 | Essential oil obtained from steam or vacuum distillation of the resin |
| Hinoki wood, *Chamaecyparis obtuse* (Siebold & Zucc.) Endl. | Japan | Protected by Japanese Government, IUCN red list of threatened species in 2007 | Essential oil from sawdust, waste and off-cuts. There are also root and leaf oils |
| Spanish juniper, *Juniperus turifera* L. | Morocco, Algeria, Spain and Corsica | IUCN red list of threatened species in 2007 | A rare oil used for leathery notes in perfumery |
| *Juniperus* spp. *J. oxycedrus* | Mediterranean Europe, Portugal, France, Morocco, Iran | IUCN red list of threatened species in 2007 | Cade oil is obtained from destructive distillation of the wood and branches of the shrub |
| *Kaempferia* spp., *K. galangal* and *K. rotunda* | Throughout East and South-East Asia | Becoming overexploited and facing extinction (Swapna *et al.*, 2004). On negative export list of Ministry of Trade, India | Essential oil of *K. galangal* used locally in hair washes, etc. (Selvam and Bandyopadhyay, 2005) |
| Spiked ginger-lily, *Hedychium spicatum* Smith | India, Nepal, Malaysia and Japan | Overexploited | Essential oil produced from the dried tuber roots, used in limited amounts in perfumery |
| Brazilian sassafras, *Ocotea pretiosa* (Nees) Mez. | Brazil and Paraguay | Logging seriously threatening this species | As a source of safarole now largely replaced by exploitation of *Cinnamomum parthenoxylon* in Vietnam |
| Olibanum, *Boswellia carterii* | Somalia, Oman and Yemen | Overexploited and damaged due to unskilled collection, result of clearing for agriculture. IUCN list of threatened species 2007 | Trees tapped for resin where essential oil is distilled from dry resin. An absolute and resinoid also produced. Used in fine perfumery |
| *Pinus* spp. *P. cembra* L., *P. roxburghii* Sarg., *P. merkusii, P. radiata, P. menziesii, P. pumila, P. silvestris, P. sibirica, P. eliotii, P. strobus* | Europe, Russia, USA, Canada, India, Pakistan and Himalayas | IUCN red list of threatened species in 2007 | Gum oleoresins and turpentine oil. Also twigs, needles and cones of some species distilled for pine needle oils, similar to *Abies* oils |
| *Prostanthera* spp., *P. rotundifolia, P. granitica, P. askania, P. hindii, P. junonis, P. palustres* | Tasmania, Victoria and New South Wales in Australia and New Zealand | IUCN red list of threatened species in 1997 | Very small production in Australia for aromatherapy |
| *Ravensara anisata* | Madagascar | Threatened by overproduction (damaged done by removing bark) | Essential oil distilled from bark |
| Rosewood, *Aniba* spp., *A. rosaeodora, A. fragrans, A. canelilla, A. parviflora* | Brazil, Colombia, Ecuador, French Guiana, Guyana, Peru, Suriname, Venezuela | Gross overexploitation (Margolis, 2004). IUCN red list of threatened species 2007 | The wood chips are steam distilled. Used in perfumery |
| Australian sandalwood, *Santalum spictum* | Northern and Western Australia | Exhausted through wild collection (Woodall and Robinson, 2003) | Used in perfumery but not of same quality as East Indian sandalwood |
| East African sandalwood, *Osyris lanceolata* Hochst & Steud. | Tanzania, Algeria, Zimbabwe, Swaziland | Protected species in Tanzania since 2005 | Essential oil and $CO_2$ extract used in perfumery |

Table 37.5 (*Continued*)

| Common and botanical name | Location | Status | Use and comments |
|---|---|---|---|
| East Indian sandalwood, *Santalum album* | India, Timor-Leste, some Indian Ocean islands, Indonesia, Philippines, Australia. Also introduced into China, Sri Lanka and Taiwan | IUCN red list of threatened species 2007 | Essential oil used in fine fragrance. Availability has dropped dramatically |
| New Caledonian sandalwood, *Santalum austrocaledonicum* Viell. var *austrocaledonicum* | New Caledonia | Nearly depleted, some production under control (Robson, 2004) | Used in cosmetics and fine fragrances |
| *Shorea robusta* Gaertn.f. | Nepal. Also other species in India, Thailand, Cambodia and Laos, Indonesia and Malaysia | Harvesting banned in Nepal | A gum is exuded from the stem. Sawdust and resin from the tree used to manufacture joss sticks |
| Siam wood, *Fokienia hodginsii* | China, Vietnam, Laos | IUCN red list of threatened species | Used as a wood source for furniture. Essential oil is distilled from off-cuts and sawdust |
| Spruce, *Picea albies* and *P. mariana* | *P. abies* distributed over central and northern Europe, *P. mariana* over USA and Canada | IUCN red list of threatened species 2007 | Needle and twigs distilled for essential oil used in perfumery |
| Styrax, *Liquidambar styraciflua* | USA, Belize, Guatemala, Honduras, Mexico, Nicaragua | IUCN red list of threatened species 2007 | Oleoresin produced in Honduras and vacuum distilled to produce styrax oil. Resinoid used as a fixative in perfumery |
| *Thymus* spp. | Mostly collected from the wild in Spain | Protected under Spanish law and quantities produced are declining, but does not include all species | Used as a flavour material |

[a]CITES Appendix II items can be commercially traded, subject to strict monitoring (i.e. export permit).
[b]The maximum permitted level of β-asarone in food is 0.1 mg/kg and 1 mg/kg for alcoholic drinks.

Large natural aromatic material producers focus their business development strategy on finding ways to develop product differentiation to strengthen their bargaining power over customers in the marketplace. These strategies involve developing isolates, fractions and specialty materials from essential oils, which other producers do not have. Producers try to convince the customer that these isolates have benefits or efficacies that are superior to the crude essential oil and worth paying a premium for. Other producers, *although very few have been successful*, have developed and marketed their own consumer product ranges around the world in attempts to get away from the price swings of a commodity product.

### 37.7.5 The regulatory environment

The regulatory environment is cited as the most difficult barrier to overcome in the development of a new natural aromatic material. The regulatory environment is currently undergoing great change. It is governed by a number of organizations with regional and product category jurisdictions. Many see these regulatory bodies as taking 'an over the top' approach to ensure consumer safety through material restriction, far beyond the actual threat availing society from the usage of aromatic materials in everyday life (Miller, 2008). Examples of some issues facing the industry today include the following:

- Limonene, a monoterpene constituent of many essential oils, including citrus oils, is under consideration for classification as an acute hazard to the environment. This means that products containing this ingredient in specified proportions (www.pesticides.gov.uk) will be required to follow EU hazard labelling rules. As it is proposed to designate limonene as R50/53 under the environmental regulations, this will disallow the certification of any product containing limonene in eco-certifications.
- The EU Scientific Committee on Consumer Products (SCCP) gave an opinion in April 2008 that oakmoss extract use be limited to where atranol and chloroatranol do not exceed >2 ppm in products (http://ec.europa.eu/health). This will potentially limit natural chypre and fougere accords in perfumery and promote more synthetic substitution for natural materials in this area.
- According to EU Directive 2003/15/EC, amending Council Directive 76/768/EEC (http://ec.europa.eu/consumers/sectors/cosmetics/documents/directive/), any product containing above specified limits of amyl cinnamal, benzyl alcohol, cinnamyl alcohol, citral, eugenol, hydroxyl-citronellal, isoeugenol, amylcin namyl alcohol, benzyl salicylate, cinnamal, coumarin, geraniol, hydroxyl-methylpentcyclohexenecarboxaldehyde, anisyl alcohol, benzyl cinnamate, farmesol, 2-(4-*tert*-butylbenzyl) propionaldehyde, linalool, bezyl benzoate, citronellol, hexyl

cinnam-aldehyde, D-limonene, methyl heptin carbonate, 3-methyl-4-(2,6,6-tri-methyl-2-cyclohexen-1-yl)-3-buten-2-one, oak moss extract and treemoss extract (all constituents found widely in essential oils), will require toxicity warning labels on consumer products.
- The SCCP in 2004 formed the opinion that the use of undiluted tea tree oil is unsafe and it is unsuitable for use in cosmetics (http://ec.europa.eu/health/ph_risk/committees/04_sccp/docs/sccp_o_160.pdf).

Under proposed legislation, small cosmetic cottage industry hobbyists and micro-companies will be required to register with the US Federal Drug Authority (FDA) like any other cosmetic manufacturer. Under the proposed FDA Globalization Act 2008 (www.teachsoap.com/forum/viewtopic.php?f=3&t=347&start=0), small companies will be required to pay the same fees as their larger counterparts, operate production and goods storage in separate locations, undertake the same paperwork requirements, register their formulations and advise the FDA of any changes in product formulation. If this act is passed, it will heavily curtail home and hand-made cosmetic and aromatherapy production in the USA.

The major consequence from the changing worldwide regulatory framework is that all essential oils must fulfil regulatory requirements, including existing ones because there are no 'grandfather' provisions in the EU regulations. This will dramatically increase the cost to producers who are not organized as a representative group and increase the costs for any new essential oil development.

New markets and opportunities are opening up for essential oils. The organic cosmetic market is reported to be growing at almost 10% per annum (Yeomans, 2013). No published market or growth figures exist for organically certified agricultural pesticides and fungicides, but growth is exponential according to those in the industry. There are also innovations in the household insecticide market. Companies like EcoSmart® have launched *knock-down* household insecticides utilizing essential oils as the major active ingredients (www.ecosmart.com).

The use of the term *natural* in the cosmetic industry has become misleading. Many multinational companies are moving into the organic cosmetic segment, previously served by smaller specialist niche manufacturers. There is a lot of confusion, as consumers are often given the impression that ingredients are 100% natural, when this is often far from the truth. There is currently lack of regulation in this area, which is allowing companies to portray their products as natural when they are really not. Natural does not mean sustainable, and the use of some materials may actually harm biodiversity rather than preserve it, as discussed in Section 37.7.3. The use of diesel to distil essential oils can be substantial. Ginger takes 20 h, linaloe 25 h, vetiver 30–90 h and ylang-ylang up to 48 h of distillation time. This creates a large carbon footprint, especially with diesel and bunker fuels that have to be carted long distances.

Various organic certification bodies exist to audit and appraise cosmetics, but these certification schemes, at present, are not uniform in their standards and are voluntary, without any legal sanction. Generally, these organic certification programmes require only 95% of materials in the product be of natural vegetable origin, processed through physical methods. Therefore, it is possible that the fragrance in the product is synthetic and not from a natural source. The sustainability of production is also not adequately assessed, leading to criticism of these certification schemes. Until these certification schemes are developed further, there is little incentive for manufacturers to increase their use of natural aromatic materials.

### 37.7.6 The rural crisis

As we are completing the second decade in this new century, much of the Asia-Pacific region is suffering a rural crisis. Australia was hit with a long drought for a number of years, a recession and a debt crisis, which has been slowly grinding rural regions to a halt. With tightening credit, many rural families are being forced off the land. Chronic shortages of labour are affecting harvests and production. Declining services and infrastructure are lowering the quality of life for rural families. Another aspect of Australian rural society is the inequality of educational pathways between rural and urban youth (Golding, 2001). This could potentially inhibit the ability of rural people to scan for opportunities and exploit them in the next generation.

To the north, in Papua New Guinea, economic and social inequality is increasing between the rural and urban populations. The rural population has fewer opportunities from limited economic activity. Disadvantages include remoteness from urban markets, the high cost and deterioration of transport services, poor access to services, lack of private and government investment and environmental pressure on the eco-system due to illegal activities such as logging. In Papua New Guinea, nearly half of the people living in rural areas live under the poverty line (http://acer.edu.au/documents).

Through South-East Asia, rural areas are facing similar problems to those of Australia and Papua New Guinea. Many rural areas lag behind in national development, unable to benefit from the high levels of growth in the region. Some governments are focusing on building infrastructure in urban areas due to the very rapid urban population growth, at the expense of the rural areas. Lack of rural investment is leading to higher rural unemployment rates, where a large percentage of the rural youth population is leaving to the cities for job opportunities (Baxter, 2001). This leaves an ageing farmer population in rural areas with low education levels, resources and capital. In some parts of the region, there is a feeling of powerlessness, with a lack of ideas, opportunities and matching skills, so very few are willing and able to embark upon new agriculture ventures (www.ausaid.gov.au/publications).

### 37.7.7 Synthesis of natural aromatic molecules

Many natural perfumes from the scents of flowers in the forests and jungles of the world are examined each year by flavour and fragrance companies (Hunter, 2006). This results in a thorough analysis of these oils and an investigation of the biological pathways that leads to their natural synthesis. From these data, thousands of new molecules are synthesized each year, their odour properties evaluated and tested in potential applications and the synthesis process for promising molecules patented (Hunter, 2008).

The discoveries of new aromatic compounds have led more to the development of specialty aroma chemicals than to the development of new natural aromatic materials. The newly discovered aromatic molecules become *molecular templates*, where economic routes of synthesis can be designed and developed on a commercial scale.

With advances in bio-fermentation, this trend is likely to continue, because many newly discovered natural molecules come from plant species where the target molecule exists in the plant in trace amounts and is too costly to extract. With very few exceptions, flavour and fragrance companies have been reluctant to invest and develop the production of new essential oils due to the ease of synthesis as an alternative. The process of finding new molecules and synthesizing them is shown in Figure 37.4.

The patenting of aromatic materials found in nature is the way that many companies have protected specialty aroma chemicals with novel notes they have developed. This allows companies to exclusively offer these materials to the industry on a monopoly basis. The issue of synthesis raises the debate about intellectual property. These claims and rights in many cases are based on indigenous knowledge and medicine, although this is sometimes hidden. Companies hide the original source of knowledge and true ownership through not using the correct botanical nomenclature and specifying individual chemicals from the plant, without specifically mentioning the plant, and referring to a number of plants. The effect of these patents in a number of cases is to prevent the development of a specific essential oil as a single party has been awarded the exclusive rights, preventing others from exploiting a known and public-domain natural resource.

Companies that have the analytical equipment and laboratories to compile the data needed for an expensive patent application are advantaged over individuals and community groups who are usually not organized and lack the needed resources (Pain, 2001). The *Cropwatch* website argues that these patents represent misallocations of public domain knowledge and occur because the patent authorities lack the expertise required to scrutinize these issues (Salvadori, 1995). Nevertheless, a number of these patents have been challenged and revoked by patent authorities. For example, WR Grace European Patent on Neem extracts *(Azadirachta indica* A. Juss) was revoked and a turmeric (*Curcuma long* L.) patent held by the University of Mississippi Medical Center was revoked in 1995.

### 37.7.8 Research and development

Most crops need considerable research before they can be commercialized. Research plays an important role in natural product development and maintaining the competitive advantage of an existing industry. Natural product research covers the disciplines of botany, ethno-botany, entomology, natural product chemistry, analytical chemistry, micro-propagation, agronomy, agricultural engineering, distillation engineering, biotechnology, bioprocess engineering, cosmetic chemistry, perfumery, flavours and industrial chemistry. However, the majority of research undertaken in these fields by universities and research institutions tends to be academic rather than practical research. Academic researchers tend to be more concerned about maintaining the rigour of science, judged by their peers in journals and conference proceedings, rather than research that contributes directly to the exploitation of essential oils and development of the industry (Rukangira, 2001).

According to Malcolm, the last 50 years of Australian agriculture research was largely irrelevant to the farming community (Malcolm, 1990). Public funding is decreasing, and this is forcing the role and organization of research into new paradigms. In the future, strategic alliances will need to be created, like for example the strategic alliance between the University of Tasmania and Essential Oils of Tasmania. Other commercial research and development arrangements are being implemented in places like New Zealand, where a corporate research institution (SCION) is collaborating with the existing kanuka oil industry to improve the value of production through creating specific isolates useful to industry, under a commercial arrangement (Carberry, 2001).

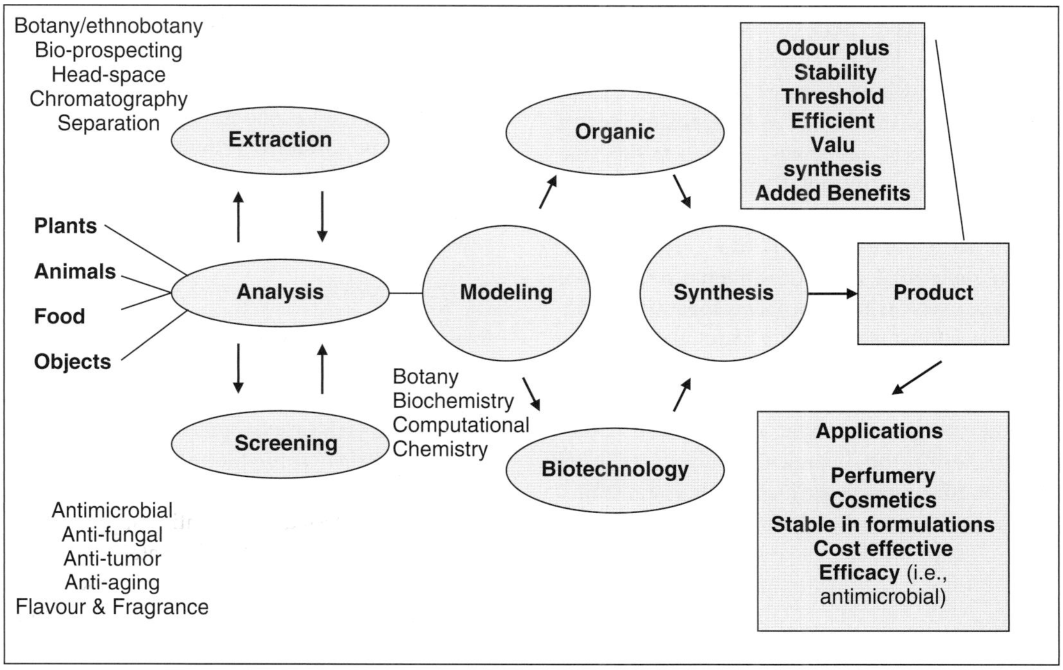


**Figure 37.4** The process of finding, analysing and synthesizing an aroma molecule based on a natural material

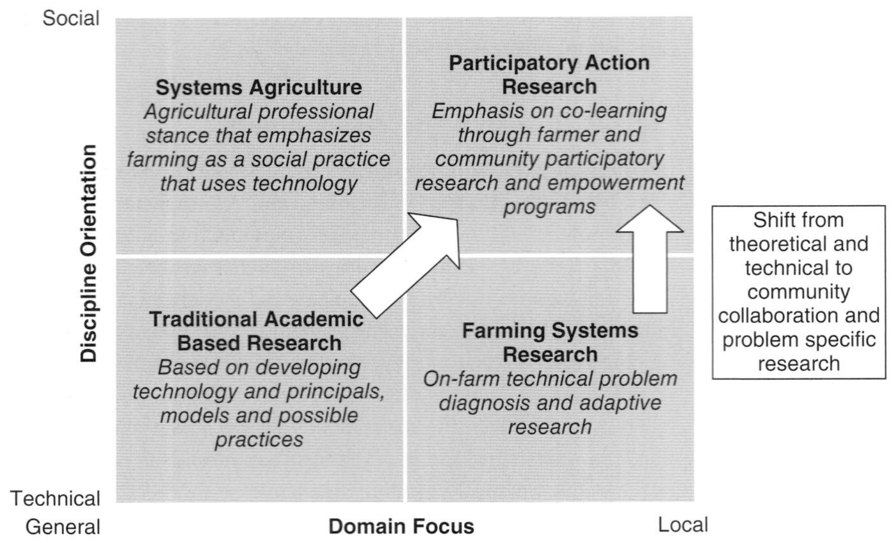


**Figure 37.5** A typology of various farming research management concepts

Future essential oil research cannot rely on public research funding to assist in the establishment of new natural products. Farmers will need to empower themselves and lead their own research in what is called *farmer-driven research* (Malcolm, 1990). This research approach escapes the irrelevant research provided by public institutions and focuses on the real needs in developing a new crop (Ecroyd *et al.*, 2008). Many research models involve the farmer in their own crop trials, without the same rigour as academic research, to achieve usable data that can be utilized by farming communities. This approach is called *participatory action research*, *farmer participatory research*, *participatory technology development* and *on-farm research*. Although these models were originally developed for Third World farming communities, they are quickly being adapted to problems of farming in developed countries (Biggs, 1995). The shift from traditional-based research to participatory action research is shown in Figure 37.5.

Research in the future is most likely to be corporate based, focused on the development of highly valued isolates and fractions of existing essential oils. This will be sponsored and dominated by research-orientated enterprises, the flavour and fragrance industry, and cosmetic and pharmaceutical companies. Growers in the Asia-Pacific region will have to rely on collaboration with other individuals (Martin and Sherington, 1997) and commercial organizations, or go down the road of participatory action research (Biggs, 1995).

From a sustainability standpoint, research without a strong component of innovation that provides some form of competitive advantage, or is based on some external unique factor like climate, may not always benefit the emerging industry. Research that does not create a strong entry barrier to others will not stop an industry declining in one country and emerging in another. Examples to support this proposition are the loss of the Australian eucalyptus industry to Europe, Africa and South America, the loss of the Australian tea tree industry to China and the move of floral oil production from Europe to Asia.

### 37.7.9 Sustainability

The concept of sustainability is elusive, difficult to define and even more difficult to know whether the enterprise is really operating in a sustainable way. Although there are now many conflicting definitions of sustainability, most include elements of both the socioeconomic and natural systems (Reijntes *et al.*, 1992). Theoretically, sustainability is about integration of the enterprise with the environment, where the environment includes natural, social, economic and strategic aspects. A sustainable enterprise is one that maintains productivity in the long term (McCown, 2001). Immediate needs should be able to be met while ecosystems and biodiversity are preserved for the future. However, it must also be accepted that everything has a limited lifecycle, and this increases the complexity of the meaning of sustainability. Sustainability needs consideration of all aspects of an enterprise. These areas include:

1. the agricultural aspects, where long-term production and preservation of resources are balanced out;
2. the surrounding ecological aspects, where the ecosystem and biodiversity are not depleted;
3. the management aspects, where the management system supports individual employees and the community;
4. the financial and economic aspects, where the benefits of new crops can be sustained without having to continually subsidize and financially support them with external funding; and
5. interactions with the socioeconomic environment and the community, where both have mutual benefits.

## 37.8 Conclusion

The future for new natural aromatic materials looks bleak. Finding new materials with very few people with the expertise and resources will be a challenge in the future. Rural communities are declining in many developing countries where the world's largest stock of biodiversity exists. Regulation is inhibiting natural product innovation and increasing the costs of commercialization to levels that threaten viability. Companies within the flavour and fragrance industry have invested too much in the roads to synthesis, and agricultural production of new materials is too risky.

The question here is, has the golden age of natural aromatic materials in the flavour and fragrance industry come and gone?

## References


Anon. (1993) A different perspective, *DCI*, September, p. 60.

Baxter, M. (2001) Enclaves or equity: the rural crisis and development in Papua New Guinea, Summary, Australian Agency for International Development, Canberra.

Biggs, S.D. (1995) Participatory technology development: reflections on current advocacy and past technology development, in *Proceedings of a Workshop on Agricultural Science for Biodiversity and Sustainability in Developing Countries* (eds S.D. Biggs, and F. Dolberg), 3–7 April, pp. 11–20.

Bolt, C. (2000) Tea-tree industry tries to cut oil glut. *Australian Financial Review*, 23 September.

Borst, A.B. (1988) An economic analysis of the US mint subsector. Thesis, Michigan State University, Department of Agricultural Economics, pp. 45–47.

Briggs, J.D. and Leigh, J.H. (1995) *Rare or Threatened Australian Plants*, revised edition, CSIRO, Canberra, p. 103.

Burfield, T. (2008) *Updated list of the threatened aromatic plants used in the aroma & cosmetics industries*, Cropwatch, October, pp. 12–14.

Carberry, P.S. (2001) Are science and industry relevance both achievable in participatory action research?, in *Proceedings of the Australian Agronomy Conference*, Australian Society of Agronomy, Hobart, Tasmania, http://www.regional.org.au/au/asa/2001/plenary/5/carberry.htm#P165_42463#P165_42463.

Ciesla, W.M. (2002), Juniper forests – a special challenge for sustainable forestry. *Forests, Trees and Livelihoods*, **12**, 195–207.

CIMAP (1997) *Indian Medicinal and Aromatic Plants Facing Genetic Erosion*, Central Institute of Medical and Aromatic Plants, Lucknow.

Ecroyd, C., Franich, R. and Meder, R. (2008) Leaf oil of *Kunzea ericoides* var. *microflora* (prostrate kanuka) isolated using supercritical carbon dioxide extraction, in *Malaysian International Conference on Essential Oils, Flavour and Fragrance Materials*, Kuala Lumpur, 28–30 October.

EEC (1991) European Council Directive on Food Flavourings 88/388/EEC, amended by 91/71/EEC.

Fenn, S. (1988) Aroma chemical usage trends in modern perfumery, in *Proceedings of the 11th International Congress of Essential Oils, Flavours and Fragrances*, Beijing.

Gallagher, M. (1993) The natural wave, *Chemical Marketing Reporter*, 12 July.

Golding, B. (2001) Great divides in learning: youth learning pathways in rural and Remote Australian towns, in *Proceedings of the ACER Research Conference 2001*.

Green, C.L. and Hone, G.A. (1992) Essential oil production in developing countries, in *Proceedings of the Third International Conference on Aromatic and Medicinal Plants*, Lyons, France, 2–4 December 1991, pp. 197–202.

Guha, R. and Pramanik, D. (2005) Know your CITES plants – II: *Nepenthes khasiana* Hook F. & *Saussurea costus* (Flac.) Lipschitz. – treasures worth saving. *Envis Newsletter*, **10** (1–2).

Harris, R. (1987) *Tanaman Minyak Atsiri*, Penebar Swadaya, Jakarta, pp. 112–134.

Hoegler, N. (2000) Plight of buchu underscores the need for international herb development. *Herbalgram*, **50**, 16.

Hunter, M.M. (1996) Malaysia: a new source of tea tree oil, in *Personal Care Ingredients Asia Exhibition & Conference*, Kuala Lumpur, 26–27 March, p. 11.

Hunter, M. (1992) A review of the comparative potentials of some essential oil crops in peninsula Malaysia, in *Proceedings of the Third Malaysian International Conference on Essential Oils, Flavours and Fragrances*, Langkawi, Malaysia.

Hunter, M. (2006) Mindset barriers against the development of Malaysian agriculture, in *Proceedings of the SME-Entrepreneurship and Global Conference 2006*, Kuala Lumpur.

Hunter, M. (2008) Revolutionary empowerment: a re-look at spirituality, cultural integrity and development, in *Proceedings of the Education and Development Conference*, Bangkok, 7–9 March.

Joulain, D. (1994) Success & failures in essential oils: a historical overview, *4th Rencontres Techniques et Economiques*, Lyons, France, 5–7 December, pp. 4–11.

Konur, O.Z. and Robertet, P. (1983) The agriculture and distillation of *Rosa damascena* in Turkey, in *Proceedings of the Ninth International Congress of Flavours, fragrances and Essential Oils*, Singapore.

Latiff, A., Ismail, G., Omar, M. *et al.* (1984) A multi-variate approach to the study of medicinal plants in Malaysia. *Journal of the Singapore National Academy of Science*, **13**, 101–103.

Lawrence, B.M. (1993) A planning scheme to evaluate new aromatic plants for the flavor and fragrance industries, in *New Crops* (eds J. Janick and J.E. Simon), John Wiley & Sons, Inc., New York, NY, pp. 620–627.

Lopes, J.N., Lopes, J.L., Vichnewski, C. *et al.* (1991) Chemical variability of *Vanillospsis erthropappa. Anais da Academia Brasileira de Ciencias*, **63**, 21–22.

Malcolm, L.R. (1990) Fifty years of farm management in Australia: survey and review. *Review of Marketing and Agricultural Economics*, **58** (1), 24–55.

Margolis, M. (2004) Jungle economics: environmentalists thought they could save the rain forest and make money at the same time. They were wrong. *Newsweek International*, 16 February.

Martin, A. and Sherington, J. (1997) Participatory research methods – implementation, effectiveness and institutional context. *Agricultural Systems*, **55** (2), 195–216.

McCown, R.L. (2001) Farming systems research and farming practice, in *Proceedings of the Australian Agronomy Conference*, Australian Society of Agronomy, Hobart, Tasmania, http://regional.org.au/au/asa/2001/plenary/4/mccown.htm.

McKay, D. (1990) Oils from India. *Manufacturing Chemist*, (July), 35–37.

Miller, J. (2008) FDA Globalization Act of 2008, Cropwatch, www.cropwatch.org/newslet13.pdf.

Ministry of Agriculture (1991) *National Agricultural Policy (1992–2010)*, Ministry of Agriculture, Kuala Lumpur, Malaysia, 1.1.7.

Pain, S. (2001) The perfume hunters. *New Scientist*, **170** (2287), 35–37.

Peter, C.M., Purata, S.E., Chibnik, M. *et al.* (2003) The life and times of *Bursera glabrifolia* (H.B.K.) Engl. in Mexico: a parable for ethnobotany. *Economic Botany*, **57** (4), 432–441.

Plowden, C. (2003) Production ecology of copaiba (*Copaifera* spp.) oleoresin in the E. Brazilian Amazon. *Economic Botany*, **57** (4), 491–501.

Robson, K. (2004) Experiences with sandalwood in plantations in the South Pacific and N. Queensland, in *Proceedings of a Workshop Prospects for High Value Hardwood in the "Dry" Tropics of N. Australia*, Mareeba, North Queensland, Australia, 12–21 October.

Rao, V.R. and Arora, R.K. (2004) Rationale for conservation of medicinal plants, in *Medicinal Plants in Research in Asia, Vol. 1: The Framework and Project Workplans* (eds P.A. Batugal, J. Kanniah, S.Y. Lee and J.T. Oliver), International Plant Genetic Resources Institute – Regional Office for Asia, the Pacific and Oceania (IPGRI-APO), Serdang, Selangor, Malaysia, p. 9.

Reijntes, C., Haverkort, B. and Waters-Bayer, A. (1992) *Farming for the Future*, Macmillan, London.

Robbins, S.R.J. (1982) *Selected Markets for the Essential Oils of Patchouli and Vetiver*, Tropical Products Institute, London.

Robbins, S.R.J. and Greenhalgh, P. (1979) *The Markets for Selected Herbaceous Essential Oils*, Tropical Products Institute, London, pp. 1–3.

Rukangira, E. (2001) The African herbal industry: constraints and challenges. *Erboristeria Domani*, **1** (August), 1–23.

Salvadori, G. (1995) Of patents and scents, in *Flavours, Fragrances and Essential Oils, Proceedings of the 13th International Congress of Flavours, Fragrances, and Essential Oils, Istanbul* (ed. K.H.C. Baser), vol. 2, AREP, Istanbul, pp. 317–325.

Selvam, A.B.D. and Bandyopadhyay, S. (2005) *Kaempferia galangal* L. – a negative listed plant with promising aromatic & medicinal values. *Enis Newsletter*, **10** (1–2).

Swapna, T.S., Binitha, M. and Manju, T.S. (2004) *In vitro* multiplication in *Kaempferia galangal* Linn. *Applied Biochemcistry and Biotechnology*, **118** (1–3), 233–241.

Verlet, N. (1993a) Essential oils: supply, demand and price determination. *Acta Horticultuea*, **344**, 9–16.

Verlet, N. (1993b) Commercial aspects, in *Volatile Oil Crops: Their Biology, Biochemistry and Production* (eds R.K.M. Hay and P.G. Waterman), Longman, Harlow, UK, pp. 137–174.

Yeomans, M. (2013) Global organic cosmetics market to grow to reach $13.2 billion by 2018, Cosmetics design-europe.com, June 27, http://www.cosmeticsdesign-europe.com/Market-Trends/Global-organic-cosmetics-market-to-reach-13.2-billion-by-2018.

Woodall, G.S. and Robinson, C.J. (2003) Natural diversity of *Santalum spicatum* host species in south-coast river systems and their incorporation into profitable and biodiverse revegetation. *Australian Journal of Botany*, **51** (6), 741–753.

## Further reading

http://www.cropwatch.org/Threatened%20Aromatic%20Spp.%20v%201.08.pdf

http://acer.edu.au/documents/RC2001_Proceedings.pdf

http://www.ausaid.gov.au/publications/pdf/enclaves_equity_summary.pdf

http://www.ecosmart.com/household/

http://ec.europa.eu/enterprise/cosmetics/doc/200315/200315_en.pdf

http://ec.europa.eu/health/ph_risk/committees/04_sccp/docs/sccp_o_131.pdf

http://ec.europa.eu/health/ph_risk/committees/04_sccp/docs/sccp_mi_002.pdf

http://energycommerce.house.gov/FDAGlobalAct-08/Dingel_60AXML.pdf

http://www.pesticides.gov.uk/uploadedfiles/Web_Assets/PSD/ EcotoxChip3BriefGuidance.pdf

Smuggling, corruption and threatened species, some press reports, http://www.cropwatch.org/cropwatch7.htm

http://www.abc.net.au /rural/vic/stories/s591725.htm

# Index



*Novel Plant Bioresources: Applications in Food, Medicine and Cosmetics*, First Edition. Edited by Ameenah Gurib-Fakim.